AF602079

Springer Monographs in Mathematics

This series publishes advanced monographs giving well-written presentations of the "state-of-the-art" in fields of mathematical research that have acquired the maturity needed for such a treatment. They are sufficiently self-contained to be accessible to more than just the intimate specialists of the subject, and sufficiently comprehensive to remain valuable references for many years. Besides the current state of knowledge in its field, an SMM volume should ideally describe its relevance to and interaction with neighbouring fields of mathematics, and give pointers to future directions of research.

More information about this series at http://www.springer.com/series/3733

Kenji Fukaya • Yong-Geun Oh • Hiroshi Ohta
Kaoru Ono

Kuranishi Structures and Virtual Fundamental Chains

Kenji Fukaya
Simons Center for Geometry and Physics
State University of New York
Stony Brook, NY, USA

Yong-Geun Oh
IBS Center for Geometry and Physics
Pohang, Gyung-Buk, Korea
Department of Mathematics, Postech
Pohang, Gyung-Buk, Korea

Hiroshi Ohta
Graduate School of Mathematics
Nagoya University
Nagoya, Japan

Kaoru Ono
Research Institute for Mathematical
Sciences
Kyoto University
Kyoto, Japan

ISSN 1439-7382 ISSN 2196-9922 (electronic)
Springer Monographs in Mathematics
ISBN 978-981-15-5561-9 ISBN 978-981-15-5562-6 (eBook)
https://doi.org/10.1007/978-981-15-5562-6

Mathematics Subject Classification: Primary 18G70, 53D37, 53D40, 53D45, 57N75, 57R18, 58D27; Secondary 14A20, 37J05, 53D42, 57N65, 57N80, 58C35

This Springer imprint is published by the registered company Springer Nature Singapore Pte Ltd.
The registered company address is: 152 Beach Road, #21-01/04 Gateway East, Singapore 189721, Singapore

Preface

The main purpose of this book is to present a complete self-contained explanation of the virtual fundamental chain technique via the Kuranishi structure introduced by Fukaya–Ono [FOn2]. This book consists of two parts and appendices (Part III). Part I studies spaces with Kuranishi structure individually. We focus on the construction of a virtual fundamental chain on each individual space equipped with Kuranishi structure, mainly employing the de Rham theory over the real coefficients. But we also include a self-contained account of the way we work out the construction over the rational coefficients, when the space with Kuranishi structure has dimension less than or equal to one. Application of this part covers the study of closed pseudo-holomorphic curves of any genus needed such as in the construction of (closed) Gromov–Witten invariants. Part II studies the system of spaces with Kuranishi structure which we call a K-system. We build an algebraic structure on certain cochain complexes arising from such a K-system. More specifically, we axiomatize two particular K-systems of moduli spaces, one appearing in the study of Floer cohomology of periodic Hamiltonian systems and the other appearing in the study of A_∞ algebras in Lagrangian Floer theory.

The exposition of this book is independent of our earlier writings such as [FOOO16] and does not assume that the readers have previous knowledge of the pseudo-holomorphic curve.

Stony Brook, NY, USA — Kenji Fukaya

Pohang, Korea — Yong-Geun Oh

Nagoya, Japan — Hiroshi Ohta

Kyoto, Japan — Kaoru Ono

Acknowledgments

The authors would like to thank Kei Irie who pointed out certain errors in the preliminary version of this manuscript. They also thank anonymous referees for their tremendous efforts in reading and checking the entire manuscript in detail and sharing important comments.

Thanks are due to Mr. Masayuki Nakamura, Springer Japan, for his sincere efforts throughout the book production process, as well as to the copyeditors for a careful proofreading of the manuscript and many helpful comments on improving our English expressions.

Kenji Fukaya is supported partially by JSPS Grant-in-Aid for Scientific Research No. 23224002, NSF Grant No. 1406423 and Simons Collaboration on Homological Mirror Symmetry, Yong-Geun Oh by the IBS project IBS-R003-D1, Hiroshi Ohta by JSPS Grant-in-Aid for Scientific Research Nos. 23340015 and 15H02054, and Kaoru Ono by JSPS Grant-in-Aid for Scientific Research Nos. 26247006, 19H00636, and NCTS, Taiwan.

Contents

Chapter 1
Introduction

The technique of virtual fundamental cycles (and chains) was introduced in the year 1996 by several groups of mathematicians [FOn2, LiTi2, LiuTi, Ru2, Sie] to provide a differential geometric foundation on the study of moduli spaces of pseudo-holomorphic curves entering on the one hand in the Gromov–Witten theory and on the other in the study of the Arnold conjecture in symplectic geometry *without the assumption of any kind of positivity* on general compact symplectic manifolds. (With suitable positivity assumption such as semi-positivity, these applications were previously established in [Ru1, RT1, RT2, MS], and in [Fl2, HS, On1] respectively.) Since then, there have been several references, such as [CLW, HWZ3, LuT, Pa1], which provide various expositions with emphasis and focus on its application to the same subjects. The central issue to deal with in these constructions is to extract some counting invariant out of the a priori singular moduli space of pseudo-holomorphic curves, especially when there occurs the phenomenon of the *negative multiple cover problem*. This problem is closely tied to that of achieving transversality of the perturbed moduli space that is equivariant under the isotropy group, i.e., in the setting of orbifolds. Under the semi-positivity assumption, a generic choice of compatible almost complex structure is all we need to make the relevant moduli space smooth and without boundary, at least in the genus 0 cases. Virtual fundamental cycle techniques were invented precisely to handle the situation where such a positivity condition is not present. This is the situation where the technique is applied to the case of moduli spaces of pseudo-holomorphic curves *without boundary*.

On the other hand, in Lagrangian Floer theory initiated by Floer [Fl1], the moduli spaces of pseudo-holomorphic maps from bordered Riemann surfaces are studied. The presence of a boundary in the moduli space of pseudo-holomorphic curves (from bordered Riemann surface) is another central issue together with the transversality problem mentioned above in this development: The presence of a boundary (of codimension 1) makes the 'invariants' constructed by 'counting the order of the moduli space' dependent on the choice of perturbations. Therefore

K. Fukaya et al., *Kuranishi Structures and Virtual Fundamental Chains*, Springer Monographs in Mathematics, https://doi.org/10.1007/978-981-15-5562-6_1

considering a single moduli space and defining a counting invariant associated thereto does not have intrinsic meaning. A good illustration of this phenomenon already occurs in Floer homology, whose definition involves a system of moduli spaces that are interrelated via an explicitly given system of gluing processes. Only after certain homological/homotopical massaging of the system of counting invariants, one does obtain an invariant that is preserved under the various geometric homotopies such as Hamiltonian isotopies of Lagrangian submanifolds. The resulting invariant is usually not a numerical one but a homological/homotopical one. In the case of pseudo-holomorphic curves from bordered Riemann surfaces, one has to work with the virtual techniques *on the level of chains* simultaneously over the system of moduli spaces. Such a process was thoroughly carried out in [FOOO3, FOOO4, FOOO5] in the construction of filtered A_∞-algebra and its associated symplectic invariants of each compact relatively spin Lagrangian submanifold. It is proved in [FOOO3, FOOO4] that this A_∞-algebra is well-defined up to filtered homotopy equivalence. This technique has been subsequently applied to the study of homological mirror symmetry on toric manifolds [FOOO7, FOOO8, FOOO10] and to the problems of symplectic topology [FOOO9, FOOO11, FOOO12, FOOO13, FOOO14, FOOO15] by the present authors and by others e.g. [AJ, CLa, CKO, FJR, Ir, Li, NNU, On2, So1, So2, Wu].

The main purpose of the present book is to provide a reference for the researchers who want to use the virtual fundamental chain technique for their research by presenting this technique in as much detail as possible. This book grew out of our earlier writing [FOOO16] of a similar nature. Besides various systematic improvements of the exposition given therein and simplification of the proof in several places, we add the following points:

(1) We present the detail of the chain level argument, especially in the de Rham version.
(2) We provide a package of the statements which arise from the virtual fundamental chain and cycle technique, in the way one can directly quote and use without referring to their proofs.
(3) We give constructions of algebraic structures from *systems of spaces with Kuranishi structures*, abbreviated as *K-systems*. We separate the topological and algebraic issues from the analytic ones of the story so that this part of the construction can be rigorously stated and proved without referring to the construction of such a system.[1]
(4) We also explain a method of working with $\mathbb{Q}$-coefficients without using a triangulation of the perturbed moduli space. This is possible if we only need to study moduli spaces of virtual dimension 0, 1 and negative. Note that this case handles all the applications appearing in [FOn2].

[1]In our main application, such a system is constructed from the moduli spaces of pseudo-holomorphic curves. However, those results we mention here are independent of the origin of such a system.

We present the material of this book in a way that it can be read without any previous knowledge of pseudo-holomorphic curves. In particular, readers' knowledge of analytic details on pseudo-holomorphic curves is not required. Occasionally we mention the moduli space of pseudo-holomorphic curves at the places where we think providing the geometric example of the construction or of the definition etc. would be helpful for readers to visualize the relevant abstract definitions. We believe that this way of writing would facilitate the readers whose background is more in topology and/or in algebra than in analysis. Actually only a basic knowledge of geometry and topology is required to read this book. In particular, only elementary knowledge is used to prove the main result of the book.

This book consists of two parts. Part I studies spaces with Kuranishi structure individually, not as a system. Item (1) and part of Items (2),(4) mentioned above compose Part I. This will be enough for most of the applications of the moduli spaces of pseudo-holomorphic curves from closed Riemann surfaces, such as the construction of (closed) Gromov–Witten invariants.

In Part II, we study an axiomatic system of spaces with Kuranishi structures which we call a K-system and build an algebraic structure on certain cochain complexes arising from such K-systems. We axiomatize two systems of the moduli spaces, one appearing in the study of Floer cohomology of periodic Hamiltonian systems and the other appearing in the study of A_∞ algebra in Lagrangian Floer theory. Item (3) and part of Items (2), (4) above compose Part II.

In the Appendices, we present various materials used in the main body of the text. In particular, we provide a self-contained exposition of orbifolds, vector bundles on them, and the covering space of an orbifold for the case of *effective orbifolds*. Our exposition of the theory of Kuranishi structures relies heavily on this theory of effective orbifolds. This enables us to bypass the higher categories such as 2-category and to take the route of a down-to-earth approach of coordinate charts. We also introduce a certain class of cornered manifolds/orbifolds, which we call *admissible* manifolds/orbifolds. Roughly speaking, the admissibility means that the corner has a collar which is preserved by the coordinate change *modulo an exponentially small* term. In Part I, only Chap. 23 of the Appendices is used. The contents of Chaps. 24, 25, 26, and 27 are used only in Part II.

Our intention is to make this book self-contained, not relying on the results of our earlier writings such as [FOOO16] of a similar nature. To achieve our goal, we reprove some parts of [FOOO16] in this book with some simplification and better exposition. In particular, the proofs of existence of good coordinate systems, CF-perturbations, multivalued perturbations are much simplified in this book. See Chaps. 11, 12, and 13.

1.1 Background: Why Virtual Fundamental Chains?

In this section, we explain, by an example, the reason why a perturbation of the almost complex structure J or various explicit geometric parameters is not enough to achieve the transversality of the moduli space of pseudo-holomorphic curves or its variant. We consider the case of the moduli space which appears in the study of Floer homology of periodic Hamiltonian systems.

Let (M, ω) be a compact symplectic manifold and $H : S^1 \times M \to \mathbb{R}$ a smooth map, a time-dependent Hamiltonian. For $t \in S^1$ we put $H_t(x) = H(t, x)$ and denote by X_{H_t} the Hamiltonian vector field of H_t.[2] A periodic solution of the Hamiltonian vector field is a map $\gamma : S^1 \to M$ such that

$$\frac{d\gamma}{dt} = X_{H_t}(\gamma(t)).$$

We consider the case when all the periodic orbits are non-degenerate, that is, the derivative of its Poincaré map at the origin does not have 1 as an eigenvalue. For two periodic orbits γ_-, γ_+ we consider the set of the maps $u : \mathbb{R} \times S^1 \to M$ such that

$$\frac{\partial u}{\partial \tau} + J\left(\frac{\partial u}{\partial t} - X_{H_t}(u)\right) = 0. \tag{1.1}$$

Here τ, t are coordinates of $\mathbb{R}$, S^1, respectively and J is an almost complex structure on M which is compatible with the symplectic structure ω. We require the asymptotic boundary condition

$$\lim_{\tau \to +\infty} u(\tau, t) = \gamma_+(t), \qquad \lim_{\tau \to -\infty} u(\tau, t) = \gamma_-(t). \tag{1.2}$$

We denote by $\overset{\circ}{\mathcal{M}}(\gamma_-, \gamma_+)$ the moduli space of the solutions u of (1.1) satisfying (1.2). Here we identify $u(\tau, t)$ with $u(\tau + \tau_0, t)$ for $\tau_0 \in \mathbb{R}$. There is a notion of a homology class of the maps satisfying (1.2) and we denote by $\overset{\circ}{\mathcal{M}}(\beta; \gamma_-, \gamma_+)$ the subset of $\overset{\circ}{\mathcal{M}}(\gamma_-, \gamma_+)$ consisting of elements of homology class β. Floer [Fl2] studied the case when the virtual dimension of $\overset{\circ}{\mathcal{M}}(\beta; \gamma_-, \gamma_+)$ is 0 and used its 'order counted with sign' to define the matrix coefficient $\langle \partial[\gamma_-], [\gamma_+] \rangle$ of the boundary operator of the chain complex defining Floer homology. We can perturb the almost complex structure J and the Hamiltonian H so that $\overset{\circ}{\mathcal{M}}(\beta; \gamma_-, \gamma_+)$ is cut out transversally, so that, in the case its virtual dimension is 0 it is a discrete set of points.

The issue is its compactness. In general the limit of a sequence of elements of $\overset{\circ}{\mathcal{M}}(\beta; \gamma_-, \gamma_+)$ can be an object represented by a map u from Σ which is a union of $\mathbb{R} \times S^1$ and sphere bubbles, where u is pseudo-holomorphic on the bubbles. We consider the case when Σ is a union $(\mathbb{R} \times S^1) \cup S^2$ and $u_*([S^2]) = \alpha \in H_2(M; \mathbb{Z})$. Then the restriction of u to $\mathbb{R} \times S^1$ is an element of $\overset{\circ}{\mathcal{M}}(\beta'; \gamma_-, \gamma_+)$ where the

[2]It is characterized by $i(X_{H_t})\omega = dH_t$.

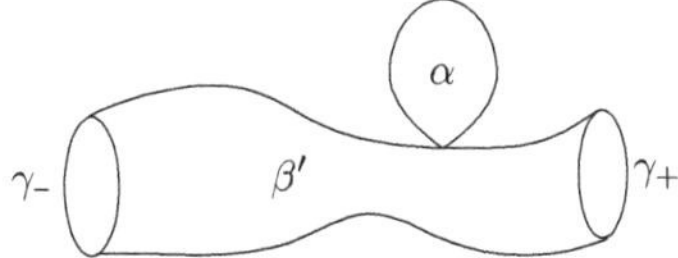

Fig. 1.1 The map $u : \Sigma \to M$

concatenation $\beta'\#\alpha$ of β' and α is β. Then we can show

$$d(\beta') + 2c_1(M) \cap \alpha = d(\beta),$$

where $d(\beta)$, $d(\beta')$ are the virtual dimensions of $\overset{\circ}{\mathcal{M}}(\beta; \gamma_-, \gamma_+)$, $\overset{\circ}{\mathcal{M}}(\beta'; \gamma_-, \gamma_+)$, respectively. ($c_1(M)$ is the first Chern class of M.) See Fig. 1.1. The map $u|_{S^2} : S^2 \to M$ represents an element of the moduli space $\overset{\circ}{\mathcal{M}}_0(\alpha)$ of pseudo-holomorphic spheres of homology class α, whose virtual dimension is

$$2c_1(M) \cap \alpha + 2n - 6,$$

which we denote by $d(\alpha)$. Here $2n$ is the (real) dimension of M. So the naive count of the dimension of the set of the configurations as in Fig. 1.1 is:

$$d(\beta') + d(\alpha) - 2n + 4 = d(\beta) - 2. \tag{1.3}$$

Here $-2n+4$ in the left hand side can be explained as follows. In Fig. 1.1 the images of $\mathbb{R} \times S^1$ and of S^2 intersect, namely $u(p) = u(q)$ with $p \in \mathbb{R} \times S^1$, $q \in S^2$. The pair (p, q) moves in a 4 dimensional space and the condition $u(p) = u(q)$ creates $2n$ equations. So the requirement 'images of $\mathbb{R} \times S^1$ and of S^2 intersect' changes the (virtual) dimension by $-2n + 4$.

In the situation to define Floer's boundary operator, we consider the case when $d(\beta) = 0$. Then (1.3) is -2, which is negative. To prove the boundary property $\partial\partial = 0$ of the boundary operator of Floer's chain complex we need to consider the case when $d(\beta) = 1$. In that case (1.3) is -1 and is still negative. Thus in the case where we can assume appropriate transversality, the configuration as in Fig. 1.1 does not appear.

However, the transversality of the configuration as in Fig. 1.1 may not hold even for a generic choice of an almost complex structure. Let us consider the case $n = 4$, $c_1(M) \cap \alpha' = -1$, $\beta = \beta''\#3\alpha'$, $\alpha = 3\alpha'$, and $d(\beta) = 0$. Then the virtual dimension of the configuration in the left hand side of Fig. 1.2 is:

$$d(\beta'') + d(\alpha') - 2n + 4 = 6 + 0 - 4 = 2.$$

In particular, it can be nonempty. Therefore the configuration of the right hand side of Fig. 1.2 consists of an at least 2-dimensional family. Here the sphere of homology class of $3\alpha'$ in the figure is a threefold cover of one of the homology classes of α' in the left hand side.

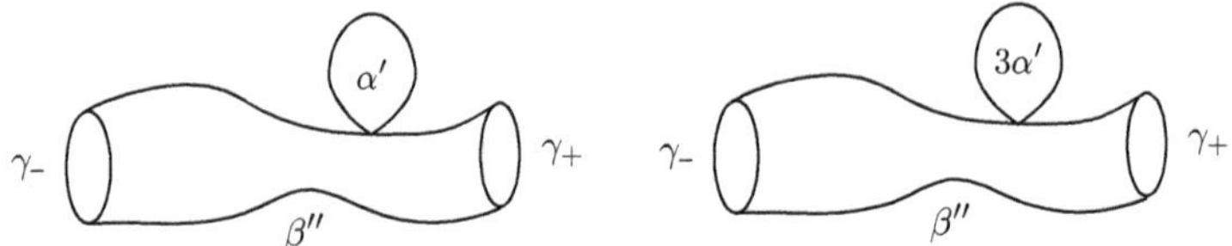

Fig. 1.2 Multiple cover sphere bubble

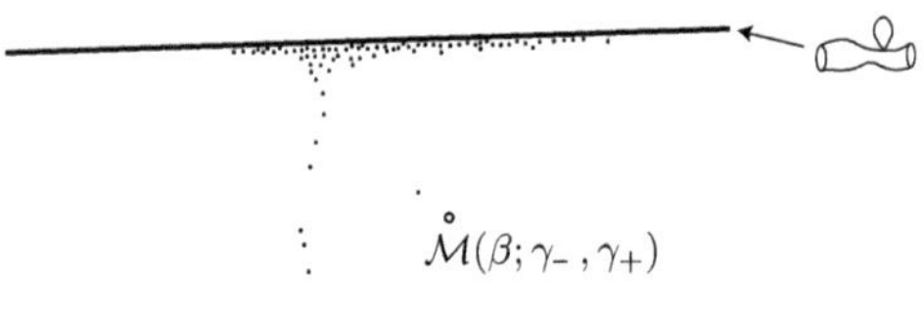

Fig. 1.3 Infinitely many points near infinity

Thus in this case a space of positive dimension appears at the 'infinity' of a 0-dimensional space. (See Fig. 1.3.)

To find a perturbation such that the perturbed moduli space of $\overset{\circ}{\mathcal{M}}(\beta;\gamma_-,\gamma_+)$ or its compactification $\mathcal{M}(\beta;\gamma_-,\gamma_+)$ is transversal, we need first to perturb it at 'infinity'. However, we cannot use the perturbation of J so that the configuration of the right hand side of Fig. 1.2 becomes transversal.

This problem is called the 'negative multiple cover problem'. An idea to resolve it is to modify (1.1) to the equation

$$\frac{\partial u}{\partial \tau} + J\left(\frac{\partial u}{\partial t} - X_{H_t}(u)\right) = s_u(z), \tag{1.4}$$

where s_u is a section of u^*TX which depends on u, so that in the limit (where we consider $u : (S^1 \times \mathbb{R}) \cup S^2 \to M$ for example) the equation becomes (1.4) on $S^1 \times \mathbb{R}$ and

$$\overline{\partial} u = s_u(z) \tag{1.5}$$

on S^2. Some versions of the method of virtual fundamental chains and cycles could be regarded as a perturbation of the moduli space via such a perturbation.

Such a perturbation had been used in several other places before 1996. For example, Donaldson in [Do1] (see also [Do2, Fur]) used a similar 'abstract' perturbation in gauge theory. On the other hand, it was in the year 1996 when the fact that abstract perturbation resolves the negative multiple cover problem was realized and abstract perturbation was applied systematically to two important problems in symplectic geometry, constructions of Gromov–Witten invariants and of Floer homology of periodic Hamiltonian systems. Since then the method of virtual fundamental chains or cycles has been developing.

To work out the perturbation as on the right hand side of (1.4), (1.5) systematically, we use the idea discovered by Kuranishi during the construction of a universal family of deformations of complex structures to construct a local finite-dimensional model of the compactified moduli space $\mathcal{M}(\beta; \gamma_-, \gamma_+)$. Namely for each element $\mathfrak{x} = [\Sigma, u] \in \mathcal{M}(\beta; \gamma_-, \gamma_+)$ we find a finite-dimensional manifold $V_{\mathfrak{x}}$, a finite-dimensional vector space $E_{\mathfrak{x}}$ and a smooth map $s_{\mathfrak{x}} : V_{\mathfrak{x}} \to E_{\mathfrak{x}}$ such that they are invariant under the finite group of automorphisms $\Gamma_{\mathfrak{x}}$ of $\mathfrak{x}$ and that the quotient space $s_{\mathfrak{x}}^{-1}(0)/\Gamma_{\mathfrak{x}}$ is homeomorphic to a neighborhood of $\mathfrak{x}$ in $\mathcal{M}(\beta; \gamma_-, \gamma_+)$. Such a description is called a Kuranishi model. A novel point in the theory of Kuranishi structures is to regard a Kuranishi model as a coordinate chart of our space $\mathcal{M}(\beta; \gamma_-, \gamma_+)$ and also to introduce the notion of coordinate changes between Kuranishi models which are defined at various points of the moduli space. In this way a certain geometric structure is defined on the moduli space $\mathcal{M}(\beta; \gamma_-, \gamma_+)$. Then a perturbation such as (1.4), (1.5) is described by the abstract language of such geometric structures.

At the time when the virtual fundamental chain and cycle technique started, it was already noticed[3] that such a method is closely related to an adaptation of the notion of a scheme or a stack to the differential topology or geometry. Actually, around the same time the construction of Gromov–Witten invariants for arbitrary smooth projective algebraic varieties was carried out by Li–Tian [LiTi1] and by Behrend–Fantechi [BF]. The method of virtual fundamental cycles in algebraic geometry was quickly established and became a standard method to study Gromov–Witten invariants in algebraic geometry. It was because in algebraic geometry, the highly sophisticated and developed language of scheme and stack had already been established and was very suitable to work out the details of the method of virtual fundamental cycles. On the other hand, in symplectic geometry, one needs to write up many parts of the theory from scratch to work out the details of the method of virtual fundamental cycles and chains. This might be a reason why it took a long time until the method of virtual fundamental chains or cycles became fully appreciated and widely used as a standard method to study pseudo-holomorphic curves among the researchers of symplectic geometry. We hope that the publication of this book helps this powerful method to be more widely applied to the study of pseudo-holomorphic curves and its applications to symplectic geometry.

1.2 The Story of Kuranishi Structures and Virtual Fundamental Chains

A Kuranishi structure is a 'generalization of the notion of a manifold with singularity'. We work entirely in the C^∞-category. In general, a singularity in the C^∞-category can be much wilder than one in the analytic/algebraic category. In the latter category, the theory of scheme provides a well-established functorial theory of spaces with singularities. Our aim is not to develop a general theory

[3] See for example [FOn1, Section 3].

of 'C^∞ analogue of scheme' but a more restricted one, which is, to define virtual fundamental chains and cycles used for the construction of some invariants, mostly related to symplectic manifolds so far. Because of this, our definitions and constructions are designed in so far as they are useful enough for that particular purpose.

Roughly speaking, a Kuranishi structure of a space X is a way to represent X locally as a zero set $s^{-1}(0)$ where s is a section of a vector bundle $\mathcal{E} \to U$ over an orbifold U. More precisely, a Kuranishi chart is a quadruple $\mathcal{U} = (U, \mathcal{E}, \psi, s)$ where s is a section of a vector bundle $\mathcal{E} \to U$ over an orbifold U and $\psi : s^{-1}(0) \to X$ is a homeomorphism to its image. (See Definition 3.5.) We call s the *Kuranishi map*. In fact, this description in the situation of various applications appears as a Kuranishi model of the local description of the moduli problem. We also need an appropriate notion of coordinate changes between them.

The main difference of the coordinate change between Kuranishi charts from that between manifold or orbifold charts lies in the fact that the coordinate change of Kuranishi charts may not be a local 'isomorphism'. This is because the dimension of the space with Kuranishi structure is, by definition, $\dim U - \operatorname{rank} \mathcal{E}$ and so the underlying orbifolds U_i of two Kuranishi charts $(U_i, \mathcal{E}_i, \psi_i, s_i)$ of the same space X may have different dimensions. (Namely $\dim U_1 \neq \dim U_2$ in general.) Recall that in the description of a manifold or orbifold structure as a geometric structure one employs the notion of a pseudo-group or a groupoid. Such a description is *not* available for Kuranishi structures since the most important axiom of groupoid (the invertibility of the morphisms) is exactly what we give up. To overcome this lack of invertibility, D. Joyce in [Jo2], for example, uses a version of localization of a category to invert the arrows appearing in the definition of coordinate changes of Kuranishi charts and employs the setting of a 2-category. On the other hand, our more down-to-earth approach is based on the observation that if one restricts to the world of *effective orbifolds*, the theory of Kuranishi structure permits the same kind of coordinate-wise description as that of manifold theory.

There are two slightly different versions of the definition of 'spaces with Kuranishi-type structures' in [FOn2]. One is what we call 'a Kuranishi structure', the other 'a good coordinate system'. In manifold theory (or in orbifold theory) we consider a maximal set of coordinate charts compatible with one another, and call this set a structure of a C^∞ manifold. Equivalently, for a manifold M we assign a (germ of a) coordinate neighborhood to each point p of M. This way of defining the manifold structure is somehow more canonical. On the other hand, to perform various operations using the given manifold structure, we usually take a locally finite cover consisting of coordinate charts of a given manifold structure. The latter is especially necessary when we use a partition of unity as in the case of defining the integration of differential forms on a manifold M. A certain amount of general topology needs to be worked out in the manifold course to prove existence of an appropriate locally finite cover extracted from the set of infinitely many charts defining the manifold structure.

The relationship between a Kuranishi structure and a good coordinate system is similar to the one between these two ways of describing a C^∞ structure of manifolds. A Kuranishi structure of X assigns a Kuranishi chart $\mathcal{U}_p$ to each point $p \in X$. So in particular, it contains uncountably many charts. On the other hand, a good coordinate system of X consists of locally finite (finite if X is compact) Kuranishi charts $\mathcal{U}_{\mathfrak{p}}$. (Here $\mathfrak{p} \in \mathfrak{P}$ and $\mathfrak{P}$ is a certain index set.)

Note that a coordinate change between Kuranishi charts is not necessarily a local isomorphism. In general, we have a coordinate change only in one direction. In the case of Kuranishi structure the coordinate change from $\mathcal{U}_q$ to $\mathcal{U}_p$ is defined only when $q \in U_p$. (Here U_p is a neighborhood of p.) We remark that $q \in U_p$ does not imply $p \in U_q$. In the case of a good coordinate system the set of Kuranishi charts is parametrized by a certain partially ordered set $\mathfrak{P}$ and the coordinate change from $\mathcal{U}_{\mathfrak{q}}$ to $\mathcal{U}_{\mathfrak{p}}$ is defined only when $\mathfrak{q} \le \mathfrak{p}$.

Let us elaborate on the notions of Kuranishi structure and good coordinate system by a simple example. Let $f : \mathbb{R}^2 \to \mathbb{R}$ be a smooth function such that

$$f(x, y) = x^2 + y^2 - 1$$

for $y \ge 0$. Here 0 is not a regular value of f on $y < 0$. We put $X = f^{-1}(0)$, which we draw as Fig. 1.4 below. We define a Kuranishi structure on X as follows:

For $p = (\cos\theta, \sin\theta) \in X$ with $\theta \in (\epsilon, \pi - \epsilon)$, we put $U_p = (\theta - \epsilon, \theta + \epsilon)$, $\mathcal{E}_p$ is the 0-bundle, $s_p = 0$, and $\psi(t) = (\cos t, \sin t)$.

For $p = (x_0, y_0) \in X$ with $y_0 < \sin\epsilon$ we put $U_p = \{(x, y) \mid y < \sin\epsilon,\ d((x, y), p) < \epsilon\}$, $\mathcal{E} = \mathbb{R} \times U_p$, $s_p : U_p \to \mathbb{R}$ is defined by $s(x, y) = f(x, y)$ and $\psi_p : s_p^{-1}(0) \to X$ is the identity map.

In both cases, ψ_p defines a homeomorphism to an open neighborhood of p. The definition of the coordinate change from $\mathcal{U}_q = (U_q, \mathcal{E}_q, \psi_q, s_q)$ to $\mathcal{U}_p = (U_p, \mathcal{E}_p, \psi_p, s_p)$ is obvious except the case $q = (\cos\theta, \sin\theta) \in X$ with $\theta \in (\epsilon, \pi - \epsilon)$ and $p = (x_0, y_0) \in X$ with $y_0 < \sin\epsilon$. In that case the coordinate change is defined as follows. We put

$$U_{pq} = \{t \in (\theta - \epsilon, \theta + \epsilon) \mid (\cos t, \sin t) \in U_p\} \subseteq U_q.$$

Fig. 1.4 The space X

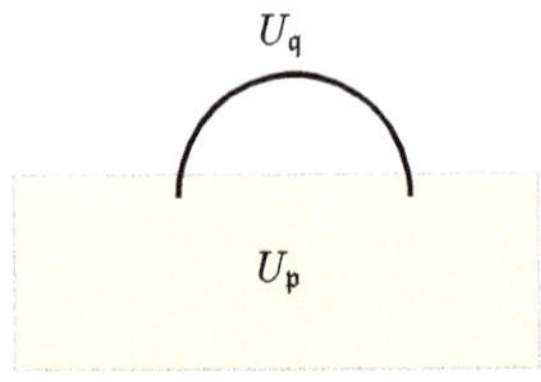

Fig. 1.5 A good coordinate system of X

We define an embedding $\varphi_{pq} : U_{pq} \to U_p$ by $\varphi_{pq}(t) = (\cos t, \sin t)$. The bundle map $\widehat{\varphi}_{pq} : \mathcal{E}_q|_{U_{pq}} \to \mathcal{E}_p$ which covers φ_{pq} is the obvious one. (Note $\mathcal{E}_q$ is the zero bundle.) They define a coordinate change $\mathcal{U}_q \to \mathcal{U}_p$ in the sense of Definition 3.6.

We observe the following features of a Kuranishi structure:

(1) It associates a Kuranishi chart $\mathcal{U}_p$ to each p. In particular, it consists of infinitely many Kuranishi charts.
(2) The dimension of U_p depends on p and the rank of vector bundles $\mathcal{E}_p$ also depends on p.
(3) The difference $\dim U_p - \operatorname{rank} \mathcal{E}_p$ is independent of p.
(4) The coordinate change map φ_{pq} is not in general a diffeomorphism.

A Kuranishi structure is a way to represent X locally as the zero set of the equation $s = 0$. The number of variables, which is the dimension of U_p and the number of equations, which is the rank of $\mathcal{E}_p$ depend on p, however the difference $\dim U_p - \operatorname{rank} \mathcal{E}_p$ is well defined and is the (virtual) dimension of X. Note in this example that U_p is a manifold. In general U_p is an orbifold.

A good coordinate system of the same space X can be taken so that it consists of two charts $\mathcal{U}_{\mathfrak{p}}, \mathcal{U}_{\mathfrak{q}}$. Here the Kuranishi chart $\mathcal{U}_{\mathfrak{q}} = (U_{\mathfrak{q}}, \mathcal{E}_{\mathfrak{q}}, \psi_{\mathfrak{q}}, s_{\mathfrak{q}})$ is defined as follows. $U_{\mathfrak{q}} = (0, \pi)$, $\mathcal{E}_{\mathfrak{q}}$ is the zero bundle, $s_{\mathfrak{q}}(t) \equiv 0$, and $\psi_{\mathfrak{q}}(t) = (\cos t, \sin t)$. The Kuranishi chart $\mathcal{U}_{\mathfrak{p}} = (U_{\mathfrak{p}}, \mathcal{E}_{\mathfrak{p}}, \psi_{\mathfrak{p}}, s_{\mathfrak{p}})$ is defined as follows. $U_{\mathfrak{p}} = \{(x, y) \mid y < \sin \epsilon\}$, $\mathcal{E}_{\mathfrak{p}}$ is the rank one trivial bundle over $U_{\mathfrak{p}}$, $s_{\mathfrak{p}}(x, y) = f(x, y)$ and $\psi_{\mathfrak{p}}$ is the identity map. We can define a coordinate change $\mathcal{U}_{\mathfrak{q}} \to \mathcal{U}_{\mathfrak{p}}$ in a similar way to the coordinate change $\mathcal{U}_q \to \mathcal{U}_p$. See Fig. 1.5.

The main reason why the notion of a Kuranishi structure was introduced is to work out the transversality issue appearing in the theory of pseudo-holomorphic curves etc. For this purpose it is important to perturb the Kuranishi map, denoted by s, so that the resulting perturbed map becomes transversal to 0. The original proof of the transversality theorem due to R. Thom [Th] constructs such perturbations by using the induction over the charts. We can follow his strategy using a partial order on the index set $\mathfrak{P}$ of Kuranishi charts for this purpose. Note that in order to perform this inductive construction it is essential to use a good coordinate system rather than a Kuranishi structure since it is hard to work with uncountably many coordinate charts composing the given Kuranishi structure. Thus the construction of a good coordinate system out of the given Kuranishi structure is an important step for the construction of an appropriate perturbation that resolves the transversality problem.

On the other hand, a Kuranishi structure has its own advantage over a good coordinate system in that it is more canonical. For example, we can define the product of two spaces with Kuranishi structure in a canonical way. However, the definition of the product of two spaces with a good coordinate system is more complicated and is less natural.[4] For our applications we need to construct virtual fundamental chains in a way that they are compatible with the fiber product and the direct product. For this purpose working with a Kuranishi structure is more appropriate. See Remark 4.26 for more detail.

Thus during the various constructions, we need to go from a Kuranishi structure to a good coordinate system and back several times. Concerning this transition from one to the other, there is one difference between the current story and manifold theory. In manifold theory, one starts with a C^∞ structure (which consists of all the compatible coordinate charts) and then picks up an appropriate locally finite cover, for example, when we define integration of differential forms. It is a fact that one can always recover the original C^∞ structure from the charts consisting of the chosen locally finite cover by enlarging the given locally finite cover to the maximal one. In the current story of Kuranishi structure, we start with a Kuranishi structure $\widehat{\mathcal{U}}$ and construct a compatible good coordinate system $\widehat{\mathcal{U}}$. We can use the resulting good coordinate system $\widehat{\mathcal{U}}$ to construct another Kuranishi structures $\widehat{\mathcal{U}^+}$. Unfortunately $\widehat{\mathcal{U}^+}$ is different from $\widehat{\mathcal{U}}$ in general. In other words, we lose certain information while we go from a Kuranishi structure to its associated good coordinate system. Some portion of the discussion in Part I is devoted to showing that this loss does not affect the construction of the virtual fundamental chain.

This technical trouble seems to reflect the fact that the notion of a Kuranishi structure is conceptually less canonical than that of a manifold. For example, we can construct a Kuranishi structure on (practically all the) moduli spaces of pseudo-holomorphic curves. However, the Kuranishi structure we obtain is not unique but depends on some choices. (We compensate this shortcoming by using an appropriate notion of cobordism.) In the case where the moduli space happens to be a manifold (and the equation defining the moduli space is Fredholm regular) the C^∞ structure on the moduli space is certainly canonical.[5] Another sign that the notion of a Kuranishi structure is not canonical enough is that we do not know how to define a morphism between two spaces with Kuranishi structure. It seems that the route taken by Joyce [Jo2, Jo4] resolves these two issues. See Sect. 1.4.5 for the reason why we nevertheless use our definitions of Kuranishi structure and good coordinate system.

In the realm of the algebro-geometric or complex analytic category, going to the limit in making all the constructions, definitions etc. as canonical as possible is often the correct point of view[6] even if it looks cumbersome at the beginning. However, because of the diversity and wildness of objects studied in the realm of differential geometry or in that of C^∞ functions, trying to realize a canonical construction in the

[4] This is mentioned also by Joyce [Jo1].

[5] Except the choice of smooth structure of gluing parameter.

[6] This is the viewpoint taken and insisted upon by Grothendieck.

ultimate level is not always a practical goal to achieve,[7] and finding a suitable place to compromise and to content ourselves with being able to achieve the particular purpose we pursue becomes an important issue.

The whole exposition of Part I is designed in the way that most of the statements and proofs find their analogs in the corresponding statements and proofs of the standard theory of manifolds. Once the right statements are given, the proofs are fairly obvious most of the time,[8] although it is not entirely so because there are also some differences between Kuranishi structure and manifold structure in certain technical points. For the researchers who have in mind the construction of virtual fundamental chains in various concrete geometric problems, a thorough understanding of cumbersome details of these technicalities should not be a part of everyone's required background.[9] It is our opinion that the whole theory now becomes nontrivial only because its presentation is heavy and lengthy. The method of making the proof 'locally trivial', which we are taking here, has been used in various branches of mathematics as an established method for building the foundation of a new theory when the theory is conceptually simple but involves certain complicated technicality in its rigorous details. We adopt the 'Bourbaki style' way in writing this book. In particular, we do our best to make explicit the assumptions we work with and the conclusions we obtain. As its consequence, there appear many definitions and statements in the text, which we acknowledge is annoying. However, it is inevitable because of the purpose of this book, that is, to provide a reference including all the technical details of the proof. We hope that in the near future many users of virtual techniques via Kuranishi structure appreciate that they do not need to know anything more than a small number of basic definitions and theorems together with the fact that the story of Kuranishi structure is mostly similar to that of smooth manifolds. Then one can safely dispose of the details such as those we provide here and use the conclusions as a 'black box'. It is the present authors' opinion that the advent of such an enlightening and agreement in the area will be important for the development of symplectic geometry and related fields. In fact, one main obstacle to its smooth development has been the nuisance of working out similar, but not precisely the same, details each time when one tries to use the moduli spaces of pseudo-holomorphic curves of various kinds in various situations.

[7]The theory of singularities of C^∞ functions is one typical example.

[8]For this reason in many places we do not need to say much about the proof.

[9]Thorough knowledge of such a technicality, of course, should be shared among the people whose interest lies also on extending the technology to the extreme of its potential border and/or using the most delicate and difficult case of the technology to obtain the sharpest results possible.

1.3 Main Results of Part I

We now provide a description of the main results of Part I.[10]

1.3.1 Elementary Material

As we explained in the previous section, we introduce two kinds of Kuranishi type structures. One is a Kuranishi structure and the other is a good coordinate system. They both are defined as a system of Kuranishi charts and coordinate changes between them. A Kuranishi chart of a space X is a 4-tuple $\mathcal{U} = (U, \mathcal{E}, \psi, s)$, where U is an orbifold, $\mathcal{E}$ is a vector bundle on U, s is its smooth section, and $\psi : s^{-1}(0) \to X$ is a homeomorphism onto an open subset of X. (Definition 3.1.)

A coordinate change from $(U_2, \mathcal{E}_2, \psi_2, s_2)$ to $(U_1, \mathcal{E}_1, \psi_1, s_1)$ consists of an open subset $U_{12} \subset U_2$, an embedding of orbifolds $\varphi_{12} : U_{12} \to U_1$, and an embedding of vector bundles $\widehat{\varphi}_{12} : \mathcal{E}_2|_{U_{12}} \to \mathcal{E}_1$ which covers φ_{12}. We require an appropriate compatibility condition of them with s_i and ψ_i.

A Kuranishi structure consists of (infinitely many) Kuranishi charts $\mathcal{U}_p = (U_p, \mathcal{E}_p, \psi_p, s_p)$ assigned to each $p \in X$ such that $p \in \psi_p(s_p^{-1}(0))$. A Kuranishi structure also assigns a coordinate change from $\mathcal{U}_q$ to $\mathcal{U}_p$ if q is sufficiently close to p, that is, $q \in \psi_p(s_p^{-1}(0))$. We require a certain cocycle condition for coordinate changes. (See Definition 3.9.)

We remark that this relation $q \in \psi_p(s_p^{-1}(0))$ is not symmetric. In other words, $q \in \psi_p(s_p^{-1}(0))$ does *not* imply $p \in \psi_q(s_q^{-1}(0))$. Note that if there is a coordinate change from $\mathcal{U}_q = (U_q, \mathcal{E}_q, \psi_q, s_q)$ to $\mathcal{U}_p = (U_p, \mathcal{E}_p, \psi_p, s_p)$, then $\dim U_q \le \dim U_p$. The fact that the dimension of U_p depends on p is an important feature of a Kuranishi structure. So this asymmetry between $q \in \psi_p(s_p^{-1}(0))$ and $p \in \psi_q(s_q^{-1}(0))$ is inevitable.

As in manifold theory it is in general very hard to work with infinitely many charts $\mathcal{U}_p$ consisting of a Kuranishi structure directly. So we use at the same time a good coordinate system. A good coordinate system consists of finitely many Kuranishi charts $\mathcal{U}_\mathfrak{p} = (U_\mathfrak{p}, \mathcal{E}_\mathfrak{p}, \psi_\mathfrak{p}, s_\mathfrak{p})$ indexed by a partially ordered set $\mathfrak{P}$. (Namely $\mathfrak{p} \in \mathfrak{P}$.) A good coordinate system assigns also a coordinate change between Kuranishi charts $\mathcal{U}_\mathfrak{p}, \mathcal{U}_\mathfrak{q}$, if $\mathfrak{q} \le \mathfrak{p}$. We require a certain cocycle condition for coordinate changes. (See Definition 3.15.)

We regard a Kuranishi structure as a basic object to study and use a good coordinate system as a tool to perform various constructions, since it seems likely that a Kuranishi structure is a less ad-hoc object than a good coordinate system. For example, if a good coordinate system is given on X then for $p \in X$ a subset $\{\mathfrak{p} \in \mathfrak{P} \mid p \in \psi_\mathfrak{p}(s_\mathfrak{p}^{-1}(0))\}$ of the partially ordered set $\mathfrak{P}$ is determined.

[10]Introduction to Part II can be seen at the beginning of Part II.

This subset changes when we replace $U_{\mathfrak{p}}$ by sightly smaller open subsets. So the process of taking open subsets of $U_{\mathfrak{p}}$ can reduce much the information it holds in a neighborhood of p. On the other hand, we can modify a Kuranishi structure, replacing U_p by an open subset U_p^0 (such that $p \in \psi_p(U_p^0 \cap s_p^{-1}(0)))$ for each $p \in X$. We call such a structure an open substructure. (See Definition 3.19.) Note that this process does not reduce the information it holds in a neighborhood of each p. Therefore replacing an open substructure of a Kuranishi structure is similar to the process of replacing a coordinate chart of a manifold by a smaller one. (This technique appears frequently in manifold theory.)

To perform various constructions on a Kuranishi structure (such as finding appropriate perturbations, integrations of differential forms, etc.) we sometimes need to construct a good coordinate system compatible with the given Kuranishi structure and then go back to a Kuranishi structure, possibly different from the given one. This process is more nontrivial than the corresponding process in manifold theory as we explained in Sect. 1.2. To describe this process precisely we will define notions of embeddings between Kuranishi structures and good coordinate systems. There are four kinds of them, which we call KK-embedding, KG-embedding, GG-embedding and GK-embedding. See Table 5.1. For example, a KG-embedding is an embedding from a Kuranishi structure to a good coordinate system. (The meaning of the other three should be clear from their names.) We also study their compositions. Certain parts of Chaps. 3 and 4 are devoted to those definitions. See Table 5.2. One technicality we need to take care of is to write down the precise definition of domains of embeddings. For example, an embedding from a Kuranishi structure $\widehat{\mathcal{U}_p} = (\{\mathcal{U}_p \mid p \in X\}, \{\Phi_{pq}\})$ to a good coordinate system $\widehat{\mathcal{U}_{\mathfrak{p}}} = (\{\mathcal{U}_{\mathfrak{p}} \mid \mathfrak{p} \in \mathfrak{P}\}, \{\Phi_{\mathfrak{p}\mathfrak{q}}\})$ assigns embeddings of Kuranishi charts, $\Phi_{\mathfrak{p}p} : \mathcal{U}_p \to \mathcal{U}_{\mathfrak{p}}$ for certain pairs $(p, \mathfrak{p})$. Also the embeddings $\Phi_{\mathfrak{p}p}$ are supposed to be compatible with coordinate changes of $\widehat{\mathcal{U}_p}$ and of $\widehat{\mathcal{U}_{\mathfrak{p}}}$. Some not-completely-trivial points to clarify are:

(1) For which pairs $(p, \mathfrak{p})$ do we require the embedding $\Phi_{\mathfrak{p}p}$ to be defined?
(2) Which subset of U_p is the domain of the embedding $\Phi_{\mathfrak{p}p}$?
(3) When do we require that the embeddings are compatible with coordinate changes?
(4) On which subset of the domain is the compatibility required?

We remark that the space X itself is the object we study and the neighborhood U_p of $p \in X$ is not necessarily contained in X. U_p contains p but the size of the neighborhood of p which U_p contains is rather ad-hoc. For this reason, there is no obvious choice for the above points (1)–(4) and hence we need to make a careful choice so that the resulting theory is consistent. We will therefore give the precise definition and the detail of the proof of composability. They are not deep but annoying. However, once we work it out we will not need to worry about it later on. The definitions are designed so that geometric intuition can be checked if necessary.

The most nontrivial and important result in Chaps. 3, 4, and 5 is the following existence theorem.

Theorem 1.1 *For any Kuranishi structure $\widehat{\mathcal{U}}$ of X there exists a good coordinate system $\overset{\triangle}{\mathcal{U}}$ of X and a KG-embedding $\widehat{\mathcal{U}} \to \overset{\triangle}{\mathcal{U}}$.*

This is a slightly simplified version of Theorem 3.35 and is proved in Chap. 11. There is another result of opposite direction, that is, for a given good coordinate system $\overset{\triangle}{\mathcal{U}}$ there exists a Kuranishi structure $\widehat{\mathcal{U}}$ and a GK-embedding $\overset{\triangle}{\mathcal{U}} \to \widehat{\mathcal{U}}$. (Proposition 5.26.) This is useful for various purposes. Its proof is easy and is given in Chap. 5.

Definitions of Kuranishi structure, good coordinate system and embeddings between them are based on the notions of an orbifold, its embeddings and vector bundle on it, which we describe in Chap. 23. The basic concepts and the mathematical contents of orbifolds were established by Satake [Sa]. However, there are a few different ways of defining an orbifold from the technical point of view. More significantly, the notion of morphisms between two orbifolds is rather delicate to define in general. We refer readers to the discussion in the book [ALR], especially its Sect. 1, about these points. In this book, we restrict ourselves to the world of *effective orbifolds* and use only embeddings as the morphisms between them. Then those delicate points disappear. For example, two morphisms are equal if and only if they coincide set-theoretically.

In Chap. 4, we define the direct product and the fiber product of spaces with Kuranishi structures. We remark that in the category theory the notion of fiber products is defined in the purely abstract language of objects and morphisms. The definition of fiber product we give in this chapter, however, is *not* the one given in the category theory.[11] In fact, we do not define the general notion of morphisms between the spaces equipped with Kuranishi structures, but define a fiber product of spaces with Kuranishi structures directly in Chap. 4. Here we consider only the fiber product over a manifold by requiring an appropriate transversality in its definition. Although our definition is not the universal one given in the category theory, it is so natural and canonical that many basic properties expected for the fiber product are fairly manifest. For example, associativity of the fiber product follows rather immediately from its definition. We do not define a fiber or a direct product of spaces with a good coordinate system here since its definition necessarily becomes more technical and complicated.

1.3.2 Multisections

In Chap. 6, we define the notion of multisections and multi-valued perturbations and their transversality.

The notion of multivalued perturbations was introduced in [FOn2]. The reason why we need such a notion to obtain a transversal perturbation can be seen from the

[11]The authors thank D. Joyce for drawing our attention to this point.

next example. Let $\mathbb{Z}_p = \mathbb{Z}/p\mathbb{Z}$ be the cyclic group of order p. For $k = 0, \ldots, p-1$ we define the action $\cdot$ of $\mathbb{Z}_p$ on $\mathbb{C}$ by

$$[m] \cdot z = e^{2\pi\sqrt{-1}mk/p} z.$$

We denote by $\mathbb{C}_p(k)$ the space $\mathbb{C}$ with this $\mathbb{Z}_p$-action. We consider a $\mathbb{Z}_5$ equivariant vector bundle:

$$\mathbb{C}_5(2) \times \mathbb{C}_5(1) \to \mathbb{C}_5(1).$$

Dividing by the $\mathbb{Z}_5$ action we obtain a vector bundle (orbibundle) $\mathcal{E}$ over $\mathbb{C}_5(1)/\mathbb{Z}_5 = U$. We consider its section, which is represented by a $\mathbb{Z}_5$ equivariant map $s : \mathbb{C}_5(1) \to \mathbb{C}_5(2) \times \mathbb{C}_5(1)$,

$$s(z) = (z^2, z).$$

This section has a zero of order 2 at the origin. Taking the $\mathbb{Z}_5$ symmetry into account we anticipate the intersection number $\mathrm{Im}(s) \cap$ zero section $= 2/5$. It is easy to see that $z \mapsto z^2 : \mathbb{C}_5(1) \to \mathbb{C}_5(2)$ cannot be perturbed in a $\mathbb{Z}_5$ equivariant way to a map transversal to 0.

A way to obtain this number $2/5$ is by using multisection. We define s_j^ϵ : $\mathbb{C}_5(1) \to \mathbb{C}_5(2) \times \mathbb{C}_5(1)$ by

$$s_j^\epsilon(z) = (z^2 + \epsilon e^{2\pi\sqrt{-1}j/5}, z)$$

$j = 0, 1, 2, 3, 4$. s_j^ϵ are not $\mathbb{Z}_5$ equivariant. However, the set $\{s_0^\epsilon, s_1^\epsilon, s_2^\epsilon, s_3^\epsilon, s_4^\epsilon\}$ is $\mathbb{Z}_5$ equivariant. In fact

$$[m] \cdot (s_j^\epsilon(z)) = s_k^\epsilon([m] \cdot z)$$

with $k \equiv 2m + j \pmod 5$ for $m = 0, 1, 2, 3, 4$. Thus on $U = \mathbb{C}_5(1)/\mathbb{Z}_5$, $\{s_0^\epsilon, s_1^\epsilon, s_2^\epsilon, s_3^\epsilon, s_4^\epsilon\}$ defines a 5-valued map, which is a special case of a 5-multisection. (In other words, it has five branches.) Note that its zero set consists of $[\sqrt{\epsilon}]$, $[e^{\pi\sqrt{-1}/5}\sqrt{\epsilon}] \in U$. The branch which becomes zero there is transversal to 0. Thus we have two transversal zeros. Since we study a 5-multisection the weighted count of zeros gives $2/5$, the expected value. See Fig. 1.6.

The above example is local. When we study the global case, the new feature appearing is that the number of branches of multisections could change when we move on an orbifold. For example, we consider an orbifold X which is S^2 as a topological space and has two orbifold singularities p and q whose neighborhoods are modeled by $\mathbb{C}_5(1)/\mathbb{Z}_5$ and $\mathbb{C}_7(1)/\mathbb{Z}_7$, respectively. See Fig. 1.7. We also consider a one-dimensional complex vector bundle (orbibundle) which is locally $(\mathbb{C}_5(2) \times \mathbb{C}_5(1))/\mathbb{Z}_5$ and $(\mathbb{C}_7(2) \times \mathbb{C}_7(1))/\mathbb{Z}_7$, around p and q. Suppose there exists a section s which is $z \mapsto [z^2, z]$ around p and $z \mapsto [z^2, z]$ around q. We perturb

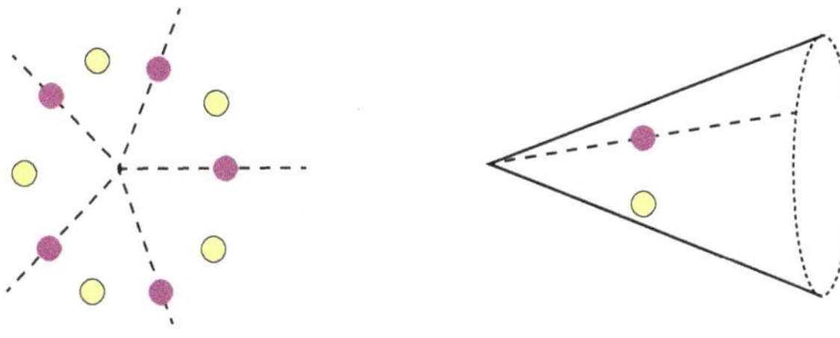

Fig. 1.6 Zero set of a multisection

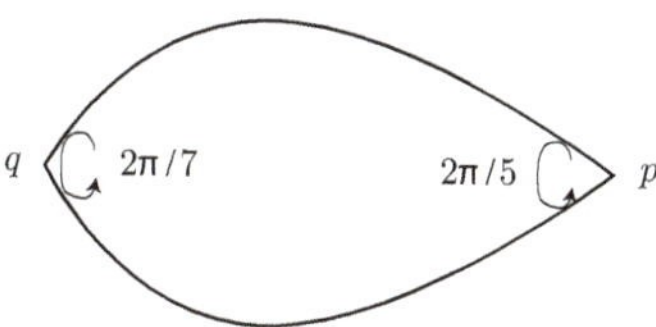

Fig. 1.7 Bad orbifold

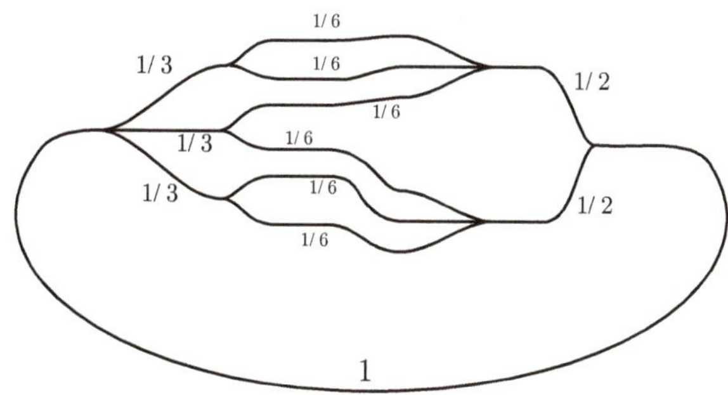

Fig. 1.8 Zero set of a multisection

s to a multisections with 5 branches around p and a multisection with 7 branches around q. To extend it to the whole X we regard both of them as multisections with 35 branches. So somewhere we need to bifurcate each one of the 5 branches to 7 branches (and then we obtain 35 branches), and also each one of the 7 branches to 5 branches.

These are examples when the dimension of the orbifold coincides with the rank of the vector bundles. In the case $\dim X - \operatorname{rank} \mathcal{E} = 1$ the zero set of a multisection is one-dimensional when the multisection is transversal to 0. However, this zero set is not a manifold in general. It bifurcates at the point where the number of branches changes. In other words, the zero set of a multisection (in the one-dimensional case) looks like Fig. 1.8. The numbers drawn in Fig. 1.8 are the weights.[12]

The main result claimed in Chap. 6 is the following existence theorem.

Theorem 1.2 *For any good coordinate system* $\widehat{\mathcal{U}}$ *of* X *there exists a multivalued perturbation that is transversal to* 0.

[12] Such a figure was observed in the year 1996 when the idea of virtual fundamental chains and cycles was incepted. (See [FOn2, Fig. 4.8].)

A multivalued perturbation is a sequence of multisections which converges to the Kuranishi map in the C^1 sense. Theorem 1.2 is a slightly simplified version of Theorem 6.23 and is proved in Chap. 13.

A technical but nontrivial result we prove in Chap. 6 is the compactness of the zero set of a multivalued perturbation, which is a part of Corollary 6.20. Corollary 6.20 also claims the fact that the zero set of a multivalued perturbation converges to the zero set of the Kuranishi map in the Hausdorff topology as our perturbation converges to the Kuranishi map. These and other related results play an important role in the proof of well-definedness of the virtual fundamental chain. (See Propositions 7.81 and 14.8.) The geometric intuition behind this well-definedness is the following. We consider the case when the good coordinate system without boundary consists of two Kuranishi charts $(U_{\mathfrak{p}}, \mathcal{E}_{\mathfrak{p}}, \psi_{\mathfrak{p}}, s_{\mathfrak{p}})$ and $(U_{\mathfrak{q}}, \mathcal{E}_{\mathfrak{q}}, \psi_{\mathfrak{q}}, s_{\mathfrak{q}})$ such that $\mathfrak{q} \le \mathfrak{p}$ and $\dim U_{\mathfrak{q}} < \dim U_{\mathfrak{p}}$. Suppose we have perturbations $\mathfrak{s}_{\mathfrak{p}}$ and $\mathfrak{s}_{\mathfrak{q}}$ of the Kuranishi maps $s_{\mathfrak{p}}$ and $s_{\mathfrak{q}}$ respectively so that they are compatible along the subset $U_{\mathfrak{pq}} \subset U_{\mathfrak{q}}$, which is embedded into $U_{\mathfrak{p}}$ by the coordinate change map. Let us consider the case when the virtual dimension of our moduli space is 0. (We also assume that $\mathfrak{s}_{\mathfrak{p}}$ and $\mathfrak{s}_{\mathfrak{q}}$ are single-valued. In the case where they are multisections we put appropriate weights.)

We want to prove that the sum of the numbers (counted with sign) of the zeros of $\mathfrak{s}_{\mathfrak{p}}$ and $\mathfrak{s}_{\mathfrak{q}}$ is independent of the system of the perturbations $(\mathfrak{s}_{\mathfrak{p}}, \mathfrak{s}_{\mathfrak{q}})$. (When there is a zero on $U_{\mathfrak{pq}}$ we do not double count it with its image in $U_{\mathfrak{p}}$.) We take a one-parameter family of systems of perturbations $(\mathfrak{s}^t_{\mathfrak{p}}, \mathfrak{s}^t_{\mathfrak{q}})$ compatible with the coordinate change which becomes $(\mathfrak{s}_{\mathfrak{p}}, \mathfrak{s}_{\mathfrak{q}})$ and $(\mathfrak{s}'_{\mathfrak{p}}, \mathfrak{s}'_{\mathfrak{q}})$ at $t = 0$ and $t = 1$, respectively.

In the case of a compact manifold without boundary we can prove the well-definedness of the number of zeros of a section of a vector bundle by using the fact that the number of the points in the boundary of a one-dimensional manifold is zero. When we try to adapt this standard argument to our case, a potential problem is the following. We consider the one-dimensional space

$$C = \bigcup_{t\in[0,1]} ((\mathfrak{s}^t_{\mathfrak{p}})^{-1}(0) \cup (\mathfrak{s}^t_{\mathfrak{q}})^{-1}(0)) \times \{t\}.$$

We regard it as a subset of the product of $U_{\mathfrak{q}} \cup U_{\mathfrak{p}}$ and $[0, 1]$, where we glue $U_{\mathfrak{q}}$ and $U_{\mathfrak{p}}$ by the coordinate change.

When we use C to show the independence of the number of the zero set (counted with sign) we prove that

$$\partial C = (\mathfrak{s}_{\mathfrak{p}}^{-1}(0) \cap \mathfrak{s}_{\mathfrak{q}}^{-1}(0)) \cup -((\mathfrak{s}'_{\mathfrak{p}})^{-1}(0) \cap (\mathfrak{s}'_{\mathfrak{q}})^{-1}(0)). \tag{1.6}$$

In the case of a single-valued section, C is a union of arcs and S^1's. So the required well-definedness follows if C is compact. In the case of a multisection, C has a bifurcation and looks like a configuration as in Fig. 1.8. In our situation C has the

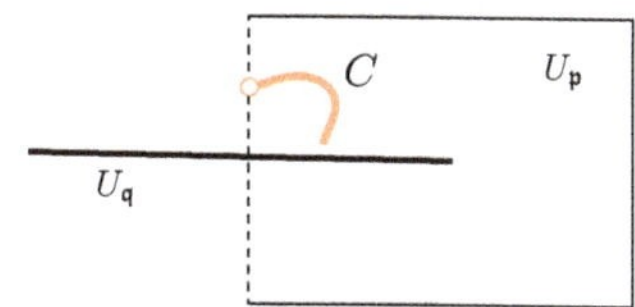

Fig. 1.9 Escape from the side

boundary as in (1.6). Thus the required well-definedness still follows from the fact that the total weight is preserved at the point where C bifurcates, if C is compact.

The issue we need to clarify to work out this argument is the compactness of C. Note that we assume that X is compact but $U_\mathfrak{p}$, $U_\mathfrak{q}$ are not compact. So a priori it could happen that C becomes noncompact, for example, when it intersects with the dotted line in Fig. 1.9 above. In such a case we cannot use C to prove the well-definedness of the weighted count of the zero set. Corollary 6.20 prevents such a phenomenon from occurring. The key observation for its proof is the fact that in a neighborhood of the image of $U_{\mathfrak{pq}}$ in $U_\mathfrak{p}$ the zero of the perturbed section still lies in the image of $U_\mathfrak{q}$, which is a submanifold of $U_\mathfrak{p}$ (Lemma 6.14). This is a consequence of the fact that the derivative of the Kuranishi map $s_\mathfrak{p}$ in the direction normal to $U_{\mathfrak{pq}}$ is invertible.

We also highlight that in Chap. 6 we work with a good coordinate system but not with a Kuranishi structure. As we mentioned before, we should perform the inductive construction of a transversal multisection with the good coordinate system but not with the Kuranishi structure itself.

1.3.3 *CF-Perturbation and Integration*

Chapter 7 is the heart of Part I where we define the integration of differential forms on a space X with a good coordinate system.

When X has a good coordinate system whose Kuranishi charts are $\{\mathcal{U}_\mathfrak{p}\}_{\mathfrak{p}\in\mathfrak{P}}$ ($\mathcal{U}_\mathfrak{p} = (U_\mathfrak{p}, \mathcal{E}_\mathfrak{p}, \psi_\mathfrak{p}, s_\mathfrak{p})$), a differential form $\widehat{h}$ on $U_\mathfrak{p}$ of this good coordinate system assigns a differential form $h_\mathfrak{p}$ to each $\mathfrak{p} \in \mathfrak{P}$ which are compatible with the coordinate changes. (See Definition 7.69.) It is impossible to integrate such forms on X in the usual sense, since X is in general a singular space and is very far from a manifold or an orbifold. The integration we define in Chap. 7 is the integration on the space obtained from X by an appropriate perturbation. Note that X is locally homeomorphic to the zero set of the Kuranishi map $s_\mathfrak{p}$. By perturbing $s_\mathfrak{p}$ to $\mathfrak{s}_\mathfrak{p}$ so that it is transversal to 0, we can integrate a differential form $\widehat{h}$. Namely we integrate $h_\mathfrak{p}$ on $\mathfrak{s}_\mathfrak{p}^{-1}(0)$. (More precisely we use an appropriate partition of unity in Sect. 7.5.) In the case where $\mathfrak{s}_\mathfrak{p}$ is a multisection (transversal to 0) its zero set is not a manifold or an orbifold. It is a set which looks like Fig. 1.8. However, we can integrate differential forms on such a space in a way similar to a manifold.

In Chap. 7 we study not only the integration but also the integration along the fiber. Then we need a continuous family of perturbations. (A multisection is a finite-set-valued perturbation. A continuous family of perturbations may be regarded as a smooth-measure-valued perturbation.)

We will define the integration along the fiber of the differential form on a good coordinate system $\widehat{\mathcal{U}} = \{\mathcal{U}_{\mathfrak{p}} \mid \mathfrak{p} \in \mathfrak{P}\}$ of X when we are given a weakly submersive map $\widehat{f} : (X, \widehat{\mathcal{U}}) \to M$. Here $\widehat{f}$ assigns a system of smooth maps $\{f_{\mathfrak{p}} : U_{\mathfrak{p}} \to M \mid \mathfrak{p} \in \mathfrak{P}\}$ which are submersions and are compatible with coordinate changes. (See Definition 3.40.) A CF-perturbation (a continuous family of perturbations) of a Kuranishi chart $\mathcal{U}_{\mathfrak{p}} = (U_{\mathfrak{p}}, \mathcal{E}_{\mathfrak{p}}, \psi_{\mathfrak{p}}, s_{\mathfrak{p}})$ is a triple $(W, \omega, \mathfrak{s}^{\epsilon})$,[13] as we will define in Definition 7.4. Here we explain an example and show how we use it to define the integration along the fiber.

Suppose $U_{\mathfrak{p}} = (\mathbb{C}_2(1) \times \mathbb{C}_2(0))/\mathbb{Z}_2 = V_{\mathfrak{p}}/\mathbb{Z}_2$, $\mathcal{E}_{\mathfrak{p}} = ((\mathbb{C}_2(1) \times \mathbb{C}_2(0)) \times (\mathbb{C}_2(1) \times \mathbb{C}_2(0)))/\mathbb{Z}_2$, $s_{\mathfrak{p}}$ is represented by $[z, w] \mapsto [z, w, z, w]$, and $\psi_{\mathfrak{p}} : s_{\mathfrak{p}}^{-1}(0) \to \{\text{point}\}$ is the obvious map. We take $M = S^2 = \mathbb{C} \cup \{\infty\}$ and let $f_{\mathfrak{p}} : U_{\mathfrak{p}} \to M$ be a smooth map $U_{\mathfrak{p}} \cong \mathbb{C}^2 \to S^2$, represented by $[z, w] \to z^2 + w$. Let $h_{\mathfrak{p}}$ be the differential 0-form, the constant function 1. Naively speaking, the zero set of $s_{\mathfrak{p}}$ is $[0]$ with multiplicity $1/2$, so the pushout (or integration along the fiber) of $h_{\mathfrak{p}}$ is $\frac{1}{2}\delta_0 dxdy$ the delta 2-form supported at $0 \in \mathbb{C} \subset M$.

We want to define an integration along the fiber as a *smooth* 2-form. So we need a perturbation. We take $W = \mathbb{C}_2(1) \times \mathbb{C}_2(0)$ and define:

$$\mathfrak{s}_{\mathfrak{p}}^{\epsilon}(z, w, \xi_1, \xi_2) = (z + \epsilon\xi_1, w + \epsilon\xi_2).$$

This is a $\mathbb{Z}_2$ equivariant map $V_{\mathfrak{p}} \times W \to \mathbb{C}_2(1) \times C_2(0)$ for each ϵ. We define a $\mathbb{Z}_2$ equivariant 4-form on ω on W as $\omega = \chi(r_1)\chi(r_2)dx_1dy_1dx_2dy_2$ where χ is a non-negative function supported on $[-1, 1]$, which is constant in a neighborhood of 0 and $r_i^2 = x_i^2 + y_i^2$, $\xi_i = x_i + \sqrt{-1}y_i$. We may choose χ so that $\int_W \omega = 1$. We thus obtain $\mathcal{S}^{\epsilon} = (W, \omega, \mathfrak{s}^{\epsilon})$.

We compose $f_{\mathfrak{p}} : \mathbb{C}_2(1) \times C_2(0) \to S^2$ with the projection $\pi : V_{\mathfrak{p}} \times W \to V_{\mathfrak{p}}$ to obtain $\tilde{f}_{\mathfrak{p}} : V_{\mathfrak{p}} \times W \to S^2$. Now the integration along the fiber of h with respect to this CF-perturbation $\mathcal{S}^{\epsilon}$ is by definition

$$f_{\mathfrak{p}}!(h; \mathcal{S}^{\epsilon}) = \frac{1}{2}(\tilde{f}_{\mathfrak{p}}|_{(\mathfrak{s}_{\mathfrak{p}}^{\epsilon})^{-1}(0)})!\pi^* h. \tag{1.7}$$

See Definition 7.11. Here $\tilde{f}_{\mathfrak{p}}|_{(\mathfrak{s}_{\mathfrak{p}}^{\epsilon})^{-1}(0)}$ is the restriction of $\tilde{f}_{\mathfrak{p}}$ to the zero set of $\mathfrak{s}_{\mathfrak{p}}^{\epsilon}$, which is a manifold.

It is easy to see that (1.7) is the 2-form $\frac{1}{2\epsilon^2}\chi'(r/\epsilon)dxdy$, where $\chi'(r)dxdy$ is a 2-form on $\mathbb{R}^2$ supported in the unit ball and integrates to 1. Thus $f_{\mathfrak{p}}!(h; \mathcal{S}^{\epsilon})$ is a smoothing of $\frac{1}{2}\delta_0 dxdy$ and converges to $\frac{1}{2}\delta_0 dxdy$ as ϵ goes to 0.

[13] Here W is a vector space, ω is a top degree compactly supported form on W and $\mathfrak{s}^{\epsilon}$ is a section of the finite-dimensional bundle on $W \times U_{\mathfrak{p}}$, which is the pullback of $\mathcal{E}_{\mathfrak{p}}$.

In the above example $s_{\mathfrak{p}}^{-1}(0)$ is a smooth manifold (a point). In the case where $s_{\mathfrak{p}}^{-1}(0)$ is singular the behavior of such $f_{\mathfrak{p}}!(h; S^{\epsilon})$ as ϵ goes to 0 is wilder. However, for each fixed positive ϵ we can define the integration along the fiber of a differential form as a smooth differential form.

To be slightly more detailed, we define the notion of a CF-perturbation in two steps. We first define such a notion on a single chart $(U_{\mathfrak{p}}, \mathcal{E}_{\mathfrak{p}}, \psi_{\mathfrak{p}}, s_{\mathfrak{p}})$, in the way we explained in the example above. We then discuss its compatibility with coordinate changes to define the notion of a CF-perturbation of a good coordinate system, which we denote by $\widehat{\mathfrak{S}} = \{\widehat{\mathfrak{S}^{\epsilon}}\}$. We define the notion of differential forms on a space with a good coordinate system. (It assigns a differential form on $U_{\mathfrak{p}}$ to each $\mathfrak{p} \in \mathfrak{P}$ which are compatible with coordinate changes. See Definition 7.69.) We use them to define the integration along the fiber

$$\widehat{h} \mapsto \widehat{f}!(\widehat{h}; \widehat{\mathfrak{S}^{\epsilon}}) \in \Omega^d(M), \tag{1.8}$$

for any sufficiently small $\epsilon > 0$. When $(X, \widehat{\mathcal{U}})$ is a manifold or an orbifold (that is, when all the obstruction bundles $\mathcal{E}_{\mathfrak{p}}$ are zero bundles), the operation (1.8) reduces to the standard integration along the fiber of a differential form. Note that $\widehat{\mathfrak{S}}$ is a one-parameter family of perturbations parametrized by $\epsilon > 0$. The integration along the fiber *does* depend on ϵ as well as the CF-perturbation. We also remark that typically (1.8) diverges as ϵ goes to 0. We firmly believe that there is *no* way of defining the integration along the fiber in a way independent of the choice of the CF-perturbation. This is related to the following most basic point of the whole story of virtual fundamental chains: In the case where we need to construct a virtual fundamental *chain* but *not a cycle*, that is, as in the case when our (moduli) space has a boundary or corners, the virtual fundamental chain depends on the choice of the perturbation. However, we can make (many) choices in a consistent way so that the resulting algebraic system is independent of such choices modulo certain homotopy equivalence. (We will discuss this point further in Part II.)

The main result of Chap. 7 is the following:

Proposition 1.3 *The integration along the fiber* (1.8) *is independent of various choices (other than the choices of the CF-perturbation and ϵ) involved. In particular, it is independent of the choice of the partition of unity we use to define the integration.*

We define the partition of unity in the current context (Definition 7.65) and prove its existence (Proposition 7.68). Proposition 1.3 is Proposition 7.81.

The following existence theorem is claimed in Chap. 7 and is proved in Chap. 12.

Theorem 1.4 *For any good coordinate system $\widehat{\mathcal{U}}$ of X there exists a CF-perturbation that is transversal to* 0.

Similar existence results hold for other kinds of transversality. See Theorem 7.51.

1.3.4 Stokes' Formula

In Chap. 8 we prove Stokes' formula for the integration along the fiber (1.8). We begin with discussing the boundary of a space with a Kuranishi structure or with a good coordinate system. We can define the notion of a boundary ∂M of a manifold with corners M, which we call the *normalized boundary*. (See Lemma-Definition 8.8.) The normalized boundary ∂M is again a manifold with corners and there is a map $\partial M \to M$ which is generically one-to-one and is a surjection to the (usual) boundary of M. Set-theoretically ∂M is not a subset of M. For example, codimension 2 corner points of M appear twice in ∂M.

Once these points are understood for the case of manifolds, it is straightforward to generalize them to a Kuranishi structure or to a good coordinate system with boundary and corners.

Suppose $(X, \widehat{\mathcal{U}})$ has a normalized boundary, $\partial(X, \widehat{\mathcal{U}}) = (\partial X, \partial\widehat{\mathcal{U}})$. Let $\widehat{f} : (X, \widehat{\mathcal{U}}) \to M$ be a weakly submersive map to a manifold. We consider its restriction to $(\partial X, \partial\widehat{\mathcal{U}})$, which is weakly submersive, and denote it by $\widehat{f_\partial}$. Suppose also we are given a CF-perturbation $\widehat{\mathfrak{S}}$ of $(X, \widehat{\mathcal{U}})$, which satisfies an appropriate transversality property. $\widehat{\mathfrak{S}}$ induces a CF-perturbation $\widehat{\mathfrak{S}_\partial}$ on the boundary. Stokes' formula for a good coordinate system now is stated as follows:

Theorem 1.5 *For any sufficiently small $\epsilon > 0$ we have*

$$d\left(\widehat{f}!(\widehat{h}; \widehat{\mathfrak{S}^\epsilon})\right) = \widehat{f}!(d\widehat{h}; \widehat{\mathfrak{S}^\epsilon}) + (-1)^{\dim(X,\widehat{\mathcal{U}})+\deg \widehat{h}}\, \widehat{f_\partial}!(\widehat{h_\partial}; \widehat{\mathfrak{S}^\epsilon_\partial}). \tag{1.9}$$

This is Theorem 8.11 and is proved in Sect. 8.2. In the case where M is a point, the integration along the fiber (1.8) is nothing but the integral of a differential form and is a real number. In that case we write $\int_{(X,\widehat{\mathcal{U}},\widehat{\mathfrak{S}^\epsilon})} h = \widehat{f}!(\widehat{h}; \widehat{\mathfrak{S}^\epsilon})$. Then (1.9) becomes

$$\int_{(X,\widehat{\mathcal{U}},\widehat{\mathfrak{S}^\epsilon})} dh = \int_{\partial(X,\widehat{\mathcal{U}},\widehat{\mathfrak{S}^\epsilon})} h. \tag{1.10}$$

This is a generalization of the standard Stokes' formula for a manifold with boundary.

In the case when a good coordinate system $(X, \widehat{\mathcal{U}})$ on X has no boundary, and $\widehat{f} : (X, \widehat{\mathcal{U}}) \to M$ is weakly submersive, the integration along the fiber of 1 (which is a differential 0-form of $(X, \widehat{\mathcal{U}})$) gives rise to a smooth differential form on M,

$$\widehat{f}!(1; \widehat{\mathfrak{S}^\epsilon}) \in \Omega^{\dim M - \dim(X,\widehat{\mathcal{U}})}(M).$$

The identity (1.9) implies that this form is closed. Moreover we have:

Theorem 1.6 *The de Rham cohomology class $[\widehat{f}!(1; \widehat{\mathfrak{S}^\epsilon})]$ is independent of the choices of $\widehat{\mathfrak{S}}$ and ϵ.*

This is a special case of Theorem 8.15. $[\widehat{f}!(1;\widehat{\mathfrak{S}^\epsilon})] \in H^{\dim M - \dim X}(M;\mathbb{R})$ is by definition the virtual fundamental class of $(X,\widehat{\mathcal{U}})$. Moreover we can show that the virtual fundamental class $[\widehat{f}!(1;\widehat{\mathfrak{S}^\epsilon})]$ does not change if we replace $\widehat{f}$: $(X,\widehat{\mathcal{U}}) \to M$ by another $\widehat{f'} : (X',\widehat{\mathcal{U}'}) \to M$, which is cobordant to $\widehat{f}$, $(X,\widehat{\mathcal{U}})$. (See Proposition 8.16.) We can use this fact to prove well-definedness of Gromov–Witten invariants.

1.3.5 Smooth Correspondence and Composition Formula

Besides Stokes' formula, an important property we use for the integration along the fiber is the composition formula. To formulate a composition formula we need to study the fiber product of CF-perturbations. Since a fiber product is better defined for a Kuranishi structure than for a good coordinate system, we rewrite the story of a CF-perturbation and the integration along the fiber with respect to a good coordinate system into the one with respect to a Kuranishi structure. We can define the notion of a CF-perturbation of a Kuranishi structure in the same way as that of a good coordinate system. The existence result of a CF-perturbation which corresponds to Theorem 1.2 is the following:

Proposition 1.7 *For any Kuranishi structure $\widehat{\mathcal{U}}$ of X there exists a Kuranishi structure $\widehat{\mathcal{U}^+}$, a KK-embedding $\widehat{\mathcal{U}} \to \widehat{\mathcal{U}^+}$ and a CF-perturbation of $\widehat{\mathcal{U}^+}$ that is transversal to* 0.[14]

Note that $\widehat{\mathcal{U}^+} \neq \widehat{\mathcal{U}}$ in Proposition 1.7. It is very difficult to construct a CF-perturbation with an appropriate transversality property on a given Kuranishi structure $\widehat{\mathcal{U}}$ of X. To prove Proposition 1.7, we first construct a good coordinate system $\widehat{\mathcal{U}}$ compatible with $\widehat{\mathcal{U}}$ and apply Theorem 1.2 to obtain a CF-perturbation $\widehat{\mathfrak{S}}$ on $\widehat{\mathcal{U}}$. Then we construct another Kuranishi structure $\widehat{\mathcal{U}^+}$ in such a way that $\widehat{\mathfrak{S}}$ induces a CF-perturbation $\widehat{\mathfrak{S}^+}$ of $(X,\widehat{\mathcal{U}^+})$. (Lemma 9.9.) Once we obtain a CF-perturbation $\widehat{\mathfrak{S}^+}$ of $(X,\widehat{\mathcal{U}^+})$ we can define the integration along the fiber for a strongly smooth map $\widehat{f} : (X,\widehat{\mathcal{U}^+}) \to M$ (see Definition 9.13 and Theorem 9.14) and can prove Stokes' formula (Theorem 9.28). This is the content of Chap. 9.

The composition formula which we prove in Chap. 10 is formulated in terms of the notion of a perturbed smooth correspondence and its composition, which we explain below.

Let M_s, M_t be compact smooth manifolds without boundary and $(X,\widehat{\mathcal{U}})$ be a space with Kuranishi structure. Suppose we are given strongly smooth maps $\widehat{f_s}$: $(X,\widehat{\mathcal{U}}) \to M_s$, $\widehat{f_t} : (X,\widehat{\mathcal{U}}) \to M_t$. We assume that $\widehat{f_t}$ is weakly submersive. We are also given a CF-perturbation $\widehat{\mathfrak{S}}$ of $(X,\widehat{\mathcal{U}})$ such that $\widehat{f_t}$ is strongly submersive with

[14] Statements on other variants of transversality also hold.

respect to $\widehat{\mathfrak{S}}$. (This means that the restriction of $f_p \circ \pi : U_p \times W \to M_t$ to the zero set $(\mathfrak{s}_p^{\epsilon})^{-1}(0)$ of $U_p \times W$ is a smooth submersion. See Definition 7.9.)

We call $\mathfrak{X} = (X, M_s, M_t, \widehat{\mathcal{U}}, \widehat{\mathfrak{S}}, \widehat{f_s}, \widehat{f_t})$ a *perturbed smooth correspondence* (Definition 10.19). To each such $\mathfrak{X}$ and sufficiently small $\epsilon > 0$, we associate a linear map $\mathrm{Corr}^{\epsilon}_{\mathfrak{X}} : \Omega^*(M_s) \to \Omega^{*+d}(M_t)$ by

$$\mathrm{Corr}^{\epsilon}_{\mathfrak{X}}(h) = \widehat{f_t}!(\widehat{f_s^*}h; \widehat{\mathfrak{S}^{\epsilon}}). \tag{1.11}$$

(See Definition 9.25.) Here $\widehat{f_s^*}$ is the pullback operation which assigns a differential form on $(X, \widehat{\mathcal{U}})$ to a differential form on M_s. The pullback is defined for an arbitrary strongly smooth map $\widehat{f_s}$. (See Definition 7.71. Here we do not need to assume weak or strong submersivity.) The integer d is given by $d = \dim M_t - \dim(X, \widehat{\mathcal{U}})$.

The composition formula claims that the assignment $\mathfrak{X} \mapsto \mathrm{Corr}^{\epsilon}_{\mathfrak{X}}$ is compatible with compositions.

For given perturbed smooth correspondences $\mathfrak{X}_{21} = (X_{21}, M_1, M_2, \widehat{\mathcal{U}_{21}}, \widehat{\mathfrak{S}_{21}}, \widehat{f_{1,21}}, \widehat{f_{2,21}})$, and $\mathfrak{X}_{32} = (X_{32}, M_2, M_3, \widehat{\mathcal{U}_{32}}, \widehat{\mathfrak{S}_{32}}, \widehat{f_{2,32}}, \widehat{f_{3,32}})$, we define their composition $\mathfrak{X}_{31} = \mathfrak{X}_{32} \circ \mathfrak{X}_{21} = (X_{31}, M_1, M_3, \widehat{\mathcal{U}_{31}}, \widehat{\mathfrak{S}_{31}}, \widehat{f_{1,31}}, \widehat{f_{3,31}})$ by taking the fiber product of Kuranishi structures $(X_{31}, \widehat{\mathcal{U}_{31}}) = (X_{21}, \widehat{\mathcal{U}_{21}}) \times_{M_2} (X_{32}, \widehat{\mathcal{U}_{32}})$.

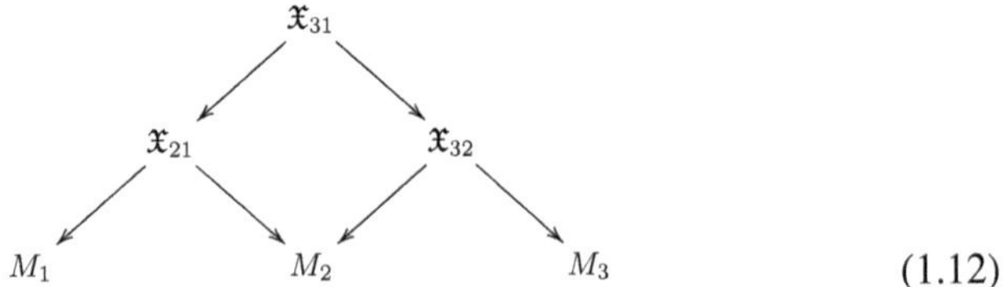

(1.12)

See Definition 10.16. Now the composition formula is stated as follows:

Theorem 1.8 *In the above situation we have*

$$\mathrm{Corr}^{\epsilon}_{\mathfrak{X}_{32} \circ \mathfrak{X}_{21}} = \mathrm{Corr}^{\epsilon}_{\mathfrak{X}_{32}} \circ \mathrm{Corr}^{\epsilon}_{\mathfrak{X}_{21}} \tag{1.13}$$

for each sufficiently small $\epsilon > 0$.

This is Theorem 10.21.

Remark 1.9 The integration along the fiber of differential forms on a space with Kuranishi structure is written in [FOOO8, Section 12]. In particular, Stokes' formula and the Composition formula were given in [FOOO8, Lemma 12.13] and [FOOO8, Lemma 12.15], respectively. Here we present them in greater details. In [FOOO8], the process of going from a Kuranishi structure to a good coordinate system and back was not written explicitly. (One reason is because the main focus of [FOOO8] lies in its application to the Lagrangian Floer theory of torus orbits of toric manifolds but not in the foundation of the general theory.) Here we provide thorough details. This theory is actually very similar to the manifold theory.

1.3.6 *Proof of Existence Theorems*

In Chap. 11, we prove the existence of a good coordinate system that is compatible with the given Kuranishi structure (Theorem 1.2 = Theorem 3.35). We also prove several variations. The proof we gave in the preprint version [FOOO19] of this book was the same as the one presented in [FOn2], which itself is more detailed in [FOOO16]. We simplify the proof by using the strategy of proving by contradiction systematically in this book. We separate the discussion on general topology issues from other parts and put it in a separate paper [FOOO17].

Chap. 12 is devoted to the proof of the existence of a CF-perturbation satisfying appropriate transversality properties. (Theorem 1.4 = Theorem 7.51.) The proof is based on the sheaf theory and proceeds as follows.

Take a compact subset $\mathcal{K}_{\mathfrak{p}} \subset \mathcal{U}_{\mathfrak{p}}$ for each $\mathfrak{p} \in \mathfrak{P}$ so that $\{\mathcal{K}_{\mathfrak{p}} \cap s_{\mathfrak{p}}^{-1}(0) \mid \mathfrak{p} \in \mathfrak{P}\}$ still covers X. (Such $\{\mathcal{K}_{\mathfrak{p}}\}$ is called a support system. See Definition 5.23.) We glue those compact spaces $\mathcal{K}_{\mathfrak{p}}$ by the coordinate change maps and obtain a compact metrizable space $|\mathcal{K}|$, which we call a hetero-dimensional compactum.

For each $x \in |\mathcal{K}|$ there exists finitely many $\mathfrak{p} \in \mathfrak{P}$ such that $x \in \mathcal{K}_{\mathfrak{p}}$. We denote by $\mathfrak{P}(x) \subseteq \mathfrak{P}$ the set of all such $\mathfrak{p}$. We consider the collection $\{\mathcal{S}_x(\mathfrak{p}) \mid \mathfrak{p} \in \mathfrak{P}(x)\}$, where $\mathcal{S}_x(\mathfrak{p})$ is a germ of a CF-perturbation of $\mathcal{U}_{\mathfrak{p}}$ at x. We require for $\mathfrak{q} < \mathfrak{p}$, $x \in \mathcal{K}_{\mathfrak{p}} \cap \mathcal{K}_{\mathfrak{p}}$ certain compatibility conditions between $\mathcal{S}_x(\mathfrak{p})$ and $\mathcal{S}_x(\mathfrak{q})$. The totality of such $\{\mathcal{S}_x(\mathfrak{p}) \mid \mathfrak{p} \in \mathfrak{P}(x)\}$ will be the stalk $(\mathcal{CF}_{\mathcal{K}})_x$ of the sheaf of CF-perturbations $\mathcal{CF}_{\mathcal{K}}$. See Definition 12.19. We can use a partition of unity for this sheaf $\mathcal{CF}_{\mathcal{K}}$ and its subsheaves of CF-perturbations satisfying various kinds of transversality conditions and use them to prove that $\mathcal{CF}_{\mathcal{K}}$ and the subsheaves are soft. The claimed existence of a CF-perturbation follows.

As far as the de Rham version is concerned, the results up to Chap. 12 provide a package we need for the case when we work on a single space with Kuranishi structure. The contents of Chaps. 13 and 14 will be used in Part II to verify that most parts of the story work when the ground field is $\mathbb{Q}$.

In Chap. 13, we prove an existence theorem of a multivalued perturbation (Theorems 1.2 and 6.23). The proof is similar to the proof of the existence of a CF-perurbation. For the proof we first define a sheaf of multivalued perturbations $\mathcal{MV}_{\mathcal{K}}$ in a similar way to $\mathcal{CF}_{\mathcal{K}}$. Then we prove its softness. In Sect. 13.3 we explain the relations of the proof of Chaps. 12 and 13 with other methods that appeared in the literature.

1.3.7 *Virtual Fundamental Chain Over* $\mathbb{Q}$

In Chap. 14, we discuss the virtual fundamental chain (over $\mathbb{Q}$) through a multivalued perturbation. The de Rham version of this section are Chaps. 7 and 8. In this chapter, only the case when the (virtual) dimension of our space $(X, \widehat{\mathcal{U}})$ with a good coordinate system is negative, 0 or 1 is studied. In the case where the dimension is

0 we define a $\mathbb{Q}$-valued virtual fundamental chain

$$[(X, \widehat{\mathcal{U}}, \widehat{\mathfrak{s}^n})] \in \mathbb{Q}$$

(Definition 14.18). It depends on the transversal multivalued perturbation $\widehat{\mathfrak{s}^n}$ and n but is independent of the other choices. Moreover we show that

$$[\partial(X, \widehat{\mathcal{U}}, \widehat{\mathfrak{s}^n})] = 0 \tag{1.14}$$

if the virtual dimension of $(X, \widehat{\mathcal{U}})$ is 1 (Theorem 14.20). This is an analogue of Stokes' formula. We show an analogue of the composition formula, Theorem 14.25, in the case where the fiber product we take is the one over a 0-dimensional space. Equation (1.14) implies the following:

Theorem 1.10 *If $(X, \widehat{\mathcal{U}})$ is a space with a 0-dimensional good coordinate system without boundary, then the virtual fundamental chain*

$$[(X, \widehat{\mathcal{U}}, \widehat{\mathfrak{s}^n})] \in \mathbb{Q}$$

is independent of the integer n and the choice of multivalued perturbation $\{\widehat{\mathfrak{s}^n}\}$.

This is Theorem 14.11. Furthermore we also prove its cobordism invariance, Proposition 14.13. We prove those results using Morse theory on a space with a good coordinate system. In the previous literature such as [FOn2] it was proved using a triangulation of the zero set of a multisection. In our situation when the virtual dimension is not greater than 1, the method using Morse theory seems simpler.

1.4 Related Works

In this section, we review some of the other related works from our point of view. We are aware of the fact that the way such a review is written depends much on the view of its authors. So we emphasize that this is a review from *our* point of view and other people have their own view, which can be very different from ours. Occasionally in this section, we assume some knowledge of the reader on the theory of pseudo-holomorphic curves, since such knowledge is required sometimes to explain the related works, but it is impossible to provide explanation of all of them here. The content of this section is not used anywhere else in this book, so readers can safely skip this section if they want.

1.4.1 Papers Appearing in the Year 1996

As we mentioned already, the technique of virtual fundamental chains and cycles in symplectic geometry was invented in 1996 by several groups of mathematicians, Fukaya–Ono [FOn2], Li–Tian [LiTi2], Liu–Tian [LiuTi], Ruan [Ru2], and Siebert [Sie]. The main applications established in 1996 are the construction of Gromov–Witten invariants with expected properties for arbitrary compact symplectic manifolds[15] and the homology version of Arnold's conjecture on the number of periodic orbits of a periodic Hamiltonian system. The differences between this method and the earlier one are the following two points:

(1) Certain notions of geometric objects are defined in an abstract context and the moduli space of pseudo-holomorphic curves or its variants is shown to be an example of such an object.
(2) The fundamental class (or chain) of the moduli space is constructed based on the abstract language of such geometric objects and such a construction is independent of the geometric origin of such an object.

In this book, we say that a method is one of virtual fundamental class or chain if it has the above two features. In this sense all of [FOn2, LiTi2, LiuTi, Ru2, Sie] are based on the virtual fundamental chain method.

There are certain differences between those five papers. One important difference is that some of them [FOn2, Ru2] are based on a finite-dimensional reduction and the others [LiTi2, LiuTi, Sie] study infinite-dimensional function spaces directly. We call the first one a finite-dimensional approach and the second one an infinite-dimensional approach.

In this book we take the finite-dimensional approach. In [FOn2] a multisection was used to define the virtual fundamental chain. In [Ru2] the Euler form of the obstruction bundle was used. The latter approach is somewhat similar to the method of CF-perturbation.

1.4.2 Issues in Developing the Theory of Virtual Fundamental Chains and Cycles

Before mentioning various related works which appeared after 1996, we explain certain issues that appear when developing the theory of virtual fundamental chains and cycles. In particular, we mention the following:

(I) Homology level argument vs. chain level argument.
(II) Morse case vs. Bott–Morse case and the choice of the chain model.

[15] For the construction of Gromov–Witten invariants there is an approach by algebraic geometry, which is somewhat similar to the symplectic geometry approach. We do not discuss it here.

(I): In the case when the moduli space we study has no boundary (for example in the sense of Kuranishi structure) its virtual fundamental chain is a cycle and its homology class is independent of the choices to be made in the definition. The study of such cases is significantly simpler than the case when the moduli space has a boundary. In the case when the moduli space has a boundary and corners, the virtual fundamental chain is not a cycle in general and so its homology class does not make sense. To obtain a well-defined invariant we need to study the whole system and construct a certain algebraic structure. We call the discussion of the former case homology level argument and the one in the latter case chain level argument.

(II): In the case of a chain level argument the moduli space often gives a smooth correspondence:

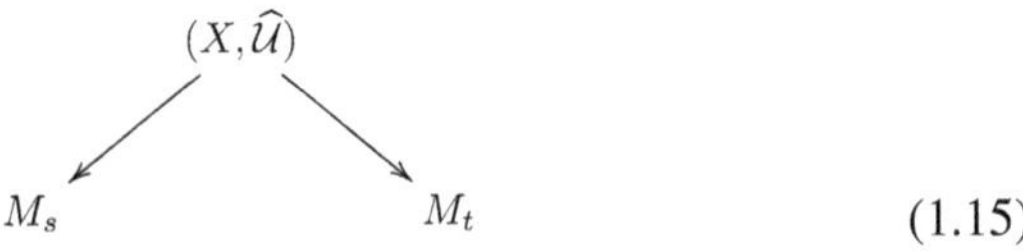

(1.15)

Here X is the moduli space with a certain structure (such as Kuranishi structure) and M_s, M_t are finite-dimensional manifolds. A typical example is Bott–Morse theory, the study of a function $f : M \to \mathbb{R}$ on a manifold such that the set of critical points of f is a smooth submanifold. In such a case M_s, M_t are critical submanifolds and X is the moduli space of gradient lines of f which join M_s to M_t. In the case when f is a Morse function in addition then M_s, M_t are zero-dimensional manifolds (finite sets). We call the case when M_s, M_t etc. appearing in the smooth correspondence has positive dimension, the Bott–Morse case and the case when M_s, M_t etc. appearing in the smooth correspondence has dimension 0, the Morse case.

The Morse case is significantly simpler than the Bott–Morse case.

The goal of the chain level argument is to obtain certain algebraic structure. In the Bott–Morse case the algebraic structure is constructed on the direct sum of certain chain complexes associated to M_s, M_t, which calculate homology. To work out the construction of such a structure we need to make a choice of the chain model we use. There are various possible choices, singular homology, de Rham cohomology, Čech cohomology, Morse homology, simplicial homology, etc. The details of the constructions depend on the choice of the chain model.

Among the two applications established in 1996, the construction of Gromov–Witten invariants is at homology level. So the choice of a chain model (of the homology group of the symplectic manifold) is not so important. The proof of the homology version of Arnold's conjecture is based on the construction of Floer homology of periodic Hamiltonian systems. The Floer's chain complex which defines the Floer homology depends on various choices involved. So this is a problem of chain level. The case where we need to define and study Floer homology of periodic Hamiltonian systems is the situation when the periodic Hamiltonian vector field has only non-degenerate periodic orbits. The set of periodic orbits are the spaces M_s, M_t in (1.15), in this case. So in the non-degenerate case they are zero-dimensional and hence we are in the Morse case. However, to calculate such a Floer homology the case when the Hamiltonian H is 0 is studied sometimes. In the case $H = 0$ the manifolds M_s, M_t in (1.15) are the symplectic manifold itself. Therefore one needs to study the Bott–Morse case.

The calculation of Floer homology of periodic Hamiltonian systems in [FOn2, LiuTi] does not use the case $H = 0$ but uses the case when H is a C^2 small time-independent Morse function. In other words, [FOn2, LiuTi] does not use the Bott–Morse case. [Ru2] (and our discussion in Part II of this book) uses the Bott–Morse case ($H = 0$ case) for the calculation.

Other issues that appear while studying the theory of virtual fundamental chains and its applications to the moduli space of pseudo-holomorphic curves, are the following:

(III) Where the virtual fundamental class lives.
(IV) Smoothness of the coordinate change map.

(III): Let us consider the situation when we are given a space with Kuranishi structure $(X, \widehat{\mathcal{U}})$ without boundary. If we are given a weakly submersive map $\widehat{f} : (X, \widehat{\mathcal{U}}) \to M$ to a manifold M then we can define a virtual fundamental class in $H_{\dim X}(M; \mathbb{R})$.[16] On the other hand, in the situation when no such map from $(X, \widehat{\mathcal{U}})$ is given, the current formulation of this book does not give a virtual fundamental class of X. (In the case when the (virtual) dimension of $(X, \widehat{\mathcal{U}})$ is zero, the virtual fundamental class is a number and so it is defined in the situation where no such map from $(X, \widehat{\mathcal{U}})$ is given.) Note that it is not reasonable to expect that the virtual fundamental class lies in, say the singular homology group $H(X; \mathbb{R})$. This is because in general the topology of the space X can be very wild (any compact subset of $\mathbb{R}^n$ can appear as a topological space X) and so its singular homology group is not a reasonable invariant.

Pardon [Pa1] made a correct choice. His virtual fundamental class lies in the dual of Čech cohomology of X. The reason why Čech cohomology is the correct choice can be explained, using our formulation of virtual fundamental class for example, as follows. We consider a good coordinate system $(X, \widehat{\mathcal{U}})$ and its support system $\mathcal{K} = \{\mathcal{K}_{\mathfrak{p}} \mid \mathfrak{p} \in \mathfrak{P}\}$ (see Definition 5.23). We can glue various $\mathcal{K}_{\mathfrak{p}}$ ($\mathfrak{p} \in \mathfrak{P}$) by coordinate changes to obtain a metric space $|\mathcal{K}|$ that contains X. It is easy to see that we can take $\mathcal{K}$ such that $|\mathcal{K}|$ is a CW-complex. Using Theorem 8.15 and Corollary 6.20, we can define the virtual fundamental class $[X, \widehat{\mathcal{U}}]$ as an element of the homology group of an arbitrary small neighborhood $B_\epsilon(X; |\mathcal{K}|)$ of X in $|\mathcal{K}|$. So

[16] See Theorem 8.15.

$$[X, \widehat{\mathcal{U}}] \in \operatorname*{proj\,lim}_{\epsilon \to 0} H_{\dim X}(B_\epsilon(X; |\mathcal{K}|); \mathbb{R}).$$

As is well-known, Čech cohomology (but not singular cohomology) commutes with projective limit.

In several works such as [LiTi2, Ru2, Sie, HWZ3] an 'ambient space' which contains X and behaves better in various homology theories is given. In those works the virtual fundamental class is a homology class of such an ambient space. Note that the authors of this book did not use an ambient space to study the virtual fundamental chain until the year 2012.

In most of the situations where virtual fundamental class is used, defining it as an element of a certain homology group of X or its 'ambient space' is not necessary. In [IP], Ionel and Parker use a certain 'localization' of the virtual fundamental class in the moduli space. For this purpose it is useful to regard it as a homology class of X.

(IV): This point actually is not directly related to the abstract theory of virtual fundamental chain. It is related to its application to the study of the moduli space of pseudo-holomorphic curves. In our definition of Kuranishi structure we require the orbifolds appearing in the definition to be of C^∞ class and also the embedding of orbifolds, which consists of a part of coordinate changes, to be of C^∞ class. For applications we need to verify that the Kuranishi structures we will construct on such moduli spaces satisfy this smoothness requirement. At an interior point of the moduli space (that is, a point of the moduli space representing a pseudo-holomorphic map from a Riemann surface without nodal points), such a smoothness result is a consequence of the standard elliptic regularity. However, at a point of the moduli space represented by a pseudo-holomorphic map from a nodal curve, there was no literature describing the details of the proof of the smoothness of the gluing map with respect to the so-called gluing parameter at the stage, for example, in the year on 2000.

Let us elaborate on this smoothness issue below. We consider a nodal curve with two irreducible components and four marked points as drawn on the left hand side of Fig. 1.10. The maps u_1, u_2 in the figure are pseudo-holomorphic maps to M from the two irreducible components respectively, which together give u.

The map u_i represents an element of the moduli space $\mathcal{M}_{0,3}(\beta_i)$ of pseudo-holomorphic spheres, where 0 means genus 0, 3 means three marked points and β_i is the homology class of u_i. The stable map (see [KM]) as drawn on the left hand side of Fig. 1.10 becomes an element of the fiber product.

$$\mathcal{M}_{0,3}(\beta_1) \times_M \mathcal{M}_{0,3}(\beta_2). \tag{1.16}$$

Here the fiber product is taken via the evaluation maps $[u_i] \mapsto \mathrm{ev}(u_i) = u_i(z_0)$ where z_0 is the nodal point. The space (1.16) is contained in the moduli space $\mathcal{M}_{0,4}(\beta_1 + \beta_2)$ of genus 0 stable maps of homology class $\beta_1 + \beta_2$ with four marked

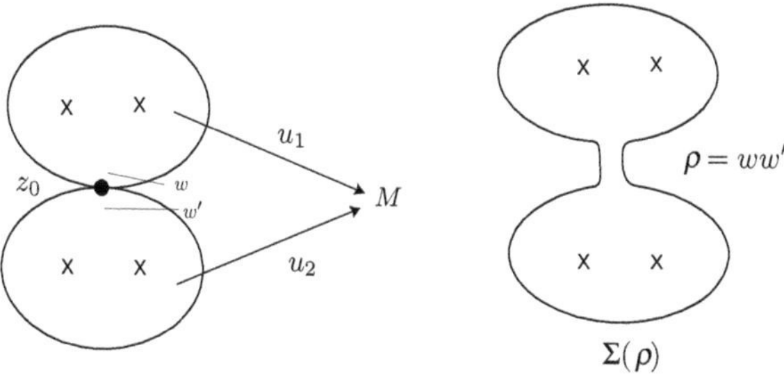

Fig. 1.10 Stable map from nodal sphere

points. We take a coordinate chart V_i of $\mathcal{M}_{0,3}(\beta_i)$ in a neighborhood of u_i. Then we can construct a coordinate chart of u in $\mathcal{M}_{0,4}(\beta_1+\beta_2)$ which is $(V_1 \times_M V_2) \times D^2(\epsilon)$ where $D^2(\epsilon)$ is the ϵ ball centered at 0 in $\mathbb{C}$. In other words, we have an embedding $\varphi : (V_1 \times_M V_2) \times D^2(\epsilon) \to \mathcal{M}_{0,4}(\beta_1 + \beta_2)$. The construction of φ is based on the gluing analysis but is not unique. The smoothness of the coordinate change issue is described as follows: Suppose we take two different choices φ_1, φ_2. The issue is whether the coordinate change map $\varphi_{12} = \varphi_1^{-1} \circ \varphi_2 : (V_1 \times_M V_2) \times D^2(\epsilon) \to (V_1' \times_M V_2') \times D^2(\epsilon')$ is smooth or not. The nontrivial part of the proof of the smoothness is that of φ_{12} with respect to the $D^2(\epsilon)$ coordinate of the domain.

Nontriviality of the smoothness proof lies in the following fact. Note that the source curve of an element $((u_1', u_2'), \rho)$ is obtained by gluing the two irreducible components S^2 in Fig. 1.10 using the parameter ρ, which we denote by $\Sigma(\rho)$. It comes with four marked points $z_{1,1}, z_{1,2}, z_{2,1}, z_{2,2}$. The map corresponding to $\varphi_1((u_1', u_2'), \rho)$ is a map $u' : \Sigma(\rho) \to M$. The source curve $\Sigma(\rho)$ depends on ρ. So to study smoothness of u' with respect to ρ, we need to identify $\Sigma(\rho)$ and $\Sigma(\rho')$. There is no canonical choice in doing so. In particular, when we take a natural family of Riemannian metrics $\Sigma(\rho)$ to define various Sobolev norms, it seems impossible to find a family of diffeomorphisms $\phi_{\rho,\rho'} : \Sigma(\rho) \to \Sigma(\rho')$ such that the operator norm of the induced map

$$(\phi_{\rho,\rho'})^* : L^p_k(\Sigma(\rho'), M) \to L^p_k(\Sigma(\rho), M)$$

is bounded and is bounded away from 0.[17]

Nevertheless one can construct a Kuranishi structure with smooth Kuranishi charts and coordinate changes. (See Sect. 1.4.6 for a relevant discussion.)

[17] When we study the interior of the moduli space $\mathcal{M}_{0,\ell}(\beta)$, we can find an appropriate local trivialization of the universal family of source curves (in the sense of fiber bundle of smooth orbifolds). So there is no such issue.

1.4.3 The Works of the Authors of This Book

During the years 1996–2006 the authors of this book had worked on the project of extending and enhancing the definition of Lagrangian Floer homology so that it is applicable to the cases as generally as possible. The books [FOOO3, FOOO4] are the outcome of that research. Lagrangian Floer theory requires study at the chain level and the manifolds M_s, M_t in (1.15) are the Lagrangian submanifold L itself or the direct product of its copies. Therefore study of the Bott–Morse case is inevitable. Our choice of the chain model in [FOOO3, FOOO4] is the singular homology. Singular homology again is used in [FOOO14] to work out the case of $\mathbb{Z}$ or $\mathbb{Z}_2$ coefficients (under certain positivity assumption of the ambient symplectic manifold). The method of using the Morse homology as a chain model is described in [FOOO5]. Before the virtual fundamental chain and cycle technique was invented, the first and second named authors studied the Morse homology model and its relation to Floer theories in [Fuk2, FOh, Oh] etc. When we start working on its applications to toric manifolds [FOOO7, FOOO8, FOOO10'] and also during the study of cyclic symmetry of Lagrangian Floer homology [Fuk4], the present authors realized that the usage of the de Rham model has various advantages. One important advantage is that while we use the de Rham model it is easiest to keep various symmetries. This is especially important for the study of Lagrangian submanifolds in toric manifolds. The theory of CF-perturbation, whose thorough details are described in this book, grew out from the study of such applications. One advantage of using our version of Kuranishi structure over others lies in the fact that it can accommodate practically *any* chain model. We believe that this flexibility will be important for the future development and application of the theory.

Recently (after 2012) we also wrote several documents which provide more systematic discussions and explain the details of various parts of the foundation of the virtual fundamental chain. In this book we start with the situation where (systems of) Kuranishi structures are already given. Three papers [FOOO18, FOOO21, FOOO22] give details of the construction of a system of Kuranishi structures on the moduli space of pseudo-holomorphic disks whose boundary lies in a Lagrangian submanifolds, which we assume in Chaps. 21 and 22 of this book. [FOOO18] is its analytic part and proves an exponential decay estimate for the gluing map and shows the way to resolve the smoothness of the coordinate change map issue (Item (IV) explained in Sect 1.4.2.) [FOOO21, FOOO22] explains the way to construct (a system of) Kuranishi structures using [FOOO18]. In those three papers we focus on the case of the moduli spaces of pseudo-holomorphic disks. However, the argument there can be easily adapted to other cases of the moduli spaces of pseudo-holomorphic curves.

1.4.4 *The Development from the Work by Li–Tian [LiTi2] and Liu-Tian [LiuTi]*

In Sect. 1.4.4 we mention several papers which might be regarded as successors of [LiTi2, LiuTi].

- Lu and Tian [LuT] describe a construction of the virtual fundamental cycle at homology level using an infinite-dimensional approach. They study stratified spaces $\mathfrak{X}$ where each stratum is an infinite-dimensional Banach orbifold. They also consider an infinite-dimensional 'bundle' $\mathfrak{E} \to \mathfrak{X}$ such that its restriction to each stratum of $\mathfrak{X}$ is a smooth infinite-dimensional orbibundle of infinite rank. Such a situation appears in Gromov–Witten theory, where $\mathfrak{X}$ is the space of the pairs (Σ, u) such that Σ is a nodal curve and $u : \Sigma \to M$ is a map which is not necessarily pseudo-holomorphic. A stratum of $\mathfrak{X}$ corresponds to the set of the pairs (Σ, u) with given combinatorial type of Σ and homology classes of the restriction of u to the irreducible components of Σ.
  They require certain conditions to such pairs $(\mathfrak{X}, \mathfrak{E})$, which is the existence of a certain class $\mathcal{A}$ of strata-wise smooth sections of $\mathfrak{E}$ with appropriate properties. One of the properties for $\mathcal{A}$ they assume (richness, see [LuT, Definition 3.8]) is that each element of a fiber of $\mathfrak{E}$ extends to a neighborhood as a section of the class $\mathcal{A}$ up to some small error term. They also assume the existence of a certain strata-wise Fredholm section S such that at each point one can find a finite number of sections s_i $(i = 1, \dots N)$ of class $\mathcal{A}$ on the local uniformization V of its neighborhood U such that the zero set of the section on $V \times \mathbb{R}^N$ defined by $(x, \xi_1, \dots, \xi_N) \mapsto S(x) + \sum_i \xi_i s_i(x)$ becomes a smooth manifold. (See [LuT, Definition 3.10].)[18] They gave a construction of the virtual fundamental cycle under those and some other conditions.
- Chen and Tian [CT] study the virtual fundamental cycles via a finite-dimensional approach. They introduce a geometric object, which they call a virtual orbifold. It is similar to a good coordinate system. Some of the differences are summarized as follows:

  (CT.1) The coordinate change of virtual orbifold is a submersion while the coordinate change in this book is an embedding. (More precisely their coordinate change is a projection of a vector bundle.) In other words, the direction of the arrow is opposite in their case and in ours.
  (CT.2) They consider the partially ordered set 2^N consisting of the subsets of a given finite set N, as the index set of the charts. The index set of the Kuranishi charts of a good coordinate system in this book is a partially ordered set, which in general is not necessarily of the form 2^N.

  In [CT, Remark 2.9] they mention a groupoid formulation of orbifolds. The definition of a virtual orbifold does not contain a notion corresponding to the

[18]It seems that under this condition one can construct a Kuranishi model of $S^{-1}(0)$ at each point.

obstruction bundle and the Kuranishi map. The integral of differential forms on virtual orbifolds is given in [CT] by using (a system of) Thom classes and Stokes' theorem is proved.
In the second half of [CT], the relation to infinite-dimensional Fredholm systems is described. There, an obstruction bundle and a similar object as a Kuranishi map appear. The obstruction bundle of [CT] is a direct sum $\bigoplus_{i \in I} \mathcal{O}_i$, where $I \subset N$ is the index of the chart. (The index set is 2^N as we mentioned in (CT.2).) They study a system of Thom classes of those obstruction bundles to define the virtual fundamental class.

Chen and Tian in [CT] study a group action on their virtual orbifold and give the localization formula. The localization formula of Gromov–Witten theory for projective algebraic manifolds with group action is established by [GP] and used much in the study of Gromov–Witten invariants and Mirror Symmetry.

- Chen, Li and Wang [CLW] study virtual fundamental cycles by an infinite-dimensional approach. They study a proper étale groupoid whose set of objects and whose set of morphisms are Banach manifolds. They define a Banach bundle on such a space in a similar way using groupoid language.
They restrict themselves to the case of the moduli space of pseudo-holomorphic spheres and also assume that the bubbles do not occur in the compactification. In such a situation they study the issue of non-differentiability of the action of the automorphism group of the source curve ($= PSL(2; \mathbb{C})$) on the Sobolev space of the maps $u : S^2 \to M$. They resolve the issue by finding a sufficiently large space of smooth functions on this Sobolev space so that such a function space is preserved by this $PSL(2; \mathbb{C})$ action.
- A recent paper [TX] by Tian and Xu contains certain new insights on virtual fundamental cycles. They use a notion similar to a good coordinate system in this book and use multisections to perturb the Kuranishi maps. However, they do not use the smoothness or differentiability of coordinate changes. They use the micro normal bundle of a topological submanifold and the notion of transversality of a C^0 map to a micro bundle and an existence theorem of transversal perturbations in the C^0 category to obtain a virtual fundamental cycle.
In [TX], this technique is applied to the moduli space of solutions of the Gauged Witten equation and a construction of the invariants of the gauged linear sigma model is given.

1.4.5 The Work by Joyce

Starting with his joint work with Akaho [AJ], Joyce has been writing a series of papers including [Jo1, Jo2, Jo3, Jo4, Jo5] on the virtual fundamental chain and its applications. Joyce's method is based on finite-dimensional reduction. Joyce developed his theory based on a certain version of derived geometry, which he calls the theory of d-manifolds. It seems to the authors that Joyce's theory is closer to the theory of scheme and stack in algebraic geometry, while the theory of Kuranishi structure in this book is closer to manifold theory in differential geometry and topology. Joyce also proposed in [Jo2], [Jo4] an alternative version of the definition of 'Kuranishi structure'.

An important advantage of his theory is that Joyce defined morphisms between d-manifolds and also between spaces equipped with his version of Kuranishi structure.

Joyce requires the cocycle condition of coordinate changes between Kuranishi charts only on the zero set of the Kuranishi map, but he requires compatibility including the derivative up to the first order. In that way Joyce succeeds in inverting the arrows of the coordinate changes of Kuranishi charts so that the trouble coming from non-invertibility of coordinate changes disappears.

However, we use our version of a Kuranishi structure and a good coordinate system in this book because of the flexibility of the choice of the chain model as mentioned above. Note that the space X which has a Kuranishi structure may have a pathological topology in general. So singular homology does not behave nicely for X. With the de Rham model, the situation gets even worse: it appears impossible to define the notion of differential forms on X. Therefore, we need to take a union of charts of X that has a positive size to work with de Rham or singular homology. In other words, we need a system of spaces $U_{\mathfrak{p}}$ which are manifolds or orbifolds and which cover X. We remark that to define the notion of differential forms of a good coordinate system or of a Kuranishi structure, we need the cocycle condition of coordinate changes in our sense, that is stronger than what Joyce assumed in [Jo4]. Namely, we also need to assume a cocycle condition at some points outside X (i.e., outside the zero set of the Kuranishi map s). For this reason, it seems that one can use neither de Rham cohomology nor singular homology directly if one uses the definition of a Kuranishi structure in the sense of [Jo4].

As far as we understand, Joyce's plan is to use a version of Kuranishi homology [Jo1] as the homology theory which makes sense in his version of Kuranishi structure.[19] We believe that this approach works. One potential issue is the Poincaré duality. In [Jo1] Joyce provides a chain level intersection pairing. The intersection 'number' in his chain level intersection pairing is not a number but is an element of a certain huge complex (whose cohomology group is $\mathbb{Q}$). Although this construction provides the same amount of information at the homology level, some nontrivial amount of work is required to use it on the chain level.

While we work on the chain level argument, sometimes we need to convert some input variables (of the algebraic operation we will obtain) to output variables. We use the (chain level) Poincaré duality for this purpose. Note that the pairing

$$(u, v) \mapsto \int_M u \wedge v \in \mathbb{R}$$

[19] Applications to the moduli space of pseudo-holomorphic curves based on such a chain model are announced but are not available yet.

works at the chain level in the de Rham theory. This identifies an element of the de Rham complex with an element in its dual. Although the dual space of the space of differential forms is the set of distributions and is different from the set of differential forms, the difference between the spaces of differential forms and of distributions is relatively small so that we can still use the chain level Poincaré duality to convert certain input variables to output variables. It seems that one needs to work more on the side of homological algebra to realize this Poincaré duality in the situation of Kuranishi homology.

Other than the points described above, Joyce's papers contain several results which are important in the study of virtual fundamental chains. For example, he pointed out the difficulty to define direct and fiber products of spaces with a good coordinate system and showed a way of resolving the difficulties. The discussion on cornered manifolds in [Jo3] is also useful for the study of virtual fundamental chains. We also like to mention that Joyce's papers utilize carefully the 2-categorical nature of the category of orbifolds to study his version of Kuranishi structure or d-manifold.

1.4.6 Polyfolds

In a series of papers including [HWZ1, HWZ2, HWZ3], H. Hofer, K. Wysocki and E. Zehnder have been developing the theory of polyfolds. Polyfold theory is a version of virtual fundamental cycles and chains in the sense we defined in Sect. 1.4.1. Polyfold theory is one of its infinite-dimensional versions. Its application [HWZ2] to the study of pseudo-holomorphic curve available at the stage of 2018 is the definition of Gromov–Witten invariants, and so is one at the homology level. The de Rham cohomology is used in [HWZ2]. Application to the case when the chain level theory is needed, especially an application to the symplectic field theory, was announced but is not yet available at this stage. In [HWZ3], certain parts of the arguments needed for the chain level construction are described based on the de Rham model. By adapting the method of CF-perturbations to their situation, it is plausible that one can study the Bott–Morse case in Polyfold theory using de Rham complex as a chain model.

The important contribution of Polyfold theory is, according to the opinion of the authors of this book, presenting the gluing analysis in an abstract language of functional analysis that they develop.

Gluing analysis is needed to study a neighborhood of the nodal stable map appearing in the compactification of the moduli space of pseudo-holomorphic curves. Starting with Taubes' work in gauge theory, a gluing analysis of this kind has been the key ingredient in the mathematical study of gauge theory and in the study of pseudo-holomorphic curves. Various approaches have been developed to work it out in various situations. Polyfold theory provides a framework to work out the gluing analysis in many situations including (most likely all the) cases of the moduli spaces of pseudo-holomorphic curves, using the abstract language enhancing usual Fredholm theory. In particular, it resolves the smoothness issue of the coordinate changes mentioned in Sect. 1.4.2 (IV). ([FOOO18] resolves this issue working in the concrete geometric situation of pseudo-holomorphic curves.)

D. Yang [Ya1, Ya2] studies the relationship between polyfolds and Kuranishi structures, so that he can associate a space with Kuranishi structure to a polyfold systematically.

For the researchers who have a stronger background in functional analysis than in algebra or geometry, Polyfold theory might be easier to read, understand and apply.

1.4.7 The Work by Pardon

In [Pa1] Pardon gave an alternative construction of Gromov–Witten invariants and an alternative proof of the homology version of Arnold's conjecture for general symplectic manifolds. He also gave a partial construction of contact homology in [Pa2] by a similar strategy. He constructed an expected chain complex and proved its well-definedness. Bao and Honda [BH] also defined a similar chain complex and proved its well-definedness, in a different way. The constructions of those authors do not give an L_∞ structure which is conjectured to exist in [EGH] and studied in [Is].

Pardon's method is a version of virtual fundamental cycle and chain technique in the sense we defined in Sect. 1.4.1. It is based on a finite-dimensional reduction. Pardon's works on Floer homology of periodic Hamiltonian systems and on contact homology contain chain level arguments. Pardon does not study the Bott–Morse case and uses Čech cohomology as the chain model.

As the title of his paper [Pa1] shows, Pardon puts more emphasis on algebra and algebraic topology than studying the space itself. The geometric object (in the sense (1) in Sect. 1.4.1) Pardon introduced is called an implicit atlas. An implicit atlas consists of objects similar to Kuranishi charts and to coordinate changes. One important difference from the story of Kuranishi structure is that Pardon does *not* perturb the Kuranishi map. Instead he directly constructs a chain complex, which is similar to the Čech complex. Pardon uses a coordinate change and a homomorphism similar to the Gysin map induced by embeddings to define the boundary operator of this complex. The virtual fundamental class, in the case of Gromov–Witten invariants, is a homology class of this complex. His construction of Floer homology of periodic Hamiltonian systems and of contact homology complex uses a similar complex.

One feature of Pardon's construction is that the smoothness of coordinate changes is never used. One reason why Pardon does not need the smoothness of coordinate changes is that he does not study the Bott–Morse case but studies the moduli spaces of virtual dimension one or less only, to define Floer homology. It is an important contribution to the theory of virtual fundamental chains that he worked out the whole construction of Floer homology of periodic Hamiltonian systems etc. in detail and in a self-contained way without using the smoothness of coordinate changes. Pardon's work made it transparent that the smoothness of coordinate changes is not needed in the construction of Floer homology of periodic Hamiltonian systems.

Various techniques of a topological nature are known to reduce the Bott–Morse case to the Morse case. For example, one can use an auxiliary Morse function on each M_s and M_t (appearing in the correspondence diagram (1.15)) and study the mixed moduli space of gradient lines and pseudo-holomorphic curves similar to those in [Fuk1, Fuk2], for example. There are similar tricks when we use simplicial homology or (in the case where the ground ring is a field) when we take a homology basis and directly work on homology groups. Using tricks of this kind it is plausible that Pardon's method may be generalized to include the Bott–Morse case.

For the researchers who have a stronger background in algebra or algebraic topology than in geometry or analysis, Pardon's paper might be easier to read, understand and apply.

It is the opinion of the authors of this book that all of these approaches to virtual fundamental chains and cycles have their own advantage. We hope that all of them will be developed so that researchers can choose the one they use according to their purpose, mathematical background or taste.

1.4.8 Other Works

In the case when the symplectic form ω satisfies $[\omega] \in H^2(M;\mathbb{Q})$ there is a method to resolve the negative multiple-cover problem explained in Sect. 1.1, *without* using the virtual fundamental chain technique. This method uses Donaldson's work [Do3] on the existence of a symplectic submanifold and was invented by Cieliebak and Mohnke [CM].

In Sect. 1.1 we explained that the perturbation of J the almost complex structure is not enough to perturb the moduli space of pseudo-holomorphic curves to be transversal. One can try to use a source-dependent J to resolve this issue. Namely we consider $u : \Sigma \to M$ such that

$$(\overline{\partial}_{J_z} u)(z) = 0. \tag{1.17}$$

To apply this argument, however, there is an issue to make sense out of the map $z \mapsto J_z$. In the case when the source Riemann surface Σ (together with the marked points attached) is unstable there exists a nontrivial automorphism $v : \Sigma \to \Sigma$. If we require a map $z \mapsto J_z$ to satisfy $J_{v(z)} = J_z$ then there is not enough freedom of J_z so that the equation (1.17) becomes Fredholm regular. Therefore we need to find a method to obtain a map $z \mapsto J_z$ which is well-defined when z is in such an unstable component. The idea by Cieliebak and Mohnke is to use a Donaldson divisor, which is a codimension 2 submanifold D of M of degree high enough. In particular, if $u : S^2 \to M$ is pseudo-holomorphic $u^{-1}(D)$ necessarily contains enough many points so that S^2 together with $u^{-1}(D)$ taken as the marked points becomes stable. In this way, Cieliebak and Mohnke defined genus 0 Gromov–Witten invariants without assuming semi-positivity.

- Comparison with [FOOO]
  - Their basic charts are essentially like ours, but w. elements $(\vec{w}, f)$ s.t. $s(\vec{w}, f) =$ "$\overline{\partial}_J f$" $\in E$. So $f, \vec{w}, \overline{\partial}_J f$ must determine a unique elt in E; while for us $\overline{\partial}_J f$ is possibly nonuniq. sum of elts in E.
  - Their sum charts do not support an action of the product group Γ_I, but just one of the stabilizer of the center point. So they build "small"sum charts with footprint generally smaller than F_I.

Fig. 1.11 McDuff's slide

Note that during the construction of Kuranishi structures we[20] used a local codimension 2 submanifold of M to make a choice of extra marked points and stabilize the domain curve. The difference between this local method and the one by Cieliebak and Mohnke lies in the fact that the codimension 2 submanifold we take is chosen locally in the moduli space to define the Kuranishi chart there, whereas Cieliebak and Mohnke take a Donaldson divisor and use it everywhere (globally) on the moduli space.

Charest and Woodward [CW1, CW2] combined this idea with that of using 'source-dependent Morse functions' to construct a filtered A_∞ structure of Lagrangian submanifolds without assuming its monotonicity. They use the Morse homology as the chain model.

McDuff and Wehrheim wrote several papers on virtual fundamental cycles at homology level via finite-dimensional reduction. We quote McDuff's view on the difference between their construction and ours from the slide (# 19) of her presentation at SCGP on March 18, 2014 (Fig. 1.11).[21]

Here is a bit more specific explanation concerning this quote. First of all their definition of a Kuranishi atlas is closer to that of a good coordinate system in our construction. For the construction of Kuranishi charts, both we and they relax the equation $\overline{\partial} f = 0$ to $\overline{\partial} f \equiv 0 \mod E(f)$. In our construction, $E(f)$ is a linear subspace of $\Gamma(f^*TX \otimes \Lambda^{01}_\Sigma)$ while in their (MW) construction, McDuff and Wehrheim consider an abstract vector space E together with a family of linear maps $I_f : E \to C^\infty(f^*TX \otimes \Lambda^{01}_\Sigma)$ (which is not necessarily injective) and then consider the equation $\overline{\partial} f = I_f(\vec{w})$ with $\vec{w} \in E$.[22] In the second sentence, the difference between the choices of the way to present locally the orbifold which is a part of a Kuranishi chart is explained.[23]

We appreciate the questioning of McDuff and Wehrheim on our construction of Kuranishi structure and virtual fundamental cycles, which ignited lively discussions on the subject in the Google group 'Kuranishi' and in the SCGP special year 2014. This book grew up in such an environment.

---

[20] [FOn2, appendix],[FOOO21] etc.

[21] http://media.scgp.stonybrook.edu/presentations/2013/20140318_McDuff.pdf

[22] A similar method was used by Ruan [Ru2, Lemma 2.5 (page 171)].

[23] Liu and Tian [LiuTi, page 8, line 3] took the same choice as McDuff and Wehrheim.

Chapter 2
Notations and Conventions

Conventions on the Way to Use Several Notations

$\widehat{\ }$ and $\overset{\triangle}{\ }$ We use 'hat' such as $\widehat{\mathcal{U}}$, $\widehat{f}$, $\widehat{\mathfrak{S}}$, $\widehat{h}$ for an object defined on a Kuranishi structure $\widehat{\mathcal{U}}$. We use 'triangle' such as $\overset{\triangle}{\mathcal{U}}$, $\overset{\triangle}{f}$, $\overset{\triangle}{\mathfrak{S}}$, $\overset{\triangle}{h}$ for an object defined on a good coordinate system $\overset{\triangle}{\mathcal{U}}$.

s and $\mathfrak{s}$ We use s (italic s, such as s_p or $s_{\mathfrak{p}}$) for the Kuranishi maps of Kuranishi charts. We use $\mathfrak{s}$ (German Fraktur $\mathfrak{s}$, such as $\mathfrak{s}_p$ or $\mathfrak{s}_{\mathfrak{p}}$) for its perturbations.

p and $\mathfrak{p}$ For a Kuranishi structure $\widehat{\mathcal{U}}$ on $Z \subseteq X$ we write $\mathcal{U}_p$ for its Kuranishi chart, where $p \in Z$. (We use an italic letter p.) For a good coordinate system $\overset{\triangle}{\mathcal{U}}$ on $Z \subseteq X$ we write $\mathcal{U}_{\mathfrak{p}}$ for its Kuranishi chart, where $\mathfrak{p} \in \mathfrak{P}$. (We use a Fraktur character $\mathfrak{p}$.) Here $\mathfrak{P}$ is a partially ordered set.

■ Indicates the end of a Situation. See Situation 6.4, for example.

M and X Usually we denote by M a smooth manifold and by X a K-space, or an orbifold unless otherwise mentioned.

Conventions on Orientation and Sign

pushout The pushout (integration along the fiber) is defined by (7.6). Namely the test differential form ρ is put on the left.

$\mathfrak{m}_k$ for DGA In this book the 0 energy part of $\mathfrak{m}_1$ is equal to the de Rham differential d (*without correcting sign*), and the 0 energy part of $\mathfrak{m}_2$ is equal to $h_1, h_2 \mapsto (-1)^{\deg h_1} h_1 \wedge h_2$, where $\wedge$ is the standard wedge product of differential forms. These two conventions are different from [FOOO3] etc. See Definition 21.29 (4), (5).

K. Fukaya et al., *Kuranishi Structures and Virtual Fundamental Chains*, Springer Monographs in Mathematics, https://doi.org/10.1007/978-981-15-5562-6_2

boundary — An open subset of $\mathbb{R}^n$ which appears in the definition of a coordinate chart of an orbifold with corners is written as $U = \overline{U} \times [0,1)^k$ where $\overline{U} \subset \mathbb{R}^{n-k}$ is open. Here $[0,1)^k$ are the *last* k coordinates.
On the other hand, as the convention of the induced orientation of the boundary ($k = 1$), we regard that $U = [0,1) \times \overline{U}$ and the orientation of $\overline{U}$ (a chart of the boundary) is given by this identification. In other words, the outer normal coordinate to the boundary is the *first* coordinate when we define the orientation of the boundary.

composition — When we compose correspondences $\mathrm{Corr}^{\epsilon}_{\tilde{\mathfrak{X}}_{21}}$ and $\mathrm{Corr}^{\epsilon}_{\tilde{\mathfrak{X}}_{32}}$, we use the fiber product $\widehat{\mathcal{U}_{21}} {}_{\widehat{f_{2,21}}} \times_{\widehat{f_{2,32}}} \widehat{\mathcal{U}_{32}}$. When $\widehat{\mathcal{U}_{21}}$ and $\widehat{\mathcal{U}_{32}}$ are oriented, we equip $\widehat{\mathcal{U}_{21}} {}_{\widehat{f_{2,21}}} \times_{\widehat{f_{2,32}}} \widehat{\mathcal{U}_{32}}$ with the fiber product orientation of $\widehat{\mathcal{U}_{32}} {}_{\widehat{f_{2,32}}} \times_{\widehat{f_{2,21}}} \widehat{\mathcal{U}_{21}}$. More generally, orientation bundles are related as follows:

$$O_{\widehat{\mathcal{U}_{31}}} = O_{\widehat{\mathcal{U}_{32}}} \otimes O_{M_2} \otimes O_{\widehat{\mathcal{U}_{21}}}.$$

Under these conventions (10.10) holds. Namely the equality holds without sign correction.

shifted degree — The shifted degree is denoted by $\deg'$. Namely $\deg' a = \deg a + 1$.

List of Notations in Part I

- Int A, $\mathring{A}$: Interior of a subset A of a topological space.
- $\overline{A}$: Closure of a subset A of a topological space. We denote the closure also by $\mathrm{Clos} A$ sometimes.
- $\mathrm{Perm}(k)$: The permutation group of order $k!$.
- $\mathrm{Supp}(h)$, $\mathrm{Supp}(f)$: The support of a differential form h, a function f, etc.
- $\varphi^{\star}\mathscr{F}$: Pullback of a sheaf $\mathscr{F}$ by a map φ.
- X: A paracompact metrizable space (Part I).
- Z: A compact subspace of X (Part I).
- $\mathcal{U} = (U, \mathcal{E}, \psi, s)$: A Kuranishi chart. Definition 3.1.
- $\mathcal{U}|_{U_0} = (U_0, \mathcal{E}|_{U_0}, \psi|_{U_0 \cap s^{-1}(0)}, s|_{U_0})$: open subchart of $\mathcal{U} = (U, \mathcal{E}, \psi, s)$. Definition 3.1.
- $\Phi = (\varphi, \widehat{\varphi})$: Embedding of Kuranishi charts. Definition 3.2.
- $o_p, o_p(q)$: Points in a Kuranishi neighborhood U_p of p. Definition 3.5.
- $\Phi_{21} = (U_{21}, \varphi_{21}, \widehat{\varphi}_{21})$: Coordinate change of Kuranishi charts from $\mathcal{U}_1$ to $\mathcal{U}_2$. Definition 3.6.
- $\widehat{\mathcal{U}} = (\{\mathcal{U}_p\}, \{\Phi_{pq}\})$: Kuranishi structure. Definition 3.9.
- $(X, \widehat{\mathcal{U}})$, $(X, Z; \widehat{\mathcal{U}})$: K-space, relative K-space. Definition 3.11.
- $\widehat{\widehat{\mathcal{U}}} = ((\mathfrak{P}, \le), \{\mathcal{U}_{\mathfrak{p}}\}, \{\Phi_{\mathfrak{pq}}\})$: Good coordinate system. Definition 3.15.

- $|\widehat{\mathcal U}|$: Definition 3.16.
- $\widehat{\Phi} : \widehat{\mathcal U} \to \widehat{\mathcal U'}$: KK-embedding. An embedding of Kuranishi structures. Definition 3.19.
- $\widehat{\Phi} : \widehat{\mathcal U} \to \widehat{\mathcal U'}$: GG-embedding. An embedding of good coordinate systems. Definition 3.24.
- $\widehat{\Phi} : \widehat{\mathcal U} \to \widehat{\mathcal U}$: KG-embedding, An embedding of a Kuranishi structure to a good coordinate system. Definition 3.30.
- $\widehat{f} : (X, Z; \widehat{\mathcal U}) \to Y$ and $\widehat{f} : (X, Z; \widehat{\mathcal U}) \to Y$: Strongly continuous map. Definitions 3.40 and 3.43.
- $(X, Z; \widehat{\mathcal U}) \times_N M$, $(X_1, Z_1; \widehat{\mathcal U_1}) \times_M (X_2, Z_2; \widehat{\mathcal U_2})$: Fiber product of Kuranishi structures. Definition 4.9.
- $S_k(X, Z; \widehat{\mathcal U})$, $S_k(X, Z; \widehat{\mathcal U})$: Corner structure stratification. Definition 4.16.
- $\mathcal S_{\mathfrak d}(X, Z; \widehat{\mathcal U})$, $\mathcal S_{\mathfrak d}(X, Z; \widehat{\mathcal U})$: Dimension stratification. Definition 5.1
- $\widehat{\mathcal U} < \widehat{\mathcal U^+}$: $\widehat{\mathcal U^+}$ being a thickening of $\widehat{\mathcal U}$. Definition 5.3.
- $\widehat{\Phi} : \widehat{\mathcal U} \to \widehat{\mathcal U}$: GK-embedding. An embedding of a good coordinate system to a Kuranishi tructure. Definition 5.6.
- $\mathcal K = \{\mathcal K_{\mathfrak p} \mid \mathfrak p \in \mathfrak P\}$: A support sytem. Definition 5.23 (1).
- $(\mathcal K^1, \mathcal K^2)$ or $(\mathcal K^-, \mathcal K^+)$: A support pair. Definition 5.23 (2).
- $\mathcal K^1 < \mathcal K^2$: $(\mathcal K^1, \mathcal K^2)$ beging a support pair. Definition 5.23 (2).
- $|\mathcal K|$: Hetero-dimensional compactum. Definition 5.23 (3).
- $\mathcal S_{\mathfrak p}(X, Z; \widehat{\mathcal U}; \mathcal K)$: Definition 5.23 (4).
- $(\mathfrak s^n_{\mathfrak p})^{-1}(0)$: The zero set of multisection. Footnote 4 at the beginning of Sect. 6.2.
- $B_\delta(A)$: Metric open ball. (6.10).
- $\mathcal S^\epsilon_x = (W_x, \omega_x, \mathfrak s^\epsilon_x)$ for each $\epsilon > 0$: Definition 7.2.
- $\mathcal S_x = (W_x, \omega_x, \{\mathfrak s^\epsilon_x\})$: CF-perturbation (= continuous family perturbation) on one orbifold chart. Definition 7.4.
- $\mathfrak S = \{(\mathfrak V_{\mathfrak r}, \mathcal S_{\mathfrak r}) \mid \mathfrak r \in \mathfrak R\}$: Representative of a CF-perturbation on Kuranishi chart $\mathcal U$. Definition 7.16.

  Here $\mathfrak V_{\mathfrak r} = (V_{\mathfrak r}, E_{\mathfrak r}, \Gamma_{\mathfrak r}, \phi_{\mathfrak r}, \widehat{\phi_{\mathfrak r}})$ is an orbifold chart of $(U, \mathcal E)$ and $\mathcal S_{\mathfrak r} = (W_{\mathfrak r}, \omega_{\mathfrak r}, \{\mathfrak s^\epsilon_{\mathfrak r}\})$ is a CF-perturbation of $\mathcal U$ on $\mathfrak V_{\mathfrak r}$.
- $\mathfrak S^\epsilon = \{(\mathfrak V_{\mathfrak r}, \mathcal S^\epsilon_{\mathfrak r}) \mid \mathfrak r \in \mathfrak R\}$ for each $\epsilon > 0$. Definition 7.16.
- $\mathcal{CF}$: Sheaf of CF-perturbations. Proposition 7.22.
- $\mathcal{CF}_{\pitchfork 0}$, $\mathcal{CF}_{f\pitchfork}$, $\mathcal{CF}_{f\pitchfork g}$: Subsheaves of $\mathcal{CF}$. Definition 7.26.
- $\mathcal{CF}^{\mathcal U^1 \triangleright \mathcal U^2}$: Sheaf of CF-perturbations on $\mathcal U^2$ which is restrictable to $\mathcal U^1$. Definition-Lemma 7.44 (1).
- $\Phi^{\mathcal{CF}}$: Restriction map of CF-perturbations, regarded as a sheaf morphism. Definition-Lemma 7.44 (4).
- $\widehat{\mathfrak S} = \{\mathfrak S_{\mathfrak p} \mid \mathfrak p \in \mathfrak P\}$: CF-perturbation of a good coordinate system. Definition 7.49.
- $\Pi((\mathfrak S^\epsilon)^{-1}(0))$: Support set of a CF-perturbation $\mathfrak S^\epsilon$. Definition 7.73.
- $\widehat{f}!(\widehat{h}; \widehat{\mathfrak S^\epsilon})$: Pushout or integration along the fiber of $\widehat{h}$ by $(\widehat{f}, \widehat{\mathfrak S^\epsilon})$ on a good coordinate system. Definition 7.79

- $\mathrm{Corr}_{(\mathfrak{X},\widehat{\mathfrak{S}^{\epsilon}})}$: Smooth correspondence associated to a good coordinate system. Definition 7.86.
- $\widehat{\mathfrak{S}}$: CF-perturbation of Kuranishi structure. Definition 9.1.
- $\widehat{f}!(\widehat{h};\widehat{\mathfrak{S}^{\epsilon}})$: Pushout or integration along the fiber of $\widehat{h}$ by $(\widehat{f},\widehat{\mathfrak{S}^{\epsilon}})$ on Kuranishi structure. Definition 9.13.
- $\mathrm{Corr}_{(\mathfrak{X},\widehat{\mathfrak{S}^{\epsilon}})}$: Smooth correspondence of Kuranishi structure. Definition 9.25.
- $\mathcal{CF}_{\mathcal{K}}$: Sheaf of CF-perturbations on a hetero-dimensional compactum. Definition 12.19.
- $\mathcal{MV}_{\mathcal{K}}$: Sheaf of multivalued perturbations on a hetero-dimensional compactum. Definition 13.3.

List of Notations in Part II

- $\mathcal{M}(\alpha_-,\alpha_+)$: Space of connecting orbits. Condition 16.1.
- $\mathcal{C} = \left(\mathfrak{A}, \mathfrak{G}, \{R_\alpha\}_{\alpha\in\mathfrak{A}}, \{o_{R_\alpha}\}_{\alpha\in\mathfrak{A}}, E, \mu, \{\mathrm{PI}_{\beta,\alpha}\}_{\beta\in\mathfrak{G},\alpha\in\mathfrak{A}}\right)$: Critical submanifold data. Definition 16.6 (1).
- $\mathcal{F} = \left(\mathcal{C}, \{\mathcal{M}(\alpha_-,\alpha_+)\}_{\alpha_\pm\in\mathfrak{A}}, (\mathrm{ev}_-,\mathrm{ev}_+), \{\mathrm{OI}_{\alpha_-,\alpha_+}\}_{\alpha_\pm\in\mathfrak{A}}, \{\mathrm{PI}_{\beta;\alpha_-,\alpha_+}\}_{\beta\in\mathfrak{G},\alpha_\pm\in\mathfrak{A}}\right)$: A linear K-system. Definition 16.6 (2).
- Λ^R_{nov}, $\Lambda^R_{0,\mathrm{nov}}$, $\Lambda^R_{+,\mathrm{nov}}$: Universal Novikov ring, and its ideal. Definition 16.11. When $R=\mathbb{R}$, we drop R from these notations.
- Λ^R, Λ^R_0, Λ^R_+: Universal Novikov ring, and its ideal. (The version without e.) Definition 16.11. When $R=\mathbb{R}$, we drop R from these notations.
- $\mathcal{N}(\alpha_1,\alpha_2)$: Interpolation space. Condition 16.17.
- $\mathfrak{N}_{ii+1} : \mathcal{F}_i \to \mathcal{F}_{i+1}$: Morphism of linear K-systems.
- $\mathcal{N}_{i+1i}$: Interpolation space of the morphism $\mathfrak{N}_{ii+1}$. Lemma-Definition 16.35. See also Remark 18.33.
- $\mathcal{FF} = (\{E^i\},\{\mathcal{F}^i\},\{\mathfrak{N}^i\})$: An inductive system of partial linear K-systems. Definition 16.36.
- $V_x^{\boxplus\tau}$: Definition 17.7.
- $\mathcal{U}_x^{\boxplus\tau} = (U_x^{\boxplus\tau},\mathcal{E}_x^{\boxplus\tau},\psi_x^{\boxplus\tau},s_x^{\boxplus\tau})$: Outer collaring, or τ-collaring, of one Kuranishi chart $\mathcal{U}_x$ at $x\in\overset{\circ}{S}_k(U)$. Lemma-Definition 17.13. See also Lemma 17.23.
- $\mathcal{S}_x^{\boxplus\tau}$: Lemma-Definition 17.14.
- $\mathcal{U}^{\boxplus\tau}$: Outer collaring or τ-collaring. Lemma-Definition 17.24.
- $X^{\boxplus\tau}$: Outer collaring, or τ-collaring, of X. Definition 17.29.
- $(X,\widehat{\mathcal{U}})^{\boxplus\tau} = (X^{\boxplus\tau},\widehat{\mathcal{U}^{\boxplus\tau}})$: Outer collaring, or τ-collaring, of K-space $(X,\widehat{\mathcal{U}})$. Lemma-Definition 17.38.
- $(X,\widehat{\mathcal{U}})^{\boxminus\tau} = (X^{\boxminus\tau},\widehat{\mathcal{U}^{\boxminus\tau}})$: Inward τ-collaring of $(X,\mathcal{U})$. Definition 17.41.
- $\mathscr{KR}(X)$: Category of Kuranishi structures. Definition 17.41.
- $\mathscr{KR}^{\boxplus\tau}(X)$: Category of τ-collared Kuranishi structures. Definition 17.41.

- $\partial_{\mathfrak{C}} U$: Normalized $\mathfrak{C}$-partial boundary of U. When we denote by $\mathfrak{C}$ a decomposition of the normalized boundary $\partial U = \partial^0 U \cup \partial^1 U$ into two disjoint unions, we write $\partial_{\mathfrak{C}} U = \partial^0 U$. Situation 18.1.
- $S_k^{\mathfrak{C}}(U)$: Definition 18.2.
- $(X, \widehat{\mathcal{U}})^{\mathfrak{C}\boxplus\tau} = (X^{\mathfrak{C}\boxplus\tau}, \widehat{\mathcal{U}^{\mathfrak{C}\boxplus\tau}})$: τ-$\mathfrak{C}$-corner trivialization, or partial outer collaring, of $(X, \widehat{\mathcal{U}})$. Definition 18.9.
- $\widehat{\mathcal{U}^{\mathrm{sm}\mathfrak{C}\boxplus\tau}}$: Definition 18.29.
- $\mathcal{N}_{12}(\alpha_1, \alpha_2) \times^{\boxplus\tau}_{R^2_{\alpha_2}} \mathcal{N}_{23}(\alpha_2, \alpha_3)$: Partially collared fiber product. Definition 18.34.
- $\widehat{\mathcal{U}_1}^{\boxplus\tau_1} < \widehat{\mathcal{U}_2}^{\boxplus\tau_2}$ as collared Kuranishi structures: Proposition 19.1.
- $\mathcal{G}(k+1, \beta)$: The set of all decorated ribbon trees $(\mathcal{T}, \beta(\cdot))$ with $(k+1)$ exterior vertices and $\sum_{\mathrm{v} \in C_{0,\mathrm{int}}(\mathcal{T})} (\beta(\mathrm{v})) = \beta$. Definition 21.1.
- $G(\mathcal{AC})$ (resp. $G(\mathcal{AC}_P)$): The discrete submonoid associated to an (resp. a P-parametrized) A_∞ correspondence $\mathcal{AC}$ (resp. $\mathcal{AC}_P$). Definition 22.1.

List of Notations in Appendices

- (V, Γ, ϕ): Orbifold chart. Definitions 23.1, 23.6.
- $(V, E, \Gamma, \phi, \widehat{\phi})$: Orbifold chart of a vector bundle. Definitions 23.18, 23.23.
- $(X, \mathcal{E})$: Orbibundle. Definition 23.21.

Part I
Abstract Theory of Kuranishi Structures, Fiber Products and Perturbations

Hereafter in Part I, the orbifold with boundary is either one in the usual sense or is an admissible orbifold in the sense we define in Chap. 25. In the latter case all the notions are ones in the admissible category. For example, vector bundle means admissible vector bundle, section means admissible section and embedding means admissible embedding. Since the whole story works in a parallel way, we do not mention this point later.

Chapter 3
Kuranishi Structures and Good Coordinate Systems

In this chapter we define the notions of a Kuranishi structure and of a good coordinate system. We also study embedding between them, which describes a relation among those structures.

3.1 Kuranishi Structures

The notion of a Kuranishi structure in this book is the same as the one in [FOOO4, Section A1] and [FOOO16], except that we include the existence of a tangent bundle in the definition of a Kuranishi structure. The notion of a good coordinate system in this book is the same as the one in [FOOO16]. We introduce some more notations which are useful to shorten the account of this book. We refer Chap. 23, for the definition of an (effective) orbifold, a vector bundle on an orbifold, and their embeddings. Our orbifold is always assumed to be *effective* unless otherwise mentioned explicitly.

Throughout Part I, X is always a separable metrizable space.

Definition 3.1 A *Kuranishi chart* of X is $\mathcal{U} = (U, \mathcal{E}, \psi, s)$ with the following properties:

(1) U is an orbifold.
(2) $\mathcal{E}$ is a vector bundle on U.
(3) s is a smooth section of $\mathcal{E}$.
(4) $\psi : s^{-1}(0) \to X$ is a homeomorphism onto an open subset of X with the induced topology.

We call U a *Kuranishi neighborhood*, $\mathcal{E}$ an *obstruction bundle*, s a *Kuranishi map* and ψ a *parametrization*.

If U' is an open subset of U, then by restricting $\mathcal{E}$, ψ and s to U', we obtain a Kuranishi chart which we write $\mathcal{U}|_{U'}$ and call it an *open subchart*.

K. Fukaya et al., *Kuranishi Structures and Virtual Fundamental Chains*, Springer Monographs in Mathematics, https://doi.org/10.1007/978-981-15-5562-6_3

The *dimension* of $\mathcal{U} = (U, \mathcal{E}, \psi, s)$ is by definition

$$\dim \mathcal{U} = \dim U - \operatorname{rank} \mathcal{E}.$$

Here rank $\mathcal{E}$ is the dimension of the fibers of $\mathcal{E} \to U$.

We say that $\mathcal{U} = (U, \mathcal{E}, \psi, s)$ is *orientable* if U and $\mathcal{E}$ are orientable.[1] An *orientation* of $\mathcal{U} = (U, \mathcal{E}, \psi, s)$ is a pair of orientations of U and of $\mathcal{E}$. See Definition 3.4. An open subchart of an oriented Kuranishi chart is oriented.

Definition 3.2 Let $\mathcal{U} = (U, \mathcal{E}, \psi, s)$, $\mathcal{U}' = (U', \mathcal{E}', \psi', s')$ be Kuranishi charts of X. An *embedding* of Kuranishi charts : $\mathcal{U} \to \mathcal{U}'$ is a pair $\Phi = (\varphi, \widehat{\varphi})$ with the following properties:

(1) $\varphi : U \to U'$ is an embedding of orbifolds. (See Definition 23.2.)
(2) $\widehat{\varphi} : \mathcal{E} \to \mathcal{E}'$ is an embedding of vector bundles over φ. (See Definition 23.21.)
(3) $\widehat{\varphi} \circ s = s' \circ \varphi$.
(4) $\psi' \circ \varphi = \psi$ holds on $s^{-1}(0)$.
(5) For each $x \in U$ with $s(x) = 0$, the derivative $D_{\varphi(x)} s'$ induces an isomorphism

$$\frac{T_{\varphi(x)} U'}{(D_x \varphi)(T_x U)} \cong \frac{\mathcal{E}'_{\varphi(x)}}{\widehat{\varphi}(\mathcal{E}_x)}. \tag{3.1}$$

In other words, the map (3.1) is the right vertical arrow of the next commutative diagram:

$$\begin{array}{ccccc}
T_x U & \xrightarrow{D_x \varphi} & T_{\varphi(x)} U' & \longrightarrow & \dfrac{T_{\varphi(x)} U'}{(D_x \varphi)(T_x U)} \\
{\scriptstyle D_x s}\downarrow & & \downarrow{\scriptstyle D_{\varphi(x)} s'} & & \downarrow \\
\widehat{\varphi}(\mathcal{E}_x) & \xrightarrow{\widehat{\varphi}} & \mathcal{E}'_{\varphi(x)} & \longrightarrow & \dfrac{\mathcal{E}'_{\varphi(x)}}{\widehat{\varphi}(\mathcal{E}_x)}
\end{array} \tag{3.2}$$

If $\dim U = \dim U'$ in addition, we call $(\varphi, \widehat{\varphi})$ an *open embedding*.

Remark 3.3 To define $D_x \varphi$ we take a coordinate (V_x, Γ_x, ϕ_x) (resp. $(V'_{\varphi(x)}, \Gamma'_{\varphi(x)}, \phi'_{\varphi(x)})$) at x (resp. $\varphi(x)$) of U (resp. U') and define $T_x U = T_x V_x$, $T_{\varphi(x)} U' = T_{\varphi(x)} V'_{\varphi(x)}$. (See Definition 23.1.) Then φ lifts locally to a smooth map $\tilde{\varphi} : V_x \to V'_{\varphi(x)}$ between manifolds. (See Definition 23.2 (2)(c).) Its derivative $D_x \varphi$ makes sense in this way.

Definition 3.4 In the situation of Definition 3.2, suppose $\mathcal{U}$ and $\mathcal{U}'$ are oriented. Then the orientations induce trivializations of $\det \mathcal{E}^* \otimes \det TU$ and of $\det \mathcal{E}'^* \otimes \det TU'$. (Here $\det \mathcal{E}$ is a real line bundle which is the determinant line bundle of $\mathcal{E}$.

[1] We might require only a weaker condition that $\det \mathcal{E}^* \otimes \det TU$ is orientable. One of the reasons we take the current choice is that then Condition 23.9 is automatically satisfied.

We define $\det TU$ etc. in the same way.) We call $\det \mathcal{E}^* \otimes \det TU$ the *orientation bundle* of U. We say $\Phi = (\varphi, \widehat{\varphi})$ is *orientation preserving* if the isomorphism

$$(\det \mathcal{E}_x)^* \otimes \det T_x U \cong (\det \mathcal{E}'_{\varphi(x)})^* \otimes \det T_{\varphi(x)} U'$$

induced by (3.1) is compatible with these trivializations.

Composition of embeddings of Kuranishi charts is again an embedding of Kuranishi charts. There is an obvious embedding of Kuranishi charts from $\mathcal{U}$ to itself, that is, the identity. We can define the notion of *isomorphism* of Kuranishi charts by using the above two facts in an obvious way.

Definition 3.5 For $A \subseteq X$, a *Kuranishi neighborhood* of A is a Kuranishi chart such that $\mathrm{Im}(\psi)$ contains A. In the case $A = \{p\}$ we call it a Kuranishi neighborhood of p or a Kuranishi chart at p.[2]

When $\mathcal{U}_p = (U_p, \mathcal{E}_p, \psi_p, s_p)$ is a Kuranishi neighborhood of p, we denote by $o_p \in U_p$ the point such that $s_p(o_p) = 0$ and $\psi_p(o_p) = p$. If $q \in \mathrm{Im}(\psi_p)$ we denote by $o_p(q) \in U_p$ the point such that $s_p(o_p(q)) = 0$ and $\psi_p(o_p(q)) = q$. Note that such o_p and $o_p(q)$ are unique and $o_p(p) = o_p$. See Definition 23.1 (1).

Definition 3.6 Let $\mathcal{U}_1 = (U_1, \mathcal{E}_1, \psi_1, s_1)$, $\mathcal{U}_2 = (U_2, \mathcal{E}_2, \psi_2, s_2)$ be Kuranishi charts of X. A *coordinate change in the weak sense* (resp. *in the strong sense*) from $\mathcal{U}_1$ to $\mathcal{U}_2$ is $\Phi_{21} = (U_{21}, \varphi_{21}, \widehat{\varphi}_{21})$ with the following properties (1) and (2) (resp. (1), (2) and (3)):

(1) U_{21} is an open subset of U_1.
(2) $(\varphi_{21}, \widehat{\varphi}_{21})$ is an embedding of Kuranishi charts : $\mathcal{U}_1|_{U_{21}} \to \mathcal{U}_2$.
(3) $\psi_1(s_1^{-1}(0) \cap U_{21}) = \mathrm{Im}(\psi_1) \cap \mathrm{Im}(\psi_2)$.

In the case where $\mathcal{U}_1$ and $\mathcal{U}_2$ are oriented, Φ_{21} is said to be *orientation preserving* if it is so as an embedding.

Remark 3.7 We use coordinate changes in the weak sense for Kuranishi structures (Definition 3.9), while we use coordinate changes in the strong sense for good coordinate systems (Definition 3.15). From now on, *coordinate changes appearing in Kuranishi structures are in the weak sense* and *coordinate changes appearing in good coordinate systems are in the strong sense.*

Convention 3.8 Hereafter in Part I, Z is assumed to be a compact subset of X, unless otherwise specified.

Definition 3.9 A *Kuranishi structure* $\widehat{\mathcal{U}}$ of $Z \subseteq X$ assigns a Kuranishi neighborhood $\mathcal{U}_p = (U_p, \mathcal{E}_p, \psi_p, s_p)$ of p in X to each $p \in Z$ and a coordinate change in the weak sense $\Phi_{pq} = (U_{pq}, \varphi_{pq}, \widehat{\varphi}_{pq}) : \mathcal{U}_q \to \mathcal{U}_p$ to each $p \in Z$, $q \in \mathrm{Im}(\psi_p) \cap Z$

---

[2]In Definition 3.1 the name, Kuranishi neighborhood, is also used for the underlying orbifold of the Kuranishi chart. Here we use it for a Kuranishi chart such that $\mathrm{Im}(\psi)$ contains a given subset of X. It should be easy to distinguish them from the context.

such that $q \in \psi_q(s_q^{-1}(0) \cap U_{pq})$ and the following holds for each $p, q \in \mathrm{Im}(\psi_p) \cap Z$, $r \in \psi_q(s_q^{-1}(0) \cap U_{pq}) \cap Z$:[3]

We put $U_{pqr} = \varphi_{qr}^{-1}(U_{pq}) \cap U_{pr}$. Then we have

$$\Phi_{pr}|_{U_{pqr}} = \Phi_{pq} \circ \Phi_{qr}|_{U_{pqr}}. \tag{3.3}$$

We also require $\Phi_{pp} = (U_p, \mathrm{id}, \mathrm{id})$.

We call Z the *support set* of our Kuranishi structure.

We also require that the dimension of $\mathcal{U}_p$ is independent of p and call it the *dimension* of $\widehat{\mathcal{U}}$.

Definition 3.10

(1) For any Kuranishi structure $\widehat{\mathcal{U}} = (\{\mathcal{U}_p\}, \{\Phi_{pq}\})$ of X we can construct a real line bundle whose fiber at p is given by $\det \mathcal{E}_p{}^* \otimes \det TU_p$ as in Definition 3.4. We call the line bundle *the orientation bundle of* $(X, \widehat{\mathcal{U}})$.

(2) We say the Kuranishi structure $(\{\mathcal{U}_p\}, \{\Phi_{pq}\})$ is *orientable* if we can choose an orientation of $\mathcal{U}_p$ for each $p \in Z$ such that all Φ_{pq} are orientation preserving. In other words, $(\{\mathcal{U}_p\}, \{\Phi_{pq}\})$ is orientable if and only if the orientation bundle is trivial. The *orientation* of an orientable Kuranishi structure is the homotopy class of the trivialization of the orientation bundle.

Definition 3.11 A *K-space* is a pair $(X, \widehat{\mathcal{U}})$ of a paracompact metrizable space X and a Kuranishi structure $\widehat{\mathcal{U}}$ of X.

A *relative K-space* is a triple $(X, Z; \widehat{\mathcal{U}})$, where $Z \subseteq X$ is a compact subspace and $\widehat{\mathcal{U}}$ is a Kuranishi structure of $Z \subseteq X$.

Remark 3.12 Various constructions in later chapters are performed by constructing objects locally and extending them inductively chart by chart. To describe such a construction in a transparent way it is convenient to consider the notion of a structure which is defined only on a part of the space of X. We introduced the notion of relative K-space for this purpose.

Remark 3.13 In [FOn2, FOOO4, FOOO16] we assumed that the orbifold appearing in the definition of Kuranishi structure is a global quotient. Namely we assumed $U = V/\Gamma$ where V is a manifold and Γ is a finite group acting on V effectively and smoothly. There is no practical difference in the definition since we can always replace U_p by a smaller open subset so that it becomes of the form $U_p = V_p/\Gamma_p$.

Remark 3.14 In [Jo1] etc. a space with Kuranishi structure is called a Kuranishi space. However, the name 'Kuranishi space' has been used for a long time for the deformation space of complex structure, which Kuranishi discovered in his celebrated work. The Kuranishi structure in our sense is much inspired by Kuranishi's

[3]Note that $\mathrm{Im}(\psi_p)$ is open in X.

work, but a space with Kuranishi structure is different from the deformation space of complex structure (Kuranishi space). So we call it a *K-space* in this book.

We call s a Kuranishi map. This is the main notion discovered by Kuranishi.

From now on when we write the Kuranishi neighborhood of p as $\mathcal{U}_p, \mathcal{U}'_p$ etc. we use the notation U_p, s_p etc. from $\mathcal{U}_p = (U_p, \mathcal{E}_p, \psi_p, s_p)$.

3.2 Good Coordinate Systems

Definition 3.15 A *good coordinate system* of $Z \subseteq X$ is

$$\widehat{\mathcal{U}} = ((\mathfrak{P}, \le), \{\mathcal{U}_{\mathfrak{p}} \mid \mathfrak{p} \in \mathfrak{P}\}, \{\Phi_{\mathfrak{p}\mathfrak{q}} \mid \mathfrak{q} \le \mathfrak{p}\})$$

such that:

(1) $(\mathfrak{P}, \le)$ is a partially ordered set. We assume $\#\mathfrak{P}$ is finite.
(2) $\mathcal{U}_{\mathfrak{p}} = (U_{\mathfrak{p}}, \mathcal{E}_{\mathfrak{p}}, \psi_{\mathfrak{p}}, s_{\mathfrak{p}})$ is a Kuranishi chart of X.
(3) $\bigcup_{\mathfrak{p}\in\mathfrak{P}} \psi_{\mathfrak{p}}(s_{\mathfrak{p}}^{-1}(0)) \supseteq Z$.
(4) Let $\mathfrak{q} \le \mathfrak{p}$. Then $\Phi_{\mathfrak{p}\mathfrak{q}} = (U_{\mathfrak{p}\mathfrak{q}}, \varphi_{\mathfrak{p}\mathfrak{q}}, \widehat{\varphi}_{\mathfrak{p}\mathfrak{q}})$ is a coordinate change in the strong sense: $\mathcal{U}_{\mathfrak{q}} \to \mathcal{U}_{\mathfrak{p}}$ in the sense of Definition 3.6.[4] We also require $\Phi_{\mathfrak{p}\mathfrak{p}} = (U_{\mathfrak{p}}, \mathrm{id}, \mathrm{id})$.
(5) If $\mathfrak{r} \le \mathfrak{q} \le \mathfrak{p}$, then by putting $U_{\mathfrak{p}\mathfrak{q}\mathfrak{r}} = \varphi_{\mathfrak{q}\mathfrak{r}}^{-1}(U_{\mathfrak{p}\mathfrak{q}}) \cap U_{\mathfrak{p}\mathfrak{r}}$ we have

$$\Phi_{\mathfrak{p}\mathfrak{r}}|_{U_{\mathfrak{p}\mathfrak{q}\mathfrak{r}}} = \Phi_{\mathfrak{p}\mathfrak{q}} \circ \Phi_{\mathfrak{q}\mathfrak{r}}|_{U_{\mathfrak{p}\mathfrak{q}\mathfrak{r}}}. \tag{3.4}$$

(6) If $\mathrm{Im}(\psi_{\mathfrak{p}}) \cap \mathrm{Im}(\psi_{\mathfrak{q}}) \ne \emptyset$, then either $\mathfrak{p} \le \mathfrak{q}$ or $\mathfrak{q} \le \mathfrak{p}$ holds.
(7) We define a relation $\sim$ on the disjoint union $\coprod_{\mathfrak{p}\in\mathfrak{P}} U_{\mathfrak{p}}$ as follows. Let $x \in U_{\mathfrak{p}}$, $y \in U_{\mathfrak{q}}$. We define $x \sim y$ if and only if one of the following holds:

    (a) $\mathfrak{p} = \mathfrak{q}$ and $x = y$.
    (b) $\mathfrak{p} < \mathfrak{q}$ and $y = \varphi_{\mathfrak{q}\mathfrak{p}}(x)$.[5]
    (c) $\mathfrak{q} < \mathfrak{p}$ and $x = \varphi_{\mathfrak{p}\mathfrak{q}}(y)$.

    Then $\sim$ is an equivalence relation.
(8) The quotient of $\coprod_{\mathfrak{p}\in\mathfrak{P}} U_{\mathfrak{p}}/\sim$ by this equivalence relation is Hausdorff with respect to the quotient topology.

In the case $Z = X$ we call it a good coordinate system of X.

In the case where $\widehat{\mathcal{U}}$ satisfies only (1)–(6), we call it *a good coordinate system in the weak sense.*

We call Z the *support set* of our good coordinate system.

---

[4] Note it may happen that $U_{\mathfrak{p}\mathfrak{q}} = \emptyset$.

[5] Here $\mathfrak{p} < \mathfrak{q}$ means $\mathfrak{p} \le \mathfrak{q}$ and $\mathfrak{p} \ne \mathfrak{q}$.

We also require that the dimension of $\mathcal{U}_{\mathfrak{p}}$ is independent of $\mathfrak{p}$ and call it the *dimension* of $\widehat{\mathcal{U}}$.

We say a good coordinate system structure $\widehat{\mathcal{U}} = ((\mathfrak{P}, \le), \{\mathcal{U}_{\mathfrak{p}} \mid \mathfrak{p} \in \mathfrak{P}\}, \{\Phi_{\mathfrak{p}\mathfrak{q}} \mid \mathfrak{q} \le \mathfrak{p}\})$ is *orientable* if we can choose an orientation of $\mathcal{U}_{\mathfrak{p}}$ such that all $\Phi_{\mathfrak{p}\mathfrak{q}}$ are orientation preserving. The notion of orientation of an orientable good coordinate system and of an oriented good coordinate system is defined in an obvious way.

Definition 3.16 We denote by $|\widehat{\mathcal{U}}|$ the quotient set of the equivalence relation in Definition 3.15 (7). (See [FOOO16, Remark 5.20] or [FOOO17, Proposition 2.11] for its topology.)

The following result is proved in [FOOO17].

Proposition 3.17 (Shrinking Lemma, [FOOO17, Theorem 2.9]) *Let* $\widehat{\mathcal{U}} = ((\mathfrak{P}, \le), \{\mathcal{U}_{\mathfrak{p}} \mid \mathfrak{p} \in \mathfrak{P}\}, \{\Phi_{\mathfrak{p}\mathfrak{q}} \mid \mathfrak{q} \le \mathfrak{p}\})$ *be a system which satisfies Definition 3.15 (1)–(6). Then there exists open subsets* $U^0_{\mathfrak{p}} \subset U_{\mathfrak{p}}$ *and* $U^0_{\mathfrak{p}\mathfrak{q}} \subset U_{\mathfrak{p}\mathfrak{q}}$ *such that*

$$((\mathfrak{P}, \le), \{\mathcal{U}_{\mathfrak{p}}|_{U^0_{\mathfrak{p}}} \mid \mathfrak{p} \in \mathfrak{P}\}, \{\Phi_{\mathfrak{p}\mathfrak{q}}|_{U^0_{\mathfrak{p}\mathfrak{q}}} \mid \mathfrak{q} \le \mathfrak{p}\})$$

satisfies Definition 3.15 *(1)–(8).*

We omit the proof and refer interested readers to [FOOO17]. We explain several related points (including history) in Sect. 3.4.

Notation 3.18 Throughout this book, we denote by $\mathcal{U}_{\mathfrak{p}}$ etc. (where the index $\mathfrak{p}$ is a German Fraktur character) a Kuranishi chart which is a member of a good coordinate system, and by $\mathcal{U}_p$ etc. (where the index p is an italic letter) a Kuranishi chart which is a member of a Kuranishi structure.

In other words, the lower case Fraktur letters $\mathfrak{p}, \mathfrak{q}$ denote the elements in the partially ordered set, the index set of the Kuranishi charts of a good coordinate system, and the italic lower case letters p, q are the points of $Z \subseteq X$ which are elements of the index set of the Kuranishi charts of a Kuranishi structure.

From now on, we denote by $\Phi_{\mathfrak{p}\mathfrak{q}} = (U_{\mathfrak{p}\mathfrak{q}}, \varphi_{\mathfrak{p}\mathfrak{q}}, \widehat{\varphi}_{\mathfrak{p}\mathfrak{q}})$ a coordinate change $\mathcal{U}_{\mathfrak{q}} \to \mathcal{U}_{\mathfrak{p}}$. The same remark applies to Φ_{pq}.

3.3 Embedding of Kuranishi Structures I

Definition 3.19 Let $\widehat{\mathcal{U}} = (\{\mathcal{U}_p\}, \{\Phi_{pq}\})$, $\widehat{\mathcal{U}'} = (\{\mathcal{U}'_p\}, \{\Phi'_{pq}\})$ be Kuranishi structures of $Z \subseteq X$. A *strict KK-embedding* from $\widehat{\mathcal{U}}$ to $\widehat{\mathcal{U}'}$ is a collection $\widehat{\Phi} = \{\Phi_p \mid p \in Z\}$ of embeddings of Kuranishi charts $\Phi_p = (\varphi_p, \widehat{\varphi}_p) : \mathcal{U}_p \to \mathcal{U}'_p$. such that the following holds:

(1) $\mathrm{Im}(\psi_p) \cap Z \subseteq \mathrm{Im}(\psi'_p) \cap Z$.
(2) $U_{pq} \subseteq \varphi_q^{-1}(U'_{pq})$ if $q \in \mathrm{Im}(\psi_p) \cap Z$.[6]
(3) $\Phi_p \circ \Phi_{pq} = \Phi'_{pq} \circ \Phi_q$ holds on U_{pq} if $q \in \mathrm{Im}(\psi_p) \cap Z$.

We say that it is an *open KK-embedding* if $\dim U_p = \dim U'_p$ for each p.

We say that $\widehat{\mathcal{U}}$ is an *open substructure* of $\widehat{\mathcal{U}'}$ if there exists an open KK-embedding $\widehat{\mathcal{U}} \to \widehat{\mathcal{U}'}$.

A *KK-embedding* from $\widehat{\mathcal{U}}$ to $\widehat{\mathcal{U}^+}$ is a strict KK-embedding $\widehat{\mathcal{U}_0} \to \widehat{\mathcal{U}^+}$ from an open substructure $\widehat{\mathcal{U}_0}$ of $\widehat{\mathcal{U}}$.

Lemma 3.20 *We can compose two strict KK-embeddings.*

Proof Let $\widehat{\mathcal{U}^j} = (\{\mathcal{U}^j_p\}, \{\Phi^j_{pq}\})$ be Kuranishi structures of $Z \subseteq X$ for $j = 1, 2, 3$ and let $\widehat{\Phi^{(j)}} = \{\Phi^{(j)}_p\} : \widehat{\mathcal{U}^j} \to \widehat{\mathcal{U}^{j+1}}$ be KK-embeddings for $j = 1, 2$. We put $\Phi_p = \Phi^{(2)}_p \circ \Phi^{(1)}_p$. We check (1)(2)(3) of Definition 3.19.

(1) If $q \in \mathrm{Im}(\psi^{(1)}_p) \cap Z$ then $q \in \mathrm{Im}(\psi^{(2)}_p) \cap Z$ since $\widehat{\Phi^{(1)}}$ is a KK-embedding.
Therefore $q \in \mathrm{Im}(\psi^{(3)}_p) \cap Z$ since $\widehat{\Phi^{(2)}}$ is a KK-embedding.
(2) $U^{(1)}_{pq} \subseteq (\varphi^{(1)}_q)^{-1}(U^{(2)}_{pq}) \subseteq (\varphi^{(1)}_q)^{-1}(\varphi^{(3)}_q)^{-1}(U^{(3)}_{pq}) = (\varphi_q)^{-1}(U^{(3)}_{pq})$.
(3) On $U^{(1)}_{pq}$ we calculate

$$\Phi_p \circ \Phi^1_{pq} = \Phi^{(2)}_p \circ \Phi^{(1)}_p \circ \Phi^1_{pq} = \Phi^{(2)}_p \circ \Phi^2_{pq} \circ \Phi^{(1)}_q = \Phi^3_{pq} \circ \Phi^{(2)}_q \circ \Phi^{(1)}_q = \Phi^3_{pq} \circ \Phi_q.$$

□

Definition 3.21 In the situation of Definition 3.19 we assume that $\widehat{\mathcal{U}}$ and $\widehat{\mathcal{U}'}$ are oriented. We say that $\widehat{\Phi} = \{\Phi_p\}$ is *orientation preserving* if each Φ_p is orientation preserving.

Remark 3.22 The notion of orientation preserving embedding can be defined for other types of embeddings (there are four types of them, see Table 5.1) in an obvious way.

Convention 3.23 Hereafter in Part I of this book we assume all the Kuranishi charts, Kuranishi structures and good coordinate systems are oriented unless otherwise mentioned explicitly. We also assume all the coordinate changes and embeddings between Kuranishi charts, Kuranishi structures and good coordinate systems are orientation preserving unless otherwise mentioned explicitly.

Definition 3.24 Let $\widehat{\mathcal{U}} = (\mathfrak{P}, \{\mathcal{U}_{\mathfrak{p}}\}, \{\Phi_{\mathfrak{p}\mathfrak{q}}\})$, $\widehat{\mathcal{U}'} = (\mathfrak{P}', \{\mathcal{U}'_{\mathfrak{p}'}\}, \{\Phi'_{\mathfrak{p}'\mathfrak{q}'}\})$ be good coordinate systems of $Z \subseteq X$. A *GG-embedding* from $\widehat{\mathcal{U}}$ to $\widehat{\mathcal{U}'}$ is a pair $\widehat{\Phi} = (\mathrm{i}, \{\Phi_{\mathfrak{p}} \mid \mathfrak{p} \in \mathfrak{P}\})$ of an order-preserving map $\mathrm{i} : \mathfrak{P} \to \mathfrak{P}'$ and a collection $\{\Phi_{\mathfrak{p}} \mid \mathfrak{p} \in \mathfrak{P}\}$ of embeddings of Kuranishi charts $\Phi_{\mathfrak{p}} = (\varphi_{\mathfrak{p}}, \widehat{\varphi}_{\mathfrak{p}}) : \mathcal{U}_{\mathfrak{p}} \to \mathcal{U}'_{\mathrm{i}(\mathfrak{p})}$ such that the following hold for each $\mathfrak{q} \leq \mathfrak{p}$:

[6]Note that under this condition Φ'_{pq} exists by Item (1).

(1) $U_{\mathfrak{pq}} = (\varphi'_{\mathfrak{i}(\mathfrak{p})\mathfrak{i}(\mathfrak{q})} \circ \varphi_{\mathfrak{q}})^{-1}(\varphi_{\mathfrak{p}}(U_{\mathfrak{p}}))$.
(2) $\Phi_{\mathfrak{p}} \circ \Phi_{\mathfrak{pq}} = \Phi'_{\mathfrak{i}(\mathfrak{p})\mathfrak{i}(\mathfrak{q})} \circ \Phi_{\mathfrak{q}}|_{U_{\mathfrak{pq}}}$.

$$\begin{array}{ccc} \mathcal{U}_{\mathfrak{q}}|_{U_{\mathfrak{pq}}} & \xrightarrow{\Phi_{\mathfrak{q}}} & \mathcal{U}'_{\mathfrak{i}(\mathfrak{q})}|_{U'_{\mathfrak{i}(\mathfrak{p})\mathfrak{i}(\mathfrak{q})}} \\ {\scriptstyle\Phi_{\mathfrak{pq}}}\downarrow & & \downarrow{\scriptstyle\Phi'_{\mathfrak{i}(\mathfrak{p})\mathfrak{i}(\mathfrak{q})}} \\ \mathcal{U}_{\mathfrak{p}} & \xrightarrow{\Phi_{\mathfrak{p}}} & \mathcal{U}'_{\mathfrak{i}(\mathfrak{p})} \end{array} \tag{3.5}$$

We say that $\widehat{\Phi}$ is a *weakly open GG-embedding* if $\dim U_{\mathfrak{p}} = \dim U'_{\mathfrak{i}(\mathfrak{p})}$ for each $\mathfrak{p}$.

We say that a weakly open embedding $\widehat{\Phi}$ is an *open GG-embedding* if $\mathfrak{i}$ is a bijection.

We say that $\widehat{\mathcal{U}}$ is an *open substructure* (resp. *weakly open substructure*) of $\widehat{\mathcal{U}'}$ if there exists an open (resp. weakly open) GG-embedding $\widehat{\mathcal{U}} \to \widehat{\mathcal{U}'}$.

We say that a GG-embedding $\widehat{\Phi}$ is an *isomorphism* if the map $\mathfrak{i}$ is a bijection and $\widehat{\varphi}_{\mathfrak{p}}$ is an isomorphism for each $\mathfrak{p}$.

Lemma 3.25 *Let $\widehat{\mathcal{U}} = (\mathfrak{P}, \{\mathcal{U}_{\mathfrak{p}}\}, \{\Phi_{\mathfrak{pq}}\})$, $\widehat{\mathcal{U}'} = (\mathfrak{P}', \{\mathcal{U}'_{\mathfrak{p}'}\}, \{\Phi'_{\mathfrak{p}'\mathfrak{q}'}\})$ be good coordinate systems of $Z \subseteq X$. Suppose that $\mathfrak{i} : \mathfrak{P} \to \mathfrak{P}'$ is an order-preserving map and that to each $\mathfrak{p} \in \mathfrak{P}$ an embedding of Kuranishi charts $\Phi_{\mathfrak{p}} = (\varphi_{\mathfrak{p}}, \widehat{\varphi}_{\mathfrak{p}}) : \mathcal{U}_{\mathfrak{p}} \to \mathcal{U}'_{\mathfrak{i}(\mathfrak{p})}$ is assigned.*

Then they define a GG-embedding if and only if the map $\coprod_{\mathfrak{p}\in\mathfrak{P}} U_{\mathfrak{p}} \to \coprod_{\mathfrak{p}'\in\mathfrak{P}'} U_{\mathfrak{p}'}$ defined by $\mathfrak{i}$ and $\{\Phi_{\mathfrak{p}}\}$ induces an injective map $|\widehat{\mathcal{U}}| \to |\widehat{\mathcal{U}'}|$.

Proof Suppose Definition 3.24 (1)(2) are satisfied. Let $x \in U_{\mathfrak{p}}$, $y \in U_{\mathfrak{q}}$ with $\mathfrak{q} < \mathfrak{p}$. Then $x \sim y$ if and only if $y \in U_{\mathfrak{pq}}$ and $x = \varphi_{\mathfrak{pq}}(y)$. Definition 3.24 (1)(2) imply $\varphi_{\mathfrak{p}}(x) \in U'_{\mathfrak{i}(\mathfrak{p})\mathfrak{i}(\mathfrak{q})}$ and $\varphi'_{\mathfrak{i}(\mathfrak{p})\mathfrak{i}(\mathfrak{q})}(\varphi_{\mathfrak{q}}(y)) = \varphi_{\mathfrak{p}}(x)$. Hence $\varphi_{\mathfrak{p}}(x) \sim \varphi_{\mathfrak{q}}(y)$.

If $\varphi_{\mathfrak{p}}(x) \sim \varphi_{\mathfrak{q}}(y)$ then y is in the right hand side of Definition 3.24 (1). Therefore $y \in U_{\mathfrak{pq}}$. Then Definition 3.24 (2) implies $x = \varphi_{\mathfrak{pq}}(y)$. We thus proved the 'only if' part. The proof of 'if' part is similar. □

Lemma 3.26 *We can compose two GG-embeddings.*

Proof This is immediate from Lemma 3.25. □

Lemma 3.27 *Let $\widehat{\mathcal{U}} = (\mathfrak{P}, \{\mathcal{U}_{\mathfrak{p}}\}, \{\Phi_{\mathfrak{pq}}\})$ be a good coordinate system of $Z \subseteq X$ and let $U^0_{\mathfrak{p}} \subseteq U_{\mathfrak{p}}$ be given open subsets such that $Z \subset \bigcup_{\mathfrak{p}\in\mathfrak{P}} \psi_{\mathfrak{p}}(s_{\mathfrak{p}}^{-1}(0) \cap U^0_{\mathfrak{p}})$. Then there exists a unique coordinate change $\Phi^0_{\mathfrak{pq}}$ such that $(\mathfrak{P}, \{\mathcal{U}_{\mathfrak{p}}|_{U^0_{\mathfrak{p}}}\}, \{\Phi^0_{\mathfrak{pq}}\})$ is an open substructure of $\widehat{\mathcal{U}}$.*

Proof Let $\Phi_{\mathfrak{pq}} = (U_{\mathfrak{pq}}, \varphi_{\mathfrak{pq}}, \widehat{\varphi}_{\mathfrak{pq}})$. We put

$$U^0_{\mathfrak{pq}} = U^0_{\mathfrak{q}} \cap \varphi_{\mathfrak{pq}}^{-1}(U^0_{\mathfrak{p}}) \tag{3.6}$$

and $\Phi^0_{\mathfrak{pq}} = (U^0_{\mathfrak{pq}}, \varphi_{\mathfrak{pq}}|_{U^0_{\mathfrak{pq}}}, \widehat{\varphi}_{\mathfrak{pq}}|_{U^0_{\mathfrak{pq}}})$. It is easy to see that $(\mathfrak{P}, \{\mathcal{U}_{\mathfrak{p}}|_{U^0_{\mathfrak{p}}}\}, \{\Phi^0_{\mathfrak{pq}}\})$ is an open substructure of $\widehat{\mathcal{U}}$.

For the uniqueness, if $(\mathfrak{P}, \{\mathcal{U}_{\mathfrak{p}}|_{U^0_{\mathfrak{p}}}\}, \{\Phi^0_{\mathfrak{pq}}\})$ is an open substructure of $\widehat{\mathcal{U}}$, then Definition 3.24 (1) implies that the domain $U^0_{\mathfrak{pq}}$ of $\Phi_{\mathfrak{pq}}$ must be as in (3.6). □

Lemma 3.28 *Let $\widehat{\mathcal{U}} = (\{\mathcal{U}_p\}, \{\Phi_{pq}\})$ be a Kuranishi structure of $Z \subseteq X$ and $U^0_p \subseteq U_p$ open subsets containing o_p. Then there exists Φ^0_{pq} such that $(\{\mathcal{U}_p|_{U^0_p}\}, \{\Phi^0_{pq}\})$ is an open substructure of $\widehat{\mathcal{U}}$.*

Remark 3.29 The uniqueness does not hold in Lemma 3.28 since there is no condition similar to Definition 3.24 (1) for Kuranishi structures. They have, however, a common open substructure.

Proof We put

$$U^0_{pq} = U^0_q \cap \varphi^{-1}_{pq}(U^0_p) \tag{3.7}$$

and $\Phi^0_{pq} = (U^0_{pq}, \varphi_{pq}|_{U^0_{pq}}, \widehat{\varphi}_{pq}|_{U^0_{pq}})$. It is easy to see that $(\{\mathcal{U}_p|_{U^0_p}\}, \{\Phi^0_{pq}\})$ is a Kuranishi structure. □

Definition 3.30 Let $\widehat{\mathcal{U}}$ be a Kuranishi structure and $\widehat{\mathcal{U}}$ a good coordinate system of $Z \subseteq X$. A *strict KG-embedding* from $\widehat{\mathcal{U}}$ to $\widehat{\mathcal{U}}$ is a collection $\{\Phi_{\mathfrak{p}p} \mid p \in Z, \mathfrak{p} \in \mathfrak{P},\ p \in \operatorname{Im}(\psi_{\mathfrak{p}})\}$ of embeddings of Kuranishi charts $\Phi_{\mathfrak{p}p} = (\varphi_{\mathfrak{p}p}, \widehat{\varphi}_{\mathfrak{p}p}) : \mathcal{U}_p \to \mathcal{U}_{\mathfrak{p}}$ with the following properties.

If $\mathfrak{p}, \mathfrak{q} \in \mathfrak{P}$, $\mathfrak{q} \le \mathfrak{p}$, $p \in \operatorname{Im}(\psi_{\mathfrak{p}}) \cap Z$, $q \in \operatorname{Im}(\psi_p) \cap \psi_{\mathfrak{q}}(U_{\mathfrak{pq}} \cap s_{\mathfrak{q}}^{-1}(0)) \cap Z$, then $\Phi_{\mathfrak{p}p} \circ \Phi_{\mathfrak{q}q} = \Phi_{\mathfrak{p}p} \circ \Phi_{pq}$ on $U_{pq} \cap \varphi^{-1}_{\mathfrak{q}q}(U_{\mathfrak{pq}})$. In other words, the following diagram commutes:

$$\begin{array}{ccc} \mathcal{U}_q|_{U_{pq} \cap \varphi^{-1}_{\mathfrak{q}q}(U_{\mathfrak{pq}})} & \xrightarrow{\Phi_{\mathfrak{q}q}} & \mathcal{U}_{\mathfrak{q}}|_{U_{\mathfrak{pq}}} \\ \Phi_{pq} \downarrow & & \downarrow \Phi_{\mathfrak{pq}} \\ \mathcal{U}_p & \xrightarrow{\Phi_{\mathfrak{p}p}} & \mathcal{U}_{\mathfrak{p}} \end{array} \tag{3.8}$$

A *KG-embedding* of $\widehat{\mathcal{U}}$ to $\widehat{\mathcal{U}}$ is by definition a strict KG-embedding of an open substructure of $\widehat{\mathcal{U}}$ to $\widehat{\mathcal{U}}$.

Remark 3.31 We remark that the condition $p \in \operatorname{Im}(\psi_{\mathfrak{p}}) \cap Z$, $q \in \operatorname{Im}(\psi_p) \cap \psi_{\mathfrak{q}}(U_{\mathfrak{pq}} \cap s_{\mathfrak{q}}^{-1}(0)) \cap Z$ we assume in Definition 3.30 is equivalent to the condition that o_q is contained in the domains of both $\varphi_{\mathfrak{pq}} \circ \varphi_{\mathfrak{q}q}$ and $\varphi_{\mathfrak{p}p} \circ \varphi_{pq}$.

Lemma 3.32 *Let $\widehat{\Phi} : \widehat{\mathcal{U}^{(1)}} \to \widehat{\mathcal{U}^{(2)}}$ be a strict KK-embedding and $\widehat{\Phi^{(2)}} : \widehat{\mathcal{U}^{(2)}} \to \widehat{\mathcal{U}}$ a strict KG-embedding. We can define the composition*

$$\widehat{\Phi^{(1)}} = \widehat{\Phi^{(2)}} \circ \widehat{\Phi} : \widehat{\mathcal{U}^{(1)}} \to \widehat{\mathcal{U}^{(2)}} \to \widehat{\mathcal{U}}$$

that is a strict KG-embedding. (See Definition 5.14 (3).)

Proof We put $\widehat{\mathcal{U}^{(j)}} = (\{\mathcal{U}_p^{(j)}\}, \{\Phi_{pq}^{(j)}\})$, $\widehat{\Phi} = \{\Phi_p\}$, (where index is an italic character), $\widehat{\mathcal{U}} = (\{\mathcal{U}_{\mathfrak{p}}\}, \{\Phi_{\mathfrak{pq}}\})$ (where index is a Fraktur character). $\widehat{\Phi^{(2)}} = \{\Phi_{\mathfrak{p}p}^{(2)}\}$.

For $p \in \mathrm{Im}(\psi_{\mathfrak{p}})$ we define

$$\Phi_{\mathfrak{p}p}^{(1)} = \Phi_{\mathfrak{p}p}^{(2)} \circ \Phi_p : \mathcal{U}_p^{(1)} \to \mathcal{U}_{\mathfrak{p}}.$$

We will check that these maps satisfy the condition of Definition 3.30.

Let $\mathfrak{p}, \mathfrak{q} \in \mathfrak{P}$, $\mathfrak{q} \le \mathfrak{p}$, $p \in \mathrm{Im}(\psi_{\mathfrak{p}}) \cap Z$, $q \in \mathrm{Im}(\psi_p^{(1)}) \cap \psi_{\mathfrak{q}}(U_{\mathfrak{pq}}) \cap Z$.

Since $q \in \mathrm{Im}\,\psi_p^{(1)} \cap Z$, Definition 3.19 (1) implies $q \in \mathrm{Im}\psi_p^{(2)} \cap Z$. By Definition 3.19 (2) we have $U_{pq}^{(1)} \subseteq \varphi_q^{-1}(U_{pq}^{(2)})$.

Let $x \in U_{pq}^{(1)} \cap (\varphi_{\mathfrak{q}q}^{(1)})^{-1}(U_{\mathfrak{pq}})$. Then $x \in \varphi_q^{-1}(U_{pq}^{(2)})$ and

$$(\varphi_p \circ \varphi_{pq}^{(1)})(x) = (\varphi_{pq}^{(2)} \circ \varphi_q)(x).$$

Moreover, $\varphi_q(x) \in U_{pq}^{(2)} \cap (\varphi_{\mathfrak{q}q}^{(2)})^{-1}(U_{\mathfrak{pq}})$ and

$$(\varphi_{\mathfrak{pq}} \circ \varphi_{\mathfrak{q}q}^{(2)})(\varphi_q(x)) = (\varphi_{\mathfrak{p}p}^{(2)} \circ \varphi_{pq}^{(2)})(\varphi_q(x)).$$

Thus

$$\begin{aligned}(\varphi_{\mathfrak{p}p}^{(1)} \circ \varphi_{pq}^{(1)})(x) &= (\varphi_{\mathfrak{p}p}^{(2)} \circ \varphi_p \circ \varphi_{pq}^{(1)})(x) = (\varphi_{\mathfrak{p}p}^{(2)} \circ \varphi_{pq}^{(2)} \circ \varphi_q)(x)\\ &= (\varphi_{\mathfrak{pq}} \circ \varphi_{\mathfrak{q}q}^{(2)} \circ \varphi_q)(x) = (\varphi_{\mathfrak{pq}} \circ \varphi_{\mathfrak{q}q}^{(1)})(x).\end{aligned}$$

We have proved the required commutativity of Diagram (3.8). □

Remark 3.33 In general we cannot compose KG-embedding and GG-embedding.[7] Let us consider the case where $\widehat{\mathcal{U}^{(1)}}, \widehat{\mathcal{U}^{(2)}}$ consist of two charts $\mathcal{U}_{\mathfrak{p}}^{(j)}, \mathcal{U}_{\mathfrak{q}}^{(j)}$, such that

$$\dim U_{\mathfrak{p}}^{(1)} = \dim U_{\mathfrak{p}}^{(2)} > \dim U_{\mathfrak{q}}^{(1)} = \dim U_{\mathfrak{q}}^{(2)}.$$

We assume $\mathfrak{i}(\mathfrak{p}) = \mathfrak{p}$, $\mathfrak{i}(\mathfrak{q}) = \mathfrak{q}$ and $\Phi_{\mathfrak{p}}$, $\Phi_{\mathfrak{q}}$ are open embeddings.

Suppose we have a Kuranishi structure $\widehat{\mathcal{U}}$ with the following property. There exists $p \in Z$ such that $\dim U_p = \dim U_{\mathfrak{p}}^{(1)}$ and

[7] There is an error in [FOOO19] on this point. Fortunately such composition was not used in [FOOO19].

$$p \in \mathrm{Im}(\psi_{\mathfrak{p}}^{(1)}), \quad p \in \mathrm{Im}(\psi_{\mathfrak{q}}^{(2)}) \cap \mathrm{Im}(\psi_{\mathfrak{p}}^{(2)}), \quad p \notin \mathrm{Im}(\psi_{\mathfrak{q}}^{(1)}).$$

Then $\Phi_{\mathfrak{p}p}^{(j)} : \mathcal{U}_p \to \mathcal{U}_{\mathfrak{p}}^{(j)}$ can exist for $j = 1, 2$. We do not need to define $\Phi_{\mathfrak{q}p}^{(1)} : \mathcal{U}_p \to \mathcal{U}_{\mathfrak{q}}^{(1)}$ because $p \notin \mathrm{Im}(\psi_{\mathfrak{q}}^{(1)})$. On the other hand, we do need to define $\Phi_{\mathfrak{q}p}^{(2)} : \mathcal{U}_p \to \mathcal{U}_{\mathfrak{q}}^{(2)}$. However, such $\Phi_{\mathfrak{q}p}^{(2)}$ cannot exist because $\dim U_p > \dim U_{\mathfrak{q}}^{(2)}$.

Thus there is a situation where KG-embedding $\widehat{\mathcal{U}} \to \widehat{\mathcal{U}^{(1)}}$ exists but $\widehat{\mathcal{U}} \to \widehat{\mathcal{U}^{(2)}}$ cannot exist.

However, the following holds.

Lemma 3.34 *If $\widehat{\mathcal{U}} \to \widehat{\mathcal{U}}$ is a KG-embedding and $\widehat{\mathcal{U}} \to \widehat{\mathcal{U}^{+}}$ is a GG-embedding, then there exist an open substructure $\widehat{\mathcal{U}^{0+}}$ of $\widehat{\mathcal{U}^{+}}$ such that a KG-embedding $\widehat{\mathcal{U}} \to \widehat{\mathcal{U}^{0+}}$ exists.*

We do not use Lemma 3.34 in this book, but we prove it in Chap. 28 for completeness.

The next result is the same as [FOn2, Lemma 6.3], [FOOO16, Theorem 7.1]. We prove it together with various addenda in Chap. 11. In particular, Theorem 3.35 is proved in Sect. 11.1.

Theorem 3.35 *For any Kuranishi structure $\widehat{\mathcal{U}}$ of $Z \subseteq X$ there exist a good coordinate system $\widehat{\mathcal{U}}$ of $Z \subseteq X$ and a KG-embedding $\widehat{\mathcal{U}} \to \widehat{\mathcal{U}}$, i.e., an open substructure $\widehat{\mathcal{U}_0}$ of $\widehat{\mathcal{U}}$ and a strict KG-embedding $\widehat{\mathcal{U}_0} \to \widehat{\mathcal{U}}$.*

Remark 3.36 According to Convention 3.23, Theorem 3.35 contains the statement that $\widehat{\mathcal{U}}$ and the KG-embedding $\widehat{\mathcal{U}} \to \widehat{\mathcal{U}}$ are oriented (provided $\widehat{\mathcal{U}}$ is oriented). We will not repeat this kind of remark later on.

Definition 3.37 A good coordinate system $\widehat{\mathcal{U}}$ is said to be *compatible* with a Kuranishi structure $\widehat{\mathcal{U}}$ if there exists a KG-embedding $\widehat{\mathcal{U}} \to \widehat{\mathcal{U}}$.

Remark 3.38

(1) The next terminology is due to Joyce [Jo1].

Definition 3.39 A good coordinate system is said to be *excellent* if $\mathfrak{P} \subset \mathbb{Z}_{\geq 0}$, $\leq$ is the standard inequality on $\mathbb{Z}_{\geq 0}$ and $\dim U_{\mathfrak{p}} = \mathfrak{p}$.

Starting with an arbitrary good coordinate system $\widehat{\mathcal{U}} = (\{U_{\mathfrak{p}}\}, \{\Phi_{\mathfrak{p}\mathfrak{q}}\})$, we can construct an excellent good coordinate system $\widehat{\mathcal{U}'}$ as follows. Suppose $\dim U_{\mathfrak{p}_1} = \dim U_{\mathfrak{p}_2}$. If $\mathrm{Im}(\psi_{\mathfrak{p}_1})$ is disjoint from $\mathrm{Im}(\psi_{\mathfrak{p}_2})$, we take its disjoint union as a new chart and remove these two charts. Suppose $\mathrm{Im}(\psi_{\mathfrak{p}_1}) \cap \mathrm{Im}(\psi_{\mathfrak{p}_2}) \neq \emptyset$. Then we may assume $\mathfrak{p}_1 < \mathfrak{p}_2$. Since an embedding between two orbifolds of the same dimension is necessarily a diffeomorphism, the coordinate change $\Phi_{\mathfrak{p}_2\mathfrak{p}_1}$ is an isomorphism. We can use this observation to construct the sum chart of $\mathcal{U}_{\mathfrak{p}_1}$ and $\mathcal{U}_{\mathfrak{p}_2}$. (See [FOOO19, Lemma 3.17].) We take it as a new chart and remove $\mathcal{U}_{\mathfrak{p}_1}$ and $\mathcal{U}_{\mathfrak{p}_2}$. It is easy to define the coordinate change between sum charts and other charts. (See [FOOO19, Lemma 3.18].)

We can continue this process finitely many times until we get an excellent coordinate system. Note there is a weakly open GG-embedding $\widehat{\mathcal{U}} \to \widehat{\mathcal{U}'}$.

(2) We note that in [FOOO16, Section 7] we introduced the notion of a mixed neighborhood. It is basically equivalent to the notion of a pair of an excellent good coordinate system $\widehat{\mathcal{U}}$ and a KG-embedding $\widehat{\mathcal{U}} \to \widehat{\mathcal{U}}$. (The only difference is that the conclusion of [FOOO16, Lemma 7.32] is not assumed in [FOOO16, Definition 7.15]. This difference is not essential at all because of [FOOO16, Lemma 7.32].) Therefore Theorem 3.35 is equivalent to [FOOO16, Theorem 7.1].

Definition 3.40 Let $\widehat{\mathcal{U}}$ be a Kuranishi structure of $Z \subseteq X$.

(1) A *strongly continuous map* $\widehat{f}$ from $(X, Z; \widehat{\mathcal{U}})$ to a topological space Y assigns a continuous map f_p from U_p to Y for each $p \in Z$ such that $f_p \circ \varphi_{pq} = f_q$ holds on U_{pq}.
(2) In the situation of (1), the map $f : Z \to Y$ defined by $f(p) = f_p(p)$ is a continuous map from Z to Y. We call $f : Z \to Y$ the *underlying continuous map* of $\widehat{f}$.
(3) We require that the underlying continuous map $f : Z \to Y$ is extended to a continuous map $f : X \to Y$ and include it to the data defining a strong continuous map.
(4) When Y is a smooth manifold, we say $\widehat{f}$ is *strongly smooth* if each of f_p is smooth.
(5) A strongly smooth map is said to be *weakly submersive* if each of f_p is a submersion.

We sometimes say that f is a strongly continuous map (resp. a strongly smooth map, a weakly submersive map) in place of $\widehat{f}$ is a strongly continuous map (resp. a strongly smooth map, weakly submersive), by an abuse of notation.

Remark 3.41 The continuity claimed in (2) follows from the next diagram:

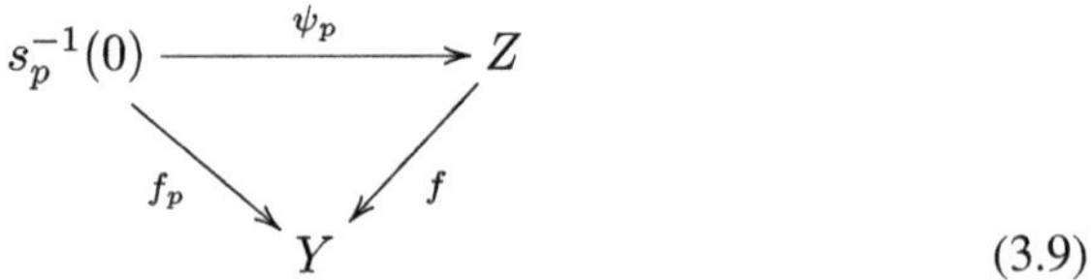

(3.9)

whose commutativity is a consequence of $f_p \circ \varphi_{pq} = f_q$.

Remark 3.42 In [Jo1], Joyce used the terminology 'strong submersion' instead of 'weak submersion' which we have been using since [FOn2]. We use the terminology 'weakly submersive' for the following two reasons:

(1) Let $V_1 \subset V_2$ be a submanifold, f_2 a map from V_2 and f_1 a restriction of f_2 to V_1. The condition that f_2 is smooth is stronger than the condition that f_1 is smooth. So we use the words strongly smooth. On the other hand, in the case where f_2 is smooth, the condition that f_2 is a submersion at each point of V_1 is weaker than the condition that f_1 is a submersion. So we use the words weak submersion.
(2) We reserve the terminology 'strongly submersive' for a different notion. See Definitions 6.21 and 6.22.

Definition 3.43 Let $\widehat{\mathcal{U}}$ be a good coordinate system of $Z \subseteq X$.

(1) A *strongly continuous map* $\widehat{f}$ from $(X, Z; \widehat{\mathcal{U}})$ to a topological space Y assigns a continuous map $f_{\mathfrak{p}}$ from $U_{\mathfrak{p}}$ to Y to each $\mathfrak{p} \in \mathfrak{P}$ such that $f_{\mathfrak{p}} \circ \varphi_{\mathfrak{p}\mathfrak{q}} = f_{\mathfrak{q}}$ holds on $U_{\mathfrak{p}\mathfrak{q}}$.
(2) In the situation of (1), the map $f : Z \to Y$ defined by $f(p) = f_{\mathfrak{p}}(o_{\mathfrak{p}}(p))$ (for $p \in \mathrm{Im}(\psi_{\mathfrak{p}}) \cap Z$) is a continuous map from Z to Y.[8] We call $f : Z \to Y$ the *underlying continuous map* of $\widehat{f}$.
(3) We require that the underlying continuous map $f : Z \to Y$ is extended to a continuous map $f : X \to Y$ and include it with the data defining a strong continuous map.
(4) When Y is a smooth manifold, we say $\widehat{f}$ is *strongly smooth* if each of $f_{\mathfrak{p}}$ is smooth.
(5) A strongly smooth map is said to be *weakly submersive* if each of $f_{\mathfrak{p}}$ is a submersion.

We sometimes say f is a strongly continuous map (resp. a strongly smooth map, weakly submersive) in place of $\widehat{f}$ is a strongly continuous map (resp. a strongly smooth map, weakly submersive), by an abuse of notation.

Definition 3.44

(1) If $\widehat{f} : (X, Z; \widehat{\mathcal{U}'}) \to Y$ is a strongly continuous map and $\widehat{\Phi} = \{\Phi_p\} : \widehat{\mathcal{U}} \to \widehat{\mathcal{U}'}$ is a KK-embeding, then $f_p \circ \varphi_p : U_p \to Y$ defines a strongly continuous map, which we call the *pullback* and write $\widehat{f} \circ \widehat{\Phi}$.
(2) Let $\widehat{\Phi}$ be a KK-embedding. If $\widehat{f}$ is strongly smooth, then so is $\widehat{f} \circ \widehat{\Phi}$. If $\widehat{\Phi}$ is an open embedding and $\widehat{f}$ is weakly submersive, then $\widehat{f} \circ \widehat{\Phi}$ is also weakly submersive.
(3) The good coordinate system version of pullback of maps can be defined in the same way as (1). A statement similar to (2) holds as well.
(4) If $\widehat{f} : (X, Z; \widehat{\mathcal{U}}) \to Y$ is a strongly continuous map from a good coordinate system and $\widehat{\Phi} : \widehat{\mathcal{U}} \to \widehat{\mathcal{U}}$ is a KG-embedding, then the pullback $\widehat{f} \circ \widehat{\Phi}$ can be defined in the same way as (1). A statement similar to (2) holds as well.

[8] The proof of continuity is the same as Remark 3.41.

We remark the following lemma. (This lemma is suggested by the referee. We do not use it in this book.)

Lemma 3.45 *Let $\widehat{f} : (X, Z; \widehat{\mathcal{U}}) \to M$ be a strongly smooth map from a relative K-space. Then there exists a Kuranishi structure $\widehat{\mathcal{U}^+}$, KK-embedding $\widehat{\mathcal{U}} \to \widehat{\mathcal{U}^+}$ and a strongly smooth map $\widehat{f^+} : (X, Z; \widehat{\mathcal{U}^+}) \to M$ such that $\widehat{f^+}$ pulls back to $\widehat{f}$ and $\widehat{f^+}$ is weakly submersive.*

Proof We take finitely many vector fields $V_1, \dots, V_N$ such that at each point $p \in M$ the vectors $V_1(p), \dots, V_N(p)$ generate T_pM. Let $\phi^t_{V_i}$ $(t \in (-\epsilon, \epsilon))$ be a one-parameter group of transformations of M generated by V_i. We put

$$F_{\vec{t}} = \phi^{t_1}_{V_1} \circ \phi^{t_2}_{V_2} \cdots \circ \phi^{t_N}_{V_N}.$$

Let $\mathcal{U}_p = (U_p, \mathcal{E}_p, \psi_p, s_p)$ be the Kuranishi chart of $\widehat{\mathcal{U}}$. We put $U^+_p = \mathcal{U}_p \times (-\epsilon, \epsilon)^N$, $\mathcal{E}^+_p = \pi^*\mathcal{E}_p \oplus \mathbb{R}^N$ (here $\pi : U^+_p \to U_p$ is the projection and $\mathbb{R}^N$ denotes the trivial bundle of rank N on U^+_p.) We then put $s^+_p(x, \vec{t}) = (s_p(x), \vec{t})$. Then $(s^+_p)^{-1}(0)$ is canonically identified with $s^{-1}_p(0)$ and $\psi_p = \psi^+_p$ under this identification. We put

$$f^+_p(x, \vec{t}) = F_{\vec{t}}(\psi_p(x)).$$

It is easy to see that they have the required properties. □

3.4 Notes on Various Versions of the Definitions

In our previous writings, slightly different definitions of Kuranishi structure and good coordinate system are used sometimes. In this section we explain the relationship between those slightly different versions and the one in this book.[9] This section is not used anywhere else in this book. Our purpose in this section is to clarify the fact that as for the previous works[10] which applied Kuranishi structure and/or good coordinate system to the moduli spaces of pseudo-holomorphic curves, its rigor or correctness is not affected by the presence of various slightly different versions of definitions or errors of our previous writings (such as those mentioned in Sects. 3.4.1, 3.4.3).

[9]There are various other versions proposed by other authors. We do not mention them unless they are directly related to the discussion here.

[10]Both by other authors and by us.

3.4.1 Tangent Bundle Condition

We first repeat the point we wrote in [FOOO2, Subsection A1.6.iv]. Definition 3.2 (5) is different from the corresponding one in [FOn2]. Namely in [FOn2] the isomorphism (3.1) is not required to be induced by the differential of the Kuranishi map but only existence of such an isomorphism for which the diagram (3.2) commutes was required. In fact, our proof of [FOn2, Theorem 6.4] implicitly uses the condition that (3.2) is induced by the differential of the Kuranishi map.

Therefore this is an error and Definition 3.2 (5) is necessary for the correct definition. [FOOO2, Example A1.64] illustrates a space X with Kuranishi structure in the sense of [FOn2] but does not have Kuranishi structure in the sense of Definition 3.2. We need to exclude X in [FOOO2, Example A1.64] so that cobordism invariance of virtual fundamental cycle holds.

In all the applications of the Kuranishi structure to the moduli problems we know, the isomorphism (3.1) is induced by the differential of the Kuranishi map. So this error does not cause any problem for applications.

3.4.2 Global Quotient

In this book, a Kuranishi chart is $(U, \mathcal{E}, s, \psi)$ where U is an orbifold. In some of the previous writing it is (V, Γ, E, s, ψ) where V is a manifold Γ is a finite group acting on it effectively and E is a vector space on which Γ acts linearly, $s : V \to E$ is a Γ equivalent map. By putting $U = V/\Gamma$, $\mathcal{E} = (V \times E)/\Gamma$ the latter induces the former. Any pair of an orbifold and its vector bundle is of the form $(V/\Gamma, (V \times E)/\Gamma)$ locally. So by replacing a Kuranishi structure with an open substructure and a good coordinate system with a weakly open substructure, we may assume that their Kuranishi charts are of the form (V, Γ, E, s, ψ).

In those previous writings, coordinate changes are assumed to be global quotients, that is, induced by an equivariant smooth embedding between manifolds. See Chap. 29, where we show we can always take an open substructure of a Kuranishi structure (or a weakly open substructure of a good coordinate system) so that coordinate changes are also global quotients.

3.4.3 Germ of Kuranishi Chart

In [FOn2] the notion of a germ of a Kuranishi chart was used for the Kuranishi structure. This notion was not appropriate and for this reason there is an error in the definition. In [FOOO2] and all of our writings thereafter we never use the notion of a germ of a Kuranishi chart. See [FOOO16, Subsection 34.1] about this point.

3.4.4 Definition 3.15 (7), (8)

Definition 3.15 (7), (8) were not included in the definition of a good coordinate system in [FOn2], [FOOO4]. Definition 3.15 (7) is equivalent to the two formulas. One is by Joyce [Jo1, Formula (32) page 38] which is $\varphi_{\mathfrak{p}\mathfrak{q}}(U_{\mathfrak{p}\mathfrak{q}}) \cap \varphi_{\mathfrak{p}\mathfrak{r}}(U_{\mathfrak{p}\mathfrak{r}}) = \varphi_{\mathfrak{p}\mathfrak{r}}(\varphi_{\mathfrak{q}\mathfrak{r}}^{-1}(U_{\mathfrak{p}\mathfrak{q}}) \cap U_{\mathfrak{p}\mathfrak{r}})$ for $\mathfrak{p} > \mathfrak{q} > \mathfrak{r}$. The other is [FOOO16, Condition 5.9], that is, $U_{\mathfrak{p}\mathfrak{r}} \cap U_{\mathfrak{q}\mathfrak{r}} = \varphi_{\mathfrak{q}\mathfrak{r}}^{-1}(U_{\mathfrak{p}\mathfrak{q}})$ for $\mathfrak{p} > \mathfrak{q} > \mathfrak{r}$. See the next item.

3.4.5 'Hausdorffness' Issue

Definition 3.15 (8) is related to so-called 'Hausdorffness issue' which was discussed much in the Google group 'Kuranishi' in the year 2012. In our construction of virtual fundamental chains in the years 1996–2011 we did not use the space which is obtained by gluing various charts $U_{\mathfrak{p}}$ by coordinate changes. Our argument was constructing perturbation on $U_{\mathfrak{q}}$ and then extending it in $U_{\mathfrak{p}}$ with $\mathfrak{q} < \mathfrak{p}$, while shrinking the domains. For this reason we did not have motivation to formulate conditions such as Definition 3.15 (7), (8), until 2012.

In the Google group 'Kuranishi', in March 2012, we were asked questions about the Hausdorffness of the space obtained by gluing various $U_{\mathfrak{p}}$'s, from people including K. Wehrheim. D. Yang also uploaded a document to the Google group 'Kuranishi'. In his document Yang proposed a 'new definition of the Kuranishi structure' (Definition 1.2.1 in that document), where he *assumed* existence of a metric space, metric background, which contains all the Kuranishi charts U_p. Then he formulated a condition which he called the maximality condition. Yang proposed a similar 'new definition of a good coordinate system' where existence of metric background and maximality condition are assumed. It is fairly obvious that Definition 3.15 (7), (8) follow from Yang's maximality condition.

Yang's formulation is very much natural and useful, for his purpose in [Ya1],[11] where he constructed a (version of) Kuranishi structure from another structure called a polyfold (see [HWZ3] etc.). A polyfold is a certain infinite-dimensional object into which moduli spaces are embedded. Therefore in the situation of [Ya1] 'metric background' is the polyfold itself he starts with and assuming its existence is very reasonable for his purpose. On the other hand, it is in general heavy and nontrivial work to obtain a polyfold containing the given moduli space and prove its nice property such as Hausdorffness. We think that, for the purpose of applications of the theory of Kuranishi structure itself, it is crucial to use the formulation of the

[11] In [Ya1] Yang used a formulation which does not explicitly use the ambient space. It is closer to the formulation of this book. One important difference from ours is that conditions similar to (7), (8) are required not only for a good coordinate system but also for a Kuranishi structure in [Ya1].

theory without relying on the existence of the ambient space (with good properties such as Hausdorffness).[12]

During the program 'Moduli Spaces of Pseudo-holomorphic curves and their applications to Symplectic Topology' and 'Workshop on Moduli Spaces of Pseudo-holomorphic Curves I: March 17–21, 2014' at Simons Center for Geometry and Physics, there was a discussion section, where the present authors discussed with people including J. Solomon, D. Yang.

After this discussion the present authors wrote and published a paper [FOOO17]. Theorem 2.9 of [FOOO17] (= Proposition 3.17), together with Theorem 3.35, clarifies the fact that we can *always* construct an 'ambient space' (which is Hausdorff) without assuming its a priori existence. In other words, if a Kuranishi structure (either the version of this book or that of [FOOO4]) is given on X we can construct a good coordinate system (of *any* version in our writing) compatible with it.

We thank all the people (including but not restricted to the people mentioned above) who contributed to the process of clearing up this 'Hausdorffness issue', by participating in the discussions at the 'Google group Kuranishi', during the Simons Center Program mentioned above, or in any other form.

[12] Actually in our recent article [FOOO21] presenting the detail of the construction of the Kuranishi structure on the moduli space of pseudo-holomorphic disks, we used an ambient *set*. While writing the article we found it is rather cumbersome and technically involved work to define a topology on this ambient set and prove its nice property. So we did not try to do so. See [FOOO21, Remark 7.9 (2)]. Proving Hausdorffness of the moduli space itself is much easier in comparison.

Chapter 4
Fiber Product of Kuranishi Structures

4.1 Fiber Product

Before studying fiber products we consider direct products. Let X_i, $i = 1, 2$ be separable metrizable spaces, $Z_i \subseteq X_i$ compact subsets, and $\widehat{\mathcal{U}_i}$ Kuranishi structures of $Z_i \subseteq X_i$. We will define a Kuranishi structure of the direct product $Z_1 \times Z_2 \subseteq X_1 \times X_2$.

Definition 4.1 For $p_i \in Z_i$ let $\mathcal{U}_{p_i} = (U_{p_i}, \mathcal{E}_{p_i}, \psi_{p_i}, s_{p_i})$ be their Kuranishi neighborhoods. Then the direct product Kuranishi neighborhood of $p = (p_1, p_2) \in Z_1 \times Z_2$ is $\mathcal{U}_p = \mathcal{U}_{p_1} \times \mathcal{U}_{p_2} = (U_p, \mathcal{E}_p, \psi_p, s_p)$ given by

$$(U_p, \mathcal{E}_p, \psi_p, s_p) = (U_{p_1} \times U_{p_2}, \mathcal{E}_{p_1} \times \mathcal{E}_{p_2}, \psi_{p_1} \times \psi_{p_2}, s_{p_1} \times s_{p_2}).$$

Here $\mathcal{E}_{p_1} \times \mathcal{E}_{p_2} = \pi_1^*(\mathcal{E}_1) \oplus \pi_2^*(\mathcal{E}_2)$ (where $\pi_i : X_1 \times X_2 \to X_i$ is the projection) and we define its section $s_{p_1} \times s_{p_2}$ by $(s_{p_1} \times s_{p_2})(x, y) = (s_{p_1}(x), s_{p_2}(y))$. This system satisfies the condition of Kuranishi neighborhood (Definition 3.5).

Suppose $q_i \in Z_i$ and $q = (q_1, q_2) \in Z_1 \times Z_2$. If $q \in \psi_p(s_p^{-1}(0))$, then it is easy to see that $q_i \in \psi_{p_i}(s_{p_i}^{-1}(0))$ for $i = 1, 2$. Therefore there exist coordinate changes $\Phi_{p_i q_i} = (\varphi_{p_i q_i}, \widehat{\varphi}_{p_i q_i}, h_{p_i q_i})$ from $\mathcal{U}_{q_i}$ to $\mathcal{U}_{p_i}$. We define

$$\begin{aligned}\Phi_{pq} &= \Phi_{p_1 q_1} \times \Phi_{p_2 q_2} = (U_{pq}, \varphi_{pq}, \widehat{\varphi}_{pq}) \\ &= (U_{p_1 q_1} \times U_{p_2 q_2}, \varphi_{p_1 q_1} \times \varphi_{p_2 q_2}, \widehat{\varphi}_{p_1 q_1} \times \widehat{\varphi}_{p_2 q_2}).\end{aligned}$$

This satisfies the condition of the coordinate change of Kuranishi charts (Definition 3.6).

Then it is also easy to show that $(\{\mathcal{U}_{p_1} \times \mathcal{U}_{p_2}\}, \{\Phi_{p_1 q_1} \times \Phi_{p_2 q_2}\})$ defines a Kuranishi structure of $Z_1 \times Z_2 \subseteq X_1 \times X_2$ in the sense of Definition 3.9. (We note that effectivity of an orbifold is preserved by the direct product.) We call this Kuranishi structure the *direct product Kuranishi structure*.

K. Fukaya et al., *Kuranishi Structures and Virtual Fundamental Chains*, Springer Monographs in Mathematics, https://doi.org/10.1007/978-981-15-5562-6_4

We can easily prove that the direct product of oriented Kuranishi structures [FOOO16, Definition 4.5] is also oriented.

Next we study fiber products. Let $(X, Z; \widehat{\mathcal{U}})$ be a relative K-space and $\widehat{f} = \{f_p\} : (X, Z; \widehat{\mathcal{U}}) \to N$ a strongly smooth map, where N is a smooth manifold of finite dimension. Let $g : M \to N$ be a smooth map between smooth manifolds. We assume M is compact. In this chapter we define a Kuranishi structure on the pair of topological spaces

$$\begin{aligned} Z \times_N M &= \{(p, q) \in Z \times M \mid f(p) = g(q)\}, \\ X \times_N M &= \{(p, q) \in X \times M \mid f(p) = g(q)\}, \end{aligned} \tag{4.1}$$

that is the fiber product. The assumption we need to require is certain transversality, which we define below.

Definition 4.2 We say $\widehat{f}$ is *weakly transversal* to g on $Z \subseteq X$ if the following holds. Let $(p, q) \in Z \times_N M$ and $\mathcal{U}_p = (U_p, \mathcal{E}_p, \psi_p, s_p)$ be a Kuranishi neighborhood of p. We then require that for each $(x, y) \in U_p \times M$ with $f_p(x) = g(y)$ we have

$$(d_x f_p)(T_x U_p) + (d_y g)(T_y M) = T_{f(x)} N. \tag{4.2}$$

Let us explain the meaning of (4.2). We take an orbifold chart $(V_p(x), \Gamma_p(x), \phi_p(x))$ of U_p at x. (Definition 23.6.) Then consider the composition of two continuous maps:

$$V_p(x) \overset{\phi_p(x)}{\longrightarrow} U_p \overset{f_p}{\longrightarrow} N.$$

By definition it is a smooth map between smooth manifolds, which we denote by $f_{p,x}$. The equality (4.2) means

$$(d_{o(x)} f_{p.x})(T_{o_p(x)} V_p(x)) + (d_y g)(T_y M) = T_{f(x)} N,$$

where $o(x) \in V_p(x)$ is the point which satisfies $\phi_p(x)(o(x)) = x$.

We define the weak transversality of a strongly smooth map $\widehat{f}$ from a space with a good coordinate system to M and $g : N \to M$ in the same way.

Example 4.3

(1) If $\widehat{f} : (X, Z; \widehat{\mathcal{U}}) \to N$ is weakly submersive in the sense of Definition 3.40, then for any $g : M \to N$, f is weakly transversal to g.
(2) If $g : M \to N$ is a submersion, then any strongly smooth map $\widehat{f} : (X, Z; \widehat{\mathcal{U}}) \to N$ is weakly transversal to g.
(3) The pullback of the map in Definition 3.44 by an open embedding preserves the weak transversality.

(4) Suppose (X_i, Z_i) $(i = 1, 2)$ have Kuranishi structures $\widehat{\mathcal{U}_i}$ and the maps $\widehat{f_i}$: $(X_i, Z_i; \widehat{\mathcal{U}_i}) \to N$ are strongly smooth. We put:

$$Z_1 \times_N Z_2 = \{(p, q) \in Z_1 \times Z_2 \mid f_1(p) = f_2(q)\}.$$

Let $(p, q) \in Z_1 \times_N Z_2$. We denote by $\mathcal{U}_p, \mathcal{U}_q$ the Kuranishi neighborhoods of p, q respectively and assume the condition

$$(d_x(f_1)_p)(T_x U_p) + (d_y(f_2)_q)(T_y U_q) = T_{(f_1)_p(x)} N \tag{4.3}$$

for each $(x, y) \in U_p \times U_q$ with $(f_1)_p(x) = (f_2)_q(y)$. (The precise meaning of (4.3) can be defined in a similar way to the case of (4.2).) It is easy to see that (4.3) is equivalent to the next condition. Consider the map

$$f = (f_1, f_2) : X_1 \times X_2 \to N \times N.$$

We use the direct product Kuranishi structure (Definition 4.1) on $Z_1 \times Z_2 \subseteq X_1 \times X_2$. Then (4.3) holds if and only if f is transversal to the diagonal embedding $N \to N \times N$ in the sense of Definition 4.2.

(5) We can generalize the situation of (4) to the case when three or more factors are involved. In fact, in the study of the moduli space of pseudo-holomorphic curves, we will encounter the situation where we consider the fiber product of various factors which are organized by a tree or a graph. See Part II Sect. 22.1 or [FOOO4, Subsection 7.1.1].

Definition 4.4 In the situation of Example 4.3 (4) we say $\widehat{f_1}$ is *weakly transversal* to $\widehat{f_2}$ if (4.3) is satisfied.

Now we assume that $\widehat{f} : (X, Z; \widehat{\mathcal{U}}) \to N$ is weakly transversal to $g : M \to N$ in the sense of Definition 4.2 and define a Kuranishi structure on the fiber product (4.1). Recalling that a Kuranishi neighborhood U_p of $p \in Z$ is assumed to be an effective orbifold, we find the following.

Lemma 4.5 *For each $(p, \mathfrak{x}) \in Z \times_N M$ the fiber product $U_p \times_N M$ is again an effective orbifold.*

Proof Let $(p, \mathfrak{x}) \in Z \times_N M$. Pick an orbifold chart (V_p, Γ_p, ϕ_p) at p. Denote by $\tilde{o}_p \in V_p$ the point, which is mapped to o_p under the map $\phi_p : V_p \to U_p$ and by $\widetilde{f_p}$ the lift of $f_p : U_p \to N$. Since $\widetilde{f_p}$ is Γ_p-invariant, we find that $K_{\tilde{o}_p} = \operatorname{Ker} d\widetilde{f_p}$ at $\tilde{o}_p$ is transversal to the tangent space $T_{\tilde{o}_p} V_p^{\Gamma_p}$ of the fixed point set $V_p^{\Gamma_p}$ by Γ_p-action. Hence the tangent space $T_{[\tilde{o}_p, x]} V_p \times_N M$ contains $K_{\tilde{o}_p}$. Since Γ_p acts trivially on $V_p^{\Gamma_p}$ and Γ_p acts effectively on V_p, Γ_p acts effectively on $K_{\tilde{o}_p}$. It implies that the fiber product $U_p \times_N M$ is an effective orbifold. □

Let $\mathcal{U}_p$ be the given Kuranishi neighborhood of p and $(p, \mathfrak{x}) \in Z \times_N M$. We define $U_{(p,\mathfrak{x})} = U_p \times_N M$. Note $U_{(p,\mathfrak{x})}$ is a smooth orbifold by Definition 4.2. The bundle $\mathcal{E}_{(p,\mathfrak{x})}$ is the pullback of $\mathcal{E}_p$ by the map $U_{(p,\mathfrak{x})} \to U_p$ that is the projection to the first factor. The section s_p induces a section $s_{(p,\mathfrak{x})}$ of $\mathcal{E}_{(p,\mathfrak{x})}$ in an obvious way. Note $s_{(p,\mathfrak{x})}^{-1}(0) = s_p^{-1}(0) \times_N M$. Therefore $\psi_p : s_p^{-1}(0) \to X$ induces

$$\psi_{(p,\mathfrak{x})} : s_{(p,\mathfrak{x})}^{-1}(0) = s_p^{-1}(0) \times_N M \to X \times_N M.$$

It is easy to see that $\psi_{(p,\mathfrak{x})}$ is a homeomorphism onto a neighborhood of $(p, \mathfrak{x})$.

In summary we have the following:

Lemma 4.6 *$\mathcal{U}_{(p,\mathfrak{x})} = (U_{(p,\mathfrak{x})}, \mathcal{E}_{(p,\mathfrak{x})}, s_{(p,\mathfrak{x})}, \psi_{(p,\mathfrak{x})})$ is a Kuranishi neighborhood of $(p, \mathfrak{x}) \in X \times_N M$.*

We next consider the coordinate change. Let $(p, \mathfrak{x}), (q, \mathfrak{y}) \in Z \times_N M$. We assume $(q, \mathfrak{y}) = \psi_{(p,\mathfrak{x})}(x, y)$ where $(x, y) \in V_{(p,\mathfrak{x})}$. By definition we have $q = \psi_p(x)$. Therefore by Definition 3.9 there exists a coordinate change $\Phi_{pq} = (U_{pq}, \varphi_{pq}, \widehat{\varphi}_{pq})$ from $\mathcal{U}_q$ to $\mathcal{U}_p$ in the sense of Definition 3.6.

Now we put:

(1) $U_{(p,\mathfrak{x}),(q,\mathfrak{y})} = U_{pq} \times_N M$.
(2) $\varphi_{(p,\mathfrak{x}),(q,\mathfrak{y})} = \varphi_{pq} \times_N \mathrm{id} : U_{pq} \times_N M \to U_p \times_N M$.
(3) $\widehat{\varphi}_{(p,\mathfrak{x}),(q,\mathfrak{y})} = \widehat{\varphi}_{pq} \times_N \mathrm{id} : \mathcal{E}_q|_{U_{pq}} \times_N M \to \mathcal{E}_p \times_N M$.

Lemma 4.7 *$\Phi_{(p,\mathfrak{x}),(q,\mathfrak{y})} = (U_{(p,\mathfrak{x}),(q,\mathfrak{y})}, \varphi_{(p,\mathfrak{x}),(q,\mathfrak{y})}, \widehat{\varphi}_{(p,\mathfrak{x}),(q,\mathfrak{y})})$ defines a coordinate change from $\mathcal{U}_{(q,\mathfrak{y})}$ to $\mathcal{U}_{(p,\mathfrak{x})}$.*

The proof is immediate from the definition.

Lemma 4.8 *Suppose $Z \subseteq X$ has a Kuranishi structure $\widehat{\mathcal{U}}$ and $\widehat{f} : (X, Z; \widehat{\mathcal{U}}) \to N$ is weakly transversal to $g : M \to N$. Then the Kuranishi neighborhoods in Lemma 4.6 together with the coordinate changes given in Lemma 4.7 define a Kuranishi structure of the fiber product of $Z \times_N M \subseteq X \times_N M$.*

The proof is again immediate from the definition.

Definition 4.9

(1) Suppose $\widehat{f} : (X, Z; \widehat{\mathcal{U}}) \to N$ is weakly transversal to $g : M \to N$. We call the Kuranishi structure obtained in Lemma 4.8 the *fiber product Kuranishi structure* and write the resulting relative K-space as

$$(X, Z; \widehat{\mathcal{U}})\ {}_f\times_g M \quad \text{or} \quad (X, Z; \widehat{\mathcal{U}}) \times_N M.$$

As for orientation, refer to Section 8.2 in [FOOO4]. Here we regard $g : M \to N$ as a weakly submersive map from a K-space.[1]

(2) Suppose $\widehat{f_i} : (X_i, Z_i; \widehat{\mathcal{U}_i}) \to N$ are strongly smooth maps. We assume that $\widehat{f_1}$ and $\widehat{f_2}$ are weakly transversal in the sense of Example 4.3 (4). Then we define the *fiber product*

$$(X_1, Z_1; \widehat{\mathcal{U}_1}) \,{}_{f_1}\!\times_{f_2} (X_2, Z_2; \widehat{\mathcal{U}_2}) \quad \text{or} \quad (X_1, Z_1; \widehat{\mathcal{U}_1}) \times_M (X_2, Z_2; \widehat{\mathcal{U}_2})$$

as the fiber product

$$\left((X_1, Z_1; \widehat{\mathcal{U}_1}) \times (X_2, Z_2; \widehat{\mathcal{U}_2})\right) \,{}_{f_1 \times f_2}\!\times_i \Delta_M.$$

Here $i : \Delta_M \to M \times M$ is the embedding of the diagonal. As for orientation, the fiber product orientation of $(X_1, Z_1; \widehat{\mathcal{U}_1}) \,{}_{f_1}\!\times_{f_2} (X_2, Z_2; \widehat{\mathcal{U}_2})$ differs from the one of $\left((X_1, Z_1; \widehat{\mathcal{U}_1}) \times (X_2, Z_2; \widehat{\mathcal{U}_2})\right) \,{}_{f_1 \times f_2}\!\times_i \Delta_M$ by $(-1)^{(\dim \widehat{\mathcal{U}_2} - \dim M) \cdot \dim M}$.

For the purpose of reference we also include other obvious statements.

Lemma 4.10 *We consider the situation of Lemma* 4.8.

(1) *If $\widehat{g} : (X, Z; \widehat{\mathcal{U}}) \to M'$ is also a strongly continuous map, then it induces a strongly continuous map $Z \times_N M \to M'$. It is weakly submersive if $(\widehat{f}, \widehat{g}) : (X, Z; \widehat{\mathcal{U}}) \to N \times M'$ is weakly submersive.*

(2) *If $\widehat{f}$ is weakly submersive, then the projection $Z \times_N M \to M$ is weakly submersive.*

Lemma 4.11 *Let $\widehat{\mathcal{U}}, \widehat{\mathcal{U}^+}$ be Kuranishi structures of $Z \subseteq X$ and $\widehat{\Phi} : \widehat{\mathcal{U}} \to \widehat{\mathcal{U}^+}$ a KK-embedding. Let $\widehat{f} : (X, Z; \widehat{\mathcal{U}^+}) \to N$ be a strongly smooth map and $g : M \to N$ a smooth map between manifolds.*

(1) *If $\widehat{f} \circ \widehat{\Phi} : (X, Z; \widehat{\mathcal{U}}) \to N$ is weakly transversal to g, then so is $\widehat{f} : (X, Z; \widehat{\mathcal{U}^+}) \to N$.*

(2) *In the situation of (1), $\widehat{\Phi}$ induces a KK-embedding*

$$\widehat{\Phi} \times_N M : \widehat{\mathcal{U}} \times_N M \to \widehat{\mathcal{U}^+} \times_N M.$$

The same conclusions hold if we replace $g : M \to N$ by a strongly smooth map from a relative K-space $g : (X', Z'; \widehat{\mathcal{U}'}) \to N$.

---

[1] Let $p : \nu_{\Gamma_g/M \times N} \to \Gamma_g$ be the normal bundle of the graph Γ_g of g. We identify $\nu_{\Gamma_g/M \times N}$ with a tubular neighborhood $U(\Gamma_g)$ of Γ_g in $M \times N$. Then N has a Kuranishi chart $(U(\Gamma_g), \nu_{\Gamma_g/M \times N}, s_{can}, \psi_{can})$, where s_{can} is the tautological section and ψ_{can} is the identification of the zero section with N. Then $pr_M|_{U(\Gamma_g)} : U(\Gamma_g) \to M$ is a submersion.

4.2 Boundaries and Corners I

So far we studied the case when our Kuranishi structures do not have a boundary or corners. Its generalization to the case when our Kuranishi structure and/or the manifold M has a boundary or corners is straightforward. However, we state them for the completeness' sake. Later we need to and will study boundaries and corners more systematically. (See Sects. 5.3, 8.1, 24.3.)

Definition 4.12 An orbifold with corners is a space locally homeomorphic to V/Γ where V is a smooth manifold with corners and Γ is a finite group acting smoothly and effectively on V. We assume the smoothness of the coordinate change as usual.

See Definition 23.12 for more precise and detailed definition.

Definition 4.13 Let M be an orbifold with corners. It has the following canonical stratification: $\{S_k(M)\}$. The stratum $S_k(M)$ is the closure of the set of the points whose neighborhoods are diffeomorphic to open neighborhoods of 0 of the space $([0,1)^k \times \mathbb{R}^{n-k})/\Gamma$. We call this stratification the *corner structure stratification* of M.

It is easy to see that $\overset{\circ}{S}_k(M) = S_k(M) \setminus S_{k+1}(M)$ carries a structure of a smooth orbifold of dimension $n-k$ without boundary or corner. However, this orbifold may not be effective. In this book, we *assume* the next condition in addition as a part of the definition of orbifold with corners.

Convention 4.14 (Corner effectivity hypothesis) When we say M is an orbifold with corners, we assume the orbifold $\overset{\circ}{S}_k(M)$ is always an *effective* orbifold in this book.

Remark 4.15 Let $M = [0,\infty)^2/\mathbb{Z}_2$ where the $\mathbb{Z}_2$-action is by exchanging the factors. M is an effective orbifold but $S_2(M) = \{0\}/\mathbb{Z}_2$ is not an effective orbifold. So this example does not satisfy the condition we assumed in Convention 4.14.

Definition 4.16 For a relative K-space $(X, Z; \widehat{\mathcal{U}})$, we put

$$\begin{aligned} S_k(X, Z; \widehat{\mathcal{U}}) &= \{p \in Z \mid o_p \in S_k(U_p)\}, \\ \overset{\circ}{S}_k(X, Z; \widehat{\mathcal{U}}) &= S_k(X, Z; \widehat{\mathcal{U}}) \setminus \bigcup_{k'>k} S_{k'}(X, Z; \widehat{\mathcal{U}}), \end{aligned} \tag{4.4}$$

where $\mathcal{U}_p = (U_p, E_p, s_p, \psi_p)$ is the Kuranishi neighborhood of p. We call this stratification the *corner structure stratification* of $\widehat{\mathcal{U}}$. We can define the corner structure stratification $\{S_k(X, Z; \widehat{\mathcal{U}})\}$ of a good coordinate system $\widehat{\mathcal{U}}$ in the same way.

Remark 4.17 By our definition, (Definition 23.12 (3)), $\varphi(\overset{\circ}{S}_k(U)) \subseteq \overset{\circ}{S}_k(U')$ holds for embedding of cornered orbifolds $\varphi : U \to U'$.

Lemma 4.18 *For any compact subset K of $\overset{\circ}{S}_k(X, Z; \widehat{\mathcal{U}}) \subseteq X$, the Kuranishi structure $\widehat{\mathcal{U}}$ induces a Kuranishi structure without boundary of dimension* $\dim(X, Z; \widehat{\mathcal{U}}) - k$ *on* $K \subseteq \overset{\circ}{S}_k(X, Z; \widehat{\mathcal{U}})$.

The same conclusion holds for a good coordinate system.

Proof We put

$$\overset{\circ}{S}_k(\mathcal{U}_p) = \left(\overset{\circ}{S}_k(\mathcal{U}_p), E_p|_{\overset{\circ}{S}_k(\mathcal{U}_p)}, s_p|_{\overset{\circ}{S}_k(\mathcal{U}_p)}, \psi_p|_{\overset{\circ}{S}_k(\mathcal{U}_p)}\right).$$

Then we define a Kuranishi neighborhood of $K \subseteq \overset{\circ}{S}_k(X, Z; \widehat{\mathcal{U}})$ at p by $\overset{\circ}{S}_k(\mathcal{U}_p)$. Suppose $q = \psi_p(o_p(q)) \in \psi_p(s_p^{-1}(0)) \cap Z$. Then $q \in S_{k'}(X, Z; \widehat{\mathcal{U}})$ if and only if $o_p(q) \in S_{k'}(U_p)$. Using this fact, we can restrict coordinate changes of $\widehat{\mathcal{U}}$ to $\overset{\circ}{S}_k(\mathcal{U}_p)$ to obtain the desired coordinate changes. The compatibility conditions follow from ones of $\widehat{\mathcal{U}}$. □

Remark 4.19

(1) In general the above Kuranishi structure on $K \subseteq \overset{\circ}{S}_k(X, Z; \widehat{\mathcal{U}})$ may not be orientable even if $\widehat{\mathcal{U}}$ is orientable.
(2) In the case $k = 1$ the above Kuranishi structure of $K \subseteq \overset{\circ}{S}_1(X, Z; \widehat{\mathcal{U}})$ is orientable if $\widehat{\mathcal{U}}$ is orientable.
(3) The Kuranishi structure induced on the normalized corner of $(X, Z; \widehat{\mathcal{U}})$ (see Definition 24.18) is not necessarily orientable even if $\widehat{\mathcal{U}}$ is orientable. It is orientable if the isotropy group of each point of the Kuranishi neighborhood U_p does not exchange the normal factors.

Definition 4.20 Let M_1 and M_2 be smooth orbifolds with corners, N a smooth orbifold without boundary or corner and $f_i : M_i \to N$ smooth maps. We say that f_1 is *transversal* to f_2 if for each k_1, k_2 the restriction $f_1 : \overset{\circ}{S}_{k_1}(M_1) \to N$ is transversal to $f_2 : \overset{\circ}{S}_{k_2}(M_2) \to N$.

We can define the transversality for the case of strongly continuous maps from relative K-spaces with corners to a manifold in the same way. The case of a good coordinate system is the same.

Lemma 4.21

(1) *Suppose that $Z \subseteq X$ has a Kuranishi structure with boundary and/or corners and $\widehat{f} : (X, Z; \widehat{\mathcal{U}}) \to N$ is weakly transversal to $g : M \to N$. Then the fiber product $Z \times_N M \subseteq X \times_N M$ has a Kuranishi structure with corners.*
(2) *If $Z_i \subseteq X_i$ has a Kuranishi structure with boundary and/or corners and $\widehat{f_i} : (X_i, Z_i; \widehat{\mathcal{U}_i}) \to N$ a strongly smooth map to a manifold. Suppose they are weakly transversal to each other. Then the fiber product $Z_1 \times_N Z_2 \subseteq X_1 \times_N X_2$ has a Kuranishi structure with corners.*

The proof is immediate from the definition.

Definition 4.22 We call the Kuranishi structure obtained in Lemma 4.21, the *fiber product Kuranishi structure.*

4.3 A Basic Property of Fiber Products

One important property of a fiber product is its associativity, which we state below.[2] We consider the following situation. Suppose (X_i, Z_i) have Kuranishi structures for $i = 1, 2, 3$ and let $\widehat{f_1} : (X_1, Z_1; \widehat{\mathcal{U}_1}) \to M_1$, $\widehat{f_2} = (\widehat{f}_{2,1}, \widehat{f}_{2,2}) : (X_2, Z_2; \widehat{\mathcal{U}_2}) \to M_1 \times M_2$, $\widehat{f_3} : (X_3, Z_3; \widehat{\mathcal{U}_3}) \to M_2$ be maps which are strongly smooth. We assume $\widehat{f_1}$ is transversal to $\widehat{f}_{2,1}$ and $\widehat{f}_{2,2}$ is transversal to $\widehat{f_3}$.

Lemma 4.23 *In the above situation, the following three conditions are equivalent:*

(1) *The map* $\widehat{f_3} : (X_3, Z_3; \widehat{\mathcal{U}_3}) \to M_2$ *is transversal to the map* $\widehat{f}'_{2,2} : (X_1, Z_1; \widehat{\mathcal{U}_1}) \times_{M_1} (X_2, Z_2; \widehat{\mathcal{U}_2}) \to M_2$, *which is induced by* $\widehat{f}_{2,2}$.
(2) *The map* $\widehat{f_1} : (X_1, Z_1; \widehat{\mathcal{U}_1}) \to M_1$ *is weakly transversal to the map* $\widehat{f}'_{2,1} : (X_2, Z_2; \widehat{\mathcal{U}_2}) \times_{M_2} (X_3, Z_3; \widehat{\mathcal{U}_3}) \to M_1$, *which is induced by* $\widehat{f}_{2,1}$.
(3) *The map*

$$(\widehat{f_1}, \widehat{f_2}, \widehat{f_3}) : (X_1, Z_1; \widehat{\mathcal{U}_1}) \times (X_2, Z_2; \widehat{\mathcal{U}_2}) \times (X_3, Z_3; \widehat{\mathcal{U}_3}) \to M_1^2 \times M_2^2 \tag{4.5}$$

is weakly transversal to

$$\Delta = \{(x_1, x_2, y_1, y_2) \in M_1 \times M_1 \times M_2 \times M_2 \mid x_1 = x_2,\ y_1 = y_2\}.$$

Here we use the direct product Kuranishi structure in the left hand side of (4.5).

In the case where any of those three equivalent conditions is satisfied, we have

$$\begin{aligned} &\left((X_1, Z_1; \widehat{\mathcal{U}_1}) \times_{M_1} (X_2, Z_2; \widehat{\mathcal{U}_2})\right) \times_{M_2} (X_3, Z_3; \widehat{\mathcal{U}_3}) \\ &\cong (X_1, Z_1; \widehat{\mathcal{U}_1}) \times_{M_1} \left((X_2, Z_2; \widehat{\mathcal{U}_2}) \times_{M_2} (X_3, Z_3; \widehat{\mathcal{U}_3})\right). \end{aligned} \tag{4.6}$$

Here the isomorphism $\cong$ in (4.6) is defined as follows.

---

[2]The fiber product in the sense of category theory is always associative if it exists. Since we do *not* study morphisms between K-spaces, the fiber product we defined is *not* the fiber product in the sense of category theory. Therefore we need to prove its associativity. However, it is obvious in our case.

Definition 4.24 Suppose $(X_1, Z_1; \widehat{\mathcal{U}_1})$ and $(X_2, Z_2; \widehat{\mathcal{U}_2})$ are relative K-spaces. Let $f : (X_1, Z_1) \to (X_2, Z_2)$ be a homeomorphism. An *isomorphism* of relative K-spaces between $(X_1, Z_1; \widehat{\mathcal{U}_1})$ and $(X_2, Z_2; \widehat{\mathcal{U}_2})$ assigns the maps f_p, $\hat{f}_p$ to each $p \in X_1$ such that the following holds. Let $\mathcal{U}^1_p, \mathcal{U}^2_{f(p)}$ be the Kuranishi charts of p, $f(p)$ in X_1, X_2, respectively.

(1) $f_p : U^1_p \to U^2_{f(p)}$ is a diffeomorphism of orbifolds.
(2) $\hat{f}_p : \mathcal{E}^1_p \to \mathcal{E}^2_{f(p)}$ is a bundle isomorphism over f_p.
(3) $s^p_2 \circ f_p = \hat{f}_p \circ s^p_1$.
(4) $\psi^2_{f(p)} \circ f_p = f \circ \psi^1_p$ on $(s^1_p)^{-1}(0)$.
(5) $f_p(o^1_p) = o^2_{f(p)}$.

Remark 4.25 This definition of isomorphism is too restrictive to be a natural notion of isomorphism between Kuranishi structures. To find a correct notion of morphisms between K-spaces and of isomorphism between them is interesting and is a highly nontrivial problem. We do not study it here since it is not necessary for our purpose. A slightly better notion is an equivalence as germs of Kuranishi structures. See [Fuk6] and [FOOO21, Definition 6.6].

The proof of Lemma 4.23 is easy and is omitted.

Remark 4.26 In the previous literature such as [FOOO4, Section A1.2] we defined a fiber product using the notion of a good coordinate system. There is one difficulty in defining the fiber product of spaces equipped with a good coordinate system, which we explain below.

For $i = 1, 2$, suppose that X_i have good coordinate systems that are defined by $\mathfrak{P}_i$, $\mathcal{U}^i_{\mathfrak{p}_i} = (U^i_{\mathfrak{p}_i}, \mathcal{E}^i_{\mathfrak{p}_i}, \psi^i_{\mathfrak{p}_i}, s^i_{\mathfrak{p}_i})$, and $\Phi_{\mathfrak{p}_i\mathfrak{q}_i} = (U^i_{\mathfrak{p}_i\mathfrak{q}_i}, \widehat{\varphi}^i_{\mathfrak{p}_i\mathfrak{q}_i}, \varphi^i_{\mathfrak{p}_i\mathfrak{q}_i})$. Let $\widehat{f}_i = \{(f_i)_{\mathfrak{p}_i}\} : (X_i, \mathcal{U}^i_{\mathfrak{p}_i}) \to Y$ be strongly smooth maps. We assume that $f_{1,\mathfrak{p}_1} : U^1_{\mathfrak{p}_1} \to Y$ is transversal to $f_{2,\mathfrak{p}_2} : U^2_{\mathfrak{p}_2} \to Y$ for each $\mathfrak{p}_1 \in \mathfrak{P}_1$ and $\mathfrak{p}_2 \in \mathfrak{P}_2$. Then we define

$$U_{(\mathfrak{p}_1,\mathfrak{p}_2)} = U^1_{\mathfrak{p}_1} \times_Y U^1_{\mathfrak{p}_2}$$

and define other objects $\mathcal{E}_{(\mathfrak{p}_1,\mathfrak{p}_2)}, \psi_{(\mathfrak{p}_1,\mathfrak{p}_2)}, s_{(\mathfrak{p}_1,\mathfrak{p}_2)}$ by taking fiber products in a similar way to define a good coordinate system.

This is written in [FOOO4, Section A1.2]. A point to take care of in this construction is as follows. (This point is mentioned in [Fuk5, Remark 10 page 165] and is discussed in detail by Joyce in [Jo1].)

Let $\mathfrak{p}_i, \mathfrak{q}_i \in \mathfrak{P}_i$ such that $\mathfrak{q}_i < \mathfrak{p}_i$. We assume that the fiber product

$$(U^1_{\mathfrak{p}_1\mathfrak{q}_1} \cap (\mathfrak{s}^1_{\mathfrak{q}_1})^{-1}(0)) \times_Y (U^2_{\mathfrak{p}_2\mathfrak{q}_2} \cap (\mathfrak{s}^2_{\mathfrak{q}_2})^{-1}(0))$$

is nonempty. Then we have

$$\psi_{(\mathfrak{q}_1,\mathfrak{p}_2)}(\mathfrak{s}^{-1}_{(\mathfrak{q}_1,\mathfrak{p}_2)}(0)) \cap \psi_{(\mathfrak{p}_1,\mathfrak{q}_2)}(\mathfrak{s}^{-1}_{(\mathfrak{p}_1,\mathfrak{q}_2)}(0)) \neq \emptyset.$$

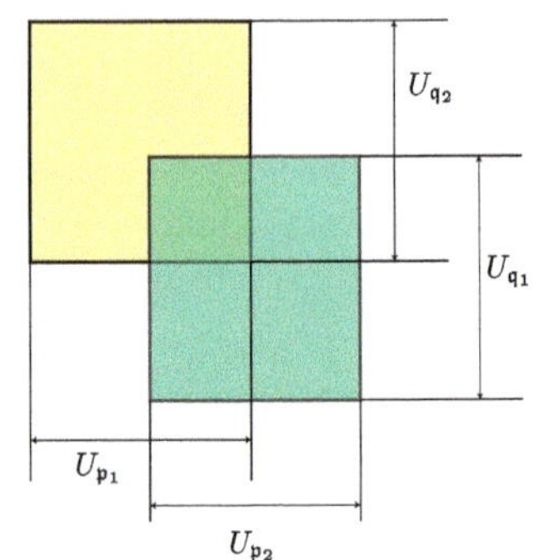

Fig. 4.1 A problem to define the product of good coordinate systems

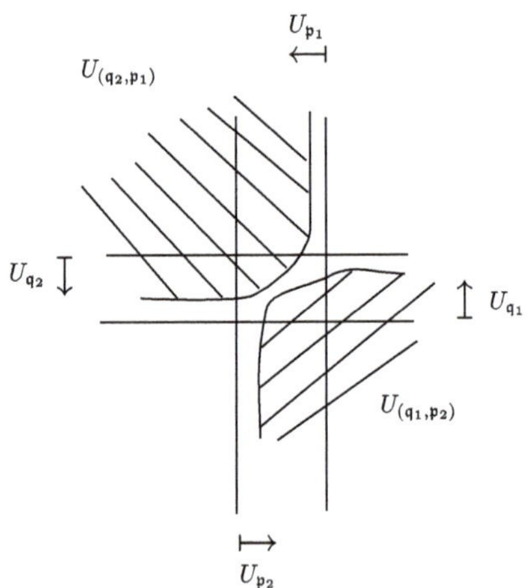

Fig. 4.2 Shrinking product chart

On the other hand, neither $(\mathfrak{q}_1, \mathfrak{p}_2) \le (\mathfrak{p}_1, \mathfrak{q}_2)$ nor $(\mathfrak{p}_1, \mathfrak{q}_2) \le (\mathfrak{q}_1, \mathfrak{p}_2)$ hold. In fact it may happen that $\dim U_{\mathfrak{q}_i} < \dim U_{\mathfrak{p}_i}$. In such a case there is no way to define $\varphi_{(\mathfrak{q}_1,\mathfrak{p}_2),(\mathfrak{p}_1,\mathfrak{q}_2)}$ or $\varphi_{(\mathfrak{p}_1,\mathfrak{q}_2),(\mathfrak{q}_1,\mathfrak{p}_2)}$. See Fig. 4.1.

Remark 4.27 Note that the same problem already occurs while we study the direct product.

We can resolve this problem by shrinking $U_{(\mathfrak{q}_1,\mathfrak{p}_2)}$ and $U_{(\mathfrak{p}_1,\mathfrak{q}_2)}$ appropriately (Fig. 4.2). Joyce [Jo1] gave a beautiful canonical way to perform this shrinking process so that the resulting fiber product is associative. (See also [Fuk5, Figure 14], which is reproduced above with minor changes.)

In the case where we have a multisection (a multivalued perturbation) on the Kuranishi structures on X_i so that the fiber product over Y is transversal on its zero set, we use the fiber product of the multisections. This is especially important when we work on the chain level. Joyce [Jo1] did not seem to discuss this point since, for his purpose in [Jo1], it is unnecessary. We have no doubt that one can incorporate the construction of multisections into Joyce's fiber product so that we can perturb the

Kuranishi maps on the good coordinate systems in a way consistent with the fiber product and also associativity of the fiber product holds together with perturbations.

However, in this book we take a slightly different way. We define the fiber product between the spaces with Kuranishi structures themselves, not those with a good coordinate system. Then the above-mentioned problem does not occur. In other words, the Kuranishi chart (of the fiber product) is defined as the fiber product of Kuranishi charts *without shrinking it* (Lemmas 4.6, 4.7, 4.8). Moreover associativity holds obviously (Lemma 4.21).

On the other hand, compatibility of the multisection with fiber product still needs to be taken care of. In fact, to find a multisection with appropriate transversality properties, we used a good coordinate system. So we need to perform a certain process to move from a good coordinate system back to a Kuranishi structure together with multisections on it. We will discuss this point in Chaps. 5, 6, 7, 8, 9, 10, 11, and 12.

The fiber product with a manifold is defined for a good coordinate system.

Definition 4.28 Let $\widehat{\mathcal{U}} = (\mathfrak{P}, \{\mathcal{U}_{\mathfrak{p}}\}, \{\Phi_{\mathfrak{pq}}\})$ be a good coordinate system of $Z \subseteq X$, $\widehat{f} : (X, \widehat{\mathcal{U}}) \to Y$ a strongly smooth map to a manifold Y and $g : M \to Y$ be a smooth map between smooth manifolds. We assume $\widehat{f}$ is weakly transversal to g.

(1) The fiber product $U_{\mathfrak{p}\ f_{\mathfrak{p}}} \times_g M$ together with various objects defines a Kuranishi chart, which we denote by $\mathcal{U}_{\mathfrak{p}\ f_{\mathfrak{p}}} \times_g M$.
(2) The coordinate change $\Phi_{\mathfrak{pq}}$ induces a coordinate change $\mathcal{U}_{\mathfrak{q}\ f_{\mathfrak{q}}} \times_g M \to \mathcal{U}_{\mathfrak{p}\ f_{\mathfrak{p}}} \times_g M$.
(3) The Kuranishi charts $U_{\mathfrak{p}\ f_{\mathfrak{p}}} \times_g M$ together with coordinate change in Item (2) define a good coordinate system of $Z_f \times_g M \subset X_f \times_g M$. We denote it by $\widehat{\mathcal{U}}_{\widehat{f}} \times_g M$.

Chapter 5
Thickening of a Kuranishi Structure

5.1 Background to Introducing the Notion of Thickening

Let X be a paracompact metrizable space, and let $\widehat{\mathcal{U}} = (\{\mathcal{U}_p\}, \{\Phi_{pq}\})$ be a Kuranishi structure on it. We consider a system of multisections[1] $\{\mathfrak{s}_p\}$ of the vector bundle $\mathcal{E}_p \to U_p$ for each p with the following property:

(*) For each p and $q \in \mathrm{Im}(\psi_p)$ the pullback of $\mathfrak{s}_p$ to U_{pq} that is a multisection of $\varphi_{pq}^*\mathcal{E}_p$ is the image of the multisection $\mathfrak{s}_q$ by the bundle embedding $\widehat{\varphi}_{pq}$.

This is a kind of obvious condition of a multisection (a multivalued perturbation) that is compatible with the Kuranishi structure $\widehat{\mathcal{U}}$. (We define such a notion precisely later in Definition 6.24.)

However, we need to note the following: Let us take a good coordinate system $\widehat{\mathcal{U}} = ((\mathfrak{P}, \le), \{\mathcal{U}_{\mathfrak{p}} \mid \mathfrak{p} \in \mathfrak{P}\}, \{\Phi_{\mathfrak{p}\mathfrak{q}} \mid \mathfrak{p}, \mathfrak{q} \in \mathfrak{P}, \mathfrak{q} \le \mathfrak{p}\})$ such that $\widehat{\mathcal{U}}$ is compatible with $\widehat{\mathcal{U}}$ in the sense of Definition 3.37 and use it to define a system of multisections $\mathfrak{s}_{\mathfrak{p}}$ on $U_{\mathfrak{p}}$. Then it is usually impossible to use $\mathfrak{s}_{\mathfrak{p}}$ to obtain a system of multisections $\mathfrak{s}_p$ of $\widehat{\mathcal{U}}$ that has property (*).

The reason is as follows. Let $p \in X$. We take $\mathfrak{p} \in \mathfrak{P}$ such that $p \in \mathrm{Im}(\psi_{\mathfrak{p}})$. By definition of the compatibility of a good coordinate system and Kuranishi structure, there exists an embedding $\Phi_{\mathfrak{p}p} : \mathcal{U}_p \to \mathcal{U}_{\mathfrak{p}}$. We put $\Phi_{\mathfrak{p}p} = (\varphi_{\mathfrak{p}p}, \widehat{\varphi}_{\mathfrak{p}p})$. Then we consider $\varphi_{\mathfrak{p}p}(o_p) \in U_{\mathfrak{p}}$. (Here $\psi_p(o_p) = p$.) By definition (see Definition 6.2),

$$\mathfrak{s}_{\mathfrak{p}}(\varphi_{\mathfrak{p}p}(o_p)) \in (\mathcal{E}_{\mathfrak{p}}|_{\varphi_{\mathfrak{p}p}(o_p)})^l. \tag{5.1}$$

[1] We will discuss multisections in Chap. 6. Here we just mention them to motivate the definition we give in this chapter. The readers who do not know the definition of a multisection can safely skip the part before Definition 5.1.

K. Fukaya et al., *Kuranishi Structures and Virtual Fundamental Chains*, Springer Monographs in Mathematics, https://doi.org/10.1007/978-981-15-5562-6_5

On the other hand, $\widehat{\varphi}_{\mathfrak{p}p}$ restricts to a linear embedding $\mathcal{E}_p|_{o_p} \to \mathcal{E}_{\mathfrak{p}}|_{\varphi_{\mathfrak{p}p}(o_p)}$. It induces

$$(\mathcal{E}_p|_{o_p})^l \to (\mathcal{E}_{\mathfrak{p}}|_{\varphi_{\mathfrak{p}p}(o_p)})^l. \tag{5.2}$$

By inspecting the construction of the multisection $\mathfrak{s}_{\mathfrak{p}}$ given in [FOn2, p. 955] or Chap. 13 of this book, we find that

$$\mathfrak{s}_{\mathfrak{p}}(\varphi_{\mathfrak{p}p}(o_p)) \notin \mathrm{Im}(5.2) \tag{5.3}$$

in general. So $\mathfrak{s}_{\mathfrak{p}}$ cannot be pulled back to a multisection of $\mathcal{E}_p$ on $U_{\mathfrak{p}p}$.

To explain the reason why (5.3) occurs, we introduce some notation, which is also used in the later chapters.

Definition 5.1 For a Kuranishi structure $\widehat{\mathcal{U}}$ of $Z \subseteq X$, we define the *dimension stratification* of Z by

$$\mathcal{S}_{\mathfrak{d}}(X, Z; \widehat{\mathcal{U}}) = \{p \in Z \mid \dim U_p \geq \mathfrak{d}\}. \tag{5.4}$$

Here $\mathfrak{d} \in \mathbb{Z}_{\geq 0}$.

When $\overset{\triangle}{\mathcal{U}}$ is a good coordinate system of $Z \subseteq X$, we define the dimension stratification of Z by

$$\mathcal{S}_{\mathfrak{d}}(X, Z; \overset{\triangle}{\mathcal{U}}) = \{p \in Z \mid \exists \mathfrak{p},\ \dim U_{\mathfrak{p}} \geq \mathfrak{d},\ p \in \mathrm{Im}\psi_{\mathfrak{p}}\}. \tag{5.5}$$

Lemma 5.2

(1) $\mathcal{S}_{\mathfrak{d}}(X, Z; \widehat{\mathcal{U}})$ *is a closed subset of* Z. $\mathcal{S}_{\mathfrak{d}}(X, Z; \overset{\triangle}{\mathcal{U}})$ *is an open subset of* Z. *Moreover, if* $\mathfrak{d}' < \mathfrak{d}$, *then*

$$\mathcal{S}_{\mathfrak{d}}(X, Z; \widehat{\mathcal{U}}) \subseteq \mathcal{S}_{\mathfrak{d}'}(X, Z; \widehat{\mathcal{U}}), \qquad \mathcal{S}_{\mathfrak{d}}(X, Z; \overset{\triangle}{\mathcal{U}}) \subseteq \mathcal{S}_{\mathfrak{d}'}(X, Z; \overset{\triangle}{\mathcal{U}}).$$

(2) *If* $\widehat{\mathcal{U}}$ *is embedded into* $\widehat{\mathcal{U}'}$ *then*

$$\mathcal{S}_{\mathfrak{d}}(X, Z; \widehat{\mathcal{U}}) \subseteq \mathcal{S}_{\mathfrak{d}}(X, Z; \widehat{\mathcal{U}'}).$$

The equality holds if and only if the embedding from $\widehat{\mathcal{U}}$ *to* $\widehat{\mathcal{U}'}$ *is an open embedding. If the good coordinate system* $\overset{\triangle}{\mathcal{U}}$ *is compatible with* $\widehat{\mathcal{U}}$, *then*

$$\mathcal{S}_{\mathfrak{d}}(X, Z; \widehat{\mathcal{U}}) \subseteq \mathcal{S}_{\mathfrak{d}}(X, Z; \overset{\triangle}{\mathcal{U}}). \tag{5.6}$$

The proof is immediate from the definition. However, we note that the equality almost never holds in (5.6). Namely $\mathcal{S}_{\mathfrak{d}}(X, Z; \overset{\triangle}{\mathcal{U}})$ contains an open neighborhood of $\mathcal{S}_{\mathfrak{d}}(X, Z; \widehat{\mathcal{U}})$. This is the reason why (5.3) occurs.

5.2 Definition of Thickening

To go around this problem we introduce the notion of thickening (Fig. 5.1).

Definition 5.3 Let $\widehat{\mathcal{U}}$ be a Kuranishi structure of $Z \subseteq X$. We say that $(\widehat{\mathcal{U}^+}, \widehat{\Phi})$ is a *strict thickening* of $\widehat{\mathcal{U}}$ if the following conditions are satisfied:

(1) $\widehat{\mathcal{U}^+}$ is a Kuranishi structure of $Z \subseteq X$ and $\widehat{\Phi} = \{(\varphi_p, \widehat{\varphi}_p)\} : \widehat{\mathcal{U}} \to \widehat{\mathcal{U}^+}$ is a strict KK-embedding.

(2) For each $p \in Z$ there exists a neighborhood O_p of p in $\psi_p((s_p)^{-1}(0)) \cap \psi_p^+((s_p^+)^{-1}(0)) \subset X$ with the following properties.

For each $q \in O_p \cap Z$ there exists a neighborhood $W_p(q)$ of $o_p(q)$[2] in U_p such that:

(a) $\varphi_p(W_p(q)) \subseteq \varphi_{pq}^+(U_{pq}^+)$.

(b) For any $x \in W_p(q)$, $y \in U_{pq}^+$ with $\varphi_p(x) = \varphi_{pq}^+(y)$, we have

$$\widehat{\varphi}_p(E_p|_x) \subseteq \widehat{\varphi}_{pq}^+(E_q^+|_y).$$

A *thickening* is by definition a strict thickening of an open substructure. We sometimes say $\widehat{\mathcal{U}^+}$ is a thickening of $\widehat{\mathcal{U}}$ by an abuse of notation.

We write $\widehat{\mathcal{U}} < \widehat{\mathcal{U}^+}$ if $\widehat{\mathcal{U}^+}$ is a thickening of $\widehat{\mathcal{U}}$.

Remark 5.4 Condition (2) above implies that

(☆) $S_{\mathfrak{d}}(X, Z; \widehat{\mathcal{U}^+})$ is a neighborhood of $S_{\mathfrak{d}}(X, Z; \widehat{\mathcal{U}})$.

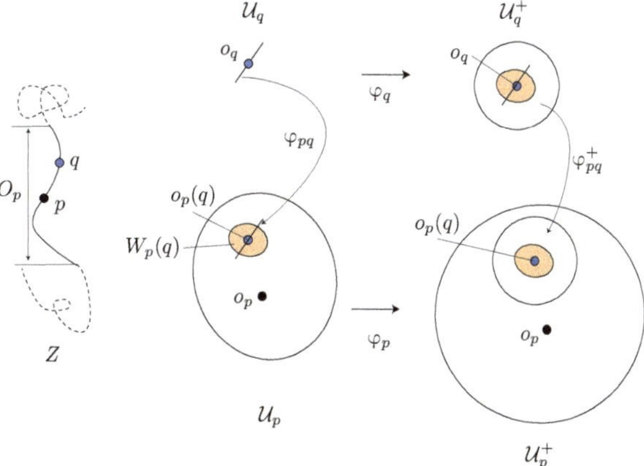

Fig. 5.1 Thickening

[2] See Definition 3.5 for this notation.

In fact if $q \in O_p \cap Z$, then

$$\dim U_q^+ \geq \dim W_p(q) = \dim U_p$$

by Condition (2)(a). In particular, $\widehat{\mathcal{U}}$ is almost never a thickening of itself, when $\dim U_p$ varies over p.

In the case $\dim U_p^+$ is strictly greater than $\dim U_p$ and $\dim U_q^+$, Condition (☆) may not imply Condition (2)(a).

A typical way to obtain a thickening of $\widehat{\mathcal{U}}$ is by constructing a good coordinate system $\widehat{\mathcal{U}}$, a Kuranishi structure $\widehat{\mathcal{U}^+}$, a KG-embedding $\widehat{\mathcal{U}} \to \widehat{\mathcal{U}}$ and a GK-embedding $\widehat{\mathcal{U}} \to \widehat{\mathcal{U}^+}$. Then $\widehat{\mathcal{U}^+}$ becomes a thickening of $\widehat{\mathcal{U}}$ by Proposition 5.11. See Sect. 5.4.

Lemma 5.5 *If $(\widehat{\mathcal{U}^+}, \widehat{\Phi})$ is a thickening of $\widehat{\mathcal{U}}$ and $(\widehat{\mathcal{U}^{++}}, \widehat{\Phi^+})$ is a thickening of $\widehat{\mathcal{U}^+}$, then $(\widehat{\mathcal{U}^{++}}, \widehat{\Phi^+} \circ \widehat{\Phi})$ is a thickening of $\widehat{\mathcal{U}}$.*

Proof Let O_p, $W_p(q)$ be as in Definition 5.3 (2) (a) for $\widehat{\Phi} : \widehat{\mathcal{U}} \to \widehat{\mathcal{U}^+}$ and let O_p^+, $W_p^+(q)$ be one for $\widehat{\Phi^+} : \widehat{\mathcal{U}^+} \to \widehat{\mathcal{U}^{++}}$. We put $O_p^{++} = O_p \cap O_p^+$ and $W_p^{++}(q) = W_p(q) \cap \varphi_p^{-1}(W_p^+(q))$. Suppose $q \in O_p^{++}$. Then we have

$$(\varphi_p^+ \circ \varphi_p)(W_p^{++}(q)) = \varphi_p^+(\varphi_p(W_p^{++}(q))) \subseteq \varphi_p^+(W_p^+(q)) \subset \varphi_p^{++}(U_{pq}^{++}).$$

Thus we have checked Definition 5.3 (2) (a). The proof of Definition 5.3 (2) (b) is similar. □

We will prove the existence of a thickening in Sect. 5.4.

5.3 Embedding of Kuranishi Structures II

Definition 5.6 Let $\widehat{\mathcal{U}} = (\mathfrak{P}, \{\mathcal{U}_{\mathfrak{p}}\}, \{\Phi_{\mathfrak{pq}}\})$ be a good coordinate system of $Z \subseteq X$ and $\widehat{\mathcal{U}^+}$ a Kuranishi structure of $Z \subseteq X$. A *GK-embedding* $\widehat{\Phi} : \widehat{\mathcal{U}} \to \widehat{\mathcal{U}^+}$ is a collection $\{(U_{\mathfrak{p}}(p), \Phi_{p\mathfrak{p}})\}$ with the following properties:

(1) $(U_{\mathfrak{p}}(p), \Phi_{p\mathfrak{p}})$ is defined for all $\mathfrak{p} \in \mathfrak{P}$ and $p \in \psi_{\mathfrak{p}}(s_{\mathfrak{p}}^{-1}(0)) \cap Z$.
(2) $U_{\mathfrak{p}}(p)$ is an open neighborhood of $o_{\mathfrak{p}}(p)$ in $U_{\mathfrak{p}}$ where $\psi_{\mathfrak{p}}(o_{\mathfrak{p}}(p)) = p$.
(3) $\Phi_{p\mathfrak{p}} : \mathcal{U}_{\mathfrak{p}}|_{U_{\mathfrak{p}}(p)} \to \mathcal{U}_p^+$ is an embedding of Kuranishi charts.
(4) Let $\mathfrak{q} \in \mathfrak{P}$, $\mathfrak{q} \leq \mathfrak{p}$, $p \in \psi_{\mathfrak{p}}(s_{\mathfrak{p}}^{-1}(0)) \cap Z$, and $q \in \psi_{\mathfrak{p}}(U_{\mathfrak{p}}(p) \cap s_{\mathfrak{p}}^{-1}(0)) \cap \psi_{\mathfrak{q}}(s_{\mathfrak{q}}^{-1}(0)) \cap Z$. Note that under this assumption

$$\psi_{\mathfrak{p}}(U_{\mathfrak{p}}(p) \cap s_{\mathfrak{p}}^{-1}(0)) \cap \psi_{\mathfrak{q}}(s_{\mathfrak{q}}^{-1}(0)) \cap Z \subseteq \psi_p^+(U_p^+ \cap (s_p^+)^{-1}(0))$$

follows from Item (1). We require:

$$\Phi_{p\mathfrak{p}} \circ \Phi_{\mathfrak{p}\mathfrak{q}} = \Phi^+_{pq} \circ \Phi_{q\mathfrak{q}}$$

on $\varphi_{\mathfrak{p}\mathfrak{q}}^{-1}(U_{\mathfrak{p}}(p)) \cap (\varphi_{q\mathfrak{q}})^{-1}(U^+_{pq})$. In other words, the next diagram commutes:

$$\begin{array}{ccc}
\mathcal{U}_{\mathfrak{q}}|_{\varphi_{\mathfrak{p}\mathfrak{q}}^{-1}(U_{\mathfrak{p}}(p))\cap(\varphi_{q\mathfrak{q}})^{-1}(U^+_{pq})} & \xrightarrow{\Phi_{q\mathfrak{q}}} & \mathcal{U}^+_q|_{U^+_{pq}} \\
\Phi_{\mathfrak{p}\mathfrak{q}}\big\downarrow & & \big\downarrow\Phi^+_{pq} \\
\mathcal{U}_{\mathfrak{p}}|_{U_{\mathfrak{p}}(p)} & \xrightarrow{\Phi_{p\mathfrak{p}}} & \mathcal{U}^+_p
\end{array} \tag{5.7}$$

Remark 5.7

(1) The case $p = q$ (but $\mathfrak{p} \neq \mathfrak{q}$) is included in Definition 5.6 (4). The case $\mathfrak{p} = \mathfrak{q}$ (but $p \neq q$) is also included.
(2) We include $U_{\mathfrak{p}}(p)$ as a part of the data to define an embedding. We sometimes need to shrink it. Such a process is included in the discussion below.

Definition-Lemma 5.8 We can pull back a strongly continuous (resp. strongly smooth) map from a Kuranishi structure to one from a good coordinate system by a GK-embedding. Weak submersivity is preserved by an open embedding.

The proof is immediate from the definition.

Shortly we will define compositions of embeddings of various types. See the table below, where KS = Kuranishi structure, GCS = good coordinate system (Tables 5.1 and 5.2).

Table 5.1 Definitions of embeddings

Source	Target	Symbol	Definition	Name	Comment
KS	KS	$\widehat{\mathcal{U}} \to \widehat{\mathcal{U}^+}$	Definition 3.19	KK-embedding	(1)
KS	GCS	$\widehat{\mathcal{U}} \to \overset{\triangle}{\mathcal{U}}$	Definition 3.30	KG-embedding	(2)
GCS	KS	$\overset{\triangle}{\mathcal{U}} \to \widehat{\mathcal{U}^+}$	Definition 5.6	GK-embedding	—
GCS	GCS	$\overset{\triangle}{\mathcal{U}} \to \overset{\triangle}{\mathcal{U}^+}$	Definition 3.24	GG-embedding	—

Comments: (1) Strict version and non-strict version exist. (Definition 5.14 (1)(2).) (2) Strict version and non-strict version exist.

Table 5.2 Compositions of embeddings 1st structure $\longrightarrow$ 2nd structure $\longrightarrow$ 3rd structure

1st structure	2nd structure	3rd structure	Definition	Comment
KS	KS	KS	Lemma 3.20	(1)
KS	KS	GCS	Lemma 3.32	(2)
KS	GCS	KS	Proposition 5.11	(3)
KS	GCS	GCS	Lemma 3.34	(4)
GCS	KS	KS	Definition-Lemma 5.10 (ii)	(5)
GCS	KS	GCS	Definition-Lemma 5.17	(6)
GCS	GCS	KS	Definition-Lemma 5.10 (i)	—
GCS	GCS	GCS	Lemma 3.26	—

Comments: (1) Strict version and non-strict version exist. Composition of non-strict version is well-defined only up to equivalence. (Definition 5.14 (3)(4).) (2) Strict version and non-strict version exist. Composition of non-strict version is well-defined only up to equivalence. (Definition 5.14 (3).) (3) Need to restrict to an open substructure. The composition becomes a thickening. (4) Need to take an open substructure for the 3rd structure. (Remark 3.33.) (5) Need to take an open substructure for the 3rd structure. (6) Need to restrict the 1st structure to a weakly open substructure.

Definition 5.9

(1) Let $\widehat{\Phi} = \{(U_{\mathfrak{p}}(p), \Phi_{\mathfrak{p}p})\} : \widehat{\mathcal{U}} \to \widehat{\mathcal{U}^+}$ be a GK-embedding. Suppose for each $\mathfrak{p} \in \mathfrak{P}$ and $p \in \psi_{\mathfrak{p}}(s_{\mathfrak{p}}^{-1}(0)) \cap Z$ we are given an open subset $U'_{\mathfrak{p}}(p)$ of $U_{\mathfrak{p}}(p)$ such that $o_{\mathfrak{p}}(p) \in U'_{\mathfrak{p}}(p)$. Then $\{(U'_{\mathfrak{p}}(p), \Phi_{\mathfrak{p}p}|_{U'_{\mathfrak{p}}(p)})\}$ is also a GK-embedding $\widehat{\mathcal{U}} \to \widehat{\mathcal{U}^+}$, which we call an *open restriction of the embedding* $\widehat{\Phi}$.

(2) Two GK-embeddings $\widehat{\mathcal{U}} \to \widehat{\mathcal{U}^+}$ are said to be *equivalent* if they have a common open restriction. This is obviously an equivalence relation.

Definition-Lemma 5.10 Let $\widehat{\Phi} : \widehat{\mathcal{U}} \to \widehat{\mathcal{U}}$ be a GK-embedding, $\widehat{\Phi_1} : \widehat{\mathcal{U}'} \to \widehat{\mathcal{U}}$ a GG-embedding and $\widehat{\Phi_2} : \widehat{\mathcal{U}} \to \widehat{\mathcal{U}^+}$ a strict KK-embedding.

(i) The composition

$$\widehat{\Phi'} := \widehat{\Phi} \circ \widehat{\Phi_1} : \widehat{\mathcal{U}'} \longrightarrow \widehat{\mathcal{U}} \longrightarrow \widehat{\mathcal{U}}$$

can be defined and is a GK-embedding.

(ii) There exists an open substructure $\widehat{\mathcal{U}^{+0}}$ of $\widehat{\mathcal{U}^+}$ such that the composition $\widehat{\Phi^+} := \widehat{\Phi_2} \circ \widehat{\Phi} : \widehat{\mathcal{U}} \to \widehat{\mathcal{U}^{+0}}$ can be defined and is a GK-embedding.

Proof We denote $\widehat{\mathcal{U}'} = \{\mathcal{U}'_{\mathfrak{p}'} \mid \mathfrak{p}' \in \mathfrak{P}'\}$, $\widehat{\mathcal{U}} = \{\mathcal{U}_{\mathfrak{p}} \mid \mathfrak{p} \in \mathfrak{P}\}$, $\widehat{\mathcal{U}} = \{\mathcal{U}_p \mid p \in Z\}$, $\widehat{\mathcal{U}^+} = \{\mathcal{U}_p^+ \mid p \in Z\}$ and $\widehat{\Phi} = \{(U_{\mathfrak{p}}(p), \Phi_{p\mathfrak{p}})\}$, $\widehat{\Phi_1} = (\mathrm{i}, \{\Phi_{1,\mathfrak{p}'}\})$, $\widehat{\Phi_2} = \{\Phi_{2,p}\}$. Here and hereafter $p, q \in Z$, $\mathfrak{p}', \mathfrak{q}' \in \mathfrak{P}'$, $\mathfrak{p}, \mathfrak{q} \in \mathfrak{P}$ and $\mathrm{i}(\mathfrak{q}') = \mathfrak{q}$, $\mathrm{i}(\mathfrak{p}') = \mathfrak{p}$.

(i) Suppose $p \in \mathrm{Im}(\psi'_{\mathfrak{p}'}) \cap Z$. It is easy to see that $p \in \mathrm{Im}(\psi_{\mathfrak{p}}) \cap Z$. So $U_{\mathfrak{p}}(p)$ is defined and is a neighborhood of $o_{\mathfrak{p}}(p) \in U_{\mathfrak{p}}$. We put

$$U'_{\mathfrak{p}'}(p) = \varphi_{1,\mathfrak{p}'}^{-1}(U_{\mathfrak{p}}(p)).$$

This is a neighborhood of $o_{\mathfrak{p}'}(p)$ in $U'_{\mathfrak{p}'}$. We define

$$\Phi'_{p\mathfrak{p}'} = \Phi_{p\mathfrak{p}} \circ \Phi_{1,\mathfrak{p}'}|_{U_{\mathfrak{p}}(p)} : \mathcal{U}_{\mathfrak{p}'}|_{U'_{\mathfrak{p}'}(p)} \to \mathcal{U}_p.$$

Definition 5.6 (1)(2)(3) are obviously satisfied. We check (4) below.

Suppose $\mathfrak{q}' \in \mathfrak{P}'$, $\mathfrak{q}' \le \mathfrak{p}'$, $q \in \psi_{\mathfrak{p}'}(U_{\mathfrak{p}'}(p) \cap s_{\mathfrak{p}'}^{-1}(0))$ and $q \in \psi_{\mathfrak{q}'}(s_{\mathfrak{q}'}^{-1}(0)) \cap Z$. They imply $q \in \psi_{\mathfrak{p}}(U_{\mathfrak{p}}(p) \cap s_{\mathfrak{p}}^{-1}(0))$, $q \in \psi_{\mathfrak{q}}(s_{\mathfrak{q}}^{-1}(0)) \cap Z$. Since $\widehat{\Phi}$ is a GK-embedding it implies $q \in \psi_p(U_p \cap s_p^{-1}(0))$.

Suppose $x \in (\varphi'_{\mathfrak{p}'\mathfrak{q}'})^{-1}(U'_{\mathfrak{p}'}(p)) \cap (\varphi'_{q\mathfrak{q}'})^{-1}(U_{pq})$. In particular, $x \in U'_{\mathfrak{p}'\mathfrak{q}'}$. Therefore by Definition 3.24 (1), $\varphi_{1,\mathfrak{q}'}(x) \in U_{\mathfrak{p}\mathfrak{q}}$. Then Definition 3.24 (2) implies:

$$(\varphi_{\mathfrak{p}\mathfrak{q}} \circ \varphi_{1,\mathfrak{q}'})(x) = (\varphi_{1,\mathfrak{p}'} \circ \varphi'_{\mathfrak{p}'\mathfrak{q}'})(x) \in U_{\mathfrak{p}}(p). \tag{5.8}$$

Therefore

$$\varphi_{1,\mathfrak{q}'}(x) \in \varphi_{\mathfrak{p}\mathfrak{q}}^{-1}(U_{\mathfrak{p}}(p)) \cap (\varphi_{q\mathfrak{q}})^{-1}(U_{pq}).$$

In fact $\varphi_{1,\mathfrak{q}'}(x) \in \varphi_{\mathfrak{p}\mathfrak{q}}^{-1}(U_{\mathfrak{p}}(p))$ is a consequence of (5.8) and $\varphi_{1,\mathfrak{q}'}(x) \in (\varphi_{q\mathfrak{q}})^{-1}(U_{pq})$ is a consequence of $x \in (\varphi'_{q\mathfrak{q}'})^{-1}(U_{pq})$.

Thus we can use (5.8) and the commutativity of Diagram (5.7) to show

$$\begin{aligned}(\varphi_{pq} \circ \varphi'_{q\mathfrak{q}'})(x) &= (\varphi_{pq} \circ \varphi_{q\mathfrak{q}} \circ \varphi_{1,\mathfrak{q}'})(x) = (\varphi_{p\mathfrak{p}} \circ \varphi_{\mathfrak{p}\mathfrak{q}} \circ \varphi_{1,\mathfrak{q}'})(x) \\ &= (\varphi_{p\mathfrak{p}} \circ \varphi_{1,\mathfrak{p}'} \circ \varphi'_{\mathfrak{p}'\mathfrak{q}'})(x) = (\varphi_{p\mathfrak{p}'} \circ \varphi'_{\mathfrak{p}'\mathfrak{q}'})(x).\end{aligned}$$

This is the orbifold embedding part of the required equality. The bundle embedding part can be proved in the same way.

(ii) The Kuranishi charts of $\widehat{\mathcal{U}^{+0}}$ are the same as those of $\widehat{\mathcal{U}^{+}}$. The coordinate changes of $\widehat{\mathcal{U}^{+0}}$ are $\Phi^{+}_{pq}|_{U^{+0}_{pq}}$ which are restrictions of the coordinate changes Φ^{+}_{pq} of $\widehat{\mathcal{U}^{+}}$ to open subsets U^{+0}_{pq} of $U^{+}_{q} = U^{+0}_{q}$. We will choose U^{+0}_{pq} later.

For $p \in \mathrm{Im}(\psi_{\mathfrak{p}})$ we define

$$U^{+}_{\mathfrak{p}}(p) = U_{\mathfrak{p}}(p), \qquad \varphi^{+}_{p\mathfrak{p}} = \varphi_{2,p} \circ \varphi_{p\mathfrak{p}} : U^{+}_{\mathfrak{p}}(p) \to U^{+}_{p}.$$

Definition 5.6 (1)(2)(3) are obviously satisfied. We check (4) below.

Suppose $\mathfrak{q} \in \mathfrak{P}$, $\mathfrak{q} \le \mathfrak{p}$, $q \in \psi_{\mathfrak{p}}(U_{\mathfrak{p}}^+(p) \cap (s_{\mathfrak{p}})^{-1}(0))$ and $q \in \psi_{\mathfrak{q}}(s_{\mathfrak{q}}^{-1}(0)) \cap Z$. Then since $\widehat{\Phi_1}$ is a GK-embedding we have $q \in \psi_p(U_p \cap (s_p)^{-1}(0))$ by Definition 5.6 (4). Therefore $q \in \psi_p^+(U_p^+ \cap (s_p^+)^{-1}(0))$ by Definition 3.19 (1). We now choose U_{pq}^{+0} small so that

$$U_{pq} \supseteq \varphi_{2,q}^{-1}(U_{pq}^{+0}).$$

Let $x \in (\varphi_{\mathfrak{p}\mathfrak{q}})^{-1}(U_{\mathfrak{p}}^+(p)) \cap (\varphi_{q\mathfrak{q}}^+)^{-1}(U_{pq}^{+0})$. Then by our choice of U_{pq}^{+0} we have $\varphi_{q\mathfrak{q}}(x) \in U_{pq}$. Hence $x \in (\varphi_{\mathfrak{p}\mathfrak{q}})^{-1}(U_{\mathfrak{p}}(p)) \cap (\varphi_{q\mathfrak{q}})^{-1}(U_{pq})$. Therefore by Definition 5.6 (3),

$$(\varphi_{pq} \circ \varphi_{q\mathfrak{q}})(x) = (\varphi_{p\mathfrak{p}} \circ \varphi_{\mathfrak{p}\mathfrak{q}})(x). \tag{5.9}$$

On the other hand, by Definition 3.19 (3),

$$(\varphi_{2,p} \circ \varphi_{pq} \circ \varphi_{q\mathfrak{q}})(x) = (\varphi_{pq}^+ \circ \varphi_{2,q} \circ \varphi_{q\mathfrak{q}})(x). \tag{5.10}$$

(5.9) and (5.10) imply

$$(\varphi_{pq}^+ \circ \varphi_{q\mathfrak{q}}^+)(x) = (\varphi_{p\mathfrak{p}}^+ \circ \varphi_{\mathfrak{p}\mathfrak{q}})(x).$$

This is the orbifold embedding part of the required equality. The bundle embedding part can be proved in the same way. □

Proposition 5.11 *If $\widehat{\Phi^1} : \widehat{\mathcal{U}} \to \widehat{\mathcal{U}}$ is a KG-embedding and $\widehat{\Phi^2} : \widehat{\mathcal{U}} \to \widehat{\mathcal{U}^+}$ is a GK-embedding, then we can find an open substructure $\widehat{\mathcal{U}_0}$ of $\widehat{\mathcal{U}}$ such that the composition of $\widehat{\mathcal{U}_0} \to \widehat{\mathcal{U}} \to \widehat{\mathcal{U}}$ and $\widehat{\mathcal{U}} \to \widehat{\mathcal{U}^+}$ is defined as a strict KK-embedding $\Phi : \widehat{\mathcal{U}_0} \to \widehat{\mathcal{U}^+}$.*[3]

Moreover $\widehat{\mathcal{U}^+}$ is a strict thickening of $\widehat{\mathcal{U}_0}$.

Proof Replacing $\widehat{\mathcal{U}}$ by its open substructure, we may assume that $\widehat{\Phi^1}$ is a strict KG-embedding.

We denote $\widehat{\mathcal{U}} = \{\mathcal{U}_{\mathfrak{p}} \mid \mathfrak{p} \in \mathfrak{P}\}$, $\widehat{\mathcal{U}} = \{\mathcal{U}_p \mid p \in Z\}$, $\widehat{\mathcal{U}}^+ = \{\mathcal{U}_p^+ \mid p \in Z\}$, $\widehat{\Phi^1} = \{\Phi_{1,p}\}$, and $\widehat{\Phi^2} = \{(U_{\mathfrak{p}}(p), \Phi_{p\mathfrak{p}})\}$.

Lemma 5.12 *For each p there exists a neighborhood U_p^1 of o_p in U_p with the following property. If $q \in \psi_p(U_p^1 \cap s_p^{-1}(0)) \cap Z$, $p \in \mathrm{Im}(\psi_{\mathfrak{p}})$ then $q \in \psi_{\mathfrak{p}}(U_{\mathfrak{p}}(p) \cap s_{\mathfrak{p}}^{-1}(0))$.*

[3]By definition, KG-embedding $\widehat{\Phi^1}$ is a strict KG-embedding from an open substructure $\widehat{\mathcal{U}'}$ of $\widehat{\mathcal{U}}$. $\widehat{\mathcal{U}_0}$ is taken as an open substructure of $\widehat{\mathcal{U}'}$. So the composition of $\widehat{\mathcal{U}_0} \to \widehat{\mathcal{U}'} \to \widehat{\mathcal{U}}$ and $\widehat{\mathcal{U}} \to \widehat{\mathcal{U}^+}$ is defined.

Proof This is a consequence of openness of $\psi_{\mathfrak{p}}(U_{\mathfrak{p}}(p) \cap s_{\mathfrak{p}}^{-1}(0))$ in X and the finiteness of $\mathfrak{P}$. □

Let $p \in Z$. We define $U_p^0 \subset U_p$ by the next formula:

$$U_p^0 = U_p^1 \cap \bigcap \varphi_{\mathfrak{q}p}^{-1}(U_{\mathfrak{q}}(p) \cap U_{\mathfrak{p}\mathfrak{q}}).$$

Here the intersection is taken over all pairs $(\mathfrak{p}, \mathfrak{q}) \in \mathfrak{P} \times \mathfrak{P}$ such that $\mathfrak{p} \geq \mathfrak{q}$, $p \in \mathrm{Im}(\psi_{\mathfrak{p}}) \cap \mathrm{Im}(\psi_{\mathfrak{q}})$. (The case $\mathfrak{p} = \mathfrak{q}$ is included. In that case $U_{\mathfrak{p}\mathfrak{p}} = U_{\mathfrak{p}}$.) We use them to define our open substructure $\widehat{\mathcal{U}_0} = (\{\widehat{\mathcal{U}_p|_{U_p^0}}\}, \{\Phi_{pq}|_{U_{pq}^0}\})$. We will define an open neighborhood $U_{pq}^0 \subset U_{pq}$ of o_q later.

First we check Definition 3.19 (1). Suppose $q \in \psi_p(U_p^0 \cap s_p^{-1}(0))$. There exists $\mathfrak{p}$ such that $p \in \mathrm{Im}(\psi_{\mathfrak{p}})$. By the definition of U_p^0 and Lemma 5.12 it implies $q \in \psi_{\mathfrak{p}}(U_{\mathfrak{p}}(p) \cap s_{\mathfrak{p}}^{-1}(0))$. Since $\Phi_{p\mathfrak{p}}^2 : \mathcal{U}_{\mathfrak{p}}|_{U_{\mathfrak{p}}(p)} \to \mathcal{U}_p^+$ is an embedding of the Kuranishi chart, $\psi_{\mathfrak{p}}(U_{\mathfrak{p}}(p) \cap s_{\mathfrak{p}}^{-1}(0)) \subset \mathrm{Im}(\psi_p^+)$. Therefore $q \in \mathrm{Im}(\psi_p^+)$ as required.

Let $p \in Z$. There exists $\mathfrak{p}$ such that $p \in \mathrm{Im}(\psi_{\mathfrak{p}})$. We consider the composition

$$\Phi_{p\mathfrak{p}}^2 \circ \Phi_{\mathfrak{p}p}^1|_{U_p^0} : \mathcal{U}_p|_{U_p^0} \to \mathcal{U}_p^+. \tag{5.11}$$

We show that (5.11) is independent of the choice of $\mathfrak{p}$. Let $\mathfrak{q}$ be another choice. We may assume $\mathfrak{q} < \mathfrak{p}$ by Definition 3.15 (6). We apply Definition 5.6 (4) for $p = q$, then we have

$$\Phi_{p\mathfrak{q}}^2 = \Phi_{p\mathfrak{p}}^2 \circ \Phi_{\mathfrak{p}\mathfrak{q}}$$

on $\varphi_{\mathfrak{p}\mathfrak{q}}^{-1}(U_{\mathfrak{p}}(p)) \cap (\varphi_{p\mathfrak{q}})^{-1}(U_p)$. By the definition of U_p^0, $\varphi_{\mathfrak{q}p}(U_p^0) \subseteq \varphi_{\mathfrak{p}\mathfrak{q}}^{-1}(U_{\mathfrak{p}}(p)) \cap (\varphi_{p\mathfrak{q}})^{-1}(U_p)$. Therefore

$$\Phi_{p\mathfrak{q}}^2 \circ \Phi_{\mathfrak{q}p}^1 = \Phi_{p\mathfrak{p}}^2 \circ \Phi_{\mathfrak{p}\mathfrak{q}} \circ \Phi_{\mathfrak{q}p}^1$$

on U_p^0. On the other hand, we apply Definition 3.30 for $\widehat{\Phi^1}$ and $p = q$ to obtain

$$\Phi_{\mathfrak{p}\mathfrak{q}} \circ \Phi_{\mathfrak{q}p}^1 = \Phi_{\mathfrak{p}p}^1$$

on $\varphi_{\mathfrak{q}p}^{-1}(U_{\mathfrak{p}\mathfrak{q}})$. Since $U_p^0 \subset \varphi_{\mathfrak{q}p}^{-1}(U_{\mathfrak{p}\mathfrak{q}})$ by definition we conclude

$$\Phi_{p\mathfrak{q}}^2 \circ \Phi_{\mathfrak{q}p}^1 = \Phi_{p\mathfrak{p}}^2 \circ \Phi_{\mathfrak{p}\mathfrak{q}} \circ \Phi_{\mathfrak{q}p}^1 = \Phi_{p\mathfrak{p}}^2 \circ \Phi_{\mathfrak{p}p}^1$$

on U_p^0 as required. We have defined $\Phi_p : \mathcal{U}_p|_{U_p^0} \to \mathcal{U}_p^+$.

We next check Definition 3.19 (2)(3). We put $\psi_p^0 = \psi_p|_{U_p^0 \cap s_p^{-1}(0)}$. Suppose

$$q \in \mathrm{Im}(\psi_p^0) \cap Z = \psi_p(U_p^0 \cap s_p^{-1}(0)) \cap Z.$$

We take $\mathfrak{p}$ such that $p \in \mathrm{Im}(\psi_{\mathfrak{p}})$. We assume also $U_{\mathfrak{p}}(p) \cap U_{\mathfrak{p}}(q) \neq \emptyset$. Then, by Lemma 5.12, $q \in \psi_{\mathfrak{p}}(U_{\mathfrak{p}}(p) \cap s_{\mathfrak{p}}^{-1}(0))$. There exists a neighborhood Ω of $o_{\mathfrak{p}}(q) = \varphi_{\mathfrak{p}q}^2(o_q) \in U_{\mathfrak{p}}(p) \cap U_{\mathfrak{p}}(q)$ such that

$$\varphi_{q\mathfrak{p}}^2(\Omega) \subset U_{pq}^+, \quad \Omega \subset U_{\mathfrak{p}}(p) \cap U_{\mathfrak{p}}(q).$$

The existence of such an Ω follows from $\varphi_{q\mathfrak{p}}^2(o_{\mathfrak{p}}(q)) = o_q^+ \in U_q^+$. (Here $o_q^+ \in U_q^+$ is the point such that $\psi_p^+(o_q^+) = q$.) Now we put

$$U_{pq}^0 = U_{pq} \cap U_q^0 \cap \varphi_{pq}^{-1}(U_p^0) \cap (\varphi_{\mathfrak{p}q}^1)^{-1}(\Omega).$$

Then

$$\varphi_q(U_{pq}^0) = (\varphi_{q\mathfrak{p}}^2 \circ \varphi_{\mathfrak{p}q}^1)(U_{pq}^0) \subseteq \varphi_{q\mathfrak{p}}^2(\Omega) \subseteq U_{pq}^+.$$

We proved Definition 3.19 (2). On U_{pq}^0 we calculate

$$\begin{aligned} \Phi_p \circ \Phi_{pq} &= \Phi_{p\mathfrak{p}}^2 \circ \Phi_{\mathfrak{p}p}^1 \circ \Phi_{pq} = \Phi_{pq}^+ \circ \Phi_{q\mathfrak{p}}^2 \circ \Phi_{\mathfrak{p}p}^1 \circ \Phi_{pq} \\ &= \Phi_{pq}^+ \circ \Phi_{q\mathfrak{p}}^2 \circ \Phi_{\mathfrak{p}q}^1 = \Phi_{pq}^+ \circ \Phi_q. \end{aligned}$$

Definition 3.19 (3) is proved.

We finally prove that $\widehat{\mathcal{U}^+}$ is a thickening of $\widehat{\mathcal{U}_0}$. We put $O_p = \psi_p(U_p^0 \cap s_p^{-1}(0)) \cap Z$. Let $p, q \in Z$, $q \in O_p$. We take $\mathfrak{p}$ such that $p \in \mathrm{Im}(\psi_{\mathfrak{p}})$. Then $q \in \psi_{\mathfrak{p}}(U_{\mathfrak{p}}(p) \cap s_{\mathfrak{p}}^{-1}(0))$ by Lemma 5.12. We take $\Omega \subset U_{\mathfrak{p}}(p) \cap U_{\mathfrak{p}}(q)$ as above and obtain U_{pq}^0. We now put

$$W_p(q) = \varphi_{pq}(U_{pq}^0).$$

Since $U_{pq}^0 \subseteq (\varphi_{\mathfrak{p}q}^1)^{-1}(\Omega)$ we have

$$\varphi_p(W_p(q)) = (\varphi_{p\mathfrak{p}}^2 \circ \varphi_{\mathfrak{p}p}^1)(W_p(q)) \subseteq \varphi_{p\mathfrak{p}}^2(\Omega) \subseteq \varphi_{pq}^+(U_{pq}^+).$$

We have proved Definition 5.3 (2)(a). The proof of (b) is similar. □

Proposition 5.11 implies that there exists a KK-embedding $\widehat{\mathcal{U}} \to \widehat{\mathcal{U}^+}$ in the situation of Proposition 5.11. This embedding is well-defined in the following sense.

Definition 5.13 Let $\widehat{\mathcal{U}}$ and $\widehat{\mathcal{U}^+}$ be Kuranishi structures. Suppose $\widehat{\mathcal{U}_{0,i}}$ $(i = 1, 2)$ are open substructures of $\widehat{\mathcal{U}}$ and $\widehat{\Phi}^{0,i} : \widehat{\mathcal{U}_{0,i}} \to \widehat{\mathcal{U}^+}$ are strict KK-embeddings. We say they are *equivalent* if there exists an open neighborhood $(U^{00})_p$ of o_p in U_p such that

$$U_p^{00} \subset U_p^{0,1} \cap U_p^{0,2}, \qquad \Phi_p^{0,1}|_{U_p^{00}} = \Phi_p^{0,2}|_{U_p^{00}}.$$

Here $U_p^{0,i}$ is the Kuranishi neighborhood of p assigned by $\widehat{\mathcal{U}_{0,i}}$.

We can define an equivalence between two KG-embeddings $\widehat{\mathcal{U}} \to \widehat{\mathcal{U}}$ in the same way.

Definition 5.14

(1) The *composition* of two strict KK-embeddings $\widehat{\mathcal{U}} \to \widehat{\mathcal{U}'}, \widehat{\mathcal{U}'} \to \widehat{\mathcal{U}''}$ is defined by Lemma 3.20.
(2) We compose two KK-embeddings $\widehat{\mathcal{U}} \to \widehat{\mathcal{U}'}, \widehat{\mathcal{U}'} \to \widehat{\mathcal{U}''}$ and obtain a KK-embedding $\widehat{\mathcal{U}} \to \widehat{\mathcal{U}''}$. The composition is well-defined up to equivalence defined in Definition 5.13.
(3) The composition of a strict KK-embedding $\widehat{\mathcal{U}} \to \widehat{\mathcal{U}'}$ and a strict KG-embedding $\widehat{\mathcal{U}'} \to \widehat{\mathcal{U}}$ is defined by Lemma 3.32.
(4) We compose a KK-embedding $\widehat{\mathcal{U}} \to \widehat{\mathcal{U}'}$ and a KG-embedding $\widehat{\mathcal{U}'} \to \widehat{\mathcal{U}}$. The composition is well-defined up to equivalence.

Lemma 5.15 *Let $\widehat{\Phi} : \widehat{\mathcal{U}} \to \widehat{\mathcal{U}^+}$ be a GK-embedding and $\widehat{\mathcal{U}_0^+}$ an open substructure of $\widehat{\mathcal{U}^+}$. Then there exists a GK-embedding $\widehat{\Phi^0} : \widehat{\mathcal{U}} \to \widehat{\mathcal{U}_0^+}$ such that the composition $\widehat{\mathcal{U}} \to \widehat{\mathcal{U}_0^+} \to \widehat{\mathcal{U}^+}$ is an open restriction of $\widehat{\Phi}$.*

Proof Let $\widehat{\Phi} = \{(U_{\mathfrak{p}}(p), \Phi_{p\mathfrak{p}})\}$. We put $U_{\mathfrak{p}}^0(p) = \varphi_{p\mathfrak{p}}^{-1}(U_{0,p}^+) \subset U_{\mathfrak{p}}(p)$. Here $U_{0,p}^+$ is the Kuranishi neighborhood of p of the Kuranishi structure $\widehat{\mathcal{U}_0^+}$. We define $\{(U_{\mathfrak{p}}^0(p), \Phi_{p\mathfrak{p}}|_{U_{\mathfrak{p}}^0(p)})\}$ and obtain the required embedding. □

Lemma 5.16 *Let $\widehat{\Phi} : \widehat{\mathcal{U}} \to \widehat{\mathcal{U}}$ be a strict KG-embedding and $\widehat{\mathcal{U}^0}$ be an open substructure of $\widehat{\mathcal{U}}$. Then there exists an open substructure $\widehat{\mathcal{U}^0}$ of $\widehat{\mathcal{U}}$ and a strict KG-embedding: $\widehat{\mathcal{U}^0} \to \widehat{\mathcal{U}^0}$.*

Proof Let the open substructure $\widehat{\mathcal{U}^0}$ consist of $\{\mathcal{U}_{\mathfrak{p}}|_{U_{\mathfrak{p}}^0}\}$. Let $\widehat{\Phi} : \widehat{\mathcal{U}} \to \widehat{\mathcal{U}}$ consist of the embeddings $\Phi_{\mathfrak{p}p} : \widehat{\mathcal{U}}_p \to \widehat{\mathcal{U}}_{\mathfrak{p}}$ with $p \in \mathrm{Im}\psi_{\mathfrak{p}}$. We put

$$U_p^0 = \bigcap_{\mathfrak{p}: p \in \mathrm{Im}(\psi_{\mathfrak{p}})} \varphi_{\mathfrak{p}p}^{-1}(U_{\mathfrak{p}}^0).$$

$\mathcal{U}_{\mathfrak{p}}|_{U_p^0}$ and restrictions of coordinate change define $\widehat{\mathcal{U}^0}$. Then the restriction of $\Phi_{\mathfrak{p}p}$ defines the required KG-embedding: $\widehat{\mathcal{U}^0} \to \widehat{\mathcal{U}^0}$. □

The composition of embeddings in another case is more nontrivial.

Definition-Lemma 5.17 Let $\widehat{\Phi} : \widehat{\mathcal{U}} \to \widehat{\mathcal{U}}$ be a GK-embedding, and $\widehat{\Phi^+} : \widehat{\mathcal{U}} \to \widehat{\mathcal{U}^+}$ a KG-embedding. Then there exists a weakly open substructure $\widehat{\mathcal{U}_0}$ of $\widehat{\mathcal{U}}$ such that

$\widehat{\mathcal{U}_0} \longrightarrow \widehat{\mathcal{U}} \xrightarrow{\widehat{\Phi}} \widehat{\mathcal{U}}$ and $\widehat{\mathcal{U}} \xrightarrow{\widehat{\Phi^+}} \widehat{\mathcal{U}^+}$ can be composed into a GG-embedding $\widehat{\mathcal{U}_0} \longrightarrow \widehat{\mathcal{U}^+}$.

We will prove Definition-Lemma 5.17 in Sect. 9.3, where we use it.

Remark 5.18 We may introduce the notion of a germ of a Kuranishi structure and use it to describe these situations. We also remark that taking an open substructure of a Kuranishi structure is an operation which one can regard as close to an 'isomorphism'. On the other hand, taking an open substructure of a good coordinate system changes the structure much. For example, the dimension of the coordinate containing $p \in Z$ does not change by the former process but does change by the latter process.

Definition 5.19 Let $\widehat{\mathcal{U}}$ be a Kuranishi structure on X and $\widehat{\mathcal{U}^+}$ be its thickening. We say a good coordinate system $\widetriangle{\mathcal{U}}$ is *in between* $\widehat{\mathcal{U}}$ and $\widehat{\mathcal{U}^+}$ and write

$$\widehat{\mathcal{U}} < \widetriangle{\mathcal{U}} < \widehat{\mathcal{U}^+},$$

if the following holds:

(1) There exist a KG-embedding $\widehat{\mathcal{U}} \to \widetriangle{\mathcal{U}}$ and a GK-embedding $\widetriangle{\mathcal{U}} \to \widehat{\mathcal{U}^+}$.
(2) The composition $\widehat{\mathcal{U}} \to \widetriangle{\mathcal{U}} \to \widehat{\mathcal{U}^+}$ is equivalent to the given embedding $\widehat{\mathcal{U}} \to \widehat{\mathcal{U}^+}$ in the sense of Definition 5.13.

Proposition 5.20 *Let $\widehat{\mathcal{U}}$ be a Kuranishi structure of $Z \subseteq X$ and $\widehat{\mathcal{U}^+}$ its thickening. Then there exists a good coordinate system $\widetriangle{\mathcal{U}}$ in between $\widehat{\mathcal{U}}$ and $\widehat{\mathcal{U}^+}$.*

We also use the following version thereof.

Proposition 5.21 *Let $\widehat{\mathcal{U}}$ be a Kuranishi structure of $Z \subseteq X$ and $\widehat{\mathcal{U}_a^+}$ $(a = 1, 2)$ thickenings of $\widehat{\mathcal{U}}$. Then there exists a good coordinate system $\widetriangle{\mathcal{U}}$ and embeddings*

$$\widehat{\mathcal{U}} \to \widetriangle{\mathcal{U}}, \qquad \widetriangle{\mathcal{U}} \to \widehat{\mathcal{U}_a^+} \quad (a = 1, 2)$$

such that their compositions $\widehat{\mathcal{U}} \to \widetriangle{\mathcal{U}} \to \widehat{\mathcal{U}_a^+}$ are equivalent to the given embedding $\widehat{\mathcal{U}} \to \widehat{\mathcal{U}_a^+}$:

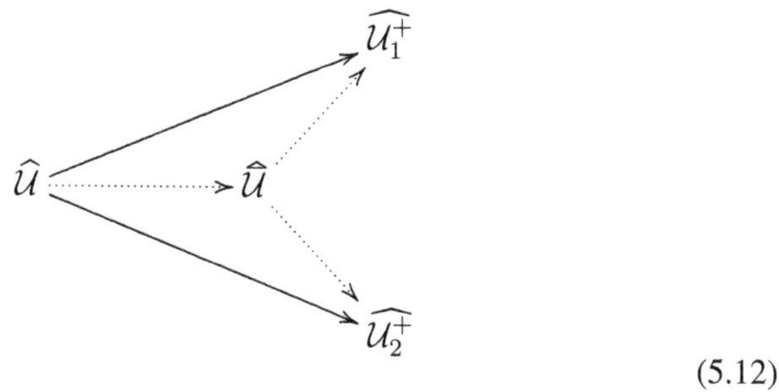

$$(5.12)$$

The proofs are almost the same as the proof of Theorem 3.35. We will prove them later in Sect. 11.2.

Remark 5.22 In [Ya1] D. Yang considered similar notions of embeddings between (his variant of) Kuranishi structure and good coordinate system. Definitions 5.3, 5.4, 5.5, 5.7, 4.1, 5.10 in [Ya1] correspond to strict KK-embedding, open KK-embedding, KK-embedding, weakly open GG-embedding, open GG-embedding, GG-embedding, respectively.

5.4 Support System and Existence of Thickening

In general the orbifold $U_{\mathfrak{p}}$ appearing in a good coordinate system is noncompact. In various situations we need to work on its compact subset. More precisely we need a system of compact subsets $\mathcal{K}_{\mathfrak{p}}$ of $U_{\mathfrak{p}}$ such that $\{\mathcal{K}_{\mathfrak{p}} \cap s_{\mathfrak{p}}^{-1}(0)\}$ covers Z. The support system which we define below is such a system.

Definition 5.23 Let $\widehat{\mathcal{U}}$ be a good coordinate system of $Z \subseteq X$.

(1) A *support system* of $\widehat{\mathcal{U}}$ is $\mathcal{K} = \{\mathcal{K}_{\mathfrak{p}} \mid \mathfrak{p} \in \mathfrak{P}\}$, where $\mathcal{K}_{\mathfrak{p}} \subset U_{\mathfrak{p}}$ is a compact subset for each $\mathfrak{p} \in \mathfrak{P}$ such that it is a closure of an open subset $\overset{\circ}{\mathcal{K}}_{\mathfrak{p}}$ of $U_{\mathfrak{p}}$, and

$$\bigcup_{\mathfrak{p} \in \mathfrak{P}} \psi_{\mathfrak{p}}(\overset{\circ}{\mathcal{K}}_{\mathfrak{p}} \cap s_{\mathfrak{p}}^{-1}(0)) \supseteq Z. \tag{5.13}$$

(2) A *support pair* $(\mathcal{K}^1, \mathcal{K}^2)$ is a pair of support systems $(\mathcal{K}^i_{\mathfrak{p}})_{\mathfrak{p} \in \mathfrak{P}}$ $i = 1, 2$ of $\widehat{\mathcal{U}}$, such that

$$\mathcal{K}^1_{\mathfrak{p}} \subset \overset{\circ}{\mathcal{K}}{}^2_{\mathfrak{p}}. \tag{5.14}$$

We write $\mathcal{K}^1 < \mathcal{K}^2$ if $(\mathcal{K}^1, \mathcal{K}^2)$ is a support pair.

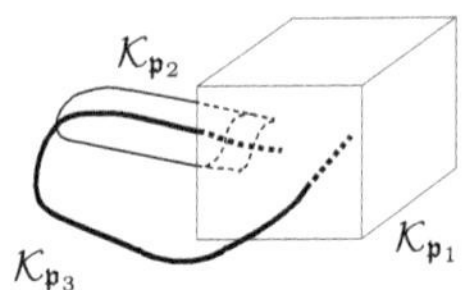

Fig. 5.2 Hetero-dimensional compactum

(3) When $\mathcal{K}$ is a support system, we define

$$|\mathcal{K}| = \left(\coprod_{\mathfrak{p} \in \mathfrak{P}} \mathcal{K}_{\mathfrak{p}} \right) / \sim .$$

Here, for $x \in \mathcal{K}_{\mathfrak{p}}$, $y \in \mathcal{K}_{\mathfrak{q}}$, the relation $x \sim y$ is defined by: $x = \varphi_{\mathfrak{p}\mathfrak{q}}(y)$ or $y = \varphi_{\mathfrak{q}\mathfrak{p}}(x)$. On $|\mathcal{K}|$, the quotient topology coincides with the induced topology from $|\widehat{\mathcal{U}}|$. Then it follows from the definition and [FOOO16, Proposition 5.17] or [FOOO17, Proposition 2.11] that the space $|\mathcal{K}|$ is metrizable.

We call $|\mathcal{K}|$ a *hetero-dimensional compactum*. If $\mathcal{K}^1 < \mathcal{K}^2$, then (5.14) induces a topological embedding $|\mathcal{K}^1| \to |\mathcal{K}^2|$, by which we identify $|\mathcal{K}^1|$ with a subset of $|\mathcal{K}^2|$ (Fig. 5.2).

(4) When $\mathcal{K}$ is a support system we define

$$\begin{aligned} \mathcal{S}_{\mathfrak{p}}(X, Z; \widehat{\mathcal{U}}; \mathcal{K}) &= \bigcup_{\mathfrak{q} \ge \mathfrak{p}} \psi_{\mathfrak{q}}(s_{\mathfrak{q}}^{-1}(0) \cap \mathcal{K}_{\mathfrak{q}}) \cap Z, \\ \mathring{\mathcal{S}}_{\mathfrak{p}}(X, Z; \widehat{\mathcal{U}}; \mathcal{K}) &= \mathcal{S}_{\mathfrak{p}}(X, Z; \widehat{\mathcal{U}}; \mathcal{K}) \setminus \bigcup_{\mathfrak{q} > \mathfrak{p}} \mathcal{S}_{\mathfrak{q}}(X; \widehat{\mathcal{U}}; \mathcal{K}). \end{aligned} \tag{5.15}$$

Remark 5.24 An idea to use the spaces whose dimension varies and embed moduli spaces of pseudo-holomorphic curves into such spaces is the method invented in the year 1996 when the idea of virtual fundamental chains was incepted.

The existence of thickening is a corollary of Proposition 5.26. We need the next definition to state it.

Definition 5.25 Let $\widehat{\mathcal{U}}$, $\widehat{\mathcal{U}^0}$ be good coordinate systems of $Z \subseteq X$. An open GG-embedding $\widehat{\Phi} : \widehat{\mathcal{U}^0} \to \widehat{\mathcal{U}}$ is said to be *relatively compact* if the subset $\varphi_{\mathfrak{p}}(U^0_{\mathfrak{p}})$ is relatively compact in $U_{\mathfrak{p}}$ for each $\mathfrak{p}$.

Proposition 5.26 *Let $\widehat{\mathcal{U}^0}$ be a relatively compact open substructure of a good coordinate system $\widehat{\mathcal{U}}$. Then there exists a Kuranishi structure $\widehat{\mathcal{U}^+}$ and a GK-embedding $\widehat{\mathcal{U}^0} \to \widehat{\mathcal{U}^+}$.*

Proof We put $\mathcal{K}^-_{\mathfrak{p}} = \overline{U^0_{\mathfrak{p}}}$. (Here $U^0_{\mathfrak{p}}$ is a Kuranishi neighborhood of the good coordinate system $\widehat{\mathcal{U}^0}$.) Since $\widehat{\mathcal{U}^0}$ is relatively compact, they are compact subsets and hence define a support system $\mathcal{K}^-$ of $\widehat{\mathcal{U}}$. We take another support system $\mathcal{K}^+$ of $\widehat{\mathcal{U}}$ such that $\mathcal{K}^- < \mathcal{K}^+$.

Lemma 5.27 *Let* $p \in Z$. *There exists a unique* $\mathfrak{p}_p \in \mathfrak{P}$ *such that* $p \in \overset{\circ}{\mathcal{S}}_{\mathfrak{p}_p}(X, Z; \widehat{\mathcal{U}}; \mathcal{K}^-)$.

Proof By definition, $\overset{\circ}{\mathcal{S}}_{\mathfrak{p}}(X, Z; \widehat{\mathcal{U}}; \mathcal{K}^-)$ are disjoint from one another for different $\mathfrak{p}$'s. By (5.13) they cover Z. This finishes the proof. □

Let $p \in \overset{\circ}{\mathcal{S}}_{\mathfrak{p}_p}(X, Z; \widehat{\mathcal{U}}; \mathcal{K}^-)$. We take an open neighborhood U_p^+ of $o_{\mathfrak{p}_p}(p) \in U_{\mathfrak{p}_p}^-$ in $\mathcal{K}_{\mathfrak{p}_p}^+$ such that

$$\psi_{\mathfrak{p}_p}(s_{\mathfrak{p}_p}^{-1}(0) \cap U_p^+) \cap \psi_{\mathfrak{p}}(s_{\mathfrak{p}}^{-1}(0) \cap \mathcal{K}_{\mathfrak{p}}^-) \neq \emptyset \Rightarrow \mathfrak{p} \le \mathfrak{p}_p. \tag{5.16}$$

Such a neighborhood exists by Condition (6) of Definition 3.15. We define

$$\mathcal{U}_p^+ = \mathcal{U}_{\mathfrak{p}_p}|_{U_p^+}. \tag{5.17}$$

We next define a coordinate change. Let $q \in \psi_{\mathfrak{p}_p}(s_{\mathfrak{p}_p}^{-1}(0) \cap U_p^+) \cap Z$. Take the unique $\mathfrak{p}_q \in \mathfrak{P}$ such that $q \in \overset{\circ}{\mathcal{S}}_{\mathfrak{p}_q}(X, Z; \widehat{\mathcal{U}}; \mathcal{K}^-)$. Since $q \in \psi_{\mathfrak{p}_p}(s_{\mathfrak{p}_p}^{-1}(0) \cap U_p^+) \cap \psi_{\mathfrak{p}_q}(s_{\mathfrak{p}_q}^{-1}(0) \cap \mathcal{K}_{\mathfrak{p}_q}^-)$, (5.16) implies $\mathfrak{p}_q \le \mathfrak{p}_p$. We put

$$U_{pq}^+ = U_q^+ \cap \varphi_{\mathfrak{p}_p\mathfrak{p}_q}^{-1}(U_p^+). \tag{5.18}$$

This is a subset of $U_q^+ \subset U_{\mathfrak{p}_q}$ and contains $o_q^+ = o_{\mathfrak{p}_q}(q)$. We define

$$\Phi_{pq}^+ = \Phi_{\mathfrak{p}_p\mathfrak{p}_q}|_{U_{pq}^+}.$$

Clearly, Φ_{pq}^+ is a coordinate change from U_q^+ to U_p^+. Using Definition 3.15 applied to $\widehat{\mathcal{U}}$, we can easily show that $\mathcal{U}_p^+$ and Φ_{pq}^+ define a Kuranishi structure. We denote it by $\widehat{\mathcal{U}^+}$.

We next define a GK-embedding $\widehat{\mathcal{U}^0} \to \widehat{\mathcal{U}^+}$. Let $p \in \psi_{\mathfrak{p}}(U_{\mathfrak{p}}^0 \cap s_{\mathfrak{p}}^{-1}(0))$. We take $\mathfrak{p}_p \in \mathfrak{P}$ such that $p \in \overset{\circ}{\mathcal{S}}_{\mathfrak{p}_p}(X, Z; \widehat{\mathcal{U}}; \mathcal{K}^-)$. By definition $\mathfrak{p} \le \mathfrak{p}_p$. We define

$$U_{\mathfrak{p}}(p) = \varphi_{\mathfrak{p}_p\mathfrak{p}}^{-1}(U_p^+) \cap U_{\mathfrak{p}}^0,$$

where U_p^+ is taken above. We define $\Phi_{p\mathfrak{p}} : \mathcal{U}_{\mathfrak{p}}|_{U_{\mathfrak{p}}(p)} \to \mathcal{U}_p^+$ by

$$\Phi_{p\mathfrak{p}} = \Phi_{\mathfrak{p}_p\mathfrak{p}}|_{U_{\mathfrak{p}}(p)}.$$

We prove that $(\{U_{\mathfrak{p}}(p)\}, \{\Phi_{p\mathfrak{p}}\})$ defines a GK-embedding $\widehat{\mathcal{U}^0} \to \widehat{\mathcal{U}^+}$. Definition 5.6 (1)(2)(3) are obvious. Suppose $\mathfrak{q} \in \mathfrak{P}$, $\mathfrak{q} \le \mathfrak{p}$, $p \in \psi_{\mathfrak{p}}(s_{\mathfrak{p}}^{-1}(0)) \cap Z$, $q \in \psi_{\mathfrak{p}}(U_{\mathfrak{p}}(p) \cap s_{\mathfrak{p}}^{-1}(0))$ and $q \in \psi_{\mathfrak{q}}(s_{\mathfrak{q}}^{-1}(0) \cap U_{\mathfrak{q}}^0) \cap Z$. Note $U_{\mathfrak{p}}(p) \subseteq \varphi_{\mathfrak{p}_p\mathfrak{p}}^{-1}(U_p^+)$. Therefore $q \in \psi_p^+(U_p^+ \cap (s_p^+)^{-1}(0))$ as required. Commutativity of Diagram (5.7) is obvious in our case since all the maps appearing there are restrictions of coordinate changes of the good coordinate system $\widehat{\mathcal{U}}$. We have thus constructed the required GK-embedding $\widehat{\mathcal{U}^0} \to \widehat{\mathcal{U}^+}$. □

Proposition 5.28 *For any Kuranishi structure $\widehat{\mathcal{U}}$ there exists its thickening.*

Proof By Theorem 3.35 we have a good coordinate system $\widehat{\mathcal{U}} = (\mathfrak{P}, \{\mathcal{U}_{\mathfrak{p}}\}, \{\Phi_{\mathfrak{p}\mathfrak{q}}\})$ compatible with $\widehat{\mathcal{U}}$. By compatibility, there exists a strict KG-embedding $\widehat{\mathcal{U}^0} \to \widehat{\mathcal{U}}$ from an open substructure $\widehat{\mathcal{U}^0}$ of $\widehat{\mathcal{U}}$. We take $\widehat{\mathcal{U}^0}$, a relatively compact open substructure of $\widehat{\mathcal{U}}$.

We apply Proposition 5.26 to show that there exist a Kuranishi structure $\widehat{\mathcal{U}^+}$ and a GK-embedding $\widehat{\mathcal{U}^0} \to \widehat{\mathcal{U}^+}$. By Lemma 5.16 there exists a strict KG-embedding $\widehat{\mathcal{U}^{00}} \to \widehat{\mathcal{U}^0}$ from some open substructure $\widehat{\mathcal{U}^{00}}$ of $\widehat{\mathcal{U}^0}$. We use Proposition 5.11 to obtain a composition $\widehat{\mathcal{U}^{000}} \to \widehat{\mathcal{U}^+}$ for some open substructure $\widehat{\mathcal{U}^{000}}$ of $\widehat{\mathcal{U}^{00}}$. Moreover $\widehat{\mathcal{U}^+}$ is a thickening of $\widehat{\mathcal{U}^{000}}$ (and so is a thickening of $\widehat{\mathcal{U}}$). The proof of Proposition 5.28 is complete. □

The following diagram visualizes the logical flow in our proof of Proposition 5.28:

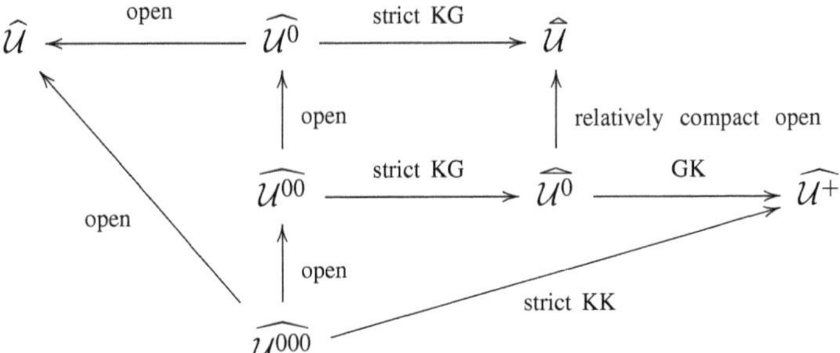

Chapter 6
Multivalued Perturbations

6.1 Multisections and Multivalued Perturbations

6.1.1 Multivalued Perturbations on an Orbifold

We next define the notion of multivalued perturbations associated to a given good coordinate system. We will slightly modify the previously given definition to make explicit certain properties which we used to study its zero set (in [FOOO16, Section 2.6] for example.)[1]

We begin with a review of multisections. We first introduce certain notations on vector bundles on orbifolds. See Chap. 23 for details.

Definition 6.1 Let U be an orbifold and $\mathcal{E}$ a vector bundle on it.

(1) (Definitions 23.1 (1) and 23.6 (1)(3)) Let $x \in U$. We call (V_x, Γ_x, ϕ_x) an *orbifold chart* of U at x if the following holds:

   (a) V_x is a smooth manifold on which a finite group Γ_x acts effectively and smoothly.
   (b) $\phi_x : V_x \to U$ is a Γ_x-equivariant map[2] which induces a diffeomorphism $\overline{\phi}_x : V_x/\Gamma_x \to U$[3] onto an open neighborhood U_x of x.
   (c) We require that there exists a unique point $o_x \in V_x$ such that o_x is a fixed point of all elements of Γ_x and $\phi_x([o_x]) = x$.

(2) (Definition 23.23 (3)) A *trivialization of our obstruction bundle* $\mathcal{E} = \widetilde{\mathcal{E}}/\Gamma_x$ *on* U_x is by definition $(E_x, \widehat{\phi}_x)$ such that:

[1] Note the definition of multisection we use here is exactly the same as one of smooth multisection in [FOn2].

[2] The Γ_x-action on U is trivial.

[3] See Definition 23.5 (1) for the definition of diffeomorphism here.

K. Fukaya et al., *Kuranishi Structures and Virtual Fundamental Chains*, Springer Monographs in Mathematics, https://doi.org/10.1007/978-981-15-5562-6_6

(a) E_x is a vector space on which Γ_x acts linearly.
(b) $\widehat{\phi}_x : V_x \times E_x \to \widetilde{\mathcal{E}}$ is a Γ_x-invariant smooth map which induces an isomorphism of vector bundles $(V_x \times E_x)/\Gamma_x \to \mathcal{E}|_{V_x/\Gamma_x}$.

(3) (Definition 23.23 (2)(4)) We call $\mathfrak{V}_x = (V_x, \Gamma_x, E_x, \phi_x, \widehat{\phi}_x)$ an *orbifold chart* of $(U, \mathcal{E})$.

Definition 6.2 Let $\mathfrak{V}_x = (V_x, \Gamma_x, E_x, \phi_x, \widehat{\phi}_x)$ be an orbifold chart of $(U, \mathcal{E})$.

(1) A *smooth ℓ-multisection of $\mathcal{E}$* on an orbifold chart $\mathfrak{V}_x$ is $\mathfrak{s} = (\mathfrak{s}_1, \dots, \mathfrak{s}_\ell)$ with the following properties:

(a) $\mathfrak{s}$ is a smooth map $V_x \to E_x^\ell$.
(b) For each $y \in V_x$ and $\gamma \in \Gamma_x$ there exists $\sigma \in \mathrm{Perm}(\ell)$ such that

$$\mathfrak{s}_{\sigma(k)}(y) = \gamma \mathfrak{s}_k(y).$$

Hereafter we simply say *ℓ-multisection* in place of smooth ℓ-multisection.
(2) Two ℓ-multisections $(\mathfrak{s}_1, \dots, \mathfrak{s}_\ell)$ and $(\mathfrak{s}'_1, \dots, \mathfrak{s}'_\ell)$ are said to be *equivalent as an ℓ-multisection on* $\mathfrak{V}_x$ if for each y there exists a permutation $\sigma \in \mathrm{Perm}(\ell)$ such that $\mathfrak{s}'_i(y) = \mathfrak{s}_{\sigma(i)}(y)$. We remark that σ may depend on y. See Example 13.13.
(3) The *ℓ'-iteration* of ℓ-multisection $\mathfrak{s}$ is the $\ell'\ell$-multisection $\mathfrak{s}'$ such that $\mathfrak{s}'_k = \mathfrak{s}_m$ for $k \equiv m \mod \ell$. We denote the ℓ'-iteration of $\mathfrak{s}$ by $\mathfrak{s}^{\times \ell'}$.
(4) Let $\mathfrak{s}_{(1)}$ be an ℓ_1-multisection and $\mathfrak{s}_{(2)}$ an ℓ_2-multisection. We say $\mathfrak{s}_{(1)}$ is *equivalent to $\mathfrak{s}_{(2)}$ as multisections* if $\mathfrak{s}_{(1)}^{\times \ell_2}$ is equivalent to $\mathfrak{s}_{(2)}^{\times \ell_1}$ as $\ell_1\ell_2$-multisections.
(5) It is easy to see that Item (4) defines an equivalence relation. We call its equivalence class a *multisection* on our orbifold chart.

Notation 6.3 Here and hereafter we use $\mathfrak{s}$ (lower case Fraktur s) such as $\mathfrak{s}^n_p$, $\mathfrak{s}^\epsilon_p$, $\mathfrak{s}^n_{\mathfrak{p}}$, $\mathfrak{s}^\epsilon_{\mathfrak{p}}$ to denote perturbations of various kinds, such as multivalued perturbations or CF-perturbations. The Kuranishi map is always denoted by s (italic lower case s), for example s_p, $s_{\mathfrak{p}}$.

Situation 6.4 Let $\mathfrak{V}_x = (V_x, \Gamma_x, E_x, \phi_x, \widehat{\phi}_x)$ and $\mathfrak{V}'_{x'} = (V'_{x'}, \Gamma'_{x'}, E'_{x'}, \phi'_{x'}, \widehat{\phi}'_{x'})$ be two orbifold charts of a vector bundle $(U, \mathcal{E})$. We assume $\phi'_{x'}(V'_{x'}) \subset \phi_x(V_x)$ and that there exist $\tilde{\varphi}_{xx'} : V'_{x'} \to V_x$, $h_{xx'} : \Gamma'_{x'} \to \Gamma_x$ such that $h_{xx'}$ is an injective group homomorphism and $\tilde{\varphi}_{xx'}$ is an $h_{xx'}$ equivariant smooth open embedding such that they induce the composition map

$$(\overline{\phi_x})^{-1} \circ \overline{\phi'_{x'}} : V'_{x'}/\Gamma'_{x'} \to V_x/\Gamma_x,$$

where $\overline{\phi_x}$ (resp. $\overline{\phi'_{x'}}$) is induced by ϕ_x (resp. $\phi'_{x'}$). In other words,

$$\phi'_{x'}(y) \equiv \phi_x(\tilde{\varphi}_{xx'}(y)) \mod \Gamma_x.$$

Moreover we assume that the composition

$$\left(\overline{\widehat{\phi}_x}\right)^{-1} \circ \overline{\widehat{\phi}'_{x'}} : (V'_{x'} \times E'_{x'})/\Gamma'_{x'} \to (V_x \times E_x)/\Gamma_x$$

is induced by a smooth map $\breve{\varphi}_{xx'} : V'_{x'} \times E'_{x'} \to E_x$ that is linear in the $E'_{x'}$ factor. In other words,

$$\widehat{\phi}'_{x'}(y, v) \equiv \widehat{\phi}_x(\tilde{\varphi}_{xx'}(y), \breve{\varphi}_{xx'}(y, v)) \mod \Gamma_x.$$

We put $\Phi_{xx'} = (h_{xx'}, \tilde{\varphi}_{xx'}, \breve{\varphi}_{xx'})$. ■

Definition 6.5 We call $\Phi_{xx'}$ a *coordinate change* from an orbifold chart $\mathfrak{V}_{x'}$ to $\mathfrak{V}_x$.

Remark 6.6 We put $\tilde{\tilde{\varphi}}_{xx'}(y, v) = (y, \breve{\varphi}_{xx'}(y, v))$. Then

$$(h_{xx'}, \tilde{\varphi}_{xx'}, \tilde{\tilde{\varphi}}_{xx'})$$

is a *local representative of an embedding of vector bundles*, id : $\mathcal{E}|_{\mathrm{Im}\overline{\widehat{\phi}'_{x'}}} \to \mathcal{E}|_{\mathrm{Im}\overline{\widehat{\phi}_x}}$ in the sense of Definition 23.27. Moreover it is a fiberwise isomorphism.

Definition 6.7 Let $\mathfrak{s}_x$ be an ℓ-multisection on $\mathfrak{V}_x$ and $\Phi_{xx'} : \mathfrak{V}_{x'} \to \mathfrak{V}_x$ a coordinate change. We define the *pullback* $\Phi^*_{xx'}\mathfrak{s}$ by

$$(\Phi^*_{xx'}\mathfrak{s})_k(y) = g_y^{-1}\mathfrak{s}_k(\tilde{\varphi}_{xx'}(y)),$$

where $g_y : E_{x'} \to E_x$ is defined by $g_y(v) = \breve{\varphi}_{xx'}(y, v)$.

Lemma 6.8 *If $\mathfrak{s}_x$ is equivalent to $\mathfrak{s}'_x$ as multisections then $\Phi^*_{xx'}\mathfrak{s}$ is equivalent to $\Phi^*_{xx'}\mathfrak{s}'$ as multisections.*

We omit the proof. (See the proof of a similar lemma, Lemma 7.8.)

Here is a notational remark.

Remark 6.9 So far we have written $\mathfrak{V}_x = (V_x, \Gamma_x, E_x, \phi_x, \widehat{\phi}_x)$. The point x plays no particular role except we assume the existence of $o_x \in V_x$ such that $\phi_x(o_x) = x$ and o_x is fixed by all the elements of Γ_x. If we change the choice of such x, the constructions so far do not change at all. So, from now on, we do not specify x in our notation of local orbifold chart but only assume existence of such x. We will write for example $\mathfrak{V}_\mathfrak{r} = (V_\mathfrak{r}, \Gamma_\mathfrak{r}, E_\mathfrak{r}, \phi_\mathfrak{r}, \widehat{\phi}_\mathfrak{r})$ instead of $\mathfrak{V}_x = (V_x, \Gamma_x, E_x, \phi_x, \widehat{\phi}_x)$.

Definition 6.10 Let U be an orbifold and $\mathcal{E}$ a vector bundle on it.

(1) A *representative of a multisection of $\mathcal{E}$ on U* is a collection $(\{\mathfrak{V}_\mathfrak{r} \mid \mathfrak{r} \in \mathfrak{R}\}, \{\mathfrak{s}_\mathfrak{r} \mid \mathfrak{r} \in \mathfrak{R}\})$ with the following properties:

   (a) $\mathfrak{V}_\mathfrak{r}$ is an orbifold chart of a vector bundle $(U, \mathcal{E})$ such that $\bigcup_{\mathfrak{r} \in \mathfrak{R}} U_\mathfrak{r} = U$.
   (b) $\mathfrak{s}_\mathfrak{r}$ is a multisection of $\mathfrak{V}_\mathfrak{r}$.

(c) For any $y \in \mathfrak{V}_{\mathfrak{r}_1} \cap \mathfrak{V}_{\mathfrak{r}_2}$, there exist an orbifold chart $\mathfrak{V}_y$ and coordinate changes $\Phi_{\mathfrak{r}_i y} : \mathfrak{V}_y \to \mathfrak{V}_{\mathfrak{r}_i}$ such that $\Phi^*_{\mathfrak{r}_1 y}\mathfrak{s}_{\mathfrak{r}_1}$ is equivalent to $\Phi^*_{\mathfrak{r}_2 y}\mathfrak{s}_{\mathfrak{r}_2}$.

(2) Let $(\{\mathfrak{V}^{(i)}_{\mathfrak{r}^{(i)}} \mid \mathfrak{r}^{(i)} \in \mathfrak{R}^{(i)}\}, \{\mathfrak{s}^{(i)}_{\mathfrak{r}^{(i)}} \mid \mathfrak{r}^{(i)} \in \mathfrak{R}^{(i)}\})$ be representatives of multisections of $\mathcal{E}$ on U for $i = 1, 2$. We say they are *equivalent* if the following holds:

For any $x \in \mathfrak{V}^{(1)}_{\mathfrak{r}^{(1)}_1} \cap \mathfrak{V}^{(2)}_{\mathfrak{r}^{(2)}_2}$, there exist an orbifold chart $\mathfrak{V}_x$ and coordinate changes $\Phi_{\mathfrak{r}_i x} : \mathfrak{V}_x \to \mathfrak{V}^{(i)}_{\mathfrak{r}^{(i)}_i}$ $(i = 1, 2)$ such that $\Phi^*_{\mathfrak{r}^{(1)}_1 x}\mathfrak{s}^{(1)}_{\mathfrak{r}^{(1)}_1}$ is equivalent to $\Phi^*_{\mathfrak{r}^{(2)}_2 x}\mathfrak{s}^{(2)}_{\mathfrak{r}^{(2)}_2}$.

An equivalence class of this equivalence relation is called a *multisection of* $(U, \mathcal{E})$.

(3) (See [FOn2, Definition 3.10].) Let $\mathfrak{s}^n$ be a sequence of multisections of $(U, \mathcal{E})$. We say that *it converges to a multisection* $\mathfrak{s}$ *in* C^k*-topology* (k is any of $0, 1, \dots, \infty$) if there exists a representative $(\{\mathfrak{V}_{\mathfrak{r}} \mid \mathfrak{r} \in \mathfrak{R}\}, \{\mathfrak{s}^n_{\mathfrak{r}} \mid \mathfrak{r} \in \mathfrak{R}\})$ of $\mathfrak{s}^n$ for sufficiently large n and $(\{\mathfrak{V}_{\mathfrak{r}} \mid \mathfrak{r} \in \mathfrak{R}\}, \{\mathfrak{s}_{\mathfrak{r}} \mid \mathfrak{r} \in \mathfrak{R}\})$ of $\mathfrak{s}$ such that $\mathfrak{s}^n_{\mathfrak{r}}$ converges to $\mathfrak{s}_{\mathfrak{r}}$ in compact C^k-topology for each $\mathfrak{r}$. We note that we assume $\mathfrak{V}_{\mathfrak{r}}$ and $\mathfrak{R}$ are independent of n.

Definition 6.11 Let $\mathfrak{s}$ be a multisection of a vector bundle $(U, \mathcal{E})$ on orbifold U and $x \in U$. We put $\mathfrak{s} = [(\{\mathfrak{V}_{\mathfrak{r}} \mid \mathfrak{r} \in \mathfrak{R}\}, \{\mathfrak{s}_{\mathfrak{r}} \mid \mathfrak{r} \in \mathfrak{R}\})]$. We take an orbifold chart $\mathfrak{V}_x$ at x. A map germ $[s]$, where $s : O_x \to E_x$, is said to be a *branch of* $\mathfrak{s}$ *at* x if the following holds:

(1) O_x is a neighborhood of o_x in V_x.
(2) Let $\mathfrak{r} \in \mathfrak{R}$ such that $x \in U_{\mathfrak{r}}$. Then there exists k such that

$$\widehat{\phi}_{\mathfrak{r}}(\tilde{\varphi}_{\mathfrak{r}x}(y), \mathfrak{s}_{\mathfrak{r},k}(\tilde{\varphi}_{\mathfrak{r}x}(y))) = \widehat{\phi}_x(y, s(y))$$

for all y in a neighborhood of x in O_x. Here $\tilde{\varphi}_{\mathfrak{r}x}$ is a part of a coordinate change $\mathfrak{V}_x \to \mathfrak{V}_{\mathfrak{r}}$.

6.1.2 *Multivalued Perturbations on a Good Coordinate System*

Let $\widehat{\mathcal{U}} = (\mathfrak{P}, \{\mathcal{U}_{\mathfrak{p}}\}, \{\Phi_{\mathfrak{pq}}\})$ be a good coordinate system of $Z \subseteq X$. Here $\mathcal{U}_{\mathfrak{p}} = (U_{\mathfrak{p}}, \mathcal{E}_{\mathfrak{p}}, \psi_{\mathfrak{p}}, s_{\mathfrak{p}})$ and $\Phi_{\mathfrak{pq}} = (U_{\mathfrak{pq}}, \varphi_{\mathfrak{pq}}, \hat{\varphi}_{\mathfrak{pq}})$.

Definition 6.12 Let $\mathcal{K} = \{\mathcal{K}_{\mathfrak{p}}\}$ be a support system of a good coordinate system $\widehat{\mathcal{U}}$. A *multisection* of $(\widehat{U}, \mathcal{K})$ is $\{\mathfrak{s}_{\mathfrak{p}} \mid \mathfrak{p} \in \mathfrak{P}\}$ such that $\mathfrak{s}_{\mathfrak{p}}$ is a multisection of $\mathcal{E}_{\mathfrak{p}}$ on a neighborhood of $\mathcal{K}_{\mathfrak{p}}$ and

$$\mathfrak{s}_{\mathfrak{p}} \circ \varphi_{\mathfrak{pq}} = \widehat{\varphi}_{\mathfrak{pq}} \circ \mathfrak{s}_{\mathfrak{q}}$$

on a neighborhood of $\mathcal{K}_{\mathfrak{q}} \cap \varphi_{\mathfrak{p},\mathfrak{q}}^{-1}(\mathcal{K}_{\mathfrak{p}})$.

We say that $\widehat{\mathfrak{s}} = \{\widehat{\mathfrak{s}^n} \mid n \in \mathbb{Z}_{\geq 0}\} = \{\mathfrak{s}_{\mathfrak{p}}^n \mid n \in \mathbb{Z}_{\geq 0}, \mathfrak{p} \in \mathfrak{P}\}$ is a *multivalued perturbation of* $(\widehat{U}, \mathcal{K})$ if $\mathfrak{s}_{\mathfrak{p}}^n$ is a multisection of $\mathcal{E}_{\mathfrak{p}}$ on a neighborhood of $\mathcal{K}_{\mathfrak{p}}$ for each $n \in \mathbb{Z}_{\geq 0}$ and $\mathfrak{p}$ and the following conditions are satisfied:

(1) $\mathfrak{s}_{\mathfrak{p}}^n \circ \varphi_{\mathfrak{p}\mathfrak{q}} = \widehat{\varphi}_{\mathfrak{p}\mathfrak{q}} \circ \mathfrak{s}_{\mathfrak{q}}^n$ on a neighborhood of $\mathcal{K}_{\mathfrak{q}} \cap \varphi_{\mathfrak{p}\mathfrak{q}}^{-1}(\mathcal{K}_{\mathfrak{p}})$.
(2) $\lim_{n\to\infty} \mathfrak{s}_{\mathfrak{p}}^n = s_{\mathfrak{p}}$ in C^1-topology on a neighborhood of $\mathcal{K}_{\mathfrak{p}}$.

A *multivalued perturbation of* $\widehat{\mathcal{U}}$ is a collection $\{\mathfrak{s}_{\mathfrak{p}}^n\}$ such that (1) and (2) hold for some support system $\mathcal{K}$.

Note that the Kuranishi map $s_{\mathfrak{p}}$, which is a single-valued section of $\mathcal{E}_{\mathfrak{p}}$, can be regarded as a multisection by Lemma 23.31. The C^1-convergence in Definition 6.12 (2) therefore is defined in Definition 6.10 (3).

Below we will elaborate on the equality in Definition 6.12 (1) further. Let $x \in \mathcal{K}_{\mathfrak{q}} \cap \varphi_{\mathfrak{p}\mathfrak{q}}^{-1}(\mathcal{K}_{\mathfrak{p}})$ and $x' = \varphi_{\mathfrak{p}\mathfrak{q}}(x) \in \mathcal{K}_{\mathfrak{p}}$. We can take orbifold charts $\mathfrak{V}_x$ of $(U_{\mathfrak{q}}, \mathcal{E}_{\mathfrak{q}})$, $\mathfrak{V}_{x'}$ of $(U_{\mathfrak{p}}, \mathcal{E}_{\mathfrak{p}})$ such that $(\varphi_{\mathfrak{p}\mathfrak{q}}, \widehat{\varphi}_{\mathfrak{p}\mathfrak{q}})$ has a local representative $(h_{\mathfrak{p}\mathfrak{q};x'x}, \tilde{\varphi}_{\mathfrak{p}\mathfrak{q};x'x}, \tilde{\widehat{\varphi}}_{\mathfrak{p}\mathfrak{q};x'x})$ with respect to these orbifold charts. (Lemma 23.26.) We define $\breve{\varphi}_{\mathfrak{p}\mathfrak{q};x'x} : V_x \times E_x \to E_{x'}$ by the relation

$$\tilde{\widehat{\varphi}}_{\mathfrak{p}\mathfrak{q};x'x}(y, v) = (\tilde{\varphi}_{\mathfrak{p}\mathfrak{q};x'x}(y), \breve{\varphi}_{\mathfrak{p}\mathfrak{q};x'x}(y, v)).$$

We may choose $\mathfrak{V}_x$ and $\mathfrak{V}_{x'}$ so small that $\mathfrak{s}_{\mathfrak{p}}^n$ and $\mathfrak{s}_{\mathfrak{q}}^n$ have representatives on the charts. Let $\mathfrak{s}_{\mathfrak{p},x'}^n$ and $\mathfrak{s}_{\mathfrak{q},x}^n$ be the representatives, which are ℓ_1 and ℓ_2 multisections, respectively. By taking an appropriate iteration we may assume $\ell_1 = \ell_2 = \ell$. We then require

$$\mathfrak{s}_{\mathfrak{p},x';k}^n(\tilde{\varphi}_{\mathfrak{p}\mathfrak{q};x'x}(y)) = \breve{\varphi}_{\mathfrak{p}\mathfrak{q};x'x}(y, \mathfrak{s}_{\mathfrak{q},x;\rho_y(k)}^n(y)) \tag{6.1}$$

for $y \in V_x$, $k = 1, \dots, \ell$, where $\rho_y \in \mathrm{Perm}(\ell)$. (6.1) is the precise form of Definition 6.12 (1).

6.2 Properties of the Zero Set of Multivalued Perturbations

In this section we prove the equality

$$\left(\bigcup_{\mathfrak{p}}((\mathfrak{s}_{\mathfrak{p}}^n)^{-1}(0) \cap \overset{\circ}{\mathcal{K}_{\mathfrak{p}}^1})\right) \cap \mathfrak{U}(Z) = \left(\bigcup_{\mathfrak{p}}((\mathfrak{s}_{\mathfrak{p}}^n)^{-1}(0) \cap \mathcal{K}_{\mathfrak{p}}^2)\right) \cap \mathfrak{U}(Z),$$

for any multivalued perturbation $\{\mathfrak{s}_{\mathfrak{p}}^n\}$ of $(\widehat{\mathcal{U}}, \mathcal{K})$ and support systems $\mathcal{K}^1$, $\mathcal{K}^2$ with $\mathcal{K}^1 < \mathcal{K}^2$, a sufficiently small neighborhood $\mathfrak{U}(Z)$ of Z and sufficiently large integer

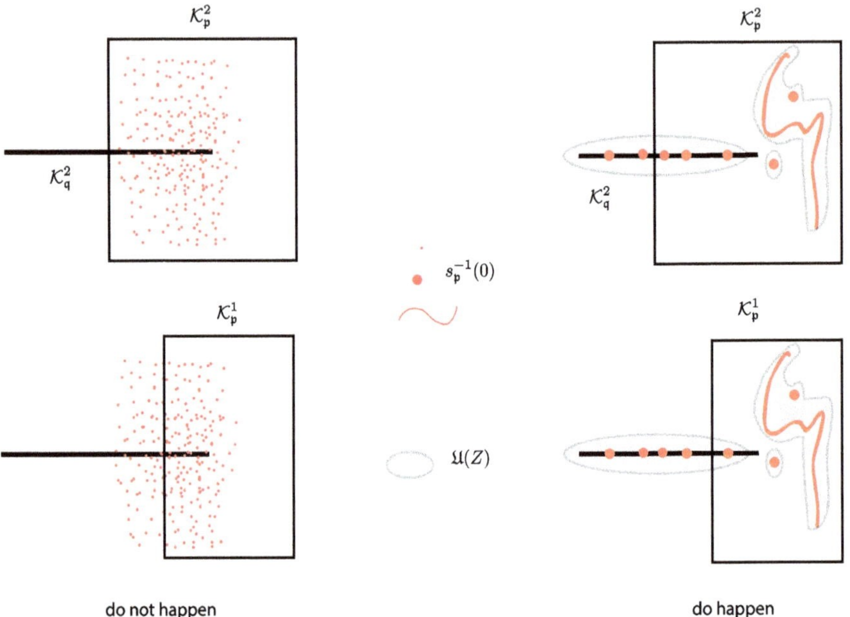

Fig. 6.1 $s_{\mathfrak{p}}^{-1}(0)$ in $|\mathcal{K}|$

n.[4] See (6.14). We also show certain related results. We use this equality as follows. The right hand side implies that this space is compact. On the other hand, if the Kuranishi structure has no boundary, $\mathfrak{s}^n$ is single valued and is transversal to 0, then the left hand side implies that this space is an orbifold. Therefore we can define the fundamental class of this space. We can use the equality in the general case of multisections to define virtual fundamental chains in a similar way.

The main idea of the proof of this equality lies in the fact that if $\mathfrak{q} < \mathfrak{p}$ then (roughly speaking) the zero of $\mathfrak{s}^n_{\mathfrak{p}}$ in a sufficiently small neighborhood of $\varphi_{\mathfrak{p}\mathfrak{q}}(U_{\mathfrak{q}})$ is actually contained in $\varphi_{\mathfrak{p}\mathfrak{q}}(U_{\mathfrak{q}})$. (Lemma 6.14. See Fig. 6.1.) This is a consequence of Definition 3.2 (5) and the C^1-convergence $\mathfrak{s}^n_{\mathfrak{p}} \to s_{\mathfrak{p}}$, To state this property precisely we need to prepare some notations.

We denote the normal bundle of our embedding $\varphi_{\mathfrak{p}\mathfrak{q}} : U_{\mathfrak{p}\mathfrak{q}} \to U_{\mathfrak{p}}$ by

$$N(\varphi_{\mathfrak{p}\mathfrak{q}}; U_{\mathfrak{p}}) := \frac{\varphi_{\mathfrak{p}\mathfrak{q}}^* TU_{\mathfrak{p}}}{TU_{\mathfrak{q}}|_{U_{\mathfrak{p}\mathfrak{q}}}}$$

It defines a vector bundle over $U_{\mathfrak{p}\mathfrak{q}}$. For a compact subset $K \subset U_{\mathfrak{q}}$ we denote by $N_K(\varphi_{\mathfrak{p}\mathfrak{q}}; U_{\mathfrak{p}})$ the restriction of this vector bundle to $K \cap U_{\mathfrak{p}\mathfrak{q}}$.

[4]Here $(\mathfrak{s}^n_{\mathfrak{p}})^{-1}(0)$ is the set of points p of $U_{\mathfrak{p}}$ such that some branch of $\mathfrak{s}^n_{\mathfrak{p}}$ vanishes at p.

Situation 6.13 We fix a Riamannian metric on $U_{\mathfrak p}$. It induces a metric on $N(\varphi_{\mathfrak{pq}}; U_{\mathfrak p})$. We denote by $N^\delta(\varphi_{\mathfrak{pq}}; U_{\mathfrak q})$ the δ-disk bundle thereof for $\delta > 0$. Using the normal exponential map of the embedding $\varphi_{\mathfrak{pq}}$, we have a diffeomorphism:

$$\mathrm{Exp} : N^\delta_K(\varphi_{\mathfrak{pq}}; U_{\mathfrak p}) \to U_{\mathfrak p}, \tag{6.2}$$

onto an open subset, which is given by $\mathrm{Exp}(x, v) = \exp_x v$ where $\exp_x$ is the exponential map of the metric given on $U_{\mathfrak p}$. (See [FOOO14, Lemma 6.5].) Here δ is a positive number depending on a compact subset K of $U_{\mathfrak q}$ and the embedding $\varphi_{\mathfrak{pq}}$. We put

$$BN_{\delta'}(K; U_{\mathfrak p}) = \bigcup_{x \in U_{\mathfrak{pq}} \cap K} \exp_x(N^{\delta'}_K(\varphi_{\mathfrak{pq}}; U_{\mathfrak p})) \subset U_{\mathfrak p} \tag{6.3}$$

for $\delta' \le \delta$. We denote by $\pi_{\delta'} : BN_{\delta'}(K; U_{\mathfrak p}) \to U_{\mathfrak{pq}} \cap K$ the composition of Exp^{-1} with the projection $N(\varphi_{\mathfrak{pq}}; U_{\mathfrak p}) \to U_{\mathfrak q}$ of the vector bundle. Note that on the image of $\varphi_{\mathfrak{pq}}$, the obstruction bundle $\mathcal{E}_{\mathfrak p}$ has a subbundle $\widetilde\varphi_{\mathfrak{pq}}(\mathcal{E}_{\mathfrak q})$. We consider a subbundle

$$\mathcal{E}_{\mathfrak q;\mathfrak p} = \pi^*_\delta(\widetilde\varphi_{\mathfrak{pq}}(\mathcal{E}_{\mathfrak q})) \subset \mathcal{E}_{\mathfrak p}$$

on $BN_\delta(K; U_{\mathfrak p})$ which restricts to the bundle $\widetilde\varphi_{\mathfrak{pq}}(\mathcal{E}_{\mathfrak q})$ on $\varphi_{\mathfrak{pq}}(K) \subset U_{\mathfrak p}$. We take the quotient bundle $\mathcal{E}_{\mathfrak p}/\mathcal{E}_{\mathfrak q;\mathfrak p}$ and consider

$$\overline{s_{\mathfrak p}} \equiv s_{\mathfrak p} \mod \mathcal{E}_{\mathfrak q;\mathfrak p},$$

that is a section of $\mathcal{E}_{\mathfrak p}/\mathcal{E}_{\mathfrak q;\mathfrak p}$. In a similar way for each branch $\mathfrak{s}^n_{\mathfrak p;k}$ of $\mathfrak{s}^n_{\mathfrak p}$ we obtain

$$\overline{\mathfrak{s}^n_{\mathfrak p;k}}(y) \in (\mathcal{E}_{\mathfrak p}/\mathcal{E}_{\mathfrak q;\mathfrak p})_y. \tag{6.4}$$

■

We denote by $E : BN_\delta(K; U_{\mathfrak p}) \to N^\delta_K(\varphi_{\mathfrak{pq}}; U_{\mathfrak q})$ the inverse of Exp on its image. Namely

$$E(y) = (x, v) \qquad \Leftrightarrow \qquad x = \pi_\delta(y),\ v = \exp_x^{-1}(y). \tag{6.5}$$

Lemma 6.14 *Suppose we are in Situation* 6.13. *There exist* $c > 0$, $\delta_0 > 0$ *and* $n_0 \in \mathbb{Z}_{\ge 0}$ *such that for* $y \in BN_{\delta_0}(K; U_{\mathfrak p})$

$$|\overline{\mathfrak{s}^n_{\mathfrak p;k}}(y)| \ge c|E(y)| \tag{6.6}$$

and

$$|\overline{s_{\mathfrak p}}(y)| \ge c|E(y)| \tag{6.7}$$

hold for any branch $\overline{\mathfrak{s}^n_{\mathfrak{p};k}}$, if $n > n_0$ and $d(\pi_\delta(y), s_{\mathfrak{q}}^{-1}(0)) < \delta_0$.

Proof We choose and fix a connection on the quotient bundle $\mathcal{E}_{\mathfrak{p}}/\mathcal{E}_{\mathfrak{q};\mathfrak{p}}$. For $y = \mathrm{Exp}(x, v) \in \varphi_{\mathfrak{p},\mathfrak{q}}(K)$ we consider the covariant derivative

$$V \mapsto D_V \overline{s_{\mathfrak{p}}} : (N_K(\varphi_{\mathfrak{p}\mathfrak{q}}; U_{\mathfrak{p}}))_y \to (\mathcal{E}_{\mathfrak{p}}/\mathcal{E}_{\mathfrak{q};\mathfrak{p}})_y. \tag{6.8}$$

By Definition 3.2 (5), the map (6.8) is an isomorphism if $y \in s_{\mathfrak{q}}^{-1}(0)$ in addition. Therefore we may choose δ_0 so that the map (6.8) is an isomorphism if $d(y, s_{\mathfrak{q}}^{-1}(0)) < \delta_0$. Then existence of δ_0, c satisfying (6.7) is an immediate consequence of the fact $s_{\mathfrak{p}}$ is smooth. The inequality (6.6) then is a consequence of C^1-convergence. □

Remark 6.15 The set-theoretical fiber of a vector bundle over an orbifold is a quotient of a vector space by a finite group acting by linear isometry. The value $\overline{s_{\mathfrak{p}}}(y)$ is well-defined as an element of vector space if we fix a local trivialization. When we do not specify the local trivialization, the value $\overline{s_{\mathfrak{p}}}(y)$ is defined as an element of a quotient of a vector space by a finite group. The left hand side of (6.7) therefore makes sense.

In the case of multisection $\overline{\mathfrak{s}^n_{\mathfrak{p};k}}(y)$, this is well-defined as an element of a vector space if we fix a local trivialization. When we change the local trivialization, it changes by the permutation of k and a finite group action. Therefore the validity of (6.6) for all branches $\mathfrak{s}^n_{\mathfrak{p};k}$ is independent of the choice of trivialization.

Remark 6.16 The way taken in [FOOO16] or in the earlier literature such as [FOn2, FOOO4] is different from that in this book and proceeds as follows. We fix the extension of the subbundle $\mathcal{E}_{\mathfrak{q};\mathfrak{p}}$ and fix the choice of the splitting $\mathcal{E}_{\mathfrak{q}} \equiv \mathcal{E}_{\mathfrak{q};\mathfrak{p}} \oplus \frac{\mathcal{E}_{\mathfrak{q}}}{\mathcal{E}_{\mathfrak{q};\mathfrak{p}}}$. We then assumed the equality

$$\Pi_{\frac{\mathcal{E}_{\mathfrak{q}}}{\mathcal{E}_{\mathfrak{q};\mathfrak{p}}}}(\mathfrak{s}^n_{\mathfrak{p},k}) = \Pi_{\frac{\mathcal{E}_{\mathfrak{q}}}{\mathcal{E}_{\mathfrak{q};\mathfrak{p}}}}(s_{\mathfrak{p}}) \tag{6.9}$$

for any branch $\mathfrak{s}^n_{\mathfrak{p},k}$ of our multisection $\mathfrak{s}^n_{\mathfrak{p}}$. (See [FOn2, (6.4.4)], for example.)[5] We did *not* assume that $\mathfrak{s}^n_{\mathfrak{p},k}$ converges to $s_{\mathfrak{p}}$ in C^1-topology but assumed only C^0-convergence. However, (6.9) implies (6.6), which is indeed what we need.

We have slightly modified the definition here, since by assuming C^1-convergence as in Definition 6.12 (2) we can prove Lemma 6.14, which we will use in this book.

We use Lemma 6.14 to prove Propositions 6.17 and 6.18 below. Hereafter, in this section we use a metric on subsets of $|\widehat{\mathcal{U}}| = \bigcup_{\mathfrak{p}\in\mathfrak{P}} U_{\mathfrak{p}}/\sim$, which we choose as follows. We start with support systems $\mathcal{K}^i$, $i = 1, 2, 3$ with $\mathcal{K}^1 < \mathcal{K}^2 < \mathcal{K}^3$. (See Definition 5.23 (2) for this notation.) The union of the images of $\mathcal{K}^3_{\mathfrak{p}}$ in $|\widehat{\mathcal{U}}|$ is denoted

[5]Actually we can make sense of the equality (6.9) without taking and fixing the splitting $\mathcal{E}_{\mathfrak{p}} \equiv \mathcal{E}_{\mathfrak{q};\mathfrak{p}} \oplus \frac{\mathcal{E}_{\mathfrak{q}}}{\mathcal{E}_{\mathfrak{q};\mathfrak{p}}}$, since the projection $\mathcal{E}_{\mathfrak{p}} \to \frac{\mathcal{E}_{\mathfrak{q}}}{\mathcal{E}_{\mathfrak{q};\mathfrak{p}}}$ is well-defined without fixing this splitting.

by $|\mathcal{K}^3|$. The quotient topology on $|\mathcal{K}^3|$ is metrizable. (See [FOOO17, Proposition 2.11].) We use this topology or its induced topology. The space X can be regarded as a subspace of $|\mathcal{K}^3|$, $|\mathcal{K}^2|$ or $|\mathcal{K}^1|$. We take a metric on a compact neighborhood of $|\mathcal{K}^3|$ in $|\widehat{\mathcal{U}}|$ and use the induced metric on various spaces appearing below. Note that all the spaces $\mathcal{K}^i_{\mathfrak{p}}$ etc. are contained in the compact neighborhood of $|\mathcal{K}^3|$ so carry this metric.

For a subset $A \subset |\mathcal{K}^3|$ we put

$$B_\delta(A) = \{x \in |\mathcal{K}^3| \mid d(x, A) < \delta\}. \tag{6.10}$$

For a point $p \in |\mathcal{K}^3|$, we define $B_\delta(p) := B_\delta(\{p\})$.

Sometimes we identify a subset $\mathcal{K}^i_{\mathfrak{p}}$ with its image in $|\mathcal{K}^3|$. Then for example, for a subset $C \subset \mathcal{K}^3_{\mathfrak{q}} \cap U_{\mathfrak{pq}}$, we identify C with $\varphi_{\mathfrak{pq}}(C)$.

Proposition 6.17 *Let $\mathcal{K}^1 < \mathcal{K}^2 < \mathcal{K}^3$ and let $\{\mathfrak{s}^n_{\mathfrak{p}}\}$ be a multivalued perturbation of $(\widehat{\mathcal{U}}, \mathcal{K}^3)$. Then there exist $\delta > 0$, $n_0 \in \mathbb{Z}_{\geq 0}$ such that for any $\mathfrak{q} \in \mathfrak{P}$ and $n > n_0$*

$$B_\delta(\mathcal{K}^1_{\mathfrak{q}} \cap Z) \cap \bigcup_{\mathfrak{p}}((\mathfrak{s}^n_{\mathfrak{p}})^{-1}(0) \cap \mathcal{K}^3_{\mathfrak{p}}) \subset \mathcal{K}^2_{\mathfrak{q}}. \tag{6.11}$$

Here $(\mathfrak{s}^n_{\mathfrak{p}})^{-1}(0)$ is the set of the points in $\mathcal{K}^3_{\mathfrak{p}}$ such that at least one of the branches of $\mathfrak{s}^n_{\mathfrak{p}}$ vanishes.

Proof The proof is by contradiction. The number n_0 is the one given in Lemma 6.14. Suppose Proposition 6.17 does not hold. Then there exist $\mathfrak{q}, \mathfrak{p}$, sequences $n_i > n_0$, $\delta_i \to 0$ and

$$x_i \in B_{\delta_i}(\mathcal{K}^1_{\mathfrak{q}} \cap Z) \cap ((\mathfrak{s}^{n_i}_{\mathfrak{p}})^{-1}(0) \cap \mathcal{K}^3_{\mathfrak{p}}) \tag{6.12}$$

such that

$$x_i \notin \mathcal{K}^2_{\mathfrak{q}}. \tag{6.13}$$

Since Z is compact, (6.12) implies that we may assume by taking a subsequence that x_i converges to $x_\infty \in \mathcal{K}^1_{\mathfrak{q}} \cap Z$. We examine two cases separately. First suppose $\mathfrak{p} \leq \mathfrak{q}$. Then

$$x_i \in \mathcal{K}^3_{\mathfrak{p}} \cap B_{\delta_i}(\mathcal{K}^1_{\mathfrak{q}})$$

implies $x_i \in \mathcal{K}^2_{\mathfrak{q}}$ by the definition of $\mathcal{K}^1 < \mathcal{K}^2$ provided i is sufficiently large. This contradicts (6.13).

Suppose $\mathfrak{p} > \mathfrak{q}$. Then, for sufficiently large i, Lemma 6.14 implies

$$|\mathfrak{s}^{n_i}_{\mathfrak{p};k}(x_i)| \geq c|E(x_i)|.$$

Since $\mathfrak{s}^{n_i}_{\mathfrak{p};k}(x_i) = 0$ for some k we have $E(x_i) = 0$. Hence $x_i \in U_{\mathfrak{q}}$. Then $\lim x_i = x_\infty \in K^1_{\mathfrak{q}}$ implies $x_i \in \mathcal{K}^2_{\mathfrak{q}}$ for sufficiently large i. This again contradicts (6.13). □

Proposition 6.18 *Let $\mathcal{K}^1, \mathcal{K}^2, \mathcal{K}^3$ be a triple of support systems of a good coordinate system $\widehat{\mathcal{U}}$ of $Z \subseteq X$ with $\mathcal{K}^1 < \mathcal{K}^2 < \mathcal{K}^3$ and $\widehat{\mathfrak{s}} = \{\mathfrak{s}^n_{\mathfrak{p}}\}$ a multivalued perturbation of $(\widehat{\mathcal{U}}, \mathcal{K}^3)$. Then there exists a neighborhood $\mathfrak{U}(Z)$ of Z in $|\mathcal{K}^2|$ and $n_0 \in \mathbb{Z}_{\geq 0}$ such that the following holds for any $n > n_0$:*

$$\left(\bigcup_{\mathfrak{p}}((\mathfrak{s}^n_{\mathfrak{p}})^{-1}(0) \cap \overset{\circ}{\mathcal{K}^1_{\mathfrak{p}}})\right) \cap \mathfrak{U}(Z) = \left(\bigcup_{\mathfrak{p}}((\mathfrak{s}^n_{\mathfrak{p}})^{-1}(0) \cap \mathcal{K}^2_{\mathfrak{p}})\right) \cap \mathfrak{U}(Z). \tag{6.14}$$

Proof The inclusion $\subseteq$ is obvious for any $\mathfrak{U}(Z)$. We will prove the inclusion of the opposite direction. The proof is by contradiction. The number n_0 is the one given in Lemma 6.14. We take $\delta_i \to 0$ and assume (6.14) does not hold for $\mathfrak{U}(Z) = B_{\delta_i}(Z)$ for any i. Then there exist $\mathfrak{p}_i$ and x_i such that

$$x_i \in (\mathfrak{s}^{n_i}_{\mathfrak{p}_i})^{-1}(0) \cap \mathcal{K}^2_{\mathfrak{p}_i} \cap B_\delta(Z) \tag{6.15}$$

but

$$x_i \notin (\mathfrak{s}^{n_i}_{\mathfrak{q}})^{-1}(0) \cap \mathcal{K}^1_{\mathfrak{q}} \tag{6.16}$$

for any $\mathfrak{q}$. By taking a subsequence we may assume $\mathfrak{p}_i = \mathfrak{p}$ is independent of i. Using (6.15) we may also assume that x_i converges to

$$x_\infty \in \mathcal{K}^2_{\mathfrak{p}} \cap Z.$$

Since $\{\overset{\circ}{\mathcal{K}^1_{\mathfrak{q}}}\}_{\mathfrak{q}}$ covers Z, there exists $\mathfrak{q}$ such that

$$x_\infty \in \overset{\circ}{\mathcal{K}^1_{\mathfrak{q}}}.$$

We first examine the case $\mathfrak{q} > \mathfrak{p}$. Then $x_i \in \mathcal{K}^2_{\mathfrak{p}}$ and $\lim x_i = x_\infty$ implies $x_i \in \overset{\circ}{\mathcal{K}^1_{\mathfrak{q}}}$, for large i. Since $\mathfrak{s}^{n_i}_{\mathfrak{p};k}(x_i) = 0$, this contradicts (6.16).

Next we consider the case $\mathfrak{p} > \mathfrak{q}$. By Lemma 6.14

$$|\mathfrak{s}^{n_i}_{\mathfrak{p};k}(x_i)| \geq c|E(x_i)|$$

for large i. Therefore $\mathfrak{s}^{n_i}_{\mathfrak{p};k}(x_i) = 0$ implies $E(x_i) = 0$. Hence $x_i \in \overset{\circ}{\mathcal{K}^1_{\mathfrak{q}}}$ for large i. This contradicts (6.16). □

Remark 6.19 Proposition 6.18 corresponds to [FOOO16, Lemma 6.6]. The proof we gave in the preprint version of this book [FOOO19, Proposition 6.30] is based on the same idea. We modify it here to give a new proof based on the argument by contradiction, which simplifies the proof.

A similar proof appeared in Irie [Ir].

Corollary 6.20 *There exist a neighborhood $\mathfrak{U}(Z)$ of Z in $|\mathcal{K}^2|$ and $n_0 \in \mathbb{Z}_{\geq 0}$ such that the space $\left(\bigcup_{\mathfrak{p}}((\mathfrak{s}^n_{\mathfrak{p}})^{-1}(0) \cap \overset{\circ}{\mathcal{K}^2_{\mathfrak{p}}})\right) \cap \mathfrak{U}(Z)$ is compact if $n > n_0$. Moreover,*

$$\lim_{n\to\infty} \left(\bigcup_{\mathfrak{p}}((\mathfrak{s}^n_{\mathfrak{p}})^{-1}(0) \cap \overset{\circ}{\mathcal{K}^2_{\mathfrak{p}}})\right) \cap \mathfrak{U}(Z) \subseteq X. \tag{6.17}$$

Here the limit is taken in Hausdorff topology.

The first claim corresponds to [FOOO16, Lemma 6.11] and the second claim corresponds to [FOOO16, Lemma 6.12]. With Proposition 6.18 at our disposal the proof is also the same as those lemmas. We reproduce them here for the reader's convenience.

Proof Proposition 6.18 implies that

$$\left(\bigcup_{\mathfrak{p}}((\mathfrak{s}^n_{\mathfrak{p}})^{-1}(0) \cap \overset{\circ}{\mathcal{K}^2_{\mathfrak{p}}})\right) \cap \mathfrak{U}(Z) = \left(\bigcup_{\mathfrak{p}}((\mathfrak{s}^n_{\mathfrak{p}})^{-1}(0) \cap \mathcal{K}^1_{\mathfrak{p}})\right) \cap \mathfrak{U}(Z). \tag{6.18}$$

We may assume that $\mathfrak{U}(Z)$ is compact. Then $\left(\bigcup_{\mathfrak{p}}((\mathfrak{s}^n_{\mathfrak{p}})^{-1}(0) \cap \mathcal{K}^1_{\mathfrak{p}})\right) \cap \mathfrak{U}(Z)$ is compact. The compactness of $\left(\bigcup_{\mathfrak{p}}((\mathfrak{s}^n_{\mathfrak{p}})^{-1}(0) \cap \overset{\circ}{\mathcal{K}^2_{\mathfrak{p}}})\right) \cap \mathfrak{U}(Z)$ follows from (6.18).

We next prove (6.17). Suppose (6.17) does not hold for any $\mathfrak{U}(Z)$. Then, using (6.18), there exist $\mathfrak{p} \in \mathfrak{P}$, $\delta > 0$, $n_i \to \infty$, and x_i such that $x_i \in (\mathfrak{s}^{n_i}_{\mathfrak{p}})^{-1}(0) \cap \mathcal{K}^1_{\mathfrak{p}} \cap \mathfrak{U}(Z)$ and $d(x_i, X) \geq \delta$ for all i. We may assume that x_i converges to x. (Note we may assume that $\mathfrak{U}(Z)$ is compact.) Then $x \in (s_{\mathfrak{p}})^{-1}(0) \cap \mathcal{K}^1_{\mathfrak{p}} \cap \mathfrak{U}(Z)$. Therefore $x \in X$. This contradicts $d(x, X) \geq \delta > 0$. □

6.3 Transversality of the Multisection

Definition 6.21 (The case of an orbifold)

(1) Let $\mathfrak{s}$ be a multisection of a vector bundle $\mathcal{E}$ on an orbifold U. We say it is *transversal to* 0 on $K \subset U$ if, for each $x \in K$ and any branch $\mathfrak{s}_k$ of $\mathfrak{s}$ at x such that $\mathfrak{s}_k(x) = 0$, $\mathfrak{s}_k$ is transversal to 0. (Note $\mathfrak{s}_k : V_x \to E_x$ is a smooth map and $(V_x, \Gamma_x, E_x, \phi_x, \widehat{\phi_x})$ is an orbifold chart of $(U, \mathcal{E})$.)

(2) In the situation of (1), let $f : U \to N$ be a smooth map to a manifold. We say f is *strongly submersive* with respect to $\mathfrak{s}$, if for any branch $\mathfrak{s}_k$ of $\mathfrak{s}$ at $x \in K$ such that $\mathfrak{s}_k(x) = 0$, the composition

$$\mathfrak{s}_k^{-1}(0) \hookrightarrow V \xrightarrow{\phi_x} U \xrightarrow{f} N \tag{6.19}$$

is a submersion.

(3) In the situation of (2), let $g : M \to N$ be a smooth map between manifolds. Suppose the multisection $\mathfrak{s}$ is transversal to 0 on K. We say $(\mathfrak{s}, f)$ is *strongly transversal to* g if $\mathfrak{s}$ is transversal to 0 and, for any branch $\mathfrak{s}_k$ of $\mathfrak{s}$ at $x \in K$ such that $\mathfrak{s}_k(x) = 0$, the composition (6.19) is transversal to g.

Definition 6.22 (The case of a good coordinate system)

(1) Let $\widehat{\mathfrak{s}}$ be a multisection of $(\widehat{\mathcal{U}}, \mathcal{K})$, where $\widehat{\mathcal{U}}$ is a good coordinate system of $Z \subseteq X$ and $\mathcal{K}$ its support system. We say it is *transversal to* 0 if for each $\mathfrak{p} \in \mathfrak{P}$, $\mathfrak{s}_{\mathfrak{p}}$ is transversal to 0 on $\mathcal{K}_{\mathfrak{p}}$.

(2) In the situation of (1), let $\widehat{f} : (X, Z; \widehat{\mathcal{U}}) \to N$ be a smooth map to a manifold. We say f is *strongly submersive* with respect to $\widehat{\mathfrak{s}}$, if for each $\mathfrak{p} \in \mathfrak{P}$, $f_{\mathfrak{p}}$ is strongly submersive with respect to $\mathfrak{s}_{\mathfrak{p}}$.

(3) In the situation of (2), let $g : M \to N$ be a smooth map between manifolds. Suppose the multisection $\widehat{\mathfrak{s}}$ is transversal to 0 on K. We say $(\widehat{\mathfrak{s}}, f)$ is *strongly transversal to* g if for each $\mathfrak{p} \in \mathfrak{P}$, $f_{\mathfrak{p}}$ is strongly transversal to g with respect to $\mathfrak{s}_{\mathfrak{p}}$.

(4) A multivalued perturbation of a good coordinate system $\widehat{\mathfrak{s}} = \{\widehat{\mathfrak{s}^n}\}$ is said to be *transversal to* 0 if $\widehat{\mathfrak{s}^n}$ is transversal to 0 for all $n = 1, 2, \ldots$.

(5) *Strong transversality of a map* $\widehat{f} : (X, Z; \widehat{\mathcal{U}}) \to N$ *on good coordinate system to* $g : M \to N$ *with respect to a multivalued perturbation* is defined in the same way.

Theorem 6.23 *Let $\widehat{\mathcal{U}}$ be a good coordinate system of $Z \subseteq X$ and $\mathcal{K}$ its support system.*

(1) *There exists a multivalued perturbation $\widehat{\mathfrak{s}} = \{\mathfrak{s}_{\mathfrak{p}}^n\}$ of $(\widehat{\mathcal{U}}, \mathcal{K})$ which is transversal to* 0.

(2) *Suppose $\widehat{f} : (X, Z; \widehat{\mathcal{U}}) \to N$ is strongly smooth and is weakly transversal to $g : M \to N$, where g is a map from a manifold M. Then we may choose $\widehat{\mathfrak{s}}$ so that $\widehat{f}$ is strongly transversal to g with respect to $\widehat{\mathfrak{s}}$.*

This theorem was actually proved during the proof of [FOOO16, Proposition 6.3]. We will prove it in Chap. 13.

6.4 Embedding of Kuranishi Structures and Multivalued Perturbations

Definition 6.24 Let $\widehat{\mathcal{U}}$ be a Kuranishi structure of $Z \subseteq X$. A *strictly compatible multivalued perturbation of* $\widehat{\mathcal{U}}$ is a collection $\widehat{\mathfrak{s}} = \{\widehat{\mathfrak{s}^n}\} = \{\mathfrak{s}^n_p\}_{p \in Z}$ such that $\mathfrak{s}^n_p$ is a multisection of E_p on U_p for each $p \in X$ and $n \in \mathbb{Z}_{\geq 0}$, which have the following properties:

(1) $\mathfrak{s}^n_p \circ \varphi_{pq} = \widehat{\varphi}_{pq} \circ \mathfrak{s}^n_q$ on U_{pq}.
(2) $\lim_{n \to \infty} \mathfrak{s}^n_p = s_p$ in C^1 sense on U_p.

The meaning of (1), (2) above is the same as in the case of Definition 6.12.

A multisection on $\widehat{\mathcal{U}}$ is defined in the same way requiring (1) but not (2).

Transversality to 0 of a multivalued perturbation and multisection of $\widehat{\mathcal{U}}$ is defined in the same way as Definition 6.22 by requiring the property on each Kuranishi chart.

Remark 6.25 We use the terminology 'strictly compatible multivalued perturbations' in Definition 6.24. The phrase 'strictly compatible' indicates that this is rather a strong condition and is hard to realize. For example, we may not expect such a perturbation exists for a given Kuranishi structure. Namely we need to replace the given Kuranishi structure by its appropriate thickening to obtain strictly compatible multivalued perturbation. (See Proposition 6.30.) Nevertheless we usually omit the phrase 'strictly compatible' and simply say multivalued perturbation.

Definition 6.26

(1) Let $\widehat{\Phi} = \{\Phi_p\} : \widehat{\mathcal{U}} \to \widehat{\mathcal{U}'}$ be a strict KK-embedding of Kuranishi structures. Let $\{\mathfrak{s}^n_p\}$ and $\{\mathfrak{s}'^n_p\}$ be multivalued perturbations of $\widehat{\mathcal{U}}$ and $\widehat{\mathcal{U}'}$, respectively. We say $\{\mathfrak{s}^n_p\}$ and $\{\mathfrak{s}'^n_p\}$ are *compatible* with $\widehat{\Phi}$ if $\mathfrak{s}'^n_p \circ \varphi_p = \widehat{\varphi}_p \circ \mathfrak{s}^n_p$.
(2) Let $\widehat{\mathcal{U}_0}$ be an open substructure of a Kuranishi structure $\widehat{\mathcal{U}}$. Let $\{\mathfrak{s}^n_p\}$ be a multivalued perturbation of $\widehat{\mathcal{U}}$. Then $\{\mathfrak{s}^n_p|_{U^0_p}\}$ is a multivalued perturbation of $\widehat{\mathcal{U}_0}$. We call it the *restriction* of $\{\mathfrak{s}^n_p\}$ and write $\{\mathfrak{s}^n_p\}|_{\widehat{\mathcal{U}^0}}$.
(3) Let $\widehat{\Phi} = \{\Phi_p\} : \widehat{\mathcal{U}} \to \widehat{\mathcal{U}'}$ be a (not necessarily strict) KK-embedding of Kuranishi structures. Let $\{\mathfrak{s}^n_p\}$ and $\{\mathfrak{s}'^n_p\}$ be multivalued perturbations of $\widehat{\mathcal{U}}$ and $\widehat{\mathcal{U}'}$, respectively. We say $\{\mathfrak{s}^n_p\}$ and $\{\mathfrak{s}'^n_p\}$ are *compatible* with $\widehat{\Phi}$ if a restriction $\{\mathfrak{s}^n_p\}|_{\widehat{\mathcal{U}_0}}$ is compatible with $\{\mathfrak{s}'^n_p\}$ with respect to a strict embedding $\widehat{\mathcal{U}_0} \to \widehat{\mathcal{U}'}$. Here $\widehat{\mathcal{U}_0}$ is an open substructure of $\widehat{\mathcal{U}}$.

Definition 6.27 Let $\widehat{\widehat{\Phi}} = \{\Phi_{\mathfrak{p}}\} : \widehat{\widehat{\mathcal{U}}} \to \widehat{\widehat{\mathcal{U}'}}$ be a GG-embedding, and $\widehat{\widehat{\mathfrak{s}}} = \{\mathfrak{s}^n_{\mathfrak{p}}\}$ and $\widehat{\widehat{\mathfrak{s}'}} = \{\mathfrak{s}'^n_{\mathfrak{p}}\}$ multivalued perturbations of $\widehat{\widehat{\mathcal{U}}}$ and $\widehat{\widehat{\mathcal{U}'}}$, respectively. We say $\{\mathfrak{s}^n_{\mathfrak{p}}\}$ and $\{\mathfrak{s}'^n_{\mathfrak{p}}\}$ are *compatible* with $\widehat{\widehat{\Phi}}$ if $\mathfrak{s}'^n_{\mathfrak{p}} \circ \varphi_{\mathfrak{p}} = \widehat{\varphi}_{\mathfrak{p}} \circ \mathfrak{s}^n_{\mathfrak{p}}$.

Definition 6.28 Let $\widehat{\Phi} = (\{U_{\mathfrak{p}}(p)\}, \{\Phi_{p\mathfrak{p}}\}) : \widehat{\mathcal{U}} \to \widehat{\mathcal{U}^+}$ be a GK-embedding. Let $\widehat{\mathfrak{s}} = \{\mathfrak{s}^n_{\mathfrak{p}}\}$ and $\widehat{\mathfrak{s}} = \{\mathfrak{s}^n_p\}$ be multivalued perturbations of $\widehat{\mathcal{U}}$ and $\widehat{\mathcal{U}^+}$, respectively. We say $\{\mathfrak{s}^n_{\mathfrak{p}}\}$ and $\{\mathfrak{s}^n_p\}$ are *compatible* with $\widehat{\Phi}$ if $\mathfrak{s}^n_p \circ \varphi_{p\mathfrak{p}} = \widehat{\varphi}_{p\mathfrak{p}} \circ \mathfrak{s}^n_{\mathfrak{p}}$ holds on $U_{\mathfrak{p}}(p)$.

There are various obvious statements about the composition of embeddings and its compatibilities with the multivalued perturbations. We leave to the interested readers to state and prove them.

The next two propositions claim that when we obtain a Kuranishi structure from good coordinate system by Proposition 5.26, we can bring various structures on the good coordinate system to the Kuranishi structure.

Definition 6.29 Let $\widehat{\Phi} : \widehat{\mathcal{U}_0} \to \widehat{\mathcal{U}}$ be a relatively compact open GG-embedding. Let $\mathcal{K} = \{\mathcal{K}_{\mathfrak{p}}\}$ and $\mathcal{K}_0 = \{\mathcal{K}_{0,\mathfrak{p}}\}$ be support systems of $\widehat{\mathcal{U}}$ and $\widehat{\mathcal{U}_0}$, respectively. We assume $\overline{\varphi_{\mathfrak{p}}(U^0_{\mathfrak{p}})} \subset \operatorname{Int} \mathcal{K}_{\mathfrak{p}}$. In particular, $\mathcal{K}_{0,\mathfrak{p}} \subset \operatorname{Int} \mathcal{K}_{\mathfrak{p}}$. Let $\widehat{\mathfrak{s}} = \{\mathfrak{s}^n_{\mathfrak{p}}\}$ be a multivalued perturbation of $(\widehat{\mathcal{U}}, \mathcal{K})$. Then $\{\mathfrak{s}^n_{\mathfrak{p}}|_{\mathcal{K}_{0,\mathfrak{p}}}\}$ is a multivalued perturbation of $(\widehat{\mathcal{U}_0}, \mathcal{K}_0)$. We denote it by $\widehat{\mathfrak{s}}|_{\widehat{\mathcal{U}_0}}$.

Proposition 6.30 *Let $\widehat{\mathcal{U}_0} \to \widehat{\mathcal{U}}$ be a relatively compact open GG-embedding of good coordinate systems of $Z \subseteq X$. Then the Kuranishi structure $\widehat{\mathcal{U}^+}$ and a GK-embedding $\widehat{\mathcal{U}_0} \to \widehat{\mathcal{U}^+}$ in Proposition* 5.26 *can be taken to have the following properties:*

(1) *In the situation of Definition* 6.29, *there exists a multivalued perturbation $\widehat{\mathfrak{s}} = \{\mathfrak{s}^n_p\}$ of $\widehat{\mathcal{U}^+}$ such that $\widehat{\mathfrak{s}}|_{\widehat{\mathcal{U}_0}}$ and $\widehat{\mathfrak{s}}$ are compatible with the embedding $\widehat{\mathcal{U}_0} \to \widehat{\mathcal{U}^+}$.*
(2) *If $\widehat{f} : (X, Z; \widehat{\mathcal{U}}) \to Y$ is a strongly continuous (resp. strongly smooth) map, then there exists $\widehat{f} : (X, Z; \widehat{\mathcal{U}^+}) \to Y$ such that $\widehat{f}|_{\widehat{\mathcal{U}_0}}$ is a pullback of $\widehat{f}$ by the embedding $\widehat{\mathcal{U}_0} \to \widehat{\mathcal{U}^+}$. If $\widehat{f}$ is strongly smooth (resp. weakly submersive) then so is $\widehat{f}$. The transversality of $\widehat{f}$ to zero and to a smooth map $M \to Y$ from a manifold M to Y is also preserved.*

Proof During the proof of Proposition 5.26 we first find $\mathfrak{p}_p \in \mathfrak{P}$ by Lemma 5.27 and an open neighborhood U^+_p of $o_{\mathfrak{p}_p}(p)$ and define the Kuranishi chart $\mathcal{U}^+_p$ by (5.17).

We define $\mathfrak{s}^n_p = \mathfrak{s}^n_{\mathfrak{p}_p}|_{U^+_p}$. Its compatibility with coordinate change follows from one of $\mathfrak{s}^n_{\mathfrak{p}}$. Thus we obtain a multivalued perturbation $\{\mathfrak{s}^n_p\}$. The compatibility of it with the GK-embedding $\widehat{\mathcal{U}_0} \to \widehat{\mathcal{U}}$ follows from the compatibility of $\mathfrak{s}^n_{\mathfrak{p}}$ with coordinate change. We have proved (1).

If $\widehat{f} = \{f_{\mathfrak{p}}\}$, then we define $f_p = f_{\mathfrak{p}_p}|_{U^+_p}$. It is easy to see that it has required properties. We have proved (2). □

Proposition 6.30 provides a way to transfer a multivalued perturbation of a good coordinate system to that of a Kuranishi structure. The next results describe the way of transferring them in the opposite direction. For this, we need one more definition.

Definition 6.31 Let $\widehat{\Phi} = \{\Phi_{\mathfrak{p}p}\} : \widehat{\mathcal{U}} \to \widehat{\mathcal{U}}$ be a KG-embedding. Let $\widehat{\mathfrak{s}} = \{\mathfrak{s}^n_p\}$ and $\widehat{\mathfrak{s}} = \{\mathfrak{s}^n_{\mathfrak{p}}\}$ be a multivalued perturbation of $\widehat{\mathcal{U}}$ and $\widehat{\mathcal{U}}$, respectively. We say $\{\mathfrak{s}^n_p\}$ and $\{\mathfrak{s}^n_{\mathfrak{p}}\}$ are *compatible* with $\widehat{\Phi}$ if they satisfy

$$\varphi_{\mathfrak{p}p} \circ \mathfrak{s}^n_p = \mathfrak{s}^n_{\mathfrak{p}} \circ \widehat{\varphi}_{\mathfrak{p}p}$$

on U_p.

Proposition 6.32 *Let $\widehat{\mathcal{U}}$ be a Kuranishi structure on $Z \subseteq X$ and $\widehat{\mathfrak{s}} = \{\mathfrak{s}^n_p\}$ a multivalued perturbation of $\widehat{\mathcal{U}}$. Then we can take a good coordinate system $\widehat{\mathcal{U}}$ and the strict KG-embedding $\widehat{\Phi} : \widehat{\mathcal{U}_0} \to \widehat{\mathcal{U}}$ in Theorem* 3.35 *so that the following holds in addition:*

(1) *There exists a multivalued perturbation $\widehat{\mathfrak{s}} = \{\mathfrak{s}^n_{\mathfrak{p}}\}$ of $\widehat{\mathcal{U}}$ such that $\widehat{\mathfrak{s}}|_{\widehat{\mathcal{U}_0}}$ and $\widehat{\mathfrak{s}}$ are compatible with the embedding $\widehat{\Phi}$.*
(2) *If $\widehat{f} : (X, Z; \widehat{\mathcal{U}}) \to Y$ is a strongly continuous map to a manifold Y, then there exists $\widehat{f} : (X, Z; \widehat{\mathcal{U}}) \to Y$ such that $\widehat{f} \circ \widehat{\Phi}$ is a pullback of $\widehat{f}$. If $\widehat{f}$ is strongly smooth (resp. weakly submersive) then so is $\widehat{f}$. The transversality of $\widehat{f}$ to zero and to a smooth map $M \to Y$ from a manifold M to Y is also preserved.*

Proposition 6.33 *Suppose we are in the situation of Propositions* 5.20 *(resp.* 5.21*) and* 6.32*. Then we can take the GK-embedding $\widehat{\Phi^+} : \widehat{\mathcal{U}} \to \widehat{\mathcal{U}^+}$ in Proposition* 5.20 *(resp. the GK-embeddings $\widehat{\Phi^+_a} : \widehat{\mathcal{U}} \to \widehat{\mathcal{U}^+_a}$ in Proposition* 5.21 *$(a = 1, 2)$) so that the following holds:*

(1) *If $\widehat{\mathfrak{s}^+}$ is a multivalued perturbation of $\widehat{\mathcal{U}^+}$ such that $\widehat{\mathfrak{s}^+}, \widehat{\mathfrak{s}}$ are compatible with the KK-embedding $\widehat{\mathcal{U}} \to \widehat{\mathcal{U}^+}$, then we may choose $\widehat{\mathfrak{s}}$ such that $\widehat{\mathfrak{s}}$, $\widehat{\mathfrak{s}^+}$ are compatible with the embedding $\widehat{\Phi^+}$. (Resp. If $\widehat{\mathfrak{s}^+_a}$ $(a = 1, 2)$ is a multivalued perturbation of $\widehat{\mathcal{U}^+_a}$ such that $\widehat{\mathfrak{s}^+_a}, \widehat{\mathfrak{s}}$ are compatible with the embedding $\widehat{\mathcal{U}} \to \widehat{\mathcal{U}^+_a}$, then we may choose $\widehat{\mathfrak{s}}$ such that $\widehat{\mathfrak{s}}$, $\widehat{\mathfrak{s}^+_a}$ are both compatible with the embedding $\widehat{\Phi^+_a} : \widehat{\mathcal{U}} \to \widehat{\mathcal{U}^+_a}$.)*
(2) *If $\widehat{f^+} : (X, Z; \widehat{\mathcal{U}^+}) \to Y$ is a strongly continuous map so that the pullback of $\widehat{f^+}$ is $\widehat{f} : (X, Z; \widehat{\mathcal{U}}) \to Y$, then we may choose $\widehat{f} : (X, Z; \widehat{\mathcal{U}}) \to Y$ such that $\widehat{f^+} \circ \widehat{\Phi^+} = \widehat{f}$. (Resp. If $\widehat{f^+_a} : (X, Z; \widehat{\mathcal{U}^+_a}) \to Y$ are strongly continuous maps such that the pullback of both $\widehat{f^+_a}$ $(a = 1, 2)$ are $\widehat{f} : (X, Z; \widehat{\mathcal{U}}) \to Y$, then we may choose $\widehat{f} : (X, Z; \widehat{\mathcal{U}}) \to Y$ such that $\widehat{f^+_a} \circ \widehat{\Phi^+_a} = \widehat{f}$.)*
(3) *In the situation of (1)(2), if $\widehat{f}$ is strongly submersive with respect to $\widehat{\mathfrak{s}}$, then $\widehat{f}$ is strongly submersive with respect to $\widehat{\mathfrak{s}}$. The transversality to $M \to Y$ is also preserved.*

Table 6.1 Object moves from one to the other

Object	GCS to K. Str.	K. Str. to GCS	K to G when < K
Structure	Proposition 5.26	Theorem 3.35	Proposition 5.20
CF-perturbation	Lemma 9.9 (1)	Lemma 9.10 (2)	Lemma 9.11 (2)
Multivalued perturbation	Proposition 6.30 (1)	Proposition 6.32 (1)	Proposition 6.33 (1)
Differential form	Lemma 9.9 (3)	Lemma 9.10 (1)	Lemma 9.11 (1)
Strongly smooth map	Lemma 6.30 (2)	Proposition 6.32 (2)	Proposition 6.33 (2)
Transversality (CF)	Lemma 9.9 (2)	Lemma 9.10 (3)	Lemma 9.11 (3)
Transversality (MV)	Lemma 6.30 (2)	Proposition 6.32 (2)	Proposition 6.33 (3)

We will prove Propositions 6.32 and 6.33 in Sect. 11.3.

We have proved and will prove several statements which claim that when we construct a good coordinate system from a Kuranishi structure, we can carry certain objects given on the Kuranishi structure to the good coordinate system we obtain, and vice versa. The above table (Table 6.1) shows a list of such statements. The proofs of those results are actually immediate from construction. We record these results explicitly for future reference. In the table GCS=Good coordinate system, K. Str.= Kuranishi structure. 'K to G when < K' means that from a Kuranishi structure to a good coordinate system when a thickening is given.

6.5 General Strategy of Construction of Virtual Fundamental Chains

In this section we summarize a general strategy we will take and show how the results of this chapter will be used in the strategy.

Let us start with a Kuranishi structure $\widehat{\mathcal{U}}$ of X. In the description of 'steps' below we sometimes replace a Kuranishi structure or a good coordinate system by its open substructure. Wo do not mention it and use the same notation for open substructure, below.

Step 1: We find a good coordinate system $\widehat{\mathcal{U}}$ such that $\widehat{\mathcal{U}} < \widehat{\mathcal{U}}$, which means that there exists a KG-embedding $\widehat{\mathcal{U}} \to \widehat{\mathcal{U}}$. (Theorem 3.35.)

Step 2: We find a multivalued perturbation $\widehat{\mathfrak{s}}$ of $\widehat{\mathcal{U}}$ that has various transversality properties. (Theorem 6.23.)

Step 3: We obtain a virtual fundamental chain associated to the perturbations of $\widehat{\mathcal{U}}$.

Step 4: We next apply Proposition 5.26 to obtain Kuranishi structure $\widehat{\mathcal{U}^+}$ such that

$$\widehat{\mathcal{U}} < \widehat{\mathcal{U}}, \quad \widehat{\mathcal{U}} < \widehat{\mathcal{U}^+}$$

and then apply Proposition 6.30 to obtain a multivalued perturbation $\widehat{\mathfrak{s}^+}$ of $\widehat{\mathcal{U}^+}$.

Step 5: We apply Theorem 3.35 to $\widehat{\mathcal{U}^+}$ and obtain a good coordinate system $\widehat{\widehat{\mathcal{U}^+}}$ such that

$$\widehat{\mathcal{U}} < \widehat{\widehat{\mathcal{U}}}, \quad \widehat{\mathcal{U}} < \widehat{\mathcal{U}^+}, \quad \widehat{\mathcal{U}^+} < \widehat{\widehat{\mathcal{U}^+}}.$$

Moreover, Proposition 6.32 implies that the multivalued perturbation $\widehat{\mathfrak{s}^+}$ induces a multivalued perturbation $\widehat{\widehat{\mathfrak{s}^+}}$ of $\widehat{\widehat{\mathcal{U}^+}}$.

Step 6: The transversality of $\widehat{\mathfrak{s}}$ implies one of $\widehat{\mathfrak{s}^+}$ and then one of $\widehat{\widehat{\mathfrak{s}^+}}$.

Step 7: We obtain a virtual fundamental chain associated to $\widehat{\widehat{\mathfrak{s}^+}}$.

Now an important statement is that the virtual fundamental chain obtained in Step 3 is the same as the virtual fundamental *chain* obtained in Step 7 for any sufficiently large n. (Roughly speaking, this is a consequence of Proposition 6.18.) Its de Rham version is Proposition 9.16. It is Lemma 14.17 in the 0-dimensional case.

This statement can be used as follows. Note the construction in Step 5 is not unique. Namely for each given $\widehat{\mathcal{U}^+}$ there are many possible choices of $\widehat{\widehat{\mathcal{U}^+}}$. However, the virtual fundamental chain associated to it is independent of such choices. Moreover it coincides with the virtual fundamental chain obtained in Step 3.

In other words, we can recover the virtual fundamental chain of $\widehat{\mathfrak{s}}$ (that is defined on $\widehat{\mathcal{U}}$ in Step 3) from $\widehat{\mathfrak{s}^+}$ that is defined on $\widehat{\mathcal{U}^+}$.

For this reason, we can forget the good coordinate system $\widehat{\mathcal{U}}$ and remember only the Kuranishi structure $\widehat{\mathcal{U}^+}$ and its perturbation $\widehat{\mathfrak{s}^+}$. Since the Kuranishi structure behaves better with fiber products than good coordinate systems, we can use this fact to make the whole construction compatible with the fiber product description of the boundaries.

In Chaps. 7, 8, 9, 10, 11, and 12, we will work out this process in the de Rham model in great detail, where we will use *CF-perturbations*, which is an abbreviation of *continuous family perturbations*, instead of multivalued perturbations. In Chaps. 13 and 14, we will work out this process for multivalued perturbations. In Chap. 14 we restrict ourselves to the case when the dimension of the K-space is not greater than 1.

In Part II, we will discuss in detail the way we make the whole construction compatible when we start with the K-system.

Chapter 7
CF-Perturbations and Integration Along the Fiber (Pushout)

7.1 Introduction to Chaps. 7, 8, 9, 10, and 12

As we mentioned in the Introduction, we study systems of K-spaces (K-systems) so that the boundary of each of its members is described by a fiber product of other members. We will obtain an algebraic structure on certain cochain complexes which realize the homology groups of certain spaces. They are the spaces over which we take fiber products between members of the system of K-spaces. To work out this process we need to make a choice of the homology theory we use. The choices are de Rham cohomology, singular homology, simplicial homology, Čech cohomology, Morse homology, Kuranishi homology (see [Jo1]) etc. In [FOOO3] we took the most standard choice, that is, the singular homology. In this book, we mainly use de Rham cohomology. There are three advantages in using de Rham cohomology. One is that it may be the shortest way to write a detailed and rigorous proof. The second is that it is the easiest way to keep as much symmetry as possible. The third is, by using de Rham cohomology, we might clarify some direct relationship to quantum field theory, (especially in regard to perturbation of the constant maps). There are certain disadvantages in using de Rham cohomology. The most serious disadvantage is that we can work only over real or complex numbers as a ground field. Certain technical points which appear when we use the singular homology will be explained elsewhere. The way of using Morse homology is discussed in [FOOO5]. Actually Morse homology is one the authors of the present book had used around 25 years ago in [Fuk2, Oh], etc. See [FOOO3, Remark 1.32].

The situation we work with is as follows.

K. Fukaya et al., *Kuranishi Structures and Virtual Fundamental Chains*, Springer Monographs in Mathematics, https://doi.org/10.1007/978-981-15-5562-6_7

Definition 7.1 (See [FOOO8, Section 12]) Let X be a compact metrizable space, and $\widehat{\mathcal{U}}$ a Kuranishi structure of X (with or without boundaries or corners). Let M_s and M_t be C^∞ manifolds. We assume $\widehat{\mathcal{U}}$, M_s and M_t are oriented.[1]

Let $\widehat{f_s} : (X;\widehat{\mathcal{U}}) \to M_s$ be a strongly smooth map and $\widehat{f_t} : (X;\widehat{\mathcal{U}}) \to M_t$ a weakly submersive strongly smooth map. We call $\mathfrak{X} = ((X;\widehat{\mathcal{U}}); \widehat{f_s}, \widehat{f_t})$ a *smooth correspondence* from M_s to M_t.

Our goal in Chaps. 7, 8, 9, and 10 is to associate a linear map

$$\mathrm{Corr}_{\mathfrak{X}} : \Omega^k(M_s) \to \Omega^{k+\ell}(M_t) \tag{7.1}$$

to a smooth correspondence $\mathfrak{X}$ and study its properties. Here $\Omega^k(M_s)$ is the set of smooth k forms on M_s and

$$\ell = \dim M_t - \dim(X,\widehat{\mathcal{U}}).$$

If $(X;\widehat{\mathcal{U}})$ is a smooth orbifold, namely if the obstruction bundles are all 0, then we can define (7.1) by

$$\mathrm{Corr}_{\mathfrak{X}}(h) = f_t!(f_s^*(h)).$$

Here $f_s^* : \Omega^k(M_s) \to \Omega^k(X)$ is the pullback of the differential form and $f_t! : \Omega^k(X) \to \Omega^{k+\ell}(M_t)$ is *the integration along the fiber*, or *pushout*, which is characterized by

$$\int_{M_t} \rho \wedge f_t!(v) = \int_X f_t^*\rho \wedge v.$$

(Note $\ell \le 0$ in this case.) The existence of such an $f_t!$ is a consequence of the fact that f_t is a proper submersion. (In our situation where $(X;\widehat{\mathcal{U}})$ is an orbifold this is a consequence of the weak submersivity of $\widehat{f_t}$.)

When the obstruction bundle is nontrivial, we need to perturb the space X so that integration along the fiber is well-defined. However, taking a *multivalued perturbation* of $\widehat{\mathcal{U}}$ discussed in Chap. 6 is not good enough for this purpose unless M_t is a point. Let us elaborate on this point below.

Suppose $\widehat{\mathfrak{s}} = \{\mathfrak{s}^n_{\mathfrak{p}}\}$ is a multivalued perturbation of $\widehat{\mathcal{U}}$ where $\widehat{\mathcal{U}}$ is a good coordinate system compatible with $\widehat{\mathcal{U}}$. If we assume that $\mathfrak{s}^n_{\mathfrak{p}}$ is transversal to 0, then in the case $\dim\widehat{\mathcal{U}} = \deg h$, we can define the number

[1] In certain situations, for example in [FOOO4, Subsection 8.8], we discussed a slightly more general case. Namely we discussed the case when $\widehat{\mathcal{U}}$, M_s and M_t are not necessarily orientable by introducing appropriate $\mathbb{Z}_2$ local systems. See Chap. 27.

$$\int_{\bigcup_{\mathfrak{p}} (\mathfrak{s}_{\mathfrak{p}}^{n})^{-1}(0)} f_s^*(h) \in \mathbb{R}.$$

However, we cannot expect that the map

$$f_t|_{\bigcup_{\mathfrak{p}} (\mathfrak{s}_{\mathfrak{p}}^{n})^{-1}(0)} : \bigcup_{\mathfrak{p}} (\mathfrak{s}_{\mathfrak{p}}^{n})^{-1}(0) \to M_t$$

is a 'submersion' in any reasonable sense. In fact, there may happen the case where $\dim \widehat{\mathcal{U}}$ is strictly smaller than $\dim M_t$. Therefore the integration along the fiber $f_t|_{\bigcup_{\mathfrak{p}} (\mathfrak{s}_{\mathfrak{p}}^{\epsilon})^{-1}(0)}!$ sends a differential form to a distributional form which may not be smooth. Thus we need to find an appropriate way to smooth it to define (7.1). The route we take here is to use *CF-perturbation*, which is an abbreviation of *continuous family perturbations*. We discussed this construction in [FOOO4, Section 7.5], [Fuk5, Section 12], [Fuk4, Section 4], [FOOO8, Section 12].

Now the outline of Chaps. 7, 8, 9, 10, and 12 is as follows. We review and describe the CF-perturbation and the integration along the fiber in greater detail and then combine them with the process of transferring various objects from a Kuranishi structure to a good coordinate system and back. More specifically, we first introduce the notion of a CF-perturbation of a single Kuranishi chart $\mathcal{U}$ in Sect. 7.2, where we find in Proposition 7.22 that the set of CF-perturbations of $\mathcal{U}$ turns out to be a sheaf $\mathcal{CF}$. We introduce several subsheaves of $\mathcal{CF}$ which satisfy various transversality conditions. Using these subsheaves we define the pushout of differential forms. Next, in Sect. 7.4, we generalize these results to the case of CF-perturbations of a good coordinate system. Then we can formulate the pushout of differential forms and smooth correspondences in a good coordinate system. We also formulate and prove Stokes' formula for a good coordinate system in Chap. 8. So far, everything here is discussed based on a good coordinate system. However, as mentioned at the end of Chap. 4, it is more convenient and natural to use the Kuranishi structure itself rather than a good coordinate system, when we study the fiber product of K-spaces. For this purpose, we start from a Kuranishi structure and use certain embeddings into a good coordinate system and/or another Kuranishi structure introduced in Chap. 5 to translate the above results based on a good coordinate system into ones based on the Kuranishi structure and study their relationship. As a result, we show in Theorem 9.14 that the pushout of differential forms for Kuranishi structures is indeed independent of the choice of the good coordinate system. After these foundational results on the pushout of differential forms are prepared, we prove a basic result about smooth correspondence, which is called *composition formula of smooth correspondence*, in Theorem 10.21. Proof of the existence of a CF-perturbation for any K-space is postponed till Chap. 12. We extend the definition of the sheaf $\mathcal{CF}$ (defined on a single Kuranishi chart) to $\mathcal{CF}_{\mathcal{K}}$, which is defined on a hetero-dimensional compactum $|\mathcal{K}|$ associated to a support system $\mathcal{K}$ of a good coordinate system. We then prove Theorem 12.24 that the sheaf

$\mathcal{CF}_{\mathcal{K}}$ of CF-perturbations, together with several subsheaves mentioned above, is soft.

Remark 7.2 The usage of the language of sheaf of CF-perturbations clarifies the fact that we can define CF-perturbations locally and glue them together when they coincide at the overlapped part.[2] This is because we take equivalence class in the definition (see Definition 7.6) rather than remembering 'how they are equivalent'. Thanks to Lemma 7.34 (3) we can safely do so.

7.2 CF-Perturbation on a Single Kuranishi Chart

We first consider the situation where we have only one Kuranishi chart. After that we will introduce a CF-perturbation of a good coordinate system in Sect. 7.4. A CF-perturbation of Kuranishi structure will be defined in Sect. 9.1.

Situation 7.3 Let $\mathcal{U} = (U, \mathcal{E}, s, \psi)$ be a Kuranishi chart of X, and $f : U \to M$ a smooth submersion to a smooth manifold M, and h a differential form on U which has compact support. Assume that U, $\mathcal{E}$ and M are oriented.■

7.2.1 *CF-Perturbation on a Kuranishi Chart Restricted to One Orbifold Chart*

Under Situation 7.3 let

$$\mathfrak{V}_x = (V_x, \Gamma_x, E_x, \phi_x, \widehat{\phi}_x)$$

be an orbifold chart of the vector bundle $(U, \mathcal{E})$. (Definition 23.18.) We assume (V_x, Γ_x, ϕ_x) is an oriented orbifold chart. (Definition 23.10 (5).) Since f is a submersion, the composition $f \circ \phi_x : V_x \to U \to M$ as continuous maps becomes a smooth submersion, which is denoted by f_x.

Definition 7.4 A *CF-perturbation (=continuous family perturbation)* of $\mathcal{U}$ on our orbifold chart $\mathfrak{V}_x = (V_x, \Gamma_x, E_x, \phi_x, \widehat{\phi}_x)$ consists of $\mathcal{S}_x = (W_x, \omega_x, \{\mathfrak{s}_x^\epsilon\})$, $0 < \epsilon \le 1$, with the following properties:

(1) W_x is an open neighborhood of 0 of a finite-dimensional oriented vector space $\widehat{W}_x$ on which Γ_x acts linearly. W_x is Γ_x-invariant.
(2) $\mathfrak{s}_x^\epsilon : V_x \times W_x \to E_x$ is a Γ_x-equivariant smooth map for each $0 < \epsilon \le 1$. $\mathfrak{s}_x^\epsilon$ depends smoothly on ϵ.
(3) For $y \in V_x$, $\xi \in W_x$ we have

[2] Such a gluing process is more involved for Kuranishi structures, for example.

$$\lim_{\epsilon \to 0} \mathfrak{s}_x^{\epsilon}(y, \xi) = s_x(y) \tag{7.2}$$

in compact C^1-topology on $V_x \times W_x$.

(4) ω_x is a smooth differential form on W_x of degree dim W_x that is Γ_x-invariant, of compact support and

$$\int_{W_x} \omega_x = 1.$$

We assume $\omega_x = |\omega_x| \mathrm{vol}_x$, where vol_x is a volume form of the oriented manifold W_x and $|\omega_x|$ is a non-negative function.

For each $0 < \epsilon \leq 1$, we denote the restriction of $\mathcal{S}_x$ at ϵ, by $\mathcal{S}_x^{\epsilon} = (W_x, \omega_x, \mathfrak{s}_x^{\epsilon})$.

Remark 7.5

(1) In Definition 7.4 (3) we regard $s_x : V_x \to E_x$ as a Γ_x-equivariant map that is a local representative of the Kuranishi map in the sense of Definition 23.33.
(2) In our earlier writings, we used a family of *multi*-sections parametrized by W_x. Here we use a family of sections parametrized by W_x on V_x such that it is Γ_x equivariant as a map from $V_x \times W_x$. We also allow W_x to have a nontrivial Γ_x action. This formulation seems simpler.
(3) We may regard $\mathfrak{s}_x^{\epsilon}$ as a local representative of a section of the vector bundle $(V_x \times W_x \times E_x)/\Gamma_x \to (V_x \times W_x)/\Gamma_x$. (Lemma 23.32.)

Definition 7.6 Let $\mathcal{S}_x^i = (W_x^i, \omega_x^i, \{\mathfrak{s}_x^{\epsilon,i}\})$ $(i = 1, 2)$ be two CF-perturbations of $\mathcal{U}$ on $\mathfrak{V}_x$.

(1) We say $\mathcal{S}_x^1$ is a *projection* of $\mathcal{S}_x^2$, if there exist a map $\Pi : \widehat{W}_x^2 \to \widehat{W}_x^1$ with the following properties:

(a) Π is a Γ_x equivariant linear projection which sends W_x^2 to W_x^1 and satisfies $\Pi!(\omega_x^2) = \omega_x^1$.
(b) For each $y \in V_x$, $\xi \in W_x^1$ and $\epsilon \in (0, 1]$, we have

$$\mathfrak{s}_x^{\epsilon,1}(y, \Pi(\xi)) = \mathfrak{s}_x^{\epsilon,2}(y, \xi).$$

(2) We say $\mathcal{S}_x^1$ is *equivalent* to $\mathcal{S}_x^2$ on $\mathfrak{V}_x$ if there exist N and $\mathcal{S}_x^{(i)}$ for $i = 0, \ldots, 2N$ with the following properties:

(a) $\mathcal{S}_x^{(i)}$ is a CF-perturbation of $\mathcal{U}$ on $\mathfrak{V}_x$.
(b) $\mathcal{S}_x^{(0)} = \mathcal{S}_x^1$, $\mathcal{S}_x^{(2N)} = \mathcal{S}_x^2$.
(c) $\mathcal{S}_x^{(2k-1)}$ and $\mathcal{S}_x^{(2k+1)}$ are both projections of $\mathcal{S}_x^{(2k)}$.

See the diagram below. It is easy to see that the relation defined in Definition 7.6 (2) is an equivalence relation.

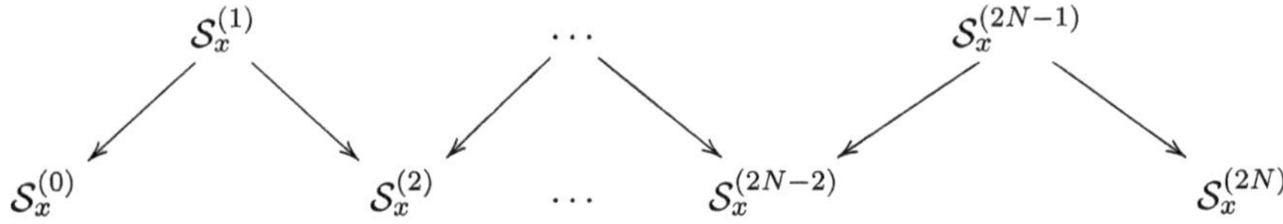

Definition 7.7 Let $\mathcal{S}_x = (W_x, \omega_x, \{\mathfrak{s}_x^\epsilon\})$ be a CF-perturbation of $\mathcal{U}$ on $\mathfrak{V}_x$. Let $\Phi_{xx'} = (h_{xx'}, \widetilde{\varphi}_{xx'}, \breve{\varphi}_{xx'})$ be a coordinate change from $\mathfrak{V}_{x'}$ to $\mathfrak{V}_x$. (See Situation 6.4.) We define the *restriction* $\Phi_{xx'}^*\mathcal{S}_x$ of $\mathcal{S}_x$ by

$$\Phi_{xx'}^*\mathcal{S}_x = (W_x, \omega_x, \{\mathfrak{s}_{x'}^{\epsilon\prime}\}).$$

Here W_x and ω_x are given. The section $\mathfrak{s}_{x'}^{\epsilon\prime}$ is defined as follows. We associate a linear isomorphism $g_y : E_{x'} \to E_x$ to each $y \in V_{x'}'$ by $\breve{\varphi}_{xx'}(y, v) = g_y(v)$. Then we put

$$\mathfrak{s}_{x'}^{\epsilon\prime}(y, \xi) = g_y^{-1}(\mathfrak{s}_x^\epsilon(\tilde{\varphi}_{xx'}(y), \xi)). \tag{7.3}$$

Lemma 7.8

(1) *If $\mathcal{S}_x^1$ is equivalent to $\mathcal{S}_x^2$, then $\Phi_{xx'}^*\mathcal{S}_x^1$ is equivalent to $\Phi_{xx'}^*\mathcal{S}_x^2$.*
(2) *The restriction $\Phi_{xx'}^*\mathcal{S}_x$ may depend on the choice of $\Phi_{xx'} = (h_{xx'}, \tilde{\varphi}_{xx'}, \breve{\varphi}_{xx'})$. However, the equivalence class of $\Phi_{xx'}^*\mathcal{S}_x$ is independent of such a choice.*

Proof (1) is obvious as far as we use the same $(h_{xx'}, \tilde{\varphi}_{xx'}, \breve{\varphi}_{xx'})$. We will prove (2). Let $(h_{xx'}^i, \tilde{\varphi}_{xx'}^i, \breve{\varphi}_{xx'}^i)$ $(i = 1, 2)$ be two choices and $\mathfrak{s}_{x'}^{\epsilon i\prime}(y, \xi)$ the restrictions obtained by these two choices for $i = 1, 2$, respectively. Then by Lemma 23.28 there exists $\gamma \in \Gamma_x$ such that

$$h_{xx'}^2(\mu) = \gamma h_{xx'}^1(\mu)\gamma^{-1}, \quad \tilde{\varphi}_{xx'}^2 = \gamma\tilde{\varphi}_{xx'}^1, \quad \breve{\varphi}_{xx'}^2 = \gamma\breve{\varphi}_{xx'}^1.$$

We put $\breve{\varphi}_{xx'}^i(y, v) = g_y^i(v)$. Then $g_y^2(v) = \gamma g_y^1(v)$. Therefore (7.3) implies

$$\begin{aligned} \mathfrak{s}_{x'}^{\epsilon 2\prime}(y, \xi) &= (g_y^2)^{-1}(\mathfrak{s}_x^{\epsilon 2\prime}(\tilde{\varphi}_{xx'}^2(y), \xi)) \\ &= (g_y^1)^{-1}\gamma^{-1}(\mathfrak{s}_x^{\epsilon 2\prime}(\gamma\tilde{\varphi}_{xx'}^1(y), \xi)) \\ &= (g_y^1)^{-1}(\mathfrak{s}_x^{\epsilon 1\prime}(\tilde{\varphi}_{xx'}^1(y), \gamma^{-1}\xi)) \\ &= \mathfrak{s}_{x'}^{\epsilon 1\prime}(y, \gamma^{-1}\xi). \end{aligned} \tag{7.4}$$

We note that γ^{-1} induces a $\Gamma_{x'}$ linear isomorphism from $(W_x, h_{xx'}^2)$ to $(W_x, h_{xx'}^1)$. (Here the $\Gamma_{x'}$ action on W_x is induced from the Γ_x action by $h_{xx'}^i$ in the case of $(W_x, h_{xx'}^i)$, $i = 1, 2$. Then the map $\xi \mapsto \gamma^{-1}\xi$ is Γ_x equivariant as a map $W_x \to W_x$, where Γ_x acts in different ways on the source and the target.) Therefore $(W_x, \omega_x, \mathfrak{s}_{x'}^{\epsilon 2\prime})$ is equivalent to $(W_x, \omega_x, \mathfrak{s}_{x'}^{\epsilon 1\prime})$. □

We next define the pushout of a differential form by using a CF-perturbation.

Definition 7.9 In Situation 7.3, let $\mathcal{S}_x = (W_x, \omega_x, \{\mathfrak{s}_x^\epsilon\})$ be a CF-perturbation of $\mathcal{U}$ on $\mathfrak{V}_x$.

(1) We say $\mathcal{S}_x$ is *transversal to* 0 if the map $\mathfrak{s}_x^\epsilon$ is transversal to 0 on a neighborhood of the support of ω_x for all $0 < \epsilon \le 1$. In particular,

$$(\mathfrak{s}_x^\epsilon)^{-1}(0) = \{(y, \xi) \in V_x \times W_x \mid \mathfrak{s}_x^\epsilon(y, \xi) = 0\}$$

is a smooth submanifold of $V_x \times W_x$ on a neighborhood of the support of ω_x.

(2) We say $f_x = f \circ \phi_x$ is *strongly submersive* with respect to $(\mathfrak{V}_x, \mathcal{S}_x)$ if $\mathcal{S}_x$ is transversal to 0 and the map

$$f_x \circ \pi_1|_{(\mathfrak{s}_x^\epsilon)^{-1}(0)} : (\mathfrak{s}_{x,k}^\epsilon)^{-1}(0) \to M \tag{7.5}$$

is a submersion on a neighborhood of the support of ω_x, for all $0 < \epsilon \le 1$. Here $\pi_1 : V_x \times W_x \to V_x$ is the projection.

(3) Let $g : N \to M$ be a smooth map between manifolds. We say f_x *is strongly transversal to* g with respect to $(\mathfrak{V}_x, \mathcal{S}_x)$ if $\mathcal{S}_x$ is transversal to 0 and the map (7.5) is weakly transversal to g, for all $0 < \epsilon \le 1$. Here $\pi_1 : V_x \times W_x \to V_x$ is the projection.

Lemma 7.10 *Suppose $\mathcal{S}_x^1$ is equivalent to $\mathcal{S}_x^2$.*

(1) *$\mathcal{S}_x^1$ is transversal to 0 if and only if $\mathcal{S}_x^2$ is transversal to 0.*
(2) *f_x is strongly submersive with respect to $(\mathfrak{V}_x, \mathcal{S}_x^1)$ if and only if f_x is strongly submersive with respect to $(\mathfrak{V}_x, \mathcal{S}_x^2)$.*
(3) *f_x is strongly transversal to $g : N \to M$ with respect to $(\mathfrak{V}_x, \mathcal{S}_x^1)$ if and only if f_x is strongly transversal to g with respect to $(\mathfrak{V}_x, \mathcal{S}_x^2)$.*

Proof It suffices to prove the lemma for the case when $\mathcal{S}_x^2$ is a projection of $\mathcal{S}_x^1$. This case follows from the fact that $\omega_x = |\omega_x| \mathrm{vol}_x$ where vol_x is a volume form of W_x and $|\omega_x|$ is a non-negative function, which is a part of Definition 7.4 (4).[3] □

We recall that a smooth differential form h on V_x/Γ_x is identified with a Γ_x-invariant smooth differential form $\tilde{h}$ on V_x. (Definition 23.10 (2).) The *support* of h is the quotient of the support of $\tilde{h}$ by Γ_x and is a closed subset of $V_x/\Gamma_x \cong U_x$. We denote it by $\mathrm{Supp}(h)$.

Definition 7.11 In Situation 7.3, let $\mathcal{S}_x = (W_x, \omega_x, \{\mathfrak{s}_x^\epsilon\})$ be a CF-perturbation of $\mathcal{U}$ on $\mathfrak{V}_x$. Let h be a smooth differential form on U_x that has compact support. Then we define a smooth differential form $f_x!(h; \mathcal{S}_x^\epsilon)$ on M for each $\epsilon > 0$ by the equation (7.6) below. We call it the *pushout* of h with respect to $f_x, \mathcal{S}_x$.

Let ρ be a smooth differential form on M. Then we require

[3] The condition $|\omega_x| \ge 0$ (Definition 7.4 (4)) is used to show $\Pi(\mathrm{Supp}(|\omega_x|)) = \mathrm{Supp}(|\overline{\omega}_x|)$, where $(\overline{W}, \overline{\omega}_x, \overline{\mathfrak{s}^\epsilon})$ is a projection of $(W, \omega_x, \mathfrak{s}^\epsilon)$.

$$\#\Gamma_x \int_M \rho \wedge f_x!(h; \mathcal{S}_x^\epsilon) = \int_{(\mathfrak{s}_x^\epsilon)^{-1}(0)} \pi_1^*(f_x^*\rho) \wedge \pi_1^*(\tilde{h}) \wedge \pi_2^*\omega_x. \tag{7.6}$$

Here π_1 (resp. π_2) is the projection of $V_x \times W_x$ to the first (resp. second) factor.

The unique existence of such $f_x!(h; \mathcal{S}_x^\epsilon)$ is an immediate consequence of the existence of pushout of a smooth form by a proper submersion.

Remark 7.12 In the left hand side of (7.6) we crucially use the fact that Γ_x is a finite group. It seems that this is the *only* place we use the finiteness of Γ_x when we use de Rham theory to realize virtual fundamental chain. One might try to use a CF-perturbation and de Rham version together with an appropriate model of equivariant cohomology to study virtual fundamental chains in the case where the isotropy group is a continuous compact group of positive dimension, such as the case of gauge theory or pseudo-holomorphic curves in a symplectic manifold acted on by a compact Lie group.

Lemma 7.13 *If $\mathcal{S}_x^1$ is equivalent to $\mathcal{S}_x^2$, then*

$$f_x!(h; \mathcal{S}_x^{1,\epsilon}) = f_x!(h; \mathcal{S}_x^{2,\epsilon}).$$

Proof It suffices to prove the equality in the case when $\mathcal{S}_x^1$ is a projection of $\mathcal{S}_x^2$. This is immediate from the definition. □

Lemma 7.14 *Suppose we are in the situation of Definition 7.7 and Situation 7.3.*

(1) *If $(\mathfrak{V}_x, \mathcal{S}_x)$ is transversal to 0, then $(\mathfrak{V}_{x'}, \Phi_{xx'}^* \mathcal{S}_x)$ is transversal to 0.*
(2) *If f_x is strongly submersive with respect to $(\mathfrak{V}_x, \mathcal{S}_x)$, then $f_{x'}$ is strongly submersive with respect to $(\mathfrak{V}_{x'}, \Phi_{xx'}^* \mathcal{S}_x)$.*
(3) *If f_x is strongly transversal to $g : N \to M$ with respect to $(\mathfrak{V}_x, \mathcal{S}_x)$, then $f_{x'}$ is strongly transversal to $g : N \to M$ with respect to $(\mathfrak{V}_{x'}, \Phi_{xx'}^* \mathcal{S}_x)$.*

The proof is immediate from the definition.

7.2.2 CF-Perturbation on a Single Kuranishi Chart

In Sect. 7.2.1, we studied locally on a single chart $U_x = V_x/\Gamma_x$. We next work globally on an orbifold U. We apply Remark 6.9 hereafter.

Remark 7.15 In Sect. 7.2.2 we consider a Kuranishi chart $\mathcal{U} = (U, \mathcal{E}, s, \psi)$. However, the parametrization ψ does not play any role in Sect. 7.2.2.

Definition 7.16 Let $\mathcal{U} = (U, \mathcal{E}, \psi, s)$ be a Kuranishi chart. A *representative of a CF-perturbation of* $\mathcal{U}$ is the following object $\mathfrak{S} = \{(\mathfrak{V}_{\mathfrak{r}}, \mathcal{S}_{\mathfrak{r}}) \mid \mathfrak{r} \in \mathfrak{R}\}$.

(1) $\{U_{\mathfrak{r}} \mid \mathfrak{r} \in \mathfrak{R}\}$ is a family of open subsets of U such that $U = \bigcup_{\mathfrak{r} \in \mathfrak{R}} U_{\mathfrak{r}}$.
(2) $\mathfrak{V}_{\mathfrak{r}} = (V_{\mathfrak{r}}, \Gamma_{\mathfrak{r}}, E_{\mathfrak{r}}, \phi_{\mathfrak{r}}, \widehat{\phi}_{\mathfrak{r}})$ is an orbifold chart of $(U, \mathcal{E})$ such that $\phi_{\mathfrak{r}}(V_{\mathfrak{r}}) = U_{\mathfrak{r}}$.

(3) $\mathcal{S}_{\mathfrak{r}} = (W_{\mathfrak{r}}, \omega_{\mathfrak{r}}, \{\mathfrak{s}^{\epsilon}_{\mathfrak{r}}\})$ is a CF-perturbation of $\mathcal{U}$ on $\mathfrak{V}_{\mathfrak{r}}$.
(4) For each $x \in U_{\mathfrak{r}_1} \cap U_{\mathfrak{r}_2}$, there exists an orbifold chart $\mathfrak{V}_{\mathfrak{r}}$ with the following properties:

(a) $x \in U_{\mathfrak{r}} \subset U_{\mathfrak{r}_1} \cap U_{\mathfrak{r}_2}$.
(b) The restriction of $\mathcal{S}_{\mathfrak{r}_1}$ to $\mathfrak{V}_{\mathfrak{r}}$ is equivalent to the restriction of $\mathcal{S}_{\mathfrak{r}_2}$ to $\mathfrak{V}_{\mathfrak{r}}$.

For each $\epsilon > 0$, we write

$$\mathfrak{S}^{\epsilon} = \{(\mathfrak{V}_{\mathfrak{r}}, \mathcal{S}^{\epsilon}_{\mathfrak{r}}) \mid \mathfrak{r} \in \mathfrak{R}\}.$$

See Definition 7.4 for the notation $\mathcal{S}^{\epsilon}_{\mathfrak{r}}$.

Definition 7.17 Let $\mathcal{U} = (U, \mathcal{E}, \psi, s)$ be as in Definition 7.16 and $\mathfrak{S}^i = \{(\mathfrak{V}^i_{\mathfrak{r}}, \mathcal{S}^i_{\mathfrak{r}}) \mid \mathfrak{r} \in \mathfrak{R}^i\}$ $(i = 1, 2)$ representatives of CF-perturbations of $\mathcal{U}$. We say that $\mathfrak{S}^2$ is *equivalent* to $\mathfrak{S}^1$ if, for each $x \in U_{\mathfrak{r}_1} \cap U_{\mathfrak{r}_2}$, there exists an orbifold chart $\mathfrak{V}_{\mathfrak{r}}$ with the following properties:

(1) $x \in U_{\mathfrak{r}} \subset U_{\mathfrak{r}_1} \cap U_{\mathfrak{r}_2}$.
(2) The restriction of $\mathcal{S}^1_{\mathfrak{r}_1}$ to $\mathfrak{V}_{\mathfrak{r}}$ is equivalent to the restriction of $\mathcal{S}^2_{\mathfrak{r}_2}$ to $\mathfrak{V}_{\mathfrak{r}}$.

Definition 7.18 Suppose we are in Situation 7.3. Let $\mathfrak{S} = \{\mathfrak{S}_{\mathfrak{r}}\}$ be a representative of a CF-perturbation of $\mathcal{U}$. Let $U' \subseteq U$ be an open subset.

Let $\mathfrak{S}_{\mathfrak{r}} = (\mathfrak{V}_{\mathfrak{r}}, \mathcal{S}_{\mathfrak{r}})$, $\mathfrak{V}_{\mathfrak{r}} = (V_{\mathfrak{r}}, \Gamma_{\mathfrak{r}}, E_{\mathfrak{r}}, \phi_{\mathfrak{r}}, \widehat{\phi}_{\mathfrak{r}})$. If $\mathrm{Im}(\phi_{\mathfrak{r}}) \cap U' = \emptyset$, then we remove $\mathfrak{r}$ from $\mathfrak{R}$. Let $\mathfrak{R}_0$ be obtained by removing all such $\mathfrak{r}$ from $\mathfrak{R}$. If $\mathrm{Im}(\phi_{\mathfrak{r}}) \cap U' \neq \emptyset$, then $\mathfrak{V}_{\mathfrak{r}}|_{\mathrm{Im}\phi_{\mathfrak{r}} \cap U'}$ is an orbifold chart of $(U, \mathcal{E})$, which we write $\mathfrak{V}_{\mathfrak{r}}|_{U'} = (V'_{\mathfrak{r}}, \Gamma'_{\mathfrak{r}}, E_{\mathfrak{r}}, \phi'_{\mathfrak{r}}, \widehat{\phi}'_{\mathfrak{r}})$. Let $\mathcal{S}_{\mathfrak{r}} = (W_{\mathfrak{r}}, \omega_{\mathfrak{r}}, \{\mathfrak{s}^{\epsilon}_{\mathfrak{r}}\})$. We define

$$\mathcal{S}_{\mathfrak{r}}|_{U'} = (W_{\mathfrak{r}}, \omega_{\mathfrak{r}}, \{\mathfrak{s}^{\epsilon}_{\mathfrak{r}}|_{V'_{\mathfrak{r}} \times W_{\mathfrak{r}}}\}). \tag{7.7}$$

Now we put:

$$\mathfrak{S}|_{U'} = \{(\mathfrak{V}_{\mathfrak{r}}|_{U'}, \mathcal{S}_{\mathfrak{r}}|_{U'}) \mid \mathfrak{r} \in \mathfrak{R}_0\}.$$

It is a representative of a CF-perturbation of $\mathcal{U}|_{U'}$, which we call the *restriction* of $\mathfrak{S}$ to U'.

Lemma 7.19 *If $\mathfrak{S}^1$ is equivalent to $\mathfrak{S}^2$, then $\mathfrak{S}^1|_{\Omega}$ is equivalent to $\mathfrak{S}^2|_{\Omega}$ for any open subset $\Omega \subseteq U$.*

The proof is immediate from the definition.

Definition 7.20 Suppose we are in Situation 7.3.

(1) A *CF-perturbation* on $\mathcal{U}$ is an equivalence class of a representative of a CF-perturbation with respect to the equivalence relation given in Definition 7.17.
(2) Let Ω be an open subset of U. We denote by $\mathcal{CF}(\Omega)$ the set of all CF-perturbations on $\mathcal{U}|_{\Omega}$.

(3) Let $\Omega_1 \subset \Omega_2 \subset U$. Then by Lemma 7.19, the restriction defined in Definition 7.18 induces a map

$$\mathfrak{i}_{\Omega_1\Omega_2} : \mathcal{CF}(\Omega_2) \to \mathcal{CF}(\Omega_1). \tag{7.8}$$

We call this map a *restriction map*.

(4) $\Omega \mapsto \mathcal{CF}(\Omega)$ together with $\mathfrak{i}_{\Omega_1\Omega_2}$ defines a presheaf. (In fact, $\mathfrak{i}_{\Omega_1\Omega_2} \circ \mathfrak{i}_{\Omega_2\Omega_3} = \mathfrak{i}_{\Omega_1\Omega_3}$ holds obviously.) The next proposition says that it is indeed a sheaf. We call it the *sheaf of CF-perturbations* on $\mathcal{U}$.

Remark 7.21 Hereafter, by a slight abuse of notation, we also use the symbol $\mathfrak{S}$ for a CF-perturbation. Namely we use this symbol not only for a representative of CF-perturbation but also for its equivalence class.

Proposition 7.22 *The presheaf $\mathcal{CF}$ is a sheaf.*

Proof This is mostly a tautology. We provide a proof for completeness' sake.

Let $\bigcup_{a\in A} \Omega_a = \Omega$ be an open cover of Ω. Suppose $\mathfrak{S}_a \in \mathcal{CF}(\Omega_a)$ and $\{(\mathfrak{V}_{a,\mathfrak{r}}, \mathcal{S}_{a,\mathfrak{r}}) \mid \mathfrak{r} \in \mathfrak{R}_a\}$ is a representative of $\mathfrak{S}_a$. We put $\Omega_{ab} = \Omega_a \cap \Omega_b$. We assume

$$\mathfrak{i}_{\Omega_{ab}\Omega_a}(\mathfrak{S}_a) = \mathfrak{i}_{\Omega_{ab}\Omega_b}(\mathfrak{S}_b). \tag{7.9}$$

To prove Proposition 7.22 it suffices to show that there exists a unique $\mathfrak{S} \in \mathcal{CF}(\Omega)$ such that

$$\mathfrak{i}_{\Omega_a\Omega}(\mathfrak{S}) = \mathfrak{S}_a. \tag{7.10}$$

Proof of uniqueness Suppose that $\mathfrak{S}, \mathfrak{S}' \in \mathcal{CF}(\Omega)$ both satisfy (7.10). Let $\{(\mathfrak{V}_\mathfrak{r}, \mathcal{S}_\mathfrak{r}) \mid \mathfrak{r} \in \mathfrak{R}\}$ and $\{(\mathfrak{V}'_{\mathfrak{r}'}, \mathcal{S}'_{\mathfrak{r}'}) \mid \mathfrak{r}' \in \mathfrak{R}'\}$ be representatives of $\mathfrak{S}$ and $\mathfrak{S}'$, respectively. We will prove that they are equivalent.

Let $x \in \Omega$. There exist $\mathfrak{r} \in \mathfrak{R}$, $\mathfrak{r}' \in \mathfrak{R}'$ such that $x \in U_\mathfrak{r} \cap U'_{\mathfrak{r}'}$, where $U_\mathfrak{r} = \mathrm{Im}(\phi_\mathfrak{r})$, $U'_{\mathfrak{r}'} = \mathrm{Im}(\phi'_{\mathfrak{r}'})$. We take a such that $x \in \Omega_a$. By (7.10), $\mathcal{S}_\mathfrak{r}|_{U_\mathfrak{r}\cap\Omega_a}$ is equivalent to $\mathcal{S}'_{\mathfrak{r}'}|_{U'_{\mathfrak{r}'}\cap\Omega_a}$. Therefore there exists an orbifold chart $\mathfrak{V}_x$ such that $U_x \subset U_\mathfrak{r} \cap U'_{\mathfrak{r}'}$ and the restriction of $\mathcal{S}_\mathfrak{r}|_{U_\mathfrak{r}\cap\Omega_a}$ to $\mathfrak{V}_x$ is equivalent to the restriction of $\mathcal{S}'_{\mathfrak{r}'}|_{U'_{\mathfrak{r}'}\cap\Omega_a}$ to $\mathfrak{V}_x$. Thus the restriction of $\mathcal{S}_\mathfrak{r}$ to $\mathfrak{V}_x$ is equivalent to the restriction of $\mathcal{S}'_{\mathfrak{r}'}$ to $\mathfrak{V}_x$.

Since this holds for any $x \in \Omega$, $\{(\mathfrak{V}_\mathfrak{r}, \mathcal{S}_\mathfrak{r}) \mid \mathfrak{r} \in \mathfrak{R}\}$ is equivalent to $\{(\mathfrak{V}'_{\mathfrak{r}'}, \mathcal{S}'_{\mathfrak{r}'}) \mid \mathfrak{r}' \in \mathfrak{R}'\}$ by definition. □

Proof of existence Let $\mathfrak{V}_{\mathfrak{r},a} = (V_{\mathfrak{r},a}, \Gamma_{\mathfrak{r},a}, E_{\mathfrak{r},a}, \phi_{\mathfrak{r},a}, \widehat{\phi}_{\mathfrak{r},a})$ and $U_{\mathfrak{r},a} = \phi_{\mathfrak{r},a}(V_{\mathfrak{r},a})$. Then $\{U_{\mathfrak{r},a} \mid a \in A,\ \mathfrak{r} \in \mathfrak{R}_a\}$ is an open cover of Ω. We put

$$\mathfrak{S} = \coprod_{a\in A} \{(\mathfrak{V}_{a,\mathfrak{r}}, \mathcal{S}_{a,\mathfrak{r}}) \mid \mathfrak{r} \in \mathfrak{R}_a\}.$$

To show $\mathfrak{S} \in \mathcal{CF}(\Omega)$ it suffices to check Definition 7.16 (4). This is a consequence of (7.9) and the definitions of the equivalence of representatives of CF-perturbations and of the restriction. We can check (7.10) also by (7.9) in the same way as in the proof of uniqueness. □

The proof of Proposition 7.22 is now complete. □

We recall that the *stalk* $\mathcal{CF}_x$ of the sheaf $\mathcal{CF}$ at $x \in U$ is by definition

$$\mathcal{CF}_x = \varinjlim_{\Omega \ni x} \mathcal{CF}(\Omega). \tag{7.11}$$

Definition-Lemma 7.23 The stalk $\mathcal{CF}_x$ is identified with the set of the equivalence classes of the equivalence relation defined in Item (2) on the set which is defined in Item (1).

(1) We consider the set $\widetilde{\mathcal{CF}_x}$ of pairs $(\mathfrak{V}_{\mathfrak{r}}, \mathfrak{S}_{\mathfrak{r}})$ where $\mathfrak{V}_{\mathfrak{r}}$ is an orbifold chart of $(U, \mathcal{E})$ at x and $\mathfrak{S}_{\mathfrak{r}}$ is a CF-perturbation on $\mathfrak{V}_{\mathfrak{r}}$.
(2) Let $(\mathfrak{V}_{\mathfrak{r}}, \mathfrak{S}_{\mathfrak{r}}), (\mathfrak{V}_{\mathfrak{r}'}, \mathfrak{S}_{\mathfrak{r}'}) \in \widetilde{\mathcal{CF}_x}$. We say that they are equivalent if there exists an orbifold chart $\mathfrak{V}_x$ at x such that $U_x \subset U_{\mathfrak{r}} \cap U_{\mathfrak{r}'}$ and the restriction of $\mathfrak{S}_{\mathfrak{r}}$ to $\mathfrak{V}_x$ equivalent to the restriction of $\mathfrak{S}_{\mathfrak{r}'}$ to $\mathfrak{V}_x$.

The proof is obvious. The next lemma is standard in the sheaf theory.

Lemma 7.24 *The set $\mathcal{CF}(\Omega)$ of global sections is identified with the following objects:*

(1) *For each $x \in \Omega$ it associates $\mathfrak{S}_x \in \mathcal{CF}_x$.*
(2) *For each $x \in \Omega$, there exists a representative $(\mathfrak{V}_{\mathfrak{r}}, \mathfrak{S}_{\mathfrak{r}})$ of $\mathfrak{S}_x$, such that for each $y \in \phi_{\mathfrak{r}}(V_{\mathfrak{r}})$ the germ $\mathfrak{S}_y$ is represented by $(\mathfrak{V}_{\mathfrak{r}}, \mathfrak{S}_{\mathfrak{r}})$.*

Definition 7.25 Let $K \subseteq U$ be a closed subset. A *CF-perturbation of* $K \subseteq U$ is an element of the inductive limit: $\varinjlim_{\Omega \supset K} \mathcal{CF}(\Omega)$. We denote the set of all CF-perturbations of $K \subseteq U$ by $\mathcal{CF}(K)$. Namely

$$\mathcal{CF}(K) = \varinjlim_{U \supset \Omega \supset K;\ \Omega \text{ open}} \mathcal{CF}(\Omega). \tag{7.12}$$

7.3 Integration Along the Fiber (Pushout) on a Single Kuranishi Chart

Definition-Lemma 7.26 Suppose we are in Situation 7.3. Let $\Omega \subset U$ be an open subset and let $\mathfrak{S} \in \mathcal{CF}(\Omega)$ be a CF-perturbation. We consider its representative $\{(\mathfrak{V}_{\mathfrak{r}}, \mathcal{S}_{\mathfrak{r}}) \mid \mathfrak{r} \in \mathfrak{R}\}$.

(1) We say that $(\mathcal{U}, \mathfrak{S})$ is *transversal to* 0 if, for each $\mathfrak{r}$, $(\mathfrak{V}_{\mathfrak{r}}, \mathcal{S}_{\mathfrak{r}})$ is transversal to 0. This is independent of the choice of representative.

(2) We say that f is *strongly submersive with respect to* $(\mathcal{U}, \mathfrak{S})$ if, for each $\mathfrak{r}$, the map f is strongly submersive with respect to $(\mathfrak{V}_{\mathfrak{r}}, \mathcal{S}_{\mathfrak{r}})$. This is independent of the choice of representative.
(3) Let $g : N \to M$ be a smooth map between manifolds. We say that the map f is *strongly transversal to* g *with respect to* $(\mathcal{U}, \mathfrak{S})$ if, for each $\mathfrak{r}$, f is strongly transversal to g with respect to $(\mathfrak{V}_{\mathfrak{r}}, \mathcal{S}_{\mathfrak{r}})$. This is independent of the choice of representative.
(4) We denote by $\mathcal{CF}_{\pitchfork 0}(\Omega)$ the set of all $\mathfrak{S} \in \mathcal{CF}(\Omega)$ transversal to 0, $\mathcal{CF}_{f\pitchfork}(\Omega)$ the set of all $\mathfrak{S} \in \mathcal{CF}(\Omega)$ such that f is strongly submersive with respect to $(\mathcal{U}, \mathfrak{S})$, and by $\mathcal{CF}_{f\pitchfork g}(\Omega)$ the set of all $\mathfrak{S} \in \mathcal{CF}(\Omega)$ such that f is strongly transversal to g with respect to $(\mathcal{U}, \mathfrak{S})$. They are subsheaves of $\mathcal{CF}$.
(5) For a closed subset $K \subseteq U$ we define $\mathcal{CF}_{\pitchfork 0}(K)$, $\mathcal{CF}_{f\pitchfork}(K)$ and $\mathcal{CF}_{f\pitchfork g}(K)$ in the same way as (7.12).

Proof The statements (1), (2), and (3) follow from Lemma 7.10. (4) is a consequence of the definition. □

Next we introduce the notion of a *member of a CF-perturbation*, which is an analogue of the notion of a branch of multisection.

Definition 7.27 Let $\mathfrak{V}_{\mathfrak{r}}$ be an orbifold chart of $(U, \mathcal{E})$ and $\mathcal{S}_{\mathfrak{r}}$ a CF-perturbation of $(\mathcal{U}, \mathfrak{V}_{\mathfrak{r}})$. Let $x \in V_{\mathfrak{r}}$. We fix $\epsilon \in (0, 1]$.

Consider the germ of a map $y \mapsto \mathfrak{s}(y)$, $O_x \to E_{\mathfrak{r}}$ (where O_x is a neighborhood of $x \in V_{\mathfrak{r}}$). We say $\mathfrak{s}$ is a *member of* $\mathcal{S}_{\mathfrak{r}}^{\epsilon}$ *at* x if there exists $\xi \in W_x$ such that the germ of $y \mapsto \mathfrak{s}_{\mathfrak{r}}^{\epsilon}(y, \xi)$ at x is $\mathfrak{s}$ and $\xi \in \operatorname{supp} \omega_x$. (See Definition 7.4 for the notation $\mathcal{S}_{\mathfrak{r}}^{\epsilon}$.)

Remark 7.28 The member of $\mathcal{S}_{\mathfrak{r}}^{\epsilon}$ at x depends on ϵ. In other words, we define the notion of the member of $\mathcal{S}^{\epsilon}$ for each $\epsilon \in (0, 1]$.

Lemma 7.29 *In the situation of Definition 7.27, let $\mathcal{S}_{\mathfrak{r}}'$ be a CF-perturbation of $(\mathcal{U}, \mathfrak{V}_{\mathfrak{r}})$ that it is equivalent to $\mathcal{S}_{\mathfrak{r}}$. Then $\mathfrak{s}$ is a member of $\mathcal{S}_{\mathfrak{r}}^{\epsilon}$ if and only if $\mathfrak{s}$ is a member of $\mathcal{S}_{\mathfrak{r}}'^{\epsilon}$.*

Proof It suffices to consider only the case when $\mathcal{S}_{\mathfrak{r}}'$ is a projection of $\mathcal{S}_{\mathfrak{r}}$. This also follows from the fact that $\omega_x = |\omega_x| \mathrm{vol}_x$ where vol_x is a volume form of W_x and $|\omega_x|$ is a non-negative function, which is a part of Definition 7.4 (4). □

Therefore we can define a member of an element of the stalk $\mathcal{CF}_x$ at x of the sheaf $\mathcal{CF}$ of CF-perturbations.

We now recall from Definition 7.4 that we denoted by $\mathcal{S}_x^{\epsilon} = (W_x, \omega_x, \mathfrak{s}_x^{\epsilon})$ the restriction of the CF-perturbation $\mathcal{S}_x = (W_x, \omega_x, \{\mathfrak{s}_x^{\epsilon}\})$ at ϵ. Recall by definition that $\{\mathfrak{s}_x^{\epsilon}\}$ is an ϵ-dependent family of parametrized sections, i.e., a map $\mathfrak{s}_x^{\epsilon} : V_x \times W_x \to E_x$.

Definition 7.30 Let $\mathfrak{S} \in \mathcal{CF}(\Omega)$ and $x \in \Omega$. A *member of* $\mathfrak{S}^{\epsilon}$ *at* x is a member of the germ of $\mathfrak{S}^{\epsilon}$ at x.

Definition 7.31 Let U be an orbifold and $K \subset U$ a compact subset and $\{U_{\mathfrak{r}}\}$ a set of finitely many open subsets such that $\cup_{\mathfrak{r}} U_{\mathfrak{r}} \supset K$. A *partition of unity* subordinate to $\{U_{\mathfrak{r}}\}$ on K assigns a smooth function $\chi_{\mathfrak{r}}$ on U to each $\mathfrak{r}$ such that:

(1) $\operatorname{supp}\chi_{\mathfrak{r}} \subset U_{\mathfrak{r}}$.
(2) $\sum_{\mathfrak{r}\in\mathfrak{R}} \chi_{\mathfrak{r}} \equiv 1$ on a neighborhood of K.

It is standard and easy to prove that a partition of unity always exists on an orbifold.

Definition 7.32 Suppose we are in Situation 7.3. Let h be a smooth differential form of compact support in U. Let $\mathfrak{S} \in \mathcal{CF}_{f\pitchfork}(\mathrm{Supp}(h))$. Let $\{\chi_{\mathfrak{r}}\}$ be a smooth partition of unity subordinate to the covering $\{U_{\mathfrak{r}}\}$ on $\mathrm{Supp}(h)$. We define the *pushout of h by f with respect to* $\mathfrak{S}^\epsilon$ by

$$f!(h;\mathfrak{S}^\epsilon) = \sum_{\mathfrak{r}\in\mathfrak{R}} f!(\chi_{\mathfrak{r}} h; \mathcal{S}^\epsilon_{\mathfrak{r}}) \tag{7.13}$$

for each $\epsilon > 0$. It is a smooth form on M of degree

$$\deg f!(h,\mathcal{S}^\epsilon) = \deg h + \dim M - \dim \mathcal{U},$$

where $\dim\mathcal{U} = \dim U - \operatorname{rank}\mathcal{E}$. The well-definedness follows from Lemma 7.34 (3).

We also call a pushout an *integration along the fiber.*

Remark 7.33 In general $f!(h;\mathfrak{S}^\epsilon)$ *depends* on ϵ. Moreover $\lim_{\epsilon\to 0} f!(h;\mathfrak{S}^\epsilon)$ typically diverges.

Lemma 7.34

(1) $f!(h_1+h_2;\mathfrak{S}^\epsilon) = f!(h_1;\mathfrak{S}^\epsilon)+f!(h_2;\mathfrak{S}^\epsilon)$ *and* $f!(ch;\mathfrak{S}^\epsilon) = cf!(h;\mathfrak{S}^\epsilon)$ *for* $c \in \mathbb{R}$.
(2) *The pushout of h defined by* (7.13) *is independent of the choice of partition of unity.*
(3) *If* $\mathfrak{S}^1$ *is equivalent to* $\mathfrak{S}^2$ *then*

$$f!(h;\mathfrak{S}^{1,\epsilon}) = f!(h;\mathfrak{S}^{2,\epsilon}). \tag{7.14}$$

Proof (1) is obvious as far as we use the same partition of unity on both sides. We will prove (2) and (3) at the same time. We take a partition of unity $\{\chi^i_{\mathfrak{r}} \mid \mathfrak{r} \in \mathfrak{R}_i\}$ subordinate to $\mathfrak{S}^i$ for $i = 1, 2$ and will prove (7.14). Here we use those partitions of unity to define the left and right hand sides of (7.14), respectively. The case $\mathfrak{S}^1 = \mathfrak{S}^2$ will be (2).

We put $h_0 = \chi_{\mathfrak{r}_0} h$. In view of (1) it suffices to prove

$$f!(h_0;\mathcal{S}^{1,\epsilon}_{\mathfrak{r}_0}) = \sum_{\mathfrak{r}\in\mathfrak{R}_2} f!(\chi^2_{\mathfrak{r}} h_0; \mathcal{S}^{2,\epsilon}_{\mathfrak{r}}). \tag{7.15}$$

To prove (7.15), it suffices to show the next equality for each $\mathfrak{r} \in \mathfrak{R}_2$.

$$f!(\chi_{\mathfrak{r}}^2 h_0; \mathcal{S}_{\mathfrak{r}_0}^{1,\epsilon}) = f!(\chi_{\mathfrak{r}}^2 h_0; \mathcal{S}_{\mathfrak{r}}^{2,\epsilon}). \tag{7.16}$$

We put $h_1 = \chi_{\mathfrak{r}}^2 h_0$. Let $K = \mathrm{Supp}(h_1)$. For each $x \in K$ there exists $\mathfrak{V}_x$ such that the restriction of $\mathcal{S}_{\mathfrak{r}_0}^{1,\epsilon}$ to U_x is equivalent to the restriction of $\mathcal{S}_{\mathfrak{r}}^{2,\epsilon}$ to U_x. We cover K by finitely many such U_{x_i}, $i = 1, \dots, N$. Let $\{\chi_i' \mid i = 1, \dots, N\}$ be a partition of unity on K subordinate to the covering $\{U_{x_i}\}$. Then we obtain:

$$\begin{aligned} f!(\chi_{\mathfrak{r}}^2 h_0; \mathcal{S}_{\mathfrak{r}_0}^{1,\epsilon}) &= \sum_{i=1}^{N} f!(\chi_i' h_1; \mathcal{S}_{\mathfrak{r}_0}^{1,\epsilon}|_{U_{x_i}}) \\ &= \sum_{i=1}^{N} f|_{U_{x_i}}!(\chi_i' h_1; \mathcal{S}_{\mathfrak{r}}^{2,\epsilon}|_{U_{x_i}}) = f!(\chi_{\mathfrak{r}}^2 h_0; \mathcal{S}_{\mathfrak{r}}^{2,\epsilon}). \end{aligned}$$

The proof of Lemma 7.34 is now complete. □

7.4 CF-Perturbations of a Good Coordinate System

To consider a CF-perturbation of a good coordinate system, we first study the pullback of a CF-perturbation by an embedding of Kuranishi charts in this subsection. Using the pullback, we then define the notion of a CF-perturbation of a good coordinate system.

7.4.1 Embedding of Kuranishi Charts and CF-Perturbations

Notation 7.35 To highlight the dependence on the Kuranishi charts, we write $\mathcal{CF}^{\mathcal{U}^1}$, $\mathcal{CF}^{\mathcal{U}^2}$ etc. in place of $\mathcal{CF}$. Namely, $\mathcal{CF}^{\mathcal{U}^2}(\Omega)$ is the set of all CF-perturbations of $\mathcal{U}^2|_\Omega$.

Situation 7.36 Let $\mathcal{U}^i = (U^i, \mathcal{E}^i, \psi^i, s^i)$ $(i = 1, 2)$ be Kuranishi charts and $\Phi_{21} = (\varphi_{21}, \widehat{\varphi}_{21}) : \mathcal{U}^1 \to \mathcal{U}^2$ an embedding of Kuranishi charts. Let $\mathfrak{S}^2 \in \mathcal{CF}^{\mathcal{U}^2}(U^2)$ and let $\{\mathfrak{S}_{\mathfrak{r}}^2\} = \{(\mathfrak{V}_{\mathfrak{r}}^2, \mathcal{S}_{\mathfrak{r}}^2)\}$ be its representative.■

We will define the pullback $\Phi_{21}^* \mathfrak{S}^2 \in \mathcal{CF}^{\mathcal{U}_1}(U_1)$. We need certain conditions for this pullback to be defined. The main property we require is that for $x \in U_1$, $y = \varphi_{21}(x)$, $\xi \in W_y$, the value $\mathfrak{s}_2^\epsilon(y, \xi)$ is contained in the image of $\hat{\varphi}_{21}(x, \cdot) : E_1 \to E_2$. We state this property as Condition 7.41 (7.17) below.

Condition 7.37 Suppose we are in Situation 7.36. We require that there exists an orbifold chart $\mathfrak{V}_{\mathfrak{r}}^1 = (V_{\mathfrak{r}}^1, \Gamma_{\mathfrak{r}}^1, E_{\mathfrak{r}}^1, \phi_{\mathfrak{r}}^1, \widehat{\phi}_{\mathfrak{r}}^1)$ such that $\phi_{\mathfrak{r}}^1(V_{\mathfrak{r}}^1) = \varphi_{21}^{-1}(U_{\mathfrak{r}}^2)$.

(Recall $U^2_{\mathfrak{r}} = \phi^2_{\mathfrak{r}}(V^2_{\mathfrak{r}})$ and $\mathfrak{V}^2_{\mathfrak{r}} = (V^2_{\mathfrak{r}}, \Gamma^2_{\mathfrak{r}}, E^2_{\mathfrak{r}}, \phi^2_{\mathfrak{r}}, \widehat{\phi}^2_{\mathfrak{r}})$.)

Lemma 7.38 *For any given $\mathfrak{S}^2$ we may choose its representative $\{(\mathfrak{V}^2_{\mathfrak{r}}, \mathcal{S}^2_{\mathfrak{r}})\}$ that satisfies Condition* 7.37.

Proof For each $x \in U_2$ there exists $\mathfrak{V}_x$ such that $\varphi_{21}^{-1}(U^2_x)$ has an orbifold chart and $\mathfrak{V}_x \subset \mathfrak{V}_{\mathfrak{r}}$ for some $\mathfrak{r}$. We cover U_2 by such $\mathfrak{V}_x$ to obtain the required $\{(\mathfrak{V}^2_{\mathfrak{r}}, \mathcal{S}^2_{\mathfrak{r}})\}$.

□

Therefore we may assume Condition 7.37. Then we can represent the orbifold embedding $(\varphi_{21}, \widehat{\varphi}_{21}) : (U^1, \mathcal{E}^1) \to (U^2, \mathcal{E}^2)$ in terms of the orbifold charts $\mathfrak{V}^1_{\mathfrak{r}}$, $\mathfrak{V}^2_{\mathfrak{r}}$ by $(h^{\mathfrak{r}}_{21}, \tilde{\varphi}^{\mathfrak{r}}_{21}, \check{\varphi}^{\mathfrak{r}}_{21})$ that have the following properties.

Property 7.39

(1) $h^{\mathfrak{r}}_{21} : \Gamma^{\mathfrak{r}}_1 \to \Gamma^{\mathfrak{r}}_2$ is an injective group homomorphism.
(2) $\tilde{\varphi}^{\mathfrak{r}}_{21} : V^{\mathfrak{r}}_1 \to V^{\mathfrak{r}}_2$ is an $h^{\mathfrak{r}}_{21}$-equivariant smooth embedding of manifolds.
(3) $h^{\mathfrak{r}}_{21}$ and $\tilde{\varphi}^{\mathfrak{r}}_{21}$ induce an orbifold embedding

$$\left(\overline{\phi^{\mathfrak{r}}_2}\right)^{-1} \circ \varphi_{21} \circ \overline{\phi^{\mathfrak{r}}_1} : V^1_{\mathfrak{r}}/\Gamma^1_{\mathfrak{r}} \to V^2_{\mathfrak{r}}/\Gamma^2_{\mathfrak{r}}.$$

(4) $\check{\varphi}^{\mathfrak{r}}_{21} : V^{\mathfrak{r}}_1 \times E^{\mathfrak{r}}_1 \to E^{\mathfrak{r}}_2$ is an $h^{\mathfrak{r}}_{21}$-equivariant smooth map such that for each $y \in V^{\mathfrak{r}}_1$ the map $v \mapsto \check{\varphi}^{\mathfrak{r}}_{21}(y, v)$ is a linear embedding $E^{\mathfrak{r}}_1 \to E^{\mathfrak{r}}_2$.
(5) $\check{\varphi}^{\mathfrak{r}}_{21}$ induces a smooth embedding of vector bundles:

$$\left(\overline{\widehat{\phi}^{\mathfrak{r}}_2}\right)^{-1} \circ \widehat{\varphi}_{21} \circ \overline{\widehat{\phi}^{\mathfrak{r}}_1} : (V^1_{\mathfrak{r}} \times E^1_{\mathfrak{r}})/\Gamma^1_{\mathfrak{r}} \to (V^2_{\mathfrak{r}} \times E^2_{\mathfrak{r}})/\Gamma^2_{\mathfrak{r}}.$$

In other words, for each $(y, v) \in V^1_{\mathfrak{r}} \times E^1_{\mathfrak{r}}$ we have

$$\widehat{\varphi}_{21}(\widehat{\phi}^{\mathfrak{r}}_1(y, v)) = \widehat{\phi}^{\mathfrak{r}}_2(\tilde{\varphi}^{\mathfrak{r}}_{21}(y), \check{\varphi}^{\mathfrak{r}}_{21}(y, v)).$$

See Lemma 23.26.

Remark 7.40 The map $(h^{\mathfrak{r}}_{21}, \tilde{\varphi}^{\mathfrak{r}}_{21}, \check{\varphi}^{\mathfrak{r}}_{21})$ satisfying (1)–(5) above is not unique.

Condition 7.41 We consider Situation 7.36 and assume Condition 7.37. We take $(h^{\mathfrak{r}}_{21}, \tilde{\varphi}^{\mathfrak{r}}_{21}, \check{\varphi}^{\mathfrak{r}}_{21})$ which satisfies Property 7.39. Let $\mathcal{S}^2_{\mathfrak{r}} = (W^2_{\mathfrak{r}}, \omega^2_{\mathfrak{r}}, \{\mathfrak{s}^{2,\epsilon}_{\mathfrak{r}}\})$. Then for each $y \in V^1_{\mathfrak{r}}$, $\xi \in W^2_{\mathfrak{r}}$, we require $\mathfrak{s}^{2,\epsilon}_{\mathfrak{r}}$ to satisfy

$$\mathfrak{s}^{2,\epsilon}_{\mathfrak{r}}(\tilde{\varphi}^{\mathfrak{r}}_{21}(y), \xi) \in \operatorname{Im}(g_y), \tag{7.17}$$

where $g_y : E^1_{\mathfrak{r}} \to E^2_{\mathfrak{r}}$ is defined by

$$\check{\varphi}^{\mathfrak{r}}_{21}(\tilde{\varphi}^{\mathfrak{r}}_{21}(y), v) = g_y(v). \tag{7.18}$$

Definition 7.42 In Situation 7.36 we say that $\{(\mathfrak{V}^2_{\mathfrak{r}}, \mathcal{S}^2_{\mathfrak{r}})\}$ is *restrictable to* $\mathcal{U}^1$ by Φ_{21} if and only if Conditions 7.37 and 7.41 are satisfied.

The *pullback* $\Phi_{21}^*\{(\mathfrak{V}_{\mathfrak{r}}^2, \mathcal{S}_{\mathfrak{r}}^2)\}$ of $\{(\mathfrak{V}_{\mathfrak{r}}^2, \mathcal{S}_{\mathfrak{r}}^2)\}$ is by definition $\{(\mathfrak{V}_{\mathfrak{r}}^1, \mathcal{S}_{\mathfrak{r}}^1) \mid \mathfrak{r} \in \mathfrak{R}_0\}$ which we define below:

(1) $\mathfrak{R}_0$ is the set of all $\mathfrak{r}$ such that $U_{\mathfrak{r}}^2 \cap \varphi_{21}(U_{21}) \neq \emptyset$.
(2) $\mathfrak{V}_{\mathfrak{r}}^1$ is then given by Condition 7.37.
(3) $\mathcal{S}_{\mathfrak{r}}^1 = (W_{\mathfrak{r}}^2, \omega_{\mathfrak{r}}^2, \{s_{\mathfrak{r}}^{1,\epsilon}\})$, where $\mathfrak{s}_{\mathfrak{r}}^{1,\epsilon}$ is defined by

$$\mathfrak{s}_{\mathfrak{r}}^{1,\epsilon}(y,\xi) = g_y^{-1}(\mathfrak{s}_{\mathfrak{r}}^{2,\epsilon}(\widetilde{\varphi}_{21}^{\mathfrak{r}}(y),\xi)). \tag{7.19}$$

Here g_y is as in (7.18). The right hand side exists because of (7.17).

It is easy to see that $\{(\mathfrak{V}_{\mathfrak{r}}^1, \mathcal{S}_{\mathfrak{r}}^1) \mid \mathfrak{r} \in \mathfrak{R}_0\}$ is a representative of a CF-perturbation of $\mathcal{U}^1$.

Lemma 7.43

(1) *If $\{(\mathfrak{V}_{\mathfrak{r}}^2, \mathcal{S}_{\mathfrak{r}}^2)\}$ is restrictable to $\mathcal{U}^1$ by Φ_{21} and $\{(\mathfrak{V}_{\mathfrak{r}'}^{\prime 2}, \mathcal{S}_{\mathfrak{r}'}^{\prime 2})\}$ is equivalent to $\{(\mathfrak{V}_{\mathfrak{r}}^2, \mathcal{S}_{\mathfrak{r}}^2)\}$, then $\{(\mathfrak{V}_{\mathfrak{r}'}^{\prime 2}, \mathcal{S}_{\mathfrak{r}'}^{\prime 2})\}$ is restrictable to $\mathcal{U}^1$ by Φ_{21}. Moreover $\Phi_{21}^*\{(\mathfrak{V}_{\mathfrak{r}}^2, \mathcal{S}_{\mathfrak{r}}^2)\}$ is equivalent to $\Phi_{21}^*\{(\mathfrak{V}_{\mathfrak{r}'}^{\prime 2}, \mathcal{S}_{\mathfrak{r}'}^{\prime 2})\}$.*
(2) *The pullback $\Phi_{21}^*\{(\mathfrak{V}_{\mathfrak{r}}^2, \mathcal{S}_{\mathfrak{r}}^2)\}$ is independent of the choice of $(h_{21}^{\mathfrak{r}}, \varphi_{21}^{\mathfrak{r}}, \widehat{\varphi}_{21}^{\mathfrak{r}})$ up to equivalence.*

Proof To prove (1) it suffices to consider only the case when $\mathcal{S}_{\mathfrak{r}}'$ is a projection of $\mathcal{S}_{\mathfrak{r}}$, which follows again from the fact that $\omega_x = |\omega_x|\mathrm{vol}_x$ where vol_x is a volume form of W_x and $|\omega_x|$ is a non-negative function, which is a part of Definition 7.4 (4). The assertion (2) follows from Lemma 23.28. □

Definition-Lemma 7.44 Suppose we are in Situation 7.36.

(1) We denote by $\mathcal{CF}^{\mathcal{U}^1 \triangleright \mathcal{U}^2}(U^2)$ the set of all elements of $\mathcal{CF}^{\mathcal{U}^2}(U^2)$ which is restrictable to $\mathcal{U}^1$. This is well-defined by Lemma 7.43 (1).
(2) The pullback

$$\Phi_{21}^* : \mathcal{CF}^{\mathcal{U}^1 \triangleright \mathcal{U}^2}(U^2) \to \mathcal{CF}^{\mathcal{U}^1}(U^1)$$

is defined via a choice of representative and an embedding, and is well-defined by Lemma 7.43 (2).
(3) $\Omega \mapsto \mathcal{CF}^{\mathcal{U}^1 \triangleright \mathcal{U}^2}(\Omega)$ is a subsheaf of $\mathcal{CF}^{\mathcal{U}^2}$.
(4) The restriction map Φ_{21}^* is induced by a sheaf morphism:

$$\Phi_{21}^{\mathcal{CF}} : \mathcal{CF}^{\mathcal{U}^1 \triangleright \mathcal{U}^2} \to \varphi_{21*}\mathcal{CF}^{\mathcal{U}^1} \tag{7.20}$$

of sheaves on U^2. Here the right hand side is the pushout sheaf.[4]

[4] In fact for any $x \in U^1$ the restriction defines a map between germs: $(\mathcal{CF}^{\mathcal{U}^1 \triangleright \mathcal{U}^2})_{\varphi_{21}(x)} \to (\mathcal{CF}^{\mathcal{U}^1})_x$, which induces the sheaf morphism (7.20).

In particular, the following diagram commutes for any open sets Ω, Ω' in U^2 with $\Omega' \subseteq \Omega$:

$$\begin{array}{ccc} \mathcal{CF}^{\mathcal{U}^1 \triangleright \mathcal{U}^2}(\Omega) & \xrightarrow{\mathfrak{i}_{\Omega'\Omega}} & \mathcal{CF}^{\mathcal{U}^1 \triangleright \mathcal{U}^2}(\Omega') \\ \Phi_{21}^* \downarrow & & \downarrow \Phi_{21}^* \\ \mathcal{CF}^{\mathcal{U}^1}(\varphi_{21}^{-1}(\Omega)) & \xrightarrow[\mathfrak{i}_{\varphi_{21}^{-1}(\Omega')\varphi_{21}^{-1}(\Omega)}]{} & \mathcal{CF}^{\mathcal{U}^1}(\varphi_{21}^{-1}(\Omega')) \end{array} \tag{7.21}$$

Note that Φ_{21}^* in (2) is equal to the map between sections over Ω induced by the sheaf morphism $\Phi_{21}^{\mathcal{CF}}$, which is sometimes denoted by $\Phi_{21}^{\mathcal{CF}}(\Omega)$.

Proof We can check the assertion directly. So we omit the proof. □

Remark 7.45 We remark that if $x \notin \mathrm{Im}\varphi_{21}$ then the stalk, $(\varphi_{21*}\mathcal{CF}^{\mathcal{U}^1})_x$ of the pushout sheaf consists of a single element. In fact if Ω is a neighborhood of x with $\Omega \cap \mathrm{Im}\varphi_{21} = \emptyset$ then

$$\varphi_{21*}\mathcal{CF}^{\mathcal{U}^1}(\Omega) = \mathcal{CF}^{\mathcal{U}^1}(\varphi_{21}^{-1}(\Omega)) = \mathcal{CF}^{\mathcal{U}^1}(\emptyset)$$

and consists of one element. (See [Go, Chapter II, Section 1.1 (p.109)].)

Lemma 7.46 *Let $\Phi_{i+1i} : \mathcal{U}^i \to \mathcal{U}^{i+1}$ $(i = 1, 2)$ be embeddings of Kuranishi charts. We put $\Phi_{31} = \Phi_{32} \circ \Phi_{21}$.*

(1) *We have*

$$(\Phi_{32}^{\mathcal{CF}})^{-1}(\varphi_{32*}\mathcal{CF}^{\mathcal{U}^1 \triangleright \mathcal{U}^2}) \subseteq \mathcal{CF}^{\mathcal{U}^1 \triangleright \mathcal{U}^3} \tag{7.22}$$

as subsheaves of $\mathcal{CF}^{\mathcal{U}^3}$.

(2) *The next diagram commutes:*

$$\begin{array}{ccc} (\Phi_{32}^{\mathcal{CF}})^{-1}(\varphi_{32*}(\mathcal{CF}^{\mathcal{U}^1 \triangleright \mathcal{U}^2})) & \xrightarrow{\Phi_{32}^{\mathcal{CF}}} & \varphi_{32*}(\mathcal{CF}^{\mathcal{U}^1 \triangleright \mathcal{U}^2}) \\ \downarrow & & \downarrow \varphi_{32*}\Phi_{21}^{\mathcal{CF}} \\ \mathcal{CF}^{\mathcal{U}^1 \triangleright \mathcal{U}^3} & \xrightarrow[\Phi_{31}^{\mathcal{CF}}]{} & \varphi_{31*}(\mathcal{CF}^{\mathcal{U}^1}) \end{array} \tag{7.23}$$

where the left vertical arrow is (7.22).

In particular, the next diagram commutes for each $\Omega \subset U^3$:

$$\begin{array}{ccc} ((\Phi_{32}^{\mathcal{CF}})^{-1}\mathcal{CF}^{\mathcal{U}^1 \triangleright \mathcal{U}^2})(\Omega) & \xrightarrow{\Phi_{32}^*} & \mathcal{CF}^{\mathcal{U}^1 \triangleright \mathcal{U}^2}(\varphi_{32}^{-1}(\Omega)) \\ \downarrow & & \downarrow \Phi_{21}^* \\ \mathcal{CF}^{\mathcal{U}^1 \triangleright \mathcal{U}^3}(\Omega) & \xrightarrow[\Phi_{31}^*]{} & \mathcal{CF}^{\mathcal{U}^1}(\varphi_{31}^{-1}(\Omega)) \end{array} \tag{7.24}$$

Proof This is a consequence of (7.19). □

Next we generalize Lemma 6.14 to the case of continuous families.

Lemma 7.47 *There exist $c > 0$, $\delta_0 > 0$ and $\epsilon_0 > 0$ such that for $y \in BN_{\delta_0}(K; U_{\mathfrak{p}})$*

$$|\mathfrak{s}^{\epsilon}(y)| \geq c|v| \tag{7.25}$$

for each $0 < \epsilon < \epsilon_0$ and member $\mathfrak{s}^{\epsilon}$ of $\mathfrak{S}_{\mathfrak{p}}^{\epsilon}$ at $y = \mathrm{Exp}(x, v)$.

The proof is the same as the proof of Lemma 6.14.

Remark 7.48 The constants ϵ_0, c, δ_0 can be taken to be independent of the choice of representative of CF-perturbations. In fact, the notion of a member is independent of the choice of representatives of CF-perturbation.

7.4.2 CF-Perturbations on Good Coordinate Systems

Definition 7.49 Let $\widehat{\mathcal{U}} = \{\mathcal{U}_{\mathfrak{p}} \mid \mathfrak{p} \in \mathfrak{P}\}$ be a good coordinate system of $Z \subseteq X$. A *CF-perturbation of* $(\widehat{\mathcal{U}}, \mathcal{K})$ is by definition $\widehat{\mathfrak{S}} = \{\mathfrak{S}_{\mathfrak{p}} \mid \mathfrak{p} \in \mathfrak{P}\}$ with the following properties:

(1) $\mathfrak{S}_{\mathfrak{p}} \in \mathcal{CF}^{\mathcal{U}_{\mathfrak{p}}}(\mathcal{K}_{\mathfrak{p}})$.
(2) If $\mathfrak{q} \leq \mathfrak{p}$ then $\mathfrak{S}_{\mathfrak{p}} \in \mathcal{CF}^{\mathcal{U}_{\mathfrak{q}} \triangleright \mathcal{U}_{\mathfrak{p}}}(\mathcal{K}_{\mathfrak{p}})$.
(3) The pullback $\Phi_{\mathfrak{pq}}^{\mathcal{CF}}(\mathfrak{S}_{\mathfrak{p}})$ is equivalent to $\mathfrak{S}_{\mathfrak{q}}$ as an element of $\mathcal{CF}^{\mathcal{U}_{\mathfrak{q}}}(\varphi_{\mathfrak{pq}}^{-1}(\mathcal{K}_{\mathfrak{p}}) \cap \mathcal{K}_{\mathfrak{q}})$.

Definition 7.50 Let $\widehat{\mathfrak{S}} = \{\mathfrak{S}_{\mathfrak{p}} \mid \mathfrak{p} \in \mathfrak{P}\}$ be a CF-perturbation of a good coordinate system $\widehat{\mathcal{U}}$ and $\mathcal{K}$ its support system.

(1) We say $\widehat{\mathfrak{S}}$ is *transversal to 0* if each of $\mathfrak{S}_{\mathfrak{p}}$ is transversal to 0.
(2) Let $\widehat{f} : (X, Z; \widehat{\mathcal{U}}) \to M$ be a strongly smooth map that is weakly submersive. We say that $\widehat{f}$ is *strongly submersive with respect to $\widehat{\mathfrak{S}}$ on $\mathcal{K}$* if for each $\mathfrak{p} \in \mathfrak{P}$ the map $f_{\mathfrak{p}}$ is strongly submersive with respect to $\mathfrak{S}_{\mathfrak{p}}$ on $\mathcal{K}_{\mathfrak{p}}$ in the sense of Definition 7.9.
(3) Let $g : N \to M$ be a smooth map between smooth manifolds. We say that $\widehat{f}$ is *strongly transversal to g with respect to $\widehat{\mathfrak{S}}$ on $\mathcal{K}$* if for each $\mathfrak{p} \in \mathfrak{P}$ the map $f_{\mathfrak{p}}$ is strongly transversal to g with respect to $\mathfrak{S}_{\mathfrak{p}}$ on $\mathcal{K}_{\mathfrak{p}}$.

Theorem 7.51 *Let $\widehat{\mathcal{U}}$ be a good coordinate system of $Z \subseteq X$ and $\mathcal{K}$ its support system.*

(1) *There exists a CF-perturbation $\widehat{\mathfrak{S}}$ of $(\widehat{\mathcal{U}}, \mathcal{K})$ transversal to* 0.
(2) *If $\widehat{f} : (X, Z; \widehat{\mathcal{U}}) \to M$ is a weakly submersive strongly smooth map, then we may take $\widehat{\mathfrak{S}}$ with respect to which $\widehat{f}$ is strongly submersive.*
(3) *If $\widehat{f} : (X, Z; \widehat{\mathcal{U}}) \to M$ is a strongly smooth map which is weakly transversal to $g : N \to M$, then we may take $\widehat{\mathfrak{S}}$ with respect to which $\widehat{f}$ is strongly transversal to g.*

The proof of Theorem 7.51 is given in Sect. 12.2.

7.4.3 Extension of a Good Coordinate System and Relative Version of the Existence Theorem of CF-Perturbations

For later use we include the relative version of Theorem 7.51, that is, Propositions 7.54, 7.59 and Lemma 7.55. To state this relative version we need some preparation.

Definition 7.52 Let X be a separable metrizable space and $Z_1, Z_2 \subseteq X$ compact subsets. We assume $Z_1 \subset \overset{\circ}{Z_2}$.

(1) For each $i = 1, 2$, let $\widehat{\mathcal{U}^i} = (\mathfrak{P}_i, \{\mathcal{U}^i_{\mathfrak{p}}\}, \{\Phi^i_{\mathfrak{pq}}\})$ be a good coordinate system of $Z_i \subseteq X$. We say $\widehat{\mathcal{U}^2}$ *strictly extends* $\widehat{\mathcal{U}^1}$ if the following holds:
   (a) $\mathfrak{P}_1 = \{\mathfrak{p} \in \mathfrak{P}_2 \mid \psi_{\mathfrak{p}}((s_{\mathfrak{p}})^{-1}(0)) \cap Z_1 \neq \emptyset\}$. The partial order of $\mathfrak{P}_1$ is the restriction of that of $\mathfrak{P}_2$.
   (b) If $\mathfrak{p} \in \mathfrak{P}_1$, then $\mathcal{U}^1_{\mathfrak{p}}$ is an open subchart of $\mathcal{U}^2_{\mathfrak{p}}$. Moreover $\mathrm{Im}(\psi^1_{\mathfrak{p}}) \cap Z_1 = \mathrm{Im}(\psi^2_{\mathfrak{p}}) \cap Z_1$.
   (c) If $\mathfrak{p}, \mathfrak{q} \in \mathfrak{P}_1$, then $\Phi^1_{\mathfrak{pq}}$ is a restriction of $\Phi^2_{\mathfrak{pq}}$.

   Note that the case $Z_2 = X$ is included in this definition.
(2) In the situation of (a), we say $\widehat{\mathcal{U}}^2$ *extends* $\widehat{\mathcal{U}^1}$ (resp. *weakly extends*) if it strictly extends an open substructure (resp. weakly open substructure) of $\widehat{\mathcal{U}^1}$.
(3) Let $\widehat{\mathcal{U}^2} = (\{\mathcal{U}^2_p \mid p \in Z_2\}, \{\Phi^2_{pq} \mid q \in \mathrm{Im}(\psi_p),\ p, q \in Z_2\})$ be a Kuranishi structure of $Z_2 \subset X$. Then $(\{\mathcal{U}^2_p \mid p \in Z_1\}, \{\Phi^2_{pq} \mid q \in \mathrm{Im}(\psi_p),\ p, q \in Z_1\})$ is a Kuranishi structure of $Z_1 \subseteq X$. We call it the *restriction* of $\widehat{\mathcal{U}^2}$ and write $\widehat{\mathcal{U}^2}|_{Z_1}$.

Lemma 7.53 *For each $i = 1, 2$ let $\widehat{\mathcal{U}^i} = (\mathfrak{P}_i, \{\mathcal{U}^i_{\mathfrak{p}}\}, \{\Phi^i_{\mathfrak{pq}}\})$ be a good coordinate system of $Z_i \subseteq X$, and $\widehat{\mathcal{U}^2}$ a Kuranishi structure of $Z_2 \subseteq X$ such that $\widehat{\mathcal{U}^2}$ is compatible with $\widehat{\mathcal{U}^2}$. Suppose that $Z_1 \subset \overset{\circ}{Z_2}$ and $\widehat{\mathcal{U}^2}$ strictly extends $\widehat{\mathcal{U}^1}$.*

Then there exists a KG-embedding $\widehat{\mathcal{U}^2}|_{Z_1} \to \widehat{\mathcal{U}^1}$ *with the following property: Let* $p \in Z_1$ *and* $\mathfrak{p} \in \mathrm{Im}(\psi^1_{\mathfrak{p}})$. *Let* $\Phi^1_{\mathfrak{p}p} : \mathcal{U}^2_p|_{U^1_p} \to \mathcal{U}^1_{\mathfrak{p}}$ *be a part of the KG-embedding* $\widehat{\mathcal{U}^2}|_{Z_1} \to \widehat{\mathcal{U}^1}$. *(Here* $\mathcal{U}^2_p|_{U^1_p}$ *is an open subchart of* $\mathcal{U}^2_p$.*) Let* $\Phi^2_{\mathfrak{p}p} : \mathcal{U}^2_p \to \mathcal{U}^2_{\mathfrak{p}}$ *be an embedding of Kuranishi charts given by the KG-embedding:* $\widehat{\mathcal{U}^2} \to \widehat{\mathcal{U}^2}$. *The next diagram commutes:*

$$\begin{array}{ccc} \mathcal{U}^1_{\mathfrak{p}} & \longrightarrow & \mathcal{U}^2_{\mathfrak{p}} \\ {\scriptstyle \Phi^1_{\mathfrak{p}p}} \uparrow & & \uparrow {\scriptstyle \Phi^2_{\mathfrak{p}p}} \\ \mathcal{U}^2_p|_{U^1_p} & \longrightarrow & \mathcal{U}^2_p \end{array} \tag{7.26}$$

Note that the horizontal arrows are embeddings as open substructures.

Moreover, the KG-embedding $\widehat{\mathcal{U}^2}|_{Z_1} \to \widehat{\mathcal{U}^1}$ *satisfying this property is unique up to equivalence in the sense of Definition* 5.13.

The same conclusion holds for the extension or the weak extension.

Proof Let $p \in Z_1$ and $\mathfrak{p} \in \mathrm{Im}(\psi^1_{\mathfrak{p}})$. By Definition 7.52 (1)(b), $\mathcal{U}^1_{\mathfrak{p}}$ is an open subchart of $\mathcal{U}^2_{\mathfrak{p}}$. In particular, $U^1_{\mathfrak{p}}$ is an open subset of $U^2_{\mathfrak{p}}$. Let $\Phi^2_{\mathfrak{p}} : \mathcal{U}^2_p \to \mathcal{U}^2_{\mathfrak{p}}$ be the embedding of Kuranishi chart given by the KG-embedding: $\widehat{\mathcal{U}^2} \to \widehat{\mathcal{U}^2}$. We put

$$U^1_p = (\varphi^2_{\mathfrak{p}p})^{-1}(U^1_{\mathfrak{p}}) \subseteq U^2_p.$$

By Lemma 3.28, there exists a Kuranishi structure of $Z_1 \subseteq X$ whose Kuranishi chart is $\mathcal{U}^2_p|_{U^1_p}$. We can define Φ^1_p by restricting Φ^2_p. The commutativity of Diagram (7.26) is obvious. □

We recall Definition 3.37 for the compatibility used in the next proposition.

Proposition 7.54 *Let X be a separable metrizable space,* $Z_1, Z_2 \subseteq X$ *be compact subsets. We assume* $Z_1 \subset \overset{\circ}{Z_2}$. *Let* $\widehat{\mathcal{U}^2}$ *be a Kuranishi structure of* $Z_2 \subseteq X$ *and* $\widehat{\mathcal{U}^1}$ *a good coordinate system of* $Z_1 \subseteq X$ *which is compatible with* $\widehat{\mathcal{U}^2}|_{Z_1}$. *Then there exists a good coordinate system* $\widehat{\mathcal{U}^2}$ *of* $Z_2 \subseteq X$ *such that:*

(1) $\widehat{\mathcal{U}^2}$ *extends* $\widehat{\mathcal{U}^1}$.
(2) $\widehat{\mathcal{U}^2}$ *is compatible with* $\widehat{\mathcal{U}^2}$.
(3) *Diagram* (7.26) *commutes.*

When a support system $\{\mathcal{K}_{\mathfrak{p}} \mid \mathfrak{p} \in \mathfrak{P}_1\}$ *of* $\widehat{\mathcal{U}^1}$ *is given, we can take an open substructure* $\widehat{\mathcal{U}^1_0}$ *of* $\widehat{\mathcal{U}^1}$ *such that* $\widehat{\mathcal{U}^2}$ *strictly extends* $\widehat{\mathcal{U}^1_0}$ *with the following additional properties:*

(4) *For any* $\mathfrak{p} \in \mathfrak{P}_1$ *we have*

$$\mathcal{K}_{\mathfrak{p}} \cap s_{\mathfrak{p}}^{-1}(0) \subset U^1_{0,\mathfrak{p}}. \tag{7.27}$$

Here $U^1_{0,\mathfrak{p}}$, the Kuranishi neighborhood of $\mathcal{U}^1_{0,\mathfrak{p}}$ which is a part of $\widehat{\mathcal{U}^1_0}$, is an open subset of $U^1_{\mathfrak{p}}$, the Kuranishi neighborhood of $\mathcal{U}^1_{\mathfrak{p}}$ which is a part of $\widehat{\mathcal{U}^1}$.

The proof will be given in Sect. 11.4.

Lemma 7.55 *Suppose we are in the situation of Proposition 7.54. We may choose $\widehat{\mathcal{U}^2}$ so that the following holds.*

Let $\widehat{\mathcal{U}^1_0}$ be an open substructure of $\widehat{\mathcal{U}^1}$ strictly extended to $\widehat{\mathcal{U}^2}$. Let $\widehat{f^1} = \{f^1_{\mathfrak{p}}\} : (X, Z_1; \widehat{\mathcal{U}^1_0}) \to M$ and $\widehat{f^2} = \{f^2_p\} : (X, Z_2; \widehat{\mathcal{U}^2}) \to M$ be strongly continuous maps. Assume that the equality $f^1_{\mathfrak{p}} \circ \varphi^1_{\mathfrak{p}p} = f^2_p$ holds on $U^1_p \subset U^2_p$ for each $p \in \mathrm{Im}\,(\psi_{\mathfrak{p}}) \cap Z_1$. Then the following hold:

(1) *There exists a strongly continuous map $\widehat{f^2} = \{f^2_{\mathfrak{p}}\} : (X, Z_2; \widehat{\mathcal{U}^2}) \to M$ such that:*
  (a) *$f^2_{\mathfrak{p}} \circ \varphi^2_{\mathfrak{p}p} = f^2_p$ for $p \in \mathrm{Im}\,(\psi_{\mathfrak{p}}) \cap Z_2$.*
  (b) *$f^2_{\mathfrak{p}} = f^1_{\mathfrak{p}}$ on $U^1_{0,\mathfrak{p}} \subset U^2_{\mathfrak{p}}$.*

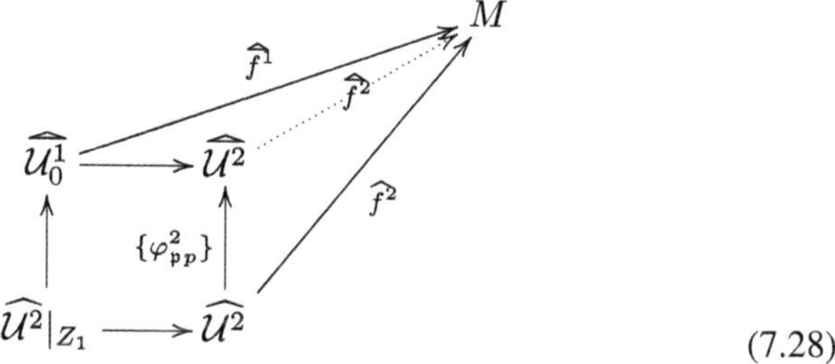

(7.28)

(2) *If $\widehat{f^1}$ and $\widehat{f^2}$ are strongly smooth (resp. weakly submersive) then so is $\widehat{f^2}$.*
(3) *If $\widehat{f^1}$ and $\widehat{f^2}$ are strongly transversal to $g : N \to M$ then so is $\widehat{f^2}$.*

The proof is given in Sect. 11.4.

Definition 7.56 Let $\widehat{\mathcal{U}^i} = (\mathfrak{P}_i, \{\mathcal{U}^i_{\mathfrak{p}}\}, \{\Phi^i_{\mathfrak{pq}}\})$ be good coordinate systems of $Z_i \subseteq X$, $i = 1, 2$ and $\mathcal{K}^i$ support systems of $\widehat{\mathcal{U}^i}$ for $i = 1, 2$.

(1) Suppose $\widehat{\mathcal{U}^2}$ strictly extends $\widehat{\mathcal{U}^1}$. We say that $\mathcal{K}^1$ is *compatible with* $\mathcal{K}^2$ if for each $\mathfrak{p} \in \mathfrak{P}_1$ we have

$$\mathcal{K}^1_{\mathfrak{p}} \subset \overset{\circ}{\mathcal{K}}{}^2_{\mathfrak{p}}.$$

(We note that $U^1_{\mathfrak{p}} \subset U^2_{\mathfrak{p}}$ by Definition 7.52 (1)(c).) In this situation we say $(\widehat{\mathcal{U}^2}, \mathcal{K}^2)$ *strictly extends* $(\widehat{\mathcal{U}^1}, \mathcal{K}^1)$.

(2) Suppose $\widehat{\mathcal{U}^2}$ extends $\widehat{\mathcal{U}^1}$. We say that $\mathcal{K}^1$ is *compatible with* $\mathcal{K}^2$ if the following holds.

By definition there exists an open substructure $\widehat{\mathcal{U}^1_0} = \{\mathcal{U}^1_{\mathfrak{p},0}\}$ of $\widehat{\mathcal{U}^1}$ such that $\widehat{\mathcal{U}^2}$ strictly extends $\widehat{\mathcal{U}^1_0}$. For each $\mathfrak{p} \in \mathfrak{P}_1$ we require

$$\mathcal{K}^1_{\mathfrak{p}} \subset U^1_{\mathfrak{p},0} \cap \overset{\circ}{\mathcal{K}^2_{\mathfrak{p}}}.$$

In this situation we say $(\widehat{\mathcal{U}^2}, \mathcal{K}^2)$ *extends* $(\widehat{\mathcal{U}^1}, \mathcal{K}^1)$.

Definition 7.57 For each $i = 1, 2$, let $\widehat{\mathcal{U}^i} = (\mathfrak{P}_i, \{\mathcal{U}^i_{\mathfrak{p}}\}, \{\Phi^i_{\mathfrak{p}\mathfrak{q}}\})$ be a good coordinate system of $Z_i \subseteq X$, $\mathcal{K}^i$ a support system of $\widehat{\mathcal{U}^i}$, and $\widehat{\mathfrak{S}^i}$ a CF-perturbation of $(\widehat{\mathcal{U}^i}, \mathcal{K}^i)$.

(1) Suppose $(\widehat{\mathcal{U}^2}, \mathcal{K}^2)$ strictly extends $(\widehat{\mathcal{U}^1}, \mathcal{K}^1)$. We say $\widehat{\mathfrak{S}^2}$ *strictly extends* $\widehat{\mathfrak{S}^1}$ if the restriction of $\mathfrak{S}^2_{\mathfrak{p}}$ to $\mathcal{K}^1_{\mathfrak{p}}$ is $\mathfrak{S}^1_{\mathfrak{p}}$ for each $\mathfrak{p} \in \mathfrak{P}_1$.
(2) Suppose $(\widehat{\mathcal{U}^2}, \mathcal{K}^2)$ extends $(\widehat{\mathcal{U}^1}, \mathcal{K}^1)$. We say $\widehat{\mathfrak{S}^2}$ *extends* $\widehat{\mathfrak{S}^1}$ if the restriction of $\mathfrak{S}^2_{\mathfrak{p}}$ to $\mathcal{K}^1_{\mathfrak{p}}$ is $\mathfrak{S}^1_{\mathfrak{p}}$ for each $\mathfrak{p} \in \mathfrak{P}_1$.

Remark 7.58 In Definition 7.57 (2) we note that $\mathcal{K}^1_{\mathfrak{p}}$ can be regarded as a support system of an open substructure of $\widehat{\mathcal{U}^1}$ by Definition 7.56 (2).

Proposition 7.59 *Suppose we are in the situation of Proposition* 7.54. *We may choose* $\widehat{\mathcal{U}^2}$ *and* $\widehat{\mathcal{U}^1_0}$ *so that the following holds.*

$\widehat{\mathcal{U}^1_0}$ *is an open substructure of* $\widehat{\mathcal{U}^1}$ *which is strictly extended to* $\widehat{\mathcal{U}^2}$ *and Proposition* 7.54 *(4) is satisfied. Let* $\mathcal{K}^1$, $\mathcal{K}^2$ *be support systems of* $\widehat{\mathcal{U}^1}$, $\widehat{\mathcal{U}^2}$ *respectively such that* $(\widehat{\mathcal{U}^2}, \mathcal{K}^2)$ *extends* $(\widehat{\mathcal{U}^1}, \mathcal{K}^1)$. *Let* $\widehat{\mathfrak{S}^1}$ *be a CF-perturbation of* $(\widehat{\mathcal{U}^1}, \mathcal{K}^1)$. *Then there exists a CF-perturbation* $\widehat{\mathfrak{S}^2}$ *of* $(\widehat{\mathcal{U}^2}, \mathcal{K}^2)$ *which extends* $\widehat{\mathfrak{S}^1}$. *Moreover the following holds:*

(1) *If* $\widehat{\mathfrak{S}^1}$ *is transversal to* 0*, so is* $\widehat{\mathfrak{S}^2}$.
(2) *Suppose we are in the situation of Lemma* 7.55 *(1) in addition. We assume* $\widehat{f^1}$ *is strongly submersive with respect to* $\widehat{\mathfrak{S}^1}$ *and* $\widehat{f^2}$ *is weakly submersive. Then* $\widehat{f^2}$ *is strongly submersive with respect to* $\widehat{\mathfrak{S}^2}$.
(3) *Suppose we are in the situation of Lemma* 7.55 *(1) in addition. We assume* $\widehat{f^1}$ *is strongly transversal to* $g : N \to M$ *with respect to* $\widehat{\mathfrak{S}^1}$ *and* $\widehat{f^2}$ *is weakly transversal to* g. *Then* $\widehat{f^2}$ *is strongly transversal to* $g : N \to M$ *with respect to* $\widehat{\mathfrak{S}^2}$.

The proof of Proposition 7.59 is given in Sect. 12.2.

We need to study a family of CF-perturbations sometimes. The following notion is useful to study such a family.

Definition 7.60 A σ-parametrized family of CF-perturbations $\{\widehat{\mathfrak{S}_\sigma} \mid \sigma \in \mathscr{A}\}$ ($\widehat{\mathfrak{S}_\sigma} = \{\mathcal{S}^\epsilon_{\sigma,\mathfrak{p}}\}$) of $(\widehat{\mathcal{U}}, \mathcal{K})$ is called a *uniform family* if the convergence in Definition 7.4 (3) is uniform. More precisely, we require the following.

For each $\mathfrak{o} > 0$ there exists $\epsilon_0(\mathfrak{o}) > 0$ such that if $0 < \epsilon < \epsilon_0(\mathfrak{o})$, then

$$|\mathfrak{s}(y) - s_{\mathfrak{p}}(y)| < \mathfrak{o}, \qquad |(D\mathfrak{s})(y) - (Ds_{\mathfrak{p}})(y)| < \mathfrak{o} \tag{7.29}$$

hold for any $\mathfrak{s}$ which is a member of $\mathcal{S}^\epsilon_{\sigma,\mathfrak{p}}$ at any point $y \in \mathcal{K}_{\mathfrak{p}}$ for any $\mathfrak{p} \in \mathfrak{P}$ and $\sigma \in \mathscr{A}$.

Lemma 7.61 *If $\{\widehat{\mathfrak{S}_\sigma} \mid \sigma \in \mathscr{A}\}$ is a uniform family of CF-perturbations of $(\widehat{\mathcal{U}}, \mathcal{K})$, then the constant ϵ_0 in Lemma 7.47 can be taken independent of σ.*

In the situation of Proposition 7.59, if $\{\widehat{\mathfrak{S}_{1,\sigma}} \mid \sigma \in \mathscr{A}\}$ is a uniform family of CF-perturbations then its extensions $\{\widehat{\mathfrak{S}_{2,\sigma}} \mid \sigma \in \mathscr{A}\}$ can be chosen to be a uniform family.

Proof The first half follows from the proof of Lemma 7.47. The second half is proved in Sect. 12.2. □

7.5 Partition of Unity Associated to a Good Coordinate System

Next we define the notion of a partition of unity on a space with a good coordinate system.

Definition 7.62 Let $\widehat{\mathcal{U}} = (\mathfrak{P}, \{\mathcal{U}_{\mathfrak{p}}\}, \{\Phi_{\mathfrak{pq}}\})$ be a good coordinate system of $Z \subseteq X$ and $\mathcal{K}$ its support system. Let Ω be an open subset of $|\mathcal{K}|$ and $f : \Omega \to \mathbb{R}$ a continuous function. We say f is *strongly smooth* if the restriction of f to $\mathcal{K}_{\mathfrak{p}} \cap \Omega$ is smooth for any $\mathfrak{p} \in \mathfrak{P}$.

We remark that the restriction to $\Omega \subset |\mathcal{K}|$ of any strongly smooth map to $\mathbb{R}$ (regarded as a manifold in the sense of Definition 3.43) is a strongly smooth function $f : \Omega \to \mathbb{R}$ in the above sense.

We take a support system $\mathcal{K}^+$ such that $(\mathcal{K}, \mathcal{K}^+)$ is a support pair. We take a metric d on the hetero-dimensional compactum $|\mathcal{K}^+|$ (see [FOOO17, Proposition 2.11]) and use it in the next definition. We remark that for an open subset Ω of $|\mathcal{K}|$ the intersection $\Omega \cap \mathcal{K}_{\mathfrak{p}}$ is open in $\mathcal{K}_{\mathfrak{p}}$ but is not necessarily open in $U_{\mathfrak{p}}$.

Definition 7.63 Let $\widehat{\mathcal{U}} = (\mathfrak{P}, \{\mathcal{U}_{\mathfrak{p}}\}, \{\Phi_{\mathfrak{pq}}\})$ be a good coordinate system of $Z \subseteq X$ and $\mathcal{K}$ its support system. For a positive number $\delta > 0$ we put

$$\mathcal{K}_{\mathfrak{p}}(2\delta) = \{x \in \mathcal{K}^+_{\mathfrak{p}} \mid d(x, \mathcal{K}_{\mathfrak{p}}) \leq 2\delta\}, \tag{7.30}$$

that is a 2δ-neighborhood of $\mathcal{K}_{\mathfrak{p}}$. We assume that $\mathcal{K}_{\mathfrak{p}}(2\delta)$ is contained in Int $\mathcal{K}_{\mathfrak{p}}^{+}$ for each $\mathfrak{p}$. We put

$$\mathcal{K}(2\delta) = \{\mathcal{K}_{\mathfrak{p}}(2\delta)\}_{\mathfrak{p}\in\mathfrak{P}},$$

which is a support system of $\widehat{\mathcal{U}}$. We define (Fig. 7.1)

$$\Omega_{\mathfrak{p}}(\mathcal{K}, \delta) = B_{\delta}(\mathcal{K}_{\mathfrak{p}}) = \{x \in |\mathcal{K}(2\delta)| \mid d(x, \mathcal{K}_{\mathfrak{p}}) < \delta\}. \tag{7.31}$$

Lemma 7.64 *If* $\mathfrak{q} < \mathfrak{p}$ *and* $\delta > 0$ *is sufficiently small, then*

$$\Omega_{\mathfrak{p}}(\mathcal{K}, \delta) \cap \mathcal{K}_{\mathfrak{q}}(2\delta) \subset \mathcal{K}_{\mathfrak{p}}(2\delta).$$

Moreover, $\Omega_{\mathfrak{p}}(\mathcal{K}, \delta) \cap \mathcal{K}_{\mathfrak{p}}(2\delta) \subset \operatorname{Int} \mathcal{K}_{\mathfrak{p}}(2\delta)$ *and* $\Omega_{\mathfrak{p}}(\mathcal{K}, \delta) \cap \mathcal{K}_{\mathfrak{p}}(2\delta)$ *is an orbifold.*

Proof The first claim is a consequence of (7.31). The second claim follows from the fact that $\operatorname{Int} \mathcal{K}_{\mathfrak{p}}(2\delta)$ is open in $\bigcup_{\mathfrak{q}\le\mathfrak{p}} \mathcal{K}_{\mathfrak{q}}(2\delta)$. □

Definition 7.65 We say $\{\chi_{\mathfrak{p}}\}$ is a *partition of unity of the quintuple* $(X, Z, \widehat{\mathcal{U}}, \mathcal{K}, \delta)$ if the following holds:

(1) $\chi_{\mathfrak{p}} : |\mathcal{K}(2\delta)| \to [0, 1]$ is a strongly smooth function.
(2) $\operatorname{supp} \chi_{\mathfrak{p}} \subseteq \Omega_{\mathfrak{p}}(\mathcal{K}, \delta)$.
(3) There exists an open neighborhood $\mathscr{U}$ of Z in $|\mathcal{K}(2\delta)|$ such that for each point $x \in \mathscr{U}$, we have

$$\sum_{\mathfrak{p}} \chi_{\mathfrak{p}}(x) = 1. \tag{7.32}$$

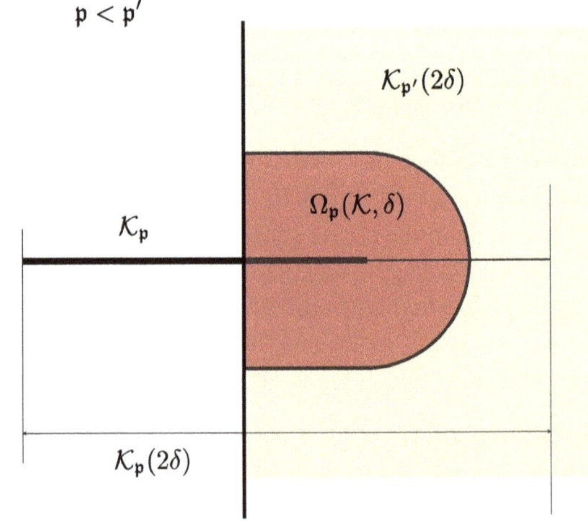

Fig. 7.1 $\Omega_{\mathfrak{p}}(\mathcal{K}, \delta)$

Remark 7.66 In our earlier writings such as [FOOO8, Section 12], we defined a partition of unity in a slightly different way. Namely we required $\chi_{\mathfrak{p}}$ to be defined on $U_{\mathfrak{p}}$ and we required

$$\chi_{\mathfrak{p}}(x) + \sum_{\mathfrak{q}>\mathfrak{p}, x\in U_{\mathfrak{q}\mathfrak{p}}} \chi_{\mathfrak{q}}(\varphi_{\mathfrak{q}\mathfrak{p}}(x)) + \sum_{\mathfrak{p}>\mathfrak{q}, x\in \mathrm{Im}\varphi_{\mathfrak{p}\mathfrak{q}}} \chi^{\delta}(\rho(x; U_{\mathfrak{p}\mathfrak{q}}))\chi_{\mathfrak{q}}(\pi(x)) = 1 \tag{7.33}$$

instead of (7.32). Here π is the projection of a tubular neighborhood of $\varphi_{\mathfrak{p}\mathfrak{q}}(U_{\mathfrak{p}\mathfrak{q}})$ in $U_{\mathfrak{p}}$, ρ is a tubular distance function of this tubular neighborhood[5] and χ^{δ} : $[0, \infty) \to [0, 1]$ is a smooth function such that it is 1 in a neighborhood of 0 and is 0 on (δ, ∞). We required (7.33) for all $x \in \mathcal{K}^1_{\mathfrak{p}}$. Formula (7.33) depends on the choice of the tubular neighborhood and we need to take a certain compatible system of tubular neighborhoods (see [Ma]) to define it. Note that the covering $|\widehat{\mathcal{U}}| = \bigcup_{\mathfrak{p}\in\mathfrak{P}} U_{\mathfrak{p}}$ is *not* an open covering. In fact, $U_{\mathfrak{p}}$ is not an open subset of $|\widehat{\mathcal{U}}|$ unless $\mathfrak{p}$ is maximal. So a partition of unity in the above sense is different from the one in the usual sense. Definition 7.65 seems to be simpler than that of our previous writings. Also it is more elementary in the sense that we do not use any compatible system of tubular neighborhoods. However, by using the third term of (7.33) we can extend the function $\chi_{\mathfrak{q}}$ to its neighborhood in $|\mathcal{K}|$ and the compatibility of tubular neighborhoods implies that it becomes a strongly smooth function. So the present definition is not very different from the earlier one.

In the rest of this section we prove the existence of a partition of unity. The proof is very similar to the corresponding one in the manifold theory. We begin with the following lemma.

Lemma 7.67 *For any open set W of $|\mathcal{K}(2\delta)|$ containing a compact subset K of Z there exists a strongly smooth function $g : |\mathcal{K}(2\delta)| \to \mathbb{R}$ that has a compact support in W and is* 1 *on a neighborhood of K.*

Proof Let $\mathcal{K}^+$ be a support system such that $\mathcal{K}(2\delta) < \mathcal{K}^+$. We take a metric d on $|\mathcal{K}^+|$. (See [FOOO17, Proposition 2.11].) For each $x \in K$ we consider

$$\epsilon_x = \inf\{d(x, \mathcal{K}_{\mathfrak{q}}(2\delta)) \mid \mathfrak{q} \in \mathfrak{P},\ x \notin \mathcal{K}_{\mathfrak{q}}(2\delta)\}.$$

Let $\mathfrak{p}_x \in \mathfrak{P}$ be the element that is maximal among the elements $\mathfrak{p} \in \mathfrak{P}$ such that $x \in \mathcal{K}_{\mathfrak{p}}(2\delta)$. Let W^+_x be an open neighborhood of x in $\mathcal{K}^+_{\mathfrak{p}_x} \cap B_{\epsilon_x/2}(x, |\mathcal{K}^+|)$. We may choose it small and may take small $\sigma_x > 0$ such that if $x \in \mathcal{K}_{\mathfrak{q}}(2\delta)$, $W^+_x \cap \overset{\circ}{\mathcal{K}}_{\mathfrak{q}}(2\delta + \sigma_x)$ is open in $U_{\mathfrak{q}}$. (Here we use the fact that $x \in \mathcal{K}_{\mathfrak{q}}(2\delta)$ implies $\mathfrak{q} \le \mathfrak{p}_x$.) Therefore $W_x = W^+_x \cap \overset{\circ}{\mathcal{K}}_{\mathfrak{p}_x}(2\delta + \sigma_x) \cap |\mathcal{K}(2\delta)|$ is open in $|\mathcal{K}(2\delta)|$. We may choose W^+_x and σ_x so small that $W_x \subset W$.

[5]The tubular distance function is, roughly speaking, the distance from the image $\varphi_{\mathfrak{p}\mathfrak{q}}(U_{\mathfrak{p}\mathfrak{q}})$. We do not give the precise definition here since we do not use this notion. See [Ma].

We can take a smooth function $f_x : W_x^+ \to [0,1]$ that has a compact support in W_x and is 1 in a neighborhood Q_x^+ of x. (This is because W_x^+ is an orbifold.) We restrict f_x to W_x and extend it by 0 to $|\mathcal{K}(2\delta)|$, which we denote by the same symbol. (Recall that $\mathfrak{p}_x$ is maximal and W_x is open in $U_{\mathfrak{p}_x}$.) Then f_x is a strongly smooth function with a compact support in W_x and equal to 1 on an open neighborhood $Q_x = Q_x^+ \cap |\mathcal{K}(2\delta)|$ of x.

We find finitely many points $x_1, \dots, x_N$ of K so that

$$\bigcup_{i=1}^{N} Q_{x_i} \supset K.$$

We put

$$f = \sum_{i=1}^{N} f_{x_i}.$$

Then the function f is a strongly smooth function on $|\mathcal{K}(2\delta)|$ with a compact support in W and satisfies $f(x) \ge 1$ if $x \in Q = \bigcup_{i=1}^{N} Q_{x_i}$. Let $\rho : \mathbb{R} \to \mathbb{R}$ be a nondecreasing smooth map such that $\rho(s) = 0$ if s is in a neighborhood of 0 and $\rho(s) = 1$ if $s \ge 1$. It is easy to see that $g = \rho \circ f$ has the required property. □

Proposition 7.68 *If $\delta > 0$ is sufficiently small, then a partition of unity of $(X, Z, \widehat{\mathcal{U}}, \mathcal{K}, \delta)$ exists.*

Proof We put

$$\mathcal{K}_{\mathfrak{p}}(-\delta) = \{x \in \mathcal{K}_{\mathfrak{p}} \mid d(x, U_{\mathfrak{p}} \setminus \mathcal{K}_{\mathfrak{p}}) \ge \delta\}.$$

It is easy to see that

$$\bigcup_{\mathfrak{p}} \mathcal{K}_{\mathfrak{p}}(-\delta) \supset Z$$

for sufficiently small $\delta > 0$. We apply Lemma 7.67 to $(K, W) = (\mathcal{K}_{\mathfrak{p}}(-\delta), \Omega_{\mathfrak{p}}(\mathcal{K}, \delta))$ to obtain $f_{\mathfrak{p}}$. Then there exists a neighborhood W' of Z such that

$$\sum_{\mathfrak{p}} f_{\mathfrak{p}}(x) \ge 1/2$$

for $x \in W'$. We apply Lemma 7.67 to $(K, W) = (Z, W')$ to obtain $f : |\mathcal{K}(2\delta)| \to [0, 1]$. Now we define

$$\chi_{\mathfrak{p}}(x) = \begin{cases} \dfrac{f(x) f_{\mathfrak{p}}(x)}{\sum_{\mathfrak{p}} f_{\mathfrak{p}}(x)} & \text{if } x \in W' \\ 0 & \text{if } x \notin W'. \end{cases} \tag{7.34}$$

Then it is easy to see that this is a partition of unity of $(X, Z, \widehat{\mathcal{U}}, \mathcal{K}, \delta)$. □

7.6 Differential Forms on a Good Coordinate System and a Kuranishi Structure

In this section we define differential forms on good coordinate systems and on Kuranishi structures, and give several basic definitions for differential forms.

Definition 7.69 Let $\widehat{\mathcal{U}} = (\mathfrak{P}, \{\mathcal{U}_{\mathfrak{p}}\}, \{\Psi_{\mathfrak{pq}}\})$ be a good coordinate system and $\mathcal{K}$ its support system. A *smooth differential k-form* $\widehat{h}$ *on* $(\widehat{\mathcal{U}}, \mathcal{K})$ by definition assigns a smooth k-form $h_{\mathfrak{p}}$ on a neighborhood of $\mathcal{K}_{\mathfrak{p}}$ for each $\mathfrak{p} \in \mathfrak{P}$ so that the next equality holds on a neighborhood of $\varphi_{\mathfrak{pq}}^{-1}(\mathcal{K}_{\mathfrak{p}}) \cap \mathcal{K}_{\mathfrak{q}}$ in $U_{\mathfrak{q}}$:

$$\varphi_{\mathfrak{pq}}^* h_{\mathfrak{p}} = h_{\mathfrak{q}}. \tag{7.35}$$

A *smooth differential k-form* $\widehat{h}$ *on* $\widehat{\mathcal{U}}$ by definition assigns a smooth differential k form $h_{\mathfrak{p}}$ on $U_{\mathfrak{p}}$ for each $\mathfrak{p}$ such that (7.35) is satisfied on $U_{\mathfrak{pq}}$.

Definition 7.70 A *differential k-form* $\widehat{h}$ *on a Kuranishi structure* $\widehat{\mathcal{U}}$ of $Z \subseteq X$ assigns a differential k-form h_p on U_p for each $p \in Z$ such that $\varphi_{pq}^* h_p = h_q$.

Definition 7.71

(1) Let $\widehat{f} = \{f_{\mathfrak{p}}\} : (X, Z; \widehat{\mathcal{U}}) \to M$ be a strongly smooth map and h a smooth differential k-form on M_s. Then $\widehat{f}^* h = \{f_{\mathfrak{p}}^* h\}$ is a smooth differential k form on $\widehat{\mathcal{U}}$, which we call the *pullback of h by* $\widehat{f} = \{f_{\mathfrak{p}}\}$ and denote by $\widehat{f}^* h$.
(2) A smooth differential 0-form is nothing but a strongly smooth function in the sense of Definition 7.62.
(3) If $\widehat{h^i} = \{h^i_{\mathfrak{p}}\}$ are smooth differential k_i-forms on $\widehat{\mathcal{U}}$ for $i = 1, 2$, then $\{h^1_{\mathfrak{p}} \wedge h^2_{\mathfrak{p}}\}$ is a smooth differential $(k_1 + k_2)$-form on $\widehat{\mathcal{U}}$. We call it the *wedge product* and denote it by $\widehat{h^1} \wedge \widehat{h^2}$.
(4) In particular, we can define a product of a smooth differential form and a strongly smooth function.
(5) The sum of smooth differential forms of the same degree is defined by taking the sum for each $\mathfrak{p} \in \mathfrak{P}$.
(6) If $\widehat{h} = \{h_{\mathfrak{p}}\}$ is a smooth differential k-form on $\widehat{\mathcal{U}}$, then $\{dh_{\mathfrak{p}}\}$ is a smooth differential $(k + 1)$-form on $\widehat{\mathcal{U}}$, which we call the *exterior derivative* of $\widehat{h}$ and denote by $d\widehat{h}$.

(7) The *support* $\mathrm{Supp}(\widehat{h})$ of $\widehat{h}$ is defined as follows:

$$\mathrm{Supp}(\widehat{h}) = \bigcup_{\mathfrak{p}} (\mathrm{Supp}(h_{\mathfrak{p}}) \cap \mathcal{K}_{\mathfrak{p}}).$$

(1)–(6) have obvious versions in the case of differential forms on a Kuranishi structure. (7) is modified as follows.

(7)' If $\widehat{h} = \{h_p \mid p \in Z\}$ is a smooth differential k-form on a Kuranishi structure $\widehat{\mathcal{U}}$ of $Z \subseteq X$, its *support* $\mathrm{Supp}(\widehat{h})$ is the set of the points $p \in Z$ such that h_p is nonzero on any neighborhood of o_p in U_p.

Note that $\mathrm{Supp}(\widehat{h})$ is a subset of Z in this case.

See Lemma 9.10 (1) about the relation between the supports of $\widehat{h}$ and of $\widehat{h}$ when a differential form $\widehat{h}$ on a good coordinate system is obtained from a differential form $\widehat{h}$ on a Kuranishi structure.

Definition 7.72 Let $\widehat{\mathcal{U}}$ be a good coordinate system of $Z \subseteq X$ and $\widehat{h} = \{h_{\mathfrak{p}}\}$ a differential form on it. We say that $\widehat{h}$ has a *compact support in* $\mathring{Z}$ if

$$\mathrm{Supp}(\widehat{h}) \cap X \subset \mathring{Z}. \tag{7.36}$$

Here the intersection in the left hand side is taken on $|\widehat{\mathcal{U}}|$.

7.7 Integration Along the Fiber (pushout) on a Good Coordinate System

To define the pushout (integration along the fiber) of a differential form using a CF-perturbation, we need a CF-perturbation version of Propositions 6.17, 6.18. To state them we introduce the notation of *support set* of a CF-perturbation.

Definition 7.73

(1) Let $\mathcal{U}$ be a Kuranishi chart, $\mathfrak{S}$ a CF-perturbation of $\mathcal{U}$ and $\{(\mathfrak{V}_{\mathfrak{r}}, \mathcal{S}_{\mathfrak{r}}) \mid \mathfrak{r} \in \mathfrak{R}\}$ its representative. For each $\epsilon > 0$ we define the *support set* $\Pi((\mathfrak{S}^{\epsilon})^{-1}(0))$ of $\mathfrak{S}$ as the set of all $x \in U$ with the following property:

There exist $\mathfrak{r} \in \mathfrak{R}$ and $y \in V_{\mathfrak{r}}$, $\xi \in W_{\mathfrak{r}}$ such that:

$$\phi_{\mathfrak{r}}([y]) = x, \qquad s^{\epsilon}_{\mathfrak{r}}(y, \xi) = 0, \qquad \xi \in \mathrm{Supp}(\omega_x).$$

This definition is independent of the choice of representative $\{(\mathfrak{V}_{\mathfrak{r}}, \mathcal{S}_{\mathfrak{r}}) \mid \mathfrak{r} \in \mathfrak{R}\}$ because of Definition 7.4 (4).[6]

(2) Let $\widehat{\mathcal{U}}$ be a good coordinate system, $\mathcal{K}$ its support system and $\widehat{\mathfrak{S}} = \{\mathfrak{S}_{\mathfrak{p}}\}$ a CF-perturbation of $(\widehat{\mathcal{U}}, \mathcal{K})$. The *support set* $\Pi((\widehat{\mathfrak{S}^{\epsilon}})^{-1}(0))$ *of* $\widehat{\mathfrak{S}}$ is defined by

$$\Pi((\widehat{\mathfrak{S}^{\epsilon}})^{-1}(0)) = \bigcup_{\mathfrak{p}\in\mathfrak{P}} \left(\mathcal{K}_{\mathfrak{p}} \cap \Pi((\mathfrak{S}^{\epsilon}_{\mathfrak{p}})^{-1}(0))\right)$$

which is a subset of $|\mathcal{K}|$.

The CF-perturbation version of Propositions 6.17 is as follows.

Lemma 7.74 *Let $\widehat{\mathcal{U}}$ a good coordinate system of $Z \subseteq X$ and $\widehat{\mathfrak{S}} = \{\mathfrak{S}_{\mathfrak{p}} \mid \mathfrak{p} \in \mathfrak{P}\}$ a CF-perturbation of $(\widehat{\mathcal{U}}, \mathcal{K})$. Let $\mathcal{K}^1 < \mathcal{K}^2 < \mathcal{K}^3$ are support systems of $\widehat{\mathcal{U}}$. Then there exist positive numbers δ_0 and ϵ_0 such that*

$$B_{\delta}(\mathcal{K}^1_{\mathfrak{q}} \cap Z) \cap \bigcup_{\mathfrak{p}\in\mathfrak{P}} (\mathcal{K}^3_{\mathfrak{p}} \cap \Pi((\widehat{\mathfrak{S}^{\epsilon}_{\mathfrak{p}}})^{-1}(0))) \subset \mathcal{K}^2_{\mathfrak{q}}$$

for any $0 < \delta < \delta_0, 0 < \epsilon < \epsilon_0$.

Proof Using Lemma 7.47, the proof is the same as the proof of Proposition 6.17. □

The next lemma is the CF-perturbation version of Proposition 6.18.

Lemma 7.75 *Let $\mathcal{K}^1, \mathcal{K}^2, \mathcal{K}^3$ be support systems of a good coordinate system $\widehat{\mathcal{U}}$ of $Z \subseteq X$ with $\mathcal{K}^1 < \mathcal{K}^2 < \mathcal{K}^3$ and $\widehat{\mathfrak{S}}$ a CF-perturbation of $(\widehat{\mathcal{U}}, \mathcal{K}^3)$. Then there exists a neighborhood $\mathfrak{U}(Z)$ of Z in $|\mathcal{K}^2|$ such that for $0 < \epsilon < \epsilon_0$*

$$\mathfrak{U}(Z) \cap \bigcup_{\mathfrak{p}} (\Pi((\widehat{\mathfrak{S}^{\epsilon}_{\mathfrak{p}}})^{-1}(0)) \cap \mathcal{K}^1_{\mathfrak{p}}) = \mathfrak{U}(Z) \cap \bigcup_{\mathfrak{p}} (\Pi((\widehat{\mathfrak{S}^{\epsilon}_{\mathfrak{p}}})^{-1}(0)) \cap \mathcal{K}^2_{\mathfrak{p}}). \tag{7.37}$$

Proof The proof is the same as the proof of Proposition 6.18. □

The next lemma is the CF-perturbation version of Corollary 6.20.

Lemma 7.76 *In the situation of Lemma 7.75, $\left(\bigcup_{\mathfrak{p}}((\widehat{\mathfrak{S}^{\epsilon}_{\mathfrak{p}}})^{-1}(0)) \cap \mathring{\mathcal{K}}^1_{\mathfrak{p}})\right) \cap \mathfrak{U}(Z)$ is compact for a sufficiently small neighborhood $\mathfrak{U}(Z)$ of Z in $|\mathcal{K}^2|$. Moreover, we have*

$$\lim_{\epsilon\to 0} \left(\bigcup_{\mathfrak{p}} (\Pi((\widehat{\mathfrak{S}^{\epsilon}_{\mathfrak{p}}})^{-1}(0) \cap \mathring{\mathcal{K}}^1_{\mathfrak{p}})\right) \cap \mathfrak{U}(Z) \subseteq X$$

[6] The condition $|\omega_x| \geq 0$ (Definition 7.4 (4)) is used to show $\Pi(\mathrm{Supp}(|\omega_x|)) = \mathrm{Supp}(|\overline{\omega}_x|)$, where $(\overline{W}, \overline{\omega}_x, \mathfrak{s}^{\epsilon})$ is a projection of $(W, \omega_x, \mathfrak{s}^{\epsilon})$.

in Hausdorff topology.

Proof The proof is the same as Corollary 6.20. □

Now to define the pushout of a differential form we consider the following situation.

Situation 7.77 Let $\widehat{\mathcal{U}} = (\mathfrak{P}, \{\mathcal{U}_{\mathfrak{p}}\}, \{\Phi_{\mathfrak{pq}}\})$ be a good coordinate system, $\mathcal{K}$ its support system, $\widehat{h} = \{h_{\mathfrak{p}}\}$ a differential form on $\widehat{\mathcal{U}}$, $\widehat{f} : (X, Z; \widehat{\mathcal{U}}) \to M$ a strongly smooth map, and $\widehat{\mathfrak{S}}$ a CF-perturbation of $(\widehat{\mathcal{U}}, \mathcal{K})$. We assume that:

(1) $\widehat{f}$ is strongly submersive with respect to $\widehat{\mathfrak{S}}$.
(2) $\mathrm{Supp}(\widehat{h}) \cap Z$ is a compact subset of $\overset{\circ}{Z}$.■

Choice 7.78 In Situation 7.77 we make the following choices:

(1) A triple of support systems $\mathcal{K}^1, \mathcal{K}^2$ with $\mathcal{K}^1 < \mathcal{K}^2 < \mathcal{K}^3 = \mathcal{K}$.
(2) We take an open neighborhood $\mathfrak{U}(X)$ of Z in $|\mathcal{K}|$ such that the conclusion of Lemma 7.75 holds.
(3) A constant $\delta > 0$ such that:
    (a) Define $\mathcal{K}^1_{\mathfrak{p}}(2\delta)$, $\Omega_{\mathfrak{p}}(\mathcal{K}, \delta)$ replacing $\mathcal{K}, \mathcal{K}^+$ by $\mathcal{K}^1, \mathcal{K}^2$ in (7.30), (7.31). Then $\Omega_{\mathfrak{p}}(\mathcal{K}, \delta)$ is relatively compact in $\mathfrak{U}(Z) \cap \mathcal{K}^1_{\mathfrak{p}}(2\delta)$.
    (b) There exists a partition of unity $\{\chi_{\mathfrak{p}}\}$ of $(X, Z, \widehat{\mathcal{U}}, \mathcal{K}^1, \delta)$. (Proposition 7.68.)
    (c) $\mathcal{K}^1(2\delta) < \mathcal{K}^2$.
    (d) δ satisfies the conclusion of Lemma 7.74.
    (e) We put

$$\delta_0 = \inf\{d(\mathcal{K}^2_{\mathfrak{p}}, \mathcal{K}^2_{\mathfrak{q}}) \mid \text{neither } \mathfrak{p} \le \mathfrak{q} \text{ nor } \mathfrak{q} \le \mathfrak{p}\},$$

    where we use the metric d of $|\mathcal{K}|$. Then $\delta < \delta_0/2$.[7]

(4) We take a partition of unity $\{\chi_{\mathfrak{p}}\}$ of $(X, Z, \widehat{\mathcal{U}}, \mathcal{K}^1, \delta)$ as in (3)(b).

Definition 7.79 In Situation 7.77, we make Choice 7.78. We define a smooth differential form $\widehat{f}!(\widehat{h}; \widehat{\mathfrak{S}^{\epsilon}})$ on the manifold M by (7.38). We call it the *pushout* or *integration along the fiber* of $\widehat{h}$ by $(\widehat{f}, \widehat{\mathfrak{S}^{\epsilon}})$:[8]

$$\widehat{f}!(\widehat{h}; \widehat{\mathfrak{S}^{\epsilon}}) = \sum_{\mathfrak{p} \in \mathfrak{P}} f_{\mathfrak{p}}!(\chi_{\mathfrak{p}} h_{\mathfrak{p}}; \mathfrak{S}^{\epsilon}_{\mathfrak{p}}|_{\mathfrak{U}(Z) \cap \mathcal{K}^1_{\mathfrak{p}}(2\delta)}). \tag{7.38}$$

We note that the restriction of $\chi_{\mathfrak{p}} h_{\mathfrak{p}}$ to $\mathcal{K}^1_{\mathfrak{p}}(2\delta)$ has compact support in $\mathrm{Int}\, \mathcal{K}^1_{\mathfrak{p}}(2\delta)$. Therefore the right hand side of (7.38) makes sense. The degree is given by

$$\deg \widehat{f}!(\widehat{h}; \widehat{\mathfrak{S}^{\epsilon}}) = \deg \widehat{h} + \dim M - \dim \widehat{\mathcal{U}}. \tag{7.39}$$

[7]The existence of such $\delta_0 > 0$ follows from Definition 3.15 (6).

[8]The integration here is taken on the *virtual fundamental chain*.

Definition 7.80 Let $F_a : (0, \epsilon_a) \to \mathscr{X}$ be a family of maps parametrized by $a \in \mathscr{B}$. We say that F_a is *independent of the choice of a in the sense of* ♠ if the following holds:

♠ For $a_1, a_2 \in \mathscr{B}$ there exists $0 < \epsilon_{a_1,a_2} < \min\{\epsilon_{a_1}, \epsilon_{a_2}\}$ such that $F_{a_1}(\epsilon) = F_{a_2}(\epsilon)$ for all $\epsilon < \epsilon_{a_1,a_2}$.

Proposition 7.81 *In Situation 7.77 the pushout* $\widehat{f}!(\widehat{h}; \widehat{\mathfrak{S}^\epsilon})$ *is independent of Choice 7.78 in the sense of* ♠.

Remark 7.82 However, $\widehat{f}!(\widehat{h}; \widehat{\mathfrak{S}^\epsilon})$ depends on the choices of ϵ and $\widehat{\mathfrak{S}}$.

Proof of Proposition 7.81 We first show the independence of $\mathfrak{U}(Z)$. Lemma 7.76 implies that if $\mathfrak{U}'(Z) \subset \mathfrak{U}(Z)$ is another open neighborhood of Z then, for sufficiently small ϵ, the value of the right hand side of (7.38) does not change if we replace $\mathfrak{U}(Z)$ by $\mathfrak{U}'(Z)$. (We use Situation 7.77 (2) also here.) This implies independence of $\mathfrak{U}(Z)$. Moreover it implies that we can always replace $\mathfrak{U}(Z)$ by a smaller open neighborhood of Z.

We next show independence of $\mathcal{K}^2$. Let $\mathcal{K}^{2\prime}$ be alternative choices of $\mathcal{K}^2$. We take $\mathcal{K}^{2\prime\prime}_{\mathfrak{p}} = \mathcal{K}^2_{\mathfrak{p}} \cup \mathcal{K}^{2\prime}_{\mathfrak{p}}$. Then $\mathcal{K}^{2\prime\prime}$ is a support system. Note in Definition 7.79, the support system $\mathcal{K}^2$ appears only when we apply Lemma 7.75 to obtain $\mathfrak{U}(Z)$ and Lemma 7.74 to obtain δ. Since we can always replace $\mathfrak{U}(Z)$ by a smaller neighborhood of Z (as far as $\epsilon > 0$ is sufficiently small) and since we do not need to change δ in Lemma 7.74 when we replace $\mathcal{K}^2$ by $\mathcal{K}^{2\prime\prime} \supset \mathcal{K}^2$, it follows that we obtain the same number in (7.38) if we replace $\mathcal{K}^2$ or $\mathcal{K}^{2\prime}$ by $\mathcal{K}^{2\prime\prime}$, as far as $\epsilon > 0$ is sufficiently small. This implies independence of $\mathcal{K}^2$.

It remains to prove the independence of $\mathcal{K}^1$ and of $\{\chi_{\mathfrak{p}}\}$, δ. We will prove the independence for this case below.

Let $\mathcal{K}^1_{\mathfrak{p}}$, $\chi_{\mathfrak{p}}$, δ and $\mathcal{K}^{1\prime}_{\mathfrak{p}}$, $\chi'_{\mathfrak{p}}$, δ' be two such choices. We take $\mathcal{K}^{1\prime\prime}_{\mathfrak{p}} = \mathcal{K}^1_{\mathfrak{p}} \cup \mathcal{K}^{1\prime}_{\mathfrak{p}}$. Then $\mathcal{K}^{1\prime\prime} < \mathcal{K}^2 < \mathcal{K}^3 = \mathcal{K}$. We can also take $\delta'' > 0$ and a partition of unity $\{\chi''_{\mathfrak{p}}\}$ of $(X, Z, \widehat{\mathcal{U}}, \mathcal{K}^{1\prime\prime}, \delta'')$. So it suffices to prove that the pushout defined by $\{\mathcal{K}^1_{\mathfrak{p}}\}$, $\{\chi_{\mathfrak{p}}\}$, δ coincides with the one defined by $\{\mathcal{K}^{1\prime\prime}_{\mathfrak{p}}\}$, $\{\chi''_{\mathfrak{p}}\}$, δ'' and that the pushout defined by $\{\mathcal{K}^{1\prime}_{\mathfrak{p}}\}$, $\{\chi'_{\mathfrak{p}}\}$, δ' coincides with the one defined by $\{\mathcal{K}^{1\prime\prime}_{\mathfrak{p}}\}$, $\{\chi''_{\mathfrak{p}}\}$, δ''. In other words, we may assume $\mathcal{K}^1_{\mathfrak{p}} \subset \mathcal{K}^{1\prime}_{\mathfrak{p}}$. We will prove the independence in this case.

We observe that

$$\widehat{f}!(\widehat{h}_1 + \widehat{h}_2; \widehat{\mathfrak{S}^\epsilon}) = \widehat{f}!(\widehat{h}_1; \widehat{\mathfrak{S}^\epsilon}) + \widehat{f}!(\widehat{h}_2; \widehat{\mathfrak{S}^\epsilon}) \tag{7.40}$$

as far as we use the same partition of unity in all these three terms. (This is a consequence of Lemma 7.34 (1).) We take $\mathfrak{p}_0 \in \mathfrak{P}$ and put

$$\widehat{h}_0 = \chi'_{\mathfrak{p}_0} \widehat{h}. \tag{7.41}$$

In view of (7.40) we find that to prove Proposition 7.81 it suffices to show the next formula:

$$f_{\mathfrak{p}_0}!((\widehat{h}_0)_{\mathfrak{p}_0}; \mathfrak{S}^{\epsilon}_{\mathfrak{p}_0}|_{\mathfrak{U}(Z)\cap\mathcal{K}^{1\prime}_{\mathfrak{p}_0}(2\delta')}) = \sum_{\mathfrak{p}} f_{\mathfrak{p}}!((\chi_{\mathfrak{p}}\widehat{h}_0)_{\mathfrak{p}}; \mathfrak{S}^{\epsilon}_{\mathfrak{p}}|_{\mathfrak{U}(Z)\cap\mathcal{K}^{1}_{\mathfrak{p}}(2\delta)}). \tag{7.42}$$

By taking $\epsilon > 0$ small, we may assume $\sum \chi_{\mathfrak{p}} = 1$ on the intersection of $\mathfrak{U}(Z)$ and the support set $\Pi((\widehat{\mathfrak{S}^{\epsilon}_{\mathfrak{p}_0}})^{-1}(0))$. (This is a consequence of Lemma 7.76, Definition 7.65 (3) and Situation 7.77 (2).) Therefore, to prove (7.42) it suffices to prove the next formula for each $\mathfrak{p}$:

$$f_{\mathfrak{p}_0}!((\chi_{\mathfrak{p}}\widehat{h}_0)_{\mathfrak{p}_0}; \mathfrak{S}^{\epsilon}_{\mathfrak{p}_0}|_{\mathfrak{U}(X)\cap\mathcal{K}^{1\prime}_{\mathfrak{p}_0}(2\delta')}) = f_{\mathfrak{p}}!((\chi_{\mathfrak{p}}\widehat{h}_0)_{\mathfrak{p}}; \mathfrak{S}^{\epsilon}_{\mathfrak{p}}|_{\mathfrak{U}(X)\cap\mathcal{K}^{1}_{\mathfrak{p}}(2\delta)}), \tag{7.43}$$

whose proof is now in order.

In the case $\mathfrak{p}_0 = \mathfrak{p}$, (7.43) follows from

$$\operatorname{Supp}(\chi_{\mathfrak{p}} h_{0,\mathfrak{p}_0}) = \operatorname{Supp}(\chi_{\mathfrak{p}_0} h_{0,\mathfrak{p}_0}) \subseteq \mathcal{K}^{1}_{\mathfrak{p}_0}(2\delta) \cap \mathcal{K}^{1\prime}_{\mathfrak{p}_0}(2\delta').$$

If neither $\mathfrak{p} \le \mathfrak{p}_0$ nor $\mathfrak{p}_0 \le \mathfrak{p}$, then both sides of (7.43) are zero because

$$\operatorname{Supp}(\chi_{\mathfrak{p}} h_{0,\mathfrak{p}_0}) \subseteq \Omega_{\mathfrak{p}_0}(\mathcal{K}^{1\prime}, \delta') \cap \Omega_{\mathfrak{p}}(\mathcal{K}^{1}, \delta) = \emptyset. \tag{7.44}$$

Note the second equality of (7.44) is a consequence of Choice 7.78 (3)(e).

We will discuss the other two cases below.

(Case 1): $\mathfrak{p} > \mathfrak{p}_0$.

By definition $\Omega_{\mathfrak{p}_0}(\mathcal{K}^{1\prime}, \delta') \subset B_{\delta'}(\mathcal{K}^{1\prime}_{\mathfrak{p}_0})$. Therefore by (7.41) the support of $\widehat{h}_0$ is in $B_{\delta'}(\mathcal{K}^{1\prime}_{\mathfrak{p}_0})$. By taking $\epsilon > 0$ small, Lemma 7.74 implies

$$\operatorname{Supp}(\widehat{h}_0) \cap \Pi((\widehat{\mathfrak{S}^{\epsilon}})^{-1}(0)) \subset \mathcal{K}^{2\prime}_{\mathfrak{p}_0} \cap B_{\delta'}(\mathcal{K}^{1\prime}_{\mathfrak{p}_0}) \subset \mathcal{K}^{1\prime}_{\mathfrak{p}_0}(2\delta').$$

Therefore

$$\operatorname{Supp}(\chi_{\mathfrak{p}}\widehat{h}_0) \cap \Pi((\widehat{\mathfrak{S}^{\epsilon}})^{-1}(0)) \cap \mathfrak{U}(Z) \subseteq \mathcal{K}^{1\prime}_{\mathfrak{p}_0}(2\delta') \cap \mathcal{K}^{1}_{\mathfrak{p}}(2\delta) \cap \mathfrak{U}(Z). \tag{7.45}$$

Then (7.43) follows from Definition 7.49 (2)(3).

(Case 2): $\mathfrak{p}_0 > \mathfrak{p}$.

By definition $\Omega_{\mathfrak{p}}(\mathcal{K}^{1}, \delta) \subset B_{\delta}(\mathcal{K}^{1}_{\mathfrak{p}})$. Therefore the support of $\chi_{\mathfrak{p}}\widehat{h}_0$ is in $B_{\delta}(\mathcal{K}^{1}_{\mathfrak{p}})$. Therefore by taking $\epsilon > 0$ small, Lemma 7.74 implies

$$\operatorname{Supp}(\chi_{\mathfrak{p}}\widehat{h}_0) \cap \Pi((\widehat{\mathfrak{S}^{\epsilon}})^{-1}(0)) \subset \mathcal{K}^{2}_{\mathfrak{p}} \cap B_{\delta}(\mathcal{K}^{1}_{\mathfrak{p}_0}) \subset \mathcal{K}^{1}_{\mathfrak{p}}(2\delta).$$

It implies (7.45). Then (7.43) follows from Definition 7.49 (2)(3).

The proof of Proposition 7.81 is complete. □

Remark 7.83 The pushout (7.38) is independent of the choice of the support system $\mathcal{K} = \mathcal{K}_3$ appearing in Situation 7.77, as long as $\widehat{\mathfrak{S}}$ and $\widehat{h}$ are defined on it. In fact $\mathcal{K}$ does not appear in the definition.

Lemma 7.84 *Let h, h_1, h_2 be differential forms on $(X, Z; \widehat{\mathcal{U}})$ and $c_1, c_2 \in \mathbb{R}$.*

(1) $\widehat{f}!(c_1\widehat{h}_1 + c_2\widehat{h}_2; \widehat{\mathfrak{S}^\epsilon}) = c_1\widehat{f}!(\widehat{h}_1; \widehat{\mathfrak{S}^\epsilon}) + c_2\widehat{f}!(\widehat{h}_2; \widehat{\mathfrak{S}^\epsilon})$.
(2) *If $\rho \in \Omega^*(M)$, then* $\widehat{f}!(\widehat{f}^*(\rho) \wedge \widehat{h}; \widehat{\mathfrak{S}^\epsilon}) = \rho \wedge \widehat{f}!(\widehat{h}; \widehat{\mathfrak{S}^\epsilon})$.

Proof Formula (1) follows from Lemma 7.34 (1). Formula (2) is immediate from the definition. □

We now define the smooth correspondence map.

Construction 7.85 Let $((X, \widehat{\mathcal{U}}), \widehat{f_s}, \widehat{f_t})$ be a smooth correspondence from M_s to M_t (Definition 7.1). We construct objects as in Situation 7.77 as follows.

We put $Z = X$. We take a good coordinate system $\widehat{\mathcal{U}}$ compatible with $\widehat{\mathcal{U}}$ such that $\widehat{f_s}$ and $\widehat{f_t}$ are pullbacks of $\widehat{f}_s : (X; \widehat{\mathcal{U}}) \to M_s$ and $\widehat{f}_t : (X; \widehat{\mathcal{U}}) \to M_t$, respectively. Moreover we may take $\widehat{f}_t$ so that it is weakly submersive. (Proposition 6.32 (2).)

We take a CF-perturbation $\widehat{\mathfrak{S}}$ of $(X; \widehat{\mathcal{U}})$ such that $\widehat{f}_t$ is strongly submersive with respect to $\widehat{\mathcal{U}}$. (Theorem 7.51 (2).)

Let $\mathcal{K}$ be a support system of $\widehat{\mathcal{U}}$. We consider the differential form $\widehat{f}_s^* h$ on $(X, \widehat{\mathcal{U}})$.

We denote the correspondence by

$$\mathfrak{X} = ((X; \widehat{\mathcal{U}}); \widehat{f_s}, \widehat{f_t}).$$

Definition 7.86 Using Construction 7.85, we define

$$\mathrm{Corr}_{(\mathfrak{X}, \widehat{\mathfrak{S}^\epsilon})}(h) = (\widehat{f}_t)!(\widehat{f}_s^* h; \widehat{\mathfrak{S}^\epsilon}). \tag{7.46}$$

We call the linear map

$$\mathrm{Corr}_{(\mathfrak{X}, \widehat{\mathfrak{S}^\epsilon})} : \Omega^*(M_s) \to \Omega^{*+\dim M_t - \dim \widehat{\mathcal{U}}}(M_t)$$

the *smooth correspondence map associated to* $\widehat{\mathfrak{X}} = ((X; \widehat{\mathcal{U}}); \widehat{f_s}, \widehat{f_t})$ and $\widehat{\mathfrak{S}^\epsilon}$.

Remark 7.87

(1) Proposition 7.81 implies that the right hand side of (7.46) is independent of various choices appearing in Definition 7.79 if $\epsilon > 0$ is sufficiently small. However, it *does* depend on $\widehat{\mathfrak{S}}$ and $\epsilon > 0$. So we keep the symbol ϵ in the notation $\widehat{\mathfrak{S}^\epsilon}$ of the left hand side of (7.46).
(2) There seems to be no way to define a smooth correspondence in the way that it becomes independent of the choices *at the chain level*. This is related to an important point of the story, that is, the construction of various structures from

a system of moduli spaces are well-defined only up to homotopy equivalence and only as a whole. (However, we note that the method of [Jo1] seems to be a way to minimize this dependence.) This is the fundamental issue which appears in *any* approach. For example, it should still exist in the infinite-dimensional approach to constructing virtual fundamental chains, such as those by Li–Tian [LiTi2], Liu–Tian [LiuTi], Siebert [Sie], Chen–Tian [CT], Chen–Li–Wang [CLW] or Hofer–Wyscoki–Zehnder [HWZ1].

(3) Dependence on the good coordinate system $\widehat{\mathcal{U}}$ and the other choices made in Construction 7.85 will be discussed in Chap. 9.

In Proposition 7.81, we have proved independence of the pushout of various choices for sufficiently small $\epsilon > 0$. On the other hand, how small ϵ must be depends on our good coordinate system and CF-perturbations on it. In certain situations appearing in applications, we need to estimate this required smallness of ϵ uniformly from below when our CF-perturbations vary in a certain family. The next proposition can be used for such a purpose.

Definition 7.88 Let $F_{\sigma,a} : (0, \epsilon_a) \to \mathscr{X}$ be a family of maps parametrized by $a \in \mathscr{B}$ and $\sigma \in \mathscr{A}$. We say that $F_{\sigma,a}$ is *uniformly independent of the choice of a in the sense of ♣* if the following holds:

♣ For $a_1, a_2 \in \mathscr{B}$ there exists $0 < \epsilon_{a_1,a_2} < \min\{\epsilon_{a_1}, \epsilon_{a_2}\}$ independent of σ such that $F_{\sigma,a_1}(\epsilon) = F_{\sigma,a_2}(\epsilon)$ for $0 < \epsilon < \epsilon_{a_1,a_2}$ and any $\sigma \in \mathscr{A}$.

Proposition 7.89 *We assume $\{\widehat{\mathfrak{S}_\sigma} \mid \sigma \in \mathscr{A}\}$ is a uniform family of CF-perturbations parametrized by $\sigma \in \mathscr{A}$ in the sense of Definition 7.60. Suppose $\widehat{f} : (X; \widehat{\mathcal{U}}) \to M$ is strongly submersive with respect to $\widehat{\mathfrak{S}_\sigma}$ for any $\sigma \in \mathscr{A}$. Then we can make Choice 7.78 in a way independent of σ.*

Moreover the pushout $\widehat{f}!(\widehat{h}; \widehat{\mathfrak{S}^\epsilon_\sigma})$ of Proposition 7.81 is uniformly independent of Choice 7.78 in the sense of ♣.

Proof From the proof of Proposition 7.81, the proof of Proposition 7.89 follows from the next lemma. □

Lemma 7.90 *Let $\{\widehat{\mathfrak{S}_\sigma} \mid \sigma \in \mathscr{A}\}$ be a uniform family of CF-perturbations. Then the following holds:*

(1) *In Lemma 7.74 the constants δ_0 and ϵ_0 can be taken independent of σ.*
(2) *In Lemma 7.75 the set $\mathfrak{U}(Z)$ and the constant ϵ_0 can be taken independent of σ.*
(3) *In Lemma 7.76 the set $\mathfrak{U}(Z)$ and the constant ϵ_0 can be taken independent of σ. Moreover*

$$\limsup_{\epsilon \to 0} \left\{ d_H \left(X, \bigcup_{\mathfrak{p}} (\Pi((\widehat{\mathfrak{S}^\epsilon_{\sigma,\mathfrak{p}}})^{-1}(0) \cap \overset{\circ}{\mathcal{K}^2_{\mathfrak{p}}} \cap \mathfrak{U}(Z)) \right) \middle| \sigma \in \mathscr{A} \right\} = 0. \tag{7.47}$$

Proof This is an immediate consequence of the proof of Lemmas 7.74, 7.75 and 7.76. □

Chapter 8
Stokes' Formula

8.1 Boundaries and Corners II

In this chapter, we state and prove Stokes' formula. We first discuss the notion of a boundary or a corner of an orbifold and of a Kuranishi structure in more detail. The discussion below is a detailed version of [FOOO4, the last paragraph of page 762]. See also [Jo1, page 11]. [Jo3] gives a systematic account on this issue.

Let U be an orbifold with boundary and corners. We defined its corner structure stratification $S_k(U)$ and $\overset{\circ}{S}_k(U)$ in Definition 4.13. Note $\overset{\circ}{S}_k(U)$ is an orbifold of dimension $\dim U - k$ without boundary. However, we also note that we *cannot* find a structure of orbifold with corners on $S_k(U)$ such that $\overset{\circ}{S}_0(S_k(U)) = \overset{\circ}{S}_k(U)$.

Example 8.1 Let $U = \mathbb{R}^2_{\ge 0}$. Then $S_1(U)$ is homeomorphic to $\mathbb{R}$ and $S_2(U)$ is one point identified with $0 \in \mathbb{R} = S_1(U)$ (Fig. 8.1).

To obtain an orbifold with corners from $S_1(U)$ we need to modify it at its boundaries and corners. Let us first consider the case of manifolds.

Lemma 8.2 *Suppose U is a manifold with corners. Then there exists a manifold with corners, denoted by ∂U, and a map $\pi : \partial U \to S_1(U)$ with the following properties:*

(1) *For each k, π induces a map*

$$\overset{\circ}{S}_k(\partial U) \to \overset{\circ}{S}_{k+1}(U). \tag{8.1}$$

(2) *The map (8.1) is a $(k+1)$-fold covering map.*
(3) *π is a smooth map $\partial U \to U$.*

K. Fukaya et al., *Kuranishi Structures and Virtual Fundamental Chains*, Springer Monographs in Mathematics, https://doi.org/10.1007/978-981-15-5562-6_8

Remark 8.3 The smoothness claimed in Lemma 8.2 (3) is defined as follows.[1] Let U be any smooth manifold with corners. We can find a smooth manifold without boundary or corner U^+ of the same dimension as U and an embedding $U \to U^+$, such that for each point $p \in U$ there exists a coordinate of U^+ at p by which U is identified with an open subset of $[0, 1)^{\dim U}$ by a diffeomorphism from U^+ onto an open subset of $\mathbb{R}^{\dim U}$. Then a map $F : U_1 \to U_2$ between two manifolds with corners is said to be smooth if F extends to F^+ that is a smooth map from a neighborhood of U_1 in U_1^+ to U_2^+.

Proof We fix a Riemannian metric on U so that each ϵ-ball $B_\epsilon(p)$ is convex. Let $p \in \overset{\circ}{S}_k(U)$. We consider $\alpha \in \pi_0(B_\epsilon(p) \cap \overset{\circ}{S}_1(U))$. The set of all such pairs (p, α) with $k \geq 1$ is our ∂U. The map $(p, \alpha) \to p$ is the map π.

By identifying U locally with $[0, \infty)^n$, it is easy to construct the structure of manifold with corners on ∂U and prove that they have the required properties. □

Definition 8.4 Let U be an orbifold. We call ∂U a *normalized boundary* of U and $\pi : \partial U \to S_1(U)$ the *normalization map.*[2]

Lemma 8.5

(1) *Let U and U' be as in Lemma* 8.2 *and $F : U \to U'$ a diffeomorphism. Then F uniquely induces a diffeomorphism*

$$F^\partial : \partial U \to \partial U'$$

such that $\pi \circ F^\partial = F \circ \pi'$.

(2) *Suppose a finite group Γ acts on U, where U is as in Lemma* 8.2. *Suppose also that each connected component of $\overset{\circ}{S}_k(U)/\Gamma$ is an effective orbifold for each k. Then Γ acts on ∂U so that π is Γ equivariant and each connected component of $\overset{\circ}{S}_k(\partial U)/\Gamma$ is an effective orbifold for each k.*

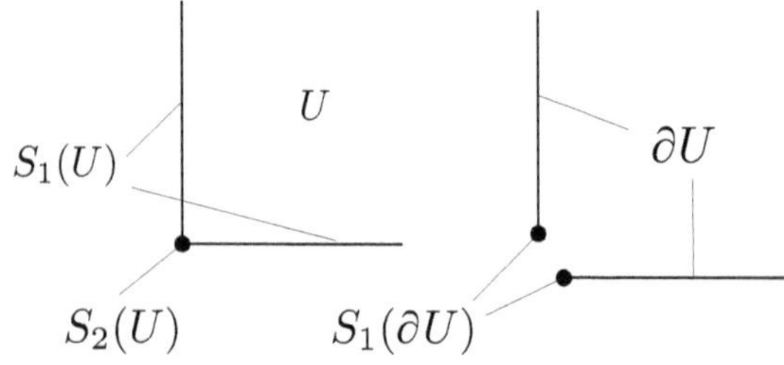

Fig. 8.1 Normalized boundary

[1] This is a standard definition going back to Whitney.

[2] In Sect. 24.3 the normalized boundary ∂U is written as $\widehat{S}_1(U)$. More generally, the notion of a normalized corner of codimension k, denoted by $\widehat{S}_k(U)$, is introduced in Definition 24.18 (see the proof of Proposition 24.17).

(3) *Let U be as in* Lemma 8.2 *and U' its open subset. Then there exists an open embedding $\partial U' \to \partial U$ which commutes with π.*

Proof (1) is immediate from the construction. Then the uniqueness implies (2) and (3). □

Now we consider the case of an orbifold U. We cover U by orbifold charts $\{(V_i, \Gamma_i, \phi_i)\}$. We apply Lemma 8.2 to V_i and obtain ∂V_i. Then Γ_i action on V_i induces one on ∂V_i by Lemma 8.5 (2). We thus obtain orbifolds $\partial V_i / \Gamma_i$. Using Lemma 8.5 (1) and (3) we can glue $\partial V_i / \Gamma_i$ for various i and obtain ∂U. We obtain also a map $\pi : \partial U \to S_1(U)$. It induces a map $\overset{\circ}{S}_k(\partial U) \to \overset{\circ}{S}_{k+1}(U)$.

Remark 8.6

(1) We note that the map $\overset{\circ}{S}_k(\partial U) \to \overset{\circ}{S}_{k+1}(U)$ is a $(k+1)$-fold orbifold covering of orbifolds in the sense we will define in Chap. 24. See Lemma 24.16.
(2) In particular, $\overset{\circ}{S}_0(\partial U) \to \overset{\circ}{S}_1(U)$ is a diffeomorphism of orbifolds.
(3) We also note that $\overset{\circ}{S}_k(\partial U) \to \overset{\circ}{S}_{k+1}(U)$ is not necessarily a $k+1$ to 1 map set-theoretically. The following is a counterexample. Let

$$U = (\mathbb{R}^2_{\geq 0} \times \mathbb{R})/\mathbb{Z}_2,$$

where the action is $(a, b, c) \mapsto (b, a, -c)$. Then $\partial U \cong \mathbb{R}_{\geq 0} \times \mathbb{R}$, $S_1(\partial U) \cong \mathbb{R}$, $S_2(U) \cong \mathbb{R}/\mathbb{Z}_2$ and the map π is the canonical projection $\mathbb{R} \to \mathbb{R}/\mathbb{Z}_2$ on $S_1(\partial U)$. So it is generically 2 to 1 but is 1 to 1 at 0.

Next we consider the case of a Kuranishi structure. We recall the following notation from Definition 4.16:

$$S_k(X, Z; \widehat{\mathcal{U}}) = \{p \in Z \mid o_p \in S_k(U_p)\}, \qquad \overset{\circ}{S}_k(X, Z; \widehat{\mathcal{U}}) = \{p \in Z \mid o_p \in \overset{\circ}{S}_k(U_p)\},$$

where $\widehat{\mathcal{U}}$ is a Kuranishi structure of $Z \subseteq X$ and

$$S_k(X, Z; \widehat{\mathcal{U}}) = \{p \in Z \mid \exists \mathfrak{p}\, \exists x \in S_k(U_{\mathfrak{p}}), \text{st } s_{\mathfrak{p}}(x) = 0, \psi_{\mathfrak{p}}(x) = p\},$$

$$\overset{\circ}{S}_k(X, Z; \widehat{\mathcal{U}}) = \{p \in Z \mid \exists \mathfrak{p}\, \exists x \in \overset{\circ}{S}_k(U_{\mathfrak{p}}), \text{st } s_{\mathfrak{p}}(x) = 0, \psi_{\mathfrak{p}}(x) = p\},$$

where $\widehat{\mathcal{U}}$ is a good coordinate system of $Z \subseteq X$.

Remark 8.7 We can rewrite the set $S_k(X, Z; \widehat{\mathcal{U}})$ as

$$S_k(X, Z; \widehat{\mathcal{U}}) = \{p \in Z \mid \forall \mathfrak{p} \forall x \in U_{\mathfrak{p}}, \text{“}s_{\mathfrak{p}}(x) = 0, \psi_{\mathfrak{p}}(x) = p \Rightarrow x \in S_k(U_{\mathfrak{p}})\text{”}\}.$$

A similar remark applies to $\overset{\circ}{S}_k(X, Z; \widehat{\mathcal{U}})$.

Lemma-Definition 8.8

(1) *Any compact subset of the space* $\overset{\circ}{S}_k(X, Z; \widehat{\mathcal{U}})$ *(resp.* $\overset{\circ}{S}_k(X, Z; \widetriangle{\mathcal{U}})$*) has a Kuranishi structure without boundary (resp. a good coordinate system without boundary) and of dimension* $\dim(X, Z; \widehat{\mathcal{U}}) - k$ *(resp.* $\dim(X, Z; \widetriangle{\mathcal{U}}) - k$*).*

(2) *There exist a space* ∂X *with a map to* X *such that* $\partial Z = \partial X \times_X Z$*, a relative K-space with corners* $\partial(X, Z; \widehat{\mathcal{U}})$ *(resp.* $\partial(X, Z; \widetriangle{\mathcal{U}})$*) whose underlying topological spaces are* $(\partial X, \partial Z)$ *and a continuous map between their underlying topological spaces* $\pi : \partial Z \to S_1(X, Z; \widehat{\mathcal{U}})$ *(resp.* $\pi : \partial Z \to S_1(X, Z; \widetriangle{\mathcal{U}})$*) such that the following holds. We call* $\partial(X, Z; \widehat{\mathcal{U}})$*,* $\partial(X, Z; \widetriangle{\mathcal{U}})$ *the normalized boundary of* $(X, Z; \widehat{\mathcal{U}})$*,* $(X, Z; \widetriangle{\mathcal{U}})$*, respectively.*

   (a) *If* $\pi(\tilde{p}) = p$*,* $\tilde{p} \in \partial Z$*,* $p \in Z$*, then the Kuranishi neighborhood of* $\tilde{p}$ *is obtained by restricting* $\mathcal{U}_p$ *to* ∂U_p*, which is as in Definition 8.4.*[3]

   (b) *The coordinate charts of* $\partial(X, Z; \widetriangle{\mathcal{U}})$ *are obtained by restricting* $\mathcal{U}_{\mathfrak{p}}$ *to* $\partial U_{\mathfrak{p}}$*.*

   (c) *The coordinate change of* $\partial(X, Z; \widehat{\mathcal{U}})$ *(resp.* $\partial(X, Z; \widetriangle{\mathcal{U}})$*) is obtained by restricting one of* $\partial \mathcal{U}_p$ *(resp.* $\partial \mathcal{U}_{\mathfrak{p}}$*).*

   (d) *The restriction of* π *induces a map*

$$\overset{\circ}{S}_0(\partial(X, Z; \widehat{\mathcal{U}})) \to \overset{\circ}{S}_1(X, Z; \widehat{\mathcal{U}})$$

   that is an isomorphism of Kuranishi structures.

   (e) *The restriction of* π *induces a map*

$$\overset{\circ}{S}_0(\partial(X, Z; \widetriangle{\mathcal{U}})) \to \overset{\circ}{S}_1(X, Z; \widetriangle{\mathcal{U}})$$

   that is an isomorphism of good coordinate systems.

   (f) *In the case of the Kuranishi structure and* $Z \neq X$*, we need to replace* $\widehat{\mathcal{U}}$ *by its open substructure.*

(3) *Various kinds of embeddings between Kuranishi structures and/or good coordinate systems induce embeddings of their normalized boundaries.*

Proof We first prove (2). Let $\mathcal{U} = (U, \mathcal{E}, s, \psi)$ be a Kuranishi chart. We restrict $\mathcal{E}$ and s to ∂U and obtain $\partial U, \mathcal{E}^{\partial}, s^{\partial}$. We will define underlying topological spaces $\partial X, \partial Z$, parametrization ψ^{∂} and the coordinate change.

Let $\Phi_{21} = (U_{21}, \varphi_{21}, \widehat{\varphi}_{21}) : \mathcal{U}_1 \to \mathcal{U}_2$ be a coordinate change of Kuranishi charts. We remark that we require the following condition for an embedding of orbifolds $\varphi_{21} : U_1 \to U_2$:

$$\varphi_{21}(S_k(U_1)) \subset S_k(U_2), \qquad \overset{\circ}{S}_k(U_1) = \varphi_{21}^{-1}(\overset{\circ}{S}_k(U_2)). \tag{8.2}$$

[3]Here and hereafter we use the term 'restriction' of objects on U_p to ∂U_p even though the map $\partial U_p \to U_p$ from the normalized boundary ∂U_p may not be injective.

We can generalize Lemma 8.5 (3) to the case when U_1, U_2 are orbifolds. Moreover by the condition (8.2) we can generalize Lemma 8.5 (1) to the embedding of orbifolds with corners. Thus φ_{21} induces $\varphi_{21}^{\partial} : \partial U_1 \to \partial U_2$. In the same way, $\widehat{\varphi}_{21} : \mathcal{E}_1 \to \mathcal{E}_2$ induces $\widehat{\varphi}_{21}^{\partial} : \mathcal{E}_1^{\partial} \to \mathcal{E}_2^{\partial}$.

Thus the data consisting of the coordinate change of the Kuranishi charts given as in (2) (a),(b) are defined by (2) (c), except the underlying topological space ∂X, ∂Z and parametrizations ψ.

Below we will construct ∂X, ∂Z and ψ.

We first consider the case of a good coordinate system and $X = Z$. Let $\widehat{\mathcal{U}} = (\mathfrak{P}, \{\mathcal{U}_{\mathfrak{p}}\}, \{\Phi_{\mathfrak{pq}}\})$. We consider $\partial U_{\mathfrak{p}}$ and $\varphi_{\mathfrak{pq}}^{\partial}$, defined as above. We glue the spaces $\partial U_{\mathfrak{p}}$ by $\varphi_{\mathfrak{pq}}^{\partial}$ and obtain a topological space $|\partial\widehat{\mathcal{U}}|$. The zero sets of the Kuranish maps $s_{\mathfrak{p}}^{\partial}$ on $\partial U_{\mathfrak{p}}$ are glued to define a subspace of $|\partial\widehat{\mathcal{U}}|$, which we define to be ∂X. ∂X is Hausdorff and metrizable.[4] We define $\psi_{\mathfrak{p}}^{\partial} : (s_{\mathfrak{p}}^{\partial})^{-1}(0) \to \partial X$ by mapping a point of $(s_{\mathfrak{p}}^{\partial})^{-1}(0)$ to its equivalence class. Then $\mathcal{U}_{\mathfrak{p}}^{\partial} = (\partial U_{\mathfrak{p}}, \mathcal{E}_{\mathfrak{p}}^{\partial}, s_{\mathfrak{p}}^{\partial}, \psi_{\mathfrak{p}}^{\partial})$ is a Kuranishi chart of ∂X. We put $\partial U_{\mathfrak{pq}} = U_{\mathfrak{pq}} \cap \partial U_{\mathfrak{q}}$. Then $\Phi_{\mathfrak{pq}}^{\partial} = (\partial U_{\mathfrak{pq}}, \varphi_{21}^{\partial}, \widehat{\varphi}_{21}^{\partial})$ is a coordinate change $\mathcal{U}_{\mathfrak{p}}^{\partial} \to \mathcal{U}_{\mathfrak{q}}^{\partial}$. Thus we obtain a good coordinate system $\partial(X, Z; \widehat{\mathcal{U}})$ in the case $Z = X$.

Next we consider the case of a good coordinate system but $X \neq Z$. We glue the zero sets of a Kuranish map on $\partial U_{\mathfrak{p}}$ in the same way as above to obtain a topological space ∂X. (See Remark 8.9.) We define the subset $\partial Z \subset \partial X$ by

$$\partial Z = \bigcup_{\mathfrak{p} \in \mathfrak{P}} \{x \in \partial U_{\mathfrak{p}} \mid s_{\mathfrak{p}}^{\partial}(x) = 0, \psi_{\mathfrak{p}}(\pi(x)) \in Z\}.$$

Here we identify $\partial U_{\mathfrak{p}}$ with its image in $|\partial\widehat{\mathcal{U}}|$ and the union is taken in $|\partial\widehat{\mathcal{U}}|$. The rest of the proof is the same as the case of $X = Z$.

Finally we consider the case of a Kuranishi structure. We take a good coordinate system $\widehat{\mathcal{U}}$ compatible with $\widehat{\mathcal{U}}$ and use the case of a good coordinate system to define ∂X and ∂Z.

It now remains to define the parametrization $\psi_p^{\partial} : (s_p^{\partial})^{-1}(0) \to \partial X$. Let $\mathcal{U}_p$ be a Kuranishi neighborhood of p which is a part of the data of $\widehat{\mathcal{U}}$. In the case when the embedding $\widehat{\mathcal{U}} \to \widehat{\mathcal{U}}$ is strict, $(s_p^{\partial})^{-1}(0) \subset U_{\mathfrak{p}}$ for some $\mathfrak{p} \in \mathfrak{P}$. Therefore we obtain ψ_p^{∂} by restricting the parametrization map $\psi_{\mathfrak{p}}$ of the good coordinate system $\partial\widehat{\mathcal{U}}$.

In the case $Z \neq X$ we replace $\widehat{\mathcal{U}}$ by its open substructure $\widehat{\mathcal{U}_0}$ such that there exists a strict embedding $\widehat{\mathcal{U}_0} \to \widehat{\mathcal{U}}$ and apply the above construction.

Suppose $Z = X$. We define $\psi_p^{\partial} : (s_p^{\partial})^{-1}(0) \to \partial X$ (without taking open substructure) as follows. Let $x \in (s_p^{\partial})^{-1}(0) \subset s_p^{-1}(0)$ and $q = \psi_p(x) \in X$. We have $o_q \in U_q$ such that $\varphi_{pq}(o_q) = x$. Equation (8.2) implies $o_q \in \partial U_q$. Moreover $o_q \in \partial U_{0,q}$. (Here $U_{0,q}$ is the Kuranishi neighborhood of the open substructure $\widehat{\mathcal{U}_0}$.)

[4]To find a topology which is metrizable, we consider a support system $\mathcal{K}$ and use the fact that $\partial X = \bigcup \mathcal{K}_{\mathfrak{p}} \cap (s_{\mathfrak{p}}^{\partial})^{-1}(0)$ and [FOOO17, Proposition 2.11].

Therefore o_q may be regarded as an element of $\partial U_{\mathfrak{p}}$ for some $\mathfrak{p} \in \mathfrak{P}$. We define $\psi_p(x)$ to be the equivalence class of $o_q \in \partial U_{\mathfrak{p}}$, which is an element of ∂X.

Therefore the proof of the statement (2) is complete.

The statement (1) can be proved in the same way and the statement (3) is obvious from the definition. □

Remark 8.9 Here is a technical remark about the way to define underlying topological space ∂X in Lemma-Definition 8.8.

(1) Let $\widehat{\mathcal{U}}$ be a Kuranishi structure of $Z \subseteq X$. Then the parametrization $\psi_p : s_p^{-1}(0) \to X$ has X as a target space. So $\widehat{\mathcal{U}}$ is *not* a Kuranishi structure of Z itself.
(2) The data consisting of $\widehat{\mathcal{U}}$ contain enough information to determine which points of Z lie in the boundary. However, the data do not contain such information for the points of $X \setminus Z$ which are far away from Z.

    The space ∂X defined in the proof of Lemma-Definition 8.8 consists of points corresponding to the 'boundary points' of X that are sufficiently close to Z. Since the image of ψ_p, $p \in Z$ lies in a neighborhood of Z, we need only a neighborhood of Z in X to define the Kuranishi structure of $Z \subseteq X$. This is the reason why it suffices to define ∂X in a neighborhood of Z.
(3) On the other hand, as a consequence of (2), the topological space ∂X is not canonically determined from $(X, Z; \widehat{\mathcal{U}})$. For example, the following phenomenon happens. Let $\widehat{\mathcal{U}}$ be a Kuranishi structure of $Z_2 \subseteq X$ and $Z_1 \subset \overset{\circ}{Z}_2$. We restrict $\mathcal{U}$ to Z_1 to obtain $\mathcal{U}|_{Z_1}$. We consider $\partial(X, Z_1; \widehat{\mathcal{U}}|_{Z_1})$ and $\partial(X, Z_2; \widehat{\mathcal{U}})$. Let $(\partial_1 X, \partial Z_1)$ and $(\partial_2 X, \partial Z_2)$ be their underlying topological spaces. Then $\partial_1 X$ may not be the same as $\partial_2 X$.

There is no such issue in the absolute case $Z = X$. In the applications we know the case $Z \neq X$ appears only together with a means of defining ∂X given.

We note that all the arguments of Chap. 7 work for Kuranishi structures or good coordinate systems with boundary or corners.

The next lemma describes how to restrict CF-perturbations etc. to the normalized boundary.

Lemma 8.10 *Let $\widehat{\mathcal{U}}$ be a good coordinate system of $Z \subseteq X$, $\mathcal{K}^1, \mathcal{K}^2, \mathcal{K}^3$ a triple of support systems of $\widehat{\mathcal{U}}$ with $\mathcal{K}^1 < \mathcal{K}^2 < \mathcal{K}^3$, and $\widehat{\mathfrak{S}^\epsilon}$ a CF-perturbation of $(\widehat{\mathcal{U}}, \mathcal{K}^3)$.*

(1) *$\{\partial U_{\mathfrak{p}} \cap \mathcal{K}^i_{\mathfrak{p}}\}$ is a support system of $\partial(X, Z; \widehat{\mathcal{U}})$, which we denote by $\mathcal{K}^i_\partial$. $\mathcal{K}^1_\partial, \mathcal{K}^2_\partial, \mathcal{K}^3_\partial$ are support systems with $\mathcal{K}^1_\partial < \mathcal{K}^2_\partial < \mathcal{K}^3_\partial$. Here $\partial U_{\mathfrak{p}} \cap \mathcal{K}^i_{\mathfrak{p}} = \pi^{-1}(\mathcal{K}^i_{\mathfrak{p}}) \subset \partial U_{\mathfrak{p}}$.*

(2) (a) *For each $\mathfrak{p}$ there exists $\mathfrak{S}^\partial_{\mathfrak{p}}$ that is a CF-perturbation of $\mathcal{K}^3_{\partial,\mathfrak{p}} \subset \partial U_{\mathfrak{p}}$.*
    (b) *The restriction of $\mathfrak{S}^\partial_{\mathfrak{p}}$ to $\overset{\circ}{S}_0(\partial U_{\mathfrak{p}})$ is identified with the restriction of $\mathfrak{S}_{\mathfrak{p}}$ to $\overset{\circ}{S}_1(U_{\mathfrak{p}})$ by the diffeomorphism in Lemma 8.8 (2)(e).*
    (c) *The collection $\{\mathfrak{S}^\partial_{\mathfrak{p}}\}$ is a CF-perturbation of $(\partial(X, Z; \widehat{\mathcal{U}}), \mathcal{K}^3_\partial)$, which we denote by $\widehat{\mathfrak{S}^\partial}$.*

   (d) *If* $\widehat{\mathfrak{S}}$ *varies in a uniform family (in the sense of Definition* 7.60*) then* $\widehat{\mathfrak{S}^{\partial}}$ *varies in a uniform family.*

(3) (a) *A strongly continuous map* $\widehat{f} : (X, \widehat{\mathcal{U}}) \to M$ *induces a strongly continuous map* $\widehat{f_\partial} : \partial(X, Z; \widehat{\mathcal{U}}) \to M$.
   (b) *The restriction of* $\widehat{f_\partial}$ *to* $\overset{\circ}{S}_0(\partial(X, \widehat{\mathcal{U}}))$ *coincides with the restriction of* $\widehat{f}$ *to* $\overset{\circ}{S}_1(X, \widehat{\mathcal{U}})$.
   (c) *If* $\widehat{f}$ *is strongly smooth (resp. weakly submersive), so is* $\widehat{f_\partial}$.

(4) (a) *If* $\widehat{\mathfrak{S}}$ *is transversal to* 0*, so is* $\widehat{\mathfrak{S}^{\partial}}$.
   (b) *If* $\widehat{f}$ *is strongly submersive with respect to* $\widehat{\mathfrak{S}}$ *then* $\widehat{f_\partial}$ *is strongly submersive with respect to* $\widehat{\mathfrak{S}^{\partial}}$.
   (c) *Let* $g : N \to M$ *be a smooth map between smooth manifolds. If* $\widehat{f}$ *is strongly smooth and weakly transversal to* g *then so is* $\widehat{f_\partial}$.
   (d) $\widehat{f}$ *is strongly transversal to* g *with respect to* $\widehat{\mathfrak{S}}$ *then* $\widehat{f_\partial}$ *is strongly transversal to* g *with respect to* $\widehat{\mathfrak{S}^{\partial}}$.

(5) (a) *A differential form* $\widehat{h}$ *on* $(X, Z; \widehat{\mathcal{U}})$ *induces a differential form on* $\partial(X, Z; \widehat{\mathcal{U}})$*, which we write* $\widehat{h_\partial}$.
   (b) *The restriction of* $\widehat{h_\partial}$ *to* $\overset{\circ}{S}_0(\partial(X, Z; \widehat{\mathcal{U}}))$ *coincides with the restriction of* $\widehat{h}$ *to* $\overset{\circ}{S}_1(X, Z; \widehat{\mathcal{U}})$.
   (c) *In particular, a strongly continuous function on* $(X, Z, \widehat{\mathcal{U}})$ *induces one on* $\partial(X, Z; \widehat{\mathcal{U}})$*, such that (b) above applies.*

(6) *If* $\{\chi_{\mathfrak{p}}\}$ *is a partition of unity of* $(X, Z, \widehat{\mathcal{U}}, \mathcal{K}^2, \delta)$ *then* $\{(\chi_{\mathfrak{p}})_\partial\}$ *is a partition of unity of* $(\partial X, \partial Z, \partial\widehat{\mathcal{U}}, \mathcal{K}^2_\partial, \delta)$*. Here* $(\chi_{\mathfrak{p}})_\partial$ *is one induced from* $\{\chi_{\mathfrak{p}}\}$ *as in (5) (c) above.*

(7) *Statements similar to (2), (4)(a)(d), with CF-perturbation replaced by multivalued perturbation, hold.*

Proof In the case when U is an orbifold with corners, various forms of transversality or submersivity are defined by requiring the conditions not only for $\overset{\circ}{S}_0(U)$ (the interior point) but also for all $\overset{\circ}{S}_k(U)$. Once we observe this point, all the statements are obvious from the definition. □

8.2 Stokes' Formula for a Good Coordinate System

Now we are ready to state and prove Stokes' formula.

Theorem 8.11 (Stokes' formula, [FOOO8, Lemma 12.13]) *Assume that we are in the situation of Lemma* 8.10 *except (4)(c), (5)(c) and (7). Then, for each sufficiently small* $\epsilon > 0$*, we have*

$$d\left(\widehat{f}!(\widehat{h};\widehat{\mathfrak{S}^\epsilon})\right) = \widehat{f}!(d\widehat{h};\widehat{\mathfrak{S}^\epsilon}) + (-1)^{\dim(X,\widehat{\mathcal{U}})+\deg\widehat{h}}\widehat{f}_\partial!(\widehat{h}_\partial;\widehat{\mathfrak{S}^\epsilon_\partial}). \tag{8.3}$$

Proof Let $\{\chi_\mathfrak{p}\}$ be a partition of unity of $(X, Z, \widehat{\mathcal{U}}, \mathcal{K}^2, \delta)$. Let $(\chi_\mathfrak{p})_\partial$ and $\mathcal{K}^1_\partial$ be defined by Lemma 8.10. We put $h_0 = \chi_\mathfrak{p} h_\mathfrak{p}$. It suffices to show

$$\begin{aligned} &d\left(f_\mathfrak{p}!(h_0;\mathfrak{S}^\epsilon_\mathfrak{p})|_{\mathfrak{U}(Z)\cap\mathcal{K}^1_\mathfrak{p}(2\delta)}\right) \\ &= f_\mathfrak{p}!(dh_0;\mathfrak{S}^\epsilon_\mathfrak{p}|_{\mathfrak{U}(Z)\cap\mathcal{K}^1_\mathfrak{p}(2\delta)}) + (-1)^{\dim(X,\widehat{\mathcal{U}})+\deg\widehat{h}} f^\partial_\mathfrak{p}!(h_0;\mathfrak{S}^{\partial,\epsilon}_\mathfrak{p}|_{\mathfrak{U}(Z)\cap\mathcal{K}^1_{\partial,\mathfrak{p}}(2\delta)}), \end{aligned} \tag{8.4}$$

where $\widehat{\mathfrak{S}^{\partial,\epsilon}} = \{\mathfrak{S}^{\partial,\epsilon}_\mathfrak{p} \mid \mathfrak{p} \in \mathfrak{P}\}$. Let $\mathfrak{S}_\mathfrak{p} = \{(\mathfrak{V}_\mathfrak{r}, \mathcal{S}^\mathfrak{p}_\mathfrak{r}) \mid \mathfrak{r} \in \mathfrak{R}\}$ and $\{\chi_\mathfrak{r}\}$ a partition of unity subordinate to $\{U_\mathfrak{r}\}$. We put $h_1 = \chi_\mathfrak{r} h_0$ and $f_\mathfrak{r} = f_\mathfrak{p}|_{U_\mathfrak{r}}$. To prove (8.4), it suffices to prove:

$$\begin{aligned} &d\left(f_\mathfrak{r}!(h_1;\mathcal{S}^\epsilon_\mathfrak{r})|_{\mathfrak{U}(Z)\cap U_\mathfrak{r}}\right) \\ &= f_\mathfrak{r}!(dh_1;\mathcal{S}^\epsilon_\mathfrak{r})|_{\mathfrak{U}(Z)\cap U_\mathfrak{r}}) + (-1)^{\dim(X,\widehat{\mathcal{U}})+\deg\widehat{h}} f^\partial_\mathfrak{r}!(h_1;\mathcal{S}^{\partial,\epsilon}_\mathfrak{r}|_{\mathfrak{U}(Z)\cap\partial U_\mathfrak{r}}), \end{aligned} \tag{8.5}$$

where $\widehat{\mathfrak{S}^\epsilon_{\partial,\mathfrak{p}}} = \{(\partial\mathfrak{V}_\mathfrak{r}, \mathcal{S}^{\partial,\epsilon}_\mathfrak{r})\}$. Equation (8.5) follows from the next lemma.

Lemma 8.12 *Let Ω be an open neighborhood of 0 in $[0,1)^m \times \mathbb{R}^{n-m}$ and $f : \Omega \to M$ a smooth map. Let h be a smooth differential k-form on Ω with compact support and ρ a differential $(n-k-1)$-form on M. Then we have*

$$\int_\Omega f^* d\rho \wedge h = (-1)^{n-k}\int_\Omega f^*\rho \wedge dh + \int_{\Omega\cap\partial([0,1)^m\times\mathbb{R}^{n-m})} f^*\rho\wedge h$$

Lemma 8.12 is an immediate consequence of the usual Stokes' formula. Thus the proof of Theorem 8.11 is complete. □

Using Stokes' formula we can immediately prove the following basic properties of smooth correspondence.

Corollary 8.13 *In Definition 7.1 we apply Construction 7.85. Let* $\mathrm{Corr}_{((X,\widehat{\mathcal{U}}),\widehat{\mathfrak{S}^\epsilon})} : \Omega^k(M_s) \to \Omega^{\ell+k}(M_t)$ *be the map obtained by Definition 7.86. (Here $\ell = \dim M_t - \dim(X,\widehat{\mathcal{U}})$.) We define the boundary by*

$$\partial(X,\widehat{\mathcal{U}}) = (\partial(X,\widehat{\mathcal{U}}), \widehat{f}_s|_{\partial(X,\widehat{\mathcal{U}})}, \widehat{f}_t|_{\partial(X,\widehat{\mathcal{U}})}).$$

$\widehat{\mathfrak{S}^\epsilon}$ *induces a CF-perturbation $\widehat{\mathfrak{S}^{\partial,\epsilon}}$ of it as in Lemma 8.10 (4). $\partial(X,\widehat{\mathcal{U}})$ and $\widehat{\mathfrak{S}^{\partial,\epsilon}}$ define a map* $\mathrm{Corr}_{(\partial(X,\widehat{\mathcal{U}}),\widehat{\mathfrak{S}^{\partial,\epsilon}})} : \Omega^k(M_s) \to \Omega^{\ell+k+1}(M_t)$*. Then for any sufficiently small $\epsilon > 0$, we have*

$$d \circ \mathrm{Corr}_{((X,\widehat{\mathcal{U}}),\widehat{\mathfrak{S}^{\epsilon}})} - \mathrm{Corr}_{((X,\widehat{\mathcal{U}}),\widehat{\mathfrak{S}^{\epsilon}})} \circ d = (-1)^{\dim(X,\widehat{\mathcal{U}})+\deg(\cdot)} \mathrm{Corr}_{(\partial(X,\widehat{\mathcal{U}}),\widehat{\mathfrak{S}^{\partial,\epsilon}})}. \tag{8.6}$$

In particular, $\mathrm{Corr}_{((X,\widehat{\mathcal{U}}),\widehat{\mathfrak{S}^{\epsilon}})}$ *is a cochain map if* $\widehat{\mathcal{U}}$ *is a Kuranishi structure without boundary.*

Proof This is immediate from Theorem 8.11. □

Lemma 8.14 *We assume that* $\widehat{\mathfrak{S}_{\sigma}}$ *is a uniform family in the sense of Definition* 7.60. *Then the positive number* ϵ *in Theorem* 8.11 *and Corollary* 8.13 *can be taken independent of* σ.

The proof is the same as that of Lemma 7.89.

8.3 Well-Definedness of Virtual Fundamental Cycle

We use Corollary 8.13 to prove well-definedness of virtual cohomology class, and well-definedness of the smooth correspondence *in the cohomology level*, when a Kuranishi structure has no boundary.

Theorem 8.15 *Let* $((X;\widehat{\mathcal{U}}); \widehat{f_s}, \widehat{f_t})$ *be a smooth correspondence from* M_s *to* M_t *as defined in Definition* 7.1. *We assume that our Kuranishi structure on* X *has no boundary. Then the map* $\mathrm{Corr}_{(\mathfrak{X},\widehat{\mathfrak{S}^{\epsilon}})} : \Omega^k(M_s) \to \Omega^{\ell+k}(M_t)$ *defined in Definition* 7.86 *is a cochain map.*

Moreover, provided ϵ *is sufficiently small, the map* $\mathrm{Corr}_{(\mathfrak{X},\widehat{\mathfrak{S}^{\epsilon}})}$ *is independent of the choices of our good coordinate system* $\widehat{\mathcal{U}}$ *and CF-perturbation* $\widehat{\mathfrak{S}}$ *and of* $\epsilon > 0$, *up to cochain homotopy.*

Proof The first half is a repetition of Corollary 8.13. We will prove the independence of the definition up to cochain homotopy below. Let $\widehat{\mathcal{U}}, \widehat{\mathcal{U}'}$ be two choices of good coordinate system and $\widehat{\mathfrak{S}}, \widehat{\mathfrak{S}'}$ CF-perturbations of $(X;\widehat{\mathcal{U}})$, $(X;\widehat{\mathcal{U}'})$ respectively. During the proof of Proposition 8.15, we do not need to make a specific choice of support system because Proposition 7.81 (see also Remark 7.87) shows the map $\mathrm{Corr}_{(\mathfrak{X},\widehat{\mathfrak{S}^{\epsilon}})}$ is independent thereof.

We put a direct product Kuranishi structure on $[0,1]\times X$. We identify $X = \{0\}\times X$. Then the good coordinate system $\widehat{\mathcal{U}}$ induces $[0,1/3)\times\widehat{\mathcal{U}}$ on $[0,1/3)\times X$ such that $\partial([0,1/3)\times X, [0,1/3)\times\widehat{\mathcal{U}})$ is isomorphic to $-(X;\widehat{\mathcal{U}})$ with opposite orientation. Similarly we have a good coordinate system $(2/3,1]\times\widehat{\mathcal{U}'}$ on $(2/3,1]\times X$ such that $\partial((2/3,1]\times X; (2/3,1]\times\widehat{\mathcal{U}'})$ is isomorphic to $(X;\widehat{\mathcal{U}'})$. Here the notion of isomorphism of a good coordinate system is defined in Definition 3.24. Then, by Proposition 7.54, there exists a good coordinate system $[0,1]\times\widehat{\mathcal{U}'}$ such that

$$\partial([0,1]\times X; [0,1]\times\widehat{\mathcal{U}'}) = -(X;\widehat{\mathcal{U}}) \cup (X,\widehat{\mathcal{U}'}). \tag{8.7}$$

We next consider two choices of CF-perturbations, which we denote by $\widehat{\mathfrak{S}}$ and $\widehat{\mathfrak{S}'}$. We assume that $\widehat{f_t}$ is strongly submersive with respect to both of them. We define $[0, 1/3) \times \widehat{\mathfrak{S}}$ and $(2/3, 1] \times \widehat{\mathfrak{S}'}$ as follows.

We consider Situation 7.3. Let $\mathfrak{V}_x = (V_x, \Gamma_x, E_x, \phi_x, \widehat{\phi}_x)$ be an orbifold chart of $(U, \mathcal{E})$ and $\mathcal{S}_x = (W_x, \omega_x, \mathfrak{s}^\epsilon_x)$ a CF-perturbation on it. (Definition 7.4.) Suppose $(f_t)_x$ is strongly submersive with respect to $\mathcal{S}_x$. We take $[0, 1/3) \times \mathfrak{V}_x = ([0, 1/3) \times V_x, \Gamma_x, [0, 1/3) \times E_x), \mathrm{id} \times \phi_x, \mathrm{id} \times \widehat{\phi}_x$ that is an orbifold chart of $([0, 1/3) \times U, [0, 1/3) \times \mathcal{E}$. In an obvious way $\mathcal{S}_x$ induces a CF-perturbation of it, with respect to which $(f_t)_x \circ \pi$ is strongly submersive. Here $\pi : [0, 1/3) \times X \to X$ is the projection. We denote it by $[0, 1/3) \times \mathcal{S}_x$. (See Definition 10.2 for detail.)

We perform this construction of multiplying $[0, 1/3)$ for each chart (once for each orbifold chart and once for each Kuranishi chart) then it is fairly obvious that they are compatible with various coordinate changes. Thus we obtain $[0, 1/3) \times \widehat{\mathfrak{S}}$ that is a CF-perturbation of $[0, 1/3) \times X$. We obtain $(2/3, 1] \times \widehat{\mathfrak{S}'}$ in the same way.

Now we use Proposition 7.59 with $Z_1 = \{0, 1\} \times X$, $Z_2 = [0, 1] \times X$. Then we obtain a CF-perturbation $\widehat{\mathfrak{S}^{[0,1]}}$ of $([0, 1] \times X, [0, 1] \times \widehat{\mathcal{U}}])$ such that its restriction to $\{0\} \times X$ and $\{1\} \times X$ are $\widehat{\mathfrak{S}}$ and $\widehat{\mathfrak{S}'}$, respectively.

Now we use Corollary 8.13 and (8.7) to show:

$$\begin{aligned} & d \circ \mathrm{Corr}_{([0,1]\times\mathfrak{X}, \widehat{\mathfrak{S}^{[0,1]\epsilon}})} - \mathrm{Corr}_{([0,1]\times\mathfrak{X}, \widehat{\mathfrak{S}^{[0,1]\epsilon}})} \circ d \\ & = (-1)^{\dim(X,\widehat{\mathcal{U}})+1+\deg(\cdot)} (\mathrm{Corr}_{(\mathfrak{X},\widehat{\mathfrak{S}'^{\epsilon}})} \cdot - \mathrm{Corr}_{(\mathfrak{X},\widehat{\mathfrak{S}^{\epsilon}})}). \end{aligned} \tag{8.8}$$

Thus we have proved that there exists ϵ_0 such that if $0 < \epsilon < \epsilon_0$ then $\mathrm{Corr}_{(\mathfrak{X},\widehat{\mathfrak{S}^{\epsilon}})}$ is cochain homotopic to $\mathrm{Corr}_{(\mathfrak{X},\widehat{\mathfrak{S}'^{\epsilon}})}$.

We next prove that $\mathrm{Corr}_{(\mathfrak{X},\widehat{\mathfrak{S}^{\epsilon}})}$ is independent of ϵ up to cochain homotopy. We consider the family $\{\widehat{\mathfrak{S}^{c\epsilon}} \mid c \in (0, 1]\}$. This is a family of CF-perturbations parametrized by $\mathscr{A} = (0, 1]$. It is easy to see that this is a uniform family. Therefore in the same way as above (using the second half of Lemma 7.61) we obtain a uniform family of CF-perturbations $\widehat{\mathfrak{S}_c^{[0,1]}}$ of $([0, 1] \times X, [0, 1] \times \widehat{\mathcal{U}})$ such that their restrictions to $\{0\} \times X$ and $\{1\} \times X$ are $\widehat{\mathfrak{S}}$ and $\widehat{\mathfrak{S}_c}$, respectively. Therefore there exists ϵ_0 independent of c such that $\mathrm{Corr}_{(\mathfrak{X},\widehat{\mathfrak{S}^{\epsilon}})}$ is cochain homotopic to $\mathrm{Corr}_{(\mathfrak{X},\widehat{\mathfrak{S}^{c\epsilon}})}$ for $0 < \epsilon < \epsilon_0$. Therefore for $0 < \epsilon, \epsilon' < \epsilon_0$, the cochain map $\mathrm{Corr}_{(\mathfrak{X},\widehat{\mathfrak{S}^{\epsilon}})}$ is cochain homotopic to $\mathrm{Corr}_{(\mathfrak{X},\widehat{\mathfrak{S}^{\epsilon'}})}$.

The proof of Theorem 8.15 is complete. □

Theorem 8.15 implies that the correspondence map $\mathrm{Corr}_{(\mathfrak{X},\widehat{\mathfrak{S}^{\epsilon}})}$ on differential forms descends to a map on cohomology which is independent of the choices of $\widehat{\mathcal{U}}$ and $\widehat{\mathfrak{S}^{\epsilon}}$. We write the cohomology class as $[\mathrm{Corr}_{\mathfrak{X}}(h)] \in H(M_t)$ for any closed differential form h on M_s by removing $\widehat{\mathfrak{S}^{\epsilon}}$ from the notation.

In Theorem 8.15 we fixed our Kuranishi structure $\widehat{\mathcal{U}}$ on X. In fact, we can prove the same conclusion under a milder assumption.

Proposition 8.16 *Let $\mathfrak{X}_i = ((X_i, \widehat{\mathcal{U}^i}), \widehat{f_s^i}, \widehat{f_t^i})$ be smooth correspondence from M_s to M_t such that $\partial X_i = \emptyset$. Here $i = 1, 2$ and M_s, M_t are independent of i. We assume that there exists a smooth correspondence $\mathfrak{Y} = ((Y, \widehat{\mathcal{U}}), \widehat{f_s}, \widehat{f_t})$ from M_s to M_t with boundary (but without corners) such that*

$$\partial\mathfrak{Y} = -\mathfrak{X}_1 \cup \mathfrak{X}_2.$$

Here $-\mathfrak{X}_1$ is the smooth correspondence $\mathfrak{X}_1$ with opposite orientation. Then we have

$$[\mathrm{Corr}_{\mathfrak{X}_1}(h)] = [\mathrm{Corr}_{\mathfrak{X}_2}(h)] \in H(M_t), \tag{8.9}$$

where h is a closed differential form on M_s.

Proof We take a good coordinate system $\widehat{\mathcal{U}}$ of Y and a KG-embedding $(Y, \widehat{\mathcal{U}}) \to (Y, \widehat{\mathcal{U}})$. (Theorem 3.35.) $\widehat{f_t}$ is pulled back from $\widehat{f_t} : (Y, \widehat{\mathcal{U}}) \to M_t$. $\widehat{f_s}$ is also pulled back from $\widehat{f_s}$ (Proposition 6.32 (2).) We also obtain a CF-perturbation $\widehat{\mathfrak{S}}$ of $\widehat{\mathcal{U}}$ with respect to which $\widehat{f}$ is strongly submersive (Theorem 7.51 (2)). They restrict to $(X_i, \widehat{\mathcal{U}_i})$, $\widehat{\mathfrak{S}_i}$ and $\widehat{f_t^i} : (Y, \widehat{\mathcal{U}}) \to M_t$, $\widehat{f_s^i} : (Y, \widehat{\mathcal{U}}) \to M_s$.

We note that $\widehat{f_t^i}$ is strongly submersive with respect to $\widehat{\mathfrak{S}_i}$. (This is the consequence of the definition of strong transversality. Namely we require the transversality on each stratum of corner structure stratification (Definition 4.20).)

By Proposition 8.15 we can use $(X_i, \widehat{\mathcal{U}_i})$, $\widehat{\mathfrak{S}_i}$, $\widehat{f_t^i}$, $\widehat{f_s^i}$ to define smooth correspondence $\mathrm{Corr}_{\mathfrak{X}_i}$ (in the cohomology level.)

The proposition now follows from Corollary 8.13 and (8.7) applied to $(Y, \widehat{\mathcal{U}})$, $\widehat{\mathfrak{S}}$, $\widehat{f_t}$, $\widehat{f_t}$. Namely we can calculate in the same way as (8.8). □

Remark 8.17 The proofs of Theorem 8.15 and Proposition 8.16 (Formulas (8.8), (8.9)) are a prototype of the proofs of various similar equalities which appear in our construction of structures and proof of its independence. We will apply a similar method in a more complicated situation in Part II systematically.

Remark 8.18 In [FOOO22, Section 9] we consider an equivalence relation between Kuranishi structures, which is generated by the relation $\sim$ where $\widehat{\mathcal{U}} \sim \widehat{\mathcal{U}'}$ if there is a KK-embedding $\widehat{\mathcal{U}} \to \widehat{\mathcal{U}'}$. We then show that two Kuranishi structures that are equivalent in the above sense are equivalent in the sense of isotopy (see [FOOO22, Definition 9.13]), which in turn implies being cobordant. It is also proved that all the Kuranishi structures obtained from the data which we called an obstruction bundle data are equivalent to each other. So we can apply Proposition 8.16 in this situation.

According to [Ya1], the above-mentioned equivalence relation (in the variant of Kuranishi structure discussed in [Ya1]) can be defined in a much simpler way, that is, one can define $\widehat{\mathcal{U}} \sim \widehat{\mathcal{U}'}$ if they are embedded into a common Kuranishi structure.

Chapter 9
From Good Coordinate Systems to Kuranishi Structures and Back with CF-Perturbations

As we explained at the end of Chap. 4, it is more canonical to define the notion of fiber products of spaces with Kuranishi structure than to define that of fiber products of spaces with a good coordinate system. On the other hand, in Chap. 7, we gave the definition of CF-perturbation and of the pushout of differential forms via good coordinate systems. In this chapter, we describe how we go from a good coordinate system to a Kuranishi structure and back together with CF-perturbations on them, and prove in Theorem 9.14 that we can define the pushout via the Kuranishi structure itself in such a way that the outcome is independent of the auxiliary choice of good coordinate system.

9.1 CF-Perturbations and Embedding of Kuranishi Structures

Definition 9.1 Let $\widehat{\mathcal{U}}$ be a Kuranishi structure on $Z \subseteq X$. A *CF-perturbation* $\widehat{\mathfrak{S}}$ *of* $\widehat{\mathcal{U}}$ assigns $\mathfrak{S}_p$ for each $p \in Z$ with the following properties:

(1) $\mathfrak{S}_p$ is a CF-perturbation of $\mathcal{U}_p$.
(2) If $q \in \text{Im}(\psi_p) \cap Z$, then $\mathfrak{S}_p$ is restrictable to $\mathcal{U}_q$. Namely

$$\mathfrak{S}_p \in \mathcal{CF}^{\mathcal{U}_q \triangleright \mathcal{U}_p}(U_p). \tag{9.1}$$

(3) If $q \in \text{Im}(\psi_p) \cap Z$, then $\mathfrak{S}_p, \mathfrak{S}_q$ are compatible with Φ_{pq}. Namely

$$\Phi_{pq}^*(\mathfrak{S}_p) = \mathfrak{S}_q|_{U_{pq}} \in \mathcal{CF}^{\mathcal{U}_q}(U_{pq}). \tag{9.2}$$

Definition 9.2 Suppose we are in the situation of Definition 9.1. Let $\widehat{f} : (X, Z; \widehat{\mathcal{U}}) \to M$ be a strongly smooth map to a smooth manifold M.

K. Fukaya et al., *Kuranishi Structures and Virtual Fundamental Chains*, Springer Monographs in Mathematics, https://doi.org/10.1007/978-981-15-5562-6_9

(1) We say $\widehat{\mathfrak{S}}$ is strictly *transversal to* 0 if each $\mathfrak{S}_p$ is transversal to 0. We say $\widehat{\mathfrak{S}}$ is *transversal to* 0 if its restriction to an open substructure is strictly so.
(2) We say $\widehat{f}$ is *strictly strongly submersive with respect to* $\widehat{\mathfrak{S}}$ if each of f_p is strongly submersive with respect to $\mathfrak{S}_p$. We say $\widehat{f}$ is *strongly submersive with respect to* $\widehat{\mathfrak{S}}$ if its restriction to an open substructure is strictly so.
(3) We say $\widehat{f}$ is strictly *strongly transversal to* $g : N \to M$ *with respect to* $\widehat{\mathfrak{S}}$ if each of f_p is strongly transversal to $g : N \to M$ with respect to $\mathfrak{S}_p$. We say $\widehat{f}$ is *strongly transversal to* $g : N \to M$ *with respect to* $\widehat{\mathfrak{S}}$ if its restriction to an open substructure is strictly so.

Definition 9.3 Let $Z \subset Z' \subset X$, $\widehat{\mathcal{U}}$ be a Kuranishi structure of $Z' \subset X$ and $\widehat{\mathfrak{S}} = \{\mathfrak{S}_p \mid p \in Z'\}$ a CF-perturbation of $\widehat{\mathcal{U}}$. Then $\{\mathfrak{S}_p \mid p \in Z\}$ is a CF-perturbation of $\widehat{\mathcal{U}}|_Z$, which we call a *restriction of* $\widehat{\mathfrak{S}}$ and denote by $\widehat{\mathfrak{S}}|_Z$.

We next define compatibility of CF-perturbations with various embeddings of Kuranishi structures and/or good coordinate systems and prove the counterparts of several lemmas in Chap. 6 corresponding to the current context of CF-perturbations.

Definition 9.4 Let $\widehat{\mathcal{U}}$ and $\widehat{\mathcal{U}^+}$ be Kuranishi structures of $Z \subseteq X$, $\widehat{\mathcal{U}}$ and $\widehat{\mathcal{U}^+}$ good coordinate systems of $Z \subseteq X$. Let $\mathcal{K}$ and $\mathcal{K}^+$ be support systems of $\widehat{\mathcal{U}}$ and $\widehat{\mathcal{U}^+}$, respectively. Let $\widehat{\mathfrak{S}}$, $\widehat{\mathfrak{S}^+}$, $\widehat{\mathfrak{S}}$, $\widehat{\mathfrak{S}^+}$ be CF-perturbations of $\widehat{\mathcal{U}}$, $\widehat{\mathcal{U}^+}$, $(\widehat{\mathcal{U}}, \mathcal{K})$, $(\widehat{\mathcal{U}^+}, \mathcal{K}^+)$, respectively.

(1) (Compatibility of CF-perturbation with strict KK-embedding) Let $\widehat{\Phi} : \widehat{\mathcal{U}} \to \widehat{\mathcal{U}^+}$ be a strict KK-embedding. We say $\widehat{\mathfrak{S}}$, $\widehat{\mathfrak{S}^+}$ are *compatible* with $\widehat{\Phi}$ if the following holds for each p:

    (a) $\mathfrak{S}_p^+ \in \mathcal{CF}^{\mathcal{U}_p \triangleright \mathcal{U}_p^+}(U_p)$. Here we use the embedding Φ_p to define the subsheaf $\mathcal{CF}^{\mathcal{U}_p \triangleright \mathcal{U}_p^+}$.
    (b) $\Phi_p^*(\mathfrak{S}_p^+) = \mathfrak{S}_p \in \mathcal{CF}^{\mathcal{U}_p}(U_p)$.

(2) (Compatibility of CF-perturbation with KK-embedding) Let $\widehat{\Phi} : \widehat{\mathcal{U}} \to \widehat{\mathcal{U}^+}$ be a KK-embedding. We say $\widehat{\mathfrak{S}}$, $\widehat{\mathfrak{S}^+}$ are *compatible* with $\widehat{\Phi}$ if there exist an open substructure $\widehat{\mathcal{U}_0}$, a CF-perturbation $\widehat{\mathfrak{S}_0}$ of $\widehat{\mathcal{U}_0}$ and a strict KK-embedding $\widehat{\Phi_0} : \widehat{\mathcal{U}_0} \to \widehat{\mathcal{U}^+}$ such that $\widehat{\mathfrak{S}_0}$, $\widehat{\mathfrak{S}^+}$ are compatible with $\widehat{\Phi_0}$ and $\widehat{\mathfrak{S}_0}$, $\widehat{\mathfrak{S}}$ are compatible with the open embedding $\widehat{\mathcal{U}_0} \to \widehat{\mathcal{U}}$.
(3) (Compatibility of support system with GG-embedding) Let $\widehat{\Phi} = (\{\Phi_{\mathfrak{p}}\}, \mathfrak{i}) : \widehat{\mathcal{U}} \to \widehat{\mathcal{U}^+}$ be a GG-embedding. We say that $\mathcal{K}, \mathcal{K}^+$ is *compatible* with $\widehat{\Phi}$ if $\varphi_{\mathfrak{p}}(\mathcal{K}_{\mathfrak{p}}) \subset \mathcal{K}^+_{\mathfrak{i}(\mathfrak{p})}$ for each $\mathfrak{p} \in \mathfrak{P}$.
(4) (Compatibility of CF-perturbation with GG-embedding) In the situation of (3), we say $\widehat{\mathfrak{S}}$, $\widehat{\mathfrak{S}^+}$ are *compatible* with $\widehat{\Phi}$ if the following holds for each $\mathfrak{p} \in \mathfrak{P}$:

    (a) $\mathfrak{S}^+_{\mathfrak{i}(\mathfrak{p})} \in \mathcal{CF}^{\mathcal{U}_{\mathfrak{p}} \triangleright \mathcal{U}^+_{\mathfrak{i}(\mathfrak{p})}}(\mathcal{K}_{\mathfrak{i}(\mathfrak{p})})$. Here we use the embedding $\Phi_{\mathfrak{p}} : \mathcal{U}_{\mathfrak{p}} \to \mathcal{U}^+_{\mathfrak{i}(\mathfrak{p})}$ to define the subsheaf $\mathcal{CF}^{\mathcal{U}_{\mathfrak{p}} \triangleright \mathcal{U}^+_{\mathfrak{i}(\mathfrak{p})}}$.

(b) $\Phi_{\mathfrak{p}}^*(\mathfrak{S}_{\mathfrak{i}(\mathfrak{p})}^+) = \mathfrak{S}_{\mathfrak{p}} \in \mathcal{CF}^{\mathcal{U}_{\mathfrak{p}}}(\mathcal{K}_{\mathfrak{p}})$.

(5) (Compatibility of CF-perturbation with strict KG-embedding) Let $\widehat{\Phi} : \widehat{\mathcal{U}} \to \widehat{\mathcal{U}}$ be a strict KG-embedding. We say $\widehat{\mathfrak{S}}, \widehat{\widehat{\mathfrak{S}}}$ are *compatible* with $\widehat{\Phi}$ if the following holds for each $\mathfrak{p}$ and $p \in \psi_{\mathfrak{p}}(\mathcal{K}_{\mathfrak{p}} \cap s_{\mathfrak{p}}^{-1}(0)) \cap Z$:

(a) $\mathfrak{S}_{\mathfrak{p}} \in \mathcal{CF}^{\mathcal{U}_p \triangleright \mathcal{U}_{\mathfrak{p}}}(\mathcal{K}_{\mathfrak{p}})$. Here we use the embedding $\Phi_{\mathfrak{p}p} : \mathcal{U}_p \to \mathcal{U}_{\mathfrak{p}}$ to define the subsheaf $\mathcal{CF}^{\mathcal{U}_p \triangleright \mathcal{U}_{\mathfrak{p}}}$.
(b) $\Phi_p^*(\mathfrak{S}_{\mathfrak{p}}) = \mathfrak{S}_p \in \mathcal{CF}^{\mathcal{U}_p}(U_p)$.

(6) (Compatibility of CF-perturbation with KG-embedding) In the case when $\widehat{\Phi} : \widehat{\mathcal{U}} \to \widehat{\mathcal{U}}$ is a KG-embedding, we can define compatibility of $\widehat{\mathfrak{S}}, \widehat{\widehat{\mathfrak{S}}}$ with $\widehat{\Phi}$ in the same way as Item (2) (using Items (1) and (5)).
(7) (Compatibility of CF-perturbation with strict GK-embedding) Let $\widehat{\Phi} = (\{U_{\mathfrak{p}}(p)\}, \{\Phi_{p\mathfrak{p}}\}) : \widehat{\mathcal{U}} \to \widehat{\mathcal{U}}$ be a GK-embedding. We say $\widehat{\widehat{\mathfrak{S}}}, \widehat{\mathfrak{S}}$ are *compatible* with $\widehat{\Phi}$ if the following holds for each $\mathfrak{p}$ and $p \in \psi_{\mathfrak{p}}(\mathcal{K}_{\mathfrak{p}} \cap s_{\mathfrak{p}}^{-1}(0)) \cap Z$:

(a) $\mathfrak{S}_p \in \mathcal{CF}^{\mathcal{U}_{\mathfrak{p}} \triangleright \mathcal{U}_p}(U_p)$. Here we use the embedding $\Phi_{p\mathfrak{p}} : \mathcal{U}_{\mathfrak{p}}|_{U_{\mathfrak{p}}(p)} \to \mathcal{U}_p$ to define the subsheaf $\mathcal{U}_{\mathfrak{p}} \triangleright \mathcal{U}_p$.
(b) $\Phi_{p\mathfrak{p}}^* \mathfrak{S}_p = \widehat{\mathfrak{S}}_{\mathfrak{p}}|_{U_{\mathfrak{p}}(p)} \in \mathcal{CF}^{\mathcal{U}_p}(U_{\mathfrak{p}}(p))$.

With these definitions of compatibility, we now prove the compatibilities relevant to various embeddings. We begin with the discussion on support systems.

Lemma 9.5 *Let $\widehat{\Phi} = (\{\Phi_{\mathfrak{p}}\}, \mathfrak{i}) : \widehat{\mathcal{U}} \to \widehat{\mathcal{U}^+}$ be a GG-embedding.*

(1) *If $\mathcal{K}$ is a support system of $\widehat{\mathcal{U}}$, then there exist a support system $\mathcal{K}^+$ of $\widehat{\mathcal{U}^+}$ such that $\mathcal{K}$, $\mathcal{K}^+$ are compatible with $\widehat{\Phi}$.*
(2) *If $\mathcal{K}_i$ $(i = 1, \dots, m)$ are support systems of $\widehat{\mathcal{U}}$ with $\mathcal{K}_i < \mathcal{K}_{i+1}$ then there exist support systems $\mathcal{K}_i^+$ $(i = 1, \dots, m)$ of $\widehat{\mathcal{U}^+}$ such that $\mathcal{K}_i$, $\mathcal{K}_i^+$ are compatible with $\widehat{\Phi}$ and $\mathcal{K}_i^+ < \mathcal{K}_{i+1}^+$.*

Proof (1) Let $\mathcal{K} = (\mathcal{K}_{\mathfrak{p}})$. Let $\mathcal{K}_{0,\mathfrak{p}_+}$ be a closure of a sufficiently small neighborhood of $\bigcup_{\substack{\mathfrak{p} \in \mathfrak{P} \\ \mathfrak{i}(\mathfrak{p}) = \mathfrak{p}_+}} \varphi_{\mathfrak{p}}(\mathcal{K}_{\mathfrak{p}})$ of $U_{p_+}^+$ for $\mathfrak{p}_+ \in \mathfrak{P}_+$. It is easy to see that $\mathcal{K}_0^+ = (\mathcal{K}_{0,\mathfrak{p}_+})$ is a support system of $\widehat{\mathcal{U}^+}$. Any $\mathcal{K}^+ > \mathcal{K}_0^+$ has the required properties. The proof of (2) is similar by using an upward induction on i. □

Lemma 9.6 *Let $\widehat{\Phi} : \widehat{\mathcal{U}} \to \widehat{\mathcal{U}^+}$ be a weakly open GG-embedding and $\mathcal{K}$, $\mathcal{K}^+$ support systems of $\widehat{\mathcal{U}}$, $\widehat{\mathcal{U}^+}$, respectively, which are compatible with $\widehat{\Phi}$. Then for any CF-perturbation $\widehat{\mathfrak{S}^+}$ of $(\widehat{\mathcal{U}^+}, \mathcal{K}^+)$, there exists a unique CF-perturbation $\widehat{\mathfrak{S}}$ of $(\widehat{\mathcal{U}}, \mathcal{K})$ such that $\widehat{\mathfrak{S}^+}$ and $\widehat{\mathfrak{S}}$ are compatible with $\widehat{\Phi}$.*

Proof For any $\mathfrak{p} \in \mathfrak{P}$ we restrict $\mathfrak{S}_{\mathfrak{i}(\mathfrak{p})}^+$ to $\mathcal{U}_{\mathfrak{p}}$ to obtain $\mathfrak{S}_{\mathfrak{p}}$. We thus obtain $\widehat{\mathfrak{S}}$. Since normal bundles are trivial in the case of weakly open embedding, the compatibility is automatic. □

Lemma 9.7 *Various transversality or submersivity of the target of an* **open** *KK-embedding imply those of the source. The same holds for a weakly open GG-embedding.*

Proof This is an easy consequence of the definition. □

Lemma 9.8 *The notion of compatibility of CF-perturbations to embeddings is preserved under the composition of embeddings of various kinds.*

The proof is obvious.

The next lemma is a CF-perturbation version of Proposition 6.30. It claims that when a CF-perturbation is given on a good coordinate system then we can bring it to the Kuranishi structure we construct from the given good coordinate system.

Lemma 9.9 *In the situation of Proposition* 6.30, *let* $\mathcal{K}_0$, $\mathcal{K}$ *be support systems of* $\widehat{\mathcal{U}_0}, \widehat{\mathcal{U}}$, *respectively, that are compatible with the open embedding* $\widehat{\mathcal{U}_0} \to \widehat{\mathcal{U}}$. *Let* $\widehat{\mathfrak{S}}$ *be a CF-perturbation of* $(\widehat{\mathcal{U}}, \mathcal{K})$, *which restricts to a CF-perturbation* $\widehat{\mathfrak{S}_0}$ *of* $(\widehat{\mathcal{U}_0}, \mathcal{K}_0)$. *Then the following holds:*

(1) *There exists a CF-perturbation* $\widehat{\mathfrak{S}}$ *of* $\widehat{\mathcal{U}^+}$ *such that* $\widehat{\mathfrak{S}_0}$ *and* $\widehat{\mathfrak{S}}$ *are compatible with the GK-embedding* $\widehat{\mathcal{U}_0} \to \widehat{\mathcal{U}^+}$.
(2) *In the situation of Proposition* 6.30 *(2), if* $\widehat{f}$ *is strongly submersive with respect to* $\widehat{\mathfrak{S}}$, *then* $\widehat{f}$ *is strongly submersive with respect to* $\widehat{\mathfrak{S}}$. *The strong transversality to* $g : M \to Y$ *is also preserved.*
(3) *If* $\widehat{h}$ *is a differential form on* $\widehat{\mathcal{U}}$, *then there exists a differential form* $\widehat{h}$ *on* $\widehat{\mathfrak{S}}$ *such that the restriction* $\widehat{h}$ *to* $\widehat{\mathcal{U}_0}$ *is a pullback of* $\widehat{h}$.

Proof The proof of Lemma 9.9 is the same as that of Proposition 6.30. In fact, the Kuranishi chart of $\widehat{\mathcal{U}^+}$ is a restriction of a Kuranishi chart of $\widehat{\mathcal{U}_0}$. Since $U_{0,\mathfrak{p}} \subset \mathcal{K}_{\mathfrak{p}}$ (see Proposition 6.30 (1)), we can restrict $\widehat{\mathfrak{S}}$ to the Kuranishi charts of $\widehat{\mathcal{U}^+}$. □

We next state the CF-perturbation versions of Propositions 6.32 and 6.33. It claims that when CF-perturbation is given on a Kuranishi structure then we can bring it to the good coordinate system we construct from the given Kuranishi structure.

In Lemmas 9.10 and 9.11 we do not specify support systems of good coordinate systems for the CF-perturbations. We take one but do not mention them.

Lemma 9.10 *Let* $\widehat{\mathcal{U}}$ *be a Kuranishi structure on* $Z \subseteq X$. *Then we can take a good coordinate system* $\widehat{\mathcal{U}}$ *and the strict KG-embedding* $\widehat{\Phi} : \widehat{\mathcal{U}_0} \to \widehat{\mathcal{U}}$ *in Theorem* 3.35 *so that the following holds in addition:*

(1) *If* $\widehat{h}$ *is a differential form on* $\widehat{\mathcal{U}}$, *then there exists a differential form* $\widehat{h}$ *on* $\widehat{\mathcal{U}}$ *such that* $\widehat{\Phi}^*(\widehat{h}) = \widehat{h}|_{\widehat{\mathcal{U}_0}}$. *If* $\widehat{h}$ *has a compact support in* $\overset{\circ}{Z}$, *then* $\widehat{h}$ *has a compact support in* $|\widehat{\mathcal{U}}|$ *and* $\mathrm{Supp}(\widehat{h}) \cap Z \subset \overset{\circ}{Z}$.
(2) *If* $\widehat{\mathfrak{S}}$ *is a CF-perturbation of* $\widehat{\mathcal{U}}$, *then there exists a CF-perturbation* $\widehat{\mathfrak{S}}$ *of* $\widehat{\mathcal{U}}$ *such that* $\widehat{\mathfrak{S}}|_{\widehat{\mathcal{U}_0}}$ *and* $\widehat{\mathfrak{S}}$ *are compatible with the KG-embedding* $\widehat{\Phi}$.
(3) *In the situation of (2) the following holds:*

(a) *If* $\widehat{\mathfrak{S}}$ *is transversal to* 0 *then so is* $\widehat{\widehat{\mathfrak{S}}}$.
(b) *If* $\widehat{f}$ *is strongly submersive with respect to* $\widehat{\mathfrak{S}}$, *then* $\widehat{\widehat{f}}$ *is strongly submersive with respect to* $\widehat{\widehat{\mathfrak{S}}}$.
(c) *If* $\widehat{f}$ *is strongly transversal to* $g : M \to Y$ *with respect to* $\widehat{\mathfrak{S}}$ *then* $\widehat{\widehat{f}}$ *is strongly transversal to* g *with respect to* $\widehat{\widehat{\mathfrak{S}}}$.

Lemma 9.11 *Suppose we are in the situation of Propositions* 5.20 *(resp. Proposition* 5.21*) and* 6.32 *(2). Then we can take the GK-embedding* $\widehat{\Phi^+} : \widehat{\widehat{\mathcal{U}}} \to \widehat{\mathcal{U}^+}$ *in Proposition* 5.20 *(resp. the GK-embeddings* $\widehat{\Phi_a^+} : \widehat{\widehat{\mathcal{U}}} \to \widehat{\mathcal{U}_a^+}$ *in Proposition* 5.21 *$(a = 1, 2)$) so that the following holds:*

(1) *In the situation of Lemma* 9.10 *(1) we assume in addition that there exists a differential form* $\widehat{h^+}$ *on* $\widehat{\mathcal{U}^+}$ *such that its pullback to* $\widehat{\mathcal{U}}$ *is* $\widehat{h}$. *Then the pullback of* $\widehat{h^+}$ *to* $\widehat{\widehat{\mathcal{U}}}$ *is* $\widehat{\widehat{h}}$.
(2) *If* $\widehat{\mathfrak{S}^+}$ *is a CF-perturbation of* $\widehat{\mathcal{U}^+}$ *such that* $\widehat{\mathfrak{S}^+}$, $\widehat{\mathfrak{S}}$ *are compatible with the embedding* $\widehat{\mathcal{U}} \to \widehat{\mathcal{U}^+}$, *then we may choose* $\widehat{\widehat{\mathfrak{S}}}$ *such that* $\widehat{\widehat{\mathfrak{S}}}$, $\widehat{\mathfrak{S}^+}$ *are compatible with the embedding* $\widehat{\Phi^+}$. *(Respectively, if* $\widehat{\mathfrak{S}_a^+}$ $(a = 1, 2)$ *is a CF-perturbation of* $\widehat{\mathcal{U}_a^+}$ *such that* $\widehat{\mathfrak{S}_a^+}$, $\widehat{\mathfrak{S}}$ *are compatible with the embedding* $\widehat{\mathcal{U}} \to \widehat{\mathcal{U}^+}$, *then we may choose* $\widehat{\widehat{\mathfrak{S}}}$ *such that* $\widehat{\widehat{\mathfrak{S}}}$, $\widehat{\mathfrak{S}_a^+}$ *are both compatible with the embedding* $\widehat{\Phi_a^+}$.*)*
(3) *In the situation of (2) the following holds:*

(a) *If* $\widehat{\mathfrak{S}}$ *is transversal to* 0, *so is* $\widehat{\widehat{\mathfrak{S}}}$.
(b) *In the situation of Proposition* 6.33 *suppose* Y *is a manifold* M. *Then if* $\widehat{f}$ *is strongly submersive with respect to* $\widehat{\mathfrak{S}}$, *then* $\widehat{\widehat{f}}$ *is strongly submersive with respect to* $\widehat{\widehat{\mathfrak{S}}}$.
(c) *In the situation of Proposition* 6.33 *suppose* M *is a manifold. Then if* $\widehat{f}$ *is strongly transversal to* $g : N \to M$ *with respect to* $\widehat{\mathfrak{S}}$, *then* $\widehat{\widehat{f}}$ *is strongly transversal to* $g : N \to M$ *with respect to* $\widehat{\widehat{\mathfrak{S}}}$.

The proofs of Lemmas 9.10 and 9.11 are given in Sect. 11.3.

9.2 Integration Along the Fiber (pushout) for Kuranishi Structures

Situation 9.12 Let $\widehat{\mathcal{U}}$ be a Kuranishi structure on X and $\widehat{\mathfrak{S}}$ a CF-perturbation of $(X, Z; \widehat{\mathcal{U}})$. Let $\widehat{f} : (X, Z; \widehat{\mathcal{U}}) \to M$ be a strongly smooth map that is strongly submersive with respect to $\widehat{\mathfrak{S}}$. Let $\widehat{h}$ be a differential form on $\widehat{\mathcal{U}}$. By Lemma 9.10, we obtain $\widehat{\widehat{\mathcal{U}}}$, $\widehat{\Phi}$, $\widehat{\widehat{\mathfrak{S}}}$, $\widehat{\widehat{f}}$, $\widehat{\widehat{h}}$.■

Definition 9.13 In Situation 9.12, we define the *pushout*, or the *integration along the fiber* $\widehat{f}!(\widehat{h}; \widehat{\mathfrak{S}^\epsilon})$ by

$$\widehat{f}!(\widehat{h};\widehat{\mathfrak{S}^{\epsilon}}) = \widehat{f}!\left(\widehat{h};\widehat{\mathfrak{S}^{\epsilon}}\right). \tag{9.3}$$

Here the right hand side is defined in Definition 7.79. Hereafter we mostly use the terminology 'pushout' in this book.

Theorem 9.14 *The right hand side of* (9.3) *is independent of choices of* $\widehat{\mathcal{U}}$, $\widehat{\Phi}$, $\widehat{\mathfrak{S}}$, $\widehat{f}$, $\widehat{h}$ *in the sense of* ♠ *of Definition* 7.80, *but depends only on* $\widehat{\mathcal{U}}$, $\widehat{\mathfrak{S}}$, $\widehat{f}$, $\widehat{h}$ *and* ϵ.

The proof uses Proposition 9.16. To state it we consider the following situation.

Situation 9.15 Let $\widehat{\mathcal{U}}$, $\widehat{\mathcal{U}^+}$ be good coordinate systems of $Z \subseteq X$, $\widehat{\Phi} : \widehat{\mathcal{U}} \to \widehat{\mathcal{U}^+}$ a GG-embedding, and $\mathcal{K}$, $\mathcal{K}^+$ the respective support systems of $\widehat{\mathcal{U}}$, $\widehat{\mathcal{U}^+}$ compatible with $\widehat{\Phi}$. Let $\widehat{\mathfrak{S}}$, $\widehat{\mathfrak{S}^+}$ be CF-perturbations of $(\widehat{\mathcal{U}}, \mathcal{K})$, $(\widehat{\mathcal{U}^+}, \mathcal{K}^+)$, respectively. Let $\widehat{h^+}$ be a differential form on $\widehat{\mathcal{U}^+}$ which has a compact support in $\overset{\circ}{Z}$ and $\widehat{f^+} : (X, Z; \widehat{\mathcal{U}^+}) \to M$ a strongly smooth map. We put $\widehat{h} = \widehat{\Phi}^*\widehat{h^+}$ and $\widehat{f} = \widehat{f^+} \circ \widehat{\Phi} : (X, Z; \widehat{\mathcal{U}}) \to M$.

We assume that $\widehat{f}$ is strongly submersive with respect to $\widehat{\mathfrak{S}}$ and $\widehat{f^+}$ is strongly submersive with respect to $\widehat{\mathfrak{S}^+}$. ■

Proposition 9.16 *In Situation* 9.15 *we have*

$$\widehat{f}!\left(\widehat{h};\widehat{\mathfrak{S}^{\epsilon}}\right) = \widehat{f^+}!\left(\widehat{h^+};\widehat{\mathfrak{S}^{+\epsilon}}\right) \tag{9.4}$$

for each sufficiently small $\epsilon > 0$.

Proof of Proposition 9.16 ⇒ Theorem 9.14 We use Definition-Lemma 5.17 also in this proof.

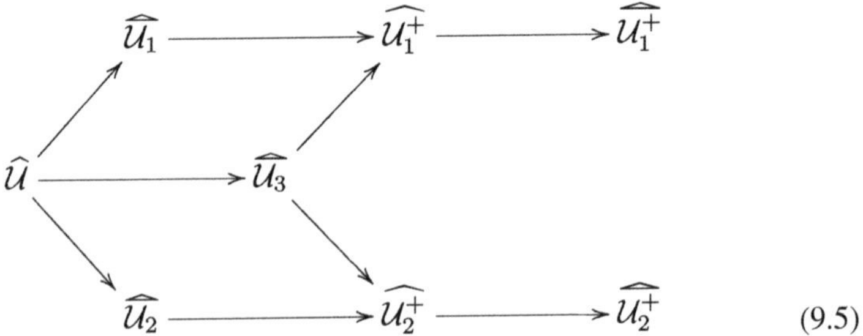

$$\tag{9.5}$$

Let $\widehat{\mathcal{U}_i}$, $\widehat{\Phi_i}$, $\widehat{\mathfrak{S}_i}$, $\widehat{f_i}$, $\widehat{h_i}$, $i = 1, 2$ be two such choices. We replace $\widehat{\mathcal{U}_i}$ by its proper open substructure such that $\widehat{\mathfrak{S}_i}$ is supported in a support system of this substructure. Then by Lemma 9.9, we have $\widehat{\mathcal{U}_i^+}$, $\widehat{\mathfrak{S}_i^+}$, $\widehat{f_i^+}$, $\widehat{h_i^+}$, $i = 1, 2$ and GK-embeddings $\widehat{\Phi_i^+} : \widehat{\mathcal{U}_i} \to \widehat{\mathcal{U}_i^+}$ with which various objects are compatible.

By Lemma 9.11 we obtain $\widehat{\mathcal{U}_3}$, $\widehat{\Phi_3}$, $\widehat{\mathfrak{S}_3}$, $\widehat{f_3}$, $\widehat{h_3}$ and GK-embeddings $\widehat{\Phi_i^{+-}} : \widehat{\mathcal{U}_3} \to \widehat{\mathcal{U}_i^+}$ with which various objects are compatible.

By Lemma 9.10, we obtain $\widehat{\mathcal{U}_i^+}$, $\widehat{\Phi_i^+}$, $\widehat{\mathfrak{S}_i^+}$, $\widehat{f_i^+}$, $\widehat{h_i^+}$ and KG-embeddings $\widehat{\Phi_i^+}$: $\widehat{\mathcal{U}_i^+} \to \widehat{\mathcal{U}_i^+}$ with which various objects are compatible.

Now we claim

$$\widehat{f_1}!\left(\widehat{h_1}; \widehat{\mathfrak{S}_1^\epsilon}\right) = \widehat{f_1^+}!\left(\widehat{h_1^+}; \widehat{\mathfrak{S}_1^{+\epsilon}}\right). \tag{9.6}$$

In fact by Definition-Lemma 5.17 there exists a weakly open substructure $\widehat{\mathcal{U}_{0,1}}$ of $\widehat{\mathcal{U}_1}$ and a GG-embedding $\widehat{\mathcal{U}_{0,1}} \to \widehat{\mathcal{U}_1^+}$. By Lemma 9.6 we can restrict $\widehat{\mathfrak{S}_1}$ to $\widehat{\mathfrak{S}_{0,1}}$, as well as other objects. Strong submersivity is preserved by Lemma 9.7. Moreover we can choose the pair of restricted CF-perturbations so that it is compatible with the GG-embedding $\widehat{\mathcal{U}_{0,1}} \to \widehat{\mathcal{U}_1^+}$. (See Lemma 9.24.) Therefore by Proposition 9.16 we find

$$\widehat{f_1}!\left(\widehat{h_1}; \widehat{\mathfrak{S}_1^\epsilon}\right) = \widehat{f_{0,1}}!\left(\widehat{h_{0,1}}; \widehat{\mathfrak{S}_{0,1}^\epsilon}\right) = \widehat{f_1^+}!\left(\widehat{h_1^+}; \widehat{\mathfrak{S}_1^{+\epsilon}}\right).$$

Here $\widehat{h_{0,1}}$ is the pullback of $\widehat{h_1}$ to $\widehat{\mathcal{U}_{0,1}}$.

We have thus proved (9.6). Using the same argument three more times, we obtain

$$\begin{aligned}\widehat{f_1}!\left(\widehat{h_1}; \widehat{\mathfrak{S}_1^\epsilon}\right) &= \widehat{f_1^+}!\left(\widehat{h_1^+}; \widehat{\mathfrak{S}_1^{+\epsilon}}\right) = \widehat{f_3}!\left(\widehat{h_3}; \widehat{\mathfrak{S}_3^\epsilon}\right) \\ &= \widehat{f_2^+}!\left(\widehat{h_2^+}; \widehat{\mathfrak{S}_2^{+\epsilon}}\right) = \widehat{f_2}!\left(\widehat{h_2}; \widehat{\mathfrak{S}_2^\epsilon}\right).\end{aligned}$$

We have thus proved the required independence of $\widehat{\Phi}$, $\widehat{\mathfrak{S}}$, $\widehat{f}$, $\widehat{h}$. □

9.3 Composition of GK-and KG-Embeddings: Proof of Definition-Lemma 5.17

In this section we prove Definition-Lemma 5.17. Definition-Lemma 5.17 claims that we can compose GK-and KG-embeddings and obtain a GG-embedding after replacing the good coordinate system of the domain by its weakly open substructure.

Recalling the notation for GK-embedding in Definition 5.6, we put

$$\widehat{\Phi} = \{(U_{\mathfrak{p}}(p), \Phi_{p\mathfrak{p}})\} : \widehat{\mathcal{U}} \to \widehat{\mathcal{U}}$$

and a KG-embedding $\widehat{\Phi^+} : \widehat{\mathcal{U}} \to \widehat{\mathcal{U}^+}$. We may assume $\widehat{\Phi^+}$ is strict. (Replace $\widehat{\mathcal{U}}$ by its open substructure.) We take a support system $\mathcal{K}$ of $\widehat{\mathcal{U}}$ and $\mathcal{K}^+$ of $\widehat{\mathcal{U}^+}$, respectively.

The idea of the proof of Definition-Lemma 5.17 is as follows. We first observe that we can cover a compact subset of $U_{\mathfrak{p}}$ by finitely many open subsets $U_{\mathfrak{p}}(p)$.

We can construct a weakly open substructure of $\widehat{\mathcal{U}}$ by restricting the charts $\mathcal{U}_{\mathfrak{p}}$ to $U_{\mathfrak{p}}(p)$.[1] Then by taking $U_{\mathfrak{p}}(p)$ sufficiently small we can compose the maps

$$U_{\mathfrak{p}}(p) \to U_p \to U_{\mathfrak{q}}^+, \tag{9.7}$$

where $p \in \psi_{\mathfrak{q}}^+(U_{\mathfrak{q}} \cap (s_{\mathfrak{q}}^+)^{-1}(0))$. We will use this composition to obtain a required GG-embedding. To work out this idea we need to find an appropriate partially ordered set $\mathfrak{P}_0$ that parametrizes the Kuranishi charts of the weakly open substructure $\widehat{\mathcal{U}^0}$ of $\widehat{\mathcal{U}}$. We also need to find an order-preserving map $\mathfrak{P}_0 \to \mathfrak{P}^+$, where $\mathfrak{P}^+$ is the partially ordered set appearing in $\widehat{\mathcal{U}^+}$, so that the composition (9.7) defines a required GG-embedding. We will use Lemma 9.17 to obtain the partially ordered set $\mathfrak{P}_0$. (In this section we write $\widehat{\mathcal{U}^0}$ in place of $\widehat{\mathcal{U}_0}$.)

For each given $\mathfrak{p} \in \mathfrak{P}$, $\mathfrak{q} \in \mathfrak{P}^+$ we define

$$Z_{\mathfrak{p}\mathfrak{q}} = (\mathcal{K}_{\mathfrak{p}} \cap Z) \cap (\mathcal{K}_{\mathfrak{q}}^+ \cap Z). \tag{9.8}$$

Here and hereafter the set-theoretical symbols such as equality and the intersection in (9.8) etc. are regarded as those between the subsets of $|\widehat{\mathcal{U}^+}|$.

Lemma 9.17 *There exist a finite subset $A_{\mathfrak{p}\mathfrak{q}}$ of $Z_{\mathfrak{p}\mathfrak{q}}$ for each $\mathfrak{p} \in \mathfrak{P}$, $\mathfrak{q} \in \mathfrak{P}^+$ and a subset $U_{(\mathfrak{p},p)}$ of $U_{\mathfrak{p}}$ for each $\mathfrak{p}$ and $p \in A_{\mathfrak{p}\mathfrak{q}}$ such that they have the following properties:*

(1) *$U_{(\mathfrak{p},p)}$ is an open subset of $U_{\mathfrak{p}}$ and $p \in U_{(\mathfrak{p},p)}$.*
(2) *$U_{(\mathfrak{p},p)} \subset U_{\mathfrak{p}}(p)$.*
(3) *If $p \in A_{\mathfrak{p}\mathfrak{q}}$, $p' \in A_{\mathfrak{p}'\mathfrak{q}'}$, $\mathfrak{p} \le \mathfrak{p}'$ and*

$$\varphi_{\mathfrak{p}'\mathfrak{p}}^{-1}(U_{(\mathfrak{p}',p')}) \cap U_{(\mathfrak{p},p)} \neq \emptyset,$$

then $\mathfrak{q} \le \mathfrak{q}'$.
(4) *For each $\mathfrak{p}_0 \in \mathfrak{P}$, $\mathfrak{q}_0 \in \mathfrak{P}^+$ we have*

$$\bigcup_{\substack{\mathfrak{p},\mathfrak{q}:\mathfrak{p}_0 \le \mathfrak{p}, \mathfrak{q}_0 \le \mathfrak{q}, \\ p \in A_{\mathfrak{p}\mathfrak{q}}}} \left(U_{(\mathfrak{p},p)} \cap Z\right) \supseteq Z_{\mathfrak{p}_0\mathfrak{q}_0}.$$

(5) *If $(\mathfrak{p},\mathfrak{q}) \neq (\mathfrak{p}',\mathfrak{q}')$ then $A_{\mathfrak{p}\mathfrak{q}} \cap A_{\mathfrak{p}'\mathfrak{q}'} = \emptyset$.*
(6) *In the situation of (3) we have $U_{(\mathfrak{p},p)} \subset U_{\mathfrak{p}'\mathfrak{p}}$ and $\varphi_{\mathfrak{p}'\mathfrak{p}}(U_{(\mathfrak{p},p)}) \subset U_{\mathfrak{p}'}(p')$.*

Proof of Lemma 9.17 ⇒ Definition-Lemma 5.17 We put

[1] $U_{\mathfrak{p}}(p)$ is the domain of Φ_{pp}.

$$\mathfrak{P}_0 = \bigcup_{(\mathfrak{p},\mathfrak{q})\in\mathfrak{P}\times\mathfrak{P}^+} A_{\mathfrak{p}\mathfrak{q}} \times \{(\mathfrak{p},\mathfrak{q})\}.$$

We define a partial order on $\mathfrak{P}\times\mathfrak{P}^+$ such that $(\mathfrak{p},\mathfrak{q}) \le (\mathfrak{p}',\mathfrak{q}')$ if and only if '$\mathfrak{p} \le \mathfrak{p}'$' and '$\mathfrak{q} \le \mathfrak{q}'$'. (Note if $\mathfrak{p} < \mathfrak{p}'$ and $\mathfrak{q} > \mathfrak{q}'$, neither $(\mathfrak{p},\mathfrak{q}) \le (\mathfrak{p}',\mathfrak{q}')$ nor $(\mathfrak{p},\mathfrak{q}) \ge (\mathfrak{p}',\mathfrak{q}')$ hold.) We choose any linear order on $A_{\mathfrak{p}\mathfrak{q}}$ and define a partial order on $\mathfrak{P}_0$ as the following lexicographic order:

$$(x,(\mathfrak{p},\mathfrak{q})) \le (x',(\mathfrak{p}',\mathfrak{q}')) \quad \text{if and only if} \quad \begin{cases} (\mathfrak{p},\mathfrak{q}) < (\mathfrak{p}',\mathfrak{q}') \\ \text{or } (\mathfrak{p},\mathfrak{q}) = (\mathfrak{p}',\mathfrak{q}'), \, x \le x'. \end{cases}$$

We define

$$U^0_{(x,(\mathfrak{p},\mathfrak{q}))} = U_{(\mathfrak{p},x)}, \quad \mathcal{U}^0_{(x,(\mathfrak{p},\mathfrak{q}))} = \mathcal{U}_{\mathfrak{p}}|_{U^0_{(x,(\mathfrak{p},\mathfrak{q}))}}.$$

We define coordinate changes between them by restricting those of $\widehat{\mathcal{U}}$. We thus obtain a good coordinate system $\widehat{\mathcal{U}^0}$. Note we use Lemma 9.17 (3) to check Definition 3.15 (6). We use Lemma 9.17 (4) to show

$$\bigcup_{(x,(\mathfrak{p},\mathfrak{q}))} \psi_{\mathfrak{p}}(s_{\mathfrak{p}}^{-1}(0) \cap U^0_{(x,(\mathfrak{p},\mathfrak{q}))}) \supseteq Z.$$

We next define a weakly open embedding $\widehat{\mathcal{U}^0} \to \widehat{\mathcal{U}}$. We first define a map $\mathfrak{P}_0 \to \mathfrak{P}$ by sending $(x,(\mathfrak{p},\mathfrak{q})) \mapsto \mathfrak{p}$. This is order preserving. We also have an open embedding of Kuranishi charts $\mathcal{U}^0_{(x,(\mathfrak{p},\mathfrak{q}))} = \mathcal{U}_{\mathfrak{p}}|_{U^0_{(x,(\mathfrak{p},\mathfrak{q}))}} \to \mathcal{U}_{\mathfrak{p}}$. They obviously commute with coordinate change.

We next define the embedding $\widehat{\mathcal{U}^0} \to \widehat{\mathcal{U}^+}$ that will be the composition of $\widehat{\mathcal{U}^0} \to \widehat{\mathcal{U}} \to \widehat{\mathcal{U}}$ and $\widehat{\mathcal{U}} \to \widehat{\mathcal{U}^+}$. We define a map $\mathfrak{P}_0 \to \mathfrak{P}^+$ by sending $(x,(\mathfrak{p},\mathfrak{q})) \mapsto \mathfrak{q}$. This is an order-preserving map. We next define a map $\varphi_{\mathfrak{q}(x,(\mathfrak{p},\mathfrak{q}))} : \mathcal{U}^0_{(x,(\mathfrak{p},\mathfrak{q}))} \to \mathcal{U}^+_{\mathfrak{q}}$ as the composition of

$$\mathcal{U}^0_{(x,(\mathfrak{p},\mathfrak{q}))} \to \mathcal{U}_{\mathfrak{p}}|_{U_{\mathfrak{p}}(x)} \to \mathcal{U}_x \to \mathcal{U}^+_{\mathfrak{q}}.$$

Here the first map $\mathcal{U}^0_{(x,(\mathfrak{p},\mathfrak{q}))} \to \mathcal{U}_{\mathfrak{p}}|_{U_{\mathfrak{p}}(x)}$ is an open embedding which exists by Lemma 9.17 (2). The second map $\mathcal{U}_{\mathfrak{p}}|_{U_{\mathfrak{p}}(x)} \to \mathcal{U}_x$ is a part of the GK-embedding $\widehat{\mathcal{U}} \to \widehat{\mathcal{U}}$. The third map $\mathcal{U}_x \to \mathcal{U}^+_{\mathfrak{q}}$ is a part of the KG-embedding $\widehat{\mathcal{U}} \to \widehat{\mathcal{U}^+}$.

We now show that the maps $\varphi_{\mathfrak{q}(x,(\mathfrak{p},\mathfrak{q}))}$ together with their lifts to obstruction bundles define a GG-embedding.

Suppose $(x,(\mathfrak{p},\mathfrak{q})) < (x',(\mathfrak{p}',\mathfrak{q}'))$. If $U^0_{(x',(\mathfrak{p}',\mathfrak{q}'))(x,(\mathfrak{p},\mathfrak{q}))} = \varphi^{-1}_{\mathfrak{p}'\mathfrak{p}}(U_{(\mathfrak{p}',x')}) \cap U_{(\mathfrak{p},x)}$ is an empty set, there is nothing to show. If this set is not an empty set, we can apply Lemma 9.17 (6) to obtain $U_{(\mathfrak{p},x)} \subset U_{\mathfrak{p}'\mathfrak{p}}$ and $\varphi_{\mathfrak{p}'\mathfrak{p}}(U_{(\mathfrak{p},x)}) \subset U_{\mathfrak{p}'}(x')$. Note $\varphi_{\mathfrak{q}'(x',(\mathfrak{p}',\mathfrak{q}'))} : \mathcal{U}^0_{(x',(\mathfrak{p}',\mathfrak{q}'))} \to \mathcal{U}^+_{\mathfrak{q}'}$ is extended to $U_{\mathfrak{p}'}(x')$ and

$$\varphi_{\mathfrak{q}'(x',(\mathfrak{p}',\mathfrak{q}'))} \circ \varphi_{\mathfrak{p}'\mathfrak{p}} = \varphi^+_{\mathfrak{q}'\mathfrak{q}} \circ \varphi_{\mathfrak{q}(x,(\mathfrak{p},\mathfrak{q}))} \tag{9.9}$$

on $U_{(\mathfrak{p},x)}$. We remark that the composition in the right hand side is always defined, since we take $U_{\mathfrak{p}}(p)$ small so that the composition (9.7) is defined. Therefore

$$\varphi^{-1}_{\mathfrak{p}'\mathfrak{p}}(U_{(\mathfrak{p}',x')}) \cap U_{(\mathfrak{p},x)} = (\varphi^+_{\mathfrak{q}'\mathfrak{q}} \circ \varphi_{\mathfrak{q}(x,(\mathfrak{p},\mathfrak{q}))})^{-1}(\varphi_{\mathfrak{q}'(x',(\mathfrak{p}',\mathfrak{q}'))}(U_{(\mathfrak{p}',x')})).$$

This is the condition Definition 3.24 (1). Then Definition 3.24 (2) is (9.9). We thus proved that the maps $\varphi_{\mathfrak{q}(x,(\mathfrak{p},\mathfrak{q}))}$ together with their lifts to obstruction bundles define a GG-embedding.

Thus to complete the proof of Definition-Lemma 5.17 it remains to prove Lemma 9.17. □

We use the following during the proof of Lemma 9.17.

Definition 9.18 Let $(\mathfrak{P}, \le)$ be a partially ordered set. A subset $\mathfrak{I} \subseteq \mathfrak{P}$ is said to be an *ideal* if $\mathfrak{p} \in \mathfrak{I}$, $\mathfrak{p}' \ge \mathfrak{p}$ implies $\mathfrak{p}' \in \mathfrak{I}$.

Proof of Lemma 9.17 Let $\mathfrak{I} \subset \mathfrak{P} \times \mathfrak{P}^+$ be an ideal. We will prove the following by induction on $\#\mathfrak{I}$.

Sublemma 9.19 *For each $(\mathfrak{p}, \mathfrak{q}) \in \mathfrak{I}$ there exist a finite subset $A_{\mathfrak{p}\mathfrak{q}}$ of $Z_{\mathfrak{p}\mathfrak{q}}$ and a subset $U_{(\mathfrak{p},p)}$ of $U_{\mathfrak{p}}$ for each $p \in A_{\mathfrak{p}\mathfrak{q}}$ such that they satisfy (1)(2)(5) of Lemma 9.17 and the following conditions (3)', (4)' and (6)':*

(3)' (a) *If $(\mathfrak{p}, \mathfrak{q}), (\mathfrak{p}', \mathfrak{q}') \in \mathfrak{I}$, then Lemma 9.17 (3) holds.*
(b) *If $(\mathfrak{p}, \mathfrak{q}) \in \mathfrak{I}$, $p \in A_{\mathfrak{p}\mathfrak{q}}$ and $(\mathfrak{p}', \mathfrak{q}') \in \mathfrak{P} \times \mathfrak{P}^+$ satisfies*

$$\overline{U_{(\mathfrak{p},p)}} \cap Z_{\mathfrak{p}'\mathfrak{q}'} \neq \emptyset,$$

then $(\mathfrak{p}, \mathfrak{q}) \ge (\mathfrak{p}', \mathfrak{q}')$.

(4)' *For each $(\mathfrak{p}_0, \mathfrak{q}_0) \in \mathfrak{I}$ we have*

$$\bigcup_{\substack{\mathfrak{p},\mathfrak{q}:(\mathfrak{p}_0,\mathfrak{q}_0)\le(\mathfrak{p},\mathfrak{q}),\\ p\in A_{\mathfrak{p}\mathfrak{q}}}} (U_{(\mathfrak{p},p)} \cap Z) \supseteq Z_{\mathfrak{p}_0\mathfrak{q}_0}.$$

(6)' *(6) holds in the situation of (3)' (a).*

Note that we do not assume $(\mathfrak{p}', \mathfrak{q}') \in \mathfrak{I}$ in Sublemma 9.19 (3)' (b).

Proof The case $\mathfrak{I} = \emptyset$ is trivial. Suppose Sublemma 9.19 is proved for all $\mathfrak{I}'$ with $\#\mathfrak{I}' < \#\mathfrak{I}$. We will prove the case of $\mathfrak{I}$.

Let $(\mathfrak{p}_1, \mathfrak{q}_1)$ be a minimal element of $\mathfrak{I}$. Then $\mathfrak{I}_- = \mathfrak{I} \setminus \{(\mathfrak{p}_1, \mathfrak{q}_1)\}$ is an ideal of $\mathfrak{P} \times \mathfrak{P}^+$. By the induction hypothesis, we obtain $A_{\mathfrak{p}\mathfrak{q}}$ for $(\mathfrak{p}, \mathfrak{q}) \in \mathfrak{I}_-$ and $U_{(\mathfrak{p},p)}$

for each $p \in A_{\mathfrak{pq}}$, $(\mathfrak{p},\mathfrak{q}) \in \mathfrak{I}_-$. By the induction hypothesis again, Sublemma 9.19 (4)', the set

$$O = \bigcup_{\substack{(\mathfrak{p},\mathfrak{q})\in\mathfrak{I}:(\mathfrak{p}_1,\mathfrak{q}_1)<(\mathfrak{p},\mathfrak{q})\\ p\in A_{\mathfrak{pq}}}} \left(\overline{U_{(\mathfrak{p},p)}} \cap Z_{\mathfrak{p}_1\mathfrak{q}_1}\right)$$

is an open neighborhood of

$$L = \left(\bigcup_{(\mathfrak{p},\mathfrak{q})\in\mathfrak{I}:(\mathfrak{p}_1,\mathfrak{q}_1)<(\mathfrak{p},\mathfrak{q})} Z_{\mathfrak{pq}}\right) \cap Z_{\mathfrak{p}_1\mathfrak{q}_1}$$

in $Z_{\mathfrak{p}_1\mathfrak{q}_1}$.

Subsublemma 9.20 *If $x \in Z_{\mathfrak{p}_1\mathfrak{q}_1} \setminus O$ and $x \in Z_{\mathfrak{pq}}$, then $(\mathfrak{p},\mathfrak{q}) \le (\mathfrak{p}_1,\mathfrak{q}_1)$.*

Note that we do *not* assume $(\mathfrak{p},\mathfrak{q}) \in \mathfrak{I}$.

Proof Since

$$x \in \mathcal{K}_{\mathfrak{p}} \cap \mathcal{K}_{\mathfrak{p}_1} \cap \mathcal{K}^+_{\mathfrak{q}} \cap \mathcal{K}^+_{\mathfrak{q}_1} \cap Z,$$

Definition 3.15 (6) implies that both '$\mathfrak{p} \le \mathfrak{p}_1$ or $\mathfrak{p} \ge \mathfrak{p}_1$' and '$\mathfrak{q} \le \mathfrak{q}_1$ or $\mathfrak{q} \ge \mathfrak{q}_1$' hold. We will prove that $(\mathfrak{p},\mathfrak{q}) \le (\mathfrak{p}_1,\mathfrak{q}_1)$, that is, $\mathfrak{p} \le \mathfrak{p}_1$ and $\mathfrak{q} \le \mathfrak{q}_1$.

Suppose to the contrary that $\mathfrak{p} > \mathfrak{p}_1$. Then $x \in \mathcal{K}_{\mathfrak{p}} \cap \mathcal{K}^+_{\mathfrak{q}_1} \cap Z$ and $(\mathfrak{p},\mathfrak{q}_1) > (\mathfrak{p}_1,\mathfrak{q}_1)$. This contradicts $x \notin O$. We can find a contradiction from $\mathfrak{q} > \mathfrak{q}_1$ in a similar way. Therefore we obtain $(\mathfrak{p},\mathfrak{q}) \le (\mathfrak{p}_1,\mathfrak{q}_1)$. □

Subsublemma 9.21 *For each $x \in Z_{\mathfrak{p}_1\mathfrak{q}_1} \setminus O$, there exists its neighborhood W_x in $U_{\mathfrak{p}_1}$ with the following properties:*

1. *$x \in W_x$ and W_x is open in $U_{\mathfrak{p}_1}$.*
2. *$W_x \subset U_{\mathfrak{p}_1}(x)$.*
3. *If $W_x \cap Z_{\mathfrak{pq}} \neq \emptyset$ then $(\mathfrak{p},\mathfrak{q}) \le (\mathfrak{p}_1,\mathfrak{q}_1)$.*
4. *If $\mathfrak{p} \ge \mathfrak{p}_1$, $p \in A_{\mathfrak{pq}}$, $(\mathfrak{p},\mathfrak{q}) \in \mathfrak{I}_-$ and $W_x \cap \varphi^{-1}_{\mathfrak{pp}_1}(U_{(\mathfrak{p},p)}) \neq \emptyset$, then $\mathfrak{q} \ge \mathfrak{q}_1$.*
5. *In the situation of (4), $W_x \subseteq U_{\mathfrak{pp}_1}$ and $\varphi_{\mathfrak{pp}_1}(W_x) \subseteq U_{\mathfrak{p}}(p)$.*

Proof It is easy to see that (1)(2) hold for a sufficiently small neighborhood W_x of x. Since $Z_{\mathfrak{pq}}$ is a closed set, Subsublemma 9.20 implies that (3) holds for a sufficiently small neighborhood W_x of x.

We next prove that (4) holds for a sufficiently small neighborhood W_x of x. Suppose $\mathfrak{p} \ge \mathfrak{p}_1$, $p \in A_{\mathfrak{pq}}$, $(\mathfrak{p},\mathfrak{q}) \in \mathfrak{I}_-$ and $x \in \overline{U_{(\mathfrak{p},p)}}$. Then $x \in \overline{U_{(\mathfrak{p},p)}} \cap Z_{\mathfrak{p}_1\mathfrak{q}_1}$. We apply the induction hypothesis Sublemma 9.19 (3)' (b) to $\mathfrak{I}_-$ and $(\mathfrak{p}',\mathfrak{q}') = (\mathfrak{p}_1,\mathfrak{q}_1)$. We then find $(\mathfrak{p},\mathfrak{q}) \ge (\mathfrak{p}_1,\mathfrak{q}_1)$. In particular, $\mathfrak{q} \ge \mathfrak{q}_1$. Thus we can take a sufficiently small neighborhood W_x so that

$$\mathfrak{p} \ge \mathfrak{p}_1,\ p \in A_{\mathfrak{pq}},\ (\mathfrak{p},\mathfrak{q}) \in \mathfrak{I}_-,\ W_x \cap \overline{\varphi^{-1}_{\mathfrak{p}p}(U_{(\mathfrak{p},p)})} \neq \emptyset \Rightarrow \mathfrak{q} \ge \mathfrak{q}_1.$$

Since W_x is open, the condition $W_x \cap \varphi_{\mathfrak{p}\mathfrak{p}_1}^{-1}(\overline{U_{(\mathfrak{p},q)}}) \neq \emptyset$ is equivalent to the condition $W_x \cap \varphi_{\mathfrak{p}\mathfrak{p}_1}^{-1}(U_{(\mathfrak{p},q)}) \neq \emptyset$. Thus we have proved (4).

Note $x \in \overline{U_{(\mathfrak{p},p)}} \cap Z_{\mathfrak{p}_1\mathfrak{q}_1} \subset U_{\mathfrak{p}}(p) \cap s_{\mathfrak{p}_1}^{-1}(0) \cap \mathcal{K}_{\mathfrak{p}_1}$ in the situation of (4). Since $\varphi_{\mathfrak{p}\mathfrak{p}_1}$ is a coordinate change *in the strong sense*, it implies $x \in U_{\mathfrak{p}\mathfrak{p}_1}$ and $\varphi_{\mathfrak{p}\mathfrak{p}_1}(x) \in U_{\mathfrak{p}}(p)$. Therefore, we can prove (5) by taking W_x small. □

We take an open neighborhood W_x^0 of x such that $\overline{W_x^0} \subset W_x$. We take a finite subset $A_{\mathfrak{p}_1\mathfrak{q}_1} \subset Z_{\mathfrak{p}_1\mathfrak{q}_1} \setminus O$ such that

$$Z_{\mathfrak{p}_1\mathfrak{q}_1} \setminus O \subset \bigcup_{x \in A_{\mathfrak{p}_1\mathfrak{q}_1}} W_x^0. \tag{9.10}$$

Lemma 9.17 (5) is obvious from the definition.

For $x \in A_{\mathfrak{p}_1\mathfrak{q}_1}$, we put

$$U_{(\mathfrak{p}_1,x)} = W_x^0. \tag{9.11}$$

Subsublemma 9.22 *There exists an open neighborhood $U'_{(\mathfrak{p},p)}$ of p for $(\mathfrak{p},\mathfrak{q}) \in \mathfrak{I}_-$ and $p \in A_{\mathfrak{p}\mathfrak{q}}$ such that the following holds:*

(1) $U'_{(\mathfrak{p},p)} \subset U_{(\mathfrak{p},p)}$.
(2) *Sublemma* 9.19 *(1)(2)(4)' hold for* $U'_{(\mathfrak{p},p)}$.
(3) *If* $p \in A_{\mathfrak{p}\mathfrak{q}}$, $(\mathfrak{p},\mathfrak{q}) \in \mathfrak{I}_-$, $\mathfrak{p}_1 \geq \mathfrak{p}$, $x \in A_{\mathfrak{p}_1\mathfrak{q}_1}$, *then*

$$\varphi_{\mathfrak{p}_1\mathfrak{p}}^{-1}(U_{(\mathfrak{p}_1,x)}) \cap U'_{(\mathfrak{p},p)} = \emptyset.$$

Proof We take

$$U'_{(\mathfrak{p},p)} = U_{(\mathfrak{p},p)} \setminus \bigcup_{x \in A_{\mathfrak{p}_1\mathfrak{q}_1}} \overline{U_{(\mathfrak{p}_1,x)}}. \tag{9.12}$$

Here we regard $U_{(\mathfrak{p},p)}$ and $\overline{U_{(\mathfrak{p}_1,x)}}$ as subsets of $|\widehat{\mathcal{U}}|$. (1) and (3) are immediate. We will prove (2). By Subsublemma 9.21 (3) and (9.11), we have

$$Z_{\mathfrak{p}\mathfrak{q}} \cap \overline{U_{(\mathfrak{p}_1,x)}} = \emptyset, \tag{9.13}$$

for each $(\mathfrak{p},\mathfrak{q}) \in \mathfrak{I}_-$, $x \in A_{\mathfrak{p}_1\mathfrak{q}_1}$. Therefore $p \in A_{\mathfrak{p}\mathfrak{q}}$, $(\mathfrak{p},\mathfrak{q}) \in \mathfrak{I}_-$ imply $p \notin \overline{U_{(\mathfrak{p}_1,x)}}$. Hence $p \in U'_{(\mathfrak{p},p)} \subset U_{(\mathfrak{p},p)}$. This implies that Sublemma 9.19 (1)(2) hold for $U'_{(\mathfrak{p},p)}$. Sublemma 9.19 (4)' is a consequence of (9.13) and (9.12). □

Hereafter we write $U_{(\mathfrak{p},p)}$ in place of $U'_{(\mathfrak{p},p)}$.

We will prove that they have the properties claimed in Sublemma 9.19. Sublemma 9.19 (1),(2),(4)' follow from Subsublemma 9.22 (2).

Proof of Sublemma 9.19 (3)' (a) Suppose $(\mathfrak{p}, \mathfrak{q}), (\mathfrak{p}', \mathfrak{q}') \in \mathfrak{I}$, $p \in A_{\mathfrak{pq}}$, $p' \in A_{\mathfrak{p}'\mathfrak{q}'}$, $\mathfrak{p} \le \mathfrak{p}'$ and $\varphi_{\mathfrak{p}'\mathfrak{p}}^{-1}(U_{(\mathfrak{p}',p')}) \cap U_{(\mathfrak{p},p)} \ne \emptyset$. We will prove $\mathfrak{q} \le \mathfrak{q}'$.

The case $(\mathfrak{p}, \mathfrak{q}), (\mathfrak{p}', \mathfrak{q}') \in \mathfrak{I}_-$ follows from the induction hypothesis.

Suppose $(\mathfrak{p}', \mathfrak{q}') = (\mathfrak{p}_1, \mathfrak{q}_1)$. Then $\varphi_{\mathfrak{p}_1\mathfrak{p}}^{-1}(U_{(\mathfrak{p}_1,p')}) \cap U_{(\mathfrak{p},p)} \ne \emptyset$. Subsublemma 9.22 (3) implies that $(\mathfrak{p}, \mathfrak{q}) \notin \mathfrak{I}_-$. Therefore $(\mathfrak{p}, \mathfrak{q}) = (\mathfrak{p}_1, \mathfrak{q}_1)$. Hence $\mathfrak{q} \le \mathfrak{q}'$ as required.

We next assume $(\mathfrak{p}_1, \mathfrak{q}_1) = (\mathfrak{p}, \mathfrak{q})$. Then Subsublemma 9.21 (4) implies $\mathfrak{q}_1 \le \mathfrak{q}'$ as required. □

Proof of Sublemma 9.19 (3)' (b) Suppose $(\mathfrak{p}, \mathfrak{q}) \in \mathfrak{I}$, $p \in A_{\mathfrak{pq}}$ and $\overline{U_{(\mathfrak{p},p)}} \cap Z_{\mathfrak{p}'\mathfrak{q}'} \ne \emptyset$. We will prove $(\mathfrak{p}, \mathfrak{q}) \ge (\mathfrak{p}', \mathfrak{q}')$.

The case $(\mathfrak{p}, \mathfrak{q}) \in \mathfrak{I}_-$ follows from the induction hypothesis. Suppose $(\mathfrak{p}, \mathfrak{q}) = (\mathfrak{p}_1, \mathfrak{q}_1)$. Then $\overline{U_{(\mathfrak{p}_1,p)}} \cap Z_{\mathfrak{p}'\mathfrak{q}'} \ne \emptyset$. Note $\overline{U}_{(\mathfrak{p}_1,x)} \subset \overline{W}_x^0 \subset W_x$. Therefore Subsublemma 9.21 (3) implies $(\mathfrak{p}', \mathfrak{q}') \le (\mathfrak{p}_1, \mathfrak{q}_1)$, as required. □

Proof of Sublemma 9.19 (6)' In the case $(\mathfrak{p}, \mathfrak{q}), (\mathfrak{p}', \mathfrak{q}') \in \mathfrak{I}_-$ this is a consequence of the induction hypothesis. If $(\mathfrak{p}, \mathfrak{q}) \in \mathfrak{I}_-$ and $(\mathfrak{p}', \mathfrak{q}') = (\mathfrak{p}_1, \mathfrak{q}_1)$ this is a consequence of Sublemma 9.21 (5). □

Therefore the proof of Sublemma 9.19 is now complete. □

Lemma 9.17 is the case $\mathfrak{I} = \mathfrak{P} \times \mathfrak{P}^+$ of Sublemma 9.19. □

During the proof of Theorem 9.14 we use the relationship between the GG-embedding (obtained by Definition-Lemma 5.17) and CF-perturbations. We explain this point in more detail below.

Situation 9.23 Let $\widehat{\mathcal{U}}, \widehat{\mathcal{U}^+}$ be good coordinate systems and $\widehat{\mathcal{U}}$ a Kuranishi structure. We consider GK-embedding $\widehat{\Phi} : \widehat{\mathcal{U}} \to \widehat{\mathcal{U}}$ and KG-embedding $\widehat{\Phi^+} : \widehat{\mathcal{U}} \to \widehat{\mathcal{U}^+}$. Then we obtain a GG-embedding $\widehat{\mathcal{U}^0} \to \widehat{\mathcal{U}^+}$. (This is Definition-Lemma 5.17.)

Suppose in addition that we are given support systems $\mathcal{K}'$, $\mathcal{K}'^+$ of $\widehat{\mathcal{U}}$, $\widehat{\mathcal{U}^+}$ respectively and CF-perturbations $\widehat{\mathfrak{S}}$, $\widehat{\mathfrak{S}^+}$ of $(\widehat{\mathcal{U}}, \mathcal{K}')$, $(\widehat{\mathcal{U}^+}, \mathcal{K}'^+)$, respectively. Moreover we assume there exists a CF-perturbation $\widehat{\mathfrak{S}}$ of $\widehat{\mathcal{U}}$ such that $\widehat{\mathfrak{S}}$, $\widehat{\mathfrak{S}^+}$, $\widehat{\mathfrak{S}}$ are compatible with $\widehat{\Phi}$, $\widehat{\Phi^+}$.■

Lemma 9.24 *In Situation 9.23 we may choose $\widehat{\mathcal{U}^0}$ and GG-embedding $\widehat{\mathcal{U}^0} \to \widehat{\mathcal{U}^+}$ so that there exist a support system $\mathcal{K}^0$ of $\widehat{\mathcal{U}^0}$ and CF-perturbations $\widehat{\mathfrak{S}^0}$ of $(\widehat{\mathcal{U}^0}, \mathcal{K}^0)$ with the following properties:*

(1) *$(\mathcal{K}^0, \mathcal{K}'^+)$ is compatible with $\widehat{\mathcal{U}^0} \to \widehat{\mathcal{U}^+}$. Moreover $\widehat{\mathfrak{S}^0}$, $\widehat{\mathfrak{S}^+}$ is compatible with $\widehat{\mathcal{U}^0} \to \widehat{\mathcal{U}^+}$.*
(2) *$(\mathcal{K}^0, \mathcal{K}')$ is compatible with $\widehat{\mathcal{U}^0} \to \widehat{\mathcal{U}}$. Moreover $\widehat{\mathfrak{S}^0}$, $\widehat{\mathfrak{S}}$ is compatible with $\widehat{\mathcal{U}^0} \to \widehat{\mathcal{U}}$.*

Proof At the beginning of the proof of Definition-Lemma 5.17 we have chosen support systems $\mathcal{K}$ and $\mathcal{K}^+$. We choose them so that $\mathcal{K} < \mathcal{K}'$ and $\mathcal{K}^+ < \mathcal{K}'^+$.

In Subsublemma 9.21 we have chosen $W_x \subset U_{\mathfrak{p}_1}(x) \subset U_{\mathfrak{p}_1}$ for $x \in Z_{\mathfrak{p}_1\mathfrak{q}_1} \subseteq \mathcal{K}_{\mathfrak{p}_1} \cap \mathcal{K}^+_{\mathfrak{q}_1}$. We may choose it so small that:

(i) $W_x \subset \operatorname{Int} \mathcal{K}'_{\mathfrak{p}_1}$.
(ii) $(\varphi_{\mathfrak{q}_1 x} \circ \varphi_{x\mathfrak{p}_1})(W_x) \subset \operatorname{Int} \mathcal{K}'^+_{\mathfrak{q}_1}$.

Then for $(x, (\mathfrak{p}, \mathfrak{q})) \in \mathfrak{P}_0$ the Kuranishi neighborhood $U_{(x,(\mathfrak{p},\mathfrak{q}))}$ is

$$U^0_{(x,(\mathfrak{p},\mathfrak{q}))} = U_{(\mathfrak{p},x)} \subseteq W^0_x \subseteq \operatorname{Int} \mathcal{K}'_{\mathfrak{p}}.$$

Therefore for any choice of support system $\mathcal{K}^0$ of $\widehat{\mathcal{U}^0}$, $(\mathcal{K}^0, \mathcal{K}')$ is compatible with $\widehat{\mathcal{U}^0} \to \widehat{\mathcal{U}}$. Since $\widehat{\mathcal{U}^0}$ is a weakly open substructure we can restrict $\widehat{\mathfrak{S}}$ to obtain $\widehat{\mathfrak{S}^0}$. Thus (2) is satisfied. (1) follows from (ii). □

9.4 GG-Embedding and Integration: Proof of Proposition 9.16

Proof of Proposition 9.16 Let $(\mathcal{K}_1^+, \mathcal{K}_2^+)$ (resp. $(\mathcal{K}_1, \mathcal{K}_2)$) be a support pair of $\widehat{\mathcal{U}^+}$ (resp. $\widehat{\mathcal{U}}$). We may choose them so that $\varphi_{\mathfrak{p}}(\mathcal{K}^i_{\mathfrak{p}}) \subseteq \mathcal{K}^{i+}_{\mathfrak{i}(\mathfrak{p})}$ and $\mathcal{K}_2^+ < \mathcal{K}^+ = \mathcal{K}_3^+$ and $\mathcal{K}_2 < \mathcal{K} = \mathcal{K}_3$. Here and hereafter we write $\mathcal{K}^i_{\mathfrak{p}}$ for $\mathcal{K}_i = \{\mathcal{K}^i_{\mathfrak{p}} \mid \mathfrak{p} \in \mathfrak{P}\}$.

We will choose δ_+, δ and $\mathfrak{U}(Z)$, later. Let $\{\chi^+_{\mathfrak{p}^+}\}$ (resp. $\{\chi_{\mathfrak{p}}\}$) be a partition of unity of $(X, Z, \widehat{\mathcal{U}^+}, \mathcal{K}^+, \delta_+)$ (resp. $(X, Z, \widehat{\mathcal{U}}, \mathcal{K}, \delta)$). By inspecting the proof of Proposition 7.68, we can take $\chi_{\mathfrak{p}}$ so that it is not only a strongly smooth function on $|\mathcal{K}_2|$ but also one on $|\mathcal{K}_2^+|$.

For a given $\mathfrak{p}_0^+ \in \mathfrak{P}^+$ we set $h_0 = \chi^+_{\mathfrak{p}_0^+} h_{\mathfrak{p}_0^+}$. To prove Proposition 9.16 it suffices to show

$$f_{\mathfrak{p}_0^+}!\left(h_0; \mathfrak{S}^{+\epsilon}_{\mathfrak{p}_0^+}|_{\mathfrak{U}(Z)\cap \mathcal{K}^{+1}_{\mathfrak{p}_0^+}(2\delta_+)}\right) = \sum_{\mathfrak{p}\in\mathfrak{P}} f_{\mathfrak{p}}!((\chi_{\mathfrak{p}} h_0)_{\mathfrak{p}}; \mathfrak{S}^{\epsilon}_{\mathfrak{p}}|_{\mathfrak{U}(Z)\cap\mathcal{K}^1_{\mathfrak{p}}(2\delta)}), \tag{9.14}$$

for each $\mathfrak{p}_0 \in \mathfrak{P}$. By taking $\epsilon > 0$ sufficiently small, we may assume $\sum \chi_{\mathfrak{p}} = 1$ on $\mathfrak{U}(Z) \cap \Pi((\mathfrak{S}^{+\epsilon}_{\mathfrak{p}_0^+})^{-1}(0))$. (This is a consequence of Lemma 7.74 and Definition 7.65 (3). Note that the differential form $(\chi_{\mathfrak{p}} h_0)_{\mathfrak{p}_0}$ is defined since the function $\chi_{\mathfrak{p}}$ is strongly smooth on $|\mathcal{K}_2^+|$.) Therefore to prove (9.14) it suffices to show

$$f_{\mathfrak{p}_0^+}!\left((\chi_{\mathfrak{p}} h_0)_{\mathfrak{p}_0^+}; \mathfrak{S}^{+\epsilon}_{\mathfrak{p}_0^+}|_{\mathfrak{U}(Z)\cap \mathcal{K}^{+1}_{\mathfrak{p}_0^+}(2\delta_+)}\right) = f_{\mathfrak{p}}!((\chi_{\mathfrak{p}} h_0)_{\mathfrak{p}}; \mathfrak{S}^{\epsilon}_{\mathfrak{p}}|_{\mathfrak{U}(Z)\cap\mathcal{K}^1_{\mathfrak{p}}(2\delta)}) \tag{9.15}$$

for each $\mathfrak{p}$. We will prove (9.15) below. There are three cases to consider.

(Case 1) Neither $\mathfrak{i}(\mathfrak{p}) \leq \mathfrak{p}_0^+$ nor $\mathfrak{i}(\mathfrak{p}) \geq \mathfrak{p}_0^+$: In this case we have

$$\mathcal{K}^{1+}_{\mathfrak{p}_0^+} \cap \mathcal{K}^1_{\mathfrak{p}} \subset \mathcal{K}^{1+}_{\mathfrak{p}_0^+} \cap \mathcal{K}^{1+}_{\mathrm{i}(\mathfrak{p})} = \emptyset.$$

Therefore in the same way as the proof of (7.44), we can choose δ,δ_+ so small that

$$\Omega^{1+}_{\mathfrak{p}_0^+}(\mathcal{K}^{1+}, \delta_+) \cap \Omega^1_{\mathfrak{p}}(\mathcal{K}^1, \delta) = \emptyset.$$

Then both sides of (9.15) are zero.

(Case 2) $\mathrm{i}(\mathfrak{p}) \le \mathfrak{p}_0^+$: We consider the embedding

$$\mathcal{U}_{\mathfrak{p}} \overset{\Phi_{\mathfrak{p}}}{\longrightarrow} \mathcal{U}^+_{\mathrm{i}(\mathfrak{p})} \overset{\Phi_{\mathfrak{p}_0^+\mathrm{i}(\mathfrak{p})}}{\longrightarrow} \mathcal{U}^+_{\mathfrak{p}_0}.$$

In the same way as the proof of (7.45) we can choose $\mathfrak{U}(Z)$ so small (depending on δ and δ_+) that

$$\mathrm{Supp}(\chi_{\mathfrak{p}}\widehat{h}_0) \cap \Pi((\widehat{\mathfrak{S}^{+\epsilon}})^{-1}(0)) \cap \mathfrak{U}(Z) \subset \mathcal{K}^{1+}_{\mathfrak{p}_0^+}(2\delta^+) \cap \mathcal{K}^1_{\mathfrak{p}}(2\delta) \cap \mathfrak{U}(Z). \tag{9.16}$$

Then (9.15) follows.

(Case 3) $\mathrm{i}(\mathfrak{p}) \ge \mathfrak{p}_0^+$: Since $\Phi_{\mathfrak{p}} : \mathcal{U}_{\mathfrak{p}} \to \mathcal{U}^+_{\mathrm{i}(\mathfrak{p})}$ is an embedding, in the same way as (Case 2) we can choose $\mathfrak{U}(Z)$ so small (depending δ, δ_+) that

$$f_{\mathrm{i}(\mathfrak{p})}!\left((\chi_{\mathfrak{p}} h_0)_{\mathrm{i}(\mathfrak{p})}; \mathfrak{S}^{+\epsilon}_{\mathrm{i}(\mathfrak{p})}|_{\mathfrak{U}(Z)\cap\mathcal{K}^{+1}_{\mathrm{i}(\mathfrak{p})}(2\delta_+)}\right) = f_{\mathfrak{p}}!((\chi_{\mathfrak{p}} h_0)_{\mathfrak{p}}; \mathfrak{S}^{\epsilon}_{\mathfrak{p}}|_{\mathfrak{U}(Z)\cap\mathcal{K}^1_{\mathfrak{p}}(2\delta)}).$$

In the same way as the proof of Proposition 7.81 Case 2, we can choose $\mathfrak{U}(Z)$ so small (depending δ, δ_+) that

$$f_{\mathrm{i}(\mathfrak{p})}!\left((\chi_{\mathfrak{p}} h_0)_{\mathrm{i}(\mathfrak{p})}; \mathfrak{S}^{+\epsilon}_{\mathrm{i}(\mathfrak{p})}|_{\mathfrak{U}(Z)\cap\mathcal{K}^{+1}_{\mathrm{i}(\mathfrak{p})}(2\delta_+)}\right) = f_{\mathfrak{p}_0^+}!\left((\chi_{\mathfrak{p}} h_0)_{\mathfrak{p}_0^+}; \mathfrak{S}^{+\epsilon}_{\mathfrak{p}_0^+}|_{\mathfrak{U}(Z)\cap\mathcal{K}^{+1}_{\mathfrak{p}_0^+}(2\delta_+)}\right).$$

Equation (9.15) follows from these two formulas.

Thus the proof of Proposition 9.16 is complete. □

9.5 CF-Perturbations of Correspondences

Definition 9.25 Let $(X, \widehat{\mathcal{U}}, \widehat{f_s}, \widehat{f_t})$ be a smooth correspondence (Definition 7.1). Let $\widehat{\mathfrak{S}}$ be a CF-perturbation of $\widehat{\mathcal{U}}$ such that $\widehat{f_t}$ is strongly submersive with respect to $\widehat{\mathfrak{S}}$.

We call such $\widehat{\mathfrak{S}}$ a *CF-perturbation of smooth correspondence* $\mathfrak{X}$. We then define

$$\mathrm{Corr}_{(\mathfrak{X},\widehat{\mathfrak{S}^\epsilon})} : \Omega^k(M_s) \to \Omega^{k+\ell}(M_t) \tag{9.17}$$

by

$$\mathrm{Corr}_{(\mathfrak{X},\widehat{\mathfrak{S}^\epsilon})}(h) = \widehat{f_t}!((\widehat{f_s})^* h; \widehat{\mathfrak{S}^\epsilon}). \tag{9.18}$$

This is well-defined ϵ-wise by Theorem 9.14. We call the linear map $\mathrm{Corr}_{(\mathfrak{X},\widehat{\mathfrak{S}^\epsilon})}$ a *smooth correspondence map* of Kuranishi structure and ℓ the *degree* of smooth correspondence $\mathfrak{X}$ and write it $\deg \mathfrak{X}$.

The next lemma says that for any smooth correspondence we can always thicken the Kuranishi structure so that the assumptions of Definition 9.25 are satisfied.

Lemma 9.26 *For each smooth correspondence $(X, \widehat{\mathcal{U}}, \widehat{f_s}, \widehat{f_t})$ there exist $\widehat{\mathcal{U}^+}, \widehat{\mathfrak{S}^+}, \widehat{f_s^+}, \widehat{f_t^+}$ with the following properties:*

(1) *$(X, \widehat{\mathcal{U}^+}, \widehat{f_s^+}, \widehat{f_t^+})$ is a smooth correspondence and $\widehat{\mathfrak{S}^+}$ a CF-perturbation of smooth correspondence such that $\widehat{f_t^+}$ is strongly submersive with respect to $\widehat{\mathfrak{S}^+}$.*
(2) *$\widehat{\mathcal{U}^+}$ is a thickening of $\widehat{\mathcal{U}}$.*
(3) *Let $\widehat{\Phi} : \widehat{\mathcal{U}} \to \widehat{\mathcal{U}^+}$ be the KK-embedding. Then $\widehat{f_s^+}$ and $\widehat{f_t^+}$ induce $\widehat{f_s}$ and $\widehat{f_t}$ by $\widehat{\Phi}$.*

Proof This is an immediate consequence of Lemmas 9.9 and 9.10. □

9.6 Stokes' Formula for a Kuranishi Structure

We have Stokes' formula in Theorem 8.11, which is the formula for a good coordinate system. In this section we translate it into the one for a Kuranishi structure.

Situation 9.27 Let $\widehat{\mathcal{U}}$ be a Kuranishi structure of $Z \subseteq X$, $\widehat{\mathfrak{S}}$ its CF-perturbation, $\widehat{f} : (X, Z; \widehat{\mathcal{U}}) \to M$ a strongly submersive map with respect to $\widehat{\mathfrak{S}}$, and $\widehat{h}$ a differential form on $(X, Z; \widehat{\mathcal{U}})$.

Let $\partial(X, Z, \widehat{\mathcal{U}}, \widehat{\mathfrak{S}}) = (\partial X, \partial Z, \widehat{\mathcal{U}_\partial}, \widehat{\mathfrak{S}_\partial})$, where $(\partial X, \partial Z; \widehat{\mathcal{U}_\partial})$ is the normalized boundary of $(X, Z; \widehat{\mathcal{U}})$ on which $\widehat{\mathfrak{S}}$ induces a CF-perturbation $\widehat{\mathfrak{S}_\partial}$ by Lemma 8.10 (2). Since $\widehat{f}$ induces a map $\widehat{f_\partial} : (\partial X, \partial Z; \widehat{\mathcal{U}_\partial}) \to M$, which is strongly submersive with respect to $\widehat{\mathfrak{S}_\partial}$ if $\widehat{f}$ is strongly submersive with respect to $\widehat{\mathfrak{S}}$ (Lemma 8.10 (4)).

Let $\widehat{h_\partial}$ be the restriction of $\widehat{h}$ to $(\partial X, \partial Z; \widehat{\mathcal{U}_\partial})$.■

Theorem 9.28 (Stokes' formula for a Kuranishi structure) *In Situation 9.27 we have the next formula for each sufficiently small $\epsilon > 0$:*

$$d\left(\widehat{f}!(\widehat{h};\widehat{\mathfrak{S}^{\epsilon}})\right) = \widehat{f}!(d\widehat{h};\widehat{\mathfrak{S}^{\epsilon}}) + (-1)^{\dim(X,\widehat{\mathcal{U}})+\deg\widehat{h}}\,\widehat{f_\partial}!(\widehat{h_\partial};\widehat{\mathfrak{S}^{\epsilon}_\partial}). \tag{9.19}$$

Proof Recall that each term of (9.19) was defined in Definition 9.13 by choosing a good coordinate system $\widehat{\mathcal{U}}$ compatible with the Kuranishi structure $\widehat{\mathcal{U}}$. By Lemmas 9.9 and 9.10, there exist a good coordinate system $\widehat{\mathcal{U}}$ and a KG-embedding $\widehat{\mathcal{U}} \to \widehat{\mathcal{U}}$. Moreover there exists a CF-perturbation $\widehat{\mathfrak{S}}$ of $\widehat{\mathcal{U}}$ such that $\widehat{\mathfrak{S}}$, $\widehat{\mathfrak{S}}$ are compatible with the KG-embedding $\widehat{\mathcal{U}} \to \widehat{\mathcal{U}}$. Futhermore there exist a strongly smooth map $\widehat{f} : (X, Z; \widehat{\mathcal{U}}) \to M$ and a differential form $\widehat{h}$, which are pulled back to $\widehat{f}$ and $\widehat{h}$, by the KG-embedding $\widehat{\mathcal{U}} \to \widehat{\mathcal{U}}$. Then $\widehat{\mathcal{U}}$, $\widehat{\mathfrak{S}}$, $\widehat{f}$ and $\widehat{h}$, induce $\widehat{\mathcal{U}_\partial}$, $\widehat{\mathfrak{S}_\partial}$, $\widehat{f_\partial}$ and $\widehat{h_\partial}$ on the boundary, respectively, which are compatible with corresponding objects on $(\partial X, \partial Z; \widehat{\mathcal{U}_\partial})$. Thus Theorem 9.28 follows by applying Theorem 8.11 to $\widehat{\mathcal{U}}, \widehat{\mathfrak{S}}, \widehat{f}, \widehat{h}$, and $\widehat{\mathcal{U}_\partial}, \widehat{\mathfrak{S}_\partial}, \widehat{f_\partial}, \widehat{h_\partial}$. □

The next corollary is an immediate consequence of Theorem 9.28.

Corollary 9.29 *In the situation of Definition 9.25 we have the next formula for each sufficiently small $\epsilon > 0$:*

$$d \circ \mathrm{Corr}_{(\mathfrak{X},\widehat{\mathfrak{S}^{\epsilon}})} = \mathrm{Corr}_{(\mathfrak{X},\widehat{\mathfrak{S}^{\epsilon}})} \circ d + (-1)^{\dim(X,\widehat{\mathcal{U}})+\deg(\cdot)}\mathrm{Corr}_{\partial(\mathfrak{X},\widehat{\mathfrak{S}^{\epsilon}})}.$$

9.7 Uniformity of CF-Perturbations on a Kuranishi Structure

In this section, we collect various facts that we use to show the existence of a uniform bound for the constants ϵ that appear in Theorem 9.14 etc..

Definition 9.30 Let $\widehat{\mathcal{U}}$ be a Kuranishi structure on $Z \subseteq X$ and $\widehat{\mathfrak{S}_\sigma}$ be a $\sigma \in \mathscr{A}$ parametrized family of CF-perturbations. We say that $\widehat{\mathfrak{S}_\sigma}$ is a *uniform family* if the convergence in Definition 7.4 is uniform. More precisely, we require the following.

For each $\mathfrak{o} > 0$, $p \in Z$ and a compact subset K_p of U_p, there exists $\epsilon_0 > 0$ such that if $0 < \epsilon < \epsilon_0$, $p \in Z$ then

$$|\mathfrak{s}(y) - s_p(y)| < \mathfrak{o}, \qquad |(D\mathfrak{s})(y) - (Ds_p)(y)| < \mathfrak{o}, \tag{9.20}$$

hold for any $\mathfrak{s}$ which is any member of $\mathfrak{S}^{\epsilon,p}_\sigma$ at any point $y \in K_p$ and $\sigma \in \mathscr{A}$.

Remark 9.31 We remark that in Definitions 9.30, the positive number ϵ_0 depends on $\mathfrak{o}$, p and K_p. It is independent of σ.

In the case of a good coordinate system, we have only a finite number of Kuranishi charts $\mathcal{U}_{\mathfrak{p}}$ and perturbations are defined on a compact set $\mathcal{K}_{\mathfrak{p}}$. Therefore whether ϵ_0 is $\mathfrak{p}$-dependent or not does not matter and we did not need to take a compact set such as K_p. In the case of a Kuranishi structure we have infinitely many Kuranishi charts $\mathcal{K}_p$. We allow ϵ_0 to depend on p and K_p. Note that to define a norm appearing in (9.20) we need to choose several objects. For example, we can define the norm if we fix a Riemannian metric on U_p and a connection of the vector bundle $\mathcal{E}_p$. However, since we allow ϵ_0 to depend on p and K_p, Definitions 9.30 are independent of such choices.

Lemma 9.32

(1) *In the situation of Lemma* 9.9 *if* $\widehat{\mathfrak{S}}$ *varies in a uniform family then* $\widehat{\mathfrak{S}}$ *varies in a uniform family.*
(2) *In the situation of Lemma* 9.10 *(2), if* $\widehat{\mathfrak{S}}$ *varies in a uniform family then* $\widehat{\widehat{\mathfrak{S}}}$ *varies in a uniform family.*
(3) *In the situation of Lemma* 9.11*, if* $\widehat{\mathfrak{S}^+}$, $\widehat{\mathfrak{S}}$ *vary in a uniform family (resp.* $\widehat{\mathfrak{S}_a^+}$ $(a = 1, 2)$ *,* $\widehat{\mathfrak{S}}$ *vary in a uniform family) then* $\widehat{\mathfrak{S}}$ *varies in a uniform family.*

(1) follows easily from the proof of Lemma 9.9 and Proposition 6.30, respectively. The proof of (2)(3) will be given at the end of Sect. 11.3.

Proposition 9.33 *In the situation of Theorem* 9.14 *suppose* $\widehat{\mathfrak{S}_\sigma}$ *varies in a uniform family. (We require that* $\widehat{\mathcal{U}}$, $\widehat{\Phi}$ *are independent of the parameter* σ*.) We also assume that* $\widehat{f}$ *is strongly submersive with respect to* $\widehat{\mathfrak{S}_\sigma}$ *for any* σ.

Then the pushout $\widehat{f}!(\widehat{h}; \widehat{\mathfrak{S}_\sigma})$ *is uniformly independent of the choices in the sense of* ♣ *in Definition* 7.88*. We may also choose the constant* ϵ *in Theorem* 9.28 *independent of* σ.

Proof Using Lemma 9.32 the proof goes in the same way as that of Proposition 7.89. □

Remark 9.34 We can choose ϵ_0 independent of $\widehat{f}$ and $\widehat{h}$.

Chapter 10
Composition Formula of Smooth Correspondences

The purpose of this chapter is to prove [FOOO8, Lemma 12.15] = Theorem 10.21, where a fiber product of Kuranishi structures is used as a way to define the composition of smooth correspondences. For this purpose we work out the plan described in Sect. 6.5 in the de Rham model.

10.1 Direct Product and CF-Perturbation

Firstly, we begin with defining the direct product of CF-perturbations.

Situation 10.1 For each $i = 1, 2$, $\mathcal{U}_i = (U_i, \mathcal{E}_i, s_i, \psi_i)$ is a Kuranishi chart of X, $x_i \in U_i$, $\mathfrak{V}^i_{x_i} = (V^i_{x_i}, \Gamma^i_{x_i}, E^i_{x_i}, \psi^i_{x_i}, \widehat{\psi}^i_{x_i})$ is an orbifold chart of $(U_i, \mathcal{E}_i)$ as in Definition 7.4. Let $\mathcal{S}^i_{x_i} = (W^i_{x_i}, \omega^i_{x_i}, \mathfrak{s}^{i\epsilon}_{x_i})$ be a CF-perturbation of $\mathcal{U}_i$ on $\mathfrak{V}^i_{x_i}$.■

Definition 10.2 In Situation 10.1, we define the *direct product of* $\mathcal{S}^1_{x_1}$ *and* $\mathcal{S}^2_{x_2}$ by

$$\mathcal{S}^1_{x_1} \times \mathcal{S}^2_{x_2} = (W^1_{x_1} \times W^2_{x_2}, \omega^1_{x_1} \times \omega^2_{x_2}, \mathfrak{s}^{1\epsilon}_{x_1} \times \mathfrak{s}^{2\epsilon}_{x_2}),$$

where

$$(\mathfrak{s}^{1\epsilon}_{x_1} \times \mathfrak{s}^{2\epsilon}_{x_2})(y_1, y_2, \xi_1, \xi_2) = (\mathfrak{s}^{1\epsilon}_{x_1}(y_1, \xi_1), \mathfrak{s}^{2\epsilon}_{x_2}(y_2, \xi_2))$$

for $y_i \in V^i_{x_i}$ $\xi_i \in W^i_{x_i}$.

Lemma 10.3

(1) $\mathcal{S}^1_{x_1} \times \mathcal{S}^2_{x_2}$ *is a CF-perturbation of* $\mathcal{U}_1 \times \mathcal{U}_2$.
(2) *If* $\mathcal{S}^i_{x_i}$ *are equivalent to* $\mathcal{S}^{i\prime}_{x_i}$ *for* $i = 1, 2$, *then* $\mathcal{S}^1_{x_1} \times \mathcal{S}^2_{x_2}$ *is equivalent to* $\mathcal{S}^{1\prime}_{x_1} \times \mathcal{S}^{2\prime}_{x_2}$.
(3) *Let* $\Phi^i : \mathfrak{V}^{i\prime}_{x'_i} \to \mathfrak{V}^i_{x_i}$ *be an embedding of orbifold chart. Then*

K. Fukaya et al., *Kuranishi Structures and Virtual Fundamental Chains*, Springer Monographs in Mathematics, https://doi.org/10.1007/978-981-15-5562-6_10

$$(\Phi^1)^* \mathcal{S}^1_{x_1} \times (\Phi^2)^* \mathcal{S}^2_{x_2}$$

is equivalent to

$$(\Phi^1 \times \Phi^2)^* (\mathcal{S}^1_{x_1} \times \mathcal{S}^2_{x_2}).$$

This is a direct consequence of definitions.

Lemma-Definition 10.4 *Suppose we are in Situation* 10.1.

(1) *Let* $\mathfrak{S}^i = \{(\mathfrak{V}^i_{\mathfrak{r}_i}, \mathcal{S}^i_{\mathfrak{r}_i}) \mid \mathfrak{r}_i \in \mathfrak{R}_i\}$ *be representatives of CF-perturbations of* $\mathcal{U}^i$ *for* $i = 1, 2$. *Then*

$$\{(\mathfrak{V}^1_{\mathfrak{r}_1} \times \mathfrak{V}^2_{\mathfrak{r}_2}, \mathcal{S}^1_{\mathfrak{r}_1} \times \mathcal{S}^2_{\mathfrak{r}_2}) \mid (\mathfrak{r}_1, \mathfrak{r}_2) \in \mathfrak{R}_1 \times \mathfrak{R}_2\}$$

is a representative of a CF-perturbation of $\mathcal{U}^1 \times \mathcal{U}^2$. *We call it the* direct product *and write* $\mathfrak{S}^1 \times \mathfrak{S}^2$.

(2) *If* $\mathfrak{S}^i$ *is equivalent to* $\mathfrak{S}^{i\prime}$, *then* $\mathfrak{S}^1 \times \mathfrak{S}^2$ *is equivalent to* $\mathfrak{S}^{1\prime} \times \mathfrak{S}^{2\prime}$.

(3) *Therefore we can define direct product of CF-perturbations.*

(4) *The direct product defines a sheaf morphism*

$$\pi_1^\star \mathcal{CF}^{\mathcal{U}_1} \times \pi_2^\star \mathcal{CF}^{\mathcal{U}_2} \to \mathcal{CF}^{\mathcal{U}_1 \times \mathcal{U}_2}, \tag{10.1}$$

where $\pi_i : U_1 \times U_2 \to U_i$ *are projections.*

In this book, we use $\star$ to denote the pullback sheaf to distinguish it from the pullback map.

Proof This is an immediate consequence of Lemma 10.3. □

Lemma 10.5 *Let* $\Phi^i : \mathcal{U}^i \to \mathcal{U}^{i+}$ *be embeddings of Kuranishi charts and* $\mathfrak{S}^i$, $\mathfrak{S}^{i+}$ *CF-perturbations of* $\mathcal{U}^i$, $\mathcal{U}^{i+}$, *for* $i = 1, 2$, *respectively.*

(1) *If* $\mathfrak{S}^{i+}$ *is restrictable to* $\mathcal{U}^i$ *for* $i = 1, 2$, *then* $\mathfrak{S}^{1+} \times \mathfrak{S}^{2+}$ *is restrictable* $\mathcal{U}^1 \times \mathcal{U}^2$.

(2) *If* $\mathfrak{S}^{i+}$, $\mathfrak{S}^i$ *are compatible with* Φ^i *for* $i = 1, 2$, *then* $\mathfrak{S}^{1+} \times \mathfrak{S}^{2+}$ *and* $\mathfrak{S}^1 \times \mathfrak{S}^2$ *are compatible with* $\Phi^1 \times \Phi^2$.

(3) *The next diagram commutes:*

$$\begin{array}{ccc}
\pi_1^\star \mathcal{CF}^{\mathcal{U}^{1+}} \times \pi_2^\star \mathcal{CF}^{\mathcal{U}^{2+}} & \xrightarrow{(10.1)} & \mathcal{CF}^{\mathcal{U}^{1+} \times \mathcal{U}^{2+}} \\
\uparrow & & \uparrow \\
\pi_1^\star \mathcal{CF}^{\mathcal{U}^1 \triangleright \mathcal{U}^{1+}} \times \pi_2^\star \mathcal{CF}^{\mathcal{U}^2 \triangleright \mathcal{U}^{2+}} & \longrightarrow & \mathcal{CF}^{(\mathcal{U}^1 \times \mathcal{U}^2) \triangleright (\mathcal{U}^{1+} \times \mathcal{U}^{2+})} \\
\downarrow{\scriptstyle \pi_1^\star (\Phi^1)^{\mathcal{CF}} \times \pi_2^\star (\Phi^2)^{\mathcal{CF}}} & & \downarrow{\scriptstyle (\Phi^1 \times \Phi^2)^{\mathcal{CF}}} \\
\pi_1^\star \Phi^1_* \mathcal{CF}^{\mathcal{U}^1} \times \pi_2^\star \Phi^2_* \mathcal{CF}^{\mathcal{U}^2} & \xrightarrow{\sharp} & (\Phi^1 \times \Phi^2)_* \mathcal{CF}^{\mathcal{U}^1 \times \mathcal{U}^2}
\end{array}$$

The map $\sharp$ in the last line is defined as follows.

If $p \in (U_1^+ \times U_2^+) \setminus \mathrm{Im}(\Phi^1 \times \Phi^2)$ then the stalk $((\Phi^1 \times \Phi^2)_* \mathcal{CF}^{\mathcal{U}^1 \times \mathcal{U}^2})_p$ consists of one point. (See Remark 7.45.) Then there exists a unique map from the stalk at p of the left hand side to the stalk of the right hand side.

If $p = (\Phi^1(p_1), \Phi^2(p_2)) \in \mathrm{Im}(\Phi^1 \times \Phi^2)$ then the stalks at p of the left and the right hand sides are both $(\mathcal{CF}^{\mathcal{U}^1})_{p_1} \times (\mathcal{CF}^{\mathcal{U}^2})_{p_2}$. Therefore there exists a canonical isomorphism between these two stalks.

The lemma is a direct consequence of the definitions.

Lemma-Definition 10.6 *Let $\widehat{\mathcal{U}^i} = (\{\mathcal{U}^i_{p_i}\}, \{\Phi^i_{p_i q_i}\})$ be Kuranishi structures of $Z_i \subseteq X_i$ for $i = 1, 2$, and $\widehat{\mathcal{U}^1} \times \widehat{\mathcal{U}^2}$ the direct product Kuranishi structure on $Z_1 \times Z_2 \subseteq X_1 \times X_2$. Let $\widehat{\mathfrak{S}^i} = \{\mathfrak{S}^i_{p_i}\}$ be CF-perturbations of $\mathcal{U}^i_{p_i}$. Then $\{\mathfrak{S}^1_{p_1} \times \mathfrak{S}^2_{p_2}\}$ defines a CF-perturbation of $\widehat{\mathcal{U}^1} \times \widehat{\mathcal{U}^2}$. We call it the* direct product *of CF perturbations and denote it by $\widehat{\mathfrak{S}^1} \times \widehat{\mathfrak{S}^2}$.*

Proof This is an immediate consequence of Lemma 10.5. □

We have thus defined the direct product of CF-perturbations.

Remark 10.7 We have defined the notion of the direct product of CF-perturbations of Kuranishi structures, but *not* one of good coordinate systems. The reason is explained at the end of Chap. 4.

10.2 Fiber Product and CF-Perturbation

We next discuss the case of fiber products.

Definition 10.8 Let $\mathcal{U} = (U, \mathcal{E}, s, \psi)$ be a Kuranishi chart of X. For $x \in U$ let $\mathfrak{V}_x$ be an orbifold chart of $(U, \mathcal{E})$ and $\mathcal{S}_x = (W_x, \omega_x, \{\mathfrak{s}^\epsilon_x\})$ a CF-perturbation of $\mathcal{U}$ on $\mathfrak{V}_x$. Let $f : U \to M$ be a smooth map to a manifold M and $g : N \to M$ a smooth map from a manifold N. Suppose f is strongly transversal to g with respect to $\mathcal{S}_x$ in the sense of Definition 7.9 (3). Then we take the fiber product $X\,{}_f\times_g N$, fiber product Kuranishi chart $(\mathfrak{V}_x)\,{}_f\times_g N$ and $(\mathcal{S}_x)\,{}_f\times_g N = (W_x, \omega_x, \{(\mathfrak{s}^\epsilon_x)\,{}_f\times_g N\})$. Here

$$((\mathfrak{s}^\epsilon_x)\,{}_f\times_g N) : ((V_x)\,{}_f\times_g N) \times W_x \to E_x$$

is defined by

$$((\mathfrak{s}^\epsilon_x)\,{}_f\times_g N)((y, z), \xi) = \mathfrak{s}^\epsilon_x(y, \xi).$$

We call $(\mathcal{S}_x)\,{}_f\times_g N$ the *fiber product CF-perturbation*. It is a CF-perturbation of $\mathcal{U}\,{}_f\times_g N$.

Lemma 10.9

(1) *If* $\mathcal{S}_x$ *is equivalent to* $\mathcal{S}'_x$ *and* f *is strongly transversal to* g *with respect to* $\mathcal{S}_x$, f *is strongly transversal to* g *with respect to* $\mathcal{S}'_x$. *Moreover* $(\mathcal{S}_x)_f \times_g N$ *is equivalent to* $(\mathcal{S}'_x)_f \times_g N$.
(2) *Let* $\Phi : \mathfrak{V}'_{x'} \to \mathfrak{V}_x$ *be an embedding of orbifold charts. If* $f \circ \Phi$ *is strongly transversal to* g *with respect to* $\Phi^*\mathfrak{S}_x$, f *is strongly transversal to* g *with respect to* $\mathfrak{S}_x$. *Moreover*

$$(\Phi \times \mathrm{id})^*((\mathcal{S}_x)_f \times_g N)$$

is equivalent to

$$(\Phi^*(\mathcal{S}_x))_f \times_g N.$$

This is a direct consequence of the definitions.

Lemma-Definition 10.10 *Let* $\mathcal{U} = (U, \mathcal{E}, s, \psi)$ *be a Kuranishi chart of* X, *and* $\mathfrak{S} = \{(\mathfrak{V}_{\mathfrak{r}}, \mathcal{S}_{\mathfrak{r}}) \mid \mathfrak{r} \in \mathfrak{R}\}$ *a representative of a CF-perturbation of* $\mathcal{U}$. *Let* $f : U \to M$ *be a smooth map to a manifold* M *and* $g : N \to M$ *a smooth map from a manifold* N.

(1) *If* f *is strongly transversal to* g *with respect to* $\mathfrak{S}$ *in the sense of Definition-Lemma 7.26 (3), then*

$$\mathfrak{S}_f \times_g N = \{((\mathfrak{V}_{\mathfrak{r}})_f \times_g N, (\mathcal{S}_{\mathfrak{r}})_f \times_g N) \mid \mathfrak{r} \in \mathfrak{R}\}$$

is a CF-perturbation of $\mathcal{U}_f \times_g N$.
(2) *If* $\mathfrak{S}$ *is equivalent to* $\mathfrak{S}'$, *then* $\mathfrak{S}_f \times_g N$ *is equivalent to* $\mathfrak{S}'_f \times_g N$.
(3) *Therefore we can define a* fiber product of CF-perturbations *with a map* $g : N \to M$ *when* $\widehat{f}$ *is strongly transversal to* g.

Proof This is an immediate consequence of Lemma 10.9. □

Lemma 10.11 *Let* $\Phi : \mathcal{U} \to \mathcal{U}^+$ *be an embedding of Kuranishi charts and* $\mathfrak{S}$, $\mathfrak{S}^+$ *CF-perturbations of* $\mathcal{U}$, $\mathcal{U}^+$, *respectively. Suppose* $\mathfrak{S}$, $\mathfrak{S}^+$ *are compatible with* Φ. *Let* $f^+ : \mathcal{U}^+ \to M$ *be a strongly smooth map to a manifold* M, $f = f^+ \circ \varphi : \mathcal{U} \to M$, *and* $g : N \to M$ *a smooth map from a manifold* N. *We assume* f, f^+ *are strongly transversal to* g *with respect to* $\mathfrak{S}$, $\mathfrak{S}^+$, *respectively.*

Then $\mathfrak{S}^+{}_{f^+} \times_g N$, $\mathfrak{S}_f \times_g N$ *are compatible with* $\Phi \times \mathrm{id} : \mathcal{U}_f \times_g N \to \mathcal{U}^+{}_{f^+} \times_g N$.

This is a direct consequence of the definitions.

Lemma-Definition 10.12 *Let* $\widehat{\mathcal{U}} = (\{\mathcal{U}_p\}, \{\Phi_{pq}\})$ *be a Kuranishi structure of* $Z \subseteq X$ *and* $\widehat{\mathfrak{S}} = \{\mathfrak{S}_p\}$ *a CF-perturbation of* $\widehat{\mathcal{U}}$. *Suppose that a strongly smooth map* $\widehat{f} : (X, \widehat{\mathcal{U}}) \to M$ *is strongly transversal to a smooth map* $g : N \to M$ *with respect to* $\widehat{\mathfrak{S}}$ *in the sense of Definition 7.50 (3).*

Then $\{(\mathfrak{S}_p)_f \times_g N\}$ *is a CF-perturbation. We call it a* fiber product CF-perturbations *and write* $(\mathfrak{S}_p)_f \times_g N$.

Lemma-Definition 10.12 is a consequence of Lemma 10.11.

Definition 10.13 Let $\widehat{\mathcal{U}^i} = (\{\mathcal{U}^i_{p_i}\}, \{\Phi^i_{p_i q_i}\})$ be Kuranishi structures of $Z_i \subseteq X_i$ and $\widehat{\mathcal{U}^1} \times \widehat{\mathcal{U}^2}$ the direct product Kuranishi structure on $Z_1 \times Z_2 \subseteq X_1 \times X_2$. Let $\widehat{\mathfrak{S}^i} = \{\mathfrak{S}^i_{p_i}\}$ be CF-perturbations of $\mathcal{U}^i_{p_i}$ and $\widehat{\mathfrak{S}^1} \times \widehat{\mathfrak{S}^2}$ their direct product. Let $\widehat{f^i} : (X_i, \widehat{\mathcal{U}^i}) \to M$ be strongly smooth maps to a manifold M.

(1) We say that $\widehat{f^1}$ is *strongly transversal to* $\widehat{f^2}$ *with respect to* $\widehat{\mathfrak{S}^1}, \widehat{\mathfrak{S}^2}$, if and only if

$$(\widehat{f^1}, \widehat{f^2}) : (X_1 \times X_2, \widehat{\mathcal{U}^1} \times \widehat{\mathcal{U}^2}) \to M \times M$$

is strongly transversal to the diagonal $\Delta_M = \{(x,x) \mid x \in M\}$, with respect to the direct product $\widehat{\mathfrak{S}^1} \times \widehat{\mathfrak{S}^2}$ in the sense of Definition 7.50 (3).

(2) In the situation of (1), we define the *fiber product of CF-perturbations* by

$$(\widehat{\mathfrak{S}^1})_{\widehat{f^1}} \times_{\widehat{f^2}} (\widehat{\mathfrak{S}^2}) = (\widehat{\mathfrak{S}^1} \times \widehat{\mathfrak{S}^2})_{(\widehat{f^1}, \widehat{f^2})} \times_{M \times M} \Delta_M.$$

Here the right hand side is defined by Lemma-Definition 10.12.

Lemma 10.14

(1) *Suppose we are in the situation of Definition* 10.13. *If* $\widehat{f^1}$ *is strongly submersive with respect to* $\widehat{\mathfrak{S}^1}$, *then* $\widehat{f^1}$ *is strongly transversal to any* $\widehat{f^2}$ *with respect to* $\widehat{\mathfrak{S}^1}$ *and* $\widehat{\mathfrak{S}^2}$, *provided* $\widehat{\mathfrak{S}^2}$ *is transversal to* 0.

(2) *In the situation of Definition* 10.13, *we assume* $\widehat{f^1}$ *is strongly submersive with respect to* $\widehat{\mathfrak{S}^1}$. *Let* $\widehat{f^3} : (X_2, \widehat{\mathcal{U}^2}) \to N$ *be another strongly smooth map such that* $\widehat{f^3}$ *is strongly submersive with respect to* $\widehat{\mathfrak{S}^2}$.

Then

$$\widetilde{\widehat{f^3}} : (X_1 \times X_2, \widehat{\mathcal{U}^1} \times \widehat{\mathcal{U}^2}) \to N$$

is strongly submersive with respect to $(\widehat{\mathfrak{S}^1})_{\widehat{f^1}} \times_{\widehat{f^2}} (\widehat{\mathfrak{S}^2})$. *Here* $\widetilde{\widehat{f^3}}$ *is the map induced from* $\widehat{f^3}$.

(3) *The fiber product of uniform families of CF-perturbations becomes a uniform family.*

Proof It suffices to prove the corresponding statement on a single orbifold chart. Namely for each $i = 1, 2$ we consider $\mathfrak{V}^i_{\mathfrak{r}_i}$ an orbifold chart of a Kuranishi neighborhood of $\widehat{\mathcal{U}}^i$, a CF-perturbation $\mathcal{S}^i_{\mathfrak{r}_i}$ of it, and maps $f^i_{\mathfrak{r}_i} : U^i_{\mathfrak{r}_i} \to M$, $f^3_{\mathfrak{r}_2} : U^2_{\mathfrak{r}_2} \to N$. We will prove this case below.

Proof of (1): By assumption

$$f^1_{\mathfrak{r}_1}|_{(\mathcal{S}^{1\epsilon}_{\mathfrak{r}_1})^{-1}(0)} : (\mathcal{S}^{1\epsilon}_{\mathfrak{r}_1})^{-1}(0) \to M$$

is a submersion. Therefore it is transversal to

$$f^2_{\mathfrak{r}_2}|_{(\mathcal{S}^{2\epsilon}_{\mathfrak{r}_2})^{-1}(0)} : (\mathcal{S}^{2\epsilon}_{\mathfrak{r}_2})^{-1}(0) \to M$$

as required.

Proof of (2): Let $(y_i, \xi_i) \in (\mathcal{S}^{i\epsilon}_{\mathfrak{r}_i})^{-1}(0)$. Here $y_i \in V^i_{\mathfrak{r}_i}$, $\xi_i \in W^i_{\mathfrak{r}_i}$, where $\mathfrak{V}^i_{x_i} = (V^i_{\mathfrak{r}_i}, \Gamma^i_{x_i}, E^i_{\mathfrak{r}_i}, \psi^i_{\mathfrak{r}_i}, \widehat{\psi}^i_{\mathfrak{r}_i})$, $\mathcal{S}^{i\epsilon}_{\mathfrak{r}_i} = (W^i_{\mathfrak{r}_i}, \omega^i_{\mathfrak{r}_i}, \mathfrak{s}^{i\epsilon}_{\mathfrak{r}_i})$.

Suppose $f^1_{\mathfrak{r}_1}(y_1) = f^2_{\mathfrak{r}_2}(y_2) = z$ and $f^3_{\mathfrak{r}_2}(y_2) = w$. We consider

$$(d_{y_1} f^1_{\mathfrak{r}_1} \oplus d_{y_2} f^2_{\mathfrak{r}_2}) \oplus d_{y_2} f^3_{\mathfrak{r}_2} : T_{(y_1,\xi_1)}(\mathcal{S}^{1\epsilon}_{\mathfrak{r}_1})^{-1}(0) \oplus T_{(y_2,\xi_2)}(\mathcal{S}^{2\epsilon}_{\mathfrak{r}_2})^{-1}(0) \\ \to T_z M \oplus T_z M \oplus T_w N.$$

Let $\mathfrak{v} \in T_w N$. Then there exists $\tilde{\mathfrak{v}_2} \in T_{(y_2,\xi_2)}(\mathcal{S}^{2\epsilon}_{\mathfrak{r}_2})^{-1}(0)$ such that

$$(d_{y_2} f^3_{\mathfrak{r}_2})(\tilde{\mathfrak{v}_2}) = \mathfrak{v}. \tag{10.2}$$

Then there exists $\tilde{\mathfrak{v}_1} \in T_{(y_1,\xi_1)}(\mathcal{S}^{1\epsilon}_{\mathfrak{r}_1})^{-1}(0)$ such that

$$(d_{y_1} f^1_{\mathfrak{r}_1})(\tilde{\mathfrak{v}_1}) = (d_{y_2} f^2_{\mathfrak{r}_2})(\tilde{\mathfrak{v}_2}). \tag{10.3}$$

Equation (10.3) implies that

$$(\tilde{\mathfrak{v}_1}, \tilde{\mathfrak{v}_2}) \in T_{((y_1),(y_2))}((\mathcal{S}^{1\epsilon}_{\mathfrak{r}_1})^{-1}(0) \ {}_{f^1}\times_{f^2} (\mathcal{S}^{2\epsilon}_{\mathfrak{r}_1})^{-1}(0))$$

and (10.2) implies that

$$(d_{((y_1),(y_2))}\overline{f^3})(\tilde{\mathfrak{v}_1}, \tilde{\mathfrak{v}_2}) = \mathfrak{v}.$$

Here $\overline{f^3} : (\mathcal{S}^{1\epsilon}_{\mathfrak{r}_1})^{-1}(0) \ {}_{f^1}\times_{f^2} (\mathcal{S}^{2\epsilon}_{\mathfrak{r}_1})^{-1}(0) \to N$ is a local representative of $\widetilde{\widetilde{f^3}}$. We have thus proved the required submersivity.

The proof of (3) is obvious from the definition. □

10.3 Composition of Smooth Correspondences

In this section we define composition of smooth correspondences and its perturbation. Let us consider the following situation.

Situation 10.15 Let $(X_{21}, \widehat{\mathcal{U}_{21}})$, $(X_{32}, \widehat{\mathcal{U}_{32}})$ be K-spaces and M_i $(i = 1, 2, 3)$ smooth manifolds. Let

$$\widehat{f_{i,ji}} : (X_{ji}, \widehat{\mathcal{U}_{ji}}) \to M_i, \qquad \widehat{f_{j,ji}} : (X_{ji}, \widehat{\mathcal{U}_{ji}}) \to M_j$$

be strongly smooth maps for $(i, j) = (1, 2)$ or $(2, 3)$. We assume $\widehat{f_{2,21}}$ and $\widehat{f_{3,32}}$ are weakly submersive. These facts imply that

$$\mathfrak{X}_{i+1i} = ((X_{i+1i}, \widehat{\mathcal{U}_{i+1i}}), \widehat{f_{i,i+1i}}, \widehat{f_{i+1,i+1i}})$$

is a smooth correspondence from M_i to M_{i+1} for $i = 1, 2$. (Lemma 10.14 (2).) Let $\widehat{\mathfrak{S}_{i+1i}}$ be a CF- perturbation of $(X_{i+1i}, \widehat{\mathcal{U}_{i+1i}})$ for each $i = 1, 2$. We assume that $\widehat{f_{i+1,i+1i}}$ is strongly submersive with respect to $\widehat{\mathfrak{S}_{i+1i}}$ for each $i = 1, 2$.■

Definition 10.16 In Situation 10.15, we put

$$X_{31} = X_{21} \times_{M_2} X_{32} = \{(x_{21}, x_{32}) \in X_{21} \times X_{32} \mid f_{2,21}(x_{21}) = f_{2,32}(x_{32})\}. \tag{10.4}$$

We put the fiber product Kuranishi structure

$$\widehat{\mathcal{U}_{31}} = \widehat{\mathcal{U}_{21}} \times_{M_2} \widehat{\mathcal{U}_{32}} \tag{10.5}$$

on X_{31} and define

$$\widehat{f_{1,31}} : (X_{31}, \widehat{\mathcal{U}_{31}}) \to M_1, \qquad \widehat{f_{3,31}} : (X_{31}, \widehat{\mathcal{U}_{31}}) \to M_3 \tag{10.6}$$

as the compositions

$$(X_{31}, \widehat{\mathcal{U}_{31}}) \to (X_{21}, \widehat{\mathcal{U}_{21}}) \to M_1, \qquad (X_{31}, \widehat{\mathcal{U}_{31}}) \to (X_{32}, \widehat{\mathcal{U}_{32}}) \to M_3,$$

where the first arrows are obvious projections. We write

$$\mathfrak{X}_{21} \times_{M_2} \mathfrak{X}_{32} = ((X_{31}, \widehat{\mathcal{U}_{31}}), \widehat{f_{1,31}}, \widehat{f_{3,31}})$$

and call it the *composition of smooth correspondences* $\mathfrak{X}_{21}$ and $\mathfrak{X}_{32}$. We also denote it by $\mathfrak{X}_{32} \circ \mathfrak{X}_{21}$.

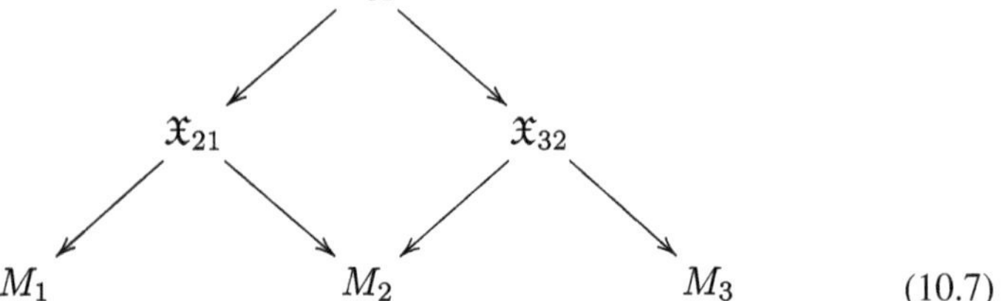

(10.7)

Remark 10.17 Note that we did not define the 'maps' $\mathfrak{X}_{31} \to \mathfrak{X}_{21}$, $\mathfrak{X}_{31} \to \mathfrak{X}_{32}$ in Diagram (10.7). This is because we never defined the notion of morphism between K-spaces in this book. However, the maps $\mathfrak{X}_{31} \to M_1$ and $\mathfrak{X}_{31} \to M_3$ are defined by composing the map $f_{21,p}$ or $f_{32,q}$ and the projection on each chart.

Lemma 10.18

(1) *The fiber product* (10.5) *is well-defined.*
(2) *The map* $(X_{31}, \widehat{\mathcal{U}_{31}}) \to M_3$ *is weakly submersive.*

Proof (1) By assumption, $\widehat{f_{2,21}}$ is weakly submersive. This implies well-definedness of (10.5).

(2) Let $(p, q) \in X_{31}$, i.e., $p \in X_{21}$, $q \in X_{32}$ $f_{2,21}(p) = f_{2,32}(q)$. We put $x = f_{2,21}(p) = f_{3;32}(q)$ and $y \in f_{3,32}(q)$. By assumption,

$$d_{o_p}(f_{2,21})_p : T_{o_p}U_p \to T_xM_2, \quad d_{o_q}(f_{3,32})_q : T_{o_q}U_q \to T_yM_3$$

are surjective. Let $v_3 \in T_yM_3$. There exists $\tilde{v}_3 \in T_{o_q}U_q$ such that $(d_{o_q}(f_{3,32})_q)(\tilde{v}_3) = v_3$. There exists $\tilde{v}_2 \in T_{o_p}U_p$ such that $(d_{o_p}(f_{2,21})_p)(\tilde{v}_2) = (d_{o_q}(f_{2,32})_q)(\tilde{v}_3)$. Then $(\tilde{v}_2, \tilde{v}_3) \in T_{o_{(p,q)}}U_{(p,q)}$ and $(d_{o_{(p,q)}}(f_{3,31})_{(p,q)})(\tilde{v}_2, \tilde{v}_3) = v_3$ as required. □

Definition 10.19

(1) Let $\mathfrak{X} = ((X, \widehat{\mathcal{U}}), \widehat{f_s}, \widehat{f_t})$ be a smooth correspondence and $\widehat{\mathfrak{S}}$ a CF-perturbation of $(X, \widehat{\mathcal{U}})$. We say that $(X, \widehat{\mathcal{U}}, \widehat{\mathfrak{S}}, \widehat{f_s}, \widehat{f_t})$ is a *perturbed smooth correspondence* if $\widehat{f_t}$ is strongly submersive with respect to $\widehat{\mathfrak{S}}$.
(2) A perturbed smooth correspondence, denoted by $\tilde{\mathfrak{X}} = (\mathfrak{X}, \widehat{\mathfrak{S}}) = (X, \widehat{\mathcal{U}}, \widehat{\mathfrak{S}}, \widehat{f_s}, \widehat{f_t})$, from M_s to M_t induces a linear map $\Omega^*(M_s) \to \Omega^{*+\deg \mathfrak{X}}(M_t)$ by (9.17) for each $\epsilon > 0$. We write it as $\mathrm{Corr}^{\epsilon}_{\tilde{\mathfrak{X}}}$.
(3) In Situation 10.15, let $\widehat{\mathfrak{S}_{i+1i}}$ be a CF-perturbation of $(X_{i+1i}, \widehat{\mathcal{U}_{i+1i}})$ for each $i = 1, 2$. Suppose $\tilde{\mathfrak{X}}_{i+1i} = (X_{i+1i}, \widehat{\mathcal{U}_{i+1i}}, \widehat{\mathfrak{S}_{i+1i}}, \widehat{f_{i,i+1i}}, \widehat{f_{i+1,i+1i}})$ is a perturbed smooth correspondence for each $i = 1, 2$. Then by Lemma 10.14,

$$\begin{aligned}(X_{21}\ {}_{f_{2,21}}\times_{f_{2,32}} X_{32}, \widehat{\mathcal{U}_{21}}\ {}_{\widehat{f_{2,21}}}\times_{\widehat{f_{2,32}}} \widehat{\mathcal{U}_{32}},\\ (\widehat{\mathfrak{S}_{21}})\ {}_{\widehat{f_{2,21}}}\times_{\widehat{f_{2,32}}} (\widehat{\mathfrak{S}_{32}}), \widehat{f_{1,31}}, \widehat{f_{3,31}})\end{aligned} \tag{10.8}$$

is a perturbed smooth correspondence from M_1 to M_3. We call (10.8) the *composition* of $\tilde{\mathfrak{X}}_{21}$ and $\tilde{\mathfrak{X}}_{32}$ and write $\tilde{\mathfrak{X}}_{32} \circ \tilde{\mathfrak{X}}_{21}$. Here,

$$\begin{aligned} \widehat{f_{1,31}} &: (X_{21}\ {}_{f_{2,21}}\times_{f_{2,32}} X_{32}, \widehat{\mathcal{U}_{21}}\ {}_{\widehat{f_{2,21}}}\times_{\widehat{f_{2,32}}} \widehat{\mathcal{U}_{32}}) \to M_1, \\ \widehat{f_{3,31}} &: (X_{21}\ {}_{f_{2,21}}\times_{f_{2,32}} X_{32}, \widehat{\mathcal{U}_{21}}\ {}_{\widehat{f_{2,21}}}\times_{\widehat{f_{2,32}}} \widehat{\mathcal{U}_{32}}) \to M_3 \end{aligned} \tag{10.9}$$

are maps as in (10.6).

10.4 Composition Formula

Convention 10.20 When we compose $\mathrm{Corr}^{\epsilon}_{\tilde{\mathfrak{X}}_{21}}$ and $\mathrm{Corr}^{\epsilon}_{\tilde{\mathfrak{X}}_{32}}$, we use the fiber product $\widehat{\mathcal{U}_{21}}\ {}_{\widehat{f_{2,21}}}\times_{\widehat{f_{2,32}}} \widehat{\mathcal{U}_{32}}$. When $\widehat{\mathcal{U}_{21}}$ and $\widehat{\mathcal{U}_{32}}$ are oriented, we equip $\widehat{\mathcal{U}_{21}}\ {}_{\widehat{f_{2,21}}}\times_{\widehat{f_{2,32}}} \widehat{\mathcal{U}_{32}}$ with the fiber product orientation of $\widehat{\mathcal{U}_{32}}\ {}_{\widehat{f_{2,32}}}\times_{\widehat{f_{2,21}}} \widehat{\mathcal{U}_{21}}$. More generally, orientation bundles are related as follows:

$$O_{\widehat{\mathcal{U}_{31}}} = O_{\widehat{\mathcal{U}_{32}}} \otimes O_{M_2} \otimes O_{\widehat{\mathcal{U}_{21}}}.$$

Under these conventions (10.10) below holds. Namely the equality holds without sign correction.

The main result of this chapter is the following.

Theorem 10.21 (Composition formula, [FOOO8, Lemma 12.15]) *Suppose that* $\tilde{\mathfrak{X}}_{i+1i} = (X_{i+1i}, \widehat{\mathcal{U}_{i+1i}}, \widehat{\mathfrak{S}_{i+1i}}, \widehat{f_{i,i+1i}}, \widehat{f_{i+1,i+1i}})$ *are perturbed smooth correspondences for* $i = 1, 2$. *Then*

$$\mathrm{Corr}^{\epsilon}_{\tilde{\mathfrak{X}}_{32}\circ\tilde{\mathfrak{X}}_{21}} = \mathrm{Corr}^{\epsilon}_{\tilde{\mathfrak{X}}_{32}} \circ \mathrm{Corr}^{\epsilon}_{\tilde{\mathfrak{X}}_{21}} \tag{10.10}$$

for each sufficiently small $\epsilon > 0$.

Remark 10.22 Note that $\mathrm{Corr}^{\epsilon}_{\tilde{\mathfrak{X}}_{**}}$ depends on the positive number ϵ.

Proof Let h_1 and h_3 be differential forms on M_1 and M_3, respectively. It suffices to show the next formula:

$$\int_{M_3} h_3 \wedge \mathrm{Corr}^{\epsilon}_{\tilde{\mathfrak{X}}_{32}\circ\tilde{\mathfrak{X}}_{21}}(h_1) = \int_{M_3} h_3 \wedge \mathrm{Corr}^{\epsilon}_{\tilde{\mathfrak{X}}_{32}}(\mathrm{Corr}^{\epsilon}_{\tilde{\mathfrak{X}}_{21}}(h_1)). \tag{10.11}$$

We use the following notation.

Definition 10.23 In Situation 9.12, we consider the case when M is a point and put:

$$\int_{(X,Z,\widehat{\mathcal{U}},\widehat{\mathfrak{S}^\epsilon})} \widehat{h} = \widehat{f}!(\widehat{h};\widehat{\mathfrak{S}^\epsilon}).$$

We call it the *integral of* $\widehat{h}$ *over* $(X,Z,\widehat{\mathcal{U}},\widehat{\mathfrak{S}^\epsilon})$. It is a real number depending on $(X,Z,\widehat{\mathcal{U}},\widehat{\mathfrak{S}})$, ϵ and $\widehat{h}$. We also define

$$\int_{(X,\mathcal{U},\mathfrak{S}^\epsilon)} h = f!(h;\mathfrak{S}^\epsilon).$$

Here $\mathcal{U}$ is a Kuranishi chart of X, h is a differential form of compact support on U, $\mathfrak{S}^\epsilon$ is a CF-perturbation of $\mathcal{U}$ on the support of h and $f : U \to$ a point is a trivial map, such that f is strongly submersive with respect to $\mathfrak{S}^\epsilon$. (Note the strong submersivity for the trivial map is nothing but transversality of the CF-perturbation to 0.) We call it the *integral of* h *over* $(\widehat{\mathcal{U}},\widehat{\mathfrak{S}^\epsilon})$.

In the notation above we omit Z if $Z = X$.

We remark that we have

$$\int_{(X,\widehat{\mathcal{U}},\widehat{\mathfrak{S}^\epsilon})} \hat{f}^*h' \wedge \hat{h} = \int_M h' \wedge f!(\hat{h};\widehat{\mathfrak{S}^\epsilon}) \tag{10.12}$$

if $\widehat{f} : (X,\widehat{\mathcal{U}}) \to M$ is strongly submersive with respect to $\widehat{\mathfrak{S}^\epsilon}$ and $\hat{h}$ (resp. h') be a differential form on $(X,\widehat{\mathcal{U}})$ (resp. M.) (10.12) is immediate from the definition.

Using (10.12), the right hand side of (10.11) is

$$\begin{aligned}
&\int_{(X_{32},\widehat{\mathcal{U}_{32}},\widehat{\mathfrak{S}^\epsilon_{32}})} (\widehat{f_{3,32}})^*h_3 \wedge (\widehat{f_{2,32}})^*(\mathrm{Corr}_{\tilde{\mathfrak{X}}_{21}}(h_1)) \\
&= \int_{(X_{32},\widehat{\mathcal{U}_{32}},\widehat{\mathfrak{S}^\epsilon_{32}})} (\widehat{f_{3,32}})^*h_3 \wedge (\widehat{f_{2,32}})^*(\widehat{f_{2,21}}!((\widehat{f_{1,21}})^*(h_1);\widehat{\mathfrak{S}^\epsilon_{21}})).
\end{aligned} \tag{10.13}$$

On the other hand, the left hand side of (10.11) is

$$\int_{(X_{31},\widehat{\mathcal{U}_{31}},\widehat{\mathfrak{S}^\epsilon_{31}})} (\widehat{f_{3,31}})^*(h_3) \wedge (\widehat{f_{1,31}})^*(h_1). \tag{10.14}$$

Therefore (10.11) follows from the next proposition. In fact we can prove (10.13) = (10.14) by applying Proposition 10.24 to $\widehat{h_1} = (\widehat{f_{3,32}})^*h_3$ and $\widehat{h_2} = (\widehat{f_{1,21}})^*h_1$.

Proposition 10.24 *For* $i = 1, 2$, *let* $\widehat{\mathcal{U}_i}$ *be Kuranishi structures of* $Z_i \subseteq X_i$, $\widehat{\mathfrak{S}_i}$ *their CF-perturbations,* $\widehat{h_i}$ *smooth differential forms on* $(X_i,\widehat{\mathcal{U}_i})$ *which have compact support in* $\mathring{Z}_i$, *and* $\widehat{f_i} : (X_i, Z_i; \widehat{\mathcal{U}_i}) \to M$ *strongly smooth maps. Suppose that* $\widehat{f_2}$ *is strongly submersive with respect to* $\widehat{\mathfrak{S}_2}$ *and* $\widehat{\mathfrak{S}_1}$ *is transversal to* 0 *and denote* *by* $(X, Z, \widehat{\mathcal{U}}, \widehat{\mathfrak{S}})$ *the fiber product*

$$(X_1, Z_1, \widehat{\mathcal{U}_1}, \widehat{\mathfrak{S}_1}) \;{}_{\widehat{f_1}}\times_{\widehat{f_2}}\; (X_2, Z_2, \widehat{\mathcal{U}_2}, \widehat{\mathfrak{S}_2})$$

over M *with the induced smooth form on it, which we denote by* $\widehat{h_1}\,{}_{\widehat{f_1}}\wedge_{\widehat{f_2}}\,\widehat{h_2}$. *We equip* $(X_1, Z_1, \widehat{\mathcal{U}_1}, \widehat{\mathfrak{S}_1}) \;{}_{\widehat{f_1}}\times_{\widehat{f_2}}\; (X_2, Z_2, \widehat{\mathcal{U}_2}, \widehat{\mathfrak{S}_2})$ *with the fiber product orientation.*[1] *Then*

$$\int_{(X,Z,\widehat{\mathcal{U}},\widehat{\mathfrak{S}^{\epsilon}})} \widehat{h_1}\,{}_{\hat{f_1}}\wedge_{\hat{f_2}}\,\widehat{h_2} = \int_{(X_1,Z_1,\widehat{\mathcal{U}_1},\widehat{\mathfrak{S}_1^{\epsilon}})} \widehat{h_1} \wedge (\widehat{f_1})^*(\widehat{f_2}!(\widehat{h_2}; \widehat{\mathfrak{S}_2^{\epsilon}})). \qquad (10.15)$$

Remark 10.25 We may regard Formula (10.15) as a version of Fubini formula.

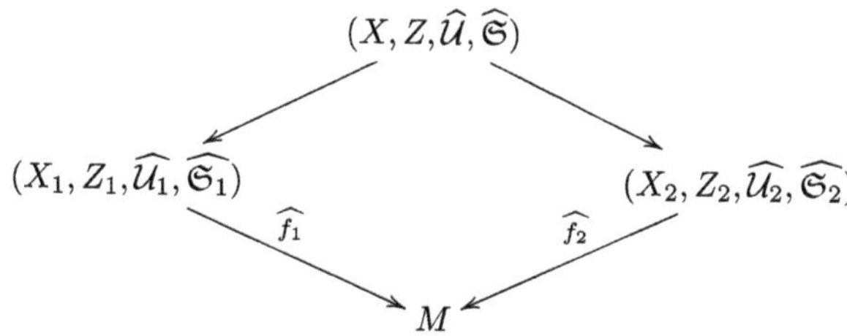

We remark that the two upper arrows in the above diagram are not defined in this book. We do not use it, however.

Hereafter we write $\hat{h}_1 \wedge \hat{h}_2$ instead of $\hat{h}_1\,{}_{\hat{f_1}}\wedge_{\hat{f_2}}\,\hat{h}_2$ when no confusion occurs.

Proof of Proposition 10.24 For $i = 1, 2$ let $\widehat{\widehat{\mathcal{U}_i}}$ be good coordinate systems of X_i and $\widehat{\mathcal{U}_i} \to \widehat{\widehat{\mathcal{U}_i}}$ KG-embeddings. We may assume that there exist CF-perturbations $\widehat{\widehat{\mathfrak{S}_i}}$ of $(X_i, Z_i; \widehat{\widehat{\mathcal{U}_i}})$ such that $\widehat{\widehat{\mathfrak{S}_i}}$, $\widehat{\mathfrak{S}_i}$ are compatible with the KG-embeddings and $\widehat{h_i}$, $\widehat{f_i}$ are pullbacks of differential forms on $\widehat{\widehat{\mathcal{U}_i}}$ and of strongly smooth maps on $\widehat{\widehat{\mathcal{U}_i}}$, which we also denote by $\widehat{h_i}$, $\widehat{f_i}$, respectively. (Theorem 3.35, Proposition 6.32 (2), Theorem 7.51.)

Let $\mathcal{K}_i$ be support systems of $\widehat{\widehat{\mathcal{U}_i}}$ and take $\delta_i > 0$, for $i = 1, 2$. Let $\{\chi^i_{\mathfrak{p}_i}\}$ be partitions of unity of $(X_i, Z_i, \widehat{\widehat{\mathcal{U}_i}}, \mathcal{K}_i, \delta_i)$. The functions $\chi^i_{\mathfrak{p}_i}$ can be regarded as strongly smooth maps $(X_i, Z_i; \widehat{\widehat{\mathcal{U}_i}}) \to \mathbb{R}$. Therefore they induce strongly smooth maps $(X_i, Z_i; \widehat{\mathcal{U}_i}) \to \mathbb{R}$, which we also denote by $\chi^i_{\mathfrak{p}_i}$. Then they induce strongly smooth functions on the fiber product $(X, Z; \widehat{\mathcal{U}})$.

[1] Here we do not use the fiber product orientation as $(X_2, Z_2, \widehat{\mathcal{U}_2}, \widehat{\mathfrak{S}_2}) \;{}_{\widehat{f_2}}\times_{\widehat{f_1}}\; (X_1, Z_1, \widehat{\mathcal{U}_1}, \widehat{\mathfrak{S}_1})$.

To prove (10.15) it suffices to prove the next formula for each $\mathfrak{p}_1$, $\mathfrak{p}_2$ and ϵ:

$$\begin{aligned}&\int_{(X,Z,\widehat{\mathcal{U}},\widehat{\mathfrak{S}^\epsilon})} \chi^1_{\mathfrak{p}_1}\widehat{h_1} \wedge \chi^2_{\mathfrak{p}_2}\widehat{h_2} \\ &= \int_{(\mathcal{U}_{1,\mathfrak{p}_1},\mathfrak{S}^\epsilon_{1,\mathfrak{p}_1})} \chi^1_{\mathfrak{p}_1} h_{1,\mathfrak{p}_1} \wedge f^*_{1,\mathfrak{p}_1}(f_{2,\mathfrak{p}_2}!(\chi^2_{\mathfrak{p}_2} h_{2,\mathfrak{p}_2};\mathfrak{S}^\epsilon_{2,\mathfrak{p}_2})).\end{aligned} \tag{10.16}$$

We will use the following result to reduce the proof of Proposition 10.24 to the case of a good coordinate system with one chart.

Proposition 10.26 *Let $Z_{(1)}$ and $Z_{(2)}$ be compact subsets of X such that $Z_{(1)} \subset \overset{\circ}{Z}_{(2)}$. Let $\widehat{\mathcal{U}}$ be a Kuranishi structure of $Z_{(2)} \subseteq X$ and $\widehat{h}$ a differential form on $\widehat{\mathcal{U}}$. Let $\widehat{f_{(2)}} : (X, Z_{(2)}; \widehat{\mathcal{U}}) \to M$ be a strongly smooth map and $\widehat{\mathfrak{S}_{(2)}}$ a CF-perturbation of $\widehat{\mathcal{U}}$. Denote by $\widehat{\mathfrak{S}_{(1)}}$ the restriction of $\widehat{\mathfrak{S}_{(2)}}$ to $\widehat{\mathcal{U}}|_{Z_{(1)}}$ (Definition 9.3). Suppose:*

(1) *$\widehat{f_{(2)}}$ is strongly submersive with respect to $\widehat{\mathfrak{S}_{(2)}}$.*
(2) *$\widehat{h}$ has compact support in $\overset{\circ}{Z}_{(1)}$.*

We denote by $\widehat{h}|_{Z_{(1)}}$ the differential form on $\widehat{\mathcal{U}}|_{Z_{(1)}}$ induced by $\widehat{h}$ via Condition (2). Then

$$\widehat{f_{(2)}}!(\widehat{h};\widehat{\mathfrak{S}^\epsilon_{(2)}}) = \widehat{f_{(1)}}!(\widehat{h}|_{Z_{(1)}};\widehat{\mathfrak{S}^\epsilon_{(1)}}), \tag{10.17}$$

where $\widehat{f_{(1)}}$ is the restriction of $\widehat{f_{(2)}}$ to $(X, Z_{(1)}; \widehat{\mathcal{U}}|_{Z_{(1)}})$.

Proof By using differential forms ρ on M it suffices to consider the case $\deg h = \dim \mathcal{U}$ and prove the next formula (see Lemma 7.84 (2)):

$$\int_{(X,Z_{(2)},\widehat{\mathcal{U}},\widehat{\mathfrak{S}^\epsilon_{(2)}})} \widehat{h} = \int_{(X,Z_{(1)},\widehat{\mathcal{U}}|_{Z_{(1)}},\widehat{\mathfrak{S}^\epsilon_{(1)}})} \widehat{h}. \tag{10.18}$$

We take good coordinate systems $\widehat{\mathcal{U}_{(i)}}$ of $Z_{(i)} \subseteq X$ and their support systems $\mathcal{K}^{(1)}$, $\mathcal{K}^{(2)}$, such that:

(1) $\widehat{\mathcal{U}_{(1)}}$ is compatible with $\widehat{\mathcal{U}}|_{Z_{(1)}}$ and $\widehat{\mathcal{U}_{(2)}}$ is compatible with $\widehat{\mathcal{U}}$.
(2) $(\widehat{\mathcal{U}_{(2)}}, \mathcal{K}^{(2)})$ strictly extends $(\widehat{\mathcal{U}_{(1)}}, \mathcal{K}^{(1)})$.
(3) There are CF-perturbations $\widehat{\mathfrak{S}_{(i)}}$, differential forms $\widehat{h_{(i)}}$ on $(\widehat{\mathcal{U}_{(i)}}, \mathcal{K}^{(i)})$ which are compatible with each other and are compatible with corresponding objects on $\widehat{\mathcal{U}_{(1)}}$ and on $\widehat{\mathcal{U}}|_{Z_{(1)}}$.
(4) $\widehat{\mathfrak{S}_i}$ are transversal to 0.

The existence of such objects is proved as follows. The existence of $\widehat{\mathcal{U}_{(1)}}$ is a consequence of Theorem 3.35. Then, by Propositions 7.54, $\widehat{\mathcal{U}_{(2)}}$ which extends $\widehat{\mathcal{U}_{(1)}}$ is obtained. CF-perturbations $\widehat{\mathfrak{S}_1}$, $\widehat{\mathfrak{S}_2}$ and support systems $\mathcal{K}^{(1)}$, $\mathcal{K}^{(2)}$ which satisfy (3)(4) are then obtained by Proposition 7.59. We replace $\widehat{\mathcal{U}_{(1)}}$ by its open substructure and also $\mathcal{K}^{(1)}_{\mathfrak{p}}$ by its subset, so that $(\widehat{\mathcal{U}_{(2)}}, \mathcal{K}^{(2)})$ strictly extends $(\widehat{\mathcal{U}_{(1)}}, \mathcal{K}^{(1)})$.

Let $\widehat{\mathcal{U}_{(i)}} = (\mathfrak{P}_{(i)}, \{\mathcal{U}_{(i),\mathfrak{p}}\}, \{\Phi_{(i),\mathfrak{pq}}\})$. By Definition 7.52 (1)(a), $\mathfrak{P}_{(1)} = \{\mathfrak{p} \in \mathfrak{P}_{(2)} \mid \mathrm{Im}(\psi_{(2),\mathfrak{p}}) \cap Z_{(1)} \neq \emptyset\}$ and $\mathcal{U}_{(1),\mathfrak{p}}$ is an open subchart of $\mathcal{U}_{(2),\mathfrak{p}}$ for $\mathfrak{p} \in \mathfrak{P}_{(1)} \subset \mathfrak{P}_{(2)}$. We may change support systems $\mathcal{K}^{(2)}$ and find a neighborhood $\mathfrak{U}_1(Z_{(1)})$ of $Z_{(1)}$ in $|\widehat{\mathcal{U}_{(2)}}|$ so that the following holds:

(a) If $\mathfrak{p} \in \mathfrak{P}_{(1)} \subset \mathfrak{P}_{(2)}$ then $\mathcal{K}^{(1)}_{\mathfrak{p}} \cap \mathfrak{U}_1(Z_{(1)}) = \mathcal{K}^{(2)}_{\mathfrak{p}} \cap \mathfrak{U}_1(Z_{(1)})$.
(b) If $\mathfrak{p} \in \mathfrak{P}_{(2)} \setminus \mathfrak{P}_{(1)}$ then $\mathfrak{U}_1(Z_{(1)}) \cap \mathcal{K}^{(2)}_{\mathfrak{p}} = \emptyset$.

To prove this claim, we take $\mathfrak{U}_1^+(Z_{(1)})$ so that it is covered by $\{\mathcal{K}^{(1)}_{\mathfrak{p}} \cap s_{\mathfrak{p}}^{-1}(0) \mid \mathfrak{p} \in \mathfrak{P}_{(1)}\}$. Choose $\mathfrak{U}_1(Z_{(1)})$ which is relatively compact in $\mathfrak{U}_1^+(Z_{(1)})$. For $\mathfrak{p} \in \mathfrak{P}_{(1)}$ we replace $\mathcal{K}^{(2)}_{\mathfrak{p}}$ by

$$\left(\mathcal{K}^{(2)}_{\mathfrak{p}} \setminus \mathfrak{U}_1(Z_{(1)})\right) \cup \mathcal{K}^{(1)}_{\mathfrak{p}}.$$

Then (a) is satisfied. For $\mathfrak{p} \in \mathfrak{P}_{(2)} \setminus \mathfrak{P}_{(1)}$ we replace $\mathcal{K}^{(2)}_{\mathfrak{p}}$ by

$$\mathcal{K}^{(2)}_{\mathfrak{p}} \setminus \mathfrak{U}_1(Z_{(1)}).$$

Then (b) is satisfied.

We next take support pairs $(\mathcal{K}^{1,(i)}, \mathcal{K}^{2,(i)})$ of $\widehat{\mathcal{U}^{(i)}}$ such that $\mathcal{K}^{1,(i)} < \mathcal{K}^{2,(i)} < \mathcal{K}^{(i)}$. $(i = 1, 2.)$ In the same way as above, we may assume that there exists a neighborhood $\mathfrak{U}_2(Z_{(1)})$ of $Z_{(1)}$ in $|\widehat{\mathcal{U}_{(2)}}|$ such that:

(a)' If $\mathfrak{p} \in \mathfrak{P}_{(1)} \subset \mathfrak{P}_{(2)}$ then $\mathcal{K}^{j,(1)}_{\mathfrak{p}} \cap \mathfrak{U}_2(Z_{(1)}) = \mathcal{K}^{j,(2)}_{\mathfrak{p}} \cap \mathfrak{U}_2(Z_{(1)})$ for $j = 1, 2$.

(An analogue (b)' follows from (b) since $\mathcal{K}^{j,(i)} < \mathcal{K}^{(i)}$.)

We can take $\delta_{(i)}$ and partitions of unity $\{\chi^{(i)}_{\mathfrak{p}}\}$ of $(X, Z_{(i)}, \widehat{\mathcal{U}_{(i)}}, \mathcal{K}^{1,(i)}, \delta_{(i)})$ and a neighborhood $\mathfrak{U}_3(Z_{(1)})$ of $Z_{(1)}$ in $|\widehat{\mathcal{U}_{(2)}}|$ for $i = 1, 2$, such that:

(c) If $\mathfrak{p} \in \mathfrak{P}_{(1)} \subset \mathfrak{P}_{(2)}$ then $\chi^{(1)}_{\mathfrak{p}} = \chi^{(2)}_{\mathfrak{p}}$ on $\mathfrak{U}_3(Z_{(1)})$.

By definition we have

$$\int_{(X, Z_{(i)}, \widehat{\mathcal{U}_{(i)}}, \widehat{\mathfrak{S}^{\epsilon}_{(i)}})} \widehat{h} = \sum_{\mathfrak{p} \in \mathfrak{P}_{(i)}} \int_{\mathcal{K}^{1,(i)}_{\mathfrak{p}}(2\delta_{(i)}) \cap \mathfrak{U}(Z_{(i)})} \chi^{(i)}_{\mathfrak{p}} h_{\mathfrak{p}} \tag{10.19}$$

for sufficiently small neighborhoods $\mathfrak{U}(Z_{(i)})$ of $Z_{(i)}$ in $|\widehat{\mathcal{U}_{(i)}}|$. By (a)', (c) above and using the fact that $\widehat{h}$ has a compact support in $\mathring{Z}_{(1)}$, (10.19) implies (10.18). In fact, if $\mathfrak{p} \in \mathfrak{P}_{(2)} \setminus \mathfrak{P}_{(1)}$ then (b) implies that

$$\int_{\mathcal{K}^{1,(2)}_{\mathfrak{p}}(2\delta_{(2)})\cap\mathfrak{U}(Z_{(2)})} \chi^{(2)}_{\mathfrak{p}} h_{\mathfrak{p}} = 0$$

and if $\mathfrak{p} \in \mathfrak{P}_{(1)}$ then (a)', (c) imply

$$\int_{\mathcal{K}^{1,(1)}_{\mathfrak{p}}(2\delta_{(1)})\cap\mathfrak{U}(Z_{(1)})} \chi^{(1)}_{\mathfrak{p}} h_{\mathfrak{p}} = \int_{\mathcal{K}^{1,(2)}_{\mathfrak{p}}(2\delta_{(2)})\cap\mathfrak{U}(Z_{(2)})} \chi^{(2)}_{\mathfrak{p}} h_{\mathfrak{p}}.$$

Thus the proof of Proposition 10.26 is complete. □

We continue to proceed with the proof of Proposition 10.24. We consider the fiber product Kuranishi chart $\mathcal{U}_{1,\mathfrak{p}_1}\ {}_{f_{1,\mathfrak{p}_1}}\times_{f_{2,\mathfrak{p}_2}} \mathcal{U}_{2,\mathfrak{p}_2}$ and the fiber product CF-perturbation $\mathfrak{S}_{1,\mathfrak{p}_1}\ {}_{f_{1,\mathfrak{p}_1}}\times_{f_{2,\mathfrak{p}_2}} \mathfrak{S}_{2,\mathfrak{p}_2}$ of it.

We apply Proposition 10.26 to $\widehat{h} = \chi^1_{\mathfrak{p}_1}\widehat{h}_1 \wedge \chi^2_{\mathfrak{p}_2}\widehat{h}_2$, $Z_{(2)} = Z$ and

$$Z_{(1)} = (\psi_{1,\mathfrak{p}_1}(U_{1,\mathfrak{p}_1} \cap s^{-1}_{1,\mathfrak{p}_1}(0)))\ {}_{f_1}\times_{f_2} (\psi_{2,\mathfrak{p}_2}(U_{2,\mathfrak{p}_2} \cap s^{-1}_{2,\mathfrak{p}_2}(0))).$$

Note there exists a good coordinate system consisting of a single Kuranishi chart $\mathcal{U}_{1,\mathfrak{p}_1}\ {}_{f_{1,\mathfrak{p}_1}}\times_{f_{2,\mathfrak{p}_2}} \mathcal{U}_{2,\mathfrak{p}_2}$ of $Z_{(1)} \subseteq X_1 \times_M X_2$. Therefore the left hand side of (10.16) is equal to

$$\int_{(\mathcal{U}_{1,\mathfrak{p}_1}\ {}_{f_{1,\mathfrak{p}_1}}\times_{f_{2,\mathfrak{p}_2}} \mathcal{U}_{2,\mathfrak{p}_2}, \mathfrak{S}^{\epsilon}_{1,\mathfrak{p}_1}\ {}_{f_{1,\mathfrak{p}_1}}\times_{f_{2,\mathfrak{p}_2}} \mathfrak{S}^{\epsilon}_{2,\mathfrak{p}_2})} \chi^1_{\mathfrak{p}_1}\widehat{h}_1 \wedge \chi^2_{\mathfrak{p}_2}\widehat{h}_2.$$

Thus to prove (10.16) it suffices to prove the next lemma.

Lemma 10.27 *Let $\mathcal{U}_i$ be Kuranishi charts of X_i, $f_i : U_i \to M$ smooth maps, h_i differential forms on U_i and $\mathfrak{S}_i$ CF-perturbations of $\mathcal{U}_i$, for $i = 1, 2$. We assume that f_2 is strongly submersive with respect to $\mathfrak{S}_2$ and $\mathfrak{S}_1$ is transversal to 0. Then*

$$\int_{(\mathcal{U}_1\ {}_{f_1}\times_{f_2} \mathcal{U}_2, \mathfrak{S}^{\epsilon}_1\ {}_{f_1}\times_{f_2} \mathfrak{S}^{\epsilon}_2)} h_1 \wedge h_2 = \int_{(\mathcal{U}_1, \mathfrak{S}^{\epsilon}_1)} h_1 \wedge f_1^*(f_2!(h_2; \mathfrak{S}^{\epsilon}_2)). \tag{10.20}$$

Proof of Lemma 10.27 Let $\mathfrak{S}_i = (\{\mathfrak{W}^i_{\mathfrak{r}_i}\}, \{\mathcal{S}^i_{\mathfrak{r}_i}\})$ and $\{\chi^i_{\mathfrak{r}_i}\}$ be smooth partitions of unity of orbifolds U_i subordinate to its open cover $\{U^i_{\mathfrak{r}_i}\}$. Then $\{\chi^1_{\mathfrak{r}_1}\chi^2_{\mathfrak{r}_2} \mid \mathfrak{r}_1 \in \mathfrak{R}_1, \mathfrak{r}_2 \in \mathfrak{R}_2\}$ is a partition of unity subordinate to the covering $\{U_1\ {}_{f_1}\times_{f_2} U_2 \mid \mathfrak{r}_1 \in \mathfrak{R}_1, \mathfrak{r}_2 \in \mathfrak{R}_2\}$. Therefore, to prove (10.20), it suffices to prove

$$\begin{aligned}&\int_{(\mathcal{U}^1_{\mathfrak{r}_1}\ {}_{f_1}\times_{f_2} \mathcal{U}^2_{\mathfrak{r}_2}, \mathcal{S}^{1\epsilon}_{\mathfrak{r}_1}\ {}_{f_1}\times_{f_2} \mathcal{S}^{2\epsilon}_{\mathfrak{r}_2})} \chi^1_{\mathfrak{r}_1} h_1 \wedge \chi^2_{\mathfrak{r}_2} h_2 \\ &= \int_{(\mathcal{U}^1_{\mathfrak{r}_1}, \mathcal{S}^{1\epsilon}_{\mathfrak{r}_1})} \chi^1_{\mathfrak{r}_1}\chi^2_{\mathfrak{r}_2} h_1 \wedge f_1^*(f_2!(h_2; \mathcal{S}^{2\epsilon}_{\mathfrak{r}_2})).\end{aligned} \tag{10.21}$$

Equation (10.21) is an immediate consequence of the next lemma.

Lemma 10.28 *For $i = 1, 2$ let N_i be smooth manifolds and $f_i : N_i \to M$ smooth maps and h_i smooth differential forms on N_i of compact support. Suppose that f_2 is a submersion. Then we have*

$$\int_{N_1\ {}_{f_1}\times_{f_2} N_2} h_1 \wedge h_2 = \int_{N_1} h_1 \wedge f_1^*(f_2!(h_2)). \tag{10.22}$$

Proof By using partitions of unity, it suffices to prove the lemma in the following special case: $N_2 = M \times \mathbb{R}^b$, $f_2 : M \times \mathbb{R}^b \to M$ is the projection to the first factor, and M, N_1, N_2 are open subsets of Euclidean spaces. We prove this case below.

In this case $N_1\ {}_{f_1}\times_{f_2} N_2 = N_1 \times \mathbb{R}^b$ (as oriented manifolds). Let $(x_1, \dots, x_m)$ be a coordinate of M, $(y_1, \dots, y_a)$ a coordinate of N_1 and $(z_1, \dots, z_b)$ a coordinate of $\mathbb{R}^b$. Then $(x_1, \dots, x_m, z_1, \dots, z_b)$ is a coordinate of N_2, and $(y_1, \dots, y_a, z_1, \dots, z_b)$ is a coordinate of $N_1\ {}_{f_1}\times_{f_2} N_2$.

We may write

$$h_1 = \sum_I g_{1,I}(y_1, \dots, y_a) dy_I$$

$$h_2 = h_2' + \sum_J g_{2,J}(x_1, \dots, x_m, z_1, \dots, z_b) dx_J \wedge dz_1 \wedge \cdots \wedge dz_b$$

for certain smooth functions $g_{1,I}, g_{2,J}$. We write them as $g_{1,I}(y), g_{2,J}(x, z)$ for simplicity. Here $I \subseteq \{1, \dots, a\}$, $J \subseteq \{1, \dots, m\}$ and $dx_I = dx_{i_1} \wedge \cdots \wedge dx_{i_{|I|}}$ for $I = \{i_1, \dots, i_{|I|}\}$ with $i_1 < \cdots < i_{|I|}$ and dx_J is defined in a similar way. The differential form h_2' is the sum of the terms which do not contain at least one of dz_1, $\dots, dz_b$.

For $I \subseteq \{1, \dots, a\}$, $J \subseteq \{1, \dots, m\}$ with $|I| = |J|$, we define $\alpha_{IJ} : N_1 \to \mathbb{R}$ by

$$f_1^*(dx_J) = \sum_I \alpha_{IJ} dy_I.$$

Then the left hand side of (10.22) is given by

$$\sum_{I,J} \int_{N_1\ {}_{f_1}\times_{f_2} N_2} \epsilon_I \alpha_{I^cJ}(y) g_{1,I}(y) g_{2,J}(f_1(y), z) dy dz, \tag{10.23}$$

where $I^c = \{1, \dots, a\} \setminus I$. The sign ϵ_I is the signature of the permutation $(1, \dots, a) \to (II^c)$.

On the other hand, the right hand side of (10.22) is given by

$$\sum_{I,J} \int_{N_1} \epsilon_I \alpha_{I^cJ}(y) g_{1,I}(y) \left(\int_{z \in \mathbb{R}^b} g_{2,J}(f_1(y), z) dz \right) dy. \tag{10.24}$$

Therefore Lemma 10.28 is an immediate consequence of Fubini's theorem. □

This completes the proof of Lemma 10.27. □

This also completes the proof of Proposition 10.24. □

Therefore the proof of Theorem 10.21 is now complete. □

We finally remark the following.

Lemma 10.29 *When CF-perturbations $\tilde{\mathfrak{X}}_{32}$, $\tilde{\mathfrak{X}}_{21}$ vary in a uniform family, we can choose ϵ in Theorem* 10.21 *in the way independent of the CF-perturbations in that family.*

The proof goes in the same way as the proof of Proposition 7.89. So we omit it.

Chapter 11
Construction of Good Coordinate Systems

In this chapter we prove Theorem 3.35 together with various addenda and variants.

11.1 Construction of Good Coordinate Systems: The Absolute Case

This section will be occupied with the proof of Theorem 3.35. Theorem 3.35 claims existence of good coordinate system compatible to the given Kuranishi structure in the sense of Definition 3.15.

Proof of Theorem 3.35 Let $\widehat{\mathcal{U}}$ be a Kuranishi structure of $Z \subseteq X$. We use the dimension stratification $\mathcal{S}_{\mathfrak{d}}(X, Z; \widehat{\mathcal{U}})$ as in Definition 5.1, for the inductive construction of a good coordinate system. Namely we construct a good coordinate system of $\mathcal{S}_{\mathfrak{d}}(X, Z; \widehat{\mathcal{U}})$ by a downward induction on $\mathfrak{d}$. The main part of the proof is the proof of Proposition 11.3 below. We use this proposition to add Kuranishi charts one by one and obtain the required good coordinate system. We first describe the situation under which Proposition 11.3 is stated.

Situation 11.1 Let $\mathfrak{d} \in \mathbb{Z}_{\geq 0}$, Z_0 a compact subset of

$$\mathcal{S}_{\mathfrak{d}}(X, Z; \widehat{\mathcal{U}}) \setminus \bigcup_{\mathfrak{d}' > \mathfrak{d}} \mathcal{S}_{\mathfrak{d}'}(X, Z; \widehat{\mathcal{U}}),$$

and Z_1 a compact subset of $\mathcal{S}_{\mathfrak{d}}(X, Z; \widehat{\mathcal{U}})$. We assume that Z_1 contains an open neighborhood of

$$\bigcup_{\mathfrak{d}' > \mathfrak{d}} \mathcal{S}_{\mathfrak{d}'}(X, Z; \widehat{\mathcal{U}})$$

K. Fukaya et al., *Kuranishi Structures and Virtual Fundamental Chains*, Springer Monographs in Mathematics, https://doi.org/10.1007/978-981-15-5562-6_11

in $\mathcal{S}_{\mathfrak{d}}(X, Z; \widehat{\mathcal{U}})$.

We also assume that we have a good coordinate system $\widehat{\mathcal{U}} = (\mathfrak{P}, \{\mathcal{U}_{\mathfrak{p}}\}, \{\Phi_{\mathfrak{pq}}\})$ of $Z_1^+ \subseteq X$, where Z_1^+ is a compact neighborhood of Z_1 in X, and a strict KG-embedding

$$\widehat{\Phi^1} = \{\Phi^1_{\mathfrak{p}p} \mid p \in \mathrm{Im}(\psi_{\mathfrak{p}}) \cap Z_1^+\} : \widehat{\mathcal{U}}|_{Z_1^+} \to \widehat{\mathcal{U}},$$

where $\widehat{\mathcal{U}}|_{Z_1^+}$ is the restriction of $\widehat{\mathcal{U}}$, which is defined in Definition 7.52 (3). We assume $\dim U_{\mathfrak{p}} \geq \mathfrak{d}$ for all $\mathfrak{p} \in \mathfrak{P}$.

Let $\mathcal{U}_{\mathfrak{p}_0} = (U_{\mathfrak{p}_0}, E_{\mathfrak{p}_0}, s_{\mathfrak{p}_0}, \psi_{\mathfrak{p}_0})$ be a Kuranishi neighborhood of Z_0^+[1] such that $\dim U_{\mathfrak{p}_0} = \mathfrak{d}$. Here Z_0^+ is a compact neighborhood of Z_0 in X (Fig. 11.1). We regard $\mathcal{U}_{\mathfrak{p}_0}$ as a good coordinate system $\widehat{\mathcal{U}_{\mathfrak{p}_0}}$ of $X_0^+ \subset X$ that consists of a single Kuranishi chart and suppose that we are given a strict KG-embedding

$$\widehat{\Phi^0} = \{\Phi^0_{\mathfrak{p}_0 p} \mid p \in \mathrm{Im}(\psi_{\mathfrak{p}_0}) \cap Z_0^+\} : \widehat{\mathcal{U}}|_{Z_0^+} \to \widehat{\mathcal{U}_{\mathfrak{p}_0}}.$$

We put

$$Z_+ = Z_1 \cup Z_0. \tag{11.1}$$

■

Remark 11.2 We note $Z_+ \subseteq \mathcal{S}_{\mathfrak{d}}(X, Z; \widehat{\mathcal{U}})$. But in general Z_0^+, Z_1^+ are not subsets of $\mathcal{S}_{\mathfrak{d}}(X, Z; \widehat{\mathcal{U}})$.

Proposition 11.3 *In Situation* 11.1, *there exists a good coordinate system* $\widehat{\mathcal{U}^+} = (\mathfrak{P}^+, \{\mathcal{U}^+_{\mathfrak{p}}\}, \{\Phi^+_{\mathfrak{pq}}\})$ *of* $Z_+^+ \subseteq X$ *with the following properties. Here* Z_+^+ *is a compact neighborhood of* Z_+ *in* X.

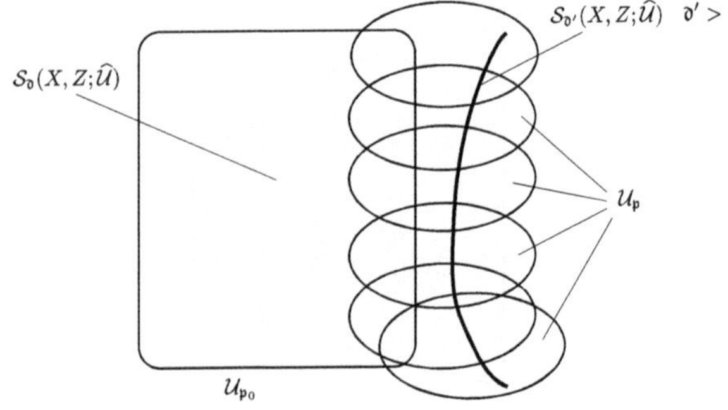

Fig. 11.1 Situation 11.1

[1] In particular, $Z_0^+ \subset \mathrm{Im}\psi_{\mathfrak{p}_0}$.

(1) $\mathfrak{P}^+ = \mathfrak{P} \cup \{\mathfrak{p}_0\}$. *The partial order on* $\mathfrak{P}^+$ *is the same as that of* $\mathfrak{P}$ *between two elements of* $\mathfrak{P}$ *and* $\mathfrak{p} > \mathfrak{p}_0$ *for any* $\mathfrak{p} \in \mathfrak{P}$.

(2) *If* $\mathfrak{p} \in \mathfrak{P}$ *then* $\mathcal{U}_{\mathfrak{p}}^+$ *is an open subchart* $\mathcal{U}_{\mathfrak{p}}|_{U_{\mathfrak{p}}^+}$ *where* $U_{\mathfrak{p}}^+$ *is an open subset of* $U_{\mathfrak{p}}$. *We have*

$$\bigcup_{\mathfrak{p} \in \mathfrak{P}} \psi_{\mathfrak{p}}^+((s_{\mathfrak{p}}^+)^{-1}(0)) \supset Z_1. \tag{11.2}$$

(3) $\mathcal{U}_{\mathfrak{p}_0}^+$ *is a restriction of* $\mathcal{U}_{\mathfrak{p}_0}$ *to an open subset* $U_{\mathfrak{p}_0}^+$ *of* $U_{\mathfrak{p}_0}$. *We have*

$$\psi_{\mathfrak{p}_0}^+((s_{\mathfrak{p}_0}^+)^{-1}(0)) \supset Z_0.$$

(4) *The coordinate change* $\Phi_{\mathfrak{p}\mathfrak{q}}^+$ *is the restriction of* $\Phi_{\mathfrak{p}\mathfrak{q}}$ *to* $U_{\mathfrak{q}}^+ \cap \varphi_{\mathfrak{p}\mathfrak{q}}^{-1}(U_{\mathfrak{p}}^+)$, *if* $\mathfrak{p}, \mathfrak{q} \in \mathfrak{P}$.

(5) *There exists an open substructure* $\widehat{\mathcal{U}^0}$ *of* $\widehat{\mathcal{U}}$ *and a strict KG-embedding* $\widehat{\Phi^+} = \{\Phi_{\mathfrak{p}p}^+ \mid p \in \mathrm{Im}(\psi_{\mathfrak{p}}^+) \cap Z_+^+\} : \widehat{\mathcal{U}^0} \to \widehat{\mathcal{U}^+}$ *with the following properties:*

(a) *If* $\mathfrak{p} \in \mathfrak{P}$ *and* $p \in \mathrm{Im}(\psi_{\mathfrak{p}}^+) \cap Z_+^+$, *then we have the commutative diagram:*

$$\begin{array}{ccc} \mathcal{U}_p^0 & \xrightarrow{\Phi_{\mathfrak{p}p}^+} & \mathcal{U}_{\mathfrak{p}}^+ \\ \downarrow & & \downarrow \\ \mathcal{U}_p & \xrightarrow{\Phi_{\mathfrak{p}p}^1} & \mathcal{U}_{\mathfrak{p}} \end{array} \tag{11.3}$$

where the vertical arrows are embeddings as open subcharts. (The commutativity of diagram means that the maps coincide when both sides are defined.)

(b) *If* $p \in \mathrm{Im}(\psi_{\mathfrak{p}_0}^+) \cap Z_+^+$, *then we have the commutative diagram:*

$$\begin{array}{ccc} \mathcal{U}_p^0 & \xrightarrow{\Phi_{\mathfrak{p}_0 p}^+} & \mathcal{U}_{\mathfrak{p}_0}^+ \\ \downarrow & & \downarrow \\ \mathcal{U}_p & \xrightarrow{\Phi_{\mathfrak{p}_0 p}^0} & \mathcal{U}_{\mathfrak{p}_0} \end{array} \tag{11.4}$$

where the vertical arrows are embeddings as open subcharts. (The commutativity of diagram means that the maps coincide when both sides are defined.)

Remark 11.4 Note that the good coordinate system $\widehat{\mathcal{U}^+}$ has one more Kuranishi chart than $\widehat{\mathcal{U}}$. Moreover $\widehat{\mathcal{U}^+}$ is a good coordinate system of a neighborhood of Z_+

which contains Z_1. $\widehat{\mathcal{U}}$ is a good coordinate system of a neighborhood of Z_1. This is the reason why we use the symbol $+$ in $\widehat{\mathcal{U}^+}$.

On the other hand, each Kuranishi chart of $\widehat{\mathcal{U}^+}$ is an open subchart either of $\widehat{\mathcal{U}}$ or of $\widehat{\mathcal{U}_{\mathfrak{p}_0}}$. In other words, each Kuranishi chart of $\widehat{\mathcal{U}^+}$ is *smaller* than that of $\widehat{\mathcal{U}}$ or of $\widehat{\mathcal{U}_{\mathfrak{p}_0}}$.

Proof of Proposition 11.3 ⇒ Theorem 3.35 We will construct a good coordinate system of a closed subset $\mathcal{S}_{\mathfrak{d}}(X, Z; \widehat{\mathcal{U}})$ of Z by a downward induction on $\mathfrak{d}$.

Let $\widehat{\mathcal{U}_{\mathfrak{d}+1}} = (\mathfrak{P}, \{\mathcal{U}_{\mathfrak{p}}\}, \{\Phi_{\mathfrak{p}\mathfrak{q}}\})$ be a good coordinate system of a compact neighborhood of $\mathcal{S}_{\mathfrak{d}+1}(X, Z; \widehat{\mathcal{U}})$. We assume $\dim U_{\mathfrak{p}} \geq \mathfrak{d} + 1$ for all $\mathfrak{p} \in \mathfrak{P}$, as a part of the induction hypothesis. We will construct a good coordinate system of $\mathcal{S}_{\mathfrak{d}}(X, Z; \widehat{\mathcal{U}})$.

We take a support system $\mathcal{K}_{\mathfrak{p}}$ of $\widehat{\mathcal{U}_{\mathfrak{d}+1}}$ and put

$$B = \mathcal{S}_{\mathfrak{d}}(X, Z; \widehat{\mathcal{U}}) \setminus \bigcup_{\mathfrak{p} \in \mathfrak{P}} \psi_{\mathfrak{p}}(s_{\mathfrak{p}}^{-1}(0) \cap \operatorname{Int} \mathcal{K}_{\mathfrak{p}}).$$

Then B is a compact subset of $\mathcal{S}_{\mathfrak{d}}(X, Z; \widehat{\mathcal{U}}) \setminus \bigcup_{\mathfrak{d}' > \mathfrak{d}} \mathcal{S}_{\mathfrak{d}'}(X, Z; \widehat{\mathcal{U}})$. We take a finite number of points $p_1, \ldots, p_N \in B$ such that

$$\bigcup_{i=1}^{N} \psi_{p_i}(s_{p_i}^{-1}(0)) \supset B.$$

We take compact subsets $K_{p_i} \subset U_{p_i}$ such that

$$\bigcup_{i=1}^{N} \psi_{p_i}(s_{p_i}^{-1}(0) \cap K_{p_i}) \supset B.$$

We put $Z_i = \psi_{p_i}(s_{p_i}^{-1}(0))$ and

$$Z(0) = \bigcup_{\mathfrak{p} \in \mathfrak{P}} \psi_{\mathfrak{p}}(s_{\mathfrak{p}}^{-1}(0) \cap \mathcal{K}_{\mathfrak{p}}) \cap Z.$$

Now using Proposition 11.3 we can construct a good coordinate system on a compact neighborhood of

$$Z(0) \cup \bigcup_{i=1}^{n} Z_i$$

by an induction on n. We thus obtain a good coordinate system on $\mathcal{S}_{\mathfrak{d}}(X, Z; \widehat{\mathcal{U}})$.

Now the downward induction on $\mathfrak{d}$ is complete and we obtain a good coordinate system on $\mathcal{S}_0(X, Z; \widehat{\mathcal{U}}) = Z$. Therefore to prove Theorem 3.35 it remains to prove Proposition 11.3. □

Proof of Proposition 11.3 We take a support system $\mathcal{K}$ of $\widehat{\mathcal{U}}$. By definition we have

$$\bigcup_{\mathfrak{p}} \psi_{\mathfrak{p}}(\overset{\circ}{\mathcal{K}_{\mathfrak{p}}} \cap s_{\mathfrak{p}}^{-1}(0)) \supset Z_1^+.$$

We take a compact subset $\mathcal{K}_{\mathfrak{p}_0}$ of $U_{\mathfrak{p}_0}$ such that

$$\psi_{\mathfrak{p}_0}(\overset{\circ}{\mathcal{K}_{\mathfrak{p}_0}} \cap s_{\mathfrak{p}_0}^{-1}(0)) \supset Z_0^+.$$

Since $\widehat{\Phi^1} : \widehat{\mathcal{U}}|_{Z_1^+} \to \widehat{\mathcal{U}}$ and $\widehat{\Phi^0} : \widehat{\mathcal{U}}|_{Z_0^+} \to \widehat{\mathcal{U}_{\mathfrak{p}_0}}$ are *strict* KG-embeddings, they have the following properties.

Property 11.5

(1) If $q \in \psi_{\mathfrak{p}}(\mathcal{K}_{\mathfrak{p}} \cap s_{\mathfrak{p}}^{-1}(0))$, then $\Phi^1_{\mathfrak{p}q} = (\varphi^1_{\mathfrak{p}q}, \widehat{\varphi^1_{\mathfrak{p}q}})$ is defined and is an embedding $\Phi^1_{\mathfrak{p}q} : \mathcal{U}_q \to \mathcal{U}_{\mathfrak{p}}$. We denote the domain of $\varphi^1_{\mathfrak{p}q}$ by $U^{1+}_{\mathfrak{p}q}$.

(2) If $q \in \psi_{\mathfrak{p}_0}(\mathcal{K}_{\mathfrak{p}_0} \cap s_{\mathfrak{p}_0}^{-1}(0))$, then $\Phi^0_{\mathfrak{p}_0 q} = (\varphi^0_{\mathfrak{p}_0 q}, \widehat{\varphi^0_{\mathfrak{p}_0 q}})$ is defined and is an embedding $\Phi^0_{\mathfrak{p}_0 q} : \mathcal{U}_q \to \mathcal{U}_{\mathfrak{p}_0}$. We denote the domain of $\varphi^0_{\mathfrak{p}_0 q}$ by $U^{0+}_{\mathfrak{p}_0 q}$.

We take $U^1_{\mathfrak{p}q}$ (resp. $U^0_{\mathfrak{p}_0 q}$), a neighborhood of o_q, which is relatively compact in $U^{1+}_{\mathfrak{p}q}$ (resp. $U^{0+}_{\mathfrak{p}q}$).

For each $\mathfrak{p} \in \mathfrak{P}$ we put

$$Z_{\mathfrak{p}\mathfrak{p}_0} = \psi_{\mathfrak{p}_0}(\mathcal{K}_{\mathfrak{p}_0} \cap s_{\mathfrak{p}_0}^{-1}(0)) \cap \mathcal{S}_{\mathfrak{d}}(X, Z; \widehat{\mathcal{U}}) \cap \psi_{\mathfrak{p}}(\mathcal{K}_{\mathfrak{p}} \cap s_{\mathfrak{p}}^{-1}(0)). \tag{11.5}$$

We remark that $Z_{\mathfrak{p}\mathfrak{p}_0} \subseteq \mathrm{Im}(\psi_{\mathfrak{p}_0}) \cap \mathrm{Im}(\psi_{\mathfrak{p}})$.

$Z_{\mathfrak{p}\mathfrak{p}_0} \cap \mathcal{S}_{\mathfrak{d}+1}(X, Z; \widehat{\mathcal{U}}) = \emptyset$ since $\mathrm{Im}\psi_{\mathfrak{p}_0} \cap \mathcal{S}_{\mathfrak{d}+1}(X, Z; \widehat{\mathcal{U}}) = \emptyset$. We use compactness of $Z_{\mathfrak{p}\mathfrak{p}_0}$ to find a finite number of points $q_1^{\mathfrak{p}}, \dots, q_{N(\mathfrak{p})}^{\mathfrak{p}} \in \psi_{\mathfrak{p}_0}(\mathcal{K}_{\mathfrak{p}_0} \cap s_{\mathfrak{p}_0}^{-1}(0)) \cap \mathcal{S}_{\mathfrak{d}}(X, Z; \widehat{\mathcal{U}}) \cap \psi_{\mathfrak{p}}(\mathcal{K}_{\mathfrak{p}} \cap s_{\mathfrak{p}}^{-1}(0))$ such that

$$\bigcup_{i=1}^{N(\mathfrak{p})} \psi_{q_i^{\mathfrak{p}}}(s_{q_i^{\mathfrak{p}}}^{-1}(0) \cap U^1_{\mathfrak{p}q_i^{\mathfrak{p}}} \cap U^0_{\mathfrak{p}_0 q_i^{\mathfrak{p}}}) \supset Z_{\mathfrak{p}\mathfrak{p}_0}. \tag{11.6}$$

(Fig. 11.2) Let $\tilde{Z}_{\mathfrak{p}\mathfrak{p}_0}$ be the compact subset of $s_{\mathfrak{p}_0}^{-1}(0) \subset U_{\mathfrak{p}_0}$ such that

$$\psi_{\mathfrak{p}_0}(\tilde{Z}_{\mathfrak{p}\mathfrak{p}_0}) = Z_{\mathfrak{p}\mathfrak{p}_0}.$$

Fig. 11.2 $U^0_{q_i^{\mathfrak p}}$

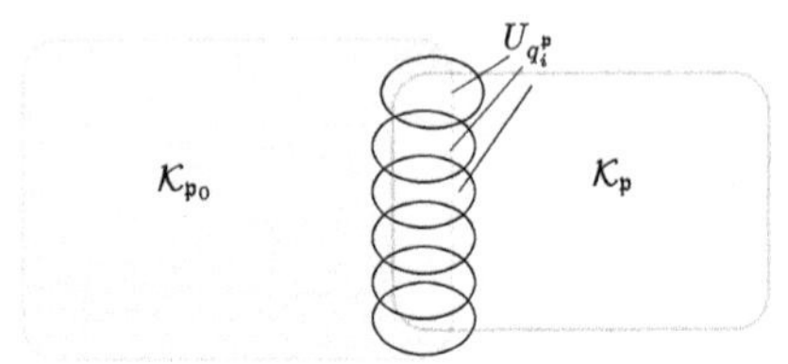

Then it follows from (11.6), the domain of $\varphi^0_{\mathfrak{p}_0 q} = U^0_{\mathfrak{p}_0 q}$, and the embedding property of $\psi_{\mathfrak{p}_0}$, ψ_q that:

$$\bigcup_{i=1}^{N(\mathfrak p)} \varphi^0_{\mathfrak{p}_0 q_i^{\mathfrak p}}(U^1_{\mathfrak{p} q_i^{\mathfrak p}} \cap U^0_{\mathfrak{p}_0 q_i^{\mathfrak p}}) \supset \tilde{Z}_{\mathfrak{p}\mathfrak{p}_0}. \tag{11.7}$$

We fix a metric d on $U_{\mathfrak{p}_0}$ and consider the ϵ neighborhood $B^{\mathfrak{p}_0}_\epsilon(\tilde{Z}_{\mathfrak{p}\mathfrak{p}_0})$ of $\tilde{Z}_{\mathfrak{p}\mathfrak{p}_0}$. Our first goal is to define

$$\varphi_{\mathfrak{p}\mathfrak{p}_0} : B^{\mathfrak{p}_0}_\epsilon(\tilde{Z}_{\mathfrak{p}\mathfrak{p}_0}) \to U_{\mathfrak p}.$$

We take $\epsilon > 0$ small such that $B_\epsilon(\tilde{Z}_{\mathfrak{p}\mathfrak{p}_0}) \subset$ the left hand side of (11.7). Let $x \in B^{\mathfrak{p}_0}_\epsilon(\tilde{Z}_{\mathfrak{p}\mathfrak{p}_0})$. Then by (11.7) there exist i and $x_i \in U^0_{\mathfrak{p}_0 q_i^{\mathfrak p}} \cap U^1_{\mathfrak{p} q_i^{\mathfrak p}}$ such that $x = \varphi^0_{\mathfrak{p}_0 q_i^{\mathfrak p}}(x_i)$. We will define $\varphi_{\mathfrak{p}\mathfrak{p}_0}(x) = \varphi^1_{\mathfrak{p} q_i^{\mathfrak p}}(x_i)$. We show that $\varphi^1_{\mathfrak{p} q_i^{\mathfrak p}}(x_i)$ is independent of i in the next lemma.

Lemma 11.6 *There exists $\epsilon_0 > 0$ such that the following holds for all $0 < \epsilon \le \epsilon_0$. Let $x \in B^{\mathfrak{p}_0}_\epsilon(\tilde{Z}_{\mathfrak{p}\mathfrak{p}_0})$. Suppose*

$$x = \varphi^0_{\mathfrak{p}_0 q_i^{\mathfrak p}}(x_i) = \varphi^0_{\mathfrak{p}_0 q_j^{\mathfrak p}}(x_j)$$

with $x_i \in U^0_{\mathfrak{p}_0 q_i^{\mathfrak p}} \cap U^1_{\mathfrak{p} q_i^{\mathfrak p}}$, $x_j \in U^0_{\mathfrak{p}_0 q_j^{\mathfrak p}} \cap U^1_{\mathfrak{p} q_j^{\mathfrak p}}$. Then

$$\varphi^1_{\mathfrak{p} q_i^{\mathfrak p}}(x_i) = \varphi^1_{\mathfrak{p} q_j^{\mathfrak p}}(x_j).$$

Proof If no such ϵ_0 exist, then there exist a sequence $\epsilon_n \to 0$ and $x_{n,i} \in U^0_{\mathfrak{p}_0 q_i^{\mathfrak p}} \cap U^1_{\mathfrak{p} q_i^{\mathfrak p}}$, $x_{n,j} \in U^0_{\mathfrak{p}_0 q_j^{\mathfrak p}} \cap U^1_{\mathfrak{p} q_j^{\mathfrak p}}$ such that

$$x_n = \varphi^0_{\mathfrak{p}_0 q_i^{\mathfrak p}}(x_{n,i}) = \varphi^0_{\mathfrak{p}_0 q_j^{\mathfrak p}}(x_{n,j}) \in B^{\mathfrak{p}_0}_{\epsilon_n}(\tilde{Z}_{\mathfrak{p}\mathfrak{p}_0})$$

but

$$\varphi^1_{\mathfrak{p}q_i^{\mathfrak{p}}}(x_{n,i}) \neq \varphi^1_{\mathfrak{p}q_j^{\mathfrak{p}}}(x_{n,j}). \tag{11.8}$$

(We can take i, j independent of n since $i, j \in \{1, \dots, N(\mathfrak{p})\}$, a finite set.) By taking a subsequence we may assume

$$\lim_{n\to\infty} x_n = r \in \tilde{Z}_{\mathfrak{p}\mathfrak{p}_0}.$$

(Fig. 11.3) We remark that $\varphi^0_{\mathfrak{p}_0 r} : U^0_{\mathfrak{p}_0 r} \to U_{\mathfrak{p}_0}$ is an *open* embedding. This is because both have dimension $\mathfrak{d}$. Therefore for large n there exists $y_n \in U^0_{\mathfrak{p}_0 r}$ such that

$$x_n = \varphi^0_{\mathfrak{p}_0 r}(y_n).$$

Note y_n converges to o_r as n goes to infinity. We also remark that $r \in \varphi_{\mathfrak{p}_0 q_i^{\mathfrak{p}}}(U^0_{\mathfrak{p}_0 q_i^{\mathfrak{p}}}) \subset \varphi_{\mathfrak{p}_0 q_i^{\mathfrak{p}}}(U^{0+}_{\mathfrak{p}_0 q_i^{\mathfrak{p}}})$. Therefore using also $s_{\mathfrak{p}_0}(r) = 0$ we obtain $\varphi_{q_i^{\mathfrak{p}} r}(o_r) \in U^{0+}_{\mathfrak{p}_0 q_i^{\mathfrak{p}}}$.[2] We thus conclude that for large n the element $\varphi^0_{\mathfrak{p}_0 q_i^{\mathfrak{p}}}(\varphi_{q_i^{\mathfrak{p}} r}(y_n))$ is defined and

$$\varphi^0_{\mathfrak{p}_0 q_i^{\mathfrak{p}}}(\varphi_{q_i^{\mathfrak{p}} r}(y_n)) = \varphi_{\mathfrak{p}_0 r}(y_n) = x_n = \varphi^0_{\mathfrak{p}_0 q_i^{\mathfrak{p}}}(x_{n,i}).$$

Thus we have

$$\varphi_{q_i^{\mathfrak{p}} r}(y_n) = x_{n,i}. \tag{11.9}$$

In the same way we obtain

$$\varphi_{q_j^{\mathfrak{p}} r}(y_n) = x_{n,j}.$$

Now we calculate

$$\varphi^1_{\mathfrak{p}q_i^{\mathfrak{p}}}(x_{n,i}) = \varphi^1_{\mathfrak{p}q_i^{\mathfrak{p}}}(\varphi_{q_i^{\mathfrak{p}} r}(y_n)) = \varphi^1_{\mathfrak{p}r}(y_n) = \varphi^1_{\mathfrak{p}q_j^{\mathfrak{p}}}(\varphi_{q_j^{\mathfrak{p}} r}(y_n)) = \varphi^1_{\mathfrak{p}q_j^{\mathfrak{p}}}(x_{n,j}).$$

This contradicts (11.8). □

We now define $\varphi_{\mathfrak{p}\mathfrak{p}_0} : B^{\mathfrak{p}_0}_{\epsilon_0}(\tilde{Z}_{\mathfrak{p}\mathfrak{p}_0}) \to U_{\mathfrak{p}}$ by

$$\varphi_{\mathfrak{p}\mathfrak{p}_0}(x) = \varphi^1_{\mathfrak{p}q_i^{\mathfrak{p}}}(x_i), \tag{11.10}$$

[2] Here we use the compatibility of parametrization with coordinate change, Definition 3.2 (4).

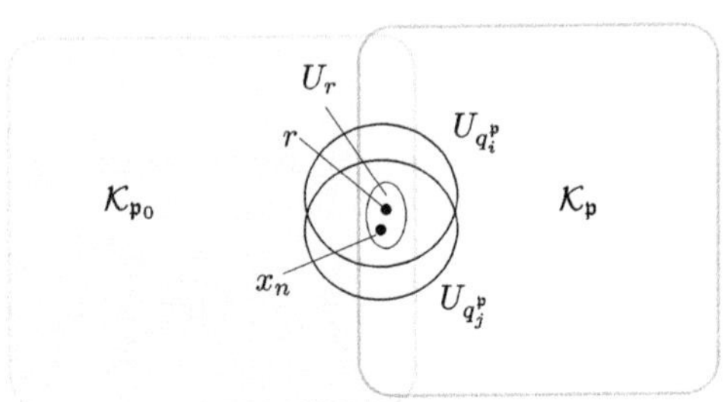

Fig. 11.3 Lemma 11.6

where $x = \varphi^0_{\mathfrak{p}_0 q_i^{\mathfrak p}}(x_i)$ with $x_i \in U^0_{\mathfrak{p}_0 q_i^{\mathfrak p}} \cap U^1_{\mathfrak{p} q_i^{\mathfrak p}}$. Using the fact that $\varphi^0_{\mathfrak{p}_0 q_i^{\mathfrak p}}$ is an open embedding we can easily show that $\varphi_{\mathfrak{p}\mathfrak{p}_0}$ is an embedding of orbifolds.

We next define $\widehat{\varphi_{\mathfrak{p}\mathfrak{p}_0}} : \mathcal{E}_{\mathfrak{p}_0}|_{B^{\mathfrak{p}_0}_{\epsilon_0}(\tilde Z_{\mathfrak{p}\mathfrak{p}_0})} \to \mathcal{E}_{\mathfrak{p}_0}$ by

$$\widehat{\varphi_{\mathfrak{p}\mathfrak{p}_0}}(v) = \widehat{\varphi^1_{\mathfrak{p} q_i^{\mathfrak p}}}(v_i), \tag{11.11}$$

where $v = \widehat{\varphi^0_{\mathfrak{p}_0 q_i^{\mathfrak p}}}(v_i)$ with $\pi(v_i) \in U^0_{\mathfrak{p}_0 q_i^{\mathfrak p}} \cap U^1_{\mathfrak{p} q_i^{\mathfrak p}}$. Well-definedness of $\widehat{\varphi_{\mathfrak{p}\mathfrak{p}_0}}$ can be proved in the same way as Lemma 11.6.

The compatibility of $(\varphi_{\mathfrak{p}\mathfrak{p}_0}, \widehat{\varphi_{\mathfrak{p}\mathfrak{p}_0}})$ with Kuranishi map and parametrization map is immediate from the definition.

We next check that $(\varphi_{\mathfrak{p}\mathfrak{p}_0}, \widehat{\varphi_{\mathfrak{p}\mathfrak{p}_0}})$, $(\varphi_{\mathfrak{q}\mathfrak{p}_0}, \widehat{\varphi_{\mathfrak{q}\mathfrak{p}_0}})$ are compatible with the coordinate change $\varphi_{\mathfrak{p}\mathfrak{q}}$ of $\widehat{\mathcal{U}}$. Note $U_{\mathfrak{p}\mathfrak{q}}$ is the domain of $\varphi_{\mathfrak{p}\mathfrak{q}}$.

We fix a metric on each $U_{\mathfrak p}$ ($\mathfrak{p} \in \mathfrak{P}$) and put

$$\begin{aligned} U^{\epsilon}_{\mathfrak p} &= B^{\mathfrak p}_{\epsilon}(\mathcal{K}_{\mathfrak p}) \subset U_{\mathfrak p} \\ U^{\epsilon}_{\mathfrak{p}\mathfrak{q}} &= U^{\epsilon}_{\mathfrak q} \cap (\varphi_{\mathfrak{p}\mathfrak{q}})^{-1}(U^{\epsilon}_{\mathfrak p}) \subset U_{\mathfrak{p}\mathfrak{q}}. \end{aligned} \tag{11.12}$$

Here and hereafter $B^{\mathfrak p}_{\epsilon}(A)$ denotes the ϵ neighborhood of a subset $A \subseteq U_{\mathfrak p}$ with respect to the metric we take on $U_{\mathfrak p}$.

Lemma 11.7 *Let ϵ_0 be as in Lemma* 11.6. *Then there exists a positive number $\epsilon_1 < \epsilon_0$ such that the following holds for $0 < \epsilon < \epsilon_1$.*

If $x \in (\varphi_{\mathfrak{q}\mathfrak{p}_0})^{-1}(U^{\epsilon}_{\mathfrak{p}\mathfrak{q}}) \cap B^{\mathfrak{p}_0}_{\epsilon}(\tilde Z_{\mathfrak{q}\mathfrak{p}_0}) \cap B^{\mathfrak{p}_0}_{\epsilon}(\tilde Z_{\mathfrak{p}\mathfrak{p}_0})$ then

$$\varphi_{\mathfrak{p}\mathfrak{q}}(\varphi_{\mathfrak{q}\mathfrak{p}_0}(x)) = \varphi_{\mathfrak{p}\mathfrak{p}_0}(x). \tag{11.13}$$

Proof The proof is again by contradiction. Suppose (11.13) does not hold. Then there exist a sequence $\epsilon_n \to 0$ and $x_n \in B^{\mathfrak{p}_0}_{\epsilon_n}(\tilde Z_{\mathfrak{q}\mathfrak{p}_0}) \cap B^{\mathfrak{p}_0}_{\epsilon_n}(\tilde Z_{\mathfrak{p}\mathfrak{p}_0})$ such that

$$\varphi_{\mathfrak{q}\mathfrak{p}_0}(x_n) \in U^{\epsilon_n}_{\mathfrak{p}\mathfrak{q}}$$

but

$$\varphi_{\mathfrak{pq}}(\varphi_{\mathfrak{qp}_0}(x_n)) \neq \varphi_{\mathfrak{pp}_0}(x_n).$$

We may assume x_n converges to $r \in \tilde{Z}_{\mathfrak{qp}_0} \cap \tilde{Z}_{\mathfrak{pp}_0}$. By the definition of $\varphi_{\mathfrak{qp}_0}$ and $\varphi_{\mathfrak{pp}_0}$, we have i, j and $x_{n,i} \in U^0_{\mathfrak{p}_0 q_i^{\mathfrak{q}}} \cap U^1_{\mathfrak{q} q_i^{\mathfrak{q}}}$, $x_{n,j} \in U^0_{\mathfrak{p}_0 q_j^{\mathfrak{p}}} \cap U^1_{\mathfrak{p} q_j^{\mathfrak{p}}}$ such that

$$x_n = \varphi^0_{\mathfrak{p}_0 q_i^{\mathfrak{q}}}(x_{n,i}), \qquad x_n = \varphi^0_{\mathfrak{p}_0 q_j^{\mathfrak{p}}}(x_{n,j}),$$

and

$$\varphi_{\mathfrak{qp}_0}(x_n) = \varphi^1_{\mathfrak{q} q_i^{\mathfrak{q}}}(x_{n,i}), \qquad \varphi_{\mathfrak{pp}_0}(x_n) = \varphi^1_{\mathfrak{p} q_j^{\mathfrak{p}}}(x_{n,j}).$$

See (11.10). In the same way as the proof of Lemma 11.6 we can find $y_n \in U^0_{\mathfrak{p}_0 r}$ such that

$$\varphi_{q_i^{\mathfrak{q}} r}(y_n) = x_{n,i}, \qquad \varphi_{q_j^{\mathfrak{p}} r}(y_n) = x_{n,j}.$$

(See the proof of (11.9).) Therefore

$$\begin{aligned}\varphi^1_{\mathfrak{pq}}(\varphi_{\mathfrak{qp}_0}(x_n)) &= \varphi^1_{\mathfrak{pq}}(\varphi^1_{\mathfrak{q} q_i^{\mathfrak{q}}}(x_{n,i})) = \varphi^1_{\mathfrak{pq}}(\varphi^1_{\mathfrak{q} q_i^{\mathfrak{q}}}(\varphi_{q_i^{\mathfrak{q}} r}(y_n))) = \varphi^1_{\mathfrak{pq}}(\varphi^1_{\mathfrak{q} r}(y_n)) \\ &= \varphi^1_{\mathfrak{p} r}(y_n) = \varphi^1_{\mathfrak{p} q_j^{\mathfrak{p}}}(\varphi_{q_j^{\mathfrak{p}} r}(y_n)) = \varphi^1_{\mathfrak{p} q_j^{\mathfrak{p}}}(x_{n,j}) = \varphi_{\mathfrak{pp}_0}(x_n),\end{aligned}$$

for large n. This is a contradiction. □

We can show a similar equality

$$\widehat{\varphi}_{\mathfrak{pq}} \circ \widehat{\varphi}_{\mathfrak{qp}_0} = \widehat{\varphi}_{\mathfrak{pp}_0}$$

in the same way.

We take ϵ_1 as in Lemma 11.7 and positive numbers $\epsilon, \delta < \epsilon_1$. We define

$$\begin{aligned} U^{\epsilon}_{\mathfrak{p}} &= B^{\mathfrak{p}}_{\epsilon}(\mathcal{K}_{\mathfrak{p}}), \qquad U^{\epsilon}_{\mathfrak{pq}} = U^{\epsilon}_{\mathfrak{q}} \cap (\varphi_{\mathfrak{pq}})^{-1}(U^{\epsilon}_{\mathfrak{p}}), \\ U^{\epsilon}_{\mathfrak{p}_0} &= B^{\mathfrak{p}_0}_{\epsilon}(\mathcal{K}_{\mathfrak{p}_0} \cap \psi^{-1}_{\mathfrak{p}_0}(S_{\mathfrak{d}}(X, Z; \widehat{\mathcal{U}}))), \\ U^{\epsilon,\delta}_{\mathfrak{pp}_0} &= \varphi^{-1}_{\mathfrak{pp}_0}(U^{\epsilon}_{\mathfrak{p}}) \cap U^{\epsilon}_{\mathfrak{p}_0} \cap B^{\mathfrak{p}_0}_{\delta}(\tilde{Z}_{\mathfrak{pp}_0}). \end{aligned} \tag{11.14}$$

Here $\mathfrak{p}, \mathfrak{q} \in \mathfrak{P}$.

The next lemma shows that we can obtain a coordinate change in the *strong sense*[3] by restricting $\varphi_{\mathfrak{pp}_0}$ to an appropriate open subset.

[3]See Definition 3.6 (3).

Lemma 11.8 *There exists a positive number $\epsilon_2(\delta)$ depending on $\delta > 0$, such that if $0 < \epsilon < \epsilon_2(\delta)$ then*

$$\psi_{\mathfrak{p}_0}(U^{\epsilon,\delta}_{\mathfrak{p}\mathfrak{p}_0} \cap s_{\mathfrak{p}_0}^{-1}(0)) = \psi_{\mathfrak{p}_0}(U^{\epsilon}_{\mathfrak{p}_0} \cap s_{\mathfrak{p}_0}^{-1}(0)) \cap \psi_{\mathfrak{p}}(U^{\epsilon}_{\mathfrak{p}} \cap s_{\mathfrak{p}}^{-1}(0)). \tag{11.15}$$

Proof The inclusion $\subseteq$ is a consequence of the definition and the compatibility of $\varphi_{\mathfrak{p}\mathfrak{p}_0}$ with Kuranishi maps and parametrization maps.

Suppose $\supseteq$ does not hold. Then there exist a sequence $\epsilon_n \to 0$ and $x_n \in U^{\epsilon_n}_{\mathfrak{p}_0} \cap s_{\mathfrak{p}_0}^{-1}(0)$, $y_n \in U^{\epsilon_n}_{\mathfrak{p}} \cap s_{\mathfrak{p}}^{-1}(0)$ such that $\psi_{\mathfrak{p}_0}(x_n) = \psi_{\mathfrak{p}}(y_n)$ but

$$x_n \notin U^{\epsilon_n,\delta}_{\mathfrak{p}\mathfrak{p}_0} \cap s_{\mathfrak{p}_0}^{-1}(0). \tag{11.16}$$

Here $\delta > 0$ is independent of n. By taking a subsequence we may assume that there exist x, y such that

$$\lim_{n\to\infty} x_n = x \in \mathcal{K}_{\mathfrak{p}_0} \cap \psi_{\mathfrak{p}_0}^{-1}(S_{\mathfrak{d}}(X, Z; \widehat{\mathcal{U}})), \qquad \lim_{n\to\infty} y_n = y \in \mathcal{K}_{\mathfrak{p}}.$$

This is a consequence of (11.14). Using also $\psi_{\mathfrak{p}_0}(x) = \psi_{\mathfrak{p}}(y)$ we have

$$x \in \tilde{Z}_{\mathfrak{p}\mathfrak{p}_0}.$$

Therefore $x_n \in B^{\mathfrak{p}_0}_{\delta}(\tilde{Z}_{\mathfrak{p}\mathfrak{p}_0})$ for large n.

In particular, it implies that $\varphi_{\mathfrak{p}\mathfrak{p}_0}(x_n)$ is defined. (See (11.10).) Then since $\psi_{\mathfrak{p}_0}(x_n) = \psi_{\mathfrak{p}}(y_n)$,

$$\varphi_{\mathfrak{p}\mathfrak{p}_0}(x_n) = y_n.$$

Therefore

$$x_n \in U^{\epsilon_n}_{\mathfrak{p}_0} \cap \varphi_{\mathfrak{p}\mathfrak{p}_0}^{-1}(U^{\epsilon}_{\mathfrak{p}}),$$

for large n. Thus by (11.14) we derive $x_n \in U^{\epsilon,\delta}_{\mathfrak{p}\mathfrak{p}_0} \cap s_{\mathfrak{p}_0}^{-1}(0)$. This is a contradiction. □

For $0 < \delta < \epsilon_0$ and $0 < \epsilon < \min\{\epsilon_1, \epsilon_2(\delta)\}$, we take Kuranishi charts

$$\{\mathcal{U}_{\mathfrak{p}}|_{U^{\epsilon}_{\mathfrak{p}}} \mid \mathfrak{p} \in \mathfrak{P}\} \cup \{\mathcal{U}_{\mathfrak{p}_0}|_{U^{\epsilon}_{\mathfrak{p}_0}}\} \tag{11.17}$$

together with coordinate changes

$$\{\Phi_{\mathfrak{p}\mathfrak{q}}|_{U^{\epsilon}_{\mathfrak{p}\mathfrak{q}}} \mid \mathfrak{p}, \mathfrak{q} \in \mathfrak{P}\} \cup \{(\varphi_{\mathfrak{p}\mathfrak{p}_0}, \widehat{\varphi}_{\mathfrak{p}\mathfrak{p}_0})|_{U^{\epsilon,\delta}_{\mathfrak{p}\mathfrak{p}_0}} \mid \mathfrak{p} \in \mathfrak{P}\}. \tag{11.18}$$

We claim that they satisfy the axiom of good coordinate system (Definition 3.15) for Z_+, except (7) and (8). In fact Definition 3.15 (1)(2)(3) are obvious from the definition. Definition 3.15 (4) (the fact (11.18) are coordinate changes in the strong sense) is immediate for $\mathfrak{p}, \mathfrak{q} \in \mathfrak{P}$. (In other words, it is the consequence for the corresponding fact for $\widehat{\mathcal{U}}$.) For $\mathfrak{p}, \mathfrak{p}_0$ this is a consequence of Lemma 11.8. Definition 3.15 (5) for $\mathfrak{p}, \mathfrak{q}, \mathfrak{r} \in \mathfrak{P}$ is the consequence for the corresponding fact for $\widehat{\mathcal{U}}$. Definition 3.15 (5) for $\mathfrak{p}, \mathfrak{q}, \mathfrak{p}_0$ (where $\mathfrak{p}, \mathfrak{q} \in \mathfrak{P}$) is a consequence of Lemma 11.7. Definition 3.15 (6) is obvious.

Now we use Proposition 3.17, Shrinking Lemma [FOOO17, Theorem 2.7] to shrink them so that Definition 3.15 (7) and (8) also hold. They satisfy the conclusion of Proposition 11.3 except (5), compatibility of good coordinate system with given Kuranishi structure.

To complete the proof of Proposition 11.3, it remains to prove Proposition 11.3 (5). We will prove that this condition is satisfied if ϵ and δ are sufficiently small. Since our Kuranishi charts are open subcharts of $\mathcal{U}_{\mathfrak{p}_0}$ or of $\mathcal{U}_{\mathfrak{p}}$, the embeddings $\Phi^+_{\mathfrak{p}_0 p}$, $\Phi^+_{\mathfrak{p} p}$ such that Diagrams (11.3), (11.4) commute are given. Moreover they are compatible with $\varphi_{\mathfrak{p}\mathfrak{q}}$ for $\mathfrak{p}, \mathfrak{q} \in \mathfrak{P}$ by hypothesis. (We use Lemma 5.16 here.) So it remains to prove the compatibility with $\Phi_{\mathfrak{p}\mathfrak{p}_0}$. Namely it suffices to show the next lemma. We may assume $\epsilon_2(\delta)$ in Lemma 11.8 satisfies $\lim_{\delta \to 0} \epsilon_2(\delta) = 0$.

Lemma 11.9 *There exists $\delta_0 > 0$ such that if $0 < \delta < \delta_0$ and $0 < \epsilon < \epsilon_1, \epsilon_2(\delta)$ the following holds for any $p \in U^{\epsilon,\delta}_{\mathfrak{p}\mathfrak{p}_0} \cap s^{-1}_{\mathfrak{p}_0}(0)$.*

The equality

$$\varphi_{\mathfrak{p}\mathfrak{p}_0} \circ \varphi^0_{\mathfrak{p}_0 p} = \varphi^1_{\mathfrak{p} p}$$

holds in a neighborhood of o_p in U_p. The same holds for bundle maps.

Proof Suppose the lemma does not hold. Then there exist a sequence $\delta_n \to 0$ $\epsilon_n < \epsilon_1, \epsilon_2(\delta_n)$ and $p_n \in U^{\epsilon_n,\delta_n}_{\mathfrak{p}\mathfrak{p}_0} \cap s^{-1}_{\mathfrak{p}_0}(0)$ such that

$$\varphi_{\mathfrak{p}\mathfrak{p}_0} \circ \varphi^0_{\mathfrak{p}_0 p_n} \neq \varphi^1_{\mathfrak{p} p_n}$$

in any small neighborhood of o_{p_n} in U_{p_n}. We may assume that p_n converges to $r \in \tilde{Z}_{\mathfrak{p}\mathfrak{p}_0}$. (This is because $U^{\epsilon_n,\delta_n}_{\mathfrak{p}\mathfrak{p}_0} \subseteq B^{\mathfrak{p}_0}_{\delta_n}(\tilde{Z}_{\mathfrak{p}\mathfrak{p}_0})$.) There exists i such that $r \in \varphi_{\mathfrak{p}_0 q^{\mathfrak{p}}_i}(U^0_{\mathfrak{p}_0 q^{\mathfrak{p}}_i} \cap U^1_{\mathfrak{p} q^{\mathfrak{p}}_i})$. By definition we have

$$\varphi_{\mathfrak{p}\mathfrak{p}_0} \circ \varphi^0_{\mathfrak{p}_0 q^{\mathfrak{p}}_i} = \varphi^1_{\mathfrak{p} q^{\mathfrak{p}}_i}$$

on $U^0_{\mathfrak{p}_0 q^{\mathfrak{p}}_i} \cap U^1_{\mathfrak{p} q^{\mathfrak{p}}_i}$. Note $\varphi^0_{\mathfrak{p}_0 q^{\mathfrak{p}}_i}(U^0_{\mathfrak{p}_0 q^{\mathfrak{p}}_i} \cap U^1_{\mathfrak{p}_0 q^{\mathfrak{p}}_i})$ is an open subset of $U_{\mathfrak{p}_0}$. Therefore for sufficiently large n a neighborhood $\Omega(p_n; \mathfrak{p}_0)$ of p_n in $U_{\mathfrak{p}_0}$ is contained in $\varphi^0_{\mathfrak{p}_0 q^{\mathfrak{p}}_i}(U^0_{\mathfrak{p}_0 q^{\mathfrak{p}}_i} \cap U^1_{\mathfrak{p}_0 q^{\mathfrak{p}}_i})$. Namely there exists $\tilde{\Omega}(p_n; q^{\mathfrak{p}}_i) \subset U^0_{\mathfrak{p}_0 q^{\mathfrak{p}}_i} \cap U^1_{\mathfrak{p} q^{\mathfrak{p}}_i}$ such that

$$\varphi^0_{\mathfrak{p}_0 q^{\mathfrak{p}}_i}(\tilde{\Omega}(p_n; q^{\mathfrak{p}}_i)) = \Omega(p_n; \mathfrak{p}_0).$$

Since p_n goes to r, for sufficiently large n we may take $\Omega(p_n; \mathfrak{p}_0)$ small so that there exists an open subset $\tilde\Omega(p_n; r) \subset U_{q_i^{\mathfrak{p}} r}$ such that

$$\tilde\Omega(p_n; q_i^{\mathfrak{p}}) = \varphi_{q_i^{\mathfrak{p}} r}(\tilde\Omega(p_n; r)).$$

Since $\lim_{n\to\infty} p_n = r$, there also exists an open neighborhood Ω_{p_n} of o_{p_n} in $U^0_{\mathfrak{p}_0 p_n} \cap U^1_{\mathfrak{p} p_n}$ such that the restrictions of coordinate changes

$$\varphi_{r p_n} : \Omega_{p_n} \to \tilde\Omega(p_n; r), \qquad \varphi_{q_i^{\mathfrak{p}} p_n} : \Omega_{p_n} \to \tilde\Omega(p_n; q_i^{\mathfrak{p}})$$

are defined.[4] Then

$$\varphi^0_{\mathfrak{p}_0 p_n} = \varphi^0_{\mathfrak{p}_0 r} \circ \varphi_{r p_n}, \qquad \varphi^1_{\mathfrak{p} p_n} = \varphi^1_{\mathfrak{p} r} \circ \varphi_{r p_n}$$

on Ω_{p_n}.

Now for $x \in \Omega_{p_n}$ we have

$$\begin{aligned}(\varphi_{\mathfrak{p}\mathfrak{p}_0} \circ \varphi^0_{\mathfrak{p}_0 p_n})(x) &= (\varphi_{\mathfrak{p}\mathfrak{p}_0} \circ \varphi^0_{\mathfrak{p}_0 r} \circ \varphi_{r p_n})(x) = (\varphi_{\mathfrak{p}\mathfrak{p}_0} \circ \varphi^0_{\mathfrak{p}_0 q_i^{\mathfrak{p}}} \circ \varphi_{q_i^{\mathfrak{p}} r} \circ \varphi_{r p_n})(x) \\ &= (\varphi^1_{\mathfrak{p} q_i^{\mathfrak{p}}} \circ \varphi_{q_i^{\mathfrak{p}} r} \circ \varphi_{r p_n})(x) = (\varphi^1_{\mathfrak{p} r} \circ \varphi_{r p_n})(x) = \varphi^1_{\mathfrak{p} p_n}(x).\end{aligned}$$

This is a contradiction. □

The proof of Proposition 11.3 is now complete. □

The proof of Theorem 3.35 is complete. □

Remark 11.10

(1) During the proof of Theorem 3.35 we crucially use the fact that equality between embeddings of orbifolds is a local property. (It implies that the equality between embeddings of Kuranishi charts is also a local property). Thanks to this fact, we can check various equalities by looking at finer and finer charts. For this main idea of the proof to work, we assumed our orbifold to be effective and the maps between them to be embeddings. Without this restriction the argument will become more cumbersome and lengthy. (However, we believe that one can prove somewhat similar results without assuming effectivity of orbifolds. Joyce may have done it in [Jo2].)

(2) The proof of Theorem 3.35 in [FOOO16] is basically the same as one we gave in the preprint version [FOOO19] of this book.

Our proof of this chapter is somewhat different from the preprint version and is based on the proof by contradiction. This change makes the proof simpler. In

[4] We remark that $\dim U_r = \dim U_{q_i^{\mathfrak{p}}} = U_{\mathfrak{p}_0} = \mathfrak{d}$ but $\dim U_{p_n}$ may be smaller than $\mathfrak{d}$.

other words, the proof in this section is simplified compared to its earlier version in the same way that the final version of [FOOO17] is different from its earlier version.

In Sects. 11.2, 11.3 and 11.4, we will prove several variants of Theorem 3.35.

11.2 Construction of Good Coordinate Systems: When Thickening Is Given

This section will be occupied with the proofs of Propositions 5.20 and 5.21.

Proof of Propositions 5.20 and 5.21 We will prove Proposition 5.21. Proposition 5.20 then follows by putting $\widehat{\mathcal{U}^+} = \widehat{\mathcal{U}_1^+} = \widehat{\mathcal{U}_2^+}$.

In Proposition 5.21 we start with the situation when two thickenings $\widehat{\mathcal{U}_1^+}, \widehat{\mathcal{U}_2^+}$ are given and will construct a good coordinate system $\widehat{\mathcal{U}}$ such that $\widehat{\mathcal{U}} < \widehat{\mathcal{U}} < \widehat{\mathcal{U}_a^+}$ for $a = 1, 2$. Namely we prove that there exist GK-embeddings $\widehat{\mathcal{U}} \to \widehat{\mathcal{U}_a^+}$ such that their compositions with $\widehat{\mathcal{U}} \to \widehat{\mathcal{U}}$ are $\widehat{\mathcal{U}} \to \widehat{\mathcal{U}_a^+}$ given by the fact that $\widehat{\mathcal{U}_a^+}$ is a thickening.

For the proof, we use the same induction scheme as Proposition 11.3 and in each inductive step we will construct GK-embeddings to $\widehat{\mathcal{U}_a^+}$. Namely we prove the following proposition.

Proposition 11.11 *Under Situation* 11.1, *we assume in addition that for each* $a = 1, 2$ *there exists a Kuranishi structure* $\widehat{\mathcal{U}_a^+}$ *of* $Z \subseteq X$ *with the following properties (see Fig. 11.4):*

(a) *There exists a strict KK-embedding* $\widehat{\Phi_a^2} : \widehat{\mathcal{U}} \to \widehat{\mathcal{U}_a^+}$.
(b) *There exists a GK-embedding* $\widehat{\Phi_a^3} : \widehat{\mathcal{U}} \to \widehat{\mathcal{U}_a^+}|_{Z_1^+}$ *such that the composition* $\widehat{\Phi_a^3} \circ \widehat{\Phi^1} : \widehat{\mathcal{U}}|_{Z_1^+} \to \widehat{\mathcal{U}} \to \widehat{\mathcal{U}_a^+}|_{Z_1^+}$ *is an open restriction of* $\widehat{\Phi_a^2}|_{Z_1^+}$.
(c) *There exists a GK-embedding* $\widehat{\Phi_a^4} : \widehat{\mathcal{U}_{\mathfrak{p}_0}} \to \widehat{\mathcal{U}_a^+}|_{Z_0^+}$ *such that the composition* $\widehat{\Phi_a^4} \circ \widehat{\Phi^0} : \widehat{\mathcal{U}}|_{Z_0^+} \to \widehat{\mathcal{U}_{\mathfrak{p}_0}} \to \widehat{\mathcal{U}_a^+}|_{Z_0^+}$ *is an open restriction of* $\widehat{\Phi_a^2}|_{Z_0^+}$.

Then there exists a good coordinate system $\widehat{\mathcal{U}^+}$ *such as in the conclusion of Proposition* 11.3 *so that the following holds in addition, if we replace* $\widehat{\mathcal{U}_a^+}$ *by its open substructure,* $\widehat{\mathcal{U}_a^{+\prime}}$ *(see Fig. 11.5):*

(1) *There exists a GK-embedding* $\widehat{\Phi_a^5} : \widehat{\mathcal{U}^+} \to \widehat{\mathcal{U}_a^+}$.

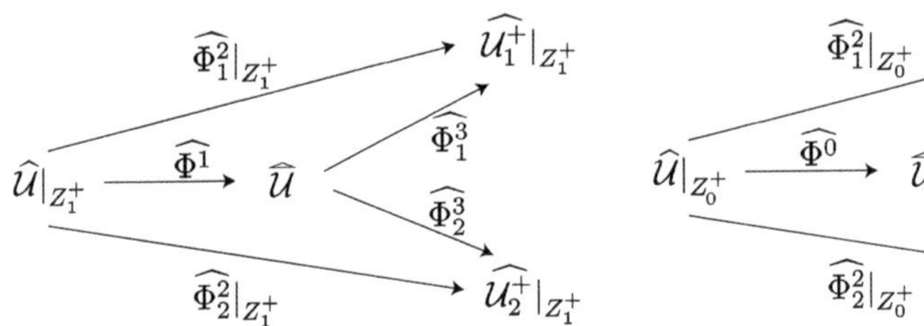

Fig. 11.4 Proposition 11.11, Assumption

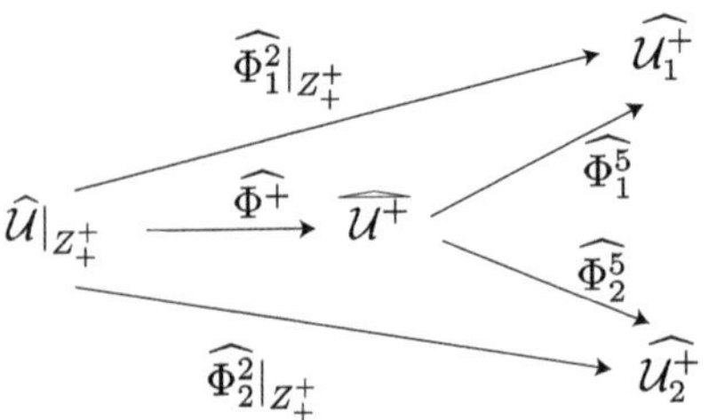

Fig. 11.5 Proposition 11.11, Conclusion

(2) *The composition* $\widehat{\Phi_a^5} \circ \widehat{\Phi^+} : \widehat{\mathcal{U}}|_{Z_+^+} \to \widehat{\mathcal{U}^+} \to \widehat{\mathcal{U}_a^+}$ *is an open restriction of* $\widehat{\Phi_a^2}|_{Z_+^+}$.

(3) *If* $\mathfrak{p} \in \mathfrak{P}$ *then* $\Phi_{a,\mathfrak{p}}^5 : \mathcal{U}_{\mathfrak{p}}|_{U_{a,\mathfrak{p}}^5(p)} \to \mathcal{U}_{a,p}^+$ *is an open restriction of* $\Phi_{a,\mathfrak{p}}^3$.

(4) *The embedding* $\Phi_{a,\mathfrak{p}_0}^5 : \mathcal{U}_{\mathfrak{p}_0}|_{U_{a,\mathfrak{p}_0}^5(p)} \to \mathcal{U}_{a,\mathfrak{p}_0}^+$ *is an open restriction of* $\Phi_{a,\mathfrak{p}_0}^4$.

Proof We will define $U_{a,\mathfrak{p}}^5(p)$, $U_{a,\mathfrak{p}_0}^5(p)$ to be sufficiently small neighborhoods of p in $U_{\mathfrak{p}}^+$, $U_{\mathfrak{p}_0}^+$, later and define $\Phi_{a,\mathfrak{p}}^5$ and $\Phi_{a,\mathfrak{p}_0}^5$ as open restrictions of $\Phi_{a,\mathfrak{p}}^3$, $\Phi_{a,\mathfrak{p}_0}^4$, respectively. Then most of the required properties follow from the assumptions. In fact (1)–(3) of Definition 5.6 (of GK-embedding) is obvious. Proposition 11.11 (2)(3)(4) follow from (b) and (c). (4) of Definition 5.6 also follows from the assumption (that is, the fact that $\widehat{\Phi_a^3}$ and $\widehat{\Phi_a^4}$ are GK embeddings) and Lemma 5.15, if $\mathfrak{p}, \mathfrak{q} \in \mathfrak{P}$ or $\mathfrak{p} = \mathfrak{q} = \mathfrak{p}_0$. Therefore it suffices to show (4) of Definition 5.6, in the case where $\mathfrak{p} \in \mathfrak{P}$ and $\mathfrak{q} = \mathfrak{p}_0$. Let

$$\begin{aligned} &p \in U_{\mathfrak{p}}^+ \subseteq U_{\mathfrak{p}}^\epsilon = B_\epsilon^{\mathfrak{p}}(\mathcal{K}_{\mathfrak{p}}), \\ &q \in U_{\mathfrak{p}_0}^+ \subseteq B_\epsilon^{\mathfrak{p}_0}((\mathcal{K}_{\mathfrak{p}_0} \cap s_{\mathfrak{p}_0}^{-1}) \cap \mathcal{S}_{\mathfrak{d}}(X, Z; \widehat{\mathcal{U}})) \subset U_{\mathfrak{p}_0}^0 \end{aligned} \tag{11.19}$$

(Note $U_{\mathfrak{p}}^+ \subset U_{\mathfrak{p}}^\epsilon$, $U_{\mathfrak{p}_0}^+ \subset U_{\mathfrak{p}_0}^\epsilon$. $U_{\mathfrak{p}}^\epsilon$ and $U_{\mathfrak{p}_0}^\epsilon$ are defined in (11.14).) We assume

$$s_{\mathfrak{p}}(p) = 0, \qquad s_{\mathfrak{p}_0}(q) = 0. \tag{11.20}$$

We choose $U^{+\prime}_{a;p}$, the Kuranishi neighborhood of an open substructure of $\widehat{\mathcal{U}^{+\prime}_a}$, such that

$$\mathrm{Diam}(\psi_p(s_p^{-1}(0) \cap U^{+\prime}_{a;p})) < \epsilon. \tag{11.21}$$

Here Diam is the diameter of a subset of the metric space X. We then choose $U^5_{a,\mathfrak{p}}(p)$ so that (11.22) below implies (11.23) (this is a part of the requirement of Definition 5.6 (4)):

$$q \in \psi_{\mathfrak{p}}(U^5_{a,\mathfrak{p}}(p) \cap s_{\mathfrak{p}}^{-1}(0)), \tag{11.22}$$

$$q \in \psi^+_{a;p}(U^{+\prime}_{a;p} \cap (s^+_{a:p})^{-1}(0)). \tag{11.23}$$

In (11.23) we identify p, q with $\psi_{\mathfrak{p}}(p), \psi_{\mathfrak{p}_0}(q) \in Z$.

It suffices to prove the commutativity of Diagram (5.7) when (11.19)–(11.23) are satisfied. Diagram (5.7) in our case is the following:

$$\begin{array}{ccc}
\mathcal{U}_{\mathfrak{p}_0}|_{\varphi^{-1}_{\mathfrak{p}\mathfrak{p}_0}(U^5_{a,\mathfrak{p}}(p)) \cap (\varphi^4_{a,q\mathfrak{p}_0})^{-1}(U^{+\prime}_{a,pq}) \cap U^5_{a,\mathfrak{p}_0}(q)} & \xrightarrow{\Phi^4_{a,q\mathfrak{p}_0}} & \mathcal{U}^+_{a,q}|_{U^{+\prime}_{a,pq}} \\
\Phi_{\mathfrak{p}\mathfrak{p}_0}\Big\downarrow & & \Big\downarrow \Phi^+_{a,pq} \\
\mathcal{U}_{a,\mathfrak{p}}|_{U^5_{a,\mathfrak{p}}(p)} & \xrightarrow{\Phi^3_{a,p\mathfrak{p}}} & \mathcal{U}^+_{a,p}
\end{array} \tag{11.24}$$

In fact, the domain of the coordinate change $\Phi^5_{a,q\mathfrak{p}_0}$ is $U^5_{a,\mathfrak{p}_0}(q)$ which satisfies

$$(\varphi^5_{a,q\mathfrak{p}_0})^{-1}(U^{+\prime}_{a,pq}) = (\varphi^4_{a,q\mathfrak{p}_0})^{-1}(U^{+\prime}_{a,pq}) \cap U^5_{a,\mathfrak{p}_0}(q).$$

Lemma 11.12 *There exists $\epsilon_3 > 0$ such that the following holds. Suppose p, q satisfy* (11.19)–(11.23). *Then, if $0 < \epsilon < \epsilon_3$ and $U^5_{a,\mathfrak{p}_0}(q) := U^5_{a,\mathfrak{p}_0}(q; p)$ is chosen to be a sufficiently small neighborhood of q, the Diagram* (11.24) *commutes.*

Note $U^5_{a,\mathfrak{p}_0}(q; p)$ may depend on p.

Proof We only prove the commutativity of the orbifold embedding part. The vector bundle embedding part can be proved in the same way. The proof is by contradiction. Suppose the lemma does not hold. Then there exist a sequence of positive numbers $\epsilon_n \to 0$ and

$$\begin{aligned}
&p_n \in U^+_{\mathfrak{p}} \subseteq U^{\epsilon_n}_{\mathfrak{p}} = B^{\mathfrak{p}}_{\epsilon_n}(\mathcal{K}_{\mathfrak{p}}) \\
&q_n \in B_{\epsilon_n}((\mathcal{K}_{\mathfrak{p}_0} \cap s^{-1}_{\mathfrak{p}_0}(0)) \cap \mathcal{S}_{\mathfrak{d}}(X, Z; \widehat{\mathcal{U}})) \\
&s_{\mathfrak{p}}(p_n) = 0
\end{aligned}$$

and

$$\begin{aligned} q_n &\in \psi^+_{a;p_n}(U^{+\prime}_{a;p_n} \cap (s^+_{a:p_n})^{-1}(0)) \\ q_n &\in \psi_{\mathfrak p}(U^5_{a,\mathfrak p}(p_n) \cap s_{\mathfrak p}^{-1}(0)). \end{aligned} \tag{11.25}$$

such that the equality

$$\varphi^3_{a,p_n\mathfrak p} \circ \varphi_{\mathfrak p\mathfrak p_0} = \varphi^+_{a,p_n q_n} \circ \varphi^4_{a,q_n\mathfrak p_0}$$

fails to hold on any small neighborhood of q_n in $U^0_{\mathfrak p_0}$. By taking a subsequence we may assume that q_n converges to $r \in (\mathcal K_{\mathfrak p_0} \cap s_{\mathfrak p_0}^{-1}) \cap \mathcal S_{\mathfrak d}(X, Z; \widehat{\mathcal U})$ and p_n converges to $p_\infty \in \mathcal K_{\mathfrak p}$.

Equations (11.21) and (11.25) imply $d(q_n, p_n) < \epsilon_n \to 0$. Therefore $r = p_\infty \in Z_{\mathfrak p\mathfrak p_0}$.

We take $q_i^{\mathfrak p}$ (as in (11.6)) such that $r \in U^0_{\mathfrak p_0 q_i^{\mathfrak p}}$. Then in the same way as the proof of Lemma 11.9, we can show that there exists a neighborhood Ω_{q_n} of o_{q_n} in U_{q_n} on which we can calculate as follows:

$$\begin{aligned} \varphi^+_{a,p_n q_n} \circ \varphi^4_{a,q_n\mathfrak p_0} \circ \varphi^0_{\mathfrak p_0 r} \circ \varphi_{r q_n} &= \varphi^+_{a,p_n q_n} \circ \varphi^4_{a,q_n\mathfrak p_0} \circ \varphi^0_{\mathfrak p_0 q_n} \\ &= \varphi^+_{a,p_n q_n} \circ \varphi^2_{a,q_n} = \varphi^2_{a,p_n} \circ \varphi_{p_n q_n}. \end{aligned}$$

In fact, the first equality follows from the fact that $\widehat{\Phi^0}$ is a KG-embedding, the second equality follows from Assumption (c) and the third equality follows from the fact that $\widehat{\Phi^2_a}$ is a KK-embedding. More precisely we need to check that o_{q_n} is contained in the domain of the maps appearing in the equality. This follows from the fact $\lim_{n\to\infty} q_n = r$ except for the maps $\varphi^+_{a,p_n q_n}$, $\varphi_{p_n q_n}$ and φ^2_{a,p_n}. In those three cases it is a consequence of (11.25).

There also exists a neighborhood Ω_{q_n} of o_{q_n} in U_{q_n} on which we can calculate as follows:

$$\begin{aligned} \varphi^3_{a,p_n\mathfrak p} \circ \varphi_{\mathfrak p\mathfrak p_0} \circ \varphi^0_{\mathfrak p_0 r} \circ \varphi_{r q_n} &= \varphi^3_{a,p_n\mathfrak p} \circ \varphi_{\mathfrak p\mathfrak p_0} \circ \varphi^0_{\mathfrak p_0 q_i^{\mathfrak p}} \circ \varphi_{q_i^{\mathfrak p} r} \circ \varphi_{r q_n} \\ &= \varphi^3_{a,p_n\mathfrak p} \circ \varphi^1_{\mathfrak p q_i^{\mathfrak p}} \circ \varphi_{q_i^{\mathfrak p} r} \circ \varphi_{r q_n} = \varphi^3_{a,p_n\mathfrak p} \circ \varphi^1_{\mathfrak p r} \circ \varphi_{r q_n} \\ &= \varphi^3_{a,p_n\mathfrak p} \circ \varphi^1_{\mathfrak p q_n} = \varphi^3_{a,p_n\mathfrak p} \circ \varphi^1_{\mathfrak p p_n} \circ \varphi_{p_n q_n} = \varphi^2_{a,p_n} \circ \varphi_{p_n q_n}. \end{aligned}$$

Here the first equality is the cocycle condition for a Kuranishi structure $\widehat{\mathcal U}$, the second equality is the definition of $\varphi_{\mathfrak p\mathfrak p_0}$ (Formula (11.10).), the third, fourth and fifth equalities follow from the fact that $\widehat{\Phi^1}$ is a KG-embedding, and the sixth equality follows from assumption (b). We also need to check that o_{q_n} is contained in the domain of the seven maps appearing in the formula. This follows from the fact

$\lim_{n\to\infty} q_n = r$ except for the three maps $\varphi^3_{a,p_n\mathfrak{p}}$, $\varphi_{p_nq_n}$ and φ^2_{a,p_n}. In those three cases it is a consequence of (11.25).

Therefore $\varphi^+_{a,pq_n} \circ \varphi^4_{a,q_n\mathfrak{p}_0} = \varphi^3_{a,p\mathfrak{p}} \circ \varphi_{\mathfrak{p}\mathfrak{p}_0}$ must hold on a neighborhood $(\varphi^0_{\mathfrak{p}_0 r} \circ \varphi_{rq_n})(\Omega(q_n))$ of q_n. This is a contradiction. □

To complete the proof of Proposition 11.11 it suffices to show that the neighborhood $U^5_{a,\mathfrak{p}_0}(q;p)$ of q in Lemma 11.12 can be taken to be independent of p. We prove it in the following way. We first take finitely many elements $p_c \in \mathcal{K}_\mathfrak{p} \cap s_\mathfrak{p}^{-1}(0)$ and $U^5_{a,\mathfrak{p}_0}(q;p_c)$ such that $\{\psi^+_{a;p_c}(U^{+\prime}_{a;p_c} \cap (s^+_{a:p_c})^{-1}(0)) \cap \psi_\mathfrak{p}(U^5_{a,\mathfrak{p}}(p_c) \cap s_\mathfrak{p}^{-1}(0))\}$ covers $\psi_\mathfrak{p}(\mathcal{K}_\mathfrak{p} \cap s_\mathfrak{p}^{-1}(0))$. Then we take

$$U^5_{a,\mathfrak{p}_0}(q)' = \bigcap_c U^5_{a,\mathfrak{p}_0}(q;p_c).$$

Here the intersection is taken over all c such that

$$q \in \psi^+_{a;p_c}(U^{+\prime}_{a;p_c} \cap (s^+_{a:p_c})^{-1}(0)) \cap \psi_\mathfrak{p}(U^5_{a,\mathfrak{p}}(p_c) \cap s_\mathfrak{p}^{-1}(0)).$$

Hence, for this choice of $U^5_{a,\mathfrak{p}_0}(q)'$, the conclusion of Lemma 11.12 holds for any $p = p_c$. For $p \in \mathcal{K}_\mathfrak{p} \cap s_\mathfrak{p}^{-1}(0)$, we replace $U^{+\prime}_{a,p}$, $U^5_{a,\mathfrak{p}}(p)$ by a smaller neighborhood so that

$$\psi_\mathfrak{p}(p) \in \psi^+_{a;p_c}(U^{+\prime}_{a;p_c} \cap (s^+_{a:p_c})^{-1}(0)) \cap \psi_\mathfrak{p}(U^5_{a,\mathfrak{p}}(p_c) \cap s_\mathfrak{p}^{-1}(0)) \tag{11.26}$$

implies

$$\begin{aligned} &\psi^+_{a;p}(U^{+\prime}_{a;p} \cap (s^+_{a:p})^{-1}(0)) \subset \psi^+_{a;p_c}(U^{+\prime}_{a;p_c} \cap (s^+_{a:p_c})^{-1}(0)) \\ &\psi_\mathfrak{p}(U^5_{a,\mathfrak{p}}(p) \cap s_\mathfrak{p}^{-1}(0)) \subset \psi_\mathfrak{p}(U^5_{a,\mathfrak{p}}(p_c) \cap s_\mathfrak{p}^{-1}(0)). \end{aligned} \tag{11.27}$$

We can still require that (11.22) implies (11.23) after this replacement.

Now suppose p, q satisfy (11.19)–(11.23). We take p_c such that (11.26) holds. Then by (11.27), p_c, q satisfy (11.19)–(11.23). Therefore Diagram (11.24) for $p = p_c$ commutes. We then can use the commutativity of Diagram (11.28) below to prove the commutativity of (11.24) for p, q by taking $U^5_{a,\mathfrak{p}_0}(q)$ sufficiently small:

$$\begin{array}{ccc} \mathcal{U}_\mathfrak{p}|_{U^5_{a,\mathfrak{p}}(p)} & \xrightarrow{\Phi^3_{a,p\mathfrak{p}}} & \mathcal{U}^+_{a,p}|_{U^{+\prime}_{a,p_cp}} \\ \subseteq\downarrow & & \downarrow \Phi^+_{a,p_cp} \\ \mathcal{U}_\mathfrak{p}|_{U^5_{a,\mathfrak{p}}(p_c)} & \xrightarrow{\Phi^3_{a,p_c\mathfrak{p}}} & \mathcal{U}^+_{a,p_c} \end{array} \tag{11.28}$$

In fact we can calculate

$$\Phi^+_{a;p_c p} \circ \Phi^+_{a;pq} \circ \Phi^4_{a;q\mathfrak{p}_0} = \Phi^+_{a;p_c q} \circ \Phi^4_{a;q\mathfrak{p}_0} = \Phi^3_{a;p_c\mathfrak{p}} \circ \Phi^+_{a;\mathfrak{p}\mathfrak{p}_0}$$
$$= \Phi^+_{a;p_c p} \circ \Phi^3_{a;p\mathfrak{p}} \circ \Phi^+_{a;\mathfrak{p}\mathfrak{p}_0}.$$

The first equality is the compatibility of the coordinate changes of $\widehat{\mathcal{U}^+}$. The second equality is commutativity of Diagram (11.24) for $p = p_c$. The third equality is commutativity of Diagram (11.28).

The commutativity of (11.28) is a consequence of the fact that $\widehat{\Phi^3_a}$ is a GK-embedding. The proof of Proposition 11.11 is complete. □

The rest of the proof of Proposition 5.21 is mostly the same as the proof of Theorem 3.35, using Proposition 11.11 in addition to Proposition 11.3. We use the same notation as in the last part of the proof of Theorem 3.35. We construct a good coordinate system of a compact neighborhood of

$$Z \cup \bigcup_{i=1}^{n} Z_i,$$

together with its GK-embedding to $\widehat{\mathcal{U}_a^+}$ by induction. Suppose we have a good coordinate system of a compact neighborhood of $Z \cup \bigcup_{i=1}^{n-1} Z_i$ together with its embedding to $\widehat{\mathcal{U}_a^+}$. To apply Propositions 11.3 and 11.11, we need to find a good coordinate system on Z_n together with the GK-embedding to $\widehat{\mathcal{U}_a^+}$. We use the following lemma for this purpose.

Lemma 11.13 *Let $\widehat{\mathcal{U}}$ be a Kuranishi structure and $\widehat{\mathcal{U}_a^+}$ its thickenings for $a = 1, 2$. Then for each $p \in \mathcal{S}_{\mathfrak{d}}(X, Z; \widehat{\mathcal{U}})$ there exists a good coordinate system $\widehat{\mathcal{U}_{\mathfrak{p}_0}}$ of a compact neighborhood of p in X with the following properties:*

(1) *$\widehat{\mathcal{U}_{\mathfrak{p}_0}}$ consists of a single Kuranishi chart $\mathcal{U}_{\mathfrak{p}_0}$ such that $\dim U_{\mathfrak{p}_0} = \mathfrak{d}$ and $\mathcal{U}_{\mathfrak{p}_0}$ is a restriction to an open set of a Kuranishi chart of a point p of $\widehat{\mathcal{U}}$.*
(2) *For each $a = 1, 2$, there exists a GK-embedding $\widehat{\mathcal{U}_{\mathfrak{p}_0}} \to \widehat{\mathcal{U}_a^+}|_{Z_0}$, where Z_0 is a compact neighborhood of p in Z.*
(3) *The composition $\widehat{\mathcal{U}}|_{Z_0} \to \widehat{\mathcal{U}_{\mathfrak{p}_0}} \to \widehat{\mathcal{U}^+}|_{Z_0}$ is the given KK-embedding.*

Proof Let $\mathcal{U}_p$ (resp. $\mathcal{U}^+_{a;p}$) be the Kuranishi chart of p induced by the Kuranishi structure $\widehat{\mathcal{U}}$ (resp. $\widehat{\mathcal{U}_a^+}$). Let $O_{a;p}$ be the neighborhood of p in X as in Definition 5.3. (Here we use thickness of $\widehat{\mathcal{U}_a^+}$.) We put $O_p = O_{1;p} \cap O_{2;p}$. We take an open neighborhood $U_{\mathfrak{p}_0}$ of o_p in U_p such that

$$\psi_p(U_{\mathfrak{p}_0} \cap s_p^{-1}(0)) \subset O_p, \tag{11.29}$$

and set $\mathcal{U}_{\mathfrak{p}_0} = \mathcal{U}_p|_{U_{\mathfrak{p}_0}}$. It is obvious that $\mathcal{U}_{\mathfrak{p}_0}$ satisfies (1). Let us prove (2). The subset Z_0 is any compact neighborhood of p contained in O_p.

Let $q \in \psi(U_{\mathfrak{p}_0} \cap s_{\mathfrak{p}_0}^{-1}(0)) \cap Z_0$. We take $W_{a;p}(q) \subset U_p$ as in Definition 5.3 and put $U(q) = W_{1;p}(q) \cap W_{2;p}(q)$. Then by Definition 5.3 (2) (a) we have

$$\varphi_{a;p}(U(q)) \subset \varphi^+_{a;pq}(U^+_{a;pq}).$$

Since $\varphi^+_{a;pq}$ is injective, we have a set-theoretical map $\varphi_{a;q\mathfrak{p}_0} : U(q) \to U^+_{a;pq}$ such that $\varphi^+_{a;pq} \circ \varphi_{a;q\mathfrak{p}_0} = \varphi_{a;p}$. Since $\varphi^+_{a;pq}$ and $\varphi_{a;p}$ are embeddings between orbifolds, the map $\varphi_{a;q\mathfrak{p}_0}$ is an embedding of orbifolds. We can use Definition 5.3 (2) (b) in the same way to obtain an embedding of vector bundles $\widehat{\varphi}_{a;q\mathfrak{p}_0} : E_{\mathfrak{p}_0} \to E^+_{a;q}$ such that it covers $\varphi_{a;q\mathfrak{p}_0}$ and satisfies $\widehat{\varphi}^+_{a;pq} \circ \widehat{\varphi}_{a;q\mathfrak{p}_0} = \widehat{\varphi}_{a;p}$. We thus obtain $\Phi_{a;q\mathfrak{p}_0} = (U(q), \widehat{\varphi}_{a;q\mathfrak{p}_0})$ for each $q \in \psi(U_{\mathfrak{p}_0} \cap s_{\mathfrak{p}_0}^{-1}(0))$.[5] It is easy to see that $\Phi_{a;q\mathfrak{p}_0}$ defines an embedding of a good coordinate system to a Kuranishi structure. The proof of Lemma 11.13 is complete. □

Remark 11.14 The proof of this lemma is the place where we use the assumption that $\widehat{\mathcal{U}^+_a}$ is a thickening of $\widehat{\mathcal{U}}$.

Since $\mathcal{U}_{\mathfrak{p}_0}$ is an open subchart of a Kuranishi chart of $\widehat{\mathcal{U}}$ there exists a KG-embedding $\widehat{\mathcal{U}} \to \widehat{\mathcal{U}_{\mathfrak{p}_0}}$. Moreover the composition $\widehat{\mathcal{U}} \to \widehat{\mathcal{U}_{\mathfrak{p}_0}} \to \widehat{\mathcal{U}^+_a}$ coincides with the given KK-embedding on a neighborhood of p. In other words, Assumption (c) of Proposition 11.11 is satisfied. Therefore we apply Proposition 11.11 inductively to complete the proof of Proposition 5.21 in the same way as the proof of Theorem 3.35. □

11.3 KG-Embeddings and Compatible Perturbations

The present section will be occupied with the proofs of Propositions 6.32, 6.33 and Lemmas 9.10, 9.11, 9.32. They claim that, if various objects such as multivalued perturbation, CF-perturbation, differential form etc. are given on the Kuranishi structure we start with then they induce the same types of objects on the good coordinate system which we obtain by Theorem 3.35. We can prove it by observing that the Kuranishi chart of the good coordinate system we obtain is a restriction of the one of the Kuranishi structure we start with and the coordinate change of the good coordinate system we obtain is locally a composition of restrictions of the coordinate changes of the Kuranishi structure we start with or the inverse of its (invertible) coordinate changes. In this sense the proof is actually a kind of tautology.

Proof of Proposition 6.32 and Lemma 9.10 We use the same induction scheme as in the proof of Theorem 3.35. The next lemma claims that in the situation of

[5] $\Phi_{a;q\mathfrak{p}_0}$ is an embedding of Kuranishi chart in the strong sense (Definition 5.6 (3)) since the embedding $\varphi_{a;q,\mathfrak{p}_0}$ is defined on the whole $U(q)$.

Proposition 11.3 we obtain various objects on the resulting good coordinate system provided that we are given objects of the same kind on the good coordinate systems and Kuranishi structures we start with.

Lemma 11.15 *We consider the situation of Proposition* 11.3 *and Situation* 11.1.

(1) (Multivalued perturbation) *Suppose there exist multivalued perturbations* $\widehat{\mathfrak{s}} = \{\mathfrak{s}^n_{\mathfrak{p}}\}$ *of* $\widehat{\mathcal{U}}$, $\widehat{\mathfrak{s}_{\mathfrak{p}_0}} = \{\mathfrak{s}^n_{\mathfrak{p}_0}\}$ *of* $\widehat{\mathcal{U}_{\mathfrak{p}_0}}$, *and* $\widehat{\mathfrak{s}} = \{\mathfrak{s}^n_p\}$ *of* $\widehat{\mathcal{U}}$. *We assume that they are compatible with the KG-embeddings* $\widehat{\Phi^0} : \widehat{\mathcal{U}}|_{Z_0^+} \to \widehat{\mathcal{U}_{\mathfrak{p}_0}}$ *and* $\widehat{\Phi^1} : \widehat{\mathcal{U}}|_{Z_1^+} \to \widehat{\mathcal{U}}$.

Then there exist multivalued perturbations $\widehat{\mathfrak{s}^+} = \{\mathfrak{s}^{n+}_{\mathfrak{p}}\}$ *of* $\widehat{\mathcal{U}^+}$ *with the following properties:*

(a) $\widehat{\mathfrak{s}^+}$, $\widehat{\mathfrak{s}}$ *are compatible with the strict KG-embedding* $\widehat{\Phi^+} : \widehat{\mathcal{U}_0}|_{Z_+^+} \to \widehat{\mathcal{U}^+}$, *where* $\widehat{\mathcal{U}_0}$ *is an open substructure of* $\widehat{\mathcal{U}}$.
(b) *If* $\mathfrak{p} \in \mathfrak{P}$, *then* $\mathfrak{s}^{n+}_{\mathfrak{p}}$ *is the restriction of* $\mathfrak{s}^n_{\mathfrak{p}}$ *to* $\mathcal{U}^+_{\mathfrak{p}}$.
(c) *In the case of* $\mathfrak{p}_0$, $\mathfrak{s}^{n+}_{\mathfrak{p}_0}$ *is the restriction of* $\mathfrak{s}^n_{\mathfrak{p}_0}$ *to* $\mathcal{U}^+_{\mathfrak{p}_0}$.
(d) *If* $\widehat{\mathfrak{s}}$, $\widehat{\mathfrak{s}}$ *are transversal to* 0, *then so is* $\widehat{\mathfrak{s}^+}$.

(2) (CF-perturbation) *Suppose there exist CF-perturbations* $\widehat{\mathfrak{S}} = \{\mathfrak{S}^{\epsilon}_{\mathfrak{p}}\}$ *of* $\widehat{\mathcal{U}}$, $\widehat{\mathfrak{S}_{\mathfrak{p}_0}} = \{\mathfrak{S}^{\epsilon}_{\mathfrak{p}_0}\}$ *of* $\widehat{\mathcal{U}_{\mathfrak{p}_0}}$, *and* $\widehat{\mathfrak{S}} = \{\mathfrak{S}^{\epsilon}_p\}$ *of* $\widehat{\mathcal{U}}$. *We assume that they are compatible with the KG-embeddings* $\widehat{\Phi^0} : \widehat{\mathcal{U}}|_{Z_0^+} \to \widehat{\mathcal{U}_{\mathfrak{p}_0}}$ *and* $\widehat{\Phi^1} : \widehat{\mathcal{U}}|_{Z_1^+} \to \widehat{\mathcal{U}}$.

Then there exists a CF-perturbation $\widehat{\mathfrak{S}^+} = \{\mathfrak{S}^{\epsilon+}_{\mathfrak{p}}\}$ *of* $\widehat{\mathcal{U}^+}$ *with the following properties:*

(a) $\widehat{\mathfrak{S}}$, $\widehat{\mathfrak{S}^+}$ *are compatible with the strict KG-embedding* $\widehat{\Phi^+} : \widehat{\mathcal{U}_0} \to \widehat{\mathcal{U}^+}$, *where* $\widehat{\mathcal{U}_0}$ *is an open substructure of* $\widehat{\mathcal{U}}$.
(b) *If* $\mathfrak{p} \in \mathfrak{P}$, *then* $\mathfrak{S}^{\epsilon+}_{\mathfrak{p}}$ *is the restriction of* $\mathfrak{S}^{\epsilon}_{\mathfrak{p}}$ *to* $\mathcal{U}^+_{\mathfrak{p}}$.
(c) *In the case of* $\mathfrak{p}_0$, $\mathfrak{S}^{\epsilon+}_{\mathfrak{p}_0}$ *is the restriction of* $\mathfrak{S}^{\epsilon}$ *to* $\mathcal{U}^+_{\mathfrak{p}_0}$.
(d) *If* $\widehat{\mathfrak{S}}$, $\widehat{\mathfrak{S}}$ *are transversal to* 0, *then so is* $\widehat{\mathfrak{S}^+}$.

(3) (Differential form and strongly smooth maps) *Suppose there exist differential forms (resp. strongly continuous maps to a manifold M)* $\widehat{h} = \{h_{\mathfrak{p}}\}$ *(resp.* $\widehat{f} = \{f_{\mathfrak{p}}\}$*) on* $\widehat{\mathcal{U}}$, $\widehat{h_{\mathfrak{p}_0}}$ *(resp.* $\widehat{f_{\mathfrak{p}_0}}$*) on* $\widehat{\mathcal{U}_{\mathfrak{p}_0}}$, *and* $\widehat{h} = \{h_p\}$ *(resp.* $\widehat{f} = \{f_p\}$*) on* $\widehat{\mathcal{U}}$. *We assume* $(\widehat{\Phi^0})^*(\widehat{h_{\mathfrak{p}_0}}) = \widehat{h}|_{Z_0^+}$ *(resp.* $\widehat{f_{\mathfrak{p}_0}} \circ \widehat{\Phi^0} = \widehat{f}|_{Z_0^+}$*) and* $(\widehat{\Phi^1})^*(\widehat{h}) = \widehat{h}|_{Z_1^+}$ *(resp.* $\widehat{f} \circ \widehat{\Phi^1} = \widehat{f}|_{Z_1^+}$*).*

Then there exists a differential form (resp. strongly continuous map to M) $\widehat{h^+} = \{h^+_{\mathfrak{p}}\}$ *(resp.* $\widehat{f^+} = \{f^+_{\mathfrak{p}}\}$ *) on* $\widehat{\mathcal{U}^+}$ *with the following properties:*

(a) $(\widehat{\Phi^+})^*(\widehat{h^+}) = \widehat{h}|_{\widehat{\mathcal{U}_0}}$ *(resp.* $\widehat{f^+} \circ \widehat{\Phi^+} = \widehat{f}|_{\widehat{\mathcal{U}_0}}$*) holds. Here* $\widehat{\mathcal{U}_0}$ *is an open substructure of* $\widehat{\mathcal{U}}$ *and* $\widehat{\Phi^+} : \widehat{\mathcal{U}_0} \to \widehat{\mathcal{U}^+}$ *is a strict KG-embedding.*
(b) *If* $\mathfrak{p} \in \mathfrak{P}$, *then* $h^+_{\mathfrak{p}}$ *(resp.* $f^+_{\mathfrak{p}}$*) is a restriction of* $h_{\mathfrak{p}}$ *(resp.* $f_{\mathfrak{p}}$ *) to* $\mathcal{U}^+_{\mathfrak{p}}$.
(c) *In the case of* $\mathfrak{p}_0$, $h^+_{\mathfrak{p}_0}$ *(resp.* $f^+_{\mathfrak{p}_0}$*) is a restriction of* $h_{\mathfrak{p}_0}$ *(resp.* $f_{\mathfrak{p}_0}$*) to* $\mathcal{U}^+_{\mathfrak{p}_0}$.

(4) (Strong transversality of maps with respect to a multivalued perturbation) *Suppose we are in the situation of (1). Let* $\widehat{f}$, $\widehat{\mathfrak{f}}$ *be as in (3) and* $g : N \to M$ *a smooth map between smooth manifolds such that* $\widehat{f}$, $\widehat{\mathfrak{f}}$ *are strongly transversal to* g *with respect to* $\widehat{\mathfrak{s}}$ *and* $\widehat{\mathfrak{s}}$*, respectively. Then* $\widehat{f^+}$ *is strongly transversal to* g *with respect to* $\widehat{\mathfrak{s}^+}$.

(5) (Strong transversality of maps with respect to CF-perturbation) *Suppose we are in the situation of (2). Let* $\widehat{f}$, $\widehat{\mathfrak{f}}$ *be as in (3).*

   (a) *If* $\widehat{f}$, $\widehat{\mathfrak{f}}$ *are strongly submersive with respect to* $\widehat{\mathfrak{S}}$ *and* $\widehat{\mathfrak{S}}$*, respectively, then* $\widehat{f^+}$ *is strongly submersive with respect to* $\widehat{\mathfrak{S}^+}$.

   (b) *Let* $g : N \to M$ *be a smooth map such that* $\widehat{f}$, $\widehat{\mathfrak{f}}$ *are strongly transversal to* g *with respect to* $\widehat{\mathfrak{S}}$ *and* $\widehat{\mathfrak{S}}$*, respectively. Then* $\widehat{f^+}$ *is strongly transversal to* g *with respect to* $\widehat{\mathfrak{S}^+}$.

Proof We will prove (1),(4). The proofs of (2),(3) and (5) are entirely similar.

We note that (1) (b) and (c) uniquely determine $\{\mathfrak{s}^{n+}_{\mathfrak{p}}\}$. Its compatibility with the coordinate change $\Phi^+_{\mathfrak{p}\mathfrak{q}}$ for $\mathfrak{p}, \mathfrak{q} \in \mathfrak{P}$ follows from the assumption, that is, the compatibility of $\{\mathfrak{s}^n_{\mathfrak{p}}\}$ with $\Phi_{\mathfrak{p}\mathfrak{q}}$. Therefore to complete the proof, it suffices to check the following three points:

(I) $\mathfrak{s}^{n+}_{\mathfrak{p}}$, $\mathfrak{s}^{n+}_{\mathfrak{p}_0}$ are compatible with $\Phi^+_{\mathfrak{p}\mathfrak{p}_0}$.
(II) Statement (1)(a) holds.
(III) Statements (1)(d) and (4) hold.

Proof of (I) Let $y \in U^+_{\mathfrak{p}\mathfrak{p}_0}$. Since $U^+_{\mathfrak{p}\mathfrak{p}_0} \subset B^{\mathfrak{p}_0}_{\delta}(\tilde{Z}_{\mathfrak{p}\mathfrak{p}_0}) \subset B^{\mathfrak{p}_0}_{\epsilon}(\tilde{Z}_{\mathfrak{p}\mathfrak{p}_0})$, (11.7) implies that there exist $q^{\mathfrak{p}}_i$ and $\tilde{y} \in U^0_{\mathfrak{p}_0 q^{\mathfrak{p}}_i}$ such that $\varphi^0_{\mathfrak{p}_0 q^{\mathfrak{p}}_i}(\tilde{y}) = y$. We can take a representative $(\mathfrak{s}^n_{\mathfrak{p}_0,k})_{k=1,\dots,\ell}$ of $\mathfrak{s}^n_{\mathfrak{p}_0}$ (resp. $(\mathfrak{s}^n_{q^{\mathfrak{p}}_i,k})_{k=1,\dots,\ell}$ of $\mathfrak{s}^n_{q^{\mathfrak{p}}_i}$) on a neighborhood U_y of y (resp. $U_{\tilde{y}}$ of $\tilde{y}$) such that

$$\mathfrak{s}^n_{\mathfrak{p}_0,k}(\varphi^0_{\mathfrak{p}_0 q^{\mathfrak{p}}_i}(\tilde{z})) = \widehat{\varphi^0_{\mathfrak{p}_0 q^{\mathfrak{p}}_i}}(\mathfrak{s}^n_{q^{\mathfrak{p}}_i,k}(\tilde{z})) \tag{11.30}$$

holds for any $\tilde{z} \in U_{\tilde{y}}$. This is a consequence of the compatibility of $\{\mathfrak{s}^n_{\mathfrak{p}_0}\}$ and $\{\mathfrak{s}^n_p\}$ with $\widehat{\Phi^0}$.

We put $\overline{y} = \varphi^+_{\mathfrak{p}\mathfrak{p}_0}(y) \in U^+_{\mathfrak{p}}$. Then we have $\overline{y} = \varphi^1_{\mathfrak{p}q^{\mathfrak{p}}_i}(\tilde{y})$. (This is the definition (11.10) of $\varphi^+_{\mathfrak{p}\mathfrak{p}_0}$.) We can take a representative $(\mathfrak{s}^n_{\mathfrak{p},k})_{k=1,\dots,\ell}$ of $\mathfrak{s}^n_{\mathfrak{p}}$ such that

$$\mathfrak{s}^n_{\mathfrak{p},k}(\varphi^1_{\mathfrak{p}q^{\mathfrak{p}}_i}(\tilde{z})) = \widehat{\varphi^1_{\mathfrak{p}q^{\mathfrak{p}}_i}}(\mathfrak{s}^n_{q^{\mathfrak{p}}_i,k}(\tilde{z})) = (\widehat{\varphi^+_{\mathfrak{p}\mathfrak{p}_0}} \circ \widehat{\varphi^0_{\mathfrak{p}_0 q^{\mathfrak{p}}_i}})(\mathfrak{s}^n_{q^{\mathfrak{p}}_i,k}(\tilde{z})) \tag{11.31}$$

holds for any $\tilde{z} \in U_{\tilde{y}}$. Here the first equality is a consequence of the compatibility of $\{\mathfrak{s}^n_{\mathfrak{p}}\}$ and $\{\mathfrak{s}^n_p\}$ with $\widehat{\Phi^1}$ and the second equality is the definition of $\widehat{\varphi^+_{\mathfrak{p}\mathfrak{p}_0}}$.

We also have

$$\varphi^{1}_{\mathfrak{p}q_i^{\mathfrak{p}}}(\tilde{z}) = \varphi^{+}_{\mathfrak{p}\mathfrak{p}_0}(\varphi^{0}_{\mathfrak{p}_0 q_i^{\mathfrak{p}}}(\tilde{z})) \tag{11.32}$$

by definition of $\varphi^{+}_{\mathfrak{p}\mathfrak{p}_0}$, which is (11.10). Then (11.30), (11.31), (11.32) imply

$$\mathfrak{s}^{n}_{\mathfrak{p},k}(\varphi^{+}_{\mathfrak{p}\mathfrak{p}_0}(z)) = \widehat{\varphi^{+}_{\mathfrak{p}\mathfrak{p}_0}}(\mathfrak{s}^{n}_{\mathfrak{p}_0,k}(z))$$

for all $z \in \varphi^{0}_{\mathfrak{p}_0 q_i^{\mathfrak{p}}}(U^{0}_{\mathfrak{p}_0 q_i^{\mathfrak{p}}})$. Since $\varphi^{0}_{\mathfrak{p}_0 q_i^{\mathfrak{p}}}(U^{0}_{\mathfrak{p}_0 q_i^{\mathfrak{p}}})$ is a neighborhood of y, this implies (I). □

Proof of (II) Suppose $p \in \mathrm{Im}(\psi^{+}_{\mathfrak{p}}) \cap Z_1^{+}$, $\mathfrak{p} \in \mathfrak{P}$. We need to prove $(\Phi^{+}_{\mathfrak{p}p})^{*}(\mathfrak{s}^{n+}_{\mathfrak{p}}) = \mathfrak{s}^{n}_{p}$, where $\Phi^{+}_{\mathfrak{p}p} : \mathcal{U}_{0,p} \to \mathcal{U}_{\mathfrak{p}}$. This is a consequence of the fact that $\{\mathfrak{s}^{n}_{\mathfrak{p}}\}$, $\{\mathfrak{s}^{n}_{p}\}$ are compatible with the embedding $\widehat{\Phi^{1}} : \widehat{\mathcal{U}}|_{Z_1^{+}} \to \widehat{\mathcal{U}}$. We can prove the case of $\mathfrak{p} = \mathfrak{p}_0$ in the same way. □

Proof of (III) This is an immediate consequence of the fact that transversality to 0, strong submersivity, and transversality to a map, is preserved under the restriction to open subsets. □

The proof of Lemma 11.15 is complete. □

Using Lemma 11.15, we complete the proof of Proposition 6.32 and Lemma 9.10, as follows. During the proof of Theorem 3.35 we constructed Kuranishi charts consisting of the good coordinate system inductively. When various objects are given on the Kuranishi structure we start with, we can use Lemma 11.15 to obtain corresponding objects on the obtained chart so that it is compatible with coordinate changes. Therefore when we complete the construction of a good coordinate system by induction, the objects we look for are obtained at the same time. □

Proof of Proposition 6.33 and Lemmas 9.11, 9.32 Proposition 6.33 and Lemmas 9.11, 9.32 are variants of Proposition 6.32 and Lemma 9.10 where we are given also a thickening of the Kuranishi structure we start with. In other words, it is also a variant of Propositions 5.20 and 5.21. The proof then is a combination of the proofs of Proposition 6.32, Lemma 9.10 and Propositions 5.20 and 5.21. The inductive step, which corresponds to Lemma 11.15 and Proposition 11.11, is the next lemma.

Lemma 11.16 *Suppose we are in the situation of Proposition* 11.11.

(1) *We suppose that the assumptions of Lemma* 11.15 *(1) are satisfied, in addition. Moreover we assume that there exists a multivalued perturbation* $\widehat{\mathfrak{s}^{+}_{a}} = \{\mathfrak{s}^{n+}_{a;p}\}$ *of* $\widehat{\mathcal{U}^{+}_{a}}$ *such that:*

(i) $\widehat{\mathfrak{s}_a^+}, \widehat{\mathfrak{s}}$ *are compatible with the KK-embedding* $\widehat{\Phi_a^2} : \widehat{\mathcal{U}} \to \widehat{\mathcal{U}_a^+}$.

(ii) $\widehat{\mathfrak{s}_a^+}, \widehat{\mathfrak{s}}$ *are compatible with the GK-embedding* $\widehat{\Phi_a^3} : \widehat{\mathcal{U}} \to \widehat{\mathcal{U}_a^+}|_{Z_1^+}$.

(iii) $\widehat{\mathfrak{s}_a^+}, \widehat{\mathfrak{s}_{\mathfrak{p}_0}}$ *are compatible with the GK-embedding* $\widehat{\Phi_a^4} : \widehat{\mathcal{U}_{\mathfrak{p}_0}} \to \widehat{\mathcal{U}_a^+}|_{Z_0^+}$.

Then we can take the multivalued perturbation $\widehat{\mathfrak{s}^+} = \{\mathfrak{s}_p^{n+}\}$ *of* $\widehat{\mathcal{U}^+}$ *as in Lemma* 11.15 *(1) such that* $\widehat{\mathfrak{s}_a^+}, \widehat{\mathfrak{s}^+}$ *are compatible with the GK-embedding* $\widehat{\Phi_a^5} : \widehat{\mathcal{U}^+} \to \widehat{\mathcal{U}_a^+}$.

(2) *A statement similar to (1) for the CF-perturbations holds.*

(3) *A statements similar to (1) for the differential forms and for strongly continuous maps hold.*

We omit the precise statement for (2)(3) above. We believe that it is not difficult to find it for the reader.

Proof Let $p \in \psi_{\mathfrak{p}}^+((s_{\mathfrak{p}}^+)^{-1}(0)) \cap Z$. It suffices to show that $\widehat{\mathfrak{s}_a^+}, \widehat{\mathfrak{s}^+}$ are compatible with the embedding $\Phi_{a;p\mathfrak{p}}^5 : \mathcal{U}_{\mathfrak{p}}^+|_{U_{\mathfrak{p}}^+(p)} \to \mathcal{U}_{a;p}^+$. In the case $\mathfrak{p} \in \mathfrak{P}$ this is a consequence of Lemma 11.15 (1)(b) and (ii). In the case $\mathfrak{p} = \mathfrak{p}_0$ this is a consequence of Lemma 11.15 (1)(c) and (iii). We thus proved (1). The proofs of (2) (3) are entirely similar. □

Using Lemma 11.16, we can prove Proposition 6.33 and Lemma 9.11 in the same way as the proof of Proposition 6.32.

We finally prove Lemma 9.32 (2)(3). They claim that if we start with a uniform family of CF-perturbations on the Kuranishi structure, then the family of CF-perturbations we obtain on the good coordinate system is also uniform. Recall that uniformity means that the C^1 convergence of CF-perturbations is uniform with respect to the parameters in the family.

This follows from the fact that the CF-perturbations we obtain on the good coordinate system are locally a restriction of CF-perturbations on the Kuranishi structure we start with.

We need to discuss the following point. CF-perturbation on a Kuranishi structure consists of $\mathfrak{S}_p^\epsilon$ for each $p \in Z$. In other words, there are infinitely many of them. We did *not* require the convergence of the members of $\mathfrak{S}_p^\epsilon$ to the Kuranishi map s_p to be uniform with respect to p. Moreover the C^1 convergence is compact C^1 convergence on U_p. We remark, however, that when we construct a good coordinate system we use only a finite number of Kuranishi charts among those infinitely many charts $\mathcal{U}_p$. We used a proof by contradiction so potentially infinitely many Kuranishi charts are involved. However, the proof by contradiction is used only to *show* certain properties of the Kuranishi charts and coordinate changes between them. In other words, all the Kuranishi charts of the good coordinate system are open restrictions of one of the finitely many Kuranishi charts of the Kuranishi structure we start with. Moreover we may start with an open substructure of the given Kuranishi structure such that each Kuranishi chart is a relatively compact subset of one of the given

Kuranishi structures. The uniformity of the family of CF-perturbations we obtain is a consequence of these observations. □

11.4 Extension of Good Coordinate Systems: The Relative Case

The present section will be occupied with the proofs of Proposition 7.54 and Lemma 7.55. These propositions claim the extendability of objects given on one compact subset $Z_1 \subset X$ to one on $Z_2 \subset X$ such that Z_1 has a relatively compact neighborhood in Z_2.

Proof of Proposition 7.54 In the statement of Proposition 7.54 we used the symbols Z_1, Z_2 for compact subsets of X. In the proof below we use the symbols $\mathcal{Z}_{(1)}, \mathcal{Z}_{(2)}$ for the compact subsets Z_1, Z_2 in Proposition 7.54 to distinguish them from Z_0, Z_1 that appear in Proposition 11.3.

To prove Proposition 7.54 we use the same induction scheme as the proof of Theorem 3.35. We will modify the statement of Proposition 11.3 to Lemma 11.18 below. We begin with modifying Situation 11.1.

In Proposition 7.54 we considered $\widehat{\mathcal{U}^1}$. We write $\widehat{\mathcal{U}^{(1)}}$ hereafter in this section in place of $\widehat{\mathcal{U}^1}$. (We also write $\widehat{\mathcal{U}^{(2)}}$ hereafter in this section in place of $\widehat{\mathcal{U}^2}$.) Let $\widehat{\mathcal{U}^{(1)}} = (\mathfrak{P}(\mathcal{Z}_{(1)}), \{\mathcal{U}^{(1)}_{\mathfrak{p}'}\}, \{\Phi^{(1)}_{\mathfrak{p}'\mathfrak{q}'}\})$. (We denote elements of $\mathfrak{P}(\mathcal{Z}_{(1)})$ by $\mathfrak{p}'$, $\mathfrak{q}'$, that is, by lower case Fraktur characters with prime.)

Let $(\mathcal{K}^{(1)}, \mathcal{K}^{(1)+})$ be a support pair of $\widehat{\mathcal{U}^{(1)}}$. We put

$$Z_{\mathfrak{p}'} = \psi^{(1)}_{\mathfrak{p}'}((s^{(1)}_{\mathfrak{p}'})^{-1}(0) \cap \mathcal{K}^{(1)}_{\mathfrak{p}'}) \cap \mathcal{Z}_{(1)}.$$

Let $\mathfrak{U}(Z_{\mathfrak{p}'})$ be a relatively compact open neighborhood of $Z_{\mathfrak{p}'}$ in $\psi^{(1)}_{\mathfrak{p}'}((s^{(1)}_{\mathfrak{p}'})^{-1}(0) \cap \operatorname{Int} \mathcal{K}^{(1)+}_{\mathfrak{p}'})$. In Proposition 7.54 a Kuranishi structure $\widehat{\mathcal{U}^2}$ is given as a part of the assumption. During the proof, we will write $\widehat{\mathcal{U}}$ in place of $\widehat{\mathcal{U}^2}$.

Situation 11.17 Let $\mathfrak{d} \in \mathbb{Z}_{\geq 0}$, and let Z_0 be a compact subset of

$$\mathcal{S}_{\mathfrak{d}}(X, \mathcal{Z}_{(2)}; \widehat{\mathcal{U}}) \setminus \bigcup_{\mathfrak{d}' > \mathfrak{d}} \mathcal{S}_{\mathfrak{d}'}(X, \mathcal{Z}_{(2)}; \widehat{\mathcal{U}}) \setminus \bigcup_{\mathfrak{p}' \in \mathfrak{P}(\mathcal{Z}_{(1)})} Z_{\mathfrak{p}'},$$

and Z_1 a compact subset of

$$\mathcal{S}_{\mathfrak{d}}(X, \mathcal{Z}_{(2)}; \widehat{\mathcal{U}}) \cup \bigcup_{\mathfrak{p}' \in \mathfrak{P}(\mathcal{Z}_{(1)})} \mathfrak{U}(Z_{\mathfrak{p}'}).$$

We assume that Z_1 contains an open neighborhood of $\mathcal{Z}_{(1)} \cup \bigcup_{\mathfrak{d}'>\mathfrak{d}} \mathcal{S}_{\mathfrak{d}'}(X, \mathcal{Z}_{(2)}; \widehat{\mathcal{U}})$ in $\mathcal{S}_{\mathfrak{d}}(X, \mathcal{Z}_{(2)}; \widehat{\mathcal{U}})$. We put

$$Z_+ = Z_0 \cup Z_1.$$

Let $\widehat{\mathcal{U}} = (\mathfrak{P}, \{\mathcal{U}_{\mathfrak{p}}\}, \{\Phi_{\mathfrak{pq}}\})$ be a good coordinate system on a compact neighborhood Z_1^+ of Z_1. We assume that $\mathfrak{P}$ is written as

$$\mathfrak{P} = \mathfrak{P}(\mathcal{Z}_{(1)}) \cup \mathfrak{P}_0$$

(disjoint union) and the inclusions $\mathfrak{P}(\mathcal{Z}_{(1)}) \to \mathfrak{P}$, $\mathfrak{P}_0 \to \mathfrak{P}$ preserve the partial order. Moreover we assume that, for $\mathfrak{p}' \in \mathfrak{P}(\mathcal{Z}_{(1)})$, the Kuranishi chart $\mathcal{U}_{\mathfrak{p}'}$ of $\widehat{\mathcal{U}}$ is an open subchart of the Kuranishi chart $\mathcal{U}^{(1)}_{\mathfrak{p}'}$ of $\widehat{\mathcal{U}^{(1)}}$ and

$$U_{\mathfrak{p}'} \cap \mathcal{Z}_{(1)} = U^{(1)}_{\mathfrak{p}'} \cap \mathcal{Z}_{(1)}.$$[6]

Furthermore we assume $\dim U_{\mathfrak{p}} \geq \mathfrak{d}$ for $\mathfrak{p} \in \mathfrak{P}_0$.

Let $\widehat{\Phi^1} = \{\Phi^1_{\mathfrak{p}p} \mid p \in \mathrm{Im}(\psi_{\mathfrak{p}}) \cap Z_1^+\} : \widehat{\mathcal{U}}|_{Z_1^+} \to \widehat{\mathcal{U}}$ be a strict KG-embedding such that, for $\mathfrak{p}' \in \mathfrak{P}(\mathcal{Z}_{(1)})$, the embedding $\Phi^1_{\mathfrak{p}'p}$ is an open restriction of one that is a part of the given KG-embedding $\widehat{\mathcal{U}}|_{\mathcal{Z}_{(1)}} \to \widehat{\mathcal{U}^{(1)}}$. Let Z_0^+ be a compact neighborhood of Z_0 in $\mathcal{S}_{\mathfrak{d}}(X, \mathcal{Z}_{(2)}; \widehat{\mathcal{U}})$ and $\mathcal{U}_{\mathfrak{p}_0} = (U_{\mathfrak{p}_0}, E_{\mathfrak{p}_0}, s_{\mathfrak{p}_0}, \psi_{\mathfrak{p}_0})$ a Kuranishi neighborhood of Z_0^+ such that $\dim U_{\mathfrak{p}_0} = \mathfrak{d}$. We regard $\mathcal{U}_{\mathfrak{p}_0}$ as a good coordinate system $\widehat{\mathcal{U}_{\mathfrak{p}_0}}$ that consists of a single Kuranishi chart and suppose that we are given a strict KG-embedding $\Phi^0 = \{\Phi^0_{\mathfrak{p}_0} \mid p \in \mathrm{Im}(\psi) \cap Z_0\} : \widehat{\mathcal{U}}|_{Z_0^+} \to \widehat{\mathcal{U}_{\mathfrak{p}_0}}$.

We put $\mathfrak{P}^+ = \mathfrak{P} \cup \{\mathfrak{p}_0\}$, so $\mathfrak{P}^+ \supset \mathfrak{P}(\mathcal{Z}_{(1)})$. ■

Lemma 11.18 *In Situation* 11.17 *there exists a good coordinate system* $\widehat{\mathcal{U}^+}$ *of* Z_+^+ *satisfying conclusions (2)–(5) of Proposition* 11.3. *Here* Z_+^+ *is a compact neighborhood of* Z_+ *in* X.

Moreover the following holds:

(1)' *If* $\mathfrak{p} \in \mathfrak{P}_0$ *then* $\mathfrak{p}_0 < \mathfrak{p}$. *If* $\mathfrak{p} \in \mathfrak{P}(\mathcal{Z}_{(1)})$ *then* $\mathfrak{p} < \mathfrak{p}_0$ *does not occur.*

(6) *If* $\mathfrak{p}' \in \mathfrak{P}(\mathcal{Z}_{(1)}) \subset \mathfrak{P}^+$, *then the Kuranishi chart* $\mathcal{U}^+_{\mathfrak{p}'}$ *of* $\widehat{\mathcal{U}^+}$ *is an open subchart of the Kuranishi chart* $\mathcal{U}^{(1)}_{\mathfrak{p}'}$ *of* $\widehat{\mathcal{U}^{(1)}}$ *and*

$$U^+_{\mathfrak{p}'} \cap \mathcal{Z}_{(1)} = U^{(1)}_{\mathfrak{p}'} \cap \mathcal{Z}_{(1)}.$$

Proof We put

[6]Compare Definition 7.52 (1)(b).

$$\mathfrak{P}_{\geq \mathfrak{d}} = \mathfrak{P} \setminus \{\mathfrak{p}' \in \mathfrak{P}(\mathcal{Z}_{(1)}) \mid \dim U_{\mathfrak{p}'} < \mathfrak{d}\}.$$

Then $\widehat{\mathcal{U}_{\geq \mathfrak{d}}} = (\mathfrak{P}_{\geq \mathfrak{d}}, \{\mathcal{U}_{\mathfrak{p}} \mid \mathfrak{p} \in \mathfrak{P}_{\geq \mathfrak{d}}\}, \{\Phi_{\mathfrak{p}\mathfrak{q}} \mid \mathfrak{p}, \mathfrak{q} \in \mathfrak{P}_{\geq \mathfrak{d}}, \mathfrak{p} \geq \mathfrak{q}\})$ is a good coordinate system of any compact subset of $\mathcal{Z}_{(2)} \cap \bigcup_{\mathfrak{p}' \in \mathfrak{P}_{\geq \mathfrak{d}}} \mathrm{Im}(\psi_{\mathfrak{p}'})$.

We take $\mathfrak{U}'(Z_{\mathfrak{p}'})$ which is an open neighborhood of $Z_{\mathfrak{p}'}$ and is relatively compact in $\mathfrak{U}(Z_{\mathfrak{p}'})$. We put

$$Z_1' = Z_1 \setminus \bigcup_{\mathfrak{p}' \in \mathfrak{P}(\mathcal{Z}_{(1)}), \dim U_{\mathfrak{p}'} < \mathfrak{d}} \mathfrak{U}'(Z_{\mathfrak{p}'}).$$

We observe that we are then in Situation 11.1, where $\widehat{\mathcal{U}_{\geq \mathfrak{d}}}$ (resp. Z_1') plays the role of $\widehat{\mathcal{U}}$ (resp. Z_1) in Situation 11.1. We apply Proposition 11.3 to our situation and obtain $\widehat{\mathcal{U}^{+\prime}}$. Note that the union of the sets of Kuranishi charts of $\widehat{\mathcal{U}^{+\prime}}$ and $\{\mathcal{U}^{(1)}_{\mathfrak{p}'} \mid \mathfrak{p}' \in \mathfrak{P}(\mathcal{Z}_{(1)}), \dim U_{\mathfrak{p}'} < \mathfrak{d}\}$ has most of the properties we need to prove. The only point to take care of is that, for $\mathfrak{p}' \in \mathfrak{P}(\mathcal{Z}_{(1)})$ with $\dim U_{\mathfrak{p}'} < \mathfrak{d}$, neither the coordinate change $\Phi^+_{\mathfrak{p}_0\mathfrak{p}'}$ nor $\Phi^+_{\mathfrak{p}'\mathfrak{p}_0}$ is defined.

Let $\mathfrak{P}^{+\prime}$ be the partially ordered set appearing in $\widehat{\mathcal{U}^{+\prime}}$. Then $\mathfrak{p}_0 \in \mathfrak{P}^{+\prime}$ and $\mathcal{U}^{+\prime}_{\mathfrak{p}_0}$ is a Kuranishi chart that is an open subchart of $\mathcal{U}_{\mathfrak{p}_0}$. To take care of the point mentioned above we shrink $\mathcal{U}^{+\prime}_{\mathfrak{p}_0}$ to $\mathcal{U}^{+}_{\mathfrak{p}_0}$ so that these two coordinates will not intersect, as follows.

Sublemma 11.19 *There exists an open subset $U^+_{\mathfrak{p}_0}$ of $U^{+\prime}_{\mathfrak{p}_0}$ such that the following holds:*

(1) *If $\mathfrak{p}' \in \mathfrak{P}(\mathcal{Z}_{(1)})$, $\dim U_{\mathfrak{p}'} < \mathfrak{d}$ then*

$$\psi^+_{\mathfrak{p}_0}(s^{-1}_{\mathfrak{p}_0}(0) \cap U^+_{\mathfrak{p}_0}) \cap \mathfrak{U}'(Z_{\mathfrak{p}'}) = \emptyset.$$

(2) $\psi_{\mathfrak{p}_0}(s^{-1}_{\mathfrak{p}_0}(0) \cap U^{+\prime}_{\mathfrak{p}_0}) \cap \mathcal{S}_{\mathfrak{d}}(X, \mathcal{Z}_{(1)}; \widehat{\mathcal{U}}) = \psi_{\mathfrak{p}_0}(s^{-1}_{\mathfrak{p}_0}(0) \cap U^{+}_{\mathfrak{p}_0}) \cap \mathcal{S}_{\mathfrak{d}}(X, \mathcal{Z}_{(1)}; \widehat{\mathcal{U}})$.

Proof By definition, we have $Z_{\mathfrak{p}'} \cap \mathcal{S}_{\mathfrak{d}}(X, \mathcal{Z}_{(1)}; \widehat{\mathcal{U}}) = \emptyset$ for $\mathfrak{p}' \in \mathfrak{P}(\mathcal{Z}_{(1)})$, $\dim U_{\mathfrak{p}'} < \mathfrak{d}$. Therefore we may choose $\mathfrak{U}(Z_{\mathfrak{p}'})$ so that

$$\mathfrak{U}(Z_{\mathfrak{p}'}) \cap \mathcal{S}_{\mathfrak{d}}(X, \mathcal{Z}_{(1)}; \widehat{\mathcal{U}}) = \emptyset$$

for such $\mathfrak{p}'$. In fact, $\mathcal{S}_{\mathfrak{d}}(X, \mathcal{Z}_{(1)}; \widehat{\mathcal{U}})$ is a closed set. Since $\mathfrak{U}'(Z_{\mathfrak{p}'})$ is relatively compact in $\mathfrak{U}(Z_{\mathfrak{p}'})$, we have

$$\overline{\mathfrak{U}'(Z_{\mathfrak{p}'})} \cap \mathcal{S}_{\mathfrak{d}}(X, \mathcal{Z}_{(1)}; \widehat{\mathcal{U}}) = \emptyset.$$

Sublemma 11.19 is an immediate consequence of this fact. □

We now put $\mathcal{U}^+_{\mathfrak{p}_0} = \mathcal{U}^{+\prime}_{\mathfrak{p}_0}|_{U^+_{\mathfrak{p}_0}}$. For $\mathfrak{p}' \in \mathfrak{P}(\mathcal{Z}_{(1)})$ with $\dim U_{\mathfrak{p}'} < \mathfrak{d}$, we take an open subset $U^+_{\mathfrak{p}'} \subset U^{(1)}_{\mathfrak{p}'}$ such that

$$\mathfrak{U}'(Z_{\mathfrak{p}'}) = \psi^{(1)}_{\mathfrak{p}'}((s^{(1)}_{\mathfrak{p}'})^{-1}(0) \cap U^+_{\mathfrak{p}'}).$$

Then we put $\mathcal{U}^+_{\mathfrak{p}'} = \mathcal{U}^{+\prime}_{\mathfrak{p}'}|_{U^+_{\mathfrak{p}'}}$. For $\mathfrak{p} \in \mathfrak{P}^{+\prime} \setminus \{\mathfrak{p}_0\}$ we put $\mathcal{U}^+_{\mathfrak{p}} = \mathcal{U}^{+\prime}_{\mathfrak{p}}$. We define a partial order $\le$ on $\mathfrak{P}_+ = \mathfrak{P}^{+\prime} \cup \mathfrak{P}(\mathcal{Z}_{(1)})$ such that $\le$ coincides with the partial orders on $\mathfrak{P}^{+\prime}$ and on $\mathfrak{P}(\mathcal{Z}_{(1)})$. Moreover we define $\le$ so that for $\mathfrak{p}' \in \mathfrak{P}^{+\prime}$ with $\dim U_{\mathfrak{p}'} < \mathfrak{d}$, neither $\mathfrak{p}' \le \mathfrak{p}_0$ nor $\mathfrak{p}' \ge \mathfrak{p}_0$.

We can define coordinate change among them by restricting the coordinate change of either $\widehat{\mathcal{U}^{+\prime}}$ or of $\widehat{\mathcal{U}}$. Sublemma 11.19 (1) implies that these two cases exhaust the cases we need to define coordinate change. The proof of Lemma 11.18 is complete. □

Using Lemma 11.18 we employ the same method as the last step of the proof of Theorem 3.35 to complete the proof of Proposition 7.54 (1), (2), (3).

We finally prove Proposition 7.54 (4). Suppose a support system $\{\mathcal{K}_{\mathfrak{p}} \mid \mathfrak{p} \in \mathfrak{P}(\mathcal{Z}_{(1)})\}$ of $\widehat{\mathcal{U}^1}$ $(= \widehat{\mathcal{U}^{(1)}})$ is given. We need to check (7.27). Note that to obtain $U^1_{0,\mathfrak{p}}$ we shrink $U^1_{\mathfrak{p}}$ several times. There are two places where we shrink $U^1_{\mathfrak{p}}$. One uses formula (11.14). We remark that here the member of the support system $\mathcal{K}_{\mathfrak{p}}$ appearing in (11.14) can be taken to be the same as $\mathcal{K}_{\mathfrak{p}}$ appearing in (7.27). Therefore (7.27) holds after this shrinking. Another place we shrink is while we apply Proposition 3.17, Shrinking Lemma [FOOO17, Theorem 2.9]. In [FOOO17] we shrink using the following formula:

$$U^{\delta}_{\mathfrak{p}} = \{x \in U^1_{\mathfrak{p}} \mid d_{\mathfrak{p}}(x, \psi^{-1}_{\mathfrak{p}}(K_{\mathfrak{p}})) < \delta\}.$$

This is [FOOO17, (3.7)]. Here $\{K_{\mathfrak{p}}\}$ is a set of compact subsets of Z contained in the image of $\psi_{\mathfrak{p}}$ which cover Z. So we may take $K_{\mathfrak{p}} = \psi_{\mathfrak{p}}(s^{-1}_{\mathfrak{p}}(0) \cap \mathcal{K}_{\mathfrak{p}})$. Therefore, (7.27) holds after this shrinking also. We thus proved Proposition 7.54 (4). □

Proof of Lemma 7.55 Using Lemma 11.18, we can prove it in the same way as in Sect. 11.3. □

Chapter 12
Construction of CF-Perturbations

In this chapter, we prove Theorem 7.51, that is, the existence of CF-perturbations with respect to which a given weakly submersive map becomes strongly submersive. We also prove its relative version, Proposition 7.59. For this purpose we use the language of sheaf theory to prove Proposition 12.2 for a single Kuranishi chart and Theorem 12.24 for a general case.

12.1 Construction of CF-Perturbations on a Single Chart

We first study the case of a single Kuranishi chart. The main result of Sect. 12.1 is Proposition 12.2 below. We recall the following well-known definition.

Definition 12.1 A sheaf (of sets) $\mathscr{F}$ on a topological space V is said to be *soft* if the restriction map

$$\mathscr{F}(V) \to \mathscr{F}(K)$$

is surjective for any closed subset K of V. (We note $\mathscr{F}(K) = \varinjlim_{W \supset K, \mathrm{open}} \mathscr{F}(W)$.)

Proposition 12.2 *Let $\mathcal{U} = (U, \mathcal{E}, \psi, s)$ be a Kuranishi chart of X and M a smooth manifold.*

(1) *The sheaf $\mathcal{CF}$ in Proposition 7.22 is soft.*
(2) *The sheaf $\mathcal{CF}_{\pitchfork 0}$ in Lemma-Definition 7.26 is soft.*
(3) *Let $f : U \to M$ be a smooth submersion. Then, the sheaf $\mathcal{CF}_{f \pitchfork}$ in Lemma-Definition 7.26 is soft.*
(4) *Let $f : U \to M$ be a smooth map and $g : N \to M$ a smooth map between manifolds. We assume that f is transversal to g. Then the sheaf $\mathcal{CF}_{f \pitchfork g}$ in Lemma-Definition 7.26 is soft.*

K. Fukaya et al., *Kuranishi Structures and Virtual Fundamental Chains*, Springer Monographs in Mathematics, https://doi.org/10.1007/978-981-15-5562-6_12

During the proof we introduce another sheaf $\mathcal{CF}_{\pitchfork\pitchfork 0}$, the sheaf of strongly transversal CF-perturbations, in Definition 12.9 and prove its softness also.

The rest of this section will be occupied with the proof of this proposition.

Proof We first prove (1). We use a partition of unity to glue sections of $\mathcal{CF}$. Note that our sheaf $\mathcal{CF}$ is a sheaf of sets. Nevertheless we can apply a partition of unity thereto, as we will discuss below.

Situation 12.3 Let $\mathcal{U} = (U, \mathcal{E}, \psi, s)$ be a Kuranishi chart of X, A a subset of U, $\{U_{\mathfrak{r}} \mid \mathfrak{r} \in \mathfrak{R}\}$ a locally finite open cover of A in U, and $\{\chi_{\mathfrak{r}}\}$ a smooth partition of unity subordinate to this covering. In other words, $\chi_{\mathfrak{r}} : U \to [0, 1]$ is a smooth function of U which has compact support in $U_{\mathfrak{r}}$, and

$$\sum_{\mathfrak{r}\in\mathfrak{R}} \chi_{\mathfrak{r}}(x) = 1$$

for $x \in A$. We assume that an element $\mathfrak{S}_{\mathfrak{r}} \in \mathcal{CF}(U_{\mathfrak{r}})$ is given for each $\mathfrak{r} \in \mathfrak{R}$. ■

In Situation 12.3 we will define the sum

$$\sum_{\mathfrak{r}} \chi_{\mathfrak{r}}\mathfrak{S}_{\mathfrak{r}} \in \mathcal{CF}(A), \tag{12.1}$$

below. For $x \in A$, let $\mathfrak{V}_x = (V_x, \Gamma_x, E_x, \phi_x, \widehat{\phi}_x)$ be an orbifold chart of $(U, \mathcal{E})$ at x. We may assume that, for each $\mathfrak{r}$ with $x \in U_{\mathfrak{r}}$, we are given a representative $\mathcal{S}_{\mathfrak{r}}$ of the CF-perturbation $\mathfrak{S}_{\mathfrak{r}}$ on a neighborhood of x. It consists of $\mathfrak{V}_{\mathfrak{r}} = (V_{\mathfrak{r}}, \Gamma_{\mathfrak{r}}, E_{\mathfrak{r}}, \phi_{\mathfrak{r}}, \widehat{\phi}_{\mathfrak{r}})$ and $(W_{\mathfrak{r}}, \omega_{\mathfrak{r}}, \{\mathfrak{s}^{\epsilon}_{\mathfrak{r}} \mid \epsilon\})$, where $\mathfrak{V}_{\mathfrak{r}} = (V_{\mathfrak{r}}, \Gamma_{\mathfrak{r}}, E_{\mathfrak{r}}, \phi_{\mathfrak{r}}, \widehat{\phi}_{\mathfrak{r}})$ is an orbifold chart of $(U, \mathcal{E})$ at x and $(W_{\mathfrak{r}}, \omega_{\mathfrak{r}}, \{\mathfrak{s}^{\epsilon}_{\mathfrak{r}} \mid \epsilon\})$ is as in Definition 7.4.

For each $x \in A$, we put

$$\mathfrak{R}(x) = \{\mathfrak{r} \in \mathfrak{R} \mid x \in \mathrm{Supp}(\chi_{\mathfrak{r}})\}. \tag{12.2}$$

By shrinking V_x if necessary we may assume $\phi_x(V_x) \subset U_{\mathfrak{r}}$ for each $\mathfrak{r} \in \mathfrak{R}(x)$ and $\chi_{\mathfrak{r}} \equiv 0$ on $\phi_x(V_x)$ for each $\mathfrak{r} \notin \mathfrak{R}(x)$.

Furthermore we may choose the orbifold chart $\mathfrak{V}_x$ so that there exist

$$\begin{aligned} h_{\mathfrak{r}x} &: \Gamma_x \to \Gamma_{\mathfrak{r}}, \\ \widetilde{\varphi}_{\mathfrak{r}x} &: V_x \to V_{\mathfrak{r}}, \\ \breve{\varphi}_{\mathfrak{r}x} &: V_x \times E_x \to E_{\mathfrak{r}} \end{aligned} \tag{12.3}$$

as in Situation 6.4, for each $\mathfrak{r} \in \mathfrak{R}(x)$. (See Lemma 23.26.)

Definition 12.4 We define

$$W_x = \prod_{\mathfrak{r}\in\mathfrak{R}(x)} W_{\mathfrak{r}}, \quad \omega_x = \prod_{\mathfrak{r}\in\mathfrak{R}(x)} \omega_{\mathfrak{r}}.$$

We define $\mathfrak{s}_x^\epsilon : V_x \times W_x \to E_x$ by the following formula:

$$\mathfrak{s}_x^\epsilon(y, (\xi_\mathfrak{r})_{\mathfrak{r}\in\mathfrak{R}(x)}) = s_x(y) + \sum_{\mathfrak{r}\in\mathfrak{R}(x)} \chi_\mathfrak{r}(\psi_x(y)) g_{\mathfrak{r},y}^{-1}(\mathfrak{s}_\mathfrak{r}^\epsilon(\widetilde{\varphi}_{\mathfrak{r}x}(y), \xi_\mathfrak{r}) - s_\mathfrak{r}(\widetilde{\varphi}_{\mathfrak{r}x}(y))). \tag{12.4}$$

Here $s_x : V_x \to E_x$ and $s_\mathfrak{r} : V_\mathfrak{r} \to E_\mathfrak{r}$ are the local expressions of the Kuranishi map (Definition 23.33) and $g_{\mathfrak{r},y} : E_x \to E_\mathfrak{r}$ is defined by $\breve{\varphi}_{\mathfrak{r}x}(y, \xi) = g_{\mathfrak{r},y}(\xi)$.

We put $\mathcal{S}_x = (W_x, \omega_x, \{\mathfrak{s}_x^\epsilon\})$.

Lemma 12.5

(1) *$\mathcal{S}_x$ is a CF-perturbation of $\mathcal{U}$ on $\mathfrak{V}_x$.*
(2) *The germ $[\mathcal{S}_x] \in \mathcal{CF}_x$ represented by $\mathcal{S}_x$ depends only on $\{\chi_\mathfrak{r}\}$, $\{\mathcal{S}_\mathfrak{r}\}$, x but is independent of the choices of orbifold chart $\mathfrak{V}_x$, the coordinate changes* (12.3), *and the representatives of $\{\mathcal{S}_\mathfrak{r}\}$.*
(3) *$x \mapsto [\mathcal{S}_x] \in \mathcal{CF}_x$ defines a (global) section of the sheaf $\mathcal{CF}$.*

Proof Statement (1) is an immediate consequence of the construction.

We prove Statement (2). We first prove independence of the coordinate changes (12.3). Let $(h'_{\mathfrak{r}x}, \widetilde{\varphi}'_{\mathfrak{r}x}, \breve{\varphi}'_{\mathfrak{r}x})$ be an alternative choice. Then there exists $\gamma_\mathfrak{r} \in \Gamma_\mathfrak{r}$ such that $h'_{\mathfrak{r}x} = \gamma_\mathfrak{r} h_{\mathfrak{r}x} \gamma_\mathfrak{r}^{-1}$, $\widetilde{\varphi}'_{\mathfrak{r}x} = \gamma_\mathfrak{r}\widetilde{\varphi}_{\mathfrak{r}x}$, $\breve{\varphi}'_{\mathfrak{r}x} = \gamma_\mathfrak{r}\breve{\varphi}_{\mathfrak{r}x}$. The third equality implies $g'_{\mathfrak{r},y} = \gamma_\mathfrak{r} g_{\mathfrak{r},y}$ by Lemma 23.28. Let $\mathfrak{s}_x^{\epsilon\prime}$ be obtained from this alternative choice. Then we have

$$\begin{aligned} &\mathfrak{s}_x^{\epsilon\prime}(y, \xi) \\ &= s_x(y) + \sum_{\mathfrak{r}\in\mathfrak{R}(x)} \chi_\mathfrak{r}(\psi_x(y))(g'_{\mathfrak{r},y})^{-1}(\mathfrak{s}_\mathfrak{r}^{\epsilon\prime}(\widetilde{\varphi}'_{\mathfrak{r}x}(y), \xi_\mathfrak{r}) - s_\mathfrak{r}(\widetilde{\varphi}'_{\mathfrak{r}x}(y))) \\ &= s_x(y) + \sum_{\mathfrak{r}\in\mathfrak{R}(x)} \chi_\mathfrak{r}(\psi_x(y)) g_{\mathfrak{r},y}^{-1}(\mathfrak{s}_\mathfrak{r}^\epsilon(\widetilde{\varphi}_{\mathfrak{r}x}(y), \gamma_\mathfrak{r}^{-1}\xi_\mathfrak{r}) - s_\mathfrak{r}(\widetilde{\varphi}_{\mathfrak{r}x}(y))). \end{aligned} \tag{12.5}$$

Here we use $\Gamma_\mathfrak{r}$ equivariance of $\mathfrak{s}_\mathfrak{r}^\epsilon$ and of $s_\mathfrak{r}$.

We define a Γ_x action on $W_\mathfrak{r}$ by $\mu \cdot \xi = h_{\mathfrak{r}x}(\mu)\xi$. We denote $W_\mathfrak{r}$ equipped with this action by $W_\mathfrak{r}^{h_{\mathfrak{r}x}}$. The notation $W_\mathfrak{r}^{h'_{\mathfrak{r}x}}$ is defined in a similar way. Its product in Definition 12.4 is denoted by W_x^h and $W_x^{h'}$, respectively.

Then $\xi_\mathfrak{r} \mapsto \gamma_\mathfrak{r}^{-1}\xi_\mathfrak{r}$ (resp. $(\xi_\mathfrak{r}) \mapsto (\gamma_\mathfrak{r}^{-1}\xi_\mathfrak{r})$) is a Γ_x equivariant linear map : $W_\mathfrak{r}^{h_{\mathfrak{r}x}} \to W_\mathfrak{r}^{h'_{\mathfrak{r}x}}$ (resp. $W_x^h \to W_x^{h'}$). Therefore (12.5) implies that the equivalence class $[\mathcal{S}_x]$ is independent of the choices of the coordinate changes (12.3).

Secondly, we prove independence of the representative of $\mathcal{S}_\mathfrak{r}$. We consider one of $\mathfrak{r}_0 \in \mathfrak{R}(x)$ and take an alternative choice $\mathcal{S}'_{\mathfrak{r}_0}$ of $\mathcal{S}_{\mathfrak{r}_0}$. It suffices to consider the case when other $\mathcal{S}_\mathfrak{r}$'s for $\mathfrak{r} \neq \mathfrak{r}_0$ are the same for both. We may also assume that $\mathcal{S}'_{\mathfrak{r}_0}$ is a projection of $\mathcal{S}_{\mathfrak{r}_0}$. Then it is immediate from the definition that the CF-perturbation $\mathcal{S}'_x$ obtained via the CF-perturbation $\mathcal{S}'_{\mathfrak{r}_0}$ is a projection of the CF-perturbation $\mathcal{S}_x$ obtained via $\mathcal{S}_{\mathfrak{r}_0}$. We have thus proved the independence of the representative of $\mathcal{S}_\mathfrak{r}$.

Thirdly, we prove independence of the orbifold chart $\mathfrak{V}_x = (V_x, \Gamma_x, E_x, \phi_x, \widehat{\phi}_x)$. Let $\mathfrak{V}'_x = (V'_x, \Gamma'_x, E'_x, \phi'_x, \widehat{\phi}'_x)$ be another orbifold chart and suppose we obtain $\mathcal{S}'_x$ when we work on $\mathfrak{V}'_x$.

By shrinking V'_x if necessary we may assume that there exists a coordinate change $(h_x, \widetilde{\varphi}_x, \breve{\varphi}_x)$ from the chart $\mathfrak{V}'_x$ to $\mathfrak{V}_x$. Let $(h_{\mathfrak{r}x}, \widetilde{\varphi}_{\mathfrak{r}x}, \breve{\varphi}_{\mathfrak{r}x})$ be the coordinate change as in (12.3). Then by putting

$$h'_{\mathfrak{r}x} = h_{\mathfrak{r}x} \circ h_x, \quad \widetilde{\varphi}'_{\mathfrak{r}x} = \widetilde{\varphi}_{\mathfrak{r}x} \circ \widetilde{\varphi}_x, \quad \breve{\varphi}'_{\mathfrak{r}x} = \breve{\varphi}_{\mathfrak{r}x} \circ \breve{\varphi}_x,$$

$(h'_{\mathfrak{r}x}, \widetilde{\varphi}'_{\mathfrak{r}x}, \breve{\varphi}'_{\mathfrak{r}x})$ becomes a coordinate change from $\mathfrak{V}'_x$ to $\mathfrak{V}_{\mathfrak{r}}$ as in (12.3).

Then

$$\begin{aligned}
&\mathfrak{s}^{\epsilon\prime}_x(y, \xi) \\
&= s_x(y) + \sum_{\mathfrak{r}\in\mathfrak{R}(x)} \chi_{\mathfrak{r}}(\psi'_x(y))(g'_{\mathfrak{r},y})^{-1}(\mathfrak{s}^{\epsilon\prime}_{\mathfrak{r}}(\widetilde{\varphi}'_{\mathfrak{r}x}(y), \xi_{\mathfrak{r}}) - s_{\mathfrak{r}}(\widetilde{\varphi}'_{\mathfrak{r}x}(y))) \\
&= s_x(y) + \sum_{\mathfrak{r}\in\mathfrak{R}(x)} \chi_{\mathfrak{r}}(\psi_x(\widetilde{\varphi}_x(y)))g^{-1}_{\mathfrak{r},y}(\mathfrak{s}^{\epsilon}_{\mathfrak{r}}(\widetilde{\varphi}_{\mathfrak{r}x}(\widetilde{\varphi}_x(y)), \gamma_{\mathfrak{r}}^{-1}\xi_{\mathfrak{r}}) - s_{\mathfrak{r}}(\widetilde{\varphi}_{\mathfrak{r}x}(\widetilde{\varphi}_x(y)))) \\
&= \mathfrak{s}^{\epsilon}_x(\widetilde{\varphi}_x(y), \xi).
\end{aligned}$$

This implies the required independence of the chart $\mathfrak{V}_x$. The proof of Statement (2) is complete.

We now prove Statement (3). Let $\mathcal{S}_x = (W_x, \omega_x, \{\mathfrak{s}^{\epsilon}_x\})$ be as above. Suppose $y \in s_x^{-1}(0) \cap U_x$ and $y = \phi_x(\tilde{y})$. We define $\Gamma_{\tilde{y}} = \{\gamma \in \Gamma_x \mid \gamma\tilde{y} = \tilde{y}\}$ and take a $\Gamma_{\tilde{y}}$-invariant neighborhood V_y of $\tilde{y}$. Then $\mathfrak{V}_y = (V_y, \Gamma_y, E_y, \phi_x|_{V_y}, \widehat{\phi}_x|_{V_y})$ is an orbifold chart of $(U, \mathcal{E})$ at y. It is easy to see that $\mathcal{S}_x|_{V_y} = (W_x, \omega_x, \{\mathfrak{s}^{\epsilon}_x|_{V_y\times E_x}\})$ is a CF-perturbation on $\mathfrak{V}_y$.

Sublemma 12.6 *$\mathcal{S}_x|_{V_y}$ is equivalent to $\mathcal{S}_y$ in the sense of Definition* 7.6.

Proof We consider $\mathfrak{R}(y) = \{\mathfrak{r} \in \mathfrak{R} \mid y \in \mathrm{Supp}(\chi_{\mathfrak{r}})\}$. Since we chose $U_x \subseteq U_{\mathfrak{r}}$ such that $U_x \cap \mathrm{Supp}(\chi_{\mathfrak{r}}) = \emptyset$ for $\mathfrak{r} \notin \mathfrak{R}(x)$, we have $\mathfrak{R}(y) \subseteq \mathfrak{R}(x)$. Therefore there exists an obvious projection

$$\pi : \widehat{W}_x = \prod_{\mathfrak{r}\in\mathfrak{R}(x)} W_{\mathfrak{r}} \to \widehat{W}_y = \prod_{\mathfrak{r}\in\mathfrak{R}(y)} W_{\mathfrak{r}}.$$

It is easy to see that $\pi!(\omega_x) = \omega_y$.

We may choose V_y so small that for $z \in V_y$ and $\mathfrak{r} \in \mathfrak{R}(x) \setminus \mathfrak{R}(y)$ we have $\chi_{\mathfrak{r}}(z) = 0$. Therefore by definition

$$\mathfrak{s}^{\epsilon}_x(\tilde{\varphi}_x(z), \xi) = \mathfrak{s}^{\epsilon}_y(\tilde{\varphi}_y(z), \pi(\xi))$$

for $z \in V_y$. Thus $\mathcal{S}_y$ is a projection of $\mathcal{S}_x|_{V_y}$. □

Statement (3) follows from Sublemma 12.6 and Lemma 7.24. □

Definition 12.7 We denote by

$$\sum_{\mathfrak{r}} \chi_{\mathfrak{r}} \mathfrak{S}_{\mathfrak{r}}$$

the element $x \mapsto [\mathcal{S}_x] \in \mathcal{CF}_x$ of $\mathcal{CF}(A)$ obtained by Lemma 12.5.

Remark 12.8 Suppose $\mathfrak{S} \in \mathcal{CF}(U)$ and $\{U_{\mathfrak{r}} \mid \mathfrak{r} \in \mathfrak{R}\}$ is a locally finite cover of U. We can define an element of $\mathcal{CF}(U)$ by

$$\sum_{\mathfrak{r}} \chi_{\mathfrak{r}} \mathfrak{S}|_{U_{\mathfrak{r}}}$$

as above. (Here $\mathcal{CF}|_{U_{\mathfrak{r}}} \in \mathcal{CF}(U_{\mathfrak{r}})$ is the restriction of $\mathfrak{S}$.) However, this section is in general *different* from the originally given $\mathfrak{S} \in \mathcal{CF}(U)$. In particular, $\mathcal{CF}$ is *not* a fine sheaf.

Proposition 12.2 (1) follows easily from Definition 12.7 and the results we proved above. We next prove Proposition 12.2 (2)(3)(4). We begin with the next definition.

Definition 12.9 Let $\mathcal{U} = (U, E, \psi, s)$ be a Kuranishi chart of X and $\mathfrak{S}_x \in \mathcal{CF}_x$ a germ of the sheaf $\mathcal{CF}$ at $x \in U$.

We say $\mathfrak{S}_x$ is *strongly transversal* if its representative $(W_x, \omega_x, \{\mathfrak{s}_x^\epsilon\})$ (which is defined on the orbifold chart $\mathfrak{V}_x = (V_x, \Gamma_x, E_x, \phi_x, \widehat{\phi}_x)$ (Definition 6.1(3))) has the following properties:

(1) For all $\epsilon \in (0, 1]$, the map $\mathfrak{s}_x^\epsilon : V_x \times W_x \to E_x$ is transversal to $c \in E_x$ on a neighborhood of $\{o_x\} \times \mathrm{Supp}(\omega_x)$ for any $c \in E_x$. (Here $o_x \in V_x$ is the point such that $\phi_x(o_x) = x$.)
(2) For $\xi \in \mathrm{Supp}(\omega_x)$ and $c = \mathfrak{s}_x^\epsilon(o_x, \xi)$ the projection

$$T_{(o_x,\xi)}(\mathfrak{s}_x^\epsilon)^{-1}(c) \to T_{o_x} V_x$$

is surjective.

We write $(\mathcal{CF}_{\pitchfork 0})_x$ for the set of all germs of the sheaf $\mathcal{CF}$ at $x \in U$ that is strongly transversal.

It is easy to see that there exists a subsheaf of $\mathcal{CF}$ whose stalk at x is $(\mathcal{CF}_{\pitchfork 0})_x$. We denote this sheaf by $\mathcal{CF}_{\pitchfork 0}$.

Remark 12.10 It is easy to see that the above properties (1)(2) are independent of the choice of the representative $(W_x, \omega_x, \{\mathfrak{s}_x^\epsilon\})$ and of the orbifold chart $(V_x, \Gamma_x, E_x, \phi_x, \widehat{\phi}_x)$ but depend only on the germ $\mathfrak{S}_x$.

Lemma 12.11 *The stalk* $(\mathcal{CF}_{\pitchfork 0})_x$ *is nonempty.*

Proof Let $\mathfrak{V}_x = (V_x, \Gamma_x, E_x, \phi_x, \widehat{\phi}_x)$ be an orbifold chart of $(U, \mathcal{E})$ at x. We put $\widehat{W}_x = E_x$ and W_x is a sufficiently small Γ_x-invariant neighborhood of 0 in $\widehat{W}_x$ and ω_x is a Γ_x-invariant differential form of compact support on W_x of degree $\dim W_x$ with $\int \omega_x = 1$. We define

$$\mathfrak{s}^\epsilon(x, \xi) = s(x) + \epsilon\xi. \tag{12.6}$$

It is easy to see that $(W_x, \omega_x, \{\mathfrak{s}_x^\epsilon \mid \epsilon\})$ is a CF-perturbation on $\mathfrak{V}_x$. Moreover it is easy to show that the projection $(\mathfrak{s}^\epsilon)^{-1}(c) \to V_x$ is a submersion for any c. Lemma 12.11 follows. □

Lemma 12.12 *Let $\mathcal{U} = (U, \mathcal{E}, \psi, s)$ be a Kuranishi chart of X and M a smooth manifold.*

(1) $(\mathcal{CF}_{\pitchfork\pitchfork 0})_x \subseteq (\mathcal{CF}_{\pitchfork})_x$.
(2) *Suppose $f : U \to M$ is a smooth submersion at x, then,* $(\mathcal{CF}_{\pitchfork\pitchfork 0})_x \subseteq (\mathcal{CF}_{f\pitchfork})_x$.
(3) *Suppose $g : N \to M$ is a smooth map between manifolds and f is transversal to g, then we have* $(\mathcal{CF}_{\pitchfork\pitchfork 0})_x \subseteq (\mathcal{CF}_{f\pitchfork g})_x$.

This is immediate from the definition.

Corollary 12.13

(1) *In the situation of Lemma* 12.12 *(1), the stalk* $(\mathcal{CF}_{\pitchfork 0})_x$ *is nonempty for any* $x \in U$.
(2) *In the situation of Lemma* 12.12 *(2), the stalk* $(\mathcal{CF}_{f\pitchfork})_x$ *is nonempty.*
(3) *In the situation of Lemma* 12.12 *(3), the stalk* $(\mathcal{CF}_{f\pitchfork g})_x$ *is nonempty.*

This is an immediate consequence of Lemmas 12.11 and 12.12.

To prove Proposition 12.2 (2)(3)(4), we need one more result (Proposition 12.14 below.)

Proposition 12.14 *Suppose we are in* Situation 12.3. *We put* $\mathfrak{R} = \{\mathfrak{r}_0\} \cup \mathfrak{R}'$. *Let* $\mathfrak{S}_\mathfrak{r} \in \mathcal{CF}(U_\mathfrak{r})$ *be given for each* $\mathfrak{r} \in \mathfrak{R}$. *Suppose # is one of* $\pitchfork\pitchfork 0$, $\pitchfork 0$, $f \pitchfork$, $f \pitchfork g$. *Then the following holds.*

If $\mathfrak{S}_{\mathfrak{r}_0} \in \mathcal{CF}_{\#}(U_{\mathfrak{r}_0})$ *and* $\mathfrak{S}_\mathfrak{r} \in \mathcal{CF}_{\pitchfork\pitchfork 0}(U_\mathfrak{r})$ *for all* $\mathfrak{r} \in \mathfrak{R}'$, *then*

$$\mathfrak{S} = \sum_{\mathfrak{r}\in\mathfrak{R}} \chi_\mathfrak{r}\mathfrak{S}_\mathfrak{r} \in \mathcal{CF}_{\#}(U).$$

Here for $\# = f \pitchfork$ *we assume that* $f : U \to M$ *is a smooth submersion at* x. *For* $\# = f \pitchfork g$ *we assume that* $g : N \to M$ *is a smooth map between manifolds and* f *is transversal to* g.

Proof To prove Proposition 12.14 we rewrite the strong transversality as follows. Let $\mathcal{S}_x = (W_x, \omega_x, \{\mathfrak{s}_x^\epsilon\})$ be a representative of a germ $\mathcal{CF}_x$ which is defined on an orbifold chart $\mathfrak{V}_x = (V_x, \Gamma_x, E_x, \phi_x, \widehat{\phi}_x)$ of $(U, \mathcal{E})$.

Lemma 12.15 *$\mathcal{S}_x$ is strongly transversal if and only if the derivative*

$$\nabla^W_{(x,\xi)} \mathfrak{s}^\epsilon_x : T_\xi W_x \to T_c E_x$$

in the W_x direction is surjective for all ξ in the support of ω_x. Here $c = \mathfrak{s}^\epsilon_x(o_x, \xi)$.

Proof We consider the following commutative diagram where all the horizontal and vertical lines are exact:

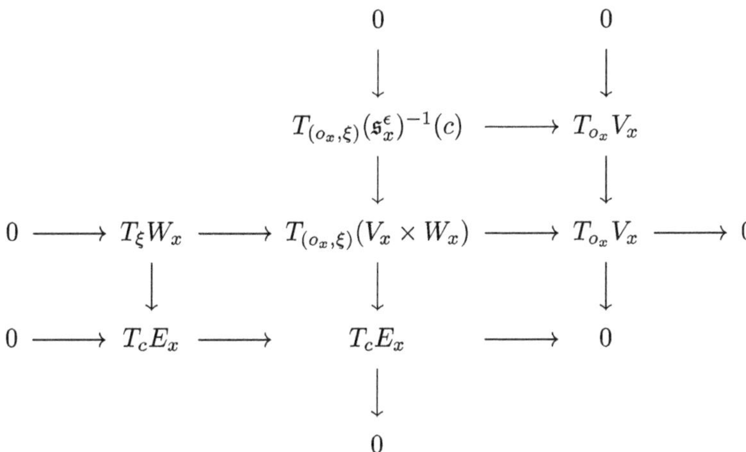

The required strong transversality is nothing but the surjectivity of the horizontal map of the top row : $T_{(o_x,\xi)}(\mathfrak{s}^\epsilon_x)^{-1}(c) \to T_{o_x} V_x$. The map $\nabla^W_{(x,\xi)} \mathfrak{s}^\epsilon_x : T_\xi W_x \to T_c E_x$ is the vertical map in the left most column. The equivalence of surjectivity of two maps is a consequence of simple diagram chase. □

The next lemma is half of the proof of Proposition 12.14.

Lemma 12.16 *Suppose we are in the situation of Proposition* 12.14 *and $\chi_{\mathfrak{r}_1}(x) \neq 0$ for some $\mathfrak{r}_1 \in \mathfrak{R}'$. Then the germ $\mathfrak{S}_x$ of $\mathfrak{S}$ at x is strongly transversal.*

Proof A representative of $\mathfrak{S}_x$ is $(W_x, \omega_x, \mathfrak{s}^\epsilon_x)$ where

$$\mathfrak{s}^\epsilon_x(y, (\xi_\mathfrak{r})_{\mathfrak{r}\in\mathfrak{R}(x)}) = s_x(y) + \sum_{\mathfrak{r}\in\mathfrak{R}(x)} \chi_\mathfrak{r}(\psi_x(y)) g^{-1}_{\mathfrak{r},y}(\mathfrak{s}^\epsilon_\mathfrak{r}(\widetilde{\varphi}_{\mathfrak{r}x}(y), \xi_\mathfrak{r}) - s_\mathfrak{r}(\widetilde{\varphi}_{\mathfrak{r}x}(y))). \tag{12.7}$$

Here $\mathfrak{R}(x) \subseteq \mathfrak{R}$ and $\mathfrak{r}_1 \in \mathfrak{R}(x)$. The derivative of $\mathfrak{s}^\epsilon_x$ in $W_{\mathfrak{r}_1}$ direction is

$$\chi_{\mathfrak{r}_1}(\psi_x(y)) g^{-1}_{\mathfrak{r}_1,y}(\nabla^{W_{\mathfrak{r}_1}} \mathfrak{s}^\epsilon_{\mathfrak{r}_1}|_{(\widetilde{\varphi}_{\mathfrak{r}_1 x}(y), \xi_{\mathfrak{r}_1})}). \tag{12.8}$$

By Lemma 12.15 the derivative $\nabla^{W_{\mathfrak{r}_1}} \mathfrak{s}^\epsilon_{\mathfrak{r}_1}$ is surjective (to $T_c E_x$). Therefore (12.8) is surjective. Therefore by Lemma 12.15 $\mathfrak{S}_x$ is strongly transversal. □

Now we are ready to complete the proof of Proposition 12.14. By Lemma 12.16 the germ $\mathfrak{S}_x$ has the property claimed in Proposition 12.14 unless $\chi_{\mathfrak{r}}(x) = 0$ for all $\mathfrak{r} \in \mathfrak{R}'$. We may also assume that $\mathfrak{s}^\epsilon(x, (\xi_{\mathfrak{r}})_{\mathfrak{r}\in\{\mathfrak{r}_0\}\cup\mathfrak{R}(x)}) = 0$. We consider such a point x. A representative of $\mathfrak{S}_x$ is $(W_x, \omega_x, \mathfrak{s}_x^\epsilon)$ where

$$
\begin{aligned}
&\mathfrak{s}_x^\epsilon(y, (\xi_{\mathfrak{r}})_{\mathfrak{r}\in\{\mathfrak{r}_0\}\cup\mathfrak{R}(x)}) \\
&= s_x(y) + \chi_{\mathfrak{r}_0}(\psi_x(y)) g_{\mathfrak{r}_0,y}^{-1}(\mathfrak{s}_{\mathfrak{r}_0}^\epsilon(\tilde{\varphi}_{\mathfrak{r}_0 x}(y), \xi_{\mathfrak{r}}) - s_{\mathfrak{r}}(\tilde{\varphi}_{\mathfrak{r}x}(y)) \\
&\qquad + \sum_{\mathfrak{r}\in\mathfrak{R}(x)} \chi_{\mathfrak{r}}(\psi_x(y)) g_{\mathfrak{r},y}^{-1}(\mathfrak{s}_{\mathfrak{r}}^\epsilon(\tilde{\varphi}_{\mathfrak{r}x}(y), \xi_{\mathfrak{r}}) - s_{\mathfrak{r}}(\tilde{\varphi}_{\mathfrak{r}x}(y)).
\end{aligned}
\tag{12.9}
$$

Here $\mathfrak{R}(x) \subseteq \mathfrak{R}'$. We remark that $\chi_{\mathfrak{r}_0}(x) = 1$ and takes the maximum there. (Note $\chi_{\mathfrak{r}}$ is a smooth map to $[0, 1]$.) Therefore the first derivative at x of $\chi_{\mathfrak{r}_0}$ is zero. In a similar way we can show that the first derivatives at x of $\chi_{\mathfrak{r}}$ are all zero.

We also remark that $\mathfrak{s}_{\mathfrak{r}_0}^\epsilon(x, \xi_{\mathfrak{r}_0}) = 0$. Therefore

$$
T_{(x,\xi_{\mathfrak{r}_0})}(\mathfrak{s}_{\mathfrak{r}_0}^\epsilon)^{-1}(0) \times \prod_{\mathfrak{r}\in\mathfrak{R}(x)} T_{\xi_{\mathfrak{r}}} W_{\mathfrak{r}} \subseteq T_{(x,(\xi_{\mathfrak{r}})_{\mathfrak{r}\in\{\mathfrak{r}_0\}\cup\mathfrak{R}(x)})}(\mathfrak{s}_x^\epsilon)^{-1}(0). \tag{12.10}
$$

Equation (12.10) implies that if $\mathfrak{S}_{\mathfrak{r}_0}$ has Property (i) at x then $\mathfrak{S} = \sum_{\mathfrak{r}} \chi_{\mathfrak{r}} \mathfrak{S}_{\mathfrak{r}}$ has the same property (i) at x, where $\chi_{\mathfrak{r}_0}(x) = 1$.

This fact together with Lemmas 12.13 and 12.16 imply Proposition 12.14. □

We are now in the position to complete the proof of Proposition 12.2 (2)(3)(4). Let $K \subset U$ be a closed subset and $\mathfrak{S}_K \in \mathcal{CF}(K)$. By definition there exists an open neighborhood $U_{\mathfrak{r}_0}$ of K such that $\mathfrak{S}_K$ is a restriction of $\mathfrak{S}_{\mathfrak{r}_0} \in \mathcal{CF}(U_{\mathfrak{r}_0})$. We take an index set $\mathfrak{R}'$ and an open covering

$$
U = U_{\mathfrak{r}_0} \cup \bigcup_{\mathfrak{r}\in\mathfrak{R}'} U_{\mathfrak{r}} \tag{12.11}
$$

with the following properties:

(a) The covering (12.11) is locally finite.
(b) $\mathcal{CF}_{\pitchfork\pitchfork 0}(U_{\mathfrak{r}}) \neq \emptyset$ for $\mathfrak{r} \in \mathfrak{R}'$.
(c) $K \cap U_{\mathfrak{r}} = \emptyset$ for $\mathfrak{r} \in \mathfrak{R}'$.

The existence of such a covering is a consequence of paracompactness of U and Lemma 12.11.

Let $\mathfrak{S}_{\mathfrak{r}} \in \mathcal{CF}_{\pitchfork\pitchfork 0}(U_{\mathfrak{r}})$ and $\chi_{\mathfrak{r}}$ a partition of unity subordinate to the covering (12.11). We put

$$
\mathfrak{S} = \sum_{\mathfrak{r}\in\{\mathfrak{r}_0\}\cup\mathfrak{R}'} \chi_{\mathfrak{r}} \mathfrak{S}_{\mathfrak{r}}.
$$

Property (c) implies that $\mathfrak{S}$ restricts to $\mathfrak{S}_K$. To prove the softness of $\mathcal{CF}_{\pitchfork 0}$, we may assume $\mathfrak{S}_{\mathfrak{r}_0} \in \mathcal{CF}_{\pitchfork 0}(U_{\mathfrak{r}_0})$. Then by Proposition 12.14, $\mathfrak{S} \in \mathcal{CF}_{\pitchfork 0}(U)$.

The proof of softness of $\mathcal{CF}_{f\pitchfork}$ and of $\mathcal{CF}_{f\pitchfork g}$ is the same. The proof of Proposition 12.2 is complete. □

Remark 12.17 The same argument also proves softness of $\mathcal{CF}_{\pitchfork\pitchfork 0}$.

12.2 Sheaf $\mathcal{CF}_{\mathcal{K}}$ of CF-Perturbations on Hetero-Dimensional Compactum

In Definition 7.20 we defined a sheaf of CF-perturbations on a single Kuranishi chart. In this section we extend the definition to a good coordinate system. Let $\widehat{\mathcal{U}} = ((\mathfrak{P}, \le), \{\mathcal{U}_{\mathfrak{p}} \mid \mathfrak{p} \in \mathfrak{P}\}, \{\Phi_{\mathfrak{p}\mathfrak{q}} \mid \mathfrak{q} \le \mathfrak{p}\})$ be a good coordinate system of $Z \subseteq X$ (Definition 3.15) and $\mathcal{K} = \{\mathcal{K}_{\mathfrak{p}} \mid \mathfrak{p} \in \mathfrak{P}\}$ its support system. Using the equivalence relation Definition 3.15 (7) on the disjoint union $\coprod \mathcal{K}_{\mathfrak{p}}$ we obtain a metric space $|\mathcal{K}|$, which we call a hetero-dimensional compactum. We will define a sheaf of CF-perturbations on $|\mathcal{K}|$.

We recall the following. Suppose Y and $\mathscr{A}$ are topological spaces and $\pi : \mathscr{A} \to Y$ is a continuous map. Then $(\mathscr{A}, \pi)$ determines a sheaf (of set) on Y if the following condition holds.

Condition 12.18 For each $p \in \mathscr{A}$ there exists its neighborhood $U(p)$ in $\mathscr{A}$ such that the restriction of π induces a homeomorphism from $U(p)$ to its image $\subset Y$ which is open.

$\mathscr{A}$ is called the étale space of the sheaf $(\mathscr{A}, \pi)$. We remark that in general $\mathscr{A}$ is not Hausdorff. Such a pair $(\mathscr{A}, \pi)$ determines a presheaf by

$$U \mapsto \mathscr{A}(U) = \{s : U \to \mathscr{A} \mid s \text{ is continuous, } \pi \circ s \text{ is the identity}\}.$$

It has the properties we checked in Proposition 7.22. The germ of this presheaf at p is $\pi^{-1}(p) \subset \mathscr{A}$. (See for example [Go, Chapter II Section 1].)

Let $x \in |\mathcal{K}|$ we put

$$\mathfrak{P}(x) = \{\mathfrak{p} \in \mathfrak{P} \mid x \in \mathcal{K}_{\mathfrak{p}}\}. \tag{12.12}$$

By Definition 3.15 (6), $(\mathfrak{P}(x), \le)$ is a totally ordered set. We denote by $\mathfrak{p}(x)$ the maximum element of $\mathfrak{P}(x)$.

Definition 12.19

(1) We denote by $(\mathcal{CF}_{\mathcal{K}})_x$ the set of all $\mathfrak{S}_x \in \mathcal{CF}_x^{\mathcal{U}_{\mathfrak{p}(x)}}$ (see Notation 7.35) such that

$$\mathfrak{S}_x \in \mathcal{CF}_x^{\mathcal{U}_{\mathfrak{p}} \triangleright \mathcal{U}_{\mathfrak{p}(x)}} \tag{12.13}$$

for any $\mathfrak{p} \in \mathfrak{P}(x)$. (See Definition-Lemma 7.44.) In other words, $(\mathcal{CF}_{\mathcal{K}})_x$ is the set of all germs of CF-perturbations of $U_{\mathfrak{p}(x)}$ at x which is restrictable to all $U_{\mathfrak{p}}$, $\mathfrak{p} \in \mathfrak{P}(x)$.

(2) We define the set $|\mathcal{CF}_{\mathcal{K}}|$ by

$$|\mathcal{CF}_{\mathcal{K}}| = \bigcup_{x \in |\mathcal{K}|} \{x\} \times (\mathcal{CF}_{\mathcal{K}})_x. \tag{12.14}$$

(3) We define the projection $\pi : |\mathcal{CF}_{\mathcal{K}}| \to |\mathcal{K}|$ by sending $(x, \mathfrak{S}_x)$ to x.

(4) If $\widehat{f} : (X, Z; \widehat{\mathcal{U}}) \to Y$ is a strongly smooth map to a manifold Y (resp. and $g : M \to Y$ is a smooth map), we say $\widehat{f}$ is strongly transversal to 0 (resp. to g) with respect to $\mathfrak{S}_x \in (\mathcal{CF}_{\mathcal{K}})_x$ if $f_{\mathfrak{p}}$ is strongly transversal to 0 (resp. to g) with respect to $\Phi^*_{\mathfrak{p}(x)\mathfrak{p}}(\mathfrak{S}_x)$ for any $\mathfrak{p} \in \mathfrak{P}(x)$.

We denote the set of all $\mathfrak{S}_x \in (\mathcal{CF}_{\mathcal{K}})_x$ such that $\widehat{f}$ is strongly transversal to 0 (resp. to g) with respect to $\mathfrak{S}_x$ by $(\mathcal{CF}_{\pitchfork f,\mathcal{K}})_x$ (resp. $(\mathcal{CF}_{g\pitchfork f,\mathcal{K}})_x$).

(5) We define subsets $|\mathcal{CF}_{\pitchfork 0,\mathcal{K}}|$, $|\mathcal{CF}_{\pitchfork f,\mathcal{K}}|$, $|\mathcal{CF}_{g\pitchfork f,\mathcal{K}}|$ of $|\mathcal{CF}_{\mathcal{K}}|$ by using item (4) in an obvious way.

We next topologize these sets. Let $\mathfrak{S}_x \in (\mathcal{CF}_{\mathcal{K}})_x$. We take its representative $(W_x, \omega_x, \{\mathfrak{s}^{\epsilon}_x \mid \epsilon\})$ on an orbifold chart $(V_x, \Gamma_x, E_x, \phi_x, \hat{\phi}_x)$ of $\mathcal{E}_{\mathfrak{p}(x)} \to U_{\mathfrak{p}(x)}$. We take V_x so small that $\phi_x(V_x) \subset U_{\mathfrak{p}(x)}$ is open in $|\mathcal{K}|$. Let $y \in |\mathcal{K}| \cap V_x$. By the above choice $\mathfrak{p}(y) \le \mathfrak{p}(x)$. (Note $\mathfrak{p}(y) \ne \mathfrak{p}(x)$ in general. See Fig. 12.1.) We may choose V_x so small that

$$[\Phi^*_{\mathfrak{p}(x)\mathfrak{p}(y)}(W_x, \omega_x, \{\mathfrak{s}^{\epsilon}_x \mid \epsilon\})] \in \mathcal{CF}^{\mathcal{U}_{\mathfrak{p}(y)}}_y \tag{12.15}$$

is defined. By an abuse of notation we write the left hand side as $\Phi^*_{\mathfrak{p}(x)\mathfrak{p}(y)}\mathfrak{S}_x$. If we take the representative $(W_x, \omega_x, \{\mathfrak{s}^{\epsilon}_x \mid \epsilon\})$ of $\mathfrak{S}_x$ then the left hand side of (12.15) is induced by a representative of $\mathfrak{S}_x$ for $y \in V_x$. By Definition-Lemma 7.44 (4), we have

$$\Phi^*_{\mathfrak{p}(x)\mathfrak{p}(y)}\mathfrak{S}_x \in (\mathcal{CF}_{\mathcal{K}})_y \tag{12.16}$$

if y is sufficiently close to x.

Moreover, if # is one of $\pitchfork 0$, $\pitchfork f$, $g \pitchfork f$ then $\mathfrak{S}_x \in (\mathcal{CF}_{\#,\mathcal{K}})_x$ implies $\Phi^*_{\mathfrak{p}(x)\mathfrak{p}(y)}\mathfrak{S}_x \in (\mathcal{CF}_{\#,\mathcal{K}})_y$ if y is sufficiently close to x.

We thus obtained a neighborhood U_x of x in $|\mathcal{K}|$ and a map $\mathfrak{I}_x : U_x \to |\mathcal{CF}_{\mathcal{K}}|$ by:

$$\mathfrak{I}_x(y) = \Phi^*_{\mathfrak{p}(x)\mathfrak{p}(y)}\mathfrak{S}_x. \tag{12.17}$$

This map is injective. When we change the representative $(W_x, \omega_x, \{\mathfrak{s}^{\epsilon}_x \mid \epsilon\})$ the map $\mathfrak{I}_x$ does not change on a small neighborhood of x.

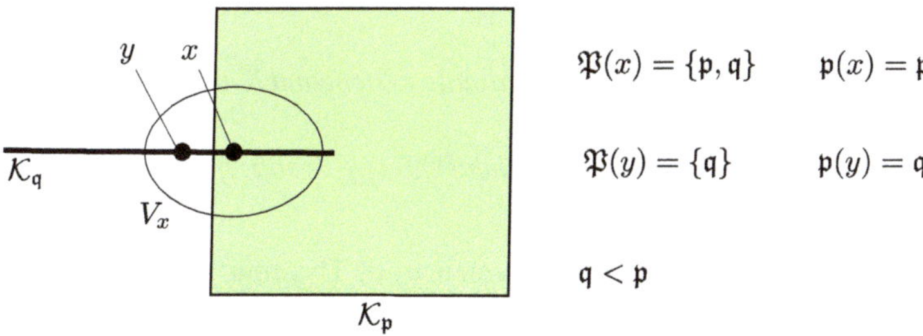

Fig. 12.1 $\mathfrak{p}(x) \neq \mathfrak{p}(y)$

Definition 12.20 We define a topology on $|\mathcal{CF}_{\mathcal{K}}|$ as follows. A subset $O \subset |\mathcal{CF}_{\mathcal{K}}|$ is open if for each $y \in O$ there exists a neighborhood O' of $x = \pi(y) \in |\mathcal{K}|$ such that $\mathfrak{I}_x(O') \subset O$.

The topologies of $|\mathcal{CF}_{\pitchfork 0,\mathcal{K}}|$, $|\mathcal{CF}_{\pitchfork f,\mathcal{K}}|$, $|\mathcal{CF}_{g\pitchfork f,\mathcal{K}}|$ are defined in the same way.

It is easy to check the axiom of topology.

Lemma 12.21 *$\pi : |\mathcal{CF}_{\mathcal{K}}| \to |\mathcal{K}|$ satisfies Condition* 12.18.
The same holds for $|\mathcal{CF}_{\pitchfork 0,\mathcal{K}}|$, $|\mathcal{CF}_{\pitchfork f,\mathcal{K}}|$, $|\mathcal{CF}_{g\pitchfork f,\mathcal{K}}|$.

The proof is standard and is left to the reader. We have thus defined a sheaf of sets $\mathcal{CF}_{\mathcal{K}}$ on $|\mathcal{K}|$ and its subsheaves $\mathcal{CF}_{\pitchfork 0,\mathcal{K}}$, $\mathcal{CF}_{\pitchfork f,\mathcal{K}}$, $\mathcal{CF}_{g\pitchfork f,\mathcal{K}}$ on $|\mathcal{K}|$.

Remark 12.22 The definition of the sheaf $\mathcal{CF}_{\mathcal{K}}$ involves not only the data of the support system $\mathcal{K}$ but also a part of data of the good coordinate system $\widehat{\mathcal{U}}$.

Lemma 12.23 *The set of global continuous sections $\mathcal{CF}_{\mathcal{K}}(|\mathcal{K}|)$ of the sheaf $\mathcal{CF}_{\mathcal{K}}$ is canonically identified with the set of CF-perturbations on $|\mathcal{K}|$.*

A global continuous section of $\mathcal{CF}_{\mathcal{K}}$ is one of the subsheaf $\mathcal{CF}_{\pitchfork 0,\mathcal{K}}$ (resp. $\mathcal{CF}_{\pitchfork f,\mathcal{K}}$, $\mathcal{CF}_{g\pitchfork f,\mathcal{K}}$ on $|\mathcal{K}|$) if the corresponding CF-perturbation is transversal to 0 *(resp. f is strongly transversal to* 0 *with respect to it, f is strongly transversal to g with respect to it.)*

Proof Let $\mathfrak{S}$ be a global section of $\mathcal{CF}_{\mathcal{K}}$. For $x \in \mathcal{K}_{\mathfrak{p}}$ we obtain

$$\Phi^*_{\mathfrak{p}(x)\mathfrak{p}}\mathfrak{S}_x, \tag{12.18}$$

where $\mathfrak{S}_x$ is the germ of $\mathfrak{S}$ at x. Equation (12.18) is defined, since $\mathfrak{p} \leq \mathfrak{p}(x)$. It is easy to see that $x \mapsto \Phi^*_{\mathfrak{p}(x)\mathfrak{p}}\mathfrak{S}_x$ defines an element of $\mathcal{CF}^{\mathcal{U}_{\mathfrak{p}}}(\mathcal{K}_{\mathfrak{p}})$ and hence a CF-perturbation on $\mathcal{K}_{\mathfrak{p}}$, which we denote by $\mathfrak{S}_{\mathfrak{p}}$. For $\mathfrak{q} \leq \mathfrak{p}$ the equality

$$\Phi^*_{\mathfrak{p}\mathfrak{q}}\mathfrak{S}_{\mathfrak{p}} = \mathfrak{S}_{\mathfrak{q}}$$

is immediate from the definition and Definition-Lemma 7.46 (2). We thus obtain a CF-perturbation on $\mathcal{K}$. The construction of the opposite direction is similar.

The second half of the lemma is immediate from the definitions. □

Now the main result of this chapter is:

Theorem 12.24 *Let $\widehat{\mathcal{U}}$ be a good coordinate system and $\mathcal{K}$ its support system. Then, the sheaves $\mathcal{CF}_{\mathcal{K}}$, $\mathcal{CF}_{\pitchfork 0,\mathcal{K}}$ are soft.*

If $\widehat{f}$ is weakly submersive then the sheaf $\mathcal{CF}_{\pitchfork f,\mathcal{K}}$ is soft. If $\widehat{f}$ is weakly transversal to g then $\mathcal{CF}_{g \pitchfork f,\mathcal{K}}$ is soft.

Theorem 7.51 are immediate consequences of Theorem 12.24.

Proof The proof is similar to the proof of Proposition 12.2. The next lemma is a replacement of Lemma 12.11.

Lemma 12.25 *For any $x \in |\mathcal{K}|$ there exists $\mathfrak{S}_x \in (\mathcal{CF}_{\mathcal{K}})_x$ such that the following holds in addition.*

Let $\mathfrak{p}_-(x)$ be the minimum element of $\mathfrak{P}(x)$. We consider the restriction $\mathfrak{S}_x|_{U_{\mathfrak{p}_-(x)}}$ of $\mathfrak{S}_x$ to $\mathcal{K}_{\mathfrak{p}_-(x)}$. Note $\mathfrak{S}_x|_{U_{\mathfrak{p}_-(x)}} \in (\mathcal{CF}^{\mathcal{U}_{\mathfrak{p}_-(x)}}_{\pitchfork 0})_x$. We require

(*)

$$\mathfrak{S}_x|_{U_{\mathfrak{p}_-(x)}} \in (\mathcal{CF}^{\mathcal{U}_{\mathfrak{p}_-(x)}}_{\pitchfork\pitchfork 0})_x,$$

that is, $\mathfrak{S}_x|_{U_{\mathfrak{p}_-(x)}}$ is strongly transversal to 0.

Remark 12.26 Suppose $\mathfrak{p} \in \mathfrak{P}(x)$, $\mathfrak{p} > \mathfrak{p}_-(x)$. Then it is impossible that the restriction $\mathfrak{S}_x|_{U_\mathfrak{p}}$ to $U_\mathfrak{p}$ is strongly transversal to 0. In fact, let $(W_x, \omega_x, \{\mathfrak{s}^\epsilon_x \mid \epsilon\})$ be a representative of $\mathfrak{S}_x$ and $(V_\mathfrak{p}, \Gamma_\mathfrak{p}, \phi_\mathfrak{p}, E_\mathfrak{p})$ (resp. $(V_{\mathfrak{p}_-(x)}, \Gamma_\mathfrak{p}, \phi_{\mathfrak{p}_-(x)}, E_{\mathfrak{p}_-(x)})$ a coordinate of $\mathcal{U}_\mathfrak{p}$ (resp. $\mathcal{U}_{\mathfrak{p}(x)}$) at x. By the assumption (12.13) the map $\mathfrak{s}^\epsilon_x$ sends $V_{\mathfrak{p}(x)} \times W_x$ to $E_{\mathfrak{p}_-(x)}$. In particular, $\mathfrak{s}^\epsilon_x(x, w) \in E_{\mathfrak{p}_-(x)}$. Therefore it is impossible that

$$D\mathfrak{s}^\epsilon_x|_{T_w W_x} : T_w W_x \to T_{\mathfrak{s}^\epsilon_x(x,w)} E_\mathfrak{p}$$

is surjective.

Proof of Lemma 12.25 Let $\mathfrak{P}(x) = \{\mathfrak{p}_1, \mathfrak{p}_2, \ldots, \mathfrak{p}_m\}$ with $\mathfrak{p}_- = \mathfrak{p}_1 < \mathfrak{p}_2 < \cdots < \mathfrak{p}_m = \mathfrak{p}(x)$. Lemma 12.11 implies that there exists $\mathfrak{S}_1 \in (\mathcal{CF}^{\mathcal{U}_{\mathfrak{p}_1}}_{\pitchfork\pitchfork 0})_x$.

We will construct, by induction on k, germs $\mathfrak{S}_k \in (\mathcal{CF}^{\mathcal{U}_{\mathfrak{p}_k}}_{\pitchfork 0})_x$ such that $\mathfrak{S}_k \in \mathcal{CF}^{\mathcal{U}_{\mathfrak{p}_i} \triangleright \mathcal{U}_{\mathfrak{p}_k}}$ for all $i < k$.

Suppose we are given $\mathfrak{S}_k = (W, \omega, \mathfrak{s}^\epsilon_k)$. We take neighborhoods $U_k \subset U_{\mathfrak{p}_k}$ and $U_{k+1} \subset U_{\mathfrak{p}_{k+1}}$ of x for which there exists a projection $\pi : U_{k+1} \to U_k$ such that $\pi \circ \varphi_{\mathfrak{p}_{k+1}\mathfrak{p}_k}$ is the identity map. Moreover we may assume that there exists an embedding of vector bundles $I_k : \pi^*\mathcal{E}_{\mathfrak{p}_k} \to \mathcal{E}_{\mathfrak{p}_{k+1}}$ so that its restriction to $\varphi_{\mathfrak{p}_{k+1}\mathfrak{p}_k}(U_k)$ coincides with $\hat{\varphi}_{\mathfrak{p}_{k+1}\mathfrak{p}_k}$. The existence of such π and I_k is obvious, since we need them only in a small neighborhood of p. Now we put

$$\mathfrak{s}^\epsilon_{k+1}(x, w) = s_{\mathfrak{p}_{k+1}}(x) + I_k(\mathfrak{s}^\epsilon_k(\pi(x), w) - s_{\mathfrak{p}_k}(x)).$$

Here $s_{\mathfrak{p}_{k+1}}$ and $s_{\mathfrak{p}_k}$ are Kuranishi maps. Then $\mathfrak{S}_{k+1} = (W, \omega, \mathfrak{s}^{\epsilon}_{k+1}) \in (\mathcal{CF}^{\mathcal{U}_{\mathfrak{p}_{k+1}}})_x$. It is easy to see $\Phi^*_{\mathfrak{p}_{k+1}\mathfrak{p}_k}(\mathfrak{S}_{k+1}) = \mathfrak{S}_k$. Therefore the induction hypothesis and Lemma 7.46 (1) imply $\mathfrak{S}_{k+1} \in \mathcal{CF}^{\mathcal{U}_{\mathfrak{p}_i} \triangleright \mathcal{U}_{\mathfrak{p}_{k+1}}}$ for all $i < k+1$.

The condition (*) is satisfied since $\mathfrak{S}_1$ is strongly transversal to 0 at x. □

Lemma 12.27 *In the situation of Lemma* 12.25 *the following holds:*

(1) $\mathfrak{S}_x \in (\mathcal{CF}_{\pitchfork 0,\mathcal{K}})_x$.
(2) *If* $\widehat{f}$ *is weakly submersive then* $\mathfrak{S}_x \in (\mathcal{CF}_{\pitchfork f,\mathcal{K}})_x$.
(3) *If* $\widehat{f}$ *is weakly transversal to* g *then* $\mathfrak{S}_x \in (\mathcal{CF}_{g \pitchfork f,\mathcal{K}})_x$.

Proof We use the notation introduced during the proof of Lemma 12.25. For $1 < k \le m$ we consider the next commutative diagram:

$$\begin{array}{ccccccccc}
0 & \longrightarrow & T_{(o_x,\xi)}(V_1 \times W) & \longrightarrow & T_{(o_x,\xi)}(V_k \times W) & \longrightarrow & \dfrac{T_{o_x}V_k}{T_{o_x}V_1} & \longrightarrow & 0 \\
 & & \downarrow D_{o_x,\xi}\mathfrak{s}^{\epsilon}_1 & & \downarrow D_{o_x,\xi}\mathfrak{s}^{\epsilon}_k & & \downarrow & & \\
0 & \longrightarrow & T_cE_{\mathfrak{p}_-(x)} & \longrightarrow & T_cE_{\mathfrak{p}_k} & \longrightarrow & \dfrac{T_cE_{\mathfrak{p}_k}}{T_cE_{\mathfrak{p}_-(x)}} & \longrightarrow & 0
\end{array} \tag{12.19}$$

The horizontal lines are exact. The first vertical arrow is surjective by (*). The third vertical line is an isomorphism by Definition 3.2 (5). Therefore the second vertical line is surjective. This proves (1).

To prove (2) we replace Diagram (12.19) with the next diagram and discuss it in the same way.

$$\begin{array}{ccccccccc}
0 & \longrightarrow & T_{(o_x,\xi)}(V_1 \times W) & \longrightarrow & T_{(o_x,\xi)}(V_k \times W) & \longrightarrow & \dfrac{T_{o_x}V_k}{T_{o_x}V_1} & \longrightarrow & 0 \\
 & & \downarrow D_{o_x,\xi}\mathfrak{s}^{\epsilon}_1 \times D_x f_{\mathfrak{p}_-(x)} & & \downarrow D_{o_x,\xi} \times D_x f_{\mathfrak{p}_k(x)}\mathfrak{s}^{\epsilon}_k & & \downarrow & & \\
0 & \longrightarrow & T_cE_{\mathfrak{p}_-(x)} \oplus T_{f(x)}Y & \longrightarrow & T_cE_{\mathfrak{p}_k} \oplus T_{f(x)}Y & \longrightarrow & \dfrac{T_cE_{\mathfrak{p}_k}}{T_cE_{\mathfrak{p}_-(x)}} & \longrightarrow & 0
\end{array}$$

The proof of (3) is similar. □

We take another support system $\mathcal{K}^+$ of $\widehat{\mathcal{U}}$ such that $\mathcal{K} < \mathcal{K}^+$. (Definition 5.23 (2).) Let # be one of $\pitchfork 0$, $\pitchfork f$, $g \pitchfork f$. Let K be a compact set and $\mathfrak{S}_K \in \mathcal{CF}_{\#,\mathcal{K}}(K)$. There exists an open neighborhood $U_{\mathfrak{r}_0}$ of K in $|\mathcal{K}^+|$ such that $\mathfrak{S}_K$ is a restriction of $\mathfrak{S}_{\mathfrak{r}_0} \in \mathcal{CF}_{\#,\mathcal{K}}(U_{\mathfrak{r}_0} \cap |\mathcal{K}|)$. For $x \in |\mathcal{K}| \setminus K$ we take $\mathfrak{S}_x$ as in Lemma 12.27 and its representative, $(W_x, \omega_x, \{\mathfrak{s}^{\epsilon}_x\})$. Here the section $\mathfrak{s}^{\epsilon}_x$ is defined on $V_x \times W_x$ and (V_x, Γ_x, ϕ_x) is an orbifold chart of $U_{\mathfrak{p}(x)}$ at x. Note that $U_x = \phi_x(V_x) \subset U_{\mathfrak{p}(x)}$ is an open neighborhood of x in $|\mathcal{K}^+|$. Note also that Lemma 12.27 claims certain transversality at $(o_x, \xi) \in V_x \times W_x$. We take V_x small so that a similar transversality holds on all $V_x \times W_x$. We stratify V_x and W_x as in the proof of Lemma 12.25 and write the strata as $V_{x,k}$, $W_{x,k}$. Now we consider the next diagram at $y \in V_{x,k}$:

$$\begin{array}{ccccccccc}
0 & \longrightarrow & T_\xi(W_x) & \longrightarrow & T_{(y,\xi)}(V_{x,k}\times W_x) & \longrightarrow & T_yV_{x,k} & \longrightarrow & 0\\
 & & \downarrow D_{(y,\xi)}\mathfrak{s}^\epsilon_1 & & \downarrow D_{(y,\xi)}\mathfrak{s}^\epsilon_k & & \downarrow & & \\
0 & \longrightarrow & I'_k(\pi(y),E_{\mathfrak{p}_-}) & \longrightarrow & T_cE_{\mathfrak{p}_k} & \longrightarrow & \frac{T_cE_{\mathfrak{p}_k}}{I'_k(\pi(y),E_{\mathfrak{p}_-})} & \longrightarrow & 0
\end{array} \tag{12.20}$$

Here $\pi : V_{x,k} \to V_{x,1}$ is a Γ_x equivariant projection and $I'_k = I_{k-1} \circ \cdots \circ I_1 : \pi^*\mathcal{E}_1 \to \mathcal{E}_k$ is the composition of the embedding appearing in the proof of Lemma 12.25.

The first vertical arrow is surjective. By taking V_x small the following $(\star)$ holds for $\delta_x > 0$, which may depend on x:

$(\star)$ If the C^1 distance between $\mathfrak{s}^\epsilon_k$ and the Kuranishi map is smaller than δ_x on $V_x \times W_x$ then the third vertical arrow $T_yV_{x,k} \to \frac{T_cE_{\mathfrak{p}_k}}{I'_k(\pi(y),E_{\mathfrak{p}_-})}$ is surjective.[1]

Now we take finitely many points $x_\mathfrak{r}$, $\mathfrak{r} \in \mathfrak{R}'$ in $|\mathcal{K}| \setminus K$ such that $U_\mathfrak{r} = \phi_{x_\mathfrak{r}}(V_{x_\mathfrak{r}}) \subset U_{\mathfrak{p}(x_\mathfrak{r})}$ satisfies:

$$U_{\mathfrak{r}_0} \cup \bigcup_{\mathfrak{r}\in\mathfrak{R}'} U_\mathfrak{r} \supset |\mathcal{K}|. \tag{12.21}$$

Let $\mathfrak{S}_\mathfrak{r} = (W_\mathfrak{r}, \omega_\mathfrak{r}, \{\mathfrak{s}^\epsilon_\mathfrak{r}\}) = (W_{x_\mathfrak{r}}, \omega_{x_\mathfrak{r}}, \{\mathfrak{s}^\epsilon_{x_\mathfrak{r}}\})$ be the representative as above. We may assume:

(a) The covering (12.21) is a finite covering.
(b) The first vertical line of (12.20) is surjective for $y \in U_\mathfrak{r} \times W_\mathfrak{r}$.
(c) $(\star)$ holds.
(d) $K \cap U_\mathfrak{r} = \emptyset$ for each $\mathfrak{r} \in \mathfrak{R}'$.

Lemma 12.28 *There exists a partition of unity subordinate to the open covering* (12.21). *Namely there exist strongly smooth functions* $\chi_\mathfrak{r} : |\mathcal{K}^+| \to [0,1]$ *for* $\mathfrak{r} \in \{\mathfrak{r}_0\} \cup \mathfrak{R}'$ *such that:*

(1) *The support of* $\chi_\mathfrak{r}$ *is a compact subset of* $U_\mathfrak{r}$.
(2) $\displaystyle\sum_{\mathfrak{r}\in\{\mathfrak{r}_0\}\cup\mathfrak{R}'} \chi_\mathfrak{r} = 1$ *on* $|\mathcal{K}|$.

Proof Note that the partition of unity appearing in Lemma 12.28 is slightly different from the one in Definition 7.65. In fact (12.21) is an open covering. We can prove this lemma in the same way as the standard proof of the existence of partition of unity using Lemma 7.67. □

We continue the proof of Theorem 12.24. Let $x \in |\mathcal{K}|$. We consider the set $\mathfrak{R}(x) = \{\mathfrak{r} \in \{\mathfrak{r}_0\} \cup \mathfrak{R}' \mid \chi_\mathfrak{r}(x) \neq 0\}$. Then $\Phi^*_{\mathfrak{p}(\mathfrak{r})\mathfrak{p}(x)}\mathfrak{S}_\mathfrak{r}$ is defined by the definition

[1] We take a Riemannian metric of V_x etc. (depending on x) so that the C^1 distance makes sense.

of $\mathcal{CF}_{\mathcal{K}}$. Let $(W_{\mathfrak{r}}, \omega_{\mathfrak{r}}, \mathfrak{s}_{\mathfrak{r}}^{\epsilon})$ be a representative of $\Phi_{\mathfrak{p}(\mathfrak{r})\mathfrak{p}(x)}^{*}\mathfrak{S}_{\mathfrak{r}}$. Then we define $W_x = \prod_{\mathfrak{r}\in\mathfrak{R}(x)} W_{\mathfrak{r}}$, $\omega_x = \prod_{\mathfrak{r}\in\mathfrak{R}(x)} \omega_{\mathfrak{r}}$ and

$$\begin{aligned}\mathfrak{s}_x^{\epsilon}(y, (\xi_{\mathfrak{r}})_{\mathfrak{r}\in\mathfrak{R}(x)}) = \; & s_x(y) + \chi_{\mathfrak{r}_0}(y)(\mathfrak{s}_{\mathfrak{r}_0}^{\epsilon}(y, \xi_0) - s_x(y)) \\ & + \sigma \sum_{\mathfrak{r}\in\mathfrak{R}(x)} \chi_{\mathfrak{r}}(y)(\mathfrak{s}_{\mathfrak{r}}^{\epsilon}(y, \xi_{\mathfrak{r}}) - s_x(y)). \end{aligned} \tag{12.22}$$

Here $\mathfrak{s}_{\mathfrak{r}_0}^{\epsilon_0}$ in the second term is the given CF-perturbation in a neighborhood of K (See Formula (12.9)), and $\sigma > 0$ is a sufficiently small positive constant determined later. When $\sigma = 1$, this is the same formula as (12.9) in the case where $\widehat{\mathcal{U}}$ consists of a single Kuranishi chart. In the same way as Lemma 12.5, we can show that $x \mapsto [W_x, \omega_x, \mathfrak{s}_x^{\epsilon}]$ defines a section of the sheaf $\mathcal{CF}_{\mathcal{K}}$.

Now we prove that $x \mapsto [W_x, \omega_x, \mathfrak{s}_x^{\epsilon}]$ is a section of $\mathcal{CF}_{\pitchfork 0,\mathcal{K}}$. In the case where $\chi_{\mathfrak{r}}(x) = 0$ for all $\mathfrak{r} \in \mathfrak{R}'$ the proof is the same as the proof of Lemma 12.5. Suppose $\chi_{\mathfrak{r}}(x) \neq 0$ for $\mathfrak{r} \in \mathfrak{R}'$. We consider the diagram

$$\begin{array}{ccccccccc} 0 & \longrightarrow & T_{\xi}(W_x) & \longrightarrow & T_{(y,\xi)}(V_{x,k} \times W_x) & \longrightarrow & T_y V_{x,k} & \longrightarrow & 0 \\ & & \downarrow D_{(y,\xi)}\mathfrak{s}_x^{\epsilon} & & \downarrow D_{(y,\xi)}\mathfrak{s}_x^{\epsilon} & & \downarrow & & \\ 0 & \longrightarrow & I_k'(\pi(y), E_{\mathfrak{p}_-}) & \longrightarrow & T_c E_{\mathfrak{p}_k} & \longrightarrow & \dfrac{T_c E_{\mathfrak{p}_k}}{I_k'(\pi(y), E_{\mathfrak{p}_-})} & \longrightarrow & 0 \end{array} \tag{12.23}$$

The first vertical arrow is surjective. In fact its restriction to $T_{\xi_{\mathfrak{r}}}(W_{\mathfrak{r}})$ is surjective. We take σ so small that the third vertical arrow is surjective. Therefore $x \mapsto [W_x, \omega_x, \mathfrak{s}_x^{\epsilon}]$ is transversal to 0 at x also.

In the same way, we can prove that $x \mapsto [W_x, \omega_x, \mathfrak{s}_x^{\epsilon}]$ is a section of $\mathcal{CF}_{\pitchfork f,\mathcal{K}}$ if $\widehat{f}$ is weakly submersive and that $x \mapsto [W_x, \omega_x, \mathfrak{s}_x^{\epsilon}]$ is a section of $\mathcal{CF}_{g\pitchfork f,\mathcal{K}}$ if $\widehat{f}$ is weakly transversal to g. In fact for the former we replace Diagram (12.23) by

$$\begin{array}{ccccccccc} 0 & \longrightarrow & T_{\xi}(W_x) & \longrightarrow & T_{(y,\xi)}(V_{x,k} \times W_x) & \longrightarrow & T_y V_{x,k} & \longrightarrow & 0 \\ & & \downarrow D_{(y,\xi)}\mathfrak{s}_x^{\epsilon} & & \downarrow D_{(y,\xi)}\mathfrak{s}_x^{\epsilon} \oplus D_y f_{\mathfrak{p}(x)} & & \downarrow & & \\ 0 & \longrightarrow & I_k'(\pi(y), E_{\mathfrak{p}_-}) & \longrightarrow & T_c E_{\mathfrak{p}_k} \oplus T_z(Y) & \longrightarrow & \dfrac{T_c E_{\mathfrak{p}_k}}{I_k'(\pi(y), E_{\mathfrak{p}_-})} \oplus T_z(Y) & \longrightarrow & 0 \end{array}$$

where $z = f_{\mathfrak{p}(x)}(y)$ and for the latter we replace it by

$$\begin{array}{ccccccccc} 0 & \longrightarrow & T_{\xi}(W_x) & \longrightarrow & T_{(y,\xi)}(V_{x,k} \times W_x) \oplus T_a M & \longrightarrow & T_y V_{x,k} \oplus T_a M & \longrightarrow & 0 \\ & & \downarrow D_{(y,\xi)}\mathfrak{s}_x^{\epsilon} \oplus D_y f & & \downarrow (D_{(y,\xi)}\mathfrak{s}_x^{\epsilon} \oplus D_y f_{\mathfrak{p}(x)}, 0 \oplus d_a g) & & \downarrow & & \\ 0 & \longrightarrow & I_k'(\pi(y), E_{\mathfrak{p}_-}) & \longrightarrow & T_c E_{\mathfrak{p}_k} \oplus T_z(Y) & \longrightarrow & \dfrac{T_c E_{\mathfrak{p}_k}}{I_k'(\pi(y), E_{\mathfrak{p}_-})} \oplus T_z(Y) & \longrightarrow & 0 \end{array}$$

where $g(a) = z$. We also modify ($\star$) appropriately. The proof of Theorem 12.24 is complete. $\square$

We next use Theorem 12.24 to prove the relative version Proposition 7.59 of Theorem 7.51. We need a few lemmas.

Lemma 12.29 *Suppose we are in the situation of Definition* 7.56. *We assume that* $\mathcal{K}^1$ *is compatible with* $\mathcal{K}^2$. *Then*

$$|\mathcal{K}^1| \subset |\mathcal{K}^2|. \tag{12.24}$$

Moreover the sheaf $\mathcal{CF}_{\mathcal{K}_1}$ *is a subsheaf of* $\mathcal{CF}_{\mathcal{K}_2}|_{|\mathcal{K}^1|}$.

Proof Equation (12.24) is an immediate consequence of the definition. For $x \in |\mathcal{K}^1|$ we define

$$\mathfrak{P}_i(x) = \{\mathfrak{p} \in \mathfrak{P}_i \mid x \in \mathcal{K}^i_{\mathfrak{p}}\}.$$

Here $\mathfrak{P}_i$ is the partially ordered set appearing in the definition of the good coordinate system $\widehat{\mathcal{U}_i}$. By Definitions 7.52 and 7.56, $\mathfrak{P}_1 \subseteq \mathfrak{P}_2$. The compatibility of $\mathcal{K}^1$ with $\mathcal{K}^2$ implies that $\mathfrak{P}_1(x) \subseteq \mathfrak{P}_2(x)$. This fact easily implies that $\mathcal{CF}_{\mathcal{K}_1}$ is a subsheaf of $\mathcal{CF}_{\mathcal{K}_2}|_{|\mathcal{K}^1|}$. □

Lemma 12.30 *Let* $\widehat{\mathcal{U}} = (\{\mathcal{U}_{\mathfrak{p}} \mid \mathfrak{p} \in \mathfrak{P}\}, \{\Phi_{\mathfrak{p}\mathfrak{q}}\})$ *be a good coordinate system of* $Z \subseteq X$ *and* $\mathcal{K}$ *its support system. Let* $\widehat{\mathcal{U}_0}$ *be an open substructure of* $\widehat{\mathcal{U}}$. *Assume*

$$\mathcal{K}_{\mathfrak{p}} \cap \psi_{\mathfrak{p}}^{-1}(0) \subset U_{0,\mathfrak{p}} \tag{12.25}$$

for all $\mathfrak{p} \in \mathfrak{P}$. *Then there exists a compact neighborhood* $\mathfrak{U}(Z)$ *of* Z *in* $|\mathcal{K}|$ *such that*

$$\mathfrak{U}(Z) \cap \mathcal{K}_{\mathfrak{p}} \subseteq U_{0,\mathfrak{p}} \cap \mathcal{K}_{\mathfrak{p}}$$

for all $\mathfrak{p} \in \mathfrak{P}$.

Proof If the lemma is false there exist $\epsilon_n \to 0$ and $x_n \in U_{\mathfrak{p}}$ such that $d(x_n, \mathfrak{U}(Z) \cap \mathcal{K}_{\mathfrak{p}}) < \epsilon_n$ but $x_n \notin U_{0,\mathfrak{p}} \cap \mathcal{K}_{\mathfrak{p}}$. By taking a subsequence if necessary we may assume that x_n converges to $x \in \mathcal{K}_{\mathfrak{p}} \cap Z \subseteq \mathcal{K}_{\mathfrak{p}} \cap \psi_{\mathfrak{p}}^{-1}(0)$. Therefore by assumption $x \in U_{0,\mathfrak{p}}$. Hence $x_n \in \mathcal{K}_{\mathfrak{p}} \cap U_{0,\mathfrak{p}}$ for large n. This is a contradiction. □

Proof of Proposition 7.59 We use the notation of Proposition 7.59. By Proposition 7.54 (4) we can apply Lemma 12.30 to obtain a compact neighborhood $\mathfrak{U}(Z_1)$ of $Z_1 \subseteq |\mathcal{K}_1|$ such that

$$\mathfrak{U}(Z_1) \cap \mathcal{K}_{\mathfrak{p}} \subseteq U_{0,\mathfrak{p}} \cap \mathcal{K}^1_{\mathfrak{p}} \tag{12.26}$$

for any $\mathfrak{p} \in \mathfrak{P}_1$. We define a support system $\mathcal{K}^1_0$ of $\widehat{\mathcal{U}^1_0}$ by

$$\mathcal{K}^1_{0,\mathfrak{p}} = \mathfrak{U}(Z_1) \cap \mathcal{K}^1_{\mathfrak{p}} \subset U_{0,\mathfrak{p}}. \tag{12.27}$$

Then $|\mathcal{K}_0^1| = |\mathcal{K}^1| \cap \mathfrak{U}(Z_1)$. Moreover, for $x \in |\mathcal{K}_0^1|$ we have $\{\mathfrak{p} \in \mathfrak{P}_1 \mid x \in \mathcal{K}_{0,\mathfrak{p}}^1\} = \{\mathfrak{p} \in \mathfrak{P}_1 \mid x \in \mathcal{K}_{\mathfrak{p}}^1\}$. Therefore $\mathcal{CF}_{\mathcal{K}_1^0}(|\mathcal{K}_0^1|) = \mathcal{CF}_{\mathcal{K}_1}(|\mathcal{K}_0^1|)$. On the other hand, by Lemma 12.29 $\mathcal{CF}_{\mathcal{K}_1^0} \subseteq \mathcal{CF}_{\mathcal{K}_2}|_{|\mathcal{K}_0^1|}$. Therefore $\widehat{\mathfrak{S}^1} \in \mathcal{CF}_{\mathcal{K}_1}(|\mathcal{K}^1|)$ induces an element $(\widehat{\mathfrak{S}^1})' \in \mathcal{CF}_{\mathcal{K}_2}(|\mathcal{K}_0^1|)$. Now we use Theorem 12.24 to obtain $\widehat{\mathfrak{S}^2} \in \mathcal{CF}_{\mathcal{K}_2}(|\mathcal{K}^2|)$ whose restriction is $(\widehat{\mathfrak{S}^1})'$. It is easy to see that, replacing $\widehat{\mathcal{U}_0^1}$ by its open substructure $\widehat{\mathcal{U}_{00}^1}$, $\widehat{\mathfrak{S}^1}$ extends $\widehat{\mathfrak{S}^2}$. In fact we choose the Kuranishi charts constituting $\widehat{\mathcal{U}_{00}^1}$ as $\mathcal{U}_{0,\mathfrak{p}}^1|_{U_{00,\mathfrak{p}}^1}$. Here the open subset $U_{00,\mathfrak{p}}^1 \subset U_{0,\mathfrak{p}}^1$ is defined by $U_{00,\mathfrak{p}}^1 = U_{0,\mathfrak{p}}^1 \cap B_\epsilon(\mathfrak{U}(Z_1))$. We take ϵ small so that $B_\epsilon(\mathfrak{U}(Z_1)) \cap \mathcal{K}_{\mathfrak{p}}^1 \subset U_{0,\mathfrak{p}}^1$.

We finally prove the second half of Lemma 7.61. Suppose we start with a uniform family of CF-perturbations. It gives a family of elements $\widehat{\mathfrak{S}_\sigma} \in \mathcal{CF}_{\mathcal{K}}(|\mathcal{K}|)$. Then in (12.22) we may take $(W_{\mathfrak{r}}, \omega_{\mathfrak{r}}, \mathfrak{s}_{\mathfrak{r}}^\epsilon)$ to be independent of $\mathfrak{r}$ if $\mathfrak{r} \neq \mathfrak{r}_0$. Moreover, by the definition of uniformity, $W_{\mathfrak{r}_0}, \omega_{\mathfrak{r}_0}$ can be taken to be independent of σ. Therefore (12.22) implies the uniformity of the CF-perturbation we obtain. □

Chapter 13
Construction of Multivalued Perturbations

In this chapter, we discuss multivalued perturbations, especially their existence result, Theorem 6.23. This result will be used in Chaps. 14 and 20. One of the advantages of using multivalued perturbations is that it enables us to work with $\mathbb{Q}$ coefficients. In the construction based on de Rham theory we can work only over $\mathbb{R}$ or $\mathbb{C}$. For many applications, it is enough to work over $\mathbb{R}$ or $\mathbb{C}$, for which we do not need to use the results of Chaps. 13 and 14. The discussion of this chapter is largely parallel to that of CF-perturbation given in Chap. 12.

13.1 Sheaf $\mathcal{MV}_{\mathcal{K}}$ of Multivalued Perturbations

Let $\widehat{\mathcal{U}} = ((\mathfrak{P}, \le), \{\mathcal{U}_{\mathfrak{p}} \mid \mathfrak{p} \in \mathfrak{P}\}, \{\Phi_{\mathfrak{pq}} \mid \mathfrak{q} \le \mathfrak{p}\})$ be a good coordinate system of $Z \subseteq X$ (Definition 3.15) and $\mathcal{K} = \{\mathcal{K}_{\mathfrak{p}} \mid \mathfrak{p} \in \mathfrak{P}\}$ its support system. Using the equivalence relation defined in Definition 3.15 (7) on the disjoint union $\coprod \mathcal{K}_{\mathfrak{p}}$, we obtain a metric space, hetero-dimensional compactum, $|\mathcal{K}|$. This section is similar to Sect. 12.2. We will define and study the sheaf of multivalued perturbations on $|\mathcal{K}|$.

Definition 13.1 Let $\mathcal{U}_{\mathfrak{p}}$ be a Kuranishi chart and $x \in U_{\mathfrak{p}}$. A *germ of multivalued perturbation on* $\mathcal{U}_{\mathfrak{p}}$ *at* x is an equivalence class of $(U, \{\mathfrak{s}^n\})$, where:

(1) U is an open neighborhood of x in $U_{\mathfrak{p}}$.
(2) $\{\mathfrak{s}^n\}$ is a multivalued perturbation of $\widehat{\mathcal{U}_{\mathfrak{p}}}|_{U_p}$.

We say $(U, \{\mathfrak{s}^n\})$ is *equivalent* to $(U', \{\mathfrak{s}'^n\})$ if there exists an open neighborhood U'' of x such that $(U'', \{\mathfrak{s}^n\})$ coincides with $(U'', \{\mathfrak{s}'^n\})$ as multisections in the sense of Definition 6.2 (4) for any n.

We denote by $(\mathcal{MV}^{\mathcal{U}_{\mathfrak{p}}})_x$ the set of all germs of multivalued perturbations on $\mathcal{U}_{\mathfrak{p}}$ at x.

K. Fukaya et al., *Kuranishi Structures and Virtual Fundamental Chains*, Springer Monographs in Mathematics, https://doi.org/10.1007/978-981-15-5562-6_13

Definition 13.2 Let $\Phi_{\mathfrak{pq}} = (U_{\mathfrak{pq}}, \varphi_{\mathfrak{pq}}, \hat{\varphi}_{\mathfrak{pq}}) : \mathcal{U}_{\mathfrak{q}} \to \mathcal{U}_{\mathfrak{p}}$ be a coordinate change. Let $x = \varphi_{\mathfrak{pq}}(y)$ with $y \in U_{\mathfrak{pq}}$.

An element $[U_x, \{\mathfrak{s}^n_{\mathfrak{p}}\}] \in (\mathcal{MV}^{\mathcal{U}_{\mathfrak{p}}})_x$ is said to be *restrictable* to $\mathcal{U}_{\mathfrak{q}}$ if there exists an element $[U_y, \{\mathfrak{s}^n_{\mathfrak{q}}\}] \in (\mathcal{MV}^{\mathcal{U}_{\mathfrak{q}}})_y$ such that we can choose their representatives so that

$$\mathfrak{s}^n_{\mathfrak{p}} \circ \varphi_{\mathfrak{pq}} = \widehat{\varphi}_{\mathfrak{pq}} \circ \mathfrak{s}^n_{\mathfrak{q}}.$$

We write

$$[U_y, \{\mathfrak{s}^n_{\mathfrak{q}}\}] = \Phi^*_{\mathfrak{pq}}[U_x, \{\mathfrak{s}^n_{\mathfrak{p}}\}]$$

and call it the *pullback* of $[U_y, \{\mathfrak{s}^n_{\mathfrak{q}}\}]$ to $\mathcal{U}_{\mathfrak{q}}$.

Hereafter we sometimes write $\mathfrak{s}^n$ instead of $[U, \{\mathfrak{s}^n\}]$ by an abuse of notation.

Definition 13.3

(1) Let $x \in |\mathcal{K}|$. A *germ of multivalued perturbation on* $|\mathcal{K}|$ *at* x is an element $\mathfrak{s}^n$ of $(\mathcal{MV}^{\mathcal{U}_{\mathfrak{p}(x)}})_x$ such that it is restrictable to $\mathcal{U}_{\mathfrak{p}}$ for any $\mathfrak{p} \in \mathfrak{P}(x)$.[1] Here we recall that

$$\mathfrak{p}(x) \in \mathfrak{P}(x)$$

is the maximal element in the ordered set $(\mathfrak{P}(x), \leq)$.

(2) We say an element $\mathfrak{s}^n \in (\mathcal{MV}^{\mathcal{U}_{\mathfrak{p}(x)}})_x$ is *transversal to* 0 if, for all $n = 1, 2, \ldots,$ $\mathfrak{s}^n$ is transversal to 0 as a multivalued perturbation of $\mathcal{U}_{\mathfrak{p}}|_U$ and if all of its restrictions $\Phi^*_{\mathfrak{p}(x)\mathfrak{p}}\mathfrak{s}^n$ are transversal to 0 in a sufficiently small neighborhood of $x_{\mathfrak{p}}$. Here $x = \varphi_{\mathfrak{p}(x)\mathfrak{p}}(x_{\mathfrak{p}})$

(3) If $\widehat{f} : (X, Z; \widehat{\mathcal{U}}) \to N$ is a strongly smooth map to a manifold N and $g : M \to N$ is a smooth map between manifolds, we say $\widehat{f}$ *is strongly transversal to* g *with respect to* $\mathfrak{s}^n$ if $f_{\mathfrak{p}}$ is strongly transversal to g with respect to $\Phi^*_{\mathfrak{p}(x)\mathfrak{p}}\mathfrak{s}^n$ for any $\mathfrak{p} \in \mathfrak{P}(x)$ and n, and a sufficiently small neighborhood of $x_{\mathfrak{p}}$.

(4) We write $(\mathcal{MV}^{\mathcal{U}_{\mathfrak{p}(x)}}_{\mathcal{K}})_x$ (resp. $(\mathcal{MV}^{\mathcal{U}_{\mathfrak{p}(x)}}_{\pitchfork 0,\mathcal{K}})_x$, $(\mathcal{MV}^{\mathcal{U}_{\mathfrak{p}(x)}}_{g \pitchfork f,\mathcal{K}})_x$) the set of all germs of multivalued perturbations on $|\mathcal{K}|$ at x (resp. the set of all germs of multivalued perturbation on $|\mathcal{K}|$ at x which is transversal to 0, which is transversal to g).

In the same way as Definition 12.19 and Lemma 12.21 we can define a topology on

$$|\mathcal{MV}_{\mathcal{K}}| = \bigcup_{x \in |\mathcal{K}|} \{x\} \times (\mathcal{MV}_{\mathcal{K}})_x, \tag{13.1}$$

[1] $\mathfrak{P}(x)$ is defined by (12.12).

so that, together with the obvious projection $\pi : |\mathcal{MV}_{\mathcal{K}}| \to |\mathcal{K}|$, it defines a sheaf on $|\mathcal{K}|$. We denote this sheaf by $\mathcal{MV}_{\mathcal{K}}$. Then $(\mathcal{MV}^{\mathcal{U}_{\mathfrak{p}(x)}}_{\pitchfork 0,\mathcal{K}})_x$ (resp. $(\mathcal{MV}^{\mathcal{U}_{\mathfrak{p}(x)}}_{g\pitchfork f,\mathcal{K}})_x$) defines its subsheaf $\mathcal{MV}_{\pitchfork 0,\mathcal{K}}$ (resp. $\mathcal{MV}_{g\pitchfork f,\mathcal{K}}$). In the same way as in Lemma 12.23, we can show that the set of global continuous sections $\mathcal{MV}_{\mathcal{K}}(|\mathcal{K}|)$ can be canonically identified with the set of multivalued perturbations on $(\widehat{\mathcal{U}}, \mathcal{K})$ and its subset $\mathcal{MV}_{\pitchfork 0,\mathcal{K}}(|\mathcal{K}|$ (resp. $\mathcal{MV}_{g\pitchfork f,\mathcal{K}}(|\mathcal{K}|)$) corresponds to the set of the multivalued perturbations which are transversal to 0 (resp. to g).

Remark 13.4 As with the case of the sheaf $\mathcal{CF}_{\mathcal{K}}$ in Sect. 12.2, we should note that the definition of the sheaf $\mathcal{MV}_{\mathcal{K}}$ involves not only the data of the support system $\mathcal{K}$ but also a part of data of the good coordinate system $\widehat{\mathcal{U}}$.

Now the main result in this chapter is stated as follows.

Theorem 13.5 *Let* $\widehat{\mathcal{U}} = ((\mathfrak{P}, \leq), \{\mathcal{U}_{\mathfrak{p}} \mid \mathfrak{p} \in \mathfrak{P}\}, \{\Phi_{\mathfrak{pq}} \mid \mathfrak{q} \leq \mathfrak{p}\})$ *be a good coordinate system of* $Z \subseteq X$. *We take a support system* $\mathcal{K} = \{\mathcal{K}_{\mathfrak{p}} \mid \mathfrak{p} \in \mathfrak{P}\}$ *of* $\widehat{\mathcal{U}}$. *Then the sheaves* $\mathcal{MV}_{\mathcal{K}}$, $\mathcal{MV}_{\pitchfork 0,\mathcal{K}}$ *are soft. Moreover, if* $\widehat{f} : (X, Z; \widehat{\mathcal{U}}) \to N$ *is a strongly smooth map to a smooth manifold* N *and transversal to a smooth map* $g : M \to N$ *between smooth manifolds, then the sheaf* $\mathcal{MV}_{g\pitchfork f,\mathcal{K}}$ *is soft.*

We prove Theorem 13.5 in Sect. 13.2. Theorem 6.23 is immediate from Theorem 13.5. We next explain its relative version. The next definition is a multivalued perturbation version of Definition 7.57.

Definition 13.6 For each $i = 1, 2$, let $\widehat{\mathcal{U}^i} = (\mathfrak{P}_i, \{\mathcal{U}^i_{\mathfrak{p}}\}, \{\Phi^i_{\mathfrak{pq}}\})$ be a good coordinate system of $Z_i \subseteq X$, $\mathcal{K}^i$ a support system of $\widehat{\mathcal{U}^i}$, and $\widehat{\mathfrak{s}^i}$ a multivalued perturbation of $(\widehat{\mathcal{U}^i}, \mathcal{K}^i)$.

(1) Suppose $(\widehat{\mathcal{U}^2}, \mathcal{K}^2)$ strictly extends $(\widehat{\mathcal{U}^1}, \mathcal{K}^1)$. (Definition 7.56). We say $\widehat{\mathfrak{s}^2}$ *strictly extends* $\widehat{\mathfrak{s}^1}$ if the restriction of $\mathfrak{S}^2_{\mathfrak{p}}$ to $\mathcal{K}^1_{\mathfrak{p}}$ is $\mathfrak{s}^1_{\mathfrak{p}}$ for each $\mathfrak{p} \in \mathfrak{P}_1$.
(2) Suppose $(\widehat{\mathcal{U}^2}, \mathcal{K}^2)$ extends $(\widehat{\mathcal{U}^1}, \mathcal{K}^1)$. We say $\widehat{\mathfrak{s}^2}$ *extends* $\widehat{\mathfrak{s}^1}$ if the restriction of $\mathfrak{s}^2_{\mathfrak{p}}$ to $\mathcal{K}^1_{\mathfrak{p}}$ is $\mathfrak{s}^1_{\mathfrak{p}}$ for each $\mathfrak{p} \in \mathfrak{P}_1$.

Proposition 13.7 *Suppose we are in the situation of Proposition 7.54. We may choose* $\widehat{\mathcal{U}^2}$ *such that the following holds.*

Let $\mathcal{U}^1_0$ *be an open substructure of* $\widehat{\mathcal{U}^1}$ *strictly extended to* $\widehat{\mathcal{U}^2}$ *so that Proposition 7.54 (4) is satisfied. Let* $\mathcal{K}^1, \mathcal{K}^2$ *be support systems of* $\widehat{\mathcal{U}^1}, \widehat{\mathcal{U}^2}$ *respectively such that* $(\widehat{\mathcal{U}^2}, \mathcal{K}^2)$ *extends* $(\widehat{\mathcal{U}^1}, \mathcal{K}^1)$. *Let* $\widehat{\mathfrak{s}^1}$ *be a multivalued perturbation of* $(\widehat{\mathcal{U}^1}, \mathcal{K}^1)$. *Then there exists a multivalued perturbation* $\widehat{\mathfrak{s}^2}$ *of* $(\widehat{\mathcal{U}^2}, \mathcal{K}^2)$ *which extends* $\widehat{\mathfrak{s}^1}$. *Moreover the following holds:*

(1) *If* $\widehat{\mathfrak{s}^1}$ *is transversal to* 0, *so is* $\widehat{\mathfrak{s}^2}$.
(2) *Suppose we are in the situation of Lemma 7.55 (1) in addition. We assume* $\widehat{f^1}$ *is strongly transversal to* $g : N \to M$ *with respect to* $\widehat{\mathfrak{s}^1}$ *and* $\widehat{f^2}$ *is weakly transversal to* g. *Then* $\widehat{f^2}$ *is strongly transversal to* $g : N \to M$ *with respect to* $\widehat{\mathfrak{s}^2}$.

Proof We can prove a multivalued perturbation analogue of Lemma 12.29 in the same way. Then using Lemma 12.30, Theorem 13.5 implies Proposition 13.7 in the same way as the last part of Sect. 12.2. □

13.2 Construction of Multivalued Perturbations

In this section we prove Theorem 13.5. The main idea of the proof is to use CF-perturbation constructed in Lemma 12.25 and partition of unity to find a parametrized family of multivalued perturbations which is transversal. The parameter space of this family is taken to be huge but is finite-dimensional. Then we use Sard's theorem to show that for a generic value of the parameter we obtain a required multivalued perturbation. The detail follows.

Proof of Theorem 13.5 Let # be one of 'absence', $\pitchfork 0$ and $g \pitchfork f$. Let

$$K \subset |\mathcal{K}|$$

be a compact subset and $\widehat{\mathfrak{s}_K} = \{\mathfrak{s}^n_K\} \in \mathcal{MV}_{\#,\mathcal{K}}(K)$. We will find $\widehat{\mathfrak{s}_K} \in \mathcal{MV}_{\#,\mathcal{K}}(|\mathcal{K}|)$ whose restriction to K is $\widehat{\mathfrak{s}_K}$.

We take a support system $\mathcal{K}^+$ of the good coordinate system

$$(\widehat{\mathcal{U}}((\mathfrak{P}, \leq), \{\mathcal{U}_{\mathfrak{p}} \mid \mathfrak{p} \in \mathfrak{P}\}, \{\Phi_{\mathfrak{p}\mathfrak{q}} \mid \mathfrak{q} \leq \mathfrak{p}\})$$

with $\mathcal{K} < \mathcal{K}^+$ in the sense of Definition 5.23 (2). (Recall that in particular, it implies $|\mathcal{K}| \subset |\mathcal{K}^+|$.) We take an open subset U_0 of $|\mathcal{K}^+|$ such that

$$K \subset U_0 \subset |\mathcal{K}^+|$$

and $\widehat{\mathfrak{s}_K}$ is represented by a section on U_0. We will fix our choice of U_0 later as in (13.5).

Under the situation above, we next take a set of open subsets $\{U_{\mathfrak{r}}\}_{\mathfrak{r}\in\mathfrak{R}'}$ of $|\mathcal{K}^+|$ with the following properties.

Property 13.8

(1) $U_{\mathfrak{r}} \cap K = \emptyset$ for any $\mathfrak{r} \in \mathfrak{R}'$. Here $\mathfrak{R}'$ is a finite set.
(2) $U_0 \cup \bigcup_{\mathfrak{r}\in\mathfrak{R}'} U_{\mathfrak{r}}$ is an open covering of $|\mathcal{K}|$.
(3) There exists $\mathfrak{p}(\mathfrak{r}) \in \mathfrak{P}$ such that $U_{\mathfrak{r}}$ is an open subset of $U_{\mathfrak{p}(\mathfrak{r})}$. Moreover, if $\mathcal{K}_{\mathfrak{p}} \cap U_{\mathfrak{r}} \neq \emptyset$, then $\mathfrak{p} \leq \mathfrak{p}(\mathfrak{r})$.
(4) $\mathfrak{V}_{\mathfrak{r}} = (V_{\mathfrak{r}}, \Gamma_{\mathfrak{r}}, E_{\mathfrak{r}}, \phi_{\mathfrak{r}}, \widehat{\phi}_{\mathfrak{r}})$ is an orbifold chart of $U_{\mathfrak{p}(\mathfrak{r})}$ such that $U_{\mathfrak{r}} = \mathrm{Im}(\phi_{\mathfrak{r}})$.

(5) There exists a CF-perturbation on $U_{\mathfrak{r}}$ satisfying (*) in Lemma 12.25, which is represented by $\mathfrak{S}_{\mathfrak{r}} = (W_{\mathfrak{r}}, \omega_{\mathfrak{r}}, \{\mathfrak{s}^{\epsilon}_{\mathfrak{r}}\})$ on $U_{\mathfrak{r}}$. Here $\mathcal{S}_{\mathfrak{r}}$ is a CF-perturbation with respect to the chart $\mathfrak{V}_{\mathfrak{r}}$ as in Lemma 12.25. $U_{\mathfrak{r}}$ is sufficiently small as we will specify during the proof of Lemma 13.11.

The existence of such $\{U_{\mathfrak{r}}\}_{\mathfrak{r}\in\mathfrak{R}'}$ etc. is a consequence of compactness of $|\mathcal{K}|$ and Lemma 12.25.

We put $\mathfrak{R} = \{0\} \cup \mathfrak{R}'$. Property 13.8 (2) says that $\{U_{\mathfrak{r}}\}_{\mathfrak{r}\in\mathfrak{R}}$ is an open covering of $|\mathcal{K}|$. Thus we can apply Lemma 12.28 to obtain a smooth partition of unity $\{\chi_{\mathfrak{r}} \mid \mathfrak{r} \in \mathfrak{R}\}$ subordinate to this open covering.

We may assume that the given $\Gamma_{\mathfrak{r}}$ action on $W_{\mathfrak{r}}$ is effective, by replacing $W_{\mathfrak{r}}$ with the product $W_{\mathfrak{r}} \times W'$ if necessary, where W' is a faithful representation of the finite group $\Gamma_{\mathfrak{r}}$. Then we define $\mathfrak{s}^{\epsilon}_{\mathfrak{r}}$ to be the pullback of a section defined on $W_{\mathfrak{r}}$ by the projection $W_{\mathfrak{r}} \times W' \to W_{\mathfrak{r}}$ and define our $\omega_{\mathfrak{r}}$ on $W_{\mathfrak{r}} \times W'$ to be the product form of a smooth $\Gamma_{\mathfrak{r}}$-invariant top degree form with compact support on W' that integrates to 1 and the given smooth top degree form on $W_{\mathfrak{r}}$.

For each $\mathfrak{r}$ we take an open subset $W^0_{\mathfrak{r}}$ of $\mathrm{Int}\,\mathrm{Supp}(\omega_{\mathfrak{r}}) \subset W_{\mathfrak{r}}$ such that

$$\gamma W^0_{\mathfrak{r}} \cap W^0_{\mathfrak{r}} = \emptyset, \tag{13.2}$$

if $\gamma \in \Gamma_{\mathfrak{r}} \setminus \{1\}$ (we use effectivity of the $\Gamma_{\mathfrak{r}}$ action here) and put

$$W_0 = \prod_{\mathfrak{r}\in\mathfrak{R}'} W^0_{\mathfrak{r}}. \tag{13.3}$$

Now let $x \in K$. We take an orbifold chart $\mathfrak{V}_{(x)} = (V_{(x)}, \Gamma_{(x)}, E_{(x)}, \phi_{(x)}, \widehat{\phi}_{(x)})$ of $U_{\mathfrak{p}(x)}$ such that if $\phi_{(x)}(V_{(x)}) \cap \mathcal{K}_{\mathfrak{p}} \neq \emptyset$ then $\mathfrak{p} \le \mathfrak{p}(x)$. (In particular, it implies that $U_{(x)} \cap |\mathcal{K}| = \phi_{(x)}(V_{(x)}) \cap |\mathcal{K}|$ is open in $|\mathcal{K}|$, since $\mathfrak{p}(x)$ is the maximum in $\mathfrak{P}(x)$.) We take a representative of $\mathfrak{s}^n_K$ in the orbifold chart $\mathfrak{V}_{(x)}$ at x and denote it by $\mathfrak{s}^n_{(x)} = (\mathfrak{s}^n_{(x),1}, \dots, \mathfrak{s}^n_{(x),\ell(x)})$. We take a finite number of points $x_1, \dots, x_k$ of K so that $\bigcup_{a=1}^k U_{(x_a)} \supset K$ and the sections $\mathfrak{s}^n_{(x_a)}$ $(a = 1, \dots, k)$. We assume that they are transversal to 0 on $\bigcup_{a=1}^k U_{(x_a)}$ if $\# = \pitchfork 0$, and we assume that $f_{\mathfrak{p}(x_a)}$ is strongly transversal to g with respect to $\mathfrak{s}^n_{(x_a)}$ if $\# = g \pitchfork f$. This is possible by the openness of transversality condition. We set $\mathfrak{V}_a = \mathfrak{V}_{(x_a)}$, $U_a = U_{(x_a)}$, $U^0_a = U^0_{(x_a)}$, $(\mathfrak{s}^{a,n}_1, \dots, \mathfrak{s}^{a,n}_{\ell_a}) = (\mathfrak{s}^n_{(x_a),1}, \dots, \mathfrak{s}^n_{(x_a),\ell(x_a)})$, and $\mathfrak{p}(a) = \mathfrak{p}(x_a)$. Then we define

$$\mathfrak{U}(K) = \bigcup_{a=1}^{k} U_a, \tag{13.4}$$

and fix an open neighborhood U_0 of K to be

$$U_0 := \mathfrak{U}(K). \tag{13.5}$$

We also take a relatively compact subset U_a^0 of U_a for each a such that

$$\mathfrak{U}_0(K) = \bigcup_{a=1}^{k} U_a^0 \supset K. \tag{13.6}$$

We may assume

$$\chi_0(x) = 1, \qquad \text{for any } x \in \mathfrak{U}_0(K).$$

In this way, for a given support system $\mathcal{K}$ and a compact subset $K \subset |\mathcal{K}|$, we take and fix open subsets

$$U_0, U_1, \cdots, U_k$$

of $|\mathcal{K}^+|$ as above.

Now for each $x \in |\mathcal{K}|$ we take its open neighborhood U_x with the following properties.

Property 13.9

(1) (a) $U_x = \tilde{U}_x \cap |\mathcal{K}|$. Here $\tilde{U}_x$ is an open neighborhood of x in $\mathcal{K}^+_{\mathfrak{p}(x)}$.
 (b) If $U_x \cap \mathcal{K}_{\mathfrak{p}} \neq \emptyset$, then $\mathfrak{p} \in \mathfrak{P}(x)$.
 (c) There exists an orbifold chart $\mathfrak{V}_x = (V_x, \Gamma_x, E_x, \phi_x, \widehat{\phi}_x)$ of $(U_{\mathfrak{p}(x)}, \mathcal{E}_{\mathfrak{p}(x)})$ at x such that $\tilde{U}_x = \phi_x(V_x)$.

(2) If $x \in \mathfrak{U}_0(K)$, then $U_x \subset \mathfrak{U}_0(K)$.

(3) If $x \in U_{\mathfrak{r}}$ for $\mathfrak{r} \in \mathfrak{R}'$, then the following holds:

 (a) $U_x \subset U_{\mathfrak{r}}$.
 (b) There exists an orbifold chart $\mathfrak{V}_{\mathfrak{r},\mathfrak{p}(x)} = (V_{\mathfrak{r},\mathfrak{p}(x)}, \Gamma_{\mathfrak{r}}, E_{\mathfrak{r}}, \phi_{\mathfrak{r},\mathfrak{p}(x)}, \widehat{\phi}_{\mathfrak{r},\mathfrak{p}(x)})$ of $U_{\mathfrak{r}} \cap U_{\mathfrak{p}(x)}$ such that $\Phi^*_{\mathfrak{p}(\mathfrak{r})\mathfrak{p}(x)}\mathcal{S}_{\mathfrak{r}}$ is represented by $(W_{\mathfrak{r}}, \omega_{\mathfrak{r}}, \{\mathfrak{s}^{\epsilon}_{\mathfrak{r}\mathfrak{p}(x)}\})$. Note that $W_{\mathfrak{r}}, \omega_{\mathfrak{r}}$ are those appearing in the representative of $\mathcal{S}_{\mathfrak{r}}$, which we took and fixed in Property 13.8 (5). $\mathfrak{s}^{\epsilon}_{\mathfrak{r}\mathfrak{p}(x)}$ is the pullback of $\mathfrak{s}^{\epsilon}_{\mathfrak{r}}$ by the embedding $U_{\mathfrak{p}(\mathfrak{r})\mathfrak{p}(x)} \to U_{\mathfrak{p}(\mathfrak{r})}$.
 (c) There exists $(h_{\mathfrak{r}x}, \tilde{\varphi}_{\mathfrak{r}x}, \breve{\varphi}_{\mathfrak{r}x}) : \mathfrak{V}_x \to \mathfrak{V}_{\mathfrak{r},\mathfrak{p}(x)}$ as in Property 7.39.

(4) If $x \notin U_{\mathfrak{r}}$, then $\overline{U}_x \cap \mathrm{Supp}(\chi_{\mathfrak{r}}) = \emptyset$.

(5) If $x \in \mathfrak{U}(K)$, then the following holds:

 (a) $U_x \subset U_a$ for some $a \in \{1, \ldots, k\}$.
 (b) There exists an orbifold chart $\mathfrak{V}_{a,\mathfrak{p}(x)} = (V_{a,\mathfrak{p}(x)}, \Gamma_a, E_a, \phi_{a,\mathfrak{p}(x)}, \widehat{\phi}_{a,\mathfrak{p}(x)})$ of $U_a \cap U_{\mathfrak{p}(x)}$ such that $\Phi^*_{\mathfrak{p}(a)\mathfrak{p}(x)}\mathfrak{s}^{a,n}$ is represented by $(\mathfrak{s}^{a,n}_{\mathfrak{p}(x),1}, \ldots, \mathfrak{s}^{a,n}_{\mathfrak{p}(x),\ell_a})$. Note that ℓ_a is the same in the representative of $\mathfrak{s}^{a,n}$, which we took and fixed as above. $\mathfrak{s}^{a,n}_{\mathfrak{p}(x),i}$ is the pullback of $\mathfrak{s}^{a,n}_i$ by the embedding $U_{\mathfrak{p}(a)\mathfrak{p}(x)} \to U_{\mathfrak{p}(a)}$.
 (c) There exists $(h_{ax}, \tilde{\varphi}_{ax}, \breve{\varphi}_{ax}) : \mathfrak{V}_x \to \mathfrak{V}_a$ as in Property 7.39.

The existence of such objects is mostly obvious from the definitions. Here are some remarks on Item (3). In $\mathfrak{V}_{\mathfrak{r},\mathfrak{p}(x)}$, the isotopy group is taken to be the same as $\Gamma_{\mathfrak{r}}$. Note that we take $U_{\mathfrak{r}}$ such that $U_{\mathfrak{r}} \cap U_{\mathfrak{p}}$ is empty unless $\mathfrak{p} \in \mathfrak{P}(x_{\mathfrak{r}})$. (Here we defined $U_{\mathfrak{r}}$ as a neighborhood of $x_{\mathfrak{r}}$.) So by the definition of embedding of orbifolds the isotropy group of $x_{\mathfrak{r}}$ in $U_{\mathfrak{p}(x)}$ is the same as $\Gamma_{\mathfrak{r}} = \Gamma_{x_{\mathfrak{r}}}$. By the same reasoning, $E_{\mathfrak{r}} = E_{x_{\mathfrak{r}}}$ is the same vector space as that appearing in $\mathfrak{V}_{\mathfrak{r},\mathfrak{p}(x)}$. We can also observe that $V_{\mathfrak{r},\mathfrak{p}(x)}$ may be taken to be a submanifold of $V_{\mathfrak{r}}$ and that $\tilde{\varphi}_{\mathfrak{r}x}$ is the canonical embedding. In such a case $\mathfrak{s}^{\epsilon}_{\mathfrak{r}\mathfrak{p}(x)}$ is a restriction of $\mathfrak{s}^{\epsilon}_{\mathfrak{r}}$. This also explains the reason why we take the same $W_{\mathfrak{r}}$, $\omega_{\mathfrak{r}}$ as those appearing in the representative of $\mathcal{S}_{\mathfrak{r}}$, which we took and fixed. Similar remarks apply to Item (5).

With preparation made above, for each $x \in |\mathcal{K}|$ and $\vec{\xi} = (\xi_{\mathfrak{r}})_{\mathfrak{r}\in\mathfrak{R}} \in W_0$ we are ready to define a multivalued perturbation $\{\mathfrak{s}^{n,\vec{\xi}}_x\}$ on $\mathfrak{V}_x$ as follows. We take a sequence $\epsilon_n > 0$ with $\lim_{n\to\infty} \epsilon_n = 0$ and fix it throughout the proof. (For example, we may take $\epsilon_n = 1/n$.)

(Case 1) $x \in \mathfrak{U}_0(K)$.

Then $U_x \subset \mathfrak{U}_0(K)$ by Property 13.9 (2). We take $a \in \{1, \dots, k\}$ such that $x \in U^0_a$. By Property 13.9 (5) we can pull back $(\mathfrak{s}^{a,n}_1, \dots, \mathfrak{s}^{a,n}_{\ell_a})$ to $\mathfrak{V}_x$. This pullback is our $\mathfrak{s}^{n,\vec{\xi}}_x$ in this case. (This is independent of $\vec{\xi}$.)

(Case 2) $x \in \mathfrak{U}(K) \setminus \mathfrak{U}_0(K)$.

Take $a \in \{1, \dots k\}$ such that $U_x \subset U_a$. By Property 13.9 (5) we have $(h_{ax}, \tilde{\varphi}_{ax}, \breve{\varphi}_{ax})$. Put

$$\mathfrak{R}(x) = \{\mathfrak{r} \in \mathfrak{R}' \mid x \in \mathrm{Supp}(\chi_{\mathfrak{r}})\}. \tag{13.7}$$

Property 13.9 (4) yields $U_x \subset U_{\mathfrak{r}}$ for $\mathfrak{r} \in \mathfrak{R}(x)$. We take $(h_{\mathfrak{r}x}, \tilde{\varphi}_{\mathfrak{r}x}, \breve{\varphi}_{\mathfrak{r}x})$ as in Property 13.9 (3) for each $\mathfrak{r} \in \mathfrak{R}(x)$. We put

$$I = \{1, \dots, \ell_a\} \times \prod_{\mathfrak{r}\in\mathfrak{R}(x)} \Gamma_{\mathfrak{r}}.$$

Then we denote $i = (j(i), (\gamma_{\mathfrak{r}}(i))) \in I$ and define $\mathfrak{s}^{n,\vec{\xi}}_{x,i} : V_x \to E_x$ for each $i \in I$ by Formula (13.8):

$$\begin{aligned}
&\mathfrak{s}^{n,\vec{\xi}}_{x,i}(y) \\
&= s_x(y) + \chi_0([y]) g^{-1}_{a,y} \left(\mathfrak{s}^{a,n}_{\mathfrak{p}(x),j(i)}(\tilde{\varphi}_{ax}(y)) - s_{a\mathfrak{p}(x)}(\tilde{\varphi}_{ax}(y)) \right) \\
&\quad + \sigma \sum_{\mathfrak{r}\in\mathfrak{R}(x)} \chi_{\mathfrak{r}}([y]) g^{-1}_{\mathfrak{r},y} \left(\mathfrak{s}^{\epsilon_n}_{\mathfrak{r},\mathfrak{p}(x)}(\tilde{\varphi}_{\mathfrak{r}x}(y), \gamma_{\mathfrak{r}}(i)^{-1}\xi_{\mathfrak{r}}) - s_{\mathfrak{r}\mathfrak{p}(x)}(\tilde{\varphi}_{\mathfrak{r}x}(y)) \right).
\end{aligned} \tag{13.8}$$

Explanation of the notations appearing in Formula (13.8) is in order:

- $s_{a\mathfrak{p}(x)} : V_{a,\mathfrak{p}(x)} \to E_a$ is the representative of the Kuranishi map.
- $g_{a,y} : E_x \to E_a$ is defined by $\breve{\varphi}_{ax}(y, \eta) = g_{a,y}(\eta)$.

- $s_{\mathfrak{r}\mathfrak{p}(x)} : V_{\mathfrak{r},\mathfrak{p}(x)} \to E_{\mathfrak{r}}$ is the representative of the Kuranishi map.
- $g_{\mathfrak{r},y} : E_x \to E_{\mathfrak{r}}$ is defined by $\breve{\varphi}_{\mathfrak{r}x}(y,\eta) = g_{\mathfrak{r},y}(\eta)$.
- $[y] \in V_{a,\mathfrak{p}(x)}/\Gamma_{a,\mathfrak{p}(x)}$ is the equivalence class of y, which is regarded as an element of $U_{\mathfrak{p}(x)} \subset |\mathcal{K}|$.
- σ is a sufficiently small positive number as we will specify during the proof of Lemma 13.11.

(Case 3) $x \in |\mathcal{K}| \setminus \mathfrak{U}(K)$.

We define $\mathfrak{R}(x)$ by (13.7). As in Case 2 above we have $U_x \subset U_{\mathfrak{r}}$ for $\mathfrak{r} \in \mathfrak{R}(x)$. Then we take $(h_{\mathfrak{r}x}, \tilde{\varphi}_{\mathfrak{r}x}, \breve{\varphi}_{\mathfrak{r}x})$ as in Property 13.9 (3) for each $\mathfrak{r} \in \mathfrak{R}(x)$ and put

$$I = \prod_{\mathfrak{r}\in\mathfrak{R}(z)} \Gamma_{\mathfrak{r}}.$$

Then we define $\mathfrak{s}^{n,\vec{\xi}}_{x,i} : V_x \to E_x$ for each $i \in I$ with $i = (\gamma_{\tau}(i))_{\tau\in\mathfrak{R}(x)}$ by

$$\mathfrak{s}^{n,\vec{\xi}}_{x,i}(y)=s_x(y)+\sigma \sum_{\mathfrak{r}\in\mathfrak{R}(x)} \chi_{\mathfrak{r}}([y])g^{-1}_{\mathfrak{r},y}\left(\mathfrak{s}^{\epsilon_n}_{\mathfrak{r},\mathfrak{p}(x)}(\tilde{\varphi}_{\mathfrak{r}x}(y), \gamma_{\mathfrak{r}}(i)^{-1}\xi_{\mathfrak{r}})-s_{\mathfrak{r}}(\tilde{\varphi}_{\mathfrak{r}x}(y))\right). \tag{13.9}$$

Here the notations are the same as in (13.8).

Lemma 13.10 *The collection* $(\mathfrak{s}^{n,\vec{\xi}}_{x,i})_{i\in I}$ *defines a multivalued perturbation on* $\mathfrak{V}_x$. *Moreover it satisfies the following properties:*

(1) *In Case 1, it is independent of the choice of a and* $(h_{ax}, \tilde{\varphi}_{ax}, \breve{\varphi}_{ax})$.
(2) *In Case 2, it is independent of the choice of* $(h_{\mathfrak{r}x}, \tilde{\varphi}_{\mathfrak{r}x}, \breve{\varphi}_{\mathfrak{r}x})$, *and a,* $(h_{ax}, \tilde{\varphi}_{ax}, \breve{\varphi}_{ax})$.
(3) *In Case 3, it is independent of the choice of* $(h_{\mathfrak{r}x}, \tilde{\varphi}_{\mathfrak{r}x}, \breve{\varphi}_{\mathfrak{r}x})$.
(4) *Let* $x' \in U_x$. *Then the pullback* $\Phi^*_{\mathfrak{p}(x)\mathfrak{p}(x')}((\mathfrak{s}^{n,\vec{\xi}}_{x,i})_{i\in I})$ *of* $(\mathfrak{s}^{n,\vec{\xi}}_{x,i})_{i\in I}$ *to* $U_x \cap U_{x'}$ *is equivalent to* $(\mathfrak{s}^{n,\vec{\xi}}_{x',i})_{i\in I}$. *Here* $\Phi_{\mathfrak{p}(x)\mathfrak{p}(x')} : \mathcal{U}_{\mathfrak{p}(x')} \to \mathcal{U}_{\mathfrak{p}(x)}$ *is the embedding of the Kuranishi chart which is a part of the data in a good coordinate system.*

Proof In Cases 2 and 3 we will show that $(\mathfrak{s}^{\epsilon,\vec{\xi}}_{x,i}(y))_{i\in I}$ is a permutation of $(\gamma\mathfrak{s}^{n,\vec{\xi}}_{x,i}(\gamma^{-1}y))_{i\in I}$ for $\gamma \in \Gamma_x$. We calculate

$$\gamma g^{-1}_{\mathfrak{r},\gamma^{-1}y}(\xi_{\mathfrak{r}}) = \gamma\tilde{\varphi}_{\mathfrak{r}x}(\gamma^{-1}y, \xi) = \tilde{\varphi}_{\mathfrak{r}x}(y, \gamma\xi) = g^{-1}_{\mathfrak{r},y}(\gamma\xi_{\mathfrak{r}}). \tag{13.10}$$

This implies that the third term of (13.8) and the second term of (13.9) are invariant under γ action modulo permutation of the indices in I. The second term of (13.8) is invariant under γ action modulo permutation of $1, \ldots, a$ since $(\mathfrak{s}^{a,n}_{\mathfrak{p}(x),1}, \ldots, \mathfrak{s}^{a,n}_{\mathfrak{p}(x),\ell_a})$ is a multisection. We have thus proved that $(\mathfrak{s}^{n,\vec{\xi}}_{x,i})_{i\in I}$ is a multisection. (In Case 1 this fact is obvious.)

Statement (1) is a consequence of the definition of multivalued perturbation, that is the well-definedness of $\mathfrak{s}_K$.

To prove Statement (2), we observe that different choices of $(h_{\mathfrak{r}x}, \tilde{\varphi}_{\mathfrak{r}x}, \breve{\varphi}_{\mathfrak{r}x})$ are related to one another by the action of $\gamma \in \Gamma_{\mathfrak{r}}$. (Lemma 23.28). Then using (13.10) we can show that the third term of (13.8) changes by the permutation of $\gamma_{\mathfrak{r}}$. The first and the second terms of (13.8) do not change.

By changing $(h_{ax}, \tilde{\varphi}_{ax}, \breve{\varphi}_{ax})$ the second term of (13.8) changes by the permutation of $j(i)$ and the third term of (13.8) does not change.

The proof of (3) is easier.

To prove (4) it suffices to consider the case $U_{x'} \subset U_x$. If Case 1 is applied to both x and x' it is a consequence of the compatibility of the restriction and pullback of the multisection.

If Case 3 is applied to both, then $\mathfrak{R}(x') \subseteq \mathfrak{R}(x)$. In this case the right hand side of (13.9) is independent of the $\Gamma_{\mathfrak{r}}$ factor for $\mathfrak{r} \notin \mathfrak{R}(x')$ on $U_{x'}$. Therefore the pullback of $(\mathfrak{s}^{n,\vec{\xi}}_{x,i})_{i \in I}$ to $U_x \cap U_{x'}$ is a permutation of the $\prod_{\mathfrak{r} \in \mathfrak{R}(x') \setminus \mathfrak{R}(x)} \#\Gamma_{\mathfrak{r}}$ iteration of the restriction of $(\mathfrak{s}^{\epsilon,\vec{\xi}}_{x',i})_{i \in I}$ to $U_x \cap U_{x'}$.

When Case 2 is applied to both, we can prove (4) by combining the above two cases.

What remains to be proved is the case when Case 2 is applied to x and Case 1 or Case 3 is applied to x'. If Case 1 is applied to x', then χ_0 becomes 1 on $U_{x'}$. Therefore the required equivalence follows from the well-definedness of the restriction of the multisection. If Case 3 is applied to x', then χ_0 becomes 0 on $U_{x'}$. Therefore the second term of (13.8) vanishes. So the pullback of $(\mathfrak{s}^{n,\vec{\xi}}_{x,i})_{i \in I}$ to $U_{x'}$ is a permutation of the $\ell_a \prod_{\mathfrak{r} \in \mathfrak{R}(x') \setminus \mathfrak{R}(x)} \#\Gamma_{\mathfrak{r}}$ iteration of the restriction of $(\mathfrak{s}^{n,\vec{\xi}}_{x',i})_{i \in I}$ to $U_{x'}$. □

Lemma 13.11 *We may choose $U_{\mathfrak{r}}$ and $\sigma > 0$ small so that the following holds.*

For each $n \in \mathbb{Z}_+$, the set of $\vec{\xi} \in W_0$ such that $(\mathfrak{s}^{n,\vec{\xi}}_{x,i})_{i \in I}$ is transversal to 0 for all $x \in |\mathcal{K}|$ is dense in W_0.

In addition, if $g : N \to M$ is a smooth map between smooth manifolds and f is transversal to g, then the set of $\vec{\xi}$ such that f is strongly transversal to g with respect to $(\mathfrak{s}^{n,\vec{\xi}}_{x,i})_{i \in I}$ is dense in W_0.

Proof It suffices to show the conclusion on a neighborhood U_x of each fixed $x \in |\mathcal{K}|$, since we can cover $|\mathcal{K}|$ by finitely many such U_x's. We first consider the case $\# =\pitchfork 0$.

In Case 1 this follows from the assumption that $\mathfrak{s}_K$ is transversal to 0.

In Case 2, we use the same argument as the proof of Lemma 12.25 and the last step of the proof of Theorem 12.24 (at the end of Sect. 12.2), to prove that the set

$$\{(y, \vec{\xi}) \mid \mathfrak{s}^{n,\vec{\xi}}_{x,i}(y) = 0\}$$

is a smooth submanifold of $V_x \times W_0$, if $U_{\mathfrak{r}}$ and $\sigma > 0$ are sufficiently small. Therefore we can prove the lemma by applying Sard's theorem to its projection to W_0. (We use (13.2) here.)

We consider Case 3. Suppose $\chi_0([x]) \neq 1$. Then we can shrink the domain U_x if necessary and assume $\chi_{\mathfrak{r}}([y]) \neq 0$ on U_x. Then by the same argument as Case 2 we can show the required transversality for $\vec{\xi}$ contained in a residual subset of W_0.

Suppose $\chi_0([x]) = 1$. Then all the functions $\chi_{\mathfrak{r}}$ together with its first derivative are small in a neighborhood of x. Moreover the first derivative of χ_0 is small in a neighborhood of x. We also remark that $\mathfrak{s}_K$ is transversal to 0 at x. Therefore by using the openness of transversality, we can shrink the neighborhood U_x so that $(\mathfrak{s}^{n,\vec{\xi}}_{x,i})_{\in I}$ remains to be transversal to 0 on U_x for any ξ contained in a compact subset of W_0.

The proof of the case $\# = g \pitchfork f$ is similar. In Case 2 above we consider the set

$$\{(y, \vec{\xi}, z) \in V_x \times W_0 \times N \mid \mathfrak{s}^{n,\vec{\xi}}_{x,i}(y) = 0, \ f(y) = g(z)\}.$$

This is a smooth submanifold of $V_x \times W_0 \times N$. Therefore we have the required transversality result in this case by applying Sard's theorem to the projection from this manifold to W_0. The rest of the proof is entirely the same. □

Remark 13.12 Note that the multivalued perturbation $\widehat{\mathfrak{s}^n}$ is a sequence. Namely it consists of countably many multisections. We use this fact to apply Baire's category theorem in this proof.

The proof of Theorem 13.5 is complete. □

13.3 Extending Multivalued Perturbations from One Chart to Another: Remarks

We discussed construction of multivalued perturbations (or multisections) in this chapter. There are several methods appearing for such construction in the literature. There are similar issues for the CF-perturbation.

Those methods are classified into two types:

(I) Work on a space such as the hetero-dimensional compactum $|\mathcal{K}|$ and construct multivalued perturbations directly on this space at once. We define a sheaf of multivalued perturbations on $|\mathcal{K}|$ and use the sheaf theory for such a construction. This is the method we introduce in this book.

(II) First we construct a multivalued perturbation $\mathfrak{s}^{\epsilon}_{\mathfrak{q}}$ on one chart $\mathcal{U}_{\mathfrak{q}}$. Then when $\mathfrak{q} < \mathfrak{p}$ we extend the multivalued perturbation on $U_{\mathfrak{q}}$ to one on $U_{\mathfrak{p}}$ in two steps:

(Step 1) We extend the multivalued perturbation $\mathfrak{s}^{\epsilon}_{\mathfrak{q}}$ on the image $\varphi_{\mathfrak{p}\mathfrak{q}}(U_{\mathfrak{p}\mathfrak{q}})$ to its tubular neighborhood.

(Step 2) We work on an obifold $U_{\mathfrak{p}}$ and use the relative version of the construction on one chart to extend the multivalued perturbation on the tubular neighborhood to the whole $U_{\mathfrak{p}}$.

The actual construction is more complicated, since for given $\mathfrak{p}$ there are several $\mathfrak{q}_i$ with $\mathfrak{q}_i < \mathfrak{p}$. So one needs to extend $\mathfrak{s}^{\epsilon}_{\mathfrak{q}_i}$ to the tubular neighborhood of $\varphi_{\mathfrak{p}\mathfrak{q}_i}(U_{\mathfrak{p}\mathfrak{q}_i})$ in $U_{\mathfrak{p}}$ so that they are consistent on the overlapped part. We use a certain double induction argument to work it out. This strategy appeared in [FOn2, page 955] and since then it has been used in various works, including the previous preprint version [FOOO19] of this book, but not in this book.

Historically, in [Th], R. Thom proved his famous transversality theorem inductively chart by chart. We followed Thom's argument in our previous writings. In more recent textbooks on manifold or elementary differential topology, the transversality theorem is proved in the following way: first take a sufficiently huge parameter space so that the version of transversality theorem with this extended parameter is easily proved: then use Sard's theorem to show the transversality for a generic value of the parameter. In other words, method (I) is close to the proof of transversality theorem in modern textbooks and method (II) is closer to its original proof by Thom.

We next explain a certain delicate point we encounter when we try to extend a multisection or multivalued perturbation defined on one chart $\mathcal{U}_q|_{U_{\mathfrak{p}\mathfrak{q}}}$ to another chart $\mathcal{U}_{\mathfrak{p}}$ into which $\mathcal{U}_q|_{U_{\mathfrak{p}\mathfrak{q}}}$ is embedded, that is, (Step 1) in method (II).

First observe that if we consider a sufficiently small chart of $\mathcal{U}_q|_{U_{\mathfrak{p}\mathfrak{q}}}$, we can easily extend a multisection to its neighborhood in $\mathcal{U}_{\mathfrak{p}}$. Then we can construct an extension globally by using a partition of unity. To work out the second step (which uses a partition of unity) in detail, we need more careful argument. In fact to extend a multivalued perturbation via a partition of unity we need to sum up only the extension of the *same* branch. This point was mentioned in [FOn2, page 955 lines 20–24] and [FOOO16, Remark 6.4]. We elaborate on it below.

Let $U_{\mathfrak{p}\mathfrak{q}} \subset U_{\mathfrak{p}}$ be an embedded submanifold. Suppose $U_{\mathfrak{p}\mathfrak{q}}$ is expressed as the union $U_{\mathfrak{p}\mathfrak{q}} = U_1 \cup U_2$ of two open subsets U_1, U_2 and we are given a multisection $\mathfrak{s}$ on $U_{\mathfrak{p}\mathfrak{q}}$. We also assume that $\mathfrak{s}|_{U_j}$ is extended to its neighborhood in $U_{\mathfrak{p}}$ and we denote this extension by $\tilde{\mathfrak{s}}_j$. (We assume that the number of branches of $\tilde{\mathfrak{s}}_j$ is the same as one of $\tilde{\mathfrak{s}}$ around each point of $U_1 \cap U_2$.) We try to glue $\tilde{\mathfrak{s}}_1$ and $\tilde{\mathfrak{s}}_2$ to obtain an extension of $\mathfrak{s}$ to a neighborhood of $U_{\mathfrak{p}\mathfrak{q}}$ in $U_{\mathfrak{p}}$.

Let $p \in U_1 \cap U_2$. We represent $\mathfrak{s}$ in a neighborhood of p as $(s_1, \dots, s_\ell)$, where $s_1, \dots, s_\ell$ are branches of $\mathfrak{s}$ in the neighborhood. We might say that extension $\tilde{\mathfrak{s}}_j$ gives $(s_{j,1}, \dots, s_{j,\ell})$ and try to glue them as

$$s_i(p) = \chi(\pi(y))s_{1,i}(y) + (1 - \chi(\pi(y)))s_{2,i}(y), \tag{13.11}$$

where χ is a function on $U_{\mathfrak{p}\mathfrak{q}}$ such that $\mathrm{Supp}\,(\chi) \subset U_1$, $\mathrm{Supp}\,(1 - \chi) \subset U_2$, and π is the projection of the tubular neighborhood of $U_{\mathfrak{p}\mathfrak{q}}$ in $U_{\mathfrak{p}}$.

However, there is a certain difficulty to defining s_i by (13.11). We first observe the following example.

Example 13.13 In Definition 6.2 (2) we allow the permutation σ to *depend* on y which lies in a neighborhood of x. For this reason, the notion of a branch of a multisection should be studied rather carefully. Here is an example: We define

$$e_{\epsilon_1\epsilon_2}(t) = \begin{cases} \epsilon_1 e^{-1/|t|} & t \le 0, \\ \epsilon_2 e^{-1/|t|} & t \ge 0. \end{cases}$$

Here ϵ_i is either plus or minus. Four functions e_{++}, e_{--}, e_{-+}, e_{+-} are all smooth functions on $\mathbb{R}$. We define 2-multisections $\mathfrak{s}$ and $\mathfrak{s}'$ as follows:

$$\mathfrak{s} = (e_{++}, e_{--}), \qquad \mathfrak{s}' = (e_{+-}, e_{-+}).$$

They are two multisections of the trivial line bundle on $\mathbb{R}$. It is easy to see that $\mathfrak{s}$ is equivalent to $\mathfrak{s}'$ in the sense of Definition 6.2 (2). (This definition coincides with [FOn2].) However, it is impossible to choose σ appearing in Definition 6.2 (2) in a way independent of y.

As Example 13.13 shows, the way that we take a representative $(s_1, \ldots, s_\ell)$ of $\mathfrak{s}$ is not unique. (The uniqueness modulo (locally well-defined) permutation also fails.) Therefore, when we extended $\mathfrak{s}$ to $\tilde{\mathfrak{s}}_j$, we might have taken a representative different from $(s_1, \ldots, s_\ell)$ in a neighborhood of p. So to add $s_{1,i}(p)$ and $s_{2,i}(p)$ does not make sense. Note that the representative $(s_{j,1}(p), \ldots, s_{j,\ell}(p))$ does make sense at each p (modulo permutation).

A similar problem appears when we extend a given multivalued (of CF) perturbation on the boundary to the interior. (This point will be discussed in Chap. 17. There we introduce another method.)

There are (at least) two methods to resolve this problem:

(IIa) To make precise what one means by 'to sum up only the same branch'.

In Definition 6.2 (2), $\mathfrak{s}'_i(y) = \mathfrak{s}'_{\sigma(i)}(y)$, the permutation σ is allowed to be y-dependent. This point is much related to Example 13.13. We can slightly modify the definition of a multivalued perturbation, which we call lifted multivalued perturbation, so that we do not allow σ to be y-dependent. Then we can actually justify the formula (13.11) for such a multivalued perturbation. We remark that all the multivalued perturbations that appeared in the geometric applications we know are lifted. The definition of lifted multivalued perturbation is given in [TF, Definition 2.3.5].

(IIb) In the preprint version [FOOO19] of this book, we defined the notion of (a system of) bundle extension data and used it to extend multivalued perturbation from one chart to the next chart, while we do not need to use the notion of bundle extension data in the current version of this book. The notion of 'bundle extension data' is basically the same as the notion of a level 1 structure, which appeared in D. Yang's paper [Ya1].

Bundle extension data is a pair of the projection $\pi_{\mathfrak{q}\mathfrak{p}} : \Omega_{\mathfrak{q}\mathfrak{p}} \to U_{\mathfrak{p}\mathfrak{q}}$ and an embedding of the bundles $\widetilde{\varphi}_{\mathfrak{p}\mathfrak{q}} : \pi^*_{\mathfrak{q}\mathfrak{p}}\mathcal{E}_{\mathfrak{q}} \to \mathcal{E}_{\mathfrak{p}}$. We require that the composition of $\pi_{\mathfrak{qp}}$ with the embedding $\varphi_{\mathfrak{pq}} : U_{\mathfrak{pq}} \to U_{\mathfrak{p}}$ is the identity map and $\widetilde{\varphi}_{\mathfrak{pq}}$ is the canonical inclusion there. $\pi_{\mathfrak{qp}}$ becomes the projection of the normal bundle when we identify a neighborhood $\Omega_{\mathfrak{qp}}$ of $\varphi_{\mathfrak{pq}}(U_{\mathfrak{pq}})$ with the tubular neighborhood and $\widetilde{\varphi}_{\mathfrak{pq}}$ determines the extension of the bundle $\mathcal{E}_{\mathfrak{q}}$ on $\varphi_{\mathfrak{pq}}(U_{\mathfrak{pq}})$ to its neighborhood $\Omega_{\mathfrak{qp}}$.

Then instead of using a partition of unity as in formula (13.11), we can define extension at once by

$$\mathfrak{s}_{\mathfrak{p},i}(y) = s_{p}(y) + \widetilde{\varphi}_{\mathfrak{pq}}(\mathfrak{s}_{\mathfrak{q},i}(\pi_{\mathfrak{qp}}(y)) - s_{q}(\pi_{\mathfrak{qp}}(y))).$$

Since there are several $\mathfrak{q}_i$ with $\mathfrak{q}_i < \mathfrak{p}$ we need certain consistency between the given various bundle extension data. We refer Yang's paper [Ya1, Section 7] and Sections 13,14 and 20 of the preprint version [FOOO19] of this book for details of the discussion about bundle extension data.

In this book we use the definition of a multisection which is exactly the same as the one in [FOn2]. We call it a liftable multisection. We want to do so since the definition of [FOn2] had been used by various people including ourselves. Note that the proofs we gave in this book (both this final version and the previous preprint version) work both for liftable multivalued perturbations and lifted multivalued perturbations. Namely the methods (I) and (IIb) both work both for liftable and for lifted multivalued perturbations. (IIa) works for lifted multivalued perturbations.

A similar problem occurs when we extend multivalued perturbations from a boundary to its neighborhood. See Chaps. 17 and 20 for such a problem. The method written in Chaps. 17 and 20 (using the *outer collar*, that is, the collar which we put outside) works both for liftable multivalued perturbations and for lifted multivalued perturbations (and for CF-perturbations). We can also use a variant of the idea to define and use a system of bundle extension data.[2] It works for both liftable multivalued perturbations and lifted multivalued perturbations. If we use lifted multisections, we can work out this extension problem by method (IIa), that is, by first extending locally and then using a formula similar to (13.11).

[2] We do not try to write the detail of this version of the proof in this book.

Chapter 14
Zero-and One-Dimensional Cases via Multivalued Perturbation

In Chaps. 7, 8, 9 and 10, we discussed smooth correspondence and defined virtual fundamental chains based on de Rham theory and CF-perturbations. In this chapter, we discuss another method based on multivalued perturbations. Here we restrict ourselves to the case when the dimension of K-spaces of our interest is 1, 0 or negative, and define a virtual fundamental chain over $\mathbb{Q}$ in the 0-dimensional case. In spite of this restriction, the argument of this chapter is enough for the purpose, for example, to prove all the results stated in [FOn2]. We recall that in [FOn2] we originally used a triangulation of the zero set of a multisection to define a virtual fundamental chain. In this chapter we present a different way from [FOn2]. Namely, we use Morse theory in place of triangulation. This change will make the relevant argument simpler and shorter for this restricted case. The triangulation of the zero set of multisection is discussed in [TF].

14.1 Virtual Fundamental Chain for a Good Coordinate System with Multivalued Perturbation

We start with the following:

Lemma 14.1 *Let $\widehat{\mathcal{U}}$ be a good coordinate system (which may or may not have a boundary or corners), $\mathcal{K}$ its support system, and $\widehat{\mathfrak{s}} = \{\mathfrak{s}^n_{\mathfrak{p}} \mid \mathfrak{p} \in \mathfrak{P}\}$ a multivalued perturbation of $(\widehat{\mathcal{U}}, \mathcal{K})$ (Definition 6.12). We assume that $\widehat{\mathfrak{s}}$ is transversal to 0. Then there exists a natural number n_0 with the following properties:*

(1) *If the dimension of $(X, \widehat{\mathcal{U}})$ is negative, then $(\mathfrak{s}^n_{\mathfrak{p}})^{-1}(0) \cap |\mathcal{K}| = \emptyset$ for $n \geq n_0$.*
(2) *If the dimension of $(X, \widehat{\mathcal{U}})$ is 0, then $(\mathfrak{s}^n_{\mathfrak{p}})^{-1}(0) \cap |\partial\widehat{\mathcal{U}}| \cap |\mathcal{K}| = \emptyset$ for $n \geq n_0$. Moreover there exists a neighborhood $\mathfrak{U}(X)$ of X in $|\widehat{\mathcal{U}}| \cap |\mathcal{K}|$ such that the intersection $(\mathfrak{s}^n_{\mathfrak{p}})^{-1}(0) \cap \mathfrak{U}(X)$ is a finite set for any $n \geq n_0$.*

K. Fukaya et al., *Kuranishi Structures and Virtual Fundamental Chains*, Springer Monographs in Mathematics, https://doi.org/10.1007/978-981-15-5562-6_14

Proof (1) is obvious. Using the fact that the dimension of the boundary of $(X, \widehat{\mathcal{U}})$ is negative, we have $(\mathfrak{s}^{\epsilon}_{\mathfrak{p}})^{-1}(0) \cap |\partial\widehat{\mathcal{U}}| \cap |\mathcal{K}| = \emptyset$ in (2) also. The finiteness of the order of the set $(\mathfrak{s}^{\epsilon}_{\mathfrak{p}})^{-1}(0) \cap \mathfrak{U}(X) \cap |\mathcal{K}|$ is a consequence of its compactness, Corollary 6.20. □

Now we consider the following situation.

Situation 14.2 Let $\widehat{\mathcal{U}}$ be a good coordinate system (which may or may not have a boundary or corners), and assume a support system $\mathcal{K}$ thereof is given. Let $\widehat{\mathfrak{s}} = \{\mathfrak{s}^n_{\mathfrak{p}} \mid \mathfrak{p} \in \mathfrak{P}\}$ be a multivalued perturbation of $(\widehat{\mathcal{U}}, \mathcal{K})$. We assume:

(1) $(X, \widehat{\mathcal{U}})$ is oriented,
(2) $\widehat{\mathfrak{s}}$ is transversal to 0.

Consider another support systems $\mathcal{K}^1, \mathcal{K}^2$ with $\mathcal{K}^1 < \mathcal{K}^2 < \mathcal{K}^3 = \mathcal{K}$. We take a neighborhood $\mathfrak{U}(X)$ of X as in Corollary 6.20. We denote the choices of $\mathcal{K}^1, \mathcal{K}^2, \mathfrak{U}(X)$ by Ξ. ■

Definition 14.3 In Situation 14.2, we assume $\dim(X, \widehat{\mathcal{U}}) = 0$. We consider $p \in U_{\mathfrak{p}} \cap \mathfrak{U}(X) \cap |\mathcal{K}^1|$ such that $\mathfrak{s}^n_{\mathfrak{p}}(p) = 0$. (This means that there is a branch of $\mathfrak{s}^n_{\mathfrak{p}}$ that vanishes at p.) Let $\mathfrak{V} = (V, \Gamma, E, \psi, \hat{\psi})$ be an orbifold chart of $(U_{\mathfrak{p}}, E_{\mathfrak{p}})$ at p. We take a representative $(\mathfrak{s}^n_{\mathfrak{p},1}, \dots, \mathfrak{s}^n_{\mathfrak{p},\ell})$ of $\mathfrak{s}^n_{\mathfrak{p}}$ on $\mathfrak{V}$. Let $\tilde{p} \in V$ such that $[\tilde{p}] = p$.

(1) For $i = 1, \dots, \ell$ we put:

$$\epsilon_{p,i} = \begin{cases} 0 & \text{if } \mathfrak{s}^n_{\mathfrak{p},i}(\tilde{p}) \neq 0. \\ +1 & \text{if } \mathfrak{s}^n_{\mathfrak{p},i}(\tilde{p}) = 0 \text{ and (14.1) below is orientation preserving.} \\ -1 & \text{if } \mathfrak{s}^n_{\mathfrak{p},i}(\tilde{p}) = 0 \text{ and (14.1) below is orientation reversing.} \end{cases}$$

In the current case of virtual dimension 0, the transversality hypothesis implies that the derivative

$$D_{\tilde{p}}\mathfrak{s}^n_{\mathfrak{p},i} : T_{\tilde{p}}V \to E_{\tilde{p}} \tag{14.1}$$

becomes an isomorphism at every point $\tilde{p}$ satisfying $\mathfrak{s}^n_{\mathfrak{p},i}(\tilde{p}) = 0$.

(2) The *multiplicity* m_p of $(\mathfrak{s}^n_{\mathfrak{p}})^{-1}(0)$ at p is a rational number and is defined by

$$m_p = \frac{1}{\ell \# \Gamma} \sum_{i=1}^{\ell} \epsilon_{p,i}.$$

Lemma 14.4

(1) *The multiplicity m_p in Definition* 14.3 *is independent of the choice of representative $(\mathfrak{s}^n_{\mathfrak{p},1}, \dots, \mathfrak{s}^n_{\mathfrak{p},\ell})$.*

(2) *If* $q \in U_{\mathfrak{p}\mathfrak{q}}$ *and* $p = \varphi_{\mathfrak{p}\mathfrak{q}}(q)$, *then the multiplicity at* p *is equal to the multiplicity at* q.

Proof This is immediate from the definition. □

Definition 14.5 In Situation 14.2 we assume $\dim(X, \widehat{\mathcal{U}}) = 0$. For each n, we define the *virtual fundamental chain* $[(X, \widehat{\mathcal{U}}, \Xi, \widehat{\mathfrak{s}^n})]$ of $(X, \widehat{\mathcal{U}}, \Xi, \widehat{\mathfrak{s}^n})$ by

$$[(X, \widehat{\mathcal{U}}, \Xi, \widehat{\mathfrak{s}^n})] = \sum_{p \in \mathfrak{U}(X) \cap |\mathcal{K}^1| \cap \bigcup_{\mathfrak{p} \in \mathfrak{P}} (\mathfrak{s}^n_{\mathfrak{p}})^{-1}(0)} m_p. \tag{14.2}$$

This is a rational number. Here the sum in the right hand side of (14.2) is defined as follows. Let us consider the *disjoint* union

$$\bigcup_{\mathfrak{p} \in \mathfrak{P}} (\mathfrak{U}(X) \cap \mathcal{K}^1_{\mathfrak{p}} \cap (\mathfrak{s}^n_{\mathfrak{p}})^{-1}(0)) \times \{\mathfrak{p}\}.$$

We define a relation $\sim$ on it by $(p, \mathfrak{p}) \sim (q, \mathfrak{q})$ if $\mathfrak{p} \le \mathfrak{q}$, $q = \varphi_{\mathfrak{q}\mathfrak{p}}(p)$ or $\mathfrak{q} \le \mathfrak{p}$, $p = \varphi_{\mathfrak{p}\mathfrak{q}}(q)$. This is an equivalence relation by Definition 3.15 (7). The set of the equivalence classes is denoted by $\mathfrak{U}(X) \cap |\mathcal{K}^1| \cap \bigcup_{\mathfrak{p} \in \mathfrak{P}} (\mathfrak{s}^n_{\mathfrak{p}})^{-1}(0)$. By Lemma 14.4 (2) the multiplicity m_p is a well-defined function on this set.

Remark 14.6 We note that in the case where $(X, \widehat{\mathcal{U}})$ has a boundary, the number $[(X, \widehat{\mathcal{U}}, \Xi, \widehat{\mathfrak{s}^n})]$ depends on the choice of the multivalued perturbation $\widehat{\mathfrak{s}} = \{\mathfrak{s}^n_{\mathfrak{p}}\}$. It also depends on n.

Definition 14.7 Let $F_a : \{n \in \mathbb{Z} \mid n \ge n_a\} \to \mathscr{X}$ be a family of maps parametrized by $a \in \mathscr{B}$. We say that F_a is *independent of the choice of* a *in the sense of* ◇ if the following holds:

◇ For $a_1, a_2 \in \mathscr{B}$ there exists $n_0 > \max\{n_{a_1}, n_{a_2}\}$ such that $F_{a_1}(n) = F_{a_2}(n)$ for all $n > n_0$.

The next proposition is a multivalued perturbation version of Proposition 7.81.

Proposition 14.8 *The number* $[(X, \widehat{\mathcal{U}}, \Xi, \widehat{\mathfrak{s}^n})]$ *as in Definition* 14.5 (14.2) *is independent of* Ξ *in the sense of* ◇.

Proof Let $\mathfrak{U}'(X)$ be an alternative choice of $\mathfrak{U}(X)$. By Corollary 6.20 we have

$$\mathfrak{U}(X) \cap |\mathcal{K}^1| \cap \bigcup_{\mathfrak{p} \in \mathfrak{P}} (\mathfrak{s}^n_{\mathfrak{p}})^{-1}(0) = \mathfrak{U}'(X) \cap |\mathcal{K}^1| \cap \bigcup_{\mathfrak{p} \in \mathfrak{P}} (\mathfrak{s}^n_{\mathfrak{p}})^{-1}(0).$$

Independence of the multiplicity m_p can be proved in the same way as the argument in Definition 14.5.

By Proposition 6.18 we have

$$\mathfrak{U}(X) \cap |\mathcal{K}^1| \cap \bigcup_{\mathfrak{p}\in\mathfrak{P}} (\mathfrak{s}_{\mathfrak{p}}^n)^{-1}(0) = \mathfrak{U}(X) \cap |\mathcal{K}^2| \cap \bigcup_{\mathfrak{p}\in\mathfrak{P}} (\mathfrak{s}_{\mathfrak{p}}^n)^{-1}(0).$$

Therefore (14.2) does not change when we take alternative $\mathcal{K}^1$ if we do not change $\mathcal{K}^2$. Moreover (14.2) does not change when we take alternative $\mathcal{K}^2$ if we do not change $\mathcal{K}^1$. □

Convention 14.9 Since $[(X, \widehat{\mathcal{U}}, \Xi, \widehat{\mathfrak{s}^n})]$ is independent of Ξ by Proposition 14.8, we will write it as $[(X, \widehat{\mathcal{U}}, \widehat{\mathfrak{s}^n})]$ hereafter.

The main result of this chapter is the following. This theorem is a multivalued perturbation version of Theorem 8.11.

Theorem 14.10 *Let $(X, \widehat{\mathcal{U}}, \widehat{\mathfrak{s}})$ be as in Situation 14.2. We assume $\dim(X, \widehat{\mathcal{U}}) = 1$. We consider its normalized boundary $\partial(X, \widehat{\mathcal{U}}) = (\partial X, \partial\widehat{\mathcal{U}})$, where $\widehat{\mathfrak{s}}$ induces a multivalued perturbation $\widehat{\mathfrak{s}_\partial}$ thereof and $(\partial X, \partial\widehat{\mathcal{U}}, \widehat{\mathfrak{s}_\partial^n})$ is as in Situation 14.2 with $\dim(\partial X, \partial\widehat{\mathcal{U}}) = 0$. Then the following formula holds for $n > n_0$:*

$$[(\partial X, \partial\widehat{\mathcal{U}}, \widehat{\mathfrak{s}_\partial^n})] = 0.$$

Theorem 14.10 is proved in Sects. 14.3 and 14.4.

We apply Theorem 14.10 to prove the next theorem. In particular, we can use it to prove that Gromov–Witten invariants are independent of the choice of multivalued perturbations.

Theorem 14.11

(1) *If $\widehat{\mathcal{U}}$ is an oriented and 0-dimensional good coordinate system without boundary of X and $\{\widehat{\mathfrak{s}^n}\}$ is its multivalued perturbation which is transversal to 0, then there exists a natural number $n_0(\widehat{\mathfrak{s}^n})$ such that then the rational number $[(X, \widehat{\mathcal{U}}, \widehat{\mathfrak{s}^n})]$ is independent of n for $n > n_0(\widehat{\mathfrak{s}^n})$.*

(2) *If $\widehat{\mathfrak{s}^{(k)n}}$ $k = 1, 2$ are two multivalued perturbations which are transversal to 0, then for $n > \max\{n(\widehat{\mathfrak{s}^{(k),n}}) \mid k = 1, 2\}$ we have*

$$[(X, \widehat{\mathcal{U}}, \widehat{\mathfrak{s}^{(1)n}})] = [(X, \widehat{\mathcal{U}}, \widehat{\mathfrak{s}^{(2)n}})]. \tag{14.3}$$

Proof We first show the following.

(*) In the situation of (2) there exists n_0 such that if $n > n_0$ then (14.3) holds.

Note at this stage we did not prove the independence of these numbers on n yet.

We consider the direct product of the good coordinate system $[0, 1] \times \widehat{\mathcal{U}}$ on $[0, 1] \times X$ so such that its restriction to $\{0\} \times \widehat{\mathcal{U}}$ (resp. $\{1\} \times \widehat{\mathcal{U}}$) is $\widehat{\mathfrak{s}^{(0)n}}$ (resp. $\widehat{\mathfrak{s}^{(1)n}}$) and such that it is transversal to 0. Moreover, we can take it so that it is constant in $[0, 1]$ direction in a neighborhood of $\partial[0, 1] \times \widehat{\mathcal{U}}$. (See the proof of

Theorem 8.15.) We apply Theorem 14.10 to $([0,1] \times X, [0,1] \times \widehat{\mathcal{U}}, \widehat{\mathfrak{s}^n})$ to obtain $[X, \widehat{\mathcal{U}}, \widehat{\mathfrak{s}^{(0)n}}] = [X, \widehat{\mathcal{U}}, \widehat{\mathfrak{s}^{(1)n}}]$. We have thus proved (*).

We next observe that $\{\widehat{\mathfrak{s}^{n+1}}\}$ is a multivalued perturbation if $\{\widehat{\mathfrak{s}^n}\}$ is so and $\{\widehat{\mathfrak{s}^{n+1}}\}$ is transversal to 0 if $\{\widehat{\mathfrak{s}^n}\}$ is so. Hence we can choose multivalued perturbations on $([0,1] \times X, [0,1] \times \widehat{\mathcal{U}}, \widehat{\mathfrak{s}^n})$ which is transversal to 0 and interpolates $\{\widehat{\mathfrak{s}^{n+1}}\}$ and $\{\widehat{\mathfrak{s}^n}\}$. Therefore we use Theorem 14.10 in the same way as the above proof of (*) to show that there exists a number n_0 such that if $n > n_0$ $[X, \widehat{\mathcal{U}}, \widehat{\mathfrak{s}^n}] = [X, \widehat{\mathcal{U}}, \widehat{\mathfrak{s}^{n+1}}]$. This implies (1).

Now (1) and (*) imply (2). □

Definition 14.12 In the situation of (2) we call the rational number $[(X, \widehat{\mathcal{U}}, \widehat{\mathfrak{s}^n})]$ the *virtual fundamental class* of $(X, \widehat{\mathcal{U}})$ and write $[(X, \widehat{\mathcal{U}})] \in \mathbb{Q}$. (Note that we assumed that $(X, \widehat{\mathcal{U}})$ has no boundary.)

We can also prove the following analogue of Proposition 8.16

Proposition 14.13 *Let $\mathfrak{X}_i = (X_i, \widehat{\mathcal{U}^i})$ be spaces with a good coordinate systems without boundary of dimension 0. Suppose that there exists a K-space $\mathfrak{Y} = (Y, \widehat{\mathcal{U}})$ with boundary (but without corners) such that*

$$\partial \mathfrak{Y} = -\mathfrak{X}_1 \cup \mathfrak{X}_2.$$

Here $-\mathfrak{X}_1$ is the smooth correspondence $\mathfrak{X}_1$ with opposite orientation. Then we have

$$[(X_1, \widehat{\mathcal{U}^1})] = [(X_2, \widehat{\mathcal{U}^2})]. \tag{14.4}$$

Proof Using Theorem 14.10 instead of Stokes' formula, the proof of Proposition 14.13 goes in the same way as the proof of Proposition 8.16. □

Remark 14.14 Let $(X, \widehat{\mathcal{U}})$ be a space with a good coordinate system of dimension 0 and without boundary. We take its support system $\mathcal{K}$. We can define a C^1 difference between the Kuranishi map $\{s_{\mathfrak{p}}\}$ and multisection $\{\mathfrak{s}_{\mathfrak{p}}\}$ on a neighborhood of $\mathcal{K}_{\mathfrak{p}}$. This C^1 distance up to equivalence is independent of choices (such as Riemannian metric of $U_{\mathfrak{p}}$ and connection of $\mathcal{E}_{\mathfrak{p}}$) since we have only a finite number of charts and $\mathcal{K}_{\mathfrak{p}}$ is compact. Now we can prove the existence of ϵ with the following properties. Suppose the C^1 distance between $\{\mathfrak{s}_{\mathfrak{p}}\}$ and the Kuranishi map $\{s_{\mathfrak{p}}\}$ is smaller than ϵ. Assume also that $\{\mathfrak{s}_{\mathfrak{p}}\}$ is transversal to 0. We can define a virtual fundamental chain $[(X, \widehat{\mathcal{U}}, \widehat{\mathfrak{s}})]$ by counting the zero set of $\widehat{\mathfrak{s}}$ with weight in the same way as Definition 14.3. Then this number is equal to the number defined in Definition 14.12.

14.2 Virtual Fundamental Chain of 0-Dimensional K-Space with Multivalued Perturbation

In this section we translate the discussion of the previous section into the situation where we have a Kuranishi structure instead of a good coordinate system. We also show the counterpart of the composition formula for the case with mutlivalued perturbations instead of CF-perturbations.

In Definition 14.5 we defined a virtual fundamental chain ($\in \mathbb{Q}$) of $(X, \widehat{\mathcal{U}}, \widehat{\mathfrak{s}})$ where $\widehat{\mathcal{U}}$ is a good coordinate system of X and $\widehat{\mathfrak{s}}$ is its multivalued perturbation transversal to 0. We adopt this story and proceed in the same way as in Sect. 9.2 to define a virtual fundamental chain ($\in \mathbb{Q}$) for a 0-dimensional K-space.

Situation 14.15

(1) We consider a triple $(X, \widehat{\mathcal{U}}, \widehat{\mathfrak{s}})$ such that $(X, \widehat{\mathcal{U}})$ is a K-space and $\widehat{\mathfrak{s}}$ is a multivalued perturbation of $\widehat{\mathcal{U}}$.
(2) We consider a triple $(X, \widehat{\mathcal{U}}, \widehat{\mathfrak{s}})$ such that $(X, \widehat{\mathcal{U}})$ is a good coordinate system and $\widehat{\mathfrak{s}}$ is a multivalued perturbation of $\widehat{\mathcal{U}}$.

■

Definition 14.16

(1) Let $(X, \mathfrak{X}_i) = (X, \widehat{\mathcal{U}_i}, \widehat{\mathfrak{s}_i})$ $(i = 1, 2)$ be as in Situation 14.15 (1). We say $\mathfrak{X}_2$ is a *thickening* of $\mathfrak{X}_1$ and write $(X, \widehat{\mathcal{U}_1}, \widehat{\mathfrak{s}_1}) < (X, \widehat{\mathcal{U}_2}, \widehat{\mathfrak{s}_2})$ if there exists a KK-embedding of Kuranishi structures $\widehat{\mathcal{U}_1} \to \widehat{\mathcal{U}_2}$ by which $\widehat{\mathcal{U}_2}$ is a thickening of $\widehat{\mathcal{U}_1}$. (See Definition 5.3.) $\widehat{\mathfrak{s}_2}, \widehat{\mathfrak{s}_1}$ are compatible with this embedding.
(2) In the case where one or both of the objects are good coordinate systems, we can define the notion of thickening such as $(X, \widehat{\mathcal{U}_1}, \widehat{\mathfrak{s}_1}) < (X, \widehat{\widehat{\mathcal{U}_2}}, \widehat{\widehat{\mathfrak{s}_2}})$, $(X, \widehat{\widehat{\mathcal{U}_1}}, \widehat{\widehat{\mathfrak{s}_1}}) < (X, \widehat{\mathcal{U}_2}, \widehat{\mathfrak{s}_2})$, $(X, \widehat{\widehat{\mathcal{U}_1}}, \widehat{\widehat{\mathfrak{s}_1}}) < (X, \widehat{\widehat{\mathcal{U}_2}}, \widehat{\widehat{\mathfrak{s}_2}})$, in the same way.

Below we will define a virtual fundamental chain of a triple $(X, \widehat{\mathcal{U}}, \widehat{\mathfrak{s}})$ as in Situation 14.15 (1) provided that $\widehat{\mathfrak{s}}$ is transversal to 0. We begin with the following lemma.

Lemma 14.17 *Let* $(X, \widehat{\widehat{\mathcal{U}_i}}, \widehat{\widehat{\mathfrak{s}_i}})$ $i = 1, 2$ *be as in Situation* 14.15 *(2) such that* $\dim(X, \widehat{\widehat{\mathcal{U}_i}}) = 0$. *We assume*

$$(X, \widehat{\widehat{\mathcal{U}_1}}, \widehat{\widehat{\mathfrak{s}_1}}) < (X, \widehat{\widehat{\mathcal{U}_2}}, \widehat{\widehat{\mathfrak{s}_2}})$$

and $\widehat{\widehat{\mathfrak{s}_1}}$ *and* $\widehat{\widehat{\mathfrak{s}_2}}$ *are transversal to* 0. *Then there exists* $n_0 > 0$ *such that the following holds for* $n \geq n_0$*:*

$$[(X, \widehat{\widehat{\mathcal{U}_1}}, \widehat{\widehat{\mathfrak{s}_1^n}})] = [(X, \widehat{\widehat{\mathcal{U}_2}}, \widehat{\widehat{\mathfrak{s}_2^n}})].$$

Proof The proof is entirely the same as the proof of Proposition 9.16. □

Definition 14.18 Let $(X, \widehat{\mathcal{U}}, \widehat{\mathfrak{s}})$ be as in Situation 14.15 (1) and $\dim(X, \widehat{\mathcal{U}}) = 0$. We assume that $\widehat{\mathfrak{s}}$ is transversal to 0. Using Proposition 6.32, we take $(X, \widehat{\widehat{\mathcal{U}}}, \widehat{\widehat{\mathfrak{s}}})$ such that

$$(X, \widehat{\mathcal{U}}, \widehat{\mathfrak{s}}) < (X, \widehat{\widehat{\mathcal{U}}}, \widehat{\widehat{\mathfrak{s}}}).$$

Then, for sufficiently large n, we define the *virtual fundamental chain of* $(X, \widehat{\mathcal{U}}, \widehat{\mathfrak{s}^n})$ by

$$[(X, \widehat{\mathcal{U}}, \widehat{\mathfrak{s}^n})] = [(X, \widehat{\widehat{\mathcal{U}}}, \widehat{\widehat{\mathfrak{s}^n}})] \in \mathbb{Q}, \tag{14.5}$$

Lemma 14.19 *The right hand side of* (14.5) *is independent of* $(X, \widehat{\widehat{\mathcal{U}}}, \widehat{\widehat{\mathfrak{s}'^n}})$ *but depends only on* $(X, \widehat{\mathcal{U}}, \widehat{\mathfrak{s}^n})$ *in the sense of* $\diamondsuit$. *(It depends on n.)*

Proof The proof is entirely the same as the proof of Theorem 9.14. □

Now we state the multivalued-perturbation-on-K-space version of Stokes' formula.

Theorem 14.20 *Let* $(X, \widehat{\mathcal{U}}, \widehat{\mathfrak{s}^n})$ *be as in Situation* 14.15 *(1) and* $\dim(X, \widehat{\mathcal{U}}) = 1$. *We assume that* $\widehat{\mathfrak{s}^n}$ *is transversal to* 0. *Then we have*

$$[\partial(X, \widehat{\mathcal{U}}, \widehat{\mathfrak{s}^n})] = 0$$

for sufficiently large n. Here $\partial(X, \widehat{\mathcal{U}}, \widehat{\mathfrak{s}^n})$ *is the normalized boundary* $\partial(X, \widehat{\mathcal{U}})$ *together with the restriction of* $\widehat{\mathfrak{s}^n}$ *to the boundary.*

Proof This immediately follows from Proposition 14.10 and the definition. □

In the same way as Theorem 14.11, we can prove the following.

Theorem 14.21

(1) *Let* $(X, \widehat{\mathcal{U}})$ *be an oriented and* 0*-dimensional K-space without boundary and* $\widehat{\mathfrak{s}^n}$ *its multivalued perturbation which is transversal to* 0. *Then there exists a natural number* $n(\widehat{\mathfrak{s}^n})$ *such that the rational number* $[(X, \widehat{\mathcal{U}}, \widehat{\mathfrak{s}^n})]$ *is independent of n for* $n > n(\widehat{\mathfrak{s}^n})$.
(2) *Let* $\widehat{\mathfrak{s}^{(k)n}}$ $k = 1, 2$ *be two multivalued perturbations which are transversal to* 0. *Then for* $n > \max\{n(\widehat{\mathfrak{s}^{(k),n}}) \mid k = 1, 2\}$ *we have*

$$[(X, \widehat{\mathcal{U}}, \widehat{\mathfrak{s}^{(1)n}})] = [(X, \widehat{\mathcal{U}}, \widehat{\mathfrak{s}^{(2)n}})]. \tag{14.6}$$

We can also prove the K-space version of Proposition 14.13.

To state the analogue of the composition formula we begin with explaining the situation and introducing the notation.

Definition 14.22 Suppose X_1, X_2 have Kuranishi structures $\widehat{\mathcal{U}_1}$, $\widehat{\mathcal{U}_2}$ respectively, and $\widehat{f_i} : (X_i, \widehat{\mathcal{U}_i}) \to R$ is strongly smooth, for $i = 1, 2$. We assume that R is a 0-dimensional compact manifold, that is nothing but a finite set. Then they are automatically transversal and we can define the fiber product $(X_1 \times_R X_2, \widehat{\mathcal{U}_1} \times_R \widehat{\mathcal{U}_2})$. We call this fiber product the *direct-like product.*

The next lemma is trivial to prove.

Lemma-Definition 14.23 *Suppose we are in the situation of Definition* 14.22.

(1) *If* $\widehat{\mathfrak{s}_i}$ *(resp.* $\mathfrak{s}_i$*) is a multivalued perturbation (resp. multisections) of* $\widehat{\mathcal{U}_i}$ *for* $i = 1, 2$, *then they induce a multivalued perturbation (resp. multisection) of their fiber product in a canonical way. We call it the fiber product of the multivalued perturbation (resp. multisection) and write* $\widehat{\mathfrak{s}_1} \times_R \widehat{\mathfrak{s}_2}$ *(resp.* $\mathfrak{s}_1 \times_R \mathfrak{s}_2$*).*
(2) *If* $\widehat{\mathfrak{s}_i}$ *(resp.* $\mathfrak{s}_i$*) is transversal to* 0, *then its direct-like product is transversal to* 0.

Situation 14.24 Suppose we are in Situation 14.22. We assume that we are given the triples $(X_i, \widehat{\mathcal{U}_i}, \widehat{\mathfrak{s}_i})$ for $i = 1, 2$ of multivalued perturbations $\widehat{\mathfrak{s}_i}$.

For $r \in R$ we put $f_i^{-1}(r) = \{x \in X_i \mid f_i(x) = r\}$. They carry Kuranishi structures and multivalued perturbations $\widehat{\mathcal{U}_i}, \widehat{\mathfrak{s}_i}$. Thus we obtain triples $f_i^{-1}(r) \cap (X_i, \widehat{\mathcal{U}_i}, \widehat{\mathfrak{s}_i})$ as in Situation 14.15 (1). ■

The next proposition is an analogue of the composition formula.

Theorem 14.25 *In Situation* 14.24 *we assume in addition that for* $i = 1, 2$*:*

(1) $\dim(X_i, \widehat{\mathcal{U}_i}) = 0$,
(2) $\widehat{\mathfrak{s}_i}$ *is transversal to* 0.

Then the direct-like product

$$(X_1, \widehat{\mathcal{U}_1}, \widehat{\mathfrak{s}_1}) \times_R (X_2, \widehat{\mathcal{U}_2}, \widehat{\mathfrak{s}_2})$$

is 0*-dimensional and transversal to* 0. *Moreover we have*

$$\begin{aligned} &[(X_1, \widehat{\mathcal{U}_1}, \widehat{\mathfrak{s}_1^n}) \times_R (X_2, \widehat{\mathcal{U}_2}, \widehat{\mathfrak{s}_2^n})] \\ &= \sum_{r \in R} [f_1^{-1}(r) \cap (X_1, \widehat{\mathcal{U}_1}, \widehat{\mathfrak{s}_1^n})][f_2^{-1}(r) \cap (X_2, \widehat{\mathcal{U}_2}, \widehat{\mathfrak{s}_2^n})], \end{aligned} \tag{14.7}$$

for sufficiently large n.

Proof The proof is the same as the proof of Proposition 10.24.[1] □

We also note the following.

Lemma 14.26 *In the situation of Theorem* 14.25 *we replace (1) by the following assumption:*

(1)' $\dim(X_1, \widehat{\mathcal{U}_1}) = -\dim(X_2, \widehat{\mathcal{U}_2}) \neq 0$.

[1]It is actually easier than that.

Except for this point, we assume the same conditions as in Theorem 14.25. *Then for sufficiently large n we have*

$$[(X_1, \widehat{\mathcal{U}_1}, \widehat{\mathfrak{s}_1^n}) \times_R (X_2, \widehat{\mathcal{U}_2}, \widehat{\mathfrak{s}_2^n})] = 0.$$

Proof We may assume $\dim(X_1, \widehat{\mathcal{U}_1}) < 0$. Then by Lemma 14.1, $(\mathfrak{s}_1^n)^{-1}(0)$ is an empty set. Therefore $(\mathfrak{s}_1^n \times_R \mathfrak{s}_2^n)^{-1}(0)$ is an empty set. The lemma follows. □

14.3 A Simple Morse Theory on a Space with a Good Coordinate System

In this section and the next, we prove Theorem 14.10. We can prove this theorem by taking an appropriate perturbation so that its zero set has a triangulation. This is the proof given in [FOn2, Theorem 6.2]. (Theorem 14.10 is a special case of [FOn2, Theorem 6.2] where Y is a point.) Here we give an alternative proof without using triangulation. The proof is based on Morse theory on a space with a good coordinate system.

Let $(\mathcal{K}', \mathcal{K})$ be a support pair of $\widehat{\mathcal{U}}$. Note that $\widehat{\mathfrak{s}}$ is a multivalued perturbation of $(\widehat{\mathcal{U}}, \mathcal{K})$. We consider a strongly smooth function $f : U(|\mathcal{K}|) \to [0, \infty)$ of a neighborhood $U(|\mathcal{K}|)$ of $|\mathcal{K}|$ in $|\widehat{\mathcal{U}}|$ such that

$$f^{-1}(0) = |\partial \widehat{\mathcal{U}}| \cap U(|\mathcal{K}|). \tag{14.8}$$

Definition 14.27 We say that f is *normally positive at the boundary* if the following holds:

(1) Suppose $p \in \overset{\circ}{S}_1 U_{\mathfrak{p}}$ and identify its neighborhood with $(W \times [0, 1))/\Gamma$ where p corresponds to $(p_0, 0)$ and $p_0 \in W$ is an interior point. Let $\vec{v} \in T_{(p_0,0)}(W \times [0, 1))$ such that $\vec{v} = (\vec{v}_0, v)$ with $v > 0$. Then $\vec{v}(f) > 0$.
(2) If $p \in U(|\mathcal{K}|) \setminus |\partial \widehat{\mathcal{U}}|$ then $f(p) > 0$.
(3) There exists $\epsilon_0 > 0$ such that if $\epsilon < \epsilon_0$ is positive then ϵ is not a critical value of $f|_{U_{\mathfrak{p}}}$ for any $\mathfrak{p}$.

Lemma 14.28 *There exists a strongly smooth function f as above that is normally positive at the boundary and satisfies* (14.8).

Proof We take $\mathcal{K}^+$ such that $(\mathcal{K}, \mathcal{K}^+)$ is a support pair. Let $x \in |\mathcal{K}|$. We then take a maximal $\mathfrak{p}$ such that $x \in \mathcal{K}_{\mathfrak{p}}$. We take a sufficiently small neighborhood Ω_x of x in $|\widehat{\mathcal{U}}|$ such that $\Omega_x \cap \mathcal{K}_{\mathfrak{q}} \neq \emptyset$ implies $\mathfrak{q} \le \mathfrak{p}$. Then we may slightly shrink Ω_x so that $\Omega_x \cap |\mathcal{K}|$ is contained in $\Omega_x \cap \mathcal{K}_{\mathfrak{p}}^+$. Note that $\Omega_x \cap \mathcal{K}_{\mathfrak{p}}^+$ is an orbifold with corners. Therefore a neighborhood of x in it is identified with a point $(x_0, (0, \dots, 0))$ in

$(V_x \times [0,1)^{k_x})/\Gamma_x$. Here $x \in \overset{\circ}{S}_{k_x}(\mathcal{K}_{\mathfrak{p}}^+)$. We can choose this coordinate so that the Γ_x action on $V_x \times [0,1)^{k_x}$ is given in the form as

$$(y, (t_1, \ldots, t_{k_x})) \mapsto (\gamma(y, (t_1, \ldots, t_{k_x})), (t_{\sigma(1)}, \ldots, t_{\sigma(k_x)})),$$

that is, the action on $[0,1)^{k_x}$ factor is given by permuting its components. We define a function f_x on $\Omega_x \cap \mathcal{K}_{\mathfrak{p}}^+$ by

$$f_x(y, (t_1, \ldots, t_{k_x})) = \begin{cases} t_1 t_2 \cdots t_{k_x} & \text{if } k_x > 0, \\ 1 & \text{if } k_x = 0. \end{cases} \tag{14.9}$$

We may regard it as a strongly smooth function on a neighborhood of $W_x = \Omega_x \cap |\mathcal{K}|$. We may assume that if $x \in \overset{\circ}{S}_{k_x}(\mathcal{K}_{\mathfrak{p}}^+)$ then $W_x \cap \overset{\circ}{S}_{k_x+1}(\mathcal{K}_{\mathfrak{p}}) = \emptyset$ for any $\mathfrak{p}$. Let $W_{0,x}$ be a relatively compact neighborhood of x in W_x. We take finitely many points x_i, $i = 1, \ldots, N$ of $|\mathcal{K}|$ such that

$$\bigcup_{i=1}^{N} W_{0,x_i} \supseteq |\mathcal{K}|.$$

Then there exist strongly smooth functions χ_i, $i = 1, \ldots, N$ on a neighborhood of $|\mathcal{K}|$ to $[0,1]$ such that:

(1) The support of χ_i is in W_{0,x_i}.
(2) $\sum_{i=1}^{N} \chi_i \equiv 1$.

We can prove the existence of such χ_i in the same way as in the proof of Proposition 7.68 by using Lemma 7.67. We put

$$f = \sum_{i=1}^{N} \chi_i f_{x_i}.$$

By (14.9) this function has the required properties. □

Definition 14.29 Let f be a strongly smooth function defined on a neighborhood of $|\mathcal{K}|$. We say that a point $p \in |\mathcal{K}|$ is a *critical point* of f if there exists $\mathfrak{p}$ such that $p \in \mathcal{K}_{\mathfrak{p}}$ and p is a critical point of the restriction of f to a neighborhood of p in $U_{\mathfrak{p}}$. We denote by $\mathrm{Crit}(f)$ the set of all critical points of f.

Definition 14.30 Let f be a strongly smooth function defined on a neighborhood of $|\mathcal{K}|$. We say that f is a *Morse function* if the restriction of f to each $U_{\mathfrak{p}}$ is a Morse function. (Namely its Hessian at any critical point is non-degenerate.)

Remark 14.31 Suppose $f : |\mathcal{K}| \to \mathbb{R}$ is a strongly smooth function and $\mathcal{K}'$ is a support system with $\mathcal{K}' < \mathcal{K}$, then the restriction $f|_{|\mathcal{K}'|} : |\mathcal{K}'| \to \mathbb{R}$ is strongly smooth. If f is Morse, so is $f|_{|\mathcal{K}'|}$.

Let $\mathcal{K}''$ be a support system with $\mathcal{K}' < \mathcal{K}'' < \mathcal{K}$. We use the following:

Lemma 14.32 *For each n, there exist finitely many orbifold charts (V_i, Γ_i, ψ_i) of $U_{\mathfrak{p}_i}$ ($\mathfrak{p}_i \in \mathfrak{P}$) and smooth embeddings $h_i : [0, 1] \to V_i$ such that*

$$\mathfrak{U}(X) \cap \bigcup_{\mathfrak{p}\in\mathfrak{P}} (\mathfrak{s}^n_{\mathfrak{p}})^{-1}(0) \cap \mathcal{K}''_{\mathfrak{p}} \subset \bigcup_{i=1}^{N} [(\psi_i \circ h_i)((0, 1))]. \tag{14.10}$$

Here $\mathfrak{U}(X)$ is as in Corollary 6.20.

Proof Since all the branches of $\mathfrak{s}^n_{\mathfrak{p}}$ are transversal to 0 and the (virtual) dimension is 1, locally the zero set of $\mathfrak{s}^n_{\mathfrak{p}}$ is a one-dimensional manifold. Then the lemma follows from compactness of the left hand side. (Corollary 6.20.) □

Proposition 14.33 *Suppose that* $\dim U_{\mathfrak{p}} \geq 2$ *for each $\mathfrak{p}$. Then there exists a strongly smooth function f on a neighborhood $\mathfrak{U}(X)$ of X in $|\mathcal{K}|$ such that:*

(1) *f is normally positive at the boundary.*
(2) *f satisfies* (14.8).
(3) *f is a Morse function.*
(4) *The composition $f \circ \psi_i \circ h_i : (0, 1) \to \mathbb{R}$ is a Morse function for each h_i : $[0, 1] \to V_i$ given in Lemma* 14.32.

Proof The proof of the lemma is a minor modification of a standard argument. We give a proof below for completeness' sake. We will use certain results concerning denseness of the set of Morse functions, which we will prove in Sect. 14.4.

We make a choice of Fréchet space we work with. Take a support system $\mathcal{K}^+$ such that $(\mathcal{K}, \mathcal{K}^+)$ is a support pair and define the set

$$C^\infty(\mathcal{K}^+) = \{(f_{\mathfrak{p}})_{\mathfrak{p}\in\mathfrak{P}} \in \prod_{\mathfrak{p}\in\mathfrak{P}} C^\infty(\mathcal{K}^+_{\mathfrak{p}}) \mid f_{\mathfrak{p}} \circ \varphi_{\mathfrak{p}\mathfrak{q}} = f_{\mathfrak{q}}, \text{on } \varphi^{-1}_{\mathfrak{p}\mathfrak{q}}(\mathcal{K}^+_{\mathfrak{p}}) \cap \mathcal{K}^+_{\mathfrak{q}}\}. \tag{14.11}$$

Here $C^\infty(\mathcal{K}^+_{\mathfrak{p}})$ is the space of C^∞ functions of an orbifold $U_{\mathfrak{p}}$ defined on its compact subset $\mathcal{K}^+_{\mathfrak{p}}$ and so is a Fréchet space with respect to the C^∞ topology. Then the set $C^\infty(\mathcal{K}^+)$ is a closed subspace of a finite product of the Fréchet spaces and so is a Fréchet space.

Let $f_0 \in C^\infty(\mathcal{K}^+_{\mathfrak{p}})$ be a function satisfying (1)(2) above. (The existence of such an f_0 follows from Lemma 14.28.) We can take another neighborhood $\mathfrak{U}_0$ of ∂X in $|\mathcal{K}|$ such that $\text{Crit} f_0 \cap \mathfrak{U}_0 = \emptyset$. We take a neighborhood $\mathfrak{U}_1$ of ∂X in $|\mathcal{K}|$ such that $\overline{\mathfrak{U}_1} \subset \mathfrak{U}_0$. Let $C^\infty(\mathcal{K}^+)_0$ be the set of all $f \in C^\infty(\mathcal{K}^+)$ that vanish on $\overline{\mathfrak{U}_1}$. Then $C^\infty(\mathcal{K}^+)_0$ itself is a Fréchet space.

Lemma 14.34 *For each $p \in X \setminus \mathfrak{U}_0$ there exists a compact neighborhood $\mathfrak{U}_p$ of p such that the set of $g \in C^\infty(\mathcal{K}^+)_0$ satisfying conditions (1)(2) below is a dense subset of $C^\infty(\mathcal{K}^+)_0$.*

(1) *$f_0 + g$ is a Morse function on $\mathfrak{U}_p$.*
(2) *The composition $(f_0 + g) \circ \psi_i \circ h_i : (0, 1) \to \mathbb{R}$ is a Morse function on $h_i^{-1}(\psi_i^{-1}(\mathfrak{U}_p))$ for each $h_i : [0, 1] \to V_i$ in Lemma* 14.32.

Proof Let $\mathfrak{p} \in \mathfrak{P}$ be the maximal element of $\{\mathfrak{p} \mid p \in |\mathcal{K}_\mathfrak{p}|\}$. (Maximal element exists because of Definition 3.15 (6).) Let Ω_p be a neighborhood of p in $\operatorname{Int}\mathcal{K}_\mathfrak{p}^+$. We may choose Ω_p sufficiently small so that the following $(*)$ holds:

(*) If $\mathcal{K}_\mathfrak{q} \cap \Omega_p \neq \emptyset$, then $\mathfrak{q} \le \mathfrak{p}$ and $K_\mathfrak{q}^+ \cap \Omega_p$ is an open subset of $U_\mathfrak{q}$.

Sublemma 14.35 *For each $\mathfrak{q} \le \mathfrak{p}$ the set of $g \in C^\infty(\mathcal{K}^+)_0$ satisfying conditions (a)(b) below is an open dense subset of $C^\infty(\mathcal{K}^+)_0$ for each $n = 1, 2, \ldots$.*

(a) *The restriction of $f_0 + g$ to $\mathcal{K}_\mathfrak{q} \cap \Omega_p$ is a Morse function on $\mathcal{K}_\mathfrak{q} \subset U_\mathfrak{q}$.*
(b) *The composition $(f_0 + g) \circ h_i : (0, 1) \to \mathbb{R}$ is a Morse function on $h_i^{-1}(\psi_i^{-1}(\Omega_p))$ for each $h_i : [0, 1] \to V_i$ in Lemma* 14.32.

Proof This is an immediate consequence of Propositions 14.41 and 14.47, which we will prove in Sect. 14.4. □

Lemma 14.34 follows from Sublemma 14.35 and Baire's category theorem. We note that we have used the fact that we consider only countably many n here. □

Proposition 14.33 easily follows from Lemma 14.34. □

Now we go back to the proof of Theorem 14.10. We take a strongly smooth function f that satisfies (1)–(4) of Proposition 14.33. For each n we take the maps $h_i : [0, 1] \to U_{\mathfrak{p}_i}$ as in Lemma 14.32. We take $0 < a_i < b_i < 1$ such that

$$\mathfrak{U}(X) \cap \bigcup_{\mathfrak{p} \in \mathfrak{P}} (\mathfrak{s}_\mathfrak{p}^n)^{-1}(0) \cap \mathcal{K}_\mathfrak{p}'' \subset \bigcup_{i=1}^{N} [(\psi_i \circ h_i)([a_i, b_i])] \tag{14.12}$$

and consider the union

$$S_n = \bigcup_{i=1}^{N} (f \circ \psi_i \circ h_i)(\mathrm{Crit}(f \circ \psi_i \circ h_i) \cap [a_i, b_i]).$$

$$S_0 = f(\mathrm{Crit}(f) \cap \mathcal{K}_\mathfrak{p}''), \qquad S = S_0 \cup \bigcup_{n \in \mathbb{Z}_+} S_n.$$

Here $\mathrm{Crit}(f \circ \psi_i \circ h_i)$ is the critical point set of the function $f \circ \psi_i \circ h_i : [0, 1] \to [0, \infty)$. Proposition 14.33 implies that S_n is a finite subset of $[0, \infty)$. (Here S_n depends on n since the set of curves $\{h_i\}$ depends on n.)

Lemma-Definition 14.36 *Suppose $s \notin S_0$. We consider $X^s = X \cap f^{-1}(s)$ and $U_\mathfrak{p}^s = f^{-1}(s) \cap \operatorname{Int}\mathcal{K}_\mathfrak{p}$. Then restricting $\mathcal{U}_\mathfrak{p}$ to $U_\mathfrak{p}^s$ and restricting the coordinate* *changes thereto, we obtain a good coordinate system on* X^s *of dimension* 0. *We write it as* $\widehat{\mathcal{U}}|_{\{U^s_{\mathfrak{p}}\}}$. *Moreover, the restriction of* $\mathfrak{s}^n_{\mathfrak{p}}$ *defines a multisection on* $\widehat{\mathcal{U}}|_{\{U^s_{\mathfrak{p}}\}}$. *The sequence of multisections* $\mathfrak{s}^n_{\mathfrak{p}}$ *defines a multivalued perturbation, which we denote by* $\widehat{\mathfrak{s}^{n,s}} = \{\mathfrak{s}^{n,s}_{\mathfrak{p}}\}$. *The multivalued perturbation* $\widehat{\mathfrak{s}^{n,s}}$ *is transversal to* 0 *if* $s \notin S_n$.

The proof is obvious from the definition. For $s \in [0, \infty) \setminus S$ we consider,

$$[(X^s, \widehat{\mathcal{U}}|_{\{U^s_{\mathfrak{p}}\}}, \widehat{\mathfrak{s}^{n,s}})] \in \mathbb{Q} \tag{14.13}$$

Lemma 14.37 *There exists* n_0 *such that we can define the virtual fundamental class* (14.13) *for* $n \geq n_0$, $s \in [0, \infty) \setminus S$. *It is independent of the choices in the sense of* $\diamondsuit$ *(Definition* 14.7*).*

Proof For each fixed s this is a consequence of Proposition 14.8. It remains to prove here the independence of n_0 on s. We consider support systems $\mathcal{K}_1 < \mathcal{K}_2 = \mathcal{K}' < \mathcal{K}_3 = \mathcal{K}''$ of $(X, \widehat{\mathcal{U}})$. They induce support systems $\mathcal{K}^{j,s}$ of $(X_s, \widehat{\mathcal{U}}|_{\{U^s_{\mathfrak{p}}\}})$ for $j = 1, 2, 3$, $s \in [0, \infty) \setminus S_0$ by $\mathcal{K}^{j,s}_{\mathfrak{p}} = \mathcal{K}^j_{\mathfrak{p}} \cap f_{\mathfrak{p}}^{-1}(s)$. We observe that if the conclusions of Propositions 6.17, 6.18 and Corollary 6.20 hold for $(X, \widehat{\mathcal{U}})$, $\mathcal{K}_j$ with δ and n_0, then the same conclusions hold for $(X_s, \widehat{\mathcal{U}}|_{\{U^s_{\mathfrak{p}}\}})$, $\mathcal{K}^{j,s}$ with the same δ and n_0.

Our s-parametrized family of multivalued perturbations are defined on the spaces $(X_s, \widehat{\mathcal{U}}|_{\{U^s_{\mathfrak{p}}\}})$, which is s-dependent. However, by the proof of Proposition 14.8, the numbers δ and n_0 are chosen so that Propositions 6.17, 6.18 and Corollary 6.20 hold. Therefore, we obtain the required independence of n_0 on s. □

Lemma 14.38 *For each* $s_0 \in (0, \infty)$ *and* $n > n_0$, *there exists a positive number* δ *such that* $[(X^s, \widehat{\mathcal{U}}|_{\{U^s_{\mathfrak{p}}\}}, \widehat{\mathfrak{s}^{n,s}})]$ *is independent of* $s \in (s_0 - \delta, s_0 + \delta) \setminus S$.

Lemma 14.39 *There exists* $\delta > 0$ *such that* $S_n \cap (0, \delta) = \emptyset$ *and*

$$[(X^s, \widehat{\mathcal{U}}|_{\{U^s_{\mathfrak{p}}\}}, \widehat{\mathfrak{s}^{n,s}})] = [(\partial X, \partial\widehat{\mathcal{U}}, \widehat{\mathfrak{s}^n_{\partial}})]. \tag{14.14}$$

Proof of Lemma 14.38 Let

$$\{p_i \mid i = 1, \dots, I\} = \mathfrak{U}(X) \cap |\widehat{\mathcal{U}}|_{\{U^{s_0}_{\mathfrak{p}}\}}| \cap \bigcup_{\mathfrak{p}} (\mathfrak{s}^{n,s_0}_{\mathfrak{p}})^{-1}(0).$$

The right hand side is a finite set by Proposition 14.33 (4). For each p_i we take $\mathfrak{p}_i$ with $p_i \in \operatorname{Int}\mathcal{K}_{\mathfrak{p}_i}$, and a representative $(\mathfrak{s}^n_{\mathfrak{p}_i,j})_{j=1,\dots,\ell_i}$ of $\mathfrak{s}^n_{\mathfrak{p}_i}$ on an orbifold chart $\mathfrak{V}_{p_i} = (V_{p_i}, \Gamma_{p_i}, E_{p_i}, \psi_{p_i}, \hat\psi_{p_i})$ of $(U_{\mathfrak{p}_i}, \mathcal{E}_{\mathfrak{p}_i})$ at p_i. Then for all sufficiently small $\delta > 0$ and $s \in (s_0 - \delta, s_0 + \delta) \setminus S_n$, we have the following. When we put $\hat f_i = f \circ \psi_{p_i} : V_{p_i} \to \mathbb{R}$:

(1) s is a regular value of $\hat f_i$,
(2) $\hat f_i^{-1}(s)$ intersects transversally to $(\mathfrak{s}^n_{\mathfrak{p}_i,j})^{-1}(0)$.

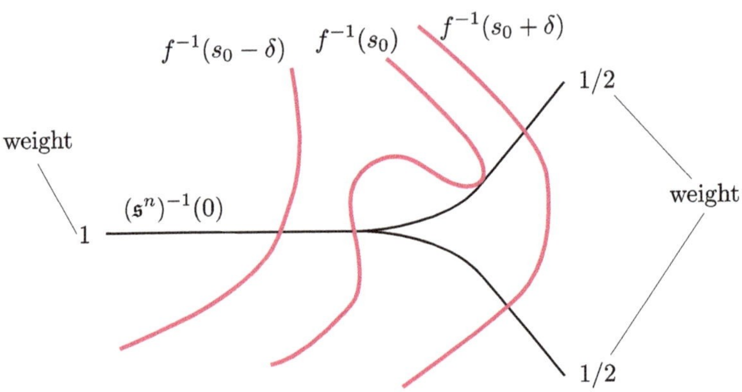

Fig. 14.1 Invariance of intersection number

Moreover, we can orient $(\mathfrak{s}^n_{\mathfrak{p}_i,j})^{-1}(0)$ for each i, j so that

$$[(X^s, \widehat{\mathcal{U}}|_{\{U^s_{\mathfrak{p}}\}}, \widehat{\mathfrak{s}^{n,s}})] = \sum_{i=1}^{I}\sum_{j=1}^{\ell_i} \frac{1}{\ell_i \# \Gamma_{\Gamma_{p_i}}} \hat{f}_i^{-1}(s) \cdot (\mathfrak{s}^n_{\mathfrak{p}_i,j})^{-1}(0).$$

Here $\frac{1}{\ell_i \# \Gamma_{\Gamma_{p_i}}} \hat{f}_i^{-1}(s) \cdot (\mathfrak{s}^n_{\mathfrak{p}_i,j})^{-1}(0)$ in the right hand side is the intersection number, that is, the order of the intersection counted with sign. We use compactness of $\mathfrak{U}(X) \cap \bigcup_{\mathfrak{p}} (\mathfrak{s}^n_{\mathfrak{p}})^{-1}(0)$ to show that the intersection number $\hat{f}_i^{-1}(s) \cdot (\mathfrak{s}^{\epsilon}_{\mathfrak{p}_i,j})^{-1}(0)$ is independent of $s \in (s_0-\delta, s_0+\delta) \setminus S$. Thus Lemma 14.38 follows. Note that we use Lemma 14.37 during this argument (Fig. 14.1). □

Proof of Lemma 14.39 The existence of δ with $S_n \cap [0,\delta) = \emptyset$ is an immediate consequence of Definition 14.27 (3). The formula (14.14) can be proved in the same way as the proof of Lemma 14.38. □

Now we are ready to complete the proof of Theorem 14.10. Lemma 14.38 implies that $[(X^s, \widehat{\mathcal{U}}|_{\{U^s_{\mathfrak{p}}\}}, \widehat{\mathfrak{s}^{n,s}})]$ is independent of $s \in (0,\infty) \setminus S$. (Note that we did not assume $s_0 \notin S$ in Lemma 14.38.) Therefore Lemma 14.39 implies that it is equal to $[(\partial X, \partial\widehat{\mathcal{U}}, \widehat{\mathfrak{s}^n_{\partial}})]$. On the other hand, by compactness of X we find that X^s is empty for sufficiently large s. Hence $[(\partial X, \partial\widehat{\mathcal{U}}, \widehat{\mathfrak{s}^n_{\partial}})] = 0$. The proof of Theorem 14.10 is complete. □

14.4 Denseness of the Set of Morse Functions on an Orbifold

In this section we review the proof of the denseness of the set of Morse functions on an orbifold. We consider the case of one orbifold chart. The case when we have several orbifold charts is the same but we do not need it. All the results of this section should be well-known. We include them here only for completeness' sake.

Situation 14.40 Let V be a manifold on which a finite group Γ acts effectively. We denote by $C^k_\Gamma(V)$ the set of all Γ-invariant C^k functions on V. ■

We take and fix a Γ-invariant Riemannian metric on V, which we use in the proof of some of the lemmas below.

Proposition 14.41 *Suppose we are in Situation 14.40. The set of all Γ-invariant smooth Morse functions on V is a countable intersection of open dense subsets in $C^\infty_\Gamma(V)$ in C^2 topology.*

Proof Let K be a compact subset of V. It suffices to prove that the set of all the functions in $C^2_\Gamma(V)$ which are Morse on K is open and dense. The openness is obvious. We will prove that it is also dense.

For $p \in V$ we put

$$\Gamma_p = \{\gamma \in \Gamma \mid \gamma p = p\},$$

and define

$$\begin{aligned} \overset{\circ}{V}(n) &= \{p \in V \mid \#\Gamma_p = n\}, \\ V(n) &= \{p \in V \mid \#\Gamma_p \geq n\}. \end{aligned} \tag{14.15}$$

Note that $\overset{\circ}{V}(n)/\Gamma$ is a smooth manifold.

Lemma 14.42 *Let $p \in \overset{\circ}{V}(n)$. Then p is a critical point of f if and only if p is a critical point of $f|_{\overset{\circ}{V}(n)}$.*

Proof This is a consequence of the fact that the directional derivative $X[f]$ is zero if $X \in T_pV$ is perpendicular to $\overset{\circ}{V}(n)$. This fact follows from the Γ invariance of f. □

We define the following sets:

$$A(n) = \{f \in C^\infty_\Gamma(V) \mid \text{all the critical points of } f \text{ on } V(n) \cap K \text{ are Morse}\},$$

$$B(n) = A(n+1) \cap \{f \in C^\infty_\Gamma(V) \mid \text{the restriction of } f \text{ to } \overset{\circ}{V}(n) \cap K \text{ is Morse}\}.$$

They are open sets.

Lemma 14.43 *If $A(n+1)$ is dense, then $B(n)$ is dense.*

Proof Let W be a relatively compact open subset of $K \cap \overset{\circ}{V}(n)$. We define a C^1-map $F : W \times C^2_\Gamma(\overline{W}) \to T^*\overset{\circ}{V}(n)$ by

$$F(x, f) = D_x f \in T^* \overset{\circ}{V}(n). \tag{14.16}$$

(Since f is Γ-invariant and $x \in \overset{\circ}{V}(n)$ it follows that $D_x f \in T^* \overset{\circ}{V}(n)$.) It is easy to see that F is transversal to the zero section $\overset{\circ}{V}(n) \subset T^* \overset{\circ}{V}(n)$. Here we identify $\overset{\circ}{V}(n)$ with the zero section of $T^* \overset{\circ}{V}(n)$. We put

$$\mathfrak{W} = \{(x, f) \in W \times C^2_\Gamma(\overline{W}) \mid F(x, f) \in \overset{\circ}{V}(n) \subset T^* \overset{\circ}{V}(n)\}.$$

Then $\mathfrak{W}$ is a sub-Banach manifold of the Banach manifold $W \times C^2_\Gamma(\overline{W})$ and the restriction of the projection

$$\mathrm{pr} : \mathfrak{W} \to C^2_\Gamma(\overline{W})$$

is a Fredholm map. Hence by the Sard–Smale theorem the regular value of pr is dense.

Sublemma 14.44 *If f is a regular value of* pr, *then $f|_{\overset{\circ}{V}(n)}$ is Morse on W.*

Proof Let $x \in W$ be a critical point of f. Then $(x, f) \in \mathfrak{W}$. We consider the following commutative diagram where all the vertical and horizontal lines are exact:

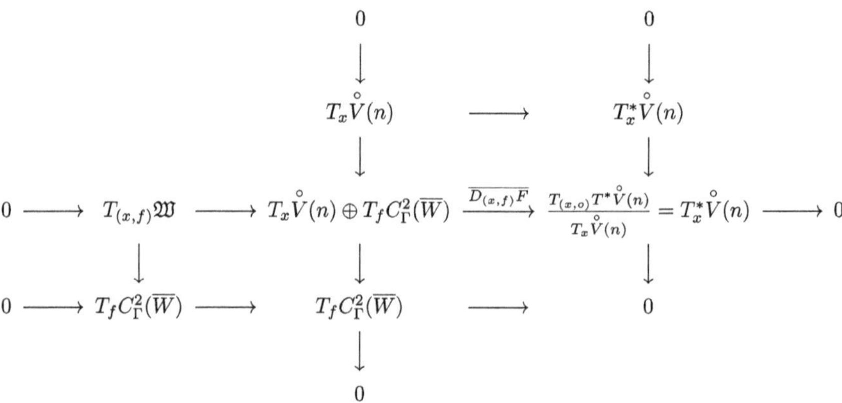

Here $\overline{D_{(x,f)}F}$ is the composition of $D_{(x,f)}F : T_x \overset{\circ}{V}(n) \oplus T_f C^2_\Gamma(\overline{W}) \to T_{(x,o)} T^* \overset{\circ}{V}(n)$ and the projection. Since f is a regular value the first vertical line is surjective. We can use it to show that the first horizontal line $T_x \overset{\circ}{V}(n) \to T^*_x \overset{\circ}{V}(n)$ is surjective by a simple diagram chase. This map is identified with the Hessian at x of $f|_{\overset{\circ}{V}(n)}$. The sublemma follows. $\square$

We observe that if $f \in A(n+1)$ then the set of critical points in $\overset{\circ}{V}(n) \cap K$ is compact. (This is because it does not have accumulation points on $V(n+1) \cap K$.) Therefore Lemma 14.43 follows from Sublemma 14.44 and the Sard–Smale theorem. □

Lemma 14.45 *If $B(n)$ is dense, then $A(n)$ is dense.*

Proof Let $f \in B(n)$. We remark that the set of critical points of $f|_{\overset{\circ}{V}(n)}$ on $\overset{\circ}{V}(n) \cap K$ is a finite set. This is because $f|_{\overset{\circ}{V}(n)}$ is a Morse function on $\overset{\circ}{V}(n) \cap K$ and $f|_{\overset{\circ}{V}(n)}$ does not have accumulation points on $V(n+1) \cap K$. Let $p_1, \dots, p_m$ be the critical points of $f|_{\overset{\circ}{V}(n)}$ on $\overset{\circ}{V}(n) \cap K$. Note that the Hessian of f at those points is non-degenerate on $T_{p_i}\overset{\circ}{V}(n)$ but may be degenerate in the normal direction to $\overset{\circ}{V}(n)$. We choose functions χ_i and V_i with the following properties:

(1) V_i is a neighborhood of p_i.
(2) The support of χ_i is in V_i
(3) $\chi_i \equiv 1$ in a neighborhood of p_i.
(4) $\overline{V}_i$ $(i = 1, \dots, m)$ are disjoint.
(5) $\overline{V}_i \cap V(n+1) = \emptyset$.
(6) If $\gamma p_i = p_j$, $\gamma \in \Gamma$, then $\gamma V_i = V_j$ and $\chi_j \circ \gamma = \chi_i$.

We use our Γ-invariant Riemannian metric and put

$$f_n(x) = d(x, \overset{\circ}{V}(n))^2$$

We choose V_i small so that $\chi_i f_n$ is a smooth function. (We use Item (5) above here.) Now

$$f_\epsilon = f + \epsilon \sum_{i=1}^{m} \chi_i f_n \tag{14.17}$$

is a Morse function for sufficiently small $\epsilon > 0$. Item (6) implies that this function is Γ-invariant. Therefore $f_\epsilon \in A(n)$. Moreover, f_ϵ converges to f as $\epsilon \to 0$. □

Lemmas 14.43 and 14.45 imply that $A(1)$ is dense. The proof of Proposition 14.41 is complete. □

Remark 14.46 The set of functions whose gradient flow is Morse–Smale is *not* dense in general in the case of orbifolds. This is an important point where Morse theory of orbifolds is different from that of manifolds. In fact we can use virtual fundamental chain techniques to work out the theory of Morse homology for orbifolds.

Proposition 14.47 *Suppose we are in Situation* 14.40. *Let* $h : [0, 1] \to V$ *be a smooth embedding. Then the set of all the functions* f *in* $C^\infty_\Gamma(V)$ *such that* $f \circ h$ *is a Morse function on* $(0, 1)$ *is a countable intersection of open dense subsets.*

Proof The proof is similar to Proposition 14.41. Let $0 < c < 1/2$. We put

$$\overset{\circ}{T}(n, c) = \{t \in [c, 1 - c] \mid h(t) \in \overset{\circ}{V}(n)\},$$
$$T(n, c) = \{t \in [c, 1 - c] \mid h(t) \in V(n)\},$$
$$S(n, c) = \{t_0 \in \overset{\circ}{T}(n, c) \mid (dh/dt)(t_0) \in T_{h(t_0)}\overset{\circ}{V}(n)\}.$$

We also put

$$C(n, c) = \{f \in C^\infty_\Gamma(V) \mid \text{all critical points of } f \circ h \text{ on } T(n, c) \text{ are Morse}\},$$

$$D(n, c) = C(n + 1, c) \cap \left\{ f \in C^\infty_\Gamma(V) \,\middle|\, \begin{array}{l} \text{all critical points of } f \circ h \text{ on } S(n, c) \\ \text{are Morse} \end{array} \right\}.$$

Lemma 14.48 *If* $C(n + 1, c)$ *is dense then* $D(n, c)$ *is dense.*

Proof We fix c and will prove $D(n, c)$ is dense. We take $\epsilon > 0$ such that $c - \epsilon > 0$. Let U_1 be a sufficiently small neighborhood of $V(n + 1)$ which we choose later. We take U_2 a neighborhood $V(n) \setminus U_1$ and $\pi : U_2 \to \overset{\circ}{V}(n)$ be the projection of normal bundle. We may assume U_1, U_2 and π are Γ equivariant. We take a compact subset Z of $U_2 \cap \overset{\circ}{V}(n)$. We take a set $S \subset [c-\epsilon, 1-c+\epsilon]$ with the following properties:

(1) $\mathrm{Int}S \supset S(n, c) \cap h^{-1}(Z)$.
(2) S is a finite disjoint union of closed intervals.
(3) The composition $\pi \circ h$ is an embedding on $S(n, c)$.

Note that the differential of $\pi \circ h$ is injective on $S(n, c) \cap h^{-1}(Z)$. Therefore we can take such an S. We define a map $G : \mathrm{Int}S \times C^2_\Gamma(K) \to \mathbb{R}$ by

$$G(t_0, f) = \frac{d(f \circ \pi \circ h)}{dt}(t_0). \tag{14.18}$$

Using the fact that $\pi \circ h$ is an embedding and $\overset{\circ}{V}(n)/\Gamma$ is a smooth manifold we can easily show that G is transversal to 0. We put $\mathfrak{N} = G^{-1}(0) \subset \mathrm{Int}S \times C^2_\Gamma(K)$. It is a Banach submanifold. The restriction of the projection $\mathfrak{N} \to C^2_\Gamma(K)$ is a Fredholm map. In the same way as Sublemma 14.44 we can show that if $f \in C^2_\Gamma(K)$ is a regular value of $\mathfrak{N} \to C^2_\Gamma(K)$ then $f \circ \pi \circ h$ is a Morse function on $\mathrm{Int}\, S$. We remark that at the point of $S(n, c)$ the Hessian of $f \circ \pi \circ h$ is the same as the Hessian of $f \circ h$. Thus by the Sard–Smale theorem we find that the set of $f \in C^2_\Gamma(K)$ such that all the critical points of $f \circ h$ on $S(n, c) \cap h^{-1}(Z)$ are Morse, is dense.

Sublemma 14.49 *Let $f \in C(n+1, c)$. Then there exists a compact set $P \subset \overset{\circ}{T}(n, c)$ such that there is no critical point of $f \circ h$ on $\overset{\circ}{T}(n, c) \setminus P$.*

Proof Suppose to the contrary that there is a sequence $t_i \in \overset{\circ}{T}(n, c)$ of $f \circ h$ and no subsequence of t_i converges to an element of $\overset{\circ}{T}(n, c)$. We may assume that t_i converges. Then the limit t should be an element of $T(n+1, c)$. Since $f \in C(n+1, c)$ the function $f \circ h$ is Morse at t. Therefore there is no critical point of $f \circ h$ other than p in a neighborhood of t. This is a contradiction. □

Now let $f \in C(n+1, c)$. We choose U_1, U_2 and Z such that $Z \supset h(P)$ and P is as in Sublemma 14.49. Then there exists a sequence of functions f_i such that f_i converges to f and all the critical points of $f_i \circ h$ on $S(n, c) \cap h^{-1}(Z)$ are Morse. Therefore $f_i \in D(n, c)$ for all sufficiently large i. The proof of Lemma 14.48 is complete. □

Lemma 14.50 *If $D(n, c)$ is dense, then $C(n, c)$ is dense.*

Proof Let $f \in D(n, c)$.

Sublemma 14.51 *The set*

$$Q = \{t \in \overset{\circ}{T}(n, c) \mid t \text{ is a critical point of } f \circ h\}$$

is a finite set.

Proof Suppose Q is an infinite set and $t_i \in Q$ is its infinitely many points. We may also assume that t_i converges to t. If $t \in \overset{\circ}{T}(n, c) \setminus S(n, c)$ then h is transversal to $\overset{\circ}{V}(n)$ at t. Therefore $h(t_i) \notin \overset{\circ}{V}(n)$ for large i. This contradicts $t_i \in Q$. Therefore $t \in S(n, c) \cup T(n+1, c)$. Since $f \in D(n, c)$ the composition $f \circ h$ is Morse at t. Therefore there is no critical point of $f \circ h$ other than t in a neighborhood of t. This contradicts $\lim t_i = t$. □

Let $Q \setminus S(n, c) = \{t_1, \ldots, t_n\}$ and we put $p_i = h(t_i) \in \overset{\circ}{T}(n, c)$. We define V_i and χ_i in the same way as the proof of Lemma 14.45 and define f_ϵ in the same way as (14.17). It is easy to see that $f_\epsilon \in C(n, c)$ and f_ϵ converges to f as $\epsilon \to 0$. The proof of Lemma 14.50 is complete. □

By Lemmas 14.48 and 14.50 we have proved that $C(1, c)$ is dense for all $c \in (0, 1/2)$. The proof of Proposition 14.47 is complete. □

Remark 14.52 We do not assume h to be Γ-invariant in Proposition 14.47.

14.5 Similarity and Difference Between CF-Perturbation and Multivalued Perturbation: Remarks

As we mentioned before, the stories of multivalued perturbations and CF-perturbations are mostly parallel. However, there are certain minor differences between the two notions which we explain below.

We defined a multivalued perturbation as an $n \in \mathbb{Z}_+$ parametrized family and a CF-perturbation as a family parametrized by $\epsilon \in (0, 1]$. In other words, the parameter space for the case of CF-perturbation is uncountable, while the parameter space is countable for the case of multivalued perturbation. In many parts of the story of multivalued perturbation we can consider $\widehat{\mathfrak{s}} = \{\widehat{\mathfrak{s}^\epsilon}\}$ in place of $\{\widehat{\mathfrak{s}^n}\}$. However, we note that we need a countable family of objects when we apply Baire's category theorem. See the end of the proof of Lemma 14.34.

When we discuss transversality of $\widehat{\mathfrak{s}} = \{\widehat{\mathfrak{s}^\epsilon}\}$, we may consider one of the following two versions:

(1) Fix $\epsilon > 0$ and define the transversality of $\widehat{\mathfrak{s}^\epsilon}$ as a multisection, for each ϵ.
(2) We consider the whole family $\widehat{\mathfrak{s}}$ and define the transversality of $\widehat{\mathfrak{s}} = \{\widehat{\mathfrak{s}^\epsilon}\}$ as a multisection on $(X, \widehat{\mathcal{U}}) \times (0, 1]$.

We say that $\widehat{\mathfrak{s}} = \{\widehat{\mathfrak{s}^\epsilon}\}$ is *transversal to* 0 *as a family* if $\widehat{\mathfrak{s}}$ is transversal to 0 in the sense of (2). Sard's theorem implies that if $\widehat{\mathfrak{s}} = \{\widehat{\mathfrak{s}^\epsilon}\}$ is transversal to 0 as a family, then *for generic* ϵ the multisection $\widehat{\mathfrak{s}^\epsilon}$ is transversal to 0 in the sense of (1). The transversality we need to define a virtual fundamental chain is one in the sense of (1).

Note when we define its transversality we require the transversality of $\widehat{\mathfrak{s}^\epsilon}$ for each fixed ϵ. So the transversality of CF-perturbation is one in the sense of (1). During the proof of existence of transversal CF-perturbation in Chap. 12 we *never* used Baire's category theorem. We proved the transversality for uncountably many $\widehat{\mathfrak{S}^\epsilon}$, by taking the auxiliary parameter space W huge.

The point we elaborate on above is related to the ϵ-dependence of the virtual fundamental chain $[(X, \widehat{\mathcal{U}}, \widehat{\mathfrak{s}^\epsilon})]$ of a 0-dimensional good coordinate system as follows. Note that this phenomenon occurs only when $(X, \widehat{\mathcal{U}})$ has a boundary.

Suppose a multivalued perturbation $\widehat{\mathfrak{s}^\epsilon}$ is transversal to 0 in the sense of (2) above. Then for a sufficiently small generic ϵ, $\widehat{\mathfrak{s}^\epsilon}$ is transversal to 0 at ϵ. So we can define the rational number $[(X, \widehat{\mathcal{U}}, \widehat{\mathfrak{s}^\epsilon})]$. On the other hand, there is a discrete subset $S \subset (0, 1]$ such that if $\epsilon_1 \in S$ then $\widehat{\mathfrak{s}^{\epsilon_1}}$ may not be transversal to 0. In particular, the zero set of $\widehat{\mathfrak{s}^{\epsilon_1}}$ may intersect with the boundary $\partial(X, \widehat{\mathcal{U}})$. Therefore

$$\lim_{\epsilon \uparrow \epsilon_1} [(X, \widehat{\mathcal{U}}, \widehat{\mathfrak{s}^\epsilon})] \neq \lim_{\epsilon \downarrow \epsilon_1} [(X, \widehat{\mathcal{U}}, \widehat{\mathfrak{s}^\epsilon})],$$

in general. In other words, wall crossing may occur at ϵ_1. For this reason the virtual fundamental chain $[(X, \widehat{\mathcal{U}}, \widehat{\mathfrak{s}^\epsilon})]$ depends on ϵ.

When we use CF-perturbations, a similar phenomenon happens and the integration along the fiber depends on ϵ. However, this phenomenon appears in a slightly different way. Suppose we consider the zero-dimensional case. Then the integration along the fiber (for the function 1 and the map from X to a point) $[(X, \widehat{\mathcal{U}}, \widehat{\mathfrak{S}^\epsilon})]$ defines a real number for each ϵ. This is defined for all sufficiently small ϵ but is not a constant function of ϵ.

The number we obtain using the CF-perturbation is a real number and so changes continuously as ϵ varies. (It is easy to see from the definition that $[(X, \widehat{\mathcal{U}}, \widehat{\mathfrak{S}^\epsilon})]$ is a smooth function of ϵ.) On the other hand, the number we obtain using the multivalued perturbation is a rational number. So it jumps. Therefore it is impossible to obtain the virtual fundamental chain $[(X, \widehat{\mathcal{U}}, \widehat{\mathfrak{s}^\epsilon})]$, for all ϵ in the case where $\widehat{\mathfrak{s}}$ is a multivalued perturbation.

Part II
System of K-Spaces and Smooth Correspondences

Throughout Part II, an orbifold with corners means an admissible orbifold with corners in the sense of Sect. 25.1. So all the notions related to the orbifold with corners are the ones in the admissible category.

Chapter 15
Introduction to Part II

In Part I, we described the foundation of the theory of Kuranishi structures, good coordinate systems, and CF-perturbations (also multivalued perturbations), and we defined the integration along the fiber (pushout) of a strongly submersive map with respect to a CF-perturbation and also proved Stokes' formula. Using these ingredients, we have established the notion of smooth correspondence and proved the composition formula. In particular, this provides us with a way to obtain a virtual fundamental chain for each K-space (space with Kuranishi structure). Thus Part I is the story for each *single K-space*. On the other hand, in Part II we are going to study a *system of K-spaces*. In actual geometric applications, there are cases for which it is not enough to study each single K-space individually, but is necessary to study a system of K-spaces satisfying certain *compatibility conditions*, especially *compatibility conditions along the boundary and corner*. The compatibility conditions we describe depend on the situation we consider in their detail. Here we have two geometric examples in mind. One is the Floer cohomology for periodic Hamiltonian systems which is established by [FOn2, LiuTi] for general closed symplectic manifolds, and the other is the A_∞ algebra associated to a Lagrangian submanifold and the Floer cohomology for the Lagrangian intersection established by [FOOO3, FOOO4]. Since this book intends to provide a 'package' of the statements appearing in the actual argument above, we begin with axiomatizing the properties and conditions in a purely abstract setting, motivated by these two geometric examples. In this way we discuss two kinds of systems of K-spaces in Part II: One is a *linear K-system* containing the Floer theory for periodic Hamiltonian systems as a typical example, and the other is a *tree-like K-system* covering the theory of the A_∞ algebra associated to a Lagrangian submanifold.

We emphasize that we discuss those two cases as a prototype of the applications of the results of this book. In fact if we are interested only in defining Floer cohomology of periodic Hamiltonian systems and proving its basic properties, certainly there is a shorter proof than those given in Chaps. 16, 17, 18, 19, and 20. In this book we give a general proof which can be used in similar situations with

K. Fukaya et al., *Kuranishi Structures and Virtual Fundamental Chains*, Springer Monographs in Mathematics, https://doi.org/10.1007/978-981-15-5562-6_15

minimal change. For this reason we will try to avoid using special features of the particular situation we work with but use only the arguments which are general enough. A similar argument works in most of the other cases where the method of pseudo-holomorphic curves is applied. We do not know a case for which this method does not work. We confirm that it also works at least in the following situations. (The references quoted below form a non-exhaustive list of the papers which (partially) work out such applications via the Kuranishi structure or its cousins.)

(1) Constructing and proving basic properties of the Gromov–Witten invariant. ([FOn2, Section 23].)
(2) Studying several Lagrangian submanifolds and constructing an A_∞ category (Fukaya category). ([AFOOO, Fuk9] and also [Fuk3, Fuk7, FOOO6].)
(3) The family version of (2) and equivariant versions of (1)(2). ([Fuk8].)
(4) Including immersed Lagrangian submanifolds. ([AJ, Fuk9].)
(5) Using the Lagrangian correspondence to construct an A_∞ functor. ([Fuk9].)
(6) Including bulk deformations into the Lagrangian Floer theory to define open-closed, closed-open maps, and proving their basic properties. ([AFOOO], [FOOO3, Chapter 3], [FOOO4, Chapter 7], [FOOO22].)
(7) Studying the moduli space of pseudo-holomorphic maps from a bordered Riemann surface of arbitrary genus with Lagrangian boundary condition to construct an IBL-infinity structure.
(8) Including the non-compact case in defining and studying symplectic homology, wrapped Floer homology etc. ([BH, DF1, DF2, Is, Pa2].)
(9) In the case of symplectic manifolds with contact type boundary, using closed Reeb orbits or Reeb chords to establish the foundation of symplectic field theory and its version with Legendrian submanifolds.
(10) Lifting the Lagrangian Floer theory to loop space. ([Ir].)

The purpose of Part II is summarized as follows: If we are given a system of K-spaces satisfying the axiom we provide in this book as an input, then we prove that we can derive certain algebraic structures from the system of K-spaces as an output. This is a ‘package’ producing an algebraic structure from a geometric input. The problem is that the resulting algebraic structure itself depends on the various choices made in the course of the construction, in general. We will specify in which sense the algebraic structure is invariant and prove the invariance in this book. Although these typical examples of K-systems arise from the moduli spaces of pseudo-holomorphic curves, the authors expect that this kind of axiomatization and framework will be applicable and useful for other problems arising from other situations in the future. Certain part of the proof of Part II are simplified compared to the proof in the preprint version ([FOOO20]).

15.1 Outline of the Story of Linear K-Systems

In Chaps. 16, 17, 18, 19, and 20 we study systems of K-spaces, which axiomatize the situation appearing during the construction of Floer cohomology of periodic Hamiltonian systems.

15.1.1 *Floer Cohomology of Periodic Hamiltonian Systems*

We first review the construction of Floer cohomology of periodic Hamiltonian systems in [Fl2].

Let $H : S^1 \times M \to \mathbb{R}$ be a real-valued smooth function. For each $t \in S^1$ we obtain a function $H_t : M \to \mathbb{R}$ by

$$H_t(x) = H(t, x).$$

We denote its Hamiltonian vector field by X_{H_t} defined by $dH_t = \omega(X_{H_t}, \cdot)$. A periodic solution of the periodic Hamiltonian system generated by H is by definition a smooth map $\ell : S^1 \to M$ that satisfies the equation:

$$\frac{d\ell}{dt} = X_{H_t} \circ \ell. \tag{15.1}$$

Let $\mathrm{Per}(H)$ be the set of all solutions of (15.1). We will work in the Bott–Morse situation so we assume that $\mathrm{Per}(H)$ is a smooth manifold satisfying certain non-degeneracy conditions. (See [FOOO13, Subsection 1.2 (2)] for example.)

Remark 15.1 To define Floer cohomology, it is enough to discuss the Morse case, that is, the case when H satisfies certain non-degeneracy condition so that $\mathrm{Per}(H)$ is discrete. However, to calculate Floer cohomology, it is useful to include the Bott–Morse case, especially the case with $H \equiv 0$.

We decompose $\mathrm{Per}(H)$ into connected components and write

$$\mathrm{Per}(H) = \bigcup_{\overline{\alpha} \in \overline{\mathfrak{A}}} \overline{R}_{\overline{\alpha}},$$

for an indexing set $\overline{\mathfrak{A}}$. For $\overline{\alpha}_-, \overline{\alpha}_+ \in \overline{\mathfrak{A}}$ we consider the set of solutions of the following equation (Floer's equation) for a map $u : \mathbb{R} \times S^1 \to M$

$$\frac{\partial u}{\partial \tau} + J\left(\frac{\partial u}{\partial t} - X_{H_t}(u)\right) = 0 \tag{15.2}$$

together with the following asymptotic boundary condition.

Condition 15.2 There exist $\gamma_{-\infty} \in \overline{R}_{\overline{\alpha}_-}$ and $\gamma_{+\infty} \in \overline{R}_{\overline{\alpha}_+}$ such that

$$\lim_{\tau \to -\infty} u(\tau, t) = \gamma_{-\infty}(t),$$
$$\lim_{\tau \to +\infty} u(\tau, t) = \gamma_{+\infty}(t). \tag{15.3}$$

We denote by $\overset{\circ}{\widetilde{\mathcal{M}}}(\overline{R}_{\overline{\alpha}_-}, \overline{R}_{\overline{\alpha}_+})$ the set of solutions of (15.2) satisfying Condition 15.2. We can define an $\mathbb{R}$ action on $\overset{\circ}{\widetilde{\mathcal{M}}}(\overline{R}_{\overline{\alpha}_-}, \overline{R}_{\overline{\alpha}_+})$ by $(\tau_0 \cdot u)(\tau, t) = u(\tau + \tau_0, t)$, and denote the associated quotient space by $\overset{\circ}{\mathcal{M}}(\overline{R}_{\overline{\alpha}_-}, \overline{R}_{\overline{\alpha}_+})$. We decompose it into

$$\overset{\circ}{\mathcal{M}}(\overline{R}_{\overline{\alpha}_-}, \overline{R}_{\overline{\alpha}_+}) = \bigcup_{\beta} \overset{\circ}{\mathcal{M}}(\overline{R}_{\overline{\alpha}_-}, \overline{R}_{\overline{\alpha}_+}; \beta),$$

according to the homology class β of u. In place of this decomposition we proceed as follows. We denote by $\mathfrak{A}$ the set of pairs $(\overline{\alpha}, [w])$ where $\overline{\alpha} \in \overline{\mathfrak{A}}$ and $[w]$ is the homology class of a disk map w bounding $\ell \in \overline{R}_{\overline{\alpha}}$.[1] For each $\alpha \in \mathfrak{A}$ we define R_α as the set of pairs consisting of an element ℓ of $\overline{R}_{\overline{\alpha}}$ and an equivalence class $[w]$ of the homology class of disk w bounding ℓ.

Then let $\overset{\circ}{\mathcal{M}}(R_{\alpha_-}, R_{\alpha_+})$ be the union of $\overset{\circ}{\mathcal{M}}(\overline{R}_{\overline{\alpha}_-}, \overline{R}_{\overline{\alpha}_+}; \beta)$ over β with $[w^-]\#[\beta] = [w^+]$ (where $\alpha_\pm = (\overline{\alpha}_\pm, [w^\pm])$ and denote the union by $\mathcal{M}^{\text{reg}}(H; \alpha_-, \alpha_+)$. Using the notion of a stable map, we can compactify each of $\mathcal{M}^{\text{reg}}(H; \alpha_-, \alpha_+)$ and denote the compactification by $\mathcal{M}(H; \alpha_-, \alpha_+)$. (See [FOn2, Definition 19.9].)

We define asymptotic evaluation maps $\mathcal{M}^{\text{reg}}(H; \alpha_-, \alpha_+) \to \overline{R}_{\alpha_\pm}$ by

$$\text{ev}_-([u]) = \gamma_{-\infty}, \qquad \text{ev}_+([u]) = \gamma_{+\infty},$$

where $\gamma_{-\infty}, \gamma_{+\infty}$ are as in (15.3). They induce maps

$$\text{ev}_\pm : \mathcal{M}(H; \alpha_-, \alpha_+) \to R_{\alpha_\pm},$$

which we call the *evaluation maps at infinity*. Using the non-degeneracy condition of R_α, we can show that $\mathcal{M}(H; \alpha_-, \alpha_+)$ carries a Kuranishi structure with corners and the evaluation map

$$(\text{ev}_-, \text{ev}_+) : \mathcal{M}(H; \alpha_-, \alpha_+) \to R_{\alpha_-} \times R_{\alpha_+} \tag{15.4}$$

[1] The equivalence class is defined by using the symplectic area and the Maslov index.

is a strongly smooth and weakly submersive map. Moreover we can find the following isomorphism of K-spaces:

$$\partial \mathcal{M}(H; \alpha_-, \alpha_+) = \bigcup_{\alpha} \mathcal{M}(H; \alpha_-, \alpha) \; {}_{\mathrm{ev}_+}\times_{\mathrm{ev}_-} \mathcal{M}(H; \alpha, \alpha_+). \tag{15.5}$$

When we regard the left hand side as the *normalized* boundary, then the right hand side becomes the *disjoint* union.

See [FOOO16, Part 5] for the construction of such a K-system.

15.1.2 Periodic Hamiltonian Systems and Axiom of Linear K-Systems

As we explained in Sect. 15.1.1, when we are given a time-dependent Hamiltonian H, we obtain a system consisting of a set of smooth manifolds R_α and a set of K-spaces $\mathcal{M}(H; \alpha_-, \alpha_+)$, together with evaluation maps (15.4). The axiom of linear K-systems, which we present in Chap. 16 Condition 16.1, spells out the properties of such a system which extract Floer cohomology.

Condition 16.1 (III)(IV) require the existence of a set of manifolds $\{R_\alpha \mid \alpha \in \mathfrak{A}\}$, a set of K-spaces $\{\mathcal{M}(H; \alpha_-, \alpha_+) \mid \alpha_-, \alpha_+ \in \mathfrak{A}\}$ and evaluation maps (15.4) indexed by a countable set $\mathfrak{A}$. In this abstract situation we call $\mathcal{M}(H; \alpha_-, \alpha_+)$ the *space of connecting orbits*.

Condition 16.1 (VI) (and (I)) requires that we can associate the Maslov index $\mu(\alpha)$ to each R_α which determines the dimension of $\mathcal{M}(H; \alpha_-, \alpha_+)$.

It is well-known that the energy

$$\int_{\mathbb{R}\times S^1} \left\| \frac{\partial u}{\partial \tau} \right\|^2 + \left\| \frac{\partial u}{\partial t} - X_{H_t}(u) \right\|^2 d\tau dt \tag{15.6}$$

of the solution u of (15.2) is a difference of the values of certain action functional at the asymptotic boundary values α_-, α_+. Moreover the energy is non-negative and is zero only when $\partial u / \partial \tau = 0$. Condition 16.1 (V) (and (I)) is an axiomatization of this property.

We note that, in our Bott–Morse situation, we need to introduce an appropriate $O(1)$-principal bundle o_{R_α} to our critical submanifold to define Floer cohomology. Namely the contribution of R_α to the Floer cohomology is the cohomology group of R_α with the coefficients twisted by this local system. (See [FOOO4, Subsection 8.8].) Then the orientation local system of the moduli space $\mathcal{M}(H; \alpha_-, \alpha_+)$ is related to the orientations of $R_{\alpha_\pm}$ and to $o_{R_{\alpha_\pm}}$ in an appropriate way. Condition 16.1 (VII) is an axiomatization of this property.

Note that an element of R_α is a pair $(\ell, [w])$ where ℓ is a periodic orbit of our Hamiltonian system and $[w]$ is a homology class of disks bounding ℓ. For an element $\beta \in H_2(M; \mathbb{Z})$ represented by a sphere, we can glue w with a representative of β to obtain another disk. By this operation we obtain another $R_{\beta\#\alpha}$. It is easy to see $\mathcal{M}(H;\alpha_-,\alpha_+) \cong \mathcal{M}(H;\beta\#\alpha_-,\beta\#\alpha_+)$. Condition 16.1 (VIII) is an axiomatization of this property.

One important property of the moduli space of pseudo-holomorphic curves or the solutions of Floer's equation, is Gromov compactness. It claims the compactness of the union of the moduli spaces whose elements have energy smaller than a fixed number. Condition 16.1 (IX) is an axiomatization of this property.

The boundary of our moduli space $\mathcal{M}(H;\alpha_-,\alpha_+)$ is described as (15.5). Moreover this isomorphism is not only one between topological spaces but also one between oriented K-spaces. Condition 16.1 (X) is an axiomatization of this property.

Our moduli space $\mathcal{M}(H;\alpha_-,\alpha_+)$ has not only a boundary but also corners in general. Its codimension k (normalized) corner $\widehat{S}_k\mathcal{M}(H;\alpha_-,\alpha_+)$ is described as the disjoint union of the fiber products

$$\begin{aligned}\mathcal{M}(H;\alpha_-,\alpha_1)\times_{R_{\alpha_1}}\mathcal{M}(H;\alpha_1,\alpha_2)\times_{R_{\alpha_2}}\cdots\\ \times_{R_{\alpha_{k-1}}}\mathcal{M}(H;\alpha_{k-1},\alpha_k)\times_{R_{\alpha_k}}\mathcal{M}(H;\alpha_k,\alpha_+),\end{aligned}$$

where $\alpha_1,\ldots,\alpha_k\in\mathfrak{A}$. Condition 16.1 (XI) is an axiomatization of this property.

We explain Condition 16.1 (XII) in Sect. 15.1.4.

A system satisfying Condition 16.1 is called a *linear K-system* (Definition 16.6 (2)). Now the main result of Chaps. 16, 17, 18, 19, and 20 is as follows. Suppose we are given a linear K-system. We consider a direct sum of $\mathbb{R}$ vector spaces

$$\bigoplus_{\alpha\in\mathfrak{A}}\Omega(R_\alpha;o_{R_\alpha}). \tag{15.7}$$

Using (15.7) and the energy filtration we define (in Definitions 16.8 and 16.12) a module

$$CF(\mathcal{C};\Lambda_{0,\mathrm{nov}})$$

over the universal Novikov ring $\Lambda_{0,\mathrm{nov}}$. (See Definition 16.11 for the definition of $\Lambda_{0,\mathrm{nov}}$.) Here $\mathcal{C}$ denotes the totality of the part of the data of our linear K-system which is related to $\{R_\alpha\}$. We call it critical submanifold data. (See Definition 16.6 (1).)

Theorem 15.3 *To each linear K-system, we can associate a cochain complex, the Floer cochain complex, which we denote by* $(CF(\mathcal{C};\Lambda_{0,\mathrm{nov}}),d)$. *This complex is independent of the choices up to cochain homotopy equivalence.*

This is a simplified version of Theorem 16.9.

15.1.3 Construction of Floer Cochain Complex

We will prove Theorem 15.3 (or Theorem 16.9) in detail in Chap. 19. The proof we present there is written in a way so that it serves as a prototype of the proof of various similar results, and can be adapted easily to the proof of similar results.

The coboundary operator d in Theorem 15.3 is a sum of the exterior differential $d_0 : \Omega(R_\alpha; o_{R_\alpha}) \to \Omega(R_\alpha; o_{R_\alpha})$ and the operator $d_{\alpha_-,\alpha_+} : \Omega(R_{\alpha_-}; o_{R_{\alpha_-}}) \to \Omega(R_{\alpha_+}; o_{R_{\alpha_+}})$ obtained by the smooth correspondence

$$R_{a_-} \xleftarrow{ev_-} \mathcal{M}(H; a_-, a_+) \xrightarrow{ev_+} R_{a_+} \tag{15.8}$$

Then (15.5) together with Stokes' formula (Proposition 9.28) and Composition formula (Theorem 10.21) 'implies'

$$d_0 \circ d_{\alpha_-,\alpha_+} + d_{\alpha_-,\alpha_+} \circ d_0 + \sum_\alpha d_{\alpha_-,\alpha} \circ d_{\alpha,\alpha_+} = 0.$$

This will imply that $d = d_0 + \sum d_{\alpha_-,\alpha_+}$ satisfies $d \circ d = 0$. Thus we obtain Floer cohomology.

More precisely speaking, to define the operator d_{α_-,α_+} from (15.8) as smooth correspondence map (Definition 7.86) we need to take and fix a CF-perturbation on $\mathcal{M}(H; \alpha_-, \alpha_+)$.

To apply Stokes' formula, our CF-perturbations of various moduli spaces must be compatible with the isomorphism (15.5). Namely we need to show the next statement.

Proposition 15.4 (slightly imprecise statement) *For each given linear K-system and sufficiently small $\epsilon > 0$, there exists a system of CF-perturbations $\widehat{\mathfrak{S}^\epsilon_{\alpha_-,\alpha_+}}$ on $\mathcal{M}(H; \alpha_-, \alpha_+)$ such that:*

(1) *$\widehat{\mathfrak{S}^\epsilon_{\alpha_-,\alpha_+}}$ is transversal to 0 and ev_+ is strongly submersive with respect to this CF-perturbation.*
(2) *The restriction of $\widehat{\mathfrak{S}^\epsilon_{\alpha_-,\alpha_+}}$ to the boundary is equivalent to the fiber product of $\widehat{\mathfrak{S}^\epsilon_{\alpha_-,\alpha}}$ and $\widehat{\mathfrak{S}^\epsilon_{\alpha,\alpha_+}}$ via the isomorphism* (15.5).

This is a slightly imprecise statement and is *not* the statement we will prove. The precise statement we will prove is Proposition 19.1. The main differences between Propositions 15.4 and 19.1 are the following:

(a) The CF-perturbation $\widehat{\mathfrak{S}^\epsilon_{\alpha_-,\alpha_+}}$ is not defined on the Kuranishi structure of $\mathcal{M}(H; \alpha_-, \alpha_+)$ itself, which is given by the axiom of linear K-system, but defined on its thickening.
(b) We replace $\mathcal{M}(H; \alpha_-, \alpha_+)$ by $\mathcal{M}(H; \alpha_-, \alpha_+)^{\boxplus\tau_0}$, which is a K-space obtained by putting a τ_0-collar on the space $\mathcal{M}(H; \alpha_-, \alpha_+)$ *outside*.

(c) We fix E_0 and construct CF-perturbations for only finitely many moduli spaces, that is, the moduli spaces consisting of the elements of energy (15.6) $\le E_0$.

The reason for Item (a) is that, to construct a CF-perturbation we first need to construct a good coordinate system and then go back to the Kuranishi structure. We explained this point already in Sect. 1.3.

The reason for Item (b) is more technical. We perform various operations in a neighborhood of the boundary and corner of our K-space. Those constructions are easier to carry out if the charts of our K-space have collars. We will use the collar to extend the Kuranishi structure on the boundary which is a thickening of the given one, to the interior. (However, see Remark 15.5 (1).) The existence of the collar on a given cornered manifold or orbifold is a fairly standard fact in differential topology. In the case of Kuranishi structures or good coordinate systems, attaching collars to all the charts so that they respect the coordinate changes is rather cumbersome. (This is because the way to put a collar on a given cornered orbifold is not unique.) We take a short-cut of putting the collar 'outside' rather than 'inside'. The process of putting the collar outside, which we call *outer collaring*, is described in detail in Chap. 17. See Sect. 17.1 for more detailed explanation on this point.

The reason for Item (c) is that it is difficult to perturb infinitely moduli spaces simultaneously. We go back to this point in Sect. 15.1.7.

Remark 15.5

(1) Recall that in this book we start from a purely abstract setting of a K-space with boundary and corners, which is not necessarily arising from a particular geometric situation like the moduli space of pseudo-holomorphic curves. In the geometric setting studied in [FOn2, FOOO3, FOOO4], an extension of the Kuranishi structure to a small neighborhood of ∂X in X is given from its construction. In fact, we start from a Kuranishi structure $\partial\widehat{\mathcal{U}}$ on the boundary (which we obtain from geometry or analysis) and construct a good coordinate system $\widehat{\mathcal{U}_\partial}$ and use it to find $\widehat{\mathcal{U}_\partial^+}$ and its perturbation. We need to extend $\widehat{\mathcal{U}_\partial^+}$ and its perturbation to a neighborhood of ∂X. In this situation, the Kuranishi charts of $\widehat{\mathcal{U}_\partial^+}$ are obtained as open subcharts of certain Kuranishi charts of $\partial\widehat{\mathcal{U}}$. (See the proof of Proposition 3.35, Lemma 6.30.) Therefore it can indeed be extended using the extension of $\partial\widehat{\mathcal{U}}$ directly.

(2) In the proof of well-definedness of the virtual fundamental chain, which corresponds to the well-definedness of the Gromov–Witten invariant, given in Part I of this book, 'outer collaring' is not necessary. This is because we only need to apply Stokes' formula and do not need the chain level argument. See Theorem 8.15, Proposition 8.16 and their proofs.

15.1.4 Corner Compatibility Conditions

The proof of Proposition 15.4 (or Proposition 19.1) is by induction over the energy level. We consider the isomorphism (15.5):

$$\partial \mathcal{M}(H;\alpha_-,\alpha_+) = \bigcup_{\alpha} \mathcal{M}(H;\alpha_-,\alpha) \;{}_{\mathrm{ev}_+}\!\times_{\mathrm{ev}_-} \mathcal{M}(H;\alpha,\alpha_+).$$

We observe that the energy of the elements of the moduli spaces appearing in the right hand side is strictly smaller than that appearing in the left hand side. So by induction hypothesis the CF-perturbation of the right hand side is already given. Therefore the statement we need to work out this induction is something like the following (*):

(*) Let $(X, \widehat{\mathcal{U}})$ be a K-space with corners. Suppose a CF-perturbation $\widehat{\mathfrak{S}^{\epsilon}_{\partial}}$ is given on the normalized boundary $\partial(X, \widehat{\mathcal{U}})$, satisfying certain transversality properties. Then we can find a CF-perturbation $\widehat{\mathfrak{S}^{\epsilon}}$ on $(X, \widehat{\mathcal{U}})$ which has the same transversality property and whose restriction to the boundary coincides with $\widehat{\mathfrak{S}^{\epsilon}_{\partial}}$.

However, we note that the statement (*), as it is, does *not* hold. In fact, since $(X, \widehat{\mathcal{U}})$ has not only a boundary but also corners, we need to assume certain compatibility conditions for $\widehat{\mathfrak{S}^{\epsilon}_{\partial}}$ at the corners. Let us elaborate on this point below.

We remark that we use the *normalized* corner of an orbifold (or Kuranishi structure) with corners. Typically a point on the corner $\widehat{S}_2U$ of an orbifold U corresponds to two points on the *normalized* boundary. In other words, we have a double cover

$$\pi : \partial\partial U \to \widehat{S}_2U. \tag{15.9}$$

Suppose we are given a CF-perturbation $\widehat{\mathfrak{S}^{\epsilon}_{\partial}}$ on the normalized boundary ∂U. The compatibility condition we need to assume for $\widehat{\mathfrak{S}^{\epsilon}_{\partial}}$ is that if $\pi(x) = \pi(y)$ then the perturbation $\widehat{\mathfrak{S}^{\epsilon}_{\partial}}$ at x coincides with $\widehat{\mathfrak{S}^{\epsilon}_{\partial}}$ at y. Namely, we need to require the next condition:

($\star$) There exists a CF-perturbation $\widehat{\mathfrak{S}^{\epsilon}_{\widehat{S}_2U}}$ on $\widehat{S}_2U$, whose pullback to $\partial\partial U$ is equivalent to the restriction of $\widehat{\mathfrak{S}^{\epsilon}_{\partial}}$ to $\partial\partial U$.

We can state a similar condition for a Kuranishi structure on X. We note that to precisely state the condition ($\star$) we first need to clarify the relationship between the Kuranishi structure on $\partial\partial X$ and the one on $\widehat{S}_2X$.

In the situation of our application where $X = \mathcal{M}(H;\alpha_-,\alpha_+)$, we have isomorphisms

$$
\begin{aligned}
&\partial\partial\mathcal{M}(H;\alpha_-,\alpha_+)\\
&\cong \partial\left(\bigcup_{\alpha}\mathcal{M}(H;\alpha_-,\alpha)\ {}_{\mathrm{ev}_+}\times_{\mathrm{ev}_-}\mathcal{M}(H;\alpha,\alpha_+)\right)\\
&\cong \bigcup_{\alpha}\partial(\mathcal{M}(H;\alpha_-,\alpha))\ {}_{\mathrm{ev}_+}\times_{\mathrm{ev}_-}\mathcal{M}(H;\alpha,\alpha_+)\\
&\quad\cup\bigcup_{\alpha}\mathcal{M}(H;\alpha_-,\alpha)\ {}_{\mathrm{ev}_+}\times_{\mathrm{ev}_-}\partial(\mathcal{M}(H;\alpha,\alpha_+))\\
&\cong \bigcup_{\alpha_1,\alpha_2}(\mathcal{M}(H;\alpha_-,\alpha_1)\ {}_{\mathrm{ev}_+}\times_{\mathrm{ev}_-}(\mathcal{M}(H;\alpha_1,\alpha_2))\ {}_{\mathrm{ev}_+}\times_{\mathrm{ev}_-}\mathcal{M}(H;\alpha_2,\alpha_+)\\
&\quad\cup\bigcup_{\alpha_1,\alpha_2}\mathcal{M}(H;\alpha_-,\alpha_1)\ {}_{\mathrm{ev}_+}\times_{\mathrm{ev}_-}((\mathcal{M}(H;\alpha_1,\alpha_2)\ {}_{\mathrm{ev}_+}\times_{\mathrm{ev}_-}\mathcal{M}(H;\alpha_2,\alpha_+)).
\end{aligned}
$$

On the other hand, by Condition 16.1 (XI) we assumed:

$$
\begin{aligned}
&\widehat{S}_2\mathcal{M}(H;\alpha_-,\alpha_+)\\
&\cong\bigcup_{\alpha_1,\alpha_2}\mathcal{M}(H;\alpha_-,\alpha_1)\ {}_{\mathrm{ev}_+}\times_{\mathrm{ev}_-}\mathcal{M}(H;\alpha_1,\alpha_2)\ {}_{\mathrm{ev}_+}\times_{\mathrm{ev}_-}\mathcal{M}(H;\alpha_2,\alpha_+).
\end{aligned}
$$

By these isomorphisms we obtain a double cover

$$
\pi' : \partial\partial\mathcal{M}(H;\alpha_-,\alpha_+)\to\widehat{S}_2\mathcal{M}(H;\alpha_-,\alpha_+).
$$

Condition 16.1 (XII) (the second corner compatibility condition) requires that this double cover π' coincides with the double cover π given in (15.9).

Remark 15.6

(1) The condition $\pi = \pi'$ (Condition 16.1 (XII)) is *not* automatic and we need to *assume* it as a part of the axiom of a linear K-system. In fact, we can define the covering map π in a canonical way for an arbitrary K-space X. On the other hand, the covering map π' depends on the choice of the isomorphism (15.5) and similar isomorphisms for the corner (Condition 16.1 (XI)). Note that in our axiomatization only the *existence* of the isomorphism (15.5) is required. The isomorphism such as (15.5) is not unique. In fact, we can change it by composing any automorphism of $\partial\mathcal{M}(H;\alpha_-,\alpha_+)$. If we change the isomorphism (15.5) then the identity $\pi = \pi'$ will no longer hold.

    In other words, Condition 16.1 (XII) is one on the consistency between various choices of the isomorphisms (15.5) and similar isomorphisms for the corner.

(2) In our geometric situation, we define the isomorphism (15.5) using geometric description of the boundary of our moduli space $\mathcal{M}(H;\alpha_-,\alpha_+)$. Then the condition $\pi = \pi'$ is fairly obvious. In this book, we need to state this condition

explicitly because our purpose here is to formulate the precise conditions for our system of K-spaces under which we can define the Floer cohomology in a way that is independent of the geometric origin of such a system of K-spaces.

Actually, we need to require consistency at the corner of arbitrary codimension using the covering space

$$\pi_{m,\ell} : \widehat{S}_m(\widehat{S}_\ell X) \to \widehat{S}_{m+\ell} X, \tag{15.10}$$

which exists for any K-space X with corners. (See Proposition 24.17.) If we assume the corner compatibility condition, the property $(\star)$ and its analogue for higher codimensional corners can be shown inductively. The inductive step of this induction can be stated as follows.

Proposition 15.7 (Slightly imprecise statement) *Let $(X, \widehat{\mathcal{U}})$ be a K-space with corners. Suppose for each k we have a CF-pertubation $\widehat{\mathfrak{S}_k}$ on $\widehat{S}_k(X, \widehat{\mathcal{U}})$ with the following properties.*

For each m and ℓ, the following two CF-perturbations on $\widehat{S}_m(\widehat{S}_\ell X)$ are equivalent to each other:

(1) *The restriction of $\widehat{\mathfrak{S}_\ell}$ to $\widehat{S}_m(\widehat{S}_\ell X)$.*
(2) *The pullback of $\widehat{\mathfrak{S}_{m+\ell}}$ by the covering map* (15.10).

Then there exists a CF-perturbation $\widehat{\mathfrak{S}}$ on $(X, \widehat{\mathcal{U}})$ such that its restriction to $\widehat{S}_k(X, \widehat{\mathcal{U}})$ coincides with $\widehat{\mathfrak{S}_k}$ for each k.

This is a simplified statement and is *not* the statement we will prove in Chap. 17. The statement we will prove is Proposition 17.81. The differences between Propositions 15.7 and 17.81 are the following:

(a) The CF-perturbation we start with is not given on $\widehat{S}_k(X, \widehat{\mathcal{U}})$ itself but is given on a thickening of $\widehat{S}_k(X, \widehat{\mathcal{U}})$. The CF-perturbation we obtain is defined on a thickening of $(X, \widehat{\mathcal{U}})$.
(b) We replace X by $X^{\boxplus \tau_0}$, which is a K-space obtained from X by outer collaring.
(c) We assume that $\widehat{\mathfrak{S}_k}$ satisfies an appropriate transversality property and will find a CF-perturbation $\widehat{\mathfrak{S}}$ satisfying the same transversality property.

The reason for (a) is that we need to go once to a good coordinate system and come back to construct a CF-perturbation. The reason for (b) is explained in detail in Sect. 17.1. We actually need to construct a system of CF-perturbations satisfying certain transversality properties. This is the reason for (c).

15.1.5 *Well-Definedness of Floer Cohomology and Morphism of Linear K-Systems*

The most important property of Floer cohomology of periodic Hamiltonian systems is its invariance under the choice of Hamiltonians. Our story contains axiomatization of this equivalence. For this purpose we introduce the notion of morphisms between two linear K-systems. To explain the relevant axiom we consider the case of a linear K-system arising from the periodic Hamiltonian system and the associated Floer equation. Let $H^i : S^1 \times M \to \mathbb{R}$ be a periodic Hamiltonian function for $i = 1, 2$. Using the set of critical points we obtain a set of manifolds $\{R^i_\alpha \mid \alpha \in \mathfrak{A}_i\}$, $i = 1, 2$ and we obtain a set, the compactified moduli space, $\mathcal{M}(H^i; \alpha_-, \alpha_+)$ of solutions of Floer's equation (15.2) for $H = H^i$, $\alpha_\pm \in \mathfrak{A}_i$. They define cochain complexes $(CF(\mathcal{C}^i; \Lambda_{0,\mathrm{nov}}), d^i)$ and their Floer cohomologies by Theorem 15.3 for $i = 1, 2$.

The well-established method to prove the independence of the Floer cohomology of periodic Hamiltonian systems under the choice of Hamiltonian is to use the moduli space of the equation (15.12) below. (This method was invented by Floer [Fl2].) We take a function $\mathcal{H} : \mathbb{R} \times S^1 \times M \to \mathbb{R}$ such that

$$\mathcal{H}(\tau, t, x) = \begin{cases} H^1(t, x) & \text{if } \tau < -C, \\ H^2(t, x) & \text{if } \tau > C, \end{cases} \tag{15.11}$$

where C is a sufficiently large fixed number. We put $H_{\tau,t}(x) = \mathcal{H}(\tau, t, x)$ and consider the equation

$$\frac{\partial u}{\partial \tau} + J\left(\frac{\partial u}{\partial t} - X_{H_{\tau,t}}(u)\right) = 0 \tag{15.12}$$

with the asymptotic boundary condition for $\tau \to \pm\infty$ given by $R^1_{\alpha_-}$, $R^2_{\alpha'_+}$, respectively. We denote the compactified moduli space of the solution of (15.12) with this boundary condition by $\mathcal{M}(\mathcal{H}; \alpha_-, \alpha'_+)$. We will define the notion of morphism of linear K-systems in Definition 16.19 and Condition 16.17 (compatibility along the boundary). The notion of the *interpolation space* $\mathcal{N}(\alpha_-, \alpha'_+)$ appearing in the definition of morphisms is the axiomatization of the properties of this moduli space $\mathcal{M}(\mathcal{H}; \alpha_-, \alpha'_+)$. For example, (16.25) corresponds to the property of the boundary of $\mathcal{M}(\mathcal{H}; \alpha_-, \alpha'_+)$, that is,

$$\begin{aligned} \partial \mathcal{M}(\mathcal{H}; \alpha_-, \alpha'_+) \cong & \bigcup_{\alpha \in \mathfrak{A}_1} \mathcal{M}(H^1; \alpha_-, \alpha) \times_{R^1_\alpha} \mathcal{M}(\mathcal{H}; \alpha, \alpha'_+) \\ & \cup \bigcup_{\alpha' \in \mathfrak{A}_2} \mathcal{M}(\mathcal{H}; \alpha_-, \alpha') \times_{R^2_{\alpha'}} \mathcal{M}(H^2; \alpha', \alpha'_+). \end{aligned} \tag{15.13}$$

Thus the set of K-spaces $\{\mathcal{M}(\mathcal{H};\alpha_-,\alpha'_+) \mid \alpha_- \in \mathfrak{A}_1, \alpha'_+ \in \mathfrak{A}_2\}$ together with various other data defines a morphism from the linear K-system associated to H^1 to the linear K-system associated to H^2.

In the study of Floer cohomology of periodic Hamiltonian systems the moduli space $\mathcal{M}(\mathcal{H};\alpha_-,\alpha'_+)$ is used to define a cochain map from Floer's cochain complex associated to H^1 to Floer's cochain complex associated to H^2. We can carry out this construction by using the properties spelled out in Definition 16.19 and Condition 16.17 only and prove the next result.

Theorem 15.8 *If $\mathfrak{N}$ is a morphism from one linear K-system $\mathcal{F}_1$ to another linear K-system $\mathcal{F}_2$, then $\mathfrak{N}$ induces a cochain map*

$$\mathfrak{N}_* : (CF(\mathcal{C}^1;\Lambda_{\rm nov}), d^1) \to (CF(\mathcal{C}^2;\Lambda_{\rm nov}), d^2).$$

Here $(CF(\mathcal{C}^i;\Lambda_{\rm nov}), d^i)$ is the cochain complex associated to $\mathcal{F}_i$ by Theorem 15.3.

The cochain map $\mathfrak{N}_*$ depends on various choices. However, it is independent of the choices up to cochain homotopy.

Theorem 15.8 is Theorem 16.31 (1). The proof is similar to the proof of Theorem 15.3 and is given in Sect. 19.6.

We define the notion of composition of morphisms in Chap. 18 and show that $\mathfrak{N} \mapsto \mathfrak{N}_*$ is functorial with respect to the composition of the morphisms in Sect. 19.4. When the interpolation spaces of the morphism $\mathfrak{N}_{i+1i} : \mathcal{F}_i \to \mathcal{F}_{i+1}$ are given by $\mathcal{N}_{ii+1}(\alpha_i,\alpha_{i+1})$ for $i = 1, 2$, the interpolation space of the composition $\mathfrak{N}_{32} \circ \mathfrak{N}_{21} : \mathcal{F}_1 \to \mathcal{F}_3$ is the K-space $\mathcal{N}_{13}(\alpha_1,\alpha_3)$ obtained, roughly speaking, by gluing the K-spaces

$$\mathcal{N}_{12}(\alpha_1,\alpha_2) \times_{R^2_{\alpha_2}} \mathcal{N}_{23}(\alpha_2,\alpha_3) \tag{15.14}$$

for various α_2 along the boundaries and corners. We need to smooth a part of the corners of this fiber product to glue them. For this purpose, we need the definition of corner smoothing of K-spaces with corners. We will discuss it in Chap. 18. See Sect. 18.2 for an issue of corner smoothing of K-spaces. To define corner smoothing in a canonical way we use the collar (obtained by outer collaring). So more precisely we use

$$\bigcup_{\alpha_2\in\mathfrak{A}_2} \mathcal{N}_{12}(\alpha_1,\alpha_2) \times^{\boxplus\tau}_{R^2_{\alpha_2}} \mathcal{N}_{23}(\alpha_2,\alpha_3) \tag{15.15}$$

in place of (15.14). See Definition 18.34 for the definition of (15.15).

We also define the notion of homotopy and homotopy of homotopies etc. of morphisms and show that homotopy between morphisms $\mathfrak{N}$ and $\mathfrak{N}'$ induces a cochain homotopy between $\mathfrak{N}_*$ and $\mathfrak{N}'_*$.

15.1.6 Identity Morphism

To make the assignment $\mathfrak{N} \mapsto \mathfrak{N}_*$ functorial, we need the notion of the identity morphism. In the second half of Chap. 18 we define and prove a basic property of the identity morphism of linear K-system.

In the geometric situation of the linear K-system arising from a periodic Hamiltonian system, the morphisms between such linear K-systems are defined by using the moduli space of the solutions of equation (15.12), for a two-parameter family of functions $\mathcal{H} : \mathbb{R} \times S^1 \times M \to \mathbb{R}$. To obtain the identity morphism of linear K-system associated to $H : S^1 \times M \to \mathbb{R}$, we consider the case of $\mathcal{H}$ such that $\mathcal{H}(\tau, t, x) = H(t, x)$. In other words, we use τ independent $\mathcal{H}$. However, note that the moduli space of solutions of (15.12) for this τ independent $\mathcal{H}$ is *different* from that of Floer's equation (15.1). Namely

$$\mathcal{M}(\mathcal{H}; \alpha_-, \alpha_+) \neq \mathcal{M}(H; \alpha_-, \alpha_+)$$

in the case $\mathcal{H}(\tau, t, x) = H(t, x)$ for all τ. Indeed, the dimensions are different. To define $\mathcal{M}(H; \alpha_-, \alpha_+)$ we divide our space by the $\mathbb{R}$ action given by translation in the $\mathbb{R}$-direction with the coordinate τ. Since $\mathcal{H}$ happens to be τ independent, our equation (15.12) is invariant under this $\mathbb{R}$ action, too. However, by definition, $\mathcal{M}(\mathcal{H}; \alpha_-, \alpha_+)$ is a special case of general $\mathcal{H}$. For general $\mathcal{H}$, (15.12) is *not* invariant under $\mathbb{R}$ action. *Before compactifiation* we can identify

$$\overset{\circ}{\mathcal{M}}(H; \alpha_-, \alpha_+) \times \mathbb{R} = \overset{\circ}{\mathcal{M}}(\mathcal{H}; \alpha_-, \alpha_+),$$

when $\mathcal{H}(\tau, t, x) = H(t, x)$ and is τ independent. However, the relationship between compactified moduli spaces $\mathcal{M}(H; \alpha_-, \alpha_+)$ and $\mathcal{M}(\mathcal{H}; \alpha_-, \alpha_+)$ is not so simple.

We describe in Sect. 18.10 a way to obtain $\mathcal{M}(\mathcal{H}; \alpha_-, \alpha_+)$ (in the case where $\mathcal{H}(\tau, t, x)$ is τ independent) from $\overset{\circ}{\mathcal{M}}(H; \alpha_-, \alpha_+)$, in an abstract setting. In other words, we start with the spaces of connecting orbits $\mathcal{M}(\alpha_-, \alpha_+) = \mathcal{M}(H; \alpha_-, \alpha_+)$ of a liner K-system $\mathcal{F}$ and define the interpolation spaces of the identity morphism $\mathcal{ID} : \mathcal{F} \to \mathcal{F}$. We also show that identity morphism is a 'homotopy unit'. Namely we show in Sect. 18.10 that the composition of the identity morphism $\mathcal{ID}$ with other morphism $\mathfrak{N}$ is homotopic to $\mathfrak{N}$. (Proposition 18.60.)

To construct the identity morphism and prove its homotopy-unitality, we imitate the proof of the corresponding results in the case of periodic Hamiltonian systems, and rewrite it so that it works in the purely abstract setting of linear K-systems without any specific geometric origin. Although we explain the geometric origin of the construction of Sect. 18.10 in Sect. 18.11, the discussion of Sect. 18.11 is *not* used in Sect. 18.10 or any other part in proving the main results of this book. We expect that those explanations will be useful for readers who know Floer cohomology of periodic Hamiltonian systems to understand the contents of Sect. 18.10.

In Chap. 19, we use the identity morphism to prove the second half of Theorem 15.3, that is, independence of the Floer cochain complex $(CF(\mathcal{C}; \Lambda_{0,\mathrm{nov}}), d)$ of the choices up to cochain homotopy equivalence, as follows. We first consider the case when our linear K-system is obtained from a Hamiltonian $H : S^1 \times M \to \mathbb{R}$ by using Floer's equation (15.2). We fix a choice of an almost complex structure and the Kuranishi structure on $\mathcal{M}(H; \alpha_-, \alpha_+)$, which are the choices we made to define a linear K-system. We then take two different systems of CF-perturbations thereon, which we denote by $\widehat{\mathfrak{S}^{i,\epsilon}_{\alpha_-,\alpha_+}}$, and obtain two different cochain complexes which we denote by $(CF(\mathcal{C}; \Lambda_{0,\mathrm{nov}}), d^i)$, for $i = 1, 2$, respectively. We want to prove that they are cochain homotopy equivalent.

In this case the interpolation spaces of the identity morphism are $\mathcal{M}(\mathcal{H}; \alpha_-, \alpha_+)$ (for various $\alpha_\pm$) with $\mathcal{H}(\tau, t, x) = H(t, x)$ for all τ. Its boundary is described by (15.13). In our situation it becomes:

$$\begin{aligned} \partial \mathcal{M}(\mathcal{H}; \alpha_-, \alpha_+) \cong & \bigcup_{\alpha \in \mathfrak{A}} \mathcal{M}(H; \alpha_-, \alpha) \times_{R_\alpha} \mathcal{M}(\mathcal{H}; \alpha, \alpha_+) \\ & \cup \bigcup_{\alpha' \in \mathfrak{A}} \mathcal{M}(\mathcal{H}; \alpha_-, \alpha') \times_{R_{\alpha'}} \mathcal{M}(H; \alpha', \alpha_+). \end{aligned} \tag{15.16}$$

Now we consider a CF-perturbation $\widehat{\mathfrak{S}^{1,\epsilon}_{\alpha_-,\alpha}}$ on $\mathcal{M}(H; \alpha_-, \alpha)$, which is the first factor of the first term of the right hand side, and another CF-perturbation $\widehat{\mathfrak{S}^{2,\epsilon}_{\alpha',\alpha_+}}$ on $\mathcal{M}(H; \alpha', \alpha_+)$, which is the second factor of second term of the right hand side. We then take a system of CF-perturbations on various $\mathcal{M}(\mathcal{H}; \alpha_-, \alpha_+)$ so that these CF-perturbations together with $\widehat{\mathfrak{S}^{1,\epsilon}_{\alpha_-,\alpha}}$, $\widehat{\mathfrak{S}^{2,\epsilon}_{\alpha',\alpha_+}}$ are compatible with the isomorphism (15.16). (To show the existence of such a system of CF-perturbations, we need to examine all the corners of arbitrary codimension and check the compatibility at the corners. We can do so by induction using a similar argument as explained in Sect. 15.1.4.) Then the correspondence by $\mathcal{M}(\mathcal{H}; \alpha_-, \alpha_+)$ together with this system of CF-perturbations defines a cochain map from $(CF(\mathcal{C}; \Lambda_{0,\mathrm{nov}}), d^1)$ to $(CF(\mathcal{C}; \Lambda_{0,\mathrm{nov}}), d^2)$. This is a consequence of Stokes' formula (Proposition 9.28) and Composition formula (Theorem 10.21).

This cochain map is actually an isomorphism since it is the identity map modulo T^ϵ for some $\epsilon > 0$.

In the case of linear K-systems, which may not come from a particular geometric construction, we can proceed in the same way using the identity morphism, to prove that $(CF(\mathcal{C}; \Lambda_{0,\mathrm{nov}}), d^1)$ is cochain homotopy equivalent to $(CF(\mathcal{C}; \Lambda_{0,\mathrm{nov}}), d^2)$.

15.1.7 Homotopy Limit

We note that to construct a system of CF-perturbations for all the spaces of connecting orbits appearing in a linear K-system, we need to find infinitely many CF-perturbations simultaneously. There is an issue in doing so. We explained this issue in detail in [FOOO4, Subsection 7.2.3]. The method to resolve it is the same as [FOOO4, Section 7.2]. The algebraic part of this method is summarized as follows. For $E > 0$ a pair (C, d) of a free Λ_0 module C and $d : C \to C$ is said to be a *partial cochain complex of energy cut level* E if $d \circ d \equiv 0 \mod T^E$. Let $(C_1, d), (C_2, d)$ be partial cochain complexes of energy cut level E. A Λ_0 module homomorphism $\varphi : C_1 \to C_2$ is said to be a *partial cochain map of energy cut level* E if $\varphi \circ d \equiv d \circ \varphi \mod T^E$. We also note that if (C, d) is a partial cochain complex of energy cut level E' and if $E < E'$ then (C, d) is a partial cochain complex of energy cut level E.

Lemma 15.9 *Let (C_i, d) be a gapped partial cochain complex of energy cut level E_i for $i = 1, 2$ with $E_1 < E_2$. Let $\varphi : C_1 \to C_2$ be a gapped partial cochain map of energy cut level E_1. We assume $\overline{\varphi} : C_1/\Lambda_{+,\mathrm{nov}}C_1 \to C_2/\Lambda_{+,\mathrm{nov}}C_2$ is an isomorphism.*

Then there exist $d^+ : C_1 \to C_1$ and $\varphi^+ : C_1 \to C_2$ such that:

(1) *(C_1, d^+) is a gapped partial cochain complex of energy cut level E_2.*
(2) *$\varphi^+ : (C_1, d^+) \to (C_2, d)$ is a gapped partial cochain map of energy cut level E_2.*
(3) *$d^+ \equiv d \mod T^{E_1}$ and $\varphi^+ \equiv \varphi \mod T^{E_1}$.*

See Definition 16.12 for the definition of gappedness. Lemma 15.9 is Lemma 19.13. We use Lemma 15.9 to construct the cochain complex $(CF(\mathcal{C}; \Lambda_{0,\mathrm{nov}}), d)$ appearing in Theorem 15.3 as follows. We take $0 < E_1 < E_2 < \dots$ with $E_i \to \infty$. We use the argument outlined in Sects. 15.1.3 and 15.1.4 using the finitely many moduli spaces (consisting of elements of energy $< E_i$) to construct a gapped partial cochain complex $(CF(\mathcal{C}; \Lambda_{0,\mathrm{nov}}), d^i)$ of energy cut level E_i for each i. We next use the argument outlined in Sects. 15.1.5 and 15.1.6 to find a gapped partial cochain map $\varphi_i : (CF(\mathcal{C}; \Lambda_{0,\mathrm{nov}}), d^i) \to (CF(\mathcal{C}; \Lambda_{0,\mathrm{nov}}), d^{i+1})$ of energy cut level E_i for each i. Now we use Lemma 15.9 inductively to obtain $d^i_k : CF(\mathcal{C}) \to CF(\mathcal{C})$ for $k > i$ and $\varphi_{i,k} : (CF(\mathcal{C}), d^i_k) \to (CF(\mathcal{C}), d^{i+1}_k)$ such that:

(1) $(CF(\mathcal{C}), d^i_k)$ is a gapped partial cochain complex of energy cut level E_k.
(2) $\varphi_{i,k}$ is a gapped partial cochain map of energy cut level E_k.
(3) $d^i_k \equiv d^i_{k+1} \mod T^{E_k}$, $\varphi_{i,k} \equiv \varphi_{i,k+1} \mod T^{E_k}$.

Then $\lim_{k\to\infty} d^1_k : CF(\mathcal{C}) \to CF(\mathcal{C})$ becomes the required coboundary operator.

To construct cochain maps, we use a similar argument using the partial homotopy instead of partial cochain map. To construct a cochain homotopy between cochain maps, we also use a similar argument using partial homotopy of homotopies. Algebraic lemmas we use in place of Lemma 15.9 or Lemma 19.13 are Propositions 19.32 and 19.41.

15.1.8 Story in the Case with Rational Coefficients

In Chap. 20 we consider the case when all the spaces R_α are 0-dimensional and prove that we can use a Novikov ring whose ground ring is $\mathbb{Q}$ in that case. The proof is based on the results of Chaps. 13 and 14.

15.2 Outline of the Story of Tree-Like K-Systems

In Chaps. 21 and 22 we study systems of K-spaces, which axiomatize the situation appearing during the construction of the filtered A_∞ algebra associated to a Lagrangian submanifold ([FOOO1, FOOO2, FOOO3, FOOO4].)

15.2.1 Moduli Space of Pseudo-holomorphic Disks: Review

In this section, we review basic properties of the moduli space of pseudo-holomorphic disks to motivate the definitions.

Let M be a symplectic manifold and L its Lagrangian submanifold. We assume that M is compact or tame at infinity and L is compact, oriented and relatively spin. We have the Maslov index group homomorphism $\mu : H_2(M, L; \mathbb{Z}) \to 2\mathbb{Z}$ and the energy group homomorphism $E : H_2(M, L; \mathbb{Z}) \to \mathbb{R}$ defined by

$$E(\beta) = \int_{D^2} u^*\omega$$

with $[u] = \beta$. For $\beta \in H_2(M, L; \mathbb{Z})$ we consider the moduli space $\overset{\circ}{\mathcal{M}}_{k+1}(\beta)$[2] consisting of $((D^2, \vec{z}), u)$ such that:

(1) $u : (D^2, \partial D^2) \to (M, L)$ is pseudo-holomorphic.
(2) $\vec{z} = (z_0, \ldots, z_k)$ are $k+1$ marked points of the boundary ∂D^2.
(3) $z_i \neq z_j$ if $i \neq j$.
(4) $(z_0, \ldots, z_k)$ respects the counterclockwise cyclic order of ∂D^2.

We define evaluation maps

$$\mathrm{ev} = (\mathrm{ev}_0, \ldots, \mathrm{ev}_k) : \overset{\circ}{\mathcal{M}}_{k+1}(\beta) \longrightarrow L^{k+1}$$

[2]Though it is better to write it as $\overset{\circ}{\mathcal{M}}_{k+1}(L; \beta)$, we omit L for the simplicity of notation.

by $\mathrm{ev}_i((D^2, \vec{z}), u) = u(z_i)$. Then $\overset{\circ}{\mathcal{M}}_{k+1}(\beta)$ has a compactification $\mathcal{M}_{k+1}(\beta)$ to which ev_i is extended. Moreover $\mathcal{M}_{k+1}(\beta)$ has an oriented Kuranish structure with corners of dimension

$$\dim \mathcal{M}_{k+1}(\beta) = \mu(\beta) + k - 2.$$

The normalized boundary of $\mathcal{M}_{k+1}(\beta)$ is a disjoint union of the fiber products:

$$\mathcal{M}_{k_1+1}(\beta_1) \ {}_{\mathrm{ev}_0}\times_{\mathrm{ev}_i} \mathcal{M}_{k_2+1}(\beta_2),$$

where $\beta_1 + \beta_2 = \beta$, $k_1 + k_2 = k + 1$, $i = 1, \dots, k_2$.

These facts are proved in [FOOO4, Subsection 7-1], [FOOO22].

15.2.2 *Axiom of Tree-Like K-System and the Construction of the Filtered A_∞ Algebra*

Axioms of the *tree-like K-system over* L or the A_∞ *correspondence over* L are stated as Conditions 21.6 and Definition 21.9 and are obtained by axiomatizing the properties of the system of the moduli spaces $\mathcal{M}_{k+1}(\beta)$ and the evaluation maps ev_i, which are described in Sect. 15.2.1. The axiomatization of various structures is parallel to that of the linear K-system and we do not repeat it. The main result we obtain is the next theorem: For a closed oriented manifold L we denote by $\Omega(L)$ the de Rham complex of L. We put

$$\Omega(L; \Lambda_0) = \Omega(L)\widehat{\otimes}\Lambda_0.$$

See Definition 16.11 for the coefficient ring Λ_0. Here $\widehat{\otimes}$ denotes the completion of the algebraic tensor product. Namely, an element of $\Omega(L; \Lambda_0)$ is a formal sum

$$\sum_{i=0}^{\infty} T^{\lambda_i} h_i,$$

where $\lambda_i \in \mathbb{R}_{\geq 0}$ with $\lambda_1 < \lambda_2 < \dots$, $\lim_{i\to\infty} \lambda_i = +\infty$ and $h_i \in \Omega(L)$.

Theorem 15.10 *Suppose* $(\mathcal{M}_{k+1}(\beta), \mathrm{ev}, \mu, E)$ *is a tree-like K-system over L. Then we can associate a filtered* A_∞ *structure* $\{\mathfrak{m}_k \mid k = 0, 1, 2 \dots\}$ *on* $\Omega(L; \Lambda_0)$.

The filtered A_∞ *algebra* $(\Omega(L; \Lambda_0), \{\mathfrak{m}_k \mid k = 0, 1, 2 \dots\})$ *is independent of the various choices up to homotopy equivalence.*

This is the de Rham version of half of [FOOO3, Theorem A]. ([FOOO3, Theorem A] also contains the part of constructing a tree-like K-system arising from a geometric situation described in Sect. 15.2.1.) It is a consequence of Theorem 21.35

(1) if we put

$$\mathfrak{m}_k = \sum_\beta T^{E(\beta)} \mathfrak{m}_{k,\beta}. \tag{15.17}$$

We recall that the filtered A_∞ structure assigns the operations

$$\mathfrak{m}_k : \underbrace{\Omega(L; \Lambda_0)[1] \widehat{\otimes} \dots \widehat{\otimes} \Omega(L; \Lambda_0)[1]}_{k \text{ times}} \to \Omega(L; \Lambda_0)[1] \tag{15.18}$$

$k = 0, 1, 2, \dots$, that satisfy the A_∞ relation

$$\sum_{k_1+k_2=k+1} \sum_{i=1}^{k-k_2+1} (-1)^* \mathfrak{m}_{k_1}(x_1, \dots, \mathfrak{m}_{k_2}(x_i, \dots, x_{i+k_2-1}), \dots, x_k) = 0. \tag{15.19}$$

Here [1] is the degree +1 shift functor. When we define $\mathfrak{m}_k$ by (15.17), the formula (15.19) follows from (21.25). See [FOOO3, Definition 3.2.20] for the definition of filtered A_∞ algebra and [FOOO3, Definition 4.2.42] for the definition of homotopy equivalence of filtered A_∞ algebras. Roughly speaking, the A_∞ operation

$$\mathfrak{m}_{k,\beta} : \underbrace{\Omega(L)[1] \otimes \cdots \otimes \Omega(L)[1]}_{k \text{ times}} \longrightarrow \Omega(L)[1]$$

is defined by

$$\mathfrak{m}_{k,\beta}(h_1, \dots, h_k) = \mathrm{ev}_0!(\mathrm{ev}_1^* h_1 \wedge \cdots \wedge \mathrm{ev}_k^* h_k) \tag{15.20}$$

using the smooth correspondence

$$\begin{array}{ccc} & \mathcal{M}_{k+1}(\beta) & \\ {}^{(\mathrm{ev}_1,\dots,\mathrm{ev}_k)}\swarrow & & \searrow^{\mathrm{ev}_0} \\ L^k & & L \end{array} \tag{15.21}$$

See (22.17) for the precise definition.[3] The integration along the fiber $\mathrm{ev}_0!$ in the formula (15.20) is defined by using an appropriate system of CF-perturbations

[3] Strictly speaking, in (22.17) we define a *partial* A_∞ algebra structure (see Definition 21.22) which depends on a parameter $\epsilon > 0$. We then use a 'homotopy limit' in a way similar to that of the construction of A_∞ algebra explained in Sect. 15.1.7.

$\widehat{\mathfrak{S}}_{k+1}(\beta)$ on the K-spaces $\mathcal{M}_{k+1}(\beta)$, the latter of which is the main part of the data defining a tree-like K-system.

The formula (15.19) (or (21.25)) is obtained from Stokes' formula (Proposition 9.28) and the Composition formula (Theorem 10.21) via the isomorphism

$$\partial \mathcal{M}_{k+1}(\beta) = \bigcup_{k_1+k_2=k} \bigcup_{i=1,\dots,k_2} \bigcup_{\beta_1+\beta_2=\beta} \mathcal{M}_{k_1+1}(\beta_1) \,{}_{\mathrm{ev}_0}\times_{\mathrm{ev}_i} \mathcal{M}_{k_2+1}(\beta_2), \tag{15.22}$$

which is a part of the axiom of a tree-like K-system, Condition 21.6 (IX).

For this argument to work, we need to choose a system of CF-perturbations $\widehat{\mathfrak{S}}_{k+1}(\beta)$ so that it is compatible with the isomorphism (15.22). Proposition 22.3 is the precise statement which claims the existence of such a system.

Construction of such a system of CF-perturbations is parallel to that of a linear K-system. We use an induction over k and $E(\beta)$ to construct $\widehat{\mathfrak{S}}_{k+1}(\beta)$. The inductive step of this construction uses Proposition 15.7 (or its precise version Proposition 17.81). To inductively verify the assumptions of Proposition 15.7 we need to construct our system $\widehat{\mathfrak{S}}_{k+1}(\beta)$ to be compatible not only along the boundary but also at the corners. Therefore we need to assume the compatibility of Kuranishi structures on $\mathcal{M}_{k+1}(\beta)$ at the corners. This is the corner compatible conditions Condition 21.6 (X) and (XI).

15.2.3 Bifurcation Method and Pseudo-isotopy

As explained in Sect. 15.2.2, the construction of a filtered A_∞ structure from the tree-like K-system given in this book is mostly similar to the construction of Floer's cochain complex from a linear K-system.

The difference between two constructions lies in the morphism part of the construction. In the case of linear K-systems we defined a morphism between two such K-systems and associated to the morphism a cochain map between their Floer's cochain complexes. (In particular, using the identity morphism we proved independence of the resulting cochain complex under the various choices we make, modulo cochain homotopy equivalence.) In the situation of tree-like K-systems we define the notion of pseudo-isotopy between two tree-like K-systems and of filtered A_∞ algebras. Then we show that a pseudo-isotopy between two tree-like K-systems induces a pseudo-isotopy between the A_∞ algebras. It is easy to show that two filtered A_∞ algebras are homotopy equivalent if they are pseudo-isotopic. (See [Fuk4, Theorem 8.2].)

In our geometric situation of Lagrangian Floer theory, a pseudo-isotopy of the tree-like K-system is obtained as follows. In the situation of Sect. 15.2.1, we consider two compatible almost complex structures J_1, J_2 on M. Then we obtain the moduli spaces of pseudo-holomorhic disks $\mathcal{M}_{k+1}(\beta; J_i)$ for $i = 1, 2$. For each $i = 1, 2$ we fix some choices to define a system of Kuranishi structures on $\mathcal{M}_{k+1}(\beta; J_i)$ so that it defines a tree-like K-system. We denote these choices by Ξ_i and the K-space obtained via these choices $\mathcal{M}_{k+1}(\beta; J_i; \Xi_i)$.

Now we consider a one-parameter family of compatible almost complex structures $\{J_t \mid t \in [1, 2]\}$ which joins J_1 to J_2. We consider the moduli space

$$\mathcal{M}_{k+1}(\beta; [1, 2]) = \bigcup_{t\in[1,2]} \{t\} \times \mathcal{M}_{k+1}(\beta; J_t). \tag{15.23}$$

Here $\mathcal{M}_{k+1}(\beta; J_t)$ is the moduli space of J_t holomorphic disks with boundary condition L, homology class β, and $k + 1$ marked points. We can find a system of Kuranishi structures on it such that its restriction to the part $t = 1$ (resp. $t = 2$) coincides with Ξ_1 (resp. Ξ_2.) We have the evaluation maps

$$\mathrm{ev} = (\mathrm{ev}_0, \dots, \mathrm{ev}_k) : \mathcal{M}_{k+1}(\beta; [1, 2]) \to L^{k+1}$$

and

$$\mathrm{ev}_{[1,2]} : \mathcal{M}_{k+1}(\beta; [1, 2]) \to [1, 2].$$

We axiomatize the properties of the system consisting of $\mathcal{M}_{k+1}(\beta; [1, 2])$, the evaluation maps, etc. and define the notion of a $[1, 2]$-parameterized family of A_∞ correspondences. (See Condition 21.11 and Definition 21.15. We define a more general notion of P-parameterized A_∞ correspondence in Definition 21.13.)

We now make choices of CF-perturbations etc. on $\mathcal{M}_{k+1}(\beta; J_i; \Xi_i)$ $i = 1, 2$ and we use them to construct the filtered A_∞ structures.

In the same way as in (15.20) we use the evaluation maps ev and $\mathrm{ev}_{[1,2]}$ together with our CF-perturbations to define operators

$$\mathfrak{m}_{k,\beta} : \Omega([1, 2] \times L)^{\otimes k} \to \Omega([1, 2] \times L).$$

Then $\mathfrak{m}_k = \sum_{k,\beta} T^{E(\beta)} \mathfrak{m}_{k,\beta}$ satisfies the A_∞ relation (15.19). The system of operators $\mathfrak{m}_{k,\beta}$ on $\Omega([1, 2] \times L)$ which satisfies the A_∞ relation and some additional properties is called a *pseudo-isotopy of filtered A_∞ algebras*. See Definition 21.26.

Thus we will prove the following:

Theorem 15.11 *A pseudo-isotopy of A_∞ correspondences induces a pseudo-isotopy of filtered A_∞ algebras.*

Theorem 15.11 is Theorem 21.35 (3).

We note that Theorem 15.11 implies the second half of Theorem 15.10 as follows. Let $(\mathcal{M}_{k+1}(\beta), \mathrm{ev}, \mu, E)$ be an A_∞ correspondence. We can define a pseudo-isotopy of this A_∞ correspondence with itself by taking

$$\mathcal{M}_{k+1}(\beta; [1, 2]) = [1, 2] \times \mathcal{M}_{k+1}(\beta)$$

etc. Then we apply Theorem 15.11 to show that the filtered A_∞ algebras obtained by two different CF-perturbations from $(\mathcal{M}_{k+1}(\beta), \mathrm{ev}, \mu, E)$ are pseudo-isotopic to each other.

Remark 15.12 Here we use the construction of a pseudo-isotopy from an A_∞ correspondence to itself for the construction of a pseudo-isotopy of A_∞ algebras in the way similar to how we use the identity morphism in Sect. 15.1.6 for the construction of a cochain map between Floer cochain complexes. We would like to mention that construction of a pseudo-isotopy from an A_∞ correspondence is much easier than the construction of the identity morphism.

For the actual proof of Theorems 15.10 and 15.11 we need to use a 'homotopy limit' argument similar to those explained in Sect. 15.1.7. We need the notion of pseudo-isotopy of pseudo-isotopies etc. for this purpose. The algebraic lemma corresponding to Lemma 15.9 is Propositions 22.8 and 22.13.

15.2.4 Bifurcation Method and Self-Gluing

In this book we use the morphism of K-systems to prove independence of the Floer's cochain complex associated to a given linear K-system of the choices we make during the construction. On the other hand, we use the pseudo-isotopy to prove independence of the filtered A_∞ structure associated to a tree-like K-system of the choices. Actually we can also use morphism for the tree-like K-system and pseudo-isotopy for the linear K-system. We use two different methods in order to demonstrate both of these two methods. We may call the method using morphism the 'cobordism method' and the method using pseudo-isotopy the 'bifurcation method'. The difference between those two methods is explained also in [FOOO4, Subsection 7.2.14].

The cobordism method is used in the Lagrangian Floer theory in [FOOO3]. An axiomatization of morphism of tree-like K-systems is given in [Fuk5]. The bifurcation method is used in Lagrangian Floer theory in, for example, [AFOOO, AJ, Fuk4]. Each of these two methods has certain advantages and disadvantages.

One advantage of the bifurcation method is that usually it is shorter and simpler, when applicable. See for example, Remark 15.12.

On the other hand, we cannot prove independence of Floer cohomology of periodic Hamiltonian systems under the change of Hamiltonian functions by the bifurcation method. This is because we need to study the situation where the sets of critical points are different.

Michael Hutchings [H2][4] and Paul Seidel [Se, Remark 10.14] mentioned some issues in proving invariance of Floer cohomology via the bifurcation method. We explain below how those issues had been resolved in our previous writings.

We discuss the case of Morse–Novikov cohomology. Let M be a compact Riemannian manifold and h a closed 1-form given locally by the exterior derivative of a Morse function. Let $R(h)$ be the set of the critical points of h. For $p, q \in R(h)$ we consider the compactified moduli space of gradient trajectories of h joining p to q and denote it by $\mathcal{M}(h; p, q)$. (We identify two gradient trajectories ℓ, ℓ' as an element of $\mathcal{M}(h; p, q)$ if $\ell(\tau) = \ell'(\tau + \tau_0)$ for some $\tau_0 \in \mathbb{R}$.) We take its subset $\mathcal{M}(h; p, q; E)$ consisting of the gradient trajectories of h with level change E. The matrix element of the coboundary operator of the Morse–Novikov complex is the sum of the signed counts of the order of $\mathcal{M}(h; p, q; E)$ augmented by the weight T^E. (We assume that the gradient vector field of h is Morse–Smale.) The proof that it defines a cochain complex is the same as the case of the Morse complex, which is similar to the one we explained in Sect. 15.1.5, as was observed by Novikov.

The issue is the way to prove the independence of the cohomology of the cochain complex associated to h when we vary h. Suppose we have two closed 1-forms h and h' whose de Rham cohomology classes in $H^1(M)$ coincide. For simplicity we assume $R(h) = R(h')$. We take a one-parameter family h_t such that $R(h) = R(h_t)$ and $h_0 = h, h_1 = h'$. We consider

$$\mathcal{M}(h_*; p, q; E) = \bigcup_{t \in [0,1]} \{t\} \times \mathcal{M}(h_t; p, q; E) \tag{15.24}$$

and try to use the moduli space to show this independence ((15.24) is similar to (15.23). So the method we explain below is a bifurcation method.)

Note that the virtual dimension of $\mathcal{M}(h; p, p; E)$ is -1. Therefore the virtual dimension of $\mathcal{M}(h_*; p, p; E)$ is 0. So there may be a discrete subset of $t_i \in [0, 1]$ at which $\mathcal{M}(h_t; p, p; E)$ is nonempty. Now since $\mathcal{M}(h_{t_1}; p, p; E)$ is nonempty, $\mathcal{M}(h_{t_1}; p, p; 2E)$ contains an object obtained by concatenating an element of $\mathcal{M}(h_{t_1}; p, p; E)$ with itself. In other words, if $[\ell] \in \mathcal{M}(h_{t_1}; p, p; E)$ then

$$([\ell], [\ell]) \in \mathcal{M}(h_{t_1}; p, p; E) \times \mathcal{M}(h_{t_1}; p, p; E) \subset \mathcal{M}(h_{t_1}; p, p; 2E).$$

By continuing this process we obtain an element of $\mathcal{M}(h_{t_1}; p, p; kE)$ for any k. So we have an element of the strata of arbitrary negative dimension. We explain three different ways to resolve this issue.

The process to obtain an object in the right hand side from the one in the left hand is called self-gluing since two of the factors in the left hand side are the same

[4] Actually Hutchings' concern is not so much about the proof of independence of Morse–Novikov cohomology of the choices but rather the explicit form of the cochain homotopy equivalence, between two Morse–Novikov complexes before and after wall crossing. The discussion below clarifies the way to prove the independence of Morse–Novikov cohomology, but to find the explicit form of the cochain homotopy equivalence we need to study more. See [H1].

moduli space. We remark that a similar point was discussed in [FOOO4, Subsection 7.2.1].

1. Instead of the moduli spaces (15.24) we can use a different moduli space given below. Take a non-decreasing function $\chi : \mathbb{R} \to [0, 1]$ such that $\chi(\tau) = 0$ for small τ and $\chi(\tau) = 1$ for large τ. We consider the 'nonautonomous' equation

$$\frac{d\ell}{d\tau}(\tau) = \operatorname{grad} h_{\chi(\tau)} \tag{15.25}$$

such that $\ell(+\infty) = q$ and $\ell(-\infty) = p$. Let $\mathcal{M}(h_{\chi(\tau)}; p, q; E)$ be the set of solutions of this equation with energy E. Contrary to the definition of a coboundary operator there is no translational symmetry. By counting the order of $\mathcal{M}(h_{\chi(\tau)}; p, q; E)$ we obtain a cochain map from the Morse–Novikov complex of h to one of h'. The standard argument shows that it becomes a cochain homotopy equivalence [Fl2].

This is the standard approach for the proof of the well-definedness of Morse–Novikov cohomology. We note that there is no self-gluing issue in this approach since the equation (15.25) is not invariant under this self-gluing construction.

In other words, when using the cobordism method the issue of self-gluing never occurs.

2. We next explain how the usage of the bifurcation method together with the de Rham model resolves the issue of 'self-gluing'.

We consider the same moduli space $\mathcal{M}(h_*; p, q; E)$ as (15.24). We define the 'evaluation maps' to the interval [0, 1]. Namely we send $\{t\} \times \mathcal{M}(h_t; p, q; E)$ to $t \in [0, 1]$. We usually consider the situation where both projections pr_s and pr_t exist. The former is the source projection and the latter is the target projection. In our setting they both are the same map defined above. We have a diagram:

$$[0,1] \xleftarrow{\mathrm{pr}_s} \mathcal{M}(h_*; p, q; E) \xrightarrow{\mathrm{pr}_t} [0,1]$$

The 'pseudo-isotopy' of the Morse–Novikov complex is a cochain complex defined on

$$\left(\bigoplus_{p \in R(h)} \Omega([0,1]) \otimes [p] \right) \widehat{\otimes} \Lambda_0$$

(where $\Omega([0, 1])$ is the de Rham complex of the interval) with the Novikov ring coefficient. We take the interval for each $p \in R(h)$ and denote it by $[0, 1]_p$. So the 'pseudo-isotopy' of the Morse–Novikov complex is defined on

$$CF(h_*) = \Omega\left(\coprod_{p \in R(h)} [0,1]_p \right) \widehat{\otimes} \Lambda_0. \tag{15.26}$$

The above diagram is regarded as

$$[0,1]_p \xleftarrow{\mathrm{pr}_s} \mathcal{M}(h_*; p, q; E) \xrightarrow{\mathrm{pr}_t} [0,1]_q. \tag{15.27}$$

It 'defines' a map $d_{p,q;E} : \Omega([0,1]_p) \to \Omega([0,1]_q)$ by

$$d_{p,q;E}(u) = (\mathrm{pr}_t)_!(\mathrm{pr}_s^*(u)). \tag{15.28}$$

Note that the pullback of the differential form is defined under rather mild assumptions. However, the pushout or an integration along the fiber $(\mathrm{pr}_t)_!$ is harder to define. This point is indeed related to the self-gluing issue as follows.

Suppose $\mathcal{M}(h_*; p, p; E)$ is transversal. It consists of finitely many points. Let us assume that it consists of a single point $\mathfrak{p}$ and $t_0 = \mathrm{pr}_s(\mathfrak{p}) = \mathrm{pr}_t(\mathfrak{p})$. We take $1 \in \Omega^0([0,1])$. Then

$$d_{p,p;E}(1) = (\mathrm{pr}_t)_!(\mathrm{pr}_s^*(1))$$

is the delta form $\delta_{t_0} dt$ supported at t_0. Now we remark that the pullback

$$\mathrm{pr}_s^*(\delta_{t_0} dt)$$

is not defined. The standard condition for distribution to be pulled back is *not* satisfied in this case. This point is related to the fact the fiber product

$$\mathcal{M}(h_*; p, p; E) \,{}_{\mathrm{pr}_t}\times_{\mathrm{pr}_s} \mathcal{M}(h_*; p, p; E)$$

is not transversal.

This discussion clarifies that the reason why the problem of self-gluing occurs lies in the fact that pr_t is not a submersion. We cannot expect submersivity because of the dimensional reason. We remark that we are in a Bott–Morse situation here even in the case when the set of critical points of h_t is a discrete set for each h_t, in the case when we study a one-parameter family of Morse forms. Therefore the way of resolving this issue is the same as that of a Bott–Morse situation. This issue can be taken care of both in the context of the de Rham and the singular homology models. Let us first explain the case of de Rham model.

The problem here is that pr_t is not a submersion. The solution to this problem is to use a CF-perturbation. As a simplified version of the CF perturbation, we take a family of perturbations (globally) parameterized by a finite-dimensional space, say W. For $w \in W$ we have perturbed moduli space $\mathcal{M}(h_*; p, p; E; w)$. We put

$$\mathcal{M}(h_*; p, p; E; W) = \bigcup_{w \in W} \mathcal{M}(h_*; p, p; E; w) \times \{w\}.$$

By taking the dimension of W sufficiently large we may assume that the map

$$\mathrm{pr}_t : \mathcal{M}(h_*; p, p; E; W) \to [0, 1] \tag{15.29}$$

is a submersion. (The space W depends on p, E.) Let $\mathrm{pr}_W : \mathcal{M}(h_*; p, p; E; W) \to W$ be the projection. We take a differential form χ_W of degree $\dim W$ and with compact support such that $\int_W \chi_W = 1$. Now we define

$$d_{p,p;E}(u) = (\mathrm{pr}_t)_!(\mathrm{pr}_s^*(u) \wedge \mathrm{pr}_W^* \chi_W).$$

Since (15.29) is a submersion this is always well defined. In this way we can define a cochain complex on (15.26) by

$$\delta = d + \sum T^E d_{p,p;E},$$

where T is a formal parameter (Novikov parameter). We can also show $\delta \circ \delta = 0$. We use $(CF(h_*), \delta)$ to prove that $CF(h)$ is cochain homotopy equivalent to $CF(h')$ as follows.

By considering embeddings $\{0\} \to [0, 1]$ and $\{1\} \to [0, 1]$ we have a map

$$CF(h_*) \to CF(h), \qquad CF(h_*) \to CF(h').$$

We can show that they are cochain maps. Moreover, using the fact that de Rham cohomology of $[0, 1]$ is $\mathbb{R}$ we can show that they are cochain homotopy equivalences. Thus we find that $CF(h)$ is cochain homotopic to $CF(h')$. This is a baby version of the proof of independence of filtered A_∞ structure of the almost complex structure etc. using the pseudo-isotopy, which we present in Chaps. 21 and 22.

In this formulation, we have an equality

$$d_{p,p;E_1} \circ d_{p,p;E_2} = 0.$$

This is because $d_{p,p;E}$ increases the degree of the differential form by 1 and the de Rham complex of $[0, 1]$ has elements only in the 0-th and the 1-st degree. So the self-gluing problem does not occur.

3. We finally prove the independence of Morse–Novikov homology using the singular homology model by the bifurcation method.

Let P be a smooth singular chain of $[0, 1]_p$. That is, it is a pair of a simplex and a smooth map from a simplex to $[0, 1]_p$.

The analogue of (15.28) in singular homology is as follows. We take fiber product

$$P \times_{\mathrm{pr}_s} \mathcal{M}(h_*; p, q; E) \tag{15.30}$$

and take its triangulation. Using the map pr_t we regard it as an element of singular chain complex of $[0, 1]_q$. So we consider

$$CF(h_*)^s = \bigoplus_{p \in R(h)} S([0,1]_p) \widehat{\otimes} \Lambda_0.$$

Here $S([0,1]_p)$ is the smooth singular chain complex of the interval $[0,1]_p$. By (15.30), we 'obtain'

$$d_{p,q;E} : S([0,1]_p) \to S([0,1]_q)$$

and the boundary operator on $CF(h_*)^s$.

We remark, however, that the fiber product (15.30) may not be transversal. Also there is no way to perturb $\mathcal{M}(h_*; p, q; E)$ so that (15.30) is transversal for *all* P.

The idea to resolve this issue, which appeared in [FOOO4, Proposition 7.2.35 etc.], is the following: We first take the fiber product (15.30) and then perturb it. In other words, our perturbation *depends not only on* $\mathcal{M}(h_*; p, q; E)$ *but also on the singular chain* P.[5]

Since we can make pr_s a submersion on each Kuranishi chart (this is the definition of weak submersivity!), we can consider the fiber product (15.30) which carries a Kuranishi structure. Then we define

$$\mathcal{M}(h_*; p, q; P) := P \times_{\mathrm{pr}_s} \mathcal{M}(h_*; p, q; E),$$

which is a space with Kuranishi structure. We take a system of perturbations of all of them and triangulations of their zero sets so that the following holds:

(1) (See [FOOO4, Compatibility Condition 7.2.38].) On $\mathcal{M}(h_*; p, q; \partial P) \subset \partial \mathcal{M}(h_*; p, q; P)$ the perturbations and triangulations are compatible.
(2) (See [FOOO4, Compatibility Condition 7.2.44].) On

$$\begin{aligned} &P \times_{\mathrm{pr}_s} \mathcal{M}(h_*; p, r; E_1)\ {}_{\mathrm{pr}_t}\!\times_{\mathrm{pr}_s} \mathcal{M}(h_*; r, q; E_2) \\ &\subset \partial \mathcal{M}(h_*; p, q; E_1 + E_2; P) \end{aligned} \tag{15.31}$$

the perturbations and triangulations are compatible.

The meaning of (1) is clear. Let us explain the meaning of (2). We consider the fiber product $\mathcal{M}(h_*; p, r; P) = P \times_{\mathrm{pr}_s} \mathcal{M}(h_*; p, r; E)$. The perturbation (multisection) and a triangulation of the perturbed space (the zero set of the multisection) are given for this space. We regard the triangulated space of the perturbed moduli space as a singular chain $\sum Q_i$ of $[0,1]_r$. Then the left hand side is the union of

[5] Our exposition below is a slightly improved version of the one that appeared in [FOOO14]. In [FOOO4] we took a countably generated subcomplex of a smooth singular chain complex. (We use Baire's category theorem uncountably many times in [FOOO14] so that we do not need to take a countably generated subcomplex as we did in [FOOO4].) Here we take singular chain complex itself, that is the way of [FOOO14].

$$Q_i \times_{\mathrm{pr}_s} \mathcal{M}(h_*; r, q; E) = \mathcal{M}(h_*; r, q; E; Q_i).$$

The perturbation and a triangulation of its zero set are also given. We require that the restriction of the perturbation of $\mathcal{M}(h_*; p, q; E; P)$ and the triangulation of its zero set coincide with the ones which are combination of $\mathcal{M}(h_*; r, q; E; Q_i)$ and of $\mathcal{M}(h_*; p, r; P)$.

Let us elaborate on the last point more. At each point in (15.31) the obstruction bundle is a direct sum of those of $\mathcal{M}(h_*; p, r; E_1)$ and of $\mathcal{M}(h_*; r, q; E_2)$. We require that the first component of the perturbation is one for $\mathcal{M}(h_*; p, r; P)$ and the second component of the perturbation is one for $\mathcal{M}(h_*; r, q; E; Q_i)$.

This is the meaning of the compatibility (2). Construction of the perturbation and a triangulation satisfying (1)(2) are given by an induction over E and $\dim P$. This is the way of obtaining $d_{p,q;E} : S([0,1]_p) \to S([0,1]_q)$ and the way taken in [FOOO4, Chapter 7].

We elaborate on this construction a bit more explicitly and show how it resolves the issue of self-gluing. We consider the case of $\mathcal{M}(h_*; p, p; E)$ that is zero-dimensional. Suppose for simplicity that it consists of one point and its t coordinate is t_0. We consider a 0-chain $P(t_1) = \{t_1\} \in [0,1]_p$. If $t_1 \neq t_0$ then $P(t_1) \times_{\mathrm{pr}_s} \mathcal{M}(h_*; p, p; E)$ is transversal and is the empty set. If $t_1 = t_0$ then $P(t_0) \times_{\mathrm{pr}_s} \mathcal{M}(h_*; p, p; E)$ is *not* transversal and we need to perturb it. After perturbation it becomes empty again.

Next we consider a 1-chain $P(a,b) = [a,b] \subset [0,1]_p$. If $a, b \neq t_0$, then $P(a,b) \times_{\mathrm{pr}_s} \mathcal{M}(h_*; p, p; E)$ is transversal. It is an empty set if $t_0 \notin [a,b]$ and is one point if $t_0 \in [a,b]$. If $a = t_0$, then $P(t_0, b) \times_{\mathrm{pr}_s} \mathcal{M}(h_*; p, p; E)$ is *not* transversal. We already fixed perturbation of $P(t_0) \times_{\mathrm{pr}_s} \mathcal{M}(h_*; p, p; E) \subset \partial P(t_0, b) \times_{\mathrm{pr}_s} \mathcal{M}(h_*; p, p; E)$. We extend it to obtain a perturbation of $P(t_0, b) \times_{\mathrm{pr}_s} \mathcal{M}(h_*; p, p; E)$. Whether it becomes an empty set or a one-point set depends on the choice of the perturbation of $P(t_0) \times_{\mathrm{pr}_s} \mathcal{M}(h_*; p, p; E)$.

Now we consider the self-gluing. We take the fiber product

$$[0,1]_p \times_{\mathrm{pr}_s} \mathcal{M}(h_*; p, p; E)_{\mathrm{pr}_t} \times_{\mathrm{pr}_s} \mathcal{M}(h_*; p, p; E). \tag{15.32}$$

We do not perturb $[0,1]_p \times_{\mathrm{pr}_s} \mathcal{M}(h_*; p, p; E)$ and this consists of a single point which is mapped to t_0 by pr_t. So the second fiber product is not transversal. However, we have already fixed the perturbation of $P(t_0) \times_{\mathrm{pr}_s} \mathcal{M}(h_*; p, p; E)$ and by this perturbation (15.32) becomes the empty set. In other words, by this perturbation the perturbed zero set does *not* hit the corner. This is how the self-gluing issue is resolved and the way we handled the Bott–Morse situation in [FOOO4].

Note that Akaho and Joyce [AJ] used the bifurcation method to show the well-definedness of the A_∞ structure using the singular homology. We think the way they adopt is basically the same as we described here.

15.3 Discussion Deferred to the Appendices

15.3.1 *Orbifolds and Covering Space of Orbifolds/K-Spaces*

Chap. 23 is a review of the notion of orbifolds and vector bundles on them. We consider effective orbifolds only and use embeddings only as morphisms. In this way we can avoid several delicate issues arising in the discussion of orbifolds. If we go beyond those cases, we need to work with the framework of the 2-category to have a proper notion of morphisms. We also use the language of chart and coordinate change, which is closer to the standard definition of manifolds. It is well-known that there is an alternative way using the language of groupoids. (See for example [ALR].) Using the groupoid language is somewhat similar to the way taken in algebraic geometry to define the notion of stacks. One advantage of using the groupoid language is that the discussion then becomes closer to the 'coordinate free' exposition. We remark that in our definition of Kuranishi structure using the coordinates (Kuranishi charts) and the coordinate changes is inevitable. No 'coordinate free' definition of Kuranishi structures is known.[6] So we think that the coordinate description of orbifolds is more natural for the study of Kuranishi structures. In other approaches to Kuranishi-like structures such as Joyce's, which is closer to that of algebraic geometry, the groupoid description of orbifolds seems to be more natural.

In Chap. 24 we discuss the covering space of an orbifold and a K-space, and define the covering space

$$\widehat{S}_m(\widehat{S}_\ell X) \to \widehat{S}_{m+\ell} X, \tag{15.33}$$

which is important for the formulation of the corner compatibilty condition. See Sect. 15.1.4. We define the notion of a covering space of orbifolds in Sect. 24.1 and generalize it to the case of K-spaces in Sect. 24.1. Then the covering space (15.33) is defined in Sect. 24.3.

15.3.2 *Admissibility of Orbifolds and of Kuranishi Structures*

In Chap. 25, we discuss the notion of admissible orbifolds and admissible Kuranishi structures. The admissibility we study here is the property of the coordinate change etc. with respect to the coordinates normal to the boundary or to the corner. Admissibility is used in the discussion of Chap. 17 for outer collaring. We explain how it is used there briefly below.

[6]The one in [Jo4] may be regarded to be closer to a 'coordinate free' definition. His definition of a Kuranishi structure there is different from the one in this book.

Consider the case of an n-dimensional manifold X with boundary ∂X. (For the simplicity of exposition we assume X has a boundary but no corner.) Let $p \in \partial X$. We take its coordinate chart and so we have a diffeomorphism ψ_p from $[V_p] \times [0, 1)$ to a neighborhood U_p of p in X. Here $[V_p]$ is an open subset of $\mathbb{R}^{n-1}$. The space $X^{\boxplus 1}$ is obtained by taking $[V_p] \times [-1, 1)$ for each p. We glue them as follows. Let $q \in \partial X$ and we take ψ_q, $[V_q] \times [0, 1)$, U_q as above. Let $V_{pq} = \psi_q^{-1}(U_p)$ which is an open subset of $[V_q] \times [0, 1)$. The coordinate change is

$$\varphi_{pq} = \psi_p^{-1} \circ \psi_q : V_{pq} \to [V_p] \times [0, 1).$$

We extend it to

$$\varphi_{pq}^{\boxplus 1} : V_{pq}^{\boxplus 1} \to [V_p] \times [-1, 1)$$

as follows. We put $[V_{pq}] = V_{pq} \cap ([V_q] \times \{0\})$. The restriction of φ_{pq} to $[V_{pq}]$ defines a map $\overline{\varphi}_{pq} : [V_{pq}] \to [V_p]$. (Here we identify $[V_p] = [V_p] \times \{0\}$.) We put

$$V_{pq}^{\boxplus 1} = V_{pq} \cup ([V_{pq}] \times [-1, 0]),$$

where we glue two spaces in the right hand side at $[V_{pq}] \times \{0\}$. The map $\varphi_{pq}^{\boxplus 1}$ is defined by

$$\varphi_{pq}^{\boxplus 1}(x, t) = \begin{cases} \varphi_{pq}(x, t) & \text{if } (x, t) \in V_{pq}, \\ (\overline{\varphi}_{pq}(x), t) & \text{if } (x, t) \in [V_{pq}] \times [-1, 0]. \end{cases}$$

It is easy to see that these maps $\varphi_{pq}^{\boxplus 1}$ satisfy an appropriate cocycle condition. Then we can glue various charts by these maps $\varphi_{pq}^{\boxplus 1}$ to obtain a space $X^{\boxplus 1}$. It is also clear that $X^{\boxplus 1}$ is a topological manifold since $\varphi_{pq}^{\boxplus 1}$ is a homeomorphism to its image.

However, in general, $\varphi_{pq}^{\boxplus 1}$ is *not* differentiable at $[V_{pq}] \times \{0\}$. So in general there is no obvious smooth structure on $X^{\boxplus 1}$.

We introduce the notion of the admissibility of manifolds (orbifolds, K-spaces) so that if we start from an admissible manifold then the coordinate change $\varphi_{pq}^{\boxplus 1}$ becomes smooth. Roughly speaking, the admissibility means that we are given distinguished system of coordinates so that the coordinate change φ_{pq} between those coordinates has the following additional properties:

(*) We put $\varphi_{pq}(x, t) = (y(x, t), s(x, t))$ then

$$s(x, t) - t, \qquad \frac{\partial}{\partial t} y(x, t)$$

together with all of their derivatives go to zero as $t \to 0$.

(More precisely, we assume certain exponential decay.) It is easy to see that (*) implies that $\varphi_{pq}^{\boxplus 1}$ is smooth. So $X^{\boxplus 1}$ becomes a smooth manifold provided X is an admissible manifold and we use admissible coordinates to define $X^{\boxplus 1}$.

We can generalize the admissibility to the case of an orbifold with corners. See Definitions 25.12 and 25.14. Then we can define admissibility of various notions on an admissible orbifold, for example, admissibility of a vector bundle, a section thereof, and a smooth map to another manifold (without boundary). We can also define admissibility of an embedding of an admissible orbifold to another admissible orbifold. We can use them to define admissibility of Kuranishi charts and coordinate changes. Then we can define the notion of an *admissible Kuranishi structure*. (Definition 25.36.) On an admissible Kuranishi structure we can define the notion of admissible CF-perturbations. It is obvious that the story of Part I can be worked out in the admissible category. Using admissibility we can extend a vector bundle $\mathcal{E}$ on X to a vector bundle $\mathcal{E}^{\boxplus 1}$ on $X^{\boxplus 1}$. Also an admissible section of $\mathcal{E}$ is canonically extended to a smooth section of $\mathcal{E}^{\boxplus 1}$.

This is the way we put a collar on the outside of a K-space X in Chap. 17 and obtain $X^{\boxplus 1}$. We can also extend various admissible objects of a K-space X to collared objects of $X^{\boxplus 1}$. Thus the operation $X \mapsto X^{\boxplus 1}$ from admissible objects to collared objects is completely canonical and functorial. We write those constructions in Chap. 17 in detail for completeness. However, we emphasize that this construction is indeed straightforward.

We note that for *any* orbifold X with corners there exists a system of charts by which X becomes an admissible orbifold. In the case of a manifold with boundary, there exists a coordinate system such that the coordinate change $\varphi_{pq} = \psi_p^{-1} \circ \psi_q : V_{pq} \to [V_p] \times [0,1)$ preserves the second factor $[0,1]$ and that the first factor ($[V_p]$ factor) of $\varphi_{pq}(x,t)$ depends only on x. This statement is nothing but the existence of the collar 'inside' a manifold X with boundary. The existence of the collar of any manifold or orbifold with corners is a classical fact which is easy to prove. So there is not much reason to put a collar 'outside' in the case of an orbifold. However, in the case of Kuranishi structures, it is cumbersome to give a detailed proof of the existence of a Kuranishi structure so that all the coordinate changes respect the collar. (As we mentioned before, this is because the way to put a collar on a manifold is not canonical.) The short-cut we take is to use the admissible structure to put the collar outside which is canonical.

To apply this story to our geometric situation such as the case of the moduli space of pseudo-holomorphic curves, we need to establish the existence of the admissible structure for such moduli spaces. This point is related to the exponential decay estimate of the gluing analysis in the following way.

The boundary or corner of the moduli space of pseudo-holomorphic curves appears typically at the infinity of the moduli space and the coordinate normal to the boundary or the corner is the gluing parameter resolving the associated nodal curve. Let us consider the case of moduli space of pseudo-holomorphic disks and consider the configuration of the three disks as in Fig. 15.1a below. This curve has two boundary nodes written p and q in the figure and one boundary marked

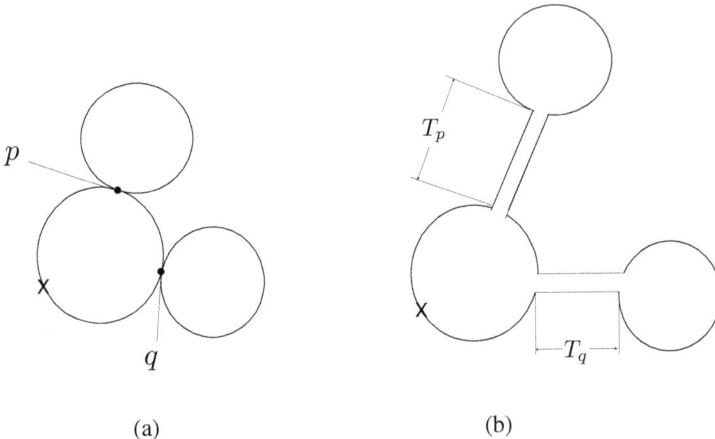

Fig. 15.1 Bordered curve consisting of three disks (**a**) and its smoothing (**b**)

point. The parameter space to resolve these singularities involves two real numbers. We write them as T_p and T_q. They are the length T_p (resp. T_q) of the neck $[0, T_p] \times [0, 1]$ associated to p (resp. $[0, T_q] \times [0, 1]$ associated to q). See Fig. 15.1b. So $T_p, T_q \in (C, \infty]$ for some large positive number C.

We study pseudo-holomorphic maps from the smoothed curves. This is the gluing construction of pseudo-holomorphic curves. We can perform the gluing in one of the following three different ways:

(1) We first glue pseudo-holomorphic maps at p and then at q.
(2) We first glue pseudo-holomorphic maps at q and then at p.
(3) We glue pseudo-holomorphic maps at the two points p, q at the same time.

Let $\mathcal{M}_1(\beta)$ be the compactified moduli space of pseudo-holomorphic disks with one marked point, of homology class $\beta \in H_2(M, L)$ and boundary condition given by a certain Lagrangian submanifold $L \subset M$. A pseudo-holomorphic map to (M, L) from the bordered nodal curve drawn in Fig. 15.1a defines an element of the compactified moduli space $\mathcal{M}_1(\beta)$. Suppose we are given such a pseudo-holomorphic curve u. We assume it is Fredholm regular, for simplicity. Let $[V]$ be the intersection of its neighborhood in $\mathcal{M}_1(\beta)$ and the stratum consisting of elements whose source curves are still singular with two boundary nodes. In other words, $[V]$ is a neighborhood of this element in the codimension 2 stratum of $\mathcal{M}_1(\beta)$.

Then any one of the above three gluing constructions gives a parameterization map from an open subset of $[V] \times (C, \infty]^2$ onto an open subset of $\mathcal{M}_1(\beta)$. Let us write those maps as ψ_1, ψ_2, ψ_3, respectively. ψ_i is a map from $[V] \times (C, \infty]^2$ to $\mathcal{M}_1(\beta)$. The coordinate change $\varphi_{ij} : [V'] \times (C', \infty]^2 \to [V] \times (C, \infty]^2$ is the composition $\psi_i^{-1} \circ \psi_j$. The issue is whether T_p and T_q coordinates (of $(C', \infty]^2$ factor) are preserved by the coordinate change φ_{ij}. (In other words, the issue is whether $\varphi_{ij}(x, T_q, T_p) = (y, T_q, T_p)$ or not.) In fact, it is unlikely that the coordinates T_p, T_q are preserved by this coordinate change φ_{ij}. We elaborate on this point at the end of this Sect. 15.3.2.

This is related to the construction of the collar of the resulting Kuranishi structure. Namely if T_p, T_q coordinates happen to be preserved by the coordinate change then we can use this geometric coordinate itself as a collaring of the corner. In other words, the neighborhood of the corner (which we denote by $S_2\mathcal{M}_1(\alpha)$) in $\mathcal{M}_1(\beta)$ is diffeomorphic to $S_2\mathcal{M}_1(\beta) \times [0, \epsilon)^2$ and the coordinates of the factor $[0, \epsilon)^2$ can be taken, for example, as $(1/T_p, 1/T_q)$.

However, it seems not so easy to find a gluing construction such that φ_{ij} etc. preserves T_p and T_q coordinates.

On the other hand, there is no need at all to obtain the collar of the corner *directly* by the analytic construction of the chart. For the case of a single orbifold, existence of the collar can be proved by a standard argument. In the situation of Kuranishi structures things are a bit more complicated, since we need to find collars for various Kuranishi charts which are preserved by the coordinate change. Though we can find such a system of collars for a good coordinate system after an appropriate shrinking, its proof is a bit cumbersome to write down in detail.

Our short-cut is to put a collar outside, and for this purpose we need to find an admissible coordinate system.

For this purpose it suffices to prove that φ_{ij} respects gluing parameter ($[0, \epsilon)^2$) modulo an error term which is exponentially small in T_p, T_q. Proving this is easier than proving that φ_{ij} strictly preserves the gluing parameter. We can prove this property as follows. We first observe that although the three different ways of gluing (1)(2)(3) give the different maps $[V] \times (C, \infty]^2 \to \mathcal{M}_1(\beta)$, it is easy to construct pre-gluing for which the above three processes (1)(2)(3) give the same result. This is because pre-gluing is a simple process via partition of unity and the pre-gluing on one neck region does not affect the other neck region. The exponential decay estimate of the gluing construction (see [FOOO18] for the detail of the proof) then implies that actual parameterization map obtained by gluing is close to the pre-gluing map modulo an error that is exponentially small in T_p, T_q. Therefore the coordinate change φ_{ij} is admissible.

Remark 15.13 Finding a collar related to the gluing parameter is a matter different from finding a collar appearing in the study of homotopy of almost complex structures etc. The latter appears, for example, when one proves independence of the Gromov–Witten invariant of the choice of almost complex structure. In the latter problem we can take a homotopy J_t between two almost complex structures J and J' so that $J_t = J$ for $t \in [0, \epsilon]$ and $J_t = J'$ for $t \in [1-\epsilon, 1]$. So existence of a collar is *trivial* to prove in that situation. The situation is different for the gluing parameter. (We also note that to prove independence of the Gromov–Witten invariant of the choice of almost complex structure we do *not* need to use the collar at all. See the proof of Proposition 8.16.)

We finally explain the reason why it is unlikely that the gluing parameters T_p, T_q are preserved by the coordinate change. Note our source curve is unstable. (Among the three irreducible components, two are unstable.) So following [FOn2, appendix] or [FOOO21] we add two extra marked points $\vec{w} = (w_1, w_2)$ on unstable irreducible components. We assume the map u is an immersion at those marked points and take codimension 2 submanifolds N_1, N_2 such that u intersects with N_i transversally at w_i. We now glue the source curves at two nodal points and obtain the source curve $\Sigma(T_p, T_q)$ as in Fig. 15.1b. We use a bump function to obtain a map $u'_{T_p,T_q} : (\Sigma(T_p, T_q), \partial\Sigma(T_p, T_q)) \to (M, L)$. Note that at this stage u'_{T_p,T_q} intersects with N_i transversally at w_i. We then modify u'_{T_p,T_q} to a pseudo-holomorphic curve u_{T_p,T_q}. According to three different ways (1), (2) and (3) to perform this process the resulting pseudo-holomorphic maps are slightly different. So we obtain $u^{(j)}_{T_p,T_q}$, $j = 1, 2, 3$. Note that $u^{(j)}_{T_p,T_q}(w_i)$ may not be contained in N_i. We take nearby points $w_i^{(j)}$ such that $u^{(j)}_{T_p,T_q}(w_i) \in N_i$. Suppose for simplicity the element in $\mathcal{M}_1(\beta)$ represented by u from the source curve Fig. 15.1a is isolated. (Namely we assume the corner locus $[V]$ is one point.) Then we can show that for T_p, T_q and $j = 2, 3$ there exist unique $T_p^{(j)}$, $T_q^{(j)}$ such that

$$(\Sigma(T_p, T_q), z_1, w_1^{(1)}, w_2^{(1)}) \cong (\Sigma(T_p^{(j)}, T_q^{(j)}), z_1, w_1^{(j)}, w_2^{(j)}),$$

where z_1 is the unique boundary marked point and the isomorphism is one of marked bordered Riemann surfaces. Then

$$\varphi_{j1}(T_p, T_q) = (T_p^{(j)}, T_q^{(j)}).$$

Since $w_i^{(1)} \neq w_i^{(j)}$ $(j = 2, 3)$ in general, φ_{j1} does not preserve the gluing parameter.

15.3.3 Stratified Submersion

When we consider a map f from a manifold (an orbifold, a K-space) with corners X to another manifold M without boundary or corners, we say that f is a submersion if its restrictions to all the corners $S_k X$ are submersions. It implies that the pushout $f_! h$ of all the smooth forms h on X by f is a smooth form on M.

When we study a family of K-systems parameterized by a manifold with corners P, we need to discuss the submersivity of a map from a manifold (an orbifold, a K-space) with corners X to a manifold with corners P.

We use the notion of stratified submersion for such a purpose. In Chap. 26 we define such a notion and discuss the pushout of a differential form to a manifold with corners.

15.3.4 Integration Along the Fiber and Local System

As we mentioned before, in certain situations, (for example when we study the Floer cohomology of periodic Hamiltonian systems in the Bott–Morse situation) we need to introduce certain $O(1)$ principal bundles on the space R_α, to control the orientation problem of the moduli space of pseudo-holomorphic curves. In Part I the integration along the fiber is defined in the situation when our K-space is oriented. We need to extend it slightly to include the case when the target and source spaces come with $O(1)$ principal bundles, the K-space (which gives smooth correspondence) may not be oriented, and the target and source spaces R_α may not be oriented, but their orientation local systems and the $O(1)$ principal bundles we put on R_α are related in some particular way. In Chap. 27 we discuss such a generalization.

Chapter 16
Linear K-Systems: Floer Cohomology I – Statement

16.1 Axiom of Linear K-Systems

We axiomatize the properties enjoyed by the system of moduli spaces of solutions of Floer's equation.

Condition 16.1 We consider the following objects:

(I) $\mathfrak{G}$ is an additive group, and $E : \mathfrak{G} \to \mathbb{R}$ and $\mu : \mathfrak{G} \to \mathbb{Z}$ are group homomorphisms. We call $E(\beta)$ the *energy* of β and $\mu(\beta)$ the *Maslov index* of β.

(II) $\mathfrak{A}$ is a set on which $\mathfrak{G}$ acts freely. We assume that the quotient set $\mathfrak{A}/\mathfrak{G}$ is a finite set. $E : \mathfrak{A} \to \mathbb{R}$ and $\mu : \mathfrak{A} \to \mathbb{Z}$ are maps such that

$$E(\beta \cdot \alpha) = E(\alpha) + E(\beta), \quad \mu(\beta \cdot \alpha) = \mu(\alpha) + \mu(\beta)$$

for any $\alpha \in \mathfrak{A}$, $\beta \in \mathfrak{G}$. We also call E the *energy* and μ the *Maslov index*.

(III) **(Critical submanifold)** For any $\alpha \in \mathfrak{A}$ we have a finite-dimensional compact manifold R_α (without boundary), which we call a *critical submanifold*.

(IV) **(Connecting orbit)** For any $\alpha_-, \alpha_+ \in \mathfrak{A}$, we have a K-space with corners $\mathcal{M}(\alpha_-, \alpha_+)$ and strongly smooth maps[1]

$$(\mathrm{ev}_-, \mathrm{ev}_+) : \mathcal{M}(\alpha_-, \alpha_+) \to R_{\alpha_-} \times R_{\alpha_+}.$$

We assume that ev_+ is weakly submersive. We call $\mathcal{M}(\alpha_-, \alpha_+)$ the *space of connecting orbits* and $\mathrm{ev}_\pm$ the *evaluation maps at infinity*.

[1] In the geometric situation when $\mathcal{M}(\alpha_-, \alpha_+)$ is the moduli space of the solution of Floer's equation (15.2) (see Sect. 15.1), the maps ev_-, ev_+ are defined as the limits $\tau \to -\infty$, $\tau \to +\infty$, respectively.

K. Fukaya et al., *Kuranishi Structures and Virtual Fundamental Chains*, Springer Monographs in Mathematics, https://doi.org/10.1007/978-981-15-5562-6_16

(V) **(Positivity of energy)** We assume $\mathcal{M}(\alpha_-, \alpha_+) = \emptyset$ if $E(\alpha_-) \geq E(\alpha_+)$.[2]

(VI) **(Dimension)** The dimension of the space of connecting orbits is given by

$$\dim \mathcal{M}(\alpha_-, \alpha_+) = \mu(\alpha_+) - \mu(\alpha_-) - 1 + \dim R_{\alpha_+}. \tag{16.1}$$

(VII) **(Orientation)**

(i) For any $\alpha \in \mathfrak{A}$, a principal $O(1)$ bundle o_{R_α} on R_α is given. We call it an *orientation system of the critical submanifold*.

(ii) For any $\alpha_1, \alpha_2 \in \mathfrak{A}$, an isomorphism

$$\mathrm{OI}_{\alpha_-,\alpha_+} : \mathrm{ev}_+^*(\det T R_{\alpha_+}) \otimes \mathrm{ev}_+^*(o_{R_{\alpha_+}}) \cong o_{\mathcal{M}(\alpha_-,\alpha_+)} \otimes \mathrm{ev}_-^*(o_{R_{\alpha_-}}) \tag{16.2}$$

of principal $O(1)$ bundles is given.[3] Here $o_{\mathcal{M}(R_{\alpha_-}, R_{\alpha_+})}$ is an orientation bundle of the K-space $\mathcal{M}(\alpha_-, \alpha_+)$ in the sense of Definition 3.10. We call the isomorphism (16.2) the *orientation isomorphism*.[4] (See Chap. 27. Otherwise the reader may consider only the case when all the spaces R_α and $\mathcal{M}(\alpha_-, \alpha_+)$ are oriented and $o_{R_{\alpha_\pm}}$ are trivial.)[5] More precisely, we fix a choice of homotopy class of the isomorphism (16.2).

(VIII) **(Periodicity)**

(i) For any $\beta \in \mathfrak{G}$ a diffeomorphism

$$\mathrm{PI}_{\beta;\alpha} : R_\alpha \to R_{\beta\alpha} \tag{16.3}$$

is given such that the equality

$$\mathrm{PI}_{\beta_2;\beta_1\alpha} \circ \mathrm{PI}_{\beta_1;\alpha} = \mathrm{PI}_{\beta_2\beta_1;\alpha}$$

holds.

---

[2]We note that $\mathcal{M}(\alpha, \alpha) = \emptyset$ in particular.

[3]In the case of the linear K-system obtained from periodic Hamiltonian systems, we take $o_R = \Theta_R^-$ for each critical submanifold R, where Θ_R^- is defined as the determinant of the index bundle of a certain family of elliptic operators. See [FOOO4, Definition 8.8.2] for the precise definition. Then [FOOO4, Proposition 8.8.6] yields the isomorphism (16.2). Note the moduli space written as $\mathcal{M}(\alpha_-, \alpha_+)$ in this book was written as $\mathcal{M}(R_{\alpha_+}, R_{\alpha_-})$ in [FOOO4]. See Remark 16.2 (1). On the other hand, in [FOOO4, Proposition 8.8.7] we take $o_R = \det TR \otimes \Theta_R^-$. Note that we use singular chains in [FOOO4], while we use differential forms in the current book. So the choices of o_R are slightly different.

[4]See [FOOO4, Section 8.8].

[5]Note that o_{R_α} may not coincide with the principal $O(1)$ bundle giving an orientation of R_α. For example, o_{R_α} may be nontrivial even in the case when R_α is orientable. On the other hand, if $\mathcal{M}(\alpha_-, \alpha_+)$ is orientable then $o_{\mathcal{M}(\alpha_-,\alpha_+)}$ is trivial.

(ii) Moreover, an isomorphism

$$\mathrm{PI}_{\beta;\alpha_-,\alpha_+} : \mathcal{M}(\alpha_-,\alpha_+) \to \mathcal{M}(\beta\alpha_-,\beta\alpha_+) \tag{16.4}$$

of K-spaces in the sense of Definition 4.24 is given such that the equality

$$\mathrm{PI}_{\beta_2;\beta_1\alpha_-,\beta_1\alpha_+} \circ \mathrm{PI}_{\beta_1;\alpha_-,\alpha_+} = \mathrm{PI}_{\beta_2\beta_1;\alpha_-,\alpha_+}$$

holds. The diagram below commutes:

$$\begin{array}{ccc}
\mathcal{M}(\alpha_-,\alpha_+) & \xrightarrow{\mathrm{PI}_{\beta;\alpha_-,\alpha_+}} & \mathcal{M}(\beta\alpha_-,\beta\alpha_+) \\
{\scriptstyle (\mathrm{ev}_-,\mathrm{ev}_+)}\downarrow & & \downarrow{\scriptstyle (\mathrm{ev}_-,\mathrm{ev}_+)} \\
R_{\alpha_-}\times R_{\alpha_+} & \xrightarrow{(\mathrm{PI}_{\beta;\alpha_-},\mathrm{PI}_{\beta;\alpha_+})} & R_{\beta\alpha_-}\times R_{\beta\alpha_+}
\end{array} \tag{16.5}$$

We call $\mathrm{PI}_{\beta;\alpha}$, $\mathrm{PI}_{\beta;\alpha_-,\alpha_+}$ the *periodicity isomorphisms*. The periodicity isomorphism preserves o_{R_α} and commutes with the orientation isomorphism.

(IX) **(Gromov compactness)** For any $E_0 \geq 0$ and $\alpha_- \in \mathfrak{A}$ the set

$$\{\alpha_+ \in \mathfrak{A} \mid \mathcal{M}(\alpha_-,\alpha_+) \neq \emptyset,\ E(\alpha_+) \leq E_0 + E(\alpha_-)\} \tag{16.6}$$

is a finite set.

(X) **(Boundary compatibility isomorphism)** There exists an isomorphism between the normalized boundary of the space of connecting orbits and a disjoint union of the fiber products:[6]

$$\partial\mathcal{M}(\alpha_-,\alpha_+) \cong \coprod_{\alpha} \left(\mathcal{M}(\alpha_-,\alpha)\ {}_{\mathrm{ev}_+}\times_{\mathrm{ev}_-}\ \mathcal{M}(\alpha,\alpha_+)\right). \tag{16.7}$$

Here $\cong$ means an isomorphism of K-spaces. We call (16.7) the *boundary compatibility isomorphism*. When we compare the orientations between the two sides, we swap the order of the factors in the fiber product in the right hand side of (16.7) and put the sign as follows:

$$\partial\mathcal{M}(\alpha_-,\alpha_+) \cong \coprod_{\alpha} (-1)^{\dim \mathcal{M}(\alpha,\alpha_+)} \left(\mathcal{M}(\alpha,\alpha_+)\ {}_{\mathrm{ev}_-}\times_{\mathrm{ev}_+}\ \mathcal{M}(\alpha_-,\alpha)\right). \tag{16.8}$$

[6]The right hand side is a finite union by Condition (IX).

See Remark 16.2 for this isomorphism. The boundary compatibility isomorphism commutes with the periodicity and orientation isomorphisms and is compatible with various evaluation maps. Namely the restriction of ev_- (resp. ev_+) of the left hand side of (16.7) coincides with ev_- (resp. ev_+) of the factor $\mathcal{M}(\alpha_-, \alpha)$ (resp. $\mathcal{M}(\alpha, \alpha_+)$) of the right hand side.

Remark 16.2

(1) The sign in (16.8) is consistent with that of [FOOO4, p.728] (line 6 of the proof of Theorem 8.8.10 (3)). Here we should take into account the following two points about the notation and convention. The first point is about notation. In [FOOO4, Chapter 8] we write $\mathcal{M}(\alpha, \beta)$ for $\lim_{\tau \to -\infty} u(\tau, t) \in R_\beta$, $\lim_{\tau \to +\infty} u(\tau, t) \in R_\alpha$ for the discussion on orientations of moduli spaces of connecting orbits. Actually, this notation is different from those used in other chapters of [FOOO3, FOOO4], as we note at the beginning of Section 8.7 and p. 723 in [FOOO4]. The order of the position of α, β in the notation $\mathcal{M}(\alpha, \beta)$ used in this book is opposite to the one used in [FOOO4, Chapter 8]. This is just a difference of notation.

The second point is the order of factors in the fiber product. This is not only a difference in conventions but also really affects the orientation on the fiber product. In [FOOO4, Chapter 8], e.g. Section 8.3, we use the evaluation map at the 0-th marked point of *the second factor* when taking the fiber product, while we are using the evaluation map at the 0-th marked point of *the first factor* in this book. See (16.7), for example. (As we note in Convention 8.3.1 and in the third line of [FOOO4, p. 699], we regard the 0-th marked point z_0 as $+1$ in the unit disk in $\mathbb{C}$ which corresponds to $+\infty$ in the strip $\mathbb{R} \times [0, 1]$.) In (16.8) we follow the convention used in [FOOO4, Chapter 8]. That is why we swap the factors in the right hand side in (16.7). Now by taking the difference of notation mentioned here into account, the formula in line 6 of the proof of [FOOO4, Theorem 8.8.10 (3), p. 728] can be rewritten with the notation in this book as

$$\begin{aligned}&\partial \mathcal{M}_{k+\ell+2}(R_{h_1}, R_{h_3}) \supset (-1)^{d_0} \mathcal{M}_{k+2}(R_{h_2}, R_{h_3}) \times_{R_{h_2}} \mathcal{M}_{\ell+2}(R_{h_1}, R_{h_2}),\\ &d_0 = k\ell + k(\mu(h_2) - \mu(h_1)) + (\mu(h_3) - \mu(h_2)) + k - 1 + \dim R_{h_3}.\end{aligned} \tag{16.9}$$

Applying this formula for the case $\alpha_- = h_1, \alpha = h_2, \alpha_+ = h_3$ and $k = \ell = 0$, we obtain (16.8) by the dimension formula (16.1).

We remark that we do not need to study compatibility of orientations at the corner (e.g. at (XI) (XII)). We do not swap the order of factors in fiber products at (XI) (XII) etc.

(2) In general, since we find

$$X_1 \times_Y X_2 = (-1)^{(\dim X_1 - \dim Y)(\dim X_2 - \dim Y)} X_2 \times_Y X_1, \tag{16.10}$$

we can rewrite (16.8) as

$$\partial \mathcal{M}(\alpha_-, \alpha_+) \cong \coprod_{\alpha} (-1)^{\epsilon} \left(\mathcal{M}(\alpha_-, \alpha) \ {}_{\mathrm{ev}_+}\times_{\mathrm{ev}_-} \ \mathcal{M}(\alpha, \alpha_+)\right), \tag{16.11}$$

where

$$\begin{aligned}
\epsilon &= \dim \mathcal{M}(\alpha, \alpha_+) + (\dim \mathcal{M}(\alpha_-, \alpha) - \dim R_\alpha)(\dim \mathcal{M}(\alpha, \alpha_+) - \dim R_\alpha) \\
&= (\mu(\alpha) - \mu(\alpha_-)) \left(\mu(\alpha_+) - \mu(\alpha) - 1 + \dim R_{\alpha_+} - \dim R_\alpha\right) + \dim R_\alpha \\
&= (\mu(\alpha) - \mu(\alpha_-)) \dim \mathcal{M}(\alpha, \alpha_+) - (\mu(\alpha) - \mu(\alpha_-) - 1) \dim R_\alpha.
\end{aligned}$$

(XI) **(Corner compatibility isomorphism)** Let $\widehat{S}_k(\mathcal{M}(\alpha_-, \alpha_+))$ be the normalized corner of the K-space $\mathcal{M}(\alpha_-, \alpha_+)$ in the sense of Definition 24.18. Then there exists an isomorphism:

$$\begin{aligned}
&\widehat{S}_k(\mathcal{M}(\alpha_-, \alpha_+)) \\
&\cong \coprod_{\alpha_1, \dots, \alpha_k \in \mathfrak{A}} \left(\mathcal{M}(\alpha_-, \alpha_1) \ {}_{\mathrm{ev}_+}\times_{R_{\alpha_1}} \cdots {}_{R_{\alpha_k}}\times_{\mathrm{ev}_-} \mathcal{M}(\alpha_k, \alpha_+)\right).
\end{aligned} \tag{16.12}$$

Here the right hand side is a disjoint union. This is an isomorphism of K-spaces, which we call a *corner compatibility isomorphism*. It commutes with the periodicity isomorphisms.[7] It is compatible with various evaluation maps. Namely the restriction of ev_- (resp. ev_+) of the left hand side of (16.12) coincides with ev_- (resp. ev_+) of the factor $\mathcal{M}(\alpha_-, \alpha_1)$ (resp. $\mathcal{M}(\alpha_k, \alpha_+)$) of the right hand side.

(XII) **(Consistency between corner compatibility isomorphisms)** The isomorphism in (XI) satisfies the compatibility condition given in Condition 16.3 below.

We call Conditions (XI)(XII) above the *corner compatibility conditions*.

To describe Condition 16.3 we need some notation. By (24.6) (the covering map associated to the corner structure) and (16.12) (the corner compatibility isomorphism) we have

$$\begin{aligned}
&\widehat{S}_\ell(\widehat{S}_k(\mathcal{M}(\alpha_-, \alpha_+))) \\
&\cong \coprod_{\alpha_1, \dots, \alpha_k \in \mathfrak{A}} \coprod_{\ell_0 + \dots + \ell_k = \ell} \left(\widehat{S}_{\ell_0}\mathcal{M}(\alpha_-, \alpha_1) \ {}_{\mathrm{ev}_+}\times_{R_{\alpha_1}} \cdots {}_{R_{\alpha_k}}\times_{\mathrm{ev}_-} \widehat{S}_{\ell_k}\mathcal{M}(\alpha_k, \alpha_+)\right).
\end{aligned}$$

We apply the corner compatibility isomorphism (16.12) to each of the fiber product factors of the right hand side and obtain

[7] We do not assume any compatibility of the orientation isomorphism *at the corner*, because what we will use is Stokes' formula where the boundary but not the corner appears.

$$
\begin{aligned}
&\widehat{S}_\ell(\widehat{S}_k(\mathcal{M}(\alpha_-,\alpha_+)))\\
&\cong \coprod_{\alpha_1,\dots,\alpha_k\in\mathfrak{A}} \coprod_{\ell_0+\dots+\ell_k=\ell} \coprod_{\alpha_{0,1},\dots,\alpha_{0,\ell_0}} \cdots \coprod_{\alpha_{k,1},\dots,\alpha_{k,\ell_k}}\\
&\Big(\mathcal{M}(\alpha_-,\alpha_{0,1}) \,{}_{\mathrm{ev}_+}\times_{R_{\alpha_{0,1}}} \cdots {}_{R_{\alpha_{0,\ell_0}}}\times_{\mathrm{ev}_-} \mathcal{M}(\alpha_{0,\ell_0},\alpha_1)\Big)\\
&{}_{\mathrm{ev}_+}\times_{R_{\alpha_1}} \Big(\mathcal{M}(\alpha_1,\alpha_{1,1}) \,{}_{\mathrm{ev}_+}\times_{R_{\alpha_{1,1}}} \cdots {}_{R_{1,\alpha_{\ell_k}}}\times_{\mathrm{ev}_-} \mathcal{M}(\alpha_{1,\ell_k},\alpha_2)\Big)\\
&\cdots\\
&{}_{\mathrm{ev}_+}\times_{R_{\alpha_k}} \Big(\mathcal{M}(\alpha_k,\alpha_{k,1}) \,{}_{\mathrm{ev}_+}\times_{R_{\alpha_{k,1}}} \cdots {}_{R_{k,\alpha_{\ell_k}}}\times_{\mathrm{ev}_-} \mathcal{M}(\alpha_{k,\ell_k},\alpha_+)\Big).
\end{aligned}
\tag{16.13}
$$

By applying the corner compatibility isomorphism to $k+\ell$ in place of k we obtain

$$
\begin{aligned}
&\widehat{S}_{k+\ell}(\mathcal{M}(\alpha_-,\alpha_+))\\
&\cong \coprod_{\alpha_1,\dots,\alpha_{k+\ell}\in\mathfrak{A}} \Big(\mathcal{M}(\alpha_-,\alpha_1) \,{}_{\mathrm{ev}_+}\times_{R_{\alpha_1}} \cdots {}_{R_{\alpha_{k+\ell}}}\times_{\mathrm{ev}_-} \mathcal{M}(\alpha_{k+\ell},\alpha_+)\Big).
\end{aligned}
\tag{16.14}
$$

It is easy to observe that each summand of (16.13) appears in (16.14) and vice versa.

Condition 16.3 The covering map $\pi_{\ell,k} : \widehat{S}_\ell(\widehat{S}_k(\mathcal{M}(\alpha_-,\alpha_+))) \to \widehat{S}_{k+\ell}(\mathcal{M}(\alpha_-,\alpha_+))$ in Proposition 24.17 restricts to the identity map on each summand of (16.13).

Remark 16.4 It is easy to see that each summand in (16.14) appears exactly $(k+\ell)!/k!\ell!$ times in (16.13). This is the covering index of the map $\pi_{\ell,k}$.

Remark 16.5 In general, in the case of orbifolds, the map $\overset{\circ}{S}_{k-1}(\partial U) \to \overset{\circ}{S}_k(U)$ is not a k to 1 map set-theoretically. (See Remark 8.6 (3).) However, it is so in our case since the isotropy group acts trivially on the part $[0,1)^k$ (which is the normal direction to the stratum).

Definition 16.6

(1) Let $\mathfrak{A}$, $\mathfrak{G}$, R_α, o_{R_α}, E, μ, $\mathrm{PI}_{\beta,\alpha}$ be the objects as in Conditions 16.1 (I), (II), (III), (VII)–(i), (VIII)–(i).[8] We denote them by

$$
\mathcal{C} = \Big(\mathfrak{A}, \mathfrak{G}, \{R_\alpha\}_{\alpha\in\mathfrak{A}}, \{o_{R_\alpha}\}_{\alpha\in\mathfrak{A}}, E, \mu, \{\mathrm{PI}_{\beta,\alpha}\}_{\beta\in\mathfrak{G},\alpha\in\mathfrak{A}}\Big)
$$

and call $\mathcal{C}$ *critical submanifold data*.

[8]Namely, we collect the same data as in Conditions 16.1 as far as critical submanifolds are concerned.

(2) Together with critical submanifold data $\mathcal{C}$, a *linear system of spaces with Kuranishi structures*, abbreviated as a *linear K-system*, $\mathcal{F}$[9] consists of objects

$$\mathcal{F} = \left(\mathcal{C}, \{\mathcal{M}(\alpha_-,\alpha_+)\}_{\alpha_\pm\in\mathfrak{A}}, (\mathrm{ev}_-,\mathrm{ev}_+), \{\mathrm{OI}_{\alpha_-,\alpha_+}\}_{\alpha_\pm\in\mathfrak{A}}, \{\mathrm{PI}_{\beta;\alpha_-,\alpha_+}\}_{\beta\in\mathfrak{G},\alpha_\pm\in\mathfrak{A}}\right)$$

which satisfy Condition 16.1.

(3) Let $E_0 > 0$. A *partial linear K-system* consists of the same objects as in Condition 16.1 except the following points. We call E_0 its *energy cut level.*

(a) The K-space (the space of connecting orbits) $\mathcal{M}(\alpha_-,\alpha_+)$ is defined only when $E(\alpha_+) - E(\alpha_-) \le E_0$.
(b) Periodicity and orientation isomorphisms of the space of connecting orbits are defined only among those satisfying $E(\alpha_+) - E(\alpha_-) \le E_0$.
(c) The boundary compatibility isomorphism given in Condition 16.1 (X) is defined only when the left hand side of (16.8) is defined.
(d) The corner compatibility isomorphism given in Condition 16.1 (XI) is defined only for $\mathcal{M}(\alpha_-,\alpha_+)$ with $E(\alpha_+) - E(\alpha_-) \le E_0$. Its consistency Condition 16.1 (XII) is required only in that case.

Remark 16.7

(1) We define the notion of a partial linear K-system to take care of the 'running out problem' which we discussed in [FOOO4, Section 7.2.3]. (See Chap. 19 for the way we will use it.) We can use a similar argument also to define symplectic homology (see [FH, CFH, BO]) in the case when M is noncompact but convex at infinity, in complete generality. One difference between the current context and the context of symplectic homology lies in the finiteness statement given in Condition 16.1 (II).
(2) To construct Floer cohomology in the situation where only partial linear K-systems are given, we need to define the notion of *inductive systems of partial linear K-systems*. To define the notion of such an inductive system, we need the notion of morphism of linear K-systems. We will define it in Definition 16.19.

16.2 Floer Cohomology Associated to a Linear K-System

To state our main theorem on the linear K-systems, we need to prepare some notations.

Definition 16.8 Let $\mathcal{C}$ be critical submanifold data as in Definition 16.6.

(1) We put

[9] We use $\mathcal{F}$ to denote a linear K-system. Here F stands for Floer.

$$CF(\mathcal{C})^0 = \bigoplus_{\alpha \in \mathfrak{A}} \Omega(R_\alpha; o_{R_\alpha}), \tag{16.15}$$

where $\Omega(R_\alpha; o_{R_\alpha})$ is the de Rham complex of R_α twisted by the principal $O(1)$ bundle o_{R_α}. (See Chap. 27.)

(2) $CF(\mathcal{C})^0$ is a graded filtered $\mathbb{R}$ vector space. Its grading is given so that the degree d part $CF^d(\mathcal{C})^0$ is to be

$$CF^d(\mathcal{C})^0 = \bigoplus_{\alpha \in \mathfrak{A}} \Omega^{d-\mu(\alpha)}(R_\alpha; o_{R_\alpha}),$$

and its filtration $\mathfrak{F}CF^d(\mathcal{C})^0$ is given by

$$\mathfrak{F}^\lambda CF(\mathcal{C})^0 = \bigoplus_{\substack{\alpha \in \mathfrak{A} \\ E(\alpha) \geq \lambda}} \Omega(R_\alpha; o_{R_\alpha}).$$

Here $\lambda \in \mathbb{R}$.

(3) The completion of $CF(\mathcal{C})^0$ with respect to the filtration $\mathfrak{F}^\lambda CF(\mathcal{C})^0$ is denoted by $CF(\mathcal{C})$. It is a graded filtered $\mathbb{R}$ vector space which is complete. Its element corresponds to an infinite formal sum $\sum_{i=1}^{\infty} h_i$, where:

(a) $h_i \in \Omega(R_{\alpha_i}; o_{R_{\alpha_i}})$,
(b) $\alpha_i \in \mathfrak{A}$,
(c) If $i \leq j$ then $E(\alpha_i) \leq E(\alpha_j)$,
(d) $\lim_{i \to \infty} E(\alpha_i) = \infty$ unless $\{\alpha_i\}$ is a finite set.

(4) For each $\beta \in \mathfrak{G}$, the inverse of the periodicity diffeomorphism $\mathrm{PI}_{\beta;\alpha}$ induces an isomorphism $\Omega(R_\alpha; o_{R_\alpha}) \to \Omega(R_{\beta\alpha}; o_{R_{\beta\alpha}})$. Its direct sum extends to the above-mentioned completion which we call the *periodicity isomorphism* and write

$$\mathrm{PI}_\beta^* : CF(\mathcal{C}) \to CF(\mathcal{C}).$$

It is of degree $\mu(\beta)$ and satisfies

$$\mathrm{PI}_\beta^*(\mathfrak{F}^\lambda CF(\mathcal{C})) \subset \mathfrak{F}^{\lambda + E(\beta)} CF(\mathcal{C}). \tag{16.16}$$

(5) We define $d_0 : CF(\mathcal{C}) \to CF(\mathcal{C})$ by

$$d_0 \left(\sum_{i=1}^{\infty} h_i \right) = \sum_{i=1}^{\infty} d_{dR} h_i \tag{16.17}$$

where $h_i \in \Omega^{\deg h_i}(R_{\alpha_i}; o_{R_{\alpha_i}})$ and d_{dR} on the right hand side is the de Rham differential. The map d_0 has degree 1 and preserves the filtration.

We call $CF(\mathcal{C})$ the *Floer cochain module associated to* $\mathcal{C}$.

Theorem 16.9 *Suppose we are in the situation of Definition* 16.8.

(1) *For any linear K-system* $\mathcal{F}$, *we can define a map* $d : CF(\mathcal{C}) \to CF(\mathcal{C})$ *such that:*

(a) $d \circ d = 0$.
(b) d *has degree* 1 *and preserves the filtration.*
(c) d *commutes with the periodicity isomorphism* PI^*_β.
(d) *There exists* $\epsilon > 0$ *such that*

$$(d - d_0)(\mathfrak{F}^\lambda CF(\mathcal{C})) \subset \mathfrak{F}^{\lambda+\epsilon} CF(\mathcal{C}).$$

(2) *The definition of the map* d *in (1) involves various choices related to the associated Kuranishi structure and* d *depends on them. However, it is independent of such choices in the following sense.*

If d_1, d_2 *are obtained from two different choices, there exists* $\psi : CF(\mathcal{C}) \to CF(\mathcal{C})$ *with the following properties:*

(a) $d_2 \circ \psi = \psi \circ d_1$.
(b) ψ *has degree* 0 *and preserves the filtration.*
(c) ψ *commutes with the periodicity isomorphism* PI^*_β.
(d) *There exists* $\epsilon > 0$ *such that:*

$$(\psi - \mathrm{id})(\mathfrak{F}^\lambda CF(\mathcal{C})) \subset \mathfrak{F}^{\lambda+\epsilon} CF(\mathcal{C})$$

where id *is the identity map.*

(e) *In particular,* ψ *induces an isomorphism in cohomologies:*

$$H(CF(\mathcal{C}), d_1) \to H(CF(\mathcal{C}), d_2).$$

(f) *The cochain map* ψ *itself depends on various choices. However, it is independent of the choices up to cochain homotopy. Namely if* ψ_i $i = 1, 2$ *are obtained from the different choices, there exists a map* K *of degree* -1 *such that*

$$\psi_1 - \psi_2 = d_2 \circ K + K \circ d_1.$$

Moreover there exists $\epsilon > 0$ *such that*

$$K(\mathfrak{F}^\lambda CF(\mathcal{C})) \subset \mathfrak{F}^{\lambda+\epsilon} CF(\mathcal{C}).$$

The proof will be given in Chap. 19.

Definition 16.10 We call $(CF(\mathcal{C}), d)$ the *Floer cochain complex* associated to the linear K-system $\mathcal{F}$ and denote it by $CF(\mathcal{F})$.

In the independence statement such as Theorem 16.9 (2) the critical submanifolds R_α are fixed. For most of the important applications of Floer cohomology we need to prove certain independence statements in the situation where the critical submanifolds vary. In that case we need to take an appropriate Novikov ring as the relevant coefficient ring. We introduce the following universal Novikov ring for this purpose.

Definition 16.11 Let R be a commutative ring with unit:

(1) The set $\Lambda^R_{\rm nov}$ consists of the formal sums

$$\mathfrak{x} = \sum_{i=1}^{\infty} P_i(e) T^{\lambda_i}, \tag{16.18}$$

where:

(a) $P_i(e) \in R[e^{1/2}, e^{-1/2}]$.[10]
(b) T is a formal parameter.
(c) $\lambda_i \in \mathbb{R}$ and $\lambda_i < \lambda_j$ for $i < j$.
(d) $\lim_{i\to\infty} \lambda_i = \infty$ unless (16.18) is a finite sum.

(2) We define the norm of the element (16.18) by

$$\|\mathfrak{x}\| = \exp\left(-\inf\{\lambda_i \mid P_i(e) \neq 0\}\right).$$

We put $\|0\| = 0$. We define a filtration on $\Lambda^R_{\rm nov}$ by

$$\mathfrak{F}^\lambda \Lambda^R_{\rm nov} = \{\mathfrak{x} \in \Lambda^R_{\rm nov} \mid \|\mathfrak{x}\| \le e^{-\lambda}\}.$$

(3) Let us consider the subset of $\Lambda^R_{\rm nov}$ consisting of formal sums (16.18) for which $P_i(e) = 0$ except finitely many indices i. We define a ring structure on it in an obvious way. It is also an $\mathbb{R}$-vector space and $\|\mathfrak{x}\|$ defines a norm, with respect to which $\Lambda^R_{\rm nov}$ is the completion. The ring structure extends to $\Lambda^R_{\rm nov}$ and $\Lambda^R_{\rm nov}$ becomes a complete normed ring. We call $\Lambda^R_{\rm nov}$ the *universal Novikov ring* with *ground ring* R.
(4) The subset consisting of elements $\mathfrak{x} \in \Lambda^R_{\rm nov}$ with $\|\mathfrak{x}\| \le 1$ is a subring, which we write $\Lambda^R_{0,\rm nov}$ and call it also the *universal Novikov ring*. Moreover, we put $\Lambda^R_{+,\rm nov} = \{\mathfrak{x} \in \Lambda^R_{\rm nov} \mid \|\mathfrak{x}\| < 1\}$.
(5) We put $\deg T = 0$, $\deg e = 2$ and the degree of elements of R to be 0. Then $\Lambda^R_{0,\rm nov}$ and $\Lambda^R_{\rm nov}$ are graded rings.
(6) In the case $R = \mathbb{R}$ we omit $\mathbb{R}$ and write $\Lambda_{\rm nov}$, $\Lambda_{0,\rm nov}$ and $\Lambda_{+,\rm nov}$.

[10] In Item (5) we put $\deg e = 2$. The Novikov ring $\Lambda^R_{\rm nov}$ here is the same as the one introduced in [FOOO3], where the indeterminate e has degree 2.

(7) When we do not include the indeterminate e, we write Λ^R, Λ_0^R and Λ_+^R in place of Λ_{nov}^R, $\Lambda_{0,\text{nov}}^R$ and $\Lambda_{+,\text{nov}}^R$. When $R = \mathbb{R}$, we also drop $\mathbb{R}$ from these notations. Λ^R becomes a field if R is so. We call it the *universal Novikov field*.

We now start from the cochain complex obtained in Theorem 16.9 and obtain a cochain complex over the universal Novikov ring $\Lambda_{0,\text{nov}}$.

Definition 16.12 Let $\mathcal{C}$ be critical submanifold data as in Definition 16.6.

(1) We consider $CF(\mathcal{C})^0$ given in (16.15) and take an algebraic tensor product with Λ_{nov} over $\mathbb{R}$, that is, $CF(\mathcal{C})^0 \otimes_{\mathbb{R}} \Lambda_{\text{nov}}$. It is a Λ_{nov} module. The Λ_{nov} module $CF(\mathcal{C})^0 \otimes_{\mathbb{R}} \Lambda_{\text{nov}}$ is graded by $\deg(x \otimes \mathfrak{x}) = \deg x + \deg \mathfrak{x}$, where $x \in CF(\mathcal{C})^0$ and $\mathfrak{x} \in \Lambda_{\text{nov}}$ are elements of pure degree. The Λ_{nov} module $CF(\mathcal{C})^0 \otimes_{\mathbb{R}} \Lambda_{\text{nov}}$ is filtered by

$$\mathfrak{F}^{\lambda}(CF(\mathcal{C})^0 \otimes_{\mathbb{R}} \Lambda_{\text{nov}}) = \bigcup_{\lambda_1+\lambda_2 \geq \lambda} \mathfrak{F}^{\lambda_1} CF(\mathcal{C})^0 \otimes_{\mathbb{R}} \mathfrak{F}^{\lambda_2} \Lambda_{\text{nov}}.$$

(2) For $\beta \in \mathfrak{G}$ we define a Λ_{nov} module homomorphism $\widehat{\text{PI}}_{\beta}^{*} : CF(\mathcal{C})^0 \otimes_{\mathbb{R}} \Lambda_{\text{nov}} \to CF(\mathcal{C})^0 \otimes_{\mathbb{R}} \Lambda_{\text{nov}}$ by

$$\widehat{\text{PI}}_{\beta}^{*}(x \otimes \mathfrak{x}) = \text{PI}_{\beta}^{*}(x) \otimes T^{-E(\beta)} e^{-\mu(\beta)/2} \mathfrak{x}.$$

It preserves the degree and the filtration.

(3) We consider the closed Λ_{nov} submodule of $CF(\mathcal{C})^0 \otimes_{\mathbb{R}} \Lambda_{\text{nov}}$ generated by elements of the form $\widehat{\text{PI}}_{\beta}^{*}(x \otimes \mathfrak{x}) - x \otimes \mathfrak{x}$ for $x \otimes \mathfrak{x} \in CF(\mathcal{C})^0 \otimes_{\mathbb{R}} \Lambda_{\text{nov}}$ and $\beta \in \mathfrak{G}$. We denote by $(CF(\mathcal{C})^0 \otimes_{\mathbb{R}} \Lambda_{\text{nov}})/\sim$ the quotient module by this submodule. It is filtered and graded.
(4) We denote the completion of $(CF(\mathcal{C})^0 \otimes_{\mathbb{R}} \Lambda_{\text{nov}})/\sim$ with respect to the filtration by $CF(\mathcal{C}; \Lambda_{\text{nov}})$. It is a filtered and graded Λ_{nov} module. We write its filtration by $\mathfrak{F}^{\lambda} CF(\mathcal{C}; \Lambda_{\text{nov}})$.
(5) We put $CF(\mathcal{C}; \Lambda_{0,\text{nov}}) = \mathfrak{F}^0 CF(\mathcal{C}; \Lambda_{\text{nov}})$. It is a $\Lambda_{0,\text{nov}}$ module.
(6) A subset $G \subset \mathbb{R}_{\geq 0} \times \mathbb{Z}$ is called a *discrete submonoid* if the following holds. Denote by $E : G \to \mathbb{R}_{\geq 0}$ and $\mu : G \to \mathbb{Z}$ the natural projections.

    (a) If $\beta_1, \beta_2 \in G$, then $\beta_1 + \beta_2 \in G$. $(0, 0) \in G$.
    (b) The image $E(G) \subset \mathbb{R}_{\geq 0}$ is discrete.
    (c) For each $E_0 \in \mathbb{R}_{\geq 0}$ the inverse image $G \cap E^{-1}([0, E_0])$ is a finite set.

(7) A $\Lambda_{0,\text{nov}}$-module homomorphism $\psi : CF(\mathcal{C}_1; \Lambda_{0,\text{nov}}) \to CF(\mathcal{C}_2; \Lambda_{0,\text{nov}})$[11] is called *G-gapped*, if ψ is decomposed into the following form:

[11] Here the case $\mathcal{C}_1 = \mathcal{C}_2$ is also included.

$$\psi = \sum_{\beta \in G} \psi_\beta T^{E(\beta)} e^{\mu(\beta)/2}, \tag{16.19}$$

where $\psi_\beta : CF(\mathcal{C}_1) \to \mathcal{C}F(\mathcal{C}_2)$ is an $\mathbb{R}$-linear map. In addition, if ψ is a cochain map, it is called a *G-gapped cochain map*. We simply say that ψ is *gapped* if it is G-gapped for some discrete submonoid G.

(8) If an operator $d : CF(\mathcal{C}; \Lambda_{0,\text{nov}}) \to CF(\mathcal{C}; \Lambda_{0,\text{nov}})$ satisfying $d \circ d = 0$ is gapped, we call $(CF(\mathcal{C}; \Lambda_{0,\text{nov}}), d)$ a *gapped cochain complex.*

Lemma 16.13 *$CF(\mathcal{C}; \Lambda_{0,\text{nov}})$ is a free $\Lambda_{0,\text{nov}}$ module.*

Proof This is a consequence of Condition 16.1 (II) and its definition. (See also [FOOO13, Lemma 2.4].) □

Corollary 16.14 *Let $\mathcal{C}$ be the critical submanifold data given in Definition* 16.6.

(1) *In the situation of Theorem* 16.9 *(1), the operator d on $CF(\mathcal{C})$ induces an operator: $CF(\mathcal{C}; \Lambda_{0,\text{nov}}) \to CF(\mathcal{C}; \Lambda_{0,\text{nov}})$, which we also denote by d. It has the same properties as in Theorem* 16.9 *(1) (a)–(d). Moreover, it is gapped in the sense of Definition* 16.12.
(2) *In the situation of Theorem* 16.9 *(2), the cochain map ψ induces a cochain map $\psi : (CF(\mathcal{C}, \Lambda_{0,\text{nov}}), d_1) \to (CF(\mathcal{C}, \Lambda_{0,\text{nov}}), d_2)$. It has the same properties as in Theorem* 16.9 *(2) (a)–(f). Moreover, it is a gapped cochain map.*

Proof

(1) This is immediate from Theorem 16.9 (1) (b)(c)(d). The gappedness follows from the compactness axiom (IX) in Condition 16.1.
(2) This is immediate from Theorem 16.9 (2) (b)(c)(d). From the construction of ψ given in Sect. 19.6, we can see the gappedness. Here the compactness axiom Condition 16.17 (IX) for morphisms of linear K-systems is used. □

Definition 16.15 In the situation of Theorem 16.9 (1), we call the cohomology of $(CF(\mathcal{C}; \Lambda_{\text{nov}}), d)$ the *Floer cohomology* of linear K-system $\mathcal{F}$. It is independent of the choices we make during the definition by Corollary 16.14 (2). We call the cochain complex $(CF(\mathcal{C}; \Lambda_{\text{nov}}), d)$ the *Floer cochain complex over the universal Novikov ring.*

16.3 Morphism of Linear K-Systems

We next define morphisms between linear K-systems.

Situation 16.16

(1) For each $i = 1, 2$, let

$$\mathcal{C}_i = \left(\mathfrak{A}_i, \mathfrak{G}_i, \{R^i_{\alpha_i}\}_{\alpha \in \mathfrak{A}_i}, \{o_{R^i_{\alpha_i}}\}_{\alpha \in \mathfrak{A}_i}, E, \mu, \{\mathrm{PI}^i_{\beta_i,\alpha_i}\}_{\beta_i \in \mathfrak{G}_i, \alpha_i \in \mathfrak{A}_i}\right)$$

be critical submanifold data and

$$\mathcal{F}_i = \Big(C_i, \{\mathcal{M}^i(\alpha_{i-}, \alpha_{i+})\}_{\alpha_{i\pm}\in\mathfrak{A}_i}, (\mathrm{ev}_-, \mathrm{ev}_+),$$
$$\{\mathrm{OI}^i_{\alpha_{i-},\alpha_{i+}}\}_{\alpha_{i\pm}\in\mathfrak{A}_i}, \{\mathrm{PI}^i_{\beta_i;\alpha_{i-},\alpha_{i+}}\}_{\beta_i\in\mathfrak{G}_i,\alpha_{i\pm}\in\mathfrak{A}_i}\Big)$$

a linear K-system. We assume $\mathfrak{G}_1 = \mathfrak{G}_2$ (together with energy E and the Maslov index μ on it) and denote it by $\mathfrak{G}$.

(2) We also consider the same objects as in (1) except we only suppose that they consist of partial linear K-systems of energy cut level E_0.■

We will define a morphism from $\mathcal{F}_1$ to $\mathcal{F}_2$. We consider the following object for this purpose.

Condition 16.17 In Situation 16.16 (1) we consider the following objects.

(I)(II)(III) Nothing to add to those from Condition 16.1.

(IV) (Interpolation space) For any $\alpha_1 \in \mathfrak{A}_1$ and $\alpha_2 \in \mathfrak{A}_2$, we have a K-space with corners $\mathcal{N}(\alpha_1, \alpha_2)$ and strongly smooth maps

$$(\mathrm{ev}_-, \mathrm{ev}_+) : \mathcal{N}(\alpha_1, \alpha_2) \to R^1_{\alpha_1} \times R^2_{\alpha_2}.$$

We assume that ev_+ is weakly submersive. We call $\mathcal{N}(\alpha_1, \alpha_2)$ the *interpolation space* and $\mathrm{ev}_\pm$ the *evaluation maps at infinity*.

(V) (Energy loss) We assume $\mathcal{N}(\alpha_1, \alpha_2) = \emptyset$ if $E(\alpha_1) \geq E(\alpha_2) + c$ for some $c \geq 0$. We call c the *energy loss*.

There is an exception in the case where $c = 0$ and $\alpha_1 = \alpha_2$. See Definition 16.20.

(VI) (Dimension) The dimension of the interpolation space is given by

$$\dim \mathcal{N}(\alpha_1, \alpha_2) = \mu(\alpha_2) - \mu(\alpha_1) + \dim R_{\alpha_2}. \tag{16.20}$$

(VII) (Orientation) For any $\alpha_1 \in \mathfrak{A}_1$ and $\alpha_2 \in \mathfrak{A}_2$, an isomorphism

$$\mathrm{OI}_{\alpha_1,\alpha_2} : \mathrm{ev}_+^*(\det T R^2_{\alpha_2}) \otimes \mathrm{ev}_+^*(o_{R^2_{\alpha_2}}) \cong o_{\mathcal{N}(\alpha_1,\alpha_2)} \otimes \mathrm{ev}_-^*(o_{R^1_{\alpha_1}}) \tag{16.21}$$

of principal $O(1)$ bundles is given. Here $o_{\mathcal{N}(\alpha_1,\alpha_2)}$ is the orientation bundle which gives an orientation of K-space $\mathcal{N}(\alpha_1, \alpha_2)$. We call $\mathrm{OI}_{\alpha_1,\alpha_2}$ an *orientation isomorphism*.

(VIII) (Periodicity) For any $\beta \in \mathfrak{G}$ an isomorphism

$$\mathrm{PI}_{\beta;\alpha_1,\alpha_2} : \mathcal{N}(\alpha_1, \alpha_2) \to \mathcal{N}(\beta\alpha_1, \beta\alpha_2) \tag{16.22}$$

of K-spaces is given such that the equality

$$\mathrm{PI}_{\beta_2;\beta_1\alpha_1,\beta_1\alpha_2} \circ \mathrm{PI}_{\beta_1;\alpha_1,\alpha_2} = \mathrm{PI}_{\beta_2\beta_1;\alpha_1,\alpha_2}$$

holds and the following diagram commutes:

$$\begin{array}{ccc} \mathcal{N}(\alpha_1,\alpha_2) & \xrightarrow{\mathrm{PI}_{\beta;\alpha_1,\alpha_2}} & \mathcal{N}(\beta\alpha_1,\beta\alpha_2) \\ {\scriptstyle (\mathrm{ev}_1,\mathrm{ev}_2)}\downarrow & & \downarrow{\scriptstyle (\mathrm{ev}_1,\mathrm{ev}_2)} \\ R^1_{\alpha_1}\times R^2_{\alpha_2} & \xrightarrow{(\mathrm{PI}_{\beta;\alpha_1},\mathrm{PI}_{\beta;\alpha_2})} & R^1_{\beta\alpha_1}\times R^2_{\beta\alpha_2} \end{array} \tag{16.23}$$

We call $\mathrm{PI}_{\beta;\alpha_1,\alpha_2}$ the *periodicity isomorphism*. The periodicity isomorphism commutes with the orientation isomorphism.

(IX) (Gromov compactness) For any $E_0 \geq 0$ and $\alpha_1 \in \mathfrak{A}_1$ the set

$$\{\alpha_2 \in \mathfrak{A}_2 \mid \mathcal{N}(\alpha_1,\alpha_2) \neq \emptyset,\ E(\alpha_2) \leq E_0 + E(\alpha_1)\} \tag{16.24}$$

is a finite set.

(X) (Boundary compatibility isomorphism) There exists an isomorphism between the normalized boundary of the interpolation space and a disjoint union of the fiber products as follows:

$$\begin{aligned} &\partial\mathcal{N}(\alpha_1,\alpha_2) \\ \cong &\coprod_{\alpha_1'\in\mathfrak{A}_1}(-1)^{\dim\mathcal{N}(\alpha_1,\alpha_2)}\left(\mathcal{N}(\alpha_1',\alpha_2)\ {}_{\mathrm{ev}_-}\times_{\mathrm{ev}_+}\ \mathcal{M}^1(\alpha_1,\alpha_1')\right) \\ &\sqcup \coprod_{\alpha_2'\in\mathfrak{A}_2}(-1)^{\dim\mathcal{M}^2(\alpha_2',\alpha_2)}\left(\mathcal{M}^2(\alpha_2',\alpha_2)\ {}_{\mathrm{ev}_-}\times_{\mathrm{ev}_+}\ \mathcal{N}(\alpha_1,\alpha_2')\right), \end{aligned} \tag{16.25}$$

where the right hand side is the disjoint union.[12] Here $\cong$ means the isomorphism of K-spaces. We call (16.25) the boundary compatibility isomorphism. The boundary compatibility isomorphism commutes with the periodicity and the orientation isomorphisms and respects various evaluation maps. The sign in the first term of the right hand side is $\dim\mathcal{N}(\alpha_1,\alpha_2)$. The moduli space $\mathcal{M}^1(\alpha_1,\alpha_1')$ (more precisely, $\overset{\circ}{\mathcal{M}}{}^1(\alpha_1,\alpha_1')$ before taking stable map compactification) is the quotient of the space of connecting orbits by the translation on the domain. In our convention, we have

$$\overset{\circ}{\mathcal{M}}{}^1(\alpha_1,\alpha_1') \cong \mathcal{M}^1(\alpha_1,\alpha_1')\times\mathbb{R}$$

as oriented K-spaces. The translation on the domain can be seen as the source of gluing parameter. This is the reason why we have signs in the first and second terms above.

[12] The right hand side is a finite sum by Condition (IX).

(XI) (Corner compatibility isomorphism) There exists an isomorphism between the normalized corner $\widehat{S}_k(\mathcal{N}(\alpha_1, \alpha_2))$ and a disjoint union of

$$\begin{aligned} &\mathcal{M}^1(\alpha_1, \alpha_{1,1}) \times_{R^1_{\alpha_{1,1}}} \cdots \times_{R^1_{\alpha_{1,k_1-1}}} \mathcal{M}^1(\alpha_{1,k_1-1}, \alpha_{1,k_1}) \\ &\times_{R^1_{\alpha_{1,k_1}}} \mathcal{N}(\alpha_{1,k_1}, \alpha_{2,1}) \\ &\times_{R^1_{\alpha_{2,1}}} \mathcal{M}^2(\alpha_{2,1}, \alpha_{2,2}) \times_{R^2_{\alpha_{2,2}}} \cdots \times_{R^2_{\alpha_{2,k_2}}} \mathcal{M}^2(\alpha_{2,k_2}, \alpha_2), \end{aligned} \tag{16.26}$$

where $k_1 + k_2 = k$, $\alpha_{1,i} \in \mathfrak{A}_1$, $\alpha_{2,i} \in \mathfrak{A}_2$. This is an isomorphism of K-spaces, which we call corner compatibility isomorphism. Corner compatibility isomorphism commutes with the periodicity isomorphisms and respects various evaluation maps.

(XII) (Consistency between corner compatibility isomorphisms) The corner compatibility isomorphisms given in (XI) have the property given in Condition 16.18 below.

Condition 16.18 Condition 16.17 (XI) and Condition 16.1 (XI) (together with (24.6)) define an isomorphism from $\widehat{S}_\ell(\widehat{S}_k(\mathcal{N}(\alpha_1, \alpha_2)))$ to the disjoint union of the summand (16.26) with k replaced by $k + \ell$. Then the map $\pi_{\ell,k} : \widehat{S}_\ell(\widehat{S}_k(\mathcal{N}(\alpha_1, \alpha_2))) \to \widehat{S}_{\ell+k}(\mathcal{N}(\alpha_1, \alpha_2))$ in Proposition 24.17 becomes the identity map on each component under those isomorphisms.

Definition 16.19 Let us consider Situation 16.16.

(1) A *morphism of linear K-systems* from $\mathcal{F}_1$ to $\mathcal{F}_2$ consists of the objects as in Condition 16.17. We say c the *energy loss* of our morphism.
(2) Let $E_i > 0$ and let $\mathcal{F}_i$ be a partial linear K-system with energy cut levels E_i for each $i = 1, 2$. Suppose $E_1 \geq E_2 + c$ with $c \geq 0$. A *morphism of partial linear K-systems* from $\mathcal{F}_1$ to $\mathcal{F}_2$ consists of the same objects as those in Condition 16.17 except the following:
    (a) The K-space, the interpolation space, $\mathcal{N}(\alpha_1, \alpha_2)$ is defined only when $-c \leq E(\alpha_2) - E(\alpha_1) \leq E_2$.
    (b) The periodicity isomorphisms of the interpolation spaces are defined only among those satisfying $-c \leq E(\alpha_2) - E(\alpha_1) \leq E_2$.
    (c) The boundary compatibility isomorphism in Condition 16.17 (X) is given only when the left hand side of (16.25) is defined.
    (d) The corner compatibility isomorphism in Condition 16.17 (XI) is given only for $\mathcal{N}(\alpha_1, \alpha_2)$ with $-c \leq E(\alpha_2) - E(\alpha_1) \leq E_2$. The consistency between corner compatibility isomorphisms in Condition 16.17 (XII) is required only in that case.

Definition 16.20 Suppose that the indexing set $\mathfrak{A}_1$ of the critical submanifold of $\mathcal{F}_1$ is identified with the indexing set $\mathfrak{A}_2$ of the critical submanifold of $\mathcal{F}_2$. A *morphism with energy loss* 0 *and congruent to an isomorphism* from $\mathcal{F}_1$ to $\mathcal{F}_2$ is a collection of objects as in Condition 16.17 where we replace (V) by (V)' below and $c = 0$.

(V)' For $\alpha = \alpha_1 = \alpha_2 \in \mathfrak{A}_1 = \mathfrak{A}_2$, we require $\mathcal{N}(\alpha, \alpha) = R_\alpha$.[13] We also require the maps $\mathrm{ev}_\pm : R_\alpha \to R_\alpha$ to be identity maps. Moreover, if $\alpha_1 \neq \alpha_2$ and $E(\alpha_1) = E(\alpha_2)$, we require $\mathcal{N}(\alpha_1, \alpha_2) = \emptyset$.

In the case when $\mathcal{F}_1 = \mathcal{F}_2$ holds in addition, we call such a morphism a *morphism with energy loss* 0 *and congruent to the identity morphism.*

16.4 Homotopy and Higher Homotopy of Morphisms of Linear K-Systems

To establish basic properties of Floer cohomology associated to a linear K-system or an inductive system of partial linear K-systems, we will use the notion of homotopy between the morphisms and also higher homotopy such as homotopy of homotopies etc. To include the most general case, we define the notion of morphisms parameterized by a manifold with corners P.

Condition 16.21 Let $\mathcal{C}_i, \mathcal{F}_i$ $(i = 1, 2)$ be critical submanifold data and linear K-systems as in Situation 16.16, respectively. We assume $\mathfrak{G}_1 = \mathfrak{G}_2$ and denote the common group by $\mathfrak{G}$. Let P be a smooth manifold with boundary or corner. We consider the following objects:

(I), (II), (III) Nothing to add.

(IV) (Interpolation space) For any $\alpha_1 \in \mathfrak{A}_1$ and $\alpha_2 \in \mathfrak{A}_2$, we have a K-space with corners $\mathcal{N}(\alpha_1, \alpha_2; P)$ and a strongly smooth map

$$(\mathrm{ev}_P, \mathrm{ev}_-, \mathrm{ev}_+) : \mathcal{N}(\alpha_1, \alpha_2; P) \to P \times R^1_{\alpha_1} \times R^2_{\alpha_2}.$$

We assume that

$$(\mathrm{ev}_P, \mathrm{ev}_+) : \mathcal{N}(\alpha_1, \alpha_2; P) \to P \times R^2_{\alpha_2}$$

is a corner stratified weak submersion. (See Definition 26.6 for the notion of corner stratified submersivity.) We call $\mathcal{N}(\alpha_1, \alpha_2; P)$ a *P-parameterized interpolation space.*

(V) (Energy loss) We assume $\mathcal{N}(\alpha_1, \alpha_2; P) = \emptyset$ if $E(\alpha_1) \geq E(\alpha_2) + c$. We call $c \geq 0$ the *energy loss.*

Exception: In the case $c = 0$ and $\alpha = \alpha_1 = \alpha_2$, we require $\mathcal{N}(\alpha, \alpha; P) = P \times R_\alpha$ instead of requiring it to be an empty set. We also require that $\mathrm{ev}_\pm : P \times R_\alpha \to R_\alpha$ and $\mathrm{ev}_P : P \times R_\alpha \to P$ are projections.

(VI) (Dimension) The dimension is given by

[13] In Condition 16.17 (V) $\mathcal{N}(\alpha_1, \alpha_2)$ is required to be the empty set if $E(\alpha_1) = E(\alpha_2) + c$.

$$\mathcal{N}(\alpha_1, \alpha_2; P) = \mu(\alpha_2) - \mu(\alpha_1) + \dim R^2_{\alpha_2} + \dim P. \tag{16.27}$$

(VII) (Orientation) P is oriented. For any $\alpha_1 \in \mathfrak{A}_1$ and $\alpha_2 \in \mathfrak{A}_2$, we are given an *orientation isomorphism*

$$\mathrm{OI}_{\alpha_1,\alpha_2} : \det TP \otimes \mathrm{ev}_+^*(\det TR^2_{\alpha_2}) \otimes \mathrm{ev}_+^*(o_{R^2_{\alpha_2}}) \cong o_{\mathcal{N}(\alpha_-,\alpha_+;P)} \otimes \mathrm{ev}_-^*(o_{R^1_{\alpha_-}}) \tag{16.28}$$

(VIII) (Periodicity) For any $\beta \in \mathfrak{G}$ an isomorphism

$$\mathrm{PI}_{\beta;\alpha_1,\alpha_2;P} : \mathcal{N}(\alpha_1, \alpha_2; P) \to \mathcal{N}(\beta\alpha_1, \beta\alpha_2; P) \tag{16.29}$$

of K-spaces is given. The equality

$$\mathrm{PI}_{\beta_2;\beta_1\alpha_1,\beta_1\alpha_2;P} \circ \mathrm{PI}_{\beta_1;\alpha_1,\alpha_2;P} = \mathrm{PI}_{\beta_2\beta_1;\alpha_1,\alpha_2;P}$$

holds. The isomorphism $\mathrm{PI}_{\beta;\alpha_1,\alpha_2;P}$ is compatible with $(\mathrm{ev}_P; \mathrm{ev}_-, \mathrm{ev}_+)$ and preserves the orientation isomorphism.

(IX) (Gromov compactness) For any $E_0 \geq 0$ and $\alpha_1 \in \mathfrak{A}_1$ the set

$$\{\alpha_2 \in \mathfrak{A}_2 \mid \mathcal{N}(\alpha_1, \alpha_2; P) \neq \emptyset,\ E(\alpha_2) \leq E_0 + E(\alpha_1)\} \tag{16.30}$$

is a finite set.

To state the next condition we note that Lemma 26.5 implies the following.

Lemma 16.22 *We assume $\mathcal{N}(\alpha_1, \alpha_2; P)$ satisfies Condition* 16.21 *(IV)–(IX).*

(1) *We can define a K-space $\mathcal{N}(\alpha_1, \alpha_2; \partial P)$ by the fiber product:*

$$\mathcal{N}(\alpha_1, \alpha_2; \partial P) := \partial P\, {}_P\times_{\mathrm{ev}_P} \mathcal{N}(\alpha_1, \alpha_2; P).$$

(2) *The periodicity and orientation isomorphisms of $\mathcal{N}(\alpha_1, \alpha_2; P)$ induce those of $\mathcal{N}(\alpha_1, \alpha_2; \partial P)$.*
(3) $(\mathrm{ev}_-, \mathrm{ev}_+, \mathrm{ev}_P)$ *of $\mathcal{N}(\alpha_1, \alpha_2; P)$ induces* $(\mathrm{ev}_-, \mathrm{ev}_+, \mathrm{ev}_{\partial P})$ *of $\mathcal{N}(\alpha_1, \alpha_2; \partial P)$.*
(4) *Objects defined in (1)–(3) above satisfy Condition* 16.21 *(I)–(IX).*

Condition 16.23

(X) (Boundary compatibility isomorphism) There exists an isomorphism between the normalized boundary of the P-parameterized interpolation space and a disjoint union of the fiber products as follows:

$$
\begin{aligned}
&\partial \mathcal{N}(\alpha_1, \alpha_2; P) \\
\cong &\coprod_{\alpha \in \mathfrak{A}_2} (-1)^{\dim \mathcal{M}^2(\alpha,\alpha_2)+\dim P(\dim \mathcal{M}^2(\alpha,\alpha_2)-\dim R_\alpha+1)} \\
&\mathcal{M}^2(\alpha, \alpha_2)\ {}_{\mathrm{ev}_-} \times_{\mathrm{ev}_+} \mathcal{N}(\alpha_1, \alpha; P) \\
&\sqcup \coprod_{\alpha \in \mathfrak{A}_1} (-1)^{\dim \mathcal{N}(\alpha_1,\alpha_2;P)} \mathcal{N}(\alpha, \alpha_2; P)\ {}_{\mathrm{ev}_-} \times_{\mathrm{ev}_+} \mathcal{M}^1(\alpha_1, \alpha) \\
&\sqcup \mathcal{N}(\alpha_1, \alpha_2; \partial P),
\end{aligned}
\tag{16.31}
$$

where the right hand side is the disjoint union. Here $\cong$ means the isomorphism of K-spaces. The isomorphism (16.31) is called the boundary compatibility isomorphism. The boundary compatibility isomorphism commutes with the orientation isomorphism and the periodicity isomorphism, which in the right hand side are obtained by taking fiber products thereof respectively. The boundary compatibility isomorphism respects various evaluation maps.

To state the next condition we note that Lemma 26.5 also implies the following:

Lemma 16.24 *We assume $\mathcal{N}(\alpha_1, \alpha_2; P)$ satisfies Condition* 16.21 *(IV)–(IX).*

(1) *We can define a K-space by the fiber product*

$$\widehat{S}_k(P)_P \times_{\mathrm{ev}_P} \mathcal{N}(\alpha_1, \alpha_2; P).$$

We denote it by $\mathcal{N}(\alpha_1, \alpha_2; \widehat{S}_k(P))$.

(2) *The periodicity isomorphism of $\mathcal{N}(\alpha_1, \alpha_2; P)$ induces one of $\mathcal{N}(\alpha_1, \alpha_2; \widehat{S}_k(P))$.*

(3) $(\mathrm{ev}_P; \mathrm{ev}_-, \mathrm{ev}_+)$, *of $\mathcal{N}(\alpha_1, \alpha_2; P)$ induces* $(\mathrm{ev}_{\widehat{S}_k(P)}; \mathrm{ev}_-, \mathrm{ev}_+)$ *of* $\mathcal{N}(\alpha_1, \alpha_2; \widehat{S}_k(P))$.

(4) *Objects defined in (1)–(3) above satisfy Condition* 16.21 *(I)–(IX).*

The next lemma is also a conclusion of Lemma 26.5.

Lemma 16.25 *Suppose we are in the situation of Lemma* 16.24.

(1) *The $(k+\ell)/k!\ell!$ fold covering map $\widehat{S}_k(\widehat{S}_\ell(P)) \to \widehat{S}_{k+\ell}(P)$ induces a $(k+\ell)!/k!\ell!$ fold covering map : $\mathcal{N}(\alpha_1, \alpha_2; \widehat{S}_k(\widehat{S}_\ell(P))) \to \mathcal{N}(\alpha_1, \alpha_2; \widehat{S}_{k+\ell}(P))$.*

(2) *The covering map in (1) commutes with the periodicity isomorphisms. It is also compatible with various evaluation maps. In particular, the following diagram commutes:*

$$
\begin{array}{ccc}
\mathcal{N}(\alpha_1, \alpha_2; \widehat{S}_k(\widehat{S}_\ell(P))) & \longrightarrow & \mathcal{N}(\alpha_1, \alpha_2; \widehat{S}_{k+\ell}(P)) \\
{\scriptstyle \mathrm{ev}_{\widehat{S}_k(\widehat{S}_\ell(P))}} \downarrow & & \downarrow {\scriptstyle \mathrm{ev}_{\widehat{S}_{k+\ell}(P)}} \\
\widehat{S}_k(\widehat{S}_\ell(P)) & \longrightarrow & \widehat{S}_{k+\ell}(P)
\end{array}
\tag{16.32}
$$

Condition 16.26

(XI) (Corner compatibility isomorphism) There exists an isomorphism between the normalized corner $\widehat{S}_k\mathcal{N}((\alpha_-, \alpha_+); P)$ and a disjoint union of fiber products:

$$\begin{aligned}&\mathcal{M}^1(\alpha_-, \alpha_1) \times_{R_{\alpha_1}} \cdots \times_{R_{\alpha_{k_1-1}}} \mathcal{M}^1(\alpha_{k_1-1}, \alpha_{k_1})\\&\times_{R_{\alpha_{k_1}}} \mathcal{N}(\alpha_{k_1}, \alpha_{k_1+1}; \widehat{S}_{k_3}(P))\\&\times_{R_{\alpha_{k_1+1}}} \mathcal{M}^2(\alpha_{k_1+1}, \alpha_{k_1+2}) \times_{R_{\alpha_{k_1+2}}} \cdots \times_{R_{\alpha_{k_1+k_2}}} \mathcal{M}^2(\alpha_{k_1+k_2}, \alpha_+).\end{aligned} \tag{16.33}$$

Here $k_1+k_2+k_3 = k$. This is an isomorphism of K-spaces and is called the corner compatibility isomorphism. The corner compatibility isomorphism commutes with the periodicity isomorphisms and respects various evaluation maps. Moreover Condition 16.28 below holds.

To state Condition 16.28 we make a digression. We consider the codimension ℓ normalized corner of the space (16.33), that is, $\widehat{S}_\ell((16.33))$. Applying (24.6) to (16.33), we get an isomorphism from the normalized corner $\widehat{S}_\ell(\widehat{S}_k(\mathcal{N}(\alpha_-, \alpha_+; P)))$ to the disjoint union of fiber products of the normalized corners of the factors of (16.33). We can identify the normalized corners of various factors of (16.33) by using (16.12) and Condition 16.26. Thus we obtain the next lemma.

Lemma 16.27 *The corner compatibility isomorphisms given in Condition* 16.26 *and in* (16.12) *canonically induce an isomorphism from* $\widehat{S}_\ell(\widehat{S}_k(\mathcal{N}(\alpha_-, \alpha_+; P)))$ *to a disjoint union of the fiber products of the following form:*

$$\begin{aligned}&\mathcal{M}^1(\alpha_-, \alpha'_1) \times_{R_{\alpha'_1}} \cdots \times_{R_{\alpha'_{k'_1-1}}} \mathcal{M}^1(\alpha'_{k'_1-1}, \alpha'_{k'_1})\\&\times_{R_{\alpha'_{k'_1}}} \mathcal{N}(\alpha'_{k'_1}, \alpha'_{k'_1+1}; \widehat{S}_{\ell'}(\widehat{S}_{k'_3}(P)))\\&\times_{R_{\alpha'_{k'_1+1}}} \mathcal{M}^2(\alpha'_{k'_1+1}, \alpha'_{k'_1+2}) \times_{R_{\alpha'_{k'_1+2}}} \cdots \times_{R_{\alpha'_{k'_1+k'_2}}} \mathcal{M}^2(\alpha'_{k'_1+k'_2}, \alpha_+).\end{aligned} \tag{16.34}$$

Here $k'_1 + k'_2 + k'_3 + \ell' = \ell + k$. *(Note that the same copies of the form* (16.34) *appear several times in* $\widehat{S}_\ell(\widehat{S}_k(\mathcal{N}(\alpha_-, \alpha_+; P)))$.*)*

Now we state the last condition.

Condition 16.28 ((XII) (Consistency between corner compatibility isomorphisms)) Under the isomorphism between (16.34) and (16.33) (with k replaced by $k + \ell$), the $(k + \ell)!/k!\ell!$-fold covering map $\widehat{S}_\ell(\widehat{S}_k(\mathcal{N}(\alpha_-, \alpha_+; P))) \to \widehat{S}_{\ell+k}(\mathcal{N}(\alpha_-, \alpha_+; P))$ obtained in Proposition 24.17 is identified on each component with the fiber product of the identity maps and of

$$\mathcal{N}(\alpha'_{k'_1}, \alpha'_{k'_1+1}; \widehat{S}_{\ell'}(\widehat{S}_{k'_3}(P))) \to \mathcal{N}(\alpha'_{k'_1}, \alpha'_{k'_1+1}; \widehat{S}_{\ell'+k'_3}(P)),$$

which is the map given in Lemma 16.25.

Definition 16.29

(1) A *P-parameterized family of morphisms* from $\mathcal{F}_1$ to $\mathcal{F}_2$ consists of objects satisfying Conditions 16.21, 16.23, 16.26, and 16.28.
(2) If a P-parameterized family of morphisms $\mathfrak{N}_P$ from $\mathcal{F}_1$ to $\mathcal{F}_2$ and a ∂P-parameterized family of morphisms $\mathfrak{N}_{\partial P}$ from $\mathcal{F}_1$ to $\mathcal{F}_2$ are related as in Condition 16.23, we call $\mathfrak{N}_{\partial P}$ the *boundary of* $\mathfrak{N}_P$ and write $\partial \mathfrak{N}_P$.
(3) Let $E_i > 0$ and $\mathcal{F}_i$ be partial linear K-systems with energy cut levels E_i ($i = 1, 2$). Suppose $E_1 \geq E_2 + c$ for some $c \geq 0$. A *P-parametrized family of morphisms of partial linear K-systems* from $\mathcal{F}_1$ to $\mathcal{F}_2$ consists of the same objects as in Condition 16.21 except the following:
    (a) The K-space, the P parameterized interpolation space, $\mathcal{N}(\alpha_1, \alpha_2; P)$ is defined only when $-c \leq E(\alpha_2) - E(\alpha_1) \leq E_2$.
    (b) The periodicity and orientation isomorphisms of the P-parameterized interpolation spaces are defined only among those which satisfy $-c \leq E(\alpha_2) - E(\alpha_1) \leq E_2$.
    (c) The boundary compatibility isomorphism in Condition 16.23 (X) and the corner compatibility isomorphism in Condition 16.26 (XI) are defined only when the left hand side of (16.31) is defined.
    (d) We require Condition 16.28 only when $\mathcal{N}(\alpha_-, \alpha_+; P)$ is defined.

Definition 16.30 Let $\mathcal{F}_1$, $\mathcal{F}_2$ be linear K-systems and $\mathfrak{N}_1$, $\mathfrak{N}_2$ two morphisms between them.

(1) *A homotopy from* $\mathfrak{N}_1$ *to* $\mathfrak{N}_2$ is a $P = [1, 2]$ parameterized family of morphisms from $\mathfrak{N}_1$ to $\mathfrak{N}_2$ such that its boundary is $-\mathfrak{N}_1 \sqcup \mathfrak{N}_2$. Here $-\mathfrak{N}_1$ is $\mathfrak{N}_1$ with the orientation systems of the critical submanifolds and orientation isomorphisms inverted and $\sqcup$ denotes the disjoint union.
(2) We say that $\mathfrak{N}_1$ is *homotopic* to $\mathfrak{N}_2$ if there exists a homotopy between them.
(3) We define partial homotopy of partial morphisms between partial linear K-systems in the same way.

Theorem 16.31 *Let $\mathcal{F}_1$, $\mathcal{F}_2$, $\mathcal{F}_3$ be linear K-systems. We make the choices mentioned in Theorem* 16.9 *(2) and obtain cochain complexes $CF(\mathcal{F}_i; \Lambda_{\text{nov}})$, $i = 1, 2, 3$.*

(1) *A morphism $\mathfrak{N} : \mathcal{F}_1 \to \mathcal{F}_2$ induces a Λ_{nov} module homomorphism*

$$\psi_{\mathfrak{N}} : CF(\mathcal{F}_1; \Lambda_{\text{nov}}) \to CF(\mathcal{F}_2; \Lambda_{\text{nov}}) \tag{16.35}$$

with the following properties:

(a) $\psi_{\mathfrak{N}} \circ d = d \circ \psi_{\mathfrak{N}}$, *where* d *in the left hand side (resp. right hand side) is the coboundary operator of* $CF(\mathcal{F}_1; \Lambda_{\text{nov}})$ *(resp.* $CF(\mathcal{F}_2; \Lambda_{\text{nov}})$*.)*

(b) *The homomorphism* $\psi_{\mathfrak{N}}$ *preserves degree and satisfies*

$$\psi_{\mathfrak{N}}\left(\mathfrak{F}^{\lambda} CF(\mathcal{F}_1; \Lambda_{\text{nov}})\right) \subset \mathfrak{F}^{\lambda-c} CF(\mathcal{F}_1; \Lambda_{\text{nov}}),$$

where c *is the energy loss of* $\mathfrak{N}$*. In particular, if* $c = 0$ *then* $\psi_{\mathfrak{N}}$ *induces a* $\Lambda_{0,\text{nov}}$ *module homomorphism:*

$$\psi_{\mathfrak{N}} : CF(\mathcal{F}_1; \Lambda_{0,\text{nov}}) \to CF(\mathcal{F}_2; \Lambda_{0,\text{nov}}).$$

(2) *Let* $\mathfrak{N}_1, \mathfrak{N}_2 : \mathcal{F}_1 \to \mathcal{F}_2$ *be morphisms. If* $\mathfrak{H}$ *is a homotopy from* $\mathfrak{N}_1$ *to* $\mathfrak{N}_2$*, then it induces a* Λ_{nov} *module homomorphism:*

$$\psi_{\mathfrak{H}} : CF(\mathcal{F}_1; \Lambda_{\text{nov}}) \to CF(\mathcal{F}_2; \Lambda_{\text{nov}}) \tag{16.36}$$

with the following properties:

(a) $\psi_{\mathfrak{H}} \circ d + d \circ \psi_{\mathfrak{H}} = \psi_{\mathfrak{N}_1} - \psi_{\mathfrak{N}_2}$.

(b) *The homomorphism* $\psi_{\mathfrak{H}}$ *decreases degree by* 1 *and satisfies*

$$\psi_{\mathfrak{H}}\left(\mathfrak{F}^{\lambda} CF(\mathcal{F}_1; \Lambda_{\text{nov}})\right) \subset \mathfrak{F}^{\lambda-c} CF(\mathcal{F}_1; \Lambda_{\text{nov}}),$$

where c *is the energy loss of* $\mathfrak{H}$*. In particular, if* $c = 0$ *then* $\psi_{\mathfrak{H}}$ *induces a* $\Lambda_{0,\text{nov}}$ *module homomorphism:*

$$\psi_{\mathfrak{H}} : CF(\mathcal{F}_1; \Lambda_{0,\text{nov}}) \to CF(\mathcal{F}_2; \Lambda_{0,\text{nov}}).$$

(3) *Let* $\mathfrak{N}_{i+1i} : \mathcal{F}_i \to \mathcal{F}_{i+1}$ *be morphisms of linear K-systems of for* $i = 1, 2$. *Let* $\mathfrak{N}_{31} : \mathcal{F}_1 \to \mathcal{F}_3$ *be the composition* $\mathfrak{N}_{32} \circ \mathfrak{N}_{21}$*. Then the composition* $\psi_{\mathfrak{N}_{23}} \circ \psi_{\mathfrak{N}_{12}}$ *is cochain homotopic to* $\psi_{\mathfrak{N}_{13}}$.

(4) *If* $\mathcal{ID}_{\mathcal{F}_i} : \mathcal{F}_i \to \mathcal{F}_i$ *is the identity morphism, the cochain map* $\psi_{\mathcal{ID}_{\mathcal{F}_i}}$ *induced by it is cochain homotopic to the identity.*

Remark 16.32 The morphism $\psi_{\mathfrak{N}}$ in Item (1) depends on the choices made for its construction. More precisely, we first make the choices mentioned in Theorem 16.9 Item (1) to define coboundary operators of $CF(\mathcal{F}_i; \Lambda_{\text{nov}})$, $i = 1, 2$. Then we make a choice (compatible with the first choices) to define the map $\psi_{\mathfrak{N}}$ which is a cochain map with respect to the coboundary operators obtained from the choices we made. Note that we can use Item (2) to show that up to cochain homotopy the cochain map $\psi_{\mathfrak{H}}$ is independent of the choices we made to define it.

We will define the notion of the identity morphism later in Lemma-Definition 18.57 of Sect. 18.10. The definition of composition of morphisms is given in Sect. 16.5 and the precise definition of the notation used there is postponed until Chap. 18, where we will discuss 'corner smoothing' systematically. The proof of Theorem 16.31 is given in Chap. 19.

Definition 16.33 We say two morphisms $\mathfrak{N}_1$ and $\mathfrak{N}_2$ are *congruent modulo* T^E if the following holds:

(1) $\mathfrak{N}_1(\alpha_1, \alpha_2) \cong \mathfrak{N}_2(\alpha_1, \alpha_2)$ for any α_1, α_2 with $E(\alpha_2) - E(\alpha_1) \leq E$.
(2) The above isomorphism is one between K-spaces. It preserves the orientation isomorphism, the periodicity isomorphism, the boundary compatibility isomorphism and the corner compatibility isomorphism.

16.5 Composition of Morphisms of Linear K-Systems

In this section we discuss composition of morphisms.

Situation 16.34 Let $C_i, \mathcal{F}_i$ $(i = 1, 2, 3)$ be critical submanifold data and linear K-systems as in Situation 16.16, respectively. We assume $\mathfrak{G}_1 = \mathfrak{G}_2 = \mathfrak{G}_3$ and denote the common group by $\mathfrak{G}$. Let $\mathcal{F}_i$ be linear K-systems for $i = 1, 2, 3$.■

Lemma-Definition 16.35 *Suppose we are in Situation* 16.34.

(1) *Let* $\mathfrak{N}_{i+1i} : \mathcal{F}_i \to \mathcal{F}_{i+1}$ *be morphisms and* $\mathcal{N}_{ii+1}(\alpha_i, \alpha_{i+1})$ *their interpolation spaces. Then we can define the* composition $\mathfrak{N}_{32} \circ \mathfrak{N}_{21} : \mathcal{F}_1 \to \mathcal{F}_3$. *The interpolation space*[14] *of* $\mathfrak{N}_{32} \circ \mathfrak{N}_{21}$ *is the union of the fiber products*

$$\bigcup_{\alpha_2 \in \mathfrak{A}_2} \mathcal{N}_{12}(\alpha_1, \alpha_2)\ {}_{\mathrm{ev}_+} \times^{\boxplus 1}_{\mathrm{ev}_-} \mathcal{N}_{23}(\alpha_2, \alpha_3). \tag{16.37}$$

The precise meaning of the union $\bigcup$ *in* (16.37) *is defined during the proof in this section and in Chap.* 18. *The symbol* $\times^{\boxplus 1}$ *will be defined in Definition* 18.34.
(2) *Let* $\mathfrak{N}^{P_i}_{i+1i} : \mathcal{F}_i \to \mathcal{F}_{i+1}$ *be* P_i*-parameterized morphisms and* $\mathcal{N}_{ii+1}(\alpha_i, \alpha_{i+1}; P_i)$ *their interpolation spaces. We can define the* composition $\mathfrak{N}^{P_2}_{32} \circ \mathfrak{N}^{P_1}_{21} : \mathcal{F}_1 \to \mathcal{F}_3$ *that is a* $P_1 \times P_2$*-parameterized morphism. Its interpolation space is*

$$\bigcup_{\alpha_2 \in \mathfrak{A}_2} \mathcal{N}_{12}(\alpha_1, \alpha_2; P_1)\ {}_{\mathrm{ev}_+} \times^{\boxplus 1}_{\mathrm{ev}_-} \mathcal{N}_{23}(\alpha_2, \alpha_3; P_2). \tag{16.38}$$

The precise meaning of the union $\bigcup$ *in* (16.38) *is defined during the proof in this section and in Chap.* 18. *The symbol* $\times^{\boxplus 1}$ *will be also defined in Definition* 18.34.
(3) *The energy loss of the morphism* $\mathfrak{N}_{32} \circ \mathfrak{N}_{21}$ *is the sum of the energy loss of* $\mathfrak{N}_{32}$ *and of* $\mathfrak{N}_{21}$. *The same holds for the* P_i *parameterized version.*

---

[14] We denote by $\mathcal{N}_{12}$ the interpolation space of the morphism $\mathfrak{N}_{21}$. See Remark 18.33.

(4) *We have* $\mathfrak{N}_{43} \circ (\mathfrak{N}_{32} \circ \mathfrak{N}_{21}) = (\mathfrak{N}_{43} \circ \mathfrak{N}_{32}) \circ \mathfrak{N}_{21}$. *The same holds for the* P_i *parameterized version.*

(5) *We have*

$$\partial(\mathfrak{N}_{32}^{P_2} \circ \mathfrak{N}_{21}^{P_1}) = (\mathfrak{N}_{32}^{\partial P_2} \circ \mathfrak{N}_{21}^{P_1}) \cup (\mathfrak{N}_{32}^{P_2} \circ \mathfrak{N}_{21}^{\partial P_1}). \tag{16.39}$$

The boundary in the left hand side is taken in the sense of Definition 16.29 *(2). The precise meaning of the union* $\cup$ *in the right hand side is defined during the proof in this section and by Definition-Lemma* 18.53.

(6) *We can generalize (1)–(5) to the case of partial linear K-systems.*

Idea of the proof Here we explain only the basic geometric idea of the proof. To give a rigorous proof, there is an issue about providing a precise definition of 'corner smoothing' in the definition of the union in (16.37) and (16.38). We postpone this point until Chap. 18.

(1) We first explain the meaning of the union in (16.37). Note that

$$\begin{aligned}
&\bigcup_{\alpha_2 \in \mathfrak{A}_2} \partial \left(\mathcal{N}_{23}(\alpha_2, \alpha_3) \,{}_{\mathrm{ev}_-}\times_{\mathrm{ev}_+} \mathcal{N}_{12}(\alpha_1, \alpha_2)\right) \\
&\supseteq \bigcup_{\alpha_2 \in \mathfrak{A}_2} \partial\mathcal{N}_{23}(\alpha_2, \alpha_3) \,{}_{\mathrm{ev}_-}\times_{\mathrm{ev}_+} \mathcal{N}_{12}(\alpha_1, \alpha_2) \\
&\qquad \cup \bigcup_{\alpha_2 \in \mathfrak{A}_2} (-1)^{\dim \mathcal{N}_{23}(\alpha_2,\alpha_3) + \dim R_{\alpha_2}} \mathcal{N}_{23}(\alpha_2, \alpha_3) \,{}_{\mathrm{ev}_-}\times_{\mathrm{ev}_+} \partial\mathcal{N}_{12}(\alpha_1, \alpha_2).
\end{aligned} \tag{16.40}$$

See [FOOO4, Lemma 8.2.3 (1)] for the sign. We also note that the fiber product

$$\mathcal{N}_{12}(\alpha_1, \alpha_2) \,{}_{\mathrm{ev}_+}\times_{\mathrm{ev}_-} \mathcal{M}^2(\alpha_2, \alpha_2') \,{}_{\mathrm{ev}_+}\times_{\mathrm{ev}_-} \mathcal{N}_{23}(\alpha_2', \alpha_3) \tag{16.41}$$

appears in both of the second and third lines of (16.40). To construct the union in (16.37), we glue several components along the codimension 1 boundaries such as (16.41). We note that those parts we glue contain certain (higher codimensional) corners.[15] See Fig. 16.1. For this purpose we use the notion of smoothing the corners of a Kuranishi structure. We will discuss the smoothing in detail in Sects. 18.4, 18.5, and 18.6. We also observe that there exists a certain K-space for which the disjoint union of the summands of (16.37) appears in its boundary. (See Proposition 18.35.) Combining them, we can put the structure of the K-space on the union (16.37). See Lemma-Definition 18.37. Thus we obtain the interpolation space of the composition

$$\mathfrak{N}_{32} \circ \mathfrak{N}_{21}.$$

[15] So the order of factors in (16.41) does not matter as we note in Remark 16.2.

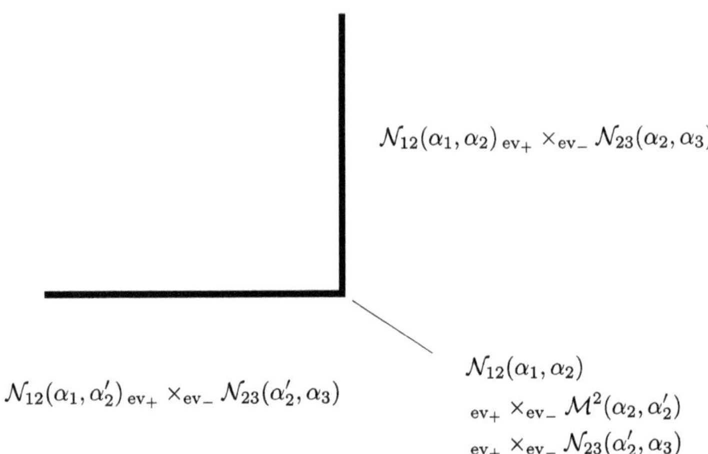

Fig. 16.1 Equation (16.41) looks like a corner

Then it is straightforward to check that this interpolation space satisfies the defining conditions of morphism.

The proof of (2) is entirely similar to the proof of (1). (See Sect. 18.9.) (3) is immediate from the definition.

We next discuss the proof of (4). Note that both sides of the equality are a union

$$\bigcup_{\alpha_2,\alpha_3} \mathcal{N}_{12}(\alpha_1, \alpha_2)\ {}_{\mathrm{ev}_+}\times_{\mathrm{ev}_-} \mathcal{N}_{23}(\alpha_2, \alpha_3)\ {}_{\mathrm{ev}_+}\times_{\mathrm{ev}_-} \mathcal{N}_{34}(\alpha_3, \alpha_4).$$

While we define this union from the disjoint union, we perform the process of corner smoothing and gluing K-spaces along the boundary twice, which we described during the proof of (1): Once at the boundary components of the form

$$\mathcal{N}_{12}(\alpha_1, \alpha_2')\ {}_{\mathrm{ev}_+}\times_{\mathrm{ev}_-} \mathcal{M}^2(\alpha_2', \alpha_2)\ {}_{\mathrm{ev}_+}\times_{\mathrm{ev}_-} \mathcal{N}_{23}(\alpha_2, \alpha_3)\ {}_{\mathrm{ev}_+}\times_{\mathrm{ev}_-} \mathcal{N}_{34}(\alpha_3, \alpha_4)$$

and once at the boundary components of the form

$$\mathcal{N}_{12}(\alpha_1, \alpha_2)\ {}_{\mathrm{ev}_+}\times_{\mathrm{ev}_-} \mathcal{N}_{23}(\alpha_2, \alpha_3')\ {}_{\mathrm{ev}_+}\times_{\mathrm{ev}_-} \mathcal{M}^3(\alpha_3', \alpha_3)\ {}_{\mathrm{ev}_+}\times_{\mathrm{ev}_-} \mathcal{N}_{34}(\alpha_3, \alpha_4).$$

Since these two components do not intersect at the interior, we can perform these two processes independently and can exchange the order of them. (4) follows. (There is an issue of showing that this isomorphism is compatible with the smooth structure we gave during the proof of (1). See Sect. 18.8.)

We next prove (5). By (16.31) applied to $\mathcal{N}_{12}(\alpha_1, \alpha_2; P_1)$ and to $\mathcal{N}_{23}(\alpha_2, \alpha_3; P_2)$, the normalized boundary of the interpolation space of $\mathfrak{N}_{32}^{P_2} \circ \mathfrak{N}_{21}^{P_1}$ is a disjoint union of the following four kinds of components:

(A) $\mathcal{N}_{12}(\alpha_1, \alpha_2; \partial P_1) \times_{R_{\alpha_2}} \mathcal{N}_{23}(\alpha_2, \alpha_3; P_2)$.
(B) $\mathcal{N}_{12}(\alpha_1, \alpha_2; P_1) \times_{R_{\alpha_2}} \mathcal{N}_{23}(\alpha_2, \alpha_3; \partial P_2)$.
(C) $\mathcal{M}^1(\alpha_1, \alpha_1') \times_{R_{\alpha_1'}} \mathcal{N}_{12}(\alpha_1', \alpha_2; P_1) \times_{R_{\alpha_2}} \mathcal{N}_{23}(\alpha_2, \alpha_3; P_2)$.
(D) $\mathcal{N}_{12}(\alpha_1, \alpha_2; P_1) \times_{R_{\alpha_2}} \mathcal{N}_{23}(\alpha_2, \alpha_3'; P_2) \times_{R_{\alpha_3'}} \mathcal{M}^3(\alpha_3', \alpha_3)$.

Note that potentially there are components of the form

$$\mathcal{N}_{12}(\alpha_1, \alpha_2'; P_1) \times_{R_{\alpha_2'}} \mathcal{M}^2(\alpha_2', \alpha_2) \times_{R_{\alpha_2}} \mathcal{N}_{23}(\alpha_2, \alpha_3; P_2) \tag{16.42}$$

in the boundary of the interpolation space of $\mathfrak{N}_{32}^{P_2} \circ \mathfrak{N}_{21}^{P_1}$. However, (16.42) appears twice in the boundary of the interpolation space of $\mathfrak{N}_{32}^{P_2} \circ \mathfrak{N}_{21}^{P_1}$: Once in

$$\partial\mathcal{N}_{12}(\alpha_1, \alpha_2; P_1) \times_{R_{\alpha_2}} \mathcal{N}_{23}(\alpha_2, \alpha_3; P_2)$$

and once in

$$\mathcal{N}_{12}(\alpha_1, \alpha_2; P_1) \times_{R_{\alpha_2}} \partial\mathcal{N}_{23}(\alpha_2, \alpha_3; P_2).$$

When we define the interpolation space of $\mathfrak{N}_{32}^{P_2} \circ \mathfrak{N}_{21}^{P_1}$, we glue

$$\mathcal{N}_{12}(\alpha_1, \alpha_2; P_1) \times_{R_{\alpha_2}} \mathcal{N}_{23}(\alpha_2, \alpha_3; P_2)$$

for various α_2 along (16.42). (See Definition-Lemma 18.53 for details.) Therefore those components of the form (16.42) do not appear in the boundary of the interpolation space of $\mathfrak{N}_{32}^{P_2} \circ \mathfrak{N}_{21}^{P_1}$. Hence the disjoint union of (A)–(D) is the normalized boundary of the interpolation space of $\mathfrak{N}_{32}^{P_2} \circ \mathfrak{N}_{21}^{P_1}$.

On the other hand, the right hand side of (16.39) is the union of the components of types (A),(B). Note that by Definition 16.29 (2) and (16.31), we have

$$\begin{aligned}&(\text{The interpolation space of } \partial(\mathfrak{N}_{32}^{P_2} \circ \mathfrak{N}_{21}^{P_1})) \cup (\mathrm{C}) \cup (\mathrm{D})\\ &= \partial\left(\text{The interpolation space of } (\mathfrak{N}_{32}^{P_2} \circ \mathfrak{N}_{21}^{P_1})\right).\end{aligned}$$

Then (5) follows from these facts.

The proof of (6) is the same as the proofs of (1)–(5). □

16.6 Inductive System of Linear K-Systems

Definition 16.36 Let $\mathcal{C}$ be critical submanifold data as in Definition 16.6.

(1) Let $E_0 < E_1$. A partial linear K-system with energy cut level E_1 induces a partial linear K-system with energy cut level E_0 by forgetting all the structures

of the former which do not appear in the definition of the latter. The same applies to the morphism, homotopy etc. We call these processes the *energy cut at* E_0.

(2) An *inductive system of partial linear K-systems*

$$\mathcal{FF} = (\{E^i\}, \{\mathcal{F}^i\}, \{\mathfrak{N}^i\})$$

consists of the following objects.[16]

(a) We are given an increasing sequence E^i of positive numbers such that $\lim_{i\to\infty} E^i = \infty$.
(b) For each i we are given a partial linear K-system with energy cut level E^i, which we denote by $\mathcal{F}^i$.
(c) The critical submanifold data

$$\mathcal{C} = \left(\mathfrak{A}, \mathfrak{G}, \{R_\alpha\}_{\alpha\in\mathfrak{A}}, \{o_{R_\alpha}\}_{\alpha\in\mathfrak{A}}, E, \mu, \{\mathrm{PI}_{\beta,\alpha}\}_{\beta\in\mathfrak{G},\alpha\in\mathfrak{A}}\right),$$

that is a part of data of $\mathcal{F}^i$, is independent of i and is given at the beginning.
(d) For each i we are given a morphism $\mathfrak{N}^i : \mathcal{F}^i \to \mathcal{F}^{i+1}|_{E^i}$, where $\mathcal{F}^{i+1}|_{E^i}$ is the partial linear K-system of energy cut level E^i that is induced from $\mathcal{F}^{i+1}$ by the energy cut.
(e) The energy loss of $\mathfrak{N}^i$ is 0.
(f) $\mathfrak{N}^i$ is congruent to the identity morphism modulo ϵ_i for some $\epsilon_i > 0$. See Definition 16.33.
(g) We assume the following *uniform Gromov compactness*. For any $E_0 \ge 0$ and $\alpha_1 \in \mathfrak{A}_1$ the set

$$\{\alpha_2 \in \mathfrak{A}_2 \mid \exists i \ \mathcal{N}^i(\alpha_1, \alpha_2) \neq \emptyset, \ E(\alpha_2) \le E_0 + E(\alpha_1)\} \tag{16.43}$$

is a finite set. Moreover

$$\{\alpha_2 \in \mathfrak{A}_2 \mid \exists i \ \mathcal{M}^i(\alpha_1, \alpha_2) \neq \emptyset, \ E(\alpha_2) \le E_0 + E(\alpha_1)\} \tag{16.44}$$

is a finite set. Here $\mathcal{N}^i(\alpha_1, \alpha_2)$ is the interpolation space of the morphism $\mathfrak{N}^i$ and $\mathcal{M}^i(\alpha_1, \alpha_2)$ is the space of connecting orbits of $\mathcal{F}^i$.

(3) Let $\mathcal{FF}_j = (\{E^i_j\}, \{\mathcal{F}^i_j\}, \{\mathfrak{N}^i_j\})$ $(j = 1, 2)$ be two inductive systems of partial linear K-systems. We assume $E^i_1 \ge E^i_2 - c$ for some $c \ge 0$. A *morphism*

$$(\{\mathfrak{N}^i_{21}\}, \{\mathfrak{H}^i\}) : \mathcal{FF}_1 \to \mathcal{FF}_2$$

with energy loss c is a pair of $\{\mathfrak{N}^i_{21}\}$ and $\{\mathfrak{H}^i\}$ with the following properties:

[16]Hereafter we use the superscript i to label the elements in the inductive system.

Fig. 16.2 Homotopy of homotopies $\mathcal{H}^i$

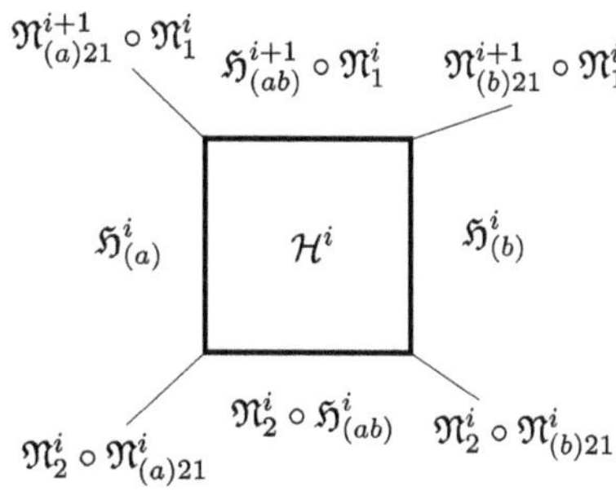

(a) $\mathfrak{N}^i_{21} : \mathcal{F}^i_1 \to \mathcal{F}^i_2$ is a morphism of energy loss c.
(b) $\mathfrak{H}^i$ is a homotopy between $\mathfrak{N}^i_2 \circ \mathfrak{N}^i_{21}$ and $\mathfrak{N}^{i+1}_{21} \circ \mathfrak{N}^i_1$. Here we regard them as morphisms with energy cut level E^i_2 and of energy loss c.

(4) In the situation of (3), let $(\{\mathfrak{N}^i_{(k)21}\}, \{\mathfrak{H}^i_{(k)}\}) : \mathcal{FF}_1 \to \mathcal{FF}_2$ be morphisms for $k = a, b$. A *homotopy* from $(\{\mathfrak{N}^i_{(a)21}\}, \{\mathfrak{H}^i_{(a)}\})$ to $(\{\mathfrak{N}^i_{(b)21}\}, \{\mathfrak{H}^i_{(b)}\})$ is the pair $(\{\mathfrak{H}^i_{(ab)}\}, \{\mathcal{H}^i\})$ with the following properties:

(a) $\mathfrak{H}^i_{(ab)}$ is a homotopy from $\mathfrak{N}^i_{(a)21}$ to $\mathfrak{N}^i_{(b)21}$.
(b) $\mathcal{H}^i$ is a $[0, 1]^2$ parameterized morphism from $\mathcal{F}^i_1$ to $\mathcal{F}^{i+1}_2$.[17]
(c) The normalized boundary $\partial\mathcal{H}^i$ is a disjoint union of the following four homotopies, see Fig. 16.2:

(i) $\mathfrak{H}^i_{(a)}$
(ii) $\mathfrak{N}^i_2 \circ \mathfrak{H}^i_{(ab)}$
(iii) $\mathfrak{H}^i_{(b)}$
(iv) $\mathfrak{H}^{i+1}_{(ab)} \circ \mathfrak{N}^i_1$

(d) For any $E_0 \geq 0$ and $\alpha_1 \in \mathfrak{A}_1$ the set

$$\{\alpha_2 \in \mathfrak{A}_2 \mid \exists i \ \mathcal{N}(\alpha_1, \alpha_2; \mathfrak{H}^i_{(ab)}) \neq \emptyset, \ E(\alpha_2) \leq E_0 + E(\alpha_1)\} \tag{16.45}$$

is a finite set. Moreover

$$\{\alpha_2 \in \mathfrak{A}_2 \mid \exists i \ \mathcal{N}(\alpha_1, \alpha_2; \mathcal{H}^i) \neq \emptyset, \ E(\alpha_2) \leq E_0 + E(\alpha_1)\} \tag{16.46}$$

is a finite set. Here $\mathcal{N}(\alpha_1, \alpha_2; \mathfrak{H}^i_{(ab)})$ (resp. $\mathcal{N}(\alpha_1, \alpha_2; \mathcal{H}^i)$) is the interpolation space which we used to define $\mathfrak{H}^i_{(ab)}$ (resp. $\mathcal{H}^i$.).

Here all the (parameterized) morphisms have energy cut level E^i_2.

[17] In other words, it is a homotopy of homotopies.

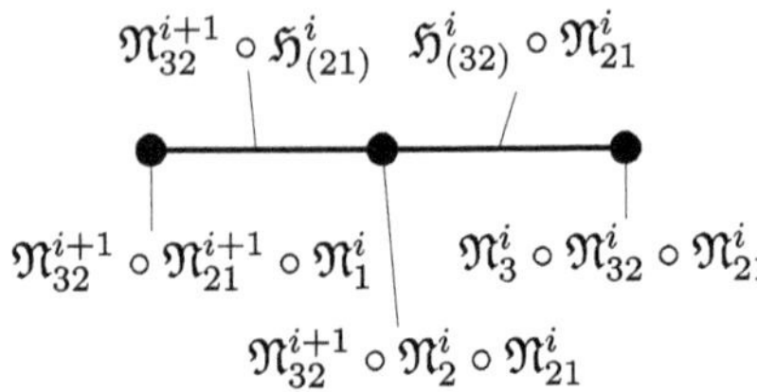

Fig. 16.3 Composition of homotopies

(5) Two morphisms of inductive systems of partial linear K-systems are said to be *homotopic* if there exists a homotopy between them.

Remark 16.37 In certain cases, especially in the study of symplectic homology (see [FH, CFH, BO]) and wrapped Floer homology (see [AS]), we need to study the case when the critical submanifold data $\mathcal{C}$ varies. We can actually study such a situation in a similar way. However, we do not try to work it out in this book.

Lemma-Definition 16.38

(1) *We can compose morphisms of inductive system of partial linear K-systems.*
(2) *Compositions of homotopic morphisms are homotopic.*
(3) *Homotopy between morphisms is an equivalence relation.*

Proof
(1) Let $(\{\mathfrak{N}^i_{j+1j}\}, \{\mathfrak{H}^i_{(j+1j)}\}) : \mathcal{FF}_j \to \mathcal{FF}_{j+1}$ be morphisms for $j = 1, 2$. We put

$$\mathfrak{N}^i_{31} = \mathfrak{N}^i_{32} \circ \mathfrak{N}^i_{21}.$$

Then $\mathfrak{H}^i_{(32)} \circ \mathfrak{N}^i_{21}$ is a homotopy from $\mathfrak{N}^i_3 \circ \mathfrak{N}^i_{31} = \mathfrak{N}^i_3 \circ \mathfrak{N}^i_{32} \circ \mathfrak{N}^i_{21}$ to $\mathfrak{N}^{i+1}_{32} \circ \mathfrak{N}^i_2 \circ \mathfrak{N}^i_{21}$ and $\mathfrak{N}^{i+1}_{32} \circ \mathfrak{H}^i_{(21)}$ is a homotopy from $\mathfrak{N}^{i+1}_{32} \circ \mathfrak{N}^i_2 \circ \mathfrak{N}^i_{21}$ to $\mathfrak{N}^{i+1}_{31} \circ \mathfrak{N}^i_1 = \mathfrak{N}^{i+1}_{32} \circ \mathfrak{N}^{i+1}_{21} \circ \mathfrak{N}^i_1$. Therefore we can construct $\mathfrak{H}^i_{(31)}$ by gluing $\mathfrak{H}^i_{(32)} \circ \mathfrak{N}^i_{21}$ and $\mathfrak{N}^{i+1}_{32} \circ \mathfrak{H}^i_{(21)}$. (See Sect. 18.9.2 for gluing process.) See Fig. 16.3.
(2) Let $(\{\mathfrak{N}^i_{k,21}\}, \{\mathfrak{H}^i_{(k,21)}\}) : \mathcal{FF}_1 \to \mathcal{FF}_2$ be morphisms for $k = a, b$ and let $(\{\mathfrak{N}^i_{32}\}, \{\mathfrak{H}^i_{(32)}\}) : \mathcal{FF}_2 \to \mathcal{FF}_3$ be a morphism. Let $(\{\mathfrak{H}^i_{(ab)}\}, \{\mathcal{H}^i\})$ be a homotopy from $(\{\mathfrak{N}^i_{a,21}\}, \{\mathfrak{H}^i_{(a,21)}\})$ to $(\{\mathfrak{N}^i_{b,21}\}, \{\mathfrak{H}^i_{(b,21)}\})$. Then we have

$$\begin{aligned} \partial(\mathfrak{N}^{i+1}_{32} \circ \mathcal{H}^i) =& \mathfrak{N}^{i+1}_{32} \circ \mathfrak{H}^{i+1}_{(ab)} \circ \mathfrak{N}^i_1 \cup \mathfrak{N}^{i+1}_{32} \circ \mathfrak{H}^i_{(b,21)} \\ & \cup \mathfrak{N}^{i+1}_{32} \circ \mathfrak{N}^i_2 \circ \mathfrak{H}^i_{(ab)} \cup \mathfrak{N}^{i+1}_{32} \circ \mathfrak{H}^i_{(a,21)} \end{aligned} \tag{16.47}$$

and

$$\begin{aligned} \partial(\mathfrak{H}^i_{(32)} \circ \mathfrak{H}^i_{(ab)}) =& \mathfrak{N}^{i+1}_{32} \circ \mathfrak{N}^i_2 \circ \mathfrak{H}^i_{(ab)} \cup \mathfrak{H}^i_{32} \circ \mathfrak{N}^i_{b,21} \\ & \cup \mathfrak{N}^i_3 \circ \mathfrak{N}^i_{32} \circ \mathfrak{H}^i_{(ab)} \cup \mathfrak{H}^i_{32} \circ \mathfrak{N}^i_{a,21}. \end{aligned} \tag{16.48}$$

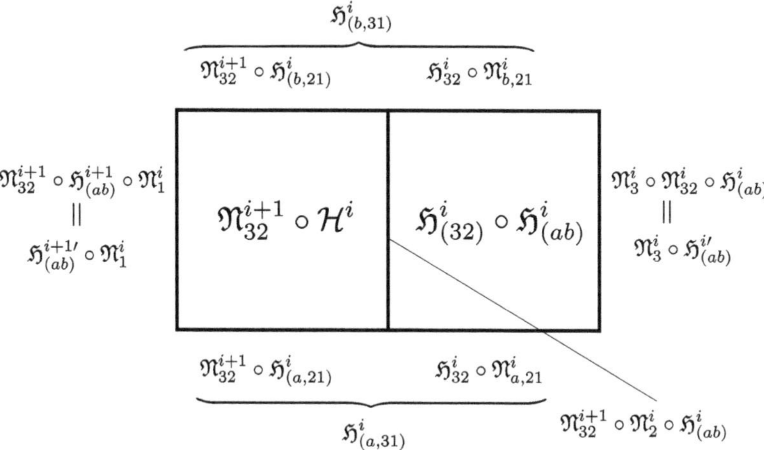

Fig. 16.4 $\mathcal{H}^{i\prime}$

We note that both are $[0, 1]^2$-parameterized families of morphisms. By the definition of composition we have

$$\mathfrak{N}^{i+1}_{32} \circ \mathfrak{H}^i_{(a,21)} \cup \mathfrak{H}^i_{32} \circ \mathfrak{N}^i_{a,21} = \mathfrak{H}^i_{(a,31)},$$
$$\mathfrak{N}^{i+1}_{32} \circ \mathfrak{H}^i_{(b,21)} \cup \mathfrak{H}^i_{32} \circ \mathfrak{N}^i_{b,21} = \mathfrak{H}^i_{(b,31)}.$$

We put

$$\mathfrak{H}^{i\prime}_{(ab)} = \mathfrak{N}^i_{32} \circ \mathfrak{H}^i_{(ab)}$$
$$\mathcal{H}^{i\prime} = (\mathfrak{N}^{i+1}_{32} \circ \mathcal{H}^i) \cup_{\mathfrak{N}^{i+1}_{32} \circ \mathfrak{N}^i_2 \circ \mathfrak{H}^i_{(ab)}} (\mathfrak{H}^i_{(32)} \circ \mathfrak{H}^i_{(ab)}).$$

See Fig. 16.4. Here in the right hand side of the second formula, we glue two $[0, 1]^2$-parameterized morphisms along one of the components of their boundaries. We can do it in the same way as in the proof of Lemma-Definition 16.35. See Sect. 18.9.2 for details. We can then easily see that

$$\partial(\mathcal{H}^{i\prime}) = \mathfrak{H}^i_{(a,31)} \cup \mathfrak{H}^i_{(b,31)} \cup (\mathfrak{N}^i_3 \circ \mathfrak{H}^i_{(ab)}) \cup (\mathfrak{H}^{i+1}_{(ab)} \circ \mathfrak{N}^i_1).$$

Namely $(\{\mathfrak{N}^i_{32}\}, \{\mathfrak{H}^i_{(32)}\}) \circ (\{\mathfrak{N}^i_{a,21}\}, \{\mathfrak{H}^i_{(a,21)}\})$ is homotopic to $(\{\mathfrak{N}^i_{32}\}, \{\mathfrak{H}^i_{(32)}\}) \circ (\{\mathfrak{N}^i_{b,21}\}, \{\mathfrak{H}^i_{(b,21)}\})$. The case of the composition of homotopies and morphisms in the opposite direction is similar.

(3) will be proved in Sect. 18.9.2. □

Theorem 16.39 *Let $\mathcal{FF} = (\{E^i\}, \{\mathcal{F}^i\}, \{\mathfrak{N}^i\})$ be an inductive system of partial linear K-systems. Note that the $\Lambda_{0,\mathrm{nov}}$ module $CF(\mathcal{F}^i; \Lambda_{0,\mathrm{nov}})$ is independent of i, which we denote by $CF(\mathcal{FF}; \Lambda_{0,\mathrm{nov}})$.*

(1) *To our inductive system of partial linear K-systems $\mathcal{FF}$, we can associate a map $d : CF(\mathcal{FF}; \Lambda_{0,\mathrm{nov}}) \to CF(\mathcal{FF}; \Lambda_{0,\mathrm{nov}})$ such that:*

(a) $d \circ d = 0$.
(b) *There exists $\epsilon > 0$ such that:*

$$(d - d_0)(\mathfrak{F}^{\lambda} CF(\mathcal{FF}; \Lambda_{0,\mathrm{nov}})) \subset \mathfrak{F}^{\lambda+\epsilon} CF(\mathcal{FF}; \Lambda_{0,\mathrm{nov}}).$$

See (16.17) *for the definition of d_0.*

(2) *The definition of the map d in (1) involves various choices and d depends on them. However, it is independent of such choices in the following sense: Suppose d_1, d_2 are obtained by two different choices. Then there exists a map $\psi : CF(\mathcal{FF}; \Lambda_{0,\mathrm{nov}}) \to CF(\mathcal{FF}; \Lambda_{0,\mathrm{nov}})$ with the following properties:*

(a) $d_2 \circ \psi = \psi \circ d_1$.
(b) *ψ is degree 0 and preserves the filtration.*
(c) *There exists $\epsilon > 0$ such that:*

$$(\psi - \mathrm{id})(\mathfrak{F}^{\lambda} CF(\mathcal{FF}; \Lambda_{0,\mathrm{nov}})) \subset \mathfrak{F}^{\lambda+\epsilon} CF(\mathcal{FF}; \Lambda_{0,\mathrm{nov}}),$$

where id *is the identity map.*
(d) *In particular, ψ induces an isomorphism on cohomologies:*

$$H(CF(\mathcal{FF}; \Lambda_{0,\mathrm{nov}}), d_1) \to H(CF(\mathcal{FF}; \Lambda_{0,\mathrm{nov}}), d_2).$$

(e) *ψ depends on various choices but it is independent of the choices up to cochain homotopy.*

(3) *A morphism $\mathfrak{N}\mathfrak{N} : \mathcal{FF}_1 \to \mathcal{FF}_2$ induces a Λ_{nov} module homomorphism*

$$\psi_{\mathfrak{N}\mathfrak{N}} : CF(\mathcal{FF}_1; \Lambda_{0,\mathrm{nov}}) \to CF(\mathcal{FF}_2; \Lambda_{0,\mathrm{nov}}) \tag{16.49}$$

with the following properties:

(a) *$\psi_{\mathfrak{N}\mathfrak{N}} \circ d = d \circ \psi_{\mathfrak{N}\mathfrak{N}}$, where d in the left hand side (resp. right hand side) is the coboundary operator of $CF(\mathcal{FF}_1; \Lambda_{0,\mathrm{nov}})$ (resp. $CF(\mathcal{FF}_2; \Lambda_{0,\mathrm{nov}})$).*
(b) *$\psi_{\mathfrak{N}\mathfrak{N}}$ preserves degree and*

$$\psi_{\mathfrak{N}\mathfrak{N}} \left(\mathfrak{F}^{\lambda} CF(\mathcal{FF}_1; \Lambda_{\mathrm{nov}})\right) \subset \mathfrak{F}^{\lambda-c} CF(\mathcal{FF}_1; \Lambda_{\mathrm{nov}}),$$

where c is the energy loss of $\mathfrak{N}\mathfrak{N}$. In particular, if $c = 0$ then $\psi_{\mathfrak{N}\mathfrak{N}}$ induces a $\Lambda_{0,\mathrm{nov}}$ module homomorphism:

$$\psi_{\mathfrak{N}\mathfrak{N}} : CF(\mathcal{F}\mathcal{F}_1; \Lambda_{0,\mathrm{nov}}) \to CF(\mathcal{F}\mathcal{F}_2; \Lambda_{0,\mathrm{nov}}).$$

(4) *Let* $\mathfrak{N}\mathfrak{N}_a$, $\mathfrak{N}\mathfrak{N}_b : \mathcal{F}\mathcal{F}_1 \to \mathcal{F}\mathcal{F}_2$ *be morphisms. If* $\mathfrak{H}\mathfrak{H}$ *is a homotopy from* $\mathfrak{N}\mathfrak{N}_a$ *to* $\mathfrak{N}\mathfrak{N}_b$*, it induces a* Λ_{nov} *module homomorphism:*

$$\psi_{\mathfrak{H}\mathfrak{H}} : CF(\mathcal{F}\mathcal{F}_1; \Lambda_{\mathrm{nov}}) \to CF(\mathcal{F}\mathcal{F}_2; \Lambda_{\mathrm{nov}}) \tag{16.50}$$

with the following properties:

(a) $\psi_{\mathfrak{H}\mathfrak{H}} \circ d + d \circ \psi_{\mathfrak{H}} = \psi_{\mathfrak{N}\mathfrak{N}_1} - \psi_{\mathfrak{N}\mathfrak{N}_2}$.
(b) $\psi_{\mathfrak{H}\mathfrak{H}}$ *decreases degree by* 1 *and*

$$\psi_{\mathfrak{H}\mathfrak{H}}\left(\mathfrak{F}^{\lambda} CF(\mathcal{F}\mathcal{F}_1; \Lambda_{\mathrm{nov}})\right) \subset \mathfrak{F}^{\lambda - c} CF(\mathcal{F}\mathcal{F}_1; \Lambda_{\mathrm{nov}}),$$

where c *is the energy loss of* $\mathfrak{H}\mathfrak{H}$*. In particular, if* $c = 0$ *then* $\psi_{\mathfrak{H}\mathfrak{H}}$ *induces a* $\Lambda_{0,\mathrm{nov}}$ *module homomorphism:*

$$\psi_{\mathfrak{H}\mathfrak{H}} : CF(\mathcal{F}\mathcal{F}_1; \Lambda_{0,\mathrm{nov}}) \to CF(\mathcal{F}\mathcal{F}_2; \Lambda_{0,\mathrm{nov}}).$$

(5) *Let* $\mathcal{F}\mathcal{F}_a$, $\mathcal{F}\mathcal{F}_b$, $\mathcal{F}\mathcal{F}_c$ *be inductive systems of partial linear K-systems and* $\mathfrak{N}\mathfrak{N}_{ba} : \mathcal{F}\mathcal{F}_a \to \mathcal{F}\mathcal{F}_b$, $\mathfrak{N}\mathfrak{N}_{cb} : \mathcal{F}\mathcal{F}_b \to \mathcal{F}\mathcal{F}_c$ *be morphisms. Then we have*

$$\psi_{\mathfrak{N}\mathfrak{N}^{cb} \circ \mathfrak{N}\mathfrak{N}^{ba}} \sim \psi_{\mathfrak{N}\mathfrak{N}^{cb}} \circ \psi_{\mathfrak{N}\mathfrak{N}^{ba}},$$

where $\sim$ *means cochain homotopic.*
(6) *The identity morphism induces a cochain map which is cochain homotopic to the identity map.*

The proof will be given in Chap. 19. The notion of identity morphism which appears in Item (6) will be defined in Sect. 18.10.

Remark 16.40 The morphism $\psi_{\mathfrak{N}\mathfrak{N}}$ in Item (3) depends on the choices made for its construction. More precisely, we first make choices mentioned in Item (2) to define coboundary operators of $CF(\mathcal{F}\mathcal{F}_i; \Lambda_{\mathrm{nov}})$, $i = 1, 2$. Then we make a choice (compatible with the first choices) to define the map $\psi_{\mathfrak{N}\mathfrak{N}}$ which is a cochain map with respect to the coboundary operators obtained from the choices we made. Note that we can use Item (4) to show that up to cochain homotopy the cochain map $\psi_{\mathfrak{N}\mathfrak{N}}$ is independent of the choices we need to make to define it.

Definition 16.41 We call the cohomology group of $(CF(\mathcal{F}\mathcal{F}; \Lambda_{0,\mathrm{nov}}), d)$ the *Floer cohomology* of an inductive system of partial linear K-systems $\mathcal{F}\mathcal{F}$ and denote the cohomology by $HF(\mathcal{F}\mathcal{F}; \Lambda_{0,\mathrm{nov}})$.

Chapter 17
Extension of a Kuranishi Structure and Its Perturbation from Boundary to Its Neighborhood

17.1 Introduction to Chap. 17

In Chap. 16 we formulated various versions of corner compatibility conditions. To prove the results stated in Chap. 16 (and which also appear elsewhere in this book and will appear in the future) we need to extend the Kuranishi structure given on the boundary ∂X satisfying corner compatibility conditions to one on X. More precisely, we will start with the situation where we have a Kuranishi structure $\widehat{\mathcal{U}}$ on X and a Kuranishi structure $\widehat{\mathcal{U}_{\partial}^{+}}$ on ∂X such that

$$\partial\widehat{\mathcal{U}} < \widehat{\mathcal{U}_{\partial}^{+}}$$

and want to find a Kuranishi structure $\widehat{\mathcal{U}^{+}}$ on X such that

$$\partial\widehat{\mathcal{U}^{+}} = \widehat{\mathcal{U}_{\partial}^{+}}, \quad \widehat{\mathcal{U}} < \widehat{\mathcal{U}^{+}}.$$

We also need a similar statement for CF-perturbations.

It is rather easy to show that if a Kuranishi structure is defined on a neighborhood Ω of a compact set K then we can extend the Kuranishi structure without changing it in a neighborhood of K. (See Lemma 17.75.) However, this statement is not enough to prove the existence of an extension in the above situation, since there we are given a Kuranishi structure on ∂X only and not on its neighborhood. In this chapter, we discuss the problem of extending a Kuranishi structure and a CF-perturbation on ∂X to its neighborhood.

Remark 17.1 If we carefully examine the whole proofs of the geometric applications appearing in previous literature such as [FOn2, FOOO3, FOOO4], we will find out that, for the Kuranishi structure $\widehat{\mathcal{U}_{\partial}^{+}}$ we actually use, an extension to its small neighborhood of ∂X in X is given from its construction. In fact, in the actual

K. Fukaya et al., *Kuranishi Structures and Virtual Fundamental Chains*, Springer Monographs in Mathematics, https://doi.org/10.1007/978-981-15-5562-6_17

situation of applications, we start from a Kuranishi structure $\partial\widehat{\mathcal{U}}$ on the boundary (which we obtain from geometry and analysis) and construct a good coordinate system $\widehat{\mathcal{U}_\partial}$ and use it to find $\widehat{\mathcal{U}_\partial^+}$ and its perturbation. We need to extend $\widehat{\mathcal{U}_\partial^+}$ and its perturbation to a neighborhood of ∂X. In this situation, the Kuranishi charts of $\widehat{\mathcal{U}_\partial^+}$ are obtained as open subcharts of certain Kuranishi charts of $\partial\widehat{\mathcal{U}}$. (See the proof of Proposition 3.35, Lemma 6.30.) Therefore it can indeed be extended using the extension of $\partial\widehat{\mathcal{U}}$, directly.

Nevertheless, the reason why we prove these extension results in this book is as follows. In this book we want to present various parts of the proofs of the whole story in a 'package' as much as possible. In other words, we want to decompose whole proofs of the geometric results into pieces, so that each piece can be stated and proved rigorously and independently from other pieces. Namely, for each divided 'package', we want to state the precise assumption and conclusion together with the proof. In this way, one can use and quote each 'package' without referring to their proofs. We want to do so because the whole story has now grown huge and becomes harder to follow all at once.

For this purpose, we want to specify and restrict the information which we 'remember' at each step of the inductive construction of the K-system and its perturbations. When we construct a system of virtual fundamental chains of, for example, $\mathcal{M}(\alpha_-, \alpha_+)$, we use, as an induction hypothesis, a certain structure on $\mathcal{M}(\alpha, \alpha')$ for $E(\alpha_-) \le E(\alpha) < E(\alpha') \le E(\alpha_+)$ (and $E(\alpha') - E(\alpha) < E(\alpha_+) - E(\alpha_-)$) and use only those structures during our construction. We need to make a careful choice of the structures we 'remember' during the inductive steps, when we proceed from one step of the induction to the next, so that the induction works. Our choice in this book is that we 'remember' a Kuranishi structure and its perturbation but forget the good coordinate system (which we use to construct the perturbation) and objects on it. We also forget the way those Kuranishi structures and various objects on them are constructed, but explicitly list all the properties we use in the next step of the induction. Therefore the relation between $\widehat{\mathcal{U}_\partial^+}$ and $\partial\widehat{\mathcal{U}}$ (except $\partial\widehat{\mathcal{U}} < \widehat{\mathcal{U}_\partial^+}$) is among the data which we forget when we proceed to the next step of the induction.

The simplest version of the extension result we mentioned above is the following:

(*) Suppose we are given a continuous function f on $\partial([0, 1)^n)$. We assume that the restriction of f to $[0, 1)^k \times \{0\} \times [0, 1)^{n-k-1}$ is smooth for any k. Then f is extended to a smooth function on $[0, 1)^n$.

The statement (*) is, of course, classical. (See [FOOO4, Lemma 7.2.121] for its proof, for example.) We can use this statement together with various techniques of manifold theory (such as partition of unity and induction on the number of coordinate charts) to prove the existence of an extension of CF-perturbations if we include enough structure into the assumptions.

Remark 17.2 One technical issue arising when applying (*) for the extension of a CF-perturbation from the boundary to the interior is the following. The boundary of a K-space does not consist of a single K-space but is rather a union of K-spaces glued along the boundary and corners. Equipping a CF-perturbation with the boundary necessarily means equipping a compatible system of CF-perturbations also with the strata of the boundary and corners that satisfy appropriate compatibility conditions. Note that a CF-perturbation is a certain equivalence class. So the compatibility conditions we require are necessarily in the form whose restrictions should be *equivalent*. Since the definition of equivalence between representatives of CF-perturbations is somewhat involved, we need some additional argument in applying (*). Outer collaring also simplifies this argument.

In this chapter we take a short-cut in the following way. Suppose M is a manifold with boundary (but has no corner). Then it is well-known that a neighborhood of ∂M is identified with $\partial M \times [0, \epsilon)$. If M has corners, we can identify a neighborhood of $\overset{\circ}{S}_k(M)$ in M with a twisted product

$$\overset{\circ}{S}_k(M) \tilde{\times} [0, 1)^{n-k}$$

that is a fiber bundle over $\overset{\circ}{S}_k(M) = S_k(M) \setminus S_{k+1}(M)$ (see Definition 4.13), whose structure group is a finite group of permutations of $(n-k)$ factors. We also require a compatibility of these structures for various k. Such a structure may be regarded as a special case of the '*system of tubular neighborhoods*' of a stratified space introduced by Mather [Ma]. Its existence is claimed and can be proved in the same way as in [Ma]. (See also [FOOO14, Proposition 8.1].)

We can include certain 'topological' objects such as vector bundle (especially obstruction bundle) and generalize the notion of collaring of the structure in a neighborhood of boundary and corner, and prove its existence without assuming extra conditions. However, we *cannot* expect that the Kuranishi map respects this trivialization. Namely the Kuranishi map in general may not be constant in the $[0, 1)^k$ factor. So we need some discussion on the way we extend perturbation to a neighborhood of ∂X.

The main idea we use in this chapter for the short-cut is summarized as follows. In the case when M is a manifold with boundary, we can attach $\partial M \times [-1, 0]$ to M by identifying $\partial M \times \{0\}$ with $\partial M \subset M$, and enhance M to a manifold $M^{\boxplus 1}$ with boundary so that a neighborhood of its boundary is *canonically* identified with $\partial M \times [-1, 0)$. Then we can extend a vector bundle $\mathcal{E}$ on M to $M^{\boxplus 1}$ so that its restriction to this neighborhood of the boundary is *canonically* identified with $\mathcal{E}|_{\partial M} \times [-1, 0)$. Then its section such as Kuranishi map can be extended to $M^{\boxplus 1}$ so that on $\partial M \times [-1, 0)$ it is constant in the $[-1, 0)$ direction. In other words, *we attach the collar '**outside**' M in place of constructing it '**inside**'*. We call this process *outer collaring*.

We can perform a similar construction in the case when M has corners, the case of orbifolds, and the case of K-spaces. Then we replace the moduli spaces such as $\mathcal{M}(\alpha_-, \alpha_+)$ by $\mathcal{M}(\alpha_-, \alpha_+)^{\boxplus 1}$ so that the fiber product description of their boundaries remains the same and their Kuranishi structures have *canonical collars* near the boundaries. We can use them to extend the Kuranishi structures and their perturbations given on the boundary to a neighborhood of the boundary.

There is a slight issue about the smoothness of the structure at the point $\partial M \times \{0\}$. The shortest and simplest way to resolve the issue is to use *exponential decay estimates* of various objects appearing there in geometric situations, which we proved in [FOOO4, Lemma A1.58], [FOOO16, Theorems 13.2 and 19.5], [FOOO18]. In the abstract setting, we impose certain *admissibility* on orbifolds, vector bundles, etc., to incorporate such an exponential decay property. See Chap. 25.

Remark 17.3

(1) In fact, we can work in the piecewise smooth category for the purpose of outer collaring. However, working in the smooth category rather than in the piecewise smooth category reduces the amount of checks we need during various constructions.
(2) It seems that for any good coordinate system with corners the outer collaring constructed in this book is isomorphic to a perturbation[1] (of the Kuranishi map) of the given good coordinate system with corners. We can prove this statement by using a collar *inside* the given good coordinate system. The existence of a collar inside is likely to be correct. However, working out its proof in the level of detailedness we intend in this book could be cumbersome. Also it seems impossible to prove a similar statement for Kuranishi structures. Therefore we use outer collaring in this book.

We will carry out this idea in detail in this chapter. It is rather lengthy since we need to write down and check various compatibilities of many objects with the outer collaring. However, the compatibilities in almost all cases are fairly obvious because almost all the constructions are canonical and functorial. In other words, the proofs are nontrivial only because they are lengthy.

Remark 17.4 An idea similar to outer collaring was mentioned in [FOOO10', Remark 4.3.89] as a method to construct compatible systems of Kuranishi structures.

17.2 Outer Collaring on One Chart

In this section, for a Kuranishi chart $\mathcal{U} = (U, \mathcal{E}, \psi, s)$, we will define in Lemma-Definition 17.13, an outer-collared Kuranishi chart $\mathcal{U}^{\boxplus\tau}$ at a point x of corners of U.

[1]Note that by construction the Kuranishi map of the outer collaring is constant in the normal direction of the collar added. So in general the outer collaring cannot be isomorphic to the original Kuranishi structure. The fact that the Kuranishi map of the outer collaring is constant in the normal direction of the collar is used in the proof of Lemma-Definition 17.14 (8) for example. However, it seems likely that we can modify the argument and can avoid the usage of this requirement.

Notation 17.5 Let U be an admissible orbifold. We consider its admissible chart (V_x, Γ_x, ϕ_x) at $x \in U$. If x lies in the interior of the codimension k corner, V is an open subset of a direct product $[V] \times [0,1)^k$, where $[V]$ is an open subset of $\mathbb{R}^{\dim U - k}$ and x is represented by $(\overline{x}, 0) \in V \subseteq [V] \times [0,1)^k$. We call $[V]$ the (codimension k) *corner locus* and $[0,1)^k$ the *normal factor* . The standard coordinates $t_1, \dots, t_k$ of $[0,1)^k$ are called *normal coordinates*.

In this chapter, we use the symbol $[V]$ for corner locus and t_i for normal coordinates, and sometimes use such notations without explicitly mentioning so.

We consider the following situation in this section.

Situation 17.6 Let $\mathcal{U} = (U, \mathcal{E}, \psi, s)$ be a Kuranishi neighborhood of X and

$$x \in \overset{\circ}{S}_k(U).$$

Let $\mathfrak{V}_x = (V_x, E_x, \Gamma_x, \phi_x, \widehat{\phi}_x)$ be an admissible orbifold chart of $(U, \mathcal{E})$ at x. Recall from Definition 23.23 (also Definition 23.18) that V_x is a manifold and

$$\phi_x : V_x \to X, \quad \widehat{\phi}_x : V_x \times E \to \mathcal{E}.$$

Let s_x be a representative of the Kuranishi map s on $\mathfrak{V}_x$. Let $o_x \in V_x$ be the base point, which corresponds to $(\overline{o}_x, (0, \dots, 0)) \in [V_x] \times [0,1)^k$. For $y \in V_x \subset [V_x] \times [0,1)^k$ we denote by $\overline{y} \in [V_x]$ the $[V_x]$-component of y. See Lemma 23.13 for this notation. ■

Definition 17.7 Under Situation 17.6, we define the *retraction map*

$$\mathcal{R}_x : [V_x] \times (-\infty, 1)^k \to [V_x] \times [0,1)^k$$

by

$$\mathcal{R}_x(\overline{y}, (t_1, \dots, t_k)) = (\overline{y}, (t'_1, \dots, t'_k)),$$

where

$$t'_i = \begin{cases} t_i & \text{if } t_i \geq 0, \\ 0 & \text{if } t_i \leq 0. \end{cases}$$

For $\tau > 0$, we define the open subset $V_x^{\boxplus\tau}$ of $[V_x] \times [-\tau, 1)^k$ associated to V_x by

$$V_x^{\boxplus\tau} := \mathcal{R}_x^{-1}(V_x) \cap ([V_x] \times [-\tau, 1)^k).$$

Here the retraction map $\mathcal{R}_x$ naturally induces a map (Fig. 17.1)

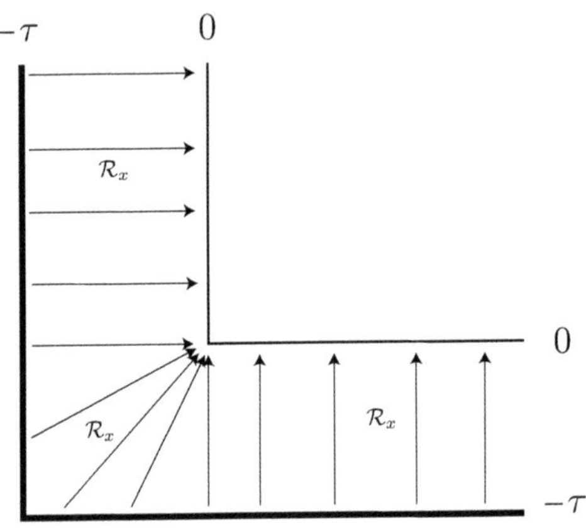

Fig. 17.1 The retraction $\mathcal{R}^x$

$$\mathcal{R}_x : V_x^{\boxplus\tau} \to V_x \subseteq [V_x] \times [0,1)^k.$$

We next extend the Γ_x action to $V_x^{\boxplus\tau}$. The extension is defined by using the retraction $\mathcal{R}_x$ and the smoothness of the action is proved by using admissibility of the action.

Definition 17.8 Let $\gamma \in \Gamma_x$. By the definition of admissible embedding (Definition 25.8 (1)(b)[2]), there exists $\sigma_\gamma \in \mathrm{Perm}(k)$ with the following properties. We define maps $\varphi_0^\gamma : [V_x] \times [0,1)^k \to [V_x]$ and $\varphi_i^\gamma : [V_x] \times [0,1)^k \to [0,1)$ requiring the components of $\gamma \cdot (\overline{y}, (t_1, \dots, t_k))$ to be

$$\gamma \cdot (\overline{y}, (t_1, \dots, t_k)) = \left(\varphi_0^\gamma(\overline{y}, (t_1, \dots, t_k)), \left(\varphi^\gamma_{\sigma_\gamma^{-1}(i)}(\overline{y}, (t_1, \dots, t_k)) \right)_{i=1}^k \right).$$

Here the left hand side is the given Γ_x action on V_x. Then we find the following:

(1) The map φ_0^γ is admissible.[3]
(2) The function $(\overline{y}, (t_1, \dots, t_k)) \mapsto \varphi_i^\gamma(\overline{y}, (t_1, \dots, t_k)) - t_i$ is exponentially small at the boundary.[4]

The existence of such $\varphi_0^\gamma, \varphi_i^\gamma$ is nothing but the definition of an admissible action. Then we define $\widehat{\varphi}_0^\gamma : [V_x] \times [-\tau, 1)^k \to [V_x]$ and $\widehat{\varphi}_i^\gamma : [V_x] \times [-\tau, 1)^k \to [-\tau, 1)$ as follows:

(A) $\widehat{\varphi}_0^\gamma = \varphi_0^\gamma \circ \mathcal{R}_x$.
(B)

[2]The map φ_{12} in Definition 25.8 corresponds to the map $(\overline{y}, (t_1, \dots, t_k)) \mapsto \gamma \cdot (\overline{y}, (t_1, \dots, t_k))$ here. The i-th normal coordinate of $\gamma \cdot (\overline{y}, (t_1, \dots, t_k))$ is denoted by $\widehat{\varphi}^\gamma_{\sigma_\gamma^{-1}(i)}(\overline{y}, (t_1, \dots, t_k))$.

[3]Definition 25.3 (1).

[4]Definition 25.3 (2).

$$\widehat{\varphi}_i^{\gamma}(\overline{y},(t_1,\ldots,t_k)) = \begin{cases} t_i & \text{if } t_i \le 0, \\ (\varphi_i^{\gamma} \circ \mathcal{R}_x)(\overline{y},(t_1,\ldots,t_k)) & \text{if } t_i \ge 0. \end{cases}$$

We now put

$$\gamma \cdot (\overline{y},(t_1,\ldots,t_k)) = \left(\widehat{\varphi}_0^{\gamma}(\overline{y},(t_1,\ldots,t_k)), \left(\widehat{\varphi}_{\sigma_\gamma^{-1}(i)}^{\gamma}(\overline{y},(t_1,\ldots,t_k)) \right)_{i=1}^{k} \right).$$

We note that $y \mapsto \gamma \cdot y$ coincides with the given γ action if $y \in V_x$.

We now prove that this defines a Γ_x-action on $V_x^{\boxplus\tau}$. We first show:

Lemma 17.9 *We have $\widehat{\varphi}_i^{\gamma}(\overline{y},(t_1,\ldots,t_k)) > 0$ if and only if $t_i > 0$. Moreover $\widehat{\varphi}_i^{\gamma}(\overline{y},(t_1,\ldots,t_k)) = 0$ if and only if $t_i = 0$.*

Proof If $t_i \le 0$ then $\widehat{\varphi}_i^{\gamma}(\overline{y},(t_1,\ldots,t_k)) = t_i \le 0$ by (B). On the other hand, if $t_i > 0$ then $\widehat{\varphi}_i^{\gamma}(\overline{y},(t_1,\ldots,t_k)) = (\varphi_i^{\gamma} \circ \mathcal{R}_x)(\overline{y},(t_1,\ldots,t_k)) > 0$ by (B). The lemma follows. □

Lemma 17.10 *We have $\gamma \cdot (\mathcal{R}_x(y)) = \mathcal{R}_x(\gamma \cdot y)$ for any $y \in V_x^{\boxplus\tau}$.*

Proof The coincidence of $[V_x]$ coordinates is an immediate consequence of the definition. We will check the coincidence of the $\sigma_\gamma(i)$-th normal coordinate.

If $t_i \le 0$, then this coordinate is 0 for both $\gamma \cdot (\mathcal{R}_x(y))$ and $\mathcal{R}_x(\gamma \cdot y)$.

If $t_i > 0$, they both are the $\sigma_\gamma(i)$-th normal coordinate of $\gamma \cdot (\mathcal{R}_x(y))$.

In fact, this is obvious for $\gamma \cdot (\mathcal{R}_x(y))$. To show this claim for $\mathcal{R}_x(\gamma \cdot y)$, we first observe that the $\sigma_\gamma(i)$-th normal coordinate of $\gamma \cdot y$ is the $\sigma_\gamma(i)$-th normal coordinate of $\gamma \cdot (\mathcal{R}_x(y))$ by (B). Then we use Lemma 17.9 to show that $\mathcal{R}_x$ does not change this coordinate. □

Lemma 17.11 *We have $\gamma \cdot (\mu \cdot y) = (\gamma\mu) \cdot y$ for any $y \in V_x^{\boxplus\tau}$ and $\gamma, \mu \in \Gamma_x$.*

Proof We calculate

$$\widehat{\varphi}_0^{\gamma}(\mu \cdot y) = \varphi_0^{\gamma}(\mathcal{R}_x(\mu \cdot y)) = \varphi_0^{\gamma}(\mu \cdot \mathcal{R}_x(y)) = \widehat{\varphi}_0^{\gamma\mu}(y).$$

Therefore the $[V_x]$ factor coincides.

Let $y = (\overline{y},(t_1,\ldots,t_k))$. Suppose $t_i > 0$. Then $\widehat{\varphi}_i^{\mu}(\overline{y},(t_1,\ldots,t_k))) > 0$ by Lemma 17.9. Therefore we obtain

$$\widehat{\varphi}_{\sigma_\mu(i)}^{\gamma}(\mu \cdot y) = \varphi_{\sigma_\mu(i)}^{\gamma}(\mathcal{R}_x(\mu \cdot y)) = \varphi_{\sigma_\mu(i)}^{\gamma}(\mu \cdot \mathcal{R}_x(y)) = \widehat{\varphi}_i^{\gamma\mu}(y).$$

Suppose $t_i \le 0$. Then $\widehat{\varphi}_i^{\mu}(\overline{y},(t_1,\ldots,t_k))) = t_i \le 0$ by Lemma 17.9. Thus we get

$$\widehat{\varphi}_{\sigma_\mu(i)}^{\gamma}(\mu \cdot y) = t_i = \widehat{\varphi}_i^{\gamma\mu}(y).$$

The proof of Lemma 17.11 is complete. □

We have thus defined a Γ_x action on $V_x^{\boxplus\tau}$.

Lemma 17.12

(1) *For any admissible map* $f : V_x \to M$ *the composition* $f \circ \mathcal{R}_x : V_x^{\boxplus\tau} \to M$ *is smooth.*
(2) *If* f *is a submersion, so is* $f \circ \mathcal{R}_x$.
(3) *The* Γ_x *action on* $V_x^{\boxplus\tau}$ *is smooth and* $\mathcal{R}_x$ *is* Γ_x *equivariant.*

Proof (1). Since the problem is local, it suffices to prove the lemma for the case $M = \mathbb{R}$. We use Lemma 25.6 to obtain f_I such that $f = \sum_I f_I$. Here $I \subset \{1, \dots, k\}$ and $f_I : [V_x] \times [0,1)^I \to \mathbb{R}$ is exponentially small near the boundary. We define a function $\widehat{f_I} : V_x^{\boxplus\tau} \to \mathbb{R}$ by

$$\widehat{f_I}(\overline{y}, (t_1, \dots, t_k)) = \begin{cases} f_I(\overline{y}, t_I) & \text{if } t_i \geq 0 \text{ for all } i \in I, \\ 0 & \text{if } t_i \leq 0 \text{ for some } i \in I. \end{cases}$$

Since $f_I : [V_x] \times [0,1)^I \to \mathbb{R}$ is exponentially small near the boundary, $\widehat{f_I}$ is smooth. It is easy to see that $f \circ \mathcal{R}_x = \sum_I \widehat{f_I}$. Therefore $f \circ \mathcal{R}_x$ is smooth.

We note that the submersivity of f by definition implies the submersivity of the restriction of f to $\overset{\circ}{S}_k V_x$. (2) is immediate from this fact and the construction.

The first half of (3) follows from the definition. The second half of (3) is Lemma 17.10. □

We define $U_x^{\boxplus\tau}$ to be the quotient $V_x^{\boxplus\tau}/\Gamma_x$ under the Γ_x action on $V_x^{\boxplus\tau}$ described above. Then Lemma 17.10 yields the retraction map

$$\mathcal{R}_x : U_x^{\boxplus\tau} \to U_x$$

induced by the one on $V_x^{\boxplus\tau}$. We denote

$$\mathcal{E}_x = (E_x \times V_x)/\Gamma_x,$$

which is a vector bundle on an orbifold $U_x = V_x/\Gamma_x$. Then we define $\mathcal{E}_x^{\boxplus\tau}$ to be the pullback

$$\mathcal{E}_x^{\boxplus\tau} = \mathcal{R}_x^*(\mathcal{E}_x) = (E_x \times V_x^{\boxplus\tau})/\Gamma_x, \tag{17.1}$$

which is a smooth vector bundle on an orbifold $U_x^{\boxplus\tau}$. The Kuranishi map s_x which is a section of $\mathcal{E}_x$ induces a section $s_x^{\boxplus\tau}$ of $\mathcal{E}_x^{\boxplus\tau}$. In the same way as in the proof of Lemma 17.12 (1) we can show that $s_x^{\boxplus\tau}$ defines a smooth section.

We define

$$(X \cap U_x)^{\boxplus\tau} := (s_x^{\boxplus\tau})^{-1}(0)/\Gamma_x, \tag{17.2}$$

which is a paracompact Hausdorff space. (We note that $X \cap U_x$ in the notation $(X \cap U_x)^{\boxplus\tau}$ does *not* stand for the set-theoretical intersection, but is just a notation.) We have a map $\psi_x^{\boxplus\tau} : (s_x^{\boxplus\tau})^{-1}(0)/\Gamma_x \to (X \cap U_x)^{\boxplus\tau}$ that is the identity map. Then from the definition we find:

Lemma-Definition 17.13 *Let $\mathcal{U} = (U, \mathcal{E}, \psi, s)$ be a Kuranishi chart of X and $x \in \overset{\circ}{S}_k(U)$ as in Situation 17.6. Then $(U_x^{\boxplus\tau} = V_x^{\boxplus\tau}/\Gamma_x, \mathcal{E}_x^{\boxplus\tau}, \psi_x^{\boxplus\tau}, s_x^{\boxplus\tau})$ is a Kuranishi chart of $(X \cap U_x)^{\boxplus\tau}$. We define:*

$$\mathcal{U}_x^{\boxplus\tau} = (U_x^{\boxplus\tau}, \mathcal{E}_x^{\boxplus\tau}, \psi_x^{\boxplus\tau}, s_x^{\boxplus\tau})$$

and call it the outer collaring *of $\mathcal{U}_x$.*

Let $\mathcal{S}_x = (W_x, \omega_x, \mathfrak{s}_x)$ be a CF-perturbation of $\mathcal{U}$ on $\mathfrak{V}_x$. (Definition 7.6.) We define $\mathfrak{s}_x^{\boxplus\tau} : V_x^{\boxplus\tau} \times W_x \to E_x$ by

$$\mathfrak{s}_x^{\boxplus\tau}(y, \xi) = \mathfrak{s}_x(\mathcal{R}_x(y), \xi). \tag{17.3}$$

In other words, the derivative of $\mathfrak{s}_x^{\boxplus\tau}(y, \xi)$ with respect to the normal coordinate t_i is zero if $t_i \leq 0$.

The next Lemma-Definition claims that various objects on U induce objects of the same type on its outer collaring $U^{\boxplus\tau}$. In other words, outer collaring is a functorial process. Such statement appears from now on repeatedly. In most of the cases they are fairly straightforward to prove.

Lemma-Definition 17.14

(1) (Boundary) *The boundary $\partial(U_x^{\boxplus\tau})$ (resp. $\partial(\mathcal{S}_x^{\boxplus\tau})$) is canonically diffeomorphic to $(\partial U_x)^{\boxplus\tau}$ (resp. $(\partial \mathcal{S}_x)^{\boxplus\tau}$).*
(2) (CF-perturbation) *The triple $(W_x, \omega_x, \mathfrak{s}_x^{\boxplus\tau})$ is a CF-perturbation of $(V_x^{\boxplus\tau}/\Gamma_x, \mathcal{E}_x^{\boxplus\tau}, s_x^{\boxplus\tau}, \psi_x^{\boxplus\tau})$. We denote it by $\mathcal{S}_x^{\boxplus\tau}$.*
(3) (Equivalence of CF-perturbations) *If $\mathcal{S}_x$ is equivalent to $\mathcal{S}'_x$, then $\mathcal{S}_x^{\boxplus\tau}$ is equivalent to $\mathcal{S}_x'^{\boxplus\tau}$.*
(4) (Strongly smooth map) *If $f : U \to M$ is a smooth map and strongly submersive on $\mathfrak{V}_x$ with respect to $\mathcal{S}_x$, then $f \circ \mathcal{R}_x$ is strongly submersive with respect to $\mathcal{S}_x^{\boxplus\tau}$. We denote it by $f_x^{\boxplus\tau} : U_x^{\boxplus\tau} \to M$.*
(5) (Transversality) *The strong transversality to $N \to M$ is also preserved.*
(6) (Multivalued perturbation) *The versions of (2)–(5) where 'CF-perturbation' is replaced by 'multivalued perturbation' also hold.*
(7) (Differential form) *If h_x is a differential form on U_x, that is, $\widetilde{h}_x$ is a Γ_x-invariant differential form on V_x, then $\widetilde{h}_x^{\boxplus\tau} := \mathcal{R}_x^* \widetilde{h}_x$ defines a differential form on $U_x^{\boxplus\tau}$. We denote it by $h_x^{\boxplus\tau}$.*
(8) (Integration along the fiber) *In the situation of (2)(4)(7), we assume that h_x is compactly supported. Then we have:*

$$(f_x^{\boxplus\tau})!(h_x^{\boxplus\tau}; \mathcal{S}_x^{\boxplus\tau}) = f_x!(h_x; \mathcal{S}_x). \tag{17.4}$$

(9) (Direct product) *Let* $\mathcal{U}' = (U', \mathcal{E}', \psi', s')$ *be a Kuranishi chart of* X'. *Then*

$$(\mathcal{U} \times \mathcal{U}')^{\boxplus\tau} \cong \mathcal{U}^{\boxplus\tau} \times \mathcal{U}'^{\boxplus\tau}.$$

Proof The proofs are mostly immediate from the definition. We only prove (7),(8) for completeness' sake. We first prove (7). The differential form h in admissible coordinates is written as

$$\begin{aligned}&\sum_{J\subseteq\{1,\dots,n-k\},\#J=\ell} h_J dx_J \\ &+ \sum_{I\subseteq\{1,\dots,k\},I\neq\emptyset}\ \sum_{J\subseteq\{1,\dots,n-k\},\#J+\#I=\ell} h_{I,J} dx_J \wedge dt_I.\end{aligned} \tag{17.5}$$

See (25.12). Here h_J, $h_{I,J}$ are admissible and $h_{I,J}(x,t) = 0$ if $t_i = 0$ for some $i \in I$. It suffices to consider the point $y = (\overline{y}; (t_1, \dots, t_k))$ such that $t_1, \dots, t_\ell < 0 \le t_{\ell+1}, \dots, t_k$. Then $\widetilde{h}_x^{\boxplus\tau}$ at y is written as

$$\begin{aligned}&\sum_{J\subseteq\{1,\dots,n-k\},\#J=\ell} h_J(\overline{y}; 0, \dots, 0, t_{\ell+1}, \dots, t_k) dx_J \\ &+ \sum_{I\subseteq\{\ell+1,\dots,k\},I\neq\emptyset}\ \sum_{J\subseteq\{1,\dots,n-k\},\#J+\#I=\ell} h_{I,J}(\overline{y}; 0, \dots, 0, t_{\ell+1}, \dots, t_k) dx_J \wedge dt_I.\end{aligned}$$

The term of (17.5) for which $\{1, \dots, \ell\} \cap I \neq \emptyset$ drops in this formula. Using the fact that $h_{I,J}$ is admissible and $h_{I,J}(x,t) = 0$ if $t_i = 0$ for some $i \in I$, the second term is smooth also when $t_i = 0$ $(i = \ell+1, \dots, k)$. The admissibility of h_J implies that the first term is smooth.

To prove (8), we recall the definition of the pushout from Definition 7.11, which results in

$$\int_M \rho \wedge f_x^{\boxplus\tau}!(h_x^{\boxplus\tau}; \mathcal{S}_x^{\boxplus\tau}) = \int_{(\mathfrak{s}_x^{\boxplus\tau})^{-1}(0)} \pi_1^*(f_x^{\boxplus\tau})^*\rho \wedge \pi_1^* h_x^{\boxplus\tau} \wedge \pi_2^*\omega_x \tag{17.6}$$

$$\int_M \rho \wedge f_x!(h_x; \mathcal{S}_x) = \int_{(\mathfrak{s}_x)^{-1}(0)} \pi_1^* f_x^* \rho \wedge \pi_1^* \widetilde{h}_x \wedge \pi_2^* \omega_x \tag{17.7}$$

for any differential form ρ on M. Here $\pi_1 : V_x^{\boxplus\tau} \times W_x \to V_x^{\boxplus\tau}$, $\pi_2 : V_x^{\boxplus\tau} \times W_x \to W_x$ are projections, $\widetilde{h}^{\boxplus\tau}$ is a Γ_x-invariant differential form on $V_x^{\boxplus\tau}$ defined by $\mathcal{R}_x^*\widetilde{h}_x$ and $f^{\boxplus\tau} = f_x \circ \mathcal{R}_x : V_x^{\boxplus\tau} \to M$ is a submersion defined by Lemma 17.12.

We compare the right hand sides of the above two integrals. First we note $\mathfrak{s}_x^{\boxplus\tau} = \mathfrak{s}_x \circ (\mathcal{R}_x \times \mathrm{id}_{W_x})$ and $\mathcal{R}_x|_{V_x} = \mathrm{id}_{V_x}$ on $\partial V_x \subset V_x \subset V_x^{\boxplus\tau}$. We decompose

$$(\mathfrak{s}_x^{\boxplus\tau})^{-1}(0) = \left((\mathfrak{s}_x^{\boxplus\tau})^{-1}(0) \cap (V_x \times W_x)\right) \cup \left((\mathfrak{s}_x^{\boxplus\tau})^{-1}(0) \setminus (V_x \times W_x)\right),$$

and its associated integral and note $\mathfrak{s}_x^{\boxplus\tau} \equiv \mathfrak{s}_x$ on $V_x \subset V_x^{\boxplus\tau}$. Then contribution of the integral (17.6) over the first part of the domain becomes (17.7).

It remains to check the contribution of (17.6) on the region $(\mathfrak{s}_x^{\boxplus\tau})^{-1}(0) \setminus (V_x \times W_x)$. In the rest of the proof, we may assume that ρ is chosen so that the degree of $\pi_1^*\widetilde{h}_x \wedge \pi_1^* f_x^* \rho \wedge \pi_2^* \omega_x$ matches the dimension of $(\mathfrak{s}_x^{-1})(0)$.

We note that the retraction $\mathcal{R}_x \times \mathrm{id}_{W_x} : V_x^{\boxplus\tau} \setminus V_x \times W_x \to \partial V_x \times W_x$ also induces a retraction of $(\mathfrak{s}_x^{\boxplus\tau})^{-1}(0) \setminus (V_x^{\boxplus\tau} \times W_x)$ to $\mathfrak{s}_x^{-1}(0) \cap (\partial V_x \times W_x)$ with one-dimensional fiber by definition of $\mathfrak{s}_x^{\boxplus\tau}$. We also note

$$(f^{\boxplus\tau})^* \rho \wedge \widetilde{h}^{\boxplus\tau} = (\mathcal{R}_x \times \mathrm{id}_{W_x})^* \eta$$

for some form η defined on ∂V_x by the definitions of $\widetilde{h}^{\boxplus\tau}$ and $f^{\boxplus\tau}$ given above. We derive

$$\begin{aligned}
&\int_{(\mathfrak{s}_x^{\boxplus\tau})^{-1}(0)\setminus(V_x\times W_x)} \pi_1^*(f^{\boxplus\tau})^*\rho \wedge \pi_1^* h^{\boxplus\tau} \wedge \pi_2^*\omega_x \\
&= \int_{(\mathfrak{s}_x^{\boxplus\tau})^{-1}(0)\setminus(V_x\times W_x)} \pi_1^*(\mathcal{R}_x \times \mathrm{id}_{W_x})^*\eta \wedge \pi_2^*\omega_x \\
&= \int_{(\mathfrak{s}_x^{\boxplus\tau})^{-1}(0)\setminus(V_x\times W_x)} (\mathcal{R}_x \times \mathrm{id}_{W_x})^*(\pi_1^*\eta \wedge \pi_2^*\omega_x) \\
&= \int_{\mathfrak{s}_x^{-1}(0)\cap(\partial V_x\times W_x)} (\pi_1^*\eta \wedge \pi_2^*\omega_x).
\end{aligned}$$

For the last equality we use

$$(\mathfrak{s}_x^{\boxplus\tau})^{-1}(0) \setminus (V_x \times W_x) = (\mathcal{R}_x \times \mathrm{id}_{W_x})^{-1}(\mathfrak{s}_x^{-1}(0) \cap (\partial V_x \times W_x))$$

by definition of $\mathfrak{s}_x^{\boxplus\tau}$. By submersion property of $\mathfrak{s}_x$, we have

$$\dim \mathfrak{s}_x^{-1}(0) \cap (\partial V_x \times W_x) = \dim \mathfrak{s}_x^{-1}(0) - 1.$$

Therefore by the degree assumption made on ρ above, the last integral vanishes. Now the proof of (17.4) is complete. □

Lemma 17.15 *We put* $\overset{\circ\circ}{S}_k(U_x^{\boxplus\tau}) = S_k(U_x^{\boxplus\tau}) \cap \mathcal{R}_x^{-1}(\overset{\circ}{S}_k(U_x))$.

(1) *The closure of* $\overset{\circ\circ}{S}_k(U_x^{\boxplus\tau})$ *in* $\overset{\circ}{S}_k(U_x^{\boxplus\tau})$ *is an orbifold with corners.*

(2) *The map* $\mathcal{R}_x$ *induces an orbifold diffeomorphism from* $\mathrm{Clos}(\overset{\circ\circ}{S}_k(U_x^{\boxplus\tau}))$ *to* $\widehat{S}_k(U_x)$.

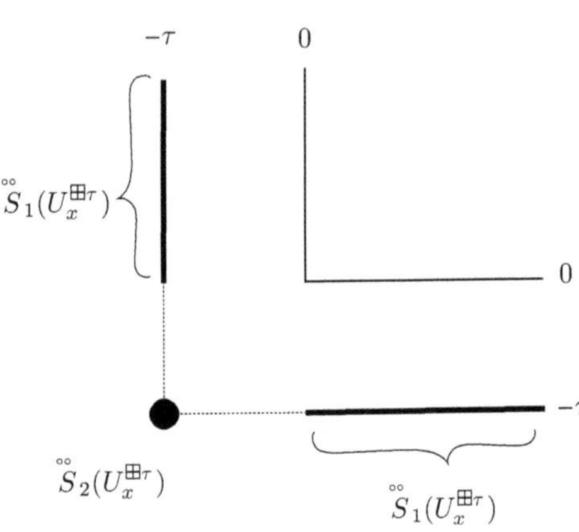

Fig. 17.2 $\overset{\circ\circ}{S}_k(U_x^{\boxplus\tau})$

Proof It suffices to prove the lemma for the case when $V = [0, 1)^k$, which is obvious. (See Fig. 17.2.) □

17.3 Outer Collaring and Embedding

To shorten the discussion we study functoriality of outer collaring for the case of coordinate change and of a local representative of an embedding simultaneously. We show that there exist functors

$$\{\text{admissible orbifold}\} \longrightarrow \{\text{collared orbifold}\},$$

$$\{\text{admissible Kuranishi charts}\} \longrightarrow \{\text{collared Kuranishi charts}\}$$

where the morphisms of those categories are embeddings. We also prove that various objects and operations are preserved by this functor. In this section we consider the following situation.

Situation 17.16 Suppose we are in Situation 17.6. Let $x' \in \overset{\circ}{S}_k(U)$ and $\mathfrak{V}_{x'} = (V_{x'}, E_{x'}, \Gamma_{x'}, \phi_{x'}, \widehat{\phi}_{x'})$ be an orbifold chart of $(U', \mathcal{E}')$ at x'. (See Definition 23.18.) Let $s_{x'}$ be a representative of the Kuranishi map s on $\mathfrak{V}_{x'}$. Let $(U', \mathcal{E}') \to (U, \mathcal{E})$ be an embedding. We take its local representative $(h_{xx'}, \varphi_{xx'}, \widehat{\varphi}_{xx'})$. ■

This includes the case when $(h_{xx'}, \varphi_{xx'}, \widehat{\varphi}_{xx'})$ represents an isomorphism, which is nothing but the case of coordinate change of an orbifold.

Lemma 17.17 *In Situation* 17.16, *there exists a unique injective map* $\mathfrak{j} : \{1, \dots, k'\} \to \{1, \dots, k\}$ *with the following properties:*

(1) *For* $\gamma' \in \Gamma_{x'}$, *we have:* $\sigma(h_{xx'}(\gamma'))(\mathfrak{j}(i)) = \mathfrak{j}(\sigma'(\gamma')(i))$.
(2) *If* $\varphi_{xx'}(\overline{y}', (t'_1, \dots, t'_{k'})) = (\overline{y}, (t_1, \dots, t_k))$ *then* $t_{\mathfrak{j}(i)} = 0$ *if and only if* $t'_i = 0$.

Proof The existence of j satisfying (2) is immediate from the fact that $\varphi_{xx'}$ preserves stratification $S_n(V_{x'})$, $S_n(V_x)$. Such a j is necessarily unique. Then (1) follows from this uniqueness. □

For $A \subset \{1, \dots, k\}$ we put

$$V_x(A) = \{(\overline{y}, (t_1, \dots, t_k)) \in V_x \mid \text{If } i \in A \text{ then } t_i = 0\}. \tag{17.8}$$

Definition 17.18 We define $\varphi_{xx'}^{\boxplus\tau} : V_{x'}^{\boxplus\tau} \to V_x^{\boxplus\tau}$ as follows:

(1) If $y' \in V_{x'}$ then $\varphi_{xx'}^{\boxplus\tau}(y') = \varphi_{xx'}(y')$.
(2) Let $y' = (\overline{y}', (t_1', \dots, t_{k'}'))$ and $\mathcal{R}_{x'}(y') \in V_{x'}(A')$, where $A' \subset \{1, \dots, k'\}$. We define $y_0 = \varphi_{xx'}(\mathcal{R}_{x'}(y'))$ and write $y_0 = (\overline{y}_0, (t_{0,1}, \dots, t_{0,k}))$. Then we define

$$\varphi_{xx'}^{\boxplus\tau}(y') = (\overline{y}_0, (t_1, \dots, t_k)),$$

where

$$t_i = \begin{cases} t_{i'}' & \text{if } i = \mathrm{j}(i'), i' \in A', \\ t_{0,i} & \text{if } i \notin \mathrm{j}(A'). \end{cases}$$

Lemma 17.19

(1) *The map* $\varphi_{xx'}^{\boxplus\tau} : V_{x'}^{\boxplus\tau} \to V_x^{\boxplus\tau}$ *is a smooth embedding of manifolds.*
(2) *The map* $\varphi_{xx'}^{\boxplus\tau} : V_{x'}^{\boxplus\tau} \to V_x^{\boxplus\tau}$ *is* $h_{xx'}$ *equivariant.*
(3) *We have* $\varphi_{xx'} \circ \mathcal{R}_{x'} = \mathcal{R}_x \circ \varphi_{xx'}^{\boxplus\tau}$.

Proof Statements (2) and (3) are obvious from the definition. We prove (1). For simplicity of notation we consider the case $\mathrm{j}(i) = i$. We study the smoothness at the point $y' = (\overline{y}', (t_1', \dots, t_{k'}'))$. We may assume without loss of generality that

$$t_1' = \cdots = t_\ell' = 0, \quad t_{\ell+1}', \dots, t_m' < 0, \quad t_{m+1}', \dots, t_{k'}' > 0.$$

Let $z' = (\overline{z}', (s_1', \dots, s_{k'}')) \in V_{x'}^{\boxplus\tau}$ be a point in a neighborhood of y'. By taking the neighborhood sufficiently small, we may assume that $s_{\ell+1}', \dots, s_m' < 0$ and $s_{m+1}', \dots, s_{k'}' > 0$. We put

$$\mathcal{R}_{x'}(z') = (\overline{z}', (s_1'', \dots, s_{k'}'')).$$

Then we have $s_{\ell+1}'', \dots, s_m'' = 0$, $s_{m+1}'' = s_{m+1}', \dots, s_{k'}'' = s_{k'}'$. Moreover, for $i \le \ell$,

$$s_i'' = \begin{cases} 0 & \text{if } s_i' \le 0, \\ s_i' & \text{if } s_i' \ge 0. \end{cases}$$

We define $\varphi_{xx'}^{\boxplus\tau}(z') = z$ and $z = (\overline{z}, (s_1, \dots, s_k))$. By definition we have

$$\overline{z} = \pi_0(\varphi_{xx'}(\overline{z}', (s_1'', \dots, s_{k'}''))),$$

where $\pi_0 : [V_x] \times [0,1)^k \to [V_x]$ is the projection. Therefore, the smooth dependence of $\overline{z}$ on z' can be proved in the same way as the proof of Lemma 17.12 (1). (Namely we use Lemma 25.10 (2).)

We also have

$$(s_{m+1}, \dots, s_k) = \pi_{m+1,\dots,k}(\varphi_{xx'}(\overline{z}', (s_1'', \dots, s_{k'}''))),$$

where $\pi_{m+1,\dots,k} : [V_x] \times [0,1)^k \to [0,1)^{k-m}$ is the projection to the last $k - m$ factors. Therefore, the smooth dependence of $(s_{k'+1}, \dots, s_k)$ on z' can be proved in the same way as in the proof of Lemma 17.12. Moreover $s_i = s_i'$ for $i = \ell + 1, \dots, m$. So the smooth dependence of s_i on z' is obvious. Finally, for $i = 1, \dots, \ell$ we have

$$s_i = \begin{cases} \pi_i(\varphi_{xx'}(\overline{z}', (s_1'', \dots, s_{k'}''))) & \text{if } s_i' \geq 0 \\ s_i' & \text{if } s_i' \leq 0. \end{cases}$$

Here $\pi_i : [V_x] \times [0,1)^k \to [0,1)$ is the projection to the i-th factor of $[0,1)^k$. Then the smooth dependence of $(s_1, \dots, s_m)$ on z' follows from Lemma 25.6.

The injectivity of $\varphi_{xx'}^{\boxplus\tau}$ can be proved easily. The smoothness of local inverse of $\varphi_{xx'}^{\boxplus\tau}$ can be proved in the same way as in the proof of smoothness of $\varphi_{xx'}^{\boxplus\tau}$. □

By Lemma 17.19 (3) the embedding of bundles $\widehat{\varphi}_{xx'} : \mathcal{E}_{x'} \to \mathcal{E}_x$ over $\varphi_{xx'}$ induces a map

$$\mathcal{R}_{x'}^* \mathcal{E}_{x'} \to \mathcal{R}_x^* \mathcal{E}_x$$

over $\varphi_{xx'}^{\boxplus\tau}$. Since $\mathcal{E}_{x'}^{\boxplus\tau} = \mathcal{R}_{x'}^* \mathcal{E}_{x'}$ and $\mathcal{E}_x^{\boxplus\tau} = \mathcal{R}_x^* \mathcal{E}_x$ by definition, we obtain

$$\widehat{\varphi}_{xx'}^{\boxplus\tau} : \mathcal{E}_{x'}^{\boxplus\tau} \to \mathcal{E}_x^{\boxplus\tau}.$$

In the same way as in the proof of Lemma 17.19 (1), we can show that $\widehat{\varphi}_{xx'}^{\boxplus\tau}$ is a smooth embedding of vector bundles. We have thus proved the next lemma.

Lemma 17.20 *Under the situation above, $(h_{xx'}, \varphi_{xx'}^{\boxplus\tau}, \widehat{\varphi}_{xx'}^{\boxplus\tau})$ is an embedding of orbifold charts. If $(h_{xx'}, \varphi_{xx'}, \widehat{\varphi}_{xx'})$ is a coordinate change, $(h_{xx'}, \varphi_{xx'}^{\boxplus\tau}, \widehat{\varphi}_{xx'}^{\boxplus\tau})$ is also a coordinate change.*

We also have the following:

Lemma 17.21 *Let $(h_{xx''}, \varphi_{xx''}, \widehat{\varphi}_{xx''})$ (resp. $(h_{x'x''}, \varphi_{x'x''}, \widehat{\varphi}_{x'x''})$) be as in Situation 17.16, where x, x' in Situation 17.16 is replaced by x, x'' (resp. x', x''). We* *define* $(h_{xx''}, \varphi_{xx''}^{\boxplus\tau}, \widehat{\varphi}_{xx''}^{\boxplus\tau})$ *(resp.* $(h_{x'x''}, \varphi_{x'x''}^{\boxplus\tau}, \widehat{\varphi}_{x'x''}^{\boxplus\tau})$*) from* $(h_{xx''}, \varphi_{xx''}, \widehat{\varphi}_{xx''})$ *(resp.* $(h_{x'x''}, \varphi_{x'x''}, \widehat{\varphi}_{x'x''})$*) in the same way as in the proof of Lemma* 17.20*. Then we have*

$$\varphi_{xx''}^{\boxplus\tau} = \varphi_{xx'}^{\boxplus\tau} \circ \varphi_{x'x''}^{\boxplus\tau}, \quad \widehat{\varphi}_{xx''}^{\boxplus\tau} = \widehat{\varphi}_{xx'}^{\boxplus\tau} \circ \widehat{\varphi}_{x'x''}^{\boxplus\tau}, \quad s_x^{\boxplus\tau} \circ \varphi_{xx'}^{\boxplus\tau} = \widehat{\varphi}_{xx'}^{\boxplus\tau} \circ s_{x'}^{\boxplus\tau}.$$

We remark that the union of $\psi_x(s_x^{-1}(0))$ for various x is an open subset of X which is homeomorphic to a subset of U. We denote this subspace by $X \cap U$. We define its outer collaring as follows.

Definition 17.22 In the Situation 17.16 we define:

$$(X \cap U)^{\boxplus\tau} = \left(\coprod_{x \in \psi(s^{-1}(0))} (s_x^{\boxplus\tau})^{-1}(0)/\Gamma_x \right) / \sim . \tag{17.9}$$

Here the equivalence relation $\sim$ in the right hand side is defined as follows. Let $y_i \in V_{x_i}^{\boxplus\tau}/\Gamma_{x_i}$ with $s_{x_i}^{\boxplus\tau}(\tilde{y}_i) = 0$, $[\tilde{y}_i] = y_i$ for $i = 1, 2$. Then $y_1 \sim y_2$ if and only if there exist $\mathfrak{V}_x$, $\tilde{y} \in V_x$, and $(h_{x_ix}, \varphi_{x_ix}, \widehat{\varphi}_{x_ix})$ as in Situation 17.16 such that $s_x^{\boxplus\tau}(\tilde{y}) = 0$ and

$$[\varphi_{x_ix}^{\boxplus\tau}(\tilde{y})] = y_i$$

in $V_{x_i}^{\boxplus\tau}/\Gamma_{x_i}$ for $i = 1, 2$.[5]

The maps $\mathcal{R}_x : (s_x^{\boxplus\tau})^{-1}(0) \to V_x$ for various x induce a map $(X \cap U)^{\boxplus\tau} \to X$, which we denote by

$$\mathcal{R} : (X \cap U)^{\boxplus\tau} \to X. \tag{17.10}$$

Lemma 17.23 *In Situation* 17.6 *we have a Kuranishi neighborhood*

$$\mathcal{U}^{\boxplus\tau} = (U^{\boxplus\tau}, \mathcal{E}^{\boxplus\tau}, \psi^{\boxplus\tau}, s^{\boxplus\tau})$$

of $(X \cap U)^{\boxplus\tau}$ *such that* $(V_x^{\boxplus\tau}/\Gamma_x, \mathcal{E}_x^{\boxplus\tau}, \psi_x^{\boxplus\tau}, s_x^{\boxplus\tau})$ *becomes its orbifold chart.*

Proof This is a consequence of Lemmas 17.19, 17.20, and 17.21. □

The next Lemma-Definition claims that various objects on $\mathcal{U}$ induce objects of the same type on its outer collaring $\mathcal{U}^{\boxplus\tau}$. In other words, outer collaring is a functorial process.

Lemma-Definition 17.24 *In the situation of Lemma* 17.23*, we call* $\mathcal{U}^{\boxplus\tau}$ *the* τ*-collaring, or* τ*-outer-collaring of* $\mathcal{U}$*.*

(1) (Boundary) $\partial(U^{\boxplus\tau})$ *is canonically diffeomorphic to* $(\partial U)^{\boxplus\tau}$*.*

---

[5] Lemma 17.21 implies that this is an equivalence relation.

(2) (CF-perturbation) *Let* $\mathfrak{S} = \{(\mathfrak{V}_{\mathfrak{r}}, \mathcal{S}_{\mathfrak{r}}) \mid \mathfrak{r} \in \mathfrak{R}\}$ *be a CF-perturbation of* $\mathcal{U}$. *Then* $\{(\mathfrak{V}_{\mathfrak{r}}^{\boxplus\tau}, \mathcal{S}_{\mathfrak{r}}^{\boxplus\tau}) \mid \mathfrak{r} \in \mathfrak{R}\}$ *is a CF-perturbation of* $\mathcal{U}^{\boxplus\tau}$. *We denote it by* $\mathfrak{S}^{\boxplus\tau}$ *and call it the* τ*-collaring of* $\mathfrak{S}$.
(3) (Equivalence of CF-perturbation) *If* $\mathfrak{S}$ *is equivalent to* $\mathfrak{S}'$, *then* $\mathfrak{S}^{\boxplus\tau}$ *is equivalent to* $\mathfrak{S}'^{\boxplus\tau}$.
(4) (Strongly smooth map) *If* $f : U \to M$ *is a smooth map and strongly submersive on* K *with respect to* $\mathfrak{S}$, *then* $f \circ \mathcal{R}$ *is strongly submersive with respect to* $\mathfrak{S}^{\boxplus\tau}$. *We denote it by* $f^{\boxplus\tau}$ *and call it* τ*-collaring of* f.
(5) (Strong transversality) *The strong transversality to a map* $N \to M$ *is also preserved.*
(6) (Multivalued perturbation) *The versions of (2)–(5) where 'CF-perturbation' is replaced by 'multivalued perturbation' also hold.*
(7) (Differential form) *If* h *is a differential form on* U, *then* $\widetilde{h}_x$ *for various* x *are glued to define a differential form on* $U^{\boxplus\tau}$. *We denote it by* $h^{\boxplus\tau}$ *and call it the* τ*-collaring of* h.
(8) (Integration along the fiber) *In the situation of (2)(3)(4)(7), we assume that* h *is compactly supported. Then we have*

$$f_!^{\boxplus\tau}(h^{\boxplus\tau}; \mathfrak{S}^{\boxplus\tau}) = f_!(h; \mathfrak{S}). \tag{17.11}$$

This follows immediately from Lemma-Definition 17.14. Also the next lemma is a straightforward generalization of Lemma 17.15 to the case of Kuranishi chart.

Lemma 17.25 *We put*

$$\overset{\circ\circ}{S}_k(U^{\boxplus\tau}) = S_k(U^{\boxplus\tau}) \cap \mathcal{R}^{-1}(\overset{\circ}{S}_k(U)).$$

(1) *The closure of* $\overset{\circ\circ}{S}_k(U^{\boxplus\tau})$ *in* $\overset{\circ}{S}_k(U^{\boxplus\tau})$ *is an orbifold with corners. We call* $\overset{\circ\circ}{S}_k(U^{\boxplus\tau})$ *a small corner of codimension* k.
(2) *The retraction map* $\mathcal{R}$ *induces an orbifold diffeomorphism from* $\mathrm{Clos}(\overset{\circ\circ}{S}_k(U^{\boxplus\tau}))$ *onto* $\widehat{S}_k(U)$.

17.4 Outer Collaring of Kuranishi Structures

In this section and the next, we study outer collaring of Kuranishi structures and good coordinate systems. For a K-space $(X, \widehat{\mathcal{U}})$ we firstly describe the underlying topological space $X^{\boxplus\tau}$ of the outer collaring of X in Definition 17.29. In the next section, we define the outer collaring $(X^{\boxplus\tau}, \widehat{\mathcal{U}^{\boxplus\tau}})$ (sometimes called τ-*collaring*) of the K-space $(X, \widehat{\mathcal{U}})$. We first consider the following situation.

Situation 17.26 Let $\mathcal{U}_i = (U_i, \mathcal{E}_i, \psi_i, s_i)$ be Kuranishi charts of X and $\Phi_{21} = (\varphi_{21}, \widehat{\varphi}_{21})$ an embedding of Kuranishi charts. We may decorate $\mathcal{U}_i$ by some of the following in addition:

(1) We are given CF-perturbations $\mathfrak{S}^i$ of $\mathcal{U}_i$ $(i = 1, 2)$ such that $\mathfrak{S}^1$, $\mathfrak{S}^2$ are compatible with Φ_{21}.
(2) We are given differential forms h_i on U_i $(i = 1, 2)$ such that $h_1 = \varphi_{21}^* h_2$.
(3) We are given smooth maps $f_i : U_i \to M$ $(i = 1, 2)$ such that $f_1 = f_2 \circ \varphi_{21}$.
(4) We are given multivalued perturbations $\mathfrak{s}^i$ of $\mathcal{U}_i$ $(i = 1, 2)$ such that $\mathfrak{s}^1$, $\mathfrak{s}^2$ are compatible with Φ_{21}.
(5) We have another Kuranishi chart $\mathcal{U}_3$ and an embedding $\Phi_{32} : \mathcal{U}_2 \to \mathcal{U}_3$. We put $\Phi_{31} = \Phi_{32} \circ \Phi_{21}$.■

Lemma 17.27 *In Situation 17.26 we have an embedding of Kuranishi charts* $\Phi_{21}^{\boxplus\tau} : \mathcal{U}_1^{\boxplus\tau} \to \mathcal{U}_2^{\boxplus\tau}$, *whose restriction to* $\mathcal{U}_1$ *coincides with* Φ_{21}. *Moreover we have the following:*

(1) *In Situation 17.26 (1),* $\mathfrak{S}^{1\boxplus\tau}$, $\mathfrak{S}^{2\boxplus\tau}$ *are compatible with* $\Phi_{21}^{\boxplus\tau}$.
(2) *In Situation 17.26 (2),* $h_1^{\boxplus\tau} = (\varphi_{21}^{\boxplus\tau})^* h_2^{\boxplus\tau}$.
(3) *In Situation 17.26 (3),* $f_1^{\boxplus\tau} = f_2^{\boxplus\tau} \circ \varphi_{21}^{\boxplus\tau}$.
(4) *In Situation 17.26 (4),* $\mathfrak{s}^{1\boxplus\tau}$ *and* $\mathfrak{s}^{2\boxplus\tau}$ *are compatible with* $\Phi_{21}^{\boxplus\tau}$.
(5) *In Situation 17.26 (5), we have* $\Phi_{31}^{\boxplus\tau} = \Phi_{32}^{\boxplus\tau} \circ \Phi_{21}^{\boxplus\tau}$.
(6) $\varphi_{21}^{\boxplus\tau} \circ \mathcal{R}_1 = \mathcal{R}_2 \circ \varphi_{21}^{\boxplus\tau}$.

Remark 17.28 Both of $\mathcal{U}_1^{\boxplus\tau}$ and $\mathcal{U}_2^{\boxplus\tau}$ are Kuranishi charts of the topological space $(X \cap U_2)^{\boxplus\tau}$.

***Proof of Lemma* 17.27** We use Lemma 23.13 to obtain objects $\{\mathfrak{V}_{\mathfrak{r}}^i \mid \mathfrak{r} \in \mathfrak{R}_i\}$ and $(h_{\mathfrak{r},21}, \varphi_{\mathfrak{r},21}, \hat{\varphi}_{\mathfrak{r},21})$ which have the properties spelled out there. Let $\mathfrak{r} \in \mathfrak{R}_1$. Then the map $\varphi_{\mathfrak{r},21} : V_{\mathfrak{r}}^1 \to V_{\mathfrak{r}}^2$ is extended to $\varphi_{\mathfrak{r},21}^{\boxplus\tau} : V_{\mathfrak{r}}^{1\boxplus\tau} \to V_{\mathfrak{r}}^{2\boxplus\tau}$ as follows. Let $(\overline{y}', (t_1', \dots, t_{d(\mathfrak{r})}')) \in V_{\mathfrak{r}}^1$ and

$$(\overline{y}'', (t_1'', \dots, t_{d(\mathfrak{r})}'')) = \varphi_{\mathfrak{r},21}(\mathcal{R}_{\mathfrak{r}}(\overline{y}', (t_1', \dots, t_{d(\mathfrak{r})}'))).$$

Then we put

$$\varphi_{\mathfrak{r},21}^{\boxplus\tau}(\overline{y}', (t_1', \dots, t_{d(\mathfrak{r})}')) = (\overline{y}, (t_1, \dots, t_{d(\mathfrak{r})})),$$

where $\overline{y}' = \overline{y}''$ and

$$t_i = \begin{cases} t_i' & \text{if } t_i' \le 0, \\ t_i'' & \text{if } t_i' \ge 0. \end{cases}$$

We define $\widehat{\varphi}_{\mathfrak{r},21}^{\boxplus\tau}$ in a similar way. Using Lemma 23.13 (4) and Lemma 23.26, it is easy to see that $(h_{\mathfrak{r},21}, \varphi_{\mathfrak{r},21}^{\boxplus\tau}, \widehat{\varphi}_{\mathfrak{r},21}^{\boxplus\tau})$ is a representative of the required embedding.

It is straightforward to check (1)–(6). □

We will glue $\mathcal{U}_p^{\boxplus\tau}$ for various p via Lemma 17.27 to obtain a Kuranishi structure $\widehat{\mathcal{U}^{\boxplus\tau}}$. (See Lemma-Definition 17.38.) Its underlying topological space is obtained as follows.

Definition 17.29 (1) Let $(X, \widehat{\mathcal{U}})$ be a K-space. We define a topological space $X^{\boxplus\tau}$ as follows. We take a disjoint union

$$\coprod_{p\in X}(X\cap U_p)^{\boxplus\tau} = \coprod_{p\in X}(s_p^{\boxplus\tau})^{-1}(0)/\Gamma_p$$

and define an equivalence relation $\sim$ as follows: Let $x_p \in (s_p^{\boxplus\tau})^{-1}(0)$ and $x_q \in (s_q^{\boxplus\tau})^{-1}(0)$. We define $[x_p] \sim [x_q]$ if there exist $r \in X$ and $x_r \in (s_p^{\boxplus\tau})^{-1}(0) \cap U_{pr}^{\boxplus\tau} \cap U_{qr}^{\boxplus\tau}$ such that

$$[x_p] = \varphi_{pr}^{\boxplus\tau}([x_r]), \quad [x_q] = \varphi_{qr}^{\boxplus\tau}([x_r]). \tag{17.12}$$

As we will see in Lemma 17.30, $\sim$ is an equivalence relation. We define $X^{\boxplus\tau}$ as the set of the equivalence classes of this equivalence relation $\sim$

$$X^{\boxplus\tau} := \coprod_{p\in X}(X\cap U_p)^{\boxplus\tau}/\!\sim\, = \left(\coprod_{p\in X}(s_p^{\boxplus\tau})^{-1}(0)/\Gamma_p\right)/\sim. \tag{17.13}$$

(2) For $k = 1, 2, \ldots$ we define

$$S_k(X^{\boxplus\tau}) := \left(\coprod_{p\in X} S_k(U_p^{\boxplus\tau}) \cap \left((s_p^{\boxplus\tau})^{-1}(0)/\Gamma_p\right)\right)/\sim. \tag{17.14}$$

The relation $\sim$ is defined on the sets $s_p^{-1}(0)/\Gamma_p$ or the enhanced sets $(s_p^{\boxplus\tau})^{-1}(0)/\Gamma_p$. So the messy process of shrinking the domain to ensure the consistency of coordinate changes is not necessary here. In fact, we show the following.

Lemma 17.30 *The relation $\sim$ in Definition 17.29 is an equivalence relation.*

Proof (1) We just check transitivity. Other properties are easier to prove. Suppose $[x_p] \sim [x_q]$ and $[x_q] \sim [x_r]$. By Definition 17.29, there exist $u, v \in X$ and $x_u \in U_u^{\boxplus\tau}$, $x_v \in U_v^{\boxplus\tau}$ such that

$$[x_p] = \varphi_{pu}^{\boxplus\tau}([x_u]),\ [x_q] = \varphi_{qu}^{\boxplus\tau}([x_u]),$$

$$[x_q] = \varphi_{qv}^{\boxplus\tau}([x_v]),\ [x_r] = \varphi_{rv}^{\boxplus\tau}([x_v]).$$

By Lemma 17.27 (6), we obtain $\psi_p(\mathcal{R}_p([x_p])) = \psi_q(\mathcal{R}_q([x_q])) = \psi_r(\mathcal{R}_r(x_r))$ from (17.12). Denote this common point by $\mathfrak{t} \in X$. We take admissible coordinate $U_{\mathfrak{t}} = V_{\mathfrak{t}}/\Gamma_{\mathfrak{t}}$ with $V_{\mathfrak{t}} = [V_{\mathfrak{t}}] \times [0,1)^d$. We note $V_{\mathfrak{t}}^{\boxplus\tau} = [V_{\mathfrak{t}}] \times [-\tau, 1)^d$. Here $[V_{\mathfrak{t}}]$ is the corner locus. Since $\mathfrak{t} \in s_p^{-1}(0) \cap s_q^{-1}(0) \cap s_r^{-1}(0) \cap s_u^{-1}(0) \cap s_v^{-1}(0)$, we have coordinate changes $\varphi_{p\mathfrak{t}}, \varphi_{q\mathfrak{t}}, \varphi_{r\mathfrak{t}}, \varphi_{u\mathfrak{t}}, \varphi_{v\mathfrak{t}}$ so that

$$\mathcal{R}_p([x_p]) = \varphi_{pu} \circ \varphi_{u\mathfrak{t}}(o_{\mathfrak{t}}) = \varphi_{p\mathfrak{t}}(o_{\mathfrak{t}}),$$

$$\mathcal{R}_q([x_q]) = \varphi_{qu} \circ \varphi_{u\mathfrak{t}}(o_{\mathfrak{t}}) = \varphi_{qv} \circ \varphi_{v\mathfrak{t}}(o_{\mathfrak{t}}) = \varphi_{q\mathfrak{t}}(o_{\mathfrak{t}}),$$

and

$$\mathcal{R}_r([x_r]) = \varphi_{rv} \circ \varphi_{v\mathfrak{t}}(o_{\mathfrak{t}}) = \varphi_{r\mathfrak{t}}(o_{\mathfrak{t}}).$$

Also note that the restrictions of $\varphi_{p\mathfrak{t}}^{\boxplus\tau}, \varphi_{q\mathfrak{t}}^{\boxplus\tau}, \varphi_{r\mathfrak{t}}^{\boxplus\tau}, \varphi_{u\mathfrak{t}}^{\boxplus\tau}$ and $\varphi_{v\mathfrak{t}}^{\boxplus\tau}$ to $\mathcal{R}_{\mathfrak{t}}^{-1}(o_{\mathfrak{t}})$ are bijections to $\mathcal{R}_p^{-1}(\mathcal{R}_p([x_p]))$, $\mathcal{R}_q^{-1}(\mathcal{R}_q([x_q]))$, $\mathcal{R}_r^{-1}(\mathcal{R}_r([x_r]))$, $\mathcal{R}_u^{-1}(\mathcal{R}_u([x_u]))$ and $\mathcal{R}_v^{-1}(\mathcal{R}_v([x_v]))$, respectively. Hence there exists $x_{\mathfrak{t}} \in V_{\mathfrak{t}}^{\boxplus\tau}$ such that $[x_q] = \varphi_{q\mathfrak{t}}^{\boxplus\tau}([x_{\mathfrak{t}}])$, $[x_u] = \varphi_{u\mathfrak{t}}^{\boxplus\tau}([x_{\mathfrak{t}}])$ and $[x_v] = \varphi_{v\mathfrak{t}}^{\boxplus\tau}([x_{\mathfrak{t}}])$. Therefore we have

$$[x_p] = \varphi_{pu}^{\boxplus}([x_u]) = \varphi_{p\mathfrak{t}}^{\boxplus\tau}([x_{\mathfrak{t}}]) \text{ and } [x_r] = \varphi_{rv}^{\boxplus\tau}([x_v]) = \varphi_{r\mathfrak{t}}^{\boxplus\tau}([x_{\mathfrak{t}}]),$$

which imply that $[x_p] \sim [x_r]$. □

Remark 17.31 In this chapter we consider the case of a Kuranishi structure on a compact metrizable space X. We may also consider the case of a Kuranishi structure of a pair $Z \subseteq X$ of a metrizable space X and its compact subspace Z. There is actually nothing new needed to do so, except the following point. To define a topological space $X^{\boxplus\tau}$ we used a Kuranishi structure on X. If we are given a Kuranishi structure of $Z \subseteq X$, we can still define $Z^{\boxplus\tau}$. However, we can define $X^{\boxplus\tau}$ only in a neighborhood of Z. The situation here is similar to the situation we met in defining the boundary $\partial(X, Z; \widehat{\mathcal{U}})$. See Remark 8.9 (2). It seems unlikely that this point becomes an important issue in the application. In fact, in all the cases we know so far appearing in the actual applications, it is enough to define $X^{\boxplus\tau}$ in a neighborhood of Z, or there is an obvious way to define $X^{\boxplus\tau}$ in the particular situations.

Definition 17.32 Using a good coordinate system $\widehat{\mathcal{U}}$ of X, we define $X^{\boxplus\tau}$ in a similar way to the case of Kuranishi structure.

17.5 Collared Kuranishi Structure

For a point $p \in X$ we can define its Kuranishi neighborhood as $\mathcal{U}_p^{\boxplus\tau}$. There is a slight issue in defining a Kuranishi neighborhood compatible with the collar structure.

Example 17.33 We consider an orbifold $X = [0,\infty)^2/\mathbb{Z}_2$, where $\mathbb{Z}_2$ acts by exchanging the factors. We want to regard it as a '1-collared orbifold'. If $p = [0, 0]$ we can take an obvious choice $[0, 1)^2/\mathbb{Z}_2$ as its 'collared neighborhood'.[6] There is an issue in the case $p = [(0, 0.5)]$. We might try to take its neighborhood such as $[0, 1) \times (0.3, 0.7)$. However, this does not work. In fact $(0.4, 0.6) \sim (0.6, 0.4)$ but $\mathbb{Z}_2$ is not contained in the isotropy group of $(0, 0.5)$. It seems impossible to find a good 'collared neighborhood' of $[(0, 0.5)]$ such that the 'length' of the collar is 1. This is a technical problem and certainly we should regard $[0,\infty)^2/\mathbb{Z}_2$ as having a collar of length ≥ 1.

It seems to the authors that the best way to define the appropriate notion of a τ-collared cornered orbifold is as follows: We do not define an orbifold chart at the points in $[0, 1)^2 \setminus \{(0, 0)\}$. The points of $[0, 1)^2 \setminus \{(0, 0)\}$ are contained in the chart at $(0, 0)$ so we do not need an orbifold chart at those points. We will define the notion of a τ-*collared Kuranishi structure* along this line below.

Remark 17.34 The above-mentioned trouble occurs only when the action of the isotropy group on the normal factor $[0, 1)^k$ is nontrivial. So it does not occur in the situation of our applications in Chaps. 16, 17, 18, 19, 20, 21, and 22. However, we present the formulation which works in more general cases. It actually appears when we will study the moduli space of pseudo-holomorphic curves from a bordered Riemann surface of arbitrary genus with arbitrary number of boundary components.

There occurs no similar issue for the definition of the τ-collared good coordinate system.

Definition 17.35 Given $\tau > 0$ let

$$X' = X^{\boxplus\tau}$$

be the τ-collaring of a certain Kuranishi structure $\widehat{\mathcal{U}}$ on X. We define:

$$\overset{\circ\circ}{S}_k(X',\widehat{\mathcal{U}}) = S_k(X') \cap \mathcal{R}^{-1}(\overset{\circ}{S}_k(X,\widehat{\mathcal{U}})), \tag{17.15}$$

where $S_k(X')$ is defined by (17.14). We also define a subset $B_\tau(\overset{\circ\circ}{S}_k(X',\widehat{\mathcal{U}})) \subset X'$ by

[6]We recall that for an orbifold chart (V, Γ, ϕ) at p we required (in Definition 23.1 (1)) that $p = \phi(o_p)$ and p is fixed under the Γ-action.

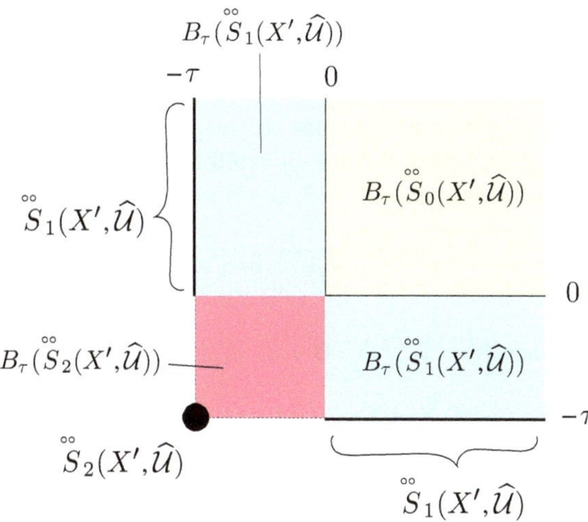

Fig. 17.3 $B_\tau(\overset{\circ\circ}{S}_k(X',\widehat{\mathcal U}))$

$$\begin{aligned}&B_\tau(\overset{\circ\circ}{S}_k(X',\widehat{\mathcal U}))\\ &= \bigcup_{p\in \overset{\circ}{S}_k(X,\widehat{\mathcal U})} \psi_p^{\boxplus\tau}\left((s_p^{\boxplus\tau})^{-1}(0)\cap\{(\overline{y},(t_1,\dots,t_k)) \mid t_i\le 0,\ i=1,\dots,k\}\right).\end{aligned} \tag{17.16}$$

Here $\overset{\circ}{S}_k(X,\widehat{\mathcal U})$ is defined in Definition 4.16.

We note that if $p'\in \overset{\circ\circ}{S}_k(X',\widehat{\mathcal U})$ then $p'=\psi_p^{\boxplus\tau}(\overline{y},(-\tau,\dots,-\tau))$ for $p=\mathcal{R}(p')$. Therefore $\overset{\circ\circ}{S}_k(X',\widehat{\mathcal U})\subset B_\tau(\overset{\circ\circ}{S}_k(X',\widehat{\mathcal U}))$. We also note that

$$\begin{aligned}B_\tau(\overset{\circ\circ}{S}_0(X',\widehat{\mathcal U})) = \overset{\circ\circ}{S}_0(X',\widehat{\mathcal U}) &= \overset{\circ}{S}_0(X,\widehat{\mathcal U}),\\ B_\tau(\overset{\circ\circ}{S}_k(X',\widehat{\mathcal U}))\cap X &= \overset{\circ}{S}_k(X,\widehat{\mathcal U}).\end{aligned} \tag{17.17}$$

See Fig. 17.3.

Lemma 17.36 *Let $X'=X^{\boxplus\tau}$ be as above. Then it has the following decomposition:*

$$X'=\coprod_k B_\tau(\overset{\circ\circ}{S}_k(X',\widehat{\mathcal U}))$$

where the right hand side is the disjoint union.

Proof Let $p' \in X'$ and put $p = \mathcal{R}(p')$. We can write $p' = \psi_p^{\boxplus\tau}(\overline{y}, (t_1, \dots, t_k))$. Without loss of generality, we may assume that $t_1, \dots, t_\ell \le 0 < t_{\ell+1}, \dots, t_k$ for some ℓ. We put $q = \psi_p^{\boxplus\tau}(\overline{y}, (0, \dots, 0, t_{\ell+1}, \dots, t_k)) \in \overset{\circ}{S}_\ell(X, \widehat{\mathcal{U}})$. We may choose the coordinate of q so that the map $\mathrm{j} : \{1, \dots, \ell\} \to \{1, \dots, k\}$ appearing in the coordinate change from an orbifold chart at q to an orbifold chart at p, which appeared in Lemma 17.17, is $\mathrm{j}(i) = i$. Then we can take $\overline{y}'$ such that

$$\psi_p^{\boxplus\tau}(\overline{y}, (0, \dots, 0, t_{\ell+1}, \dots, t_k)) = \psi_q^{\boxplus\tau}(\overline{y}', (0, \dots, 0)).$$

Then $p' = \psi_q^{\boxplus\tau}(\overline{y}', (t_1, \dots, t_\ell)) \in B_\tau(\overset{\circ\circ}{S}_\ell(X', \widehat{\mathcal{U}}))$. Moreover,

$$B_\tau(\overset{\circ\circ}{S}_k(X', \widehat{\mathcal{U}})) \cap B_\tau(\overset{\circ\circ}{S}_\ell(X', \widehat{\mathcal{U}})) = \emptyset$$

for $k \neq \ell$ is obvious from the definition. □

We next define a τ-collared version of various objects. This definition shows functoriality of outer collaring.

Definition 17.37 Suppose we are in the situation of Definition 17.35. In particular, X' is a compact metrizable space homeomorphic to $X^{\boxplus\tau}$ for a certain K-space $(X, \widehat{\mathcal{U}})$.

(1) (Kuranishi neighborhood) Let $p' \in \overset{\circ\circ}{S}_k(X', \widehat{\mathcal{U}})$. A *$\tau$-collared Kuranishi neighborhood* at p' is a Kuranishi chart $\mathcal{U}_{p'}$ of X' such that $\mathcal{U}_{p'} = (\mathcal{U}_p)^{\boxplus\tau}$ for a certain Kuranishi neighborhood $\mathcal{U}_p$ of $p = \mathcal{R}(p')$. (See Fig. 17.4.) Note that for $k = 0$, $p' = p$ and $\mathcal{U}_{p'} = \mathcal{U}_p^{\boxplus\tau} = \mathcal{U}_p$, and for $k > 0$, $p' \notin X$, the notation $\mathcal{U}_{p'}$ is non-ambiguous in meaning as well.

(2) (Coordinate change) For $p' \in \overset{\circ\circ}{S}_k(X', \widehat{\mathcal{U}})$ and $q' \in \overset{\circ\circ}{S}_\ell(X')$, let $\mathcal{U}_{p'} = \mathcal{U}_p^{\boxplus\tau} = (\mathcal{U}_p)^{\boxplus\tau}$ and $\mathcal{U}_{q'} = \mathcal{U}_q^{\boxplus\tau}$ be their τ-collared Kuranishi neighborhoods, respectively. Suppose $q' \in \psi_{p'}(s_{p'}^{-1}(0))$. (So $q \in \psi_p(s_p^{-1}(0))$.) A *$\tau$-collared coordinate change* $\Phi_{p'q'}$ from $\mathcal{U}_{q'}$ to $\mathcal{U}_{p'}$ is $\Phi_{pq}^{\boxplus\tau}$ defined by Lemma 17.27, where Φ_{pq} is a coordinate change from $\mathcal{U}_q$ to $\mathcal{U}_p$.

(3) (Kuranishi structure) A *τ-collared Kuranishi structure* $\widehat{\mathcal{U}'}$ on X' consists of the following objects:

   (a) To each $p' \in \overset{\circ\circ}{S}_k(X', \widehat{\mathcal{U}})$, $\widehat{\mathcal{U}'}$ assigns a τ-collared Kuranishi neighborhood $\mathcal{U}_{p'}$.

   (b) To each $p' \in \overset{\circ\circ}{S}_k(X', \widehat{\mathcal{U}})$ and $q' \in \overset{\circ\circ}{S}_\ell(X', \widehat{\mathcal{U}})$ with $q' \in \psi_{p'}(s_{p'}^{-1}(0))$, $\widehat{\mathcal{U}'}$ assigns a τ-collared coordinate change $\Phi_{p'q'}$.

   (c) If $p' \in \overset{\circ\circ}{S}_k(X', \widehat{\mathcal{U}})$, $q' \in \overset{\circ\circ}{S}_\ell(X', \widehat{\mathcal{U}})$, $r' \in \overset{\circ\circ}{S}_m(X', \widehat{\mathcal{U}})$ with $q' \in \psi_{p'}(s_{p'}^{-1}(0))$ and $r' \in \psi_{q'}(s_{q'}^{-1}(0))$, then we require

Fig. 17.4 $\widehat{\mathcal{U}_p^{\boxplus\tau}}$

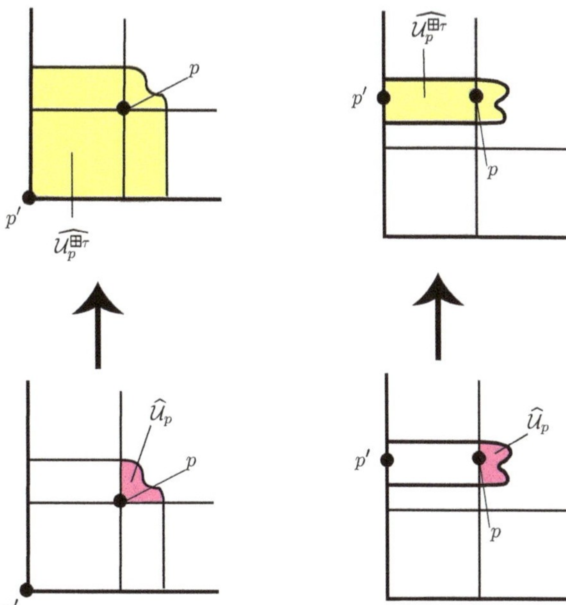

$$\Phi_{p'q'} \circ \Phi_{q'r'}|_{U_{p'q'r'}} = \Phi_{p'r'}|_{U_{p'q'r'}},$$

where $U_{p'q'r'} = U_{p'r'} \cap \varphi_{q'r'}^{-1}(U_{p'q'})$. Note that

$$B_\tau(\overset{\circ\circ}{S}_k(X',\widehat{\mathcal{U}})) \subset \bigcup_{p\in \overset{\circ}{S}_k(X,\widehat{\mathcal{U}})} (X \cap U_p)^{\boxplus\tau}.$$

So Lemma 17.36, $X = \coprod_k \overset{\circ}{S}_k(X,\widehat{\mathcal{U}})$ and the fact that $\mathcal{R} : \overset{\circ\circ}{S}_k(X',\widehat{\mathcal{U}}) \to \overset{\circ}{S}_k(X,\widehat{\mathcal{U}})$ is a homeomorphism imply that X is covered by the Kuranishi charts of $\widehat{\mathcal{U}'}$.

(4) (K-space) A *τ-collared K-space* is a pair of a compact metrizable space and its τ-collared Kuranishi structure. It is obtained from a K-space $(X,\widehat{\mathcal{U}})$ as in Lemma-Definition 17.38 below.
(5) (Various other objects) We can define the notion of a τ-collared CF-perturbation, τ-collared multivalued perturbation, τ-collared good coordinate system, τ-collared Kuranishi chart, τ-collared vector bundle, τ-collared smooth section, τ-collared embedding of various kinds, etc. in the same way. Actually those objects on $(X^{\boxplus\tau},\widehat{\mathcal{U}^{\boxplus\tau}})$ are obtained from the corresponding objects on $(X,\widehat{\mathcal{U}})$ by applying the process of outer collaring on each chart as in Lemma 17.40 below.

(6) (Uniformity) An $\mathscr{A}$-parametrized family of τ-collared CF-perturbations is said to be *uniform* if it is of the form $\{\widehat{\mathfrak{S}_\sigma^{\boxplus\tau}} \mid \sigma \in \mathscr{A}\}$ for a certain uniform family $\{\widehat{\mathfrak{S}_\sigma} \mid \sigma \in \mathscr{A}\}$ of CF-perturbations on $(X, \widehat{\mathcal{U}})$.

When $(X^{\boxplus\tau}, \widehat{\mathcal{U}'})$ is a τ-collared Kuranishi structure which is the outer collaring of $(X, \widehat{\mathcal{U}})$ then its boundary $\partial(X^{\boxplus\tau}, \widehat{\mathcal{U}'})$ is by definition the τ-collared K-space, which is the outer collaring $(\partial(X, \widehat{\mathcal{U}}))^{\boxplus\tau}$ of $\partial(X, \widehat{\mathcal{U}})$.

Lemma-Definition 17.38 *For any K-space $(X, \widehat{\mathcal{U}})$ we can assign a τ-collared K-space $(X^{\boxplus\tau}, \widehat{\mathcal{U}^{\boxplus\tau}})$ such that:*

(1) *Its underlying topological space $X^{\boxplus\tau}$ is as in Definition* 17.29.
(2) *If $p \in \overset{\circ\circ}{S}_k(X^{\boxplus\tau})$, its Kuranishi neighborhood is $\mathcal{U}^{\boxplus\tau}_{\mathcal{R}(p)}$, where $\mathcal{U}^{\boxplus\tau}_{\mathcal{R}(p)}$ is defined in Lemma* 17.23.
(3) *The coordinate changes are $\Phi^{\boxplus\tau}_{\mathcal{R}(p)\mathcal{R}(q)}$, where $\Phi^{\boxplus\tau}_{\mathcal{R}(p)\mathcal{R}(q)}$ is defined in Lemma* 17.27.

We call $(X^{\boxplus\tau}, \widehat{\mathcal{U}^{\boxplus\tau}})$ the τ-collaring (or outer collaring) of $(X, \widehat{\mathcal{U}})$. We sometimes write $(X, \widehat{\mathcal{U}})^{\boxplus\tau}$ in place of $(X^{\boxplus\tau}, \widehat{\mathcal{U}^{\boxplus\tau}})$.

Proof This is immediate from the definition. □

For a τ-collared Kuranishi structure $\widehat{\mathcal{U}^{\boxplus\tau}}$ on $X' = X^{\boxplus\tau}$ we sometimes write $\overset{\circ\circ}{S}_k((X, \widehat{\mathcal{U}})^{\boxplus\tau})$ etc. in place of $\overset{\circ\circ}{S}_k(X^{\boxplus\tau}, \widehat{\mathcal{U}^{\boxplus\tau}})$ etc.

Definition 17.39 Let $\widehat{\mathcal{U}} = ((\mathfrak{P}, \le), \{\mathcal{U}_{\mathfrak{p}} \mid \mathfrak{p} \in \mathfrak{P}\}, \{\Phi_{\mathfrak{p}\mathfrak{q}} \mid \mathfrak{q} \le \mathfrak{p}\})$ be a good coordinate system of X. We define a good coordinate system $\widehat{\mathcal{U}^{\boxplus\tau}}$ of $X^{\boxplus\tau}$ as

$$\widehat{\mathcal{U}^{\boxplus\tau}} = ((\mathfrak{P}, \le), \{\mathcal{U}_{\mathfrak{p}}^{\boxplus\tau} \mid \mathfrak{p} \in \mathfrak{P}\}, \{\Phi_{\mathfrak{p}\mathfrak{q}}^{\boxplus\tau} \mid \mathfrak{q} \le \mathfrak{p}\}).$$

We call $(X^{\boxplus\tau}, \widehat{\mathcal{U}^{\boxplus\tau}})$ the τ-*collaring* (or *outer collaring*) of $(X, \widehat{\mathcal{U}})$.

We call a good coordinate system τ-*collared* if it is isomorphic to $\widehat{\mathcal{U}^{\boxplus\tau}}$ for some $\widehat{\mathcal{U}}$.

The next lemma says that many objects defined on $(X, \widehat{\mathcal{U}})$ induce collared objects on $(X^{\boxplus\tau}, \widehat{\mathcal{U}^{\boxplus\tau}})$. The object discussed in each item of the lemma is: (1) boundary, (2) CF-perturbation, (3) strongly continuous map, (4) strong submersivity, (5) transversality, (6) CF-perturbation, (7) differential form, (8) integration along the fiber, (9) multivalued perturbation, (10) retraction, (11) KK-embedding, (12) various types of embeddings, (13) thickening, (14) covering, (15) good coordinate system. Lemma 17.40 is also regarded as functoriality of outer collaring and is actually a trivial statement to prove.

Lemma 17.40 *We consider the situation of Lemma-Definition* 17.38.

(1) *$(\partial(X, \widehat{\mathcal{U}}))^{\boxplus\tau}$ is canonically isomorphic to $\partial(X^{\boxplus\tau}, \widehat{\mathcal{U}^{\boxplus\tau}})$.*

(2) *A CF-perturbation* $\widehat{\mathfrak{S}}$ *on* $(X, \widehat{\mathcal{U}})$ *induces a* τ*-collared CF-perturbation* $\widehat{\mathfrak{S}^{\boxplus\tau}}$ *on* $(X^{\boxplus\tau}, \widehat{\mathcal{U}^{\boxplus\tau}})$.

(3) *A strongly continuous map* $\widehat{f}$ *from* $(X, \widehat{\mathcal{U}})$ *induces a* τ*-collared strongly continuous map* $\widehat{f^{\boxplus\tau}}$ *from* $(X^{\boxplus\tau}, \widehat{\mathcal{U}^{\boxplus\tau}})$. *Strong smoothness and weak submersivity are preserved.*

(4) *In the situation of (2)(3), if* $\widehat{f}$ *is strongly submersive with respect to* $\widehat{\mathfrak{S}}$, *then* $\widehat{f^{\boxplus\tau}}$ *is a* τ*-collared strongly submersive map with respect to* $\widehat{\mathfrak{S}^{\boxplus\tau}}$.

(5) *Transversality to a map* $N \to M$ *is also preserved.*

(6) *The versions of (2)(4) where 'CF-perturbation' is replaced by 'multivalued perturbation' also hold.*

(7) *A differential form* $\widehat{h}$ *on* $(X, \widehat{\mathcal{U}})$ *induces a* τ*-collared differential form* $\widehat{h^{\boxplus\tau}}$ *on* $(X^{\boxplus\tau}, \widehat{\mathcal{U}^{\boxplus\tau}})$.

(8) *In the situation of (2)(4)(7), if* $\widehat{f} : (X, \widehat{\mathcal{U}}) \to M$ *is strongly submersive with respect to* $\widehat{\mathfrak{S}}$, *then we have*

$$\widehat{f}!(\widehat{h}; \widehat{\mathfrak{S}}) = \widehat{f^{\boxplus\tau}}!(\widehat{h^{\boxplus\tau}}; \widehat{\mathfrak{S}^{\boxplus\tau}}). \tag{17.18}$$

(9) *We put* $\overset{\circ\circ}{S}_k(X^{\boxplus\tau}, \widehat{\mathcal{U}^{\boxplus\tau}}) = S_k(X^{\boxplus\tau}, \widehat{\mathcal{U}^{\boxplus\tau}}) \cap \mathcal{R}^{-1}(\overset{\circ}{S}_k(X, \widehat{\mathcal{U}}))$ *and call it the small codimension k corner. (We note that* $\overset{\circ\circ}{S}_k(X^{\boxplus\tau}, \widehat{\mathcal{U}^{\boxplus\tau}}) = \overset{\circ\circ}{S}_k(X^{\boxplus\tau}, \widehat{\mathcal{U}})$ *in* (17.15).*) Then the Kuranishi structure of* $S_k(X^{\boxplus\tau}, \widehat{\mathcal{U}^{\boxplus\tau}})$ *induces a Kuranishi structure on* $\mathrm{Clos}(\overset{\circ\circ}{S}_k(X^{\boxplus\tau}, \widehat{\mathcal{U}^{\boxplus\tau}}))$.

(10) *The restriction of the retraction map* $\mathcal{R}$ *is an underlying homeomorphism of an isomorphism between the K-spaces* $\mathrm{Clos}(\overset{\circ\circ}{S}_k(X^{\boxplus\tau}, \widehat{\mathcal{U}^{\boxplus\tau}}))$ *and* $\widehat{S}_k(X, \widehat{\mathcal{U}})$.

(11) *If there exists an embedding* $\widehat{\mathcal{U}} \to \widehat{\mathcal{U}^+}$ *of Kuranishi structures, then the space* $X^{\boxplus\tau}$ *defined by* $\widehat{\mathcal{U}}$ *is canonically homeomorphic to the one defined by* $\widehat{\mathcal{U}^+}$. *The same holds for various types of embeddings between Kuranishi structures* $\widehat{\mathcal{U}}$ *and/or good coordinate systems* $\widehat{\widehat{\mathcal{U}}}$.

(12) *Various types of embeddings between Kuranishi structures* $\widehat{\mathcal{U}}$ *and/or good coordinate systems* $\widehat{\widehat{\mathcal{U}}}$ *induce* τ*-collared embeddings between* $\widehat{\mathcal{U}^{\boxplus\tau}}$ *and/or* $\widehat{\widehat{\mathcal{U}^{\boxplus\tau}}}$. *Compatibility among various objects on them (such as CF-perturbation) is preserved by outer collaring.*

(13) *If* $\widehat{\mathcal{U}^+}$ *is a thickening of* $\widehat{\mathcal{U}}$, *then* $\widehat{\mathcal{U}^{+\boxplus\tau}}$ *is a thickening of* $\widehat{\mathcal{U}^{\boxplus\tau}}$.

(14) *If* $(\widetilde{X}, \widetilde{\widehat{\mathcal{U}}})$ *is a k-fold covering of* $(X, \widehat{\mathcal{U}})$, *then* $(\widetilde{X^{\boxplus\tau}}, \widetilde{\widehat{\mathcal{U}^{\boxplus\tau}}})$ *is a k-fold covering of* $(X^{\boxplus\tau}, \widehat{\mathcal{U}^{\boxplus\tau}})$.

(15) *The same results as (1)–(14) hold when we replace Kuranishi structures by good coordinate systems.*

Proof (1)–(8) are consequences of Lemma-Definition 17.24 (1)–(8), respectively. (9), (10) are consequences of Lemma 17.25 (1), (2), respectively. The proof of

(11) is similar to the proof of Lemma 17.30. (12) follows from Lemma 17.27. We can prove (13) by putting $O_p^{\boxplus\tau} = \mathcal{R}_{\mathcal{R}(p)}^{-1}(O_{\mathcal{R}(p)})$ and $W_p(q)^{\boxplus\tau} = \mathcal{R}_{\mathcal{R}(p)}^{-1}(W_{\mathcal{R}(p)}(\mathcal{R}(q))$, where the notations are as in Definition 5.3 (2). (14) is obvious from the definition. The proof of (15) is the same as the proof of (1)–(14). □

Definition 17.41

(1) We denote by $\mathscr{KR}(X)$ the category whose objects are Kuranishi structures of X and whose morphisms are strict KK-embeddings. We also denote by $\mathscr{KR}^{\boxplus\tau}(X)$ the category whose objects are τ-collared Kuranishi structures of $X^{\boxplus\tau}$ and whose morphisms are τ-collared strict KK-embeddings.
(2) Let $(X, \widehat{\mathcal{U}})$ be a τ-collared Kuranishi structure. We take a subset $X^{\boxminus\tau}$ of X and its Kuranishi structure $\widehat{\mathcal{U}^{\boxminus\tau}}$ such that $(X^{\boxminus\tau}, \widehat{\mathcal{U}^{\boxminus\tau}})^{\boxplus\tau} = (X, \widehat{\mathcal{U}})$.
    We call $(X^{\boxminus\tau}, \widehat{\mathcal{U}^{\boxminus\tau}})$ an *inward τ-collaring* of $(X, \widehat{\mathcal{U}})$.

Remark 17.42 We remark that for a given $(X, \widehat{\mathcal{U}})$ the K-space $(X^{\boxminus\tau}, \widehat{\mathcal{U}^{\boxminus\tau}})$ satisfying $(X^{\boxminus\tau}, \widehat{\mathcal{U}^{\boxminus\tau}})^{\boxplus\tau} = (X, \widehat{\mathcal{U}})$ may not be unique. When we say $(X, \widehat{\mathcal{U}})$ has a τ-collared Kuranishi structure, we include the choice of $(X^{\boxminus\tau}, \widehat{\mathcal{U}^{\boxminus\tau}})$ as a part of the data consisting of τ-collared Kuranishi structure. So, in this sense, the inward τ-collaring $(X^{\boxminus\tau}, \widehat{\mathcal{U}^{\boxminus\tau}})$ is determined by the τ-collared Kuranishi structure of $(X, \widehat{\mathcal{U}})$.

By this remark we have the following:

Corollary 17.43 *$\widehat{\mathcal{U}} \mapsto \widehat{\mathcal{U}^{\boxplus\tau}}$ defines an equivalence of categories $\mathscr{KR}(X) \to \mathscr{KR}^{\boxplus\tau}(X)$.*

We next define another functor

$$\mathscr{KR}^{\boxplus\tau}(X) \to \mathscr{KR}(X^{\boxplus\tau}).$$

Let $(X', \mathcal{U}') = (X, \mathcal{U})^{\boxplus\tau}$ be a τ-collared K-space. For $p \in X' = X^{\boxplus\tau}$ we take $\mathcal{R}(p) \in X$ and define a Kuranishi chart

$$\mathcal{U}_p^{\sim} := (\mathcal{U}_{\mathcal{R}(p)})^{\boxplus\tau}.$$

Here the right hand side is the τ-collaring of the Kuranishi chart $\mathcal{U}_{\mathcal{R}(p)}$ assigned to $\mathcal{U}$ and is a Kuranishi chart of $X' = X^{\boxplus\tau}$. We put $\mathcal{U}_p^{\sim} = (U_p^{\sim}, \mathcal{E}_p^{\sim}, \psi_p^{\sim}, s_p^{\sim}) = (U_{\mathcal{R}(p)}^{\boxplus\tau}, \mathcal{E}_{\mathcal{R}(p)}^{\boxplus\tau}, \psi_{\mathcal{R}(p)}^{\boxplus\tau}, s_{\mathcal{R}(p)}^{\boxplus\tau})$. Suppose $q \in \psi_p^{\sim}((s_p^{\sim})^{-1}(0))$. Since $\psi_p^{\sim} = \psi_{\mathcal{R}(p)}^{\boxplus\tau}$ we have

$$\mathcal{R}(q) \in \psi_{\mathcal{R}(p)}((s_{\mathcal{R}(p)})^{-1}(0)).$$

Therefore the coordinate change

$$\Phi_{\mathcal{R}(p)\mathcal{R}(q)} : \mathcal{U}_{\mathcal{R}(q)} \to \mathcal{U}_{\mathcal{R}(p)}$$

is defined. We define

$$\Phi^{\sim}_{pq} := (\Phi_{\mathcal{R}(p)\mathcal{R}(q)})^{\boxplus\tau} : \mathcal{U}^{\sim}_{q} \to \mathcal{U}^{\sim}_{p}.$$

It is easy to see that $(\{\mathcal{U}^{\sim}_{p}\}, \{\Phi^{\sim}_{pq}\})$ defines a Kuranishi structure on $X' = X^{\boxplus\tau}$.

Remark 17.44 Note that the Kuranishi structure $(\{\mathcal{U}^{\sim}_{p}\}, \{\Phi^{\sim}_{pq}\})$ on $X' = X^{\boxplus\tau}$ and the collared Kuranishi structure on the same space X' are mostly the same thing. The only difference is that the Kuranishi neighborhood is assigned only to an element of $\bigcup_k \overset{\circ\circ}{S}_k(X', \widehat{\mathcal{U}})$ for the collared Kuranishi structure but to all elements of X' for the Kuranishi structure $(\{\mathcal{U}^{\sim}_{p}\}, \{\Phi^{\sim}_{pq}\})$.

Definition 17.45 We define the functor $\mathscr{KR}^{\boxplus\tau}(X) \to \mathscr{KR}(X^{\boxplus\tau})$ by sending $(X', \mathcal{U}')$ to $(X', (\{\mathcal{U}^{\sim}_{p}\}, \{\Phi^{\sim}_{pq}\}))$.

It is easy to see that this is a functor. Namely a strict embedding of τ-collared Kuranishi structures $(X'_1, \mathcal{U}'_1) \to (X'_2, \mathcal{U}'_2)$ induces a strict embedding of the corresponding Kuranishi structures.

We call this functor the *forgetful functor of a τ-collared structure.*

We use Definition 17.45 to prove the following:

Definition-Lemma 17.46 For $0 < \tau' < \tau$ we can define a functor

$$\mathscr{KR}^{\boxplus\tau}(X) \to \mathscr{KR}^{\boxplus\tau'}(X^{\boxplus(\tau-\tau')}). \tag{17.19}$$

In other words, to a τ-collared Kuranishi structure on $X^{\boxplus\tau}$ we can canonically associate a τ'-collared Kuranishi structure on the same space $X^{\boxplus\tau}$.

Proof The functor (17.19) is the composition

$$\mathscr{KR}^{\boxplus\tau}(X) \to \mathscr{KR}(X) \to \mathscr{KR}^{\boxplus(\tau-\tau')}(X) \to \mathscr{KR}(X^{\boxplus(\tau-\tau')}) \to \mathscr{KR}^{\boxplus\tau'}(X^{\boxplus(\tau-\tau')}).$$

Here the first functor is the inverse of the functor in Corollary 17.43, the second functor is the functor in Corollary 17.43 (with τ replaced by $\tau - \tau'$), the third functor is the one in Definition 17.45 (with τ replaced by $\tau - \tau'$), the fourth functor is the one in Corollary 17.43 (with τ replaced by τ'). □

In the construction of this chapter, Chap. 19 etc., we need to replace a τ-collared structure by a τ'-collared structure with $\tau' < \tau$ several times. We use the next lemma also.

Lemma 17.47 *We consider one of the functors appearing in Corollary* 17.43, *Definition* 17.45 *or Definition-Lemma* 17.46. *Suppose we are given an object such as a differential form, CF-perturbation, multivalued perturbation, strongly continuous map, etc. on the source, then it induces the corresponding object on the target.*

The proof is obvious from construction.

17.6 Products of Collared Kuranishi Structures[7]

Definition 17.48 Let $(X_i, \widehat{\mathcal{U}_i})$ be τ-collared Kuranishi structures for $i = 1, 2$. We take their inward τ-collarings $(X_i^{\boxminus\tau}, \widehat{\mathcal{U}_i^{\boxminus\tau}})$.

Their *direct product* $(X_1, \widehat{\mathcal{U}_1}) \times (X_2, \widehat{\mathcal{U}_2})$ as a τ-collared Kuranishi structure is by definition

$$(X_1, \widehat{\mathcal{U}_1}) \times (X_2, \widehat{\mathcal{U}_2}) = \left((X_1^{\boxminus\tau}, \widehat{\mathcal{U}_1^{\boxminus\tau}}) \times (X_2^{\boxminus\tau}, \widehat{\mathcal{U}_2^{\boxminus\tau}})\right)^{\boxplus\tau}.$$

Lemma 17.49 *The direct product commutes with the functors appearing in Corollary* 17.43 *Definition* 17.45 *or Definition-Lemma* 17.46.

Proof $(X \times Y)^{\boxplus\tau} = X^{\boxplus\tau} \times Y^{\boxplus\tau}$ is obvious from the definition. The lemma then follows easily from Lemma-Definition 17.14 (9). □

We next explain that the results of this section on direct products are generalized to fiber products. For this purpose we first study a fiber product of collared Kuranishi structures with manifolds.

Definition 17.50 Let $(X, \widehat{\mathcal{U}})$ be a τ-collared Kuranishi structure and let $\widehat{f^{\boxminus\tau}}$: $(X, \widehat{\mathcal{U}})^{\boxminus\tau} \to M$ be a strongly smooth map to a manifold which is transversal to a smooth map $g : N \to M$ between two manifolds without boundary. $\widehat{f^{\boxminus\tau}}$ induces a τ-collared map $\widehat{f} : (X, \widehat{\mathcal{U}}) \to M$.

We define the fiber product $(X, \widehat{\mathcal{U}}) \times_{\widehat{f}} N$ as a τ-collared Kuranishi structure to be

$$(X, \widehat{\mathcal{U}}) \times_{\widehat{f}} N = \left((X, \widehat{\mathcal{U}}^{\boxminus\tau}) \times_{\widehat{f^{\boxminus\tau}}} N\right)^{\boxplus\tau}. \tag{17.20}$$

Its underlying topological space is $X \times_{f^{\boxminus\tau}} N$. The definition (17.20) gives a τ-collared Kuranishi structure on $X \times_{f^{\boxminus\tau}} N$.

Lemma 17.51 *For any* $0 < \tau' \le \tau$ *we have an isomorphism*

$$\left((X, \widehat{\mathcal{U}}) \times_{\widehat{f}} N\right)^{\boxminus\tau'} \cong (X, \widehat{\mathcal{U}})^{\boxminus\tau'} \times_{\widehat{f^{\boxminus\tau'}}} N.$$

Here $\widehat{f^{\boxminus\tau'}} : (X, \widehat{\mathcal{U}})^{\boxminus\tau'} \to M$ *is the map induced by* $\widehat{f}$ *in an obvious way and the right hand side is the fiber product of Kuranishi structure.*

[7] We thank a referee who pointed out that establishing consistency of fiber products and outer collaring should be included in the study of the main results of this book.

The proof is obvious from the definition.

Definition 17.52 Let $(X_i, \widehat{\mathcal{U}_i})$ be τ-collared Kuranishi structures for $i = 1, 2$ and $\widehat{f_i} : (X_i, \widehat{\mathcal{U}_i}) \to M$ be τ-collared strongly smooth maps. We assume $\widehat{f_1}$ is transversal to $\widehat{f_2}$, that is, $\widehat{f_1^{\boxminus\tau}} : (X_1, \widehat{\mathcal{U}_1})^{\boxminus\tau} \to M$ is transversal to $\widehat{f_2^{\boxminus\tau}} : (X_2, \widehat{\mathcal{U}_2})^{\boxminus\tau} \to M$.

Their *fiber product* $(X_1, \widehat{\mathcal{U}_1})_{\widehat{f_1}} \times_{\widehat{f_2}} (X_2, \widehat{\mathcal{U}_2})$ as a τ-collared Kuranishi structure is by definition given by

$$(X_1, \widehat{\mathcal{U}_1})_{\widehat{f_1}} \times_{\widehat{f_2}} (X_2, \widehat{\mathcal{U}_2}) = \left((X_1^{\boxminus\tau}, \widehat{\mathcal{U}_1^{\boxminus\tau}})_{\widehat{f_1^{\boxminus\tau}}} \times_{\widehat{f_2^{\boxminus\tau}}} (X_2^{\boxminus\tau}, \widehat{\mathcal{U}_2^{\boxminus\tau}}) \right)^{\boxplus\tau}.$$

Lemma 17.53 *The fiber product commutes with the functors appearing in Corollary* 17.43, *Definition* 17.45 *or Definition-Lemma* 17.46.

Proof This is a consequence of Lemmas 17.49 and 17.51. □

17.7 Extension of Collared Kuranishi Structures

The discussion of the previous sections is rather tiresome routine work. However, thanks to this routine work, we are now ready to prove Propositions 17.58 and 17.73 below, which are the extension theorems of a τ-collared Kuranishi structure and a τ-collared CF-perturbation from the boundary to its neighborhood, respectively. They claim that if various collared objects are given on the boundary of the K-space, one can extend it to a neighborhood of the boundary of its outer collaring. See Fig. 17.5. In Sect. 17.7 we prove Proposition 17.58 and in Sect. 17.8 we prove Proposition 17.73.

Remark 17.54 Let $S_k(X; \widehat{\mathcal{U}})$ be a codimension k stratum of a K-space $(X, \widehat{\mathcal{U}})$ and $\widehat{S}_k(X; \widehat{\mathcal{U}})$ a normalized codimension k corner of $(X, \widehat{\mathcal{U}})$. (See Definition 4.16 for $S_k(X; \widehat{\mathcal{U}})$ and Definition 24.18 for $\widehat{S}_k(X; \widehat{\mathcal{U}})$, respectively.) If $\widehat{\mathcal{U}} \to \widehat{\mathcal{U}^+}$ is a KK-embedding of X, the underlying topological space of $S_k(X; \widehat{\mathcal{U}})$ (resp. $\widehat{S}_k(X; \widehat{\mathcal{U}})$) is

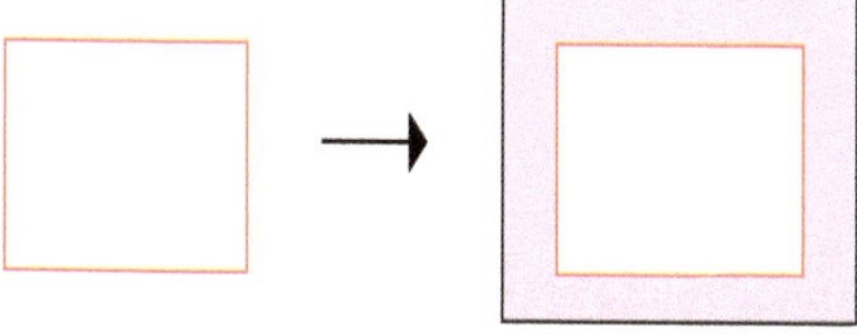

Fig. 17.5 Extension of objects to a neighborhood of the boundary of the outer collar

canonically homeomorphic to the underlying topological space of $S_k(X;\widehat{\mathcal{U}^+})$ (resp. $\widehat{S}_k(X;\widehat{\mathcal{U}^+})$.)

Hereafter we write $S_k(X)$ or $\widehat{S}_k(X)$ in place of $S_k(X;\widehat{\mathcal{U}})$, $\widehat{S}_k(X;\widehat{\mathcal{U}})$. The simplified notations $S_k(X)$, $\widehat{S}_k(X)$ stand for the underlying topological spaces, unless otherwise specified.

17.7.1 Statement

To state the extension theorem (Proposition 17.58) of a τ-collared Kuranishi structure we consider the following situation. Roughly speaking, Situation 17.55 says that we are given thickening of the restriction $\widehat{\mathcal{U}}|_{\partial Z}$ of the Kuranishi structure $\widehat{\mathcal{U}}$ to the boundary, such that the thickening given on the boundary satisfies a certain compatibility condition at the corner. We use covering given in Proposition 24.17 to formulate the compatibility condition at the corner. Proposition 17.58 claims that those Kuranishi structures induce the one in a neighborhood of the boundary of the outer collaring.

Situation 17.55 Let $(X,\widehat{\mathcal{U}})$ be a τ-collared K-space and ∂X the normalized boundary of X.

For given $\tau > 0$, we are given a τ-collared Kuranishi structure $\widehat{\mathcal{U}^+_\partial}$ of ∂X such that

$$\partial\widehat{\mathcal{U}} < \widehat{\mathcal{U}^+_\partial}. \tag{17.21}$$

We assume that $\widehat{\mathcal{U}^+_\partial}$ satisfies the following conditions:

(1) For each $k \ge 1$ there exists a τ-collared Kuranishi structure $\widehat{\mathcal{U}^+_{S_k}}$ on $\widehat{S}_k(X)$ such that $\widehat{\mathcal{U}^+_{S_1}} = \widehat{\mathcal{U}^+_\partial}$.
(2) The τ-collared Kuranishi structure on $\widehat{S}_k(\widehat{S}_\ell(X);\widehat{\mathcal{U}^+_{S_\ell}})$ is isomorphic to the $(k+\ell)!/k!\ell!$ fold covering space of $(\widehat{S}_{k+\ell}(X);\widehat{\mathcal{U}^+_{S_{k+\ell}}})$.
(3) The following diagram of K-spaces commutes:

$$\begin{array}{ccc}
\widehat{S}_{k_1}(\widehat{S}_{k_2}(\widehat{S}_{k_3}(X);\widehat{\mathcal{U}^+_{S_{k_3}}})) & \xrightarrow{\pi_{k_1,k_2}} & \widehat{S}_{k_1+k_2}(\widehat{S}_{k_3}(X);\widehat{\mathcal{U}^+_{S_{k_3}}}) \\
\downarrow & & \downarrow \\
\widehat{S}_{k_1}(\widehat{S}_{k_2+k_3}(X);\widehat{\mathcal{U}^+_{S_{k_2+k_3}}}) & \longrightarrow & (\widehat{S}_{k_1+k_2+k_3}(X);\widehat{\mathcal{U}^+_{S_{k_1+k_2+k_3}}})
\end{array} \tag{17.22}$$

Here π_{k_1,k_2} is the covering map in Proposition 24.17. The right vertical and lower horizontal arrows are covering maps in (2). The left vertical arrow is induced by the covering map $\widehat{S}_{k_2}(\widehat{S}_{k_3}(X);\widehat{\mathcal{U}^+_{S_{k_3}}}) \to (\widehat{S}_{k_2+k_3}(X);\widehat{\mathcal{U}^+_{S_{k_2+k_3}}})$ in (2).

(4) There exists a τ-collared embedding $\widehat{S}_k(X;\widehat{\mathcal{U}}) \to \widehat{\mathcal{U}^+_{S_k}}$.
(5) The following diagram of K-spaces commutes:

$$\begin{array}{ccc}\widehat{S}_k(\widehat{S}_\ell(X;\widehat{\mathcal{U}})) & \longrightarrow & \widehat{S}_k(\widehat{S}_\ell(X);\widehat{\mathcal{U}^+_{S_\ell}}) \\ \downarrow & & \downarrow \\ \widehat{S}_{k+\ell}(X;\widehat{\mathcal{U}}) & \longrightarrow & (\widehat{S}_{k+\ell}(X);\widehat{\mathcal{U}^+_{S_{k+\ell}}})\end{array} \tag{17.23}$$

Here the map in the first row is induced by the embedding $\widehat{S}_\ell(X;\widehat{\mathcal{U}}) \to \widehat{\mathcal{U}^+_{S_\ell}}$. The map in the first column is given by Proposition 24.17. The map in the second row is given by (4). The map in the second column is given by (2). ■

Remark 17.56

(1) Here we used the notion of the covering space of a K-space we discuss in Chap. 24 to formulate the compatibility condition in Situation 17.55 at the corner of general codimension. In our application in Chaps. 16, 17, 18, 19, 20, 21, and 22, the stratum $\widehat{S}_k(\widehat{S}_\ell(X);\widehat{\mathcal{U}^+_{S_k}})$ is a disjoint union of $(k+\ell)!/k!\ell!$ copies of $\widehat{S}_{k+\ell}(X;\widehat{\mathcal{U}^+_{S_{k+\ell}}})$. So the notion of the covering space of a K-space is not necessary there. The result in the generality stated here will become necessary to study the case of higher genus Lagrangian Floer theory and/or symplectic field theory, for example.
(2) If $\widehat{\mathcal{U}^+_{S_{k+\ell}}}$ is a restriction of the Kuranishi structure $\widehat{\mathcal{U}^+}$ on X to $S_{k+\ell}(X)$ such that $\widehat{\mathcal{U}} < \widehat{\mathcal{U}^+}$, then Conditions (1)–(5) above follow from Proposition 24.17. Proposition 17.58 below may be regarded as a kind of converse of this statement.

Definition 17.57 In Situation 17.55, we define

$$X_0 := X \setminus X^{\boxminus\tau}. \tag{17.24}$$

Here $X^{\boxminus\tau}$ is the inward τ-collaring of $(X,\widehat{\mathcal{U}})$. We note that X_0 is an open neighborhood of $S_1(X)$ in X.

We remark that

$$\widehat{S}_k(X_0) = \widehat{S}_k(X), \qquad \text{for } k \geq 1. \tag{17.25}$$

Now the next proposition is our main result of Sect. 17.7. We complete the proof at the end of Sect. 17.7.

Proposition 17.58 *Under Situation* 17.55, *for any* $0 < \tau' < \tau$, *there exists a* τ'-*collared Kuranishi structure* $\widehat{\mathcal{U}^+}$ *on* X_0 *with the following properties:*

(1) *The restriction of* $\widehat{\mathcal{U}^+}$ *to* $\widehat{S}_k(X)$ *is isomorphic to* $\widehat{\mathcal{U}^+_{S_k}}$ *as* τ'-*collared Kuranishi structures.*
(2) *There exists an embedding of* τ'-*collared Kuranishi structures* $\widehat{\mathcal{U}}|_{X_0} \to \widehat{\mathcal{U}^+}$.
(3) *The restriction of (2) to* $\widehat{S}_k X$ *coincides with the one induced from* (17.21) *under the identification (1).*
(4) *The isomorphism between the K-spaces* $\widehat{S}_k(X;\widehat{\mathcal{U}^+})$ *and* $(\widehat{S}_k(X);\widehat{\mathcal{U}^+_{S_k}})$ *in (1) respects the following commutative diagram of K-spaces:*

$$\begin{array}{ccc}\widehat{S}_k(\widehat{S}_\ell(X;\widehat{\mathcal{U}^+})) & \xrightarrow{\cong} & \widehat{S}_k(\widehat{S}_\ell(X);\widehat{\mathcal{U}^+_{S_\ell}}) \\ \downarrow & & \downarrow \\ \widehat{S}_{k+\ell}(X;\widehat{\mathcal{U}^+}) & \xrightarrow{\cong} & (\widehat{S}_{k+\ell}(X);\widehat{\mathcal{U}^+_{S_{k+\ell}}}) \end{array} \tag{17.26}$$

Here the second horizontal arrow is the one claimed in (1), the left vertical arrow is given by Proposition 24.17 *and the right vertical arrow is given in Situation* 17.55 *(2).*
(5) *The two embeddings* $\widehat{\mathcal{U}}|_{\widehat{S}_k(X)} \to \widehat{\mathcal{U}^+}|_{\widehat{S}_k(X)}$ *and* $\widehat{\mathcal{U}}|_{\widehat{S}_k(X)} \to \widehat{\mathcal{U}^+_{S_k}}$ *coincide via the isomorphism in (1). Here the first embedding is induced by the embedding* $\widehat{\mathcal{U}} \to \widehat{\mathcal{U}^+}$ *claimed in (2) and the second embedding is as in Situation* 17.55 *(4). They are defined on* $\widehat{S}_k(X_0) = \widehat{S}_k(X)$.

17.7.2 Extension Theorem for a Single Collared Kuranishi Chart

The main part of the proof of Proposition 17.58 is to prove the corresponding result for one Kuranishi chart. For this purpose we consider the following situation. Situation 17.59 is the single chart version of Situation 17.55.

Situation 17.59 Let $\mathcal{U}$ be a τ-collared Kuranishi chart of X and $\mathcal{U}^+_\partial$ a τ-collared Kuranishi chart of ∂X. We assume that there exists an embedding

$$\partial\mathcal{U} \to \mathcal{U}^+_\partial \tag{17.27}$$

of τ-collared Kuranishi charts and the following conditions are satisfied:

(1) For each $k \geq 1$ there exists a τ-collared Kuranishi chart $\mathcal{U}^+_{S_k}$ on $\widehat{S}_k(X)$ such that $\mathcal{U}^+_{S_1} = \mathcal{U}^+_\partial$.

(2) The orbifold $\widehat{S}_k(U^+_{S_\ell})$ is isomorphic to the $(k+\ell)!/k!\ell!$ fold covering space of $U^+_{S_{k+\ell}}$. (Note that $U^+_{S_{k+\ell}}$ is the underlying orbifold of $\mathcal{U}^+_{S_{k+\ell}}$.) Restrictions of the obstruction bundle and the Kuranishi map of $\mathcal{U}^+_{S_\ell}$ to $\widehat{S}_k(U^+_{S_\ell})$ are the pullbacks of the ones of $\mathcal{U}^+_{S_{k+\ell}}$, respectively. Thus we have an induced covering between Kuranishi charts : $\widehat{S}_k(\mathcal{U}^+_{S_\ell}) \to \mathcal{U}^+_{S_{k+\ell}}$.
(3) The following diagram commutes:

$$\begin{array}{ccc} \widehat{S}_{k_1}(\widehat{S}_{k_2}(\mathcal{U}^+_{S_{k_3}})) & \xrightarrow{\pi_{k_1,k_2}} & \widehat{S}_{k_1+k_2}(\mathcal{U}^+_{S_{k_3}}) \\ \downarrow & & \downarrow \\ \widehat{S}_{k_1}(\mathcal{U}^+_{S_{k_2+k_3}}) & \longrightarrow & \mathcal{U}^+_{S_{k_1+k_2+k_3}} \end{array} \tag{17.28}$$

Here π_{k_1,k_2} is the covering map in Proposition 24.17. The right vertical and lower horizontal arrows are the covering maps given in (2). The left vertical arrow is induced by the covering map $\widehat{S}_{k_2}(\mathcal{U}^+_{S_{k_3}}) \to \mathcal{U}^+_{S_{k_2+k_3}}$ of (2).
(4) There exists a τ-collared embedding $\widehat{S}_k(\mathcal{U}) \to \mathcal{U}^+_{S_k}$ of τ-collared Kuranishi charts.
(5) The following diagram commutes:

$$\begin{array}{ccc} \widehat{S}_k(\widehat{S}_\ell(\mathcal{U})) & \longrightarrow & \widehat{S}_k(\mathcal{U}^+_{S_\ell}) \\ \downarrow & & \downarrow \\ \widehat{S}_{k+\ell}(\mathcal{U}) & \longrightarrow & \mathcal{U}^+_{S_{k+\ell}} \end{array} \tag{17.29}$$

The maps are as in the case of Diagram (17.23). ■

The following is the extension theorem for a single collared Kuranishi chart and is the single chart version of Proposition 17.58.

Lemma 17.60 *Under Situation* 17.59, *we put*

$$U_0 := U \setminus U^{\boxminus\tau}.$$[8]

Then for any $0 < \tau' < \tau$ *there exists a* τ'*-collared Kuranishi chart* $\mathcal{U}^+$ *of* $X_0 = X \setminus X^{\boxminus\tau}$ *with the following properties:*

(1) *The restriction of* $\mathcal{U}^+$ *to* $\widehat{S}_k(U)$ *is isomorphic to* $\mathcal{U}^+_{S_k}$ *as* τ'*-collared Kuranishi charts.*
(2) *There exists an embedding of* τ'*-collared Kuranishi charts* $\mathcal{U}|_{U_0} \to \mathcal{U}^+$.

[8] We remark that $\widehat{S}_k(U) = \widehat{S}_k(U_0)$ for $k \geq 1$.

(3) *The restriction of (2) to* $\widehat{S}_k(U)$ *coincides with the one induced from Situation* 17.59 *(4) under the identification (1).*

(4) *The following diagram commutes:*

$$\begin{array}{ccc} \widehat{S}_k(\widehat{S}_\ell(\mathcal{U}^+)) & \xrightarrow{\cong} & \widehat{S}_k(\mathcal{U}^+_{S_\ell}) \\ \downarrow & & \downarrow \\ \widehat{S}_{k+\ell}(\mathcal{U}^+) & \xrightarrow{\cong} & \mathcal{U}^+_{S_{k+\ell}} \end{array} \tag{17.30}$$

Here the first horizontal arrow is induced from (1). The second horizontal arrow is (1). The left vertical arrow is given by Proposition 24.17. *The right vertical arrow is induced by a map given in Situation* 17.59 *(2).*

(5) *The embeddings* $\widehat{S}_k(\mathcal{U}) \to \widehat{S}_k(\mathcal{U}^+)$ *and* $\widehat{S}_k(\mathcal{U}) \to \mathcal{U}^+_{S_k}$ *coincide via the isomorphism in Situation* 17.59 *(4). Here the first embedding is induced by the embedding* $\mathcal{U}|_{U_0} \to \mathcal{U}^+$ *claimed in (2) and the second embedding is as in Situation* 17.59 *(4).*[9]

The proof of Lemma 17.60 occupies Sects. 17.7.3 and 17.7.4.

17.7.3 Construction of Kuranishi Chart $\mathcal{U}^+$

Let $p' \in S_1(U)$. Firstly, we will construct a Kuranishi chart $\mathcal{U}^+_{p'}$, mentioned in Lemma 17.60. We use Kuranishi neighborhoods of various $\tilde{p}' \in \widehat{S}_k(U)$ with $\pi(\tilde{p}') = p'$. The Kuranishi neighborhoods we use are those of the Kuranishi chart $\mathcal{U}^+_{S_k}$ given in Situation 17.59 (1). We will modify and glue them in a canonical way to obtain $\mathcal{U}^+_{p'}$. The detail is in order.

We first begin with describing the situation of the Kuranishi chart $\mathcal{U}$ in Situation 17.59 in more detail. Suppose $p \in \overset{\circ}{S}_k(U)$ for some k. Let $\mathfrak{V}_{\mathfrak{r}} = (V_{\mathfrak{r}}, \Gamma_{\mathfrak{r}}, \phi_{\mathfrak{r}})$ be an orbifold chart of U, which is the underlying orbifold of our Kuranishi chart $\mathcal{U}$. (Here $\mathfrak{r}$ stands for the index of orbifold charts.) Let $p \in U_{\mathfrak{r}} = V_{\mathfrak{r}}/\Gamma_{\mathfrak{r}}$ such that the base point $o_{\mathfrak{r}}$ of the chart goes to p (Definition 23.1), that is, $p = \phi_{\mathfrak{r}}(o_{\mathfrak{r}}) \in \overset{\circ}{S}_k(U)$. Then we take an admissible coordinate $V_{\mathfrak{r}} \subset [V_{\mathfrak{r}}] \times [0,1)^k$ and $o_{\mathfrak{r}} = (\overline{o}_{\mathfrak{r}}, (0,\dots,0))$. We have a representation $\sigma : \Gamma_{\mathfrak{r}} \to \mathrm{Perm}(k)$ by the definition of an admissible orbifold (see Definition 25.8 (1)(b)), where $\mathrm{Perm}(k)$ is the group of permutations of $\{1,\dots,k\}$.

Let $A \subset \{1,\dots,k\}$. We put

[9] They are defined on $\widehat{S}_k(U) = \widehat{S}_k(U_0)$.

$$\begin{aligned}\overset{\circ}{S}_A([0,1)^k) &= \{(t_1,\dots,t_k) \in [0,1)^k \mid i \in A \Rightarrow t_i = 0,\ i \notin A \Rightarrow t_i > 0\},\\ \overset{\circ}{S}_A(V_\mathfrak{r}) &= V_\mathfrak{r} \cap ([V_\mathfrak{r}] \times \overset{\circ}{S}_A([0,1)^k)),\\ \Gamma_\mathfrak{r}^A &= \{\gamma \in \Gamma_\mathfrak{r} \mid \gamma A = A\}.\end{aligned} \tag{17.31}$$

The subset A determines a point $p(A)$ of $\widehat{S}_\ell(U)$ which goes to p by the map $\widehat{S}_\ell(U) \to U$. Here $\ell = \#A$. An orbifold chart of $\widehat{S}_\ell(U)$ at $p(A)$ is given by $S_A(V_\mathfrak{r})$, which is the closure of $\overset{\circ}{S}_A(V_\mathfrak{r})$, the isotropy group $\Gamma_\mathfrak{r}^A$ and ψ_A, which is a lift of the restriction of the parametrization map ψ to $S_A(V_\mathfrak{r}) \cap s_\mathfrak{r}^{-1}(0)$. We put

$$V_\mathfrak{r}(p; A) = (S_A(V_\mathfrak{r}))^{\boxplus\tau} \times [-\tau, 0)^A. \tag{17.32}$$

Let us elaborate on (17.32). We first note that

$$S_A(V_\mathfrak{r}) \subset [V_\mathfrak{r}] \times [0,1)^{A^c},$$

where $A^c = \{1,\dots,k\} \setminus A$. (In fact, if $i \in A$ then $t_i = 0$ for an element of $S_A(V_\mathfrak{r})$.) We have a retraction map

$$\mathcal{R}_{A^c} : [V_\mathfrak{r}] \times [-\tau, 1)^{A^c} \to [V_\mathfrak{r}] \times [0,1)^{A^c},$$

which changes $t_i < 0$ to $t_i = 0$. Then we find

$$(S_A(V_\mathfrak{r}))^{\boxplus\tau} = (\mathcal{R}_{A^c})^{-1}(S_A(V_\mathfrak{r})) \subset [V_\mathfrak{r}] \times [-\tau, 1)^{A^c}.$$

Let $\Pi_{A^c} : [V_\mathfrak{r}] \times [-\tau, 1)^k \to [V_\mathfrak{r}] \times [-\tau, 1)^{A^c}$ be the projection. Then we can write

$$\begin{aligned}V_\mathfrak{r}(p; A) = \{y = (\overline{y}, (t_1,\dots,t_k)) &\in [V_\mathfrak{r}] \times [-\tau, 1)^k \mid\\ &\Pi_{A^c}(y) \in (S_A(V_\mathfrak{r}))^{\boxplus\tau}, i \in A \Rightarrow t_i \in [-\tau, 0)\}.\end{aligned} \tag{17.33}$$

Example 17.61 Figures 17.6, 17.7, 17.8, and 17.9 depict examples of $V_\mathfrak{r}$, $S_A(V_\mathfrak{r})$, $S_A(V_r)^{\boxplus\tau}$ and $V_\mathfrak{r}(p, A)$.

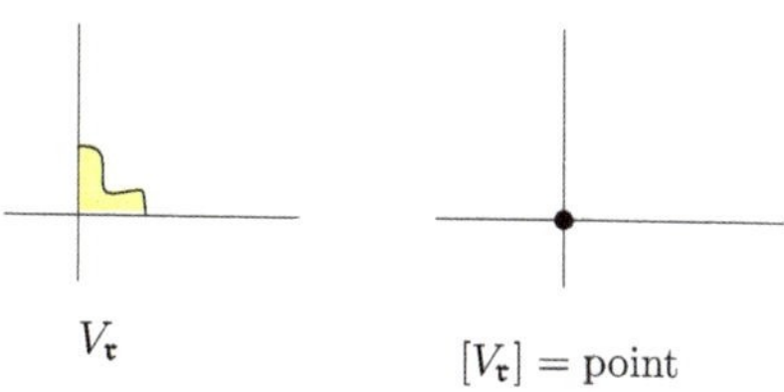

Fig. 17.6 $V_\mathfrak{r}$ and $[V_\mathfrak{r}]$

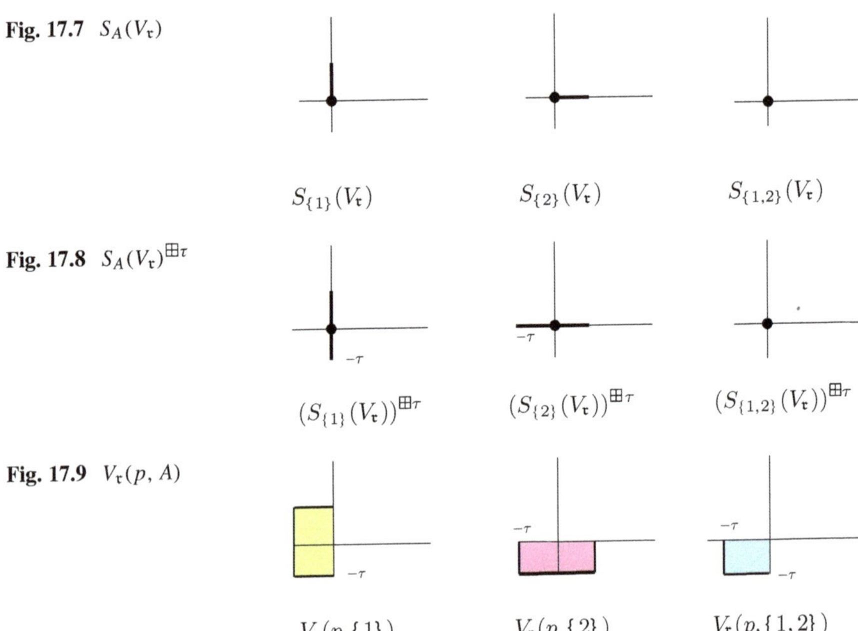

Fig. 17.7 $S_A(V_\mathfrak{r})$

Fig. 17.8 $S_A(V_\mathfrak{r})^{\boxplus\tau}$

Fig. 17.9 $V_\mathfrak{r}(p, A)$

Remark 17.62 We observe that the space $V_\mathfrak{r}(p, A)$, when it is written as (17.32), is defined by the data of $S_A(V_\mathfrak{r})$ only. In particular, it is independent of $\overset{\circ}{S}_0(V_\mathfrak{r})$. Note, for our extension $V_\mathfrak{r}^+$, we are given only data on $S_1(V_\mathfrak{r}^+)$. We will use this remark to construct $V_\mathfrak{r}^+(p, A)$ in this situation.

We next suppose that $B \supset A$ with $\#B = \ell+m$. Then the triple (p, A, B) determines a point $p(A, B) \in \widehat{S}_m(\widehat{S}_\ell(U))$. We consider the maps $\pi_m : \widehat{S}_m(\widehat{S}_\ell(U)) \to S_m(\widehat{S}_\ell(U))$ and $\pi_{m,\ell} : \widehat{S}_m(\widehat{S}_\ell(U)) \to \widehat{S}_{m+\ell}(U)$ defined in Proposition 24.17. We have $\pi_m(p(A, B)) = p(A)$ and $\pi_{m,\ell}(p(A, B)) = p(B)$.

We put $\Gamma_\mathfrak{r}^{A,B} = \Gamma_\mathfrak{r}^A \cap \Gamma_\mathfrak{r}^B$. The map $S_B(V_\mathfrak{r})/\Gamma_\mathfrak{r}^{A,B} \to S_A(V_\mathfrak{r})/\Gamma_\mathfrak{r}^A$ is a restriction of π_m and the map $S_B(V_\mathfrak{r})/\Gamma_\mathfrak{r}^{A,B} \to S_B(V_\mathfrak{r})/\Gamma_\mathfrak{r}^B$ is a restriction of $\pi_{m,\ell}$. We note

$$V_\mathfrak{r}(p; B) = \{(\overline{y}, (t_1, \dots, t_k)) \in V_\mathfrak{r}(p; A) \mid i \in B \Rightarrow t_i \in [-\tau, 0)\}.$$

Therefore we get $V_\mathfrak{r}(p; B) \subset V_\mathfrak{r}(p; A)$. (See Fig. 17.9.)

Using the expression (17.32), we can rewrite the embedding $V_\mathfrak{r}(p; B) \subset V_\mathfrak{r}(p; A)$ as follows. We remark that

$$S_{B\setminus A}(S_A(V_\mathfrak{r})) \times [0, \epsilon)^{B\setminus A} \subset S_A(V_\mathfrak{r}). \tag{17.34}$$

Here $S_{B\setminus A}(S_A(V_\mathfrak{r}))$ is a subset of $\widehat{S}_m(S_A(V_\mathfrak{r}))$. Note that $\pi_0(\widehat{S}_m(S_A(V_\mathfrak{r})))$ corresponds one-to-one to the set of B satisfying $\{1, \dots, k\} \supset B \supset A$ with

$\#B = \ell + m$. Then $S_{B\setminus A}(S_A(V_\mathfrak{r}))$ is the connected component corresponding to B under this one-to-one correspondence.

Equation (17.34) implies

$$(S_{B\setminus A}(S_A(V_\mathfrak{r})))^{\boxplus\tau} \times [-\tau, 0)^{B\setminus A} \subset (S_A(V_\mathfrak{r}))^{\boxplus\tau}. \tag{17.35}$$

The map $\pi_{m,\ell}$ induces a map

$$\pi_{B\setminus A,A} : S_{B\setminus A}(S_A(V_\mathfrak{r})) \to S_B(V_\mathfrak{r}), \tag{17.36}$$

which is an isomorphism. (The $(m+\ell)!/m!\ell!$ different choices of the points in $\pi_{m,\ell}^{-1}$(one point) correspond to $(m+\ell)!/m!\ell!$ different choices of A in the given B.) Therefore composing the inverse of (17.36) and the inclusion (17.35), we obtain

$$\begin{aligned}
&(S_B(V_\mathfrak{r}))^{\boxplus\tau} \times [-\tau, 0)^B \\
&\overset{(\pi^{\boxplus\tau}_{B\setminus A,A})^{-1}\times \mathrm{id}}{\longrightarrow} (S_{B\setminus A}(S_A V_\mathfrak{r}))^{\boxplus\tau} \times [-\tau, 0)^{B\setminus A} \times [-\tau, 0)^A \\
&\longrightarrow (S_A(V_\mathfrak{r}))^{\boxplus\tau} \times [-\tau, 0)^A.
\end{aligned} \tag{17.37}$$

It is easy to see that (17.37) coincides with the inclusion $V_\mathfrak{r}(p; B) \subset V_\mathfrak{r}(p; A)$.

Example 17.63 Figure 17.10 depicts the embedding $V_\mathfrak{r}(p; B) \subset V_\mathfrak{r}(p; A)$ in the situation of Example 17.61 and $A = \{1\}$, $B = \{1, 2\}$.

For $A \subset B \subset C$ we have $V_\mathfrak{r}(p; C) \subset V_\mathfrak{r}(p; B) \subset V_\mathfrak{r}(p; A)$. The composition of the two embeddings $V_\mathfrak{r}(p; C) \subset V_\mathfrak{r}(p; B)$ and $V_\mathfrak{r}(p; B) \subset V_\mathfrak{r}(p; A)$ coincides with $V_\mathfrak{r}(p; C) \subset V_\mathfrak{r}(p; A)$. This is equivalent to the commutativity of Diagram (17.22) with $\mathcal{U}^+$ replaced by $\mathcal{U}$. (See Sublemma 17.64.) We put

$$V_\mathfrak{r}(p) = \bigcup_{A\subseteq\{1,\dots,k\}} V_\mathfrak{r}(p; A).$$

This is a $\Gamma_\mathfrak{r}$-equivariant open subset of $\mathcal{R}^{-1}(U) \setminus U$. We may take $V_\mathfrak{r}(p)/\Gamma_\mathfrak{r}$ (together with other data) as a Kuranishi neighborhood of the point in $\mathcal{R}^{-1}(\{p\})\setminus U$.

To construct $V_\mathfrak{r}^+(p)$ we imitate the above description using only the data given on the boundary as follows.

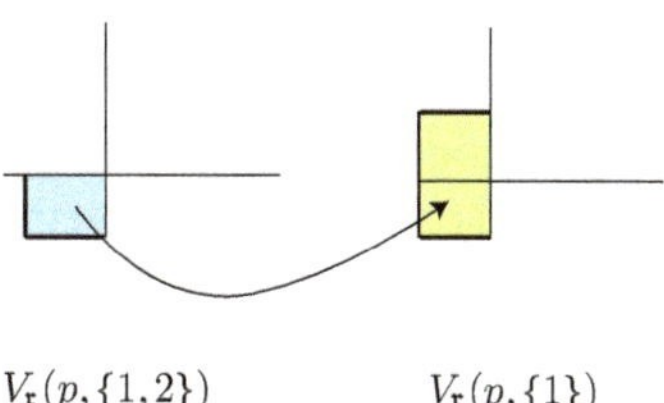

Fig. 17.10
$V_\mathfrak{r}(p; B) \subset V_\mathfrak{r}(p; A)$

The Kuranishi chart $\mathcal{U}^+_{S_\ell}$ given in Situation 17.59 induces $V^+_{\mathfrak{r},S_A}/\Gamma^A_{\mathfrak{r}}$. (It is an open subset of $U^+_{S_\ell}$. Also recall $\#A = \ell$.) We define

$$V^+_{\mathfrak{r}}(p;A) = (V^+_{\mathfrak{r},S_A})^{\boxplus\tau} \times [-\tau,0)^A. \tag{17.38}$$

For $B \supset A$ we have an embedding denoted by $h_{A,B}$:

$$h_{A,B} : (S_{B\setminus A}(V^+_{\mathfrak{r},S_A}))^{\boxplus\tau} \times [-\tau,0)^{B\setminus A} \hookrightarrow (V^+_{\mathfrak{r},S_A})^{\boxplus\tau}. \tag{17.39}$$

The covering map $\widehat{S}_m(\mathcal{U}^+_{S_\ell}) \to \mathcal{U}^+_{S_{\ell+m}}$ given in Situation 17.59 (2) induces the map

$$\pi'_{B\setminus A,A} : S_{B\setminus A}(V^+_{\mathfrak{r},S_A}) \to V^+_{\mathfrak{r},S_B}, \tag{17.40}$$

which is an isomorphism. We define

$$\phi_{AB} : V^+_{\mathfrak{r}}(p;B) \to V^+_{\mathfrak{r}}(p;A)$$

by

$$\begin{aligned} &(V^+_{\mathfrak{r},S_B})^{\boxplus\tau} \times [-\tau,0)^B \\ &\xrightarrow{(\pi'^{\boxplus\tau}_{B\setminus A,A})^{-1}\times \mathrm{id}} (S_{B\setminus A}(V^+_{\mathfrak{r},S_A}))^{\boxplus\tau} \times [-\tau,0)^{B\setminus A} \times [-\tau,0)^A \\ &\xrightarrow{h_{A,B}\times \mathrm{id}} (V^+_{\mathfrak{r},S_A})^{\boxplus\tau} \times [-\tau,0)^A. \end{aligned} \tag{17.41}$$

Sublemma 17.64 *If $C \supset B \supset A$, then $\phi_{AB} \circ \phi_{BC} = \phi_{AC}$.*

Proof The sublemma follows from the commutativity of Diagram (17.28) as follows. Recall that $h_{A,B}$ is the inclusion map in (17.39). Then the following diagram commutes:

$$\begin{array}{ccc} (V^+_{\mathfrak{r},S_B})^{\boxplus\tau} & \xleftarrow{\pi'^{\boxplus\tau}_{B\setminus A,A}} & (S_{B\setminus A}(V^+_{\mathfrak{r},S_A}))^{\boxplus\tau} \\ \uparrow h_{B,C} & & \uparrow h_{B,C} \\ (S_{C\setminus B}(V^+_{\mathfrak{r},S_B}))^{\boxplus\tau} \times [-\tau,0)^{C\setminus B} & \xleftarrow{(S_{C\setminus B}(\pi'_{B\setminus A,A}))^{\boxplus\tau}\times \mathrm{id}} & (S_{C\setminus B}(S_{B\setminus A}(V^+_{\mathfrak{r},S_A})))^{\boxplus\tau} \times [-\tau,0)^{C\setminus B} \end{array} \tag{17.42}$$

The commutativity of Diagram (17.42) is a consequence of the commutativity of Diagram (17.28) and the fact that $\pi'_{B\setminus A,A}$ is a diffeomorphism of cornered manifolds and the definition of $h_{*,*}$. We also have the following commutative diagram.

$$
\begin{array}{ccc}
(V^+_{\mathfrak{r},S_A})^{\boxplus\tau} \times [-\tau,0)^A & \xleftarrow{h_{A,B}\times \mathrm{id}_A} & (S_{B\setminus A}(V^+_{\mathfrak{r},S_A}))^{\boxplus\tau} \times [-\tau,0)^B \\
\uparrow h_{A,C}\times \mathrm{id}_A & & \uparrow h_{B,C}\times \mathrm{id}_B \\
(S_{C\setminus A}(V^+_{\mathfrak{r},S_A}))^{\boxplus\tau} \times [-\tau,0)^C & \xleftarrow{(\pi_{C\setminus B,B\setminus A})^{\boxplus\tau}\times \mathrm{id}_C} & (S_{C\setminus B}(S_{B\setminus A}(V^+_{\mathfrak{r},S_A})))^{\boxplus\tau} \times [-\tau,0)^C
\end{array}
\tag{17.43}
$$

Note that $\pi_{C\setminus B,B\setminus A}$ in Diagram (17.43) is the map in Proposition 24.17. (The map $\pi'_{B\setminus A,A}$ in Diagram 17.42 is the one in Situation 17.59 (2).) The commutativity of Diagram 17.43 is an immediate consequence of the definition of $\pi_{C\setminus B,B\setminus A}$ and $h_{*,*}$. Therefore we have

$$
\begin{aligned}
&\phi_{AB} \circ \phi_{BC} \\
&= (h_{A,B} \times \mathrm{id}) \circ ((\pi'^{\boxplus\tau}_{B\setminus A,A})^{-1} \times \mathrm{id}) \circ (h_{B,C} \times \mathrm{id}) \circ ((\pi'^{\boxplus\tau}_{C\setminus B,B})^{-1} \times \mathrm{id}) \\
&= (h_{A,B} \times \mathrm{id}) \circ (h_{B,C} \times \mathrm{id}) \circ ((S_{C\setminus B}(\pi'_{B\setminus A,A}))^{\boxplus\tau})^{-1} \times \mathrm{id}) \circ ((\pi'^{\boxplus\tau}_{C\setminus B,B})^{-1} \times \mathrm{id}) \\
&= (h_{A,C} \times \mathrm{id}) \circ ((\pi_{C\setminus B,B\setminus A})^{\boxplus\tau} \times \mathrm{id}) \\
&\qquad\qquad \circ ((S_{C\setminus B}(\pi'_{B\setminus A,A}))^{\boxplus\tau})^{-1} \times \mathrm{id}) \circ ((\pi'^{\boxplus\tau}_{C\setminus B,B})^{-1} \times \mathrm{id}) \\
&= (h_{A,C} \times \mathrm{id}) \circ ((\pi'^{\boxplus\tau}_{C\setminus A,A})^{-1} \times \mathrm{id}) \\
&= \phi_{AC}.
\end{aligned}
$$

Here the first equality is the definition. The second equality is the commutativity of Diagram (17.42). The third equality is the commutativity of Diagram (17.43). The fourth equality is the commutativity of Diagram (17.28). □

Remark 17.65 Here $\pi_{*,*}$ is the covering map induced by the corner structure and $\pi'_{*,*}$ (with prime) is the map given by assumption. The key ingredient of the proof is the compatibility of $\pi_{*,*}$ with $\pi'_{*,*}$, that is, the commutativity of Diagram (17.28).

We consider the disjoint union

$$\coprod_A V^+_{\mathfrak{r}}(p; A)$$

and define $\sim$ on it as follows. For $x \in V^+_{\mathfrak{r}}(p; A)$ and $y \in V^+_{\mathfrak{r}}(p; B)$ we say $x \sim y$ if and only if there exist C and $z \in V^+_{\mathfrak{r}}(p; C)$ such that $x = \phi_{AC}(z)$ and $y = \phi_{BC}(z)$.

Sublemma 17.66 *$\sim$ is an equivalence relation.*

Proof It suffices to prove the transitivity. Let $x = (x', (t_i)_{i\in A})$ where $x' \in (V^+_{\mathfrak{r},S_A})^{\boxplus\tau}$ and $t_i \in [-\tau, 0)$ for $i \in A$. Furthermore we write $x' = (\overline{x}, (t_i)_{i\in A^c})$. We observe that x' is in the image of $(S_{C\setminus A}(V^+_{\mathfrak{r},S_A}))^{\boxplus\tau} \times [-\tau, 0)^{C\setminus A}$ if and only if $t_i < 0$ for all $i \in C$. Therefore for each x there exists *unique* C such that:

Fig. 17.11 $V_{\mathfrak{r}}^{+}(p)$

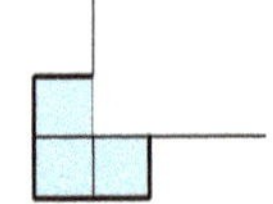

(1) $x \in \mathrm{Im}(\phi_{AC})$.
(2) If $x \in \mathrm{Im}(\phi_{AD})$ then $D \subseteq C$.

Transitivity follows from this fact, Sublemma 17.64 and the fact that ϕ_{AB} is injective. □

We define

$$V_{\mathfrak{r}}^{+}(p) = \left(\coprod_{A} V_{\mathfrak{r}}^{+}(p; A)\right) /\!\sim .$$

Example 17.67 Figure 17.11 depicts the space $V_{\mathfrak{r}}^{+}(p)$ in the situation of Example 17.61.

Sublemma 17.68 *The quotient space $V_{\mathfrak{r}}^{+}(p)$ is Hausdorff with respect to the quotient topology.*

Proof Let $x, y \in V_{\mathfrak{r}}^{+}(p)$ such that $x \neq y$. We take C_x and C_y as in (1)(2) above and take the representatives $\tilde{x} \in V_{\mathfrak{r}}^{+}(p; C_x)$ and $\tilde{y} \in V_{\mathfrak{r}}^{+}(p; C_y)$, respectively. If $C_x = C_y = C$, we can find an open set U_x, U_y in $V_{\mathfrak{r}}^{+}(p; C)$ such that $\tilde{x} \in U_x$, $\tilde{y} \in U_y$ and $U_x \cap U_y = \emptyset$. The images of U_x and U_y in $V_{\mathfrak{r}}^{+}(p)$ separate x, y.

Suppose $C_x \neq C_y$. We may assume that there exists $j \in C_x \setminus C_y$. We write $\tilde{x} = (x', (t_i^0)_{i \in C_x})$, $x' = (\overline{x}, (t_i^0)_{i \in C_x^c})$. We also write $\tilde{y} = (y', (s_i^0)_{i \in C_y})$, $y' = (\overline{y}, (s_i^0)_{i \in C_y^c})$. Then $t_j^0 < 0$ and $s_j^0 \geq 0$. Let U_x be the set of all points in $V_{\mathfrak{r}}^{+}(p; C_x)$ such that $t_j < t_j^0/2$ and U_y be the set of all points in $V_{\mathfrak{r}}^{+}(p; C_y)$ such that $s_j > t_j^0/2$. They induce disjoint open sets in $V_{\mathfrak{r}}^{+}(p)$ containing x and y respectively. □

Since ϕ_{AB}'s are open embeddings of manifolds, Sublemma 17.68 implies that $V_{\mathfrak{r}}^{+}(p)$ is a smooth manifold.

We next define an obstruction bundle and a Kuranishi map on it. Equation (17.38) shows that each $V_{\mathfrak{r}}^{+}(p; A)$ comes with an obstruction bundle and a Kuranishi map on it. We denote them by $\mathcal{V}_{\mathfrak{r},p;A}^{+}$. Moreover (17.41) implies that each ϕ_{AB} is covered by the bundle isomorphism with which Kuranishi map is compatible. Moreover the identity $\phi_{AB} \circ \phi_{BC} = \phi_{AC}$ is promoted to the identity among bundle maps. Therefore we obtain an obstruction bundle and a Kuranishi map on $V_{\mathfrak{r}}^{+}(p)$. We denote them by $\mathcal{E}_{\mathfrak{r},p}$ and $s_{\mathfrak{r},p}^{+}$. We can also define the parametrization map $\psi_{\mathfrak{r},p}^{+} : (s_{\mathfrak{r},p}^{+})^{-1}(0) \to X_0$ in an obvious way. The following is immediate from the construction.

Sublemma 17.69 *For each $\gamma \in \Gamma_p$ and A there exists $\varphi_{\gamma,A} : \mathcal{V}_{p;A}^{+} \to \mathcal{V}_{p;\gamma A}^{+}$. Moreover $\varphi_{\gamma,A} \circ \phi_{AB} = \phi_{(\gamma A)(\gamma B)} \circ \varphi_{\gamma,B}$.*

Then Sublemma 17.69 implies that

$$(V_{\mathfrak{r}}^{+}(p), \Gamma_{\mathfrak{r},p}, \mathcal{E}_{\mathfrak{r},p}, s_{\mathfrak{r},p}^{+}, \psi_{\mathfrak{r},p}^{+})$$

is a Kuranishi chart at each point of $\mathcal{R}^{-1}(p)$.

Next we define coordinate change. Let $p \in \overset{\circ}{S}_k(X)$. We will use the same notations as those used in the construction of the Kuranishi chart $(V_{\mathfrak{r}}^{+}(p), \Gamma_{\mathfrak{r},p}, \mathcal{E}_{\mathfrak{r},p}, s_{\mathfrak{r},p}^{+}, \psi_{\mathfrak{r},p}^{+})$.

We take $k' \le k$ such that $q \in \overset{\circ}{S}_{k'}(X)$ and we may assume $q \in \psi_{A_q}(s_{A_q}^{-1}(0))$ with $\#A_q = k'$. We have $\tilde{q} \in S_{A_q}(V_{\mathfrak{r}})$ which parameterizes q. Moreover, for each $A \subseteq A_q$ there exists $\tilde{q}_A \in S_A(V_{\mathfrak{r}})$, which parametrizes q.

For $A \subset A_q$ with $\#A = \ell$, the Kuranishi chart $\mathcal{U}_{S_\ell}^{+}$ gives an orbifold chart $\mathfrak{V}_{\mathfrak{o},A}^{+} = (V_{\mathfrak{o},A}^{+}, \Gamma_{\mathfrak{o}}^{A}, \psi_{\mathfrak{o},A}^{+})$ at $\tilde{q}_A$. The coordinate change of the underlying orbifold $U_{S_\ell}^{+}$ of $\mathcal{U}_{S_\ell}^{+}$ induces a group homomorphism $h_{\mathfrak{r}\mathfrak{o}}^{A} : \Gamma_{\mathfrak{o}}^{A} \to \Gamma_{\mathfrak{r}}^{A}$ and an $h_{\mathfrak{r}\mathfrak{o}}^{A}$ equivariant embedding

$$\varphi_{\mathfrak{r}\mathfrak{o}}^{A+} : V_{\mathfrak{o},A}^{+} \to V_{\mathfrak{r},A}^{+}. \tag{17.44}$$

Moreover the admissibility of our orbifolds implies that we have

$$\begin{aligned} V_{\mathfrak{o},A}^{+} &\subset [V_{\mathfrak{o},A}^{+}] \times [0,1)^{A_q \setminus A}, \\ V_{\mathfrak{r},A}^{+} &\subset [V_{\mathfrak{r},A}^{+}] \times [0,1)^{A_p \setminus A} \end{aligned} \tag{17.45}$$

and

$$\varphi_{\mathfrak{r}\mathfrak{o}}^{A+}(\overline{y}, (t_i)_{i \in A_q \setminus A}) = \left(\varphi_{\mathfrak{r}\mathfrak{o},0}^{A+}(\overline{y}, (t_i)), \quad (\varphi_{\mathfrak{r}\mathfrak{o},j}^{A+}(\overline{y}, (t_i))_{j \in A_p \setminus A}\right)$$

such that:

(1) $\varphi_{\mathfrak{r}\mathfrak{o},0}^{A+}$ is admissible.
(2) For $j \in A_q \setminus A$, $\varphi_{\mathfrak{r}\mathfrak{o},j}^{A+} - t_j$ is exponentially small near the boundary.
(3) For $j \in A_p \setminus A_q$, $\varphi_{\mathfrak{r}\mathfrak{o},j}^{A+}$ is admissible.

Below we will extend $\varphi_{\mathfrak{r}\mathfrak{o}}^{A+}$ to

$$\varphi_{\mathfrak{r}\mathfrak{o}}^{A+\boxplus\tau} : V^{+}(q, A) \to V^{+}(p, A).$$

Note that

$$\begin{aligned} V_{\mathfrak{o}}^{+}(q, A) &= (V_{\mathfrak{o},A}^{+})^{\boxplus\tau} \times [-\tau, 0)^{A}, \\ V_{\mathfrak{r}}^{+}(p, A) &= (V_{\mathfrak{r},A}^{+})^{\boxplus\tau} \times [-\tau, 0)^{A}. \end{aligned}$$

Let $y = (y', (t_i)_{i \in A}) \in V_{\mathfrak{o}}^{+}(q, A)$. We define

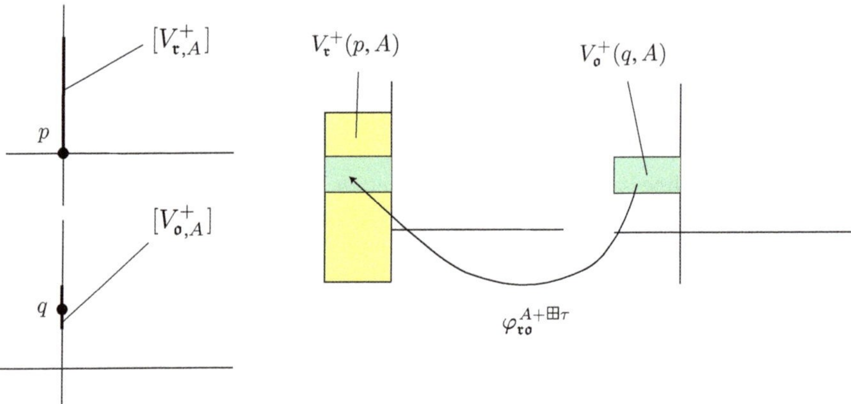

Fig. 17.12 $\varphi^{A+\boxplus\tau}_{\mathfrak{r}\mathfrak{o}}$

$$\varphi^{A+\boxplus\tau}_{\mathfrak{r}\mathfrak{o}}(y) = ((\varphi^{A+}_{\mathfrak{r}\mathfrak{o}})^{\boxplus\tau}(y'), (t_i)_{i\in A}) \in V^+_{\mathfrak{r}}(p, A). \tag{17.46}$$

Figure 17.12 depicts this map in the case $A = \{1\}$.

Sublemma 17.70 *If $A \subseteq B \subseteq A_q$ then*

$$\varphi^{A+\boxplus\tau}_{\mathfrak{r}\mathfrak{o}} \circ \phi_{AB} = \phi_{AB} \circ \varphi^{B+\boxplus\tau}_{\mathfrak{r}\mathfrak{o}}.$$

Proof This is a consequence of the fact that two maps appearing in (17.41) are compatible with the coordinate change. □

By Sublemma 17.70 we can glue $\varphi^{A+\boxplus\tau}_{\mathfrak{r}\mathfrak{o}}$ for various A to obtain a map

$$\varphi^{+\boxplus\tau}_{\mathfrak{r}\mathfrak{o}} : V^+_{\mathfrak{o}}(q) \to V^+_{\mathfrak{r}}(p)$$

and a bundle map

$$\widehat{\varphi}^{+\boxplus\tau}_{\mathfrak{r}\mathfrak{o}} : \mathcal{E}^+_{\mathfrak{o},q} \to \mathcal{E}^+_{\mathfrak{r},p}.$$

They are $h_{pq} : \Gamma_q \to \Gamma_p$ equivariant by construction. Moreover the Kuranishi maps and parameterizations ψ^+_p, ψ^+_q are compatible with it. We have thus constructed the coordinate change.

The cocycle condition among the coordinate changes follows from the cocycle condition among the coordinate changes of various $\mathcal{U}^+_{S_\ell}$.[10]

[10]Since we are constructing the space U together with its orbifold structure, we need to check the cocycle condition. It is easy to check, however.

17.7.4 Completion of the Proof of Lemma 17.60

To complete the proof of Lemma 17.60 it suffices to check that the Kuranishi chart $\mathcal{U}^+$ we obtained has the required properties.

Proof of the τ'-collaredness of $\widehat{\mathcal{U}^+}$ Each $V_\mathfrak{r}^+(p,A) = (V_{\mathfrak{r},A}^+)^{\boxplus\tau} \times [-\tau,0)^A$ is τ'-collared. The open embedding ϕ_{AB} which we used to glue them are τ'-collared. The coordinate change is obtained by gluing $\varphi_{\mathfrak{r}\mathfrak{o}}^{A+\boxplus\tau}$, which is τ'-collared. Therefore, we can construct a τ'-collared Kuranishi structure in the same way as in the proof of Lemma 17.46. □

Proof of Lemma* 17.60 *(1) Let $p' \in S_\ell(U^{\boxplus\tau} \setminus U)$ and $\mathcal{R}(p') = p \in S_\ell(U)$. We take $k \ge \ell$ such that $p \in \overset{\circ}{S}_k(U)$. We use the same notations as in the construction of $V^+(p)$. Take p'_A to be a point in the underlying topological space of $\widehat{S}_\ell(U^{\boxplus\tau} \setminus U)$ which goes to p'. We have a corresponding point $p_A \in \widehat{S}_\ell(U)$ which goes to p. The point p_A is associated to a certain subset $A \subset \{1,\dots,k\}$ with $\#A = \ell$. An orbifold neighborhood of p_A in $U_{S_\ell}^+$ is $V_{\mathfrak{r},S_A}^+/\Gamma_\mathfrak{r}^A$ by definition.

Note that $V_\mathfrak{r}^+(p;A) = V_{\mathfrak{r},S_A}^+ \times [-\tau,0)^A$ and a neighborhood of p'_A in $S_A(V_\mathfrak{r}^+(p;A))$ is $V_{\mathfrak{r},S_A}^+ \times \{(0,\dots,0)\}$ in $V_\mathfrak{r}^+(p;A)$.

Thus we have shown that $\widehat{S}_\ell(\mathcal{U}^+)$ and $\mathcal{U}_{S_\ell}^+$ are locally diffeomorphic to each other as orbifolds. This diffeomorphism is compatible with the gluing by ϕ_{AB} and by coordinate changes. So the underlying orbifolds of $\widehat{S}_\ell(\mathcal{U}^+)$ and $\mathcal{U}_{S_\ell}^+$ are diffeomorphic. Moreover it is covered by the bundle isomorphism of obstruction bundles which is compatible with coordinate change, the Kuranishi map and the parametrization map. □

Proof of Lemma* 17.60 *(2) By assumption there exists an embedding $\mathcal{U}|_{S_k(U)} \to \mathcal{U}_{S_k}^+$. (Situation 17.59 (4).) By comparing (17.32) and (17.38) it induces an embedding $V_\mathfrak{r}(p;A) \to V_\mathfrak{r}^+(p;A)$.

By the commutativity of (17.29) we have the following commutative diagram:

$$\begin{array}{ccc} V_\mathfrak{r}(p;B) & \longrightarrow & V_\mathfrak{r}^+(p;B) \\ \downarrow{\scriptstyle \phi_{AB}} & & \downarrow{\scriptstyle \phi_{AB}} \\ V_\mathfrak{r}(p;A) & \longrightarrow & V_\mathfrak{r}^+(p;A) \end{array}$$

for $B \supset A$. Therefore we have an embedding $V_\mathfrak{r}(p) \to V_\mathfrak{r}^+(p)$. It is covered by a bundle map and is Γ_p equivariant. Moreover it is compatible with the Kuranishi map. Thus this embedding $V_\mathfrak{r}(p) \to V_\mathfrak{r}^+(p)$ is promoted to an embedding of Kuranishi charts.

On the other hand, the embeddings $V_*(p;A) \to V_*^+(p;A)$ commute with the embeddings $\varphi_{\mathfrak{r}\mathfrak{o}}^{A+}$ and $\varphi_{\mathfrak{r}\mathfrak{o}}^{A}$. Therefore we can glue the embeddings $V_*(p) \to V_*^+(p)$ to obtain the required embedding. □

Proof of Lemma* 17.60 *(3) This follows from the proof of Lemma 17.60 (1),(2). □

Proof of Lemma* 17.60 *(4) This follows from the proof of Lemma 17.60 (1). □

Proof of Lemma* 17.60 *(5) This follows from the proof of Lemma 17.60 (2). □

Therefore the proof of Lemma 17.60 is now complete.

17.7.5 Proof of Proposition 17.58

Proof It suffices to construct the coordinate changes between Kuranishi charts produced in Lemma 17.60 and show that the coordinate changes are compatible with various embeddings and isomorphisms appearing in the statement of Proposition 17.58 and of Lemma 17.60. This is indeed straightforward. In fact, the Kuranishi structure in Lemma 17.60 is constructed from $\mathcal{U}^+_{S_k}$, which are Kuranishi charts of $\widehat{\mathcal{U}^+_{S_k}}$. They are compatible with the coordinate change by definition. The process to construct our Kuranishi chart from $\mathcal{U}^+_{S_k}$ is by outer collaring, $* \mapsto *^{\boxplus\tau}$, and gluing by the map in Situation 17.59 (2). The former is compatible with coordinate change as we proved in the first half of this chapter. The latter is compatible since it is induced by the corresponding map (Situation 17.55 (2)) of Kuranishi structures. □

Remark 17.71 What is written as $\mathcal{U}$ in the notation of Proposition 17.58 corresponds to $\mathcal{U}^{\boxplus\tau}$ in the notation of Lemma 17.60.

17.8 Extension of Collared CF-Perturbations

In this section we prove Proposition 17.73.

Situation 17.72 In Situation 17.55, let $\widehat{\mathfrak{S}^+_\partial}$ be a τ-collared CF-perturbation of $\widehat{\mathcal{U}^+_\partial}$. We assume the following:

(1) For each $k \geq 1$ there exists a τ-collared CF-perturbation $\widehat{\mathfrak{S}^+_{S_k}}$ of $\widehat{\mathcal{U}^+_{S_k}}$ such that $\widehat{\mathfrak{S}^+_{S_1}} = \widehat{\mathfrak{S}^+_\partial}$.

(2) The pullback of $\widehat{\mathfrak{S}^+_{S_{k+\ell}}}$ by $\pi_{k,\ell} : \widehat{S}_k(\widehat{S}_\ell(X), \widehat{\mathcal{U}^+_{S_\ell}}) \to (\widehat{S}_{k+\ell}(X), \widehat{\mathcal{U}^+_{S_{k+\ell}}})$ is equivalent to the restriction of $\widehat{\mathfrak{S}^+_\ell}$. ■

Proposition 17.73 *Suppose we are in Situation* 17.72. *Then for any* $0 < \tau' < \tau$ *there exists a* τ'*-collared CF-perturbation* $\widehat{\mathfrak{S}^+}$ *of the Kuranishi structure* $\widehat{\mathcal{U}^+}$ *obtained in Proposition* 17.58 *such that the restriction of* $\widehat{\mathfrak{S}^+}$ *to* $(\widehat{S}_k(X), \widehat{\mathcal{U}^+_{S_k}})$ *is* *equivalent to* $\widehat{\mathfrak{S}^+_{S_k}}$. *When* $\widehat{\mathfrak{S}^+_{S_k}}$ *varies in a uniform family, we may take* $\widehat{\mathfrak{S}^+}$ *to be uniform.*

Proof We first consider the situation of one chart. We use the same notations used in the construction of the chart $V^+(p; A)$ during the proof of Lemma 17.60. By assumption (Situation 17.72 (1)) we are given a CF-perturbation $\mathcal{S}^+_{\mathfrak{r},A}$ on $V^+_{\mathfrak{r},S_A}/\Gamma^A_{\mathfrak{r}}$. It induces $\mathcal{S}^{+\boxplus\tau}_{\mathfrak{r},A}$ on $(V^+_{\mathfrak{r},S_A})^{\boxplus\tau}/\Gamma^A_{\mathfrak{r}}$. We obtain $\mathcal{S}_{p,A}$ on $V^+(p; A)/\Gamma^A_{\mathfrak{r}}$ which extends it by a constant on the $[-\tau, 0)^A$ factor.

If $A' = \gamma A$ for $\gamma \in \Gamma_{\mathfrak{r}}$, then $\mathcal{S}_{p,A}$ is isomorphic to $\mathcal{S}_{p,\gamma A}$. Below we will define $\mathcal{S}^+_{p,\ell}$ on

$$\left(\bigcup_{A;\#A=\ell} V^+_{\mathfrak{r}}(p; A)\right)/\Gamma_{\mathfrak{r}}. \tag{17.47}$$

The open sets (17.47) for various ℓ cover $V^+_{\mathfrak{r}}(p)$. We also remark that

$$(V^+_{\mathfrak{r},S_A} \times [-\tau, 0)^A)/\Gamma^A_{\mathfrak{r}} \tag{17.48}$$

is diffeomorphic to an open subset of (17.47). (Note (17.48) is a subset of $V^+_{\mathfrak{r}}(p, A)/\Gamma^A_{\mathfrak{r}} = ((V^+_{\mathfrak{r},S_A})^{\boxplus\tau} \times [-\tau, 0)^A)/\Gamma^A_{\mathfrak{r}}$. The set $V^+_{\mathfrak{r}}(p, A)/\Gamma^A_{\mathfrak{r}}$ itself may not be a subset of (17.48). See Remark 17.74.) Moreover, for $B \supseteq A$,

$$(V^+_{\mathfrak{r},S_B} \times [-\tau, 0)^B)/\Gamma^B_{\mathfrak{r}} \tag{17.49}$$

becomes also an open subset of (17.47) via ϕ_{AB}. (See Fig. 17.10.)

For any element x in (17.47) there exists $B \supseteq A$ such that x is in the image of (17.49). Namely, $x = \phi_{AB}(x')$. We define the germ of $\mathcal{S}^+_{p,\ell}$ at x as the germ $\mathcal{S}_{p,B}$ at x' transformed via the coordinate change. When x moves, B can jump. The jump occurs at the point where a certain normal coordinate (a coordinate of the $[-\tau, 0)$ factor) becomes 0. However, we can use Situation 17.72 (2) to show that the germ of $\mathcal{S}^+_{p,\ell}$ at x does not jump while x moves. (In fact if $B' \supset B$. Situation 17.72 (2) implies that the restriction of $\mathcal{S}_{p,B'}$ by the map $\phi_{B'B}$ is equivalent to $\mathcal{S}_{p,B}$.) We have thus defined $\mathcal{S}^+_{p,\ell}$.

Again by Situation 17.72 (2) $\mathcal{S}^+_{p,\ell}$ is equivalent to $\mathcal{S}^+_{p,m}$ on the intersection of the domains (17.47). Thus we get a CF-perturbation $\mathfrak{S}^+_p$ on $V^+_{\mathfrak{r}}(p)/\Gamma_{\mathfrak{r}}$.

If $q \in \psi^+_p((s^+_p)^{-1}(0))$, then we have $\varphi^{A+}_{\mathfrak{r}\mathfrak{o}} : V^+_{\mathfrak{o},A} \to V^+_{\mathfrak{r},A}$, that is, (17.44). Since $\widehat{\mathfrak{S}^+_{S_\ell}}$ is a CF-perturbation, the pullback of $\mathcal{S}^+_{\mathfrak{r},A}$ by $\varphi^{A+}_{\mathfrak{r}\mathfrak{o}}$ is equivalent to $\mathcal{S}^+_{\mathfrak{o},A}$. Therefore $\mathfrak{S}^+_p$ and $\mathfrak{S}^+_q$ are glued to define a CF-perturbation on the union of domains. We have thus constructed a CF-perturbation on each of the Kuranishi charts. The compatibility with the coordinate change follows from the corresponding compatibility of $\widehat{\mathcal{U}^+_{S_k}}$'s. Thus we have obtained a CF-perturbation $\widehat{\mathfrak{S}^+}$.

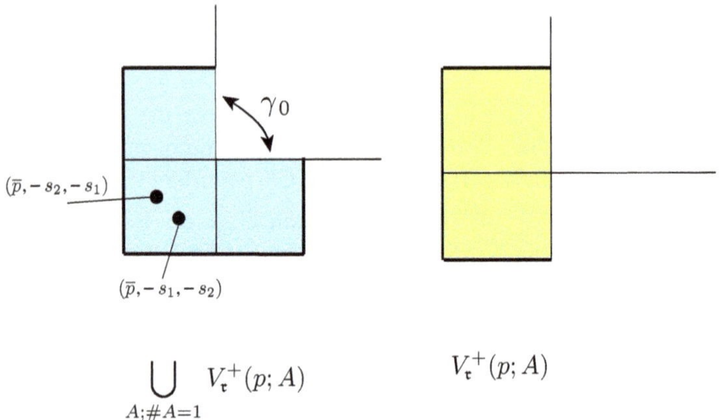

Fig. 17.13 $V_{\mathfrak{r}}^{+}(p, A)/\Gamma_{\mathfrak{r}}^{A}$ and (17.47)

The equivalence of the restriction of $\widehat{\mathfrak{S}^+}$ to $(\widehat{S}_k(X), \widehat{\mathcal{U}^+_{S_k}})$ and $\widehat{\mathfrak{S}^+_k}$ is obvious from the construction. The uniformity also follows from the construction. □

Remark 17.74 We consider the case $\ell = 1$ as in Fig. 17.13 above and suppose that there exists $\gamma_0 \in \Gamma_{\mathfrak{r}}$ which exchanges two normal coordinates. Then $(\overline{p}; -s_1, -s_2)$ is identified with $(\overline{p}; -s_2, -s_1)$, where $A = \{1\}$ and $(\overline{p}, -s_2) \in (V^+_{\mathfrak{r},S_A})^{\boxplus\tau}$. The element γ_0 is not contained in the subgroup $\Gamma_{\mathfrak{r}}^{A}$. Therefore $(\overline{p}; -s_1, -s_2)$ is *not* identified with $(\overline{p}; -s_2, -s_1)$ in $V_{\mathfrak{r}}^{+}(p, A)/\Gamma_{\mathfrak{r}}^{A}$.

17.9 Extension of Kuranishi Structures and CF-Perturbations from a Neighborhood of a Compact Set

In Sect. 17.9 we prove extension lemmas of Kuranishi structures and of CF-perturbations defined on a *neighborhood* of a compact set. In Sect. 7.4.3, we proved certain extension results for a good coordinate system and CF-perturbation thereon. Namely, we assumed that they are given on a neighborhood of $Z_1 \subseteq X$ and showed that they can be extended to a subset $Z_2 \subseteq X$ satisfying $Z_1 \subset \overset{\circ}{Z_2}$, under the assumption that the given good coordinate system on Z_1 is compatible with the restriction $\widehat{\mathcal{U}}|_{Z_1}$ of a Kuranishi structure $\widehat{\mathcal{U}}$ on Z_2. During the proof of Lemma 17.77, Lemma 17.75 will be used to show that this assumption is satisfied.

The next lemma says that a thickening $\widehat{\mathcal{U}^+_Z}$ of $\widehat{\mathcal{U}}|_Z$ given on a compact neighborhood Z of K can be extended to a thickening of $\widehat{\mathcal{U}}$ on X without changing it on K. Note that they are results about Kuranishi structures and CF-perturbations, not about τ-collared ones.

Lemma 17.75 *Let K be a compact set of X and $Z \subseteq X$ a compact neighborhood of K such that $K \subset \operatorname{Int} Z$. Suppose we are given a Kuranishi structure $\widehat{\mathcal{U}}$ on X and $\widehat{\mathcal{U}^+_Z}$ on Z. We assume*

$$\widehat{\mathcal{U}}|_Z < \widehat{\mathcal{U}^+_Z}.$$

Let Ω be a relatively compact neighborhood of K in Z such that

$$K \subset \Omega \subset \overline{\Omega} \subset \operatorname{Int} Z.$$

For each given $p \in K$:

(i) *Assume the Kuranishi neighborhood $\mathcal{U}^+_{Z,p} = (U^+_{Z,p}, \mathcal{E}^+_{Z,p}, s^+_{Z,p}, \psi^+_{Z,p})$ of $\widehat{\mathcal{U}^+_Z}$ at p satisfies*

$$\psi^+_{Z,p}((s^+_{Z,p})^{-1}(0)) \subset \Omega.$$

(ii) *Assume the Kuranishi neighborhood $\mathcal{U}_p = (U_p, \mathcal{E}_p, s_p, \psi_p)$ of $\widehat{\mathcal{U}}$ at p satisfies*

$$\psi_p((s_p)^{-1}(0)) \subset \Omega.$$

Then there exist a Kuranishi structure $\widehat{\mathcal{U}^+}$ on X and a strict KK-embedding $\widehat{\mathcal{U}} \to \widehat{\mathcal{U}^+}$ with the following properties:

(1) *For any $p \in K$ the Kuranishi neighborhood $\mathcal{U}^+_p$ of $\widehat{\mathcal{U}^+}$ at p is isomorphic to the Kuranishi neighborhood $\mathcal{U}^+_{Z,p}$ of $\widehat{\mathcal{U}^+_Z}$ at p. For any $p, q \in K$ the coordinate change between $\mathcal{U}^+_p$ and $\mathcal{U}^+_q$ coincides with the coordinate change between $\mathcal{U}^+_{Z,p}$ and $\mathcal{U}^+_{Z,q}$.*
(2) *$\widehat{\mathcal{U}^+}|_\Omega$ is an open substructure of $\widehat{\mathcal{U}^+_Z}|_\Omega$.*
(3) *$\widehat{\mathcal{U}} < \widehat{\mathcal{U}^+}$.*
(4) *The next diagram commutes:*

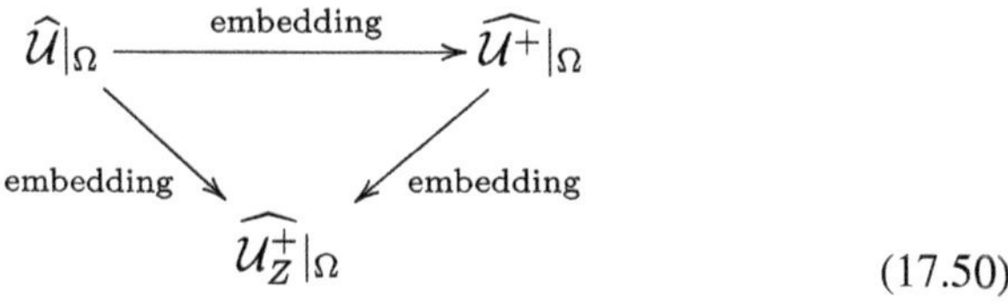

(17.50)

Here the right down arrow is the open embedding given by (2).
(5) *The embedding $\widehat{\mathcal{U}}|_K \to \widehat{\mathcal{U}^+_Z}|_K$ coincides with the embedding $\widehat{\mathcal{U}}|_K \to \widehat{\mathcal{U}^+}|_K$ via the isomorphism (1).*

Proof We take an open set $\Omega_1 \subset X$ such that

$$\overline{\Omega} \subset \Omega_1 \subset \overline{\Omega}_1 \subset \operatorname{Int} Z.$$

We next take an open set $W_1 \subset X$ such that

$$\overline{W}_1 \cap \overline{\Omega} = \emptyset, \quad \Omega_1 \cup W_1 = X.$$

We replace $\widehat{\mathcal{U}}$ and $\widehat{\mathcal{U}_Z^+}$ by their open substructures (but without changing the Kuranishi neighborhoods of the point $p \in K$) if necessary, and may assume that the following holds:

(1) If $\psi^+_{Z,p}(U^+_{Z,p} \cap (s^+_{Z,p})^{-1}(0)) \cap \overline{W}_1 \neq \emptyset$, then $\psi^+_{Z,p}(U^+_{Z,p} \cap (s^+_{Z,p})^{-1}(0)) \cap \overline{\Omega} = \emptyset$.
(2) If $\psi_p(U_p \cap s_p^{-1}(0)) \cap \overline{W}_1 \neq \emptyset$, then $\psi_p(U_p \cap s_p^{-1}(0)) \cap \overline{\Omega} = \emptyset$.

We define a Kuranishi structure $\widehat{\mathcal{U}'}$ on X as follows:

(a) If $p \in \overline{\Omega}_1$, we put $\mathcal{U}'_p = \mathcal{U}^+_{Z,p}$.
(b) If $p \notin \overline{\Omega}_1$, we put $\mathcal{U}'_p = \mathcal{U}_p|_{U_p \setminus \psi_p^{-1}(\overline{\Omega}_1)}$.

The coordinate change is defined as follows. Let $q \in \psi'_p((s'_p)^{-1}(0))$. If $p, q \in \overline{\Omega}_1$, then we define $\Phi'_{pq} = \Phi^+_{Z,pq}$. If $p, q \notin \overline{\Omega}_1$, then we define $\Phi'_{pq} = \Phi_{pq}|_{U_{pq} \setminus \psi_q^{-1}(\overline{\Omega}_1)}$. Among the other two cases $q \in \overline{\Omega}_1$, $p \notin \overline{\Omega}_1$ cannot occur (by (b)). We consider the remaining case, $q \notin \overline{\Omega}_1$, $p \in \overline{\Omega}_1$. We have an embedding $\Phi_q : \mathcal{U}_q \to \mathcal{U}^+_{Z,q}$. We compose it with the embedding of Kuranishi structure $\widehat{\mathcal{U}_Z^+}$ to obtain

$$\Phi^+_{Z,pq} \circ \Phi_q : \mathcal{U}_q|_{(\varphi_q)^{-1}(U^+_{Z,pq})} \to \mathcal{U}^+_{Z,q}|_{U^+_{Z,pq}} \to \mathcal{U}^+_{Z,p}.$$

The composition gives the coordinate change Φ'_{pq} in this case.

Note $\Phi^+_{Z,pq} \circ \Phi_q = \Phi_p \circ \Phi_{pq}$ on $(\varphi_q)^{-1}(U^+_{Z,pq})$, by the definition of embedding of Kuranishi structures. Using this fact, it is easy to see that they define a Kuranishi structure on X.

The Kuranishi structure $\widehat{\mathcal{U}'}$ has all the properties we need, except property (3). We will further modify $\widehat{\mathcal{U}'}$ to $\widehat{\mathcal{U}^+}$ for this reason. Firstly, we use Proposition 5.28 to find a Kuranishi structure $\widehat{\mathcal{U}''}$ such that

$$\widehat{\mathcal{U}'} < \widehat{\mathcal{U}''}.$$

Although there are various choices of such $\widehat{\mathcal{U}''}$, we choose one of them in the proof of Lemma 17.75. For later purposes, we will take a more specific $\widehat{\mathcal{U}''}$ in the proof of Lemma 17.77.

Next, we modify $\widehat{\mathcal{U}''}$ to $\widehat{\mathcal{U}^+}$, which has the required properties as follows. We take an open set $W_2 \subset X$ such that

$$\overline{W}_2 \cap \overline{\Omega} = \emptyset, \quad \overline{\Omega}_1 \cup W_2 = X.$$

We replace the various Kuranishi structures involved by their open substructures (but without changing the Kuranishi neighborhoods of the point $p \in K$) and may assume the following:

(I) If $\psi_p(U^+_{Z,p} \cap (s^+_{Z,p})^{-1}(0)) \cap \overline{W}_2 \neq \emptyset$, then $\psi^+_{Z,p}(U^+_{Z,p} \cap (s^+_{Z,p})^{-1}(0)) \cap \overline{\Omega} = \emptyset$
(II) If $\psi_p(U_p \cap s_p^{-1}(0)) \cap \overline{W}_2 \neq \emptyset$, then $\psi_p(U_p \cap s_p^{-1}(0)) \cap \overline{\Omega} = \emptyset$.

Now we define a Kuranishi structure $\widehat{\mathcal{U}^+}$ on X as follows:

(A) If $p \in \overline{W}_2$, we put $\mathcal{U}^+_p = \mathcal{U}''_p$.
(B) If $p \notin \overline{W}_2$, we put $\mathcal{U}^+_p = \mathcal{U}'_p|_{U'_p \setminus \psi_p^{-1}(\overline{W}_2)}$.

The coordinate change is defined as follows. Let $q \in \psi^+_p((s^+_p)^{-1}(0))$. If $p, q \in \overline{W}_2$, then we define $\Phi^+_{pq} = \Phi''_{pq}$. If $p, q \notin \overline{W}_2$, then we define $\Phi^+_{pq} = \Phi'_{pq}|_{U'_{pq} \setminus \psi_q^{-1}(\overline{W}_2)}$. Among the other two cases $q \in \overline{W}_2$, $p \notin \overline{W}_2$ cannot occur. Suppose $p \in \overline{W}_2$, $q \notin \overline{W}_2$. Then there is an embedding $\Phi_q : \mathcal{U}'_q \to \mathcal{U}''_q$. The coordinate change of $\widehat{\mathcal{U}^+}$ is given by the composition

$$\Phi''_{pq} \circ \Phi_q : \mathcal{U}'_q|_{(\varphi_q)^{-1}(U''_{pq})} \to \mathcal{U}''_q|_{U''_{pq}} \to \mathcal{U}''_p.$$

It is easy to see that this Kuranishi structure $\widehat{\mathcal{U}^+}$ has the required properties. □

We next discuss extension of CF-perturbations.

Situation 17.76 Suppose we are in the situation of Lemma 17.75. We assume the following in addition:

(1) We have a strongly smooth and weakly submersive map $\widehat{f} : (X, \widehat{\mathcal{U}}) \to M$ to a manifold M. (See Definition 3.43.)
(2) We have a CF-perturbation $\widehat{\mathfrak{S}^+_Z}$ of $\widehat{\mathcal{U}^+_Z}$.
(3) We have a strongly smooth map $\widehat{f_Z} : (Z, \widehat{\mathcal{U}^+_Z}) \to M$ which is strongly submersive with respect to $\widehat{\mathfrak{S}^+_Z}$. (See Definition 9.2.)
(4) The following diagram commutes:

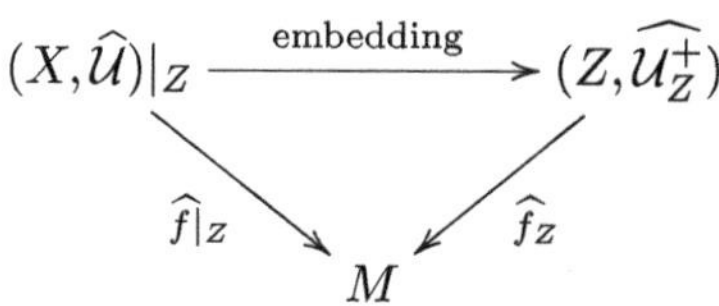

■

Lemma 17.77 *In Situation* 17.76, *we may choose the Kuranishi structure* $\widehat{\mathcal{U}^+}$ *in Lemma* 17.75 *so that the following holds in addition:*

(1) *There exists a CF-perturbation* $\widehat{\mathfrak{S}^+}$ *of* $\widehat{\mathcal{U}^+}$.
(2) *The map* $\widehat{f}$ *is extended to a strongly smooth map* $\widehat{f}^+ : (X, \widehat{\mathcal{U}^+}) \to M$.
(3) *The map* $\widehat{f}^+ : (X, \widehat{\mathcal{U}^+}) \to M$ *which extends* $\widehat{f}$ *is strongly submersive with respect to* $\widehat{\mathfrak{S}^+}$.
(4) *For any* $p \in K$, *two CF-perturbations* $\widehat{\mathfrak{S}_Z^+}$ *and* $\widehat{\mathfrak{S}^+}$ *assign the same CF-perturbation on the Kuranishi chart* $\mathcal{U}^+_{Z,p} = \mathcal{U}^+_p$.
(5) $\widehat{\mathfrak{S}_Z^+}|_\Omega$ *is a restriction of* $\widehat{\mathfrak{S}^+}|_\Omega$ *to the open substructure.*
(6) *When* $\widehat{\mathfrak{S}_Z^+}|_\Omega$ *varies in a uniform family, we may take* $\widehat{\mathfrak{S}^+}$ *to be uniform.*

Proof The lemma is a consequence of combination of results in Chaps. 3, 6, 7, and 9. Before we start the proof, we recall from Chap. 7 that we used a good coordinate system to define a CF-perturbation. Thus we also need to involve an extension of good coordinate system in the course of the proof of Lemma 17.77. Indeed, we use a good coordinate system from the given Kuranishi structure to find an extension of the given CF-perturbation, and come back from the good coordinate system to a Kuranishi structure together with the CF-perturbation. This process is described in Chap. 9. This is a rough description of the structure of the proof of Lemma 17.77 given below.

Now we start the proof. In the proof of Lemma 17.75, we took a relatively compact open subset $\Omega_1 \subset \operatorname{Int} Z$ such that

$$\overline{\Omega} \subset \Omega_1 \subset \overline{\Omega}_1 \subset \operatorname{Int} Z.$$

As in the proof, we replace $\widehat{\mathcal{U}}$ and $\widehat{\mathcal{U}_Z^+}$ by their open substructures without changing the Kuranishi neighborhoods of the point $p \in K$ if necessary, and may assume that the map $\widehat{f_Z} : (Z, \widehat{\mathcal{U}_Z^+}) \to M$ is strictly strongly submersive (Definition 9.2) with respect to $\widehat{\mathfrak{S}_Z^+}$. We put

$$Z_1 = \overline{\Omega}_1, \quad \widehat{\mathcal{U}_{Z_1}^+} = \widehat{\mathcal{U}_Z^+}|_{Z_1}.$$

Then by the definition of the Kuranishi structure $\widehat{\mathcal{U}'}$ in the proof of Lemma 17.75, we note

$$\widehat{\mathcal{U}'}|_{Z_1} = \widehat{\mathcal{U}_{Z_1}^+}. \tag{17.51}$$

We apply Theorem 3.35 to the Kuranishi structure $\widehat{\mathcal{U}_{Z_1}^+}$ to find a good coordinate system $\widetriangle{\mathcal{U}_{Z_1}^+}$ on Z_1 and a KG-embedding

$$\Phi : \widehat{\mathcal{U}_{Z_1}^+} \longrightarrow \widetriangle{\mathcal{U}_{Z_1}^+}. \tag{17.52}$$

Since we are given a CF-perturbation $\widehat{\mathfrak{S}^+_Z|_{Z_1}}$ of $\widehat{\mathcal{U}^+_{Z_1}}$, Lemma 9.10 shows that there exists a CF-perturbation $\widehat{\mathfrak{S}^+_{Z_1}}$ of $\widehat{\mathcal{U}^+_{Z_1}}$ such that $\widehat{\mathfrak{S}^+_{Z_1}}$ and $\widehat{\mathfrak{S}^+_Z|_{Z_1}}$ are compatible with the KG-embedding Φ in (17.52). In addition, Lemma 7.55 and Proposition 7.59 yield that there exists a strongly smooth map $\widehat{f_{Z_1}} : (Z_1, \widehat{\mathcal{U}^+_{Z_1}}) \to M$ such that it is strongly submersive with respect to $\widehat{\mathfrak{S}^+_{Z_1}}$ and the following diagram commutes:

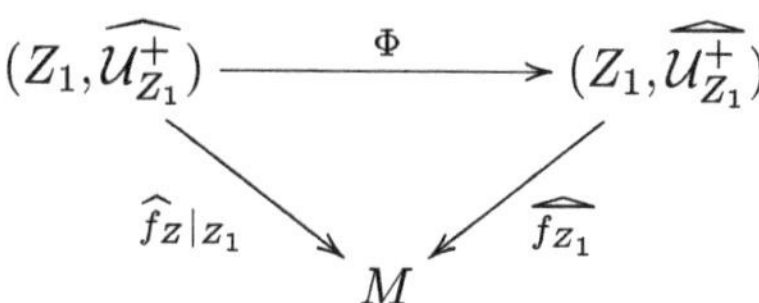

Since $\widehat{\mathcal{U}^+_{Z_1}} = \widehat{\mathcal{U}'}|_{Z_1}$ is the restriction of the Kuranishi structure $\widehat{\mathcal{U}'}$ on X to Z_1, we can apply Proposition 7.54 for the case $Z_1 = Z_1$ and $Z_2 = X$ to obtain a good coordinate system $\widehat{\widehat{\mathcal{U}'}}$ on X such that it is an extension of $\widehat{\mathcal{U}^+_{Z_1}}$, and $\widehat{\mathcal{U}'}$ and $\widehat{\widehat{\mathcal{U}'}}$ are compatible in the sense of Definition 3.37, i.e., there exists a KG-embedding

$$\widehat{\mathcal{U}'} \longrightarrow \widehat{\widehat{\mathcal{U}'}}.$$

Moreover, the CF-perturbation $\widehat{\mathfrak{S}^+_{Z_1}}$ is also extended to a CF-perturbation $\widehat{\widehat{\mathfrak{S}'}}$ of $\widehat{\widehat{\mathcal{U}'}}$ by Proposition 7.59. In addition, by Lemma 7.55 there exists a strongly smooth and strongly submersive map

$$\widehat{\widehat{f'}} : (X, \widehat{\widehat{\mathcal{U}'}}) \longrightarrow M$$

with respect to $\widehat{\widehat{\mathfrak{S}'}}$ such that the following diagram commutes:

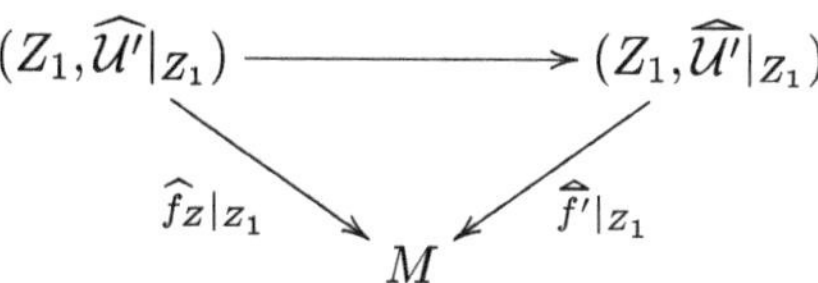

Now we go back to a Kuranishi structure from the good coordinate system. We apply Proposition 6.30 for the case $\widehat{\mathcal{U}_0} = \widehat{\mathcal{U}} = \widehat{\widehat{\mathcal{U}'}}$ to find a Kuranishi structure $\widehat{\mathcal{U}''}$ together with a GK-embedding

$$\widehat{\widehat{\mathcal{U}'}} \longrightarrow \widehat{\mathcal{U}''},$$

and a strongly smooth map

$$\widehat{f'} : (X, \widehat{\mathcal{U}''}) \longrightarrow M$$

such that $\widehat{f'}$ is a pullback of $\widehat{f'}$. By Proposition 5.11, $\widehat{\mathcal{U}''}$ is a thickening of $\widehat{\mathcal{U}'}$:

$$\widehat{\mathcal{U}'} < \widehat{\mathcal{U}''}.$$

Moreover, Lemma 9.9 shows that there exists a CF-perturbation $\widehat{\mathfrak{S}''}$ of $\widehat{\mathcal{U}''}$ such that $\widehat{f'}$ is strongly submersive with respect to $\widehat{\mathfrak{S}''}$.

Finally, in the exactly same way as in the proof of Lemma 17.75, we modify the Kuranishi structure $\widehat{\mathcal{U}''}$ obtained above to $\widehat{\mathcal{U}^+}$. Accordingly, we also have the corresponding CF-perturbation $\widehat{\mathfrak{S}^+}$ of $\widehat{\mathcal{U}^+}$ and the corresponding map $f^+ : (X, \widehat{\mathcal{U}^+}) \to M$. Then all the assertions of Lemma 17.77 follow from the construction. □

17.10 Conclusion of Chap. 17

We now combine the results of Sects. 17.7, 17.8, and 17.9 to prove results which we will use later in our applications. Roughly speaking, they claim that Kuranishi structure and its CF-perturbation, which are defined on the boundary and which satisfy certain corner compatibility conditions, can be extended to the whole space.

Proposition 17.78 *In Situation* 17.55, *there exists a* τ'*-collared Kuranishi structure* $\widehat{\mathcal{U}^{++}}$ *on* X *for any* $0 < \tau' < \tau$ *such that Items (1)–(5) of Proposition* 17.58 *hold, by replacing* $\widehat{\mathcal{U}^+}$ *by* $\widehat{\mathcal{U}^{++}}$.

Remark 17.79 The difference between Propositions 17.58 and 17.78 is that the τ'-collared Kuranishi structure $\widehat{\mathcal{U}^{++}}$ in Proposition 17.78 is defined on whole X, while the τ'-collared Kuranishi structure $\widehat{\mathcal{U}^+}$ in Proposition 17.58 is defined only on a neighborhood of the boundary.

Proof We will use Lemma 17.75 to prove Proposition 17.78. For this purpose, we will slightly modify the collared Kuranishi structure $\widehat{\mathcal{U}^+}$ produced in Proposition 17.58 to get $\widehat{\mathcal{U}^+_\Omega}$ satisfying the assumption of Lemma 17.75. The detail is in order.

We take $0 < \tau' < \tau'' < \tau''' < \tau'''' < \tau$. Let $\widehat{\mathcal{U}^+}$ be the τ''-collared Kuranishi structure produced in Proposition 17.58. (Here the role of τ' in Proposition 17.58 is played by τ''.) We put $X' = X^{\boxminus\tau'}$ and $X_- = X^{\boxminus\tau''}$. We note that $X' = X_-^{\boxplus(\tau''-\tau')}$. Since $\widehat{\mathcal{U}^+}$ be the τ''-collared there is a Kuranishi structure $\widehat{\mathcal{U}'}$ on $X_- \setminus X^{\boxminus\tau}$ such that $(X_-, \widehat{\mathcal{U}'})^{\boxplus\tau''} = (X, \widehat{\mathcal{U}^+})|_{X_0}$.

We then obtain a Kuranishi structure $\widehat{\mathcal{U}''}$ on $X' \setminus X^{\boxminus\tau}$ such that $(X', \widehat{\mathcal{U}''})^{\boxplus\tau'} = (X, \widehat{\mathcal{U}^+})|_{X_0}$ as a τ'-collared Kuranishi structure. Namely $\widehat{\mathcal{U}''}$ is the $\tau'' - \tau'$ outer collaring of $\widehat{\mathcal{U}'}$.

We shrink the Kuranishi neighborhood $\mathcal{U}''_p$ of the Kuranishi structure $\widehat{\mathcal{U}''}$ to obtain $\mathcal{U}'''_p$ and a Kuranishi structure $\widehat{\mathcal{U}'''}$ on $X' \setminus X^{\boxminus\tau}$ so that the following is satisfied:

(1) If $p \in X^{\boxminus(\tau-\tau''')}$, then $\psi_p'''((s_p''')^{-1}(0)) \subset X^{\boxminus(\tau-\tau'''')}$.
(2) If $p \in S_1(X')$, then $\partial\mathcal{U}_p''' = \partial\mathcal{U}_p''$.

Note that (2) above implies

$$(X', \widehat{\mathcal{U}'''})^{\boxplus\tau'}|_{\overline{X \setminus X'}} = (X, \widehat{\mathcal{U}^+})|_{\overline{X \setminus X'}}. \tag{17.53}$$

We put $K = \overline{X' \setminus X^{\boxminus(\tau-\tau''')}}$ and $Z = \overline{X'' \setminus X^{\boxminus(\tau-\tau'''')}}$. We now apply Lemma 17.75. Here the role of X, K, Z, $\widehat{\mathcal{U}_Z^+}$ in Lemma 17.75 is played by X', K, Z and $\widehat{\mathcal{U}'''}$, respectively.

Remark 17.80 As we mentioned at the top of Sect. 17.9, Lemma 17.75 is about genuine Kuranishi structures and not τ-collared Kuranishi structures. So here we apply Lemma 17.75 literally, not its τ-collared version.

We thus obtain a Kuranishi structure which we wrote as $\widehat{\mathcal{U}^+}$ in Lemma 17.75. We denote it here by $\widehat{\mathcal{U}'^+}$. $\widehat{\mathcal{U}'^+}$ is a Kuranishi structure on X'.

We put $\widehat{\mathcal{U}^{++}} = \widehat{\mathcal{U}'^{+\boxplus\tau'}}$. Equation (17.53), Lemma 17.75 (1) and the fact that $\widehat{\mathcal{U}^+}$ satisfies Proposition 17.58 imply that $\widehat{\mathcal{U}^{++}}$ satisfies Proposition 17.58. The proof of Proposition 17.78 is now complete. □

We next include CF-perturbations.

Proposition 17.81 *In Situation 17.72 there exists a τ'-collared CF-perturbation $\widehat{\mathfrak{S}^{++}}$ on the Kuranishi structure $\widehat{\mathcal{U}^{++}}$ obtained in Proposition 17.78 such that:*

(1) *Its restriction to $(\widehat{S}_k(X), \widehat{\mathcal{U}_{S_k}^+})$ is equivalent to $\widehat{\mathfrak{S}_k^+}$.*
(2) *If $f : (X, \widehat{\mathcal{U}}) \to M$ is weakly submersive and $f|_{\widehat{\mathcal{U}_k^+}}$ is strongly submersive with respect to $\widehat{\mathfrak{S}_k^+}$, then we may take $\widehat{\mathfrak{S}^{++}}$ such that $f : (X, \widehat{\mathcal{U}^{++}}) \to M$ is strongly submersive with respect to $\widehat{\mathfrak{S}^{++}}$.*
   Transversality to $g : N \to M$ is preserved in a similar sense.
(3) *Uniformity of CF-perturbations is preserved in this construction.*

Proof We use the notation in the proof of Proposition 17.78. We apply Proposition 17.73 to obtain a CF-perturbation $\widehat{\mathfrak{S}^+}$ on $\widehat{\mathcal{U}^+}$. Since $\widehat{\mathfrak{S}^+}$ is τ'-collared, it is induced by a CF-perturbation $\widehat{\mathfrak{S}''}$ of $\widehat{\mathcal{U}''}$. Therefore by restriction we obtain a CF-perturbation $\widehat{\mathfrak{S}_Z^+}$ on $\widehat{\mathcal{U}_Z^+}$. Here Z is one in the proof of Proposition 17.78. Thus we can apply Lemma 17.77 to obtain the required $\widehat{\mathfrak{S}^{++}}$ and $\widehat{\mathcal{U}^{++}}$. □

One minor point remains to be explained to apply the results of this section. Note that a τ-collared Kuranishi structure is not a Kuranishi structure, since some points are not assigned to its Kuranishi neighborhood. Since in Part I the 'pushout' is defined for the case of a Kuranishi structure and a good coordinate system but not for the case of a τ-collared Kuranishi structure, we need some explanation to define the notion of the 'pushout' for a τ-collared Kuranishi structure.

One can define the notion of a τ-collared good coordinate system, and prove the existence of a τ-collared good coordinate system compatible with each τ-collared Kuranishi structure, and use it to define the 'pushout'. (We note that a τ-collared good coordinate system is a special case of a good coordinate system. See the end of Remark 17.34.) It is certainly possible to proceed in that way.

Here we take a slightly different way which seems shorter, as follows. Let $\widehat{\mathcal{U}}$ be a τ-collared Kuranishi structure on $X' = X^{\boxplus\tau}$. We take a Kuranishi structure on X' obtained by Definition 17.45, which we denote by $\widehat{\mathcal{U}'}$. If $\widehat{\mathfrak{S}}$ is a τ-collared CF-perturbation on $\widehat{\mathcal{U}}$, it induces a CF-perturbation $\widehat{\mathfrak{S}'}$ on $\widehat{\mathcal{U}'}$.[11]

Definition 17.82 Let $f : (X', \widehat{\mathcal{U}}) \to M$ be a strongly smooth map which is strongly submersive with respect to $\widehat{\mathfrak{S}}$. Let h be a differential form on $(X', \widehat{\mathcal{U}})$. We define its *pushout*

$$f_!(h; \widehat{\mathfrak{S}}^{\epsilon}) = f_!(h; \widehat{\mathfrak{S}'^{\epsilon}}). \tag{17.54}$$

Here h in the right hand side is the differential form on $(X', \widehat{\mathcal{U}'})$ which is induced from h.

Lemma 17.83 *Stokes' formula (Theorem* 9.28*) and the composition formula (Theorem* 10.21*) hold for the pushout in Definition* 17.82.

Proof They are direct consequences of the corresponding results (Theorems 9.28 and 10.21). □

[11]When a τ-collared CF-perturbation $\widehat{\mathfrak{S}}$ on $\widehat{\mathcal{U}}$ varies in a uniform family, we may take the induced CF-perturbation $\widehat{\mathfrak{S}'}$ to be uniform.

Chapter 18
Corner Smoothing and Composition of Morphisms

The goal of this chapter is to define composition of morphisms of linear K-systems (Lemma-Definition 18.37) and to show that it is associative (Proposition 18.41). There are two key ingredients for the construction of composition of morphisms. One is 'partial outer collaring' and the other is 'corner smoothing'. The precise definition of 'partial outer collaring' will be given in Definition 18.9 and 'corner smoothing' will be described in detail in Sects. 18.5 and 18.6. After that, we will define composition of morphisms in Sect. 18.7 and prove its associativity in Sect. 18.8. We also discuss the identity morphism in Sect. 18.10.

18.1 Why Corner Smoothing?

We use corner smoothing for the purpose of defining a category of linear K-systems and obtain a functor

$$\{\text{linear K-system}\} \to \{\text{filtered chain complex}\}.$$

In other words, we use corner smoothing so that the construction of Chaps. 16, 17, 18, and 19 becomes fully functorial. Actually, for the application we can construct a (partial) cochain map directly from the union of the spaces (18.1). To prove that it becomes a cochain map we can use the isomorphism among the terms of the form (18.2) and the isomorphism for (18.3) so that the corresponding terms appearing in the boundary of the maps cancel each other. If we take that way we do not need to define composition of morphisms in the category of linear K-systems. The usage of identity morphism can be bypassed by various ways which were introduced to study Floer homology and its independence by various people. In this book we do our best to avoid any ad-hoc argument and work out as much canonical and functorial construction as possible.

K. Fukaya et al., *Kuranishi Structures and Virtual Fundamental Chains*, Springer Monographs in Mathematics, https://doi.org/10.1007/978-981-15-5562-6_18

We remark that if we include a certain kind of objects which are the 'union' of K-spaces 'glued' along the boundary and corners and allow such an object as a space of connecting orbits or an interpolation space in the axiom of linear K-systems, it is likely that we do not need to study corner smoothing. However, then we need to describe such an object and its perturbation, in detail. For example, we need to assume some kinds of 'corner compatibility condition' (which is related to but is different from the corner compatibility condition in the system of K-spaces we spelled out). In fact, when we glue several K-spaces along the boundaries we need to assume that gluing maps are consistent at the (higher codimensional) corners. Then the notion of CF-perturbations for such objects should be described. Studying such objects is certainly possible and could be useful for certain applications. However, working it out in detail is rather heavy and in this book we prefer to use corner smoothing.

18.2 Introduction to Chap. 18

In this section, we explain the idea of the construction of composition of morphisms and its geometric background.

Firstly, we explain the reason why we need the notion of 'partial outer collaring', or 'partially collared fiber products', instead of outer collaring or usual fiber products. Let $\mathcal{N}_{12}(\alpha_1, \alpha_2)$ and $\mathcal{N}_{23}(\alpha_2, \alpha_3)$ be interpolation spaces of morphisms $\mathfrak{N}_{12}$ and $\mathfrak{N}_{23}$ of linear K-systems, respectively. As we mentioned in Lemma-Definition 16.35, the interpolation space of the morphism $\mathfrak{N}_{32} \circ \mathfrak{N}_{21}$ is a union of fiber products

$$\bigcup_{\alpha_2} \mathcal{N}_{12}(\alpha_1, \alpha_2) \times_{R_{\alpha_2}} \mathcal{N}_{23}(\alpha_2, \alpha_3). \tag{18.1}$$

The union in (18.1) may not be a disjoint union, in general. In fact, the summands corresponding to α_2 and to α_2' may intersect at

$$\mathcal{N}_{12}(\alpha_1, \alpha_2) \times_{R_{\alpha_2}} \mathcal{M}^2(\alpha_2, \alpha_2') \times_{R_{\alpha_2'}} \mathcal{N}_{23}(\alpha_2', \alpha_3). \tag{18.2}$$

Moreover, three such summands may intersect at

$$\mathcal{N}_{12}(\alpha_1, \alpha_2) \times_{R_{\alpha_2}} \mathcal{M}^2(\alpha_2, \alpha_2') \times_{R_{\alpha_2'}} \mathcal{M}^2(\alpha_2', \alpha_2'') \times_{R_{\alpha_2''}} \mathcal{N}_{23}(\alpha_2'', \alpha_3). \tag{18.3}$$

The pattern of how the summands of (18.1) intersect one another is similar to the way the components of the boundary of certain K-spaces (or of orbifolds) intersect one another. (Namely, each of the summands of (18.1) corresponds to a codimension 1 boundary and (18.2), (18.3) correspond to codimension 2 and 3 corners, respectively.)

We can use this observation to apply a version of Proposition 17.58, that is, we 'put the collar' *outside* the union (18.1) to obtain a collared K-space so that its boundary is

$$\bigcup_{\alpha_2} \mathcal{N}_{12}(\alpha_1, \alpha_2) \times^{\boxplus\tau}_{R_{\alpha_2}} \mathcal{N}_{23}(\alpha_2, \alpha_3).$$

Here the notation $\times^{\boxplus\tau}_{R_{\alpha_2}}$ is defined in Definition 18.34, where it is called a *partially collared fiber product*. The notion of *partial outer collaring* is introduced in Sect. 18.3. See Definition 18.9. The reason why we introduce the notion of partial outer collaring is as follows. Note that the boundary or corner of K-space of the summand of (18.1) has different components from those appearing in (18.2), (18.3). In fact, a boundary of the form

$$\mathcal{M}^1(\alpha_1, \alpha_1') \times_{R_{\alpha_1'}} \mathcal{N}_{12}(\alpha_1', \alpha_2) \times_{R_{\alpha_2}} \mathcal{N}_{23}(\alpha_2, \alpha_3)$$

also appears. To study boundary components of this kind together, we introduce the notion of partial outer collaring and use it to modify Proposition 17.58 so that it can be directly applicable to our situation.

After we have done partial outer collaring, we will next discuss *corner smoothing*. Here we use the fact that a K-space X has a collar where all the objects are 'constant' in the direction of the collar. See Sect. 18.4 for the reason why this fact is useful for our construction. Then we can finally define composition of morphisms.

In the rest of this section, we explain a geometric origin of the idea that the union (18.1) looks like a boundary of a certain K-space, by considering the situation of the Floer cohomology of periodic Hamiltonian systems. (The story of linear K-systems will be applied to define and study the Floer cohomology of periodic Hamiltonian systems. Namely, we associate such a system to each periodic Hamiltonian function H. See Sect. 15.1.) Let H^1, H^2, H^3 be periodic Hamiltonian functions. To define a cochain map between Floer's cochain complexes associated to them, we use τ-dependent Hamiltonian functions interpolating them ($\tau \in \mathbb{R}$). Namely, we take $\mathcal{H}^{ij} : \mathbb{R} \times S^1 \times M \to \mathbb{R}$ such that

$$\mathcal{H}^{ij}(\tau, t, x) = \begin{cases} H^j(t, x) & \text{if } \tau \le -T_0, \\ H^i(t, x) & \text{if } \tau \ge T_0. \end{cases} \tag{18.4}$$

We then use the moduli space of the solutions of the equation

$$\frac{\partial u}{\partial \tau} + J\left(\frac{\partial u}{\partial t} - X_{H^{ij}_{\tau,t}}(u)\right) = 0, \tag{18.5}$$

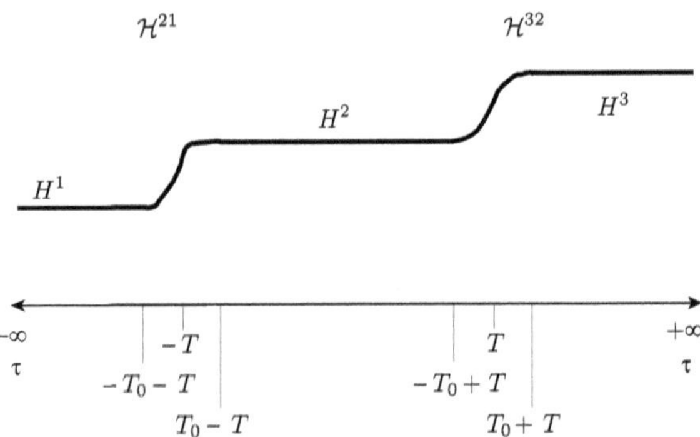

Fig. 18.1 The concatenation $\mathcal{H}^{31}$

where $H^{ij}_{\tau,t}(x) = \mathcal{H}^{ij}(\tau, t, x)$ and $X_{H^{ij}_{\tau,t}}$ is its Hamiltonian vector field. The solution space of (18.5) with an appropriate boundary condition becomes an interpolation space $\mathcal{N}^{ji}(\alpha_i, \alpha_j)$ of the morphism $\mathfrak{N}^{ji}$.

To study the relation between $\mathfrak{N}^{31}$ and the composition $\mathfrak{N}^{32} \circ \mathfrak{N}^{21}$ we use a one-parameter family of τ-dependent Hamiltonian functions $\mathcal{H}^{31,T}$ where

$$\mathcal{H}^{31,T}(\tau, t, x) = \begin{cases} H^1(t,x) & \text{if } \tau \le -T_0 - T \\ \mathcal{H}^{21}(\tau + T, t, x) & \text{if } -T_0 - T \le \tau \le T_0 - T \\ H^2(t,x) & \text{if } T_0 - T \le \tau \le T - T_0 \\ \mathcal{H}^{32}(\tau - T, t, x) & \text{if } T - T_0 \le \tau \le T + T_0 \\ H^3(t,x) & \text{if } T + T_0 \le \tau. \end{cases} \tag{18.6}$$

See Fig. 18.1.

We may choose $\mathcal{H}^{31} = \mathcal{H}^{31,2T_0}$, for example. We consider the limit as $T \to +\infty$ and the set of solutions of (18.5) with $X_{H^{ij}_{\tau,t}}$ replaced by $X_{H^{31,T}_{\tau,t}}$. The moduli space of its solutions becomes the union of fiber products (18.1). Thus the union of solution spaces for $T \in [2T_0, \infty)$ and the fiber product (18.1) gives a homotopy between the morphism defined by $\mathcal{H}^{31} = \mathcal{H}^{31,2T_0}$ and the composition whose interpolation space is (18.1). The space (18.1) itself is a part of the boundary of this cobordism.

18.3 Partial Outer Collaring of Cornered K-Spaces

In this section we explain that the story of outer collaring in Chap. 17 can be generalized to the case of partial outer collaring in a quite straightforward way. Because of the nature of this book, we repeat the statements. We believe that the readers can go through those parts very quickly since there is nothing new to do.

Situation 18.1 Let U be an admissible orbifold with corners. We decompose its normalized boundary ∂U into a *disjoint* union

$$\partial U = \partial^0 U \cup \partial^1 U,$$

where both $\partial^0 U$ and $\partial^1 U$ are open subsets in ∂U. We denote this decomposition by $\mathfrak{C}$. We also denote $\partial^0 U$ by $\partial_{\mathfrak{C}} U$. ■

Definition 18.2 In Situation 18.1 we define a closed subset $S_k^{\mathfrak{C}}(U)$ of U as follows. Let k be given. For any given $p \in U$, we take its orbifold chart (V_p, Γ_p, ϕ_p) where $V_p \subset [V_p] \times [0,1)^{k'}$ and $p = \phi(y_0, (0, \dots, 0))$. For $i = 1, 2, \dots, k'$ we put

$$\partial_i V_p = V_p \cap ([V_p] \times [0,1)^{i-1} \times \{0\} \times [0,1)^{k'-i}).$$

We require

$$\#\{i \in \{1, \dots, k'\} \mid \phi(\partial_i V_p) \subset \partial^0 U\} = k. \tag{18.7}$$

Then $\overset{\circ}{S}{}_k^{\mathfrak{C}}(U)$ is defined to be the set of all $p \in U$ such that (18.7) is satisfied. We put

$$S_k^{\mathfrak{C}}(U) = \bigcup_{\ell \geq k} \overset{\circ}{S}{}_\ell^{\mathfrak{C}}(U).$$

Convention 18.3 In the case $p \in S_k^{\mathfrak{C}}(U)$ as above, we take its orbifold chart (V_p, Γ_p, ϕ_p) as above such that $\phi(\partial_i V_p) \subset \partial^0 U$ if and only if $i = k'-k+1, \dots, k'$.

Situation 18.4 Let $(X, \widehat{\mathcal{U}})$ be an n-dimensional K-space. We assume that for each $p \in \partial X$, we have a decomposition of its Kuranishi neighborhood into a disjoint union

$$\partial U_p = \partial^0 U_p \cup \partial^1 U_p \tag{18.8}$$

such that for each coordinate change Φ_{pq} with $p, q \in \partial X$ we have

$$\varphi_{pq}(U_{pq} \cap \partial^0 U_q) \subset \partial^0 U_p, \qquad \varphi_{pq}(U_{pq} \cap \partial^1 U_q) \subset \partial^1 U_p. \tag{18.9}$$

We also denote this decomposition by $\mathfrak{C}$.■

Definition 18.5 In Situation 18.4 we define $S_k^{\mathfrak{C}}(X, \widehat{\mathcal{U}})$ as follows. If $p \notin \partial X$ then $p \in S_0^{\mathfrak{C}}(X, \widehat{\mathcal{U}})$ but $p \notin S_k^{\mathfrak{C}}(X, \widehat{\mathcal{U}})$ for $k \geq 1$. If $p \in \partial X$ then $p \in S_k^{\mathfrak{C}}(X, \widehat{\mathcal{U}})$ if and only if $o_p \in S_k^{\mathfrak{C}}(U_p)$. We define $\partial_{\mathfrak{C}} X = S_1^{\mathfrak{C}}(X, \widehat{\mathcal{U}})$. We also put

$$\overset{\circ}{S}{}^{\mathfrak{C}}_{k}(X,\widehat{\mathcal{U}}) = S^{\mathfrak{C}}_{k}(X,\widehat{\mathcal{U}}) \setminus \bigcup_{\ell>k} S^{\mathfrak{C}}_{\ell}(X,\widehat{\mathcal{U}}).$$

We can define a similar notion for a good coordinate system by modifying the above definition in an obvious way.[1]

We can generalize Proposition 24.17 without change as follows.

Proposition 18.6 *In Situation 18.4, for each k there exist a compact $(n-k)$-dimensional K-space $\widehat{S}^{\mathfrak{C}}_{k}(X,\widehat{\mathcal{U}})$ with corners, a map $\pi_k : \widehat{S}^{\mathfrak{C}}_{k}(X,\widehat{\mathcal{U}}) \to S^{\mathfrak{C}}_{k}(X,\widehat{\mathcal{U}})$, and a decomposition of $\partial \widehat{S}^{\mathfrak{C}}_{k}(X,\widehat{\mathcal{U}})$ as in Situation 18.4, (which we also denote by $\mathfrak{C}$), and a map $\pi_{\ell,k} : \widehat{S}^{\mathfrak{C}}_{\ell}(\widehat{S}^{\mathfrak{C}}_{k}(X,\widehat{\mathcal{U}})) \to \widehat{S}^{\mathfrak{C}}_{k+\ell}(X,\widehat{\mathcal{U}})$ for each ℓ, k such that they enjoy the following properties:*

(1) *The map π_k is a continuous map between underlying topological spaces.*
(2) *The interior of $\widehat{S}^{\mathfrak{C}}_{k}(X,\widehat{\mathcal{U}})$ is isomorphic to $\overset{\circ}{S}{}^{\mathfrak{C}}_{k}(X,\widehat{\mathcal{U}})$. The underlying homeomorphism of this isomorphism is the restriction of π_k.*
(3) *The map $\pi_{\ell,k}$ is an $(\ell+k)!/\ell!k!$ fold covering map of K-spaces.*
(4) *The following objects on $(X,\widehat{\mathcal{U}})$ induce those on $\widehat{S}^{\mathfrak{C}}_{k}(X,\widehat{\mathcal{U}})$. Moreover the induced objects are compatible with the covering maps $\pi_{\ell,k}$.*

  (a) *CF-perturbation.*
  (b) *Multivalued perturbation.*
  (c) *Differential form.*
  (d) *Strongly continuous map. Strongly smooth map.*
  (e) *Covering map.*

(5) *The following diagram commutes:*

$$\begin{array}{ccc}
\widehat{S}^{\mathfrak{C}}_{k_1}(\widehat{S}^{\mathfrak{C}}_{k_2}(\widehat{S}^{\mathfrak{C}}_{k_3}(X,\widehat{\mathcal{U}}))) & \xrightarrow{\pi_{k_1,k_2}} & \widehat{S}^{\mathfrak{C}}_{k_1+k_2}(\widehat{S}^{\mathfrak{C}}_{k_3}(X,\widehat{\mathcal{U}})) \\
{\scriptstyle \widehat{S}^{\mathfrak{C}}_{k_1}(\pi_{k_2,k_3})}\downarrow & & \downarrow{\scriptstyle \pi_{k_1+k_2,k_3}} \\
\widehat{S}^{\mathfrak{C}}_{k_1}(\widehat{S}^{\mathfrak{C}}_{k_2+k_3}(X,\widehat{\mathcal{U}})) & \xrightarrow{\pi_{k_1,k_2+k_3}} & \widehat{S}^{\mathfrak{C}}_{k_1+k_2+k_3}(X,\widehat{\mathcal{U}})
\end{array} \tag{18.10}$$

Here $\widehat{S}^{\mathfrak{C}}_{k_1}(\pi_{k_2,k_3})$ is the covering map induced from π_{k_2,k_3}.

(6) *For $i = 1,2$ let $f_i : (X_i,\widehat{\mathcal{U}_i}) \to M$ be a strongly smooth map. If f_1 is transversal to f_2, then*

$$\widehat{S}^{\mathfrak{C}}_{k}\left((X_1,\widehat{\mathcal{U}_1}) \times_M (X_2,\widehat{\mathcal{U}_2})\right) \cong \coprod_{k_1+k_2=k} \widehat{S}^{\mathfrak{C}}_{k_1}(X_1,\widehat{\mathcal{U}_1}) \times_M \widehat{S}^{\mathfrak{C}}_{k_2}(X_2,\widehat{\mathcal{U}_2}).$$

[1] We do not define it in detail since it is never used in this book.

Here the right hand side is the disjoint union. The decomposition of the fiber product $(X_1, \widehat{\mathcal{U}_1}) \times_M (X_2, \widehat{\mathcal{U}_2})$ *as in Situation* 18.4 *is induced from those of* $(X_1, \widehat{\mathcal{U}_1})$ *and* $(X_2, \widehat{\mathcal{U}_2})$ *as follows:*

$$\begin{aligned}&\partial^0((X_1, \widehat{\mathcal{U}_1}) \times_M (X_2, \widehat{\mathcal{U}_2}))\\&= \partial^0(X_1, \widehat{\mathcal{U}_1}) \times_M (X_2, \widehat{\mathcal{U}_2}) \cup (-1)^{\dim X_1 + \dim M}(X_1, \widehat{\mathcal{U}_1}) \times_M \partial^0(X_2, \widehat{\mathcal{U}_2}).\end{aligned}$$

(7) *(1)–(6) also hold when we replace 'Kuranishi structure' by 'good coordinate system'.*
(8) *Various kinds of embeddings of Kuranishi structures and/or good coordinate systems induce ones of* $\widehat{S}_k^{\mathfrak{C}}(X, \widehat{\mathcal{U}})$.

The proof is the same as the proof of Proposition 24.17 and so is omitted.

Next we generalize the process of outer collaring in Chap. 17 to partial outer collaring.

Suppose that we are in Situation 17.6 and a decomposition $\partial U = \partial^0 U \cup \partial^1 U$ is given as in Situation 18.1. We define a map

$$\mathcal{R}_x^{\mathfrak{C}} : [V_x] \times [0,1)^{k'-k} \times (-\infty, 1)^k \to [V_x] \times [0,1)^{k'}$$

by

$$\mathcal{R}_x^{\mathfrak{C}}(\overline{y}, (t_1, \dots, t_{k'})) = (\overline{y}, (t'_1, \dots, t'_{k'})),$$

where

$$t'_i = \begin{cases} t_i & \text{if } t_i \ge 0, \\ 0 & \text{if } t_i \le 0. \end{cases}$$

For $\tau \ge 0$, we define an open subset $V_x^{\mathfrak{C}\boxplus\tau}$ of $[V_x] \times [0,1)^{k'-k} \times [-\tau, 1)^k$ to be

$$V_x^{\mathfrak{C}\boxplus\tau} = (\mathcal{R}_x^{\mathfrak{C}})^{-1}(V_x) \cap ([V_x] \times [0,1)^{k'-k} \times [-\tau, 1)^k).$$

Then $\mathcal{R}_x^{\mathfrak{C}}$ induces a map $\mathcal{R}_x^{\mathfrak{C}} : V_x^{\mathfrak{C}\boxplus\tau} \to V_x \subset [V_x] \times [0,1)^{k'}$.

We can define a Γ_x action on $V_x^{\mathfrak{C}\boxplus\tau}$ in the same way as in Definition 17.8 and put $U_x^{\mathfrak{C}\boxplus\tau} = V_x^{\mathfrak{C}\boxplus\tau}/\Gamma_x$. We define $\mathcal{E}_x^{\mathfrak{C}\boxplus\tau}$ in the same way as in (17.1) by

$$\mathcal{E}_x^{\mathfrak{C}\boxplus\tau} = (\mathcal{R}_x^{\mathfrak{C}})^*(\mathcal{E}_x) = (E_x \times V_x^{\mathfrak{C}\boxplus\tau})/\Gamma_x.$$

The section s_x of $\mathcal{E}_x$ induces a section $s_x^{\mathfrak{C}\boxplus\tau}$ of $\mathcal{E}_x^{\mathfrak{C}\boxplus\tau}$ in an obvious way. We define

$$(X \cap V_x)^{\mathfrak{C}\boxplus\tau} = (s_x^{\mathfrak{C}\boxplus\tau})^{-1}(0)/\Gamma_x.$$

Let $\psi_x^{\mathfrak{C}\boxplus\tau} : (s_x^{\mathfrak{C}\boxplus\tau})^{-1}(0)/\Gamma_x \to (X \cap V_x)^{\mathfrak{C}\boxplus\tau}$ be the identity map. Then similarly to Lemma-Definition 17.13, we find that

$$\mathcal{U}^{\mathfrak{C}\boxplus\tau} = (V_x^{\mathfrak{C}\boxplus\tau}/\Gamma_x, \mathcal{E}_x^{\mathfrak{C}\boxplus\tau}, \psi_x^{\mathfrak{C}\boxplus\tau}, s_x^{\mathfrak{C}\boxplus\tau})$$

is a Kuranishi chart of $(X \cap V_x)^{\mathfrak{C}\boxplus\tau}$. Moreover the following objects on $\mathcal{U} = (U, \mathcal{E}, s, \psi)$ induce the corresponding objects on $\mathcal{U}^{\mathfrak{C}\boxplus\tau}$. The proof is the same as that of Lemma-Definition 17.14 so is omitted.

- CF-perturbation.
- Strongly smooth map.
- Differential form.
- Multivalued perturbation.

We put $\overset{\circ\circ}{S}_k(V_x^{\mathfrak{C}\boxplus\tau}) = S_k(V_x^{\mathfrak{C}\boxplus\tau}) \cap (\mathcal{R}_x^{\mathfrak{C}})^{-1}(\overset{\circ}{S}_k(V_x^{\mathfrak{C}\boxplus\tau}))$. Then Lemma 17.15 is generalized in an obvious way. Furthermore we have:

Lemma 18.7 *Suppose we are in Situation 17.26 and $\partial U_i = \partial^0 U_i \cup \partial^1 U_i$ for $i = 1, 2$. We denote by $\mathfrak{C}$ the decomposition of the boundary. Then $\Phi_{21} = (\varphi_{21}, \widehat{\varphi}_{21})$ induces an embedding $\Phi_{21}^{\mathfrak{C}\boxplus\tau} : \mathcal{U}_1^{\mathfrak{C}\boxplus\tau} \to \mathcal{U}_2^{\mathfrak{C}\boxplus\tau}$ of Kuranishi charts whose restriction to $\mathcal{U}_1$ coincides with Φ_{21}. Moreover we have:*

(1) *In Situation 17.26 (1), $\mathfrak{S}^{1\mathfrak{C}\boxplus\tau}$, $\mathfrak{S}^{2\mathfrak{C}\boxplus\tau}$ are compatible with $\Phi_{21}^{\mathfrak{C}\boxplus\tau}$.*
(2) *In Situation 17.26 (2), $(\varphi_{21}^{\mathfrak{C}\boxplus\tau})^*(h_2^{\mathfrak{C}\boxplus\tau}) = (h_1^{\mathfrak{C}\boxplus\tau})$.*
(3) *In Situation 17.26 (3), $f_2^{\mathfrak{C}\boxplus\tau} \circ \varphi_{21}^{\mathfrak{C}\boxplus\tau} = f_1^{\mathfrak{C}\boxplus\tau}$.*
(4) *In Situation 17.26 (4), $\mathfrak{s}^{1\mathfrak{C}\boxplus\tau}$, $\mathfrak{s}^{2\mathfrak{C}\boxplus\tau}$ are compatible with $\Phi_{21}^{\mathfrak{C}\boxplus\tau}$.*
(5) *In Situation 17.26 (5), we have $\Phi_{\mathfrak{C}31}^{\boxplus\tau} = \Phi_{32}^{\mathfrak{C}\boxplus\tau} \circ \Phi_{21}^{\mathfrak{C}\boxplus\tau}$.*
(6) $\varphi_{21}^{\mathfrak{C}\boxplus\tau} \circ \mathcal{R}_1^{\mathfrak{C}} = \mathcal{R}_2^{\mathfrak{C}} \circ \varphi_{21}^{\mathfrak{C}\boxplus\tau}$.

The proof is the same as the proof of Lemma 17.27.

Definition 18.8 Suppose we are in Situation 18.4. Consider a disjoint union

$$\coprod_{p \in X} (s_p^{\mathfrak{C}\boxplus\tau})^{-1}(0)/\Gamma_p$$

and define an equivalence relation $\sim$ on it as follows: Let $x_p \in (s_p^{\mathfrak{C}\boxplus\tau})^{-1}(0)$ and $x_q \in (s_q^{\mathfrak{C}\boxplus\tau})^{-1}(0)$. Then we define $[x_p] \sim [x_q]$ if there exist $r \in X$ and $x_r \in (s_p^{\mathfrak{C}\boxplus\tau})^{-1}(0) \cap U_{pr}^{\mathfrak{C}\boxplus\tau} \cap U_{qr}^{\mathfrak{C}\boxplus\tau}$ such that

$$[x_p] = \varphi_{pr}^{\mathfrak{C}\boxplus\tau}([x_r]), \quad [x_q] = \varphi_{qr}^{\mathfrak{C}\boxplus\tau}([x_r]). \tag{18.11}$$

The same argument as in Lemma 17.30 shows that $\sim$ is an equivalence relation. Then we define a topological space $X^{\mathfrak{C}\boxplus\tau}$ by the set of the equivalence classes of this equivalence relation $\sim$ with the quotient topology:

$$X^{\mathfrak{C}\boxplus\tau} := \left(\coprod_{p \in X} (s_p^{\mathfrak{C}\boxplus\tau})^{-1}(0)/\Gamma_p \right) / \sim .$$

In the situation of Definition 18.8 we put

$$\overset{\circ\circ}{S}_k(X^{\mathfrak{C}\boxplus\tau}) = S_k(X^{\mathfrak{C}\boxplus\tau}, \widehat{\mathcal{U}}^{\mathfrak{C}\boxplus\tau}) \cap (\mathcal{R}^{\mathfrak{C}})^{-1}(\overset{\circ}{S}_k(X, \widehat{\mathcal{U}})). \tag{18.12}$$

We define $B_\tau(\overset{\circ\circ}{S}_k(X^{\mathfrak{C}\boxplus\tau})) \subset X^{\mathfrak{C}\boxplus\tau}$ as the union of

$$\psi_p^{\mathfrak{C}\boxplus\tau}\left((s_p^{\mathfrak{C}\boxplus\tau})^{-1}(0) \cap \{(\overline{y}, (t_1, \dots, t_k)) \mid t_i \le 0,\ i = k' - k + 1, \dots, k'\}\right). \tag{18.13}$$

We can show

$$X^{\mathfrak{C}\boxplus\tau} = \coprod_k B_\tau(\overset{\circ\circ}{S}_k(X^{\mathfrak{C}\boxplus\tau})) \tag{18.14}$$

in the same way as in Lemma 17.36.

Definition 18.9

(1) Let $p' \in \overset{\circ\circ}{S}_k(X^{\mathfrak{C}\boxplus\tau})$. A *$\tau$-$\mathfrak{C}$-collared Kuranishi neighborhood* at p' is a Kuranishi chart $\mathcal{U}_{p'}$ of $X^{\mathfrak{C}\boxplus\tau}$ which is $(\mathcal{U}_p)^{\mathfrak{C}\boxplus\tau}$ for a certain Kuranishi neighborhood $\mathcal{U}_p$ of $p = \mathcal{R}^{\mathfrak{C}}(p')$.

(2) Let $p' \in \overset{\circ\circ}{S}_k(X^{\mathfrak{C}\boxplus\tau})$, $q' \in \overset{\circ\circ}{S}_\ell(X^{\mathfrak{C}\boxplus\tau})$ and let $\mathcal{U}_{p'} = (\mathcal{U}_p)^{\mathfrak{C}\boxplus\tau}$, $\mathcal{U}_{q'} = (\mathcal{U}_q)^{\mathfrak{C}\boxplus\tau}$ be their τ-$\mathfrak{C}$-collared Kuranishi neighborhoods, respectively. Suppose $q' \in \psi_{p'}(s_{p'}^{-1}(0))$. A *$\tau$-$\mathfrak{C}$-collared coordinate change* $\Phi_{p'q'}$ from $\mathcal{U}_{q'}$ to $\mathcal{U}_{p'}$ is by definition $\Phi_{pq}^{\mathfrak{C}\boxplus\tau}$, where Φ_{pq} is a coordinate change from $\mathcal{U}_q$ to $\mathcal{U}_p$.

(3) A *τ-$\mathfrak{C}$-collared Kuranishi structure* $\widehat{\mathcal{U}'}$ on $X^{\mathfrak{C}\boxplus\tau}$ is the following objects:

   (a) For each $p' \in \overset{\circ\circ}{S}_k(X^{\mathfrak{C}\boxplus\tau})$, $\widehat{\mathcal{U}'}$ assigns a τ-$\mathfrak{C}$-collared Kuranishi neighborhood $\mathcal{U}_{p'}$.

   (b) For each $p' \in \overset{\circ\circ}{S}_k(X^{\mathfrak{C}\boxplus\tau})$ and $q' \in \overset{\circ\circ}{S}_\ell(X^{\mathfrak{C}\boxplus\tau})$ with $q' \in \psi_{p'}(s_{p'}^{-1}(0))$, $\widehat{\mathcal{U}'}$ assigns a τ-$\mathfrak{C}$-collared coordinate change $\Phi_{p'q'}$.

   (c) If $p' \in \overset{\circ\circ}{S}_k(X^{\mathfrak{C}\boxplus\tau})$, $q' \in \overset{\circ\circ}{S}_\ell(X^{\mathfrak{C}\boxplus\tau})$, $r' \in \overset{\circ\circ}{S}_m(X^{\mathfrak{C}\boxplus\tau})$ with $q' \in \psi_{p'}(s_{p'}^{-1}(0))$ and $r' \in \psi_{q'}(s_{q'}^{-1}(0))$, then we require

$$\Phi_{p'q'} \circ \Phi_{q'r'}|_{U_{p'q'r'}} = \Phi_{p'r'}|_{U_{p'q'r'}}$$

   where $U_{p'q'r'} = U_{p'r'} \cap \varphi_{q'r'}^{-1}(U_{p'q'})$.

(4) A *τ-$\mathfrak{C}$-collared K-space* is a pair of paracompact Hausdorff space $X^{\mathfrak{C}\boxplus\tau}$ and its τ-$\mathfrak{C}$-collared Kuranishi structure $\widehat{\mathcal{U}^{\mathfrak{C}\boxplus\tau}}$. We call

$$(X^{\mathfrak{C}\boxplus\tau}, \widehat{\mathcal{U}^{\mathfrak{C}\boxplus\tau}})$$

the *τ-$\mathfrak{C}$-corner trivialization*, or *partial outer collaring*, of $(X, \widehat{\mathcal{U}})$. We sometimes write

$$(X, \widehat{\mathcal{U}})^{\mathfrak{C}\boxplus\tau}$$

in place of $(X^{\mathfrak{C}\boxplus\tau}, \widehat{\mathcal{U}^{\mathfrak{C}\boxplus\tau}})$.

(5) We can define the notion of a *τ-$\mathfrak{C}$-collared CF-perturbation, τ-$\mathfrak{C}$-collared multivalued perturbation, τ-$\mathfrak{C}$-collared good coordinate system, τ-$\mathfrak{C}$-collared Kuranishi chart, τ-$\mathfrak{C}$-collared vector bundle, τ-$\mathfrak{C}$-collared smooth section, τ-$\mathfrak{C}$-collared embedding* of various kinds, etc. in the same way.

The decomposition $\mathfrak{C}$ induces a decomposition of the boundary $\partial(X^{\mathfrak{C}\boxplus\tau}, \widehat{\mathcal{U}}^{\mathfrak{C}\boxplus\tau})$ in an obvious way. We also denote it by $\mathfrak{C}$.

Lemma 18.10 *Lemma* 17.40 *can be generalized to a $\mathfrak{C}$-version in an obvious way.*

Lemma 18.11 *If $(X', \widehat{\mathcal{U}'})$ is τ–$\mathfrak{C}$-collared, then for any $0 < \tau' < \tau$, X' has a τ'-$\mathfrak{C}$-collared Kuranishi structure determined in a canonical way from the τ–$\mathfrak{C}$-collared Kuranishi structure $(X', \widehat{\mathcal{U}'})$. The same holds for a CF-perturbation, multivalued perturbation and good coordinate system.*

The proof is the same as the proof of Definition-Lemma 17.46.

Situation 18.12 Let $(X, \widehat{\mathcal{U}})$ be a K-space. Suppose $\mathfrak{C}_1$, $\mathfrak{C}_2$, $\mathfrak{C}$ are decompositions of $\partial(X, \widehat{\mathcal{U}})$ as in Situation 18.4. We assume the following two conditions:

(1) $\partial_{\mathfrak{C}_1} U_p \cap \partial_{\mathfrak{C}_2} U_p = \emptyset$.
(2) $\partial_{\mathfrak{C}_1} U_p \cup \partial_{\mathfrak{C}_2} U_p = \partial_{\mathfrak{C}} U_p$.■

Lemma 18.13 *In Situation* 18.12 *we have the following canonical isomorphisms:*

$$((X, \widehat{\mathcal{U}})^{\mathfrak{C}_1\boxplus\tau})^{\mathfrak{C}_2\boxplus\tau} \cong ((X, \widehat{\mathcal{U}})^{\mathfrak{C}_2\boxplus\tau})^{\mathfrak{C}_1\boxplus\tau} \cong (X, \widehat{\mathcal{U}})^{\mathfrak{C}\boxplus\tau}.$$

Remark 18.14 The decomposition $\mathfrak{C}_2$ of $\partial(X, \widehat{\mathcal{U}})$ induces one on $\partial((X, \widehat{\mathcal{U}})^{\mathfrak{C}_1\boxplus\tau})$, which we denote by the same symbol. We used this fact in the statement of Lemma 18.13.

***Proof of Lemma* 18.13** Suppose a Kuranishi chart of $(X, \widehat{\mathcal{U}})$ is given as a quotient open subset V_p of $[V_p] \times [0,1)^{k_1} \times [0,1)^{k_2} \times [0,1)^{k_3}$ where

$$\partial_{\mathfrak{C}_1} V_p = V_p \cap \left([V_p] \times [0,1)^{k_1} \times \partial[0,1)^{k_2} \times [0,1)^{k_3}\right),$$

$$\partial_{\mathfrak{C}_2} V_p = V_p \cap \left([V_p] \times [0,1)^{k_1} \times [0,1)^{k_2} \times \partial[0,1)^{k_3}\right).$$

Then

$$\partial_{\mathfrak{C}} V_p = V_p \cap \left([V_p] \times [0,1)^{k_1} \times \partial([0,1)^{k_2} \times [0,1)^{k_3})\right).$$

Therefore we have

$$V_p^{\mathfrak{C}_1 \boxplus \tau} = \mathcal{R}^{-1}(V_p) \cap \left([V_p] \times [0,1)^{k_1} \times [-\tau,1)^{k_2} \times [0,1)^{k_3}\right)$$

and

$$V_p^{\mathfrak{C} \boxplus \tau} = \mathcal{R}^{-1}(V_p) \cap \left([V_p] \times [0,1)^{k_1} \times [-\tau,1)^{k_2} \times [-\tau,1)^{k_3})\right).$$

Lemma 18.13 follows easily. □

18.4 In Which Sense Is Corner Smoothing Canonical?

We next discuss corner smoothing. Corner smoothing of manifolds is a standard process (see, for example, [Ta]) and its generalization to orbifolds is also straightforward. An issue in generalizing the corner smoothing to Kuranishi structures lies in the way we fix a smooth structure around the corners and how much we can make the smooth structure canonical. We can go around these technicalities quite nicely, especially when we use *outer collaring*, introduced in Chap. 17, which is exactly the case, for which we do the smoothing. We discuss those issues in Sects. 18.4, 18.5, and 18.6.

We begin with the following remark. Let M be a manifold (or an orbifold) with corners. We have a structure of a manifold with boundary (but without corners) on the *same* underlying topological space. We write this manifold with boundary (but without corners) as M'. We denote by $i' : M \to M'$ the identity map as sets. Then it has the following properties:

(*) i' induces a smooth embedding $\widehat{S}_k(M) \to M'$.

Moreover, if M is admissible, we have the following:

(1) If $f : M \to \mathbb{R}$ is an admissible function, then $f \circ (i')^{-1}$ is smooth.
(2) If $\mathcal{E} \to M$ is an admissible vector bundle, the underlying continuous map $\mathcal{E} \to M$ has a structure of C^∞-vector bundle on M'. We write it $\mathcal{E}' \to M'$.
(3) If s is an admissible section of $\mathcal{E}$, the same (set-theoretical) map $M' \to \mathcal{E}'$ is a smooth section.

The proofs of (1)–(3) above are easy. Since we do not use them, we do not prove them. We next see how much the smooth structure of M' is canonical. The following statement is also standard.

Lemma 18.15 *We can construct M' from M in such a way that the differentiable manifold M' is well-defined modulo diffeomorphism. More precisely, we have the following: Suppose we obtain another M'' from M. Let $i'' : M \to M''$ be the identity* *map. Then we have a diffeomorphism* $f : M' \to M''$ *such that* $f(i'(S_k(M)) = i''(S_k(M))$, *for* $k = 0, 1, 2, \ldots$.

In other words, M' is well-defined modulo stratification preserving diffeomorphism. Here the stratification means the corner structure stratification of M. This lemma is fairly standard and its proof is omitted. It seems that it is more nontrivial to find a 'canonical' way so that the above diffeomorphism f can be taken to be the identity map. In this book, we do not try to find such a way in general situation of (admissible) orbifold, but will do so in the case of *collared orbifold*.

Before doing so, we explain a reason why the uniqueness in the sense of Lemma 18.15 is not enough for our purpose. When we generalize the process of corner smoothing to that for Kuranishi structures, we need to study the situation where we have an embedding $N \to M$ of cornered orbifolds. When we smooth the corners of N and M, we want the *same* map $N' \to M'$ to be a *smooth* embedding. This is not obvious because of the non-uniqueness of the smooth structure we put on M' and N'. However, it is still true and not difficult to prove that we *can find* smooth structures of M' and N' so that $N' \to M'$ is a smooth embedding.

On the other hand, in order to smooth the corners of Kuranishi structures we need to smooth the corners of all the Kuranishi charts, simultaneously. This now becomes a nontrivial problem. If we try to use the uniqueness in Lemma 18.15, we should include the diffeomorphism f as a part of the data in the construction. Then the compatibility of the coordinate change might be broken.[2]

We go around this issue by using the collar. When our orbifold is collared, the way we smooth the corners still involves choices. However, we can make the choice to smooth the model $[0, 1)^k$ once for all and then use that particular choice to smooth all the collared orbifolds, simultaneously. This way is so canonical that all the embeddings of Kuranishi charts become smooth embeddings automatically after smoothing the corners.

In other words, we will define a functor

$$\begin{aligned}&\{\text{Collared orbifold with or without corners}\}\\ &\longrightarrow \{\text{Collared orbifold with or without boundary but no corners}\},\end{aligned}$$

where morphisms of those categories are embeddings. Then it induces a similar functor on the category of Kuranishi structures of given spaces (where morphisms are strict KK-embeddings).

[2] It seems that we can still prove that for a given good coordinate system we can construct a corner smoothing compatible with the coordinate change for any proper open substructure of it. We need to work out rather cumbersome induction to prove it at the level of detail we intend to achieve in this book.

18.5 Corner Smoothing of $[0, \infty)^k$

In this section, we fix data which we need to smooth the corner of a partially collared orbifold and a partially collared Kuranishi structure. Namely, we fix a way to smooth the local model $[0, \infty)^k$ so that it is compatible with various k and also with the symmetry exchanging the factors. The latter is important in studying the case of orbifolds. Let $\mathrm{Perm}(k+1)$ be a group of permutations of $\{1, \dots, k+1\}$.

Definition 18.16 We define a $\mathrm{Perm}(k+1)$ action on $\mathbb{R}^k$ as follows. We regard $\mathbb{R}^k$ as a linear subspace of $\mathbb{R}^{k+1}$ defined by the equation $\sum_{i=0}^k t_i = 0$. In other words, we put

$$\mathbb{R}^k = \left\{ (t_0, \dots, t_k) \in \mathbb{R}^{k+1} \;\middle|\; \sum_{i=0}^{k} t_i = 0 \right\}.$$

The group $\mathrm{Perm}(k+1)$ acts on $\mathbb{R}^{k+1}$ by exchanging the factors. It restricts to an action of $\mathrm{Perm}(k+1)$ to $\mathbb{R}^k$. In this section, the $\mathrm{Perm}(k+1)$ action on $\mathbb{R}^k$ always means this particular action.

Below we will show the existence of a set of homeomorphisms

$$\Phi_k : [0, \infty)^k \to \mathbb{R}^{k-1} \times [0, \infty) \tag{18.15}$$

and smooth structures $\mathfrak{sm}_k$ on $[0, \infty)^k$, simultaneously, for any $k \in \mathbb{Z}_+$, which satisfy Condition 18.17.

Condition 18.17 We require Φ_k and $\mathfrak{sm}_k$ to satisfy the following conditions.

(1) Φ_k is a diffeomorphism from $([0, \infty)^k, \mathfrak{sm}_k)$ to $\mathbb{R}^{k-1} \times [0, \infty)$. Here we use the standard smooth structure on $\mathbb{R}^{k-1} \times [0, \infty)$.
(2) Let $\Phi_k(t_1, \dots, t_k) = (x, t)$. Then

$$\Phi_k(ct_1, \dots, ct_k) = (cx, ct)$$

for $c \in \mathbb{R}_+$.
(3) For $\mathbf{t} \in [0, \infty)^k \setminus 0$ there exists an open neighborhood of $\mathbf{t}$ that is isometric to $V \times [0, \epsilon)^\ell$ where $V \subset \mathbb{R}^{k-\ell}$. (Here V is an open set. When we say 'isometric', we use the Euclidian metrics on $[0, \infty)^k$, $\mathbb{R}^{k-\ell}$, $[0, \epsilon)^\ell$.) Then the map

$$\mathrm{id} \times \Phi_\ell : V \times [0, \epsilon)^\ell \to V \times \mathbb{R}^{\ell-1} \times [0, \infty)$$

is a diffeomorphism onto its image. Here we put the restriction of the smooth structure $\mathfrak{sm}_k$ to $V \times [0, \epsilon)^\ell$. (The space $V \times [0, \epsilon)^\ell$ is identified with an open subset of $[0, \infty)^k \setminus 0$ by the isometry.) The smooth structure of $V \times \mathbb{R}^{\ell-1} \times [0, \infty)$ is the standard one.

(4) The map $\Phi_k : [0, \infty)^k \to \mathbb{R}^{k-1} \times [0, \infty)$ is Perm(k) equivariant. Here Perm(k) acts on $[0, \infty)^k$ by permutation of factors, and acts on $\mathbb{R}^{k-1}$ by Definition 18.16. On the last factor $[0, \infty)$ the action is trivial.

Remark 18.18 In (3) above we require a neighborhood of $\mathbf{t}$ to be isometric to $V \times [0, \epsilon)^\ell$. The reason why we require such a rather restrictive assumption that they are isometric is that we want to specify the diffeomorphism from a neighborhood of $\mathbf{t}$ to $V \times [0, \epsilon)^\ell$. We use the Euclidian metric on $\mathbb{R}^n$ here only to make the choice of diffeomorphism (that is, isometry) as canonical as possible.

Lemma 18.19 *For any $k \in \mathbb{Z}_+$ there exist Φ_k and $\mathfrak{sm}_k$ satisfying Condition* 18.17.

Proof The proof is by induction on k. For $k = 1$, Φ_1 is the identity map and $\mathfrak{sm}_1$ is the standard smooth structure on $[0, \infty)$.

Suppose we have Φ_i, $\mathfrak{sm}_i$ for $i < k$. We observe that Condition (3) determines a smooth structure $\mathfrak{sm}_k$ on $[0, \infty)^k \setminus 0$ uniquely. Indeed, well-definedness of this structure can be checked by Condition (3) itself inductively. Moreover, by the definition of the smooth structure $\mathfrak{sm}_k$, we find that the Perm(k)-action is smooth with respect to this smooth structure. The map $(t_1, \ldots, t_k) \mapsto (ct_1, \ldots, ct_k)$ is also a diffeomorphism for this smooth structure if $c \in \mathbb{R}_+$.

Next we will construct a homeomorphism Φ_k and extend the smooth structure $\mathfrak{sm}_k$ to $[0, \infty)^k$. We choose a compact subset $S \subset [0, \infty)^k \setminus 0$ which is a smooth $(k-1)$-dimensional submanifold with corners (with respect to the standard structure of manifold with corners of $[0, \infty)^k$) such that:

(a) S is a slice of the multiplicative $\mathbb{R}_+$ action on $[0, \infty)^k \setminus 0$.
(b) S is perpendicular to all the strata $\overset{\circ}{S}_\ell([0, \infty)^k)$. We can take a tubular neighborhood of $\overset{\circ}{S}_\ell([0, \infty)^k) \cap S$ in S such that the fiber of the projection to $\overset{\circ}{S}_\ell([0, \infty)^k) \cap S$ is flat with respect to the Euclidean metric of $[0, \infty)^k$.
(c) S is invariant under the Perm(k)-action on $[0, \infty)^k$.

We can find such an S by Perm(k)-equivariantly modifying the intersection of the unit ball and $[0, \infty)^k$ around the boundary slightly.

By Condition (b) and the induction hypothesis, S is a smooth submanifold with boundary of $([0, \infty)^k \setminus 0, \mathfrak{sm}_k)$. Since we can construct S by modifying the intersection of unit ball and $[0, \infty)^k$ around the boundary slightly, there exists a Perm(k) equivariant diffeomorphism from S to

$$S' = \{(x, t) \in \mathbb{R}^{k-1} \times \mathbb{R}_{\geq 0} \mid \|x\|^2 + t^2 = 1\}.$$

Here we use the smooth structure of S induced from $([0, \infty)^k \setminus 0, \mathfrak{sm}_k)$ and the standard smooth structure on S'. See Fig. 18.2. (Note that S becomes a manifold

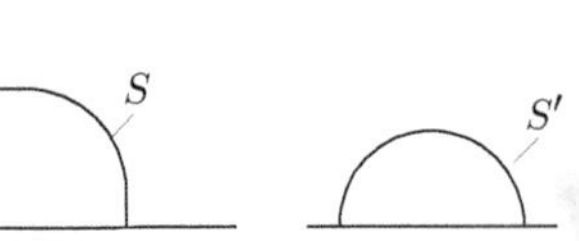

Fig. 18.2 Submanifolds S and S'

with boundary and without corners with respect to this smooth structure.) We fix this diffeomorphism. Then we can define Φ_k by extending the diffeomorphism $S \to S'$ so that Condition (2) is satisfied. By construction, Φ_k is a diffeomorphism outside the origin.

Then we extend the smooth structure $\mathfrak{sm}_k$ to the origin so that Φ_k also becomes a diffeomorphism at the origin. The proof is now complete by induction. □

Remark 18.20 During the proof we made choices of S and a diffeomorphism between S and S' for each k. The resulting smooth structure $\mathfrak{sm}_k$ *depends* on these choices in the sense that the identity map is not a diffeomorphism when we use two smooth structures obtained by different choices for the source and the target. However, since two different choices of S and the diffeomorphism $S \to S'$ are isotopic to each other, the resulting smooth structure $\mathfrak{sm}_k$ is independent of the choices in the sense of diffeomorphism. (This is a proof of Lemma 18.15 in this case.)

When we apply the construction of corner smoothing of Kuranishi structures, we sometimes need to put collars to the smoothed K-space. We use Lemma 18.22 below for this purpose. Let Φ_k and $\mathfrak{sm}_k$ be as in Condition 18.17.

Condition 18.21 For any $k \in \mathbb{Z}_+$ we consider $\mathfrak{Trans}_{k-1}$ and Ψ_k with the following properties (see Fig. 18.3):

(1) $\mathfrak{Trans}_{k-1}$ is a smooth $(k-1)$-dimensional submanifold of $[0, \infty)^k$ and is contained in $(0, \infty)^k \setminus (1, \infty)^k$.
(2) $\mathfrak{Trans}_{k-1}$ is invariant under the $\mathrm{Perm}(k)$ action on $[0, \infty)^k$.
(3) $\mathfrak{Trans}_{k-1} \cap ([0, \infty)^{k-1} \times [1, \infty)) = \mathfrak{Trans}_{k-2} \times [1, \infty)$. This is an equality as subsets of $[0, \infty)^k = [0, \infty)^{k-1} \times [0, \infty)$.
(4)

$$\Psi_k : [0, 1] \times \mathfrak{Trans}_{k-1} \to [0, \infty)^k$$

is a homeomorphism onto its image. Let $\mathfrak{U}_k$ be its image.

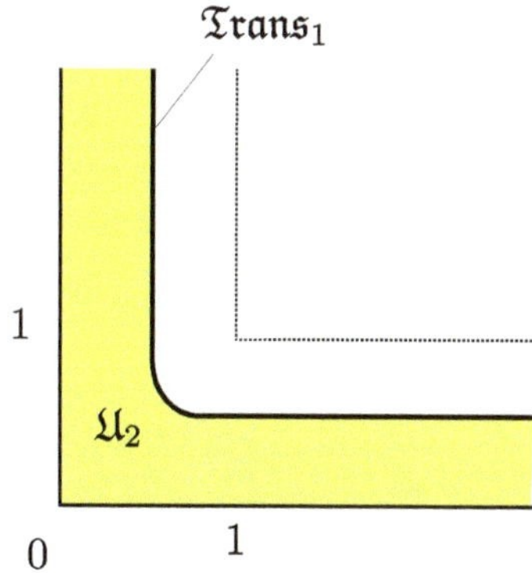

Fig. 18.3 $\mathfrak{Trans}_{k-1}$ and $\mathfrak{U}_k$

(5) Using the smooth structure $\mathfrak{sm}_k$ on $[0,\infty)^k$, the subset $\mathfrak{U}_k \subset [0,\infty)^k$ is a smooth k-dimensional submanifold with boundary and Ψ_k is a diffeomorphism. Moreover

$$\partial \mathfrak{U}_k = \partial([0,\infty)^k) \cup \mathfrak{Trans}_{k-1}$$

and the restriction of Ψ_k to $\{0\} \times \mathfrak{Trans}_{k-1}$ is a diffeomorphism onto $\partial([0,\infty)^k)$. The restriction of Ψ_k to $\{1\} \times \mathfrak{Trans}_{k-1}$ is the identity map.

(6) Ψ_k is equivariant under the $\mathrm{Perm}(k)$ action. The $\mathrm{Perm}(k)$ action on $\mathfrak{Trans}_{k-1}$ is defined in Item (2) and the action on $[0,\infty)^k$ is by permutation of factors.

(7) If $s \ge 1$, $t \in [0,1]$ and $(x_1,\dots,x_{k-1}) \in \mathfrak{Trans}_{k-2}$, then

$$\Psi_k(t,(x_1,\dots,x_{k-1},s)) = (\Psi_{k-1}(t,(x_1,\dots,x_{k-1})),s).$$

Here we use the identification in Item (3) to define the left hand side.

Lemma 18.22 *For any $k \in \mathbb{Z}_+$ there exist $\mathfrak{Trans}_{k-1}$ and Ψ_k satisfying Condition* 18.21. *Moreover, for each given $\delta > 0$, we may take them so that $\mathfrak{U}_k$ contains $[0,\infty)^k \setminus [1-\delta,\infty)^k$.*

Proof The proof is by induction. If $k = 1$, we put $\mathfrak{Trans}_0 = \{1-\delta/2\}$ and there is nothing to prove. Suppose we have $\mathfrak{Trans}_{k'-1}$, $\Psi_{k'}$ for $k' < k$. Conditions 18.21 (2)(3) determine $\mathfrak{Trans}_{k-1}$ outside $[0,1]^k$. Conditions 18.21 (6)(7) determine Ψ_k outside $[0,1]^k$. It is easy to see that we can extend them to $[0,1]^k$ and obtain $\mathfrak{Trans}_{k-1}$, Ψ_k. □

We note that Ψ_k defines a collar of $([0,\infty)^k, \mathfrak{sm}_k)$, which is a manifold with boundary (but without corners).

18.6 Corner Smoothing of Collared Orbifolds and of Kuranishi Structures

In this section we combine the story of partial outer collaring of Kuranishi structure in Sect. 18.3 with that of corner smoothing in Sects. 18.4 and 18.5.

Let U be an orbifold. We consider its normalized boundary ∂U. Let $\mathfrak{C}$ be a decomposition of ∂U as in Situation 18.1. We denote by $U^{\mathfrak{C}\boxplus\tau}$ the partial outer collaring. We will define corner smoothing of $U^{\mathfrak{C}\boxplus\tau}$ below.

Let $p \in \overset{\circ\circ}{S}_k(U^{\mathfrak{C}\boxplus\tau})$ and put $\overline{p} = \mathcal{R}^{\mathfrak{C}}(p)$. The point p has an orbifold neighborhood

$$\mathfrak{V}_p = ([V_{\overline{p}}] \times [-\tau,0)^k, \Gamma_{\overline{p}}, \phi_p). \tag{18.16}$$

Here $([V_{\overline{p}}] \times [0,1)^k, \Gamma_{\overline{p}}, \phi_{\overline{p}})$ is an orbifold neighborhood of $\overline{p}$ in U with $\overline{p} \in \overset{\circ}{S}{}^{\mathfrak{C}}_k(U)$, and $[V_{\overline{p}}]$ may have boundary or corners but the boundary of $[V_{\overline{p}}]$ does not correspond to a boundary component in $\mathfrak{C}$.

Definition 18.23 We define a smooth structure of $\mathfrak{V}_p$ in (18.16) as follows. We identify $[-\tau, 0)^k \cong [0, \tau)^k$ by the diffeomorphism $(t_1, \dots, t_k) \mapsto (t_1 + \tau, \dots, t_k + \tau)$. We use the smooth structure $\mathfrak{sm}_k$ on $[0, \tau)^k$ to obtain a smooth structure on $[-\tau, 0)^k$. Then by taking the direct product we obtain a smooth structure on $[V_{\overline{p}}] \times [-\tau, 0)^k$.

Lemma 18.24 *Consider the smooth structure on* $[V_{\overline{p}}] \times [-\tau, 0)^k$ *given as in Definition* 18.23. *Then we have:*

(1) *The* $\Gamma_{\overline{p}}$ *action on* $[V_{\overline{p}}] \times [-\tau, 0)^k$ *is smooth with respect to this smooth structure.*
(2) *If* $q \in \psi_p([V_{\overline{p}}] \times [-\tau, 0)^k)$ *and* $q \in \overset{\circ\circ}{S}_\ell(U^{\mathfrak{C}\boxplus\tau})$, *then the coordinate change from* $\mathfrak{V}_q$ *to* $\mathfrak{V}_p$ *is smooth with respect to the above smooth structure.*

Proof Item (1) is a consequence of Condition 18.17 (4) and Item (2) follows from Condition 18.17 (3). □

Definition 18.25 We obtain an atlas of an orbifold structure on $U^{\mathfrak{C}\boxplus\tau}$ by Lemma 18.24 and Definition 18.23. We call $U^{\mathfrak{C}\boxplus\tau}$ with this smooth structure the *orbifold with corners obtained by* τ*-*$\mathfrak{C}$*-partial smoothing of corners*, and when no confusion can occur, we simply call the smoothing τ*-partial smoothing of corners*. We denote it by

$$U^{\mathrm{sm}\mathfrak{C}\boxplus\tau}.$$

In the case when $\mathfrak{C}$ is the whole set of all the components of the boundary, this smooth structure has no corner.

Lemma 18.26 *If* $p \in \overset{\circ}{S}_{k+\ell}(U^{\mathfrak{C}\boxplus\tau}) \cap \overset{\circ}{S}{}^{\mathfrak{C}}_k(U^{\mathfrak{C}\boxplus\tau})$, *then* $p \in \overset{\circ}{S}_{\ell+1}(U^{\mathrm{sm}\mathfrak{C}\boxplus\tau})$.

The proof is obvious. We also have the following:

Lemma 18.27 *If* $0 < \tau' < \tau$, *the orbifold obtained by* τ*-partial smoothing of corners is* τ'*-collared.*

Proof Let (V, Γ, ϕ) be an orbifold chart of U. We may choose V so that it is an open subset of $[V] \times [0,1)^k \times [0,1)^{k'-k}$ and Convention 18.3 is satisfied. Then the corresponding orbifold chart of $U^{\mathfrak{C}\boxplus\tau}$ is

$$V = \mathcal{R}^{-1}(V) \cap \left([V] \times [-\tau, 1)^k \times [0,1)^{k'-k}\right).$$

We use the smooth structure $\mathfrak{sm}_k$ on the $[-\tau, 1)^k$ factor (which we identity with $[0, 1+\tau)^k$) and obtain the orbifold chart of $U^{\mathrm{sm}\mathfrak{C}\boxplus\tau}$.

Now we use $\mathfrak{Trans}_{k-1}$ and Ψ_k produced in Lemma 18.22. The map Ψ_k is a diffeomorphism

$$\Psi_k : \mathfrak{Trans}_{k-1} \times [0,1) \to [-\tau, 0)^k$$

on the image. The image is a neighborhood of $\partial[-\tau, 0)^k$. Then we have a smooth embedding

$$[V] \times \mathfrak{Trans}_{k-1} \times [0,1) \times [0,1)^{k'-k} \supseteq V' \to V$$

where V' is an open subset. Its image is a neighborhood of ∂V. This embedding is a diffeomorphism to its image if we use the differential structure after smoothing the corners. This gives a collar of U on this chart. Using Conditions 18.21 (6)(7), we can show that this collar is compatible with the coordinate change.

We can take $\tau' > 0$ for any $\tau > 0$ by choosing δ small in Lemma 18.22. □

In the next lemma we summarize the properties of (partial) corner smoothing.

Lemma 18.28 *Let U be an orbifold. We consider its normalized boundary ∂U. Let $\mathfrak{C}$ be a decomposition of ∂U as in Situation 18.1. We denote by $U^{\mathfrak{C}\boxplus\tau}$ the partial outer collaring. Let $0 < \tau' < \tau$.*

1. *If $\mathcal{E}$ is an admissible vector bundle on U, $\mathcal{E}^{\mathrm{sm}\mathfrak{C}\boxplus\tau}$ becomes an admissible vector bundle on $U^{\mathrm{sm}\mathfrak{C}\boxplus\tau}$. It is τ'-collared.*
2. *If $f : U \to M$ is an admissible map, $f^{\mathfrak{C}\boxplus\tau} : U^{\mathrm{sm}\mathfrak{C}\boxplus\tau} \to M$ is an admissible map from $U^{\mathrm{sm}\mathfrak{C}\boxplus\tau}$. When M has corners, we take the decomposition $\mathfrak{C}$ of the normalized boundary ∂U so that $\partial_{\mathfrak{C}} U$ is contained in the horizontal boundary as in Definition 26.10. Then the same assertion also holds.*
3. *If $f : U_1 \to U_2$ is an admissible embedding and $f(\partial_{\mathfrak{C}_1} U_1) \subset \partial_{\mathfrak{C}_2} U_2 \cap f(U_1)$, then $f^{\mathfrak{C}_1\boxplus\tau} : U_1^{\mathrm{sm}\mathfrak{C}_1\boxplus\tau} \to U_2^{\mathrm{sm}\mathfrak{C}_2\boxplus\tau}$ is an admissible embedding. It is τ'-collared.*
4. *In the situation of (3) suppose $\mathcal{E}_1 \to U_1$, $\mathcal{E}_2 \to U_2$ are admissible vector bundles and f is covered by an admissible embedding of vector bundles $\widehat{f} : \mathcal{E}_1 \to \mathcal{E}_2$. Then $\widehat{f}^{\mathfrak{C}_1\boxplus\tau} : \mathcal{E}_1^{\mathrm{sm}\mathfrak{C}_1\boxplus\tau} \to \mathcal{E}_2^{\mathrm{sm}\mathfrak{C}_2\boxplus\tau}$ is an admissible embedding of vector bundles which covers $f^{\mathfrak{C}_1\boxplus\tau}$. It is τ'-collared.*
5. *In the situation of (1) if s is an admissible section of $\mathcal{E}$, $s^{\mathfrak{C}\boxplus\tau}$ is an admissible section of $\mathcal{E}^{\mathrm{sm}\mathfrak{C}\boxplus\tau}$. It is τ'-collared.*

The proof is obvious.

Now we consider the case of a Kuranishi structure. This generalization is quite straightforward.

Definition 18.29 Let $(X, \widehat{\mathcal{U}})$ be a K-space. Suppose for each $p \in X$ we have a decomposition of ∂U_p into two unions of connected components. Let $\mathfrak{C}_p$ be the first union. We assume that for each $q \in \psi_p(s_p^{-1}(0))$ we have

$$\varphi_{pq}(\partial_{\mathfrak{C}_q} U_q \cap U_{pq}) = \varphi_{pq}(U_{pq}) \cap \partial_{\mathfrak{C}_p} U_p.$$

Let $(X^{\mathfrak{C}\boxplus\tau}, \widehat{\mathcal{U}^{\mathfrak{C}\boxplus\tau}})$ be a τ-$\mathfrak{C}$-collared K-space as in Definition 18.9. We define its *partial corner smoothing* as follows:

If $p \in \overset{\circ\circ}{S}_k(X^{\mathfrak{C}\boxplus\tau})$ and its partially τ-collared Kuranishi neighborhood is $\mathcal{U}^{\mathfrak{C}\boxplus\tau}_{\overline{p}} = (U^{\mathfrak{C}\boxplus\tau}_{\overline{p}}, \mathcal{E}^{\mathfrak{C}\boxplus\tau}_{\overline{p}}, s^{\mathfrak{C}\boxplus\tau}_{\overline{p}}, \psi^{\mathfrak{C}\boxplus\tau}_{\overline{p}})$, then we change its smooth structure by smoothing its corners. We obtain a Kuranishi chart denoted by $\mathcal{U}^{\mathrm{sm}\mathfrak{C}\boxplus\tau}_{\overline{p}}$. Then the coordinate change is induced by one of $\widehat{\mathcal{U}^{\mathfrak{C}\boxplus\tau}}$ because of Lemma 18.28. It is τ'-collared for any $0 < \tau' < \tau$. We call the Kuranishi structure obtained above the *Kuranishi structure obtained by partial corner smoothing* and denote it by

$$\widehat{\mathcal{U}^{\mathrm{sm}\mathfrak{C}\boxplus\tau}}.$$

Lemma 18.30 *In the situation of Definition* 18.29 *the following (τ-collared) object on $(X, \widehat{\mathcal{U}})$ induces the corresponding (τ'-collared) object of the Kuranishi structure obtained by partial corner smoothing:*

(1) *Strongly smooth map.*
(2) *CF-perturbation.*
(3) *Multivalued perturbation.*
(4) *Differential form.*

The proof is immediate from construction.

Lemma 18.31 *In the situation of Lemma* 18.30, *we put*

$$\partial_{\mathfrak{C}}(X, \widehat{\mathcal{U}}) = \coprod_{c \in \mathfrak{C}} \partial_c(X, \widehat{\mathcal{U}}).$$

Let $\widehat{\mathcal{U}^{\mathfrak{C}}}$ be the Kuranishi structure of $\partial_{\mathfrak{C}}(X, \widehat{\mathcal{U}})$ with smoothed corners. We consider the K-space $(\partial_{\mathfrak{C}}(X), \widehat{\mathcal{U}^{\mathfrak{C}}})$. Let h be a differential form on X and $f : X \to M$ a strongly smooth map which is strongly submersive with respect to a CF-perturbation $\mathfrak{S}^{\epsilon}$. The CF-perturbation $\mathfrak{S}^{\epsilon}$ induces one on $(\partial_{\mathfrak{C}}(X), \widehat{\mathcal{U}^{\mathfrak{C}}})$ and on $\partial_c(X, \widehat{\mathcal{U}})$. We also denote them by $\mathfrak{S}^{\epsilon}$. Then we have

$$\sum_c f!(h|_{\partial_c(X)}; \mathfrak{S}^{\epsilon}) = f!(h|_{\partial_{\mathfrak{C}}(X)}; \mathfrak{S}^{\epsilon}).$$

The proof is obvious and so omitted.

18.7 Composition of Morphisms of Linear K-Systems

We refer Conditions 16.1, 16.17, 16.21 and Definitions 16.6, 16.19, 16.30 for the various definitions and notations concerning linear K-systems, morphisms and homotopy.

Situation 18.32 Suppose we are in Situation 16.34 and $\mathfrak{N}_{i+1i}$ is a morphism from $\mathcal{F}_i$ to $\mathcal{F}_{i+1}$. We denote by

$$\mathcal{M}^i(\alpha_-, \alpha_+)$$

the space of connecting orbits of $\mathcal{F}_i$ and by

$$\mathcal{N}_{ii+1}(\alpha_-, \alpha_+)$$

the interpolation space of $\mathfrak{N}_{i+1i}$. Let $R^i_{\alpha_i}$ be a critical submanifold of $\mathcal{F}_i$ and $\alpha_i \in \mathfrak{A}_i$. ■

Remark 18.33 In the above situation we denote the interpolation space of a morphism from $\mathcal{F}_i$ to $\mathcal{F}_{i+1}$ by $\mathcal{N}_{ii+1}(*, *)$, while we denote the corresponding morphism by $\mathfrak{N}_{i+1i}$. The suffix is $ii+1$ for the former and is $i+1i$ for the latter. The suffix $i+1i$ is compatible with algebraic formulas of compositions of morphisms.

Definition 18.34 In Situation 18.32 we define the *partially collared fiber product*

$$\mathcal{N}_{12}(\alpha_1, \alpha_2) \times^{\boxplus\tau}_{R^2_{\alpha_2}} \mathcal{N}_{23}(\alpha_2, \alpha_3) \tag{18.17}$$

as follows. We consider the fiber product $\mathcal{N}_{12}(\alpha_1, \alpha_2) \times_{R^2_{\alpha_2}} \mathcal{N}_{23}(\alpha_2, \alpha_3)$ equipped with fiber product Kuranishi structure. Then we consider the decomposition $\mathfrak{C}$ of its boundary which consists of the following two kinds of components of its normalized boundary:

$$\mathcal{N}_{12}(\alpha_1, \alpha'_2) \times_{R^2_{\alpha'_2}} \mathcal{M}^2(\alpha_{\alpha'_2}, \alpha_{\alpha_2}) \times_{R^2_{\alpha_2}} \mathcal{N}_{23}(\alpha_2, \alpha_3),$$
$$\mathcal{N}_{12}(\alpha_1, \alpha_2) \times_{R^2_{\alpha_2}} \mathcal{M}^2(\alpha_{\alpha_2}, \alpha_{\alpha'_2}) \times_{R^2_{\alpha'_2}} \mathcal{N}_{23}(\alpha'_2, \alpha_3).$$

The first line is contained in $\partial\mathcal{N}_{12}(\alpha_1, \alpha_2) \times_{R^2_{\alpha_2}} \mathcal{N}_{23}(\alpha_2, \alpha_3)$ and the second line is contained in $\mathcal{N}_{12}(\alpha_1, \alpha_2) \times_{R^2_{\alpha_2}} \partial\mathcal{N}_{23}(\alpha_2, \alpha_3)$. Now we define (18.17) by

$$(\mathcal{N}_{12}(\alpha_1, \alpha_2) \times_{R^2_{\alpha_2}} \mathcal{N}_{23}(\alpha_2, \alpha_3))^{\mathfrak{C}\boxplus\tau}.$$

Proposition 18.35 *In Situation* 18.32 *there exists a* τ*-*$\mathfrak{C}$*-collared K-space*

$$\mathcal{N}_{123}(\alpha_1, \alpha_3)$$

with the following properties:

(1) *Its normalized boundary in* $\mathfrak{C}$ *is isomorphic to the disjoint union of*

$$\mathcal{N}_{12}(\alpha_1, \alpha_2) \times^{\boxplus\tau}_{R^2_{\alpha_2}} \mathcal{N}_{23}(\alpha_2, \alpha_3) \tag{18.18}$$

over various α_2.

(2) *Its normalized boundary which is not in* $\mathfrak{C}$ *is isomorphic to the disjoint union of*

$$\mathcal{M}^1(\alpha_1, \alpha'_1) \times_{R^1_{\alpha'_1}} \mathcal{N}_{123}(\alpha'_1, \alpha_3) \tag{18.19}$$

over various α'_1 *and of*

$$\mathcal{N}_{123}(\alpha_1, \alpha'_3) \times_{R_{\alpha'_3}} \mathcal{M}^3(\alpha'_3, \alpha_3) \tag{18.20}$$

over various α'_3.

(3) *The isomorphisms (1)(2) satisfy the compatibility conditions, Condition* 18.36, *at the corners.*

(4) *The evaluation maps, periodicity and orientation isomorphisms are defined on* $\mathcal{N}_{123}(\alpha_1, \alpha_3)$ *and commute with the isomorphisms in (1)(2)(3) above.*

(5) *Similar statements of Conditions* 16.17 *(V)(IX) hold.*

See Fig. 18.4. We omit describing the precise ways to modify Conditions 16.17 (V)(IX) to our situation. Since they are not hard, we leave them to the readers.

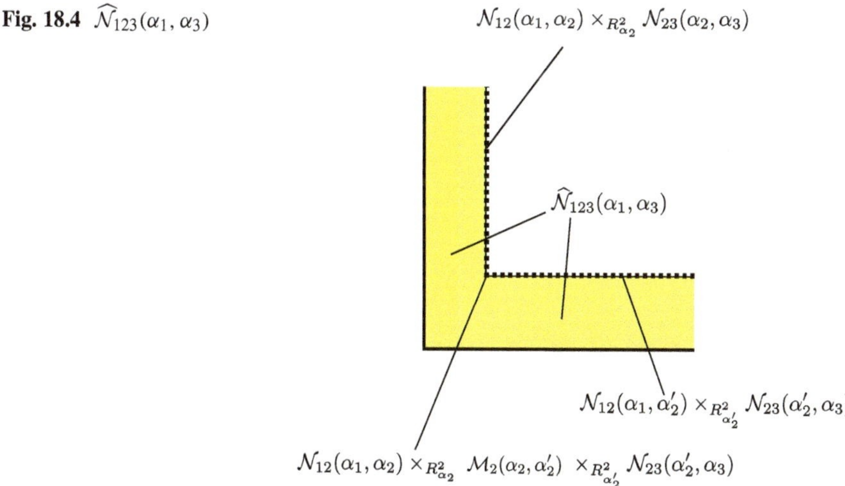

Fig. 18.4 $\widehat{\mathcal{N}}_{123}(\alpha_1, \alpha_3)$

Condition 18.36

(1) The codimension k normalized corner $\widehat{S}_k(\mathcal{N}_{123}(\alpha_1,\alpha_3))$ of $\mathcal{N}_{123}(\alpha_1,\alpha_3)$ is isomorphic to a disjoint union of the one of the two types (18.21) and (18.22):

$$\begin{aligned}&\mathcal{M}^1(\alpha_1,\alpha_1^2)\times_{R^1_{\alpha_1^2}}\cdots\times_{R^1_{\alpha_1^{k_1-1}}}\mathcal{M}^1(\alpha_1^{k_1-1},\alpha_1^{k_1})\\&\times_{R^1_{\alpha_1^{k_1}}}\widehat{S}^{\mathfrak{C}}_m\left(\mathcal{N}_{12}(\alpha_1^{k_1},\alpha_2)\times^{\boxplus\tau}_{R^2_{\alpha_2}}\mathcal{N}_{23}(\alpha_2,\alpha_3^1)\right)\\&\times_{R^3_{\alpha_3^1}}\mathcal{M}^3(\alpha_3^1,\alpha_3^2)\times_{R^3_{\alpha_3^2}}\cdots\times_{R^3_{\alpha_3^{k_3-1}}}\mathcal{M}^3(\alpha_3^{k_3-1},\alpha_3),\end{aligned}\tag{18.21}$$

and

$$\begin{aligned}&\mathcal{M}^1(\alpha_1,\alpha_1^2)\times_{R^1_{\alpha_1^2}}\cdots\times_{R^1_{\alpha_1^{k_1-1}}}\mathcal{M}^1(\alpha_1^{k_1-1},\alpha_1^{k_1})\\&\times_{R^1_{\alpha_1^{k_1}}}\mathcal{N}_{123}(\alpha_1^{k_1},\alpha_3^1)\\&\times_{R^3_{\alpha_3^1}}\mathcal{M}^3(\alpha_3^1,\alpha_3^2)\times_{R^3_{\alpha_3^2}}\cdots\times_{R^3_{\alpha_3^{k_3-1}}}\mathcal{M}^3(\alpha_3^{k_3-1},\alpha_3).\end{aligned}\tag{18.22}$$

Here in (18.21), $k_1+m-2+k_3=k$ and in (18.22), $k_1+k_3-2=k$. Note $k_1,k_3\in\mathbb{Z}_{\ge 1}$. In the case $k_1=1$ the first line of (18.21) or (18.22) is void and $\alpha_1^1=\alpha_1$. In the case $k_3=1$ the third line of (18.21) or (18.22) is void and $\alpha_3^1=\alpha_3$. We also note that (1) (2) of Proposition 18.35 is the case $k=1$ of this condition.

(2) (1) implies that $\widehat{S}_\ell(\widehat{S}_k(\mathcal{N}_{123}(\alpha_1,\alpha_3)))$ is isomorphic to the disjoint union of the spaces of type (18.21), (18.22) or

$$\begin{aligned}&\mathcal{M}^1(\alpha_1,\alpha_1^2)\times_{R^1_{\alpha_1^2}}\cdots\times_{R^1_{\alpha_1^{k_1-1}}}\mathcal{M}^1(\alpha_1^{k_1-1},\alpha_1^{k_1})\\&\times_{R^1_{\alpha_1^{k_1}}}\widehat{S}^{\mathfrak{C}}_n\left(\widehat{S}^{\mathfrak{C}}_m\left(\mathcal{N}_{12}(\alpha_1^{k_1},\alpha_2)\times^{\boxplus\tau}_{R^2_{\alpha_2}}\mathcal{N}_{23}(\alpha_2,\alpha_3^1)\right)\right)\\&\times_{R^3_{\alpha_3^1}}\mathcal{M}^3(\alpha_3^1,\alpha_3^2)\times_{R^3_{\alpha_3^2}}\cdots\times_{R^3_{\alpha_3^{k_3-1}}}\mathcal{M}^3(\alpha_3^{k_3-1},\alpha_3).\end{aligned}\tag{18.23}$$

We require that the covering map $\widehat{S}_\ell(\widehat{S}_k(\mathcal{N}_{123}(\alpha_1,\alpha_3)))\to\widehat{S}_{\ell+k}(\mathcal{N}_{123}(\alpha_1,\alpha_3)))$ is the identity map on the components of type (18.21), (18.22) and is induced by

$$\widehat{S}_n^{\mathfrak{C}}\left(\widehat{S}_m^{\mathfrak{C}}\left(\mathcal{N}_{12}(\alpha_1^{k_1},\alpha_2)\times^{\boxplus\tau}_{R^2_{\alpha_2}}\mathcal{N}_{23}(\alpha_2,\alpha_3^1)\right)\right)$$
$$\longrightarrow \widehat{S}_{n+m}^{\mathfrak{C}}\left(\mathcal{N}_{12}(\alpha_1^{k_1},\alpha_2)\times^{\boxplus\tau}_{R^2_{\alpha_2}}\mathcal{N}_{23}(\alpha_2,\alpha_3^1)\right)$$

on the components of type (18.23).

Lemma-Definition 18.37 *By Proposition* 18.35 *we can partially smooth the corner of* $\mathcal{N}_{123}(\alpha_1,\alpha_3)$. *Then we obtain a K-space which is the union of* $\mathcal{N}_{12}(\alpha_1,\alpha_2)\times^{\boxplus\tau}_{R^2_{\alpha_2}}\mathcal{N}_{23}(\alpha_2,\alpha_3)$ *over various* α_2*:*

$$\bigcup_{\alpha_2}\mathcal{N}_{12}(\alpha_1,\alpha_2)\times^{\boxplus\tau}_{R^2_{\alpha_2}}\mathcal{N}_{23}(\alpha_2,\alpha_3). \tag{18.24}$$

There is a morphism from $\mathcal{F}_1$ *to* $\mathcal{F}_3$ *whose interpolation space is* (18.24). *We call this morphism the composition of* $\mathfrak{N}_{21}$ *and* $\mathfrak{N}_{32}$ *and write* $\mathfrak{N}_{32}\circ\mathfrak{N}_{21}$.

This follows immediately from Proposition 18.35.

Remark 18.38 Before proving Proposition 18.35, we note the following. Using the moduli space of the solutions of the equation (18.5), which uses a one-parameter family of homotopies of Hamiltonians (18.6), we can find a space $\widehat{\mathcal{N}}_{123}(\alpha_1,\alpha_3)$ whose normalized boundary is the union of

$$\mathcal{N}_{12}(\alpha_1,\alpha_2)\times_{R^2_{\alpha_2}}\mathcal{N}_{23}(\alpha_2,\alpha_3), \tag{18.25}$$

and

$$\mathcal{M}^1(\alpha_1,\alpha_1')\times_{R^1_{\alpha_1'}}\widehat{\mathcal{N}}_{123}(\alpha_1',\alpha_3) \tag{18.26}$$

over various $\alpha_1'\in\mathfrak{A}_1$ and

$$\widehat{\mathcal{N}}_{123}(\alpha_1,\alpha_3')\times_{R_{\alpha_3'}}\mathcal{M}^3(\alpha_3',\alpha_3) \tag{18.27}$$

over various $\alpha_3'\in\mathfrak{A}_3$.

We take $\mathfrak{C}$ as the boundary of type (18.25). Then we can take

$$\mathcal{N}_{123}(\alpha_1,\alpha_3)=\widehat{\mathcal{N}}_{123}(\alpha_1,\alpha_3)^{\mathfrak{C}\boxplus\tau}\setminus\widehat{\mathcal{N}}_{123}(\alpha_1,\alpha_3).$$

In the proof of Proposition 18.35 below we will construct this space without using the whole $\mathcal{N}_{123}(\alpha_1,\alpha_3)$ but using only its 'boundary' which is (18.25), (18.19) and (18.20). This argument is very much similar to the proof of Proposition 17.58.

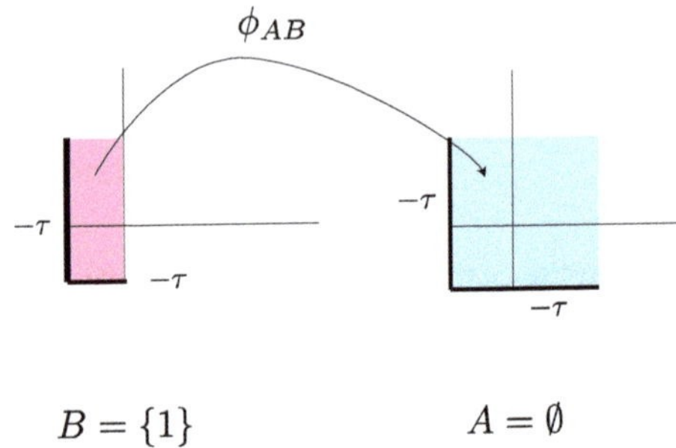

Fig. 18.5 ϕ_{AB}

***Proof of Proposition* 18.35** We begin with proving Lemma 18.39. For $A \subset \{1, \dots, k\}$ we put

$$[0,1)^A = \{(t_1, \dots, t_k) \in [0,1)^k \mid t_i = 0 \ \text{for } i \notin A\}.$$

Lemma 18.39 *For $A \subset B \subseteq \{1, \dots, k\}$ we have a smooth embedding*

$$\phi_{AB} : ([0,1)^{B^c})^{\boxplus\tau} \times [-\tau, 0)^{\#B} \to ([0,1)^{A^c})^{\boxplus\tau} \times [-\tau, 0)^{\#A},$$

where A^c, B^c are the complements of A, B in the set $\{1, \dots, k\}$ respectively (Fig. 18.5). Moreover, if $A \subset B \subset C \subseteq \{1, \dots, k\}$, we have

$$\phi_{AB} \circ \phi_{BC} = \phi_{AC}. \tag{18.28}$$

Proof Regrading $V^+_{\mathfrak{r},S_A}, V^+_{\mathfrak{r},S_B}$ in (17.41) as $[0,1)^{A^c}, [0,1)^{B^c}$, respectively, we can adopt (17.41) to define the map ϕ_{AB} above. (Use the obvious inclusion map and the local inverse of the covering map.) For example, if $A = \{1, \dots, a\}$, $B = \{1, \dots, b\}$, then we have

$$\phi_{AB}((t_{b+1}, \dots, t_k), (s_1, \dots, s_b)) = (s_{a+1}, \dots, s_b, t_{b+1}, \dots, t_k, s_1, \dots, s_a).$$

The formula (18.28) is easy to prove from this definition. □

For $\alpha_2^1, \dots, \alpha_2^k \in \mathfrak{A}_2$, we put

$$\begin{aligned} &\mathcal{N}(\alpha_1, \alpha_3; \alpha_2^1, \dots, \alpha_2^k) \\ &= \Big(\mathcal{N}_{12}(\alpha_1, \alpha_2^1) \times_{R^2_{\alpha_2^1}} \mathcal{M}^2(\alpha_2^1, \alpha_2^2) \times_{R^2_{\alpha_2^2}} \\ &\quad \cdots \times_{R^2_{\alpha_2^{k-1}}} \mathcal{M}^2(\alpha_2^{k-1}, \alpha_2^k) \times_{R^2_{\alpha_2^k}} \mathcal{N}_{23}(\alpha_2^k, \alpha_3) \Big)^{\mathfrak{C}\boxplus\tau} \times [-\tau, 0)^k. \end{aligned} \tag{18.29}$$

Here $\mathfrak{C}$ is the decomposition of the boundary such that its *complement* consists of

$$\mathcal{M}^1(\alpha_1, \alpha_1') \times_{R^1_{\alpha_1'}} \mathcal{N}_{12}(\alpha_1', \alpha_2^1) \times_{R^2_{\alpha_2^1}} \mathcal{M}^2(\alpha_2^1, \alpha_2^2) \times_{R^2_{\alpha_2^2}} \cdots \times_{R^2_{\alpha_2^{k-1}}} \mathcal{M}^2(\alpha_2^{k-1}, \alpha_2^k) \times_{R^2_{\alpha_2^k}} \mathcal{N}_{23}(\alpha_2^k, \alpha_3)$$

and

$$\mathcal{N}_{12}(\alpha_1, \alpha_2^1) \times_{R^2_{\alpha_2^1}} \mathcal{M}^2(\alpha_2^1, \alpha_2^2) \times_{R^2_{\alpha_2^2}} \cdots \times_{R^2_{\alpha_2^{k-1}}} \mathcal{M}^2(\alpha_2^{k-1}, \alpha_2^k) \times_{R^2_{\alpha_2^k}} \mathcal{N}_{23}(\alpha_2^k, \alpha_3') \times_{R^3_{\alpha_3'}} \mathcal{M}^3(\alpha_3', \alpha_3).$$

Let $A = \{i_1, \dots, i_a\} \subset \{1, \dots, k\}$ with $i_1 < i_2 < \cdots < i_a$. We define an embedding

$$\hat{\phi}_{A\{1,\dots,k\}} : \mathcal{N}(\alpha_1, \alpha_3; \alpha_2^1, \dots, \alpha_2^k) \to \mathcal{N}(\alpha_1, \alpha_3; \alpha_2^{i_1}, \dots, \alpha_2^{i_a})$$

as follows. For any $p \in \mathcal{N}(\alpha_1, \alpha_3; \alpha_2^1, \dots, \alpha_2^k)$ we have a Kuranishi neighborhood U_p of $\mathcal{N}(\alpha_1, \alpha_3)$ such that U_p has an orbifold chart at p of the form

$$V_p^{\mathfrak{C}\boxplus\tau} \times [-\tau, 0)^k. \tag{18.30}$$

On the other hand, p has a Kuranishi neighborhood U_p' of $\mathcal{N}(\alpha_1, \alpha_3; \alpha_2^{i_1}, \dots, \alpha_2^{i_a})$ such that U_p' has an orbifold chart at p of the form

$$V_p^{\mathfrak{C}\boxplus\tau} \times ([0, \delta)^{k-a})^{\boxplus\tau} \times [-\tau, 0)^a. \tag{18.31}$$

In fact

$$\mathcal{N}_{12}(\alpha_1, \alpha_2^1) \times_{R^2_{\alpha_2^1}} \mathcal{M}^2(\alpha_2^1, \alpha_2^2) \times_{R^2_{\alpha_2^2}} \cdots \times_{R^2_{\alpha_2^{k-1}}} \mathcal{M}^2(\alpha_2^{k-1}, \alpha_2^k) \times_{R^2_{\alpha_2^k}} \mathcal{N}_{23}(\alpha_2^k, \alpha_3)$$

lies in the codimension $(k-a)$-$\mathfrak{C}$-corner of

$$\mathcal{N}_{12}(\alpha_1, \alpha_2^{i_1}) \times_{R^2_{\alpha_2^{i_1}}} \mathcal{M}^2(\alpha_2^{i_1}, \alpha_2^{i_2}) \times_{R^2_{\alpha_2^{i_2}}} \cdots \times_{R^2_{\alpha_2^{i_{a-1}}}} \mathcal{M}^2(\alpha_2^{i_{a-1}}, \alpha_2^{i_a}) \times_{R^2_{\alpha_2^{i_a}}} \mathcal{N}_{23}(\alpha_2^{i_a}, \alpha_3)$$

by (16.33).[3]

Therefore the map $\hat{\phi}_{A\{1,\dots,k\}}$ in Lemma 18.39 defines an embedding from (18.30) to (18.31). See Fig. 18.6. ($A = \{2\} \subset \{1, 2, 3\}$ in Fig. 18.6.)

[3] In the geometric situation $[0, \delta)^{k-a}$ are gluing parameters.

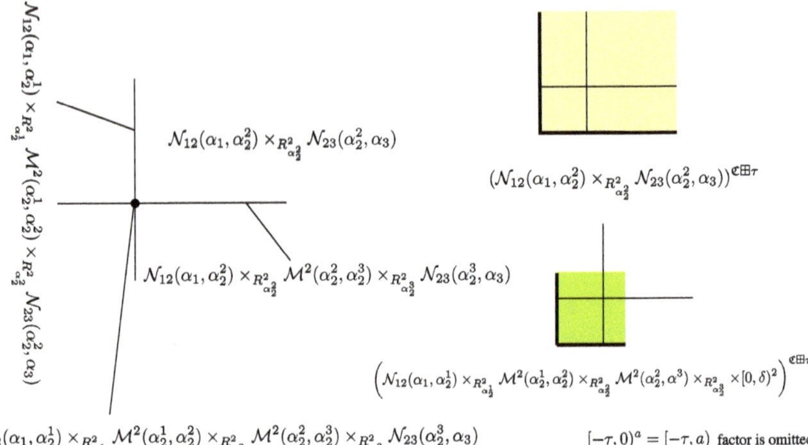

Fig. 18.6 $\hat{\phi}_{\{2\}\{1,2,3\}}$

This is compatible with the coordinate change of a Kuranishi structure and defines a required embedding. We can glue various $\mathcal{N}(\alpha_1,\alpha_3;\alpha_2^1,\dots,\alpha_2^k)$ by the embeddings $\hat{\phi}_{A\{1,\dots,k\}}$. More precisely, we will glue charts of their Kuranishi structures. We can do so in the same way as the proof of Proposition 17.58 using the compatibility of interpolation spaces and (18.28).

We have thus obtained a $\mathfrak{C}$-collared K-space $\mathcal{N}_{123}(\alpha_1,\alpha_3)$. It is easy to see from the construction that $\mathcal{N}_{123}(\alpha_1,\alpha_3)$ has the required properties. (See Fig. 18.4, which is the case $a=1$, $i_1=2$, $k=3$.) □

By construction we have the following:

Lemma 18.40 *Suppose that there exists a K-space*

$$\mathcal{N}'_{123}(\alpha_1,\alpha_3)$$

and its boundary components $\mathfrak{C}$ *such that the following holds:*

(1) $\partial_{\mathfrak{C}}\mathcal{N}'_{123}(\alpha_1,\alpha_3)$ *is a disjoint union of* $\mathcal{N}_{12}(\alpha_1,\alpha_2)\times_{R^2_{\alpha_2}}\mathcal{N}_{23}(\alpha_2,\alpha_3)$ *over* α_2.

(2) *The* k*-th normalized corner* $\widehat{S}_k^{\mathfrak{C}}(\mathcal{N}'_{123}(\alpha_1,\alpha_3))$ *is identified with the disjoint union of*

$$\mathcal{N}_{12}(\alpha_1,\alpha_2^1)\times_{R^2_{\alpha_2^1}}\mathcal{M}^2(\alpha_2^1,\alpha_2^2)\times_{R^2_{\alpha_2^2}} \cdots\times_{R^2_{\alpha_2^{k-1}}}\mathcal{M}^2(\alpha_2^{k-1},\alpha_2^k)\times_{R^2_{\alpha_2^k}}\mathcal{N}_{23}(\alpha_2^k,\alpha_3)$$

over $\alpha_2^1,\dots,\alpha_2^k$.

(3) *The map* $\hat{S}_k(\hat{S}_\ell(\mathcal{N}'_{123}(\alpha_1,\alpha_2))) \to \hat{S}_{k+\ell}(\mathcal{N}'_{123}(\alpha_1,\alpha_2))$ *is compatible with the isomorphisms (1)(2) in the sense similar to Condition* 18.36 *(2).*

Then there exists an isomorphism

$$\mathcal{N}'_{123}(\alpha_1,\alpha_2)^{\mathfrak{C}\boxplus\tau} \setminus \mathcal{N}'_{123}(\alpha_1,\alpha_3) \cong \mathcal{N}_{123}(\alpha_1,\alpha_3). \tag{18.32}$$

This isomorphism is compatible with the isomorphisms in (1)(2) above and Condition 18.36 *(2) at the corners.*

18.8 Associativity of the Composition

In this section we present the detail of the proof of the associativity of the composition.

Proposition 18.41 *Suppose we are in Situation* 18.32 *for* $i = 1, 2, 3$. *Then we have the following identity:*

$$(\mathfrak{N}_{43} \circ \mathfrak{N}_{32}) \circ \mathfrak{N}_{21} = \mathfrak{N}_{43} \circ (\mathfrak{N}_{32} \circ \mathfrak{N}_{21}). \tag{18.33}$$

Proof As we already discussed in Sect. 16.5, this equality is mostly obvious. Namely the interpolation spaces of the left hand and the right hand sides are isomorphic to each other by the associativity of the fiber product. The only issue is about the way we smooth the corners. Since corner smoothing is not completely canonical, this point is non-trivial. To clarify this tiny technicality we will use the following.[4]

Lemma 18.42 *There exists a K-space*

$$\mathcal{N}_{1234}(\alpha_1,\alpha_4)$$

with the following properties.

(1) *Its normalized boundary is isomorphic to the union of the following four types of fiber products:*

$$\mathcal{N}_{123}(\alpha_1,\alpha_3) \times_{R^3_{\alpha_3}} \mathcal{N}_{34}(\alpha_3,\alpha_4), \tag{18.34}$$

[4] The fact that the left and right hand sides induce the same cochain map in the de Rham complex by the smooth correspondence follows without comparing the smooth structures near the corner. Since the part where we smooth the corners lies in the collar, it does not contribute to the integration along the fiber (see (17.18)). So we never need the part of the proof of this proposition given in this section for applications.

$$\mathcal{N}_{12}(\alpha_1,\alpha_2) \times_{R^2_{\alpha_2}} \mathcal{N}_{234}(\alpha_2,\alpha_4), \tag{18.35}$$

$$\mathcal{M}^1(\alpha_1,\alpha_1') \times_{R^1_{\alpha_1'}} \mathcal{N}_{1234}(\alpha_1',\alpha_4), \tag{18.36}$$

$$\mathcal{N}_{1234}(\alpha_1,\alpha_4') \times_{R^4_{\alpha_4'}} \mathcal{M}^4(\alpha_4',\alpha_1'). \tag{18.37}$$

(2) *A compatibility condition similar to Condition* 18.36 *holds at the corners.*
(3) *The evaluation maps, periodicity and orientation isomorphisms are defined on* $\mathcal{N}_{1234}(\alpha_1,\alpha_4)$ *and commute with the isomorphisms in (1)(2) above.*
(4) *A condition similar to Condition* 16.17 *(V)(IX) holds.*

We omit the precise definition of the compatibility condition in Lemma 18.42 (2). Since it is not difficult, we leave it to the readers. We will prove Lemma 18.42 later in this section. In the mean time, we continue the proof of Proposition 18.41.

We consider two decompositions $\mathfrak{C}_1$, $\mathfrak{C}_2$ of the boundary of $\mathcal{N}_{1234}(\alpha_1,\alpha_4)$, where $\mathfrak{C}_1$ consists of the boundary of type (18.34) and $\mathfrak{C}_2$ consists of the boundary of type (18.35). Thus we are in the situation of Lemma 18.13. Here we use the next lemma.

Lemma 18.43 *Suppose we are in the situation of Lemma* 18.13. *Then the following constructions described in (A) and (B) give the same K-space:*

(A)(1) *We first take* τ-$\mathfrak{C}_1$*-outer collaring of the corner.*
(2) *We smooth the boundaries in* $\mathfrak{C}_1$.
(3) *We next take* τ-$\mathfrak{C}_2$*-outer collaring of the corner.*
(4) *We smooth the boundaries in* $\mathfrak{C}_2$.
(B)(1) *We first take* τ-$\mathfrak{C}_2$*-outer collaring of the corner.*
(2) *We smooth the boundaries in* $\mathfrak{C}_2$.
(3) *We next take* τ-$\mathfrak{C}_1$*-outer collaring of the corner.*
(4) *We smooth the boundaries in* $\mathfrak{C}_1$.

Proof It suffices to consider the case of $[0,1)^{k_1} \times [0,1)^{k_2}$ where $\mathfrak{C}_1$ corresponds to $\partial[0,1)^{k_1} \times [0,1)^{k_2}$ and $\mathfrak{C}_2$ corresponds to $[0,1)^{k_1} \times \partial[0,1)^{k_2}$. However, the proof for this case is straightforward from the definition. □

We note that the intersection of (18.34) and (18.35) is

$$\mathcal{N}_{12}(\alpha_1,\alpha_2) \times_{R^2_{\alpha_2}} \mathcal{N}_{23}(\alpha_2,\alpha_3) \times_{R^3_{\alpha_3}} \mathcal{N}_{34}(\alpha_3,\alpha_4). \tag{18.38}$$

The interpolation spaces of both sides of (18.33) are obtained by appropriately modifying the union of (18.38) for α_2, α_3.

We study how the processes (A) and (B) of Lemma 18.43 affect the subspace (18.38). By (A)(1), (18.38) becomes

$$\mathcal{N}_{12}(\alpha_1,\alpha_2) \times_{R^2_{\alpha_2}} \mathcal{N}_{23}(\alpha_2,\alpha_3) \times^{\boxplus\tau}_{R^3_{\alpha_3}} \mathcal{N}_{34}(\alpha_3,\alpha_4).$$

The step (A)(2) smooths the corners of the (union over α_3 of) the second fiber product factor. So it becomes

$$\mathcal{N}_{12}(\alpha_1,\alpha_2) \times_{R^2_{\alpha_2}} \mathcal{N}_{24}(\alpha_2,\alpha_4).$$

Here $\mathcal{N}_{24}(\alpha_2,\alpha_4)$ is the interpolation space of the composition $\mathfrak{N}_{43} \circ \mathfrak{N}_{32}$. Then (A)(3) changes it to

$$\mathcal{N}_{12}(\alpha_1,\alpha_2) \times^{\boxplus\tau}_{R^2_{\alpha_2}} \mathcal{N}_{24}(\alpha_2,\alpha_4).$$

Thus when (A) completed we obtain an interpolation space of

$$(\mathfrak{N}_{43} \circ \mathfrak{N}_{32}) \circ \mathfrak{N}_{21}.$$

In the same way, (B) gives an interpolation space of

$$\mathfrak{N}_{43} \circ (\mathfrak{N}_{32} \circ \mathfrak{N}_{21}).$$

Thus Proposition 18.41 follows from Lemma 18.43.

***Proof of Lemma* 18.42** We fix α_1, α_4. Let $\alpha_2^1, \dots, \alpha_2^{k_2} \in \mathfrak{A}_2$ and $\alpha_3^1, \dots, \alpha_3^{k_3} \in \mathfrak{A}_3$. We put

$$\begin{aligned}
&\mathcal{N}(\alpha_1,\alpha_4;\alpha_2^1,\dots,\alpha_2^{k_2};\alpha_3^1,\dots,\alpha_3^{k_3})\\
&= \Bigg(\mathcal{N}_{12}(\alpha_1,\alpha_2^1) \times_{R^2_{\alpha_2^1}} \mathcal{M}^2(\alpha_2^1,\alpha_2^2) \times_{R^2_{\alpha_2^2}}\\
&\quad \cdots \times_{R^2_{\alpha_2^{k_2-1}}} \mathcal{M}^2(\alpha_2^{k_2-1},\alpha_2^{k_2}) \times_{R^2_{\alpha_2^{k_2}}} \mathcal{N}_{23}(\alpha_2^{k_2},\alpha_3^1) \times_{R^3_{\alpha_3^1}} \mathcal{M}^3(\alpha_3^1,\alpha_3^2) \times_{R^3_{\alpha_3^2}}\\
&\quad \cdots \times_{R^3_{\alpha_3^{k_3-1}}} \mathcal{M}^3(\alpha_3^{k_3-1},\alpha_3^{k_3}) \times_{R^3_{\alpha_3^{k_3}}} \mathcal{N}_{34}(\alpha_3^{k_3},\alpha_4)\Bigg)^{\mathfrak{C}\boxplus\tau} \times [-\tau,0)^{k_2+k_3}.
\end{aligned}$$

Here $\mathfrak{C}$ is defined in a similar way to (18.29). We can glue them in a similar way to the proof of Proposition 18.35 to obtain the required $\mathcal{N}_{1234}(\alpha_1,\alpha_4)$. □

The proof of Proposition 18.41 is now complete. □

We note that in the geometric situation of periodic Hamiltonian systems we can prove Lemma 18.42 using the two-parameter family of moduli space of solutions of the equation

$$\frac{\partial u}{\partial \tau} + J\left(\frac{\partial u}{\partial t} - X_{H^{ST}_{\tau,t}}(u)\right) = 0, \tag{18.39}$$

where

$$\mathcal{H}\mathcal{H}^{ST}(\tau,t,x)=\begin{cases} H^1(t,x) & \text{if } \tau \le -T-T_0 \\ \mathcal{H}^{21}(\tau+T,t,x) & \text{if } -T-T_0 \le \tau \le -T+T_0 \\ H^2(t,x) & \text{if } -T+T_0 \le \tau \le -T_0 \\ \mathcal{H}^{32}(\tau,t,x) & \text{if } -T_0 \le \tau \le T_0 \\ H^3(t,x) & \text{if } T_0 \le \tau \le S-T_0 \\ \mathcal{H}^{43}(\tau-S,t,x) & \text{if } S-T_0 \le \tau \le S+T_0 \\ H^4(t,x) & \text{if } S+T_0 \le \tau, \end{cases} \tag{18.40}$$

and $\mathcal{H}^{ij}(\tau,t,x):\mathbb{R}\times S^1\times M\to\mathbb{R}$ $(i,j=1,\dots,4)$ is defined by (18.4).

18.9 Parametrized Version of Morphism: Composition and Gluing

18.9.1 Compositions of Parametrized Morphisms

Situation 18.44 Suppose we are in Situation 16.34 and $\mathfrak{N}^{P_i}_{i+1i}$ is a P_i-parametrized morphism from $\mathcal{F}_i$ to $\mathcal{F}_{i+1}$. We denote by

$$\mathcal{M}^i(\alpha_-,\alpha_+)$$

the space of connecting orbits of $\mathcal{F}_i$ and by

$$\mathcal{N}_{ii+1}(\alpha_-,\alpha_+;P_i)$$

the interpolation space of $\mathfrak{N}^{P_i}_{i+1i}$. Let $R^i_{\alpha_i}$ be a critical submanifold of $\mathcal{F}_i$ and $\alpha_i\in\mathfrak{A}_i$. ■

Definition 18.45 In Situation 18.44 we define a K-space

$$\mathcal{N}_{12}(\alpha_1,\alpha_2;P_1)\times^{\boxplus\tau}_{R^2_{\alpha_2}}\mathcal{N}_{23}(\alpha_2,\alpha_3;P_2) \tag{18.41}$$

as follows. We take the fiber product $\mathcal{N}_{12}(\alpha_1,\alpha_2;P_1)\times_{R^2_{\alpha_2}}\mathcal{N}_{23}(\alpha_2,\alpha_3;P_2)$ with a fiber product Kuranishi structure and consider the decomposition $\mathfrak{C}$ of its boundary consisting of the following two kinds of components of its normalized boundary:

$$\mathcal{N}_{12}(\alpha_1,\alpha_2';P_1)\times_{R^2_{\alpha_2'}}\mathcal{M}^2(\alpha_{\alpha_2'},\alpha_{\alpha_2})\times_{R^2_{\alpha_2}}\mathcal{N}_{23}(\alpha_2,\alpha_3;P_2),$$

$$\mathcal{N}_{12}(\alpha_1,\alpha_2;P_1)\times_{R^2_{\alpha_2}}\mathcal{M}^2(\alpha_{\alpha_2},\alpha_{\alpha_2'})\times_{R^2_{\alpha_2'}}\mathcal{N}_{23}(\alpha_2',\alpha_3;P_2).$$

The first line is contained in $\partial\mathcal{N}_{12}(\alpha_1,\alpha_2;P_1)\times_{R^2_{\alpha_2}}\mathcal{N}_{23}(\alpha_2,\alpha_3;P_2)$ and the second line is contained in $\mathcal{N}_{12}(\alpha_1,\alpha_2;P_1)\times_{R^2_{\alpha_2}}\partial\mathcal{N}_{23}(\alpha_2,\alpha_3;P_2)$. Now we define (18.41) by

$$(\mathcal{N}_{12}(\alpha_1,\alpha_2;P_1)\times_{R^2_{\alpha_2}}\mathcal{N}_{23}(\alpha_2,\alpha_3);P_2)^{\mathfrak{C}\boxplus\tau}.$$

Note that besides the evaluation maps ev_- and ev_+, (18.41) carries an evaluation map to $P_1\times P_2$, which is stratumwise weakly submersive.

Proposition 18.46 *In Situation* 18.44 *there exists a* τ*-*$\mathfrak{C}$*-collared K-space*

$$\mathcal{N}_{123}(\alpha_1,\alpha_3;P_1\times P_2)$$

with the following properties:

(1) *Its normalized boundary in* $\mathfrak{C}$ *is isomorphic to the disjoint union of*

$$\mathcal{N}_{12}(\alpha_1,\alpha_2;P_1)\times^{\boxplus\tau}_{R^2_{\alpha_2}}\mathcal{N}_{23}(\alpha_2,\alpha_3;P_2) \qquad (18.42)$$

over various α_2.

(2) *Its normalized boundary not in* $\mathfrak{C}$ *is isomorphic to the disjoint union of*

$$\mathcal{M}^1(\alpha_1,\alpha_1')\times_{R^1_{\alpha_1'}}\mathcal{N}_{123}(\alpha_1',\alpha_3;P_1\times P_2) \qquad (18.43)$$

over various α_1' *and*

$$\mathcal{N}_{123}(\alpha_1,\alpha_3';P_1\times P_2)\times_{R_{\alpha_3'}}\mathcal{M}^3(\alpha_3',\alpha_3) \qquad (18.44)$$

over various α_3', *and*

$$\mathcal{N}_{123}(\alpha_1,\alpha_3;\partial P_1\times P_2)\cup\mathcal{N}_{123}(\alpha_1,\alpha_3;P_1\times\partial P_2). \qquad (18.45)$$

(3) *Conditions similar to (3)(4)(5) of Proposition* 18.35 *hold.*

(4) *We have an evaluation map* $\mathcal{N}_{123}(\alpha_1,\alpha_3;P_1\times P_2)\to P_1\times P_2$ *which is a stratumwise weakly submersive map. It is compatible with the boundary description given in (1)(2) above.*

The proof is the same as the proof of Proposition 18.35, so is omitted.

Definition-Lemma 18.47 By Proposition 18.46 we can partially smooth the corner of $\mathcal{N}_{123}(\alpha_1, \alpha_3; P_1 \times P_2)$ to obtain a K-space given by

$$\bigcup_{\alpha_2} \mathcal{N}_{12}(\alpha_1, \alpha_2; P_1) \times^{\boxplus\tau}_{R^2_{\alpha_2}} \mathcal{N}_{23}(\alpha_2, \alpha_3; P_2). \tag{18.46}$$

There is a $(P_1 \times P_2)$-parametrized morphism from $\mathcal{F}_1$ to $\mathcal{F}_3$ whose interpolation space is given by (18.46). We call this morphism the *parametrized composition of* $\mathfrak{N}^{P_1}_{21}$ and $\mathfrak{N}^{P_2}_{32}$ and write $\mathfrak{N}^{P_1}_{32} \circ \mathfrak{N}^{P_2}_{21}$.

Lemma 18.48 *The parametrized composition is associative in the same sense as in Proposition* 18.41. *Moreover the boundary of the parametrized composition* $\mathfrak{N}^{P_1}_{32} \circ \mathfrak{N}^{P_2}_{21}$ *is the disjoint union of*

$$\mathfrak{N}^{\partial P_1}_{32} \circ \mathfrak{N}^{P_2}_{21} \qquad \text{and} \qquad \mathfrak{N}^{P_1}_{32} \circ \mathfrak{N}^{\partial P_2}_{21}.$$

Proof The proof of the first half is similar to the proof of Proposition 18.41. The second half is obvious from the construction. □

18.9.2 Gluing Parametrized Morphisms

We first review well-known obvious facts on the gluing of cornered manifolds along their boundaries.

Definition-Lemma 18.49 Let P_1, P_2 be two admissible manifolds with corners, and for each $i = 1, 2$ let $\partial_{\mathfrak{C}_i} P_i \subset \partial P_i$ be an open and closed subset of the normalized boundary ∂P_i and assume $S^{\mathfrak{C}_i}_2(P_i) = \emptyset$. (See Definition 18.2 for the notation.) Suppose we have an admissible diffeomorphism of manifolds with corners $I : \partial_{\mathfrak{C}_1} P_1 \to \partial_{\mathfrak{C}_2} P_2$.

(1) We can define a structure of admissible manifold with corners on

$$P_1 \,{}_{\mathfrak{C}_1}\!\cup_{\mathfrak{C}_2} P_2 = (P_1 \cup P_2)/\sim. \tag{18.47}$$

Here $\sim$ is defined by $x \sim I(x)$ for $x \in \partial_{\mathfrak{C}_1} P_1$.

(2) The boundary $\partial(P_1 \,{}_{\mathfrak{C}_1}\!\cup_{\mathfrak{C}_2} P_2)$ is described as follows. Let $\partial_{\mathfrak{C}^c_i} P_i$ be the complement of $\partial_{\mathfrak{C}_i} P_i$ in ∂P_i. Then $\mathfrak{C}_i$ induces a decomposition

$$\partial(\partial_{\mathfrak{C}^c_i} P_i) = \partial_{\mathfrak{C}_i}(\partial_{\mathfrak{C}^c_i} P_i) \cup \partial_{\mathfrak{C}^c_i}(\partial_{\mathfrak{C}^c_i} P_i).$$

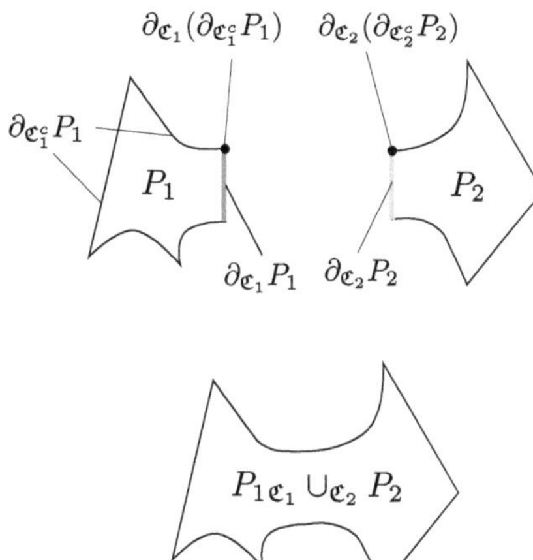

Fig. 18.7 Gluing cornered manifolds

Moreover I induces a diffeomorphism[5]

$$\partial_{\mathfrak{C}_1}(\partial_{\mathfrak{C}_1^c} P_1) \cong \partial_{\mathfrak{C}_2}(\partial_{\mathfrak{C}_2^c} P_2).$$

Now we have, see Fig. 18.7,

$$\partial(P_1 {}_{\mathfrak{C}_1}\cup_{\mathfrak{C}_2} P_2) = \partial_{\mathfrak{C}_1^c} P_1 {}_{\mathfrak{C}_1}\cup_{\mathfrak{C}_2} \partial_{\mathfrak{C}_2^c} P_2. \tag{18.48}$$

(3) Suppose that we have one of the following admissible objects on P_1 and on P_2 such that their restrictions to $\partial_{\mathfrak{C}_1} P_1$ and to $\partial_{\mathfrak{C}_2} P_2$ are isomorphic to each other (or coincide with each other) when we identify $\partial_{\mathfrak{C}_1} P_1$ with $\partial_{\mathfrak{C}_2} P_2$ under the admissible diffeomorphism I. Then we obtain a glued object on $P_1 {}_{\mathfrak{C}_1}\cup_{\mathfrak{C}_2} P_2$.

(a) Vector bundle.
(b) Section of vector bundle.
(c) Differential form.
(d) Smooth map to a manifold.

(4) In addition, suppose that we have admissible manifolds with corners P_1', P_2' and for each $i = 1, 2$ let $\partial_{\mathfrak{C}_i'} P_i'$ be an open and closed subset of the normalized boundary $\partial P_i'$. Let $I' : \partial_{\mathfrak{C}_1'} P_1' \cong \partial_{\mathfrak{C}_2'} P_2'$ be an admissible diffeomorphism as above. We also assume that for each $i = 1, 2$ we have an admissible smooth embedding $f_i : P_i' \to P_i$ of cornered orbifolds such that:

(a) $f_i^{-1}(\partial_{\mathfrak{C}_i} P_i) = \partial_{\mathfrak{C}_i'} P_i'$.

[5] In fact, since $S_2^{\mathfrak{C}_i}(P_i) = \emptyset$ the restriction of I to the boundary of the domain induces an isomorphism: $\partial_{\mathfrak{C}_1^c}(\partial_{\mathfrak{C}_1} P) \cong \partial_{\mathfrak{C}_2^c}(\partial_{\mathfrak{C}_2} P)$. Therefore using $\partial_{\mathfrak{C}_i^c}(\partial_{\mathfrak{C}_i} P) \cong \partial_{\mathfrak{C}_i}(\partial_{\mathfrak{C}_i^c} P)$ (with the opposite orientation), we obtain the isomorphism as claimed.

(b)

$$I \circ f_1 = f_2 \circ I'$$

on $\partial_{\mathfrak{C}'_1} P'_1$.

Then f_1, f_2 induce a smooth admissible embedding of cornered orbifolds

$$f_1 \; {}_{\mathfrak{C}_1}\cup_{\mathfrak{C}_2} \; f_2 : P'_1 \; {}_{\mathfrak{C}'_1}\cup_{\mathfrak{C}'_2} \; P'_2 \to P_1 \; {}_{\mathfrak{C}_1}\cup_{\mathfrak{C}_2} \; P_2.$$

(5) In the situation of (4), suppose that we are given admissible vector bundles $\mathcal{E}_i$ on P_i and $\mathcal{E}'_i$ on P'_i. We assume $\mathcal{E}_1|_{\partial_{\mathfrak{C}_1} P_1} \cong \mathcal{E}_2|_{\partial_{\mathfrak{C}_2} P_2}$ and the isomorphism covers I. We assume the same condition for $\mathcal{E}'_1$ and $\mathcal{E}'_2$. Moreover we assume that there exist admissible embeddings of vector bundles $\hat{f}_i : E'_i \to E_i$ which cover f_i for $i = 1, 2$, and assume that $\hat{f}_1$ and $\hat{f}_2$ are compatible with the isomorphisms $\mathcal{E}_1|_{\partial_{\mathfrak{C}_1} P_1} \cong \mathcal{E}_2|_{\partial_{\mathfrak{C}_2} P_2}$ and $\mathcal{E}'_1|_{\partial_{\mathfrak{C}'_1} P'_1} \cong \mathcal{E}'_2|_{\partial_{\mathfrak{C}'_2} P'_2}$.

Then we obtain an admissible embedding of vector bundles:

$$\hat{f}_1 \; {}_{\mathfrak{C}_1}\cup_{\mathfrak{C}_2} \; \hat{f}_2 : \mathcal{E}'_1 \; {}_{\mathfrak{C}'_1}\cup_{\mathfrak{C}'_2} \; \mathcal{E}'_2 \to \mathcal{E}_1 \; {}_{\mathfrak{C}_1}\cup_{\mathfrak{C}_2} \; \mathcal{E}_2.$$

(6) In addition to (1)(2)(3), suppose that we have P'_1, P'_2 and $\partial_{\mathfrak{C}'_i} P'_i$ (for $i = 1, 2$), $I' : \partial_{\mathfrak{C}'_1} P'_1 \cong \partial_{\mathfrak{C}'_2} P'_2$ as above. We also assume that for each $i = 1, 2$ we have an admissible smooth stratumwise submersion $\pi_i : P'_i \to P_i$ of cornered orbifolds such that:

(a) $\pi_i^{-1}(\partial_{\mathfrak{C}_i} P_i) = \partial_{\mathfrak{C}'_i} P'_i$.

(b)

$$I \circ \pi_1 = \pi_2 \circ I'$$

on $\partial_{\mathfrak{C}'_1} P'_1$.

Then π_1, π_2 induce an admissible smooth stratumwise submersion of cornered orbifolds

$$\pi_1 \; {}_{\mathfrak{C}_1}\cup_{\mathfrak{C}_2} \; \pi_2 : P'_1 \; {}_{\mathfrak{C}'_1}\cup_{\mathfrak{C}'_2} \; P'_2 \to P_1 \; {}_{\mathfrak{C}_1}\cup_{\mathfrak{C}_2} \; P_2.$$

Proof The proof is mostly obvious except the smoothness of various objects (such as coordinate change) at the boundary where we glue P_1 and P_2. The smoothness can be proved by using admissibility and Lemma 25.10 (2). In fact, the admissibility condition on the normal direction to the boundary of all the objects holds. So gluing along the boundary gives smooth objects. □

Situation 18.50 We work in admissible category. Let P_1, P_2, $\partial_{\mathfrak{C}_i} P_i \subset \partial P_i$ and $I : \partial_{\mathfrak{C}_1} P_1 \to \partial_{\mathfrak{C}_2} P_2$ be given as in Definition-Lemma 18.49. For $i = 1, 2$ let $(X_i, \widehat{\mathcal{U}_i})$ be a K-space and $\mathrm{ev}_{P_i} : (X_i, \widehat{\mathcal{U}_i}) \to P_i$ be a strongly smooth and stratumwise weakly submersive map. Via the map $\mathrm{ev}_{P_i} : (X_i, \widehat{\mathcal{U}_i}) \to P_i$, the decomposition $\partial P_i = \partial_{\mathfrak{C}_i} P_i \cup \partial_{\mathfrak{C}_i^c} P_i$ induces a decomposition $\partial(X_i, \widehat{\mathcal{U}_i}) = \partial_{\mathfrak{C}_i}(X_i, \widehat{\mathcal{U}_i}) \cup \partial_{\mathfrak{C}_i^c}(X_i, \widehat{\mathcal{U}_i})$ as in Situation 18.4. Namely, $\partial_{\mathfrak{C}_i}(X_i, \widehat{\mathcal{U}_i})$ is defined by taking Kuranishi charts so that they are mapped to $\partial_{\mathfrak{C}_i} P_i$ via ev_{P_i} and the coordinate change satisfies the condition as in Situation 18.4. We assume that there exists an isomorphism

$$\hat{I} : \partial_{\mathfrak{C}_1}(X_1, \widehat{\mathcal{U}_1}) \cong \partial_{\mathfrak{C}_2}(X_2, \widehat{\mathcal{U}_2}) \tag{18.49}$$

which is compatible with the admissible diffeomorphism $I : \partial_{\mathfrak{C}_1} P_1 \to \partial_{\mathfrak{C}_2} P_2$.

Then we glue X_1 and X_2 by the underlying homeomorphism of $\hat{I}$ to obtain

$$X = X_1 \ {}_{\mathfrak{C}_1}\cup_{\mathfrak{C}_2} \ X_2$$

in the same way as (18.47). ■

Definition-Lemma 18.51 In Situation 18.50 we can glue $\widehat{\mathcal{U}_1}$ and $\widehat{\mathcal{U}_2}$ to obtain a Kuranishi structure $\widehat{\mathcal{U}}$ of X. We write

$$(X, \widehat{\mathcal{U}}) = (X_1, \widehat{\mathcal{U}_1}) \ {}_{\mathfrak{C}_1}\cup_{\mathfrak{C}_2} \ (X_2, \widehat{\mathcal{U}_2}), \qquad \widehat{\mathcal{U}} = \widehat{\mathcal{U}_1} \ {}_{\mathfrak{C}_1}\cup_{\mathfrak{C}_2} \ \widehat{\mathcal{U}_2}.$$

It has the following properties:

(1) We have

$$\partial(X, \widehat{\mathcal{U}}) = \partial_{\mathfrak{C}_1^c}(X_1, \widehat{\mathcal{U}_1}) \ {}_{\mathfrak{C}_1}\cup_{\mathfrak{C}_2} \ \partial_{\mathfrak{C}_2^c}(X_2, \widehat{\mathcal{U}_2}).$$

(2) Suppose that we have one of the following objects on $(X_1, \widehat{\mathcal{U}_1})$ and on $(X_2, \widehat{\mathcal{U}_2})$ such that their restrictions to $\partial_{\mathfrak{C}_1}(X_1, \widehat{\mathcal{U}_1})$ and to $\partial_{\mathfrak{C}_2}(X_2, \widehat{\mathcal{U}_2})$ are isomorphic to each other (or coincide with each other) when we identify $\partial_{\mathfrak{C}_1}(X_1, \widehat{\mathcal{U}_1})$ with $\partial_{\mathfrak{C}_2}(X_2, \widehat{\mathcal{U}_2})$ under the diffeomorphism $\hat{I}$. Then we obtain the corresponding object on $(X, \widehat{\mathcal{U}})$:

   (a) Differential form.
   (b) Strongly smooth map to a manifold.
   (c) CF-perturbation.
   (d) Multivalued perturbation.

(3) Let h_i $(i = 1, 2)$ be differential forms on $(X_i, \widehat{\mathcal{U}_i})$, $f_i : (X_i, \widehat{\mathcal{U}_i}) \to M$ strongly smooth maps, and $\widehat{\mathfrak{S}_i}$ CF-perturbations of $(X_i, \widehat{\mathcal{U}_i})$ such that f_i is strongly submersive with respect to $\widehat{\mathfrak{S}_i}$ respectively. Suppose we can glue them in the sense of Item (2) and obtain h, f and $\widehat{\mathfrak{S}}$. Then we have

$$f!(h; \widehat{\mathfrak{S}^{\epsilon}}) = f_1!(h_1; \widehat{\mathfrak{S}_1^{\epsilon}}) + f_2!(h_2; \widehat{\mathfrak{S}_2^{\epsilon}}).$$

(4) In addition, suppose that we have $(X_i, \widehat{\mathcal{U}_i'})$ $(i = 1, 2)$ and $\hat{I}' : \partial_{\mathfrak{C}_1}(X_1, \widehat{\mathcal{U}_1'}) \cong \partial_{\mathfrak{C}_2}(X_2, \widehat{\mathcal{U}_2'})$ as above. (Note that both of $\widehat{\mathcal{U}_i'}$ and $\widehat{\mathcal{U}_i}$ are Kuranishi structures of X_i for each $i = 1, 2$.)

We also assume that we have embeddings $(X_i, \widehat{\mathcal{U}_i'}) \to (X_i, \widehat{\mathcal{U}_i})$ of Kuranishi structures for $i = 1, 2$ that are compatible with $\widehat{I}$ and $\widehat{I}'$ in an obvious sense. Then they induce an embedding of Kuranishi structures : $\widehat{\mathcal{U}_1'}\ {}_{\mathfrak{C}_1}\cup_{\mathfrak{C}_2} \widehat{\mathcal{U}_2'} \to \widehat{\mathcal{U}_1}\ {}_{\mathfrak{C}_1}\cup_{\mathfrak{C}_2} \widehat{\mathcal{U}_2}$.

The compatibility of objects (a)(b)(c)(d) in Item (2) is preserved by this process.

(5) We have a stratumwise weakly submersive map:

$$(X_1, \widehat{\mathcal{U}_1})\ {}_{\mathfrak{C}_1}\cup_{\mathfrak{C}_2} (X_2, \widehat{\mathcal{U}_2}) \to P_1\ {}_{\mathfrak{C}_1}\cup_{\mathfrak{C}_2} P_2.$$

The proof is immediate from Definition-Lemma 18.49.

Situation 18.52 We work in the admissible category. Let $\mathcal{C}_i, \mathcal{F}_i$ $(i = 1, 2)$ be critical submanifold data and linear K-systems as in Situation 16.16, respectively. We assume $\mathfrak{G}_1 = \mathfrak{G}_2$ and denote the common group by $\mathfrak{G}$. Let P_1, P_2, $\partial_{\mathfrak{C}_i} P_i \subset \partial P_i$ and $I : \partial_{\mathfrak{C}_1} P_1 \to \partial_{\mathfrak{C}_2} P_2$ be as in Definition-Lemma 18.49. Let $\mathfrak{N}^{P_i}$ be a P_i-parametrized morphism from $\mathcal{F}_1$ to $\mathcal{F}_2$. We denote by

$$\mathcal{N}_i(\alpha_-, \alpha_+; P_i)$$

the interpolation space of $\mathfrak{N}_i^{P_i}$. We obtain a $\partial_{\mathfrak{C}_i} P_i$-morphism $\mathfrak{N}^{\partial_{\mathfrak{C}_i} P_i}$ from $\mathcal{F}_1$ to $\mathcal{F}_2$, whose interpolation space is

$$\mathcal{N}_i(\alpha_-, \alpha_+; \partial_{\mathfrak{C}_i} P_i) = \partial_{\mathfrak{C}_i} P_i \times_{P_i} \mathcal{N}_i(\alpha_-, \alpha_+; P_i).$$

We assume that $\mathfrak{N}^{\partial_{\mathfrak{C}_1} P_1}$ is isomorphic to $\mathfrak{N}^{\partial_{\mathfrak{C}_2} P_2}$ under an isomorphism which covers $I : \partial_{\mathfrak{C}_1} P_1 \to \partial_{\mathfrak{C}_2} P_2$ in an obvious sense. ■

Definition-Lemma 18.53 In Situation 18.52, we can glue $\mathfrak{N}^{P_1}$ and $\mathfrak{N}^{P_2}$ to define a P-parametrized morphism $\mathfrak{N}^P$ for $P = P_1\ {}_{\mathfrak{C}_1}\cup_{\mathfrak{C}_2} P_2$ from $\mathcal{F}_1$ to $\mathcal{F}_2$ with the following properties:

(1) The interpolation space of $\mathfrak{N}^P$ is

$$\mathcal{N}_1(\alpha_-, \alpha_+; P_1)\ {}_{\mathfrak{C}_1}\cup_{\mathfrak{C}_2} \mathcal{N}_2(\alpha_-, \alpha_+; P_2).$$

(2) The boundary of $\mathfrak{N}^P$ is

$$\mathfrak{N}^{P_1}|_{\partial_{\mathfrak{C}_1^c} P_1}\ {}_{\mathfrak{C}_1}\cup_{\mathfrak{C}_2} \mathfrak{N}^{P_1}|_{\partial_{\mathfrak{C}_2^c} P_2}.$$

The proof is immediate from Definition-Lemma 18.51.

Now we return to the situation of Lemma-Definition 16.38. We recall that to prove Item (1) we glued two [0, 1]-parametrized morphisms $\mathfrak{H}^i_{(32)} \circ \mathfrak{N}^i_{21}$ and $\mathfrak{N}^i_{32} \circ \mathfrak{H}^i_{(21)}$ along a part of their boundaries, and to prove Item (2) we glued two $[0,1]^2$-parametrized morphisms $\mathfrak{N}^{i+1}_{32} \circ \mathcal{H}^i$ and $\mathfrak{H}^i_{(32)} \circ \mathfrak{H}^i_{(ab)}$ along a part of their boundaries. Using Definition-Lemma 18.53, we can perform such gluing.

Proof of Lemma-Definition* 16.38 *(3) We can prove transitivity by gluing two homotopies by using Definition-Lemma 18.53. Other properties are easier to prove, so they are omitted. □

18.10 Identity Morphism

In this section, we define the identity morphism.

Remark 18.54 Note that we *cannot* define the identity morphism by defining its interpolation space such as

$$\mathcal{N}(\alpha_-, \alpha_+) = \begin{cases} R_{\alpha_-} & \alpha_- = \alpha_+, \\ \emptyset & \alpha_- \neq \alpha_+. \end{cases} \tag{18.50}$$

In fact, for $\alpha_- \neq \alpha_+$, (16.25) and (18.50) yield

$$\begin{aligned} &\partial \mathcal{N}(\alpha_-, \alpha_+) \\ &\supseteq (\mathcal{M}(\alpha_-, \alpha_+) \times_{R_{\alpha_+}} \mathcal{N}(\alpha_+, \alpha_+)) \cup (\mathcal{N}(\alpha_-, \alpha_-) \times_{R_{\alpha_-}} \mathcal{M}(\alpha_-, \alpha_+)) \\ &= \mathcal{M}(\alpha_-, \alpha_+) \sqcup \mathcal{M}(\alpha_-, \alpha_+) \end{aligned} \tag{18.51}$$

which is not an empty set in general.

Suppose that we put two different systems of perturbations on $\mathcal{M}(\alpha, \beta)$. They give two different coboundary operations of the Floer cochain complex on $CF(C)$. We can use the identity morphism and a system of perturbations on its interpolation spaces to define a cochain map between them. This is the way we proceed in Chap. 19. While working out this proof, we put two different perturbations on the two copies on $\mathcal{M}(\alpha_-, \alpha_+)$ appearing in the right hand side of (18.51) and extend them to $\mathcal{N}(\alpha_-, \alpha_+)$. Thus the fact that $\partial \mathcal{N}(\alpha_-, \alpha_+)$ is nonempty is important for proving that Floer cohomology is independent of perturbations on the interpolation spaces of the identity morphism.

Let $\mathcal{F}$ be a linear K-system whose space of connecting orbits is given by $\mathcal{M}(\alpha_-, \alpha_+)$. We will define the identity morphism from $\mathcal{F}$ to itself. The interpolation spaces are defined as follows.

Definition 18.55

(1) We define $\overset{\circ}{\mathcal{N}}(\alpha_-, \alpha_+)$ as follows.[6]

(a) If $\alpha_- \neq \alpha_+$, then

$$\overset{\circ}{\mathcal{N}}(\alpha_-, \alpha_+) = \overset{\circ}{\mathcal{M}}(\alpha_-, \alpha_+) \times \mathbb{R}.$$

(b) If $\alpha = \alpha_- = \alpha_+$, then

$$\overset{\circ}{\mathcal{N}}(\alpha, \alpha) = R_\alpha.$$

(2) We compactify $\overset{\circ}{\mathcal{N}}(\alpha_-, \alpha_+)$ as follows. In the case $\alpha = \alpha_- = \alpha_+$, we put $\mathcal{N}(\alpha, \alpha) = \overset{\circ}{\mathcal{N}}(\alpha, \alpha)$.

In the case $\alpha_- \neq \alpha_+$, a stratum $\overset{\circ}{S}_k(\mathcal{N}(\alpha_-, \alpha_+))$ of the compactification $\mathcal{N}(\alpha_-, \alpha_+)$ is the union of the following two types of fiber products:

(a)

$$\begin{aligned} &\overset{\circ}{\mathcal{M}}(\alpha_-, \alpha_1) \times_{R_{\alpha_1}} \overset{\circ}{\mathcal{M}}(\alpha_1, \alpha_2) \times_{R_{\alpha_2}} \cdots \times_{R_{\alpha_{k'-1}}} \overset{\circ}{\mathcal{M}}(\alpha_{k'-1}, \alpha_{k'}) \\ &\times_{R_{\alpha_{k'}}} (\overset{\circ}{\mathcal{M}}(\alpha_{k'}, \alpha_{k'+1}) \times \mathbb{R}) \\ &\times_{R_{\alpha_{k'+1}}} \overset{\circ}{\mathcal{M}}(\alpha_{k'+1}, \alpha_{k'+2}) \times_{R_{\alpha_{k'+2}}} \cdots \times_{R_{\alpha_k}} \overset{\circ}{\mathcal{M}}(\alpha_k, \alpha_+). \end{aligned} \tag{18.52}$$

Here we take all possible choices of $k', \alpha_1, \ldots, \alpha_k$ with $0 \le k' \le k+1$ and

$$E(\alpha_-) < E(\alpha_1) < \cdots < E(\alpha_{k'}) < \cdots < E(\alpha_k) < E(\alpha_+).$$

(b)

$$\begin{aligned} &\overset{\circ}{\mathcal{M}}(\alpha_-, \alpha_1) \times_{R_{\alpha_1}} \overset{\circ}{\mathcal{M}}(\alpha_1, \alpha_2) \times_{R_{\alpha_2}} \cdots \times_{R_{\alpha_{k'-1}}} \overset{\circ}{\mathcal{M}}(\alpha_{k'-1}, \alpha_{k'}) \\ &\times_{R_{\alpha_{k'}}} R_{\alpha_{k'}} \\ &\times_{R_{\alpha_{k'}}} \overset{\circ}{\mathcal{M}}(\alpha_{k'}, \alpha_{k'+1}) \times_{R_{\alpha_{k'+1}}} \cdots \times_{R_{\alpha_{k-1}}} \overset{\circ}{\mathcal{M}}(\alpha_{k-1}, \alpha_+). \end{aligned} \tag{18.53}$$

Here we take all possible choices of $k', \alpha_1, \ldots, \alpha_{k-1}$ with $0 \le k' \le k$ and

---

[6]In case (a), $\overset{\circ}{\mathcal{N}}(\alpha_-, \alpha_+)$ is defined so that its quotient by $\mathbb{R}$ action is $\overset{\circ}{\mathcal{M}}(\alpha_-, \alpha_+)$. See Remark 18.59 for the geometric background to this definition.

$$E(\alpha_-) < E(\alpha_1) < \cdots < E(\alpha_{k'}) < \cdots < E(\alpha_{k-1}) < E(\alpha_+).$$

Remark 18.56

(1) In (2)(a) above, we take $\alpha_0 = \alpha_-$ and $\alpha_{k+1} = \alpha_+$. For example, in the case $k' = 0$, the stratum (18.52) becomes

$$(\overset{\circ}{\mathcal{M}}(\alpha_-, \alpha_1) \times \mathbb{R}) \times_{R_{\alpha_1}} \overset{\circ}{\mathcal{M}}(\alpha_1, \alpha_2) \times_{R_{\alpha_2}} \cdots \times_{R_{\alpha_k}} \overset{\circ}{\mathcal{M}}(\alpha_k, \alpha_+).$$

(2) Similarly in (2)(b) above, we take $\alpha_0 = \alpha_-$ and $\alpha_k = \alpha_+$. For example, in the case $k' = k$, the stratum (18.53) becomes

$$\overset{\circ}{\mathcal{M}}(\alpha_-, \alpha_1) \times_{R_{\alpha_1}} \overset{\circ}{\mathcal{M}}(\alpha_1, \alpha_2) \times_{R_{\alpha_2}} \cdots \times_{R_{\alpha_{k-1}}} \overset{\circ}{\mathcal{M}}(\alpha_{k-1}, \alpha_+) \times_{R_{\alpha_+}} R_{\alpha_+}. \tag{18.54}$$

(3) Indeed, (18.53) is isomorphic to

$$\begin{aligned}&\overset{\circ}{\mathcal{M}}(\alpha_-, \alpha_1) \times_{R_{\alpha_1}} \overset{\circ}{\mathcal{M}}(\alpha_1, \alpha_2) \times_{R_{\alpha_2}} \cdots \times_{R_{\alpha_{k'-1}}} \overset{\circ}{\mathcal{M}}(\alpha_{k'-1}, \alpha_{k'})\\ &\times_{R_{\alpha_{k'}}} \overset{\circ}{\mathcal{M}}(\alpha_{k'}, \alpha_{k'+1}) \times_{R_{\alpha_{k'+1}}} \cdots \times_{R_{\alpha_{k-1}}} \overset{\circ}{\mathcal{M}}(\alpha_{k-1}, \alpha_+).\end{aligned}$$

We write it as in (18.53) to distinguish the various components that are actually the same space and take their disjoint union. However, in the definition of $\overset{\circ}{S}_k(\mathcal{N}(\alpha_-, \alpha_+))$ above, we regard those various components written as (18.53) as *different spaces* and take their disjoint union.

(4) Similarly, the space (18.52) is independent of $k' = 0, \ldots, k+1$ up to isomorphism. However, we regard each of them for different k's as *different spaces* in the definition of $\overset{\circ}{S}_k(\mathcal{N}(\alpha_-, \alpha_+))$.

Lemma-Definition 18.57 *There exists a morphism from $\mathcal{F}$ to itself whose interpolation space has a stratification described in Definition* 18.55. *We call the morphism the identity morphism. We can also define the identity morphism for the case of partial linear K-systems in the same way.*

Before proving the lemma we give an example.

Example 18.58 Let us consider the case when there is exactly one α such that $E(\alpha_-) < E(\alpha) < E(\alpha_+)$. Then $\mathcal{N}(\alpha_-, \alpha_+)$ is stratified as follows:

$$\overset{\circ}{S}_0(\mathcal{N}(\alpha_-, \alpha_+)) = \overset{\circ}{\mathcal{M}}(\alpha_-, \alpha_+) \times \mathbb{R}.$$

$$\begin{aligned}\overset{\circ}{S}_1(\mathcal{N}(\alpha_-,\alpha_+)) =&(\overset{\circ}{\mathcal{M}}(\alpha_-,\alpha)\times\mathbb{R})\times_{R_\alpha}\overset{\circ}{\mathcal{M}}(\alpha,\alpha_+)\\ &\cup\overset{\circ}{\mathcal{M}}(\alpha_-,\alpha)\times_{R_\alpha}(\overset{\circ}{\mathcal{M}}(\alpha,\alpha_+)\times\mathbb{R})\\ &\cup R_{\alpha_-}\times_{R_{\alpha_-}}\overset{\circ}{\mathcal{M}}(\alpha_-,\alpha_+)\\ &\cup\overset{\circ}{\mathcal{M}}(\alpha_-,\alpha_+)\times_{R_{\alpha_+}}R_{\alpha_+}.\end{aligned} \tag{18.55}$$

$$\begin{aligned}\overset{\circ}{S}_2(\mathcal{N}(\alpha_-,\alpha_+)) =&R_{\alpha_-}\times_{R_{\alpha_-}}\overset{\circ}{\mathcal{M}}(\alpha_-,\alpha)\times_{R_\alpha}\overset{\circ}{\mathcal{M}}(\alpha,\alpha_+)\\ &\cup\overset{\circ}{\mathcal{M}}(\alpha_-,\alpha)\times_{R_\alpha}R_\alpha\times_{R_\alpha}\overset{\circ}{\mathcal{M}}(\alpha,\alpha_+)\\ &\cup\overset{\circ}{\mathcal{M}}(\alpha_-,\alpha)\times_{R_\alpha}\overset{\circ}{\mathcal{M}}(\alpha,\alpha_+)\times_{R_{\alpha_+}}R_{\alpha_+}.\end{aligned} \tag{18.56}$$

See Fig. 18.8. Note that in this case $\mathcal{M}(\alpha_-,\alpha_+)$ has a Kuranishi structure with boundary $\overset{\circ}{\mathcal{M}}(\alpha_-,\alpha)\times_{R_\alpha}\overset{\circ}{\mathcal{M}}(\alpha,\alpha_+)$ but without corners. The K-space $\mathcal{N}(\alpha_-,\alpha_+)$ may be regarded as $\mathcal{M}(\alpha_-,\alpha_+)\times[0,1]$. However, the stratification of $\mathcal{N}(\alpha_-,\alpha_+)$ is different from that of $\mathcal{M}(\alpha_-,\alpha_+)\times[0,1]$. Namely, we stratify $\mathcal{M}(\alpha_-,\alpha_+)\times[0,1]$ as follows:

Its codimension 0 stratum is

(0) $\mathcal{M}(\alpha_-,\alpha_+)\times[0,1]$.

Its codimension 1 strata are

(1-1) $\mathcal{M}(\alpha_-,\alpha_+)\times\{0\}$,
(1-2) $\mathcal{M}(\alpha_-,\alpha_+)\times\{1\}$,
(1-3) $\partial\mathcal{M}(\alpha_-,\alpha_+)\times[0,1/2]$,
(1-4) $\partial\mathcal{M}(\alpha_-,\alpha_+)\times[1/2,1]$.

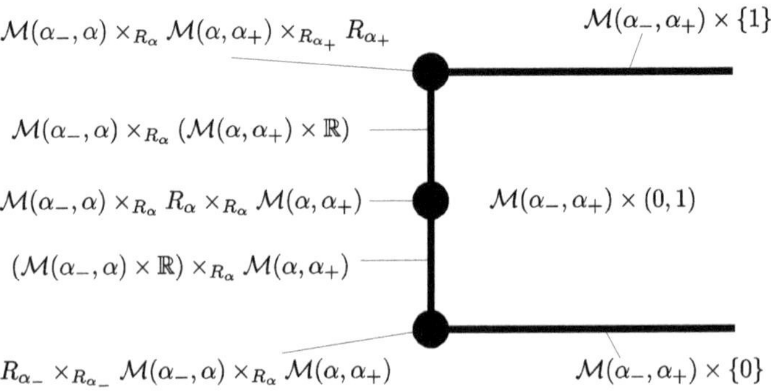

Fig. 18.8 $\mathcal{N}(\alpha_-,\alpha_+)$: Case 1

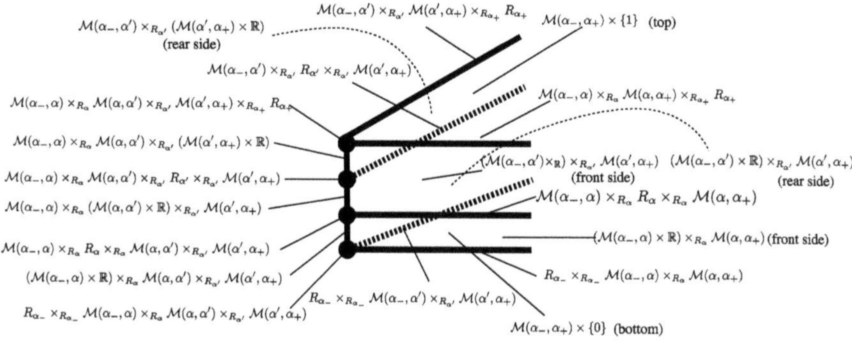

Fig. 18.9 $\mathcal{N}(\alpha_-, \alpha_+)$: Case 2

Its codimension 2 strata are

(2-1) $\partial\mathcal{M}(\alpha_-, \alpha_+) \times \{0\}$,
(2-2) $\partial\mathcal{M}(\alpha_-, \alpha_+) \times \{1/2\}$,
(2-3) $\partial\mathcal{M}(\alpha_-, \alpha_+) \times \{1\}$.

The strata (1-1), (1-2), (1-3), (1-4) correspond to the 1st, 2nd, 3rd, 4th terms of (18.55), respectively. The strata (2-1), (2-2), (2-3) correspond to the 1st, 2nd, 3rd terms of (18.56), respectively. Note that $\mathcal{N}(\alpha_-, \alpha_+)$ has an $\mathbb{R}$-action preserving the stratification. Namely, on the stratum where there is a (0, 1) factor, we identify it with $\mathbb{R}$ and the $\mathbb{R}$ action is the translation on it. The other strata are in the fixed point set of this action.

The quotient space $\mathcal{N}(\alpha_-, \alpha_+)/\mathbb{R}$ is similar to $\mathcal{M}(\alpha_-, \alpha_+)$ but is different therefrom. In the case of our example, $\overset{\circ}{S}_0(\mathcal{N}(\alpha_-, \alpha_+))/\mathbb{R} = \overset{\circ}{S}_0(\mathcal{M}(\alpha_-, \alpha_+)) = \overset{\circ}{\mathcal{M}}(\alpha_-, \alpha_+)$. However, $\overset{\circ}{S}_1(\mathcal{N}(\alpha_-, \alpha_+))/\mathbb{R}$ is the union of the disjoint union of two copies of $\overset{\circ}{S}_0(\mathcal{M}(\alpha_-, \alpha_+))$ and the disjoint union of two copies of $\overset{\circ}{S}_1(\mathcal{M}(\alpha_-, \alpha_+))$. Moreover, $\overset{\circ}{S}_2(\mathcal{N}(\alpha_-, \alpha_+))/\mathbb{R}$ is the disjoint union of three copies of $\overset{\circ}{S}_1(\mathcal{M}(\alpha_-, \alpha_+))$. Then the quotient space $\mathcal{N}(\alpha_-, \alpha_+)/\mathbb{R}$ with quotient topology is non-Hausdorff. At any rate, we do not use the quotient space in this book at all.

When there exist exactly two α and α' with $E(\alpha_-) < E(\alpha) < E(\alpha') < E(\alpha_+)$, the stratification of $\mathcal{N}(\alpha_-, \alpha_+)$ is drawn in Fig. 18.9. We note that in this figure all the codimension 2 strata (the vertices in Fig. 18.9) are contained in exactly three edges. So its neighborhood can be identified with a corner point.

***Proof of Lemma-Definition* 18.57** We first explain how we glue the strata and define the topology on $\mathcal{N}(\alpha_-, \alpha_+)$. (See Sect. 18.11.2 for a geometric origin of the identification (18.57).)

Let E be the energy homomorphism. We will identity

$$\mathcal{N}(\alpha_-, \alpha_+) \cong \mathcal{M}(\alpha_-, \alpha_+) \times [E(\alpha_-), E(\alpha_+)] \tag{18.57}$$

as follows. We identify the factor $\mathbb{R}$ in (18.52) with the open interval $(E(\alpha_{k'}), E(\alpha_{k'+1}))$. The fiber product of other factors in (18.52) is

$$\overset{\circ}{\mathcal{M}}(\alpha_-, \alpha_1) \times_{R_{\alpha_1}} \overset{\circ}{\mathcal{M}}(\alpha_1, \alpha_2) \times_{R_{\alpha_2}} \cdots \times_{R_{\alpha_k}} \overset{\circ}{\mathcal{M}}(\alpha_k, \alpha_+) \tag{18.58}$$

and is a component of $\overset{\circ}{S}_k(\mathcal{M}(\alpha_-, \alpha_+))$. Thus (18.52) is identified with a subset of

$$\overset{\circ}{S}_k(\mathcal{M}(\alpha_-, \alpha_+)) \times (E(\alpha_{k'}), E(\alpha_{k'+1}))$$

that is a subset of (18.57).

We next consider the fiber product (18.53). This space is isomorphic to (18.58) and so is a component of $\overset{\circ}{S}_k(\mathcal{M}(\alpha_-, \alpha_+))$. We set its $(E(\alpha_-), E(\alpha_+))$ factor as $E(\alpha_{k'})$. Thus (18.53) is a subset of

$$\overset{\circ}{S}_k(\mathcal{M}(\alpha_-, \alpha_+)) \times \{E(\alpha_{k'})\}$$

that is a subset of (18.57).

We have thus specified the way we embed all the strata of $\mathcal{N}(\alpha_-, \alpha_+)$ into (18.57). It is easy to see that they are disjoint from one another and their union gives (18.57). Thus we identify $\mathcal{N}(\alpha_-, \alpha_+)$ with (18.57) and define the topology on $\mathcal{N}(\alpha_-, \alpha_+)$ using this identification.

As we observed in Example 18.58, the space $\mathcal{N}(\alpha_-, \alpha_+)$ is homeomorphic to $\mathcal{M}(\alpha_-, \alpha_+) \times [0, 1]$ but its corner structure stratification is different from the direct product. We next define a Kuranishi structure with corners on $\mathcal{N}(\alpha_-, \alpha_+)$ compatible with the stratification given by (18.52) and (18.53). We will define a Kuranishi neighborhood of each point of $\mathcal{N}(\alpha_-, \alpha_+)$ below.

Firstly, let $\hat{p}$ be a point of (18.52). We write $\hat{p} = (p, s_0)$ where p is an element of (18.58) and $s_0 \in (E(\alpha_{k'}), E(\alpha_{k'+1}))$. Let $\mathcal{U}_p$ be a Kuranishi neighborhood of p. For simplicity of the discussion, we assume that U_p consists of a single orbifold chart V_p/Γ_p. (In the general case we can perform the construction below for each orbifold chart.) Then V_p is an open set of $[V_p] \times [0, 1)^k$. We put $\mathcal{U}_p \times (s_0 - \epsilon, s_0 + \epsilon)$ as our Kuranishi neighborhood of $\hat{p}$. In fact, the parametrization map $s_{\hat{p}}^{-1}(0) \to \mathcal{N}(\alpha_-, \alpha_+)$ is defined as follows. Let $(q, s) \in s_{\hat{p}}^{-1}(0)$. Then q parametrizes a certain point of $\mathcal{M}(\alpha_-, \alpha_+)$. Hence (q, s) defines a point in (18.57). Therefore we obtain $\psi_{\hat{p}}(q, s) \in \mathcal{N}(\alpha_-, \alpha_+)$ by our identification.

Next we consider the case $\hat{p}$ is in (18.53). First we suppose $k' = k$. Then (18.53) coincides with (18.54) that is identified with a subset of

$$\overset{\circ}{S}_k(\alpha_-, \alpha_+) \times \{E(\alpha_+)\} \subset \overset{\circ}{S}_k(\alpha_-, \alpha_+) \times \partial[E(\alpha_-), E(\alpha_+)].$$

Fig. 18.10 $(s, t) \mapsto (\sigma, \rho)$

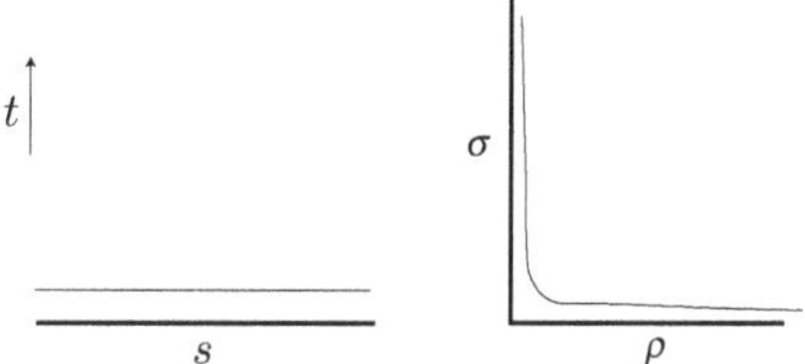

We take $\mathcal{U}_p \times (E(\alpha_+) - \epsilon, E(\alpha_+)]$ as our Kuranishi neighborhood $\mathcal{U}_{\hat{p}}$. Here $\mathcal{U}_p$ is a Kuranishi neighborhood of p in $\mathcal{M}(\alpha_-, \alpha_+)$. The definition of the parametrization map is similar to the first case. The case $k' = 0$ is similar to the above case.

The case $k' \neq k$ and $k' \neq 0$ is the most involved, which we discuss now. We will construct a Kuranishi neighborhood of $\hat{p} = (p, E(\alpha_{k'}))$. Let $\mathcal{U}_p$ be the Kuranishi neighborhood of p in (18.58). Put $p = (p_0, p_1, \ldots, p_k)$ according to this fiber product and take a Kuranishi neighborhood $\mathcal{U}_{p_i}$ of p_i in $\overset{\circ}{\mathcal{M}}(\alpha_i, \alpha_{i+1})$. Then the orbifold chart of $\mathcal{U}_p$ is given by

$$U_{p_0} \times_{R_{\alpha_1}} \times \cdots \times_{R_{\alpha_k}} U_{p_k} \times [0, \epsilon)^k.$$

We write its element as $(x_0, \ldots, x_k; (t_1, \ldots, t_k))$. We note that the triple $(x_{k'-1}, x_{k'}, t_{k'})$ parametrizes a Kuranishi neighborhood of $(p_{k'-1}, p_{k'})$ in $\mathcal{M}(\alpha_{k'-1}, \alpha_{k'+1})$.

For $s' \in (E(\alpha_{k'}) - \epsilon, E(\alpha_{k'}) + \epsilon)$ we write $(s, t) = (s' - E(\alpha_{k'}), t_{k'})$ and change variables from (s, t) to $(\rho, \sigma) \in [0, 1)^2$ by

$$s = \rho - \sigma, \qquad t = \rho\sigma.$$

In other words (Fig. 18.10),

$$\rho = \frac{s + \sqrt{s^2 + 4t}}{2}, \qquad \sigma = \frac{-s + \sqrt{s^2 + 4t}}{2}. \tag{18.59}$$

We use $x_0, \ldots, x_k, t_1, \ldots, t_{k'-1}, t_{k'+1}, \ldots, t_k$ and ρ, σ as the coordinates of

$$U_{p_0} \times_{R_{\alpha_1}} \times \cdots \times_{R_{\alpha_k}} U_{p_k} \times [0, \epsilon)^k \times (E(\alpha_{k'}) - \epsilon, E(\alpha_{k'}) + \epsilon). \tag{18.60}$$

This gives a smooth structure on (18.60). In fact, the point $\hat{p}$ lies in the codimension $k + 1$ corner in *this smooth structure* which is different from the *direct product smooth structure* on (18.60). (We note that $\hat{p}$ lies in the codimension k corner with respect to the direct product smooth structure.) In the case of Example 18.58 the space drawn in Fig. 18.8 becomes as in Fig. 18.11 under this new smooth structure. We call this process *bending*.

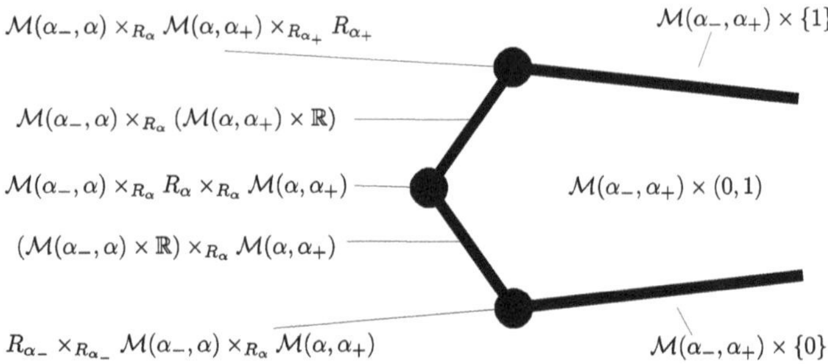

Fig. 18.11 Bending of $\mathcal{N}(\alpha_-,\alpha_+)$

To complete the definition of the Kuranishi structure of $\overset{\circ}{\mathcal{N}}(\alpha_-,\alpha_+)$ it suffices to show that the coordinate change is admissible in this last case. (Admissibility of the coordinate change is trivial in other cases.) We will check it below.

We denote by y the totality of the coordinates $x_0,\dots,x_k,t_1,\dots,t_{k'-1},t_{k'+1},\dots,t_l$ Then y,t,s or y,ρ,σ are the coordinates of our Kuranishi neighborhood. Let y',t',s' or y',ρ',σ' be the other coordinates. Then the coordinate changes among them are given by

$$y'=\varphi(y,t),\quad t'=\psi(y,t),\quad s'=s,$$

where φ and ψ are smooth and satisfy

$$\left\|\frac{\partial y'}{\partial t}\right\|_{C^k}\le C_ke^{-c_k/t},\quad \|t'-t\|_{C^k}\le C_ke^{-c_k/t}$$

for some $c_k>0$, $C_k>0$. (See Definition 25.8, Lemma 25.10 (2) and (25.9).) We note that y' and t' are independent of s. Then using (18.59) it is easy to check

$$\left\|\frac{\partial y'}{\partial \rho}\right\|_{C^k}\le C_ke^{-c_k/\rho},\quad \left\|\frac{\partial y'}{\partial \sigma}\right\|_{C^k}\le C_ke^{-c_k/\sigma}$$

and

$$\|\rho'-\rho\|_{C^k}\le C_ke^{-c_k/\rho},\quad \|\sigma'-\sigma\|_{C^k}\le C_ke^{-c_k/\sigma}.$$

This implies admissibility of the coordinate changes with respect to our new coordinates. Thus we have constructed a Kuranishi structure on $\mathcal{N}(\alpha_-,\alpha_+)$.

We define evaluation maps $\mathrm{ev}_\pm$ on $\mathcal{N}(\alpha_-,\alpha_+)$ by taking one on the factor where α_- or α_+ appears. The periodicity isomorphism and orientation isomorphism are induced by ones on $\mathcal{M}(\alpha_-,\alpha_+)$ in an obvious way.

We note that in (18.52) the factor $\overset{\circ}{\mathcal{M}}(\alpha_{k'}, \alpha_{k'+1}) \times \mathbb{R}$ can be identified with $\overset{\circ}{\mathcal{N}}(\alpha_{k'}, \alpha_{k'+1})$ and in (18.53) the factor $R_{\alpha_{k'}}$ can be identified with $\overset{\circ}{\mathcal{N}}(\alpha_{k'}, \alpha_{k'})$. The fact that $\mathcal{N}(\alpha_-, \alpha_+)$ satisfies Conditions 16.23 and 16.28 is a consequence of the compatibility condition at the boundary and the corner of $\mathcal{M}(\alpha_-, \alpha_+)$. The proof of Lemma-Definition 18.57 is now complete. □

Remark 18.59 In the situation where we obtain a linear K-system from the Floer equation of a periodic Hamiltonian system, morphisms between them are obtained by studying the two-parameter family of Hamiltonians $\mathcal{H} : \mathbb{R} \times S^1 \times M \to \mathbb{R}$, $H_{\tau,t}(x) = \mathcal{H}(\tau, t, x)$. In that case the interpolation space is the compactified moduli space of solutions of the equation

$$\frac{\partial u}{\partial \tau} + J\left(\frac{\partial u}{\partial t} - X_{H_{\tau,t}}(u)\right) = 0. \tag{18.61}$$

See [FOOO13, Section 9]. In the case when $H_{\tau,t} = H_t$ is τ-independent, the morphism we obtain becomes the identity morphism defined above. (See Sect. 18.11.1.)

The next proposition shows that the identity morphism $\mathfrak{ID}$ is a homotopy unit with respect to the product given by the composition of morphisms.

Proposition 18.60 *For any morphism $\mathfrak{N}$, the compositions $\mathfrak{N} \circ \mathfrak{ID}$ and $\mathfrak{ID} \circ \mathfrak{N}$ are both homotopic to $\mathfrak{N}$.*

Proof We will prove $\mathfrak{N} \circ \mathfrak{ID} \sim \mathfrak{N}$. The proof of $\mathfrak{ID} \circ \mathfrak{N} \sim \mathfrak{N}$ is similar. Let $\mathcal{N}(\alpha_-, \alpha_+)$ be an interpolation space of $\mathfrak{N}$. By the definition of the composition $\mathfrak{N} \circ \mathfrak{ID}$, its interpolation space is decomposed into the following two types of fiber products:

(1)

$$\begin{aligned}
&\overset{\circ}{\mathcal{M}}{}^1(\alpha_-, \alpha_1) \times_{R_{\alpha_1}} \overset{\circ}{\mathcal{M}}{}^1(\alpha_1, \alpha_2) \times_{R_{\alpha_2}} \cdots \times_{R_{\alpha_{k'_1-1}}} \overset{\circ}{\mathcal{M}}{}^1(\alpha_{k'_1-1}, \alpha_{k'_1}) \\
&\times_{R_{\alpha_{k'_1}}} (\overset{\circ}{\mathcal{M}}{}^1(\alpha_{k'_1}, \alpha_{k'_1+1}) \times \mathbb{R}) \\
&\times_{R_{\alpha_{k'_1+1}}} \overset{\circ}{\mathcal{M}}{}^1(\alpha_{k'_1+1}, \alpha_{k'_1+2}) \times_{R_{\alpha_{k'_1+2}}} \cdots \times_{R_{\alpha_{k_1-1}}} \overset{\circ}{\mathcal{M}}{}^1(\alpha_{k_1-1}, \alpha_{k_1}) \\
&\times_{R_{\alpha_{k_1}}} \overset{\circ}{\mathcal{N}}(\alpha_{k_1}, \alpha'_1) \\
&\times_{R_{\alpha'_1}} \overset{\circ}{\mathcal{M}}{}^2(\alpha'_1, \alpha'_2) \times_{R_{\alpha'_2}} \cdots \times_{R_{\alpha'_{k_2}}} \overset{\circ}{\mathcal{M}}{}^2(\alpha'_{k_2}, \alpha'_+).
\end{aligned} \tag{18.62}$$

(2)

$$\begin{aligned}
&\overset{\circ}{\mathcal{M}}{}^1(\alpha_-, \alpha_1) \times_{R_{\alpha_1}} \overset{\circ}{\mathcal{M}}{}^1(\alpha_1, \alpha_2) \times_{R_{\alpha_2}} \cdots \times_{R_{\alpha_{k'_1-1}}} \overset{\circ}{\mathcal{M}}{}^1(\alpha_{k'_1-1}, \alpha_{k'_1}) \\
&\times_{R_{\alpha_{k'_1}}} R_{\alpha_{k'_1}} \\
&\times_{R_{\alpha_{k'_1}}} \overset{\circ}{\mathcal{M}}{}^1(\alpha_{k'_1}, \alpha_{k'_1+1}) \times_{R_{\alpha_{k'_1+1}}} \cdots \times_{R_{\alpha_{k_1-1}}} \overset{\circ}{\mathcal{M}}{}^1(\alpha_{k_1-1}, \alpha_{k_1}) \qquad (18.63) \\
&\times_{R_{\alpha_{k_1}}} \overset{\circ}{\mathcal{N}}(\alpha_{k_1}, \alpha'_1) \\
&\times_{R_{\alpha'_1}} \overset{\circ}{\mathcal{M}}{}^2(\alpha'_1, \alpha'_2) \times_{R_{\alpha'_2}} \cdots \times_{R_{\alpha'_{k_2}}} \overset{\circ}{\mathcal{M}}{}^2(\alpha'_{k_2}, \alpha'_+).
\end{aligned}$$

Note that the process of gluing, described in the proof of Lemma-Definition 16.35, is included in this description. Namely, the fiber products (18.62), (18.63) appear only once in this decomposition. We also note that $\mathcal{N}(\alpha_-, \alpha'_+)$, that is an interpolation space of $\mathfrak{N}$, is decomposed into

$$\begin{aligned}
&\overset{\circ}{\mathcal{M}}{}^1(\alpha_-, \alpha_1) \times_{R_{\alpha_1}} \cdots \times_{R_{\alpha_{k_1-1}}} \overset{\circ}{\mathcal{M}}{}^1(\alpha_{k_1-1}, \alpha_{k_1}) \\
&\times_{R_{\alpha_{k_1}}} \overset{\circ}{\mathcal{N}}(\alpha_{k_1}, \alpha'_1) \qquad (18.64) \\
&\times_{R_{\alpha'_1}} \overset{\circ}{\mathcal{M}}{}^2(\alpha'_1, \alpha'_2) \times_{R_{\alpha'_2}} \cdots \times_{R_{\alpha'_{k_2}}} \overset{\circ}{\mathcal{M}}{}^2(\alpha'_{k_2}, \alpha'_+).
\end{aligned}$$

By Definition 16.30 of homotopy, it suffices to find a K-space whose boundary is a union of (18.62), (18.63), (18.64) and the obvious component $\mathcal{N}(\alpha_-, \alpha_+; \partial[0, 1])$ corresponding to the last line in (16.31). (We omit the last obvious component here.)

The interpolation space of our homotopy between $\mathfrak{N} \circ \mathfrak{ID}$ and $\mathfrak{N}$ is stratified so that its strata consist of (18.62), (18.63), (18.64) and the following spaces (18.65), (18.66):

$$\begin{aligned}
&\overset{\circ}{\mathcal{M}}{}^1(\alpha_-, \alpha_1) \times_{R_{\alpha_1}} \cdots \times_{R_{\alpha_{k_1-1}}} \overset{\circ}{\mathcal{M}}{}^1(\alpha_{k_1-1}, \alpha_{k_1}) \\
&\times_{R_{\alpha_{k_1}}} \overset{\circ}{\mathcal{N}}(\alpha_{k_1}, \alpha'_1) \times (0, 1) \qquad (18.65) \\
&\times_{R_{\alpha'_1}} \overset{\circ}{\mathcal{M}}{}^2(\alpha'_1, \alpha'_2) \times_{R_{\alpha'_2}} \cdots \times_{R_{\alpha'_{k_2}}} \overset{\circ}{\mathcal{M}}{}^2(\alpha'_{k_2}, \alpha'_+),
\end{aligned}$$

$$\begin{aligned}
&\overset{\circ}{\mathcal{M}}{}^1(\alpha_-, \alpha_1) \times_{R_{\alpha_1}} \cdots \times_{R_{\alpha_{k_1-1}}} \overset{\circ}{\mathcal{M}}{}^1(\alpha_{k_1-1}, \alpha_{k_1}) \\
&\times_{R_{\alpha_{k_1}}} \overset{\circ}{\mathcal{N}}(\alpha_{k_1}, \alpha'_1) \times \{0\} \\
&\times_{R_{\alpha'_1}} \overset{\circ}{\mathcal{M}}{}^2(\alpha'_1, \alpha'_2) \times_{R_{\alpha'_2}} \cdots \times_{R_{\alpha'_{k_2}}} \overset{\circ}{\mathcal{M}}{}^2(\alpha'_{k_2}, \alpha'_+).
\end{aligned} \tag{18.66}$$

We remark that when $t \in (0, 1)$ in the second line of (18.65) goes to 1 the element goes to an element in (18.65) becomes an element of (18.64). This is the reason why $\{1\}$ is not included in the second line of (18.66).

We claim that the union of (18.62), (18.63), (18.64) and (18.65), (18.66) above has a Kuranishi structure with corners. The proof follows.

We first define a topology on the disjoint union $\mathcal{N}(\alpha_-, \alpha'_+; [0, 1])$ of the spaces (18.62), (18.63), (18.64), (18.65), and (18.66). Let c be the energy loss of $\mathfrak{N}$. We identify the underlying topological space of $\mathcal{N}(\alpha_-, \alpha'_+; [0, 1])$ with

$$\mathcal{N}(\alpha_-, \alpha'_+) \times [E(\alpha_-), E(\alpha'_+) + c]. \tag{18.67}$$

See Sect. 18.11.4 for the geometric origin of this identification. We identify (18.62), (18.63), (18.64), (18.65), and (18.66) to a subset of (18.67) as follows.
(The case of (18.62)): We identify the $\mathbb{R}$ factor with $(E(\alpha_{k'_1}), E(\alpha'_1))$. The fiber product of the other factors is

$$\begin{aligned}
&\overset{\circ}{\mathcal{M}}{}^1(\alpha_-, \alpha_1) \times_{R_{\alpha_1}} \overset{\circ}{\mathcal{M}}{}^1(\alpha_1, \alpha_2) \times_{R_{\alpha_2}} \cdots \times_{R_{\alpha_{k_1-1}}} \overset{\circ}{\mathcal{M}}{}^1(\alpha_{k'_1-1}, \alpha_{k'_1}) \\
&\times_{R_{\alpha_{k_1}}} \overset{\circ}{\mathcal{N}}(\alpha_{k_1}, \alpha'_1) \\
&\times_{R_{\alpha'_1}} \overset{\circ}{\mathcal{M}}{}^2(\alpha'_1, \alpha'_2) \times_{R_{\alpha'_2}} \cdots \times_{R_{\alpha'_{k_2}}} \overset{\circ}{\mathcal{M}}{}^2(\alpha'_{k_2}, \alpha'_+),
\end{aligned} \tag{18.68}$$

which is a subset of $\mathcal{N}(\alpha_-, \alpha'_+)$. Thus (18.62) is identified with the subset of (18.67).
(The case of (18.63)): (18.63) is the same as (18.68) and so is a subset of $\mathcal{N}(\alpha_-, \alpha'_+)$. We take $E(\alpha'_{k_1})$ as the $[E(\alpha_-), E(\alpha'_+) + c]$ factor and regard (18.63) as a subset of (18.67).
(The case of (18.64)): (18.64) is again the same as (18.68) and so is a subset of $\mathcal{N}(\alpha_-, \alpha'_+)$. We take $E(\alpha'_+)+c$ as the $[E(\alpha_-), E(\alpha'_+)+c]$ factor and regard (18.64) as a subset of (18.67).
(The case of (18.65) $\cup$ (18.66)): We identify $[0, 1)$ with $[E(\alpha_{k_1}), E(\alpha'_+) + c)$. The fiber product of the other factors is the same as (18.68). Thus (18.65) is identified with a subset of (18.67).

Thus we have identified the union $\mathcal{N}(\alpha_-, \alpha'_+; [0,1])$ of (18.62), (18.63), (18.64), (18.65), and (18.66) with (18.67). Using this identification we define the topology of $\mathcal{N}(\alpha_-, \alpha'_+; [0,1])$.

In a way similar to the proof of Lemma-Definition 18.57 we define a Kuranishi structure on $\mathcal{N}(\alpha_-, \alpha'_+; [0,1])$ as follows. We first observe that codimension 1 strata of $\mathcal{N}(\alpha_-, \alpha'_+; [0,1])$ are one of the following four types of fiber products:

$$(\overset{\circ}{\mathcal{M}}{}^1(\alpha_-, \alpha_1) \times \mathbb{R}) \times_{R_{\alpha_1}} \overset{\circ}{\mathcal{N}}(\alpha_1, \alpha'_+), \tag{18.69}$$

$$\overset{\circ}{\mathcal{M}}{}^1(\alpha_-, \alpha_1) \times_{R_{\alpha_1}} (\overset{\circ}{\mathcal{N}}(\alpha_1, \alpha'_+) \times (0,1)), \tag{18.70}$$

$$\overset{\circ}{\mathcal{N}}(\alpha_-, \alpha'_+) \times \{0\}, \tag{18.71}$$

$$\overset{\circ}{\mathcal{N}}(\alpha_-, \alpha'_+) \times \{1\}. \tag{18.72}$$

The $\mathbb{R}$ factor in (18.69) is identified with $(E(\alpha_-), E(\alpha_1))$. The $(0,1)$ factor in (18.70) is identified with $(E(\alpha_1), E(\alpha'_+)+c)$. They are glued where these factors are identified with $E(\alpha_1)$. The union is identified with

$$(\overset{\circ}{\mathcal{M}}{}^1(\alpha_-, \alpha_1) \times_{R_{\alpha_1}} \overset{\circ}{\mathcal{N}}(\alpha_1, \alpha'_+)) \times (E(\alpha_-), E(\alpha'_+) + c). \tag{18.73}$$

However, we regard the subset

$$(\overset{\circ}{\mathcal{M}}{}^1(\alpha_-, \alpha_1) \times_{R_{\alpha_1}} \overset{\circ}{\mathcal{N}}(\alpha_1, \alpha'_+)) \times \{E(\alpha_1)\} \tag{18.74}$$

of (18.73) now as a codimension 2 stratum, that is, a corner. In other words, we bend the space (18.73) at (18.74). This bending is performed in the same way as the argument of case $k' \neq 0, k$ in the proof of Lemma-Definition 18.57.

The point $\{0\}$ in (18.71) is identified with $E(\alpha_-)$. Therefore the closures of (18.71) and of (18.73) intersect at the codimension 2 stratum

$$(\overset{\circ}{\mathcal{M}}{}^1(\alpha_-, \alpha_1) \times_{R_{\alpha_1}} \overset{\circ}{\mathcal{N}}(\alpha_1, \alpha'_+)) \times \{E(\alpha_-)\}.$$

We smooth this corner in the way we explain Sect. 18.7. In fact, this smoothing occurs during the definition of the composition $\mathfrak{N} \circ \mathfrak{ID}$. (See Definition 18.34.)

The point $\{1\}$ in (18.72) is identified with $E(\alpha'_+)$. Therefore, the closures of (18.72) and of (18.73) intersect at the codimension 2 stratum

$$(\overset{\circ}{\mathcal{M}}{}^1(\alpha_-, \alpha_1) \times_{R_{\alpha_1}} \overset{\circ}{\mathcal{N}}(\alpha_1, \alpha'_+)) \times \{E(\alpha'_+) + c\}.$$

We do not smooth this corner.

We can perform an appropriate bending or smoothing and check the consistency at the corner of higher codimension by using the compatibility conditions of $\mathcal{M}^i(\alpha_1, \alpha_2)$, $(i = 1, 2)$ and of $\mathcal{N}(\alpha, \alpha')$.

We define a map $\mathcal{N}(\alpha_-, \alpha'_+; [0, 1]) \to [0, 1]$ appearing in Condition 16.21 (IV), so that the inverse image of $\{0\}$ is the closure of the union of the strata (18.69) and (18.71). The inverse image of $\{1\}$ is the closure of the union of the strata (18.72). We note that this map is *not* diffeomorphic to the projection

$$\begin{aligned}\mathcal{N}(\alpha_-, \alpha'_+; [0, 1]) &\cong \mathcal{N}(\alpha_-, \alpha'_+) \times [E(\alpha_-), E(\alpha'_+) + c] \\ &\to [E(\alpha_-), E(\alpha'_+) + c] \to [0, 1],\end{aligned}$$

where the first identification is one we used to define the topology of $\mathcal{N}(\alpha_-, \alpha'_+; [0, 1])$ by identifying it with (18.67).

We are now ready to wrap up the proof of Proposition 18.60. So far we have defined a K-space $\mathcal{N}(\alpha_-, \alpha'_+; [0, 1])$. It has most of the properties of the interpolation space of the homotopy. In other words, its boundary has mostly the required properties, except the boundary is related to the composition $\mathfrak{N} \circ \mathfrak{ID}$ *before* equipping outer collaring with the boundary and smoothing of the corners take place. Therefore to obtain the required interpolation space of homotopy we perform partial outer collaring and corner smoothing at those collared corners. (See Sect. 18.7.) Namely the required interpolation space of the homotopy is

$$\mathcal{N}(\alpha_-, \alpha'_+; [0, 1])^{\mathfrak{C}\boxplus\tau},$$

where $\mathfrak{C}$ is the part of the boundary corresponding to $\{1\} \in [0, 1]$ in the factor $[0, 1]$. Thus we obtain an interpolation space of the required $[0, 1]$ parametrized morphism and complete the proof of Proposition 18.60. □

Example 18.61 We consider the case when $\partial\mathcal{N}(\alpha_-, \alpha_+)$ has only one component $\mathcal{M}(\alpha_-, \alpha) \times_{R_\alpha} \mathcal{N}(\alpha, \alpha_+)$. Then the top stratum of our homotopy is $\mathcal{N}(\alpha_-, \alpha_+) \times [0, 1]$. There are four codimension 1 strata: $\mathcal{N}(\alpha_-, \alpha_+)$, $R_{\alpha_-} \times_{R_{\alpha_-}} \mathcal{N}(\alpha_-, \alpha_+)$, $(\mathcal{M}(\alpha_-, \alpha) \times \mathbb{R}) \times_{R_\alpha} \mathcal{N}(\alpha, \alpha_+)$, and $\mathcal{M}(\alpha_-, \alpha) \times_{R_\alpha} (\mathcal{N}(\alpha, \alpha_+) \times [0, 1])$. The first one corresponds to the morphism $\mathfrak{N}$. The union of the second and the third corresponds to the morphism $\mathfrak{N} \circ \mathfrak{ID}$. (See Fig. 18.12.) Then we change its corner structure to obtain the space as in Fig. 18.13.

Example 18.62 Let us consider the case when there are α_1 and α_2 such that

$$\partial\mathcal{N}(\alpha_-, \alpha_+) = \mathcal{M}(\alpha_-, \alpha_1) \times_{R_{\alpha_1}} \mathcal{N}(\alpha_1, \alpha_+) \cup \mathcal{M}(\alpha_-, \alpha_2) \times_{R_{\alpha_2}} \mathcal{N}(\alpha_2, \alpha_+)$$

and the two components in the right hand side are glued at the corner $\mathcal{M}(\alpha_-, \alpha_1) \times_{R_{\alpha_1}} \times \mathcal{N}(\alpha_1, \alpha_2) \times_{R_{\alpha_2}} \mathcal{N}(\alpha_2, \alpha_+)$. Then our homotopy has one top stratum $\mathcal{N}(\alpha, \alpha_+) \times (0, 1)$ and six codimension 1 strata. Those six strata are

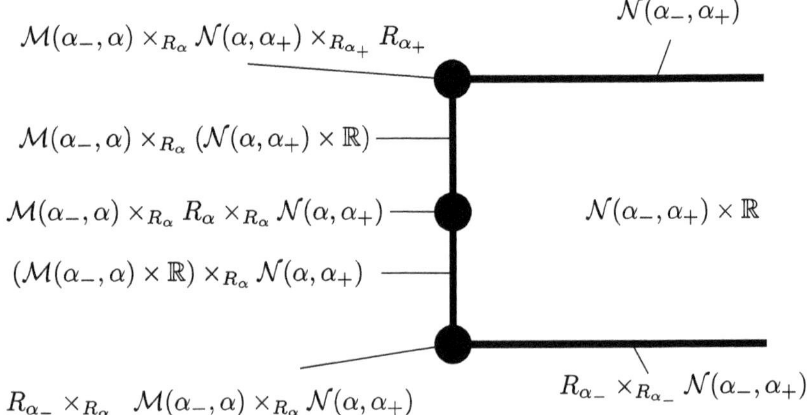

Fig. 18.12 Homotopy between $\mathfrak{N} \circ \mathfrak{ID}$ and $\mathfrak{N}$: Case 1

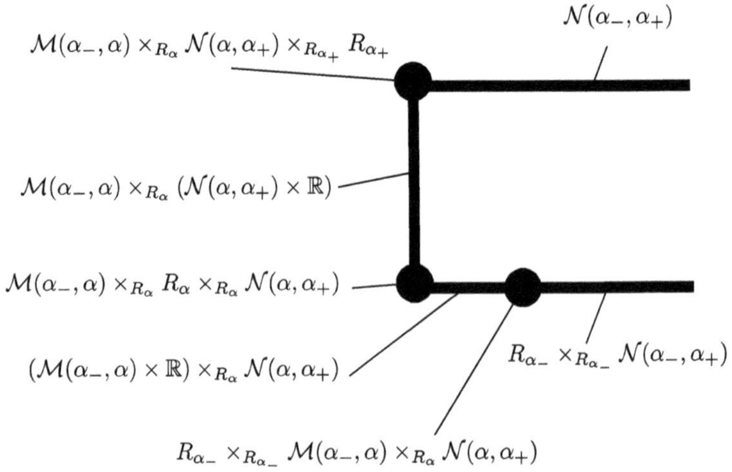

Fig. 18.13 Bending and smoothing of Fig. 18.12

$$R_{\alpha_-} \times_{R_{\alpha_-}} \mathcal{N}(\alpha_-, \alpha_+),$$
$$(\mathcal{M}(\alpha_-, \alpha_1) \times \mathbb{R}) \times_{R_{\alpha_1}} \mathcal{N}(\alpha_1, \alpha_+),$$
$$(\mathcal{M}(\alpha_-, \alpha_2) \times \mathbb{R}) \times_{R_{\alpha_2}} \mathcal{N}(\alpha_2, \alpha_+),$$
$$\mathcal{M}(\alpha_-, \alpha_1) \times_{R_{\alpha_1}} (\mathcal{N}(\alpha_1, \alpha_+) \times (0, 1)),$$
$$\mathcal{M}(\alpha_-, \alpha_2) \times_{R_{\alpha_1}} (\mathcal{N}(\alpha_2, \alpha_+) \times (0, 1)),$$
$$\mathcal{N}(\alpha_-, \alpha_+).$$

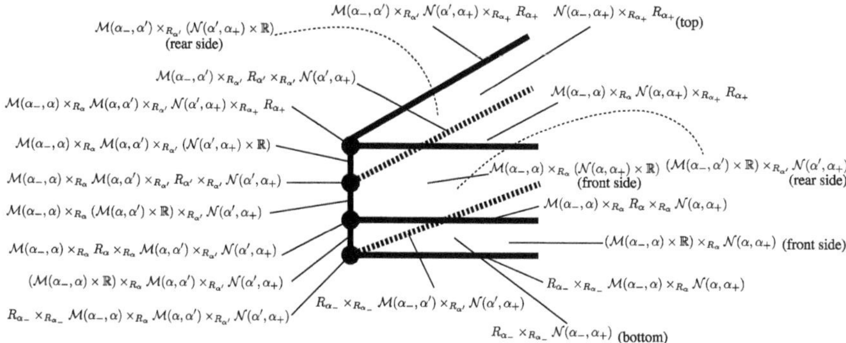

Fig. 18.14 Homotopy between $\mathfrak{N}\circ\mathfrak{ID}$ and $\mathfrak{N}$: Case 2

The first three strata constitute $\mathfrak{N}\circ\mathfrak{ID}$ and the last stratum is $\mathfrak{N}$. See Fig. 18.14. We note that each vertex in the figure is contained in exactly three edges. So this configuration is one of a manifold with boundary. (However, note that the boundary in the interior of three strata constituting $\mathfrak{N}\circ\mathfrak{ID}$ are smooth during the definition of the composition.)

Remark 18.63 The proof of Proposition 18.60 seems rather complicated. We can prove Proposition 18.60 for the case appearing in our geometric application in a more intuitive way. See Sect. 18.11.

In the rest of this section, we describe an alternative way to define the identity morphism $\mathfrak{ID}$ and a homotopy between $\mathfrak{N}\circ\mathfrak{ID}$ and $\mathfrak{N}$. The method we explain below is similar to that of [FOOO3, Subsection 4.6.1] in defining an A_∞ homomorphism, where we used *time-ordered fiber products*. We put

$$C_k(\mathbb{R}) = \{(\tau_1,\dots,\tau_k)\in(\mathbb{R}\cup\{\pm\infty\})^k \mid \tau_1\le\cdots\le\tau_k\}.$$

When $\alpha_-\neq\alpha_+$, we consider the union of

$$\mathcal{M}(\alpha_-,\alpha_1)\times_{R_{\alpha_1}}\mathcal{M}(\alpha_1,\alpha_2)\times_{R_{\alpha_2}}\cdots\times_{R_{\alpha_{k-1}}}\mathcal{M}(\alpha_{k-1},\alpha_+)\times C_k(\mathbb{R}) \tag{18.75}$$

over various k. When $\alpha_- = \alpha_+ = \alpha$, we consider R_α instead of (18.75). We glue them as follows. For the point $(\tau_1,\dots,\tau_k)\in C_k(\mathbb{R})$ with $\tau_i=\tau_{i+1}$, we put

$$(\tau_1,\dots,\tau_i,\tau_{i+2},\dots,\tau_k) = (\tau'_1,\dots,\tau'_{k-1}).$$

We also consider the embedding

$$\begin{aligned}I_i : &\mathcal{M}(\alpha_-,\alpha_1)\times_{R_{\alpha_1}}\cdots\times_{R_{\alpha_{k-1}}}\mathcal{M}(\alpha_{k-1},\alpha_+)\\ &\subset\mathcal{M}(\alpha_-,\alpha_1)\times_{R_{\alpha_1}}\cdots\mathcal{M}(\alpha_i,\alpha_{i+2})\cdots\times_{R_{\alpha_{k-1}}}\mathcal{M}(\alpha_{k-1},\alpha_+),\end{aligned}$$

which is induced by the inclusion map

$$\mathcal{M}(\alpha_i, \alpha_{i+1}) \times_{R_{\alpha_{i+1}}} \mathcal{M}(\alpha_{i+1}, \alpha_{i+2}) \subset \partial\mathcal{M}(\alpha_i, \alpha_{i+2}) \subset \mathcal{M}(\alpha_i, \alpha_{i+2}).$$

We now identify

$$(\mathfrak{x}, (\tau_1, \dots, \tau_k)) \sim (I_i(\mathfrak{x}), (\tau'_1, \dots, \tau'_{k-1})).$$

Under this identification we obtain a K-space with corners, which we take as an interpolation space $\mathcal{ID}(\alpha_-, \alpha_+)$ of our morphism $\mathfrak{ID}$. (We actually need to smooth the corners for this purpose.)

Note that we consider the part where $\tau_i = -\infty$, $i = 1, \dots, k$ to find

$$\begin{aligned}&\widehat{S}_k(\mathcal{ID}(\alpha_-, \alpha_+))\\ &\supset \mathcal{M}(\alpha_-, \alpha_1) \times_{R_{\alpha_1}} \cdots \times_{R_{\alpha_{k-1}}} \mathcal{M}(\alpha_{k-1}, \alpha_k) \times_{R_{\alpha_k}} \mathcal{ID}(\alpha_k, \alpha_+).\end{aligned}$$

We also have a similar embedding in the case when $\tau_i = +\infty$ and consider both of them. Then we find that $\mathfrak{ID}$ has the required structure at the corners.

To construct a homotopy from $\mathfrak{N} \circ \mathfrak{ID}$ to $\mathfrak{N}$ we consider the union of

$$\mathcal{M}(\alpha_-, \alpha_1) \times_{R_{\alpha_1}} \mathcal{M}(\alpha_1, \alpha_2) \times_{R_{\alpha_2}} \cdots \times_{R_{\alpha_{k-1}}} \mathcal{N}(\alpha_{k-1}, \alpha_+) \times C_k(\mathbb{R}) \tag{18.76}$$

over various k and identify them in a similar way. Here $\tau_k \in \mathbb{R} \cup \{\pm\infty\} \cong [1, 2]$ defines a map to $[1, 2]$. We can prove that this space is an interpolation space of a $[1, 2]$-parametrized family of morphisms and gives a homotopy from $\mathfrak{N}$ to $\mathfrak{N} \circ \mathfrak{ID}$.

Remark 18.64 The second method explained above seems to be shorter. (To work out the detail of the second map, we need to glue several K-spaces to obtain the required K-space. We omit this detail.) Here, however, we present the first one in detail since it is directly tied to the geometric situation appearing in the Floer theory of periodic Hamiltonian systems, as we mentioned in Remark 18.59 and explained in Sect. 18.11.

The identity morphism obtained by the second method also appears in the geometric situation in a different case. For example, when we consider two Lagrangian submanifolds and Floer cohomology of their intersection. Let us assume there is no pseudo-holomorphic disk whose boundary lies in one of those two Lagrangian submanifolds. Then we can define Lagrangian intersection Floer theory as Floer did. ([Fl1]. See also [FOOO3, Chapter 2].) When we prove the independence of the choice of compatible almost complex structures, we consider a one-parameter family of moduli spaces $\mathcal{M}(\alpha_1, \alpha_2; J_\tau)$, where α_1, α_2 are connected components of the intersections. From this we can construct a similar moduli space as (18.75), which provides a required cochain homotopy. If the family J_τ is the trivial family, it boils down to the identity morphism constructed by the second method.

Remark 18.65 There is an alternative way to prove well-definedness of Floer cohomology in the situation when R_α does *not* vary. We call it the *bifurcation method* in [FOOO4, Subsection 7.2.14]. The method of using morphisms which is taken in this section corresponds to the one which we called the *cobordism method* there. There is a way to translate the bifurcation method into the cobordism method, which is explained in [Fuk4, Proposition 9.1, Remark 12.3]. When we start from the bifurcation method and translate it into the cobordism method, the time-ordered fiber product appears. Then we end up with the same isomorphism as one we obtain by the alternative proof.

18.11 Geometric Origin of the Definition of the Identity Morphism

In this section we explain the geometric background to our definition of the identity morphism given in Sect. 18.10. We assume the reader is familiar with the construction of Floer cohomology of periodic Hamiltonian systems such as the one given in [Fl2]. Since the content of this section is never used in the proof of the results of this book, the reader can skip this section if he/she prefers.

18.11.1 Interpolation Space of the Identity Morphism

As we mentioned in Remark 18.59, when we study Floer cohomology of periodic Hamiltonian systems on a symplectic manifold M, we define a morphism between two linear K-systems associated to $H : S^1 \times M \to \mathbb{R}$ and to $H' : S^1 \times M \to \mathbb{R}$ by considering the homotopy

$$H_{\tau,t}(x) = \mathcal{H}(\tau, t, x) : \mathbb{R} \times S^1 \times M \to \mathbb{R},$$

where $\mathcal{H}(\tau, t, x) = H(t, x)$ for τ sufficiently small and $\mathcal{H}(\tau, t, x) = H'(t, x)$ for τ sufficiently large, and the interpolation space of the morphism is the compactified moduli space of the solutions of the equation (18.61). Note that (18.61) is *not* invariant under translation of the $\tau \in \mathbb{R}$ direction. We denote by $\mathcal{M}(\mathcal{H}; [\gamma_-, w_-], [\gamma_-, w_+])$ the moduli space of solutions of (18.61) with asymptotic boundary condition

$$\lim_{\tau \to -\infty} u(\tau, t) = \gamma_-(t), \qquad \lim_{\tau \to +\infty} u(\tau, t) = \gamma_+(t).$$

(See Condition 15.2.) Moreover we assume $[w_- \# u] = [w_+]$. Here γ_-, γ_+ are the periodic orbits of the periodic Hamiltonian system associated to H and H'

respectively, and $w_- : (D^2, \partial D^2) \to (M, \gamma_-)$, $w_+ : (D^2, \partial D^2) \to (M, \gamma_+)$ are disks bounding them. We denote by $[\gamma_\pm, w_\pm]$ its homology class.

We consider the Bott–Morse case. Namely, the set of $[\gamma_-, w_-]$ etc. consists of (possibly positive dimensional) smooth manifolds $\{R_\alpha\}_\alpha$. We put

$$\mathcal{M}(\mathcal{H}; \alpha_-, \alpha'_+) := \bigcup_{[\gamma_-, w_-]\in R_{\alpha_-}, [\gamma_+, w_+]\in R_{\alpha'_+}} \mathcal{M}(\mathcal{H}; [\gamma_-, w_-], [\gamma_-, w_+]),$$

that is the interpolation space of the morphisms. Then the codimension k corner of $\mathcal{M}(\mathcal{H}; \alpha_-, \alpha'_+)$ is described by the union of

$$\begin{aligned} &\mathcal{M}(H; \alpha_-, \alpha_1) \times_{R_{\alpha_1}} \cdots \times_{R_{\alpha_{k_1-1}}} \mathcal{M}(H; \alpha_{k_1-1}, \alpha_{k_1}) \\ &\times_{R_{\alpha_{k_1}}} \mathcal{M}(\mathcal{H}; \alpha_{k_1}, \alpha'_1) \\ &\times_{R_{\alpha'_1}} \mathcal{M}(H'; \alpha'_1, \alpha'_2) \times_{R_{\alpha'_2}} \cdots \times_{R_{\alpha'_{k_2}}} \mathcal{M}(H'; \alpha'_{k_2}, \alpha'_+) \end{aligned} \tag{18.77}$$

with $k_1 + k_2 = k$. Here the moduli space $\mathcal{M}(H; \alpha_i, \alpha_{i+1})$ is the set of solutions of the Floer equation

$$\frac{\partial u}{\partial \tau} + J\left(\frac{\partial u}{\partial t} - X_{H_t}(u)\right) = 0 \tag{18.78}$$

with asymptotic boundary conditions given by R_{α_i}, $R_{\alpha_{i+1}}$. We note that we divide the set of solutions by the $\mathbb{R}$ action defined by translation in the τ direction to obtain $\mathcal{M}(H; \alpha_i, \alpha_{i+1})$. The space $\mathcal{M}(H'; \alpha'_i, \alpha'_{i+1})$ is defined in a similar way, replacing H by H'.

To define the identity morphism, we consider the case when $H = H'$ and take the trivial homotopy. Namely,

$$\mathcal{H}_{\text{tri}}(\tau, t, x) \equiv H(t, x).$$

Then the equation (18.61) is exactly the same as (18.78). However, there is one important difference between them: When we consider the interpolation space of the morphism $\mathcal{M}(\mathcal{H}_{\text{tri}}; \alpha_-, \alpha_+)$ with $\mathcal{H}_{\text{tri}}(\tau, t, x) = H(t, x)$, we do *not* divide the moduli space by $\mathbb{R}$ action, even though in the case there is an $\mathbb{R}$ action. In other words, we have an isomorphism:

$$\overset{\circ}{\mathcal{M}}(\mathcal{H}_{\text{tri}}; \alpha_-, \alpha_+) = \overset{\circ}{\mathcal{M}}(H; \alpha_-, \alpha_+) \times \mathbb{R}.$$

Thus (18.77) becomes the closure of (18.52).

Another point we need to consider on $\mathcal{M}(\mathcal{H}_{\text{tri}}; R_{\alpha_-}, R_{\alpha_+})$ is that the case when $R = R_{\alpha_-} = R_{\alpha_+}$ can occur. In this case, the set of solutions of (18.61) = (18.78) consists of maps, each of which is constant in the $\tau \in \mathbb{R}$ direction and is an element of R in the $t \in S^1$ direction. When we consider $\mathcal{M}(H; R, R)$, we did not regard them as the elements. Hence this space is empty. This is because these elements are unstable because of the $\mathbb{R}$ invariance.

However, when we consider $\mathcal{M}(\mathcal{H}_{\text{tri}}; R_{\alpha_-}, R_{\alpha_+})$ with $R = R_{\alpha_-} = R_{\alpha_+}$, this element u is included. This is because we do not regard the $\mathbb{R}$ translation as a symmetry. In other words, we have an isomorphism

$$\mathcal{M}(\mathcal{H}_{\text{tri}}; R, R) = R$$

in this case. Thus (18.77) becomes the closure of (18.53).

18.11.2 Identification of the Interpolation Space of the Identity Morphism with Direct Product

Next we explain a geometric origin of the identification

$$\mathcal{N}(\alpha_-, \alpha_+) \cong \mathcal{M}(\alpha_-, \alpha_+) \times [E(\alpha_-), E(\alpha_+)], \tag{18.79}$$

that is (18.57). In our geometric situation

$$\mathcal{M}(\alpha_-, \alpha_+) = \mathcal{M}(R_{\alpha_-}, R_{\alpha_+})$$

and $E(\alpha)$ is the value of the action functional

$$\mathcal{A}_H([\gamma, w]) = \int w^*\omega + \int_{S^1} H(t, \gamma(t))dt$$

at $[\gamma, w] \in R_\alpha$. (Note that the sign here is opposite to that in [FOOO13, (1.2)].) Then our interpolation space $\mathcal{N}(\alpha_-, \alpha_+)$ is a compactification of $\overset{\circ}{\mathcal{N}}(\alpha_-, \alpha_+) = \overset{\circ}{\mathcal{M}}(R_{\alpha_-}, R_{\alpha_+}) \times \mathbb{R}$. An element of $\overset{\circ}{\mathcal{M}}(R_{\alpha_-}, R_{\alpha_+})$ that is the complement of the boundary of $\mathcal{M}(R_{\alpha_-}, R_{\alpha_+})$ may be regarded as a map $u : \mathbb{R} \times S^1 \to M$ satisfying (18.78) with the asymptotic boundary condition specified by R_{α_-} and R_{α_+}. Since we put an $\mathbb{R}$ factor in the definition of $\mathcal{N}(\alpha_-, \alpha_+)$, its element is a map u itself but not an equivalence class by the $\mathbb{R}$ action induced by the translation on the $\mathbb{R}$ direction. Therefore the loop $\gamma_\tau : S^1 \to M$, $\gamma_\tau(t) = u(\tau, t)$ is well-defined for each τ. Let $[\gamma_-, w_-]$ be an element of R_{α_-} such that $\gamma_-(t) = \lim_{\tau\to-\infty} u(\tau, t)$. We consider the concatenation of w_- and the restriction of u to $(-\infty, \tau] \times S^1$ and denote it by w_τ. We put

$$E(u) = \mathcal{A}_H([\gamma_0, w_0]). \tag{18.80}$$

Using the $\mathbb{R}$ action $(\tau_0 \cdot u)(\tau, t) = u(\tau + \tau_0, t)$, we have

$$E(\tau_0 \cdot u) = \mathcal{A}_H([\gamma_{\tau_0}, w_{\tau_0}]).$$

Therefore $E(\tau_0 \cdot u)$ is an increasing function of $\tau_0 \in \mathbb{R}$ and

$$\lim_{\tau_0 \to -\infty} E(\tau_0 \cdot u) = E(\alpha_-), \qquad \lim_{\tau_0 \to +\infty} E(\tau_0 \cdot u) = E(\alpha_+).$$

Therefore using this function E we obtain a homeomorphism

$$\overset{\circ}{\mathcal{N}}(\alpha_-, \alpha_+) \cong \overset{\circ}{\mathcal{M}}(R_{\alpha_-}, R_{\alpha_+}) \times (E(\alpha_-), E(\alpha_+)).$$

It is easy to see that this homeomorphism extends to (18.79).

18.11.3 Interpolation Space of the Homotopy

We next describe a geometric origin of the homotopy $\mathfrak{N} \circ \mathfrak{ID} \sim \mathfrak{N}$ given in the proof of Proposition 18.60.

We consider the case when the interpolation space of $\mathfrak{N}$ is $\mathcal{M}(\mathcal{H}; R_-, R_+)$ where $\mathcal{H}$ is a homotopy from H to H'. This moduli space is the set of solutions of the equation (18.61) with asymptotic boundary condition given by R_- and R_+. Note that $\mathcal{H}(\tau, t, x)$ is τ-dependent. We write this $\mathcal{H}$ as $\mathcal{H}^{32}$.

On the other hand, the interpolation space of the morphism $\mathfrak{ID}$ is by definition the set of solutions of (18.61) with $\mathcal{H}_{\rm tri}(\tau, t, x) = H(t, x)$, which is τ-independent. We write this $\mathcal{H}_{\rm tri}$ as $\mathcal{H}^{21}$.

Then the composition $\mathfrak{N} \circ \mathfrak{ID}$ is obtained by using the two-parameter family of Hamiltonians concatenating $\mathcal{H}^{32}$ and $\mathcal{H}^{21}$ as in (18.6). Let $\mathcal{H}^{31,T}$ be obtained by this concatenation. More precisely, the interpolation space of $\mathfrak{N} \circ \mathfrak{ID}$ appears when we take the limit of $\mathcal{H}^{31,T}$ as $T \to \infty$. Note that in our case $\mathcal{H}^{31,T}$ is

$$\mathcal{H}^{31,T}(\tau, t, x) = \begin{cases} H(t, x) & \text{if } \tau \le -T_0 - T \\ \mathcal{H}^{21}(\tau + T, t, x) = H(t, x) & \text{if } -T_0 - T \le \tau \le T_0 - T \\ H^2(t, x) = H(t, x) & \text{if } T_0 - T \le \tau \le T - T_0 \\ \mathcal{H}^{32}(\tau - T, t, x) = H_{\tau - T, t}(x) & \text{if } T - T_0 \le \tau \le T + T_0 \\ H^3(t, x) = H'(t, x) & \text{if } T + T_0 \le \tau. \end{cases} \tag{18.81}$$

See Fig. 18.15. Thus actually we have

$$\mathcal{H}^{31,T}(\tau, t, x) = H_{\tau - T, t}(x).$$

Therefore the set of solutions of (18.5), with respect to (τ, t)-dependent Hamiltonian $H^{31,T}$, that is:

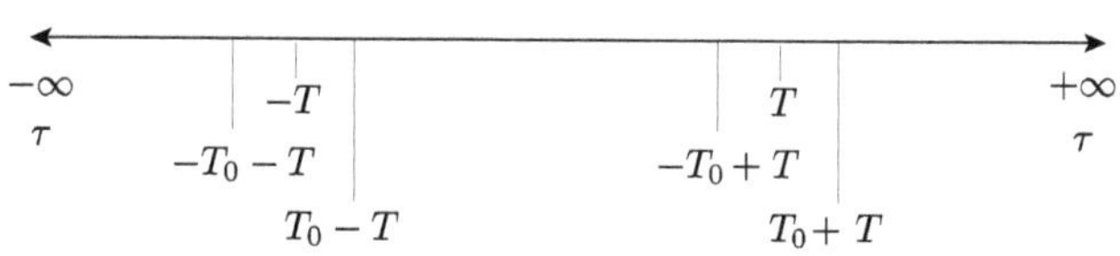

Fig. 18.15 $\mathcal{H}^{31,T}(\tau,t,x)$

$$\frac{\partial u}{\partial \tau} + J\left(\frac{\partial u}{\partial t} - X_{H_{\tau-T,t}}(u)\right) = 0, \tag{18.82}$$

is indeed independent of T up to the canonical isomorphism. Moreover, if $T = 0$, this equation is exactly the same as one we used to define $\mathfrak{N}$.

We consider the union of the set of solutions of (18.82) for $T \in [0,\infty)$. Namely,

$$\bigcup_{T\in[0,\infty)} \overset{\circ}{\mathcal{M}}(\mathcal{H}^{31,T};\alpha_-,\alpha'_+) \times \{T\}. \tag{18.83}$$

The interpolation space of the homotopy is a compactification of (18.83). Except the part $T = \infty$, the compactification is a product of $[0,\infty)$ and the compactification of the moduli space of solutions of (18.82) for fixed $T = 0$, which is $\mathcal{N}(\alpha_-,\alpha'_+) \times [0,\infty)$.[7]

However, there is rather a delicate issue related to the (source) $\mathbb{R}$ action at the limit as $T \to \infty$. Recall that $\mathcal{H}^{21}$ is τ-independent. As we already mentioned several times, we do not use this symmetry to divide our moduli space. In order to clarify this point, we take and fix a marked point and regard our moduli space as a map from a marked cylinder. (We take some marked point $(-T, 1/2) \in (-\infty,0] \times S^1$.) The coordinate $T \in [0,\infty)$ in turn becomes the $[0,\infty)$ factor of $\mathcal{N}(\alpha_-,\alpha'_+) \times [0,\infty)$. We note that $\mathcal{N}(\alpha_-,\alpha'_+)$ is the set of solutions of

$$\frac{\partial u}{\partial \tau} + J\left(\frac{\partial u}{\partial t} - X_{H_{\tau,t}}(u)\right) = 0. \tag{18.84}$$

Then we can identify $\mathcal{N}(\alpha_-,\alpha'_+) \times [0,\infty)$ with the set of $(u,(-T,1/2))$ where u solves the equation (18.84) and $T \in [0,\infty)$. In other words, putting the marked

[7]Note $\mathcal{N}(\alpha_-,\alpha'_+) = \mathcal{M}(\mathcal{H}^{31,T};\alpha_-,\alpha'_+)$.

point $(-T, 1/2)$ corresponds to shifting $\mathcal{H}^{31,0}$ to $\mathcal{H}^{31,T}$. Then by a standard gluing analysis we can describe its limit as $T \to \infty$ by the union of

$$\begin{aligned}
&\mathcal{M}(H; \alpha_-, \alpha_1) \times_{R_{\alpha_1}} \cdots \times_{R_{\alpha_{k_1'-2}}} \mathcal{M}(H; \alpha_{k_1'-2}, \alpha_{k_1'-1}) \\
&\times_{R_{\alpha_{k_1'-1}}} \mathcal{M}(H; \alpha_{k_1'-1}, \alpha_{k_1'}) \times \mathbb{R} \\
&\times_{R_{\alpha_{k_1'}}} \mathcal{M}(H; \alpha_{k_1'}, \alpha_{k_1'+1}) \times_{R_{\alpha_{k_1'+1}}} \cdots \times_{R_{k_1-1}} \mathcal{M}(H; \alpha_{k_1-1}, \alpha_{k_1}) \\
&\times_{R_{\alpha_{k_1}}} \mathcal{M}(\mathcal{H}; \alpha_{k_1}, \alpha_1') \\
&\times_{R_{\alpha_1'}} \mathcal{M}(H'; \alpha_1', \alpha_2') \times_{R_{\alpha_2'}} \cdots \times_{\alpha_{k_2}'} \mathcal{M}(H'; \alpha_{k_2}', \alpha_+').
\end{aligned} \tag{18.85}$$

See Fig. 18.16, that is, the case $k_1 = 1$, $k_2' = 0$. An object in the compactification of (18.83) corresponding to the limit as $T \to \infty$ is obtained by gluing maps appearing in (18.85). In fact, since an element of $\mathcal{M}(\mathcal{H}^{31,T}; \alpha_-, \alpha_+')$ comes with a marked point, the limit is assigned with a marked point on one of the factors. We put the $\mathbb{R}$ factor to the factor on which the marked point lies. In (18.85) it lies on the map representing an element of $\mathcal{M}(H; \alpha_{k_1'-1}, \alpha_{k_1'})$.

The space (18.85) is the closure of (18.62). By the same reasoning as we discussed in Sect. 18.11.2, the limit as $T \to \infty$ also contains a component

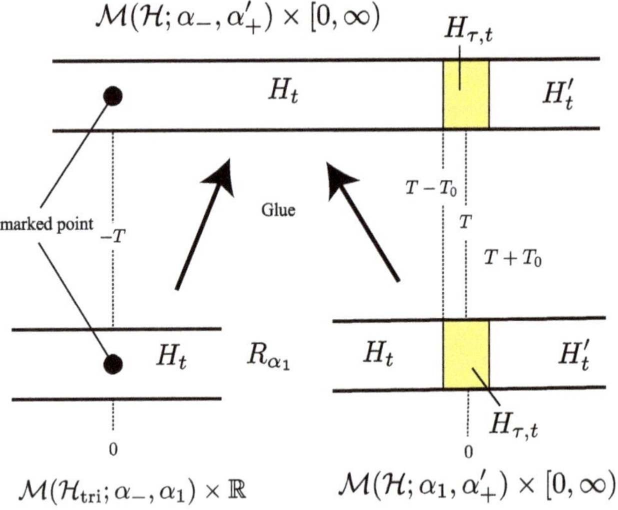

Fig. 18.16 Boundary of $\mathcal{M}(\mathcal{H}; \alpha_-, \alpha_+') \times [0, \infty)$

$$\begin{aligned}
&\mathcal{M}(H;\alpha_-,\alpha_1)\times_{R_{\alpha_1}}\cdots\times_{\alpha_{k_1-1}}\mathcal{M}(H;\alpha_{k_1'-1},\alpha_{k_1'})\\
&\times_{R_{\alpha_{k_1'}}} R_{\alpha_{k_1'}}\\
&\times_{R_{\alpha_{k_1'}}}\mathcal{M}(H;\alpha_{k_1'},\alpha_{k_1'+1})\times_{\alpha_{k_1+1}}\cdots\times_{R_{\alpha_{k_1-1}}}\mathcal{M}(H;\alpha_{k_1-1},\alpha_{k_1}) \qquad (18.86)\\
&\times_{R_{\alpha_{k_1}}}\mathcal{M}(\mathcal{H};\alpha_{k_1},\alpha_1')\\
&\times_{R_{\alpha_1'}}\mathcal{M}(H';\alpha_1',\alpha_2')\times_{R_{\alpha_2'}}\cdots\times_{R_{\alpha_{k_2}'}}\mathcal{M}(H';\alpha_{k_2}',\alpha_+').
\end{aligned}$$

Namely, this space (18.86) corresponds to the case when the marked point lies in the component corresponding to a map $u : \mathbb{R}\times S^1 \to M$ which is constant in the $\mathbb{R}$ direction and represents an element of $R_{\alpha_{k_1'}}$ on each $\{\tau\}\times S^1$. The space (18.86) coincides with the closure of (18.63).

Therefore the compactification of (18.83) coincides with the interpolation space of the homotopy constructed in the proof of Proposition 18.60.

18.11.4 *Identification of the Interpolation Space of the Homotopy with Direct Product*

Finally, we explain a geometric origin of the identification

$$\mathcal{N}(\alpha_-,\alpha_+';[0,1]) \cong \mathcal{N}(\alpha_-,\alpha_+')\times[E(\alpha_-),E(\alpha_+')+c]. \qquad (18.87)$$

The discussion is similar to that in Sect. 18.11.2 but is slightly more involved. We denote by $\overset{\circ}{\mathcal{N}}(\alpha_-,\alpha_+';[0,1])$ the set of pairs $(u,-T)$, where $u : \mathbb{R}\times S^1 \to M$ is a map solving the equation (18.84) and satisfying the asymptotic boundary conditions as $\tau\to\pm\infty$ given by R_{α_-} and $R_{\alpha_+'}$, and $(-T,1/2)$ is the marked point in $(-\infty,0]\times S^1$. As we discussed in Sect. 18.11.3, $\mathcal{N}(\alpha_-,\alpha_+';[0,1])$ is a compactification of this space.

We first define a map

$$E : \overset{\circ}{\mathcal{N}}(\alpha_-,\alpha_+';[0,1]) \to \mathbb{R}$$

and modify it to E' later. For $(u,T)\in\overset{\circ}{\mathcal{N}}(\alpha_-,\alpha_+';[0,1])$ we put $\gamma_\tau(t)=u(\tau,t)$ and

$$\gamma_- = \lim_{\tau\to-\infty}\gamma_\tau.$$

The asymptotic boundary condition we assumed for $\overset{\circ}{\mathcal{N}}(\alpha_-, \alpha'_+; [0, 1])$ implies that there exists w_- such that $[\gamma_-, w_-] \in R_{\alpha_-}$. We denote by w_T the concatenation of w_- and the restriction of u to $(-\infty, -T) \times S^1$ and define

$$E(u, -T) = \int_{D^2} w_T^* \omega + \int_{t \in S^1} H_{\tau,t}(\gamma_T(t))dt \in \mathbb{R}. \tag{18.88}$$

We note that

$$\lim_{T \to \infty} E(u, -T) = \mathcal{A}_H([\gamma_-, w_-]) = E(\alpha_-). \tag{18.89}$$

On the other hand, contrary to the situation of Sect. 18.11.2, the map

$$(-\infty, 0] \ni -T \mapsto E(u, -T) \in \mathbb{R}$$

may not be an increasing function in general, although it is an increasing function for sufficiently large T. Moreover there is no obvious relation between $E(u, 0)$ and $E(\alpha'_+)$. Since the energy loss is c, the inequality $E(\alpha_-) < E(\alpha'_+) + c$ does hold, but $E(\alpha_-) < E(\alpha'_+)$ may not hold in general. Thus we modify $E(u, T)$ to $E'(u, T)$ with the following properties:

(1) $E'(u, -T) = E(u, -T)$ if T is sufficiently large.
(2) $-T \mapsto E'(u, -T)$ is strictly increasing.
(3) $E'(u, 0) = E(\alpha'_+) + c$.

More explicitly, we can define E' as follows. We first take a sufficiently large number $T_1 > 0$ with the following properties:

(a) $E(u, -T) \le E(\alpha_+) + c$ holds for each $u \in \mathcal{N}(\alpha_-, \alpha'_+)$ and $T \ge T_1$.
(b) $H_{\tau,t} = H_t$ if $\tau < -T_1$.

Then we define

$$E'(u, -T) = \begin{cases} E(u, -T) & \text{if } T \ge T_1 \\ \frac{T_1 - T}{T_1}(E(\alpha_+) + c - E(u, -T_1)) + E(u, -T_1) & \text{if } T \in [0, T_1]. \end{cases} \tag{18.90}$$

Item (b) implies that the function $-T \mapsto E(u, -T)$ is increasing for $T > T_1$. Item (a) implies that $-T \mapsto E'(u, -T)$ is an increasing function for $T \in [0, T_1]$. We have thus verified (2). (1) and (3) are obvious from the definition. We can easily extend it to $\mathcal{N}(\alpha_-, \alpha'_+; [0, 1])$. In fact, the value E' is in the interval $(E(\alpha_{k'_1 - 1}), E(\alpha_{k'_1}))$ on (18.85) and is $E(\alpha_{k'_1})$ on (18.86). This is a consequence of Item (1). Then using E' as the second factor, we obtain the identification (18.87).

Chapter 19
Linear K-Systems: Floer Cohomology II – Proof

The purpose of this chapter is to prove the theorems we claimed in Chap. 16.

19.1 Construction of Cochain Complexes

Let us start with a linear K-system or a partial linear K-system of energy cut level E_0 as in Definition 16.6. We set $E_0 = +\infty$ for the case of a linear K-system and E_0 to be the energy cut level for the case of a partial linear K-system. We consider the set

$$\mathfrak{E}_{\le E_0} = \{E(\alpha_+) - E(\alpha_-) \mid \mathcal{M}(\alpha_-, \alpha_+) \ne \emptyset,\ E(\alpha_+) - E(\alpha_-) \le E_0\}. \tag{19.1}$$

This is a discrete set by Condition 16.1 (IX). We put

$$\mathfrak{E}_{\le E_0} = \{E^1_{\mathfrak{E}}, E^2_{\mathfrak{E}}, \ldots\}$$

with $E^1_{\mathfrak{E}} < E^2_{\mathfrak{E}} < \ldots$. We will use the results of Chap. 17 to prove the next proposition by induction on k of the energy cut level $E^k_{\mathfrak{E}}$. In other words, the statement of Proposition 19.1, which provides the precise induction hypothesis we work with, is the main idea of its proof. We also remark that Proposition 19.1 is the detailed and precise version of Proposition 15.4 in the introduction. Put $\tau_0 = 1$.

Proposition 19.1 *For each $0 < \tau < 1$, there exist a τ-collared Kuranishi structure $\widehat{\mathcal{U}^+}(\alpha_-, \alpha_+)$ of $\mathcal{M}(\alpha_-, \alpha_+)^{\boxplus \tau_0}$ and a CF-perturbation $\widehat{\mathfrak{S}^+}(\alpha_-, \alpha_+)$ of $\widehat{\mathcal{U}^+}(\alpha_-, \alpha_+)$ for every α_-, α_+ with $E(\alpha_+) - E(\alpha_-) \le E^k_{\mathfrak{E}}$ and they enjoy the following properties:*

K. Fukaya et al., *Kuranishi Structures and Virtual Fundamental Chains*, Springer Monographs in Mathematics, https://doi.org/10.1007/978-981-15-5562-6_19

(1) *Let* $\widehat{\mathcal{U}}(\alpha_-,\alpha_+)$ *be the Kuranishi structure on* $\mathcal{M}(\alpha_-,\alpha_+)$ *given in Condition* 16.1. *Then* $\widehat{\mathcal{U}}(\alpha_-,\alpha_+)^{\boxplus\tau_0} < \widehat{\mathcal{U}^+}(\alpha_-,\alpha_+)$ *as oriented, collared Kuranishi structures.*[1]

(2) *The CF-perturbation* $\widehat{\mathfrak{S}^+}(\alpha_-,\alpha_+)$ *is transversal to* 0. *Moreover*[2] $\mathrm{ev}_+ : \mathcal{M}(\alpha_-,\alpha_+)^{\boxplus\tau_0} \to R_{\alpha_+}$ *is strongly submersive with respect to* $\widehat{\mathfrak{S}^+}(\alpha_-,\alpha_+)$.

(3) *We have the following isomorphism, called the periodicity isomorphism,*

$$\widehat{\mathcal{U}^+}(\alpha_-,\alpha_+) \longrightarrow \widehat{\mathcal{U}^+}(\beta\alpha_-,\beta\alpha_+)$$

for any $\beta \in \mathfrak{G}$. *It is compatible with the periodicity isomorphism in Condition* 16.1 *(VIII) via the embedding in (1). The pullback of* $\widehat{\mathfrak{S}^+}(\beta\alpha_-,\beta\alpha_+)$ *by this isomorphism is equivalent to* $\widehat{\mathfrak{S}^+}(\alpha_-,\alpha_+)$.

(4) *There exists an isomorphism of* τ*-collared K-spaces:*

$$\begin{aligned}&\partial(\mathcal{M}(\alpha_-,\alpha_+)^{\boxplus\tau_0},\widehat{\mathcal{U}^+}(\alpha_-,\alpha_+))\\ &=\coprod_{\alpha}(-1)^{\dim\mathcal{M}(\alpha,\alpha_+)}(\mathcal{M}(\alpha,\alpha_+)^{\boxplus\tau_0},\widehat{\mathcal{U}^+}(\alpha,\alpha_+))\\ &\qquad {}_{\mathrm{ev}_-}\times_{\mathrm{ev}_+}(\mathcal{M}(\alpha_-,\alpha)^{\boxplus\tau_0},\widehat{\mathcal{U}^+}(\alpha_-,\alpha)).\end{aligned} \tag{19.2}$$

The isomorphism (19.2) *is compatible with the boundary compatibility isomorphism:*

$$\partial\widehat{\mathcal{U}}(\alpha_-,\alpha_+)^{\boxplus\tau_0}=\coprod_{\alpha}(-1)^{\dim\widehat{\mathcal{U}}(\alpha,\alpha_+)}(\widehat{\mathcal{U}}(\alpha,\alpha_+)^{\boxplus\tau_0})_{\mathrm{ev}_-}\times_{\mathrm{ev}_+}(\widehat{\mathcal{U}}(\alpha_-,\alpha)^{\boxplus\tau_0})$$

in Condition 16.1 *(X) via the KK-embedding in (1).*

(5) *The pullback of* $\widehat{\mathfrak{S}^+}(\alpha_-,\alpha_+)$ *by the isomorphism* (19.2) *is equivalent to the fiber product*

$$\widehat{\mathfrak{S}^+}(\alpha,\alpha_+)_{\mathrm{ev}_-}\times_{\mathrm{ev}_+}\widehat{\mathfrak{S}^+}(\alpha_-,\alpha).$$

This fiber product is well-defined by (2).

(6) *There exists an isomorphism of* τ*-collared K-spaces:*

---

[1] Let $\widehat{\mathcal{U}_1}^{\boxplus\tau_1}, \widehat{\mathcal{U}_2}^{\boxplus\tau_2}$ be two collared Kuranishi structures. We define the relation $\widehat{\mathcal{U}_1}^{\boxplus\tau_1} < \widehat{\mathcal{U}_2}^{\boxplus\tau_2}$ *as collared Kuranishi structures* if there exist $\tau' < \tau_1, \tau_2$ and Kuranishi structures $\widehat{\mathcal{U}'_1}, \widehat{\mathcal{U}'_2}$ such that $\widehat{\mathcal{U}'_1} < \widehat{\mathcal{U}'_2}$ and $\widehat{\mathcal{U}'_1}^{\boxplus\tau'} = \widehat{\mathcal{U}_1}^{\boxplus\tau_1}$, $\widehat{\mathcal{U}'_2}^{\boxplus\tau'} = \widehat{\mathcal{U}_2}^{\boxplus\tau_2}$. Here we regard $\widehat{\mathcal{U}_1}^{\boxplus\tau_1}$ and $\widehat{\mathcal{U}_2}^{\boxplus\tau_2}$ as τ' collared Kuranishi structures by using the functor in Definition-Lemma 17.46.

[2] According to Lemma 17.40 (3), we should write $\mathrm{ev}_+^{\boxplus\tau_0}$ but not ev_+. However, to simplify the notation we drop ${}^{\boxplus\tau_0}$ from the notation of evaluation maps if no confusion can occur.

$$\begin{aligned}&\widehat{S}_k(\mathcal{M}(\alpha_-,\alpha_+)^{\boxplus\tau_0},\widehat{\mathcal{U}^+}(\alpha_-,\alpha_+))\\ &\cong \coprod_{\alpha_1,\dots,\alpha_k\in\mathfrak{A}} \Big((\mathcal{M}(\alpha_-,\alpha_1)^{\boxplus\tau_0},\widehat{\mathcal{U}^+}(\alpha_-,\alpha_1))\ {}_{\mathrm{ev}_+}\times_{R_{\alpha_1}}\cdots\\ &\qquad\cdots{}_{R_{\alpha_k}}\times_{\mathrm{ev}_-}(\mathcal{M}(\alpha_k,\alpha_+)^{\boxplus\tau_0},\widehat{\mathcal{U}^+}(\alpha_k,\alpha_+))\Big).\end{aligned} \tag{19.3}$$

The isomorphism (19.3) *is compatible with the corner compatibility isomorphism*

$$\begin{aligned}&\widehat{S}_k(\widehat{\mathcal{U}}(\alpha_-,\alpha_+)^{\boxplus\tau_0})\\ &\cong \coprod_{\alpha_1,\dots,\alpha_k\in\mathfrak{A}} \Big(\widehat{\mathcal{U}}(\alpha_-,\alpha_1)^{\boxplus\tau_0}{}_{\mathrm{ev}_+}\times_{R_{\alpha_1}}\cdots{}_{R_{\alpha_k}}\times_{\mathrm{ev}_-}\widehat{\mathcal{U}}(\alpha_k,\alpha_+)^{\boxplus\tau_0}\Big)\end{aligned}$$

in Condition 16.1 *(XI) via the KK-embedding in (1).*

(7) *The isomorphism* (19.3) *implies that* $\widehat{S}_\ell(\widehat{S}_k(\mathcal{M}(\alpha_-,\alpha_+)^{\boxplus\tau_0},\widehat{\mathcal{U}^+}(\alpha_-,\alpha_+)))$ *is decomposed into similar fiber products as* (19.3) *where k is replaced by* $\ell+k$*. On the other hand,* $\widehat{S}_{\ell+k}(\mathcal{M}(\alpha_-,\alpha_+)^{\boxplus\tau_0},\widehat{\mathcal{U}^+}(\alpha_-,\alpha_+))$ *is decomposed into similar fiber products as* (19.3) *where k is replaced by* $\ell+k$*. The map*

$$\begin{aligned}\pi_{\ell,k}:&\widehat{S}_\ell(\widehat{S}_k(\mathcal{M}(\alpha_-,\alpha_+)^{\boxplus\tau_0},\widehat{\mathcal{U}^+}(\alpha_-,\alpha_+)))\\ &\to \widehat{S}_{\ell+k}(\mathcal{M}(\alpha_-,\alpha_+)^{\boxplus\tau_0},\widehat{\mathcal{U}^+}(\alpha_-,\alpha_+))\end{aligned}$$

in Proposition 24.17 *becomes an identity map on each component under this identification.*

(8) *The pullback of* $\widehat{\mathfrak{S}^+}(\alpha_-,\alpha_+)$ *by the isomorphism* (19.3) *is equivalent to the fiber product:*

$$\widehat{\mathfrak{S}^+}(\alpha_-,\alpha_1)\ {}_{\mathrm{ev}_+}\times_{R_{\alpha_1}}\cdots{}_{R_{\alpha_k}}\times_{\mathrm{ev}_-}\widehat{\mathfrak{S}^+}(\alpha_k,\alpha_+). \tag{19.4}$$

This fiber product is well-defined by (2). This equivalence is compatible with the covering map $\pi_{\ell,k}$*.*

Proof Using the results of Sect. 17.10, we can prove the proposition in a straightforward way as follows. We take τ^+ with

$$\tau<\tau^+<1=\tau_0.$$

Suppose we constructed $\widehat{\mathcal{U}^+}(\alpha_-,\alpha_+)$ and $\widehat{\mathfrak{S}^+}(\alpha_-,\alpha_+)$ satisfying (1)–(8) above for any α_-,α_+ with $E(\alpha_+)-E(\alpha_-)<E^k_{\mathfrak{C}}$ and for τ replaced by τ^+. We consider α_-, α_+ with $E(\alpha_+)-E(\alpha_-)=E^k_{\mathfrak{C}}$ and $\mathcal{M}(\alpha_-,\alpha_+)^{\boxplus\tau_0}$.

To apply Propositions 17.58 and 17.78, we check that we are in Situation 17.55. The space X in Situation 17.55 is $\mathcal{M}(\alpha_-,\alpha_+)^{\boxplus\tau_0}$ and τ in Situation 17.55 is

τ^+ here. The τ^+-collared Kuranishi structure $\widehat{\mathcal{U}}$ in Situation 17.55 is the one obtained from $\widehat{\mathcal{U}}(\alpha_-,\alpha_+)^{\boxplus\tau_0}$ by using Lemma 17.46. The τ^+-collared Kuranishi structure $\widehat{\mathcal{U}^+_{S_k}}$ in Situation 17.55 is the one obtained by using Lemma 17.46 from the Kuranishi structure in the right hand side of (19.3), which is given by the induction hypothesis. We write it as $\widehat{\mathcal{U}^+_{S_k}}$. Then from Lemma 17.53 and the definition that $\widehat{S}_k(\widehat{\mathcal{U}^+_{S_\ell}})$ coincides with the $(k+\ell)!/k!\ell!$ disjoint union of $\widehat{\mathcal{U}^+_{S_{k+\ell}}}$. Therefore Situation 17.55 (2) holds. The commutativity of Diagram (17.22) holds because every map in (17.22) is the identity map via the isomorphism (19.3). We next construct the embedding $\widehat{S}_k(\mathcal{M}(\alpha_-,\alpha_+)^{\boxplus\tau_0},\widehat{\mathcal{U}}(\alpha_-,\alpha_+)^{\boxplus\tau_0}) \to \widehat{\mathcal{U}^+_{S_k}}$. By the induction hypothesis the embedding $\widehat{\mathcal{U}}(\alpha,\alpha')^{\boxplus\tau_0} \to \widehat{\mathcal{U}^+}(\alpha,\alpha')$ for $E(\alpha')-E(\alpha) < E^k_{\mathfrak{C}}$ is given. By using Condition 16.1 (XI) we have

$$\begin{aligned}&\widehat{S}_k(\widehat{\mathcal{U}}(\alpha_-,\alpha_1)^{\boxplus\tau_0})\\ &\cong \coprod_{\alpha_1,\dots,\alpha_k\in\mathfrak{A}} \left(\widehat{\mathcal{U}}(\alpha_-,\alpha_1)^{\boxplus\tau_0}{}_{\mathrm{ev}_+}\times_{R_{\alpha_1}}\cdots{}_{R_{\alpha_k}}\times_{\mathrm{ev}_-}\widehat{\mathcal{U}}(\alpha_k,\alpha_+)^{\boxplus\tau_0}\right).\end{aligned}$$

Therefore, the right hand side is embedded in

$$\begin{aligned}\widehat{\mathcal{U}^+_{S_k}} \cong \coprod_{\alpha_1,\dots,\alpha_k\in\mathfrak{A}} &\Big((\mathcal{M}(\alpha_-,\alpha_1)^{\boxplus\tau_0},\widehat{\mathcal{U}^+}(\alpha_-,\alpha_1))\ {}_{\mathrm{ev}_+}\times_{R_{\alpha_1}}\cdots\\ &\cdots{}_{R_{\alpha_k}}\times_{\mathrm{ev}_-}(\mathcal{M}(\alpha_k,\alpha_+)^{\boxplus\tau_0},\widehat{\mathcal{U}^+}(\alpha_k,\alpha_+))\Big).\end{aligned}$$

Using Lemma 17.53, this isomorphism holds as the isomorphism of τ^+-collared Kuranishi structures. So Situation 17.55 (4) is satisfied. The commutativity of Diagram (17.22) and Diagram (17.23) follows from the fact that all the maps in Diagrams (17.22), (17.23) become the identity maps via the isomorphism (19.3), which is a part of the induction hypothesis (Proposition 19.1 (7)). We have thus checked the assumption of Propositions 17.58 and 17.78. The Kuranishi structure we obtain in Proposition 17.78 is our Kuranishi structure $\widehat{\mathcal{U}^+}(\alpha_-,\alpha_+)$.

We next consider CF-perturbations. We check that we are in Situation 17.72. We define $\widehat{\mathfrak{S}^+_{S_k}}$ in Situation 17.72 by the right hand side of (19.4). Situation 17.72 (2) can be checked easily in our case by using an inductive hypothesis and Lemma 17.47. We can thus apply Propositions 17.73 and 17.81. The CF-perturbations we obtain by Proposition 17.81 is $\widehat{\mathfrak{S}^+}(\alpha_-,\alpha_+)$.

Now we will check that $\widehat{\mathcal{U}^+}(\alpha_-,\alpha_+)$ and $\widehat{\mathfrak{S}^+}(\alpha_-,\alpha_+)$ satisfy Proposition 19.1 (1)–(8).

(1) Since $\tau < \tau^+$, the τ-collaredness of $\widehat{\mathcal{U}^+}(\alpha_-,\alpha_+)$ is a consequence of Proposition 17.58 and $\widehat{\mathcal{U}}(\alpha_-,\alpha_+)^{\boxplus\tau} < \widehat{\mathcal{U}^+}(\alpha_-,\alpha_+)$ also follows from Proposition 17.58.

(2) is a consequence of Proposition 17.81.

(3) We apply Proposition 17.58 to each $\mathfrak{G}$ equivalence class of the pair (α_-, α_+). Then for other $(\beta\alpha_-, \beta\alpha_+)$ we define $\widehat{\mathcal{U}^+}(\alpha_-, \alpha_+)$ so that it is identified with (α_-, α_+) by using existence of the periodicity isomorphism on the boundary. Then existence of the periodicity isomorphism for α_-, α_+ with $E(\alpha_+) - E(\alpha_-) = E^k_{\mathfrak{C}}$ is immediate from the definition. The compatibility of the periodicity isomorphism with CF-perturbations can be proved in the same way.

(4),(6),(7) This is a consequence of Proposition 17.58. Namely, it is a consequence of Proposition 17.58 (1) and the induction hypothesis.

(5),(8) This is a consequence of Proposition 17.81 (1) and the induction hypothesis. Hence the proof of Proposition 19.1 is now complete. □

Remark 19.2 In the case of partial linear K-systems, the induction to prove Proposition 19.1 stops in finite steps. In the case of linear K-systems, the number of inductive steps is countably infinitely many.

We next rewrite the geometric conclusion of Proposition 19.1 into algebraic structures.

Definition 19.3 In the situation of Proposition 19.1, we define

$$\mathfrak{m}^{\epsilon}_{1;\alpha_+,\alpha_-} : \Omega(R_{\alpha_-}; o_{R_{\alpha_-}}) \longrightarrow \Omega(R_{\alpha_+}; o_{R_{\alpha_+}}) \tag{19.5}$$

by

$$\mathfrak{m}^{\epsilon}_{1;\alpha_+,\alpha_-}(h) = (-1)^{\mu(\alpha_+)-\mu(\alpha_-)+\deg h} \mathrm{ev}_+!(\mathrm{ev}_-^* h; \widehat{\mathfrak{S}^{+\epsilon}}(\alpha_-, \alpha_+)). \tag{19.6}$$

Here the right hand side is defined by Definition 17.82 on the K-space

$$(\mathcal{M}(\alpha_-, \alpha_+)^{\boxplus\tau_0}, \widehat{\mathcal{U}^+}(\alpha_-, \alpha_+)).$$

See Theorem 27.1 for the correspondence coupled with local systems. Note that Condition 16.1 (VII) is compatible with the conventions (27.1), (27.2).

The degree of the map $\mathfrak{m}^{\epsilon}_{1;\alpha_+,\alpha_-}$ is

$$\dim R_{\alpha_+} - \dim \mathcal{M}(\alpha_-, \alpha_+) = 1 - \mu(\alpha_+) + \mu(\alpha_-)$$

by Condition 16.1 (VI) and Definition 7.79. Therefore after the degree shift in Definition 16.8 (2) its degree becomes +1.

Remark 19.4 In the situation of a linear K-system where infinitely many different K-spaces are involved, we need to make a careful choice of $\epsilon > 0$. Note that the well-definedness of pushout (Theorem 9.14) says that for each K-space and its CF-perturbation the pushout is well-defined for sufficiently small $\epsilon > 0$. In the situation of a linear K-system the infimum of such an ϵ over all $\alpha_\pm$ of $\widehat{\mathfrak{S}^{+\epsilon}}(\alpha_-, \alpha_+)$ may be

0.[3] (It might be possible to prove that we can take the same ϵ for all $\widehat{\mathfrak{S}^{+\epsilon}}(\alpha_-,\alpha_+)$ at the same time. However, it is cumbersome to formulate the condition without referring to the construction itself. For this reason, we do not try to prove or use such a uniformity in this book.) More precisely speaking, we have the following:

($\flat$) For any energy cut level E_0 there exists $\epsilon_0(E_0) > 0$ such that the operator $\mathfrak{m}^{\epsilon}_{1;\alpha_+,\alpha_-}$ is defined when $0 < E(\alpha_+) - E(\alpha_-) \le E_0$ and $\epsilon < \epsilon_0(E_0)$.

Hereafter when the relationship between energy cut level and ϵ (which is the parameter of the approximation) appears in this way, we write *in the sense of* ($\flat$) in place of repeating the above sentence over again.

Lemma 19.5 *The operators* $\mathfrak{m}^{\epsilon}_{1;\alpha_+,\alpha_-}$ *satisfy the following equality in the sense of* ($\flat$)*:*

$$d_0 \circ \mathfrak{m}^{\epsilon}_{1;\alpha_+,\alpha_-} + \mathfrak{m}^{\epsilon}_{1;\alpha_+,\alpha_-} \circ d_0 + \sum_{\alpha; E(\alpha_-)<E(\alpha)<E(\alpha_+)} \mathfrak{m}^{\epsilon}_{1;\alpha_+,\alpha} \circ \mathfrak{m}^{\epsilon}_{1;\alpha,\alpha_-} = 0. \tag{19.7}$$

Here and hereafter we denote

$$d_0(h) = d_{dR}(h) \tag{19.8}$$

for a differential form $h \in \Omega(R_\alpha)$, where d_{dR} denotes the de Rham differential on R_α. This sign arises from d_0 being the classical part of $\mathfrak{m}_1$ in a filtered A_∞ structure. See (16.17).

Proof By Stokes' formula and the definition (19.6) we have

$$(d_0 \circ \mathfrak{m}^{\epsilon}_{1;\alpha_+,\alpha_-} + \mathfrak{m}^{\epsilon}_{1;\alpha_+,\alpha_-} \circ d_0)(h) = (-1)^{\dim R_+ + 1} \mathrm{ev}_+!(\mathrm{ev}_-^* h; \partial\widehat{\mathfrak{S}^{+\epsilon}}(\alpha_-,\alpha_+)). \tag{19.9}$$

By Proposition 19.1 (5) and the composition formula, the right hand side of (19.9) is

$$-\sum_{\alpha; E(\alpha_-)<E(\alpha)<E(\alpha_+)} (\mathfrak{m}^{\epsilon}_{1;\alpha_+,\alpha} \circ \mathfrak{m}^{\epsilon}_{1;\alpha,\alpha_-})(h).$$

The lemma follows. □

Lemma 19.5 implies that

$$d_0 + \sum \mathfrak{m}^{\epsilon}_{1;\alpha_+,\alpha_-}$$

[3] This is a version of the 'running out' problem discussed in [FOOO4, Subsection 7.2.8]. The way we resolve it in Sect. 19.2 of this book is the same as that in [FOOO4].

is a 'coboundary operator modulo higher order term'. Since we need to take care of the point mentioned in Remark 19.4, we have to stop at some energy cut level. So we still need to do some more work to prove Theorem 16.9 (1).

19.2 Construction of Cochain Maps

We next consider morphism of linear K-systems or of partial linear K-systems.

Situation 19.6 Suppose we are in Situation 16.16. We assume that for each $i = 1, 2$ we have a CF-perturbation $\widehat{\mathfrak{S}^+}(i; \alpha_{i-}, \alpha_{i+})$ of a τ-collared Kuranishi structure $\widehat{\mathcal{U}^+}(i; \alpha_{i-}, \alpha_{i+})$ on $\mathcal{M}^i(\alpha_{i-}, \alpha_{i+})^{\boxplus\tau_0}$ such that they satisfy the conclusion of Proposition 19.1. Here $\mathcal{M}^i(\alpha_{i-}, \alpha_{i+})$ is as in Condition 16.17. (From now on, we write $\alpha_\pm$ in place of $\alpha_{i\pm}$ if no confusion can occur.)

Let $0 < \tau' < \tau < \tau_0 = 1$ where τ_0, τ are as in Proposition 19.1. ■

Proposition 19.7 *Suppose we are in Situation* 19.6 *and are given a morphism of partial linear K-systems of energy cut level* E_0 $(\in \mathbb{R}_{\ge 0} \cup \{\infty\})$ *and energy loss* c *from* $\mathcal{F}_1$ *to* $\mathcal{F}_2$. *Let* $\mathcal{N}(\alpha_1, \alpha_2)$ *be its interpolation space where* $E(\alpha_2) - E(\alpha_1) \le E_0$.

Then for each τ' *with* $0 < \tau' < \tau < \tau_0 = 1$ *as above, there exist a* τ'*-collared Kuranishi structure* $\widehat{\mathcal{U}^+}(\mathrm{mor}; \alpha_1, \alpha_2)$ *on* $\mathcal{N}(\alpha_1, \alpha_2)^{\boxplus\tau_0}$ *and a* τ'*-collared CF-perturbation* $\widehat{\mathfrak{S}^+}(\mathrm{mor}; \alpha_1, \alpha_2)$ *of* $\widehat{\mathcal{U}^+}(\mathrm{mor}; \alpha_1, \alpha_2)$ *such that they have the following properties:*

(1) *Let* $\widehat{\mathcal{U}}(\mathrm{mor}; \alpha_1, \alpha_2)$ *be the Kuranishi structure of* $\mathcal{N}(\alpha_1, \alpha_2)$ *in Situation* 16.17 *(IV). Then* $\widehat{\mathcal{U}}(\mathrm{mor}; \alpha_1, \alpha_2)^{\boxplus\tau_0} < \widehat{\mathcal{U}^+}(\mathrm{mor}; \alpha_1, \alpha_2)$ *as oriented, collared Kuranishi structures.*
(2) *The CF-perturbation* $\widehat{\mathfrak{S}^+}(\mathrm{mor}; \alpha_1, \alpha_2)$ *is transversal to* 0. *Moreover*[4] $\mathrm{ev}_+ : \mathcal{N}(\alpha_1, \alpha_2)^{\boxplus\tau_0} \to R_{\alpha_2}$ *is strongly submersive with respect to* $\widehat{\mathfrak{S}^+}(\mathrm{mor}; \alpha_1, \alpha_2)$.
(3) *We have the following periodicity isomorphism,*

$$\widehat{\mathcal{U}^+}(\mathrm{mor}; \alpha_1, \alpha_2) \longrightarrow \widehat{\mathcal{U}^+}(\mathrm{mor}; \beta\alpha_1, \beta\alpha_2),$$

which is compatible with the isomorphism in Condition 16.17 *(VIII) via the embedding given in (1). The pullback of* $\widehat{\mathfrak{S}^+}(\mathrm{mor}; \beta\alpha_1, \beta\alpha_2)$ *by this isomorphism is* $\widehat{\mathfrak{S}^+}(\mathrm{mor}; \alpha_1, \alpha_2)$.
(4) *There is an isomorphism of* τ'*-collared K-spaces*

[4] As we note in the footnote of Proposition 19.1 (2), we simply write ev_+ in place of $\mathrm{ev}_+^{\boxplus\tau_0}$.

$$
\begin{aligned}
&\partial(\mathcal{N}(\alpha_1,\alpha_2)^{\boxplus\tau_0}, \widehat{\mathcal{U}^+}(\text{mor};\alpha_1,\alpha_2)) \\
\cong &\coprod_{\alpha_1'\in\mathfrak{A}_1} (-1)^{\dim\mathcal{N}(\alpha_1,\alpha_2)}\Big((\mathcal{N}(\alpha_1',\alpha_2)^{\boxplus\tau_0}, \widehat{\mathcal{U}^+}(\text{mor};\alpha_1',\alpha_2)) \\
&\qquad {}_{\text{ev}_-}\times_{\text{ev}_+} (\mathcal{M}^1(\alpha_1,\alpha_1')^{\boxplus\tau_0}, \widehat{\mathcal{U}^+}(1;\alpha_1,\alpha_1'))\Big) \\
\sqcup &\coprod_{\alpha_2'\in\mathfrak{A}_2} (-1)^{\dim\mathcal{M}^2(\alpha_2',\alpha_2)}\Big((\mathcal{M}^2(\alpha_2',\alpha_2)^{\boxplus\tau_0}, \widehat{\mathcal{U}^+}(2;\alpha_2',\alpha_2)) \\
&\qquad {}_{\text{ev}_-}\times_{\text{ev}_+} (\mathcal{N}(\alpha_1,\alpha_2')^{\boxplus\tau_0}, \widehat{\mathcal{U}^+}(\text{mor};\alpha_1,\alpha_2'))\Big).
\end{aligned}
\tag{19.10}
$$

Here the first union is taken over $\alpha_1' \in \mathfrak{A}_1$ *with* $E(\alpha_1) < E(\alpha_1') \le E(\alpha_2) + c$ *and the second union is taken over* $\alpha_2' \in \mathfrak{A}_2$ *with* $E(\alpha_1) - c \le E(\alpha_2') < E(\alpha_2)$. *The number c is the energy loss of the given morphism.*

The isomorphism (19.10) *is compatible with the boundary compatibility isomorphism in Condition* 16.17 *(X) via the KK-embedding in (1).*

(5) *The pullback of* $\widehat{\mathfrak{S}^+}(\text{mor};\alpha_1,\alpha_2)$ *by the isomorphism* (19.10) *is equivalent to the fiber product of* $\widehat{\mathfrak{S}^+}(1;\alpha_1,\alpha_1')$, $\widehat{\mathfrak{S}^+}(\text{mor};\alpha_1',\alpha_2)$ *and of* $\widehat{\mathfrak{S}^+}(\text{mor};\alpha_1,\alpha_2')$, $\widehat{\mathfrak{S}^+}(1;\alpha_2',\alpha_2)$.

(6) *The normalized corner* $\widehat{S}_k(\mathcal{N}(\alpha_1,\alpha_2)^{\boxplus\tau_0}, \widehat{\mathcal{U}^+}(\text{mor};\alpha_1,\alpha_2))$ *is the disjoint union of*

$$
\begin{aligned}
&\widehat{\mathcal{U}^+}(1;\alpha_1,\alpha_{1,1}) \times_{R^1_{\alpha_{1,1}}} \cdots \times_{R^1_{\alpha_{1,k_1-1}}} \widehat{\mathcal{U}^+}(1;\alpha_{1,k_1-1},\alpha_{1,k_1}) \\
&\times_{R^1_{\alpha_{1,k_1}}} \widehat{\mathcal{U}^+}(\text{mor};\alpha_{1,k_1},\alpha_{2,1}) \\
&\times_{R^1_{\alpha_{2,1}}} \widehat{\mathcal{U}^+}(2;\alpha_2,\alpha_{2,1}) \times_{R^2_{\alpha_{2,1}}} \cdots \times_{R^2_{\alpha_{2,k_2-1}}} \widehat{\mathcal{U}^+}(2;\alpha_{2,k_2-1},\alpha_{2,k_2})
\end{aligned}
\tag{19.11}
$$

where $k_1 + k_2 = k$, $\alpha_{1,i} \in \mathfrak{A}_1$, $\alpha_{2,i} \in \mathfrak{A}_2$.

This isomorphism is compatible with the corner compatibility isomorphism in Condition 16.17 *(XI) via the KK-embedding in (1).*

(7) (19.3) *and* (19.11) *imply that* $\widehat{S}_\ell(\widehat{S}_k(\mathcal{N}(\alpha_1,\alpha_2)^{\boxplus\tau_0}, \widehat{\mathcal{U}^+}(\text{mor};\alpha_1,\alpha_2)))$ *is a fiber product similar to* (19.11) *with k replaced by* $k+\ell$.

Moreover $\widehat{S}_{k+\ell}(\mathcal{N}(\alpha_1,\alpha_2)^{\boxplus\tau_0}, \widehat{\mathcal{U}^+}(\text{mor};\alpha_1,\alpha_2))$ *is also a fiber product similar to* (19.11) *with k replaced by* $k+\ell$. *The map*

$$
\begin{aligned}
\pi_{\ell,k} :&\widehat{S}_\ell(\widehat{S}_k(\mathcal{N}(\alpha_1,\alpha_2)^{\boxplus\tau_0}, \widehat{\mathcal{U}^+}(\text{mor};\alpha_1,\alpha_2))) \\
&\to \widehat{S}_{k+\ell}(\mathcal{N}(\alpha_1,\alpha_2)^{\boxplus\tau_0}, \widehat{\mathcal{U}^+}(\text{mor};\alpha_1,\alpha_2))
\end{aligned}
$$

in Proposition 24.17 *becomes the identity map on components via those identifications.*

(8) *The pullback of the restriction of* $\widehat{\mathfrak{S}^+}(\text{mor};\alpha_1,\alpha_2)$ *to* $\widehat{S}_k(\widehat{\mathcal{U}^+}(\text{mor};\alpha_1,\alpha_2))$ *by the isomorphism in (6) is equivalent to the fiber product of* $\widehat{\mathfrak{S}^+}(\text{mor};*,*)$, $\widehat{\mathfrak{S}^+}(1;*,*)$, $\widehat{\mathfrak{S}^+}(2;*,*)$ *along* (19.11).

(9) *Suppose we are given a uniform family* $\widehat{\mathfrak{S}^+}(i;\alpha_-,\alpha_+)^\sigma$ *of CF-perturbations as in Condition 19.6. Then the* σ*-parametrized family of CF-perturbations* $\widehat{\mathfrak{S}^+}(\text{mor};\alpha_1,\alpha_2)^\sigma$ *satisfying (5)(6)(7) for each* σ *is also uniform.*

Remark 19.8 Note that in Formula (19.11) we simplify the notation and omit the symbol of the underlying topological space. Namely, we write $\widehat{\mathcal{U}^+}(1;\alpha_1,\alpha_{1,1})$ in place of $(\mathcal{M}^1(\alpha_1,\alpha_{1,1})^{\boxplus\tau_0},\widehat{\mathcal{U}^+}(1;\alpha_1,\alpha_{1,1}))$. From now on, we use this kind of simplified notation when no confusion can occur.

Proof The proof is by induction on k. It is entirely similar to the proof of Proposition 19.1. So we omit it. □

We next rewrite the geometric conclusion of Proposition 19.7 in algebraic language.

Definition 19.9 In the situation of Proposition 19.7, we define

$$\psi^{\epsilon}_{\alpha_2,\alpha_1} : \Omega(R_{\alpha_1};o_{R_{\alpha_1}}) \longrightarrow \Omega(R_{\alpha_2};o_{R_{\alpha_2}}) \tag{19.12}$$

by

$$\psi^{\epsilon}_{\alpha_2,\alpha_1}(h) = (-1)^{\mu(\alpha_2)-\mu(\alpha_1)}\text{ev}_+!(\text{ev}_-^* h;\widehat{\mathfrak{S}^{+\epsilon}}(\text{mor};\alpha_1,\alpha_2)). \tag{19.13}$$

Here the right hand side is defined by Definition 17.82 on the K-space

$$(\mathcal{N}(\alpha_1,\alpha_2)^{\boxplus\tau_0},\widehat{\mathcal{U}^+}(\text{mor};\alpha_1,\alpha_2)).$$

By Condition 16.17 (VI) and Definition 7.79, the degree of $\psi^{\epsilon}_{\alpha_1,\alpha_2}$ is $\mu(\alpha_1)-\mu(\alpha_2)$. Therefore, after the degree shift as in Definition 16.8 (2), its degree becomes 0. If the energy loss of our morphism $\mathfrak{N}$ is c, the family $\{\psi^{\epsilon}_{\alpha_1,\alpha_2}\}$ of maps induces

$$\mathfrak{F}^{\lambda}CF(\mathcal{F}_1) \longrightarrow \mathfrak{F}^{\lambda-c}CF(\mathcal{F}_2),$$

where the filtration $\mathfrak{F}^{\lambda}$ is defined in Definition 16.8 (2)(3).

Lemma 19.10 *The operators* $\{\psi^{\epsilon}_{\alpha_2,\alpha_1}\}$ *satisfy the following equality in the sense of* $(\mathfrak{b})$*:*

$$\begin{aligned}&d_0\circ\psi^{\epsilon}_{\alpha_2,\alpha_1}-\psi^{\epsilon}_{\alpha_2,\alpha_1}\circ d_0\\&+\sum_{\alpha_2'}\mathfrak{m}^{2,\epsilon}_{1;\alpha_2,\alpha_2'}\circ\psi^{\epsilon}_{\alpha_2',\alpha_1}-\sum_{\alpha_1'}\psi^{\epsilon}_{\alpha_2,\alpha_1'}\circ\mathfrak{m}^{1,\epsilon}_{1;\alpha_1',\alpha_1}=0.\end{aligned} \tag{19.14}$$

Here the first sum in the second line is taken over $\alpha_2' \in \mathfrak{A}_2$ *with* $E(\alpha_1) - c \leq E(\alpha_2') < E(\alpha_2)$ *and the second sum in the second line is taken over* $\alpha_1' \in \mathfrak{A}_1$ *with* $E(\alpha_1) < E(\alpha_1') \leq E(\alpha_2) + c$. *Here* $c \geq 0$ *is the energy loss of our morphism.*

Proof By Stokes' formula the sum $d_0 \circ \psi^{\epsilon}_{\alpha_2,\alpha_1} - \psi^{\epsilon}_{\alpha_2,\alpha_1} \circ d_0$ is equal to the correspondence induced by the boundary of $\widehat{\mathfrak{S}^+}(\text{mor}; \alpha_1, \alpha_2)$. By Proposition 19.7 (4) and the composition formula this is equal to the second line of (19.14). □

19.3 Proof of Theorem 16.9 (1) and Theorem 16.39 (1)

In this section we will prove Theorem 16.9 (1). We will also prove Theorem 16.39 (1) at the same time.

Situation 19.11 We study a partial linear K-system $\mathcal{F}^i$.

We define $\mathfrak{E}$ as a set of $E \in \mathbb{R}$ such that one of the following holds:

(1) There exist α_-, α_+ and i such that $\mathcal{N}^i(\alpha_-, \alpha_+) \neq \emptyset$ and $E(\alpha_+) - E(\alpha_-) = E$.
(2) There exist α_-, α_+ and i such that $\mathcal{M}^i(\alpha_-, \alpha_+) \neq \emptyset$ and $E(\alpha_+) - E(\alpha_-) = E$.

Then $\mathfrak{E}$ is a discrete set by Definition 16.36 (2)(g). We put

$$\mathfrak{E} = \{E^1_{\mathfrak{E}}, E^2_{\mathfrak{E}}, \dots\}$$

such that $0 < E^1_{\mathfrak{E}} < E^2_{\mathfrak{E}} < \dots$. Note that the graded and filtered vector space $CF(\mathcal{F}^i)$ as in Definition 16.8 is independent of i. We denote it by $CF(\mathcal{F})$. ■

We observe that for each fixed i the set $\mathfrak{E} \cap [0, E_0]$ can be strictly bigger than the set $\mathfrak{E}_{\leq E_0}$ in (19.1). Nevertheless we can replace $\mathfrak{E}_{\leq E_0}$ by $\mathfrak{E} \cap [0, E_0]$ in Proposition 19.1 etc., by putting $\mathfrak{m}^{\epsilon}_{1;\alpha_+,\alpha_-} = 0$ when $\mathcal{M}^i(\alpha_-, \alpha_+) = \emptyset$.

In Situation 19.11, we take $i \in \mathbb{Z}_+$ and study the relationship between the operators $\mathfrak{m}^{i,\epsilon}_{1;\alpha_+,\alpha_-}$ and $\mathfrak{m}^{i+1,\epsilon}_{1;\alpha_+,\alpha_-}$ defined by applying Proposition 19.1 and Lemma 19.5 to partial linear K-systems $\mathcal{F}^i$ and $\mathcal{F}^{i+1}$. The next definition will be applied with $\mathfrak{m}^1_{1;\alpha_+,\alpha_-}$ and $\mathfrak{m}^2_{1;\alpha_+,\alpha_-}$ replaced by $\mathfrak{m}^{i,\epsilon}_{1;\alpha_+,\alpha_-}$ and $\mathfrak{m}^{i+1,\epsilon}_{1;\alpha_+,\alpha_-}$ respectively.

Definition 19.12 Suppose we are in Situation 19.11.

(1) A *partial cochain complex structure on* $CF(\mathcal{F})$ *of energy cut level* E_0 assigns $\mathfrak{m}_{1;\alpha_+,\alpha_-}$ to each α_+, α_- with $0 < E(\alpha_+) - E(\alpha_-) \leq E_0$ such that (19.7) is satisfied.
(2) Let $(CF(\mathcal{F}), \{\mathfrak{m}^j_{1;\alpha_+,\alpha_-}\})$ be a partial cochain complex structure on $CF(\mathcal{F})$ of energy cut level E_0 for $j = 1, 2$. A *partial cochain map of energy cut level* E *with energy loss 0* from $(CF(\mathcal{F}), \{\mathfrak{m}^1_{1;\alpha_+,\alpha_-}\})$ to $(CF(\mathcal{F}), \{\mathfrak{m}^2_{1;\alpha_+,\alpha_-}\})$ assigns a map ψ_{α_2,α_1} to each α_1, α_2 with $0 \leq E(\alpha_2) - E(\alpha_1) \leq E_0$ such that (19.14) is satisfied: $\psi_{\alpha_2,\alpha_1} = 0$ whenever $\alpha_1 \neq \alpha_2$ with $E(\alpha_1) = E(\alpha_2)$; $\psi_{\alpha_2,\alpha_1} =$ identity for $\alpha_1 = \alpha_2$.

(3) If $(CF(\mathcal{F}), \{\mathfrak{m}_{1;\alpha_+,\alpha_-}\})$ is a partial cochain complex structure on $CF(\mathcal{F})$ of energy cut level E_0 and $E_0' < E_0$, then by forgetting a part of the operations $\mathfrak{m}_{1;\alpha_+,\alpha_-}$ we may regard $(CF(\mathcal{F}), \{\mathfrak{m}_{1;\alpha_+,\alpha_-}\})$ as a partial cochain complex structure on $CF(\mathcal{F})$ of energy cut level E_0'. We call it the *reduction by energy cut at* E_0'.

We can define a reduction by energy cut at E_0' of a partial cochain map of energy cut level E_0 with energy loss 0 in the same way.

(4) Let $(CF(\mathcal{F}), \{\mathfrak{m}_{1;\alpha_+,\alpha_-}\})$ be a partial cochain complex structure of energy cut level E'. A partial cochain complex structure of energy cut level E is said to be its *promotion* if its reduction by energy cut at E_0 is $(CF(\mathcal{F}), \{\mathfrak{m}_{1;\alpha_+,\alpha_-}\})$.

A *promotion* of a partial cochain map is defined in the same way.

The next lemma is a baby version of [FOOO4, Lemma 7.2.72].

Lemma 19.13 *Let $(CF(\mathcal{F}), \{\mathfrak{m}^j_{1;\alpha_+,\alpha_-}\})$ be partial cochain complexes of energy cut level $E^{k_j}_{\mathfrak{C}}$ for $j = 1, 2$. Suppose $k_1 < k_2$. Let $\{\psi_{\alpha_2,\alpha_1}\}$ be a partial cochain map of energy cut level $E^{k_1}_{\mathfrak{C}}$ from $(CF(\mathcal{F}), \{\mathfrak{m}^1_{1;\alpha_+,\alpha_-}\})$ to a reduction by energy cut at $E^{k_1}_{\mathfrak{C}}$ of $(CF(\mathcal{F}), \{\mathfrak{m}^2_{1;\alpha_+,\alpha_-}\})$. Then there exists a promotion of $(CF(\mathcal{F}), \{\mathfrak{m}^1_{1;\alpha_+,\alpha_-}\})$ to energy cut level $E^{k_2}_{\mathfrak{C}}$ and a promotion of $\{\psi_{\alpha_2,\alpha_1}\}$ to the energy cut level $E^{k_2}_{\mathfrak{C}}$ from this promoted partial cochain complex structure.*

Proof By an obvious induction argument it suffices to prove the case $k_2 = k_1 + 1$. Suppose $E(\alpha_+) - E(\alpha_-) = E^{k_2}_{\mathfrak{C}}$. We define a linear map

$$o(\alpha_+, \alpha_-) : \Omega(R_{\alpha_-}; o_{R_{\alpha_-}}) \longrightarrow \Omega(R_{\alpha_+}; o_{R_{\alpha_+}})$$

by

$$o(\alpha_+, \alpha_-) = \sum_{\alpha; E(\alpha_-) < E(\alpha) < E(\alpha_+)} (\mathfrak{m}^1_{1;\alpha_+,\alpha} \circ \mathfrak{m}^1_{1;\alpha,\alpha_-}).$$

We will prove that $o(\alpha_+, \alpha_-)$ is a d_0-coboundary.

Notation 19.14 We use the following notation. For an $\mathbb{R}$ linear map $F : CF(\mathcal{F}) \to CF(\mathcal{F})$, F_{α_+,α_-} denotes the $\mathrm{Hom}_{\mathbb{R}}(\Omega(R_{\alpha_-}, o_{R_{\alpha_-}}), \Omega(R_{\alpha_+}, o_{R_{\alpha_+}}))$ component of F.

We put $\psi_{\alpha,\alpha} = \mathrm{id}$. We define $\hat{d}^j : CF(\mathcal{F}) \to CF(\mathcal{F})$ $(j = 1, 2)$ and $\widehat{\psi} : CF(\mathcal{F}) \to CF(\mathcal{F})$ by

$$\hat{d}^j = d_0 \oplus \bigoplus_{\substack{\alpha_1,\alpha_2 \\ E(\alpha_2)-E(\alpha_1) \le E^{k_1}_{\mathfrak{C}}}} \mathfrak{m}^j_{1;\alpha_2,\alpha_1}, \quad \widehat{\psi} = \bigoplus_{\substack{\alpha_1,\alpha_2 \\ E(\alpha_2)-E(\alpha_1) \le E^{k_1}_{\mathfrak{C}}}} \psi_{\alpha_2,\alpha_1}. \tag{19.15}$$

We have

$$(\hat{d}^1 \circ \hat{d}^1)_{\alpha_+,\alpha_-} = o(\alpha_+,\alpha_-), \tag{19.16}$$

if $E(\alpha_+) - E(\alpha_-) = E^{k_2}_{\mathfrak{E}}$. On the other hand, we have

$$(\hat{d}^1 \circ \hat{d}^1)_{\alpha_2,\alpha_1} = 0 \tag{19.17}$$

if $E(\alpha_2) - E(\alpha_1) \le E^{k_1}_{\mathfrak{E}}$. We note

$$(\hat{d}^1 \circ \hat{d}^1) \circ \hat{d}^1 - \hat{d}^1 \circ (\hat{d}^1 \circ \hat{d}^1) = 0. \tag{19.18}$$

Then (19.17) implies

$$((\hat{d}^1 \circ \hat{d}^1) \circ \hat{d}^1)_{\alpha_+,\alpha_-} = (\hat{d}^1 \circ \hat{d}^1)_{\alpha_+,\alpha_-} \circ d_0,$$
$$(\hat{d}^1 \circ (\hat{d}^1 \circ \hat{d}^1))_{\alpha_+,\alpha_-} = d_0 \circ (\hat{d}^1 \circ \hat{d}^1)_{\alpha_+,\alpha_-}.$$

Therefore (19.16) and (19.18) imply

$$d_0 \circ o(\alpha_+,\alpha_-) - o(\alpha_+,\alpha_-) \circ d_0 = 0.$$

Namely, $o(\alpha_+,\alpha_-)$ is a cocycle as a map between cochain complexes.

We continue to show that $o(\alpha_+,\alpha_-)$ is a coboundary. When $E(\alpha_+) - E(\alpha_-) = E^{k_2}_{\mathfrak{E}}$, we put

$$b(\alpha_+,\alpha_-) = (\hat{d}^2 \circ \widehat{\psi} - \widehat{\psi} \circ \hat{d}^1)_{\alpha_+,\alpha_-}.$$

Since $\widehat{\psi}$ is assumed to be a cochain map of energy cut level $E^{k_1}_{\mathfrak{E}}$, we have

$$(\hat{d}^2 \circ \widehat{\psi} - \widehat{\psi} \circ \hat{d}^1)_{\alpha_2,\alpha_1} = 0$$

if $E(\alpha_2) - E(\alpha_1) \le E^{k_1}_{\mathfrak{E}}$. Therefore we find

$$\begin{aligned}&(\hat{d}^2 \circ (\hat{d}^2 \circ \widehat{\psi} - \widehat{\psi} \circ \hat{d}^1) + (\hat{d}^2 \circ \widehat{\psi} - \widehat{\psi} \circ \hat{d}^1) \circ \hat{d}^1)_{\alpha_+,\alpha_-}\\ &= d_0 \circ b(\alpha_+,\alpha_-) + b(\alpha_+,\alpha_-) \circ d_0.\end{aligned} \tag{19.19}$$

On the other hand, an obvious calculation leads to

$$\begin{aligned}&(\hat{d}^2 \circ (\hat{d}^2 \circ \widehat{\psi} - \widehat{\psi} \circ \hat{d}^1) + (\hat{d}^2 \circ \widehat{\psi} - \widehat{\psi} \circ \hat{d}^1) \circ \hat{d}^1)_{\alpha_+,\alpha_-}\\ &= ((\hat{d}^2 \circ \hat{d}^2) \circ \widehat{\psi} - \widehat{\psi} \circ (\hat{d}^1 \circ \hat{d}^1))_{\alpha_+,\alpha_-}\\ &= (\hat{d}^2 \circ \hat{d}^2)_{\alpha_+,\alpha_-} \circ \widehat{\psi}_{\alpha_-,\alpha_-} - \widehat{\psi}_{\alpha_+,\alpha_+} \circ (\hat{d}^1 \circ \hat{d}^1)_{\alpha_+,\alpha_-}\\ &= (\hat{d}^2 \circ \hat{d}^2)_{\alpha_+,\alpha_-} - o(\alpha_+,\alpha_-).\end{aligned} \tag{19.20}$$

We observe that the equation $\hat{d}^2 \circ \hat{d}^2 \equiv 0 \mod T^{E^{k_2}_{\mathfrak{E}}}$ gives rise to

$$d_0 \circ \mathfrak{m}^2_{1;\alpha_+,\alpha_-} + \mathfrak{m}^2_{1;\alpha_+,\alpha_-} \circ d_0 + (\hat{d}^2 \circ \hat{d}^2)_{\alpha_+,\alpha_-} = 0, \tag{19.21}$$

since the energy cut level of $(CF(\mathcal{F}), \{\mathfrak{m}^2_{1;\alpha_+,\alpha_-}\})$ is $E^{k_2}_{\mathfrak{E}}$. Combination of (19.19), (19.20), (19.21) implies

$$o(\alpha_+,\alpha_-) = -d_0 \circ (\mathfrak{m}^2_{1;\alpha_+,\alpha_-} + b(\alpha_+,\alpha_-)) - (\mathfrak{m}^2_{1;\alpha_+,\alpha_-} + b(\alpha_+,\alpha_-)) \circ d_0. \tag{19.22}$$

Namely, $o(\alpha_+,\alpha_-)$ is a d_0-coboundary. Therefore there exists $\mathfrak{m}^1_{1;\alpha_+,\alpha_-}$ such that

$$d_0 \circ \mathfrak{m}^1_{1;\alpha_+,\alpha_-} + \mathfrak{m}^1_{1;\alpha_+,\alpha_-} \circ d_0 + o(\alpha_+,\alpha_-) = 0. \tag{19.23}$$

Hence we have a promotion of $(CF(\mathcal{F}), \{\mathfrak{m}^1_{1;\alpha_+,\alpha_-}\})$ to the energy level $E^{k_2}_{\mathfrak{E}}$.

We note that the choice of $\mathfrak{m}^1_{1;\alpha_+,\alpha_-}$ satisfying (19.23) is not unique and can be changed by adding a d_0-cocycle. We will use this freedom in the next part of the proof.

We next promote ψ. For this purpose it suffices to find ψ_{α_+,α_-} for $E(\alpha_+) - E(\alpha_-) = E^{k_2}_{\mathfrak{E}} - c$ such that

$$\begin{aligned} d_0 \circ \psi_{\alpha_+,\alpha_-} - \psi_{\alpha_+,\alpha_-} \circ d_0 + \mathfrak{m}^2_{1;\alpha_+,\alpha_-} \circ \psi_{\alpha_-,\alpha_-} \\ - \psi_{\alpha_+,\alpha_+} \circ \mathfrak{m}^1_{1;\alpha_+,\alpha_-} + b(\alpha_+,\alpha_-) = 0. \end{aligned} \tag{19.24}$$

We will prove it below. We put

$$\begin{aligned} o'(\alpha_+,\alpha_-) &= \mathfrak{m}^2_{1;\alpha_+,\alpha_-} \circ \psi_{\alpha_-,\alpha_-} - \psi_{\alpha_+,\alpha_+} \circ \mathfrak{m}^1_{1;\alpha_+,\alpha_-} + b(\alpha_+,\alpha_-) \\ &= \mathfrak{m}^2_{1;\alpha_+,\alpha_-} - \mathfrak{m}^1_{1;\alpha_+,\alpha_-} + b(\alpha_+,\alpha_-). \end{aligned}$$

The identities (19.22) and (19.23) imply

$$d_0 \circ o'(\alpha_+,\alpha_-) + o'(\alpha_+,\alpha_-) \circ d_0 = 0.$$

Thus $o'(\alpha_+,\alpha_-)$ is a d_0-cocycle. Using the freedom of the choice of $\mathfrak{m}^1_{1;\alpha_+,\alpha_-}$ we mentioned above, we may assume that $o'(\alpha_+,\alpha_-)$ is a d_0-coboundary. Hence we can find ψ_{α_+,α_-} satisfying (19.24). The proof of Lemma 19.13 is complete. □

Proof of Theorem* 16.9 *(1) and Theorem* 16.39 *(1) Below we will prove Theorem 16.39 (1) and indicate modifications needed to prove Theorem 16.9 (1).

Suppose we are in Situation 19.11. (We note that for the proof of Theorem 16.9 (1) we consider the situation where $\mathcal{F}^i = \mathcal{F}$ and the morphisms $\mathfrak{N}^i$ appearing in Definition 16.36 (2)(d) are the identity morphisms for all i.)

We assume that the energy cut level of $\mathcal{F}^i$ is $E_{\mathfrak{E}}^{k_i}$ such that $k_i < k_{i+1}$. (It implies $\lim_{i\to\infty} E_{\mathfrak{E}}^{k_i} = \infty$ by the discreteness of $\mathfrak{E}$.) We apply Proposition 19.1 and Lemma 19.5 to $\mathcal{F}^i$ for each i. Then we obtain $\epsilon_{0,i}$ such that for each $\epsilon < \epsilon_{0,i}$ and i we obtain a partial cochain complex of energy cut level $E_{\mathfrak{E}}^{k_i}$, which we denote by $(CF(\mathcal{F}), \{\mathfrak{m}_{1;\alpha_+,\alpha_-}^{i,\epsilon}\})$. Note that the operator $\mathfrak{m}_{1;\alpha_+,\alpha_-}^{i,\epsilon}$ is defined as the correspondence by a Kuranishi structure and a CF-perturbation. We write them as $(\widehat{\mathcal{U}^+}(i;\alpha_-,\alpha_+), \widehat{\mathfrak{S}^+}(i;\alpha_-,\alpha_+))$ respectively. We define

$$CF(\mathcal{F}^i;\epsilon) = (CF(\mathcal{F}), \{\mathfrak{m}_{1;\alpha_+,\alpha_-}^{i,\epsilon}\}). \tag{19.25}$$

This is well-defined if $\epsilon < \epsilon_{0,i}$, which is a partial cochain complex of energy cut level $E_{\mathfrak{E}}^{k_i}$ by Lemma 19.5.

We next consider the morphisms $\mathfrak{N}^i : \mathcal{F}^i \to \mathcal{F}^{i+1}$. Their interpolation spaces are denoted by $\mathcal{N}(i;\alpha_-,\alpha_+)$. We apply Proposition 19.7 to obtain a τ_{i+1}-collared Kuranishi structure on it, which we write $\widehat{\mathcal{U}^+}(\text{mor}, i;\alpha_-,\alpha_+)$. We note that here we regard $\widehat{\mathcal{U}^+}(i;\alpha_-,\alpha_+)$ as a τ_{i+1}-collared Kuranishi structure. This is possible since $\tau_{i+1} < \tau_i$ and $\widehat{\mathcal{U}^+}(i;\alpha_-,\alpha_+)$ is τ_i-collared.

We next take $\rho_i > 0$ such that

$$\rho_i \le \min\{\epsilon_{k,i+1}/\epsilon_{k,i} \mid k = 0, 1, 2, \dots\}, \tag{19.26}$$

where $\epsilon_{k,i+1}, \epsilon_{k,i}, k = 1, 2, \dots$ will be defined later.

Remark 19.15 There appear only a finite number of k's. Note that $\epsilon_{0,i}$ is already defined above and $\epsilon_{1,i}, \epsilon_{2,i}$ and $\epsilon_{3,i}$ will be chosen after Lemma 19.16, Lemma 19.22 and Lemma 19.29 below, respectively. Other $\epsilon_{4,i}$ etc. will be taken during the proof of Theorem 16.39 (3) in Sect. 19.6 for studying homotopy of homotopies.

We consider a CF-perturbation $\epsilon \mapsto \widehat{\mathfrak{S}^{+\rho_i\epsilon}}(i+1;\alpha_-,\alpha_+)$ of $\widehat{\mathcal{U}^+}(i+1;\alpha_-,\alpha_+)$, which is defined as follows. Note that $\widehat{\mathfrak{S}^+}(i+1;\alpha_-,\alpha_+)$ is a CF-perturbation of $\widehat{\mathcal{U}^+}(i+1;\alpha_-,\alpha_+)$. Its local representative on the Kuranishi charts is $(W_{\mathfrak{r}}, \omega_{\mathfrak{r}}, \mathfrak{s}_{\mathfrak{r}}^{\epsilon})$. (See Definition 7.4.) We replace $\mathfrak{s}_{\mathfrak{r}}^{\epsilon}$ by $\mathfrak{s}_{\mathfrak{r}}^{\epsilon\rho_i}$ but do not change anything else. It is easy to find that it is compatible with the coordinate change etc. and defines a CF-perturbation, which we denoted by $\epsilon \mapsto \widehat{\mathfrak{S}^{+\rho_i\epsilon}}(i+1;\alpha_-,\alpha_+)$ above. Hereafter we write it as $\widehat{\mathfrak{S}^{+\rho_i\cdot}}(i+1;\alpha_-,\alpha_+)$. By definition we have

$$f!(h;(\widehat{\mathfrak{S}^{+\rho_i\cdot}}(i+1;\alpha_-,\alpha_+)^{\epsilon}) = f!(h;(\widehat{\mathfrak{S}^{+\rho_i\epsilon}}(i+1;\alpha_-,\alpha_+)) \tag{19.27}$$

when the integrations along the fibers of both sides are defined.

Now we apply Proposition 19.7 to obtain $\widehat{\mathfrak{S}^{+i}}$ on $\widehat{\mathcal{U}^+}(\text{mor};\alpha_-,\alpha_+)$. Here we use the CF-perturbations $\widehat{\mathfrak{S}^+}(i;\alpha_-,\alpha_+)$ on the space of connecting orbits of $\mathcal{F}^i$ and $\widehat{\mathfrak{S}^{+\rho_i\cdot}}(i+1;\alpha_-,\alpha_+)$ on the space of connecting orbits of $\mathcal{F}^{i+1}$. (We put $\rho_i\cdot$

for the second one.) Then we obtain from Proposition 19.7 a CF-perturbation on $\widehat{\mathcal{U}^+}(\text{mor}; \alpha_-, \alpha_+)$. We denote it by $\widehat{\mathfrak{S}^{+i}}(\text{mor}, i, \sigma_i; \alpha_-, \alpha_+)$.

Lemma 19.16 *The family* $\widehat{\mathfrak{S}^{+i}}(\text{mor}, i, \rho_i; \alpha_-, \alpha_+)$ *for* $\rho_i \in (0, \epsilon_{1,i+1}/\epsilon_{1,i}]$ *is a uniform family in the sense of Definition* 9.30.

Proof This is immediate from Proposition 19.7 (9). □

By Lemmas 7.89 and 19.16 we can find $\epsilon_{1,i}$ such that if $\epsilon < \epsilon_{1,i}$ the integration along the fiber is defined by using the CF-perturbation $\widehat{\mathfrak{S}^{+i}}(\text{mor}, i, \rho_i; \alpha_-, \alpha_+)^\epsilon$ for $\epsilon < \epsilon_{1,i+1}$.

Remark 19.17 The fact that we can take ϵ in a way independent of ρ_i is crucial here. Otherwise, the process of defining those numbers would become circular. Namely, while working in this step of induction, we do not know in advance how small $\epsilon_{0,i+1}$, $\epsilon_{1,i+1}$ must be. So we cannot estimate ρ_i *from below* at this stage.

See also Remark 20.7.

Thus for $\epsilon < \min(\epsilon_{0,i}, \epsilon_{1,i})$ we obtain $\psi^i_{\alpha_+,\alpha_-}$ by (19.13). We use (19.27) together with Lemma 19.10 to prove that $\{\psi^i_{\alpha_+,\alpha_-}\} : CF(\mathcal{F}^i; \epsilon) \to CF(\mathcal{F}^{i+1}; \epsilon)$ is a partial cochain map of energy cut level $E^{k_i}_{\mathfrak{C}}$. (Here we reduce the energy cut level of $CF(\mathcal{F}^{i+1}; \epsilon)$ to $E^{k_i}_{\mathfrak{C}}$.)

Now we use Lemma 19.13 inductively to promote $CF(\mathcal{F}^{i+1}; \epsilon)$ to a partial cochain complex of energy cut level $E^{k_n}_{\mathfrak{C}}$ for each n and $\{\psi^i_{\alpha_+,\alpha_-}\}$ to a partial cochain map of energy cut level $E^{k_n}_{\mathfrak{C}}$ for each n.

To prove Theorem 16.9 (1), we regard $\mathcal{F} = \mathcal{F}^i$ and use the identity morphism $\mathcal{F} \to \mathcal{F}$ as $\mathfrak{N}^i : \mathcal{F}^i \to \mathcal{F}^{i+1}$ for all i. Then Theorem 16.39 (1) implies Theorem 16.9 (1).

The proofs of Theorem 16.9 (1) and Theorem 16.39 (1) are now complete. □

19.4 Composition of Morphisms and of Induced Cochain Maps

In this section we show that the composition of morphisms (defined in Lemma-Definition 16.35 and in Sect. 18.7) induces the composition of the partial cochain maps given by Definition 19.9. Since the partial cochain map in Definition 19.9 depends on the choice of the perturbation, we need to state it a bit carefully.

We consider a situation that is similar to but is slightly different from Situation 16.16. Namely:

Situation 19.18

(1) For $j = 1, 2, 3$, let

$$\mathcal{C}_j = \left(\mathfrak{A}_j, \mathfrak{G}_j, \{R^j_{\alpha_j}\}_{\alpha\in\mathfrak{A}_j}, \{o_{R^j_{\alpha_j}}\}_{\alpha\in\mathfrak{A}_j}, E, \mu, \{\mathrm{PI}^j_{\beta_j,\alpha_j}\}_{\beta_j\in\mathfrak{G}_j,\alpha_j\in\mathfrak{A}_j}\right)$$

be critical submanifold data and

$$\mathcal{F}_j = \Big(\mathcal{C}_j, \{\mathcal{M}^j(\alpha_{j-},\alpha_{j+})\}_{\alpha_{j\pm}\in\mathfrak{A}_j}, (\mathrm{ev}_-,\mathrm{ev}_+), \{\mathrm{OI}^j_{\alpha_{j-},\alpha_{j+}}\}_{\alpha_{j\pm}\in\mathfrak{A}_j}, \{\mathrm{PI}^j_{\beta_j;\alpha_{j-},\alpha_{j+}}\}_{\beta_j\in\mathfrak{G}_j,\alpha_{j\pm}\in\mathfrak{A}_j}\Big)$$

linear K-systems. We assume $\mathfrak{G}_1 = \mathfrak{G}_2 = \mathfrak{G}_3$ (together with energy E and the Maslov index μ on it) and denote it by $\mathfrak{G}$.

(2) The same as (1) except the assumption that they consist of partial linear K-systems of energy cut level E_0. ■

Situation 19.19 Suppose we are in Situation 19.18 (2).

We assume that for each $j = 1, 2, 3$ we have $\widehat{\mathfrak{S}^+}(j;\alpha_{j-},\alpha_{j+})$ of $\widehat{\mathcal{U}^+}(j;\alpha_{j-},\alpha_{j+})$ on $\mathcal{M}^j(\alpha_{j-},\alpha_{j+})^{\boxplus\tau_0}$ satisfying the conclusions of Proposition 19.1. Here $\mathcal{M}^j(\alpha_{j-},\alpha_{j+})$ is as in Situation 19.18. (From now on, we write $\alpha_\pm$ in place of $\alpha_{j\pm}$ if no confusion can occur.) We assume that we have a partial morphism $\mathfrak{N}_{j+1j} : \mathcal{F}_j \to \mathcal{F}_{j+1}$ for $j = 1, 2$, whose interpolation space is $\mathcal{N}_{jj+1}(\alpha_j,\alpha_{j+1})$.

Furthermore, we have $\widehat{\mathfrak{S}^+}(\mathrm{mor}; j+1, j; \alpha_j, \alpha_{j+1})$ and $\widehat{\mathcal{U}^+}(\mathrm{mor}; j+1, j; \alpha_j, \alpha_{j+1})$ on $\mathcal{N}_{jj+1}(\alpha_j,\alpha_{j+1})$ satisfying the conclusions of Proposition 19.7. ■

We obtain a composition $\mathfrak{N}_{31} = \mathfrak{N}_{32}\circ\mathfrak{N}_{21}$ by Lemma-Definition 16.35 and Lemma-Definition 18.37. By (19.21) we have

$$\begin{aligned}&\widehat{\mathcal{U}}(\mathrm{mor}; 3, 1; \alpha_1, \alpha_3)\\ &= \bigcup_{\alpha_2\in\mathfrak{A}_2} \widehat{\mathcal{U}}(\mathrm{mor}; 2, 1; \alpha_1, \alpha_2) \times^{\boxplus\tau}_{R_{\alpha_2}} \widehat{\mathcal{U}}(\mathrm{mor}; 3, 2; \alpha_2, \alpha_3).\end{aligned} \tag{19.28}$$

Here the summand of the right hand side is defined by Definition 18.34. Therefore

$$\begin{aligned}&\widehat{\mathcal{U}}(\mathrm{mor}; 3, 1; \alpha_1, \alpha_3)^{\boxplus\tau}\\ &= \bigcup_{\alpha_2\in\mathfrak{A}_2} \widehat{\mathcal{U}}(\mathrm{mor}; 2, 1; \alpha_1, \alpha_2)^{\boxplus\tau} \times_{R_{\alpha_2}} \widehat{\mathcal{U}}(\mathrm{mor}; 3, 2; \alpha_2, \alpha_3)^{\boxplus\tau}.\end{aligned} \tag{19.29}$$

(Note we take an appropriate corner smoothing in the right hand side.) On its underlying topological space

$$\bigcup_{\alpha_2\in\mathfrak{A}_2} \mathcal{N}_{12}(\alpha_1,\alpha_2)^{\boxplus\tau} \times_{R_{\alpha_2}} \mathcal{N}_{23}(\alpha_2,\alpha_3)^{\boxplus\tau}, \tag{19.30}$$

we will define a Kuranishi structure

$$\begin{aligned}&\widehat{\mathcal{U}^+}(\mathrm{mor}; 3, 1; \alpha_1, \alpha_3)\\ &= \bigcup_{\alpha_2 \in \mathfrak{A}_2} \widehat{\mathcal{U}^+}(\mathrm{mor}; 2, 1; \alpha_1, \alpha_2) \times_{R_{\alpha_2}} \widehat{\mathcal{U}^+}(\mathrm{mor}; 3, 2; \alpha_2, \alpha_3)\end{aligned} \tag{19.31}$$

as follows.

Remark 19.20 The fiber product appearing in each summand of (19.31) obviously gives a Kuranishi structure on each summand of (19.30). Below we explain how we smooth the corners to obtain a Kuranishi structure of the union.

Note that the Kuranishi structure $\widehat{\mathcal{U}^+}(\mathrm{mor}; 2, 1; \alpha_1, \alpha_2)$ is τ'-collared. We take a Kuranishi structure $\widehat{\mathcal{U}^+}(\mathrm{mor}; 2, 1; \alpha_1, \alpha_2)^{\boxminus(\tau-\tau')}$ such that

$$\left(\widehat{\mathcal{U}^+}(\mathrm{mor}; 2, 1; \alpha_1, \alpha_2)^{\boxminus(\tau-\tau')}\right)^{\boxplus\tau'} = \widehat{\mathcal{U}^+}(\mathrm{mor}; 2, 1; \alpha_1, \alpha_2).$$

We define $\widehat{\mathcal{U}^+}(\mathrm{mor}; 3, 2; \alpha_2, \alpha_3)^{\boxminus(\tau-\tau')}$ in the same way. We then consider a Kuranishi structure

$$\widehat{\mathcal{U}^+}(\mathrm{mor}; 2, 1; \alpha_1, \alpha_2)^{\boxminus(\tau-\tau')} \times_{R_{\alpha_2}} \widehat{\mathcal{U}^+}(\mathrm{mor}; 3, 2; \alpha_2, \alpha_3)^{\boxminus(\tau-\tau')} \tag{19.32}$$

on

$$\mathcal{N}(\mathrm{mor}; 2, 1; \alpha_1, \alpha_2)^{\boxplus\tau'} \times_{R_{\alpha_2}} \mathcal{N}(\mathrm{mor}; 3, 2; \alpha_2, \alpha_3)^{\boxplus\tau'}.$$

See Fig. 19.1.

On the other hand, let $\mathcal{N}_{123}(\alpha_1, \alpha_3)$ be as in Proposition 18.35. That is, after corner smoothing, its boundary contains $(\mathcal{N}(\alpha_1, \alpha_3), \widehat{\mathcal{U}}(\mathrm{mor}; 3, 1; \alpha_1, \alpha_3))$. We denote the Kuranishi structure of $\mathcal{N}_{123}(\alpha_1, \alpha_3)$ by $\widehat{\mathcal{U}}_{123}(\alpha_1, \alpha_3)$.

We consider the complement $\mathfrak{C}^c$ of the boundary components $\mathfrak{C}$ of $\mathcal{N}_{123}(\alpha_1, \alpha_3)$ appearing in Proposition 18.35. Namely $\mathfrak{C}$ consists of

$$\bigcup \mathcal{N}_{12}(\alpha_1, \alpha_2)^{\boxplus\tau} \times_{R_{\alpha_2}} \mathcal{N}_{23}(\alpha_2, \alpha_3)^{\boxplus\tau}.$$

The K-space $\mathcal{N}_{123}(\alpha_1, \alpha_3)$ is τ'-$\mathfrak{C}$-collared. We take $\mathcal{N}_{123}(\alpha_1, \alpha_3)^{\mathfrak{C}\boxminus(\tau-\tau')}$ and consider the topological space (see Fig. 19.2)

$$\left(\mathcal{N}_{123}(\alpha_1, \alpha_3) \setminus \mathcal{N}_{123}(\alpha_1, \alpha_3)^{\mathfrak{C}\boxminus(\tau-\tau')}\right)^{\mathfrak{C}^c\boxplus\tau'}. \tag{19.33}$$

Then $\widehat{\mathcal{U}}_{123}(\alpha_1, \alpha_3)$ induces a Kuranishi structure on it. We write the resulting Kuranishi structure as $\widehat{\mathcal{U}}_{123}(\alpha_1, \alpha_3)^{\mathfrak{C}\boxminus(\tau-\tau'), \mathfrak{C}^c\boxplus\tau'}$.

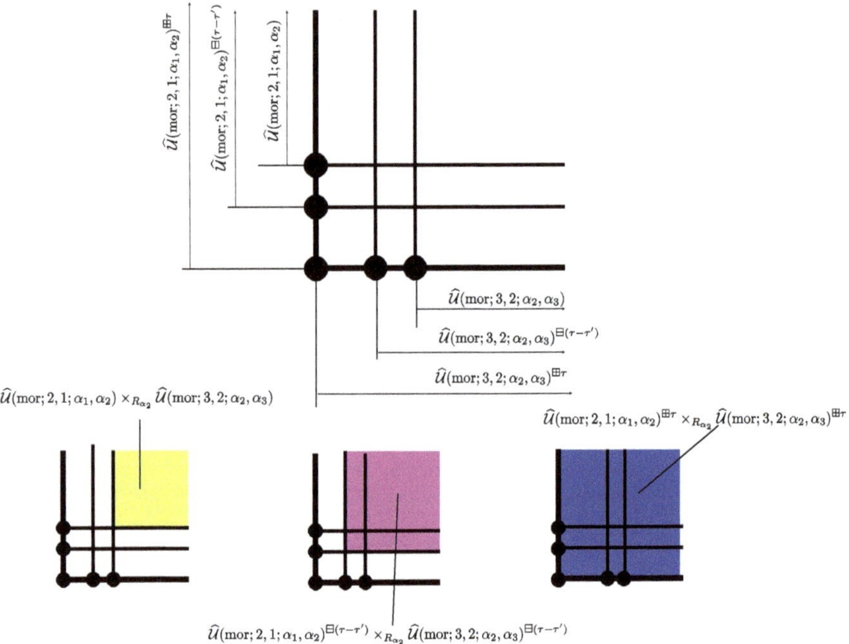

Fig. 19.1 $\mathcal{N}(\text{mor};2,1;\alpha_1,\alpha_2)^{\boxplus\tau}\times_{R_{\alpha_2}}\mathcal{N}(\text{mor};3,2;\alpha_2,\alpha_3)^{\boxplus\tau}$

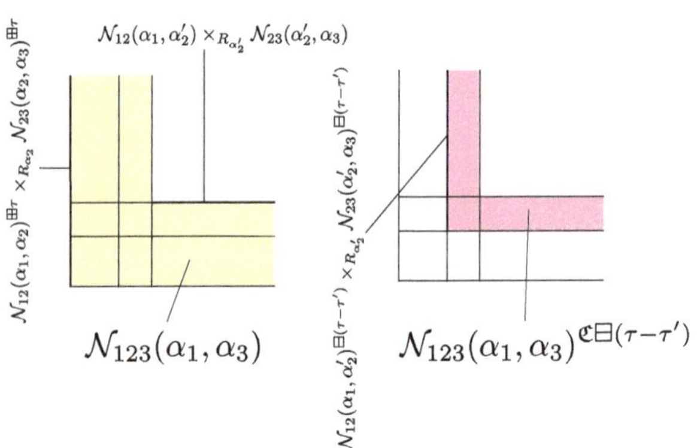

Fig. 19.2 $\mathcal{N}_{123}(\alpha_1,\alpha_3)^{\mathfrak{C}\boxminus(\tau-\tau')}$

Let $0<\tau''<\tau'$. Starting from (19.32) we apply Proposition 18.35 to obtain a τ''-$\mathfrak{C}$-collared Kuranishi structure on the topological space (19.33), which we denote by $\widehat{\mathcal{U}_{123}^{+}}(\alpha_1,\alpha_3)$. The Kuranishi structure $\widehat{\mathcal{U}}_{123}(\alpha_1,\alpha_3)^{\mathfrak{C}\boxminus(\tau-\tau'),\mathfrak{C}^c\boxplus\tau'}$ is embedded in $\widehat{\mathcal{U}_{123}^{+}}(\alpha_1,\alpha_3)$.

The τ''-$\mathfrak{C}$-partial corner smoothing of $\widehat{\mathcal{U}^{+}_{123}}(\alpha_1,\alpha_3)$ defines our Kuranishi structure $\widehat{\mathcal{U}^{+}}(\text{mor}; 3, 1; \alpha_1, \alpha_3)$ of $\mathcal{N}_{13}(\alpha_1,\alpha_3)^{\boxplus\tau}$ that appears in (19.31), where $\mathcal{N}_{13}(\alpha_1,\alpha_3)$ is the interpolation space of the morphism $\mathfrak{N}_{31}$. We put

$$\begin{aligned}&\widehat{\mathfrak{S}^{+}}(\text{mor}; 3, 1; \alpha_1, \alpha_3)\\ &= \bigcup_{\alpha_2\in\mathfrak{A}_2} \widehat{\mathfrak{S}^{+}}(\text{mor}; 2, 1; \alpha_1, \alpha_2) \times_{R_{\alpha_2}} \widehat{\mathfrak{S}^{+}}(\text{mor}; 3, 2; \alpha_2, \alpha_3).\end{aligned} \tag{19.34}$$

Here each summand of the right hand side of (19.34) gives a CF-perturbation of each summand in (19.31). Using the collaredness, we can glue them to obtain a CF-perturbation on $\widehat{\mathcal{U}^{+}}(\text{mor}; 3, 1; \alpha_1, \alpha_3)$. (In other words, they are automatically smooth on the part where they are glued.)

Lemma 19.21 *The Kuranishi structure* $\widehat{\mathcal{U}^{+}}(\text{mor}; 3, 1; \alpha_1, \alpha_3)$ *and the CF-perturbation* $\widehat{\mathfrak{S}^{+}}(\text{mor}; 3, 1; \alpha_1, \alpha_3)$ *satisfy the conclusion of Proposition* 19.7.

Proof By construction, the corners contained in

$$\widehat{\mathcal{U}^{+}}(\text{mor}; 2, 1; \alpha_1, \alpha_2) \times_{R_{\alpha_2}} \widehat{\mathcal{U}^{+}}(2; \alpha_2, \alpha_2') \times_{R_{\alpha_2'}} \widehat{\mathcal{U}^{+}}(\text{mor}; 3, 2; \alpha_2', \alpha_3)$$

and

$$\widehat{\mathcal{U}}(\text{mor}; 2, 1; \alpha_1, \alpha_2)^{\boxplus\tau} \times_{R_{\alpha_2}} \widehat{\mathcal{U}}(2; \alpha_2, \alpha_2')^{\boxplus\tau} \times_{R_{\alpha_2'}} \widehat{\mathcal{U}}(\text{mor}; 3, 2; \alpha_2', \alpha_3)^{\boxplus\tau}$$

are smooth. Therefore the boundary of $\widehat{\mathcal{U}^{+}}(\text{mor}; 3, 1; \alpha_1, \alpha_3)$ consists of

$$\widehat{\mathcal{U}^{+}}(1; \alpha_1, \alpha_1') \times_{R_{\alpha_1'}} \widehat{\mathcal{U}^{+}}(\text{mor}; 2, 1; \alpha_1', \alpha_2) \times_{R_{\alpha_2}} \widehat{\mathcal{U}^{+}}(\text{mor}; 3, 2; \alpha_2, \alpha_3)$$

and

$$\widehat{\mathcal{U}^{+}}(\text{mor}; 2, 1; \alpha_1, \alpha_2) \times_{R_{\alpha_2}} \widehat{\mathcal{U}^{+}}(\text{mor}; 3, 2; \alpha_2, \alpha_3') \times_{R_{\alpha_3'}} \widehat{\mathcal{U}^{+}}(3; \alpha_3', \alpha_3).$$

This is Proposition 19.7 (4). Since $\widehat{\mathcal{U}}_{123}(\alpha_1,\alpha_3)^{\mathfrak{C}\boxminus(\tau-\tau'),\mathfrak{C}^c\boxplus\tau'}$ is embedded in $\widehat{\mathcal{U}^{+}_{123}}(\alpha_1,\alpha_3)$, the K-space $\widehat{\mathcal{U}}(\text{mor}; 3, 1; \alpha_1, \alpha_3)^{\boxplus\tau}$ is embedded in $\widehat{\mathcal{U}^{+}}(\text{mor}; 3, 1; \alpha_1, \alpha_3)$. This is Proposition 19.7 (1). The rest of the proof is obvious. □

Let (i, i') be one of $(3, 2)$, $(2, 1)$, $(3, 1)$. We use the pair $\widehat{\mathcal{U}^{+}}(\text{mor}; i', i; \alpha_i, \alpha_{i'})$ and $\widehat{\mathfrak{S}^{+}}(\text{mor}; i', i; \alpha_i, \alpha_{i'})$ to apply Definition 19.9. We then obtain:

$$\psi^{i'i;\epsilon}_{\alpha_{i'},\alpha_i} : \Omega(R_{\alpha_i}; o_{R_{\alpha_i}}) \longrightarrow \Omega(R_{\alpha_{i'}}; o_{R_{\alpha_{i'}}}).$$

Lemma 19.22

$$\psi^{31;\epsilon}_{\alpha_3,\alpha_1} = \sum_{\alpha_2 \in \mathfrak{A}_2} \psi^{32;\epsilon}_{\alpha_3,\alpha_2} \circ \psi^{21;\epsilon}_{\alpha_2,\alpha_1} \tag{19.35}$$

in the sense of (♭). *(See Remark* 19.4 *for the meaning of* (♭).*)*

Proof This is immediate from the composition formula and Lemma 18.31. □

We will choose $\epsilon_{2,i}$ so that *(19.35)* holds for $\epsilon < \epsilon_{2,i}$. We remark that the suffix i in $\epsilon_{2,i}$ here is not related to the suffix i in α_i etc. above. See Remark 19.15 for $\epsilon_{2,i}$.

Corollary 19.23 *Suppose we are in the situation of Theorem 16.31 (3). Let E be an arbitrary positive number. Then we can make the choices to define* $\mathfrak{N}_{13}$ *so that* $\mathfrak{N}_{13} \equiv \mathfrak{N}_{23} \circ \mathfrak{N}_{12} \mod T^E$ *holds.*

19.5 Construction of Homotopy

In this section we start from a homotopy of (partial) morphisms of linear K-systems and construct a (partial) cochain homotopy. We can study higher homotopy in the same way. Since the definition of parametrized morphism is a bit heavy, we discuss the case of homotopy first in this section. The general case of higher homotopy will be discussed in Sect. 19.7.

Situation 19.24 Let $\mathcal{C}_i, \mathcal{F}_i$ $(i = 1, 2)$ be critical submanifold data and linear K-systems as in Situation 16.16, respectively. We assume $\mathfrak{G}_1 = \mathfrak{G}_2$ and denote the common group by $\mathfrak{G}$. Suppose also we are given partial linear K-systems $\mathcal{F}_i$ of energy cut level E_0 for $i = 1, 2$ and a τ-collared Kuranishi structure $\widehat{\mathcal{U}^+}(i; \alpha_{i-}, \alpha_{i+})$ on $\mathcal{M}^i(\alpha_{i-}, \alpha_{i+})^{\boxplus \tau_0}$ equipped with CF-perturbations $\widehat{\mathfrak{S}^+}(i; \alpha_{i-}, \alpha_{i+})$ for $i = 1, 2$ which satisfy the conclusion of Proposition 19.1. Here $0 < \tau < \tau_0 = 1$ as in Proposition 19.1 and $\mathcal{M}^i(\alpha_{i-}, \alpha_{i+})$ are as in Condition 16.17. (From now on, we write $\alpha_\pm$ in place of $\alpha_{i\pm}$ if no confusion can occur.) ■

Situation 19.25 Suppose we are in Situation 19.24.

(1) For $j = 1, 2$, we are given partial morphisms $\mathfrak{N}_j : \mathcal{F}_1 \to \mathcal{F}_2$ of energy cut level E_0 and energy loss c. We denote by $\mathcal{N}(j; \alpha_1, \alpha_2)$ its interpolation space.
(2) Suppose we are given a homotopy $\mathfrak{H}$ between partial morphisms $\mathfrak{N}_1$ and $\mathfrak{N}_2$. By its definition, it is a $[1, 2]$-parametrized family of partial morphisms from $\mathcal{F}_1$ to $\mathcal{F}_2$. Suppose its energy cut level is E_0 and energy loss is c. We denote its interpolation space by $\mathcal{N}(\alpha_1, \alpha_2; [1, 2])$. ■

Situation 19.26 Suppose we are in Situation 19.25. Let $0 < \tau' < \tau < \tau_0 = 1$.

Suppose also that, for $j = 1, 2$, we are given a τ'-collared Kuranishi structure and a τ'-collared CF-perturbation on $\mathcal{N}(j; \alpha_1, \alpha_2)^{\boxplus \tau_0}$ which satisfy the conclusions of Proposition 19.7. We denote them by $\widehat{\mathcal{U}^+}(\text{mor}, j; \alpha_1, \alpha_2)$, $\widehat{\mathfrak{S}^+}(\text{mor}, j; \alpha_1, \alpha_2)$. ■

Before we state the main result, we explicitly write the boundary and corner compatibility conditions for the case of $[1,2]^{\boxplus\tau_0}$-parametrized morphism below. (Note $[1,2]^{\boxplus\tau_0} = [1-\tau_0, 2+\tau_0]$.) The compatibility condition at the boundary (Condition 16.23) is as follows:

$$\begin{aligned}
&\partial\mathcal{N}(\alpha_1,\alpha_2;[1,2])^{\boxplus\tau_0}\\
&= \coprod_{\alpha_1'\in\mathfrak{A}_1} (-1)^{\dim\mathcal{N}(\alpha_1,\alpha_2;[1,2])}\mathcal{N}(\alpha_1',\alpha_2;[1,2])^{\boxplus\tau_0} \times_{R_{\alpha_1'}} \mathcal{M}^1(\alpha_1,\alpha_1')^{\boxplus\tau_0}\\
&\sqcup \coprod_{\alpha_2'\in\mathfrak{A}_2} (-1)^{\dim R_{\alpha_2'}+1}\mathcal{M}^2(\alpha_2',\alpha_2)^{\boxplus\tau_0} \times_{R_{\alpha_2'}} \mathcal{N}(\alpha_1,\alpha_2';[1,2])^{\boxplus\tau_0}\\
&\sqcup \mathcal{N}(2;\alpha_1,\alpha_2)^{\boxplus\tau_0} \sqcup -\mathcal{N}(1;\alpha_1,\alpha_2)^{\boxplus\tau_0}.
\end{aligned} \tag{19.36}$$

The first of the corner compatibility conditions (Condition 16.26) says that the normalized corner $\widehat{S}_k(\mathcal{N}(\alpha_1,\alpha_2;[1,2]))$ is the disjoint union of the following two types of fiber products:

$$\begin{aligned}
&\mathcal{M}^1(\alpha_-,\alpha_1)^{\boxplus\tau_0} \times_{R_{\alpha_1}} \cdots \times_{R_{\alpha_{k_1-1}}} \mathcal{M}^1(\alpha_{k_1-1},\alpha_{k_1})^{\boxplus\tau_0}\\
&\times_{R_{\alpha_{k_1}}} \mathcal{N}(\alpha_{k_1},\alpha_{k_1+1};[1,2])^{\boxplus\tau_0}\\
&\times_{R_{\alpha_{k_1+1}}} \mathcal{M}^2(\alpha_{k_1+1},\alpha_{k_1+2})^{\boxplus\tau_0} \times_{R_{\alpha_{k_1+2}}} \cdots \times_{R_{\alpha_{k_1+k_2}}} \mathcal{M}^2(\alpha_{k_1+k_2},\alpha_+)^{\boxplus\tau_0},
\end{aligned} \tag{19.37}$$

with $k_1+k_2=k$, and

$$\begin{aligned}
&\mathcal{M}^1(\alpha_-,\alpha_1)^{\boxplus\tau_0} \times_{R_{\alpha_1}} \cdots \times_{R_{\alpha_{k_1-1}}} \mathcal{M}^1(\alpha_{k_1-1},\alpha_{k_1})^{\boxplus\tau_0}\\
&\times_{R_{\alpha_{k_1}}} \mathcal{N}(j;\alpha_{k_1},\alpha_{k_1+1})^{\boxplus\tau_0}\\
&\times_{R_{\alpha_{k_1+1}}} \mathcal{M}^2(\alpha_{k_1+1},\alpha_{k_1+2})^{\boxplus\tau_0} \times_{R_{\alpha_{k_1+2}}} \cdots \times_{R_{\alpha_{k_1+k_2}}} \mathcal{M}^2(\alpha_{k_1+k_2},\alpha_+)^{\boxplus\tau_0},
\end{aligned} \tag{19.38}$$

with $k_1+k_2=k-1$, $j=1,2$.

The second of the corner compatibility conditions (Condition 16.28) says the following: Consider the ℓ-th normalized corner of the K-spaces (19.37). According to the descriptions of $\widehat{S}_n(\mathcal{N}(\alpha_1,\alpha_2;[1,2])^{\boxplus\tau_0})$, $\mathcal{N}(j;\alpha_1,\alpha_2)^{\boxplus\tau_0}$ and $\mathcal{M}^i(\alpha_1,\alpha_2)^{\boxplus\tau_0}$ we gave above, we see that the ℓ-th normalized corner of (19.37) or (19.38) is given by the disjoint union of the same type of fiber products as (19.37) or (19.38). Condition 16.28 requires that this description coincides with the description of $\widehat{S}_{k+\ell}(\mathcal{N}(\alpha_1,\alpha_2;[1,2])^{\boxplus\tau_0})$.

Proposition 19.27 *Suppose we are in Situations* 19.24, 19.25, 19.26 *and* $\tau'' < \tau'$. *Then for any* α_1,α_2 *with* $E(\alpha_2)-E(\alpha_1)\le E_0-c$, *there exist* $\widehat{\mathcal{U}^+}(\alpha_1,\alpha_2;[1,2])$ *and* $\widehat{\mathfrak{S}^+}(\alpha_1,\alpha_2;[1,2])$ *such that they enjoy the following properties:*

(1) $\widehat{\mathcal{U}^+}(\alpha_1, \alpha_2; [1,2])$ *is a* τ''*-collared Kuranishi structure of* $\mathcal{N}(\alpha_1, \alpha_2; [1,2])^{\boxplus\tau}$ *and* $\widehat{\mathfrak{S}^+}(\alpha_1, \alpha_2; [1,2])$ *is its* τ''*–collared CF-perturbation.*

(2) $\widehat{\mathfrak{S}^+}(\alpha_1, \alpha_2; [1,2])$ *is transversal to* 0*. Moreover an evaluation map*

$$(\mathrm{ev}_+, \mathrm{ev}_{[1,2]}) : (\mathcal{N}(\alpha_1, \alpha_2; [1,2])^{\boxplus\tau}, \widehat{\mathcal{U}^+}(\alpha_1, \alpha_2; [1,2])) \longrightarrow R_{\alpha_2} \times [1,2]^{\boxplus\tau_0}$$

is given and is strongly stratumwise submersive with respect to $\widehat{\mathfrak{S}^+}(\alpha_1, \alpha_2; [1,2])$.

(3) *We have periodicity isomorphisms among* $\widehat{\mathcal{U}^+}(\alpha_1, \alpha_2; [1,2])$*'s that are compatible with* $\widehat{\mathfrak{S}^+}(\alpha_1, \alpha_2; [1,2])$.

(4) *There exists an embedding of* τ''*-collared Kuranishi structures from* $\mathcal{N}(\alpha_1, \alpha_2; [1,2])^{\boxplus\tau}$ *to the* τ''*- collared Kuranishi structure* $\widehat{\mathcal{U}^+}(\alpha_1, \alpha_2; [1,2])$*. Here for the source* $\mathcal{N}(\alpha_1, \alpha_2; [1,2])^{\boxplus\tau}$ *we use the outer collaring of the Kuranishi structure induced by that of* $\mathcal{N}(\alpha_1, \alpha_2; [1,2])$ *which is given by the definition of* $[1,2]$*-parametrized interpolation space. This outer collaring gives a* τ''*-collared Kuranishi structure. The KK-embedding respects orientation isomorphisms, periodicity isomorphisms, and evaluation maps.*

(5) *There is an isomorphism of* τ''*-collared K-spaces*

$$\begin{aligned}
&\partial\widehat{\mathcal{U}^+}(\alpha_1, \alpha_2; [1,2]) \\
&= \coprod_{\alpha_1' \in \mathfrak{A}_1} (-1)^{\dim \widehat{\mathcal{U}^+}(\alpha_1,\alpha_2;[1,2])} \widehat{\mathcal{U}^+}(\alpha_1', \alpha_2; [1,2]) \times_{R_{\alpha_1'}} \widehat{\mathcal{U}^+}(1; \alpha_1, \alpha_1') \\
&\sqcup \coprod_{\alpha_2' \in \mathfrak{A}_2} (-1)^{\dim R_{\alpha_2'}+1} \widehat{\mathcal{U}^+}(2; \alpha_2', \alpha_2) \times_{R_{\alpha_2'}} \widehat{\mathcal{U}^+}(\alpha_1, \alpha_2'; [1,2]) \\
&\sqcup \widehat{\mathcal{U}^+}(\mathrm{mor}, 2; \alpha_1, \alpha_2) \sqcup -\widehat{\mathcal{U}^+}(\mathrm{mor}; 1; \alpha_1, \alpha_2).
\end{aligned} \tag{19.39}$$

The isomorphism (19.39) *is compatible with the isomorphism* (19.36) *via the KK-embedding (4). It is also compatible with the periodicity isomorphism and the evaluation maps.*

(6) *The pullback of* $\widehat{\mathfrak{S}^+}(\alpha_1, \alpha_2; [1,2])$ *by the isomorphism* (19.39) *is equivalent to the fiber product of* $\widehat{\mathfrak{S}^+}(j; \alpha, \alpha')$*, and* $\widehat{\mathfrak{S}^+}(\alpha, \alpha'; [1,2])$ *on the first two summands. It is isomorphic to* $\widehat{\mathfrak{S}^+}(\mathrm{mor}, 1; \alpha, \alpha')$ *(resp.* $\widehat{\mathfrak{S}^+}(\mathrm{mor}, 2; \alpha, \alpha')$ *) on the 4th (resp. 3rd) summand.*

(7) *The normalized corner* $\widehat{S}_k(\widehat{\mathcal{U}^+}(\alpha_1, \alpha_2; [1,2]))$ *is a disjoint union of the following two types of fiber products:*

$$\widehat{\mathcal{U}^+}(1;\alpha_-,\alpha_1) \times_{R_{\alpha_1}} \cdots \times_{R_{\alpha_{k_1-1}}} \widehat{\mathcal{U}^+}(1;\alpha_{k_1-1},\alpha_{k_1})$$
$$\times_{R_{\alpha_{k_1}}} \widehat{\mathcal{U}^+}(\alpha_{k_1},\alpha_{k_1+1};[1,2])$$
$$\times_{R_{\alpha_{k_1+1}}} \widehat{\mathcal{U}^+}(2;\alpha_{k_1+1},\alpha_{k_1+2}) \times_{R_{\alpha_{k_1+2}}} \cdots \times_{R_{\alpha_{k_1+k_2}}} \widehat{\mathcal{U}^+}(2;\alpha_{k_1+k_2},\alpha_+) \tag{19.40}$$

and

$$\widehat{\mathcal{U}^+}(1;\alpha_-,\alpha_1) \times_{R_{\alpha_1}} \cdots \times_{R_{\alpha_{k_1-1}}} \widehat{\mathcal{U}^+}(1;\alpha_{k_1-1},\alpha_{k_1})$$
$$\times_{R_{\alpha_{k_1}}} \widehat{\mathcal{U}^+}(\mathrm{mor};j;\alpha_{k_1},\alpha_{k_1+1})$$
$$\times_{R_{\alpha_{k_1+1}}} \widehat{\mathcal{U}^+}(2;\alpha_{k_1+1},\alpha_{k_1+2}) \times_{R_{\alpha_{k_1+2}}} \cdots \times_{R_{\alpha_{k_1+k_2}}} \widehat{\mathcal{U}^+}(2;\alpha_{k_1+k_2},\alpha_+). \tag{19.41}$$

This isomorphism is compatible with the isomorphism (19.37), (19.38) *via the KK-embedding (4). It is also compatible with the periodicity isomorphism and the evaluation maps.*

(8) *The pullback of the restriction of* $\widehat{\mathfrak{S}^+}(\alpha_1,\alpha_2;[1,2])$ *to* $\widehat{S}_k(\widehat{\mathcal{U}^+}(\alpha_1,\alpha_2;[1,2]))$ *by the isomorphism in (7) is equivalent to the fiber product of* $\widehat{\mathfrak{S}^+}(\mathrm{mor};j;*,*)$, $\widehat{\mathfrak{S}^+}(1;*,*)$, $\widehat{\mathfrak{S}^+}(2;*,*)$.

(9) *The isomorphism of (7) is compatible with the covering map*

$$\widehat{S}_\ell(\widehat{S}_k(\widehat{\mathcal{U}^+}(\alpha_1,\alpha_2;[1,2]))) \longrightarrow \widehat{S}_{k+\ell}(\widehat{\mathcal{U}^+}(\alpha_1,\alpha_2;[1,2])).$$

(The precise meaning of this compatibility is the same as the case of

$$\mathcal{N}(\alpha_1,\alpha_2;[1,2]),$$

which we explained right before this proposition.)

(10) *If we start from a uniform family of* $\widehat{\mathfrak{S}^+}(i;\alpha_-,\alpha_+)$ *and* $\widehat{\mathfrak{S}^+}(\mathrm{mor},j;\alpha_1,\alpha_2)$, *then we can take the family of* $\widehat{\mathfrak{S}^+}(\alpha_1,\alpha_2;[1,2])$ *to be uniform.*

Proof We prove Proposition 19.27 with E_0 replaced by $E^n_{\mathfrak{E}}$, by induction on n. The corner compatibility conditions Conditions 16.26 and Conditions 16.28 are written in such a way that they imply the assumptions of Proposition 17.58, which is Situation 17.55 (especially its (1)(2)), after appropriately modifying the τ-collared Kuranishi structures, in the same way as the proof of Proposition 19.1 (using Lemma 17.53). □

We rewrite the geometric conclusion of Proposition 19.27 into algebraic language.

Definition 19.28 In the situation of Proposition 19.27, we define

$$\mathfrak{h}^{\epsilon}_{\alpha_2,\alpha_1} : \Omega(R_{\alpha_1};o_{R_{\alpha_1}}) \longrightarrow \Omega(R_{\alpha_2};o_{R_{\alpha_2}}) \tag{19.42}$$

by

$$\mathfrak{h}^{\epsilon}_{\alpha_2,\alpha_1}(h) = (-1)^{\dim R_{\alpha_2}+\deg h}\mathrm{ev}_{+}!(\mathrm{ev}_{-}^{*}h; \widehat{\mathfrak{S}^{+\epsilon}}(\alpha_1,\alpha_2;[1,2])). \tag{19.43}$$

Here the right hand side is defined by Definition 17.82 on the K-space

$$(\mathcal{N}(\alpha_1,\alpha_2;[1,2])^{\boxplus\tau_0}, \widehat{\mathcal{U}^{+}}(\alpha_1,\alpha_2;[1,2])).$$

By Condition 16.21 (VI) and Definition 7.79, the degree of $\mathfrak{h}^{\epsilon}_{\alpha_2,\alpha_1}$ is $\eta(\alpha_2)-\eta(\alpha_1)-1$. Therefore after degree shift as in Definition 16.8 (2) its degree becomes -1.

If the energy loss of our homotopy is c, the family $\{\mathfrak{h}^{\epsilon}_{\alpha_2,\alpha_1}\}$ of maps induces

$$\mathfrak{F}^{\lambda}CF(\mathcal{F}_1) \to \mathfrak{F}^{\lambda-c}CF(\mathcal{F}_2),$$

where the filtration $\mathfrak{F}^{\lambda}$ is defined in Definition 16.8 (2)(3).

Suppose we are in the situation of Proposition 19.27. We have two morphisms of partial linear K-systems equipped with CF-perturbations. Namely, we have $\mathcal{N}(j;\alpha_1,\alpha_2)$, $\widehat{\mathcal{U}^{+}}(\mathrm{mor},j;\alpha_1,\alpha_2)$, $\widehat{\mathfrak{S}^{+}}(\mathrm{mor},j;\alpha_1,\alpha_2)$ for $j = 1,2$. We use Definition 19.9 for $j=1,2$ to obtain maps

$$\psi^{j,\epsilon}_{\alpha_2,\alpha_1} : \Omega(R_{\alpha_1}; o_{R_{\alpha_1}}) \to \Omega(R_{\alpha_2}; o_{R_{\alpha_2}}).$$

Lemma 19.29 *The linear maps $\{\mathfrak{h}^{\epsilon}_{\alpha_1,\alpha_2}\}$ satisfy the following equality in the sense of $(\flat)$ in Remark 19.4:*

$$\begin{aligned}
&d_0 \circ \mathfrak{h}^{\epsilon}_{\alpha_2,\alpha_1} + \mathfrak{h}_{\alpha_2,\alpha_1} \circ d_0 \\
&= -\sum_{\alpha_1'} \mathfrak{h}^{\epsilon}_{\alpha_2,\alpha_1'} \circ \mathfrak{m}^{1,\epsilon}_{1;\alpha_1',\alpha_1} - \sum_{\alpha_2'} \mathfrak{m}^{2,\epsilon}_{1;\alpha_2,\alpha_2'} \circ \mathfrak{h}^{2,\epsilon}_{\alpha_2',\alpha_1} - (\psi^{2,\epsilon}_{\alpha_2,\alpha_1} - \psi^{1,\epsilon}_{\alpha_2,\alpha_1}).
\end{aligned} \tag{19.44}$$

Here the first sum in the second line is taken over $\alpha_1' \in \mathfrak{A}_1$ with $E(\alpha_1) < E(\alpha_1') \le E(\alpha_2)+c$ and the second sum in the second line is taken over $\alpha_2' \in \mathfrak{A}_2$ with $E(\alpha_1)-c \le E(\alpha_2') < E(\alpha_2)$. The number c is the energy loss of our morphism.

We will define $\epsilon_{3,i}$ so that (19.44) holds for $\epsilon < \epsilon_{3,i}$.

Proof The proof is similar to the proof of Lemmas 19.5, 19.10. By Stokes' formula the left hand side is obtained from $\partial\widehat{\mathfrak{S}^{+\epsilon}}(\alpha_1,\alpha_2;[1,2])$ in the same way as (19.43). We can decompose the boundary $\partial\widehat{\mathfrak{S}^{+\epsilon}}(\alpha_1,\alpha_2;[1,2])$ into a disjoint union by Proposition 19.27. Then we use the composition formula to obtain the right hand side of (19.44). Namely the 1,2,3,4-th unions of (19.39) correspond to the 1,2,3,4-th terms of (19.44), respectively. □

19.6 Proof of Theorem 16.9 (2)(except (f)), Theorem 16.31 (1) and Theorem 16.39 (2)(except (e)), (3)

In this section we use the result of Sect. 19.5 to prove Theorem 16.9 (2) (a)–(e). We also prove Theorem 16.31 (1) and Theorem 16.39 (2)(3) (except (2)(e)) at the same time.

To prove Theorem 16.31 (1) we need to take the 'projective limit' in a similar way to the argument of Sect. 19.3. Proposition 19.32 below is its algebraic part. It is similar to Lemma 19.13 and is a baby version of [FOOO4, Lemma 7.2.129].

Definition 19.30 Suppose we are given two partial linear K-systems $\mathcal{F}_i$ $(i = 1, 2)$. Also suppose we are given $(CF(\mathcal{F}_i), \{\mathfrak{m}^i_{1;\alpha_+,\alpha_-}\})$, a partial cochain complex structure on $CF(\mathcal{F}_i)$ of energy cut level $E_{(i)}$ for $i = 1, 2$. We assume $0 \le c < E_0 \le E_{(1)}$ and $E_0 \le E_{(2)} - c$.

(1) A *partial cochain map* $\mathcal{F}_1 \to \mathcal{F}_2$ *of energy cut level* E_0 *and energy loss* c is a family $\widehat{\psi} = \{\psi_{\alpha_2,\alpha_1}\}$ consisting of the following objects ψ_{α_2,α_1}:
If $\alpha_i \in \mathfrak{A}_i$ and $E(\alpha_2) \le E(\alpha_1) + E_0 - c$, we have an $\mathbb{R}$ linear map

$$\psi_{\alpha_2,\alpha_1} : \Omega(R_{\alpha_1}; o_{R_{\alpha_1}}) \longrightarrow \Omega(R_{\alpha_2}; o_{R_{\alpha_2}}).$$

We require that it satisfies (19.14) for $E(\alpha_2) \le E(\alpha_1) + E_0 - c$.

(2) For $j = 1, 2$, let $\{\psi_{j;\alpha_2,\alpha_1}\}$ be partial cochain maps of energy cut level E_0 with energy loss c. A *partial cochain homotopy of energy cut level* E_0 *with energy loss* c *from* $\widehat{\psi}_1 = \{\psi_{1;\alpha_2,\alpha_1}\}$ *to* $\widehat{\psi}_2 = \{\psi_{2;\alpha_2,\alpha_1}\}$ is a family $\widehat{\mathfrak{h}} = \{\mathfrak{h}_{\alpha_2,\alpha_1}\}$ consisting of the following objects $\mathfrak{h}_{\alpha_2,\alpha_1}$:
If $\alpha_i \in \mathfrak{A}_i$ and $E(\alpha_2) \le E(\alpha_1) + E_0 - c$, we have an $\mathbb{R}$ linear map

$$\mathfrak{h}_{\alpha_2,\alpha_1} : \Omega(R_{\alpha_1}; o_{R_{\alpha_1}}) \longrightarrow \Omega(R_{\alpha_2}; o_{R_{\alpha_2}}).$$

We require that it satisfies (19.44) for $E(\alpha_2) \le E(\alpha_1) + E_0 - c$.

(3) For $i = 1, 2$, let $\widehat{\psi}_{i+1i} = \{\psi_{\alpha_{i+1},\alpha_i}\}$ be partial cochain maps of energy loss c_i. Its energy cut level is E_0 for $i = 1$ and $E_0 - c_1$ for $i = 2$. We define the *composition* $\widehat{\psi}_{31} = \widehat{\psi}_{32} \circ \widehat{\psi}_{21}$ by

$$(\psi_{31})_{\alpha_3\alpha_1} = \sum_{\alpha_2 \in \mathfrak{A}_2} (\psi_{32})_{\alpha_3\alpha_2} \circ (\psi_{21})_{\alpha_2\alpha_1}.$$

Then $\widehat{\psi}_{31}$ is a partial cochain map of energy cut level E_0 and energy loss $c_1 + c_2$.

(4) Consider E_0' such that $c < E_0' < E_0$. In the situation of (1) we forget all of ψ_{α_2,α_1} for $E(\alpha_2) \le E(\alpha_1) + E_0' - c$. We then obtain a partial cochain map $\mathcal{F}_1 \to \mathcal{F}_2$ of energy cut level E_0'. We call it the *energy cut of* $\widehat{\psi}$ *at energy cut level* E_0'.
The *energy cut of* $\widehat{\mathfrak{h}}$ *at energy cut level* E_0' is defined in the same way.

(5) Let $\widehat{\psi} : \mathcal{F}_1 \to \mathcal{F}_2$ be a partial cochain map of energy cut level E_0 and energy loss c and let $c < E_0' < E_0$. If $\widehat{\psi}'$ is an energy cut of $\widehat{\psi}$ at energy cut level E_0', we call $\widehat{\psi}$ a *promotion of* $\widehat{\psi}'$ *to the energy cut level* E_0. A promotion of a partial cochain homotopy is defined in the same way.

Lemma-Definition 19.31 *Two partial cochain maps are said to be cochain homotopic if there exists a partial cochain homotopy between them. This is an equivalence relation.*

Proof If $\widehat{\mathfrak{h}}^j = \{\mathfrak{h}^j_{\alpha_2\alpha_1}\}$ is a partial cochain homotopy from $\widehat{\psi}_j$ to $\widehat{\psi}_{j+1}$ for $j = 1, 2$, then $\{\mathfrak{h}^1_{\alpha_2\alpha_1} + \mathfrak{h}^2_{\alpha_2\alpha_1}\}$ is a partial cochain homotopy from $\widehat{\psi}_1$ to $\widehat{\psi}_3$. The other part of the proof is obvious. □

Proposition 19.32 *Let* $(CF(\mathcal{F}_i), \{\mathfrak{m}^i_{1;\alpha_+\alpha_-}\})$ *and* $(CF(\mathcal{F}_i'), \{\mathfrak{m}^{i\prime}_{1;\alpha_+\alpha_-}\})$ *be partial cochain complexes of energy cut level* E_2 *for* $i = 1, 2$. *We take* $E_1 < E_2$. *Suppose we have a diagram*

$$\begin{array}{ccc}
(CF(\mathcal{F}_1'), \{\mathfrak{m}^{1\prime}_{1;\alpha_+,\alpha_-}\}) & \xrightarrow{\widehat{\psi}_{21}'} & (CF(\mathcal{F}_2'), \{\mathfrak{m}^{2\prime}_{1;\alpha_+,\alpha_-}\}) \\
{\scriptstyle \widehat{\psi}_1}\uparrow & & \uparrow{\scriptstyle \widehat{\psi}_2} \\
(CF(\mathcal{F}_1), \{\mathfrak{m}^{1}_{1;\alpha_+,\alpha_-}\}) & \xrightarrow{\widehat{\psi}_{21}} & (CF(\mathcal{F}_2), \{\mathfrak{m}^{2}_{1;\alpha_+,\alpha_-}\})
\end{array} \tag{19.45}$$

such that:

(i) $\widehat{\psi}_{21}$, $\widehat{\psi}_{21}'$ *are partial cochain maps of energy cut level* E_2 *and energy loss* 0. *We assume that* $\widehat{\psi}_{21}$, $\widehat{\psi}_{21}'$ *induce isomorphisms modulo* T^{ϵ} *for a sufficiently small* $\epsilon > 0$.
(ii) $\widehat{\psi}_2$ *is a partial cochain map of energy cut level* E_2 *and energy loss* c.
(iii) $\widehat{\psi}_1$ *is a partial cochain map of energy cut level* E_1 *and energy loss* c.
(iv) *Diagram* 19.45 *is homotopy commutative as partial cochain maps of energy cut level* E_1 *and energy loss* c. *Namely, there exists a partial cochain homotopy* $\widehat{\mathfrak{h}} = \{\mathfrak{h}_{\alpha_2,\alpha_1}\}$, *where* $\mathfrak{h}_{\alpha_2,\alpha_1}$ *is defined when* $E(\alpha_2) \le E(\alpha_1) + E_2 - c$, *and satisfies*

$$\begin{aligned}
& d_0 \circ \mathfrak{h}_{\alpha_2',\alpha_1} + \mathfrak{h}_{\alpha_2',\alpha_1} \circ d_0 \\
&= - \sum_{\hat\alpha_1 \in \mathfrak{A}_1} \mathfrak{h}_{\alpha_2',\hat\alpha_1} \circ \mathfrak{m}^1_{1;\hat\alpha_1,\alpha_1} - \sum_{\hat\alpha_2' \in \mathfrak{A}_2'} \mathfrak{m}^2_{1;\alpha_2',\hat\alpha_2'} \circ \mathfrak{h}^2_{\hat\alpha_2',\alpha_1} \\
&\quad + \sum_{\hat\alpha_2 \in \mathfrak{A}_2} (\hat\psi 2)_{\alpha_2'\hat\alpha_2} \circ (\hat\psi 21)_{\hat\alpha_2\alpha_1} - \sum_{\hat\alpha_1' \in \mathfrak{A}_1'} (\hat\psi_{21}')_{\alpha_2'\hat\alpha_1'} \circ (\widehat{\psi} 1)_{\hat\alpha_1',\alpha_1},
\end{aligned} \tag{19.46}$$

if $E(\alpha_2') \le E(\alpha_1) + E_1 - c$.

Then we can promote $\widehat{\psi}_1$ to a partial cochain map of energy cut level E_2 and energy loss c and $\widehat{\mathfrak{h}}$ to a partial cochain homotopy of energy cut level E_2 and energy loss c. That is, (19.46) *holds if $E(\alpha'_2) \le E(\alpha_1) + E_2 - c$.*

We use the next proposition for the proof.

Proposition 19.33 *Let $\mathscr{F}_i$ ($i = 1, 2$) be partial cochain complexes of energy cut level E_2 and D_j ($j = 1, 2$) be cochain complexes over the ground ring R, which is a field. Let $D_2 \to D_1$ be an injective cochain map which induces isomorphism in cohomology. Let $\phi_j : \mathscr{F}_1 \otimes D_j \to \mathscr{F}_2$ be a partial cochain map of energy cut at level E_j. We assume $E_1 < E_2$ and energy cut of ϕ_2 by E_1 coincides with the restriction of ϕ_1.*

Then there exists a partial cochain map $\hat{\phi}_1$ of energy cut level E_2 such that:

(1) *The energy cut of $\hat{\phi}_1$ at level E_1 coincides with ϕ_1.*
(2) *The restriction to $\mathscr{F}_1 \otimes D_2$ of $\hat{\phi}_1$ is ϕ_2.*

Proof We consider the following commutative diagram:

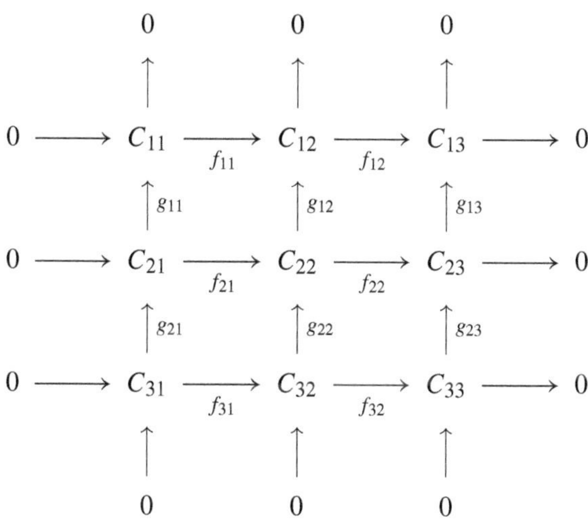

where

$$C_{11} = Hom(\mathscr{F}_1 \otimes D_1/D_2, \mathscr{F}_2) \otimes \Lambda_0/T^{E_1}\Lambda_0,$$

$$C_{12} = Hom(\mathscr{F}_1 \otimes D_1, \mathscr{F}_2) \otimes \Lambda_0/T^{E_1}\Lambda_0,$$

$$C_{13} = Hom(\mathscr{F}_1 \otimes D_2, \mathscr{F}_2) \otimes \Lambda_0/T^{E_1}\Lambda_0,$$

$$C_{21} = Hom(\mathscr{F}_1 \otimes D_1/D_2, \mathscr{F}_2) \otimes \Lambda_0/T^{E_2}\Lambda_0,$$

$$C_{22} = Hom(\mathscr{F}_1 \otimes D_1, \mathscr{F}_2) \otimes \Lambda_0/T^{E_2}\Lambda_0,$$

$$C_{23} = Hom(\mathscr{F}_1 \otimes D_2, \mathscr{F}_2) \otimes \Lambda_0/T^{E_2}\Lambda_0,$$

$$C_{31} = Hom(\mathscr{F}_1 \otimes D_1/D_2, \mathscr{F}_2) \otimes T^{E_1}\Lambda_0/T^{E_2}\Lambda_0,$$

$$C_{32} = Hom(\mathscr{F}_1 \otimes D_1, \mathscr{F}_2) \otimes T^{E_1}\Lambda_0/T^{E_2}\Lambda_0,$$

$$C_{33} = Hom(\mathscr{F}_1 \otimes D_2, \mathscr{F}_2) \otimes T^{E_1}\Lambda_0/T^{E_2}\Lambda_0.$$

Here the vertical and horizontal lines are all exact and C_{11}, C_{21}, C_{31} are acyclic.

By assumption we have cycles $x_1 \in C_{12}$, $x_2 \in C_{23}$ with $f_{12}(x_1) = g_{13}(x_2)$. We look for a cycle $y \in C_{22}$, such that $g_{12}(y) = x_1$, $f_{22}(y) = x_2$. We can find such y by a diagram chase. □

***Proof of Proposition* 19.32** We take $\mathscr{F}_i = (CF(\mathcal{F}_i), \{\mathfrak{m}^i_{1;\alpha_+,\alpha_-}\})$. D_1 is a simplicial chain complex associated to an interval $[1, 2]$ and D_2 is its subcomplex corresponding to the vertex $\{1\}$. Let v_1, v_2 be the generator of D_1 corresponding to $\{1\}$, $\{2\}$ and e the generator corresponding to $[1, 2]$. We put

$$\phi_1(x \otimes \mathrm{e}) = \widehat{\mathfrak{h}}(x), \quad \phi_1(x \otimes \mathrm{v}_2) = (\widehat{\psi'}_{21} \circ \widehat{\psi}_1)(x), \quad \phi_1(x \otimes \mathrm{v}_1) = (\widehat{\psi}_2 \circ \widehat{\psi}_{21})(x)$$

and $\phi_2(x \otimes \mathrm{v}_1) = (\widehat{\psi}_2 \circ \widehat{\psi}_{21})(x)$. See Fig. 19.3. Thus we can apply Proposition 19.33 to promote $\widehat{\mathfrak{h}}$ and $\widehat{\psi'}_{21} \circ \widehat{\psi}_1$ to energy cut level E_2. Since $\widehat{\psi'}_{21}$ is an isomorphism we can promote $\widehat{\psi}_1$ to energy cut level E_2. □

***Proof of Theorem* 16.39 (3)** Let $\mathcal{FF}_j$ $(j = a, b)$ be inductive systems of partial linear K-systems which consist of partial linear K-systems $\mathcal{F}^i_j$, $i = 1, 2, \dots$ and

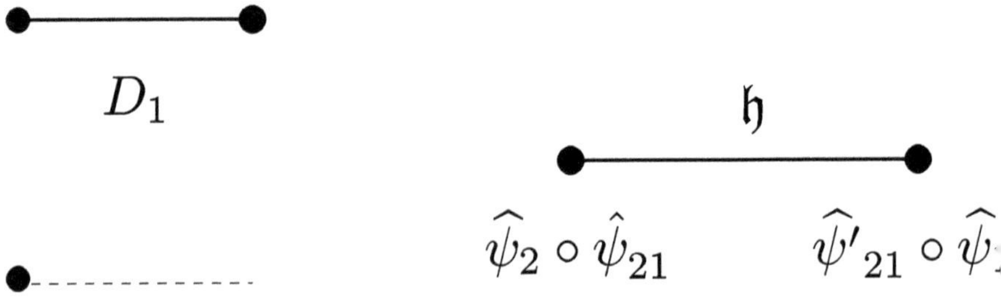

Fig. 19.3 D_1, D_2, ϕ_1, ϕ_2: 1

partial morphisms $\mathfrak{N}_j^{i+1i} : \mathcal{F}_j^i \to \mathcal{F}_j^{i+1}$. The morphism $\mathcal{FF}_a \to \mathcal{FF}_b$ consists of morphisms $\mathfrak{NN}^i : \mathcal{F}_a^i \to \mathcal{F}_b^i$ and homotopy. Namely, we have a homotopy commutative diagram of partial morphisms:

$$\begin{array}{ccc} \mathcal{F}_b^i & \xrightarrow{\mathfrak{N}_b^{i+1i}} & \mathcal{F}_b^{i+1} \\ {\scriptstyle \mathfrak{NN}^i}\uparrow & & \uparrow{\scriptstyle \mathfrak{NN}^{i+1}} \\ \mathcal{F}_a^i & \xrightarrow{\mathfrak{N}_a^{i+1i}} & \mathcal{F}_a^{i+1} \end{array} \tag{19.47}$$

In the proof of Theorem 16.39 (1) given in Sect. 19.3, we made a choice of

$$\widehat{\mathcal{U}^+}(j,i;\alpha_-,\alpha_+), \quad \widehat{\mathfrak{S}^+}(j,i;\alpha_-,\alpha_+)$$

that are Kuranishi structures of the spaces of connecting orbits and their CF-perturbations to define partial cochain complex $(CF(\mathcal{F}_j^i), \{\mathfrak{m}_{1;\alpha_+\alpha_-}^{ji;\epsilon}\})$ and this partial cochain complex is defined from $\mathcal{F}_j^i$. We also made a choice of

$$\widehat{\mathcal{U}^+}(\text{mor}; j,i,i+1;\alpha_-,\alpha_+), \quad \widehat{\mathfrak{S}^+}(\text{mor}; j,i,i+1;\alpha_-,\alpha_+;\rho_j^{ii+1})$$

that are Kuranishi structures of the spaces of interpolation spaces of $\mathfrak{N}_j^{i+1i}$ and their CF-perturbations. Here the parameter $\rho = \rho_j^{ii+1} \in (0,1]$ is taken as follows. We need to fix CF-perturbations of $\widehat{\mathcal{U}^+}(j,i;\alpha_-,\alpha_+)$ and of $\widehat{\mathcal{U}^+}(j,i+1;\alpha_-,\alpha_+)$ with which our CF-perturbations on the interpolation spaces are compatible. On one of those Kuranishi structures, we take $\widehat{\mathfrak{S}^+}(j,i;\alpha_-,\alpha_+)$. On the other Kuranishi structure, we take $\epsilon \mapsto \widehat{\mathfrak{S}^+}(j,i+1;\alpha_-,\alpha_+)^{\rho\epsilon}$. The parameter ρ appears here. Using these choices, we obtain partial cochain maps at the horizontal arrows in Diagram (19.47).

Next we apply Proposition 19.7 to the vertical arrows of Diagram (19.47). Then we can choose

$$\widehat{\mathcal{U}^+}(\text{mor}; ab,i;\alpha_-,\alpha_+), \quad \widehat{\mathfrak{S}^+}(\text{mor}; ab,i;\alpha_-,\alpha_+;\rho_{ab}^i),$$

and

$$\widehat{\mathcal{U}^+}(\text{mor}; ab,i+1;\alpha_-,\alpha_+), \quad \widehat{\mathfrak{S}^+}(\text{mor}; ab,i+1;\alpha_-,\alpha_+;\rho_{ab}^{i+1}),$$

that are Kuranishi structures and CF-perturbations of the interpolation spaces of $\mathfrak{NN}^i$, $\mathfrak{NN}^{i+1}$, respectively.

For each i we take $\epsilon_{bi} \le \epsilon_{ai} \le \epsilon_{4,i}$ such that $\epsilon_{ai+1} \le \epsilon_{ai}$, $\epsilon_{bi+1} \le \epsilon_{bi}$. (We specify our choice of $\epsilon_{4,i}$ later.) We take $\rho_j^{ii+1} = \epsilon_{ji+1}/\epsilon_{ji}$. We put

$$CF^i_j := \left(CF(\mathcal{F}^i_j), \{\mathfrak{m}^{ji;\epsilon}_{1;\alpha_+\alpha_-}\}\right). \tag{19.48}$$

Then we have a diagram of partial cochain complexes:

$$\begin{array}{ccc} CF^i_b & \longrightarrow & CF^{i+1}_b \\ \uparrow & & \uparrow \\ CF^i_a & \longrightarrow & CF^{i+1}_a \end{array} \tag{19.49}$$

By construction, the vertical arrows are partial cochain maps of energy cut level E^i and E^{i+1}, respectively.

We now use Proposition 19.27 to obtain Kuranishi structures and CF-perturbations on the interpolation spaces of the homotopy of Diagram (19.47) which are compatible with the choices we had made for the interpolation spaces of the arrows of Diagram (19.47) and the space of connecting orbits. (We had made the choice of them already as explained above.) We can take $\epsilon_{4,i}$ so small that this choice of Kuranishi structures and CF-perturbations determines a cochain homotopy which makes Diagram (19.49) commutative up to cochain homotopy. (This is a consequence of Lemma 19.29.) Recall from Remark 19.15 that the numbers $\epsilon_{0,i}, \epsilon_{1,i}, \epsilon_{2,i}, \epsilon_{3,i}$ in (19.26) are already chosen. We will need $\epsilon_{4,i}, \epsilon_{5,i}$ etc. for homotopy of homotopies etc. (However, we need only finitely many of them.)

Therefore by Proposition 19.32 we can promote Diagram (19.49) to a homotopy commutative diagram of the partial cochain maps of energy cut level E_{i+1}.

The rest of the proof is purely algebraic. We now consider the following diagram:

$$\begin{array}{ccccccccc} CF^1_b & \longrightarrow & \cdots & \longrightarrow & CF^i_b & \longrightarrow & CF^{i+1}_b & \longrightarrow & \cdots \\ \uparrow & & & & \uparrow & & \uparrow & & \\ CF^1_a & \longrightarrow & \cdots & \longrightarrow & CF^i_a & \longrightarrow & CF^{i+1}_a & \longrightarrow & \cdots \end{array} \tag{19.50}$$

Note that our construction of cochain complex $CF(\mathcal{F}_a)$ is done by promoting the horizontal lines inductively to those of energy cut level E_k and taking the limit $k \to \infty$. We do so for both of the horizontal lines. Then we continue to promote the vertical lines so that the whole diagram becomes homotopy commutative. Thus we obtain a cochain map $CF(\mathcal{F}_a) \to CF(\mathcal{F}_b)$. This finishes the proof of Theorem 16.39 (3). □

Proof of Theorem* 16.39 *(2) (a)–(d) When we proved Theorem 16.39 (1) in Sect. 19.3, we made the following choices:

(1) For each i the pairs $\left(\widehat{\mathcal{U}^+}(i;\alpha_-,\alpha_+), \widehat{\mathfrak{S}^+}(i;\alpha_-,\alpha_+)\right)$ for various α_-, α_+ that are a system of Kuranishi structures of the spaces of connecting orbits and their CF-perturbations define a partial cochain complex

$$(CF(\mathcal{F}^i), \{\mathfrak{m}^{i;\epsilon}_{1;\alpha_+\alpha_-}\})$$

and this partial cochain complex is defined from the partial linear K-system $\mathcal{F}^i$ with energy cut level $E^{k_i}_{\mathfrak{E}}$.

(2) For each i the pairs

$$\left(\widehat{\mathcal{U}^+}(\text{mor}; i, i+1; \alpha_-, \alpha_+), \widehat{\mathfrak{S}^+}(\text{mor}; i, i+1; \alpha_-, \alpha_+; \rho^{ii+1}_j)\right)$$

for various α_-, α_+ are a system of Kuranishi structures of the interpolation spaces of $\mathfrak{N}^{i+1i}$ and their CF-perturbations. Here the parameter $\rho = \rho^{ii+1}_j \in (0, 1]$ is as explained during the proof of Theorem 16.39 (3).

(3) The small numbers ϵ_i.

We then define $\left(CF(\mathcal{F}^i), \{\mathfrak{m}^{i;\epsilon_i}_{1;\alpha_+\alpha_-}\}\right)$ by using the choices (1),(3). Moreover, using the choices (2), (3), we define partial cochain maps

$$\widehat{\psi}^i \;:\; \left(CF(\mathcal{F}^i), \{\mathfrak{m}^{i;\epsilon_i}_{1;\alpha_+\alpha_-}\}\right) \longrightarrow \left(CF(\mathcal{F}^{i+1}), \{\mathfrak{m}^{i+1;\epsilon_{i+1}}_{1;\alpha_+\alpha_-}\}\right). \tag{19.51}$$

Finally, using (19.51) and an algebraic result (Lemma 19.13), we promote partial cochain complexes $\left(CF(\mathcal{F}^i), \{\mathfrak{m}^{i;\epsilon_i}_{1;\alpha_+\alpha_-}\}\right)$ and partial cochain maps $\widehat{\psi}^i$ to cochain complexes and cochain maps. This algebraic process also involves choices.

We will prove that the resulting cochain complex is independent of those choices up to cochain homotopy equivalence. Let

$$\left(\widehat{\mathcal{U}^+}(ji; \alpha_-, \alpha_+), \widehat{\mathfrak{S}^+}(ji; \alpha_-, \alpha_+)\right)$$

and

$$\left(\widehat{\mathcal{U}^+}(\text{mor}, j; i, i+1; \alpha_-, \alpha_+), \widehat{\mathfrak{S}^+}(\text{mor}, j; i, i+1; \alpha_-, \alpha_+; \rho^{ii+1}_j)\right),$$

and ϵ_{ji} be two choices, where $j = a, b$.

We consider the next diagram, which defines a morphism $\mathcal{FF} \to \mathcal{FF}$:

$$\begin{array}{ccccccccc}
\mathcal{F}^1 & \xrightarrow{\mathfrak{N}^{21}} & \cdots & \xrightarrow{\mathfrak{N}^{ii-1}} & \mathcal{F}^i & \xrightarrow{\mathfrak{N}^{i+1i}} & \mathcal{F}^{i+1} & \xrightarrow{\mathfrak{N}^{i+2i+1}} & \cdots \\
\uparrow \mathfrak{ID} & & & & \uparrow \mathfrak{ID} & & \uparrow \mathfrak{ID} & & \\
\mathcal{F}^1 & \xrightarrow{\mathfrak{N}^{21}} & \cdots & \xrightarrow{\mathfrak{N}^{ii-1}} & \mathcal{F}^i & \xrightarrow{\mathfrak{N}^{i+1i}} & \mathcal{F}^{i+1} & \xrightarrow{\mathfrak{N}^{i+2i+1}} & \cdots
\end{array} \tag{19.52}$$

The homotopy commutativity of Diagram (19.52) follows from Proposition 18.60.

Now we apply Theorem 16.39 (3). Namely, we apply the choice we made for $j = b$ for the first line and the choice we made for $j = a$ for the second line. We also use the particular way to promote the inductive system of partial cochain complexes which we used for choices $j = b$ and $j = a$. (We obtain those cochain complexes from the first and second lines of Diagram (19.52).)

Now we apply Proposition 19.32 and obtain a cochain map $CF_a \to CF_b$. Here CF_a (resp. CF_b) is the cochain complex we obtain by this promotion of the second line (resp. first line). The proof of Theorem 16.39 (2) (except (e)) is complete. □

Proof of Theorem* 16.9 (2) *(a)–(e) This is nothing but a special case of Theorem 16.39 (2) where $\mathfrak{N}^{i+1i}$ is the identity morphism. □

Proof of Theorem* 16.31 *(1) We can use Proposition 18.60 to obtain the following homotopy commutative diagram:

$$\begin{array}{ccccccccc}
\mathcal{F}_2 & \xrightarrow{\mathfrak{ID}} & \cdots & \xrightarrow{\mathfrak{ID}} & \mathcal{F}_2 & \xrightarrow{\mathfrak{ID}} & \mathcal{F}_2 & \xrightarrow{\mathfrak{ID}} & \cdots \\
\uparrow \mathfrak{N} & & & & \uparrow \mathfrak{N} & & \uparrow \mathfrak{N} & & \\
\mathcal{F}_1 & \xrightarrow{\mathfrak{ID}} & \cdots & \xrightarrow{\mathfrak{ID}} & \mathcal{F}_1 & \xrightarrow{\mathfrak{ID}} & \mathcal{F}_1 & \xrightarrow{\mathfrak{ID}} & \cdots
\end{array} \tag{19.53}$$

In fact, we have $\mathfrak{ID} \circ \mathfrak{N} \sim \mathfrak{N} \sim \mathfrak{N} \circ \mathfrak{ID}$, by Proposition 18.60.

Following the proof of Theorem 16.9 (1), we make the following choices:

(1) For each $j = 1, 2$ we regard $\mathcal{F}_j$ as a partial linear K-system of energy cut level E^i. We write it as $\mathcal{F}_j^i$.
(2) We take $\widehat{\mathcal{U}^+}(j, i; \alpha_-, \alpha_+)$ and $\widehat{\mathfrak{S}^+}(j, i; \alpha_-, \alpha_+)$ that are a Kuranishi structure of the space of connection orbits of $\mathcal{F}_j^i$ and its CF-perturbation, respectively. They satisfy the conclusion of Proposition 19.1.
(3) We take $\widehat{\mathcal{U}^+}(\mathrm{mor}, j, ii+1; \alpha_-, \alpha_+)$ and $\widehat{\mathfrak{S}^+}(\mathrm{mor}, j, ii+1; \alpha_-, \alpha_+)$ that are a Kuranishi structure of the interpolation space of $\mathfrak{ID} : \mathcal{F}_j^i \to \mathcal{F}_j^{i+1}$ and its CF-perturbation, respectively. They satisfy the conclusion of Proposition 19.7.
(4) We also take $\widehat{\mathcal{U}^+}(\mathrm{mor}, 12, i; \alpha_1, \alpha_2)$ and $\widehat{\mathfrak{S}^+}(\mathrm{mor}, 12, i; \alpha_1, \alpha_2)$ that are a Kuranishi structure of the interpolation space of $\mathfrak{N} : \mathcal{F}_1^i \to \mathcal{F}_2^i$ and its CF-perturbation, respectively. They satisfy the conclusion of Proposition 19.7.
(5) We take $\widehat{\mathcal{U}^+}(\mathrm{mor}, i; \alpha_-, \alpha_+; [0, 1])$ and $\widehat{\mathfrak{S}^+}(\mathrm{mor}, i; \alpha_-, \alpha_+; [0, 1])$ that are a Kuranishi structure of the interpolation space of the homotopy $\mathfrak{ID} \circ \mathfrak{N} \sim \mathfrak{N} \circ \mathfrak{ID} : \mathcal{F}_1^i \to \mathcal{F}_2^{i+1}$ and its CF-perturbation, respectively. The Kuranishi structure is compatible with ones in Items (1)(2)(3)(4) at the boundary.

By these choices the geometric Diagram (19.53) is converted to the algebraic diagram below:

$$\begin{array}{ccc}
\left(CF(\mathcal{F}_2), \{\mathfrak{m}_{1;\alpha_+,\alpha_-}^{2i\epsilon_{2i}}\}\right) & \xrightarrow{\widehat{\psi}_2^{i+1i}} & \left(CF(\mathcal{F}_2), \{\mathfrak{m}_{1;\alpha_+,\alpha_-}^{2i\epsilon_{2i+1}}\}\right) \\
\mathfrak{n}^i \uparrow & & \uparrow \mathfrak{n}^{i+1} \\
\left(CF(\mathcal{F}_1), \{\mathfrak{m}_{1;\alpha_+,\alpha_-}^{1i,\epsilon_{1i}}\}\right) & \xrightarrow{\widehat{\psi}_1^{i+1i}} & \left(CF(\mathcal{F}_1), \{\mathfrak{m}_{1;\alpha_+,\alpha_-}^{1i+1,\epsilon_{1i+1}}\}\right)
\end{array} \tag{19.54}$$

Here:

- $\left(CF(\mathcal{F}_j), \{\mathfrak{m}_{1;\alpha_+,\alpha_-}^{ji,\epsilon_{ji}}\}\right)$ is obtained by the choice (2) above.
- $\widehat{\psi}_j^{i+1i}$, $j = 1, 2$ is obtained by the choice (3) above.
- $\mathfrak{n}^i$ is obtained by the choice (4) above.
- By the choice (5) above we obtain a cochain homotopy between $\widehat{\psi}_2^{i+1i} \circ \mathfrak{n}^i$ and $\mathfrak{n}^{i+1} \circ \widehat{\psi}_1^{i+1i}$. We denote it by $\mathfrak{h}_i$.

Here we use Lemmas 19.21 and 19.22 to prove that the composition $\widehat{\psi}_2^{i+1i} \circ \mathfrak{n}^i$ (resp. $\mathfrak{n}^{i+1} \circ \widehat{\psi}_1^{i+1i}$) is the cochain map associated to the morphism that is the composition of the identity morphism and $\mathfrak{N}^i$ (resp. $\mathfrak{N}^i$ and the identity morphism).

We note that at this stage the energy cut level of the objects (partial cochain maps and partial cochain complexes) in the right vertical line is E^{i+1} and the energy cut level of all the other objects in (19.54) (and the partial cochain homotopy $\mathfrak{h}_i$) is E^i.

Now we start promoting the objects in Diagram (19.54). First according to the proof of Theorem 16.9 (1), we promote all the objects in the upper and lower horizontal lines in Diagram (19.54) to a cochain complex and cochain map. (Namely we promote them to the energy cut level ∞.)

Thus all the objects other than $\mathfrak{n}^i$, $\mathfrak{n}^{i+1}$ and the cochain homotopy in Diagram (19.54) are promoted to the energy cut level ∞. We now use the fact that the energy cut level of $\mathfrak{n}^{i+1}$ is E^{i+1} to promote $\mathfrak{n}^i$ and $\mathfrak{h}^i$ to the energy cut level E^{i+1}. We use Proposition 19.32 to do so.

Then by induction using Diagrams (19.54) for all i, we can promote $\mathfrak{n}^i$, $\mathfrak{n}^{i+1}$ to the energy cut level ∞ so that Diagram (19.54) commutes up to cochain homotopy. The proof of Theorem 16.31 (1) is complete. □

19.7 Construction of Higher Homotopy

In this section, we generalize Proposition 19.27 to the case of P-parametrized morphisms. Situation 19.36 we assume is lengthy. However, its meaning is simple. Let $\mathcal{N}(\alpha_1, \alpha_2; P)$ be a P parametrized morphism. We assume on the boundary ∂P we are given a system of CF-perturbations with appropriate transversality property. Proposition 19.37 claims that it can be extended to a CF-perturbation of P parametrized morphism under appropriate conditions. More precisely, we need to replace the given Kuranishi structure on $\mathcal{N}(\alpha_1, \alpha_2; P)$ with a thickening of its outer collaring. The condition for CF-perturbations to be extended, is appropriate compatibility of given CF-perturbations on the boundary and at the corners. Situation 19.36 describes such a compatibility condition precisely. The detail follows.

We first define:

Definition 19.34 Suppose we are given a P-parametrized family of morphisms of (partial) linear K-system. We consider the evaluation map

$$\mathrm{ev}_P : \mathcal{N}(\alpha_1, \alpha_2; P) \longrightarrow P,$$

from its interpolation space. The inverse image $\mathrm{ev}_P^{-1}(\partial P)$ is a part of the boundary $\partial\mathcal{N}(\alpha_1, \alpha_2; P)$, which we denote by $\partial_{\mathfrak{C}^v}$ and call the *vertical boundary*. We put

$$\partial_{\mathfrak{C}^h}\mathcal{N}(\alpha_1, \alpha_2; P) := \partial\mathcal{N}(\alpha_1, \alpha_2; P) \setminus \partial_{\mathfrak{C}^v}\mathcal{N}(\alpha_1, \alpha_2; P)$$

and call it the *horizontal boundary*.

Horizontal and vertical boundaries induce one on the outer collaring $\mathcal{N}(\alpha_1, \alpha_2; P)^{\boxplus\tau_0}$. We denote them by the same symbol.

We remark that

$$\begin{aligned} \widehat{S}_k^{\mathfrak{C}^v}(\mathcal{N}(\alpha_1, \alpha_2; P)) &= \mathcal{N}(\alpha_1, \alpha_2; \widehat{S}_k(P)), \\ \widehat{S}_k^{\mathfrak{C}^v}(\mathcal{N}(\alpha_1, \alpha_2; P)^{\boxplus\tau_0}) &= \mathcal{N}(\alpha_1, \alpha_2; \widehat{S}_k(P^{\boxplus\tau_0})). \end{aligned} \tag{19.55}$$

Notation 19.35 In Situation 19.36 and Proposition 19.37 below we call the totality of periodicity isomorphisms, orientation isomorphisms, evaluation maps, boundary compatibility isomorphisms, and corner compatibility isomorphisms, the *structure isomorphisms*.

Situation 19.36 Suppose we are given a P-parametrized family of morphisms of a (partial) linear K-system. Let E_0 be its energy cut level and c be its energy loss. Let $\mathcal{N}(\alpha_1, \alpha_2; P)$ be its interpolation spaces.

We assume that we are given a family of the pairs

$$\widehat{\mathcal{U}^+}(\alpha_1, \alpha_2; \widehat{S}_k(P)), \qquad \widehat{\mathfrak{S}^+}(\alpha_1, \alpha_2; \widehat{S}_k(P))$$

for $k \geq 1$, $\alpha_i \in \mathfrak{A}_i$ with $E(\alpha_2) - E(\alpha_1) \leq E_0 - c$ and $0 < \tau < \tau_0$, with the following properties:

(1) $\widehat{\mathcal{U}^+}(\alpha_1, \alpha_2; \widehat{S}_k(P))$ is a τ-collared Kuranishi structure on the underlying topological space of $\mathcal{N}(\alpha_1, \alpha_2; S_k(P))^{\boxplus\tau_0}$. It comes with the structure isomorphisms, by which it becomes the interpolation space of an $S_k(P)^{\boxplus\tau_0}$ parametrized morphism, for each $k \geq 1$.

(2) We assume

$$\mathcal{N}(\alpha_1, \alpha_2; S_k(P))^{\boxplus\tau_0} < \widehat{\mathcal{U}^+}(\alpha_1, \alpha_2; \widehat{S}_k(P)). \tag{19.56}$$

Here we equip $\mathcal{N}(\alpha_1, \alpha_2; P)$ with the Kuranishi structure given in Definition 19.34 and the left hand side is the outer collaring of its restriction to the (vertical) corner. This KK-embedding respects the structure isomorphisms.

(3) $\widehat{\mathfrak{S}^+}(\alpha_1, \alpha_2; \widehat{S}_k(P))$ is a τ-collared CF-perturbation of the Kuranishi structure $\widehat{\mathcal{U}^+}(\alpha_1, \alpha_2; \widehat{S}_k(P))$. It is transversal to 0 and the evaluation map

$$(\mathrm{ev}_+, \mathrm{ev}_{S_k(P)}) : \widehat{\mathcal{U}^+}(\alpha_1, \alpha_2; \widehat{S}_k(P)) \longrightarrow R_{\alpha_2} \times \widehat{S}_k(P^{\boxplus\tau_0})$$

is stratified strongly submersive with respect to $\widehat{\mathfrak{S}^+}(\alpha_1, \alpha_2; \widehat{S}_k(P))$. The CF-perturbations $\widehat{\mathfrak{S}^+}(\alpha_1, \alpha_2; \widehat{S}_k(P))$ are compatible with the boundary and corner compatibility isomorphisms in Conditions 16.23, 16.26, 16.28.

(4) We are given a $(k+\ell)!/k!\ell!$-fold covering:

$$\begin{aligned} &\widehat{S}^{\mathfrak{C}^v}_\ell(\mathcal{N}(\alpha_1, \alpha_2; S_k(P))^{\boxplus\tau_0}, \widehat{\mathcal{U}^+}(\alpha_1, \alpha_2; \widehat{S}_k(P))) \\ &\to (\mathcal{N}(\alpha_1, \alpha_2; S_{k+\ell}(P))^{\boxplus\tau_0}, \widehat{\mathcal{U}^+}(\alpha_1, \alpha_2; \widehat{S}_{k+\ell}(P))) \end{aligned} \tag{19.57}$$

of τ-collared K-spaces. (19.57) is compatible with the structure isomorphisms.[5]

(5) The covering map (19.57) is compatible with the covering map

$$S^{\mathfrak{C}^v}_\ell(S^{\mathfrak{C}^v}_k(\mathcal{N}(\alpha_1, \alpha_2; P)^{\boxplus\tau_0})) \to S^{\mathfrak{C}^v}_{\ell+k}(\mathcal{N}(\alpha_1, \alpha_2; P)^{\boxplus\tau_0}) \tag{19.58}$$

via (19.56).[6] The covering map (19.58) is given in Proposition 18.6.

(6) The restriction of CF-perturbation $\widehat{\mathfrak{S}^+}(\alpha_1, \alpha_2; \widehat{S}_k(P))$ to

$$\widehat{S}^{\mathfrak{C}^v}_\ell(\mathcal{N}(\alpha_1, \alpha_2; S_k(P))^{\boxplus\tau_0}, \widehat{\mathcal{U}^+}(\alpha_1, \alpha_2; \widehat{S}_k(P)))$$

is equivalent to the pullback of $\widehat{\mathfrak{S}^+}(\alpha_1, \alpha_2; \widehat{S}_{k+\ell}(P))$ by (19.57).

(7) The next diagram commutes. (In this diagram we omit underlying topological space from the notation.)

$$\begin{array}{ccc} \widehat{S}^{\mathfrak{C}^v}_\ell(\widehat{S}^{\mathfrak{C}^v}_m(\mathcal{U}^+(\alpha_1, \alpha_2; \widehat{S}_k(P))) & \xrightarrow{\pi_{\ell,m}} & \widehat{S}^{\mathfrak{C}^v}_{\ell+m}(\mathcal{U}^+(\alpha_1, \alpha_2; \widehat{S}_k(P)) \\ \downarrow {\scriptstyle \widehat{S}^{\mathfrak{C}^v}_\ell((19.57))} & & \downarrow {\scriptstyle (19.57)} \\ \widehat{S}^{\mathfrak{C}^v}_\ell(\mathcal{U}^+(\alpha_1, \alpha_2; \widehat{S}_{m+k}(P))) & \xrightarrow{(19.57)} & (\mathcal{U}^+(\alpha_1, \alpha_2; \widehat{S}_{\ell+m+k}(P)) \end{array} \tag{19.59}$$

Here the first horizontal arrow is the covering map in Proposition 18.6. The right vertical and second horizontal arrows are the covering map (19.57). The left vertical arrow is the restriction of the covering map (19.57) to the codimension ℓ corner.

■

[5] The domains of boundary compatibility isomorphisms or corner compatibility isomorphisms are *horizontal* boundary or corner. So they induce maps between *vertical* corners, which is the left hand side of (19.57).

[6] See (19.55).

Proposition 19.37 *Suppose we are in Situation 19.36. Let $0 < \tau' < \tau$. Then we can find a system of Kuranishi structures and CF-perturbations*

$$\widehat{\mathcal{U}^+}(\alpha_1, \alpha_2; P), \qquad \widehat{\mathfrak{S}^+}(\alpha_1, \alpha_2; P)$$

$\alpha_i \in \mathfrak{A}_i$ with $E(\alpha_2) - E(\alpha_1) \le E_0 - c$ with the following properties:

(1) *$\widehat{\mathcal{U}^+}(\alpha_1, \alpha_2; P)$ is a τ'-collared Kuranishi structure on the underlying topological space of $\mathcal{N}(\alpha_1, \alpha_2; P)^{\boxplus\tau_0}$. It comes with the structure isomorphisms, by which it becomes the interpolation space of $P^{\boxplus\tau_0}$ parametrized morphism.*

(2) *We have:*

$$\mathcal{N}(\alpha_1, \alpha_2; P)^{\boxplus\tau_0} < (\mathcal{N}(\alpha_1, \alpha_2; P)^{\boxplus\tau_0}, \widehat{\mathcal{U}^+}(\alpha_1, \alpha_2; P)). \tag{19.60}$$

Here the Kuranishi structure of $\mathcal{N}(\alpha_1, \alpha_2; P)$ in the left hand side is given in Definition 19.34. This KK-embedding respects structure isomorphisms.

(3) *$\widehat{\mathfrak{S}^+}(\alpha_1, \alpha_2; P)$ is a CF-perturbation of the Kuranishi structure $\widehat{\mathcal{U}^+}(\alpha_1, \alpha_2; P)$. It is transversal to 0 and $(\mathrm{ev}_+, \mathrm{ev}_P)$ is stratumwise strongly submersive with respect to $\widehat{\mathfrak{S}^+}(\alpha_1, \alpha_2; P)$. The CF-perturbations $\widehat{\mathfrak{S}^+}(\alpha_1, \alpha_2; P)$ for various α_1, α_2 are compatible with boundary and corner compatibility isomorphisms.*

(4) *There exists an isomorphism:*

$$\begin{aligned} &\widehat{S}_k^{\mathfrak{C}v}(\mathcal{N}(\alpha_1, \alpha_2; P)^{\boxplus\tau_0}, \widehat{\mathcal{U}^+}(\alpha_1, \alpha_2; \widehat{P})) \\ &\cong (\mathcal{N}(\alpha_1, \alpha_2; \widehat{S}_k(P))^{\boxplus\tau_0}, \widehat{\mathcal{U}^+}(\alpha_1, \alpha_2; \widehat{S}_k(P))) \end{aligned} \tag{19.61}$$

of τ'-collared K-spaces. (Note the right hand side is the Kuranishi structure given in Situation 19.36.) (19.61) is compatible with the structure isomorphisms.[7] *It is also compatible with KK-embeddings (19.56), (19.60).*

(5) *The restriction of $\widehat{\mathfrak{S}^+}(\alpha_1, \alpha_2; P)$ to $\widehat{S}_k^{\mathfrak{C}v}(\mathcal{N}(\alpha_1, \alpha_2; P)^{\boxplus\tau_0}, \widehat{\mathcal{U}^+}(\alpha_1, \alpha_2; \widehat{P}))$ is equivalent to $\widehat{\mathfrak{S}^+}(\alpha_1, \alpha_2; \widehat{S}_k(P))$ via the isomorphism (19.61).*

(6) *The next diagram commutes. (In this diagram we omit underlying topological space from the notation.)*

$$\begin{array}{ccc} \widehat{S}_m^{\mathfrak{C}v}(\widehat{S}_k^{\mathfrak{C}v}(\widehat{\mathcal{U}}^+(\alpha_1, \alpha_2; P))) & \xrightarrow{\pi_{m,k}} & \widehat{S}_{m+k}^{\mathfrak{C}v}(\widehat{\mathcal{U}}^+(\alpha_1, \alpha_2; P)) \\ \Big\downarrow{\scriptstyle \widehat{S}_m^{\mathfrak{C}v}((19.61))} & & \Big\downarrow{\scriptstyle (19.61)} \\ \widehat{S}_m^{\mathfrak{C}v}(\widehat{\mathcal{U}}^+(\alpha_1, \alpha_2; \widehat{S}_k(P))) & \xrightarrow{(19.57)} & \widehat{\mathcal{U}}^+(\alpha_1, \alpha_2; \widehat{S}_{m+k}(P)) \end{array} \tag{19.62}$$

[7] The domains of boundary compatibility isomorphisms or corner compatibility isomorphisms are *horizontal* boundary or corner. So they induce maps between *vertical* corners, which is the left hand side of (19.61).

Here the first horizontal arrow is the covering map in Proposition 18.6. *The second horizontal arrow is the covering map* (19.57). *The right vertical arrow is the covering map* (19.61). *The left vertical arrow is its restriction of the codimension m corner.*

(7) *When the τ-collared CF-perturbation $\widehat{\mathfrak{S}^+}(\alpha_1, \alpha_2; \widehat{S}_k(P))$ given in Situation* 19.36 *varies in a uniform family, we can take $\widehat{\mathfrak{S}^+}(\alpha_1, \alpha_2; P)$ to be uniform.*

Proof The proof is similar to the proof of Propositions 19.1 and 19.27. We construct $\widehat{\mathcal{U}^+}(\alpha_1, \alpha_2; P)$, $\widehat{\mathfrak{S}^+}(\alpha_1, \alpha_2; P)$ by an upward induction on $E(\alpha_2) - E(\alpha_1)$. At each inductive step we construct a Kuranishi structure and CF-perturbation which extends the given one on its boundary and corner. On the horizontal boundary they are given by induction hypothesis. On the vertical boundary they are given by Situation 19.36, the assumption. We use Propositions 17.78 and 17.81 to prove that they can be extended. To apply those propositions, we need to check Situations 17.55, 17.72, the compatibility at the iterated corner. For the iteration of two vertical corners $\widehat{S}^{\mathfrak{C}^v}_\ell \widehat{S}^{\mathfrak{C}^v}_m$ this is Situation 19.36, the assumption. (We use Lemma 17.53 in a way similar to the proof of Proposition 19.1.) For the iteration of two horizontal corners or mixture of horizontal and vertical corners, it is an induction hypothesis. □

We use (19.62) to construct $P^{\boxplus\tau_0}$ parametrized family of CF-perturbation inductively on corner structure stratification of P.

The translation to algebra is fairly immediate.

Definition 19.38 In the situation of Proposition 19.37, we define

$$\psi^{P,\epsilon}_{\alpha_2,\alpha_1} : \Omega(R_{\alpha_1}; o_{R_{\alpha_1}}) \longrightarrow \Omega(R_{\alpha_2}; o_{R_{\alpha_2}}) \tag{19.63}$$

by

$$\psi^{P,\epsilon}_{\alpha_2,\alpha_1}(h) = (-1)^{\epsilon(R_{\alpha_2}, R_{\alpha_2}; h)} \mathrm{ev}_+!(\mathrm{ev}_-^* h; \widehat{\mathfrak{S}^{+\epsilon}}(\alpha_1, \alpha_2; P)), \tag{19.64}$$

where

$$\epsilon(R_{\alpha_2}, R_{\alpha_2}; h) = (\dim P + 1)(\mu(\alpha_+) - \mu(\alpha_-)) + \dim P(\dim R_{\alpha_2} + \deg h).$$

Note that in the case $\dim P = 0$ this formula coincides with (19.13). In the case $P = [0, 1]$ this formula coincides with (19.43).

In (19.64), the right hand side is defined by Definition 17.82 on the K-spaces

$$(\mathcal{N}(\alpha_1, \alpha_2; P)^{\boxplus\tau_0}, \widehat{\mathcal{U}^+}(\mathrm{mor}; \alpha_1, \alpha_2; P)).$$

The degree of $\psi^{P,\epsilon}_{\alpha_2,\alpha_1}$ after shifting is $-\dim P$. If the energy loss of our parametrized family of morphisms is c, the map $\psi^{P,\epsilon}_{\alpha_2,\alpha_1}$ induces

$$\mathfrak{F}^{\lambda} CF(\mathcal{F}_1) \longrightarrow \mathfrak{F}^{\lambda-c} CF(\mathcal{F}_2),$$

where the filtration $\mathfrak{F}^\lambda$ is defined in Definition 16.8 (2)(3).

Lemma 19.39 *The operators* $\psi^{P,\epsilon}_{\alpha_2,\alpha_1}$ *satisfy the following equality in the sense of* ($\mathfrak{b}$)*:*

$$\begin{aligned}&(d_0 \circ \psi^{P,\epsilon}_{\alpha_2,\alpha_1} + \sum_{\alpha_2'} \mathfrak{m}^{2,\epsilon}_{1;\alpha_2,\alpha_2'} \circ \psi^{P,\epsilon}_{\alpha_2',\alpha_1})\\ &-(-1)^{\dim P}(\psi^{P,\epsilon}_{\alpha_2,\alpha_1} \circ d_0 + \sum_{\alpha_1'} \psi^{P,\epsilon}_{\alpha_2,\alpha_1'} \circ \mathfrak{m}^{1,\epsilon}_{1;\alpha_1',\alpha_1})\\ &-(-1)^{\dim P}\psi^{\partial P,\epsilon}_{\alpha_2,\alpha_1} = 0.\end{aligned} \tag{19.65}$$

Here the first sum in the second line is taken over $\alpha_1' \in \mathfrak{A}_1$ *with* $E(\alpha_1) < E(\alpha_1') \le E(\alpha_2)+c$ *and the second sum in the second line is taken over* $\alpha_2' \in \mathfrak{A}_2$ *with* $E(\alpha_1)-c \le E(\alpha_2') < E(\alpha_2)$*. The number* c *is the energy loss of our morphism.*

Proof The proof is by Stokes' formula and the composition formula and is entirely similar to the proof of Lemma 19.10. In fact (16.31) and Proposition 19.37 (1) imply that the contribution by horizontal boundary gives the 4th and the 5th terms of (19.65). Proposition 19.37 (4) implies that the contribution by vertical boundary gives the 3rd term of (19.65). □

19.8 Proof of Theorem 16.39 (2)(e), (4)–(6) and Theorem 16.9 (2)(f)

We begin with an algebraic result that is similar to Proposition 19.32 and is a baby version of [FOOO4, Theorem 7.2.212].

Situation 19.40 For $j = 1, 2$, let $(CF(\mathcal{F}^i_j), \hat d^i_j)$, $(CF(\mathcal{F}^{i+1}_j), \hat d^{i+1}_j)$ be partial cochain complexes of energy cut level E^{i+1}.

(1) For $j = 1, 2$, $\psi^{i+1i}_j : CF(\mathcal{F}^i_j) \to CF(\mathcal{F}^{i+1}_j)$ is a partial cochain map of energy cut level E^{i+1} and energy loss 0. Moreover, we assume ψ^{i+1i}_j induces an isomorphism modulo T^ϵ for small $\epsilon > 0$.
(2) For $k = a, b$ and $\ell = i, i+1$, $\mathfrak{n}^\ell_k : CF(\mathcal{F}^\ell_1) \to CF(\mathcal{F}^\ell_2)$ is a partial cochain map of energy cut level E^{i+1} and energy loss c.
(3) $\mathfrak{h}^i_{ab} : CF(\mathcal{F}^i_1) \to CF(\mathcal{F}^i_2)$ is a cochain homotopy between $\mathfrak{n}^i_a$ and $\mathfrak{n}^i_b$ of energy cut level E^i and energy loss c. $\mathfrak{h}^{i+1}_{ab} : CF(\mathcal{F}^{i+1}_1) \to CF(\mathcal{F}^{i+1}_2)$ is a cochain homotopy between $\mathfrak{n}^{i+1}_a$ and $\mathfrak{n}^{i+1}_b$ of energy cut level E^{i+1} and energy loss c. See Diagram (19.66).
(4) For $k = a, b$, $\mathfrak{h}^{i+1i}_k : CF(\mathcal{F}^i_1) \to CF(\mathcal{F}^{i+1}_2)$ is a partial cochain homotopy between $\mathfrak{n}^{i+1}_k \circ \psi^{i+1i}_1$ and $\psi^{i+1i}_2 \circ \mathfrak{n}^i_k$ of energy cut level E^{i+1} and energy loss c. ■

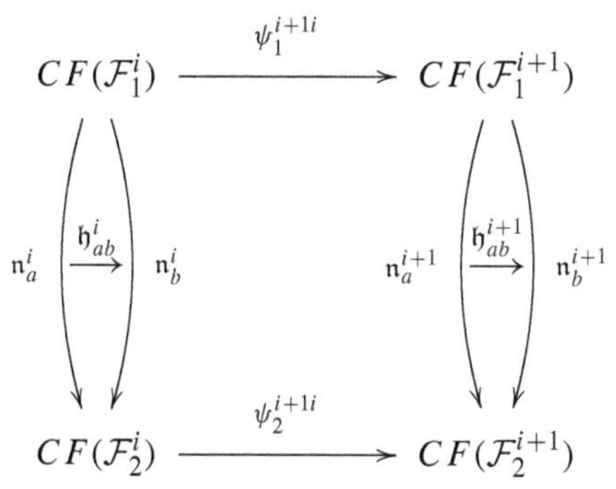

$$
\begin{array}{ccc}
CF(\mathcal{F}_1^i) & \xrightarrow{\psi_1^{i+1i}} & CF(\mathcal{F}_1^{i+1}) \\
\mathfrak{n}_a^i \Big\downarrow \overset{\mathfrak{h}_{ab}^i}{\longrightarrow} \Big\downarrow \mathfrak{n}_b^i & & \mathfrak{n}_a^{i+1} \Big\downarrow \overset{\mathfrak{h}_{ab}^{i+1}}{\longrightarrow} \Big\downarrow \mathfrak{n}_b^{i+1} \\
CF(\mathcal{F}_2^i) & \xrightarrow{\psi_2^{i+1i}} & CF(\mathcal{F}_2^{i+1})
\end{array}
\tag{19.66}
$$

Proposition 19.41 *Suppose in Situation* 19.40 *there exists (homotopy of homotopies)*

$$\mathfrak{H}_{ab}^{i+1i} : CF(\mathcal{F}_1^i) \longrightarrow CF(\mathcal{F}_2^{i+1})$$

which satisfies

$$
\begin{aligned}
&\hat{d}_2^{i+1} \circ \mathfrak{H}_{ab}^{i+1i} - \mathfrak{H}_{ab}^{i+1i} \circ \hat{d}_1^i \\
&= \mathfrak{h}_b^{i+1i} - \mathfrak{h}_a^{i+1i} + \mathfrak{h}_{ab}^{i+1} \circ \psi_1^{i+1i} - \psi_2^{i+1i} \circ \mathfrak{h}_{ab}^i
\end{aligned}
\tag{19.67}
$$

as equality of maps of energy cut level E^i and energy loss c.

Then we can promote $\mathfrak{h}_{ab}^i$ and $\mathfrak{H}_{ab}^{i+1i}$ to the energy cut level E^{i+1} so that Formula (19.67) *holds as an equality of maps of energy cut level E^{i+1} and energy loss c.*

Proof We use Proposition 19.33. We put $\mathscr{F}_1 = CF(\mathcal{F}_1^i)$, $\mathscr{F}_2 = CF(\mathcal{F}_2^{i+1})$, $E_1 = E_i - c$, $E_2 = E_{i+1} - c$. D_2 is the chain complex associated to the cell complex $[1,2]^2$ and D_1 is a subcomplex associated to $\partial([1,2]^2) \setminus ([1,2] \times \{1\})$. We define ϕ_1 and ϕ_2 by assigning various maps on each cell as drawn in Fig. 19.4. Proposition 19.33 implies that we can promote $\mathfrak{H}_{ab}^{i+1i}$ and $\psi_2^{i+1i} \circ \mathfrak{h}_{ab}^i$ to energy cut level E_{i+1}. Using the fact that ψ_2^{i+1i} is invertible, we can promote $\mathfrak{h}_{ab}^i$ also. □

Remark 19.42 In Situation 19.40 (1) we assumed that the energy zero part of the partial cochain map of energy loss 0 is the identity map. Actually we only need a milder assumption that the energy 0 part of the partial cochain map of energy loss 0 induces an injection on d_0 cohomology.

Proof of Theorem* 16.39 *(4) We are given $\mathfrak{N}\mathfrak{N}_a$, $\mathfrak{N}\mathfrak{N}_b$, that are morphisms of inductive systems $\mathcal{F}\mathcal{F}_1 \to \mathcal{F}\mathcal{F}_2$. Each of them consists of partial morphisms

$$\mathfrak{N}_k^i : \mathcal{F}_1^i \to \mathcal{F}_2^i, \quad k = a, b,$$

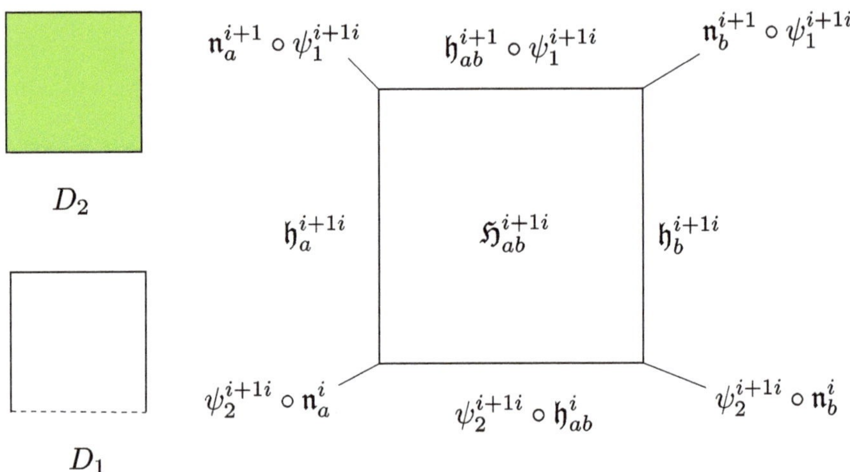

Fig. 19.4 D_1, D_2, ϕ_1, ϕ_2: 2

respectively. They induce partial cochain maps

$$\mathfrak{n}_k^i : CF(\mathcal{F}_1^i) \to CF(\mathcal{F}_2^i), \quad k = a, b$$

for each i. We are also given a homotopy $\mathfrak{H}\mathfrak{H}$ from $\mathfrak{N}\mathfrak{N}_a$ to $\mathfrak{N}\mathfrak{N}_b$. It consists of partial homotopies from $\mathfrak{N}_a^i$ to $\mathfrak{N}_b^i$. By Proposition 19.27 and Lemma 19.39, it induces a partial homotopy

$$\mathfrak{h}_{ab}^i : CF(\mathcal{F}_1^i) \to CF(\mathcal{F}_2^i)$$

from $\mathfrak{n}_a^i$ to $\mathfrak{n}_b^i$.

Since $\mathfrak{N}\mathfrak{N}_a$, $\mathfrak{N}\mathfrak{N}_b$ are morphisms of a partial linear K-system, we are given a partial homotopy from $\mathfrak{N}_2^{i+1i} \circ \mathfrak{N}_k^i$ to $\mathfrak{N}_k^{i+1} \circ \mathfrak{N}_1^{i+1i}$. Again by Proposition 19.27 and Lemma 19.39, it induces a partial homotopy

$$\mathfrak{h}_k^{i+1i} : CF(\mathcal{F}_1^i) \to CF(\mathcal{F}_2^{i+1})$$

from $\psi_2^{i+1i} \circ \mathfrak{n}_k^i$ to $\mathfrak{n}_k^{i+1} \circ \psi_1^{i+1i}$. (Note $\psi_j^{i+1i} : CF(\mathcal{F}_j^i) \to CF(\mathcal{F}_j^{i+1})$ is a cochain map induced by $\mathfrak{N}_j^{i+1i}$.) Thus we are in Situation 19.40.

Now the existence of homotopy of homotopy $\mathcal{H}^i$ which is Definition 16.36 (4) (b), (c), together with Proposition 19.37 and Lemma 19.39 implies that there exists

$$\mathfrak{H}_{ab}^{i+1i} : CF(\mathcal{F}_1^i) \to CF(\mathcal{F}_2^{i+1})$$

that satisfies (19.67). Thus we can apply Proposition 19.41 to promote our partial homotopy $\mathfrak{h}^i_{ab}$ to one of energy cut level ∞. Therefore $\mathfrak{n}^i_a$ is cochain homotopic to $\mathfrak{n}^i_b$. □

Proof of Theorem* 16.31 *(2) This is a special case of Theorem 16.39 (4) where $\mathfrak{N}^{i+1i}$ is the identity morphism. □

Proof of Theorem* 16.39 *(5) and Theorem* 16.31 *(3) We will prove Theorem 16.39 (5). Theorem 16.31 (3) is its special case.

We recall that $\mathfrak{N}_{ba} : \mathcal{F}\mathcal{F}_a \to \mathcal{F}\mathcal{F}_b$ consists of partial morphisms

$$\mathfrak{N}^i_{ba} : \mathcal{F}^i_a \to \mathcal{F}^i_b$$

and partial homotopies $\mathfrak{H}^i_{ba}$ between $\mathfrak{N}^{i+1i}_b \circ \mathfrak{N}^i_{ba}$ and $\mathfrak{N}^{i+1}_{ba} \circ \mathfrak{N}^{i+1i}_a$. (See Diagram 19.68 below.)

$$\begin{array}{ccc} \mathcal{F}^i_b & \xrightarrow{\mathfrak{N}^{i+1i}_b} & \mathcal{F}^{i+1}_b \\ {\scriptstyle \mathfrak{N}^i_{ba}}\uparrow & & \uparrow{\scriptstyle \mathfrak{N}^{i+1}_{ba}} \\ \mathcal{F}^i_a & \xrightarrow{\mathfrak{N}^{i+1i}_a} & \mathcal{F}^{i+1}_a \end{array} \tag{19.68}$$

Also $\mathfrak{N}_{cb} : \mathcal{F}\mathcal{F}_b \to \mathcal{F}\mathcal{F}_c$ consists of partial morphisms $\mathfrak{N}^i_{cb} : \mathcal{F}^i_b \to \mathcal{F}^i_c$ and partial homotopies $\mathfrak{H}^i_{cb}$ in a similar way. The definition of the composition $\mathfrak{N}_{ca} = \mathfrak{N}_{cb} \circ \mathfrak{N}_{ba}$ is given in Lemma-Definition 16.38.

Let $\mathcal{N}^i_{ab}(\alpha_a, \alpha_b)$ and $\mathcal{N}^i_{bc}(\alpha_b, \alpha_c)$ be interpolation spaces of $\mathfrak{N}^i_{ba}$ and $\mathfrak{N}^i_{cb}$, respectively. For $k = a, b, c$, let $\mathcal{N}^{i+1i}_k(\alpha_k, \alpha'_k)$ be an interpolation space of $\mathfrak{N}^{i+1i}_k$. We denote by $\mathcal{N}^i_{ab}(\alpha_a, \alpha_b; [0,1])$ and $\mathcal{N}^i_{bc}(\alpha_b, \alpha_c; [0,1])$ interpolation spaces of $\mathfrak{H}^i_{ba}$ and $\mathfrak{H}^i_{cb}$, respectively. By definition we have

$$\begin{aligned} &\partial \mathcal{N}^i_{ab}(\alpha_a, \alpha_b; [0,1]) \\ =&\bigcup_{\alpha'_a} \mathcal{N}^{i+1i}_a(\alpha_a, \alpha'_a) \times^{\boxplus\tau}_{R_{\alpha'_a}} \mathcal{N}^i_{ab}(\alpha'_a, \alpha_b) \\ &\cup \bigcup_{\alpha'_b} \mathcal{N}^i_{ab}(\alpha_a, \alpha'_b) \times^{\boxplus\tau}_{R_{\alpha'_b}} \mathcal{N}^{i+1i}_b(\alpha'_b, \alpha_b). \end{aligned} \tag{19.69}$$

Here $\times^{\boxplus\tau}_{R_{\alpha'_a}}$ and $\times^{\boxplus\tau}_{R_{\alpha'_b}}$ are as in Definition 18.34. Similarly, we have

$$\begin{aligned}
&\partial \mathcal{N}^i_{bc}(\alpha_b, \alpha_c; [0,1]) \\
&= \bigcup_{\alpha'_b} \mathcal{N}^{i+1i}_b(\alpha_b, \alpha'_b) \times^{\boxplus\tau}_{R_{\alpha'_b}} \mathcal{N}^i_{bc}(\alpha'_b, \alpha_c) \\
&\cup \bigcup_{\alpha'_c} \mathcal{N}^i_{bc}(\alpha_a, \alpha'_c) \times^{\boxplus\tau}_{R_{\alpha'_c}} \mathcal{N}^{i+1i}_c(\alpha'_c, \alpha_c).
\end{aligned} \tag{19.70}$$

The composition $\mathfrak{N}_{ca} = \mathfrak{N}_{cb} \circ \mathfrak{N}_{ba}$ consists of $\mathfrak{N}^i_{ca}$ and $\mathfrak{H}^i_{ca}$. Here the interpolation space $\mathcal{N}^i_{ac}(\alpha_a, \alpha_c)$ of $\mathfrak{N}_{ca}$ is given by

$$\mathcal{N}^i_{ac}(\alpha_a, \alpha_c) = \bigcup_{\alpha_b} \mathcal{N}^i_{ab}(\alpha_a, \alpha_b) \times^{\boxplus\tau}_{R_{\alpha_b}} \mathcal{N}^i_{bc}(\alpha_b, \alpha_c).$$

Note that the union in the above formula is different from the disjoint union and is defined as in Lemma-Definition 18.37.

The homotopy $\mathfrak{H}^i_{ca}$ is obtained by gluing $\mathfrak{H}^i_{cb} \circ \mathfrak{N}^i_{ba}$ and $\mathfrak{N}^i_{cb} \circ \mathfrak{H}^i_{ba}$ as follows. The interpolation space of $\mathfrak{H}^i_{cb} \circ \mathfrak{N}^i_{ba}$ is

$$\bigcup_{\alpha_b} \mathcal{N}^i_{ab}(\alpha_a, \alpha_b; [0,1]) \times^{\boxplus\tau}_{R_{\alpha_b}} \mathcal{N}^i_{bc}(\alpha_b, \alpha_c). \tag{19.71}$$

The interpolation space of $\mathfrak{N}^i_{cb} \circ \mathfrak{H}^i_{ba}$ is

$$\bigcup_{\alpha_b} \mathcal{N}^i_{ab}(\alpha_a, \alpha_b) \times^{\boxplus\tau}_{R_{\alpha_b}} \mathcal{N}^i_{bc}(\alpha_b, \alpha_c; [0,1]). \tag{19.72}$$

We observe that both (19.71) and (19.72) contain

$$\bigcup_{\alpha_b} \bigcup_{\alpha'_b} \mathcal{N}^i_{ab}(\alpha_a, \alpha_b) \times^{\boxplus\tau}_{R_{\alpha_b}} \mathcal{N}^{i+1i}_b(\alpha_b, \alpha'_b) \times^{\boxplus\tau}_{R_{\alpha'_b}} \mathcal{N}^i_{bc}(\alpha_b, \alpha_c) \tag{19.73}$$

in its boundary. We smooth the corners contained in (19.73) and glue (19.71) and (19.72) there. See Chap. 18.

Next we recall that while we constructed a partial cochain map

$$\psi^i_{ba} : CF^i_a \to CF^i_b$$

(see (19.48) for the notation CF^i_*), we took a thickening $\widehat{\mathcal{U}^{i,+}_{ab}}(\alpha_a, \alpha_b)$ of $\widehat{\mathcal{U}^i_{ab}}(\alpha_a, \alpha_b)$ (note that they are Kuranishi structures of $\mathcal{N}^i_{ab}(\alpha_a, \alpha_b)^{\mathfrak{C}^h \boxplus \tau}$) and a CF-perturbation $\widehat{\mathfrak{S}^i_{ab}}(\alpha_a, \alpha_b)$ of $\widehat{\mathcal{U}^{i,+}_{ab}}(\alpha_a, \alpha_b)$. (Here $\mathfrak{C}^h$ stands for the horizontal boundary as in Definition 19.34.) During the construction of a partial cochain map

$$\psi^i_{cb} : CF^i_b \to CF^i_c,$$

we took a thickening $\widehat{\mathcal{U}^{i,+}_{bc}}(\alpha_b, \alpha_c)$ (of $\widehat{\mathcal{U}^{i}_{bc}}(\alpha_b, \alpha_c)$) and its CF-perturbation $\widehat{\mathfrak{S}^i_{bc}}(\alpha_b, \alpha_c)$. During the construction of partial cochain maps

$$\psi^{i+1i}_k : CF^i_k \to CF^{i+1}_k, \quad k = a, b, c,$$

we took thickenings $\widehat{\mathcal{U}^{i+1i,+}_k}(\alpha_k, \alpha'_k)$ of $\widehat{\mathcal{U}^{i+1i}_k}(\alpha_k, \alpha'_k)$ which are Kuranishi structures on $\mathcal{N}^{i+1i}_k(\alpha_k, \alpha'_k)^{\boxplus\tau}$), and their CF-perturbations $\widehat{\mathfrak{S}^{i+1i}_k}(\alpha_k, \alpha'_k)$.

Furthermore, in the course of our construction of a partial cochain homotopy

$$\mathfrak{h}^i_{ba} \quad \text{between } \mathfrak{n}^{i+1i}_b \circ \psi^i_{ba} \text{and } \psi^{i+1}_{ba} \circ \mathfrak{n}^{i+1i}_a$$

(where $\mathfrak{n}^{i+1i}_b \circ \psi^i_{ba}$ and $\psi^{i+1}_{ba} \circ \mathfrak{n}^{i+1i}_a$ are cochain maps : $CF^i_a \to CF^{i+1}_b$), we took a thickening $\widehat{\mathcal{U}^{i,+}_{ab}}(\alpha_a, \alpha_b; [0,1])$ of $\widehat{\mathcal{U}^{i}_{ab}}(\alpha_a, \alpha_b; [0,1])$ which is a Kuranishi structure on $\mathcal{N}^i_{ab}(\alpha_a, \alpha_b; [0,1])^{\mathfrak{C}^h\boxplus\tau}$, and its CF-perturbation $\widehat{\mathfrak{S}^i_{ab}}(\alpha_a, \alpha_b; [0,1])$.

$$\begin{array}{ccccc} CF^i_a & \xrightarrow{\psi^i_{ba}} & CF^i_b & \xrightarrow{\psi^i_{cb}} & CF^i_c \\ {\scriptstyle \mathfrak{n}^{i+1i}_a}\downarrow & & {\scriptstyle \mathfrak{n}^{i+1i}_b}\downarrow & & {\scriptstyle \mathfrak{n}^{i+1i}_c}\downarrow \\ CF^{i+1}_a & \xrightarrow{\psi^{i+1}_{ba}} & CF^{i+1}_b & \xrightarrow{\psi^{i+1}_{cb}} & CF^{i+1}_c \end{array} \tag{19.74}$$

During our construction of a partial cochain homotopy

$$\mathfrak{h}^i_{cb} \quad \text{between } \mathfrak{n}^{i+1i}_c \circ \psi^i_{cb} \text{and } \psi^{i+1}_{cb} \circ \mathfrak{n}^{i+1i}_b,$$

we took a thickening $\widehat{\mathcal{U}^{i,+}_{bc}}(\alpha_b, \alpha_c; [0,1])$ of $\widehat{\mathcal{U}^{i}_{bc}}(\alpha_b, \alpha_c; [0,1])$ which is a Kuranishi structure on $\mathcal{N}^i_{bc}(\alpha_b, \alpha_c; [0,1])^{\mathfrak{C}^h\boxplus\tau}$) and its CF-perturbation $\widehat{\mathfrak{S}^i_{bc}}(\alpha_b, \alpha_c; [0,1])$.

Note that by Lemma 19.21 we may use

$$\bigcup_{\alpha_b} \widehat{\mathcal{U}^{i,+}_{ab}}(\alpha_a, \alpha_b) \times_{R_{\alpha_b}} \widehat{\mathcal{U}^{i,+}_{bc}}(\alpha_b, \alpha_c) \tag{19.75}$$

and

$$\bigcup_{\alpha_b} \widehat{\mathfrak{S}^i_{ab}}(\alpha_a, \alpha_b) \times_{R_{\alpha_b}} \widehat{\mathfrak{S}^i_{bc}}(\alpha_b, \alpha_c) \tag{19.76}$$

to define a partial cochain map

$$\psi^i_{ca} : CF^i_a \to CF^i_c.$$

Here (19.75) is a thickening of $\bigcup_{\alpha_b} \widehat{\mathcal{U}^i_{ab}}(\alpha_a, \alpha_b) \times_{R_{\alpha_b}} \widehat{\mathcal{U}^i_{bc}}(\alpha_b, \alpha_c)$ which is a Kuranishi structure on $\mathcal{N}^i_{ac}(\alpha_a, \alpha_c)^{\mathfrak{C}^h \boxplus \tau}$, and (19.76) is its CF-perturbation. Therefore composition formula and Lemma 18.31 yield

$$\psi^i_{ca} = \psi^i_{cb} \circ \psi^i_{ba}, \tag{19.77}$$

if we define ψ^i_{ca} by this particular choice. (Lemma 19.22.)

Next we consider $\mathfrak{h}^i_{ca}$. We take

$$\begin{aligned} &\left(\bigcup_{\alpha_b} \widehat{\mathcal{U}^{i,+}_{ab}}(\alpha_a, \alpha_b; [0,1]) \times_{R_{\alpha_b}} \widehat{\mathcal{U}^{i+1,+}_{bc}}(\alpha_b, \alpha_c) \right) \\ \cup &\left(\bigcup_{\alpha_b} \widehat{\mathcal{U}^{i,+}_{ab}}(\alpha_a, \alpha_b) \times_{R_{\alpha_b}} \widehat{\mathcal{U}^{i,+}_{bc}}(\alpha_b, \alpha_c; [0,1]) \right). \end{aligned} \tag{19.78}$$

Here we take a partial corner smoothing of the right hand side and glue them. Then (19.78) is a thickening of $\mathcal{N}^i_{ac}(\alpha_a, \alpha_c)^{\mathfrak{C}^h \boxplus \tau}$ and

$$\begin{aligned} &\left(\bigcup_{\alpha_b} \widehat{\mathfrak{S}^i_{ab}}(\alpha_a, \alpha_b; [0,1]) \times_{R_{\alpha_b}} \widehat{\mathfrak{S}^{i+1}_{bc}}(\alpha_b, \alpha_c) \right) \\ \cup &\left(\bigcup_{\alpha_b} \widehat{\mathfrak{S}^i_{ab}}(\alpha_a, \alpha_b) \times_{R_{\alpha_b}} \widehat{\mathfrak{S}^i_{bc}}(\alpha_b, \alpha_c; [0,1]) \right) \end{aligned} \tag{19.79}$$

is a CF-perturbation of (19.78). We use (19.78) and (19.79) to define $\mathfrak{h}^i_{ca}$. Then by composition formula and Lemma 18.31 again we find

$$\mathfrak{h}^i_{ca} = \psi^{i+1}_{cb} \circ \mathfrak{h}^i_{ba} + \mathfrak{h}^i_{cb} \circ \psi^i_{ba}. \tag{19.80}$$

Lemma 19.43

$$\hat{d} \circ \mathfrak{h}^i_{ca} + \mathfrak{h}^i_{ca} \circ \hat{d} = \psi^{i+1}_{ca} \circ \mathfrak{n}^{i+1i}_a - \mathfrak{n}^{i+1i}_c \circ \psi^i_{ca}. \tag{19.81}$$

Proof

$$
\begin{aligned}
&\hat{d} \circ \mathfrak{h}^i_{ca} + \mathfrak{h}^i_{ca} \circ \hat{d} \\
&= \psi^{i+1}_{cb} \circ \hat{d} \circ \mathfrak{h}^i_{ba} + \psi^{i+1}_{cb} \circ \mathfrak{h}^i_{ba} \circ \hat{d} \\
&\quad + \mathfrak{h}^i_{cb} \circ \hat{d} \circ \psi^i_{ba} + \hat{d} \circ \mathfrak{h}^i_{cb} \circ \psi^i_{ba} \\
&= \psi^{i+1}_{cb} \circ \psi^{i+1}_{ba} \circ \mathfrak{n}^{i+1i}_{a} - \psi^{i+1}_{cb} \circ \mathfrak{n}^{i+1i}_{b} \circ \psi^i_{ba} \\
&\quad + \psi^{i+1}_{cb} \circ \mathfrak{n}^{i+1i}_{b} \circ \psi^i_{ba} - \mathfrak{n}^{i+1i}_{c} \circ \psi^{i+1}_{cb} \circ \psi^i_{ba} \\
&= \psi^{i+1}_{ca} \circ \mathfrak{n}^{i+1i}_{a} - \mathfrak{n}^{i+1i}_{c} \circ \psi^i_{ca}.
\end{aligned}
$$

□

We note that the equalities (19.77), (19.80), (19.81) are those of energy cut level E^i. We recall that while we constructed ψ_{ba} and ψ_{cb} we promoted $\mathfrak{n}^{i+1i}_k$ $(k = a, b, c)$, ψ^i_{ba}, ψ^i_{cb}, $\mathfrak{h}^i_{ba}$ and $\mathfrak{h}^i_{cb}$ to energy cut level ∞. We now promote ψ^i_{ca} and $\mathfrak{h}^i_{ca}$ to energy cut level ∞ so that (19.77), (19.80) hold as the equalities at the energy cut level ∞. Then (19.81) holds as an equality at the energy cut level ∞.

Thus for this particular choice of promotion, the equality $\psi^i_{ca} = \psi^i_{cb} \circ \psi^i_{ba}$ holds not only up to cochain homotopy but also as a strict equality. Since ψ^i_{ca} is independent of various choices up to cochain homotopy (Theorem 16.39, which we proved in Sect. 19.6), $\psi^i_{ca} = \psi^i_{cb} \circ \psi^i_{ba}$ holds for any choice of ψ^i_{ca} up to cochain homotopy. The proof of Theorem 16.39 (5) is complete. □

We next prove a partial cochain map version of Theorem 16.31 (4).

Lemma 19.44 *In the situation of Theorem* 16.31 *(4) we have* $\mathcal{ID}_* \sim \mathrm{id} \mod T^E$ *for any* E. *Here* $\sim$ *means cochain homotopic.*

Proof This is a consequence of Theorem 16.31 (2) and Proposition 18.60. In fact, Theorem 16.31 (2) implies that $\mathcal{ID}_* \circ \mathcal{ID}_* \sim \mathcal{ID}_*$. (Here $\mathcal{ID}_*$ is the cochain map induced by the identity morphism and $\sim$ is a chain homotopy.) On the other hand, Lemma 19.45 below implies that $\mathcal{ID}_*$ is an isomorphism. Therefore $\mathcal{ID}_* \sim \mathrm{id}$. □

Lemma 19.45 *Let* $\widehat{\psi_{21}} = \{\psi_{\alpha_2\alpha_1}\} : \mathcal{F}_1 \to \mathcal{F}_2$ *be a partial cochain map of energy cut level* E_0 *and energy loss* 0. *We assume it is congruent to the isomorphism in the sense of Definition* 16.20. *Then there exists a partial cochain map* $\widehat{\psi_{12}} = \{(\psi_{12})_{\alpha_1\alpha_2}\} : \mathcal{F}_2 \to \mathcal{F}_1$ *of energy cut level* E_0 *and energy loss* 0 *such that* $\widehat{\psi_{12}} \circ \widehat{\psi_{21}}$ *and* $\widehat{\psi_{21}} \circ \widehat{\psi_{12}}$ *are identity maps.*

Proof We construct $(\psi_{12})_{\alpha_1\alpha_2}$ by induction on $E(\alpha_1) - E(\alpha_2)$. This induction is possible because the set of values of $E(\alpha_1) - E(\alpha_2)$ is a discrete set by uniform Gromov compactness Definition 16.36 (2)(g).

We start with the case when $E(\alpha_1) - E(\alpha_2) = 0$. By definition of a partial cochain map of energy loss 0 congruent to the isomorphism (Definition 16.20), we have

$$(\psi_{12})_{\alpha_1\alpha_2} = \begin{cases} 0 & \text{if } \alpha_1 \neq \alpha_2 \\ \text{id} & \text{if } \alpha_1 = \alpha_2. \end{cases}$$

In fact, the interpolation space $\mathcal{N}(\alpha_2, \alpha_1)$ is an empty set if $\alpha_1 \neq \alpha_2$ and $E(\alpha_1) - E(\alpha_2) = 0$ by Condition 16.17 (V). This implies the first equality. If $\alpha_1 = \alpha_2 = \alpha$, we have $\mathcal{N}(\alpha, \alpha) = R_\alpha$ by Definition 16.20. Moreover, $\text{ev}_\pm : \mathcal{N}(\alpha, \alpha) \to R_\alpha$ is the identity map. This implies the second equality.

Suppose $E(\alpha_1) - E(\alpha_2) = E_0$ and we have defined $(\psi_{12})_{\alpha_1'\alpha_2'}$ for $E(\alpha_1) - E(\alpha_2) < E_0$. Then the condition that $\widehat{\psi}_{21}$ is a right inverse of $\widehat{\psi}_{12}$ at the energy cut level E_0 is written as

$$(\psi_{12})_{\alpha_1\alpha_2} + \sum_{\alpha_2'} (\psi_{12})_{\alpha_1\alpha_2'} \circ (\psi_{21})_{\alpha_2'\alpha_2} = 0.$$

Since the second term is already defined we can find $(\psi_{12})_{\alpha_1\alpha_2}$ uniquely so that this condition holds. Thus we have found the left inverse by induction. We can find the right inverse in the same way. A standard fact in group theory yields that the right and left inverse coincide. It is also easy to see that partial inverse of a partial cochain map is a partial cochain map. □

Proof of Theorem* 16.9 *(2)(f) and Theorem* 16.39 *(2)(e) Theorem 16.39 (2)(e) follows from Theorem 16.39 (4) applied to the identity morphism. Theorem 16.9 (2)(f) is a special case of Theorem 16.39 (2)(e) when $\mathfrak{N}^{i+1i}$ are the identity morphisms. □

Proof of Theorem* 16.39 *(6) and Theorem* 16.31 *(4) We will prove Theorem 16.39 (6). Theorem 16.31 (4) is its special case.

We first define the identity morphism of an inductive system of linear K-systems. Let $\mathcal{FF} = (\{\mathcal{F}^i\}, \{\mathfrak{N}^i\})$ be an inductive system of linear K-systems. (Definition 16.36 (2).) We put $\mathcal{FF}_k = \mathcal{FF}$ for $k = a, b$ and $\mathfrak{N}^i_{ba} = \mathcal{ID}_{\mathcal{F}^i}$ the identity morphism of $\mathcal{F}^i$. By Proposition 18.60 we have

$$\mathfrak{N}^i \circ \mathcal{ID}_{\mathcal{F}^i} \sim \mathfrak{N}^i \sim \mathcal{ID}_{\mathcal{F}^i} \circ \mathfrak{N}^i.$$

Let $\mathfrak{H}^i_{ba}$ be this homotopy. (We take the particular choice of the homotopy which we gave during the proof of Proposition 18.60.)

Definition 19.46 We define $\mathcal{ID}_{\mathcal{FF}} = (\{\mathcal{ID}_{\mathcal{F}^i}\}, \{\mathfrak{H}^i_{ba}\})$ to be the *identity morphism* from $\mathcal{FF}$ to itself.

Theorem 16.39 (6) claims that the cochain map induced by $\mathcal{ID}_{\mathcal{FF}}$ is cochain homotopic to the identity. To prove this it suffices to show the following lemma.

Lemma 19.47 *If $\mathfrak{N}_{cb} : \mathcal{FF}_b \to \mathcal{FF}_c$ be a morphism of inductive systems of linear K-systems, then the composition $\mathfrak{N}_{cb} \circ \mathcal{ID}_{\mathcal{FF}}$ is homotopic to $\mathfrak{N}_{cb}$. The same holds for $\mathcal{ID}_{\mathcal{FF}} \circ \mathfrak{N}_{cb}$.*

Proof Write $\mathcal{FF}_c = (\{\mathcal{F}_c^i\}, \{\mathfrak{N}_c^{i+1i}\})$ and $\mathfrak{N}_{cb} = (\{\mathfrak{N}_{cb}^i\}, \{\mathfrak{H}_{cb}^i\}) : \mathcal{FF}_b \to \mathcal{FF}_c$. Here $\mathfrak{N}_k^{i+1i} : \mathcal{F}_k^i \to \mathcal{F}_k^{i+1}$ $(k = b, c)$, $\mathfrak{N}_{cb}^i : \mathcal{F}_b^i \to \mathcal{F}_c^i$ are partial morphisms of partial linear K-systems, and $\mathfrak{H}_{cb}^i$ is a homotopy between $\mathfrak{N}_{cb}^{i+1} \circ \mathfrak{N}_b^{i+1i}$ and $\mathfrak{N}_c^{i+1i} \circ \mathfrak{N}_{cb}^i$. (They are partial morphisms $: \mathcal{F}_b^i \to \mathcal{F}_c^{i+1}$.)

We denote by $\mathcal{M}(ki; \alpha_-, \alpha_+)$ the moduli space of connecting orbits for $\mathcal{F}_k^i$. $(k = a, b, c)$. Note that

$$\mathcal{M}(ai; \alpha_-, \alpha_+) = \mathcal{M}(bi; \alpha_-, \alpha_+).$$

Let $\mathcal{N}(k, ii+1; \alpha_-, \alpha_+)$ and $\mathcal{N}(bc, i; \alpha_-, \alpha'_+)$ be interpolation spaces of $\mathfrak{N}_k^{i+1i}$ and $\mathfrak{N}_{cb}^i$, respectively. Let $\mathcal{N}(bc, ii+1; \alpha_-, \alpha'_+; [1, 2])$ be the interpolation space of $\mathfrak{H}_{cb}^i$.

$$\begin{array}{ccc}
\mathcal{F}_c^i & \xrightarrow[\mathfrak{N}_c^{i+1i}]{} & \mathcal{F}_c^{i+1} \\
{\scriptstyle \mathfrak{N}_{cb}^i}\uparrow & & \uparrow{\scriptstyle \mathfrak{N}_{cb}^{i+1}} \\
\mathcal{F}_b^i & \xrightarrow[\mathfrak{N}_b^{i+1i}]{} & \mathcal{F}_b^{i+1} \\
{\scriptstyle \mathcal{ID}}\uparrow & & \uparrow{\scriptstyle \mathcal{ID}} \\
\mathcal{F}_a^i & \xrightarrow[\mathfrak{N}_a^{i+1i}]{} & \mathcal{F}_a^{i+1}
\end{array} \tag{19.82}$$

Here we also note that $\mathcal{F}_a^i = \mathcal{F}_b^i$ and $\mathfrak{N}_a^{i+1i} = \mathfrak{N}_b^{i+1i}$.

We put $\mathfrak{N}^i \circ \mathcal{ID}_{\mathcal{F}^i} = (\{\mathfrak{N}_{cb}^i \circ \mathcal{ID}\}, \{\mathfrak{H}_{ca}^i\})$. By definition the homotopy $\mathfrak{H}_{ca}^i$ is obtained as

$$\mathfrak{H}_{ca}^i = (\mathfrak{N}_{cb}^{i+1} \circ \mathfrak{H}_{ba}^i) \cup (\mathfrak{H}_{cb}^i \circ \mathcal{ID}). \tag{19.83}$$

Note that $\mathfrak{N}_{cb}^{i+1} \circ \mathfrak{H}_{ba}^i$ is a homotopy from $\mathfrak{N}_{cb}^{i+1} \circ \mathcal{ID} \circ \mathfrak{N}_a^{i+1i}$ to $\mathfrak{N}_{cb}^{i+1} \circ \mathfrak{N}_b^{i+1i} \circ \mathcal{ID}$ and $\mathfrak{H}_{cb}^i \circ \mathcal{ID}$ is a homotopy from $\mathfrak{N}_{cb}^{i+1} \circ \mathfrak{N}_b^{i+1i} \circ \mathcal{ID}$ to $\mathfrak{N}_c^{i+1i} \circ \mathfrak{N}_{cb}^i \circ \mathcal{ID}$. (See Diagram (19.82).)

Now we start the construction of the homotopy between $\mathfrak{N}_{cb} \circ \mathcal{ID}_{\mathcal{FF}}$ and $\mathfrak{N}_{cb}$. By Proposition 18.60 we have $\mathfrak{N}_{cb}^i \circ \mathcal{ID} \sim \mathfrak{N}_{cb}^i$. Let $\mathfrak{H}^i$ be this homotopy. Note that as $\mathfrak{H}^i$ we take the specific homotopy we constructed during the proof of Proposition 18.60. To prove Lemma 19.47 it suffices to construct a homotopy of homotopies $\mathcal{H}^i$ appearing in Definition 16.36 (4). Recall that $\mathcal{H}^i$ is a $[0, 1]^2$-parametrized partial morphism from $\mathcal{F}_a^i$ to $\mathcal{F}_c^{i+1}$ such that its normalized boundary $\partial \mathcal{H}^i$ is a disjoint union of the following four homotopies:

(i) $\mathfrak{H}_{cb}^i$.
(ii) $\mathfrak{H}^{i+1} \circ \mathfrak{N}_a^{i+1i}$.
(iii) $\mathfrak{H}_{ca}^i = (\mathfrak{N}_{cb}^{i+1} \circ \mathfrak{H}_{ba}^i) \cup (\mathfrak{H}_{cb}^i \circ \mathcal{ID})$.

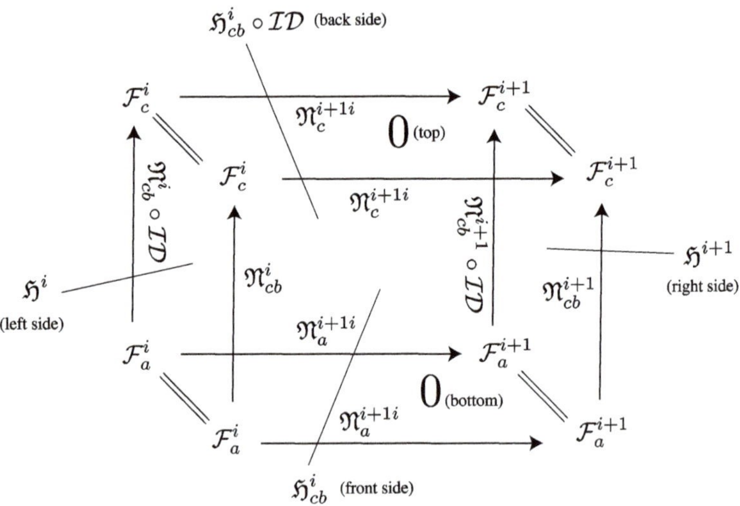

Fig. 19.5 Cubic diagram

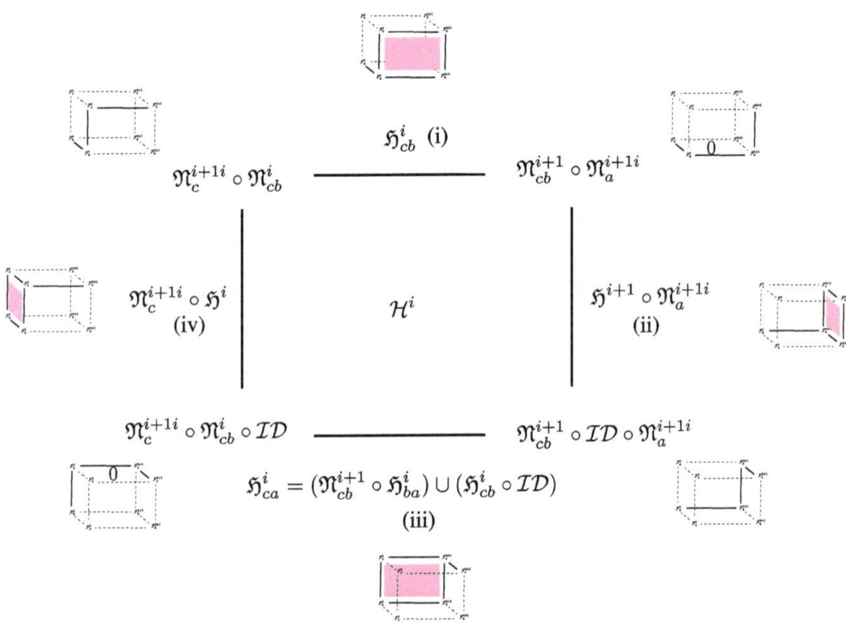

Fig. 19.6 $\mathcal{H}^i$ and its boundary

(iv) $\mathfrak{N}^{i+1i}_c \circ \mathfrak{H}^i$.

See Figs. 19.5 and 19.6. We remark that the boundary of $\mathcal{H}^i$ has four components instead of six components because two of the homomtopies which give the top and bottom faces of Fig. 19.5 are trivial. We will construct an interpolation space

$$\mathcal{N}(ac, ii+1, \alpha_-, \alpha'_+; [1,2]^2)$$

of the homotopy of homotopies $\mathcal{H}^i$ by modifying the interpolation space

$$\mathcal{N}(bc, ii+1, \alpha_-, \alpha'_+; [1,2])$$

of $\mathfrak{H}^i_{cb}$ in a way similar to the proof of Proposition 18.60 as follows: We note that the restriction of $\mathcal{N}(bc, ii+1; \alpha_-, \alpha_+; [1,2])$ to $1 \in \partial[1,2]$ and to $2 \in \partial[1,2]$ is the union of the following fiber products, respectively.

(I) $\mathcal{N}(b, ii+1; \alpha_-, \alpha) \times^{\boxplus\tau}_{R_\alpha} \mathcal{N}(bc, i+1; \alpha, \alpha'_+)$.
(II) $\mathcal{N}(bc, i; \alpha_-, \alpha') \times^{\boxplus\tau}_{R_{\alpha'}} \mathcal{N}(c, ii+1; \alpha', \alpha'_+)$.

There are two other kinds of boundary of $\mathcal{N}(bc, ii+1; \alpha_-, \alpha'_+; [1,2])$ as follows:

(III) $\mathcal{M}(b, i; \alpha_-, \alpha) \times_{R_\alpha} \mathcal{N}(bc, ii+1; \alpha, \alpha'_+; [1,2])$.
(IV) $\mathcal{N}(bc, ii+1; \alpha_-, \alpha'; [1,2]) \times_{R_{\alpha'}} \mathcal{M}(b, i; \alpha', \alpha'_+)$.

Let C be a sufficiently large positive number. We assume that it is large enough compared to the energy loss of $\mathfrak{N}^{cb}_i$. The top-dimensional stratum of our interpolation space $\mathcal{N}(ac, ii+1; \alpha_-, \alpha'_+; [1,2])$ is

(1) $\overset{\circ}{\mathcal{N}}(bc, ii+1; \alpha_-, \alpha'_+; [1,2]) \times (E(\alpha_-), E(\alpha'_+) + C)$.

This is the only stratum of top dimension. Below we list the codimension 1 strata:

(2) $R_{\alpha_-} \times \{E(\alpha_-)\}$
$\times_{R_\alpha} \overset{\circ}{\mathcal{N}}(bc, ii+1; \alpha_-, \alpha_+; [1,2])$
(3) $\overset{\circ}{\mathcal{M}}(b, i; \alpha_-, \alpha) \times (E(\alpha_-), E(\alpha))$
$\times^{\boxplus\tau}_{R_\alpha} \overset{\circ}{\mathcal{N}}(bc, ii+1; \alpha, \alpha_+; [1,2])$
(4) $\overset{\circ}{\mathcal{M}}(b, i; \alpha_-, \alpha)$
$\times_{R_\alpha} \overset{\circ}{\mathcal{N}}(bc, ii+1; \alpha, \alpha'_+; [1,2]) \times (E(\alpha), E(\alpha'_+) + C)$
(5) $\overset{\circ}{\mathcal{N}}(b, ii+1; \alpha_-, \alpha) \times (E(\alpha_-), E(\alpha))$
$\times^{\boxplus\tau}_{R_\alpha} \overset{\circ}{\mathcal{N}}(bc, i+1; \alpha, \alpha'_+)$
(6) $\overset{\circ}{\mathcal{N}}(b, ii+1; \alpha_-, \alpha)$
$\times^{\boxplus\tau}_{R_\alpha} \overset{\circ}{\mathcal{N}}(bc, i+1; \alpha, \alpha'_+) \times (E(\alpha), E(\alpha'_+) + C)$
(7) $\overset{\circ}{\mathcal{N}}(bc, ii+1; \alpha_-, \alpha'_+; [1,2]) \ {}_{R_{\alpha'_+}} \times R_{\alpha'_+} \times \{E(\alpha'_+) + C\}$
(8) $\overset{\circ}{\mathcal{N}}(bc, i; \alpha_-, \alpha') \times (E(\alpha_-), E(\alpha'_+) + C)$
$\times^{\boxplus\tau}_{R_{\alpha'}} \overset{\circ}{\mathcal{N}}(c, ii+1; \alpha', \alpha'_+)$

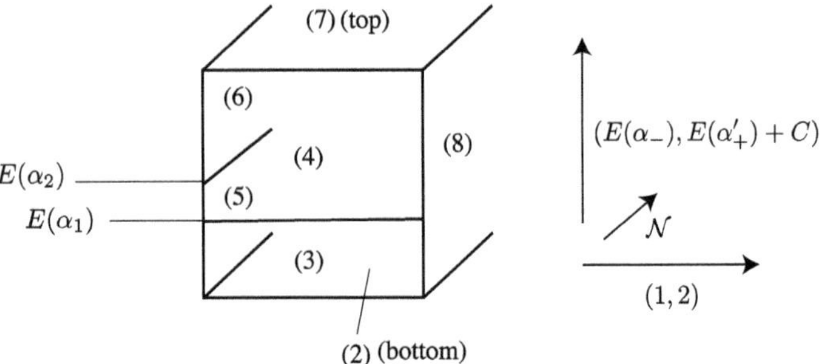

$$(3) = \overset{\circ}{\mathcal{M}}(b, i; \alpha_-, \alpha_1) \times (E(\alpha_-), E(\alpha_1)) \times^{\boxplus \tau}_{R_{\alpha_1}} \overset{\circ}{\mathcal{N}}(bc, ii+1; \alpha_1, \alpha_+; [1,2])$$

$$(5) = \overset{\circ}{\mathcal{N}}(b, ii+1; \alpha_-, \alpha_2) \times (E(\alpha_-), E(\alpha_2)) \times^{\boxplus \tau}_{R_{\alpha_2}} \overset{\circ}{\mathcal{N}}(bc, i+1; \alpha_2, \alpha'_+)$$

Fig. 19.7 Boundary components (2)–(9)

(9) $\overset{\circ}{\mathcal{N}}(bc, ii+1; \alpha_-, \alpha'; [1,2]) \times (E(\alpha_-), E(\alpha'_+) + C)$
$\times_{R_{\alpha'}} \overset{\circ}{\mathcal{M}}(c, i; \alpha', \alpha'_+)$

See Fig. 19.7. Observe that (3) ∪ (4), (5) ∪ (6), (8), (9) are products of (III), (I), (II), (IV) and the interval $(E(\alpha_-), E(\alpha'_+) + C)$, respectively.

We also find that $(2) \cup (3) = \mathfrak{H}_i^{cb} \circ \mathcal{ID}$ and $(5) = \mathfrak{N}_{i+1}^{cb} \circ \mathfrak{H}_i^{ba}$. Therefore $(2) \cup (3) \cup (5) = $ (iii). We also have (6) = (ii), (7) = (i).

We can regard (8) = (iv). Here the interval $(E(\alpha_-), E(\alpha'_+) + C)$ appearing in (8) is not $(E(\alpha_-), E(\alpha') + C)$, but it is diffeomorphic.

Moreover, (4) is nothing but

$$\overset{\circ}{\mathcal{M}}(a, i; \alpha_-, \alpha) \times_{R_\alpha} \overset{\circ}{\mathcal{N}}(ac, ii+1, \alpha, \alpha'_+; [1,2]^2)$$

and we can regard (9) as

$$\overset{\circ}{\mathcal{N}}(ac, ii+1, \alpha_-, \alpha'; [1,2]^2) \times_{R_{\alpha'}} \overset{\circ}{\mathcal{M}}(c, i+1; \alpha', \alpha).$$

Thus the boundary of $\mathcal{N}(ac, ii+1, \alpha_-, \alpha'_+; [1,2]^2)$ has the required properties.

We note that we need to smooth corners and bend a part of the boundary so that corner structure stratification of $\mathcal{N}(ac, ii+1, \alpha_-, \alpha'_+; [1,2]^2)$ becomes the correct one. In fact, since the union of (2)(3)(5) is (iv), we need to smooth the corner where they intersect.

For example, the intersection of (3) and (5) is

$$\begin{aligned}&\overset{\circ}{\mathcal{M}}(b, i; \alpha_-, \alpha_1) \times^{\boxplus\tau}_{R_{\alpha_1}} \overset{\circ}{\mathcal{N}}(b, ii+1; \alpha_1, \alpha_2) \times^{\boxplus\tau}_{R_\alpha} \overset{\circ}{\mathcal{N}}(bc, ii+1; \alpha_2, \alpha'_+; [1,2]) \\ &\times (E(\alpha_-), E(\alpha_1)).\end{aligned}$$

So we smooth the corner here. On the other hand

$$\begin{aligned}&\overset{\circ}{\mathcal{M}}(b, i; \alpha_-, \alpha_1) \times^{\boxplus\tau}_{R_{\alpha_1}} \overset{\circ}{\mathcal{N}}(b, ii+1; \alpha_1, \alpha_2) \times^{\boxplus\tau}_{R_\alpha} \overset{\circ}{\mathcal{N}}(bc, ii+1; \alpha_2, \alpha'_+; [1,2]) \\ &\times (E(\alpha_1), E(\alpha_2))\end{aligned}$$

is a part of the boundary of (5) which is not contained in the boundary of (3). We bend the boundary component

$$\begin{aligned}&\overset{\circ}{\mathcal{M}}(b, i; \alpha_-, \alpha_1) \times^{\boxplus\tau}_{R_{\alpha_1}} \overset{\circ}{\mathcal{N}}(b, ii+1; \alpha_1, \alpha_2) \times^{\boxplus\tau}_{R_\alpha} \overset{\circ}{\mathcal{N}}(bc, ii+1; \alpha_2, \alpha'_+; [1,2]) \\ &\times (E(\alpha_-), E(\alpha_2))\end{aligned}$$

at

$$\begin{aligned}&\overset{\circ}{\mathcal{M}}(b, i; \alpha_-, \alpha_1) \times^{\boxplus\tau}_{R_{\alpha_1}} \overset{\circ}{\mathcal{N}}(b, ii+1; \alpha_1, \alpha_2) \times^{\boxplus\tau}_{R_\alpha} \overset{\circ}{\mathcal{N}}(bc, ii+1; \alpha_2, \alpha'_+; [1,2]) \\ &\times \{E(\alpha_1)\}.\end{aligned}$$

See Figs. 19.8 and 19.9. This bending is included in the process we constructed the homotopy $\mathfrak{H}^i$ between $\mathfrak{N}^i_{cb}$ and $\mathcal{ID} \sim \mathfrak{N}^i_{cb}$. (See the proof of Proposition 18.60.)

We also note that in (4)(9) we take $\times_{R_\alpha}$ etc. but in other places we take $\times^{\boxplus}_{R_\alpha}$. This does *not* cause inconsistency at the intersection of (3) and (4) by the following reasoning. We first take the fiber product

$$\overset{\circ}{\mathcal{M}}(b, i; \alpha_-, \alpha) \times_{R_\alpha} \overset{\circ}{\mathcal{N}}(bc, ii+1; \alpha, \alpha'_+; [1,2]) \times (E(\alpha_-), E(\alpha'_+))$$

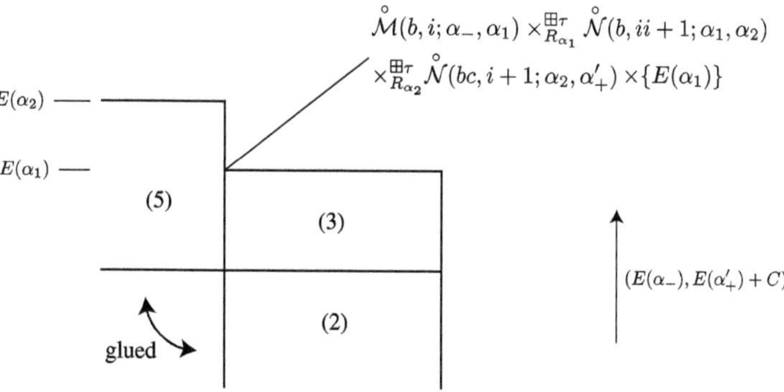

Fig. 19.8 Domain (2) + (3) + (5) developed

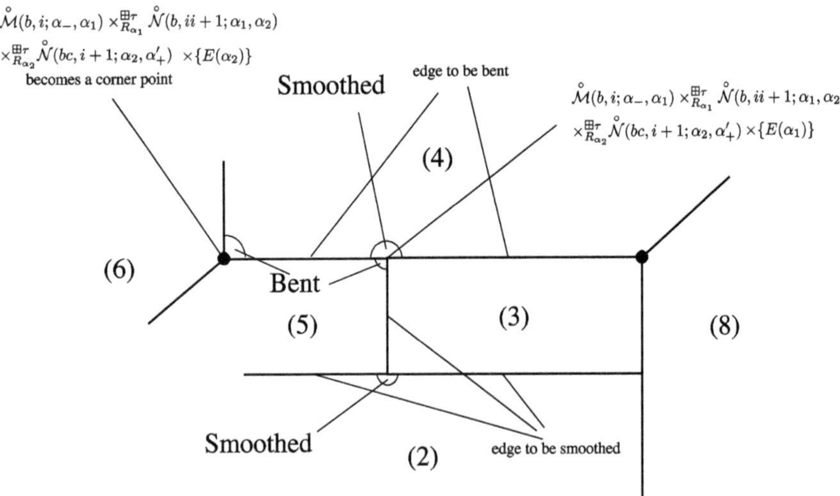

Fig. 19.9 Bent and smoothed edges and vertex around (2) + (3) + (5)

and bend it at

$$\overset{\circ}{\mathcal{M}}(b,i;\alpha_-,\alpha) \times_{R_\alpha} \overset{\circ}{\mathcal{N}}(bc,ii+1;\alpha,\alpha'_+;[1,2]) \times \{E(\alpha)\}.$$

We then partially trivialize by choosing $\mathfrak{C}$ = union of components

$$\overset{\circ}{\mathcal{M}}(b,i;\alpha_-,\alpha) \times_{R_\alpha} \overset{\circ}{\mathcal{N}}(bc,ii+1;\alpha,\alpha'_+;[1,2]) \times (E(\alpha_-),E(\alpha)).$$

Then when we consider the corner

$$\overset{\circ}{\mathcal{M}}(b,i;\alpha_-,\alpha_1) \times_{R_{\alpha_1}} \mathcal{M}(b,i;\alpha_1,\alpha_2) \times_{R_{\alpha_2}} \overset{\circ}{\mathcal{N}}(bc,ii+1;\alpha_2,\alpha'_+;[1,2]) \\ \times (E(\alpha),E(\alpha'_+)),$$

the part where the $\mathfrak{C}$-outer collaring is performed is nothing but

$$\overset{\circ}{\mathcal{M}}(b,i;\alpha_-,\alpha_1) \times_{R_{\alpha_1}} \mathcal{M}(b,i;\alpha_1,\alpha_2) \times_{R_{\alpha_2}} \overset{\circ}{\mathcal{N}}(bc,ii+1;\alpha_2,\alpha'_+;[1,2]) \\ \times (E(\alpha_2),E(\alpha'_+)).$$

This is because the $\mathfrak{C}$-outer collaring is performed at the intersection of the two boundary components which belongs to $\mathfrak{C}$. We smooth those corners to obtain the union of (3) and (4).

In the same way as in the proof of Proposition 18.60, we can describe the higher codimension boundary of our K-space $\mathcal{N}(ac,ii+1,\alpha_-,\alpha'_+;[1,2]^2)$, and can check the consistency in a straightforward way. This finishes the proof of Lemma 19.47.

□

The proof of Theorem 16.39 (6) is now complete. □

We remark that a fiber product of orbifolds with corners is always an orbifold with corners. Therefore 'polygon' appearing in the above proof is always locally modeled by $[0, 1)^k \times \mathbb{R}^m$.

Thus we have completed the proof of all the results claimed in Chap. 16.

Remark 19.48 In this section we use the homotopy of homotopies version of algebraic lemmas (Proposition 19.41) for the promotion of homotopies. We can avoid it if we use the next lemma instead.

Lemma 19.49 *Let* $\psi_1, \psi_2 : CF(\mathcal{F}) \to CF(\mathcal{F}')$ *be two gapped cochain maps. (See Definition* 16.12 *for gappedness.) Suppose* $\psi_1|_E$ *is partially cochain homotopic to* $\psi_2|_E$ *for any* E*, up to energy cut level* E*. Then* ψ_1 *is cochain homotopic to* ψ_2*.*

We can prove Lemma 19.49 in the same way as in the proof of [FOOO4, Lemma 7.2.177]. Note that Lemma 19.49 itself is valid for any ground ring R, while [FOOO4, Lemma 7.2.177] is proved only for the case when the ground ring is $\mathbb{C}$ or a finite field. (In fact, [FOOO4, Remark 7.2.181] gives a counterexample to [FOOO4, Lemma 7.2.177] for the case when the ground ring is $\mathbb{Q}$.) The reason why Lemma 19.49 is valid for any ground ring R is that the equation for a map to be a cochain map or a cochain homotopy is linear. On the other hand, the equation for a map to be an A_∞ map is nonlinear.

We choose to prove Proposition 19.41 rather than using Lemma 19.49, since it is more direct to apply Proposition 19.41 to other situations.

Chapter 20
Linear K-Systems: Floer Cohomology III – Morse Case by Multisection

In this chapter we provide the way we associate Floer cohomology with *ground ring* $\mathbb{Q}$ to a linear K-system when the critical submanifolds are 0-dimensional.

Definition 20.1 We say a (partial) linear K-system as in Condition 16.1 and Definition 16.6 is *of Morse type* if all the critical submanifolds R_α consist of finite sets.

Theorem 20.2 *For a (partial) linear K-system of Morse type, Theorems* 16.9, 16.31, 16.39 *hold with* $\mathbb{R}$ *replaced by* $\mathbb{Q}$.

The proof is done by replacing *CF-perturbations* used in Chap. 19 by *multivalued perturbations* (or multisections) in this chapter. We recall that we defined the notion of a fiber product of CF-perturbations in Chap. 10. There is a certain difficulty in defining the notion of a fiber product of multivalued perturbations. (In fact, it is not correct to assume that the evaluation map restricted to the zero set of multisection is submersive, because this assumption is not satisfied even in the case of generic perturbations by an obvious dimensional reason.) On the other hand, in the case of direct product or fiber product over 0-dimensional spaces, we can define the notion of a (fiber) product of multivalued perturbations in an obvious way. This is the main idea of the proof. To work it out in detail, we need to check carefully that the whole proof in Chap. 19 (and ones in Chap. 17 which are used in Chap. 19) can be carried out using multivalued perturbation in place of CF-perturbations. The proofs are either automatic or a straightforward analogue of the corresponding results in Chap. 17 or 19.

The contents of this chapter are not used in the other parts of this book.

K. Fukaya et al., *Kuranishi Structures and Virtual Fundamental Chains*, Springer Monographs in Mathematics, https://doi.org/10.1007/978-981-15-5562-6_20

20.1 Extension of a Multisection from Boundary to Its Neighborhood

In this section we discuss an analogue of the story in Chap. 17 for multivalued perturbations.

Lemma 20.3 *Suppose we are in the situation of Situation* 17.55. *We assume in addition that we are given a τ-collared multivalued perturbation $\widehat{\mathfrak{s}^+_{S_k}}$ of $\widehat{\mathcal{U}^+_{S_k}}$ such that the pullback of $\widehat{\mathfrak{s}^+_{S_{k+\ell}}}$ by the map $\pi_{k,\ell} : \widehat{S}_k(\widehat{S}_\ell(X), \widehat{\mathcal{U}^+_{S_\ell}}) \to (\widehat{S}_{k+\ell}(X), \widehat{\mathcal{U}^+_{S_{k+\ell}}})$ coincides with the restriction of $\widehat{\mathfrak{s}^+_{S_\ell}}$.*

Then for any $0 < \tau' < \tau$ there exists a τ'-collared multivalued perturbation $\widehat{\mathfrak{s}^+}$ on the Kuranishi structure $\widehat{\mathcal{U}^+}$ obtained in Proposition 17.58 *such that its restriction to $(\widehat{S}_k(X), \widehat{\mathcal{U}^+_{S_k}})$ coincides with $\widehat{\mathfrak{s}^+_{S_k}}$.*

Proof The proof is the same as the proof of Proposition 17.73.

Now we discuss results corresponding to one in Sect. 17.10.

Lemma 20.4 *Suppose we are in the situation of Lemma* 20.3. *We assume that $\widehat{\mathfrak{s}^+_k}$ are transversal to* 0. *Then there exists a τ'-collared multivalued perturbation $\widehat{\mathfrak{s}^{++}}$ on the Kuranishi structure $\widehat{\mathcal{U}^{++}}$ obtained in Proposition* 17.78 *such that:*

(1) *Its restriction to $(\widehat{S}_k(X), \widehat{\mathcal{U}^+_{S_k}})$ coincides with $\widehat{\mathfrak{s}^+_k}$.*
(2) *$\widehat{\mathfrak{s}^{++}}$ is transversal to* 0.

Proof The proof is the same as the proof of Proposition 17.81. □

20.2 Completion of the Proof of Theorem 20.2

***Proof of Theorem* 20.2** We first show a version of Proposition 19.1.

Proposition 20.5 *Suppose we are in the situation of Proposition* 19.1. *Moreover we assume that the linear K-system is of Morse type.*

Then for any $0 < \tau < 1$ there exists a τ-collared Kuranishi structure $\widehat{\mathcal{U}^+}(\alpha_-, \alpha_+)$ of $\mathcal{M}(\alpha_-, \alpha_+)^{\boxplus\tau_0}$, multivalued perturbations $\widehat{\mathfrak{s}^+}(\alpha_-, \alpha_+)$ of $\widehat{\mathcal{U}^+}(\alpha_-, \alpha_+)$ for every α_-, α_+ with $E(\alpha_+) - E(\alpha_-) \le E^k_{\mathfrak{C}}$, which enjoy the following properties:

(1) *The same as Proposition* 19.1 *(1).*
(2) *$\widehat{\mathfrak{s}^+}(\alpha_-, \alpha_+)$ is transversal to* 0.
(3) *The pullback of $\widehat{\mathfrak{s}^+}(\beta\alpha_-, \beta\alpha_+)$ by the periodicity isomorphism coincides with $\widehat{\mathfrak{s}^+}(\alpha_-, \alpha_+)$.*

(4) *Proposition* 19.1 *(4) holds.*

(5) *The pullback of* $\widehat{\mathfrak{s}^+}(\alpha_-, \alpha_+)$ *by the isomorphism* (19.2) *coincides with the fiber product*

$$\widehat{\mathfrak{s}^+}(\alpha_-, \alpha)_{\mathrm{ev}_+} \times_{\mathrm{ev}_-} \widehat{\mathfrak{s}^+}(\alpha, \alpha_+).$$

These fiber products are well-defined since they are direct-like products.

(6) *Proposition* 19.1 *(6) holds.*

(7) *Proposition* 19.1 *(7) holds.*

(8) *The pullback of* $\widehat{\mathfrak{s}^+}(\alpha_-, \alpha_+)$ *by the isomorphism* (19.3) *is the fiber product*

$$\widehat{\mathfrak{s}^+}(\alpha_-, \alpha_1)\ {}_{\mathrm{ev}_+} \times_{R_{\alpha_1}} \cdots {}_{R_{\alpha_k}} \times_{\mathrm{ev}_-} \widehat{\mathfrak{s}^+}(\alpha_k, \alpha_+). \tag{20.1}$$

This fiber product is well-defined because it is a direct-like product.

Proof Using the results of Sect. 20.1, the proof is the same as the proof of Proposition 19.1. □

We next rewrite Proposition 20.5 in the algebraic language. In our situation (linear K-system of Morse type) we define

$$\Omega(R_\alpha) = \bigoplus_{r \in \Omega(R_\alpha)} \mathbb{Q}[r]. \tag{20.2}$$

We next define

$$\mathfrak{m}^n_{1;\alpha_+,\alpha_-} : \Omega(R_{\alpha_-}) \to \Omega(R_{\alpha_+}) \tag{20.3}$$

in the case when $\dim \mathcal{M}(\alpha_-, \alpha_+) = 0$ by the formula

$$\begin{aligned} &\mathfrak{m}^n_{1;\alpha_+,\alpha_-}([r_-]) \\ &= \sum_{r_+ \in R_{\alpha_+}} [(ev_-, ev_+)^{-1}((r_-, r_+)) \\ &\qquad \cap (\mathcal{M}(\alpha_-, \alpha_+)^{\boxplus\tau_0}, \widehat{\mathcal{U}^+}(\alpha_-, \alpha_+), \widehat{\mathfrak{s}^{+n}}(\alpha_-, \alpha_+))][r_+]. \end{aligned} \tag{20.4}$$

Here the coefficient of $[r_+]$ in the right hand side is the virtual fundamental chain as in Definition 14.18, which is a rational number, and $r_- \in R_{\alpha_-}$.

We modify ($\mathfrak{b}$) appearing in Remark 19.4 as follows:

($\mathfrak{b}'$) For any energy cut level E_0 there exists $n_0 = n_0(E_0) > 0$ such that the operator $\mathfrak{m}^n_{1;\alpha_+,\alpha_-}$ is defined when $0 < E(\alpha_+) - E(\alpha_-) \le E_0$ and $n > n_0$.

Lemma 20.6 *The operators* $\mathfrak{m}^n_{1;\alpha_+,\alpha_-}$ *in* (20.4) *satisfy the following equality in the sense of* ($\flat'$)*:*

$$\sum_{\alpha; E(\alpha_-)<E(\alpha)<E(\alpha_+)} \mathfrak{m}^n_{1;\alpha_+,\alpha} \circ \mathfrak{m}^n_{1;\alpha,\alpha_-} = 0. \tag{20.5}$$

Proof Using Theorems 14.20, 14.25 and Proposition 20.5, the proof is the same as the proof of Lemma 19.5. □

We have thus rewritten Sect. 19.1 by using multivalued perturbations in the case of a linear K-system of Morse type. It is now obvious that we can rewrite Sects. 19.2, 19.3, 19.4, 19.5, 19.6, 19.7, and 19.8 in the same way and complete the proof of Theorem 20.2. □

Remark 20.7 However, we need to remark the following point. We defined the notion of uniformity of CF-perturbations and used it in Chap. 19. See Remark 19.17. We do not define such a notion for multivalued perturbations. We explain the reason why we do not need uniformity for multivalued perturbations. The construction of the chain complex uses the homotopy inductive limit. In our situation we use the following two facts. Let $G = \{E_1, E_2, \dots, \}$. We put

$$CF(\mathcal{C}) = \bigoplus_{\alpha} \Omega(R_\alpha).$$

(1) For each k there exists n_k such that we can construct a partial chain complex

$$(CF(\mathcal{C}), \mathfrak{m}^n_1)$$

of energy cut level E_k for $n \geq n_k$.

(2) For each k there exists n'_k such that there exists a partial chain map ψ_n between energy cut to energy level E_k of $(CF(\mathcal{C}), \mathfrak{m}^n_1)$ and of $(CF(\mathcal{C}), \mathfrak{m}^{n+1}_1)$ if $n \geq n'_k$. Moreover ψ_n is partial chain homotopy equivalence and is congruent to the identity map modulo T^ϵ.

Now we consider $(CF(\mathcal{C}), \mathfrak{m}^{n_k}_1)$ and $(CF(\mathcal{C}), \mathfrak{m}^{n_{k+1}}_1)$. We use the composition

$$\psi_{n_{k+1}-1} \circ \psi_{n_{k+1}-2} \circ \dots \circ \psi_{n_k} : (CF(\mathcal{C}), \mathfrak{m}^{n_k}_1) \to (CF(\mathcal{C}), \mathfrak{m}^{n_{k+1}}_1),$$

which is a partial chain map of energy cut level E_k and is a chain homotopy equivalence. Moreover it is congruent to the identity map modulo T^ϵ. Therefore we can use Lemma 19.13 to promote $(CF(\mathcal{C}), \mathfrak{m}^{n_k}_1)$ to energy cut level E_{k+1}. We can continue in the same way to promote $(CF(\mathcal{C}), \mathfrak{m}^{n_k}_1)$ to energy cut level ∞. By the same argument we can promote all the partial chain complexes $(CF(\mathcal{C}), \mathfrak{m}^n_1)$ to energy cut level ∞ so that they are all chain homotopy equivalent. The inductive construction of chain maps or of chain homotopies can be performed in the same way. Thus we do not need the notion of uniformity. The reason why we used the uniformity in the case of CF-perturbations is that a CF-perturbation is an ϵ-parametrized family and ϵ moves in the uncountable set $(0, 1]$. So to prove all the promotions of the boundary operators $\mathfrak{m}_1^\epsilon$ are chain homotopy equivalent to each other, we need to show that $\mathfrak{m}_1^\epsilon$ is chain homotopy equivalent to $\mathfrak{m}_1^{\epsilon'}$ for all the sufficiently small pair ϵ, ϵ'. This is the place where we use uniformity.

Chapter 21
Tree-Like K-Systems: A_∞ Structure I – Statement

In Chaps. 21 and 22 we discuss construction of a filtered A_∞ structure associated to a relatively spin Lagrangian submanifold L of a symplectic manifold. This construction had been worked out in great detail in the book [FOOO3], [FOOO4] based on singular homology, before its de Rham version was given in [FOOO8, Section 12], [FOOO10], [Fuk4].

In this book, based on de Rham cohomology, we give its detail again. We also provide a package so that the construction part of various Kuranishi structures and their usage part of obtaining a filtered A_∞ structure are clearly separated from each other. For this purpose, we take an axiomatic approach as in the case of linear K-systems developed up to the previous chapters. A similar axiomatic treatment was given in [Fuk5]. The axiom we give here is slightly different from that of [Fuk5]. In [Fuk5] we used a geometric operad (the Stasheff operad) to formulate Kuranishi A_∞ correspondence. In this book we use the *K-system organized by a tree*, sometimes called a *tree-like K-system* in short (plus certain additional data) rather than the Stasheff operad. In fact, in [Fuk5] the Stasheff operad ([MSS]) was used only to describe the combinatorial data on the way various strata (which are fiber products of moduli spaces of pseudo-holomorphic disks) are glued. Since the members of the Stasheff operad are cells, they do not carry a nontrivial homology class. So in [Fuk5] we actually did not use the geometric data of the Stasheff operad but used only its combinatorial structure (that is, the way various strata intersect).

21.1 Axiom of Tree-Like K-Systems: A_∞ Correspondence

In the rest of Part II, we assume that L is a smooth oriented closed manifold.

Definition 21.1 Let $\mathfrak{G}$ be an additive group and $\mu : \mathfrak{G} \to \mathbb{Z}$, $E : \mathfrak{G} \to \mathbb{R}$ group homomorphisms. We call $\mu(\beta)$ the *Maslov index* of β and $E(\beta)$ the *energy* of β.

A *decorated rooted ribbon tree* is $(\mathcal{T}, \beta(\cdot))$ such that:

K. Fukaya et al., *Kuranishi Structures and Virtual Fundamental Chains*, Springer Monographs in Mathematics, https://doi.org/10.1007/978-981-15-5562-6_21

(1) $\mathcal{T}$ is a connected tree. Let $C_0(\mathcal{T})$, $C_1(\mathcal{T})$ be the sets of all vertices and edges of $\mathcal{T}$, respectively.
(2) For each $\mathrm{v} \in C_0(\mathcal{T})$ we fix a cyclic order of the set of edges containing v. This is equivalent to fixing an isotopy type of an embedding of $\mathcal{T}$ to the plane $\mathbb{R}^2$. (Namely, the cyclic order of the edges is given by the orientation of the plane so that the edges are enumerated according to the counterclockwise orientation. We call it a *ribbon structure* at the vertex v.)
(3) $C_0(\mathcal{T})$ is divided into the set of *exterior vertices* $C_{0,\mathrm{ext}}(\mathcal{T})$ and the set of *interior vertices* $C_{0,\mathrm{int}}(\mathcal{T})$.
(4) We fix one element of $C_{0,\mathrm{ext}}(\mathcal{T})$, which we call the *root*.
(5) The valency of all the exterior vertices are 1.
(6) $\beta(\cdot) : C_{0,\mathrm{int}}(\mathcal{T}) \to \mathfrak{G}$ is a map. We require $E(\beta(\mathrm{v})) \geq 0$. Moreover, if $E(\beta(\mathrm{v})) = 0$, then $\beta(\mathrm{v})$ is required to be the unit.
(7) **(Stability)** For each $\mathrm{v} \in C_{0,\mathrm{int}}(\mathcal{T})$ we assume that one of the following holds:

    (a) $E(\beta(\mathrm{v})) > 0$.
    (b) The valency of v is not smaller than 3.

We denote by $\mathcal{G}(k+1, \beta)$ the set of all decorated ribbon trees $(\mathcal{T}, \beta(\cdot))$ such that:

(I) $\#C_{0,\mathrm{ext}}(\mathcal{T}) = k + 1$.
(II) $\sum_{\mathrm{v} \in C_{0,\mathrm{int}}(\mathcal{T})} (\beta(\mathrm{v})) = \beta$.

We decompose the set of edges $C_1(\mathcal{T})$ as follows. If an edge e contains an exterior vertex, we call e an *exterior edge*. Otherwise we call e an *interior edge*. We denote by $C_{1,\mathrm{int}}(\mathcal{T})$, (resp. $C_{1,\mathrm{ext}}(\mathcal{T})$) the set of all interior (resp. exterior) edges.

Next we define the fiber product of K-spaces along an element $(\mathcal{T}, \beta(\cdot))$ of $\mathcal{G}(k + 1, \beta)$. Suppose we are given K-spaces $\mathcal{M}_{k+1}(\beta)$ and maps

$$\mathrm{ev} = (\mathrm{ev}_0, \ldots, \mathrm{ev}_k) : \mathcal{M}_{k+1}(\beta) \to L^{k+1}$$

for each β and $k \in \mathbb{Z}_{\geq 0}$. Let $(\mathcal{T}, \beta(\cdot)) \in \mathcal{G}(k + 1, \beta)$. We define the K-space

$$\prod_{(\mathcal{T},\beta(\cdot))} \mathcal{M}_{k_{\mathrm{v}}+1}(\beta(\mathrm{v})) \tag{21.1}$$

as follows. We consider the direct product $\prod_{\mathrm{v} \in C_{0,\mathrm{int}}(\mathcal{T})} \mathcal{M}_{k_{\mathrm{v}}+1}(\beta(\mathrm{v}))$, where $k_{\mathrm{v}} + 1$ is the valency of the vertex v. We take two copies of L for each interior edge $\mathrm{e} \in C_{1,\mathrm{int}}(\mathcal{T})$. We define a map

$$\mathrm{ev} : \prod_{\mathrm{v} \in C_{0,\mathrm{int}}(\mathcal{T})} \mathcal{M}_{k_{\mathrm{v}}+1}(\beta(\mathrm{v})) \to \prod_{\mathrm{e} \in C_{1,\mathrm{int}}(\mathcal{T})} L^2 \tag{21.2}$$

as follows. For each $\mathrm{v} \in C_{0,\mathrm{int}}(\mathcal{T})$ we enumerate the edges containing v as

$$\mathrm{e}_{\mathrm{v},0}, \ldots, \mathrm{e}_{\mathrm{v},k_{\mathrm{v}}}$$

such that the following conditions are satisfied.

Condition 21.2

(1) The edge $\mathrm{e}_{\mathrm{v},0}$ is contained in the connected component of $\mathcal{T}\setminus\{\mathrm{v}\}$ which contains the root.
(2) $(\mathrm{e}_{\mathrm{v},0},\dots,\mathrm{e}_{\mathrm{v},k_{\mathrm{v}}})$ respects the cyclic ordering of the edges given by the ribbon structure at v.

Such an enumeration is unique. Each edge e contains two vertices. For one of them v_- we have $\mathrm{e}=\mathrm{e}_{\mathrm{v}_-,0}$. For the other vertex v_+ contained in e, we have $\mathrm{e}=\mathrm{e}_{\mathrm{v}_+,i}$ for some $i\in\{1,\dots,k_{\mathrm{v}_+}\}$. We define the e component of $\mathrm{ev}((\mathbf{x}_{\mathrm{v}})_{\mathrm{v}\in C_{0,\mathrm{int}}(\mathcal{T})})$ as $(\mathrm{ev}_0(\mathbf{x}_{\mathrm{v}_-}),\mathrm{ev}_i(\mathbf{x}_{\mathrm{v}_+}))$, where $\mathrm{e}=\mathrm{e}_{\mathrm{v}_+,i}$ and $\mathbf{x}_{\mathrm{v}}\in\mathcal{M}_{k_{\mathrm{v}}+1}(\beta(\mathrm{v}))$. (Here ev is the map in (21.2).)

Definition 21.3 The fiber product (21.1) is defined by

$$\left(\prod_{\mathrm{v}\in C_{0,\mathrm{int}}(\mathcal{T})}\mathcal{M}_{k_{\mathrm{v}}+1}(\beta(\mathrm{v}))\right)\ {}_{\mathrm{ev}}\times_{\prod_{\mathrm{e}\in C_{1,\mathrm{int}}(\mathcal{T})}L^2}\left(\prod_{\mathrm{e}\in C_{1,\mathrm{int}}(\mathcal{T})}L\right). \tag{21.3}$$

Here ev is as in (21.2) and $\prod_{\mathrm{e}\in C_{1,\mathrm{int}}(\mathcal{T})}L$ is the product of the diagonal $L\subset L^2$ and is contained in $\prod_{\mathrm{e}\in C_{1,\mathrm{int}}(\mathcal{T})}L^2$. We call (21.3) *the fiber product of* $\mathcal{M}_{k+1}(\beta)$ *along* $(\mathcal{T},\beta(\cdot))$.

Remark 21.4 The fiber product (21.3) in the sense of K-spaces may not be defined because of the transversality problem. It is defined if Condition 21.5 is satisfied:

Condition 21.5 The map $\mathrm{ev}_0:\mathcal{M}_{k+1}(\beta)\to L$ is weakly submersive.

Condition 21.6 We consider the following objects:

(I) $\mathfrak{G}$ is an additive group. (We denote the unit $0\in\mathfrak{G}$ by β_0.) $E:\mathfrak{G}\to\mathbb{R}$ and $\mu:\mathfrak{G}\to\mathbb{Z}$ are group homomorphisms. We call $E(\beta)$ the *energy* of β and $\mu(\beta)$ the *Maslov index* of β.
(II) L is a smooth oriented manifold without boundary.
(III) **(Moduli space)** For each $\beta\in\mathfrak{G}$ and $k\in\mathbb{Z}_{\geq 0}$ we have a K-space with corners $\mathcal{M}_{k+1}(\beta)$ and a strongly smooth map

$$\mathrm{ev}=(\mathrm{ev}_0,\dots,\mathrm{ev}_k):\mathcal{M}_{k+1}(\beta)\to L^{k+1}.$$

We assume that ev_0 is weakly submersive. We call $\mathcal{M}_{k+1}(\beta)$ the *moduli space of* A_∞ *operations*.
(IV) **(Non-negativity of energy)** We assume $\mathcal{M}_{k+1}(\beta)=\emptyset$ if $E(\beta)<0$.
(V) **(Energy zero part)** In the case $E(\beta)=0$, we have $\mathcal{M}_{k+1}(\beta)=\emptyset$ unless $\beta=0$ and $k\geq 2$. If $\beta=\beta_0=0$, then $\mathcal{M}_{k+1}(\beta_0)=L\times D^{k-2}$ and $\mathrm{ev}_i:\mathcal{M}_{k+1}(\beta_0)\to L$ is the projection. Here we regard D^{k-2} as a Stasheff cell which is a manifold with corners. (See [FOh, Section 10], for example.)

(VI) **(Dimension)** The dimension of the moduli space of A_∞ operations is given by

$$\dim \mathcal{M}_{k+1}(\beta) = \mu(\beta) + \dim L + k - 2. \tag{21.4}$$

(VII) **(Orientation)** $\mathcal{M}_{k+1}(\beta)$ is oriented.

(VIII) **(Gromov compactness)** For any E_0 the set

$$\{\beta \in \mathfrak{G} \mid \exists k \ \mathcal{M}_{k+1}(\beta) \neq \emptyset, \ E(\beta) \leq E_0\} \tag{21.5}$$

is a finite set.

(IX) **(Boundary compatibility isomorphism)** There exists an isomorphism between the normalized boundary of the moduli space of A_∞ operations and a disjoint union of the fiber products as follows.[1]

$$\partial \mathcal{M}_{k+1}(\beta) \cong \coprod_{\beta_1,\beta_2,k_1,k_2,i} (-1)^{\epsilon} \mathcal{M}_{k_1+1}(\beta_1) \ {}_{\mathrm{ev}_i}\times_{\mathrm{ev}_0} \mathcal{M}_{k_2+1}(\beta_2), \tag{21.6}$$

where

$$\epsilon = k_2(k_1 + i) + \dim L + i \tag{21.7}$$

and the union is taken over $\beta_1, \beta_2, k_1, k_2, i$ such that $\beta_1 + \beta_2 = \beta$, $k_1 + k_2 = k + 1$, $i = 1, \dots, k_2$ and i. (21.6) is called the *boundary compatibility isomorphism*. The boundary compatibility isomorphism is compatible with orientation and is compatible with evaluation maps in the following sense. Let $\mathbf{x}_1 \in \mathcal{M}_{k_1+1}(\beta_1)$ and $\mathbf{x}_2 \in \mathcal{M}_{k_2+1}(\beta_2)$ with $\mathrm{ev}_i(\mathbf{x}_1) = \mathrm{ev}_0(\mathbf{x}_2)$. We denote by $\mathbf{x}$ the element of $\partial \mathcal{M}_{k+1}(\beta)$ which is obtained from $(\mathbf{x}_1, \mathbf{x}_2)$ by the boundary compatibility isomorphism. Then

$$\mathrm{ev}_j(\mathbf{x}) = \begin{cases} \mathrm{ev}_j(\mathbf{x}_1) & \text{if } j = 0, \dots, i-1, \\ \mathrm{ev}_{j-i+1}(\mathbf{x}_2) & \text{if } j = i, \dots, i + k_2 - 1, \\ \mathrm{ev}_{j-k_2+1}(\mathbf{x}_1) & \text{if } j = i + k_2, \dots, k. \end{cases} \tag{21.8}$$

See Remark 21.8 below for the signs in (21.7) and (21.8).

In the case $\beta = 0$, we require that (21.15) coincides with the standard decomposition appearing at the boundary of Stasheff cell. (See [FOh, Section 10] and [MSS].)

(X) **(Corner compatibility isomorphism)** There exists an isomorphism between the normalized corner $\widehat{S}_m(\mathcal{M}_{k+1}(\beta))$ of the K-space $\mathcal{M}_{k+1}(\beta)$ in the sense of Definition 24.18 and a disjoint union of

[1] See Remark 16.2 for the sign and the order of the fiber products.

$$\prod_{(\mathcal{T},\beta(\cdot))} \mathcal{M}_{k_{\rm v}+1}(\beta({\rm v})). \tag{21.9}$$

Here the union is taken over all $(\mathcal{T},\beta(\cdot)) \in \mathcal{G}(k+1,\beta)$ such that $\#C_{1,\rm int}(\mathcal{T}) = m$. This isomorphism, which we call the *corner compatibility isomorphism*, is compatible with the evaluation map in the following sense.

Let ${\rm v}_i$ be the i-th exterior vertex of $\mathcal{T}$. The (unique) edge e containing ${\rm v}_i$ contains one interior vertex denoted by v. Suppose e is the j-th edge of v. Assume an element $(\mathbf{x}_{\rm v})_{{\rm v}\in C_{0,\rm int}(\mathcal{T})}$ of (21.9) is sent to an element $\mathbf{x}$ in $\widehat{S}_m(\mathcal{M}_{k+1}(\beta))$ by the corner compatibility isomorphism. Then we require

$$\mathrm{ev}_i(\mathbf{x}) = \mathrm{ev}_j(\mathbf{x}_{\rm v}). \tag{21.10}$$

When $\beta = 0$, we require that (21.9) coincides with the standard decomposition appearing at the corner of the Stasheff cell.

(XI) **(Consistency of the corner compatibility isomorphisms)** Condition (X) implies that

$$\widehat{S}_\ell(\widehat{S}_m(\mathcal{M}_{k+1}(\beta)))$$

is a disjoint union of $(m+\ell)!/m!\ell!$ copies of (21.9), where the union is taken over all $(\mathcal{T},\beta(\cdot)) \in \mathcal{G}(k+1,\beta)$ such that $\#C_{1,\rm int}(\mathcal{T}) = m+\ell$.

The map $\widehat{S}_\ell(\widehat{S}_m(\mathcal{M}_{k+1}(\beta))) \to \widehat{S}_{m+\ell}(\mathcal{M}_{k+1}(\beta))$ is identified with the identity map on each of the component (21.9).

We call Conditions (X)(XI) the *corner compatibility conditions*.

Remark 21.7 The version of Condition 21.5 for the case of moduli spaces involving interior marked points can be found in [FOOO22, Theorem 2.16].

Remark 21.8 The sign in (21.6) is consistent with our conventions adopted in [FOOO4]. [FOOO4, Proposition 8.3.3] is the same as (21.6) for the case $i = 1$. Also we can derive the sign in (21.6) by using [FOOO4, Proposition 8.3.3] and [FOOO4, (8.4.5)]. Namely, we have

$$\mathcal{M}_{k_1+1}(\beta_1)\ {}_{\mathrm{ev}_i}\times_{\mathrm{ev}_0} \mathcal{M}_{k_2+1}(\beta_2) = (-1)^{\delta}\mathcal{M}_{k_1+1}(\beta_1)\ {}_{\mathrm{ev}_1}\times_{\mathrm{ev}_0} \mathcal{M}_{k_2+1}(\beta_2), \tag{21.11}$$

where

$$\delta \equiv (i-1)(1+k_2) \mod 2 \tag{21.12}$$

by taking the dimension formula (21.4) into account. Moreover, we note that the convention of the order of boundary marked points after gluing described in [FOOO4, Remark 8.3.4] is nothing but the convention (21.8) for the case $i = 1$.

Definition 21.9 A *tree-like K-system*, or sometimes called an A_∞ *correspondence* over L, is a system of $(\mathcal{M}_{k+1}(\beta), \mathrm{ev}, \mu, E)$ satisfying Condition 21.6.

We refer the readers to [FOOO22, Definition 2.17] for the version of a tree-like K-system with interior marked points.

In [FOOO4, Propositions 7.1.1 and 7.1.2] and [FOOO22, Theorem 2.5] the authors of this book associated a tree-like K-system to a relatively spin Lagrangian submanifold of a compact symplectic manifold. So the story of Chaps. 21 and 22 can be applied to such situation.

We next define a notion of *partial* A_∞ *correspondence*. The moduli space $\mathcal{M}_{k+1}(\beta)$ depends on k and β. In the version of partial A_∞ correspondence which we use in this book, we include the Kuranishi structure on $\mathcal{M}_{k+1}(\beta)$ for only a finite number of the pairs (β, k). This coincides with the way taken in [FOOO4, Section 7], where we used the notion of an $A_{n,K}$ structure. In [Fuk4] the notion of an A_∞ structure modulo E_0 was used. It includes only a finite number of β's but infinitely many k's are included. In [Fuk4] the Kuranishi structure of $\mathcal{M}_{k+1}(\beta)$ such that it is compatible with the forgetful map $\mathcal{M}_{k+1}(\beta) \to \mathcal{M}_1(\beta)$ was used. This is the reason why infinitely many of k's were included in [Fuk4].

Since we postpone technical detail concerning the forgetful map to [FOOO23], we use the formulation where only finitely many k's are included in the partial structure. Since we are working in de Rham theory, it is certainly possible to include infinitely many k's at this stage. However, it seems easier to use only a finite number of moduli spaces at each step of the construction.

Definition 21.10 A *partial* A_∞ *correspondence of energy cut level* E_0 *and minimal energy* e_0 *over* L is defined in the same way as A_∞ correspondence except the following:

(1) The moduli space $\mathcal{M}_{k+1}(\beta)$ of A_∞ operations is defined only when $E(\beta) + ke_0 \le E_0$.
(2) The boundary and corner compatibility isomorphisms in Conditions 21.6 (IX)(X) are defined only when $E(\beta) + ke_0 \le E_0$. Conditions 21.6 (XI) is assumed only in that case.
(3) We assume that $\mathcal{M}_{k+1}(\beta) = \emptyset$ if $0 < E(\beta) < e_0$.

Hereafter we say $\mathcal{M}_{k+1}(\beta)$ is a (partial) A_∞ correspondence, (and omit ev, μ, E) for simplicity.

We next describe a parametrized version of Condition 21.6. See Chap. 26 for the definition of submersivity of a map to a manifold with corners.

Condition 21.11 We consider the following objects:

(I) $\mathfrak{G}$ is an additive group. (We denote the unit 0 by β_0.) $E : \mathfrak{G} \to \mathbb{R}$ and $\mu : \mathfrak{G} \to \mathbb{Z}$ are group homomorphisms. We call $E(\beta)$ the *energy* of β and $\mu(\beta)$ the *Maslov index* of β.
(II) L is a smooth oriented manifold without boundary. P is a smooth oriented manifold with corners.

(III) **(Moduli space)** For each $\beta \in \mathfrak{G}$ and $k \in \mathbb{Z}_{\ge 0}$ we have a K-space with corners $\mathcal{M}_{k+1}(\beta; P)$ and a strongly smooth map

$$\mathrm{ev} = (\mathrm{ev}_P, \mathrm{ev}_0, \dots, \mathrm{ev}_k) : \mathcal{M}_{k+1}(\beta; P) \to P \times L^{k+1}.$$

We assume that $(\mathrm{ev}_P, \mathrm{ev}_0)$ is weakly submersive stratumwisely. (See Chap. 26 for its definition.) We call $\mathcal{M}_{k+1}(\beta; P)$ the *moduli space of P-parametrized A_∞ operations.*

(IV) **(Non-negativity of energy)** We assume $\mathcal{M}_{k+1}(\beta; P) = \emptyset$ if $E(\beta) < 0$.

(V) **(Energy zero part)** In the case $E(\beta) = 0$, we have $\mathcal{M}_{k+1}(\beta; P) = \emptyset$ unless $\beta = 0$ and $k \ge 2$. If $\beta = \beta_0 = 0$, then $\mathcal{M}_{k+1}(\beta_0; P) = P \times L \times D^{k-2}$ and $\mathrm{ev}_i : \mathcal{M}_{k+1}(\beta_0; P) \to L$ is the projection. Also $\mathrm{ev}_P : \mathcal{M}_{k+1}(\beta_0; P) \to P$ is the projection. Here we again identify D^{k-2} with the Stasheff cell.

(VI) **(Dimension)** The dimension of the moduli space of P-parametrized A_∞ operations is given by

$$\dim \mathcal{M}_{k+1}(\beta; P) = \mu(\beta) + \dim L + k - 2 + \dim P. \tag{21.13}$$

(VII) **(Orientation)** $\mathcal{M}_{k+1}(\beta; P)$ is oriented.

(VIII) **(Gromov compactness)** For any E_0 the set

$$\{\beta \in \mathfrak{G} \mid \exists k\ \mathcal{M}_{k+1}(\beta; P) \neq \emptyset,\ E(\beta) \le E_0\} \tag{21.14}$$

is a finite set.

(IX) **(Boundary compatibility isomorphism)** There exists an isomorphism between the normalized boundary of the moduli space of A_∞ operations and a disjoint union of the fiber products as follows.[2]

$$\begin{aligned} &\partial \mathcal{M}_{k+1}(\beta; P) \\ \cong &\coprod_{\beta_1,\beta_2,k_1,k_2,i} (-1)^\epsilon \mathcal{M}_{k_1+1}(\beta_1; P) \ {}_{(\mathrm{ev}_P,\mathrm{ev}_i)}\times_{(\mathrm{ev}_P,\mathrm{ev}_0)} \mathcal{M}_{k_2+1}(\beta_2; P) \\ &\sqcup \left(\partial P \ {}_P\times_{\mathrm{ev}_P} \mathcal{M}_{k+1}(\beta; P)\right). \end{aligned} \tag{21.15}$$

where

$$\epsilon = k_2(k_1 + i) + \dim L + i + \dim P \tag{21.16}$$

and the union in the second line is taken over $\beta_1, \beta_2, k_1, k_2, i$ such that $\beta_1 + \beta_2 = \beta$, $k_1 + k_2 = k + 1$, $i = 1, \dots, k_2$. The fiber product in the

[2]Following our convention in [FOOO4, (8.9.1)], that the parameter space P is put on the first factor in the fiber product. So there is no extra sign contribution from the parameter space in the second line on the right hand side of (21.15).

second line is taken over $P \times L$. (21.15) is called the boundary compatibility isomorphism and is compatible with orientation. The boundary compatibility isomorphism is compatible with evaluation maps, in the same sense as (21.8). In the case $\beta = 0$, we require that the boundary compatibility isomorphism coincides with the decomposition induced from the standard decomposition appearing at the boundary of Stasheff cell.

(X) **(Corner compatibility isomorphism)** There exists an isomorphism between the normalized corner $\widehat{S}_m(\mathcal{M}_{k+1}(\beta; P))$ of the K-space $\mathcal{M}_{k+1}(\beta; P)$ and a disjoint union of

$$\prod_{(\mathcal{T},\beta(\cdot))} \mathcal{M}_{k_{\rm v}+1}(\beta({\rm v}); \widehat{S}_{m'}(P)). \tag{21.17}$$

(We will explain the fiber product (21.17) right after Condition 21.11.) Here the union is taken over all $(\mathcal{T}, \beta(\cdot)) \in \mathcal{G}(k+1, \beta)$ and $m' \in \mathbb{Z}_{\ge 0}$ such that $\#C_{1,{\rm int}}(\mathcal{T}) + m' = m$ and we put

$$\mathcal{M}_{k_{\rm v}+1}(\beta({\rm v}); \widehat{S}_{m'}(P)) = \widehat{S}_{m'}(P) \ {}_P\times_{{\rm ev}_P} \mathcal{M}_{k_{\rm v}+1}(\beta({\rm v}); P).$$

This isomorphism, which we call corner compatibility isomorphism, is compatible with the evaluation map in the same sense as (21.10). In the case $\beta = 0$, we require that the corner compatibility isomorphism coincides with the decomposition induced from the standard decomposition appearing at the corner of the Stasheff cell.

(XI) **(Consistency of corner compatibility isomorphisms)** Condition (X) implies that $\widehat{S}_\ell(\widehat{S}_m(\mathcal{M}_{k+1}(\beta; P)))$ is a disjoint union of copies

$$\prod_{(\mathcal{T},\beta(\cdot))} \mathcal{M}_{k_{\rm v}+1}(\beta({\rm v}); \widehat{S}_{\ell'}(\widehat{S}_{m'}(P))) \tag{21.18}$$

where the union is taken over all $(\mathcal{T}, \beta(\cdot)) \in \mathcal{G}(k+1, \beta)$, and m', ℓ' such that $\#C_{1,{\rm int}}(\mathcal{T}) + m' + \ell' = m + \ell$ with $m' \le m$ and $\ell' \le \ell$, and we put

$$\mathcal{M}_{k_{\rm v}+1}(\beta({\rm v}); \widehat{S}_{\ell'}(\widehat{S}_{m'}(P))) = \widehat{S}_{\ell'}(\widehat{S}_{m'}(P)) \ {}_P\times_{{\rm ev}_P} \mathcal{M}_{k_{\rm v}+1}(\beta({\rm v}); P).$$

The covering map $\widehat{S}_\ell(\widehat{S}_m(\mathcal{M}_{k+1}(\beta; P))) \to \widehat{S}_{m+\ell}(\mathcal{M}_{k+1}(\beta; P))$ is identified with the map induced from the covering map $\widehat{S}_{\ell'}(\widehat{S}_{m'}(P)) \to \widehat{S}_{\ell'+m'}(P)$ on each of the component (21.18).

Now we define the fiber product (21.17) as follows. We consider the direct product $\prod_{{\rm v}\in C_{0,{\rm int}}(\mathcal{T})} \mathcal{M}_{k_{\rm v}+1}(\beta({\rm v}); P)$ as in (21.2). We have

$${\rm ev} : \prod_{{\rm v}\in C_{0,{\rm int}}(\mathcal{T})} \mathcal{M}_{k_{\rm v}+1}(\beta({\rm v}); P) \to \prod_{{\rm e}\in C_{1,{\rm int}}(\mathcal{T})} L^2.$$

Using $\mathrm{ev}_P : \mathcal{M}_{k_{\rm v}+1}(\beta(\mathrm{v}); P) \to P$ we have

$$\mathrm{ev}_P : \prod_{\mathrm{v} \in C_{0,\mathrm{int}}(\mathcal{T})} \mathcal{M}_{k_{\rm v}+1}(\beta(\mathrm{v}); P) \to \prod_{\mathrm{e} \in C_{1,\mathrm{int}}(\mathcal{T})} P^2.$$

The fiber product (21.17) is by definition

$$\left(\prod_{\mathrm{e} \in C_{1,\mathrm{int}}(\mathcal{T})} \widehat{S_{m'}}(P) \times L \right)$$

$$\prod\nolimits_{\mathrm{e} \in C_{1,\mathrm{int}}(\mathcal{T})((\widehat{S_{m'}}(P))^2 \times L^2)} \times_{(\mathrm{ev}_P, \mathrm{ev})} \left(\prod_{\mathrm{v} \in C_{0,\mathrm{int}}(\mathcal{T})} \mathcal{M}_{k_{\rm v}+1}(\beta(\mathrm{v}); \widehat{S_{m'}}(P)) \right). \tag{21.19}$$

Lemma 21.12 *The fiber product (21.19) is well-defined.*

Proof We assumed that the map

$$(\mathrm{ev}_P, \mathrm{ev}_0) : \mathcal{M}_{k_{\rm v}+1}(\beta(\mathrm{v}); P) \to P \times L$$

is weakly submersive stratumwisely. We also note that for each $\mathrm{e} \in C_{1,\mathrm{int}}(\mathcal{T})$ there exists a unique vertex v such that e is its 0-th vertex. These two facts imply the lemma immediately. □

Definition 21.13 A *P-parametrized A_∞ correspondence* over L is a system of $(\mathcal{M}_{k+1}(\beta; P), \mathrm{ev}, \mu, E)$ satisfying Condition 21.11.

A *partial P-parametrized A_∞ correspondence of energy cut level E_0 and minimal energy e_0 over L* is defined in the same way as P-parametrized A_∞ correspondence except the following:

(1) The moduli space of P-parametrized A_∞ operations $\mathcal{M}_{k+1}(\beta; P)$ is defined only when $E(\beta) + ke_0 \le E_0$.[3]
(2) The boundary and corner compatibility isomorphisms as in Condition 21.11 (IX)(X) are defined only when $E(\beta) + ke_0 \le E_0$. Condition 21.11 (XI) is assumed only in that case.
(3) We assume $\mathcal{M}_{k+1}(\beta, P) = \emptyset$ if $0 < E(\beta) < e_0$.

Lemma 21.14 *Let $\mathcal{M}_{k+1}(\beta; P)$ be a P-parametrized A_∞ correspondence over L and $\partial_i P$ a connected component of the normalized boundary of P. Then*

$$\mathcal{M}_{k+1}(\beta; \partial_i P) = \partial_i P \ {}_P\times_{\mathrm{ev}_P} \mathcal{M}_{k+1}(\beta; P)$$

[3]It is easy to see that if $E(\beta) + ke_0 \le E_0$ and the moduli space $\mathcal{M}_{k'+1}(\beta'; P)$ appears as a factor of the boundary of $\mathcal{M}_{k+1}(\beta; P)$, then the inequality $E(\beta') + k'e_0 < E_0$ holds.

defines a $\partial_i P$-parametrized A_∞ correspondence over L. The same holds for partial P-parametrized A_∞ correspondence.

The proof is obvious.

Definition 21.15 Suppose we are given two A_∞ correspondences over L denoted by $\mathcal{M}^j_{k+1}(\beta)$ with $j = 1, 2$. Then a *pseudo-isotopy* between them is a $P = [1, 2]$ parametrized A_∞ correspondence $\mathcal{M}_{k+1}(\beta; [1, 2])$ such that for $\{1\} \subset \partial[1, 2]$ (resp. $\{2\} \subset \partial[1, 2]$) the A_∞ correspondence $\mathcal{M}_{k+1}(\beta; \{1\})$ (resp. $\mathcal{M}_{k+1}(\beta; \{2\})$) is isomorphic to $\mathcal{M}^1_{k+1}(\beta)$ (resp $\mathcal{M}^2_{k+1}(\beta)$). We define the notion of pseudo-isotopy of partial A_∞ correspondences in the same way.

Let (X, ω) be a compact symplectic manifold and J_i $(i = 1, 2)$ be two compatible almost complex structures. We consider relatively spin Lagrangian submanifold L of X. Then, by the choice J_i of almost complex structure together with other choices, we obtain a tree-like K-system on the de Rham complex of L. In [FOOO22, Theorem 2.20] the authors of this book constructed a pseudo-isotopy between two tree-like K-systems obtained from those two different choices of J_i etc. Thus we can use the story of Chaps. 21 and 22 to prove independence of filtered A_∞ algebra associated to L on the choices up to pseudo-isotopy in the sense we will define in Definition 21.25.

Remark 21.16 In this book we use the notion of pseudo-isotopy of A_∞ correspondences to prove well-definedness of the filtered A_∞ algebra induced by the A_∞ correspondence. (See Theorem 21.35 (2).) On the other hand, we can define the notion of morphism of A_∞ correspondences and use it instead to prove well-definedness. In other words, we are using the bifurcation method here but not the cobordism method. (See [FOOO4, Subsection 7.2.14] for these two methods. In a slightly different formulation, a morphism of A_∞ correspondences is defined in [Fuk5, Definition 8].) In [FOOO4] and [Fuk5] we used the cobordism method. In [AJ] and [Fuk4] the bifurcation method was used. They will give the same morphism at the end of the day as explained in [Fuk4, Remark 12.3].

Definition 21.17 An *inductive system of A_∞ correspondences over L* consists of the following objects:

(1) We are given a sequence $\{E^i\}_{i=1}^\infty$ of positive real numbers such that $E^i < E^{i+1}$ and $\lim_{i\to\infty} E^i = \infty$. We are also given $e_0 > 0$ independent of i.
(2) For each i, we are given a partial A_∞ correspondence $\mathcal{M}^i_{k+1}(\beta)$ over L of energy cut level E^i and minimal energy e_0.
(3) For each i we are given a pseudo-isotopy $\mathcal{M}_{k+1}(\beta; [i, i+1])$ between $\mathcal{M}^i_{k+1}(\beta)$ and $\mathcal{M}^{i+1}_{k+1}(\beta)$. Here the energy cut level of $\mathcal{M}_{k+1}(\beta; [i, i+1])$ is E^i and its minimal energy is e_0.
(4) We assume the following *uniform Gromov compactness*. For each $E' > 0$ the next set is of finite order:

$$\{\beta \in \mathfrak{G} \mid \exists k\, \exists i\ \mathcal{M}^i_{k+1}(\beta) \neq \emptyset,\ E(\beta) \le E'\}. \tag{21.20}$$

The next set is also of finite order for each $E' > 0$:

$$\{\beta \in \mathfrak{G} \mid \exists k \, \exists i \; \mathcal{M}^i_{k+1}(\beta; [i, i+1]) \neq \emptyset, \; E(\beta) \leq E'\}. \tag{21.21}$$

Remark 21.18 In the situation of Definition 21.17, suppose E'^i is another sequence so that $E'^i < E'^{i+1}$, $\lim_{i\to\infty} E'^i = \infty$ and $E'^i < E^i$. We forget the K-spaces $\mathcal{M}_{k+1}(\beta; [i, i+1])$ and $\mathcal{M}^i_{k+1}(\beta + ke_0)$ for $E(\beta) > E'^i$. Then we obtain another inductive system of A_∞ correspondences over L.

Let $e'_0 < e_0$. We replace E^i by $E'^i = E^i e'_0/e_0$. Then $ke'_0 + E \leq E'^i$ implies $ke_0 + E \leq E^i$. Therefore any partial A_∞ correspondence of energy cut level E^i and of minimal energy e_0 induces one of energy cut level E'^i and of minimal energy e'_0 by forgetting certain moduli spaces. In this way when we compare two inductive systems of A_∞ correspondences over L we may always assume that the numbers E^i, e_0 are common without loss of generality. We will assume it in the next definition.

Definition 21.19 Suppose for $j = 0, 1$ we are given inductive systems of A_∞ correspondences over L, denoted by $\mathcal{M}^{ji}_{k+1}(\beta)$, $\mathcal{M}^j_{k+1}(\beta; [i, i+1])$. (We take the same E^i and e_0 for $j = 0, 1$ as we explained in Remark 21.18.) A *pseudo-isotopy between these two inductive systems* consists of the following objects:

(1) For each i we are given a pseudo-isotopy $\mathcal{M}^i_{k+1}(\beta; [0, 1])$ between $\mathcal{M}^{0i}_{k+1}(\beta)$ and $\mathcal{M}^{1i}_{k+1}(\beta)$. The energy cut level and minimal energy of this pseudo-isotopy are E^i and e_0, respectively.
(2) For each i we are given a $P = [0, 1] \times [i, i+1]$ parametrized A_∞ correspondence $\mathcal{M}_{k+1}(\beta; [0, 1] \times [i, i+1])$ satisfying the following properties:
   (a) Its restriction to the boundary component $\{j\} \times [i, i+1]$ is isomorphic to $\mathcal{M}^j_{k+1}(\beta; [i, i+1])$. Here $j = 0, 1$.
   (b) Its restriction to the boundary component $[0, 1] \times \{i\}$ is isomorphic to $\mathcal{M}^{1i}_{k+1}(\beta)$.
   (c) Its restriction to the boundary component $[0, 1] \times \{i+1\}$ is isomorphic to $\mathcal{M}^{1i+1}_{k+1}(\beta)$.
   (d) The isomorphisms in (a)(b)(c) are consistent at $\mathcal{M}_{k+1}(\beta; \widehat{S}_2([0, 1] \times [i, i+1]))$.
   (e) The energy cut level of $\mathcal{M}_{k+1}(\beta; [0, 1] \times [i, i+1])$ is E^i and its minimal energy is e_0.
   (f) The isomorphisms in Items (a)(b)(c)(d) satisfy appropriate corner compatibility conditions.

(3) We assume the following *uniform Gromov compactness*. For each $E' > 0$ the next set is of finite order:

$$\{\beta \in \mathfrak{G} \mid \exists k \, \exists i \; \mathcal{M}_{k+1}(\beta; [0, 1] \times [i, i+1]) \neq \emptyset, \; E(\beta) \leq E'\}. \tag{21.22}$$

21.2 Filtered A_∞ Algebra and its Pseudo-Isotopy

In this section we review certain algebraic material in [FOOO3], [FOOO4], [Fuk4].

Definition 21.20 We call a subset $G \subset \mathbb{R}_{\ge 0} \times 2\mathbb{Z}$ a *discrete submonoid*[4] if the following holds. We denote by $E : G \to \mathbb{R}_{\ge 0}$ and $\mu : G \to 2\mathbb{Z}$ the natural projections.

(1) If $\beta_1, \beta_2 \in G$, then $\beta_1 + \beta_2 \in G$. $(0,0) \in G$.
(2) The image $E(G) \subset \mathbb{R}_{\ge 0}$ is discrete.
(3) For each $E_0 \in \mathbb{R}_{\ge 0}$ the inverse image $G \cap E^{-1}([0, E_0])$ is a finite set.

Let $\Omega(L)$ be the de Rham complex of L. We put $\Omega(L)[1]^d = \Omega^{d+1}(L)$, where $\Omega^d(L)$ is the space of degree d smooth forms. We put

$$B_k(\Omega(L)[1]) = \underbrace{\Omega(L)[1] \otimes \cdots \otimes \Omega(L)[1]}_{k \text{ times}}. \tag{21.23}$$

Let G be a discrete submonoid as in Definition 21.20.

Definition 21.21 ([FOOO3, Definition 3.2.26, Definition 3.5.6, Remark 3.5.8]) A *G-gapped filtered A_∞ algebra structure* on $\Omega(L)$ is a sequence of multilinear maps

$$\mathfrak{m}_{k,\beta} : B_k(\Omega(L)[1]) \to \Omega(L)[1] \tag{21.24}$$

for each $\beta \in G$ and $k \in \mathbb{Z}_{\ge 0}$ of degree $1 - \mu(\beta)$ with the following properties:

(1) $\mathfrak{m}_{k,\beta_0} = 0$ for $\beta_0 = (0,0)$, $k \ne 1, 2$.
(2) $\mathfrak{m}_{2,\beta_0}(h_1, h_2) = (-1)^* h_1 \wedge h_2$, where $* = \deg h_1$.[5]
(3) $\mathfrak{m}_{1,\beta_0}(h) = dh$.
(4)

$$\sum_{k_1+k_2=k+1} \sum_{\beta_1+\beta_2=\beta} \sum_{i=1}^{k-k_2+1} (-1)^* \mathfrak{m}_{k_1,\beta_1}(h_1, \dots, \mathfrak{m}_{k_2,\beta_2}(h_i, \dots, h_{i+k_2-1}), \dots, h_k) = 0 \tag{21.25}$$

holds for any $\beta \in G$ and k. The sign is given by $* = \deg' h_1 + \cdots + \deg' h_{i-1}$. Here $\deg'$ is the shifted degree by $+1$.

[4]In the situation of oriented Lagrangian submanifolds, the Maslov index is even. Thus we assume $G \subset \mathbb{R}_{\ge 0} \times 2\mathbb{Z}$, while we consider $G \subset \mathbb{R}_{\ge 0} \times \mathbb{Z}$ in Definition 16.12.

[5] The sign in Items (2), (3) here is different from the one given in [FOOO3, (3.2.5)], because the sign convention of the correspondence is different.

Definition 21.22 A *partial G-gapped filtered A_∞ algebra structure of energy cut level E and minimal energy* e_0 is a sequence of operators (21.24) for $E(\beta)+ke_0 \le E$ satisfying the same properties, except (21.25) is assumed only for $E(\beta)+ke_0 \le E$. We require $\mathfrak{m}_{k,\beta} = 0$ if $0 < E(\beta) < e_0$.

Definition 21.23 A multilinear map $F : B_k(\Omega(L)[1]) \to \Omega(L)[1]$ is said to be *continuous in C^∞ topology* if the following holds. Suppose $h_{i,a} \in \Omega(L)$ converges to $h_i \in \Omega(L)$ in C^∞ topology as $a \to \infty$, then $F(h_{1,a}, \ldots, h_{k,a})$ converges to $F(h_1, \ldots, h_k)$ in C^∞ topology as $a \to \infty$.

Definition 21.24 A (partial) G gapped filtered A_∞ algebra structure on $\Omega(L)$ is said to be *continuous* if the operations $\mathfrak{m}_{k,\beta}$ are continuous in C^∞ topology. Hereafter we assume that all the operations of (partial) A_∞ algebra structure on $\Omega(L)$ are continuous in C^∞ topology.

Definition 21.25 ([Fuk4, Definition 8.5]) For each $t \in [0, 1]$, $\beta \in G$ and $k \in \mathbb{Z}_{\ge 0}$, let $\mathfrak{m}^t_{k,\beta}$ be as in Definition 21.21 and $\mathfrak{c}^t_{k,\beta}$ a sequence of multilinear maps

$$\mathfrak{c}^t_{k,\beta} : B_k(\Omega(L)[1]) \to \Omega(L)[1] \tag{21.26}$$

of degree $-\mu(\beta)$. We say $(\{\mathfrak{m}^t_{k,\beta}\}, \{\mathfrak{c}^t_{k,\beta}\})$ is a *pseudo-isotopy of G-gapped filtered A_∞ algebra structures*, or in short, *gapped pseudo-isotopy* on $\Omega(L)$ if the following holds:

(1) $\mathfrak{m}^t_{k,\beta}$ and $\mathfrak{c}^t_{k,\beta}$ are continuous in C^∞ topology. The map sending t to $\mathfrak{m}^t_{k,\beta}$ or $\mathfrak{c}^t_{k,\beta}$ is smooth. Here we use the operator topology with respect to the C^∞ topology for $\mathfrak{m}^t_{k,\beta}$ and $\mathfrak{c}^t_{k,\beta}$ to define their smoothness.
(2) For each (but fixed) t, the set of operators $\{\mathfrak{m}^t_{k,\beta}\}$ defines a G-gapped filtered A_∞ algebra structure on $\Omega(L)$.
(3) For each $h_i \in \Omega(L)[1]$ the following equality holds:

$$\begin{aligned}
&\frac{d}{dt}\mathfrak{m}^t_{k,\beta}(h_1, \ldots, h_k) \\
&+ \sum_{k_1+k_2=k+1} \sum_{\beta_1+\beta_2=\beta} \sum_{i=1}^{k-k_2+1} (-1)^* \mathfrak{c}^t_{k_1,\beta_1}(h_1, \ldots, \mathfrak{m}^t_{k_2,\beta_2}(h_i, \ldots), \ldots, h_k) \\
&- \sum_{k_1+k_2=k+1} \sum_{\beta_1+\beta_2=\beta} \sum_{i=1}^{k-k_2+1} \mathfrak{m}^t_{k_1,\beta_1}(h_1, \ldots, \mathfrak{c}^t_{k_2,\beta_2}(h_i, \ldots), \ldots, h_k) \\
&= 0.
\end{aligned} \tag{21.27}$$

Here $* = \deg' h_1 + \cdots + \deg' h_{i-1}$.
(4) $\mathfrak{c}^t_{k,\beta} = 0$ if $E(\beta) \le 0$.

Sometimes we say that $(\{\mathfrak{m}^t_{k,\beta}\}, \{\mathfrak{c}^t_{k,\beta}\})$ is a pseudo-isotopy or *gapped pseudo-isotopy* between $\{\mathfrak{m}^0_{k,\beta}\}$ and $\{\mathfrak{m}^1_{k,\beta}\}$.

It is proved in [Fuk4] that two filtered A_∞ algebras which are pseudo-isotopic are homotopy equivalent.

Definition 21.26 We define the notion of *pseudo-isotopy of partial G-gapped filtered A_∞ algebra structures on $\Omega(L)$ of energy cut level E and minimal energy* e_0 in the same way, except we replace (2) and (3) by the following (2)' and (3)' and we further require (5) below:

(2)' We require that the set of operators $\{\mathfrak{m}^t_{k,\beta}\}$ defines partial G-gapped filtered A_∞ algebra structures on $\Omega(L)$ of energy cut level E and of minimal energy e_0.
(3)' We require (21.27) only for β, k with $E(\beta) + ke_0 \le E$.
(5) $\mathfrak{m}^t_{k,\beta} = \mathfrak{c}^t_{k,\beta} = 0$ if $0 < E(\beta) < e_0$.

We also use the notion of pesudo-isotopy of pseudo-isotopies. It seems simpler to define a more general notion of the P-parametrized family of G-gapped filtered A_∞ algebra structures on $\Omega(L)$. So we define this notion below.

Let P be a manifold with corners. We consider the totality of smooth differential forms on $P \times L$ which we write $\Omega(P \times L)$. Here $\Omega(P \times L)[1]$ is its degree shift as before. Let $t_1, \dots, t_d$ be local coordinates of P. (In this book we only consider the case $P \subset \mathbb{R}^d$ for $d = \dim P$. So we actually have canonical global coordinates.) For $h \in \Omega(P \times L)$ we put

$$h = \sum_{I \subset \{1,\dots,d\}} dt_I \wedge h_I. \tag{21.28}$$

Here $I = \{i_1, \dots, i_{|I|}\}$ $(i_1 < \cdots < i_{|I|})$, $dt_I = dt_{i_1} \wedge \cdots \wedge dt_{i_{|I|}}$ and h_I does not contain dt_i.

Definition 21.27 A multilinear map $F : B_k(\Omega(P \times L)[1]) \to \Omega(P \times L)[1]$ is said to be *pointwise in the P direction* if the following holds:

For each $I \subset \{1, \dots, d\}$ and $\mathbf{t} \in P$ there exists a continuous map

$$F^{\mathbf{t}}_{I;j_1,\dots,j_k} : B_k(\Omega(L)[1]) \to \Omega(L)[1]$$

such that

$$\begin{aligned} &F(dt_{j_1} \wedge h_1, \dots, dt_{j_k} \wedge h_k)|_{\{\mathbf{t}\}\times L} \\ &= \sum_I dt_I \wedge dt_{j_1} \wedge \cdots \wedge dt_{j_k} \wedge F^{\mathbf{t}}_{I;j_1,\dots,j_k}(h^{\mathbf{t}}_1, \dots, h^{\mathbf{t}}_k), \end{aligned} \tag{21.29}$$

where $|_{\{\mathbf{t}\}\times L}$ means the restriction to $\{\mathbf{t}\} \times L$. Moreover $F^{\mathbf{t}}_{I;j_1,\dots,j_k}$ depends smoothly on $\mathbf{t}$ with respect to the operator topology. Here $h^{\mathbf{t}}_i$ is the restriction of h_i to $\{\mathbf{t}\} \times L$.

Remark 21.28 This condition is equivalent to the following:

(*) For smooth differential forms σ_i on P, we have

$$F(\sigma_1 h_1, \ldots, \sigma_k h_k) = \pm \sigma_1 \wedge \cdots \wedge \sigma_k \wedge F(h_1, \ldots, h_k).$$

Definition 21.29 A *P-parametrized family of G-gapped filtered A_∞ algebra structures on* $\Omega(L)$ is $\{\mathfrak{m}^P_{k,\beta}\}$ satisfying the following properties:

(1) $\mathfrak{m}^P_{k,\beta} : B_k(\Omega(P \times L)[1]) \to \Omega(P \times L)[1]$ is a multilinear map of degree $1 - \mu(\beta)$.
(2) $\mathfrak{m}^P_{k,\beta}$ is pointwise in P direction if $\beta \neq \beta_0$.
(3) $\mathfrak{m}^P_{k,\beta_0} = 0$ for $k \neq 1, 2$.
(4) $\mathfrak{m}^P_{1,\beta_0}(h) = dh$. Here d is the de Rham differential.
(5) $\mathfrak{m}^P_{2,\beta_0}(h_1 \wedge h_2) = (-1)^* h_1 \wedge h_2$. Here $\wedge$ is the wedge product and $* = \deg h_1$.
(6) $\mathfrak{m}^P_{k,\beta}$ satisfies the following A_∞ relation:

$$\sum_{k_1+k_2=k+1} \sum_{\beta_1+\beta_2=\beta} \sum_{i=1}^{k-k_2+1} (-1)^* \mathfrak{m}^P_{k_1,\beta_1}(h_1, \ldots, \mathfrak{m}^P_{k_2,\beta_2}(h_i, \ldots, h_{i+k_2-1}), \ldots, h_k) = 0, \tag{21.30}$$

where $* = \deg' h_1 + \ldots + \deg' h_{i-1}$.

Definition 21.30 A *partial P-parametrized family of G-gapped filtered A_∞ algebra structures on $\Omega(L)$ of energy cut level E and of minimal energy e_0* is $\{\mathfrak{m}^P_{k,\beta}\}$ satisfying the same properties as above except the following points:

(a) $\mathfrak{m}^P_{k,\beta}$ is defined only for β, k with $E(\beta) + ke_0 \leq E$.
(b) We require the A_∞ relation (21.30) only for β, k with $E(\beta) + ke_0 \leq E$.
(c) $\mathfrak{m}^P_{k,\beta} = 0$ if $0 < E(\beta) < e_0$.

Lemma 21.31 *The notion of pseudo-isotopy of G-gapped filtered A_∞ algebra structures on $\Omega(L)$ is the same as the notion of the $P = [0, 1]$ parametrized family of G-gapped filtered A_∞ algebra structures on $\Omega(L)$. The same holds for the partial G-gapped filtered A_∞ algebra structure on $\Omega(L)$.*

Proof Let $(\{\mathfrak{m}^t_{k,\beta}\}, \{\mathfrak{c}^t_{k,\beta}\})$ be the objects as in Definition 21.25. We define $\mathfrak{m}^P_{k,\beta}$ as follows. It suffices to consider the case $\beta \neq \beta_0$.

Suppose h_i does not contain dt. Then we put

$$\mathfrak{m}^P_{k,\beta}(h_1, \ldots, h_k) = \mathfrak{m}^t_{k,\beta}(h_1, \ldots, h_k) + dt \wedge \mathfrak{c}^t_{k,\beta}(h_1, \ldots, h_k).$$

We also put

$$\mathfrak{m}^P_{k,\beta}(h_1, \ldots, dt \wedge h_i, \ldots, h_k) = (-1)^* dt \wedge \mathfrak{m}^t_{k,\beta}(h_1, \ldots, h_k),$$

where $* = \deg' h_1 + \cdots + \deg' h_{i-1} + 1$. If at least two of $\hat{h}_1, \ldots, \hat{h}_k$ contain dt, then

$$\mathfrak{m}^P_{k,\beta}(\hat{h}_1, \ldots \hat{h}_k) = 0.$$

It is straightforward to check that (21.27) is equivalent to (21.30). □

Lemma-Definition 21.32 *Let Q be a connected component of the normalized corner $\widehat{S}_k(P)$. Then a P-parametrized family of G-gapped filtered A_∞ algebra structures on $\Omega(L)$ induces a Q-parametrized family of G-gapped filtered A_∞ algebra structures on $\Omega(L)$.*

Proof This is a consequence of the following fact. If $F : B_k(\Omega(P \times L)[1]) \to \Omega(P \times L)[1]$ is pointwise in P direction, it induces $B_k(\Omega(Q \times L)[1]) \to \Omega(Q \times L)[1]$ which is pointwise in the Q direction. This fact is a consequence of the definition of pointwiseness. □

Remark 21.33 If two G-gapped filtered A_∞ algebra structures on $\Omega(L)$ are pseudo-isotopic, then the two filtered A_∞ algebras induced from those two structures are homotopy equivalent in the sense of [FOOO3, Definition 4.2.42]. This fact is proved in [Fuk4, Theorem 8.2].

21.3 Statement of the Results

In this section and hereafter we say a filtered A_∞ structure, pseudo-isotopy or P-parametrized A_∞ structure is *gapped* when it is G-gapped for some discrete submonoid $G \subset \mathbb{R}_{\ge 0} \times 2\mathbb{Z}$ in Definition 21.20.

Situation 21.34 Let $\mathfrak{G}$ be an additive group and $\mu : \mathfrak{G} \to \mathbb{Z}$, $E : \mathfrak{G} \to \mathbb{R}$ group homomorphisms.

Let L be a compact oriented smooth manifold without boundary. In addition to those, we consider one of the following situations:

(1) We are given $\mathcal{AF} = \{(\mathcal{M}_{k+1}(\beta), \mathrm{ev}) \mid \beta, k\}$, which defines an A_∞ correspondence over L. Hereafter we will not include ev in the notation for simplicity.
(2) We are given $\mathcal{AF} = \{\mathcal{M}_{k+1}(\beta) \mid \beta, k\}$, which defines a partial A_∞ correspondence over L of energy cut level E_0 and minimal energy e_0.
(3) We are given

$$\mathcal{AF}^j = \{\mathcal{M}^j_{k+1}(\beta) \mid \beta, k\}\ (j = 0, 1), \quad \mathcal{AF}^{[0,1]} = \{\mathcal{M}_{k+1}(\beta; [0, 1]) \mid \beta, k\},$$

which are two A_∞ correspondences over L and a pseudo-isotopy among them, respectively.
(4) We are given

$$\mathcal{AF}^j = \{\mathcal{M}^j_{k+1}(\beta) \mid \beta, k\} \ (j = 0, 1), \quad \mathcal{AF}^{[0,1]} = \{\mathcal{M}_{k+1}(\beta; [0, 1]) \mid \beta, k\},$$

which are two partial A_∞ correspondences over L and a pseudo-isotopy among them, respectively. Their energy cut level are E_0 and minimal energy are e_0.

(5) We are given

$$\mathcal{AF}^i = \{\mathcal{M}^i_{k+1}(\beta) \mid \beta, k\}, \quad \mathcal{AF}^{[i,i+1]} = \{\mathcal{M}_{k+1}(\beta; [i, i + 1]) \mid \beta, k\}$$

for $i = 1, 2, \ldots$ such that

$$\mathcal{IAF} = \left(\{\mathcal{AF}^i \mid i = 1, 2, \ldots\}, \{\mathcal{AF}^{[i,i+1]} \mid i = 1, 2, \ldots\}\right)$$

consists of an inductive system of A_∞ correspondences over L.

(6) For $j = 0, 1$, we are given

$$\mathcal{AF}^{ji} = \{\mathcal{M}^{ji}_{k+1}(\beta) \mid \beta, k\}, \quad \mathcal{AF}^{j,[i,i+1]} = \{\mathcal{M}^j_{k+1}(\beta; [i, i + 1]) \mid \beta, k\}$$

for $i = 1, 2, \ldots$ such that

$$\mathcal{IAF}^j = \left(\{\mathcal{AF}^{ji} \mid i = 1, 2, \ldots\}, \{\mathcal{AF}^{j,[i,i+1]} \mid i = 1, 2, \ldots\}\right)$$

consist of two inductive systems of A_∞ correspondences over L. Moreover we are given $\{\mathcal{M}^i_{k+1}(\beta; [0, 1]) \mid \beta, k\}$ and $\{\mathcal{M}_{k+1}(\beta; [0, 1] \times [i, i + 1]) \mid \beta, k\}$ which consist of a pseudo-isotopy of the inductive systems $\mathcal{IAF}^0$, $\mathcal{IAF}^1$ of A_∞ correspondences.

(7) Let $\mathcal{AF}^j = \{\mathcal{M}^j_{k+1}(\beta) \mid \beta, k\}$ define A_∞ correspondences over L for $j = 0, 1$. Let $\mathcal{AF}^{[0,1],\ell} = \{\mathcal{M}_{k+1}(\beta; [0, 1]) \mid \beta, k\}$, $\ell = a, b$ be two pseudo-isotopies from $\mathcal{AF}^1$ to $\mathcal{AF}^2$. Let

$$\mathcal{AF}^{[0,1]\times[1,2]}$$

be a $[0, 1] \times [1, 2]$-parametrized family of A_∞ correspondences over L with the following properties:

(a) On $[0, 1] \times \{1\}$ it is isomorphic to $\mathcal{AF}^{[0,1],a}$.
(b) On $[0, 1] \times \{2\}$ it is isomorphic to $\mathcal{AF}^{[0,1],b}$.
(c) Let $j = 1$ or $j = 2$ then on $\{j\} \times [1, 2]$ it is isomorphic to the direct product $\mathcal{AF}^j \times [1, 2]$.
(d) At the corner $\{0, 1\} \times \{1, 2\}$ the isomorphisms (a)(b)(c)(d) and the various isomorphisms included in the definitions of A_∞ correspondence and its pseudo-isotopies are compatible in the same sense as Condition 21.11 (X)(XI).

(8) For $j = 0, 1$, we are given

$$\mathcal{AF}^{ji} = \{\mathcal{M}^{ji}_{k+1}(\beta) \mid \beta, k\}, \quad \mathcal{AF}^{j,[i,i+1]} = \{\mathcal{M}^{j}_{k+1}(\beta; [i, i+1]) \mid \beta, k\}$$

for $i = 1, 2, \ldots$ such that

$$\mathcal{IAF}^j = \Big(\{\mathcal{AF}^{ji} \mid i = 1, 2, \ldots\}, \{\mathcal{AF}^{j,[i,i+1]} \mid i = 1, 2, \ldots\}\Big)$$

consist of inductive systems of A_∞ correspondences over L. Moreover, for $\ell = a, b$, we are given $\{\mathcal{M}^{i,\ell}_{k+1}(\beta; [0, 1]) \mid \beta, k\}$ and $\{\mathcal{M}^{\ell}_{k+1}(\beta; [0, 1] \times [i, i+1]) \mid \beta, k\}$ which consist of two pseudo-isotopies of the inductive systems $\mathcal{IAF}^0$, $\mathcal{IAF}^1$ of A_∞ correspondences.

Furthermore we assume that we have a pseudo-isotopy between these two pseudo-isotopies in the following sense: For each i we have a $[0, 1] \times [i, i+1] \times [1, 2]$-parametrized family of partial A_∞ correspondence

$$\mathcal{AF}^{[0,1]\times[i,i+1]\times[1,2]}$$

over L of energy cut level E^i and minimal energy e_0 with the following properties:

(i) On $[0, 1] \times \{i\} \times [1, 2]$, it satisfies the same condition as (7) (a)–(d) up to energy level E^i.
(ii) On $[0, 1] \times [i, i+1] \times \{c\}$ with $c = 1$ (resp. $c = 2$) it is isomorphic to $\{\mathcal{M}^a_{k+1}(\beta; [0, 1] \times [i, i+1]) \mid \beta, k\}$ (resp. $\{\mathcal{M}^b_{k+1}(\beta; [0, 1] \times [i, i+1]) \mid \beta, k\}$).
(iii) On $\{j\} \times [i, i+1] \times [1, 2]$ with $j = 0$ or $j = 1$, it is isomorphic to the direct product $\mathcal{AF}^{j,[i,i+1]} \times [1, 2]$.
(iv) Various isomorphisms in (i)(ii)(iii) above and those appearing in the definitions of A_∞ correspondences or its pseudo-isotopies are compatible at $S_m([0, 1] \times [i, i+1] \times [1, 2])$ in the same sense as Condition 21.11 (X)(XI).
(v) A similar uniform Gromov compactness as in Definition 21.19 (3) is satisfied.

■

Theorem 21.35 *Let $\mathfrak{G}$ be an additive group and $\mu : \mathfrak{G} \to \mathbb{Z}$, $E : \mathfrak{G} \to \mathbb{R}$ group homomorphisms. Let L be a compact oriented smooth manifold without boundary.*

(1) *Suppose we are in Situation 21.34 (1). We can associate a gapped filtered A_∞ structure on $\Omega(L)$. This filtered A_∞ structure is independent of the choices made for its construction up to pseudo-isotopy.*
(2) *Suppose we are in Situation 21.34 (2). We can associate a gapped partial filtered A_∞ structure on $\Omega(L)$ of energy cut level E_0 and of minimal energy e_0. This partial filtered A_∞ structure is independent of the choices made for its construction up to gapped pseudo-isotopy.*

(3) *Suppose we are in Situation 21.34 (3). We can associate a gapped pseudo-isotopy of filtered A_∞ structures on $\Omega(L)$ among the two gapped filtered A_∞ algebras which associate by (1) to $\mathcal{M}^j_{k+1}(\beta)$ for $j = 0, 1$.*

In particular, it induces a gapped homotopy equivalence between those two gapped filtered A_∞ algebras.

(4) *Suppose we are in Situation 21.34 (4). We can associate a gapped partial pseudo-isotopy of filtered A_∞ structures on $\Omega(L)$ of energy cut level E_0 and of minimal energy e_0 among the two gapped partial filtered A_∞ algebras which we associate by (2) to $\mathcal{M}^j_{k+1}(\beta)$ for $j = 0, 1$.*

(5) *Suppose we are in Situation 21.34 (5). We can associate a gapped filtered A_∞ structure on $\Omega(L)$. This gapped filtered A_∞ structure is independent of the choices made for its construction up to gapped pseudo-isotopy.*

(6) *Suppose we are in Situation 21.34 (6). We can associate a gapped pseudo-isotopy of gapped filtered A_∞ structures on $\Omega(L)$ among the two gapped filtered A_∞ algebras which we associate in (5) to $\mathcal{M}^{ji}_{k+1}(\beta)$ and $\mathcal{M}^j_{k+1}(\beta; [i, i+1])$ for $j = 0, 1$.*

In particular, it induces a gapped homotopy equivalence between those two gapped filtered A_∞ algebras.

(7) *Suppose we are in Situation 21.34 (7). By (1) we obtain two gapped filtered A_∞ algebras $(\Omega(L), \mathfrak{m}^j_k)$ for $j = 1, 2$. By (3) we obtain two gapped homotopy equivalences φ_a and φ_b from $(\Omega(L), \mathfrak{m}^1_k)$ to $(\Omega(L), \mathfrak{m}^2_k)$, corresponding to $\ell = a$ and $\ell = b$, respectively.*

Then the claim of (7) is that φ_a is gapped homotopic to φ_b.

(8) *We obtain the same conclusion as (7) in the Situation 21.34 (8).*

Chapter 22
Tree-Like K-Systems: A_∞ Structure II – Proof

In this chapter we prove Theorem 21.35. The proof is mostly parallel to the proofs given in Chap. 19, and also similar to those given in [FOOO4, Subection 7.2], [Fuk4].

22.1 Existence of CF-Perturbations

Definition 22.1 Let $\mathcal{M}_{k+1}(\beta)$ be the moduli spaces of the A_∞ operations of a (partial) A_∞ correspondence $\mathcal{AC}$. We use the notation of Condition 21.11. Consider the submonoid of $\mathbb{R}_{\ge 0} \times 2\mathbb{Z}$ generated by the subset

$$\{(E(\beta), \mu(\beta)) \mid \beta \in \mathfrak{G}, \mathcal{M}_{k+1}(\beta) \neq \emptyset\}$$

of $\mathbb{R}_{\ge 0} \times 2\mathbb{Z}$. This submonoid is discrete by Condition 21.11 (VIII). We call it *the discrete submonoid associated to* $\mathcal{AC}$, and denote it by $G(\mathcal{AC})$.

When we have a partial P-parametrized family $\mathcal{AC}_P$ of A_∞ correspondences whose moduli spaces of P-parametrized A_∞ operations are $\mathcal{M}_{k+1}(\beta; P)$, we define the *discrete submonoid* $G(\mathcal{AC}_P)$ *associated to* $\mathcal{AC}_P$ as the submonoid generated by the subset

$$\{(E(\beta), \mu(\beta)) \mid \beta \in \mathfrak{G}, \mathcal{M}_{k+1}(\beta; P) \neq \emptyset\}.$$

Definition 22.2 Consider a discrete submonoid $G \subset \mathbb{R}_{\ge 0} \times 2\mathbb{Z}$ in the sense of Definition 21.20.

(1) We put

$$e_{\min}(G) = \inf\{E(\beta) \mid \beta \in G, E(\beta) > 0\},$$

K. Fukaya et al., *Kuranishi Structures and Virtual Fundamental Chains*, Springer Monographs in Mathematics, https://doi.org/10.1007/978-981-15-5562-6_22

if the image of $E : G \to \mathbb{R}$ is not 0. Otherwise we put $e_{\min}(G) = 1$.

(2) For $E_0, e_0 > 0$ with $e_0 \le e_{\min}(G)$ we define

$$\mathcal{GK}(G; E_0, e_0) = \{(\beta, k) \mid \beta \in G, k \in \mathbb{Z}_{\ge 0},\ E(\beta) = 0 \Rightarrow k > 1 \\ E(\beta) + ke_0 \le E_0\}.$$

Note that if e_0 is a minimal energy of a G-gapped (partial) A_∞ correspondence $\mathcal{AC}$ and $G \supseteq G(\mathcal{AC})$, then $e_0 \le e_{\min}(G)$. In the case $G \ne G(\mathcal{AC})$ we put $\mathcal{M}_{k+1}(\beta) = \emptyset$ for $\beta \notin G(\mathcal{AC})$ as convention.

Proposition 22.3 *Let $\mathcal{AC}$ be a partial A_∞ correspondence of energy cut level E_0 and minimal energy e_0, and G a discrete submonoid containing $G(\mathcal{AC})$. Suppose $e_0 \le e_{\min}(G)$ and $0 < \tau < \tau_0 = 1$. We can find a system of τ-collared Kuranishi structures and CF-perturbations, $\{(\widehat{\mathcal{U}^+_{k+1}}(\beta), \widehat{\mathfrak{S}}_{k+1}(\beta)) \mid (\beta, k) \in \mathcal{GK}(G; E_0, e_0)\}$, with the following properties:*

(1) *$\widehat{\mathcal{U}^+_{k+1}}(\beta)$ is a τ-collared Kuranishi structure of $\mathcal{M}_{k+1}(\beta)^{\boxplus\tau_0}$ and*

$$(\mathcal{M}_{k+1}(\beta), \widehat{\mathcal{U}_{k+1}}(\beta))^{\boxplus\tau_0} < (\mathcal{M}_{k+1}(\beta)^{\boxplus\tau_0}, \widehat{\mathcal{U}^+_{k+1}}(\beta)), \tag{22.1}$$

where $\widehat{\mathcal{U}_{k+1}}(\beta)$ is the Kuranishi structure given in Condition 21.6 (III). Evaluation maps are extended to τ-collared strongly smooth maps on this τ-collared Kuranishi structure and its component ev_0 is weakly submersive.

(2) *$\widehat{\mathfrak{S}}_{k+1}(\beta)$ is a τ-collared CF-perturbation of the τ-collared Kuranishi structure $\widehat{\mathcal{U}^+_{k+1}}(\beta)$. It is transversal to 0 and ev_0 is strongly submersive with respect to $\widehat{\mathfrak{S}}_{k+1}(\beta)$.*

(3) *There exists an isomorphism of τ-collared K-spaces,*

$$\partial(\mathcal{M}_{k+1}(\beta)^{\boxplus\tau_0}, \widehat{\mathcal{U}^+_{k+1}}(\beta)) \cong \coprod_{\beta_1,\beta_2,k_1,k_2,i} (-1)^{\epsilon}(\mathcal{M}_{k_1+1}(\beta_1)^{\boxplus\tau_0}, \widehat{\mathcal{U}^+_{k_1+1}}(\beta_1)) \\ {}_{\mathrm{ev}_i}\times_{\mathrm{ev}_0} (\mathcal{M}_{k_2+1}(\beta_2)^{\boxplus\tau_0}, \widehat{\mathcal{U}^+_{k_2+1}}(\beta_2)), \tag{22.2}$$

where

$$\epsilon = k_2(k_1 + i) + \dim L + i. \tag{22.3}$$

The isomorphism is compatible with the evaluation maps in the sense of Condition 21.6 (IX).

(4) *The restriction of $\widehat{\mathfrak{S}}_{k+1}(\beta)$ to the boundary is equivalent (see Definition 7.6 for the definition of equivalence of CF-perturbations) to the fiber product of $\widehat{\mathfrak{S}}_{k_1+1}(\beta_1)$ and $\widehat{\mathfrak{S}}_{k_2+1}(\beta_2)$ under the isomorphism (22.2).*

(5) *There is an isomorphism*

$$\begin{aligned}&\widehat{S}_m(\mathcal{M}_{k+1}(\beta)^{\boxplus\tau_0}, \widehat{\mathcal{U}_{k+1}^+}(\beta))\\ &\cong \coprod_{\substack{(\mathcal{T},\beta(\cdot))\in\mathcal{G}(k+1,\beta)\\ \#C_{1,\mathrm{int}}(\mathcal{T})=m}} \prod_{(\mathcal{T},\beta(\cdot))} (\mathcal{M}_{k_{\mathrm{v}}+1}(\beta(\mathrm{v}))^{\boxplus\tau_0}, \widehat{\mathcal{U}_{k_{\mathrm{v}}+1}^+}(\beta(\mathrm{v})))\end{aligned} \tag{22.4}$$

of τ-collared K-spaces. Here the right hand side is the disjoint union of the fiber products defined as in Definition 21.3. *The isomorphism is compatible with the evaluation maps in the sense of Condition* 21.6 *(X).*

(6) *The isomorphism* (22.4) *is also compatible with the outer collaring of the corner compatibility isomorphism*

$$\begin{aligned}&\widehat{S}_m((\mathcal{M}_{k+1}(\beta), \widehat{\mathcal{U}_{k+1}}(\beta)))\\ &\cong \coprod_{(\mathcal{T},\beta(\cdot))\in\mathcal{G}(k+1,\beta)} \prod_{(\mathcal{T},\beta(\cdot))} (\mathcal{M}_{k_{\mathrm{v}}+1}(\beta(\mathrm{v})), \widehat{\mathcal{U}_{k_{\mathrm{v}}+1}}(\beta(\mathrm{v})))\end{aligned}$$

given in Condition 21.6 *(X) via the KK-embedding* (22.1).

In particular the isomorphism (22.2) *is compatible with the boundary compatibility isomorphism* (21.6) *(Condition* 21.6 *(IX)) via the KK-embedding* (22.1).

(7) *The restriction of $\widehat{\mathfrak{S}}_{k+1}(\beta)$ to $\widehat{S}_m(\mathcal{M}_{k+1}(\beta)^{\boxplus\tau_0})$ is equivalent to the fiber product of $\widehat{\mathfrak{S}}_{k_{\mathrm{v}}+1}(\beta(\mathrm{v}))$ under the isomorphism* (22.4).

(8) *We denote by $\mathscr{I}_{m,(22.4)}$ the isomorphism* (22.4). *We denote its restriction to the ℓ-th normalized corner $\widehat{S}_\ell(\widehat{S}_m((\mathcal{M}_{k+1}(\beta)^{\boxplus\tau_0}, \widehat{\mathcal{U}_{k+1}^+}(\beta))))$ by $\widehat{S}_\ell(\mathscr{I}_{m,(22.4)})$. The target of the map $\widehat{S}_\ell(\mathscr{I}_{m,(22.4)})$ is*

$$\coprod_{(\mathcal{T},\beta(\cdot))\in\mathcal{G}(k+1,\beta)} \coprod_{\sum \ell_{\mathrm{v}}=\ell} \prod_{(\mathcal{T},\beta(\cdot))} \widehat{S}_{\ell_{\mathrm{v}}}(\mathcal{M}_{k_{\mathrm{v}}+1}(\beta(\mathrm{v}))^{\boxplus\tau_0}, \widehat{\mathcal{U}_{k_{\mathrm{v}}+1}^+}(\beta(\mathrm{v}))).$$

We apply $\mathscr{I}_{\ell_{\mathrm{v}},(22.4)}$ to all the factors and take the disjoint union. We denote by $\coprod\prod\mathscr{I}_{(22.4)}$ the map we obtain in this way. Its target is a disjoint union of copies of fiber products

$$\prod_{(\mathcal{T},\beta(\cdot))} (\mathcal{M}_{k_{\mathrm{v}}+1}(\beta(\mathrm{v}))^{\boxplus\tau_0}, \widehat{\mathcal{U}_{k_{\mathrm{v}}+1}^+}(\beta(\mathrm{v}))), \tag{22.5}$$

where the union is taken over all $(\mathcal{T}, \beta(\cdot)) \in \mathcal{G}(k+1, \beta)$ with $\#C_{1,\mathrm{int}}(\mathcal{T}) = m + \ell$. The fiber product (22.5) *is defined as in Definition* 21.3.

Then we claim

$$\left(\coprod\prod\mathscr{I}_{(22.4)}\right) \circ \widehat{S}_\ell(\mathscr{I}_{m,(22.4)}) = \mathscr{I}_{\ell+m,(22.4)} \circ \pi_{\ell,m},$$

where

$$\pi_{\ell,m} : \widehat{S}_\ell(\widehat{S}_m(\mathcal{M}_{k+1}(\beta)^{\boxplus\tau_0}, \widehat{\mathcal{U}^+_{k+1}}(\beta))) \to \widehat{S}_{m+\ell}(\mathcal{M}_{k+1}(\beta)^{\boxplus\tau_0}, \widehat{\mathcal{U}^+_{k+1}}(\beta))$$

is the covering map in Proposition 24.17.

To prove Proposition 22.3 we use the following:

Lemma 22.4 *If* $(\mathcal{T}, \beta(\cdot)) \in \mathcal{G}(k+1, \beta)$, $(k, \beta) \in \mathcal{GK}(G; E_0, e_0)$ *and* $\mathrm{v} \in C_{0,\mathrm{int}}(\mathcal{T})$, *then* $(\beta(\mathrm{v}), k_\mathrm{v}) \in \mathcal{GK}(G; E_1, e_0)$ *with* $E_1 < E_0$.

Proof We consider $\mathcal{T} \setminus \{\mathrm{v}\}$. It has $k_\mathrm{v}+1$ connected components $\mathcal{T}_i$, $i = 0, \dots, k$. Let ℓ be the number of its connected components that do not contain exterior vertices. If $\mathcal{T}_i$ does not contain exterior vertices, then by Definition 21.1 (7) we have

$$\sum_{\mathrm{v}' \in C_{0,\mathrm{int}}(\mathcal{T}) \cap \mathcal{T}_i} E(\beta_{\mathrm{v}'}) > 0.$$

Therefore $\ell e_0 + E(\beta_\mathrm{v}) \le E_0$. The lemma follows immediately from this fact. □

Proof of Proposition 22.3. The proof is given by induction on E_0. Lemma 22.4 implies that we already obtained $\widehat{\mathcal{U}^+_{k_\mathrm{v}+1}}(\beta(\mathrm{v}))$ and $\widehat{\mathfrak{S}}_{k_\mathrm{v}+1}(\beta_\mathrm{v})$ while we construct $\widehat{\mathcal{U}^+_{k+1}}(\beta)$ and $\widehat{\mathfrak{S}}_{k+1}(\beta)$. Here k_v and $\beta(\mathrm{v})$ are as in (22.5). Therefore using Propositions 17.78, 17.81, we can prove Proposition 22.3 in the same way as the proof of Proposition 19.1. □

Next we consider the case of a P-parametrized family. Proposition 22.6 below claims the following. Suppose we are given a partial P-parametrized family $\mathcal{AC}_P$ of A_∞ correspondences. We assume that on the boundary ∂P we are given a system of CF-perturbations. Then we can extend it to a CF-perturbation defined on P. Note that we need to construct a system of thickenings of the Kuranishi structures at the same time and we need to assume compatibilities at the corners of the various given objects on the boundary. We formulate the assumption as Situation 22.5 below.

We consider the vertical boundary $\mathcal{M}_{k+1}(\beta; \partial P) \subseteq \partial\mathcal{M}_{k+1}(\beta; P)$, which we denote by $\mathfrak{C}^v$. Its complement, the horizontal boundary, is denoted by $\mathfrak{C}^h$. We define $\widehat{S}^{\mathfrak{C}^v}_m(\mathcal{M}_{k+1}(\beta; P))$ as in Proposition 18.6. We also define partial outer collaring by Definition 18.9. We remark

$$\widehat{S}^{\mathfrak{C}^v}_m(\mathcal{M}_{k+1}(\beta; P), \widehat{\mathcal{U}}_{k+1}(\beta; P)) = (\mathcal{M}_{k+1}(\beta; S_m(P)), \widehat{\mathcal{U}}_{k+1}(\beta; P)|_{\mathcal{M}_{k+1}(\beta; S_m(P))}).$$

Situation 22.5 Suppose we are given a partial P-parametrized family $\mathcal{AC}_P$ of A_∞ correspondences whose moduli spaces of P-parametrized A_∞ operations are $\mathcal{M}_{k+1}(\beta; P)$. Let E_0 be its energy cut level and e_0 its minimal energy. Let G be a discrete monoid containing $G(\mathcal{AC}_P)$.

We assume that we are given a family of the pairs

$$\widehat{\mathcal{U}^+_{k+1}}(\beta; \widehat{S}_m(P)), \quad \widehat{\mathfrak{S}}_{k+1}(\beta; \widehat{S}_m(P))$$

for $m \ge 1$ and $0 < \tau < \tau_0 = 1$ with the following properties:

(1) $\widehat{\mathcal{U}^+_{k+1}}(\beta; \widehat{S}_m(P))$ is a τ-collared Kuranishi structure of $\mathcal{M}_{k+1}(\beta; \widehat{S}_m(P))^{\boxplus\tau_0}$. It comes with evaluation maps, boundary compatibility isomorphisms, corner compatibility isomorphisms, so that we obtain a partial tree-like K-system parametrized by $\widehat{S}_m(P)^{\boxplus\tau_0}$, for which $\mathcal{M}_{k+1}(\beta; \widehat{S}_m(P))^{\boxplus\tau_0}$ is the moduli spaces of $\widehat{S}_m(P)^{\boxplus\tau_0}$-parametrized A_∞ operations.

(2)

$$\begin{aligned}
&\widehat{S}^{\mathfrak{C}v}_m(\mathcal{M}_{k+1}(\beta; \widehat{S}_m(P)), \widehat{\mathcal{U}_{k+1}}(\beta))^{\boxplus\tau_0} \\
&< (\mathcal{M}_{k+1}(\beta; \widehat{S}_m(P))^{\boxplus\tau_0}, \widehat{\mathcal{U}^+_{k+1}}(\beta; \widehat{S}_m(P))).
\end{aligned} \tag{22.6}$$

Here the left hand side is the outer collaring of the Kuranishi structure, which is the restriction to $\widehat{S}_m(P)$ of the Kuranishi structure given by Condition 21.11 (III). Orientation, evaluation maps, boundary compatibility isomorphisms, corner compatibility isomorphisms are compatible with this KK-embedding.

(3) $\widehat{\mathfrak{S}}_{k+1}(\beta; \widehat{S}_m(P))$ is a CF-perturbation of $\widehat{\mathcal{U}^+_{k+1}}(\beta; \widehat{S}_m(P))$. We assume that $\widehat{\mathfrak{S}}_{k+1}(\beta; \widehat{S}_m(P))$ is transversal to 0. The evaluation map $(\mathrm{ev}_0, \mathrm{ev}_{\widehat{S}_m(P)})$ is stratified strongly submersive with respect to $\widehat{\mathfrak{S}}_{k+1}(\beta; \widehat{S}_{\ell'+m}(P))$. The CF-perturbations $\widehat{\mathfrak{S}_{k+1}}(\beta; \widehat{S}_m(P))$ for various k, β are compatible with the boundary and corner compatibility isomorphisms.

(4) We are given a $(m+\ell)!/m!\ell!$-hold covering map:

$$\begin{aligned}
&\widehat{S}^{\mathfrak{C}v}_\ell((\mathcal{M}_{k+1}(\beta; \widehat{S}_m(P))^{\boxplus\tau_0}, \widehat{\mathcal{U}^+_{k+1}}(\beta; \widehat{S}_m(P)))) \\
&\to ((\mathcal{M}_{k+1}(\beta; \widehat{S}_{\ell+m}(P))^{\boxplus\tau_0}, \widehat{\mathcal{U}^+_{k+1}}(\beta; \widehat{S}_{\ell+m}(P))))
\end{aligned} \tag{22.7}$$

of τ-collared K-spaces. (22.7) is compatible with the evaluation maps, boundary compatibility isomorphisms, corner compatibility isomorphisms.

(5) The covering map (22.7) is also compatible with the covering map (Proposition 18.6)

$$\widehat{S}^{\mathfrak{C}v}_\ell(\widehat{S}^{\mathfrak{C}v}_m(\mathcal{M}_{k+1}(\beta; P), \widehat{\mathcal{U}_{k+1}})^{\boxplus\tau_0}) \to \widehat{S}^{\mathfrak{C}v}_{\ell+m}(\mathcal{M}_{k+1}(\beta; P), \widehat{\mathcal{U}_{k+1}})^{\boxplus\tau_0} \tag{22.8}$$

via the KK-embedding (22.6).

(6) The restriction of the CF-perturbation $\widehat{\mathfrak{S}}_{k+1}(\beta; \widehat{S}_m(P))$ to

$$\widehat{S}^{\mathfrak{C}v}_\ell(\mathcal{M}_{k+1}(\beta; \widehat{S}_m(P))^{\boxplus\tau_0})$$

is equivalent to the pullback of the CF-perturbations

$$\widehat{\mathfrak{S}}_{k+1}(\beta(\mathrm{v}); \widehat{S}_{\ell+m}(P))$$

by the map (22.7).

(7) The next diagram commutes. (In this diagram we omit underlying topological space from the notation.)

$$\begin{array}{ccc} \widehat{S}^{\mathfrak{C}v}_n(\widehat{S}^{\mathfrak{C}v}_\ell(\mathcal{U}^+_{k+1}(\beta; \widehat{S}_m(P)))) & \xrightarrow{\pi_{n,\ell}} & \widehat{S}^{\mathfrak{C}v}_{n+\ell}(\mathcal{U}^+_{k+1}(\beta; \widehat{S}_m(P)) \\ \widehat{S}^{\mathfrak{C}v}_n((22.7)) \downarrow & & \downarrow (22.7) \\ \widehat{S}^{\mathfrak{C}v}_n(\mathcal{U}^+_{k+1}(\beta; \widehat{S}_{\ell+m}(P))) & \xrightarrow{(22.7)} & (\mathcal{U}^+_{k+1}(\beta; \widehat{S}_{n+\ell+m}(P)) \end{array} \tag{22.9}$$

Here the first horizontal arrow is the covering map in Proposition 18.6. The right vertical and second horizontal arrows are the covering map (22.8). The left vertical arrow is the restriction of the covering map (22.8) to the codimension n corner.

■

Proposition 22.6 *Suppose we are in Situation 22.5 and $0 < \tau' < \tau$. Then we can find a system of Kuranishi structures and CF-perturbations*

$$\{(\widehat{\mathcal{U}^+_{k+1}}(\beta; P), \widehat{\mathfrak{S}}_{k+1}(\beta; P)) \mid (\beta, k) \in \mathcal{GK}(G; E_0, e_0)\}$$

with the following properties:

(1) $\widehat{\mathcal{U}^+_{k+1}}(\beta; P)$ *is a τ'-collared Kuranishi structure of $\mathcal{M}_{k+1}(\beta; P)^{\boxplus\tau_0}$. It comes with evaluation maps, boundary compatibility isomorphisms, corner compatibility isomorphisms so that we obtain a partial tree-like K-system parametrized by P, for which $(\mathcal{M}_{k+1}(\beta; P)^{\boxplus\tau_0}, \widehat{\mathcal{U}^+_{k+1}}(\beta; P))$ is the moduli spaces of $P^{\boxplus\tau_0}$-parametrized A_∞ operations.*

(2) *We have:*

$$(\mathcal{M}_{k+1}(\beta; P), \widehat{\mathcal{U}_{k+1}}(\beta; P))^{\boxplus\tau_0} < (\mathcal{M}_{k+1}(\beta; P)^{\boxplus\tau_0}, \widehat{\mathcal{U}^+_{k+1}}(\beta; P)). \tag{22.10}$$

By Condition 21.11 (III) a Kuranishi structure is given on $\mathcal{M}_{k+1}(\beta; P)$, which we denote by $\widehat{\mathcal{U}_{k+1}}(\beta; P)$. The left hand side is its outer collaring.

The KK-embedding in (22.10) respects orientation, evaluation maps, boundary compatibility isomorphisms, corner compatibility isomorphisms.

(3) $\widehat{\mathfrak{S}}_{k+1}(\beta; P)$ *is a CF-perturbation of the Kuranishi structure $\widehat{\mathcal{U}^+_{k+1}}(\beta; P)$, which is transversal to 0 and the map $(\mathrm{ev}_0, \mathrm{ev}_P)$ is stratified strongly submersive with respect to $\widehat{\mathfrak{S}}_{k+1}(\beta; P)$. The CF-perturbations $\widehat{\mathfrak{S}_{k+1}}(\beta; P)$ for various k, β are compatible with boundary and corner compatibility isomorphisms.*

(4) *There exists an isomorphism:*

$$\begin{aligned}&\widehat{S}^{\mathfrak{C}v}_{m}(\mathcal{M}_{k+1}(\beta;P)^{\boxplus\tau_0},\widehat{\mathcal{U}^+_{k+1}}(\beta;P))\\ &\cong(\mathcal{M}_{k+1}(\beta;\widehat{S}_m(P))^{\boxplus\tau_0},\widehat{\mathcal{U}^+_{k+1}}(\beta;\widehat{S}_m(P)))\end{aligned}\tag{22.11}$$

of τ'-collared K-spaces. (Note the right hand side is the K-space given in Situation 22.5.) The isomorphism (22.11) is compatible with the evaluation maps boundary compatibility isomorphisms, corner compatibility isomorphisms.[1] *It is also compatible with KK-embeddings (22.6), (22.10).*

(5) *The restriction of $\widehat{\mathfrak{S}}_{k+1}(\beta;P)$ to $\widehat{S}^{\mathfrak{C}v}_{m}(\mathcal{M}_{k+1}(\beta;P)^{\boxplus\tau_0},\widehat{\mathcal{U}^+_{k+1}}(\beta;P))$ is equivalent to $\widehat{\mathfrak{S}}_{k+1}(\beta;\widehat{S}_m(P))$ via the isomorphism (22.11).*

(6) *The next diagram commutes. (In this diagram we omit underlying topological space from the notation.)*

$$\begin{array}{ccc}\widehat{S}^{\mathfrak{C}v}_{\ell}(\widehat{S}^{\mathfrak{C}v}_{m}(\widehat{\mathcal{U}^+_{k+1}}(\beta;P)) & \xrightarrow{\pi_{\ell,m}} & \widehat{S}^{\mathfrak{C}v}_{\ell+m}(\widehat{\mathcal{U}^+_{k+1}}(\beta;P)\\ {\scriptstyle \widehat{S}^{\mathfrak{C}v}_{\ell}((22.11))}\downarrow & & \downarrow{\scriptstyle (22.11)}\\ \widehat{S}^{\mathfrak{C}v}_{\ell}(\widehat{\mathcal{U}^+_{k+1}}(\beta;\widehat{S}_m(P))) & \xrightarrow{(22.7)} & \widehat{\mathcal{U}^+_{k+1}}(\beta;\widehat{S}_{\ell+m}(P))\end{array}\tag{22.12}$$

Here the first horizontal arrow is the covering map in Proposition 18.6. The second horizontal arrow is the covering map (22.7). The right vertical arrow is the covering map (22.11). The left vertical arrow is its restriction of the codimension ℓ partial corner.

(7) *When the CF-perturbations $\widehat{\mathfrak{S}}_{k+1}(\beta;\widehat{S}_m(P))$ given in Situation 22.5 move in a uniform family, then the CF-perturbation $\widehat{\mathfrak{S}}_{k+1}(\beta;P)$ we obtain becomes a uniform family.*

Proof Using Lemma 22.4, Propositions 17.78, 17.81, we can prove Proposition 22.6 in the same way as the proof of Propositions 19.1, 19.27 and 22.3. In fact the items in Situation 17.55 and items in Situation 22.5 are related as in Table 22.1 below. Moreover the items in Proposition 17.58 and Proposition 22.6 are related as in Table 22.2 below. Note that the statements concerning only the vertical boundary appearing in Situation 22.5 and Proposition 22.6 correspond to the statements on the boundary appearing in Situation 17.55, Proposition 17.58 in these tables. The horizontal boundary of the former is taken care of in Situation 22.5 (1) and Proposition 22.6 (1) as a claim made on the K-spaces which are the moduli spaces of parametrized A_∞ operations. In fact, this claim contains the existence of boundary compatibility and corner compatibility isomorphisms and their consistency.

[1] The domains of boundary compatibility isomorphisms or corner compatibility isomorphisms are *horizontal* boundary or corner. So they induce maps between *vertical* corners, which is the left hand side of *(22.11)*.

Table 22.1 Situations 17.55 and 22.5

Situation 17.55 (1)	Situation 22.5 (1)
Situation 17.55 (2)	Situation 22.5 (4)
Situation 17.55 (3)	Situation 22.5 (7)
Situation 17.55 (4)	Situation 22.5 (2)
Situation 17.55 (5)	Situation 22.5 (5)
Situation 17.72 (1)	Situation 22.5 (3)
Situation 17.72 (2)	Situation 22.5 (6)

Table 22.2 Propositions 17.58 and 22.6

Proposition 17.58, second line	Proposition 22.6 (1)
Proposition 17.58(1)	Proposition 22.6 (4)
Proposition 17.58 (2)	Proposition 22.6 (2)
Proposition 17.58 (3)	Proposition 22.6 (4), last sentence
Proposition 17.58 (4)	Proposition 22.6 (6)
Proposition 17.58 (5)	Proposition 22.6 (4), last sentence
Proposition 17.73	Proposition 22.6 (3) and (5)

□

22.2 Algebraic Lemmas: Promotion Lemmas via Pseudo-Isotopy

Definition 22.7 Let $E_0' < E_0$.

(1) When $\{\mathfrak{m}_{k,\beta}\}$ is a partial G-gapped filtered A_∞ structure on $\Omega(L)$ of energy cut level E_0 and of minimal energy e_0. We forget all the operations $\mathfrak{m}_{k,\beta}$ with $E(\beta) > E_0'$ and obtain a partial G-gapped filtered A_∞ structure on $\Omega(L)$ of energy cut level E_0'. We call it the partial filtered A_∞ structure on $\Omega(L)$ obtained by the *energy cut at* E_0'.
(2) When $\{\mathfrak{m}_{k,\beta}\}$ is obtained from $\{\mathfrak{m}'_{k,\beta}\}$ by the energy cut at E_0' and $\{\mathfrak{m}'_{k,\beta}\}$ is a partial G-gapped filtered A_∞ structure on $\Omega(L)$ of energy cut level E_0, we call $\{\mathfrak{m}'_{k,\beta}\}$ a *promotion of* $\{\mathfrak{m}_{k,\beta}\}$ *to the energy cut level* E_0.
(3) We define an energy cut or a promotion of pseudo-isotopy or a P-parametrized family of partial A_∞ structures in the same way.

We will use the next proposition which says that we can extend the promotion of partial A_∞ structures via pseudo-isotopy.

Proposition 22.8 *Let G be a discrete submonoid, $e_0 \le e_{\min}(G)$ and $E_0 < E_1$. For each $j = 0, 1$ let $\{\mathfrak{m}^j_{k,\beta}\}$ be a G-gapped partial filtered A_∞ structure of energy cut level E_j and minimal energy e_0 on $\Omega(L)$. Suppose that we are given a G-gapped partial filtered A_∞ pseudo-isotopy $(\{\mathfrak{m}^t_{k,\beta}\}, \{\mathfrak{c}^t_{k,\beta}\})$ of energy cut level E_0*

and minimal energy e_0, *from* $\{\mathfrak{m}^0_{k,\beta}\}$ *to the energy cut of* $\{\mathfrak{m}^1_{k,\beta}\}$ *at* E_0. *Then we can promote* $\{\mathfrak{m}^0_{k,\beta}\}$ *to energy cut level* E_1 *and* $(\{\mathfrak{m}^t_{k,\beta}\}, \{\mathfrak{c}^t_{k,\beta}\})$ *to energy cut level* E_1.

Proof The proof is the same as the proof of [Fuk4, Theorem 8.1]. (The only difference is the following point: In [Fuk4, Theorem 8.1] partial structures are ones where we take only finitely many β's: Here we take finitely many (β, k)'s.) We repeat the proof for completeness.

We consider the set

$$\mathfrak{E} = \{E(\beta) + ke_0 \mid (\beta, k) \in G \times \mathbb{Z}_{\geq 0}\}.$$

This is a discrete set. So by applying an induction we may and will assume that

$$\#(\mathfrak{E} \cap (E_0, E_1]) = 1. \tag{22.13}$$

Let $(\beta, k) \in G \times \mathbb{Z}_{\geq 0}$ such that $E(\beta) + ke_0 \in (E_0, E_1]$. We will define $\mathfrak{m}^t_{k,\beta}$ and $\mathfrak{c}^t_{k,\beta}$ for each such (β, k).

We put $\mathfrak{c}^t_{k,\beta} = 0$. Then there exists a unique $\mathfrak{m}^t_{k,\beta}$ such that it satisfies (21.27) and $\mathfrak{m}^t_{k,\beta} = \mathfrak{m}^1_{k,\beta}$ for $t = 1$. Note that (21.27) can be regarded as an ordinary differential equation for each fixed $(h_1, \dots, h_k)$. Therefore $\mathfrak{m}^t_{k,\beta}$ depends smoothly on t and is local in the $[0, 1]$-direction.

Next we check the A_∞ relation for each fixed t in the case of (β, k). We calculate

$$\begin{aligned}
&\frac{d}{dt} \sum_{k_1+k_2=k+1} \sum_{\beta_1+\beta_2=\beta} \sum_{i=1}^{k-k_2+1} \mathfrak{m}^t_{k_1,\beta_1}(h_1, \dots, \mathfrak{m}^t_{k_2,\beta_2}(h_i, \dots), \dots, h_k) \\
&= \sum_{k_1+k_2=k+1} \sum_{\beta_1+\beta_2=\beta} \sum_{i=1}^{k-k_2+1} \frac{d\mathfrak{m}^t_{k_1,\beta_1}}{dt}(h_1, \dots, \mathfrak{m}^t_{k_2,\beta_2}(h_i, \dots), \dots, h_k) \\
&+ \sum_{k_1+k_2=k+1} \sum_{\beta_1+\beta_2=\beta} \sum_{i=1}^{k-k_2+1} \mathfrak{m}^t_{k_1,\beta_1}(h_1, \dots, \frac{d\mathfrak{m}^t_{k_2,\beta_2}}{dt}(h_i, \dots), \dots, h_k).
\end{aligned} \tag{22.14}$$

Using (21.27) and the A_∞ relation (that is, the induction hypothesis), it is easy to see that (22.14) is zero. We define $\mathfrak{m}^0_{k,\beta}$ as the case $t = 0$ of $\mathfrak{m}^t_{k,\beta}$. The proof of Proposition 22.8 is complete. □

We next study the promotion of pseudo-isotopy using pseudo-isotopy of pseudo-isotopies. The proof is similar to that of [Fuk4, Theorem 14.1]. We repeat the detail of the proof for completeness.

We first define the notion of collaredness of a $P^{\boxplus\tau}$-parametrized (partial) A_∞ structure. Here $P^{\boxplus\tau}$ is the outer collaring of our manifold with corners P. (A similar assumption appeared in [Fuk4, Assumption 14.1].)

Definition 22.9 Let $\{\mathfrak{m}_{k,\beta}^{P^{\boxplus\tau}}\}$ be a $P^{\boxplus\tau}$-parametrized partial A_∞-structure of energy cut level E_0 and minimal energy e_0. We say it is *τ-collared* if the following conditions (1) and (2) are satisfied.

Let $\mathbf{t} \in \overset{\circ\circ}{S}_k(P^{\boxplus\tau})$. Its τ-collared neighborhood is identified with $V \times [-\tau, 0)^k$. Let $(t'_1, \dots, t'_m)$ be a coordinate of V and let $(t''_1, \dots, t''_k)$ be the standard coordinate of $[-\tau, 0)^k$. A differential form on P in a neighborhood is written as $\sum f_{I'I''} dt'_{I'} \wedge dt''_{I''}$ where $dt'_{I'}$ are wedge products of dt'_i's, and $dt''_{I''}$ are wedges products of dt''_i's.

By definition, $\mathfrak{m}_{k,\beta}^{P^{\boxplus\tau}}$ is written on this neighborhood as the form

$$\mathfrak{m}_{k,\beta}^{P^{\boxplus\tau}}(h_1, \dots, h_k) = \sum_{I,I'} dt'_{I'} \wedge dt''_{I''} \wedge \mathfrak{m}_{k,\beta;I',I''}^{t',t''}(h_1, \dots, h_k).$$

Now we require:

(1) $\mathfrak{m}_{k,\beta;I',I''}^{t',t''}(h_1, \dots, h_k) = 0$ unless $I'' = \emptyset$.
(2) If $I'' = \emptyset$, $\mathfrak{m}_{k,\beta;I',\emptyset}^{t',t''}(h_1, \dots, h_k)$ is independent of $t'' \in [-\tau, 0)^k$.

We say a *P-parametrized partial A_∞ structure* $\{\mathfrak{m}_{k,\beta}^P\}$ *is collared* if there exist $\tau > 0$ and P' such that $P = P'^{\boxplus\tau}$ and $\{\mathfrak{m}_{k,\beta}^P\}$ is collared in the above sense.

Example 22.10 The case when $P = [0, 1]$ in Definition 22.9 is nothing but the case of pseudo-isotopy in Definition 21.25. In this case, $P^{\boxplus\tau} = [-\tau, 1 + \tau]$ and the τ-collaredness properties (1), (2) in Definition 22.9 imply the following properties of pseudo-isotopy $(\{\mathfrak{m}_{k,\beta}^t\}, \{\mathfrak{c}_{k,\beta}^t\})$, respectively:

(1) $\mathfrak{c}_{k,\beta}^t = 0$ for $t \in [-\tau, 0] \cup [1, 1 + \tau]$.
(2) $\frac{d}{dt}\mathfrak{m}_{k,\beta}^t = 0$ for $t \in [-\tau, 0] \cup [1, 1 + \tau]$.

Situation 22.11 Let P be a manifold with corners and $E_1 > E_0 \geq 0$, $e_0 > 0$. We assume that we are given the following objects:

(1) A $P \times [0, 1]$-parametrized collared partial A_∞ structure $\{\mathfrak{m}_{k,\beta}^{P\times[0,1]}\}$ of energy cut level E_0 and of minimal energy e_0 on $\Omega(L)$.
(2) A collared version of promotion of the restriction of $\{\mathfrak{m}_{k,\beta}^{P\times[0,1]}\}$ to $P \times \{1\}$ to energy cut level E_1.
(3) Let $\partial P = \coprod \partial_i P$ be the decomposition of the normalized boundary of P into the connected components. We also assume that a collared promotion of the restriction of $\{\mathfrak{m}_{k,\beta}^{P\times[0,1]}\}$ to $\partial_i P \times [0, 1]$ to energy cut level E_1 is given for each i.
(4) We assume that the restriction of the promotion in (2) coincides with the promotion in (3) on $\partial_i P \times \{1\}$.
(5) Suppose that the images of $\partial_i P$ and $\partial_j P$ intersect each other in P at the component $\partial_{ij} P$ of the codimension 2 corner of P. (Note that the case $i = j$ is included. In this case, $\partial_{ii} P$ is the 'self intersection' of $\partial_i P$.) Then we assume

that the promotions of the restrictions on $\partial_i P \times [0, 1]$ and on $\partial_j P \times [0, 1]$ in (3) coincide with each other on $\partial_{ij} P \times [0, 1]$.

■

Remark 22.12 In Situation 22.11 we assumed the compatibility of the promotion only at the codimension 2 corners. In this situation it automatically implies that they coincide at higher codimensional corners. This is because our assumptions are their exact coincidence and not coincidence up to certain equivalence relation.

Proposition 22.13 *In Situation* 22.11 *there exists a promotion of* $\{\mathfrak{m}_{k,\beta}^{P\times[0,1]}\}$ *to energy cut level* E_1 *such that the promotion coincides with those given in Situation* 22.11 *(2) (resp. (3)) on* $P \times \{1\}$ *(resp.* $\partial_i P \times [0, 1]$*).*

Proof We first prove Proposition 22.13 for the case $P = [-\tau, 1 + \tau]$. We regard $P \times [0, 1]$ as $P \times [-\tau, 1 + \tau] = ([0, 1]^2)^{\boxplus\tau}$ and assume that our structures are τ-collared.

We change the corner structure of $([0, 1]^2)^{\boxplus\tau}$ so that we smooth the corners at two points $(-\tau, -\tau)$, $(1+\tau, -\tau)$ and make two points $(-\tau, 1+\tau/4)$, $(1+\tau, 1+\tau/4)$ into new corners instead. We leave two other corners $(-\tau, 1 + \tau)$, $(1 + \tau, 1 + \tau)$ as corners. We then get a new cornered 2 manifold Q diffeomorphic to $[-\tau, 1 + \tau]^2$. We denote this diffeomorphism by $F : [-\tau, 1 + \tau]^2 \to Q$. The diffeomorphism F is different from the set theoretical identity map $\mathrm{id} : [-\tau, 1 + \tau]^2 \to Q$. In fact, we can take F satisfying the following properties. See Fig. 22.1.

(i) $F(-\tau, 1 + \tau/4) = (-\tau, -\tau)$ and $F(1 + \tau, 1 + \tau/4) = (1 + \tau, -\tau)$.
(ii) F is the identity on the edge $[-\tau, 1 + \tau] \times \{1 + \tau\}$.

Using the τ-collaredness, our structures give a Q-parametrized family of partial A_∞ structures. We regard it as a $[-\tau, 1 + \tau]^2$-parametrized family of partial A_∞ structures under the diffeomorphism F. By assumption, its energy cut level is E_0 and the energy cut level of its restriction to $[-\tau, 1 + \tau] \times \{1 + \tau\}$ is E_1. Therefore we can apply Proposition 22.8 to promote this $[-\tau, 1 + \tau]^2$-parametrized family to energy cut level E_1. Using collaredness again, we find that on $\partial[-\tau, 1 + \tau] \times [-\tau, 1 + \tau] \subset [-\tau, 1 + \tau]^2 \overset{F}{\cong} Q$ this promotion coincides with the structure of energy cut level E_1 given at the beginning.

Now we identify $Q \overset{\mathrm{id}}{\cong} [-\tau, 1 + \tau]^2$. At the place where we smooth corners or make new corners, we can use the τ-collaredness to show that the promotion coincides with the one originally given at the beginning. Thus we obtain the required promotion. Note that the structure obtained is τ'-collared for some $0 < \tau' < \tau$ by construction.

Thus we have proved Proposition 22.13 for the case $P = [0, 1]$. (We use only this case in this book.)

The general case can be proved in a similar way. Namely, we smooth some corners of $(P \times [0, 1])^{\boxplus\tau}$ and make certain points into new corners to obtain a cornered manifold Q so that the following holds:

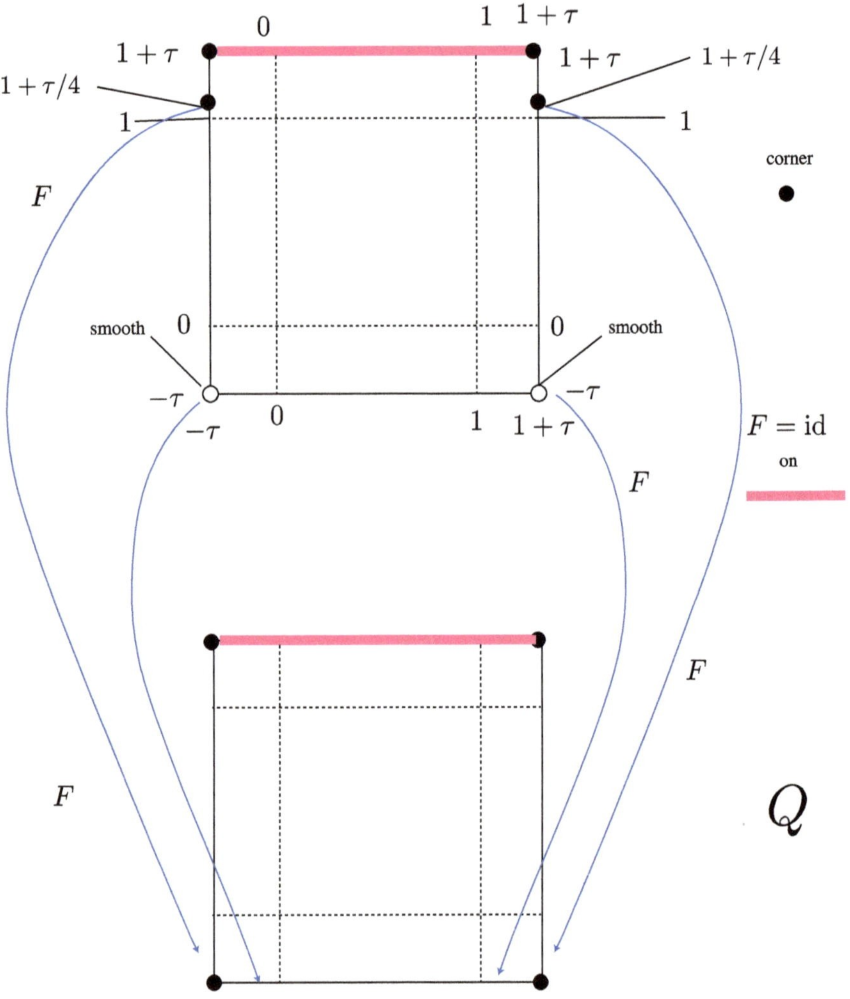

Fig. 22.1 Q and F

(1) There exists a diffeomorphism $F : (P \times [0,1])^{\boxplus\tau} \to Q$ which is the identity on $P^{\boxplus\tau} \times \{1+\tau\}$.
(2) $\{\mathfrak{m}_{k,\beta}^{P\times[0,1]}\}$ induces a Q-parameter family of partial A_∞ structures.

By the diffeomorphism in (1), the set $P^{\boxplus\tau} \times \{1+\tau\} \subset (P \times [0,1])^{\boxplus\tau} = Q$, where the last equality is the set-theoretical one, is mapped from a subset of $(\partial P^{\boxplus\tau} \times [-\tau, 1+\tau]) \cup (P^{\boxplus\tau} \times \{1+\tau\})$. Therefore the partial structure in (2) is one of energy cut level E_1 on $P^{\boxplus\tau} \times \{1+\tau\}$. On $(P \times [0,1])^{\boxplus\tau}$ it is of energy cut level E_0. We apply Proposition 22.8 to promote it to the energy cut level E_1. Using the diffeomorphism F, we regard it as a $(P \times [0,1])^{\boxplus\tau}$-parametrized structure. Using collaredness, we find that it induces a $(P \times [0,1])$-parametrized structure via the identity map. (Note that identity map $(P \times [0,1])^{\boxplus\tau} \cong Q$ is not a diffeomorphism. However, the structures are constant at the place where differentiability breaks down.)

(1) implies that on $[-\tau, 1+\tau] \times \{1+\tau\}$ the structure we obtained coincides with given one as partial A_∞ structure of energy cut level E_1.

Thus we have obtained the required promotion. □

22.3 Pointwiseness of Parametrized Family of Smooth Correspondences

In this section we prove Proposition 22.17 which reads that the operation defined as a smooth correspondence associated to a P-parametrized family of A_∞ correspondences is pointwise in the P-direction in the sense of Definition 21.27. To state the result in the way we can utilize in similar but different situations, we slightly generalize Definition 21.27. We use the notation t_J etc. of Definition 21.27 in the next definition.

Definition 22.14 Let M_s, M_t be smooth manifolds (without boundary) and P a smooth manifold with corners. A linear map $F : \Omega(P \times M_s)[1] \to \Omega(P \times M_t)[1]$ is said to be *pointwise in the P direction* if the following holds:

For each $I, J \subset \{1, \dots, d\}$ with $I \cap J = \emptyset$ and $\mathbf{t} \in P$ there exists a linear and continuous map $F^{\mathbf{t}}_{I;J} : \Omega(M_s)[1] \to \Omega(M_t)[1]$ such that

$$F(dt_J \wedge h)|_{\{\mathbf{t}\}\times M_t} = \sum_I dt_I \wedge dt_J \wedge F^{\mathbf{t}}_{I;J}(h|_{\{\mathbf{t}\}\times M_t}). \tag{22.15}$$

Moreover $F^{\mathbf{t}}_{I;J}$ depends smoothly on $\mathbf{t}$ (with respect to the operator topology) and is independent of J up to the sign.

In the case when $M_s = L^k$ and $M_t = L$, Definition 22.14 is nothing but Definition 21.27.

Situation 22.15 Let $(X, \widehat{\mathcal{U}})$ be a K-space, and let M_s, M_t be smooth manifolds without boundary and P a smooth manifold with corners. Let $f_s : (X, \widehat{\mathcal{U}}) \to M_s$, $f_t : (X, \widehat{\mathcal{U}}) \to M_t$ and $f_P : (X, \widehat{\mathcal{U}}) \to P$ be strongly smooth maps. We assume that $(f_t, f_P) : (X, \widehat{\mathcal{U}}) \to M_t \times P$ is stratumwise weakly submersive.

Let $\widehat{\mathfrak{S}}$ be a CF-perturbation of $(X, \widehat{\mathcal{U}})$. We assume that (f_t, f_P) is stratumwise strongly submersive with respect to $\widehat{\mathfrak{S}}$. ■

Definition 22.16 We call $\mathfrak{X}_P = ((X, \widehat{\mathcal{U}}), f_s, f_t, f_P)$ as in Situation 22.15 a *P parametrized family of smooth correspondences.*

Let $\widehat{\mathfrak{S}}$ be a CF-perturbation such that (f_t, f_P) is strongly submersive with respect to $\widehat{\mathfrak{S}}$. Then for any $\epsilon > 0$ we associate a linear map

$$\mathrm{Corr}_{\mathfrak{X}_P}(\cdot; \widehat{\mathfrak{S}}^{\epsilon}) \,:\, \Omega(P \times M_s)[1] \longrightarrow \Omega(P \times M_t)[1]$$

by

$$\mathrm{Corr}_{\mathfrak{X}_P}(h; \widehat{\mathfrak{S}}^{\epsilon}) = (f_P, f_t)!((f_P, f_s)^* h; \widehat{\mathfrak{S}}^{\epsilon}). \tag{22.16}$$

Then we have

Proposition 22.17 *The map* $\mathrm{Corr}_{\mathfrak{X}_P}(\cdot; \widehat{\mathfrak{S}}^{\epsilon})$ *is pointwise in* P *direction.*

Proof Let $h \in \Omega(M_s)$. We put

$$(f_P, f_t)!((f_s)^* h; \widehat{\mathfrak{S}}^{\epsilon}) = \sum_I dt_I \wedge F_I(h).$$

Let $F_I^{\mathbf{t}}(h)$ be the restriction of $F_I(h)$ to $\{\mathbf{t}\} \times M_t$. Then it is easy to see that this $F_I^{\mathbf{t}}$ satisfies (22.15) up to the sign. □

22.4 Proof of Theorem 21.35

In this section, we complete the proof of Theorem 21.35.

Proof of Theorem* 21.35 *(2) Suppose $\mathcal{AF} = \{\mathcal{M}_{k+1}(\beta) \mid \beta, k\}$ defines a partial A_∞ correspondence over L of energy cut level E_0 and minimal energy e_0. Let G be a discrete submonoid containing the discrete submonoid $G(\mathcal{AC})$ in Definition 22.1.

Remark 22.18 To prove Theorem 21.35 (2) itself, it suffices to take $G = G(\mathcal{AC})$. However, we may also take G which is strictly bigger than $G(\mathcal{AC})$. We may replace e_0 by a smaller one.

We apply Proposition 22.3 and find a system of τ-collared Kuranishi structures and CF-perturbations, $\{(\widehat{\mathcal{U}^+_{k+1}}(\beta), \widehat{\mathfrak{S}}_{k+1}(\beta)) \mid (\beta, k) \in \mathcal{GK}(G; E_0, e_0)\}$. We regard

$$\left(\left(\mathcal{M}_{k+1}(\beta)^{\boxplus \tau_0}, \widehat{\mathcal{U}^+_{k+1}}(\beta)\right); (\mathrm{ev}_1, \ldots, \mathrm{ev}_k), \mathrm{ev}_0\right)$$

as a smooth correspondence from L^k to L and write it as $\mathfrak{M}_{k+1}(\beta)$. We now define:

$$\mathfrak{m}^{\epsilon}_{k,\beta}(h_1, \ldots, h_k) := (-1)^* \mathrm{Corr}_{\mathfrak{M}_{k+1}(\beta)}\left(h_1 \times \cdots \times h_k; \widehat{\mathfrak{S}}^{\epsilon}_{k+1}(\beta)\right), \tag{22.17}$$

where

$$* = \sum_{i=1}^{k} i(\deg h_i + 1) + 1.$$

Here and hereafter we write

$$h_1 \times \cdots \times h_k := \pi_1^* h_1 \wedge \cdots \wedge \pi_k^* h_k,$$

where $\pi_i : L^k \to L$ is the i-th projection. By Stokes' formula (Theorem 9.28) we have

$$\begin{aligned}(d \circ \mathfrak{m}_{k,\beta}^{\epsilon})(h_1, \ldots, h_k) &+ (\mathfrak{m}_{k,\beta}^{\epsilon} \circ \hat{d})(h_1, \ldots, h_k) \\ &= (-1)^{*'} \mathrm{Corr}_{\partial \mathfrak{M}_{k+1}(\beta)} \left(h_1 \times \cdots \times h_k; \widehat{\mathfrak{S}}_{k+1}^{\epsilon}(\beta) \right),\end{aligned}$$

where

$$*' = * + \dim \mathcal{M}_{k+1}(\beta) - \sum_{i=1}^{k} \deg h_i.$$

Here we define $\hat{d}$ by

$$\hat{d}(h_1, \ldots, h_k) = \sum_{i=1}^{k} (-1)^{\deg' h_1 + \cdots + \deg' h_{i-1}} h_1 \wedge \cdots \wedge dh_i \wedge \cdots \wedge h_k.$$

We recall (22.2), that is,

$$\begin{aligned}\partial(\mathcal{M}_{k+1}(\beta)^{\boxplus \tau_0}, \widehat{\mathcal{U}_{k+1}^{+}}(\beta)) \cong \coprod_{\beta_1, \beta_2, k_1, k_2, i} &(-1)^{\epsilon} (\mathcal{M}_{k_1+1}(\beta_1)^{\boxplus \tau_0}, \widehat{\mathcal{U}_{k_1+1}^{+}}(\beta_1)) \\ &{}_{\mathrm{ev}_i} \times_{\mathrm{ev}_0} (\mathcal{M}_{k_2+1}(\beta_2)^{\boxplus \tau_0}, \widehat{\mathcal{U}_{k_2+1}^{+}}(\beta_2)),\end{aligned} \tag{22.18}$$

where

$$\epsilon = k_2(k_1 + i) + \dim L + i \tag{22.19}$$

We denote by $\mathfrak{M}_{k_1,k_2,i}(\beta_1, \beta_2)$ the component corresponding to $\beta_1, \beta_2, k_1, k_2, i$ in the right hand side together with evaluation maps. Note that the evaluation maps to the source of the left hand side restrict to the evaluation maps to the source of either the first or the second fiber product factor of the right hand side. So our situation is (very slightly) different from that of the composition formula Theorem 10.21. However, we can apply Proposition 10.24 instead by putting

$$(X_1, \widehat{\mathcal{U}}_1, \widehat{\mathfrak{S}}_1, \widehat{f}_1) = \left(\mathcal{M}_{k_1+1}(\beta_1)^{\boxplus\tau_0}, \widehat{\mathcal{U}^+_{k_1+1}}(\beta_1), \widehat{\mathfrak{S}}^{\epsilon}_{k_1+1}(\beta_1), \mathrm{ev}_0\right),$$
$$(X_2, \widehat{\mathcal{U}}_2, \widehat{\mathfrak{S}}_2, \widehat{f}_2) = \left(\mathcal{M}_{k_2+1}(\beta_2)^{\boxplus\tau_0}, \widehat{\mathcal{U}^+_{k_2+1}}(\beta_2), \widehat{\mathfrak{S}}^{\epsilon}_{k_2+1}(\beta_2), \mathrm{ev}_i\right),$$
$$\widehat{h}_1 = (\mathrm{ev}_0, \mathrm{ev}_1, \dots, \mathrm{ev}_{i-1}, \mathrm{ev}_{i+1}, \dots, \mathrm{ev}_{k_1})^* \\ (h_0 \times h_1 \times \cdots \times h_{i-1} \times h_{i+k_2} \times \cdots \times h_{k_1+k_2-1}),$$
$$\widehat{h}_2 = (\mathrm{ev}_1, \dots, \mathrm{ev}_{k_2})^*(h_i \times \cdots \times h_{i+k_2-1}).$$

Then (10.15) and (22.2) = (22.18) imply

$$\begin{aligned}&\int_L h_0 \wedge \mathrm{Corr}_{\partial\mathfrak{M}_{k+1}(\beta)}\left(h_1 \times \cdots \times h_k; \widehat{\mathfrak{S}}^{\epsilon}_{k+1}(\beta)\right) \\ &= (-1)^{\#} \sum_{\beta_1,\beta_2,k_1,k_2,i} \int_L h_0 \wedge \mathrm{Corr}_{\mathfrak{M}_{k_1+1}(\beta_1)}\left(\diamond; \widehat{\mathfrak{S}}^{\epsilon}_{k_1+1}(\beta_1)\right),\end{aligned} \tag{22.20}$$

where

$$\begin{aligned}\diamond =& h_1 \times \cdots \times h_{i-1} \times \\ &\times \mathrm{Corr}_{\mathfrak{M}_{k_2+1}(\beta_2)}\left(h_i \times \cdots \times h_{i+k_2-1}; \widehat{\mathfrak{S}}^{\epsilon}_{k_2+1}(\beta_2)\right) \times h_{i+k_2} \times \cdots \times h_{k_1+k_2-1}\end{aligned}$$

and

$$\# = k_2(k_1 + i) + \dim L + i + k_2 \cdot \sum_{j=i+k_2}^{k_1+k_2-1} \deg h_j.$$

(22.20) implies that $\{\mathfrak{m}^{\epsilon}_{k,\beta}\}$ defines a G-gapped partial A_∞-structure of energy cut level E_0 and minimal energy e_0.

Thus we have constructed the required partial A_∞-correspondence. Its well-definedness up to pseudo-isotopy will have been proved if Theorem 21.35 (4) is proved. □

Remark 22.19 In the formulation of this book, we do not perturb $\mathfrak{m}_{2,\beta_0}$. (Recall $\beta_0 = 0$.) Namely, $\mathfrak{m}_{2,\beta_0}$ coincides with the wedge product up to the sign and $\mathfrak{m}_{k,\beta_0} = 0$ for $k \geq 3$. We take $\widehat{\mathfrak{S}}^{\epsilon}_{2+1}(\beta_0)$ as the trivial perturbation. Note that $\mathcal{M}_{2+1}(\beta_0) = L$ and the evaluation map $\mathrm{ev}_0 : \mathcal{M}_{2+1}(\beta_0) \to L$ is the identity map (that is a submersion). So we do not need to perturb it. We also take $\mathcal{M}_{k+1}(\beta_0) = L \times D^{k-2}$, where we identify D^{k-2} with the Stasheff cell. Note that $\mathrm{ev}_0 : \mathcal{M}_{k+1}(\beta_0) \to L$ factors through the projection $L \times D^{k-2} \to L$ whose fiber is of positive dimension. Therefore this smooth correspondence induces the zero map.

We can proceed in a different way and perturb $\mathcal{M}_{2+1}(\beta_0)$ so that $\mathfrak{m}_{2,\beta_0}$ has a smooth Schwartz kernel. Then we necessarily include nonzero $\mathfrak{m}_{k+1,\beta_0}$ for $k > 2$.

We need to take such a choice of perturbation to generalize our story to the case of bordered Riemann surfaces of higher genus and/or those which have more than one boundary component, because the corresponding moduli space of constant maps is not transversal.

Proof of Theorem* 21.35 *(4) We are given two partial A_∞ correspondences over $LA\mathcal{F}^j = \{\mathcal{M}^j_{k+1}(\beta) \mid \beta, k\}(j = 0, 1)$, and a pseudo-isotopy $A\mathcal{F}^{[0,1]} = \{\mathcal{M}_{k+1}(\beta; [0,1]) \mid \beta, k\}$ between them. We assume that both of their energy cut levels are E_0 and minimal energies are e_0. Let G be a discrete submonoid containing both $G(\mathcal{AC}^j)$ for $j = 0, 1$. We can make such a choice by Remark 21.18. We also assume that G contains $G(A\mathcal{F}^{[0,1]})$ and $e_0 \le e_{\min}(G)$.

We assume that we have obtained partial filtered A_∞ structures

$$\{\mathfrak{m}^{j,\epsilon_j}_{k,\beta} \mid (k, \beta) \in \mathcal{GK}(G; E_0, e_0)\}$$

associated with the partial A_∞ correspondences $A\mathcal{F}^j$ given in the proof of Theorem 21.35 (2) above. It means that we have taken a system of

$$\{(\widehat{\mathcal{U}^{j+}_{k+1}(\beta)}, \widehat{\mathfrak{S}}^j_{k+1}(\beta)) \mid (\beta, k) \in \mathcal{GK}(G; E_0, e_0)\},$$

where $\widehat{\mathcal{U}^{j+}_{k+1}(\beta)}$ is a τ-collared Kuranishi structure on $\mathcal{M}^j_{k+1}(\beta)^{\boxplus\tau_0}$ and $\widehat{\mathfrak{S}}^j_{k+1}(\beta)$ is a CF-perturbation of $\widehat{\mathcal{U}^{j+}_{k+1}(\beta)}$ such that they satisfy (22.18). Here recall that τ and τ_0 satisfies the following inequality:

$$0 < \tau < \tau_0 = 1.$$

Now we apply Proposition 22.6 to $P = [0,1]^{\boxplus(\tau_0-\tau)}$ (then $P^{\boxplus\tau} = [0,1]^{\boxplus\tau_0}$) to obtain objects

$$\widehat{\mathcal{U}^{+}_{k+1}(\beta; [0,1])}, \quad \widehat{\mathfrak{S}}_{\rho,k+1}(\beta; [0,1])$$

with $\rho \in (0, 1]$ described below. Firstly, $\widehat{\mathcal{U}^{+}_{k+1}(\beta; [0,1])}$ is a τ-collared Kuranishi structure on $\mathcal{M}_{k+1}(\beta; [0,1])^{\boxplus\tau_0}$ with the following properties.

Property 22.20

(1) For $j = 0, 1$, the restriction of $\widehat{\mathcal{U}^{+}_{k+1}(\beta; [0,1])}$ to $\mathcal{M}^j_{k+1}(\beta)^{\boxplus\tau_0} \subset \partial(\mathcal{M}_{k+1}(\beta; [0,1])^{\boxplus\tau_0})$ is $\widehat{\mathcal{U}^{j+}_{k+1}(\beta)}$.

(2) The restriction of $\widehat{\mathcal{U}^{+}_{k+1}(\beta; [0,1])}$ to

$$\mathcal{M}_{k_1+1}(\beta_1; [0,1])^{\boxplus\tau_0} {}_{(\mathrm{ev}_0, \mathrm{ev}_{[0,1]})}\times_{(\mathrm{ev}_i, \mathrm{ev}_{[0,1]})} \mathcal{M}_{k_2+1}(\beta_2; [0,1])^{\boxplus\tau_0}$$

is $\widehat{\mathcal{U}^{+}_{k_1+1}}(\beta_1; [0,1]) \,{}_{(\mathrm{ev}_0,\mathrm{ev}_{[0,1]})}\times_{(\mathrm{ev}_i,\mathrm{ev}_{[0,1]})} \widehat{\mathcal{U}^{+}_{k_2+1}}(\beta_2; [0,1])$.

Note that $\widehat{\mathcal{U}^{j+}_{k+1}}(\beta)$ is a $[0,1]^{\boxplus\tau_0} = [-\tau_0, 1+\tau_0]$-parametrized family.

The τ-collared Kuranishi structure $\widehat{\mathcal{U}^{+}_{k+1}}(\beta; [0,1])$ also satisfies the compatibility conditions at the corner. However, we do not describe them here since they are special cases of the statement of Proposition 22.6 and we do not use them below directly.

Secondly, $\widehat{\mathfrak{S}}_{\rho,k+1}(\beta; [0,1])$ is a family of CF-perturbations of $\widehat{\mathcal{U}^{+}_{k+1}}(\beta; [0,1])$ parametrized by $\rho \in (0,1]$ with the following properties.

Property 22.21

(1) (a) The restriction of $\widehat{\mathfrak{S}}_{\rho,k+1}(\beta; [0,1])$ to $\mathcal{M}^{j}_{k+1}(\beta)^{\boxplus\tau_0} \subset \partial\mathcal{M}_{k+1}(\beta; [0,1])^{\boxplus\tau_0}$ with $j = 0$ is $\widehat{\mathfrak{S}}^{0}_{k+1}(\beta)$.

(b) The restriction of $\widehat{\mathfrak{S}}_{\rho,k+1}(\beta; [0,1])$ to $\mathcal{M}^{j}_{k+1}(\beta)^{\boxplus\tau_0} \subset \partial\mathcal{M}_{k+1}(\beta; [0,1])^{\boxplus\tau_0}$ with $j = 1$ is $\epsilon \mapsto \widehat{\mathfrak{S}}^{1\rho\epsilon}_{k+1}(\beta)$.

(2) The restriction of $\widehat{\mathfrak{S}}_{\rho,k+1}(\beta; [0,1])$ to

$$\mathcal{M}_{k_1+1}(\beta_1; [0,1])^{\boxplus\tau_0} \,{}_{(\mathrm{ev}_0,\mathrm{ev}_{[0,1]})}\times_{(\mathrm{ev}_i,\mathrm{ev}_{[0,1]})} \mathcal{M}_{k_2+1}(\beta_2; [0,1])^{\boxplus\tau_0}$$

is $\widehat{\mathfrak{S}}_{\rho,k_1+1}(\beta_1; [0,1]) \,{}_{(\mathrm{ev}_0,\mathrm{ev}_{[0,1]})}\times_{(\mathrm{ev}_i,\mathrm{ev}_{[0,1]})} \widehat{\mathfrak{S}}_{\rho,k_2+1}(\beta_2; [0,1])$.

(3) $\widehat{\mathfrak{S}}_{\rho,k+1}(\beta; [0,1])$ is transversal to 0. The map $(\mathrm{ev}_0, \mathrm{ev}_{[0,1]})$ is strongly submersive with respect to $\widehat{\mathfrak{S}}_{\rho,k+1}(\beta; [0,1])$.

(4) $\{\widehat{\mathfrak{S}}_{\rho,k+1}(\beta; [0,1]) \mid \rho \in (0,1]\}$ is a uniform family.

We regard

$$\left(\left(\mathcal{M}_{k+1}(\beta)^{\boxplus\tau_0}, \widehat{\mathcal{U}^{+}_{k+1}}(\beta; [0,1])\right); ((\mathrm{ev}_1, \mathrm{ev}_{[0,1]}), \dots, (\mathrm{ev}_k, \mathrm{ev}_{[0,1]})), (\mathrm{ev}_0, \mathrm{ev}_{[0,1]})\right)$$

together with $\widehat{\mathfrak{S}}_{\rho,k+1}(\beta; [0,1])$ as a smooth correspondence from $([0,1]^{\boxplus\tau_0} \times L)^k$ to $[0,1]^{\boxplus\tau_0} \times L$ and write it as

$$\mathfrak{M}_{\rho,k+1}(\beta; [0,1]^{\boxplus\tau_0}).$$

Now for differential forms $h_1, \dots, h_k$ on $[0,1]^{\boxplus\tau_0} \times L$ we put

$$\mathfrak{m}^{\epsilon,\rho,[0,1]^{\boxplus\tau_0}}_{k,\beta}(h_1, \dots, h_k) = (-1)^{*}\mathrm{Corr}_{\mathfrak{M}_{\rho,k+1}(\beta;[0,1]^{\boxplus\tau_0})}\left(h_1, \dots, h_k; \widehat{\mathfrak{S}}^{\epsilon}_{\rho,k+1}(\beta; [0,1])\right),$$

where $*$ is the same as in (22.17).

Using Property 22.21 (2) and Stokes' formula (Theorem 26.12) in the same way as in the proof of Theorem 21.35 (2), we can prove

$$\sum_{k_1+k_2=k+1}\sum_{\beta_1+\beta_2=\beta}\sum_{i=1}^{k-k_2+1} \tag{22.21}$$
$$(-1)^*\mathfrak{m}_{k_1,\beta_1}^{\epsilon,\rho,[0,1]^{\boxplus\tau_0}}(h_1,\ldots,\mathfrak{m}_{k_2,\beta_2}^{\epsilon,\rho,[0,1]^{\boxplus\tau_0}}(h_i,\ldots,h_{i+k_2-1}),\ldots,h_k)=0,$$

where $* = \deg' h_1 + \ldots + \deg' h_{i-1}$.

By Proposition 22.17, $\mathfrak{m}_{k,\beta}^{\epsilon,\rho,[0,1]^{\boxplus\tau_0}}$ is pointwise in the $[0,1]^{\boxplus\tau_0}$-direction. Moreover, Property 22.21 (1) (a) implies that the restriction of $\mathfrak{m}_{k,\beta}^{\epsilon,\rho,[0,1]^{\boxplus\tau_0}}(h_1,\ldots,h_k)$ to $\{-\tau\}\times L$ is $\mathfrak{m}_{k,\beta}^{0,\epsilon}(h_1,\ldots,h_k)$ and Property 22.21 (1) (b) implies that the restriction of $\mathfrak{m}_{k,\beta}^{\epsilon,\rho,[0,1]}(h_1,\ldots,h_k)$ to $\{1+\tau\}\times L$ is $\mathfrak{m}_{k,\beta}^{1,\epsilon\rho}(h_1,\ldots,h_k)$. Therefore $\mathfrak{m}_{k,\beta}^{\epsilon,\rho,[0,1]^{\boxplus\tau_0}}$ gives a partial A_∞ pseudo-isotopy between $\mathfrak{m}_{k,\beta}^{0,\epsilon}$ and $\mathfrak{m}_{k,\beta}^{1,\epsilon\rho}$. □

Remark 22.22 We introduced the parameter ρ and constructed a pseudo-isotopy $\mathfrak{m}_{k,\beta}^{\epsilon,\rho,[0,1]^{\boxplus\tau_0}}$ between $\mathfrak{m}_{k,\beta}^{0,\epsilon\rho}$ and $\mathfrak{m}_{k,\beta}^{1,\epsilon\rho}$ since we need it in the inductive construction of a filtered A_∞ structure associated to an inductive system of A_∞ correspondences. See Remark 19.17.

Proof of Theorem* 21.35 *(5) Recall from Definition 21.17 that we have a divergent sequence $\{E^i\}_i$ with

$$0<\cdots<E^i<E^{i+1}<\cdots\to+\infty$$

of energy cut levels in the definition of the inductive system. By (2) there exists a filtered A_∞ structure of energy cut level E^i and minimal energy e_0 on $\Omega(L)$ induced by

$$\mathcal{AF}^i=\{\mathcal{M}_{k+1}^i(\beta)\mid\beta,k\}.$$

We denote it by $\{\mathfrak{m}_k^i\}$. By (4) there exists a partial pseudo-isotopy from $\{\mathfrak{m}_k^i\}$ to $\{\mathfrak{m}_k^{i+1}\}$. Its energy cut level is E^i and minimal energy is e_0. It is induced by

$$\mathcal{AF}^{[i,i+1]}=\{\mathcal{M}_{k+1}(\beta;[i,i+1])\mid\beta,k\}.$$

We denote it by $\{\mathfrak{m}_k^{[i,i+1]}\}$. Then we can prove the following lemma by induction on N.

Lemma 22.23 *For each $n\le N$ we have the following:*

(1) *For $i\le n$, there exists a promotion of $\{\mathfrak{m}_k^i\}$ to the energy cut level E^n.*
(2) *For $i\le n-1$, there exists a promotion of $\{\mathfrak{m}_k^{[i,i+1]}\}$ to the energy cut level E^n. It is a pseudo-isotopy between the promotions in (1).*

Moreover, if $n'<n$ the structures in (1)(2) for n' are the energy cut at $E^{n'}$ of the structures in (1)(2) for n.

Proof This is immediate from Proposition 22.8. □

Now by mathematical induction we obtain the same conclusion as in Lemma 22.23 in the case $N = \infty$. This implies Theorem 21.35 (5). Namely, the filtered A_∞ structure associated to our inductive system of linear K-systems is $\{\mathfrak{m}_k^{[0,1],1}\}$ obtained by promotion to energy cut level ∞. □

Proof of Theorem* 21.35 *(1) Using trivial pseudo-isotopy (the direct product), this is a special case of Theorem 21.35 (5). □

Proof of Theorem* 21.35 *(6) Suppose we are in Situation 21.34 (6). We apply Lemma 22.23 to each of the two inductive systems $\mathcal{IAF}^0, \mathcal{IAF}^1$. Namely, we start with $\{\mathfrak{m}_k^{ji} \mid i = 1, 2, \dots\}$ which are partial A_∞ structures of energy cut level E^i and minimal energy e_0 on $\Omega(L)$ (where $j = 0, 1$) and with $\{\mathfrak{m}_k^{j,[i,i+1]} \mid i = 1, 2, \dots\}$ which are partial pseudo-isotopies of energy cut level E^i and minimal energy e_0 among them. Then by Lemma 22.23 and induction, we promote them to the energy cut level ∞.

Next we consider $\{\mathcal{M}_{k+1}^i(\beta; [0,1]) \mid \beta, k\}$ and $\{\mathcal{M}_{k+1}(\beta; [0,1] \times [i, i+1]) \mid \beta, k\}$ in Situation 21.34 (6).

The former defines $\{\mathfrak{m}_k^{[0,1],i} \mid i = 1, 2, \dots\}$ that is a pseudo-isotopy of energy cut level E^i and minimal energy e_0 from $\{\mathfrak{m}_k^{0i} \mid i = 1, 2, \dots\}$ to $\{\mathfrak{m}_k^{1i} \mid i = 1, 2, \dots\}$.

The latter defines $\{\mathfrak{m}_k^{[0,1],[i,i+1]} \mid i = 1, 2, \dots\}$ that is a pseudo-isotopy of pseudo isotopies. In other words, it is a ($[0,1] \times [i, i+1]$)-parametrized family of partial A_∞ algebras of energy cut level E^i and minimal energy e_0 and their restrictions to the normalized boundary are disjoint union of $\{\mathfrak{m}_k^{[0,1],i} \mid i = 1, 2, \dots\}$, $\{\mathfrak{m}_k^{[0,1],i+1} \mid i = 1, 2, \dots\}$, $\{\mathfrak{m}_k^{0,[i,i+1]} \mid i = 1, 2, \dots\}$ and $\{\mathfrak{m}_k^{1,[i,i+1]} \mid i = 1, 2, \dots\}$.

Now we apply Proposition 22.8 and the same induction argument as in the proof of Lemma 22.23 to promote $\{\mathfrak{m}_k^{[0,1],i} \mid i = 1, 2, \dots\}$ and $\{\mathfrak{m}_k^{[0,1],[i,i+1]} \mid i = 1, 2, \dots\}$ to the energy cut level ∞. Thus after promotion, $\{\mathfrak{m}_k^{[0,1],1}\}$ gives a pseudo-isotopy from the promotion of $\{\mathfrak{m}_k^{0i} \mid i = 1, 2 \dots\}$ to the promotion of $\{\mathfrak{m}_k^{1i} \mid i = 1, 2 \dots\}$ of energy cut level ∞. This is what we want to construct.

We note that there exist A_∞ homomorphisms

$$\{\mathfrak{m}_k^{[0,1],1}\} \longrightarrow \{\mathfrak{m}_k^{j1}\}, \quad j = 0, 1, \tag{22.22}$$

which are linear homotopy equivalences induced by the inclusion $L = L \times \{j\} \to L \times [0,1]$. Therefore inverting one of them and using the Whitehead theorem for A_∞ algebra [FOOO3, Theorem 4.2.45], we obtain the homotopy equivalence $\{\mathfrak{m}_k^{01}\} \to \{\mathfrak{m}_k^{11}\}$. □

Proof of Theorem* 21.35 *(3) This is a special case of Theorem 21.35 (6). □

Proof of Theorem* 21.35 *(8) For each i we use $\mathcal{AF}^{[0,1]\times[i,i+1]\times[1,2]}$ to obtain a $[0,1] \times [i, i+1] \times [1,2]$ parametrized family of partial A_∞ structures of energy cut level E^i and minimal energy e_0. On $[0,1] \times [i, i+1] \times \{1\}$ and $[0,1] \times [i, i+1] \times \{2\}$ this family restricts to the family we obtain in the above proof of Theorem 21.35 (6) applied to $\{\mathcal{M}^{i,\ell}_{k+1}(\beta; [0,1]) \mid \beta, k\}$, $\{\mathcal{M}^{\ell}_{k+1}(\beta; [0,1] \times [i, i+1]) \mid \beta, k\}$ with $\ell = a$ and $\ell = b$, respectively. Moreover it restricts to the direct product on $\{j\} \times [i, i+1] \times [1,2]$ with $j = 0$ or $j = 1$.

Now applying Proposition 22.8, we use the same induction argument as in the proof of Lemma 22.23 to obtain at the part $i = 1$ the $[0,1] \times \{1\} \times [1,2]$ parametrized family of A_∞ structures of energy cut level ∞. We denote it by $\{\mathfrak{m}_k^{[0,1]\times\{1\}\times[1,2]}\}$. We have a commutative diagram:

$$\begin{array}{ccccc}
\{\mathfrak{m}_k^{01}\} & \longleftarrow & \{\mathfrak{m}_k^{[0,1],1,b}\} & \longrightarrow & \{\mathfrak{m}_k^{11}\} \\
\uparrow & & \uparrow & & \uparrow \\
\{\mathfrak{m}_k^{01}\} \times [1,2] & \longleftarrow & \{\mathfrak{m}_k^{[0,1]\times\{1\}\times[1,2]}\} & \longrightarrow & \{\mathfrak{m}_k^{11}\} \times [1,2] \\
\downarrow & & \downarrow & & \downarrow \\
\{\mathfrak{m}_k^{01}\} & \longleftarrow & \{\mathfrak{m}_k^{[0,1],1,a}\} & \longrightarrow & \{\mathfrak{m}_k^{11}\}
\end{array} \tag{22.23}$$

The first and the third horizontal lines define homotopy equivalences we obtain in Theorem 21.35 (6) applied for $\ell = a$ and $\ell = b$ respectively. (We invert the first arrow.)

By the symbol $\{\mathfrak{m}_k^{01}\} \times [1,2]$, we denote the direct product pseudo-isotopy, that is, the isotopy $\{\mathfrak{m}_k^t, \mathfrak{c}_k^t\}$ such that $\mathfrak{m}_k^t$ is independent of t and $\mathfrak{c}_k^t$ are all zero. Then by inverting one of the arrows in the first or third vertical line we obtain identity maps. Thus the commutativity of (22.23) implies that the two homotopy equivalences obtained for $\ell = a$ and $\ell = b$ are homotopic. This is the conclusion of Theorem 21.35 (8). □

Proof of Theorem* 21.35 *(7) This is a special case of proof of Theorem 21.35 (8). □

Remark 22.24 The proof we gave here using Diagram (22.23) is similar to those in [AJ]. It is based on a way to define homotopy equivalence we took here, that is, to invert homotopy equivalence (22.22).

There is an alternative way to define homotopy equivalence, given in [Fuk4, Proof of Theorem 8.2], where we take an appropriate integration and sum over trees to construct homotopy equivalence from pseudo-isotopy directly. This method has an advantage that in the case when the pseudo-isotopy has cyclic symmetry the resulting homotopy equivalence is also cyclic. (We do not know the version of [FOOO3, Theorem 4.2.45] with cyclic symmetry.)

We can use the same method to prove Theorem 21.35 (7). Namely, we regard $\{\mathfrak{m}_k^{[0,1]\times\{1\}\times[1,2]}\}$ as the pseudo-isotopy from $\{\mathfrak{m}_k^{01}\} \times [1,2]$ to $\{\mathfrak{m}_k^{11}\} \times [1,2]$ and apply the same formula [Fuk4, Definition 9.4]. Using the fact that this is a pseudo-isotopy between direct product A_∞ algebras, we can easily check that the resulting filtered A_∞ homomorphism becomes the required homotopy equivalence.

Remark 22.25 We stop here at the stage where we prove consistency up to homotopy of homotopies. It is fairly obvious from the proof that we can prove as many higher consistencies of the homotopies as we want.

Remark 22.26 We are aware of the opinion that using iterated homotopies and homological algebra such as those we developed in this chapter is cumbersome and had better be avoided. Our opinion is different.

One of the origins of the 'homotopy everything structure' in algebraic topology is the study of the 'homotopy limit'. So the language of A_∞ structures which we are using here is very much suitable for such a discussion. Moreover, taking an inductive limit is necessary anyway to study, for example, symplectic homology.

Furthermore, the general strategy taken here does not use any of the special features of the problem and uses only the facts which are 'intuitively obvious'. For this reason it should work in almost all the situations we meet and will meet in the future. We do not know of a situation when this strategy does not work. (Once we get used to it, applying this strategy becomes a routine.) The general strategy is summarized as follows:

(1) The problem is that it is hard and sometimes impossible to perturb infinitely many moduli spaces simultaneously.
(2) Those moduli spaces are filtered by certain quantities, typically by energy.
(3) We fix a certain 'energy cut level' and perturb the (finitely many) moduli spaces only up to that level.
(4) Then we obtain a partial structure, such as the partial A_∞ structure of energy cut level E.
(5) We may take a *finite* number E as large as we want.
(6) Let $E < E'$. We obtain the structures of energy cut level E and of E'. The latter can be regarded as the structure of energy cut level E. Those two structures are not the same but are 'homotopy equivalent' in a sense of 'homotopy everything structure'.
(7) Then by a general method of homological algebra we can take the homotopy limit to obtain the desired structure.

We note that a similar technique also appears in the renormalization theory. In particular, Costello [Co] uses a similar technique.

Part III
Appendices

Chapter 23
Orbifolds and Orbibundles by Local Coordinates

In this chapter we describe the story of orbifolds as much as we need in this book. We restrict ourselves to effective orbifolds and regard only embeddings as morphisms. The category $\mathcal{OB}_{\rm ef,em}$ where objects are effective orbifolds and morphisms are embeddings among them is naturally a 1 category. Moreover it has the following property. We consider the forgetful functor

$$\mathfrak{forget} : \mathcal{OB}_{\rm ef,em} \to \mathcal{TOP},$$

where $\mathcal{TOP}$ is the category of topological spaces. Then

$$\mathfrak{forget} : \mathcal{OB}_{\rm ef,em}(c, c') \to \mathcal{TOP}(\mathfrak{forget}(c), \mathfrak{forget}(c'))$$

is injective. In other words, we can check the equality between morphisms set-theoretically. This is a nice property, which we use extensively in the main body of this book. If we go beyond this category, then we need to distinguish carefully the two notions, two morphisms are equal, two morphisms are isomorphic. It will make the argument much more complicated.[1]

We emphasize that there is nothing new in this chapter. The story of orbifolds is classical and is well-established. It has been used in various branches of mathematics since its invention by Satake [Sa] more than 50 years ago. In particular, if we restrict ourselves to effective orbifolds, the story of orbifolds is nothing more than a straightforward generalization of the theory of smooth manifolds. The only important issue is the observation that for effective orbifolds almost everything works in the same way as manifolds.

[1]We need to use covering maps of orbifolds. We will check set-theoretical equality is enough in the cases for which we use them.

K. Fukaya et al., *Kuranishi Structures and Virtual Fundamental Chains*, Springer Monographs in Mathematics, https://doi.org/10.1007/978-981-15-5562-6_23

23.1 Orbifolds and Embeddings Between Them

Definition 23.1 Let X be a paracompact Hausdorff space.

(1) An *orbifold chart of X (as a topological space)* is a triple (V, Γ, ϕ) such that V is a manifold, Γ is a finite group acting smoothly and effectively on V and $\phi : V \to X$ is a Γ equivariant continuous map[2] which induces a homeomorphism $\overline{\phi} : V/\Gamma \to X$ onto an open subset of X. We assume that there exists $o \in V$ such that $\gamma o = o$ for all $\gamma \in \Gamma$. We call o the *base point*.[3] We say (V, Γ, ϕ) is an orbifold chart *at x* if $x = \phi(o)$. We call Γ the *isotropy group*, ϕ the *local uniformization map* and $\overline{\phi}$ the *parametrization*.

(2) Let (V, Γ, ϕ) be an orbifold chart and $p \in V$. We put $\Gamma_p = \{\gamma \in \Gamma \mid \gamma p = p\}$. Let V_p be a Γ_p-invariant open neighborhood of p in V. We assume the map $\overline{\phi} : V_p/\Gamma_p \to X$ is injective. (In other words, we assume that $\gamma V_p \cap V_p \neq \emptyset$ implies $\gamma \in \Gamma_p$.) We call such a triple $(V_p, \Gamma_p, \phi|_{V_p})$ a *subchart* of (V, Γ, ϕ).

(3) Let (V_i, Γ_i, ϕ_i) $(i = 1, 2)$ be orbifold charts of X. We say that they are *compatible* if the following holds for each $p_1 \in V_1$ and $p_2 \in V_2$ with $\phi_1(p_1) = \phi_2(p_2)$:

   (a) There exists a group isomorphism $h : (\Gamma_1)_{p_1} \to (\Gamma_2)_{p_2}$.
   (b) There exists an h equivariant diffeomorphism $\tilde{\varphi} : V_{1,p_1} \to V_{2,p_2}$. Here V_{i,p_i} is a $(\Gamma_i)_{p_i}$ equivariant subset of V_i such that $(V_{i,p_i}, (\Gamma_i)_{p_i}, \phi_i|_{V_{i,p_i}})$ is a subchart of (V_i, Γ_i, ϕ_i).
   (c) $\phi_2 \circ \tilde{\varphi} = \phi_1$ on V_{1,p_1}.

(4) A *representative of an orbifold structure* on X is a set of orbifold charts $\{(V_i, \Gamma_i, \phi_i) \mid i \in I\}$ such that each two of the charts are compatible in the sense of (3) above and $\bigcup_{i\in I} \phi_i(V_i) = X$, is a locally finite open cover of X.

Definition 23.2 Suppose that X, Y are equipped with representatives of orbifold structures $\{(V_i^X, \Gamma_i^X, \phi_i^X) \mid i \in I\}$ and $\{(V_j^Y, \Gamma_j^Y, \phi_j^Y) \mid j \in J\}$, respectively. A continuous map $f : X \to Y$ is said to be an *embedding* if the following holds:

(1) f is an embedding of topological spaces.
(2) Let $p \in V_i^X$, $q \in V_j^Y$ with $f(\phi_i(p)) = \phi_j(q)$. Then we have the following:

   (a) There exists an isomorphism of groups $h_{p;ji} : (\Gamma_i^X)_p \to (\Gamma_j^Y)_q$.
   (b) There exist $V_{i,p}^X$ and $V_{j,q}^Y$ such that $(V_{i,p}^X, (\Gamma_i^X)_p, \phi_i|_{V_{i,p}^X})$ is a subchart for $i = 1, 2$. There exists an $h_{p;ji}$ equivariant embedding of manifolds $\tilde{f}_{p;ji} : V_{i,p}^X \to V_{j,q}^Y$.

[2]The Γ action on X is trivial.

[3]The existence of a base point means that Γ is an isotropy group of some point of V. This is sometimes useful for the proof.

(c) The diagram below commutes:

$$\begin{array}{ccc} V_{i,p}^X & \xrightarrow{\tilde{f}_{p;ji}} & V_{j,q}^Y \\ {\scriptstyle \phi_i}\downarrow & & \downarrow{\scriptstyle \phi_j} \\ X & \xrightarrow{f} & Y \end{array} \tag{23.1}$$

Two orbifold embeddings are said to be *equal* if they coincide set-theoretically.

Remark 23.3 Note that an embedding of effective orbifolds is a continuous map $f : X \to Y$ of underlying topological spaces, which has the properties (2) above.

When we study morphisms among ineffective orbifolds or morphisms between effective orbifolds which is not an embedding, such a morphism is a continuous map $f : X \to Y$ of underlying topological spaces *plus* certain additional data. For example, if we consider an ineffective orbifold that is a point with an action of a nontrivial finite group Γ, then the morphism from this ineffective orbifold to itself contains the datum of an automorphism of the group Γ. (Two such morphisms h_1, h_2 are equivalent if there exists an inner automorphism h such that $h_1 = h \circ h_2$.)

Lemma 23.4

(1) *The composition of embeddings is an embedding.*
(2) *The identity map is an embedding.*
(3) *If an embedding is a homeomorphism, then its inverse is also an embedding.*

The proof is easy and is left to the reader.

Definition 23.5

(1) We call an embedding of orbifolds a *diffeomorphism* if it is a homeomorphism in addition.
(2) We say that two representatives of orbifold structures on X are *equivalent* if the identity map regarded as a map between X equipped with those two representatives of orbifold structures is a diffeomorphism. This is an equivalence relation by Lemma 23.4.
(3) An equivalence class of the equivalence relation (2) is called an *orbifold structure* on X. An *orbifold* is a pair of a topological space and its orbifold structure.
(4) The condition for a map $X \to Y$ to be an embedding does not change if we replace representatives of orbifold structures to equivalent ones. So we can define the notion of an *embedding of orbifolds*.
(5) If U is an open subset of an orbifold X, then there exists a unique orbifold structure on U such that the inclusion $U \to X$ is an embedding. We call U with this orbifold structure an *open suborbifold*.

Definition 23.6

(1) Let X be an orbifold. An orbifold chart (V, Γ, ϕ) of underlying topological space X in the sense of Definition 23.1 (1) is called an *orbifold chart of an orbifold* X if the map $\overline{\phi} : V/\Gamma \to X$ induced by ϕ is an embedding of orbifolds.
(2) Hereafter when X is an orbifold, an 'orbifold chart' always means an orbifold chart of an orbifold in the sense of (1).
(3) In the case when an orbifold structure on X is given, a representative of its orbifold structure is called an *orbifold atlas*.
(4) Two orbifold charts (V_i, Γ_i, ϕ_i) are said to be *isomorphic* if there exist a group isomorphism $h : \Gamma_1 \to \Gamma_2$ and an h-equivariant diffeomorphism $\tilde{\varphi} : V_1 \to V_2$ such that $\phi_2 \circ \tilde{\varphi} = \phi_1$. The pair $(h, \tilde{\varphi})$ is called an *isomorphism* or a *coordinate change* between two orbifold charts.

Proposition 23.7 *In the situation of Definition 23.6 (4), suppose $(h, \tilde{\varphi})$ and $(h', \tilde{\varphi}')$ are both isomorphisms between two orbifold charts (V_1, Γ_1, ϕ_1) and (V_2, Γ_2, ϕ_2). Then there exists $\mu \in \Gamma_2$ such that*

$$h'(\gamma) = \mu h(\gamma)\mu^{-1}, \qquad \tilde{\varphi}'(x) = \mu\tilde{\varphi}(x). \tag{23.2}$$

Conversely, if $(h, \tilde{\varphi})$ is an isomorphism between orbifold charts, then $(h', \tilde{\varphi}')$ defined by (23.2) *is also an isomorphism between orbifold charts. In particular, any automorphism of an orbifold chart $(h, \tilde{\varphi})$ is given by $h(\gamma) = \mu\gamma\mu^{-1}$, $\tilde{\varphi}(x) = \mu x$ for some element $\mu \in \Gamma$.*

Proof The proposition immediately follows from the next lemma.

Lemma 23.8 *Let V_1, V_2 be manifolds on which finite groups Γ_1, Γ_2 act effectively and smoothly. Assume that V_1 is connected. Let $(h_i, \tilde{\varphi}_i)$ $(i = 1, 2)$ be pairs such that $h_i : \Gamma_1 \to \Gamma_2$ are injective group homomorphisms and $\tilde{\varphi}_i : V_1 \to V_2$ are h_i-equivariant embeddings of manifolds. Moreover, we assume that the induced maps $\varphi_i : V_1/\Gamma_1 \to V_2/\Gamma_2$ are embeddings of orbifolds and φ_1 coincides with φ_2 set-theoretically. Then there exists $\mu \in \Gamma_2$ such that*

$$\tilde{\varphi}_2(x) = \mu\tilde{\varphi}_1(x), \qquad h_2(\gamma) = \mu h_1(\gamma)\mu^{-1}.$$

Proof For the sake of simplicity we prove only the case when Condition 23.9 below is satisfied. Let X be an orbifold. For a point $x \in X$ we take its orbifold chart (V_x, Γ_x, ψ_x). We say $x \in \mathrm{Reg}(X)$ if $\Gamma_x = \{1\}$, and put $\mathrm{Sing}(X) = X \setminus \mathrm{Reg}(X)$.

Condition 23.9 We assume that $\dim \mathrm{Sing}(X) \le \dim X - 2$.

This condition is satisfied if X is oriented.[4] (In fact, Condition 23.9 fails only when there exists an element of Γ_x (an isotropy group of some orbifold chart) whose action is given by $(x_1, x_2, \ldots, x_n) \mapsto (-x_1, x_2, \ldots, x_n)$ for some coordinate

[4]This is one reason we assumed the orientability of Kuranishi neighborhood.

$(x_1, \ldots, x_n)$. Therefore we can always assume Condition 23.9 in the study of Kuranishi structures, by adding a trivial factor which is acted on by the induced representation of $t \mapsto -t$ to both the obstruction bundle and to the Kuranishi neighborhood.)

Let $x_0 \in \mathrm{Reg}(V_1)$. By assumption there exists a unique $\mu \in \Gamma_2$ such that $\tilde{\varphi}_2(x_0) = \mu\tilde{\varphi}_1(x_0)$. By Condition 23.9 the subset $\mathrm{Reg}(V_1)$ is connected. Therefore the above element μ is independent of $x_0 \in \mathrm{Reg}(V_1)$ by uniqueness. Since V_1^0 is dense, we conclude $\tilde{\varphi}_2(x) = \mu\tilde{\varphi}_1(x)$ for any $x \in V_1$. Now, for $\gamma \in \Gamma_1$, we calculate

$$h_1(\gamma)\tilde{\varphi}_1(x_0) = \tilde{\varphi}_1(\gamma x_0) = \mu^{-1}\tilde{\varphi}_2(\gamma x_0) = \mu^{-1}h_2(\gamma)\tilde{\varphi}_2(x_0) = \mu^{-1}h_1(\gamma)\mu\tilde{\varphi}_1(x_0).$$

Since the induced map is an embedding of orbifold, it follows that the isotropy group of $\tilde{\varphi}_1(x_0)$ is trivial. Therefore $h_1(\gamma) = \mu^{-1}h_2(\gamma)\mu$ as required. □

The proof of Proposition 23.7 is complete. □

Definition 23.10 Let X be an orbifold.

(1) A function $f : X \to \mathbb{R}$ is said to be a *smooth function* if for any orbifold chart (V, Γ, ϕ) the composition $f \circ \phi : V \to \mathbb{R}$ is smooth.
(2) A *differential form* on an orbifold X assigns a Γ-invariant differential form $h_{\mathfrak{V}}$ on V to each orbifold chart $\mathfrak{V} = (V, \Gamma, \phi)$ such that the following holds:
    (a) If (V_1, Γ_1, ϕ_1) is isomorphic to (V_2, Γ_2, ϕ_2) and $(h, \tilde{\varphi})$ is an isomorphism, then $\tilde{\varphi}^* h_{\mathfrak{V}_2} = h_{\mathfrak{V}_1}$.
    (b) If $\mathfrak{V}_p = (V_p, \Gamma_p, \phi_p)$ is a subchart of $\mathfrak{V} = (V, \Gamma, \phi)$, then $h_{\mathfrak{V}}|_{V_p} = h_{\mathfrak{V}_p}$.
(3) An n-dimensional orbifold X is said to be *orientable* if there exists a differential n-form ω such that $\omega_{\mathfrak{V}}$ never vanishes.
(4) Let ω be an n-form as in (3) and $\mathfrak{V} = (V, \Gamma, \phi)$ an orbifold chart. Then we give V an orientation so that it is compatible with $\omega_{\mathfrak{V}}$. The Γ action preserves the orientation. We call such (V, Γ, ϕ) equipped with an orientation of V, an *oriented orbifold chart*.
(5) Let $\bigcup_{i \in I} U_i = X$ be an open covering of an orbifold X. A *smooth partition of unity subordinate to the covering* $\{U_i\}$ is a set of functions $\{\chi_i \mid i \in I\}$ such that:
    (a) χ_i are smooth functions.
    (b) The support of χ_i is contained in U_i.
    (c) $\sum_{i \in I} \chi_i = 1$.
(6) Let M be a smooth manifold (without boundary) and $f : X \to M$ a continuous map. We say that f is a *smooth map* if for any smooth function $g : M \to \mathbb{R}$ the composition $g \circ f : X \to \mathbb{R}$ is a smooth function.

Lemma 23.11 *For any locally finite open covering of an orbifold X there exists a smooth partition of unity subordinate thereto.*

We omit the proof, which is an obvious analogue of the standard proof for the case of manifolds.

Definition 23.12 An *orbifold with corners* is defined in the same way. We require the following:

(1) In Definition 23.1 (1) we assume that V is a manifold with corners.
(2) Let $S_k(V)$ be the set of points which lie on the codimension k corner and $\overset{\circ}{S}_k(V) = S_k(V) \setminus \bigcup_{k'>k} S_{k'}(V)$. We require that Γ action on each connected component of $\overset{\circ}{S}_k(V)$ is effective. (Compare Condition 4.14.)
(3) For an embedding of orbifolds with corners we require that the map $\tilde{f}$ in Definition 23.2 (c) satisfies $\tilde{f}(\overset{\circ}{S}_k(V_1)) \subset \overset{\circ}{S}_k(V_2)$.

Lemma 23.13 *Let X_i $(i = 1, 2)$ be orbifolds and $\varphi_{21} : X_1 \to X_2$ an embedding. Let X_1^0 be relatively compact open subset of X_1. Then we can find an orbifold atlas $\{\mathfrak{V}^i_{\mathfrak{r}} = \{(V^i_{\mathfrak{r}}, \Gamma^i_{\mathfrak{r}}, \phi^i_{\mathfrak{r}})\} \mid \mathfrak{r} \in \mathfrak{R}_i\}$ of X_1^0 with the following properties:*

(1) $\mathfrak{R}_1 \subseteq \mathfrak{R}_2$.
(2) $V^2_{\mathfrak{r}} \cap \varphi_{21}(X_1) \neq \emptyset$ *if and only if* $\mathfrak{r} \in \mathfrak{R}_1$.
(3) *If* $\mathfrak{r} \in \mathfrak{R}_1$ *then* $\varphi_{21}^{-1}(\phi^2_{\mathfrak{r}}(V^2_{\mathfrak{r}})) = \phi^1_{\mathfrak{r}}(V^1_{\mathfrak{r}})$ *and there exists* $(h_{\mathfrak{r},21}, \tilde{\varphi}_{\mathfrak{r},21})$ *such that:*

   (a) $h_{\mathfrak{r},21} : \Gamma^1_{\mathfrak{r}} \to \Gamma^2_{\mathfrak{r}}$ *is a group isomorphism.*
   (b) $\tilde{\varphi}_{\mathfrak{r},21} : V^1_{\mathfrak{r}} \to V^2_{\mathfrak{r}}$ *is an* $h_{\mathfrak{r},21}$*-equivariant embedding of smooth manifolds.*
   (c) *The next diagram commutes:*

$$\begin{array}{ccc} V^1_{\mathfrak{r}} & \xrightarrow{\tilde{\varphi}_{\mathfrak{r},21}} & V^2_{\mathfrak{r}} \\ \phi^{\mathfrak{r}}_1 \downarrow & & \downarrow \phi^{\mathfrak{r}}_2 \\ X_1 & \xrightarrow{\varphi_{21}} & X_2 \end{array} \tag{23.3}$$

(4) *In the case when X_i has a boundary or corners we may choose our charts so that the following is satisfied:*

   (a) $V^i_{\mathfrak{r}}$ *is an open subset of* $[V^i_{\mathfrak{r}}] \times [0, 1)^{d(\mathfrak{r})}$, *where* $d(\mathfrak{r})$ *is independent of* i *and* $[V^i_{\mathfrak{r}}]$ *is a manifold without boundary.*
   (b) *There exists a point* $o^i(\mathfrak{r})$ *which is fixed by all* $\gamma \in \Gamma^i_{\mathfrak{r}}$ *such that* $[0, 1)^{d(\mathfrak{r})}$ *components of* $o^i(\mathfrak{r})$ *are all* 0.
   (c) *If we write*

$$\varphi_{\mathfrak{r},21}(\overline{y}', (t'_1, \ldots, t'_{d(\mathfrak{r})})) = (\overline{y}, (t_1, \ldots, t_{d(\mathfrak{r})})),$$

   then $t_i = 0$ *if and only if* $t'_i = 0$.

We may take our atlas that consists of refinements of the given coverings of $\overline{X_1^0}$ and X_2.

Proof For each $x \in \overline{X_1^0}$ we can find orbifold charts $\mathfrak{V}^i_x$ for $i = 1, 2$, such that $\varphi_{21}^{-1}(U^2_x) = U^1_x$, $x \in U^1_x$ and that there exists a representative $(h_{x,21}, \tilde{\varphi}_{x,21})$ of embedding $U_x^1 \to U_x^2$ that is a restriction of φ_{21}. In the case when X_i has a boundary or corners, we choose them so that (4) is also satisfied.

We cover $\overline{X_1^0}$ by finitely many such $U_{x_j}^1$. This is our choice of atlas $\{\mathfrak{V}_{\mathfrak{r}}^1 \mid \mathfrak{r} \in \mathfrak{R}_1\}$. Then the associated $\{\mathfrak{V}_{\mathfrak{r}}^2 \mid \mathfrak{r} \in \mathfrak{R}_1\}$ satisfies (3)(4) and covers $\varphi_{21}(\overline{X_1^0})$. We can extend it to $\{\mathfrak{V}_{\mathfrak{r}}^2 \mid \mathfrak{r} \in \mathfrak{R}_2\}$ so that (1)(2) are also satisfied. □

Definition 23.14 We call $(h_{\mathfrak{r},21}, \tilde{\varphi}_{\mathfrak{r},21})$ a *local representative of embedding* $\varphi_{\mathfrak{r},21}$ on the charts $\mathfrak{V}_{\mathfrak{r}}^1$, $\mathfrak{V}_{\mathfrak{r}}^2$.

Lemma 23.15 *If* $(h_{\mathfrak{r},21}, \tilde{\varphi}_{\mathfrak{r},21})$, $(h'_{\mathfrak{r},21}, \tilde{\varphi}'_{\mathfrak{r},21})$ *are local representatives of an embedding of the same charts* $\mathfrak{V}_{\mathfrak{r}}^1$, $\mathfrak{V}_{\mathfrak{r}}^2$, *then there exists* $\mu \in \Gamma_2$ *such that*

$$\tilde{\varphi}'_{\mathfrak{r},21}(x) = \mu\tilde{\varphi}_{\mathfrak{r},21}(x), \qquad h'_{\mathfrak{r},21}(\gamma) = \mu h_{\mathfrak{r},21}(\gamma)\mu^{-1}.$$

This is a consequence of Lemma 23.8.

Notation 23.16 Let U be an orbifold. We consider its chart (V_x, Γ_x, ϕ_x) at $x \in U$. If x lies in the interior of the codimension k corner, we may choose V_x which is an open subset of a direct product $[V] \times [0,1)^k$, where $[V]$ is an open subset of $\mathbb{R}^{\dim U - k}$ and x is represented by $(\overline{x}, 0) \in [V] \times [0,1)^k$. We call such chart, a chart of *product type*, $[V]$ the (codimension k) *corner locus* and $[0,1)^k$ the *normal factor*. The standard coordinates $t_1, \dots, t_k$ of $[0,1)^k$ are called *normal coordinates*.

For a chart of product type, we use the symbol $[V]$ for corner locus and t_i for normal coordinates, and sometimes we use such notations without explicitly mentioning so.

Lemma 23.17 *Let* X *be a topological space,* Y *an orbifold, and* $f : X \to Y$ *an embedding of topological spaces. Then the orbifold structure on* X *by which* f *becomes an embedding of orbifolds is unique if there exists one.*

Proof Let X_1, X_2 be orbifolds whose underlying topological spaces are both X and satisfy that $f = f_i : X_i \to Y$ are embeddings of orbifolds for $i = 1, 2$. We will prove that the identity map id : $X_1 \to X_2$ is a diffeomorphism of orbifolds. Since the condition for a homeomorphism to be a diffeomorphism of orbifolds is a local condition, it suffices to check it on a neighborhood of each point. Let $p \in X$ and $q = f(p)$. We take a representative $(h_i, \tilde{\varphi}_i)$ of the orbifold embeddings $f_i : X_i \to Y$ using the orbifold charts $\mathfrak{V}_p^i = (V_p^i, \Gamma_p^i, \phi_p^i)$ of X and $\mathfrak{V}_q = (V_q, \Gamma_q, \phi_q)$ of Y. The maps $h_i : \Gamma_p^i \to \Gamma_q^i$ are group isomorphisms. So we have a group isomorphism $h = h_2^{-1} \circ h_1 : \Gamma_p^1 \to \Gamma_p^2$. Since $\tilde{\varphi}_1(V^1)/\Gamma_p = \tilde{\varphi}_2(V^2)/\Gamma_p$ set-theoretically, we have $\tilde{\varphi}_1(V_p^1) = \tilde{\varphi}_2(V_p^2) \subset V_q$. They are smooth submanifolds since f_i are embeddings of orbifolds. Therefore $\varphi = \tilde{\varphi}_2^{-1} \circ \tilde{\varphi}_1$ is defined in a neighborhood of the base point o_p^i and is a diffeomorphism. Then $(h, \tilde{\varphi})$ is a local representative of id. □

23.2 Vector Bundles on Orbifolds

Definition 23.18 Let $(X, \mathcal{E}, \pi)$ be a pair of orbifolds X and $\mathcal{E}$ with a continuous map $\pi : \mathcal{E} \to X$ between their underlying topological spaces. Hereafter we write $(X, \mathcal{E})$ in place of $(X, \mathcal{E}, \pi)$.

(1) An *orbifold chart* of $(X, \mathcal{E})$ is a quintuple $(V, E, \Gamma, \phi, \widehat{\phi})$ with the following properties:

(a) $\mathfrak{V} = (V, \Gamma, \phi)$ is an orbifold chart of the orbifold X.
(b) E is a finite-dimensional vector space equipped with a linear Γ action.
(c) $(V \times E, \Gamma, \widehat{\phi})$ is an orbifold chart of the orbifold $\mathcal{E}$.
(d) The diagram below commutes set-theoretically:

$$\begin{array}{ccc} V \times E & \xrightarrow{\widehat{\phi}} & \mathcal{E} \\ \downarrow & & \downarrow \pi \\ V & \xrightarrow{\phi} & X \end{array} \tag{23.4}$$

Here the left vertical arrow is the projection to the first factor.

(2) In the situation of (1), let $p \in V$ and $(V_p, \Gamma_p, \phi|_{V_p})$ be a subchart of (V, Γ, ϕ) in the sense of Definition 23.1 (2). Then $(V_p, E, \Gamma_p, \phi|_{V_p}, \widehat{\phi}|_{V_p \times E})$ is an orbifold chart of $(X, \mathcal{E})$. We call it a *subchart* of $(V, E, \Gamma, \phi, \widehat{\phi})$.

(3) Let $(V^i, E^i, \Gamma^i, \phi^i, \widehat{\phi^i})$ $(i = 1, 2)$ be orbifold charts of $(X, \mathcal{E})$. We say that they are *compatible* if the following holds for each $p_1 \in V^1$ and $p_2 \in V^2$ with $\phi^1(p_1) = \phi^2(p_2)$: There exist open neighborhoods $V^i_{p_i}$ of $p_i \in V^i$ such that:

(a) There exists an isomorphism $(h, \tilde{\varphi}) : (V^1, \Gamma^1, \phi^1)|_{V^1_{p_1}} \to (V^2, \Gamma^2, \phi^2)|_{V^2_{p_2}}$ between orbifold charts of X, which are subcharts.
(b) There exists an isomorphism $(h, \tilde{\hat{\varphi}}) : (V^1 \times E^1, \Gamma^1, \phi^1)|_{V^1_{p_1} \times E^1} \to (V^2 \times E^2, \Gamma^2, \phi^2)|_{V^2_{p_2} \times E^2}$ between orbifold charts of $\mathcal{E}$, which are subcharts.
(c) For each $y \in V^1_{p_1}$ the map $E^1 \to E^2$ given by $\xi \to \pi_{E^2}\tilde{\hat{\varphi}}(y, \xi)$ is a linear isomorphism. Here $\pi_{E^2} : V^2 \times E^2 \to E^2$ is the projection.

(4) A *representative of a vector bundle structure* on $(X, \mathcal{E})$ is a set of orbifold charts $\{(V_i, E_i, \Gamma_i, \phi_i, \widehat{\phi_i}) \mid i \in I\}$ such that any two of the charts are compatible in the sense of (3) above and

$$\bigcup_{i \in I} \phi_i(V_i) = X, \quad \bigcup_{i \in I} \widehat{\phi_i}(V_i \times E_i) = \mathcal{E},$$

are locally finite open covers.

Definition 23.19 Suppose $(X^*, \mathcal{E}^*)$ $(* = a, b)$ have representatives of vector bundle structures $\{(V_i^*, E_i^*, \Gamma_i^*, \phi_i^*, \widehat{\phi}_i^*) \mid i \in I^*\}$, respectively. A pair of orbifold embeddings $(f, \widehat{f})$, $f : X^a \to X^b$, $\widehat{f} : \mathcal{E}^a \to \mathcal{E}^b$ is said to be an *embedding of vector bundles* if the following holds:

(1) Let $p \in V_i^a$, $q \in V_j^b$ with $f(\phi_i^a(p)) = \phi_j^b(q)$. Then there exist open subcharts $(V_{i,p}^a \times E_{i,p}^a, \Gamma_{i,p}^a, \widehat{\phi}_{i,p}^a)$ and $(V_{j,q}^b \times E_{j,q}^b, \Gamma_{j,q}^b \widehat{\phi}_{j,q}^b)$ and a local representative $(h_{p;i,j}, f_{p;i,j}, \widehat{f}_{p;i,j})$ of the embeddings f and $\widehat{f}$ such that for each $y \in V_i^a$ the map $\xi \mapsto \pi_{E^b}(\widehat{f}_{p;i,j}(y, \xi))$, $E_{i,p}^a \to E_{j,q}^b$ is a linear embedding. Here $\pi_{E^b} : V^b \times E^b \to E^b$ is the projection.
(2) The diagram below commutes set-theoretically:

$$
\begin{array}{ccc}
\mathcal{E}^a & \xrightarrow{\widehat{f}} & \mathcal{E}^b \\
{\scriptstyle \pi_{E^a}}\downarrow & & \downarrow{\scriptstyle \pi_{E^b}} \\
X^a & \xrightarrow{f} & X^b
\end{array}
\tag{23.5}
$$

Two orbifold embeddings of vector bundles are said to be *equal* if they coincide set-theoretically as pairs of maps.

Lemma 23.20

(1) *A composition of embeddings of vector bundles is an embedding.*
(2) *The pair of identity maps* $(\mathrm{id}, \widehat{\mathrm{id}})$ *is an embedding.*
(3) *If an embedding of vector bundles is a pair of homeomorphisms, then the pair of their inverses is also an embedding.*

The proof is easy and is omitted.

Definition 23.21 Let $(X, \mathcal{E})$ be as in Definition 23.18.

(1) An embedding of vector bundles is said to be an *isomorphism* if it is a pair of diffeomorphisms of orbifolds.
(2) We say that two representatives of a vector bundle structure on $(X, \mathcal{E})$ are *equivalent* if the pair of identity maps regarded as a self-map of vector bundle $(X, \mathcal{E})$ equipped with those two representatives of vector bundle structure is an isomorphism. This is an equivalence relation by Lemma 23.20.
(3) An equivalence class of the equivalence relation (2) is called a *vector bundle structure* on $(X, \mathcal{E})$.
(4) A pair $(X, \mathcal{E})$ together with its vector bundle structure is called a *vector bundle* on X. We call $\mathcal{E}$ the *total space*, X the *base space*, and $\pi : \mathcal{E} \to X$ the *projection*.
(5) The condition for the pair $(f, \widehat{f}) : (X^a, \mathcal{E}^a) \to (X^b, \mathcal{E}^b)$ to be an embedding depends only on the equivalence class of vector bundle structures and is independent of its representatives. This enables us to define the notion of an *embedding of vector bundles*.

(6) We say $(f, \widehat{f})$ is an embedding *over the orbifold embedding f.*

Remark 23.22

(1) We may use the terminology 'orbibundle' in place of vector bundle. We use this terminology when we want to emphasize that it is different from the vector bundle over the underlying topological space.
(2) We sometimes simply say $\mathcal{E}$ is a vector bundle on an orbifold X.

Definition 23.23

(1) Let $(X, \mathcal{E})$ be a vector bundle. We call an orbifold chart $(V, E, \Gamma, \phi, \widehat{\phi})$ in the sense of Definition 23.18 (1) of underlying pair of topological spaces $(X, \mathcal{E})$ an *orbifold chart of a vector bundle* $(X, \mathcal{E})$ if the pair of maps $(\overline{\phi}, \overline{\widehat{\phi}}) : (V/\Gamma, (V \times E)/\Gamma) \to (X, \mathcal{E})$ induced from $(\phi, \widehat{\phi})$ is an embedding of vector bundles.
(2) If $(V, E, \Gamma, \phi, \widehat{\phi})$ is an orbifold chart of a vector bundle, we call a pair $(E, \widehat{\phi})$ a *trivialization* of our vector bundle on V/Γ.
(3) Hereafter when $(X, \mathcal{E})$ is a vector bundle, its 'orbifold chart' always means an orbifold chart of a vector bundles in the sense of (1).
(4) In the case when a vector bundle structure on $(X, \mathcal{E})$ is given, a representative of this vector bundle structure is called an *orbifold atlas* of $(X, \mathcal{E})$.
(5) Two orbifold charts $(V_i, E_i, \Gamma_i, \phi_i, \widehat{\phi_i})$ of a vector bundle are said to be *isomorphic* if there exist an isomorphism $(h, \tilde{\varphi})$ of orbifold charts $(V_1, \Gamma_1, \phi_1) \to (V_2, \Gamma_2, \phi_2)$ and an isomorphism $(h, \tilde{\hat{\varphi}})$ of orbifold charts $(V_1 \times E_1, \Gamma_1, \widehat{\phi_1}) \to (V_2 \times E_2, \Gamma_2, \widehat{\phi_2})$ such that they induce an embedding of vector bundles $(\varphi, \hat{\varphi})$: $(V_1/\Gamma_1, (V_1 \times E_1)/\Gamma_1) \to (V_2/\Gamma_2, (V_2 \times E_2)/\Gamma_2)$. The triple $(h, \tilde{\varphi}, \tilde{\hat{\varphi}})$ is called an *isomorphism* or a *coordinate change* between orbifold charts of the vector bundle.
(6) We define Whitney sum, tensor product, dual, exterior power, quotient bundle by a subbundle of vector bundles on orbifold, in the same way as the standard definition in the case of usual vector bundles.

The pullback of a vector bundle for orbifold is in general a delicate matter. (See [ALR, Section 2.4].) However, in our case of interest, where morphisms are embeddings, it is rather easy to pull back.

Lemma 23.24 *Let $(X^b, \mathcal{E}^b)$ be a vector bundle over an orbifold X^b and $f : X^a \to X^b$ an embedding of orbifolds. Let $\mathcal{E}^a = X^a \times_{X^b} \mathcal{E}^b$ be the fiber product in the category of topological space. By definition of the fiber product, we have maps π : $\mathcal{E}^a \to X^a$ and $\widehat{f} : \mathcal{E}^a \to \mathcal{E}^b$. Then there exists a unique structure of vector bundle on $(X^a, \mathcal{E}^a)$ such that the projection is given the above map π and $(f, \widehat{f})$ is an embedding of vector bundles.*

Proof Let $\{\mathfrak{V}^*_{\mathfrak{r}} \mid \mathfrak{r} \in \mathfrak{R}_*\}$, $* = a, b$ be orbifold atlases, where $\mathfrak{V}^*_{\mathfrak{r}} = (V^*_{\mathfrak{r}}, \Gamma^*_{\mathfrak{r}}, \phi^*_{\mathfrak{r}})$. Let $(V^b_{\mathfrak{r}}, E^b_{\mathfrak{r}}, \Gamma^b_{\mathfrak{r}}, \phi^b_{\mathfrak{r}}, \widehat{\phi}^b_{\mathfrak{r}})$ be an orbifold atlas of the vector bundle $(X^b, \mathcal{E}^b)$. Let $(h_{\mathfrak{r},ba}, \tilde{\varphi}_{\mathfrak{r},ba})$ be a local representative of the embedding f on the charts $\mathfrak{V}^a_{\mathfrak{r}}, \mathfrak{V}^b_{\mathfrak{r}}$. We put $E^a_{\mathfrak{r}} = E^b_{\mathfrak{r}}$, on which $\Gamma^a_{\mathfrak{r}}$ acts by the isomorphism $h_{\mathfrak{r},ba}$. By definition of a fiber product, there exists uniquely a map $\widehat{\phi}^a_{\mathfrak{r}} : V^a_{\mathfrak{r}} \times E^b_{\mathfrak{r}} \to \mathcal{E}^a$ such that the next diagram commutes:

$$\begin{array}{ccccc} V^a_{\mathfrak{r}} & \xleftarrow{\mathrm{pr}_1} & V^a_{\mathfrak{r}} \times E^b_{\mathfrak{r}} & \xrightarrow{\tilde{\varphi}_{\mathfrak{r},ba} \times id} & V^b_{\mathfrak{r}} \times E^b_{\mathfrak{r}} \\ \phi^a_{\mathfrak{r}} \downarrow & & \downarrow \widehat{\phi}^a_{\mathfrak{r}} & & \downarrow \widehat{\phi}^b_{\mathfrak{r}} \\ X^a & \xleftarrow{\pi} & \mathcal{E}^a & \xrightarrow{\hat{f}} & \mathcal{E}^b \end{array} \tag{23.6}$$

In fact,

$$f \circ \phi^a_{\mathfrak{r}} \circ \mathrm{pr}_1 = \phi^b_{\mathfrak{r}} \circ \varphi_{\mathfrak{r},ba} \circ \mathrm{pr}_1 = \phi^b_{\mathfrak{r}} \circ \pi \circ (\tilde{\varphi}_{\mathfrak{r},ba} \times id) = \pi \circ \widehat{\phi}^b_{\mathfrak{r}} \circ (\tilde{\varphi}_{\mathfrak{r},ba} \times id).$$

Thus $\{(V^a_{\mathfrak{r}}, E^a_{\mathfrak{r}}, \Gamma^a_{\mathfrak{r}}, \phi^a_{\mathfrak{r}}, \widehat{\phi}^a_{\mathfrak{r}}) \mid \mathfrak{r} \in \mathfrak{R}\}$ is an atlas of the vector bundle $(X^a, \mathcal{E}^a)$. □

Definition 23.25 We call the vector bundle in Lemma 23.24 the *pullback* and write $f^*(X^b, \mathcal{E}^b)$. (Sometimes we write $f^*\mathcal{E}^b$ by an abuse of notation.)

In the case when X^a is an open subset of X^b equipped with open substructure we call it a *restriction* in place of pullback of $\mathcal{E}^b$ and write $\mathcal{E}^b|_{X^a}$ in place of $f^*\mathcal{E}^b$.

Lemma 23.26 *In the situation of Lemma 23.13 suppose in addition that $\mathcal{E}^i$ is a vector bundle over X^i and $\widehat{\varphi}_{21} : \mathcal{E}^1 \to \mathcal{E}^2$ is an embedding of vector bundles over φ_{21}. Then in addition to the conclusion of Lemma 23.13, there exists $\tilde{\widehat{\varphi}}_{\mathfrak{r};21} : V^1_{\mathfrak{r}} \times E^1_{\mathfrak{r}} \to V^2_{\mathfrak{r}} \times E^2_{\mathfrak{r}}$ that is an $h_{\mathfrak{r};21}$ equivariant embedding of manifolds with the following properties:*

(1) *The next diagram commutes:*

$$\begin{array}{ccc} V^1_{\mathfrak{r}} \times E^1_{\mathfrak{r}} & \xrightarrow{\tilde{\widehat{\varphi}}_{\mathfrak{r},21}} & V^2_{\mathfrak{r}} \times E^2_{\mathfrak{r}} \\ \widehat{\phi}^1_{\mathfrak{r}} \downarrow & & \downarrow \widehat{\phi}^2_{\mathfrak{r}} \\ \mathcal{E}^1 & \xrightarrow{\widehat{\varphi}_{21}} & \mathcal{E}^2 \end{array} \tag{23.7}$$

(2) *For each $y \in V^1_{\mathfrak{r}}$ the map $\xi \mapsto \pi_2(\tilde{\widehat{\varphi}}_{\mathfrak{r},21}(y, \xi)) : E^1_{\mathfrak{r}} \to E^2_{\mathfrak{r}}$ is a linear embedding.*

The proof is similar to the proof of Lemma 23.13 and is omitted.

Definition 23.27 We call $(h_{\mathfrak{r},21}, \tilde{\varphi}_{\mathfrak{r},21}, \tilde{\widehat{\varphi}}_{\mathfrak{r},21})$ a *local representative of embedding* $(\varphi_{21}, \widehat{\varphi}_{21})$ on the charts $(V^1 \times E^1, \Gamma^1, \widehat{\phi}^1)$, $(V^2 \times E^2, \Gamma^2, \widehat{\phi}^2)$.

Lemma 23.28 *If $(h_{\mathfrak{r},21}, \tilde{\varphi}_{\mathfrak{r},21}, \tilde{\widehat{\varphi}}_{\mathfrak{r},21})$, $(h'_{\mathfrak{r},21}, \tilde{\varphi}'_{\mathfrak{r},21}, \tilde{\widehat{\varphi}}'_{\mathfrak{r},21})$ are local representatives of an embedding of vector bundles of the same charts $(V^1 \times E^1, \Gamma^1, \widehat{\phi}^1)$, $(V^2 \times E^2, \Gamma^2, \widehat{\phi}^2)$, then there exists $\mu \in \Gamma^2$ such that*

$$\tilde{\varphi}'_{\mathfrak{r},21}(x) = \mu\tilde{\varphi}_{\mathfrak{r},21}(x), \quad \tilde{\widehat{\varphi}}'_{\mathfrak{r},21}(x, \xi) = \mu\tilde{\widehat{\varphi}}_{\mathfrak{r},21}(x, \xi) \quad h'_{\mathfrak{r},21}(\gamma) = \mu h_{\mathfrak{r},21}(\gamma)\mu^{-1}.$$

Proof This is a consequence of Lemma 23.8. □

Remark 23.29 In Situation 6.4 we introduced the notation $(h_{\mathfrak{r},21}, \tilde{\varphi}_{\mathfrak{r},21}, \breve{\varphi}_{\mathfrak{r},21})$ where $\breve{\varphi}_{\mathfrak{r},21}$ is related to $\tilde{\tilde{\varphi}}_{\mathfrak{r},21}$ by the formula

$$\tilde{\tilde{\varphi}}_{\mathfrak{r},21}(y,\xi) = (\tilde{\varphi}_{\mathfrak{r},21}(y), \breve{\varphi}_{\mathfrak{r},21}(y,\xi)).$$

Definition 23.30 Let $(X, \mathcal{E})$ be a vector bundle. A *section* of $(X, \mathcal{E})$ is an embedding of orbifolds $s : X \to \mathcal{E}$ such that the composition of s and the projection is the identity map set-theoretically.

Lemma 23.31 *Let* $\{(V_{\mathfrak{r}}, E_{\mathfrak{r}}, \Gamma_{\mathfrak{r}}, \psi_{\mathfrak{r}}, \widehat{\psi}_{\mathfrak{r}}) \mid \mathfrak{r} \in \mathfrak{R}\}$ *be an atlas of* $(X, \mathcal{E})$. *Then a section of* $(X, \mathcal{E})$ *corresponds one-to-one to the following objects:*

(1) *For each* $\mathfrak{r}$ *we have a* Γ_r *equivariant map* $s_{\mathfrak{r}} : V_{\mathfrak{r}} \to E_{\mathfrak{r}}$, *which is compatible in the sense of* (2) *below.*
(2) *Suppose* $\phi_{\mathfrak{r}_1}(x_1) = \phi_{\mathfrak{r}_2}(x_2) = p$. *Then the definition of an orbifold atlas implies that there exist subcharts* $(V_{\mathfrak{r}_i,x_i}, E_{\mathfrak{r}_i,x_i}, \Gamma_{\mathfrak{r}_i,x_i}, \phi_{\mathfrak{r}_i,x_i}, \widehat{\phi}_{\mathfrak{r}_i})$ *of the orbifold charts* $(V_{\mathfrak{r}_i}, E_{\mathfrak{r}_i}, \Gamma_{\mathfrak{r}_i}, \phi_{\mathfrak{r}_i}, \widehat{\phi}_{\mathfrak{r}_i})$ *at* $x_i \in V_{\mathfrak{r}_i}$ *for* $i = 1, 2$ *and an isomorphism of charts*

$$(h_{12}^{\mathfrak{r},p}, \tilde{\varphi}_{12}^{\mathfrak{r},p}, \tilde{\tilde{\varphi}}_{12}^{\mathfrak{r},p}) : (V_{\mathfrak{r}_2,x_2}, E_{\mathfrak{r}_2}, \Gamma_{\mathfrak{r}_2,x_2}, \phi_{\mathfrak{r}_2,x_2}, \widehat{\phi}_{\mathfrak{r}_2}) \to (V_{\mathfrak{r}_1,x_1}, E_{\mathfrak{r}_1,x_1}, \Gamma_{\mathfrak{r}_1,x_1}, \phi_{\mathfrak{r}_1,x_1}, \widehat{\phi}_{\mathfrak{r}_1}).$$

Now we require the following equality:

$$\tilde{\tilde{\varphi}}_{12}^{\mathfrak{r},p}(s_{\mathfrak{r}_1}(y,\xi)) = s_{\mathfrak{r}_2}(\tilde{\varphi}_{12}^{\mathfrak{r},p}(y),\xi). \tag{23.8}$$

Proof The proof is mostly the same as the corresponding standard result for the case of vector bundle on a manifold or on a topological space. Let $s : X \to \mathcal{E}$ be a section, which is an orbifold embedding. Let $p = \phi_{\mathfrak{r}}(x) \in \phi_{\mathfrak{r}}(V_{\mathfrak{r}})$. Then there exist a subchart $(V_{\mathfrak{r},x}, \Gamma_{\mathfrak{r},x}, \phi_{\mathfrak{r},x})$ of $\mathfrak{V}_{\mathfrak{r}}$ and a subchart $(V'_{\mathfrak{r},x}, E_{\mathfrak{r}}, \Gamma_{\mathfrak{r},x}, \phi_{\mathfrak{r},x}, \hat{\phi}_{\mathfrak{r},x})$ of $(V_{\mathfrak{r}}, E_{\mathfrak{r}}, \Gamma_{\mathfrak{r}}, \phi_{\mathfrak{r}}, \hat{\phi}_{\mathfrak{r}})$ such that a representative (h', s') of s exists on the subcharts. Since $\pi \circ s$ =identity set-theoretically, it follows that $\pi_1(s'(y)) \equiv y \mod \Gamma_{\mathfrak{r},x}$ for any $y \in V_{\mathfrak{r},x}$. We take y such that $\Gamma_y = \{1\}$. Then, there exists a *unique* $\mu \in \Gamma_{\mathfrak{r},x}$ such that $\pi_1(s'(y)) \equiv \mu y$. By continuity this μ is independent of y. (We use Condition 23.9 here.)

We replace x by $\mu^{-1}x$ and $(V_{\mathfrak{r},x}, \Gamma_{\mathfrak{r},x}, \phi_{\mathfrak{r},x})$, by $(\mu^{-1}V_{\mathfrak{r},x}, \mu^{-1}\Gamma_{\mathfrak{r},x}\mu, \phi_{\mathfrak{r},x} \circ \mu)$ and (h', s') by $(h' \circ \mathrm{conj}_{\mu}, s' \circ \mu^{-1})$. (Here $\mathrm{conj}_{\mu}(\gamma) = \mu\gamma\mu^{-1}$.) Therefore we may assume $\pi_1(s'(y)) = y$. Note that s' is h'-equivariant and π_1 is id-equivariant. Here id is the identity map $\Gamma_{\mathfrak{r},x} \to \Gamma_{\mathfrak{r},x}$. Therefore the identity map $V_{\mathfrak{r},x} \to V_{\mathfrak{r},x}$ is h' equivariant. Hence $h' = \mathrm{id}$.

In sum, we have the following. (We put $s_{\mathfrak{r},x} = s'$.) For a sufficiently small $\mathfrak{V}_{\mathfrak{r},x}$ there exists uniquely a map $s_{\mathfrak{r},x} : V_{\mathfrak{r},x} \to V_{\mathfrak{r},x} \times E_{\mathfrak{r}}$ such that:

(a) $\pi_1(s_{\mathfrak{r},x}(y)) = y$,
(b) $s_{\mathfrak{r},x}$ is equivariant with respect to the embedding $\Gamma_{\mathfrak{r},x} \to \Gamma_{\mathfrak{r}}$. (Recall $\Gamma_{\mathfrak{r},x} = \{\gamma \in \Gamma_{\mathfrak{r}} \mid \gamma x = x\}$.)
(c) $(\mathrm{id}, s_{\mathfrak{r},x})$ is a local representative of s.

We can use uniqueness of such $s_{\mathfrak{r},x}$ to glue them to obtain a map $V_{\mathfrak{r}} \to V_{\mathfrak{r}} \times E_{\mathfrak{r}}$. By (a) this map is of the form $y \mapsto (y, \mathfrak{s}_{\mathfrak{r}}(y))$ for some map $\mathfrak{s}_{\mathfrak{r}} : V_{\mathfrak{r}} \to E_{\mathfrak{r}}$. This is the map $\mathfrak{s}_{\mathfrak{r}}$ required in (1). Since $y \mapsto \gamma^{-1}\mathfrak{s}_{\mathfrak{r}}(\gamma y)$ also has the same property, the uniqueness implies that $\mathfrak{s}_{\mathfrak{r}}$ is $\Gamma_{\mathfrak{r}}$ equivariant. Equation (23.8) is also a consequence of the uniqueness.

Thus we find a map from the set of sections to the set of $(s_{\mathfrak{r}})_{\mathfrak{r}\in\mathfrak{R}}$ satisfying (1)(2). The construction of the converse map is obvious. □

The next lemma is proved during the proof of Lemma 23.31.

Lemma 23.32 *Let $(V_{\mathfrak{r}}, E_{\mathfrak{r}}, \Gamma_{\mathfrak{r}}, \phi_{\mathfrak{r}}, \widehat{\phi}_{\mathfrak{r}})$ be an orbifold chart of $(X, \mathcal{E})$ and s a section of $(X, \mathcal{E})$. Then there exists uniquely a Γ equivariant map $s_{\mathfrak{r}} : V_{\mathfrak{r}} \to E_{\mathfrak{r}}$ such that the following diagram commutes:*

$$\begin{array}{ccc} V_{\mathfrak{r}} \times E_{\mathfrak{r}} & \xrightarrow{\widehat{\phi}_{\mathfrak{r}}} & \mathcal{E}_{\mathfrak{r}} \\ {\scriptstyle \mathrm{id}\times s_{\mathfrak{r}}}\uparrow & & \uparrow{\scriptstyle s} \\ V_{\mathfrak{r}} & \xrightarrow{\phi_{\mathfrak{r}}} & X \end{array} \tag{23.9}$$

Definition 23.33 We call the system of maps $s_{\mathfrak{r}}$ the *local expression* of s in the orbifold chart $(V_{\mathfrak{r}}, E_{\mathfrak{r}}, \Gamma_{\mathfrak{r}}, \phi_{\mathfrak{r}}, \widehat{\phi}_{\mathfrak{r}})$.

Definition 23.34 Let s be a section of a vector bundle $(X, \mathcal{E})$. We say s is *transversal to 0* if the local representatives $s_{\mathfrak{r}} : V_{\mathfrak{r}} \to E_{\mathfrak{r}}$ are transversal to 0 for all $\mathfrak{r}$.

If s is transversal to 0 then it is easy to see that $s^{-1}(0) \subset X$ has a structure of orbifold such that $s^{-1}(0) \to X$ is an embedding of an orbifold.

Chapter 24
Covering Space of Effective Orbifolds and K-Spaces

24.1 Covering Space of an Orbifold

We first define the notion of a covering space of an orbifold. Let U_1, U_2 be orbifolds and let $\pi : U_1 \to U_2$ be a continuous map between their underlying topological spaces.

Definition 24.1 For $i = 1, 2$ let $x_i \in U_i$ with $\pi(x_1) = x_2$ and $\mathfrak{V}_i = (V_i, \Gamma_i, \phi_i)$ be orbifold charts of U_i at x_i. We say that $(\mathfrak{V}_1, \mathfrak{V}_2)$ is a *covering chart* if the following holds:

(1) There exists an injective group homomorphism $h_{21} : \Gamma_1 \to \Gamma_2$.
(2) There exists an h_{21}-equivariant diffeomorphism $\varphi_{21} : V_1 \to V_2$.
(3) $\phi_2 \circ \varphi_{21} = \phi_1$.

The index $[\Gamma_2 : h_{21}(\Gamma_1)]$ is called the *covering index* of the covering chart $(\mathfrak{V}_1, \mathfrak{V}_2)$.

Definition 24.2 The map $\pi : U_1 \to U_2$ is called a *covering map* if the following holds at each $x \in U_2$:

(1) The set $\pi^{-1}(x)$ is a finite set, which we write $\{\tilde{x}_1, \dots, \tilde{x}_{m_x}\}$.
(2) There exist an orbifold chart $\mathfrak{V}_x$ of U_2 at x and orbifold charts $\mathfrak{V}_{\tilde{x}_j}$ of U_1 at $\tilde{x}_j$ respectively for $j = 1, \dots, m_x$ such that $(\mathfrak{V}_{\tilde{x}_j}, \mathfrak{V}_x)$ is a covering chart. (Here $\mathfrak{V}_x$ is independent of j.) We write its covering index $n_j(x)$.
(3) $\sum_{j=1}^{m_x} n_j(x)$ is independent of x.

We call $\sum_{j=1}^{m_x} n_j(x)$ the *covering index* of π.

Remark 24.3

(1) We only define a finite covering here since we do not use an infinite covering in the present book.
(2) If Definition 24.2 (1)–(2) is satisfied and U_2 is connected, then (3) is equivalent to the following condition:

K. Fukaya et al., *Kuranishi Structures and Virtual Fundamental Chains*, Springer Monographs in Mathematics, https://doi.org/10.1007/978-981-15-5562-6_24

(3)' $\pi^{-1}(U_x) = \bigcup_{j=1}^{m_x} U_{\tilde{x}_j}$ and the right hand side is the disjoint union.

In fact, (3)' implies that $\sum_{j=1}^{m_x} n_j(x)$ is a locally constant function. We can replace (3) by (3)' and define an infinite covering in the same way.

(3) The composition of covering maps is a covering map.

Lemma 24.4 *Let $\varphi_{21} : U_1 \to U_2$ be an embedding of orbifolds and $\pi_2 : \widetilde{U}_2 \to U_2$ a covering map of orbifolds. We consider the fiber product $U_1 \times_{U_2} \widetilde{U}_2$ in the category of topological spaces. It comes with continuous maps $\pi_1 : U_1 \times_{U_2} \widetilde{U}_2 \to U_1$ and $\widetilde{\varphi}_{21} : U_1 \times_{U_2} \widetilde{U}_2 \to \widetilde{U}_2$. Then $U_1 \times_{U_2} \widetilde{U}_2$ has a structure of orbifolds such that:*

(1) *π_1 is a covering map.*
(2) *$\widetilde{\varphi}_{21}$ is an embedding of orbifolds.*

The conditions (1) (2) uniquely determine the orbifold structure on $U_1 \times_{U_2} \widetilde{U}_2$.

Proof We take the atlas $\{\mathfrak{V}^i_{\mathfrak{r}} \mid \mathfrak{r} \in \mathfrak{R}_i\}$ as in Lemma 23.13. We may choose it sufficiently fine so that for each $\mathfrak{r} \in \mathfrak{R}_2$ there exist $\mathfrak{V}^{2,\mathfrak{r}}_j$, $j = 1, \dots, m_{\mathfrak{r}}$ such that $(\mathfrak{V}^{2,\mathfrak{r}}_j, \mathfrak{V}^2_{\mathfrak{r}})$ is a covering atlas and satisfies $\pi^{-1}(U^2_{\mathfrak{r}}) = \bigcup_{j=1}^{m_{\mathfrak{r}}} U^2_{\mathfrak{r},j}$. For each $j = 1, \dots, m_{\mathfrak{r}}$, the map π determines a finite index subgroup $\Gamma^2_{j,\mathfrak{r}}$ of $\Gamma^2_{\mathfrak{r}}$ for $\mathfrak{r} \in \mathfrak{R}_2$. Note that in the case $\mathfrak{r} \in \mathfrak{R}_1$ the group $\Gamma^2_{\mathfrak{r}}$ is isomorphic to $\Gamma^1_{\mathfrak{r}}$. Then a finite subgroup $\Gamma^1_{j,\mathfrak{r}}$ of $\Gamma^1_{\mathfrak{r}}$ determines $\Gamma^2_{j,\mathfrak{r}} \subset \Gamma^2_{\mathfrak{r}}$. Therefore the collection $(V^1_{\mathfrak{r}}, \Gamma^1_{\mathfrak{r},j}, \phi^1_{\mathfrak{r},j})$ determines an orbifold chart. Here the map $\phi^1_{\mathfrak{r},j}$ is defined by

$$\phi^1_{\mathfrak{r},j}(y) = (\phi^1_{\mathfrak{r}}(y), \phi^2_{\mathfrak{r},j}(\varphi^{\mathfrak{r}}_{21}(y))),$$

where $\varphi^{\mathfrak{r}}_{21} : V^{\mathfrak{r}}_1 \to V^{\mathfrak{r}}_2$ is determined by the orbifold embedding φ_{21} and $\phi^2_{\mathfrak{r},j} : V^{\mathfrak{r}}_2 \to \widetilde{U}_2$ is a part of the orbifold chart $\mathfrak{V}^{2,\mathfrak{r}}_j$. It is easy to check that the charts $(V^1_{\mathfrak{r}}, \Gamma^1_{\mathfrak{r},j}, \phi^1_{\mathfrak{r},j})$ for various $\mathfrak{r}$ and j determine an orbifold structure on the fiber product. We can easily check (1), (2). The uniqueness is also easy to check. □

Lemma-Definition 24.5 *We can pull back a vector bundle and its section by a covering map of an orbifold.*

Proof Let $\widetilde{U} \to U$ be a covering map and $\mathcal{E} \to U$ a vector bundle. We take a local coordinate $(V \times E, \Gamma, \widehat{\phi})$ of $\mathcal{E} \to U$ at $p \in U$, where (V, Γ, ϕ) is the corresponding chart of U. We may shrink V and may assume that there is a covering chart (V, Γ_i, ϕ_i) for each $p_i \in \tilde{U}$ with $\pi(p_i) = p$. The two maps $\phi_i \circ \mathrm{pr}_V : V \times E \to \tilde{U}$ and $\widehat{\phi} : V \times E \to |\mathcal{E}|$ give rise to a map

$$\widehat{\phi}_i : V \times E \to \widetilde{U} \times_U |\mathcal{E}|.$$

It is easy to see that $(V \times E, \Gamma_i, \widehat{\phi}_i)$ gives a structure of vector bundle on $\widetilde{U} \times_U |\mathcal{E}| \to \widetilde{U}$. □

Remark 24.6 Our covering map in the sense of Definition 24.2 defines a *good map* in the sense of [ALR]. Then the above lemma is a special case of [ALR, Theorem 2.43].

Lemma 24.7 *Let X be a topological space, Y an orbifold, and let $f : X \to Y$ be a continuous map. Then the orbifold structure on X, under which f becomes a covering map of orbifolds, is unique if it exists.*

Proof Suppose we have two such structures. It suffices to show that the identity map is a diffeomorphism of orbifolds. Let $p \in Y$ and (V, Γ, ϕ) be an orbifold chart of Y at p. We may shrink V if necessary and assume that there are covering charts (V_i, Γ_i, ϕ_i) $(i = 1, 2)$ for two orbifold structures on X such that $f : X \to Y$ is a covering map. We put

$$S(V) = \{x \in V \mid \exists \gamma \in \Gamma \setminus \{1\},\ \gamma x = x\}, \quad U^{\text{reg}} = (V \setminus S(V))/\Gamma.$$

Note that $f^{-1}(U^{\text{reg}}) \to U^{\text{reg}}$ is a covering space. An element of $\gamma \in \Gamma$ is contained in the image of Γ_i if and only if the corresponding loop in U^{reg} lifts to a loop in $f^{-1}(U^{\text{reg}})$. Therefore $\Gamma_1 = \Gamma_2$ as subgroups of Γ. It is now easy to see that the identity map is a diffeomorphism at p_i. □

24.2 Covering Space of a K-Space

Using Lemma 24.4 and Lemma-Definition 24.5, it is fairly straightforward to define the notion of a covering space of a K-space and show that various objects are pulled back by a covering map. We will describe them below for completeness.

Situation 24.8 Let $\mathcal{U} = (U, \mathcal{E}, s, \psi)$ be a Kuranishi chart of $Z \subseteq X$. Suppose we are given a topological space $\widetilde{X}$ and a continuous map $\pi : \widetilde{X} \to X$. We put $\tilde{Z} = \pi^{-1}(Z)$. ■

Definition 24.9 We call a Kuranishi chart $\widetilde{\mathcal{U}} = (\widetilde{U}, \widetilde{\mathcal{E}}, \widetilde{s}, \widetilde{\psi})$ of $\tilde{X}$ a *covering chart* of $\mathcal{U} = (U, \mathcal{E}, s, \psi)$, if $\widetilde{U}$ is a covering space of the orbifold U, $\widetilde{\mathcal{E}}$ is the pullback of $\mathcal{E}$ by $\widetilde{U} \to U$, $\widetilde{s}$ is obtained from s by the pullback and $\widetilde{\psi}$, a homeomorphism from $\widetilde{s}^{-1}(0)$ to $\widetilde{X}$, satisfies $\pi \circ \widetilde{\psi} = \psi$.

Lemma-Definition 24.10 *Suppose we are in Situation 24.8. Let $\mathcal{U}' = (U', \mathcal{E}', s', \psi')$ be another Kuranishi chart and $\Phi : \mathcal{U}' \to \mathcal{U}$ an embedding of Kuranishi chart. Let $\widetilde{\mathcal{U}}$ be a covering chart of $\mathcal{U}$. Then we can define a covering chart $\widetilde{\mathcal{U}'} = (\widetilde{U'}, \widetilde{\mathcal{E}'}, \widetilde{s'}, \widetilde{\psi'})$ of $\mathcal{U}'$ and an embedding of Kuranishi chart $\widetilde{\Phi} : \widetilde{\mathcal{U}'} \to \widetilde{\mathcal{U}}$ such that*

$$\widetilde{U'} = U' {}_{\Phi}\times_U \tilde{U} \tag{24.1}$$

with the orbifold structure given by Lemma 24.4. We call $\widetilde{\mathcal{U}'}$ the pullback covering chart *of $\widetilde{\mathcal{U}}$ by Φ. In particular, if U_0 is an open subset of U, then we can define a restriction $\widetilde{\mathcal{U}'}|_{U_0}$ of $\widetilde{\mathcal{U}'}$.*

Proof We define $\widetilde{U'}$ by (24.1). Then $\widetilde{U'} \to U'$ is a covering space. Therefore by Lemma 24.5 we can pull back $\mathcal{E}$ to $\widetilde{U'}$. We define $\widetilde{\mathcal{E}'}$ by the pullback. We can then obtain $\widetilde{s}'$, $\widetilde{\psi}'$ from definition. The existence of $\widetilde{\Phi}$ is immediate from construction. □

Definition 24.11 Let $\widehat{\mathcal{U}}$ be a Kuranishi structure on $Z \subseteq X$, $\pi : \widetilde{X} \to X$ a finite-to-one continuous map, and $\widehat{Z} = \pi^{-1}(Z)$. Let $\widehat{\widetilde{\mathcal{U}}}$ be a Kuranishi structure on $\widehat{Z} \subseteq \widetilde{X}$. We say that $(\widetilde{X}, \widehat{Z}; \widehat{\widetilde{\mathcal{U}}})$ is a *covering space* of $(X, Z; \widehat{\mathcal{U}})$ if the following holds:

(1) For each $p \in Z$ we are given a covering chart $\widetilde{\mathcal{U}_p}$ of the Kuranishi chart $\mathcal{U}_p$, which is a part of the data constituting $\widehat{\mathcal{U}}$.
(2) If $q \in \psi_p(s_p^{-1}(0))$, an isomorphism between the restriction $\widetilde{\mathcal{U}_q}|_{U_{pq}}$ of the covering chart $\widetilde{\mathcal{U}_q}$ given in Item (1) and the pullback of the covering chart of $\widetilde{\mathcal{U}_p}$ by the coordinate change Φ_{pq} is given so that the next diagram commutes:

$$\begin{array}{ccc} \widetilde{\mathcal{U}_q}|_{U_{pq}} & \longrightarrow & \widetilde{\Phi}^*_{pq}\widetilde{\mathcal{U}_p} \\ \downarrow & & \downarrow \\ \mathcal{U}_q|_{U_{pq}} & \xrightarrow{=} & \mathcal{U}_q|_{U_{pq}} \end{array} \tag{24.2}$$

We remark that the chart $\widetilde{\Phi}^*_{pq}\widetilde{\mathcal{U}_p}$ is defined in Lemma-Definition 24.10. Its underlying orbifold is a covering of U_{pq}, which is an open subset of U_q.
(3) Let $p_j \in \widehat{Z}$ and $p \in Z$ with $\pi(p_j) = p$ and $\mathcal{U}_{p_j} = (U_{p_j}, \mathcal{E}_{p_j}, s_{p_j}, \psi_{p_j})$ a Kuranishi chart of $\widehat{Z}$ at p_j which is a part of the data of $(\widehat{Z}, \widehat{\widetilde{\mathcal{U}}})$. Then there exists an open embedding Φ_{p_j} of the Kuranishi chart $\mathcal{U}_{p_j}$ to the covering chart $\widetilde{\mathcal{U}_p}$ given in Item (1).
(4) For any $q \in \psi_p(s_p^{-1}(0))$, $q_j \in \psi_{p_j}(s_{p_j}^{-1}(0))$, the following diagram commutes:

$$\begin{array}{ccc} \mathcal{U}_{q_j}|_{U_{p_jq_j}} & \xrightarrow{\Phi_{p_jq_j}} & \mathcal{U}_{p_j} \\ {\scriptstyle\Phi_{q_j}}\downarrow & & \downarrow{\scriptstyle\Phi_{p_j}} \\ \widetilde{\mathcal{U}_q}|_{U_{pq}} \cong \widetilde{\Phi}^*_{pq}\widetilde{\mathcal{U}_p} & \xrightarrow{\widetilde{\Phi}_{pq}} & \widetilde{\mathcal{U}_p} \end{array} \tag{24.3}$$

Here $\Phi_{p_jq_j}$ is the coordinate change of the Kuranishi structure $(\widetilde{X}, \widehat{Z}; \widehat{\widetilde{\mathcal{U}}})$ and $\widetilde{\Phi}_{pq}$ is obtained by Lemma-Definition 24.10. We use the isomorphism in Item (2) to identify two coordinates in the lower left corner.

(5) Let $n_{p,j}$ be the covering index of the covering $\widetilde{U}_{p,j} \to U_p$. Then the number

$$\sum_{p_j \in \widetilde{X}:\pi(p_j)=p} n_{p,j} \tag{24.4}$$

is independent of p. We call it the *covering index* of the covering $(\widetilde{X}, \widehat{Z}; \widehat{\widehat{\mathcal{U}}})$ of $(X, Z; \widehat{\mathcal{U}})$.

Remark 24.12

(1) If Z is connected, Condition (5) follows from Conditions (1), (2). In fact, Conditions (1), (2) imply that the covering index of $\widetilde{U}_p \to U_p$ is locally constant. However, it does not seem to be a good idea to assume connectivity of Z in our situation, since the topology of Z can be pathological. In fact any closed subset of $\mathbb{R}^n$ can be a zero set of a smooth function and so can be our space Z.

(2) The commutativity of Diagram (24.3) means the set-theoretic equalities of the underlying topological spaces of orbifolds and of the total spaces of obstruction bundles. We can safely do so since all the maps involved are embeddings.

(3) We can define the notion of a covering space of a space equipped with a good coordinate system and can prove that for a given covering space of K-space we can construct a covering space equipped with a good coordinate system in the same way.

Lemma 24.13 *Let $(\widetilde{X}, \widetilde{Z}; \widehat{\widehat{\mathcal{U}}})$ be a covering space of $(X, Z; \widehat{\mathcal{U}})$ in the sense of Definition* 24.11.

(1) *If $\widehat{\mathfrak{S}}$ is a CF-perturbation of $\widehat{\mathcal{U}}$, it induces a CF-perturbation $\widehat{\widehat{\mathfrak{S}}}$ of $\widehat{\widehat{\mathcal{U}}}$.*

(2) *A strongly continuous (strongly smooth) map $\widehat{f}$ from $(X, \widehat{\mathcal{U}})$ can be pulled back to a strongly continuous (strongly smooth) map $\widehat{\widehat{f}}$ from $(\widetilde{X}, Z; \widehat{\widehat{\mathcal{U}}})$.*

(3) *A differential form $\widehat{h}$ on $(X, Z; \widehat{\mathcal{U}})$ can be pulled back to $\widehat{\widehat{h}}$ on $(\widetilde{X}, \widehat{\widehat{\mathcal{U}}})$.*

(4) *A strongly smooth map $\widehat{f}$ is strongly submersive with respect to $\widehat{\mathfrak{S}}$ if and only if $\widehat{\widehat{f}}$ in* (2) *is strongly submersive with respect to $\widehat{\widehat{\mathfrak{S}}}$.*

(5) *The statements such that 'CF-perturbations' in* (2) (4) *are replaced by 'multi-valued perturbations' also hold.*

(6) *Suppose we are in the situation of* (4) *and $Z = \emptyset$. Then we have*

$$n\widehat{f}!(\widehat{h}; \widehat{\mathfrak{S}}) = \widehat{\widehat{f}}!\,(\widehat{\widehat{h}}; \widehat{\widehat{\mathfrak{S}}}).$$

Here n is the covering index.

The proof is straightforward and so omitted.

Definition 24.14 Suppose we are given CF-perturbations on $(\widetilde{X}, \widehat{\widetilde{\mathcal{U}}})$ and on $(X, \widehat{\mathcal{U}})$. We say they are compatible if the former is induced by the latter via Lemma 24.13 (1). We define compatibility of a strongly smooth map, differential form or multivalued perturbation with a covering map in the same way.

24.3 Covering Spaces Associated to the Corner Structure Stratification

One of the main reasons we introduced the notion of the covering space of a K-space is that we use Proposition 24.17 to clarify the discussion of normalized boundaries.

Proposition 24.15 *If $(X, Z; \widehat{\mathcal{U}})$ is a relative K-space with corners, then $\overset{\circ}{S}_{k-1}(\partial(X, Z; \widehat{\mathcal{U}}))$ is a covering space of $\overset{\circ}{S}_k(X, Z; \widehat{\mathcal{U}})$ with covering index k.*

Proof We first prove the proposition for the case of orbifolds.

Lemma 24.16 *Let U be an orbifold with corners. The map $\overset{\circ}{S}_{k-1}(\partial U) \to \overset{\circ}{S}_k U$ which is the restriction of the map π in Lemma 8.8 (2) is a k-fold covering of orbifolds. If $\mathcal{E}$ is a vector bundle on U, then the bundle induced on $\overset{\circ}{S}_{k-1}(\partial U)$ is canonically isomorphic to the pullback of the restriction of $\mathcal{E}$ to $\overset{\circ}{S}_k U$.*

Proof Let $x \in \overset{\circ}{S}_k U$ and $\mathfrak{V}_x = (V_x, \Gamma_x, \phi_x)$ be an orbifold chart of U at x. We may assume $V_x \subset [V_x] \times [0, 1)^k$ and $o_x = (\overline{o}_x, (0, \dots, 0))$, where $[V_x]$ is a manifold without boundary. There exists a group homomorphism $\sigma : \Gamma_x \to \mathrm{Perm}(k)$ such that if $\gamma(\overline{y}, (t_1, \dots, t_k)) = (\overline{y}', (t_1', \dots, t_k'))$ then $t_k' = 0$ if and only if $t_{\sigma(\gamma)(k)} = 0$.

We consider $I \subset \{1, \dots, k\}$ a complete set of representatives of $\{1, \dots, k\}/\Gamma_x$. For each $i \in I$, we put

$$\Gamma_{x,i} = \{\gamma \in \Gamma_x \mid \sigma(\gamma)i = i\},$$
$$\partial_i V_x = \{(\overline{y}, (t_1, \dots, t_k)) \in V_x \mid t_i = 0\}.$$

The given Γ_x action on V_x induces a $\Gamma_{x,i}$ action on $\partial_i V_x$. By the definition of a normalized boundary, there exists a map $\phi_{x,i} : \partial_i V_x \to \partial U$ such that $(\partial_i V_x, \Gamma_{x,i}, \phi_{x,i})$ is an orbifold chart of ∂U at $\tilde{x}_i$. Here $\{\tilde{x}_i \mid i \in I\} = \pi^{-1}(x) \subset \overset{\circ}{S}_{k-1}(\partial U)$. An orbifold chart of $\overset{\circ}{S}_{k-1}(\partial U)$ at $\tilde{x}_i$ is

$$([V_x], \Gamma_{x,i}, \phi_{x,i}|_{[V_x]}),$$

where $[V_x]$ is identified with the subset $[V_x] \times \{0\}$ of $\partial_i V_x$.

On the other hand, an orbifold chart of $\overset{\circ}{S}_k U$ at x is $([V_x], \Gamma_x, \phi_x|_{[V_x]})$. Since $\#(\Gamma_x \cdot i) = \#(\Gamma_x/\Gamma_{x,i})$, we have

$$\sum_{i \in I} \#(\Gamma_x / \Gamma_{x,i}) = k.$$

We have thus proved that $\pi : \overset{\circ}{S}_{k-1}(\partial U) \to \overset{\circ}{S}_k(U)$ is a k-fold covering of orbifolds. We can prove the second half of the lemma using the above description of the orbifold charts. □

Now we consider the case of a Kuranishi structure. Let $\mathcal{U}_p = (U_p, \mathcal{E}_p, \psi_p, s_p)$ be a Kuranishi chart of $\widehat{\mathcal{U}}$ at $p \in \overset{\circ}{S}_k(X, Z; \widehat{\mathcal{U}})$. We put

$$\overset{\circ}{S}_k(\mathcal{U}_p) := (\overset{\circ}{S}_k(U_p), \mathcal{E}_p|_{\overset{\circ}{S}_k}, \psi_p|_{\overset{\circ}{S}_k}, s_p|_{\overset{\circ}{S}_k}),$$

which is a Kuranishi chart at p of the relative K-space $\overset{\circ}{S}_k(X, Z; \widehat{\mathcal{U}})$. Let π be the map from the underlying topological space of $\overset{\circ}{S}_{k-1}(\partial(X, Z; \widehat{\mathcal{U}}))$ to that of $\overset{\circ}{S}_k(X, Z; \widehat{\mathcal{U}})$. We use the notation in the proof of Lemma 24.16 by putting $x = o_p \in U_p$, $U = U_p$ and $\mathcal{E} = \mathcal{E}_p$. Then $\pi^{-1}(p)$ consists of $\#I$ points which we write $\widetilde{p}_i$, $i \in I$. Then $([V_x], \Gamma_{x,i}, \psi_{x,i}|_{[V_x]})$ is an orbifold chart of $\overset{\circ}{S}_{k-1}(\partial(X, Z; \widehat{\mathcal{U}}))$ at $\widetilde{p}_i$. We restrict $\overset{\circ}{S}_k(\mathcal{U}_p)$ to $[V_x]/\Gamma_x \subset U_p$ to get a Kuranishi chart of $\overset{\circ}{S}_{k-1}(\partial(X, Z; \widehat{\mathcal{U}}))$.

The second half of Lemma 24.16 implies that the pullback of the restriction of the obstruction bundle $\mathcal{E}_p$ to $[V_x]/\Gamma_x \subset U_p$ defines a vector bundle on $[V_x]/\Gamma_{x,i}$ which is isomorphic to the obstruction bundle of the Kuranishi structure $\overset{\circ}{S}_{k-1}(\partial(X, Z; \widehat{\mathcal{U}}))$. It is easy to see from construction that Kuranishi maps are preserved by this isomorphism.

We have thus constructed open substructures of $\overset{\circ}{S}_{k-1}(\partial(X, Z; \widehat{\mathcal{U}}))$ and $\overset{\circ}{S}_k(X, Z; \widehat{\mathcal{U}})$ so that the Kuranishi chart of the former is a k-fold covering of the Kuranishi chart of the latter. It is easy to see that this isomorphism is compatible with the coordinate change. Hence the proof of Proposition 24.15 is complete. □

We next generalize Proposition 24.15 to the corners of arbitrary codimension.

Proposition 24.17 *Let $(X, Z; \widehat{\mathcal{U}})$ be an n-dimensional relative K-space. Then for each k there exists an $(n-k)$-dimensional relative K-space $\widehat{S}_k(X, Z; \widehat{\mathcal{U}})$ with corners and maps $\pi_k : \widehat{S}_k(X, Z; \widehat{\mathcal{U}}) \to S_k(X, Z; \widehat{\mathcal{U}})$, $\pi_{\ell,k} : \widehat{S}_\ell(\widehat{S}_k(X, Z; \widehat{\mathcal{U}})) \to \widehat{S}_{k+\ell}(X, Z; \widehat{\mathcal{U}})$ with the following properties:*

(1) *π_k is a continuous map between underlying topological spaces.*
(2) *$\widehat{S}_1(X, Z; \widehat{\mathcal{U}})$ is the normalized boundary $\partial(X, Z; \widehat{\mathcal{U}})$.*
(3) *The interior of $\widehat{S}_k(X, Z; \widehat{\mathcal{U}})$ is isomorphic to $\overset{\circ}{S}_k(X, Z; \widehat{\mathcal{U}})$. The underlying homeomorphism of this isomorphism is the restriction of π_k.*
(4) *$\pi_{\ell,k}$ is an $(\ell + k)!/\ell!k!$ fold covering map of K-spaces.*
(5) *The following objects on $(X, Z; \widehat{\mathcal{U}})$ induce the corresponding ones on $\widehat{S}_k(X, Z; \widehat{\mathcal{U}})$. The induced objects are compatible with the covering map $\pi_{\ell,k}$.*

(a) *CF-perturbation.*
(b) *Multivalued perturbation.*
(c) *Differential form.*
(d) *Strongly continuous map. Strongly smooth map.*
(e) *Covering map.*

(6) *The following diagram commutes:*

$$\begin{array}{ccc} \widehat{S}_{k_1}(\widehat{S}_{k_2}(\widehat{S}_{k_3}(X,Z;\widehat{\mathcal{U}}))) & \xrightarrow{\pi_{k_1,k_2}} & \widehat{S}_{k_1+k_2}(\widehat{S}_{k_3}(X,Z;\widehat{\mathcal{U}})) \\ {\scriptstyle \widehat{S}_{k_1}(\pi_{k_2,k_3})}\downarrow & & \downarrow{\scriptstyle \pi_{k_1+k_2,k_3}} \\ \widehat{S}_{k_1}(\widehat{S}_{k_2+k_3}(X,Z;\widehat{\mathcal{U}})) & \xrightarrow{\pi_{k_1,k_2+k_3}} & \widehat{S}_{k_1+k_2+k_3}(X,Z;\widehat{\mathcal{U}}) \end{array} \tag{24.5}$$

Here $\widehat{S}_{k_1}(\pi_{k_2,k_3})$ is the covering map induced from π_{k_2,k_3}.

(7) *Let $f_i : (X_i, Z_i; \widehat{\mathcal{U}_i}) \to M$ be a strongly smooth map and assume that f_1 is transversal to f_2. Then*

$$\widehat{S}_k\left((X_1,Z_1;\widehat{\mathcal{U}_1}) \times_M (X_2,Z_2;\widehat{\mathcal{U}_2})\right) \cong \coprod_{k_1+k_2=k} \widehat{S}_{k_1}(X_1,Z_1;\widehat{\mathcal{U}_1}) \times_M \widehat{S}_{k_2}(X_2,Z_2;\widehat{\mathcal{U}_2}).$$

Here the right hand side is the disjoint union.

(8) *(1)–(6) also hold when we replace 'Kuranishi structure' by 'good coordinate system'.*

(9) *Various kinds of embeddings of Kuranishi structures and/or good coordinate systems induce the corresponding ones of $\widehat{S}_k(X,Z;\widehat{\mathcal{U}})$.*

We note that Proposition 24.17 (7) implies

$$\widehat{S}_\ell(X_1 \times_{M_1} \cdots \times_{M_{n-1}} X_n) = \coprod_{\ell_1+\cdots+\ell_n=\ell} (\widehat{S}_{\ell_1}(X_1) \times_{M_1} \cdots \times_{M_{n-1}} \widehat{S}_{\ell_n}(X_n)) \tag{24.6}$$

and a similar formula for relative case (for (X_i, Z_i)).

Definition 24.18 We call $\widehat{S}_k(X,Z;\widehat{\mathcal{U}})$ the *normalized (codimension k) corner* of $(X,Z;\widehat{\mathcal{U}})$. The *normalized corner $\widehat{S}_k(U)$ of an orbifold U* is defined in the same way.

***Proof of Proposition* 24.17** Let M be a manifold with corners. We first define $\widehat{S}_k(M)$. Let $x \in \overset{\circ}{S}_m(M)$. We take a chart $\mathfrak{V}_x = (V_x, \psi_x)$ of product type $(V_x \cong [V_x] \times [0,1)^m)$. Let $A \subset \{1,\dots,m\}$ with $\#A = k$. A pair (x, A) becomes an element of $\widehat{S}_k(M)$.

We next define a topology on $\widehat{S}_k(M)$. Let $y = \psi(\tilde{y})$ with $\tilde{y} \in V_x$. We write $\tilde{y} = (\tilde{y}_0, (t_1,\dots,t_m))$. If $t_i = 0$ for all $i \in A$, we consider elements $y_A = (y, A) \in \widehat{S}_k(M)$ as follows. Suppose $B = \{i \mid t_i = 0\} \supset A$. Let W be a neighborhood of $(t_i)_{i \notin B}$ in $(0,1)^{\{1,\dots,m\}\setminus B}$. Then $[V] \times W \times [0,1)^B$ together with the restriction of

ψ_x is a chart of y. Thus we have $(y, A) \in \widehat{S}_k(U)$. We say (y^a, A) above converges to (x, A) if $\tilde{y}^a$ converges to o_x, where o_x is the point such that $\psi_x(o_x) = x$.

It is easy to see that $\widehat{S}_k(M)$ with this topology becomes a manifold with corners. This construction is canonical so that it induces one of orbifolds and of Kuranishi structures. (The proof of this part is entirely similar to the case of a normalized boundary of a manifold and so is omitted.)

We next construct the covering map $\pi_{\ell,k}$. We consider the case of manifolds. Let $x \in \overset{\circ}{S}_m(M)$ and let $\mathfrak{V}_x$, A be as above. For simplicity of notation, we put $A = \{1, \dots, k\}$. Suppose $(x, A) \in S_\ell(\widehat{S}_k(M))$. It implies $m \geq k + \ell$.

By definition the neighborhood of (x, A) in $\widehat{S}_k(M)$ is (y, A), where $y \in [V] \times \{(0, \dots, 0)\} \times [0, 1)^{m-k}$. Therefore a point $\tilde{x}$ in $\widehat{S}_\ell(\widehat{S}_k(M))$ such that $\pi_\ell(\tilde{x}) = (x, A)$ corresponds one-to-one to the set $A^+ \supset A$ with $\#A^+ = \ell + k$. We put $B = A^+ \setminus A$. We thus may regard $(x, A, B) \in \widehat{S}_\ell(\widehat{S}_k(M))$. (Here $\#B = \ell$.) Now we define the map $\pi_{\ell,k} : \widehat{S}_\ell(\widehat{S}_k(M)) \to \widehat{S}_{\ell+k}(M)$ by

$$\pi_{\ell,k}(x, A, B) = (x, A \cup B).$$

Given $(x, C) \in \widehat{S}_{\ell+k}(M)$, the element in the fiber of $\pi_{\ell,k}$ corresponds one-to-one to the partition of C into $A \cup B$ where $\#A = k$ and $\#B = \ell$. We can use this fact to show that $\pi_{\ell,k}$ is a covering map of covering index $(k + \ell)!/k!\ell!$.

We have thus constructed the covering map $\pi_{\ell,k}$ in the case of manifolds. To prove the case of orbifolds and of K-spaces, it suffices to observe that this construction is canonical and so is compatible with various kinds of coordinate changes.

Once the K-space $\widehat{S}_k(X, \widehat{\mathcal{U}})$ and the covering map $\pi_{\ell,k}$ are defined as above, it is very easy to check the properties (1)–(9). □

Remark 24.19 Like some other parts of this book, Proposition 24.17 is *not* new. In particular, we would like to mention that mostly the same construction appeared in D. Joyce's paper [Jo3]. (The article [Jo3] discusses the case of manifolds. However, it is straightforward to generalize the story to the case of K-spaces.) In [Jo3] the notion of the boundary ∂X is defined, which is the same as our definition of a normalized boundary. Then the action of Perm(k) on $\underbrace{\partial \cdots \partial}_{k \text{ times}} X$ is introduced. The quotient space $\underbrace{\partial \cdots \partial}_{k \text{ times}} X/\text{Perm}(k)$ (which is denoted by $C_k X$ in [Jo3]), coincides with our $\widehat{S}_k(X)$.

24.4 Finite Group Action on a K-Space

Definition 24.20 Let X be an orbifold and G a finite group. A G action on X as a topological space is said to be a *G action on the orbifold X* if the homeomorphism $X \to X$ induced by each element of G is a diffeomorphism of orbifolds.

Two actions are said to be *the same* if they are the same as maps $G \times X \to X$, set-theoretically.

Lemma 24.21 *Let X be an orbifold on which a finite group G acts (as an orbifold). Assume that the action is effective on each connected component. Then there exists a unique orbifold structure on X/G such that X is a covering space of X/G and the natural map $X \to X/G$ is a covering map.*

Proof Let $x \in X$ and $G_x = \{g \in G \mid gx = x\}$. We take an orbifold chart $\mathfrak{V}_x = (V_x, \Gamma_x, \phi_x)$ such that $U_x (= \phi_x(V_x) \cong V_x/\Gamma_x)$ is G_x-invariant (by using a G-invariant Riemannian metric, for example). Using the effectivity of G action on each connected component, we can easily show that the G_x action on U_x is effective. For each $g \in G_x$ we obtain a map $\varphi_g : V_x \to V_x$ and a group homomorphism $h_g : \Gamma_x \to \Gamma_x$. Since $\varphi_{g_1}\varphi_{g_2}$ induces the same continuous map as $\varphi_{g_1g_2}$ between the underlying topological spaces, there exists a unique element $\gamma_{g_1g_2g_3} \in \Gamma_x$ such that

$$\varphi_{g_1}\varphi_{g_2} = \gamma_{g_1g_2}\varphi_{g_1g_2}.$$

Moreover we have

$$h_{g_1}h_{g_2} = \text{conj}_{\gamma_{g_1g_2}} h_{g_1g_2}.$$

Note that φ_g is h_g equivariant. Then we define a group structure on the direct product set $\Gamma_x \times G_x$ by

$$(\gamma_1, g_1) \circ (\gamma_2, g_2) = (\gamma_1 h_{g_1}(\gamma_2)\gamma_{g_1,g_2}, g_1g_2). \tag{24.7}$$

We define $\cdot : (\Gamma_x \times G_x) \times V_x \to V_x$ by

$$(\gamma, g) \cdot x = \gamma(\varphi_g(x)).$$

Then we observe

$$\begin{aligned}&(\gamma_1, g_1) \cdot ((\gamma_2, g_2) \cdot x) = (\gamma_1, g_1) \cdot \gamma_2(\varphi_{g_2}(x)) = \gamma_1(\varphi_{g_1}(\gamma_2(\varphi_{g_2}(x))))\\ &= \gamma_1 h_{g_1}(\gamma_2)\varphi_{g_1}(\varphi_{g_2}(x)) = \gamma_1 h_{g_1}(\gamma_2)\gamma_{g_1g_2}(\varphi_{g_1g_2}(x)) = ((\gamma_1, g_1) \circ (\gamma_2, g_2)) \cdot x.\end{aligned}$$

It follows from effectivity that $\circ$ defines a group structure. We denote this group by $\Gamma_x \tilde{\times} G_x$.

Now we define $\overline{\phi}_x : V_x \to X/G$ by the composition of $\phi_x : V_x \to X$ and the projection $X \to X/G$. Then it is easy to see that $(V_x, \Gamma_x \tilde{\times} G_x, \overline{\phi}_x)$ defines an orbifold structure on X/G. The rest of the proof is obvious. □

Now we define the definition of the action of a finite group on K-space.

Definition 24.22 Let $(X, \widehat{\mathcal{U}})$ be a K-space. An *automorphism* Φ consists of a pair $(|\Phi|, \{\Phi_p\})$ of a homeomorphism $|\Phi| : X \to X$ and an assignment $X \ni p \mapsto \Phi_p = (\varphi_p, \widehat{\varphi}_p)$ with the following properties:

(1) $\varphi_p : U_p \to U_{|\Phi|(p)}$ is a diffeomorphism of orbifolds.
(2) $\widehat{\varphi}_p : E_p \to E_{|\Phi|(p)}$ is an isomorphism of vector bundles over φ_p.
(3) $\widehat{\varphi}_p \circ s_p = s_{|\Phi|(p)} \circ \varphi_p$ holds on U_p.
(4) $|\Phi| \circ \psi_p = \psi_{|\Phi|(p)} \circ \varphi_p$ holds on $s_p^{-1}(0)$.
(5) Let $q \in \psi_p(s_p^{-1}(0))$. Suppose $\Phi_{pq} = (U_{pq}, \varphi_{pq}, \widehat{\varphi}_{pq})$ and

$$\Phi_{|\Phi|(p)|\Phi|(q)} = (U_{|\Phi|(p)|\Phi|(q)}, \varphi_{|\Phi|(p)|\Phi|(q)}, \widehat{\varphi}_{|\Phi|(p)|\Phi|(q)})$$

are the coordinate changes. Then we have the following:

(a) $\varphi_q(U_{pq}) = U_{|\Phi|(p)|\Phi|(q)}$.
(b) $\varphi_{|\Phi|(p)|\Phi|(q)} \circ \varphi_q = \varphi_p \circ \varphi_{pq}$.
(c) $\widehat{\varphi}_{|\Phi|(p)|\Phi|(q)} \circ \widehat{\varphi}_q = \widehat{\varphi}_p \circ \widehat{\varphi}_{pq}$.

We say $(|\Phi|, \{\Phi_p\})$ is *the same* as $(|\Phi'|, \{\Phi'_p\})$ if $|\Phi| = |\Phi'|$ and $\Phi_p = \Phi'_p$ for all p.

Remark 24.23

(1) The equality $\Phi_p = \Phi'_p$ has an obvious meaning. Namely, we defined the notion of two diffeomorphisms and bundle isomorphisms of orbifolds to be the same. (That is, they coincide set-theoretically.)
(2) It happens that two automorphisms Φ and Φ' with the same underlying homeomorphisms $|\Phi|$ and $|\Phi'|$ could be different.

Definition-Lemma 24.24

(1) We can compose two automorphisms of K-spaces. The composition is again an automorphism.
(2) The set of automorphisms of a given K-space $(X, \widehat{\mathcal{U}})$ is a group whose product is the composition of automorphisms. We denote this group by $\mathrm{Aut}(X, \widehat{\mathcal{U}})$.
(3) An action of a finite group G on $(X, \widehat{\mathcal{U}})$ is, by definition, a group homomorphism $G \to \mathrm{Aut}(X, \widehat{\mathcal{U}})$.
(4) A G action on $(X, \widehat{\mathcal{U}})$ induces a G action on the underlying topological space X.

Definition 24.25 An action of a finite group G on a K-space $(X, \widehat{\mathcal{U}})$ is said to be *effective* if the following holds for each $p \in X$.

We put $G_p = \{g \in G \mid gp = p\}$. Let $U_p = V_p/\Gamma_p$ be the Kuranishi neighborhood of p. By definition, G_p acts on U_p. We require that this action is effective on each connected component of U_p.

Lemma 24.26 *Suppose a finite group G acts effectively on a K-space* $(X, \widehat{\mathcal{U}})$. *Then there exists a unique Kuranishi structure on* X/G *such that the projection* $X \to X/G$ *is an underlying map of the covering map and each* $\mathcal{U}_p$ *gives a covering chart of this covering.*

Proof This follows from Lemma 24.21. □

We remark that the theory of covering space of algebraic stack is deeper and harder ([Gro]). Our story of covering space of K-space is much simpler in comparison.

Chapter 25
Admissible Kuranishi Structures

In this chapter we introduce the notion of an admissible Kuranishi structure. For this purpose we introduce the notion of an admissible orbifold, an admissible vector bundle, and various admissible objects associated to them, like an admissible section, and provide their fundamental properties. Roughly speaking, 'admissibility' in this chapter is some condition that objects in question obey certain exponential decay estimates at asymptotic ends. Here we regard boundary or corner points as the end points, so the 'exponential decay estimates at asymptotic ends' means the exponential decay estimates in the direction normal to boundary or corners.

25.1 Admissible Orbifolds

Firstly, we discuss admissibility for the case of manifolds with corners before going to the case of orbifolds with corners, because the key idea can be already seen for the case of manifolds.

Situation 25.1 Let $V \subset [V] \times [0,1)^k$ be an open subset where $[V]$ is a manifold without corners. We put $\mathbf{t} = (t_1, \dots, t_k) \in [0,1)^k$. ■

Convention 25.2

(1) We put

$$T_i = e^{1/t_i}, \qquad (\text{i.e. } t_i = \frac{1}{\log T_i}). \tag{25.1}$$

We consider the corner structure stratification of V. Then each connected component of open stratum $\overset{\circ}{S}_\ell V$ has coordinates that are the union of the coordinate of $[V]$ and $k - \ell$ of T_i's. Here $T_i \in [1, \infty)$. We consider the C^m norm of a function $f : V \to \mathbb{R}$ stratumwise (i.e., the norm of the differential

K. Fukaya et al., *Kuranishi Structures and Virtual Fundamental Chains*, Springer Monographs in Mathematics, https://doi.org/10.1007/978-981-15-5562-6_25

in the stratum direction) using the above coordinate T_i. (Namely, we use T_i and not t_i to define the C^m norm.)

(2) For a function $f : V \to \mathbb{R}$, we denote by $|f|_{C^m}$ the pointwise C^m norm in the above sense. Thus $|f|_{C^m}$ can be regarded as a non-negative real-valued function on V. To define it we use a certain Riemannian metric on $[V]$. (We use the standard metric for $T_i \in [0, \infty)$.) Since we only consider its value on a compact subset, the difference of the metric affects it only by a bounded ratio. So we do not need to care about the difference of the metric.

Definition 25.3

(1) We say a function $f : V \to \mathbb{R}$ is *admissible* if for each compact subset K and $m > 0$, there exist $\sigma(m, K) > 0$ and $C(m, K) > 0$ such that the following holds for each i:

$$\left| \frac{\partial f}{\partial T_i} \right|_{C^m} \le C(m, K) e^{-\sigma(m,K)T_i}. \tag{25.2}$$

(2) We say a function $f : V \to \mathbb{R}$ is *exponentially small near the boundary* if for each compact subset K and $m > 0$, there exist $\sigma(m, K) > 0$ and $C(m, K) > 0$ such that the following holds for each i:

$$|f|_{C^m} \le C(m, K) e^{-\sigma(m,K)T_i}. \tag{25.3}$$

Example 25.4 If $k = 1$, an admissible function $f(\overline{y}, t)$ is written as in the form

$$f(\overline{y}, t) = f_0(\overline{y}) + f_1(\overline{y}, t)$$

such that $f_1(\overline{y}, 1/\log T)$ decays in an exponential order in T. (Here $t = 1/\log T$.)

Remark 25.5 We remark that (25.2) implies that f is a smooth function on $[V] \times [0, 1)^k$. Moreover the following holds. Let $x_1, \dots, x_{n-k}$ be coordinates of $[V]$ and $t_1 \dots, t_k$ the standard coordinates of $[0, 1)^k$ and $m_1, \dots, m_n \in \mathbb{Z}_{\ge 0}$ with $m_k > 0$. Then

$$\left| \frac{\partial^{m_1+\dots+m_n} f}{\partial^{m_1} t_1 \dots \partial^{m_k} t_k \partial^{m_{k+1}} x_1 \dots \partial^{m_n} x_{n-k}} \right| \le C'(m, K) e^{-\sigma'(m,K)/t_k} \tag{25.4}$$

This is an easy consequence of the chain rule.

In a way similar to Example 25.4, we can prove the following.

Lemma 25.6 *On a subset $K \times [0, c)^k$, any admissible function f is written uniquely in the following form:*

$$f(\overline{y}, (t_1, \dots, t_k)) = \sum_{I \subseteq \{1,\dots,k\}} f_I(\overline{y}, t_I). \tag{25.5}$$

Here $t_I = (t_i)_{i\in I}$ *and* f_I *is a function on* $K \times [0,c)^I$ *which is exponentially small near the boundary.*

Proof Firstly we observe that the set of functions of the form (25.5) forms an $\mathbb{R}$ vector space. We also note that if $[V] \times [0,c)^k \to [V] \times [0,c)^I$ is a projection then the pullback of admissible functions are admissible. The same holds for 'exponentially small near the boundary' and 'of the form (25.5)'.

Sublemma 25.7 *If an admissible function is zero on the boundary, then it is exponentially small near the boundary.*

Proof We have

$$f(\overline{y}; (t_1, \ldots, t_k)) = -\int_{T_i=e^{1/t_i}}^{\infty} \frac{\partial f}{\partial T_i} dT_i.$$

This implies the sublemma. □

We will prove, by an upward induction on m, that for $\#I \le m$ there exists a function f_I on $K \times [0,c)^I$ which is exponentially small near the boundary and such that the equality

$$f(\overline{y}, (t_1, \ldots, t_k)) - \sum_{I \subseteq \{1,\ldots,k\}, \#I \le \ell} f_I(\overline{y}, t_I) = 0 \tag{25.6}$$

holds on $S_{k-\ell}(K \times [0,c)^k)$ for $\ell = 0, \ldots, m$. The lemma is the case $m = k$.

For $m = 0$ we put $f_\emptyset(\overline{y}) = f(\overline{y}; (0, \ldots, 0))$.

As an induction hypothesis, we assume that we obtained $f_{I'}$ for I' with $\#I' < m$ which is exponentially small near the boundary and such that (25.6) holds for $\ell < m$. Let $I \subset \{1, \ldots, k\}$ with $\#I = m$. We may replace f by

$$f'(\overline{y}, (t_1, \ldots, t_k)) = f(\overline{y}, (t_1, \ldots, t_k)) - \sum_{I \subseteq \{1,\ldots,k\}, \#I < m} f_I(\overline{y}, t_I)$$

and may assume that f is zero on $S_{k-m+1}(K \times [0,c)^k)$. We embed $K \times [0,c)^I$ into $K \times [0,c)^k$ by putting $t_i = 0$ for $i \notin I$. Restricting f to its image we obtain an admissible function on $K \times [0,c)^I$. Since we assumed f is zero on $S_{k-m+1}(K \times [0,c)^k)$, then $f = 0$ on $\partial(K \times [0,c)^I)$. We define $f_I = f|_{K \times [0,c)^I}$. Then f_I is exponentially small near the boundary and therefore its pullback to $K \times [0,c)^k$ is of the form (25.5). By taking

$$S_{k-m}(K \times [0,c)^k) = \bigcup_{I \subseteq \{1,\ldots,k\}, \#I = m} K \times [0,c)^I$$

into account, we can see that

$$f - \sum_{I \subseteq \{1,\ldots,k\}, \#I = m} f_I$$

is zero on $S_{k-m}(K \times [0,c)^k)$. The proof of the existence part of the lemma is complete by induction.

We next prove uniqueness. Let

$$\sum_{I \subseteq \{1,\ldots,k\}} f_I(\overline{y}, t_I) = \sum_{I \subseteq \{1,\ldots,k\}} f'_I(\overline{y}, t_I), \tag{25.7}$$

where f_I, f'_I are exponentially small near the boundary. We will prove $f_I = f'_I$ by an upward induction on $\#I$. We restrict this equality at $(0,\ldots,0)$ and obtain $f_\emptyset = f'_\emptyset$. Therefore we may assume $f_\emptyset = f'_\emptyset = 0$. Suppose $f_I = f'_I$ for $\#I < k$. Then we may assume $f_I = f'_I = 0$ for $\#I < k$. Let $\#I = k$. Then we restrict the equality to $K \times [0,1)^I$ and find $f_I = f'_I$ for this I. The proof of uniqueness is complete by induction. □

Definition 25.8

(1) Let $V_i \subset [V_i] \times [0,1)^k$ $(i = 1,2)$ be open subsets as in Situation 25.1 and let $\varphi_{21} : V_1 \to V_2$ be an embedding of manifolds. We say that φ_{21}, $\varphi_{21}(\overline{y}, (t_1,\ldots,t_k)) = (\overline{y}', (t'_1,\ldots t'_k))$, is an admissible embedding if there exists a permutation $\sigma : (1,\ldots,k) \to (1,\ldots,k)$ such that:
   (a) The coordinates of $\overline{y}' = \overline{y}'(\overline{y}, (t_1,\ldots,t_k))$ are admissible functions in the sense of Definition 25.3.
   (b) We put $T'_i = e^{1/t'_i}$, where $t'_i = t'_i(\overline{y}, (t_1,\ldots,t_k))$. Then for each i, $T'_i - T_{\sigma^{-1}(i)}$ is an admissible function in the sense of Definition 25.3 near the boundary.[1]
(2) An admissible embedding is said to be an *admissible diffeomorphism* if it is also a homeomorphism.[2]
(3) An action of a finite group Γ on $V \subset [V] \times [0,1)^k$ is said to be an *admissible action* if each element of Γ induces an admissible diffeomorphism.
(4) An orbifold chart in the sense of Definition 23.1 (1) is said to be an *admissible chart* if it is of product type (Notation 23.16) and the Γ action is admissible.

Remark 25.9

(1) In the geometric context of pseudo-holomorphic curves we took T to be the 'length' of the neck region in [FOOO4] etc. For this choice, Definition 25.8 (1)(b) is satisfied. See Sect. 25.3.

[1] When $\sigma : \{1,\ldots,k\} \to \{1,\ldots,k\}$ acts by permutation of the factors on $[0,\infty)^k$ the i-th factor of $\sigma \cdot (T_1,\ldots,T_k)$ is $T_{\sigma^{-1}(i)}$.

[2] See Lemma 25.13.

(2) The choice of the coordinate $t = 1/T$ used in [FOOO4] is different from that of (25.1). See also Sect. 25.3 for this point.

Lemma 25.10 *Suppose* $V_i = [V_i] \times [0,1)^k$ *for* $i = 1, 2$. *Let* $\varphi_{21} : V_1 \to V_2$ *be an admissible embedding as in Definition* 25.8.

(1) *An admissible embedding* φ_{21} *induces a smooth embedding*

$$\varphi_{21} : [V_1] \times [0,1)^k \to [V_2] \times [0,1)^k.$$

(2) *Denote by* φ_{21}^j *the* j*-th component of the* $[0,1)^k$ *factor of* φ_{21}. *If we put*

$$R_i = \frac{1}{t_i}, \tag{25.8}$$

then for each compact set $K \subset V$ *and* $m \geq 0$ *there exist* $C(m, K) > 0$ *and* $\sigma(m, K) > 0$ *such that*

$$\left\| \varphi_{21}^i - t_{\sigma^{-1}(i)} \right\|_{C_K^{m,R}} \leq C(m, K) e^{-\sigma(m,K) R_i}. \tag{25.9}$$

Here $\sigma^{-1}(i) \in \{1, \ldots, k\}$ *and we identify* $V \cong [V] \times (1, \infty]^k$ *using* R_i *as the coordinates of the second factor and* $C_K^{m,R}$ *stands for the* C^m *norm on* K *with respect to the* R_i *coordinates.*

Proof If $\sigma(i) = i$ for all i in addition, then it is easy to see that an admissible embedding induces a smooth embedding. So it suffices to consider the case when $\overline{y}' = \overline{y}$ and $\sigma(i) = i$. We also note that (1) follows from (2). Then by Definition 25.8 (1)(b) and (25.1) we have

$$t_i' = (\log(e^{\frac{1}{t_i}} + f_i(x, (t_1, \ldots, t_k))))^{-1}$$

for some admissible function f_i. The right hand side is equal to

$$\frac{t_i}{1 + t_i \log(1 + e^{-R_i} f_i(x, (t_1, \ldots, t_k)))}.$$

Statement (2) easily follows from this formula.[3] □

Remark 25.11 The admissible function f_i above may not be zero at $t_i = 0$. However, $\log(1 + e^{-R_i} f_i(x, (t_1, \ldots, t_k)))$ goes to 0 in exponential order as $R_i \to \infty$. Therefore we can smoothly extend φ_{21} to a *collared* neighborhood. Here is the key point that coordinate changes of Kuranishi structures can be extended

[3]Recall $T_i = e^{R_i}$. Therefore a function which decays in exponential order in T_i coordinates also decays in exponential order in R_i coordinates.

smoothly to a collared neighborhood, once we establish the exponential decay estimate (25.9) of the *coordinate changes* with respect to the R_i *coordinates*. See also Remark 25.39. On the other hand, we recall from Definition 25.3 that admissible *functions* are required that their derivatives[4] with respect to the T_i *coordinates* satisfy the exponential decay estimate.

Next, we go to the case of an orbifold with corners.

Definition 25.12

(1) Two admissible charts are said to be *compatible as admissible charts* if the diffeomorphism $\tilde{\varphi}$ in Definition 23.1 (3) is admissible.
(2) A representative of an orbifold structure (with boundary or corners) is said to be a *representative of an admissible orbifold structure* if:

   (a) Each member is an admissible chart.
   (b) Two of them are compatible as admissible charts.

(3) In Definition 23.2, suppose $\{(V_i^X, \Gamma_i^X, \phi_i^X) \mid i \in I\}$ and $\{(V_j^Y, \Gamma_j^Y, \phi_j^Y) \mid j \in J\}$ are representatives of admissible orbifold structures. Then the embedding f in Definition 23.2 (2) is said to be an *admissible embedding* if $\tilde{f}_{p;ji}$ in Definition 23.2 (2) is an admissible embedding in the sense of Definition 25.8 (1).

Lemma 25.13

(1) *Composition of admissible embeddings is an admissible embedding.*
(2) *The identity map is an admissible embedding.*
(3) *If an admissible embedding is a homeomorphism, the inverse is also an admissible embedding.*

The proof is obvious.

Definition 25.14

(1) We say an admissible embedding of orbifolds is an *admissible diffeomorphism* if it is a homeomorphism in addition.
(2) We say that two representatives of admissible orbifold structures on X are *equivalent* if the identity map regarded as a map between X equipped with those two representatives of admissible orbifold structures is an admissible diffeomorphism. This is an equivalence relation by Lemma 25.13.
(3) An equivalence class of the equivalence relation (2) is called an *admissible orbifold structure* on X.
(4) An orbifold X with an admissible orbifold structure is called an *admissible orbifold*.
(5) The condition for a map $X \to Y$ to be an admissible embedding does not change if we replace representatives of admissible orbifold structures with

[4]We assume $m > 0$ in Definition 25.3, while we assume $m \geq 0$ in (25.9).

equivalent ones. So we can define the notion of an *admissible embedding of admissible orbifolds*.

(6) If U is an open subset of an admissible orbifold X, then there exists a unique admissible orbifold structure on U such that the inclusion $U \to X$ is an admissible embedding. We call U with this admissible orbifold structure an *open admissible suborbifold*.

(7) A direct product of admissible orbifolds has a structure of an admissible orbifold in an obvious way.

Definition 25.15

(1) Let X be an admissible orbifold. An admissible orbifold chart (V, Γ, ϕ) of underlying topological space X is called an *admissible orbifold chart of orbifold* X if the map $V/\Gamma \to X$ induced by ϕ is an admissible embedding of orbifolds.

(2) Hereafter when X is an admissible orbifold, an admissible orbifold chart always means an admissible orbifold chart of orbifold X in the sense of (1).

(3) In the case when an admissible orbifold structure on X is given, a representative of its admissible orbifold structure is called an *admissible orbifold atlas*.

(4) Two admissible orbifold charts (V_i, Γ_i, ϕ_i) are said to be *isomorphic* if there exists a group isomorphism $h : \Gamma_1 \to \Gamma_2$ and an h equivariant admissible diffeomorphism $\varphi : V_1 \to V_2$ such that $\phi_2 \circ \varphi = \phi_1$. The pair (h, φ) is called an *admissible isomorphism* or *admissible coordinate change* between two admissible orbifold charts.

The proofs of the following lemmas are obvious from the definition.

Lemma 25.16 *Suppose V, V_1, V_2 are as in Situation* 25.1.

(1) *A restriction of an admissible function on V to its open subset is also admissible.*

(2) *Let $f : V_2 \to \mathbb{R}$ be an admissible function and $\varphi_{21} : V_{21} \to V_1$ an admissible embedding, then the composition $f \circ \varphi_{21} : V_1 \to \mathbb{R}$ is admissible.*

Lemma 25.17 *A subchart (in the sense of Definition* 23.1 *(2)) of an admissible chart is also admissible.*

Definition 25.18 Let X be an admissible orbifold. A function $f : X \to \mathbb{R}$ is said to be *admissible* if for all the orbifold charts (V, Γ, ϕ) of X the composition $f \circ \phi : V \to \mathbb{R}$ is admissible in the sense of Definition 25.3.

A smooth map $f : X \to M$ from an admissible orbifold to a manifold without boundary[5] is said to be admissible if for any smooth function $g : M \to \mathbb{R}$ the composition $g \circ f : X \to \mathbb{R}$ is an admissible function.

Lemma 25.19 *Let X be an admissible orbifold and $f : X \to \mathbb{R}$ a function.*

(1) *If there exists a representative $\{(V_i, \Gamma_i, \phi_i) \mid i \in I\}$ of the orbifold structure on X such that $f \circ \phi_i : V_i \to \mathbb{R}$ is admissible for any i, then f is admissible.*

[5]The case when M has a boundary and corners is discussed in Chap. 26.

(2) *The composition of an admissible embedding and an admissible function (resp. a function which is exponentially small near the boundary) is again admissible (resp. a function which is exponentially small near the boundary).*

The same statement holds for admissibility of a map to a smooth manifold.

This is a consequence of Lemma 25.16. We can also prove the next lemma easily.

Lemma 25.20 *For any locally finite open cover of admissible orbifold, there exists a partition of unity subordinate to it that consists of admissible functions.*

25.2 Admissible Vector Bundles

We next describe the admissible version of Definition 23.18.

Definition 25.21 Let $(X, \mathcal{E}, \pi)$ be a pair of admissible orbifolds X and $\mathcal{E}$ with a continuous map $\pi : \mathcal{E} \to X$ between their underlying topological spaces. Hereafter we write $(X, \mathcal{E})$ in place of $(X, \mathcal{E}, \pi)$.

(1) An *admissible orbifold chart* of $(X, \mathcal{E})$ is a quintuple $(V, E, \Gamma, \phi, \widehat{\phi})$ with the following properties:

(a) $\mathfrak{V} = (V, \Gamma, \phi)$ is an admissible orbifold chart of X.
(b) E is a finite-dimensional vector space equipped with a linear Γ action.
(c) $(V \times E, \Gamma, \widehat{\phi})$ is an admissible orbifold chart of $\mathcal{E}$.
(d) The diagram below commutes set-theoretically:

$$\begin{array}{ccc} V \times E & \xrightarrow{\widehat{\phi}} & \mathcal{E} \\ \downarrow & & \downarrow \pi \\ V & \xrightarrow{\phi} & X \end{array} \tag{25.10}$$

Here the left vertical arrow is the projection to the first factor.

(2) In the situation of (1), let $p \in V$ and $(V_p, \Gamma_p, \phi|_{V_p})$ be a subchart of (V, Γ, ϕ) in the sense of Definition 23.1 (2). Then $(V_p, E, \Gamma_p, \phi|_{V_p}, \widehat{\phi}|_{V_p \times E})$ is an admissible orbifold chart of $(X, \mathcal{E})$. We call it a *subchart* of $(V, E, \Gamma, \phi, \widehat{\phi})$.

(3) Let $(V_i, E_i, \Gamma_i, \phi_i, \widehat{\phi}_i)$ $(i = 1, 2)$ be admissible orbifold charts of $(X, \mathcal{E})$. We say that they are *compatible as admissible charts* if the following holds for each $p_1 \in V_1$ and $p_2 \in V_2$ with $\phi_1(p_1) = \phi_2(p_2)$: There exist open neighborhoods $V^i_{p_i}$ of $p_i \in V_i$ such that:

(a) There exists an isomorphism $(h, \varphi) : (V_1, \Gamma_1, \phi_1)|_{V^1_{p_1}} \to (V_2, \Gamma_2, \phi_2)|_{V^2_{p_2}}$ of admissible orbifold charts of X, which are subcharts.

(b) There exists an isomorphism $(h, \widehat{\varphi}) : (V_1 \times E_1, \Gamma_1, \widehat{\phi}_1)|_{V^1_{p_1} \times E^1} \to (V_2 \times E_2, \Gamma_2, \widehat{\phi}_2)|_{V^2_{p_2} \times E^2}$ of admissible orbifold charts of $\mathcal{E}$, which are subcharts.

(c) For each $y \in V^1_{p_1}$ the map $E_1 \to E_2$ given by $\xi \to \pi_{E_2}\widehat{\varphi}(y, \xi)$ is a linear isomorphism. Here $\pi_{E_2} : V^2_{p_2} \times E_2 \to E_2$ is the projection.

(d) Each component of the map $V^1_{p_1} \times E_1 \to E_2$, which is the composition of $\widehat{\varphi}$ and the projection $V^2_{p_2} \times E_2 \to E_2$, is an admissible function.

(4) A *representative of an admissible vector bundle structure* on $(X, \mathcal{E})$ is a set of admissible orbifold charts $\{(V_i, E_i, \Gamma_i, \phi_i, \widehat{\phi}_i) \mid i \in I\}$ such that any two of the charts are compatible in the sense of (3) above and

$$\bigcup_{i\in I} \phi_i(V_i) = X, \quad \bigcup_{i\in I} \widehat{\phi}_i(V_i \times E_i) = \mathcal{E},$$

are locally finite open coverings.

Definition 25.22 Let $(X^*, \mathcal{E}^*)$ $(* = a, b)$ have representatives of vector bundle structures $\{(V_i^*, E_i^*, \Gamma_i^*, \phi_i^*, \widehat{\phi}_i^*) \mid i \in I^*\}$, respectively. A pair $(f, \widehat{f})$ of admissible orbifold embeddings, $f : X^a \to X^b$, $\widehat{f} : \mathcal{E}^a \to \mathcal{E}^b$, is said to be an *admissible embedding of admissible vector bundles* if the following holds:

(1) Let $p \in V_i^a$, $q \in V_j^b$ with $f(\phi_i^a(p)) = \phi_j^b(q)$. Then there exist admissible open subcharts $(V^a_{i,p} \times E^a_{i,p}, \Gamma^a_{i,p}, \widehat{\phi}^a_{i,p})$ and $(V^b_{j,q} \times E^b_{j,q}, \Gamma^b_{j,q}\widehat{\phi}^b_{j,q})$ and a local representative $(h_{p;ji}, f_{p;ji}, \widehat{f}_{p;ji})$ of the embeddings f and $\widehat{f}$ such that the following holds. For each $y \in V_i^a$ the map $\xi \mapsto \pi_2(\widehat{f}_{p;ji}(y, \xi))$, $E_1^a \to E_2^b$ is a linear embedding.

(2) Each component of the map $\pi_2 \circ \widehat{f}_{p;ji} : V_1^a \times E_1^a \to E_2^b$ is admissible.

(3) The diagram below commutes set-theoretically:

$$\begin{array}{ccc} \mathcal{E}^a & \xrightarrow{\widehat{f}} & \mathcal{E}^b \\ \pi \downarrow & & \downarrow \pi \\ X^a & \xrightarrow{f} & X^b \end{array} \tag{25.11}$$

Two orbifold embeddings are said to be *equal* if they coincide set-theoretically as a pair of maps.

Lemma 25.23

(1) *The composition of admissible embeddings of vector bundles is an admissible embedding.*

(2) *The pair of identity maps* (id, $\widehat{\text{id}}$) *is an admissible embedding.*

(3) *If an admissible embedding of vector bundles is a pair of homeomorphisms, then the pair of their inverses is also an admissible embedding.*

The proof is obvious.

Definition 25.24 Let $(X, \mathcal{E})$ be as in Definition 23.18.

(1) An admissible embedding of vector bundles is said to be an *isomorphism* if it is a pair of admissible diffeomorphisms of admissible orbifolds.
(2) We say that two representatives of an admissible vector bundle structure on $(X, \mathcal{E})$ are *equivalent* if the pair of identity maps regarded as a map between $(X, \mathcal{E})$ equipped with those two representatives of vector bundle structures is an admissible embedding. This is an equivalence relation by Lemma 25.23.
(3) An equivalence class of the equivalence relation (1) is called an *admissible vector bundle structure* on $(X, \mathcal{E})$.
(4) A pair $(X, \mathcal{E})$ together with its admissible vector bundle structure is called an *admissible vector bundle* on X. We call $\mathcal{E}$ the *total space*, X the *base space*, and $\pi : \mathcal{E} \to X$ the *projection*.
(5) The condition for $(f, \widehat{f}) : (X^a, \mathcal{E}^a) \to (X^b, \mathcal{E}^b)$ to be an admissible embedding does not change if we replace representatives of admissible vector bundle structures with equivalent ones. So we can define the notion of an *admissible embedding of vector bundles*.
(6) We say $(f, \widehat{f})$ is an admissible embedding *over the admissible orbifold embedding f.*

Definition 25.25

(1) Let $(X, \mathcal{E})$ be an admissible vector bundle. We call an admissible orbifold chart $(V, E, \Gamma, \phi, \widehat{\phi})$ in the sense of Definition 25.21 (1) of underlying pair of topological spaces $(X, \mathcal{E})$ an *admissible orbifold chart of an admissible vector bundle* $(X, \mathcal{E})$ if the pair of maps $(\overline{\phi}, \overline{\widehat{\phi}}) : (V/\Gamma, (V \times E)/\Gamma) \to (X, \mathcal{E})$ induced from $(\phi, \widehat{\phi})$ is an admissible embedding of admissible vector bundles.
(2) If $(V, E, \Gamma, \phi, \widehat{\phi})$ is an admissible orbifold chart of an admissible vector bundle, we call a pair $(E, \widehat{\phi})$ a *trivialization* of our admissible vector bundle on V/Γ.
(3) Hereafter when $(X, \mathcal{E})$ is an admissible vector bundle, its 'admissible orbifold chart' always means an admissible orbifold chart of an admissible vector bundle in the sense of (1).
(4) In then case when an admissible vector bundle structure on $(X, \mathcal{E})$ is given, a representative of this admissible vector bundle structure is called an *admissible vector bundle atlas* of $(X, \mathcal{E})$.
(5) Two admissible orbifold charts $(V_i, E_i, \Gamma_i, \phi_i, \widehat{\phi}_i)$ of an admissible vector bundle are said to be *isomorphic* if there exist an isomorphism $(h, \tilde{\varphi})$ of admissible orbifold charts $(V_1, \Gamma_1, \phi_1) \to (V_2, \Gamma_2, \phi_2)$ and an admissible isomorphism $(h, \tilde{\hat{\varphi}})$ of admissible orbifold charts $(V_1 \times E_1, \Gamma_1, \widehat{\phi}_1) \to (V_2 \times E_2, \Gamma_2, \widehat{\phi}_2)$ such that they induce an admissible embedding of admissible vector bundles $(\varphi, \hat{\varphi}) : (V_1/\Gamma_1, (V_1 \times E_1)/\Gamma_1) \to (V_2/\Gamma_2, (V_2 \times E_2)/\Gamma_2)$. The triple $(h, \tilde{\varphi}, \tilde{\hat{\varphi}})$ is called an *admissible isomorphism* or *admissible coordinate change* between admissible orbifold charts of the admissible vector bundle.

Once we have established these basic notions related to admissible vector bundles as above, the next lemma obviously holds.

Lemma 25.26 *The tangent bundle of an admissible orbifold has a canonical structure of an admissible vector bundle.*

Taking the Whitney sum, tensor product, dual, exterior power, and quotient of admissible vector bundles preserves admissibility.

Lemma-Definition 25.27 *If $(X^b, \mathcal{E}^b)$ is an admissible vector bundle and $f : X^a \to X^b$ is an admissible embedding, then the pullback $f^*\mathcal{E}^b$ has a unique structure of an admissible vector bundle such that the embedding of vector bundles $(f, \widehat{f}) : (X^a, f^*\mathcal{E}^b) \to (X^b, \mathcal{E}^b)$ becomes an admissible embedding.*

We call $f^*\mathcal{E}^b$ equipped with this admissible vector bundle structure the pullback in the sense of admissible vector bundles.

The proof is straightforward, so is omitted. The following lemmas are also straightforward consequences of the definitions.

Lemma 25.28 *A covering space $\tilde{X}$ of an admissible orbifold X has a canonical structure of an admissible orbifold such that admissible functions of X are pulled back to admissible functions.*

Lemma 25.29 *The normalized boundary or corner of an admissible orbifold is admissible. The covering maps in Lemma* 24.16 *or Proposition* 24.17 *are admissible.*

Definition 25.30 An *admissible section* of an admissible vector bundle $(X, \mathcal{E})$ is an admissible embedding of orbifolds $s : X \to \mathcal{E}$ such that the composition of s and the projection is the identity map set-theoretically.

The next lemma obviously follows from the definition.

Lemma 25.31 *Let $\{(V_{\mathfrak{r}}, E_{\mathfrak{r}}, \Gamma_{\mathfrak{r}}, \phi_{\mathfrak{r}}, \widehat{\phi}_{\mathfrak{r}})\}$ be an admissible orbifold atlas of the admissible bundle $(X, \mathcal{E})$. Suppose $s_{\mathfrak{r}}$ is a representative of a section s of $\mathcal{E}$ in the sense of Definition* 23.33. *Then s is admissible if and only if $s_{\mathfrak{r}} : V_{\mathfrak{r}} \to E_{\mathfrak{r}} \cong \mathbb{R}^m$ is admissible for any $\mathfrak{r}$.*

Lemma 25.32 *Let s be an admissible section of an admissible bundles $(X, \mathcal{E})$ over an admissible orbifold. We assume that s is transversal to* 0. *Then $s^{-1}(0) \subset X$ has a structure of an admissible orbifold such that $s^{-1}(0) \to X$ is an admissible embedding.*

Proof Let $p \in s^{-1}(0)$. We take admissible coordinates $(V_{\mathfrak{r}}, E_{\mathfrak{r}}, \Gamma_{\mathfrak{r}}, \phi_{\mathfrak{r}}, \widehat{\phi}_{\mathfrak{r}})$ such that $\phi_{\mathfrak{r}}(o) = p$. Here $o = (0, \dots, 0)$. We put $V_{\mathfrak{r}} = [V_{\mathfrak{r}}] \times [0, 1)^k$ and let $x_1, \dots, x_{n-k}$ be coordinates of $[V_{\mathfrak{r}}]$ and $t_1, \dots, t_k$ the standard coordinates of $[0, 1)^k$. We identify $E_{\mathfrak{r}} = \mathbb{R}^d$. Let $s_{\mathfrak{r}} = (s_{\mathfrak{r}}^1, \dots, s_{\mathfrak{r}}^d)$ be the local representative of s. We may choose $x_1, \dots, x_k$ such that the matrix $(\partial s_{\mathfrak{r}}^i / \partial x_j)_{i,j=1}^m$ is invertible at o. Then by the implicit function theorem $s_{\mathfrak{r}}^{-1}(0)$ is written as the image of a map

$$g : \Omega \to [V_{\mathfrak{r}}] \times [0, 1)^k$$

such that Ω is a neighborhood of 0 in $\mathbb{R}^{n-k-d}$ and $g(y) = (g_1(y), y, g_2(y))$. Moreover g_1, g_2 are exponentially small near the boundary. We may choose the coordinate x_i so that g becomes $\Gamma_{\mathfrak{r}}$ equivariant. Then $\phi \circ g$ gives an orbifold chart of $s^{-1}(0)$. Using the fact that g_1, g_2 are exponentially small near the boundary we can easily show that the coordinate changes of such coordinates are admissible and $s^{-1}(0) \to X$ is an admissible embedding. □

In a similar way we can show the next lemma.

Lemma 25.33 *Let X be an admissible orbifold and $f : X \to M$ an admissible smooth map to a manifold without boundary. Let $g : N \to M$ be a smooth map between manifolds. We assume that f is transversal to g. (Definition 4.2.)*

Then the fiber product $X\,{}_f\times_g N$ has a structure of an admissible orbifold such that $X\,{}_f\times_g N \to X \times N$ is an admissible embedding of admissible orbifolds.

We can use Lemma 25.33 to define a fiber product of K spaces in the admissible category.

We next define admissibility of differential forms. We consider the dual to the tangent bundle TX of X and its ℓ-th exterior power $\Lambda^\ell X$. It is an admissible vector bundle. Let $(V_{\mathfrak{r}}, \Gamma_{\mathfrak{r}}, \phi_{\mathfrak{r}})$ be an admissible coordinate of X. We write $V_{\mathfrak{r}} = [V_{\mathfrak{r}}] \times [0,1)^k$ and let $t_1, \dots, t_k$ be the coordinate of $[0,1)^k$ and $x_1, \dots, x_{n-k}$ be the coordinate of $[V_{\mathfrak{r}}]$. Let s be a section of $\Lambda^\ell X$. On $V_{\mathfrak{r}}$ it is represented as

$$\begin{aligned} &\sum_{J \subseteq \{1,\dots,n-k\}, \#J=\ell} h_J dx_J \\ &+ \sum_{I \subseteq \{1,\dots,k\}, I \neq \emptyset} \; \sum_{J \subseteq \{1,\dots,n-k\}, \#J+\#I=\ell} h_{I,J} dx_J \wedge dt_I. \end{aligned} \tag{25.12}$$

Here, we put $I = \{i_1, \dots, i_{\#I}\}$, $J = \{j_1, \dots, j_{\#J}\}$ and define $dt_I = dt_{i_1} \wedge \cdots \wedge dt_{i_{\#I}}$, $dx_J = dx_{j_1} \wedge \cdots \wedge dx_{j_{\#J}}$.

Definition 25.34 A section s of $\Lambda^\ell X$ is called an *admissible differential form* if the local expression (25.12) has the following properties:

(1) The functions h_J and $h_{I,J}$ are admissible.
(2) $h_{I,J}(x_1, \dots, x_{n-k}, t_1, \dots, t_k) = 0$ if there exists $i \in I$ with $t_i = 0$.

We remark that this condition is more restrictive than the condition that s is an admissible section of $\Lambda^\ell X$. In fact for s to be an admissible section, Condition (2) can be weakened to the admissibility of $h_{I,J}$. Condition (2) is used in the proof of Lemma-Definition 17.14.

Lemma 25.35

(1) *If $f : X \to M$ is an admissible smooth map from an admissible orbifold to a manifold (without boundary) and h is a smooth differential form on M, then the pullback f^*h is an admissible differential form.*

(2) *The pullback of an admissible differential form by an admissible embedding of admissible orbifolds is admissible.*

Proof We use an admissible coordinate of X and a coordinate of M to write f locally as $f^j(x_1, \dots, x_{n-k}, t_1, \dots, t_k)$ $j = 1, \dots, \dim M$. Then $\partial f^j/\partial x_i$ is admissible and $\partial f^j/\partial t_i$ is exponentially small, by definition. (1) follows from this fact. (2) is obvious from the definition. □

In this way, we can translate various stories for manifolds to the admissible realm. In particular, we can define the notion of admissibility of Kuranishi structures. In fact, the whole story works just by adding the word admissible to various constructions.

Definition 25.36 A Kuranishi structure

$$\widehat{\mathcal{U}} = (\{\mathcal{U}_p = (U_p, \mathcal{E}_p, \psi_p, s_p)\}, \{\Phi_{pq} = (U_{pq}, \varphi_{pq}, \widehat{\varphi}_{pq})\})$$

is *admissible* if:

- U_p is an admissible orbifold in the sense of Definition 25.14.
- $\mathcal{E}_p$ is an admissible vector bundle in the sense of Definition 25.24.
- s_p is an admissible section of $\mathcal{E}_p$ in the sense of Definition 25.30.
- U_{pq} is an open admissible suborbifold of U_q in the sense of Definition 25.14.
- $(\varphi_{pq}, \widehat{\varphi}_{pq})$ is an admissible embedding of admissible vector bundles in the sense of Definition 25.24.

25.3 Admissibility of the Moduli Spaces of Pseudo-holomorphic Curves

In [FOOO18, Theorem 6.4, Proposition 8.27, Proposition 8.32] (see also [FOOO16, Proposition 16.11]), we proved the exponential decay of the coordinate change. It implies

$$\left\| \frac{\partial}{\partial T'_{\mathrm{e}}}(T'_{\mathrm{e}} - T_{\mathrm{e}}) \right\| \le C e^{-\delta T'_{\mathrm{e}}}. \tag{25.13}$$

Here T_{e} is the coordinate corresponding to a singular point which is resolved and T'_{e} is the coordinate corresponding to the same singular point after coordinate change.

In this section, we axiomatize the properties of the coordinate change proved in [FOOO16, FOOO18] under an abstract setting described in Situation 25.37. Then we can directly see that the results proved in [FOOO16, FOOO18] actually imply that the Kuranishi structure of the moduli space of pseudo-holomorphic curves is admissible in the sense of Definition 25.36. This is tautological, but such an axiomatization will be useful when we prove admissibility of coordinate change of a Kuranishi structure constructed in other situation.

Situation 25.37 We consider open subsets $V_1 \subset [V_1] \times (0, \infty]^k$ and $V_2 \subset [V_2] \times (0, \infty]^k$, where $[V_i]$ are open subsets of $\mathbb{R}^{n_i}$. Let $\varphi : V_1 \to V_2$ be an embedding of topological spaces such that the following holds: We consider the stratification of V_i such that $S_m V_i$ consists of the points where at least m of the coordinates of the $(0, \infty]^k$ factor are ∞. We put $\overset{\circ}{S}_m V_i = S_m V_i \setminus S_{m+1} V_i$. We assume the following:

(1) $\varphi(p) \in S_m V_2$ if and only if $p \in S_m V_1$.
(2) The restriction of φ to $\overset{\circ}{S}_m V_1$ is a smooth embedding $\overset{\circ}{S}_m V_1 \to \overset{\circ}{S}_m V_2$.
(3) We write

$$\varphi(x; T_1, \ldots, T_k) = (\overline{\varphi}(x; T_1, \ldots, T_k); T_1'(x; T_1, \ldots, T_k), \ldots, T_k'(x; T_1, \ldots, T_k)).$$

Then the following holds:

(a)

$$\left\| \frac{\partial \overline{\varphi}}{\partial T_i} \right\|_{C^k} \le C_k e^{-c_k T_i}.$$

Here C_k, c_k are positive numbers depending on k and φ, and $\| \cdot \|_{C^k}$ is the C^k norm with respect to all of x, T_i.

(b)

$$\left\| \frac{\partial T_j'}{\partial T_i} - \delta_{ij} \right\|_{C^k} \le C_k e^{-c_k T_i}. \tag{25.14}$$

Here $C_k, c_k, \| \cdot \|_{C^k}$ are the same as above and $\delta_{ij} = 0$ if $i \neq j$ and $\delta_{ii} = 1$.

We assume the above inequalities (a) and (b) hold stratumwise. Namely, in the case when a certain coordinate T_i is ∞, we require the inequality only for T_j derivatives which are not ∞. ■

Lemma 25.38 *In Situation* 25.37 *the coordinate change φ is admissible in the sense of Definition* 25.8.

Proof This is immediate from the definition. □

We can use this lemma to obtain an admissible coordinate in the geometric situation appearing in the moduli space of pseudo-holomorphic curves.

Remark 25.39 As we explained in [FOOO4, Remark A1.63], the coordinate appearing in algebraic geometry is e^{-cT_e} which decays faster than $1/T_e$. On the other hand, $1/T_e$ is the coordinate used in [FOOO4, FOOO16] etc. Here the coordinate change is smooth with respect to this coordinate $1/T_e$. In this book we take an even slower coordinate $1/\log T_e$ than $1/T_e$ (see (25.1)) so that the coordinate change is admissible.

When we use the coordinate $s_e = 1/T_e$ in place of $1/\log T_e$, Lemma 25.10 (1) still holds, while Lemma 25.10 (2) does not. In fact, $f(x, T) = 1$ is an admissible function and so

$$T' = T + 1 \tag{25.15}$$

is an admissible coordinate change. Then putting $s' = 1/T'$ and $s = 1/T$, we have $s' = \frac{1}{1+1/s} = \frac{s}{s+1}$. Hence $\frac{d^2s'}{ds^2}(0) = -2 \neq 0$. As we mentioned in Remark 25.11, Lemma 25.10 (2) is necessary to extend the coordinate change to the collared neighborhood. Indeed, in Chap. 17 of this book we extended the coordinate change from $[V]\times[0, 1)$ to $[V]\times(-1, 1)$ by taking the $(-1, 1)$ component to be the identity map on $(-1, 0)$. Lemma 25.10 (2) implies that this extended coordinate change is smooth at $t = 1/\log T = 0$.

On the other hand, in [FOOO4] we used the fact that the coordinate change is smooth in the coordinate $s = 1/T$. However, we note that we did not need Lemma 25.10 (2) in [FOOO4] because we did not use the *collared* Kuranishi neighborhood and did not need to extend the coordinate change.

When we study a *forgetful map of marked points* for the Kuranishi structures on the moduli spaces of pseudo-holomorphic curves we use exponential decay. See [FOOO4, pages 777–778]. Maps between K-spaces (e.g. forgetful map) is not studied in this book. We will discuss this point in detail in the forthcoming paper [FOOO23].

Remark 25.40 In the geometric situation, the coordinate change where $T' - T$ is positive at $T = \infty$ actually occurs. So the coordinate change of the form (25.15) should be considered. In fact, the parameter T corresponds to the 'length of the neck' region. Namely, if the neck of the source curve is $[0, 1] \times [-5T, 5T]$, the corresponding element in the (thickened) moduli space has the coordinate T. However, the value of T depends on the choice of the coordinate at the nodes. Actually, when nodes have coordinates z and w, we identify $zw = -r$ and

$$r = e^{10\pi T}.$$

See [FOOO18, Section 8] right above Figure 13. If we take a different choice of z, say $z' = ez$, then r becomes $r' = er$. So $T' = T + 1/10\pi$.

Lemma 25.38 and [FOOO18, Section 8] imply the following.

Corollary 25.41 *The Kuranishi structure on* $\mathcal{M}_{k+1,\ell}(X, L; \beta)$ *defined in [FOOO21, Theorem 7.1] is admissible.*

Proof In [FOOO18, Section 8] it is proved that the coordinate change of the obstruction bundle is admissible and the Kuranishi map is also admissible. (The former is a consequence of [FOOO18, Proposition 8.19] and the latter is proved in the course of the proof of [FOOO18, Proposition 8.31].) Therefore it turns out that the Kuranishi structure we constructed on the moduli space of pseudo-holomorphic curves is admissible in the sense of Definition 25.36.

The argument of [FOOO18, Section 8] quoted above is the case of the Kuranishi chart of a stable map $((\Sigma, \vec{z}), u)$ when the marked source curve $(\Sigma, \vec{z})$ is stable. There are cases when the pair of a marked source curve $(\Sigma, \vec{z})$ and a map $u : \Sigma \to X$ is stable but $(\Sigma, \vec{z})$ is not stable. Admissibility of the Kuranishi chart in such cases is proved in [FOOO21]. In fact Condition 25.37 (a) and (b) are consequence of [FOOO21, (9.13)]. □

Remark 25.42 There are different kinds of boundaries or corners appearing in applications. For example, to prove independence of the filtered A_∞ structure associated to a Lagrangian submanifold under the change of compatible almost complex structures chosen in the course of the construction, we take a one-parameter family of almost complex structures $\{J_s\}$ joining two almost complex structures J_0 and J_1 which we choose for the construction. Then we consider the union of moduli spaces of pseudo-holomorphic disks bounding our Lagrangian submanifold with respect to the almost complex structures J_s for $s \in [0, 1]$. In this case the part $s = 0, 1$ becomes a boundary. To prove that our Kuranishi structure is admissible at this boundary, we choose our family $\{J_s\}$ so that $J_s = J_0$ (resp. $J_s = J_1$) for $s \in [0, \epsilon]$ (resp. for $s \in [1 - \epsilon, 1]$). Then the boundary corresponding to $s = 0, 1$ has a canonical collar. Therefore admissibility is obviously satisfied for this collar.

We can study the case when we consider a homotopy of Hamiltonians or homotopy of homotopies of almost complex structures (or Hamiltonians) in the same way. The study of such boundaries is much easier than that of the boundary corresponding to the boundary node.

Chapter 26
Stratified Submersion to a Manifold with Corners

In Definition 3.44 we defined the notions of a strongly smooth map and weakly submersive map from a K-space to a manifold *without boundary or corners*. In this chapter we give the corresponding definitions for the case when the target manifold P has a boundary or corners. In Chaps. 16 and 19, we used them to define and study homotopy and/or higher homotopy of morphisms of linear K-systems. In Chaps. 21 and 22, we also used them to define and study pseudo-isotopy of filtered A_∞ structure associated to a Lagrangian submanifold.

Let P be a manifold with corners (cornered manifold). For $p \in P$ we take a coordinate (V_p, ϕ_p) of product type. Here $p \in \overset{\circ}{S}_k(P)$ and $p = \phi(y_p, (0, \dots, 0))$. In this chapter we take the coordinate of P in this form.

Let X be an orbifold with corners (cornered orbifold), and $f : X \to P$ a continuous map.

Definition 26.1 Under the situation above, we say $f : X \to P$ is a *corner stratified smooth map* if for each $q \in X$ and $p = f(q)$ we can choose coordinates $\mathfrak{V}_q = (V_q, \Gamma_q, \phi_q)$ and $\mathfrak{V}_p = (V_p, \phi_p)$ respectively with the following properties. (Note that since P is a smooth manifold, $\Gamma_p = \{\mathrm{id}\}$.)

(1) $V_p = [V_p] \times [0, 1)^k$ is as above and $V_q = [V_q] \times [0, 1)^{\ell+k}$. Here $p \in \overset{\circ}{S}_k(P)$, $q \in \overset{\circ}{S}_{k+\ell}(X)$ and $q = \phi(y_q, (0, \dots, 0))$.
(2) There exists a map $f_q : V_q \to V_p$ of the form

$$f_q(y; (s_1, \dots, s_\ell, t_1, \dots, t_k)) = (\overline{f}_q(y; (s_1, \dots, s_\ell, t_1, \dots, t_k)), (t_1, \dots, t_k))$$

such that $\overline{f}_q : V_q \to [V_p]$ is smooth.
(3) The following diagram commutes:

K. Fukaya et al., *Kuranishi Structures and Virtual Fundamental Chains*, Springer Monographs in Mathematics, https://doi.org/10.1007/978-981-15-5562-6_26

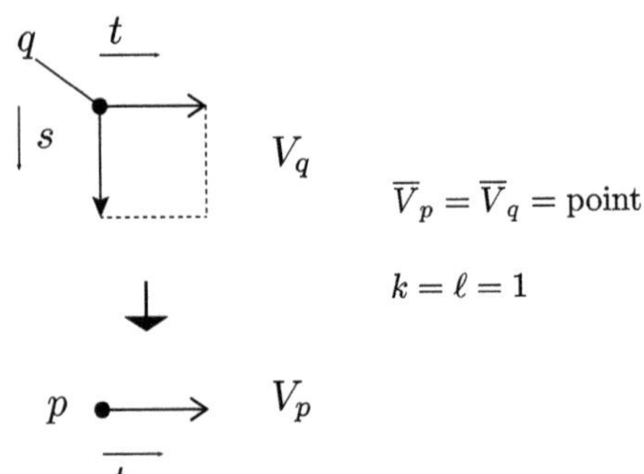

Fig. 26.1 Definition 26.1

$$\begin{array}{ccc} V_q & \xrightarrow{\phi_q} & X \\ {\scriptstyle f_q}\downarrow & & \downarrow{\scriptstyle f} \\ V_p & \xrightarrow{\phi_p} & P \end{array} \tag{26.1}$$

See Fig. 26.1. In the case when X, P are admissible, we require $\mathfrak{V}_p$, $\mathfrak{V}_q$ to be admissible charts and $\overline{f}_q$ to be admissible.

Remark 26.2 Throughout this chapter, we can work either in the category of a smooth manifold (or orbifold) with corners, or in the admissible category. We do not mention admissibility from now on in this chapter.

Definition 26.3 In the situation of Definition 26.1, we say $f : X \to P$ is a *corner stratified submersion* if the following holds.

We use the notation of Definition 26.1 (2). For any $(t_1, \dots, t_k) \in [0, 1)^k$ the map

$$(y; s_1, \dots, s_\ell) \mapsto \overline{f}_q(y; s_1, \dots, s_\ell, t_1, \dots, t_k)$$

is a submersion $[V_q] \times [0, 1)^\ell \to [V_p]$.[1]

Lemma 26.4 *Let X_1, X_2 be cornered orbifolds and let P_1, P_2, P be cornered manifolds, and R a smooth manifold without boundary.*

(1) *Let $f_i : X_i \to P \times R$ be smooth maps. Suppose f_1 is a corner stratified submersion. Then the fiber product $X_1 \times_{P\times R} X_2$ carries a structure of a cornered orbifold. In addition, if $\pi_P \circ f_2 : X_2 \to P$ is also a corner stratified submersion, then the map $X_1 \times_P X_2 \to P$ induced from f_1 and f_2 in an obvious way is a corner stratified submersion.*

[1]It means that $[V_q] \times [0, 1)^\ell \to [V_p]$ is a submersion on each stratum of $[V_q] \times [0, 1)^\ell$. For example, $y \mapsto \overline{f}_q(y; 0, \dots, 0, t_1, \dots, t_k)$ is required to be a submersion $[V_q] \to [V_p]$.

(2) *Let* $f_i : X_i \to P_i \times R$ *be smooth maps. Suppose* f_1 *is a corner stratified submersion. Then the fiber product* $X_1 \times_R X_2$ *carries a structure of a cornered orbifold. In addition, if* $\pi_{P_2} \circ f_2 : X_2 \to P_2$ *is also a corner stratified submersion, then the map* $X_1 \times_R X_2 \to P_1 \times P_2$ *induced from* f_1 *and* f_2 *in an obvious way is a corner stratified submersion.*

The proof is obvious.

Lemma 26.5 *Let* $f : X \to P$ *be a corner stratified smooth map from a cornered orbifold to a cornered manifold. Let* $\widehat{S}_k(P)$ *be the normalized corner of* P *and* $\pi : \widehat{S}_k(P) \to S_k(P) \subset P$ *the projection.*

(1) *The fiber product* $\widehat{S}_k(P) \times_P X$ *as topological space carries a structure of a cornered orbifold. The projection* $\widehat{S}_k(P) \times_P X \to \widehat{S}_k(P)$ *is a corner stratified smooth map.*
(2) *The projection* $\widehat{S}_k(P) \times_P X \to \widehat{S}_k(P)$ *is a corner stratified submersion if* $f : X \to P$ *is a corner stratified submersion.*
(3) *The map* $\widehat{S}_\ell(\widehat{S}_k(P)) \times_P X \to \widehat{S}_{\ell+k}(P) \times_P X$ *is a* $(k+\ell)!/k!\ell!$ *fold covering map.*

The proof is again obvious.

Now it is straightforward to generalize the story for an orbifold X to the case when X is a K-space.

Definition 26.6 Let $(X, \widehat{\mathcal{U}})$ be a K-space and P a manifold with corners.

(1) A strongly continuous map $\widehat{f} : (X, \widehat{\mathcal{U}}) \to P$ is said to be a *corner stratified smooth map* if $f_p : U_p \to P$ is a corner stratified smooth map for any $p \in X$.
(2) A corner stratified smooth map $\widehat{f} : (X, \widehat{\mathcal{U}}) \to P$ is said to be a *corner stratified weak submersion* if $f_p : U_p \to P$ is a corner stratified submersion for any $p \in X$.
(3) Let $\widehat{\mathfrak{S}}$ be a CF-perturbation of X. We say that a corner stratified smooth map $\widehat{f} : (X, \widehat{\mathcal{U}}) \to P$ is a *corner stratified strong submersion with respect to* $\widehat{\mathfrak{S}}$ if the following holds. Let $p \in X$ and (U_p, E_p, ψ_p, s_p) be a Kuranishi chart at $p \in X$. Let $(V_\mathfrak{r}, \Gamma_\mathfrak{r}, \phi_\mathfrak{r})$ be an orbifold chart of U_p at some point and $(W_\mathfrak{r}, \omega_\mathfrak{r}, \mathfrak{s}^\epsilon_\mathfrak{r})$ be a representative of $\widehat{\mathfrak{S}}$ in this orbifold chart. Then

$$f \circ \psi_p \circ \phi_\mathfrak{r} \circ \mathrm{pr} : (\mathfrak{s}^\epsilon_\mathfrak{r})^{-1}(0) \to P$$

is a corner stratified submersion. Here $\mathrm{pr} : V_\mathfrak{r} \times W_\mathfrak{r} \to V_\mathfrak{r}$ is the projection.

We can define a corner stratified smooth map and a corner stratified weak submersion from a space equipped with a good coordinate system in the same way.

Lemma-Definition 26.7 *Let* P *be a manifold with corners and* R *a manifold with boundary. Let* $\widehat{f} : (X, \widehat{\mathcal{U}}) \to P \times R$ *be a corner stratified strong submersion with respect to* $\widehat{\mathfrak{S}}$*. Then for any differential form* h *on* $(X, \widehat{\mathcal{U}})$ *and for each sufficiently small* $\epsilon > 0$*, we can define the pushout*

$$\widehat{f}!(h; \widehat{\mathfrak{S}^{\epsilon}})$$

which is a smooth differential form on $P \times R$, in the same way as in Theorem 9.14 using a good coordinate system compatible with the given Kuranishi structure. It is independent of the choice of the compatible good coordinate system if $\epsilon > 0$ is sufficiently small.

The proof is the same as that of Theorem 9.14 and so is omitted.

Lemma 26.8 *For each $i = 1, 2$ let $(X_i, \widehat{\mathcal{U}_i})$ be a K-space and $\widehat{f_i} : (X_i, \widehat{\mathcal{U}_i}) \to P \times R$ a corner stratified smooth map, where P is a manifold with corners and R is a manifold without boundary. We assume that $\widehat{f_1}$ is a corner stratified weak submersion and $\pi \circ \widehat{f_2} : (X_2, \widehat{\mathcal{U}_2}) \to P$ is a corner stratified weak submersion.*

1. *The fiber product $X_1 \times_{P \times R} X_2$ carries a Kuranishi structure and the map $X_1 \times_{P \times R} X_2 \to P$ induced from $\widehat{f_1}$ and $\widehat{f_2}$ in an obvious way is a corner stratified weak submersion.*
2. *Let $\widehat{\mathfrak{S}_i}$ be a CF-perturbation of $(X_i, \widehat{\mathcal{U}_i})$. We assume that $\widehat{f_1}$ is a corner stratified strong submersion with respect to $\widehat{\mathfrak{S}_1}$ and $\pi_P \circ \widehat{f_2} : (X_2, \widehat{\mathcal{U}_2}) \to P$ is a corner stratified strong submersion with respect to $\widehat{\mathfrak{S}_2}$. Then we can define the fiber product $\widehat{\mathfrak{S}_1} \times_{P \times R} \widehat{\mathfrak{S}_2}$ of CF-perturbations. The map $X_1 \times_{P \times R} X_2 \to P$ is a corner stratified strong submersion with respect to $\widehat{\mathfrak{S}_1} \times_{P \times R} \widehat{\mathfrak{S}_2}$.*

Proof This follows from Lemma 26.4 (1). □

There is a slightly different situation where we take the fiber product as follows:

Lemma 26.9 *For each $i = 1, 2$ let $(X_i, \widehat{\mathcal{U}_i})$ be a K-space and $\widehat{f_i} : (X_i, \widehat{\mathcal{U}_i}) \to R \times P_i$ a corner stratified smooth map, where P_i is a cornered manifold and R is a manifold without boundary. We assume that $\widehat{f_1}$ is a corner stratified weak submersion and $\pi_{P_2} \circ \widehat{f_2}$ is a corner stratified weak submersion.*

1. *The fiber product $X_1 \times_R X_2$ carries a Kuranishi structure and the map $X_1 \times_R X_2 \to P_1 \times P_2$ induced from $\widehat{f_1}$ and $\widehat{f_2}$ in an obvious way is a corner stratified weak submersion.*
2. *Let $\widehat{\mathfrak{S}_i}$ be a CF-perturbation of $(X_i, \widehat{\mathcal{U}_i})$. We assume that $\widehat{f_i}$ is a corner stratified strong submersion with respect to $\widehat{\mathfrak{S}_i}$. Then we can define the fiber product $\widehat{\mathfrak{S}_1} \times_R \widehat{\mathfrak{S}_2}$ of CF-perturbations. The map $X_1 \times_R X_2 \to P_1 \times P_2$ is a corner stratified strong submersion with respect to $\widehat{\mathfrak{S}_1} \times_R \widehat{\mathfrak{S}_2}$.*

Proof This follows from Lemma 26.4 (2). □

Next we discuss Stokes' formula and the composition formula under the smooth correspondence.

Definition 26.10 Let $\widehat{f} : (X, \widehat{\mathcal{U}}) \to P$ be a corner stratified weak submersion. We decompose the boundary of X into two components:

$$\partial_{\mathfrak{C}v}(X, \widehat{\mathcal{U}}) = \widehat{f}^{-1}(\partial P) \tag{26.2}$$

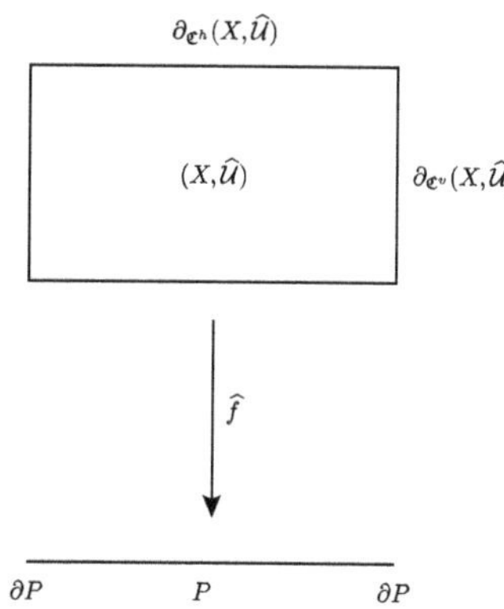

Fig. 26.2 Vertical/horizontal boundary

and

$$\partial_{\mathfrak{C}^h}(X,\widehat{\mathcal{U}}) = \partial(X,\widehat{\mathcal{U}}) \setminus \partial_{\mathfrak{C}^v}(X,\widehat{\mathcal{U}}). \tag{26.3}$$

We call (26.2) the *vertical boundary* and (26.3) the *horizontal boundary*. See Fig. 26.2. They are induced by the decomposition of the boundary satisfying (18.9) in Situation 18.4.

Lemma 26.11 *In the situation of Definition* 26.10, *the restriction of* $\widehat{f}$ *to the horizontal boundary induces a corner stratified weak submersion* $\widehat{f}|_{\partial_{\mathfrak{C}^h}(X,\widehat{\mathcal{U}})} : \partial_{\mathfrak{C}^h}(X,\widehat{\mathcal{U}}) \to P$. *If* $\widehat{f}$ *is a corner stratified strong submersion with respect to a CF-perturbation* $\widehat{\mathfrak{S}}$, *then the restriction* $\widehat{f}|_{\partial_{\mathfrak{C}^h}(X,\widehat{\mathcal{U}})}$ *is a corner stratified strong submersion with respect to* $\widehat{\mathfrak{S}}|_{\partial_{\mathfrak{C}^h}(X,\widehat{\mathcal{U}})}$.

The proof is obvious.

Theorem 26.12 *In the situation of Lemma* 26.11, *let* h *be a differential form on* $(X,\widehat{\mathcal{U}})$. *Then for each sufficiently small* $\epsilon > 0$ *we have*

$$d\widehat{f}!(h;\widehat{\mathfrak{S}}^{\epsilon}) = \widehat{f}!(dh;\widehat{\mathfrak{S}}^{\epsilon}) + (-1)^{\dim(X,\widehat{\mathcal{U}})+\deg h}\,\widehat{f}!(h|_{\partial_{\mathfrak{C}^h}(X,\widehat{\mathcal{U}})};\widehat{\mathfrak{S}}^{\epsilon}|_{\partial_{\mathfrak{C}^h}(X,\widehat{\mathcal{U}})}). \tag{26.4}$$

Proof Since both sides are smooth forms, it suffices to prove the formula pointwise on $\operatorname{Int} P$. Let $p \in \operatorname{Int} P$ and take a compact set $K \subset \operatorname{Int} P$ containing an open neighborhood of p. Using a partition of unity, we may assume without loss of generality that h is supported in $\widehat{f}^{-1}(K)$. Then we can apply Theorem 9.28 to $X \setminus \partial_{\mathfrak{C}^v}(X,\widehat{\mathcal{U}})$ to prove the equality (26.4) at p. □

Definition 26.13 Let P be a manifold with corners and let M_s, M_t be manifolds without boundary.

(1) A *P-parametrized smooth correspondence* is a system

$$\mathfrak{X} = \left((X, \widehat{\mathcal{U}}), \widehat{f_s}, \widehat{f_t}, \pi_{s,P}, \pi_{t,P}\right),$$

where $(X, \widehat{\mathcal{U}})$ is a K-space and $\widehat{f_s} : (X, \widehat{\mathcal{U}}) \to P \times M_s$ and $\widehat{f_t} : (X, \widehat{\mathcal{U}}) \to P \times M_t$ are strongly smooth maps. We assume that $\widehat{f_t}$ is a corner stratified weak submersion and satisfies

$$\pi_{s,P} \circ \widehat{f_s} = \pi_{t,P} \circ \widehat{f_t}. \tag{26.5}$$

Here $\pi_{s,P}$ and $\pi_{t,P}$ are the projections to the P factor.

(2) A *P-parametrized perturbed smooth correspondence* is $(\mathfrak{X}, \widehat{\mathfrak{S}})$ where $\mathfrak{X} = ((X, \widehat{\mathcal{U}}), \widehat{f_s}, \widehat{f_t})$ is a P-parametrized smooth correspondence and $\widehat{\mathfrak{S}}^\epsilon$ is a CF-perturbation of $(X, \widehat{\mathcal{U}})$ such that $\widehat{f_t}$ is a corner stratified strong submersion with respect to $\widehat{\mathfrak{S}}$.

(3) Let $(\mathfrak{X}, \widehat{\mathfrak{S}})$ $(\mathfrak{X} = ((X, \widehat{\mathcal{U}}), \widehat{f_s}, \widehat{f_t}, \pi_{s,P}, \pi_{t,P}))$ be a P-parametrized perturbed smooth correspondence. The restrictions of $\widehat{f_s}, \widehat{f_t}, \widehat{\mathfrak{S}}$ to the horizontal boundary $\partial_{\mathfrak{C}^h}(X, \widehat{\mathcal{U}})$ define a P-parametrized perturbed smooth correspondence. We call it the *boundary* of $(\mathfrak{X}, \widehat{\mathfrak{S}})$ and denote it by $\partial(\mathfrak{X}, \widehat{\mathfrak{S}})$.

Definition 26.14 A P-parametrized perturbed smooth correspondence $(\mathfrak{X}, \widehat{\mathfrak{S}})$ from M_s to M_t induces a continuous family of maps

$$\mathrm{Corr}_{(\mathfrak{X}, \widehat{\mathfrak{S}^\epsilon})} : \Omega^k(P \times M_s) \to \Omega^{k+\ell}(P \times M_t)$$

by

$$\mathrm{Corr}_{(\mathfrak{X}, \widehat{\mathfrak{S}^\epsilon})}(h) = \widehat{f_t}!(\widehat{f_s}^* h; \widehat{\mathfrak{S}}^\epsilon) \tag{26.6}$$

for each sufficiently small $\epsilon > 0$.

Lemma 26.15 *Suppose we are in the situation of Definition* 26.14. *If ρ is a differential form on P, then for each sufficiently small $\epsilon > 0$ we have*

$$\mathrm{Corr}_{(\mathfrak{X}, \widehat{\mathfrak{S}^\epsilon})}(\rho \wedge h) = \rho \wedge \mathrm{Corr}_{(\mathfrak{X}, \widehat{\mathfrak{S}^\epsilon})}(h).$$

Proof This is a consequence of (26.5). □

Theorem 26.12 immediately implies the following.

Proposition 26.16 *For each sufficiently small $\epsilon > 0$ we have*

$$d \circ \mathrm{Corr}_{(\mathfrak{X}, \widehat{\mathfrak{S}^\epsilon})} - \mathrm{Corr}_{(\mathfrak{X}, \widehat{\mathfrak{S}^\epsilon})} \circ d = (-1)^{\dim(\mathfrak{X}) + \deg(\cdot)} \mathrm{Corr}_{\partial(\mathfrak{X}, \widehat{\mathfrak{S}^\epsilon})}.$$

Next we discuss the composition formula.

Definition-Lemma 26.17 Let M_1, M_2, M_3 be smooth manifolds and P, P_1, P_2 manifolds with corners.

(1) Let $\mathfrak{X}_{ii+1} = (X_{ii+1}, \widehat{\mathcal{U}}_{ii+1}, \widehat{f}_{ii+1;i}, \widehat{f}_{ii+1;i+1})$ be P-parametrized smooth correspondences from M_i to M_{i+1} for $i = 1, 2$.

(a) The *composition* $\mathfrak{X}_{13} = \mathfrak{X}_{23} \circ \mathfrak{X}_{12}$ is

$$\Big((X_{12}, \widehat{\mathcal{U}}_{12}) \times_{P \times M_2} (X_{23}, \widehat{\mathcal{U}}_{23}),\ \widehat{f}_{12;1} \circ \pi,\ \widehat{f}_{12,2} \circ \pi\Big),$$

which is a P-parametrized smooth correspondence from M_1 to M_3.

(b) In addition, if $(\mathfrak{X}_{ii+1}, \widehat{\mathfrak{S}}_{ii+1})$ is a P-parametrized perturbed smooth correspondence, then together with $\widehat{\mathfrak{S}}_{23} = \widehat{\mathfrak{S}}_{12} \times_{P \times M_2} \widehat{\mathfrak{S}}_{23}$, the composition $\mathfrak{X}_{13} = \mathfrak{X}_{23} \circ \mathfrak{X}_{12}$ defines a P-parametrized perturbed smooth correspondence from M_1 to M_3. We say $(\mathfrak{X}_{13}, \widehat{\mathfrak{S}}_{13})$ is the *composition* of $(\mathfrak{X}_{12}, \widehat{\mathfrak{S}}_{12})$ and $(\mathfrak{X}_{23}, \widehat{\mathfrak{S}}_{23})$.

(2) Let $\Xi_{ii+1} = (X_{ii+1}, \widehat{\mathcal{U}}_{ii+1})$ be P_i-parametrized smooth correspondences from M_i to M_{i+1} for $i = 1, 2$.

(a) The *composition* $\mathfrak{X}_{13} = \mathfrak{X}_{23} \circ \mathfrak{X}_{12}$ is defined by $(X_{12}, \widehat{\mathcal{U}}_{12}) \times_{M_2} (X_{23}, \widehat{\mathcal{U}}_{23})$ which is a $(P_1 \times P_2)$-parametrized smooth correspondence from M_1 to M_3.

(b) In addition, if $(\mathfrak{X}_{ii+1}, \widehat{\mathfrak{S}}_{ii+1})$ is a P_i-parametrized perturbed smooth correspondence, then together with $\widehat{\mathfrak{S}}_{23} = \widehat{\mathfrak{S}}_{12} \times_{M_2} \widehat{\mathfrak{S}}_{23}$, the composition $\mathfrak{X}_{13}$ defines a $(P_1 \times P_2)$-parametrized perturbed smooth correspondence from M_1 to M_3. We say $(\Xi_{13}, \widehat{\mathfrak{S}}_{13})$ is the *composition* of $(\mathfrak{X}_{12}, \widehat{\mathfrak{S}}_{12})$ and $(\mathfrak{X}_{23}, \widehat{\mathfrak{S}}_{23})$.

Proof This is a consequence of Lemma 26.9. □

Proposition 26.18 *In the situation of Definition-Lemma* 26.17 *(1) we have*

$$\mathrm{Corr}_{(\mathfrak{X}_{23}, \widehat{\mathfrak{S}}^{\epsilon}_{23})} \circ \mathrm{Corr}_{(\mathfrak{X}_{12}, \widehat{\mathfrak{S}}^{\epsilon}_{12})} = \mathrm{Corr}_{(\mathfrak{X}_{13}, \widehat{\mathfrak{S}}^{\epsilon}_{13})} \tag{26.7}$$

for each sufficiently small $\epsilon > 0$.

We will discuss the situation of Definition-Lemma 26.17 (2) later.

Proof Let us consider the following situation.

Situation 26.19 For $i = 1, 2$, let $(X_i, \widehat{\mathcal{U}_i})$ be K-spaces, P, P_i manifolds with corners, and R a manifold without boundary. Let $\widehat{\mathfrak{S}_i}$ be CF-perturbations of $(X_i, \widehat{\mathcal{U}_i})$.

(1) $\widehat{f_i} : (X_i, \widehat{\mathcal{U}_i}) \to P \times R$ are corner stratified strongly smooth maps for $i = 1, 2$. We assume that $\pi_P \circ \widehat{f_1} : (X_1, \widehat{\mathcal{U}_1}) \to P$ and $\widehat{f_2} : (X_2, \widehat{\mathcal{U}_2}) \to P \times R$ are corner stratified weakly submersive and corner stratified strongly submersive with respect to $\widehat{\mathfrak{S}_1}, \widehat{\mathfrak{S}_2}$, respectively.

(2) $\widehat{f_i} : (X_i, \widehat{\mathcal{U}_i}) \to P_i \times R$ are corner stratified strongly smooth maps. We assume $\pi_{P_1} \circ \widehat{f_1}$ and $\widehat{f_2}$ and are corner stratified weakly submersive and corner stratified strongly submersive with respect to $\widehat{\mathfrak{S}_1}$, $\widehat{\mathfrak{S}_2}$, respectively. ■

Lemma 26.20 *In Situation 26.19 (1) we consider differential forms h_i on $(X_i, \widehat{\mathcal{U}_i})$. They define a differential form $h_1 \wedge h_2$ on the fiber product $(X_1, \widehat{\mathcal{U}_1}) \times_{P \times R} (X_2, \widehat{\mathcal{U}_2})$. Then for each sufficiently small $\epsilon > 0$ we have*

$$\begin{aligned} &\int_{((X_1,\widehat{\mathcal{U}_1}) \times_{P\times R} (X_2,\widehat{\mathcal{U}_2}), (\widehat{\mathfrak{S}_1} \times_{P\times R} \widehat{\mathfrak{S}_2})^\epsilon)} h_1 \wedge h_2 \\ &= \int_{((X_1,\widehat{\mathcal{U}_1}), \widehat{\mathfrak{S}_1})} h_1 \wedge \widehat{f_1}^* \widehat{f_2}!(h_2; (\widehat{\mathfrak{S}_2})^\epsilon). \end{aligned} \tag{26.8}$$

Proof We can use Sublemma 26.21 below in place of Lemma 10.28. Then the proof is the same as that of Proposition 10.24. □

Sublemma 26.21 *For $i = 1, 2$ let N_i, P be smooth manifolds with corners, and $f_i : N_i \to P \times M$ smooth maps, and h_i smooth differential forms on N_i of compact support. Suppose that f_2 is a corner stratified submersion. Then we have*

$$\int_{N_1 \, {}_{f_1}\times_{f_2} N_2} h_1 \wedge h_2 = \int_{N_1} h_1 \wedge f_1^*(f_2!(h_2)). \tag{26.9}$$

Proof Using a partition of unity, it suffices to prove (26.9) when $P = \overline{P} \times [0, 1)^b$, $N_1 = P \times M \times \mathbb{R}^{a_1} \times [0, 1)^{a_2}$ and $f_i : N_i \to P \times M$ is the obvious projection. We can prove this by Fubini's theorem in the same way as in Lemma 10.28. □

Lemma 26.22 *In Situation 26.19 (2) we consider differential forms h_i on $(X_i, \widehat{\mathcal{U}_i})$ $(i = 1, 2)$ and the wedge product $h_1 \wedge h_2$ on the fiber product $(X_1, \widehat{\mathcal{U}_1}) \times_R (X_2, \widehat{\mathcal{U}_2})$. Then for each sufficiently small $\epsilon > 0$ we have*

$$\begin{aligned} &\int_{((X_1,\widehat{\mathcal{U}_1}) \times_R (X_2,\widehat{\mathcal{U}_2}), (\widehat{\mathfrak{S}_1} \times_R \widehat{\mathfrak{S}_2})^\epsilon)} h_1 \wedge h_2 \\ &= \int_{((X_1,\widehat{\mathcal{U}_1}), \widehat{\mathfrak{S}_1})} h_1 \wedge (\pi_R \circ \widehat{f_1})^* (\pi_R \circ \widehat{f_2})!(h_2; (\widehat{\mathfrak{S}_2})^\epsilon). \end{aligned} \tag{26.10}$$

Proof This is a consequence of Proposition 10.24. □

The proof of Proposition 26.18 is now complete. □

To discuss Situation 26.19 (2) we slightly generalize the notion of correspondence.

Definition 26.23 Let $(\mathfrak{X}, \widehat{\mathfrak{S}}) = ((X, \widehat{\mathcal{U}}, \widehat{f_s}, \widehat{f_t}, \pi_{s,P_1}, \pi_{t,P_1}), \widehat{\mathfrak{S}})$ be a P_1-parametrized perturbed smooth correspondence from M_s to M_t. Let P_2 be a manifold with corners. We regard

$$\left(P_2 \times (X,\widehat{\mathcal{U}}), \mathrm{id}_{P_2} \times \widehat{f_s}, \mathrm{id}_{P_2} \times \widehat{f_t}, \pi_{P_2} \circ (\mathrm{id}_{P_2} \times \pi_{s,P_1}), \pi_{P_2} \circ (\mathrm{id}_{P_2} \times \pi_{t,P_1})\right)$$

as a $(P_2 \times P_1)$-parametrized smooth correspondence from M_s to M_t. Here $\pi_{P_2} : P_2 \times P_1 \to P_2$ is the projection. Then this defines a map which we denote by

$$\mathrm{Corr}_{(\mathfrak{X},\widehat{\mathfrak{S}^\epsilon}),P_2} : \Omega^k(P_2 \times P_1 \times M_s) \to \Omega^{k+\ell}(P_2 \times P_1 \times M_t) \tag{26.11}$$

for each sufficiently small $\epsilon > 0$. Here $\ell = \dim M_t - \dim(X,\widehat{\mathcal{U}})$. Similarly we define

$$\mathrm{Corr}_{(\mathfrak{X},\widehat{\mathfrak{S}^\epsilon}),P_1} : \Omega^k(P_1 \times P_2 \times M_s) \to \Omega^{k+\ell}(P_1 \times P_2 \times M_t). \tag{26.12}$$

We note that when we define these maps, we do not use the orientations on P_1, P_2. So the order of factors in the direct product does not cause the sign problem. Thus we may write as

$$\mathrm{Corr}_{(\mathfrak{X},\widehat{\mathfrak{S}^\epsilon}),P_2} : \Omega^k(P_1 \times P_2 \times M_s) \to \Omega^{k+\ell}(P_1 \times P_2 \times M_t).$$

Proposition 26.24 *In the situation of Definition-Lemma* 26.17 *(2) we have*

$$\mathrm{Corr}_{(\mathfrak{X}_{23},\widehat{\mathfrak{S}}^\epsilon_{23}),P_1} \circ \mathrm{Corr}_{(\mathfrak{X}_{12},\widehat{\mathfrak{S}}^\epsilon_{12}),P_2} = \mathrm{Corr}_{(\mathfrak{X}_{13},\widehat{\mathfrak{S}}^\epsilon_{13})} \tag{26.13}$$

for each sufficiently small $\epsilon > 0$.

Proof

Lemma 26.25 *Suppose we are in Situation* 26.19 *(2). Let h_i be differential forms on $(X_i,\widehat{\mathcal{U}_i})$ and ρ_i differential forms on P_i for $i = 1, 2$. Then we obtain a differential form*

$$(\pi_{P_1} \circ f_1)^*\rho_1 \wedge h_1 \wedge (\pi_{P_2} \circ f_2)^*\rho_2 \wedge h_2$$

on $(X_1,\widehat{\mathcal{U}_1}) \times_R (X_2,\widehat{\mathcal{U}_2})$. Moreover we have the following equality:

$$\begin{aligned}&\int_{(X_1,\widehat{\mathcal{U}_1})\times_R(X_2,\widehat{\mathcal{U}_2})} (\pi_{P_1} \circ f_1)^*\rho_1 \wedge h_1 \wedge (\pi_{P_2} \circ f_2)^*\rho_2 \wedge h_2\\ &= \int_{((X_1,\widehat{\mathcal{U}_1}),\widehat{\mathfrak{S}_1})} (\pi_R \circ f_1)^*\rho_1 \wedge h_1\\ &\qquad\qquad \wedge(\pi_R \circ f_1)^*(\pi_R \circ f_2)!((\pi_{P_2} \circ f_2)^*\rho_2 \wedge h_2; (\widehat{\mathfrak{S}_2})^\epsilon).\end{aligned} \tag{26.14}$$

Proof Applying Lemma 26.22 to $(\pi_{P_1} \circ f_1)^*\rho_1 \wedge h_1$ and $(\pi_{P_2} \circ f_2)^*\rho_2 \wedge h_2$, we obtain Lemma 26.25. □

Proposition 26.24 is a consequence of Lemma 26.25. □

Next we rewrite Lemma 26.5 to the case when X is a K-space.

Lemma 26.26 *Let $\widehat{f} : (X, \widehat{\mathcal{U}}) \to P$ be a corner stratified smooth map from a K-space to a manifold with corners. Let $\widehat{S}_k(P)$ be the normalized corner of P and $\pi : \widehat{S}_k(P) \to S_k(P) \subset P$ the projection.*

(1) *The fiber product $\widehat{S}_k(P) \times_P X$ as topological space carries a Kuranishi structure. The projection $\widehat{S}_k(P) \times_P X \to \widehat{S}_k(P)$ is a corner stratified smooth map.*
(2) *The projection $\widehat{S}_k(P) \times_P X \to \widehat{S}_k(P)$ is a corner stratified submersion if $\widehat{f} : X \to P$ is a corner stratified submersion.*
(3) *The map $\widehat{S}_\ell(\widehat{S}_k(P)) \times_P X \to \widehat{S}_{\ell+k}(P) \times_P X$ is induced by a $(k+\ell)!/k!\ell!$ fold covering map of K-spaces.*

The proof is again obvious.

Next we mention the relation to (partial) outer collaring. (See Chaps. 17 and 18.) The proof of the next lemma is straightforward and so omitted.

Lemma 26.27 *Suppose $\widehat{f} : (X, \widehat{\mathcal{U}}) \to P$ is a corner stratified submersion from a K-space to an (admissible) manifold with corners. Then $\widehat{f}$ induces a map*

$$\widehat{f}^{\boxplus\tau_0} : (X, \widehat{\mathcal{U}})^{\boxplus\tau_0} \to P^{\boxplus\tau_0}.$$

Let $\mathfrak{C}^h$ be the horizontal component of the corner of $(X, \widehat{\mathcal{U}})$. Then we obtain a map

$$\widehat{f}^{\mathfrak{C}^h\boxplus\tau_0} : (X, \widehat{\mathcal{U}})^{\mathfrak{C}^h\boxplus\tau_0} \to P.$$

In both cases, if $\widehat{f}$ is corner stratified weakly submersive (resp. corner stratified strongly submersive with respect to $\widehat{\mathfrak{S}}$), then $\widehat{f}^{\boxplus\tau_0}$ and $\widehat{f}^{\mathfrak{C}^h\boxplus\tau_0}$ are corner stratified weakly submersive (resp. corner stratified strongly submersive with respect to $\widehat{\mathfrak{S}}$).

Most of the stories of Kuranishi structure, CF-perturbation, pushout etc. can be generalized to the case when the target space has a corner, in a straightforward way. We will describe them when we need them.

Remark 26.28 In Chap. 19 etc. we have used corner stratified submersions (Definition 26.6) to define and study homotopy and higher homotopy of morphisms of linear K-systems. On the other hand, we like to mention that there is another way to define homotopy and/or higher homotopy of morphisms etc. without using corner stratified submersions to manifolds with corners. Indeed, while we were writing [FOOO2] we sometimes took this way. For example, in [FOOO2, Subsection 19.2] we take a small number $\epsilon > 0$ and consider $P = (-\epsilon, 1+\epsilon)$ instead of $P = [0, 1]$. When we construct $\mathcal{N}(\alpha_1, \alpha_2; P)$ in [FOOO2], we consider the P-parametrized version of the moduli space so that it is constant on $(-\epsilon, 0)$ and $(1, 1+\epsilon)$. This method also works rigorously. However, choosing $P = [0, 1]$ and using corner stratified submersions seem more natural.

Chapter 27
Local System and Smooth Correspondence in de Rham Theory with Twisted Coefficients

Let $\mathcal{L}$ be a local system, i.e., a flat vector bundle, on a manifold M. We denote by $(\Omega^\bullet(M;\mathcal{L}) = \Gamma(M;\bigwedge^\bullet T^*M \otimes \mathcal{L}), d = d_{\mathcal{L}})$ the de Rham complex with coefficients in $\mathcal{L}$. We recall some basic operations on the de Rham complex with twisted coefficients.

1. (Pullback.) Let $f : N \to M$ be a smooth map. Clearly, the pullback $f^*\mathcal{L}$ is a flat vector bundle and we have a cochain homomorphisms

$$f^* : \Omega^\bullet(M;\mathcal{L}) \to \Omega^\bullet(N; f^*\mathcal{L}).$$

As in the usual de Rham theory, we have $d \circ f^* = f^* \circ d$.
2. (Wedge product.) Let $\mathcal{L}_1, \mathcal{L}_2$ be flat vector bundles. Then $\mathcal{L}_1 \otimes \mathcal{L}_2$ is a flat vector bundle and we have the product

$$\wedge : \Omega^\bullet(M;\mathcal{L}_1) \otimes \Omega^\bullet(M;\mathcal{L}_2) \to \Omega^\bullet(M;\mathcal{L}_1 \otimes \mathcal{L}_2).$$

The wedge product and the differential enjoy the Leibniz' rule.
3. (Integration along fibers.) Let $\pi : N \to M$ be a proper submersion and let O_M (resp. O_N) be the flat real line bundle associated with the orientation $O(1)$-bundle of M (resp. N). We denote by $O_\pi = \pi^* O_M \otimes O_N$ be the relative orientation bundle of the submersion π. Then we have the integration along fibers

$$\pi! : \Omega^\bullet(N; O_\pi) \to \Omega^\bullet(M).$$

For a flat vector bundle $\mathcal{L}$ on M, we have the integration along fibers with twisted coefficients

$$\pi! : \Omega^\bullet(N; O_\pi \otimes \pi^*\mathcal{L}) \to \Omega^\bullet(M;\mathcal{L}).$$

K. Fukaya et al., *Kuranishi Structures and Virtual Fundamental Chains*, Springer Monographs in Mathematics, https://doi.org/10.1007/978-981-15-5562-6_27

Suppose that the boundary ∂N of N is not empty and the restriction $\pi|_{\partial N} : \partial N \to M$ is also a submersion, then we have

$$d \circ \pi! = \pi! \circ d + (-1)^{\dim N + \deg(\cdot)} (\pi|_{\partial N})!.$$

Now consider the following situation. Let $f_s : X \to M_s$ be a smooth map and $f_t : X \to M_t$ a proper submersion.

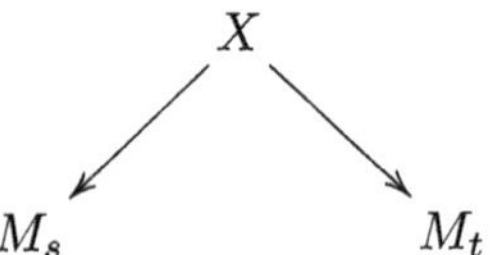

Let $\mathcal{L}_s$ (resp. $\mathcal{L}_t$) be a flat vector bundle such that $(f_t)^*\mathcal{L}_t \cong O_{f_t} \otimes (f_s)^*\mathcal{L}_s$. By composing the pullback operation by f_s and the integration along fibers of f_t, we obtain the correspondence

$$f_t! \circ f_s^* : \Omega^\bullet(M_s; \mathcal{L}_s) \to \Omega^\bullet(M_t; \mathcal{L}_t).$$

Taking these arguments into account, we can obtain the following. Let $\mathfrak{X} = ((X; \widehat{\mathcal{U}}); \widehat{f_s}, \widehat{f_t})$ be a smooth correspondence from M_s to M_t (Definition 7.1). Namely, $(X, \widehat{\mathcal{U}})$ is a K-space, $f_s : (X, \widehat{\mathcal{U}}) \to M_s$ is a strongly smooth map and $f_t : (X, \widehat{\mathcal{U}}) \to M_t$ is a weak submersion.

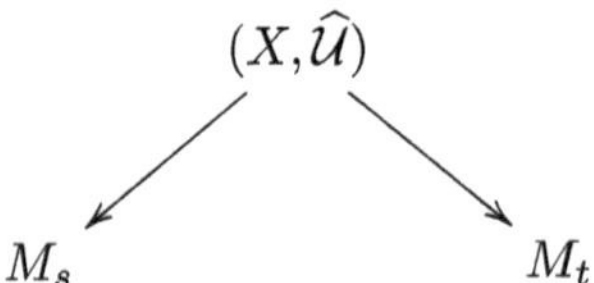

Theorem 27.1 *Let $\widehat{\mathfrak{S}} = \{\widehat{\mathfrak{S}}^\epsilon\}$ be a CF-perturbation with respect to which f_t is a strong submersion. Let $\mathcal{L}_s$ (resp. $\mathcal{L}_t$) be a flat vector bundle on M_s (resp. M_t) such that*

$$(f_t)^*\mathcal{L}_t \cong O_{f_t} \otimes (f_s)^*\mathcal{L}_s. \tag{27.1}$$

Here O_{f_t} is the flat real line bundle associated with the relative orientation $O(1)$-bundle, i.e.,

$$O_{f_t} = (f_t)^* O_{M_t}^* \otimes O_X. \tag{27.2}$$

Then for $\widehat{\mathfrak{X}} = (\mathfrak{X}, \widehat{\mathfrak{S}})$ we have the map

$$\mathrm{Corr}_{\widehat{\mathfrak{X}}}^\epsilon : \Omega^\bullet(M_s; \mathcal{L}_s) \to \Omega^\bullet(M_t; \mathcal{L}_t).$$

The following properties are fundamental.

Theorem 27.2 *If the restriction of f_t to $(\partial X, \partial\widehat{\mathcal{U}})$ is strongly submersive with respect to $\widehat{\mathfrak{S}}|_{\partial X}$, we have*

$$d\,\mathrm{Corr}^{\epsilon}_{\widehat{\mathfrak{X}}}(h) = \mathrm{Corr}^{\epsilon}_{\widehat{\mathfrak{X}}}(dh) + (-1)^{\dim(X,\widehat{\mathcal{U}})+\deg h}\mathrm{Corr}^{\epsilon}_{\partial\widehat{\mathfrak{X}}}(h)$$

for $h \in \Omega^\bullet(M_s; \mathcal{L}_s)$.

In addition to Situation 10.15, let $\mathcal{L}_i$ be a flat vector bundle on M_i, $i = 1, 2, 3$ such that

$$f_{2,21}^*\mathcal{L}_2 \cong O_{f_{2,21}} \otimes f_{1,21}^*\mathcal{L}_1, \quad f_{3,32}^*\mathcal{L}_3 \cong O_{f_{3,32}} \otimes f_{2,32}^*\mathcal{L}_2.$$

Note that $O_{f_{3,31}} \cong g_{32,31}^* O_{f_{3,32}} \otimes g_{21,31}^* O_{f_{2,21}}$. Here $g_{32,31} : \mathfrak{X}_{31} \to \mathfrak{X}_{32}$ and $g_{21,31} : \mathfrak{X}_{31} \to \mathfrak{X}_{21}$ are natural projections from the fiber product. Hence we have

$$f_{3,31}^*\mathcal{L}_3 \cong O_{f_{3,31}} \otimes f_{1,31}^*\mathcal{L}_1.$$

Then we have the following composition formula.

Theorem 27.3

$$\mathrm{Corr}^{\epsilon}_{\widehat{\mathfrak{X}_{32}}\circ\widehat{\mathfrak{X}_{21}}} = \mathrm{Corr}^{\epsilon}_{\widehat{\mathfrak{X}_{32}}} \circ \mathrm{Corr}^{\epsilon}_{\widehat{\mathfrak{X}_{21}}}.$$

Chapter 28
Composition of KG-and GG-Embeddings: Proof of Lemma 3.34

***Proof of Lemma* 3.34** We put $\widehat{\mathcal{U}} = \{\mathcal{U}_{\mathfrak{p}} \mid \mathfrak{p} \in \mathfrak{P}\}, \widehat{\mathcal{U}} = \{\mathcal{U}_p \mid p \in Z\}, \widehat{\mathcal{U}^+} = \{\mathcal{U}^+_{\mathfrak{p}} \mid \mathfrak{p} \in \mathfrak{P}^+\}$, $\widehat{\Phi} = (\mathfrak{i}, \{\Phi_{\mathfrak{p}}\})$, where $\widehat{\Phi} : \widehat{\mathcal{U}} \to \widehat{\mathcal{U}^+}$ is the given GG-embedding and $\widehat{\Phi} = \{\Phi_{\mathfrak{p}p} \mid p \in \mathrm{Im}(\psi_{\mathfrak{p}})\}$, where $\widehat{\Phi} : \widehat{\mathcal{U}} \to \widehat{\mathcal{U}}$ is the given KG-embedding. We assume it is a strict KG-embedding. We write $\Phi_{\mathfrak{p}} = (\varphi_{\mathfrak{p}}, \hat{\varphi}_{\mathfrak{p}})$ and $\Phi_{\mathfrak{p}p} = (U_{\mathfrak{p}}(p), \varphi_{\mathfrak{p}p}, \hat{\varphi}_{\mathfrak{p}p})$.

We choose a support pair $\mathcal{K} < \mathcal{K}^+$ of $\widehat{\mathcal{U}}$. For $\mathfrak{p}^+ \in \mathfrak{P}^+$ we put $\mathfrak{P}(\mathfrak{p}^+) = \mathfrak{i}^{-1}(\mathfrak{p}^+) \subseteq \mathfrak{P}$ and

$$K(\mathfrak{p}^+) = \bigcup_{\mathfrak{p}\in\mathfrak{P}(\mathfrak{p}^+)} \varphi_{\mathfrak{p}}(\mathcal{K}_{\mathfrak{p}} \cap s_{\mathfrak{p}}^{-1}(0)).$$

We remark

$$\bigcup_{\mathfrak{p}^+\in\mathfrak{P}^+} K(\mathfrak{p}^+) \supseteq Z. \tag{28.1}$$

Sublemma 28.1 *There exists a neighborhood $U(\mathfrak{p}^+)$ of $K(\mathfrak{p}^+)$ such that*

$$U(\mathfrak{p}^+) \cap s_{\mathfrak{p}^+}^{-1}(0) \subseteq \bigcup_{\mathfrak{p}\in\mathfrak{P}(\mathfrak{p}^+)} \varphi_{\mathfrak{p}}(\mathcal{K}^+_{\mathfrak{p}} \cap s_{\mathfrak{p}}^{-1}(0)).$$

Proof If there is no such neighborhood, there exists a sequence $x_n \in U^+_{\mathfrak{p}^+}$ such that $s_{\mathfrak{p}^+}(x_n) = 0$, x_n converges to $x \in K(\mathfrak{p}^+)$ but x_n is not contained in $\bigcup_{\mathfrak{p}\in\mathfrak{P}(\mathfrak{p}^+)} \varphi_{\mathfrak{p}}(\mathcal{K}^+_{\mathfrak{p}} \cap s_{\mathfrak{p}}^{-1}(0))$. Suppose $x \in \varphi_{\mathfrak{p}}(\mathcal{K}_{\mathfrak{p}} \cap s_{\mathfrak{p}}^{-1}(0))$. Using the fact that $\varphi_{\mathfrak{p}} : U_{\mathfrak{p}} \to U^+_{\mathfrak{p}^+}$ is an embedding of a Kuranishi chart we can show an analogue of Lemma 6.14 and use it to show $x_n \in \varphi_{\mathfrak{p}}(\mathcal{K}^+_{\mathfrak{p}} \cap s_{\mathfrak{p}}^{-1}(0))$ for sufficiently large n. This is a contradiction. □

K. Fukaya et al., *Kuranishi Structures and Virtual Fundamental Chains*, Springer Monographs in Mathematics, https://doi.org/10.1007/978-981-15-5562-6_28

We consider $\mathcal{U}^+_{\mathfrak{p}^+}|_{U(\mathfrak{p}^+)}$ together with the restriction of coordinate changes to obtain a good coordinate system $\widehat{\mathcal{U}^{0+}} = \{\mathcal{U}^+_{\mathfrak{p}^+}|_{U(\mathfrak{p}^+)} \mid \mathfrak{p}^+ \in \mathfrak{P}^+\}$ of $Z \subseteq X$. (Here we use (28.1).)

Let $p \in \psi_{\mathfrak{p}^+}(U(\mathfrak{p}^+) \cap s_{\mathfrak{p}^+}^{-1}(0))$. By Sublemma 28.1 there exists $\mathfrak{p} \in \mathfrak{P}(\mathfrak{p}^+)$ such that

$$p \in \psi_{\mathfrak{p}}(\mathcal{K}^+_{\mathfrak{p}} \cap s_{\mathfrak{p}}^{-1}(0)).$$

We choose a neighborhood U^0_p of o_p in $U_{\mathfrak{p}}$ such that

$$U^0_p \subset (\varphi_{\mathfrak{p}} \circ \varphi_{\mathfrak{p}p})^{-1}(U(\mathfrak{p}^+)) \tag{28.2}$$

for all such $\mathfrak{p}^+$ and $\mathfrak{p}$. It is easy to check that $\varphi_{\mathfrak{p}^+p} = \varphi_{\mathfrak{p}} \circ \varphi_{\mathfrak{p}p} : U^0_p \to U(\mathfrak{p}^+)$ is independent of the choice of $\mathfrak{p}$.

Let $p \in \psi_{\mathfrak{p}^+}(U(\mathfrak{p}^+) \cap s_{\mathfrak{p}^+}^{-1}(0))$. We define $\mathfrak{p} = \mathfrak{p}(p, \mathfrak{p}^+) \in \mathfrak{P}(\mathfrak{p}^+)$ with $p \in \psi_{\mathfrak{p}}(\mathcal{K}^+_{\mathfrak{p}} \cap s_{\mathfrak{p}}^{-1}(0))$ to be the maximum element of

$$\{\mathfrak{p} \in \mathfrak{P}(\mathfrak{p}^+) \mid p \in \psi_{\mathfrak{p}}(\mathcal{K}^+_{\mathfrak{p}} \cap s_{\mathfrak{p}}^{-1}(0))\}, \tag{28.3}$$

We may choose U^0_p such that if $\mathfrak{p} < \mathfrak{q}$, $\mathfrak{i}(\mathfrak{q}) \le \mathfrak{i}(\mathfrak{p})$ then

$$\psi_p(U^0_p \cap s_p^{-1}(0)) \cap \psi_{\mathfrak{q}}(\mathcal{K}^+_{\mathfrak{q}} \cap s_{\mathfrak{q}}^{-1}(0)) = \emptyset. \tag{28.4}$$

In fact, $\mathfrak{p} < \mathfrak{q}$, $\mathfrak{i}(\mathfrak{q}) \le \mathfrak{i}(\mathfrak{p})$ implies $p \notin \psi_{\mathfrak{q}}(\mathcal{K}^+_{\mathfrak{q}} \cap s_{\mathfrak{q}}^{-1}(0))$ by the maximality of $\mathfrak{p}$ in (28.3). (Note $\mathfrak{p} < \mathfrak{q}$, $\mathfrak{i}(\mathfrak{q}) \le \mathfrak{i}(\mathfrak{p})$ implies $\mathfrak{i}(\mathfrak{q}) = \mathfrak{i}(\mathfrak{p})$.)

We finally check the commutativity of (3.8). Let $p \in \psi_{\mathfrak{p}^+}(U(\mathfrak{p}^+) \cap s_{\mathfrak{p}^+}^{-1}(0))$, $q \in \psi_p(U^0_p \cap s_p^{-1}(0))$, $q \in \psi_{\mathfrak{q}^+}(U(\mathfrak{q}^+) \cap s_{\mathfrak{q}^+}^{-1}(0))$, $\mathfrak{q}^+ \le \mathfrak{p}^+$. Take $\mathfrak{p} = \mathfrak{p}(p, \mathfrak{p}^+) \in \mathfrak{P}(\mathfrak{p}^+)$ as above and let $\mathfrak{q} \in \mathfrak{P}(\mathfrak{q}^+)$ with $q \in \psi_{\mathfrak{q}}(\mathcal{K}^+_{\mathfrak{q}} \cap s_{\mathfrak{q}}^{-1}(0))$.

Equation (28.2) implies $q \in \psi_{\mathfrak{p}}(\mathcal{K}^+_{\mathfrak{p}} \cap s_{\mathfrak{p}}^{-1}(0))$. Therefore $\mathfrak{q} \le \mathfrak{p}$ or $\mathfrak{p} < \mathfrak{q}$. Then (28.4) implies $\mathfrak{q} \le \mathfrak{p}$.

We also remark that $q \in \psi_{\mathfrak{p}^+}(U(\mathfrak{p}^+) \cap s_{\mathfrak{p}^+}^{-1}(0))$ by (28.2).

Now the commutativity of (3.8) for $\widehat{\Phi'} : \widehat{\mathcal{U}^0} \to \widehat{\mathcal{U}^{0+}}$ which we are defining, follows from the commutativity of (3.8) for $\widehat{\Phi} : \widehat{\mathcal{U}} \to \widehat{\mathcal{U}}$ and the commutativity of (3.5) for $\widehat{\Phi} : \widehat{\mathcal{U}} \to \widehat{\mathcal{U}^+}$, by choosing $U^0_{pq} \subset \varphi_{pq}(U^0_p) \cap U^0_q$ to be a sufficiently small neighborhood of o_q in $U^0_q \cap U_{pq}$. □

Chapter 29
Global Quotients and Orbifolds

In some of our earlier writings we assumed that a Kuranishi neighborhood is not only an orbifold but also is a quotient of a manifold by a finite group action globally. We explain that there is no difference between two formulations when we apply the theory. The contents of this chapter are not used anywhere else in this book. Our purpose is to clarify the relationship of the formulation of this book with that in the previous literature.

We say an orbifold M is a *global quotient* if there exists a manifold N and a finite group Γ acting effectively on it such that M is diffeomorphic to N/Γ as orbifolds. An embedding $f : M \to M'$ between orbifolds is said to be a *global quotient* if $M = N/\Gamma$ and $M' = N'/\Gamma'$ are global quotients and there exists a group homomorphism $h : \Gamma \to \Gamma'$ and an h equivariant smooth embedding $\tilde{f} : N \to N'$ such that $(h, \tilde{f})$ induces f in an obvious sense.

Lemma 29.1 *Any Kuranishi structure has an open substructure such that all the Kuranishi neighborhoods (of its Kuranishi charts) and coordinate changes are global quotient.*

The proof is easy and is omitted. The next proposition is slightly more nontrivial.

Proposition 29.2 *Any good coordinate system has a weakly open substructure such that all the Kuranishi neighborhoods (of its Kuranishi charts) and coordinate changes are global quotient.*

We first show the following.

Lemma 29.3 *Let U be an orbifold and K its compact subset. Then there exists finitely many orbifold charts $\mathfrak{V}_i = (V_i, \Gamma_i, \phi_i)$ of U which cover K and have the following properties.*

The index set $\{i\}$ has a total order such that if $i < j$ then the coordinate change from $\mathfrak{V}_i$ to $\mathfrak{V}_j$ is a global quotient. Namely, there exist a group homomorphism $h_{ji} : \Gamma_i \to \Gamma_j$ and an open smooth embedding $\varphi_{ji} : \phi_i^{-1}(\phi_i(V_i) \cap \phi_j(V_j)) \to V_j$, such

K. Fukaya et al., *Kuranishi Structures and Virtual Fundamental Chains*, Springer Monographs in Mathematics, https://doi.org/10.1007/978-981-15-5562-6_29

that φ_{ji} is h_{ji} equivariant and φ_{ji} is a lift of the identity map $\phi_i(V_i) \cap \phi_j(V_j) \to \phi_j(V_j)$.

Proof This lemma is well-known to experts. We prove it for completeness' sake.

For $p \in U$ we take an orbifold chart (V_p, Γ_p, ϕ_p) with $o_p \in V_p$ the base point, which goes to p, and Γ_p fixing o_p. The order of Γ_p depends only on p and we put $d(p) = \#\Gamma_p$. We define $SG_d = \{p \mid d(p) \geq d\}$, which is a closed set and $\overset{\circ}{SG}_d = SG_d \setminus SG_{d+1}$ is a smooth manifold. Since K is compact we may assume $SG_d = \emptyset$ for $d > d_0$, by shrinking U if necessary. We will prove the existence of an atlas which covers $K \cap SG_d$ and such that coordinate changes among them are global quotients, by downward induction on d.

We first consider the case of $d = d_0$. We take a Riemannian metric on U and use the exponential map $\exp_p : B_\epsilon(T_p(U))/\Gamma_p \to U$. Let $i(p)$ be the injectivity radius, which is the maximum of ϵ where the exponential map exists and is injective.

Sublemma 29.4 *The injectivity radius is uniformly bounded away from zero on $SG_{d_0} \cap K$.*

Proof Otherwise there exist $p_i \in SG_{d_0}$ such that $\lim_{i\to\infty} i(p_i) = 0$. We may assume p_i converges to p_∞. Then $p_\infty \in SG_{d_0}$. Since SG_{d_0+1} is empty it follows that for large i the isotropy group of p_i is identified with the isotropy group of p_∞. We can then show $i(p_i)$ is away from 0 by using the coordinate at p_∞. This is a contradiction. □

Let ϵ_1 be the infimum in Sublemma 29.4. We take $\epsilon_{p_i} < \epsilon_1/5$ and cover $SG_{d_0} \cap K$ by finitely many metric balls $B_{\epsilon_{p_i}}(p_i)$ of radius ϵ_{p_i} centered at $p_i \in SG_{d_0} \cap K$. Then using the exponential map we obtain an orbifold chart of $B_{\epsilon_{p_i}}(p_i)$.[1]

If $B_{\epsilon_{p_i}}(p_i) \cap B_{\epsilon_{p_j}}(p_j) \neq \emptyset$ then they both are contained in the image of the exponential map at p_i. Together with $d(p_i) = d(p_j)$ it implies $\Gamma_{p_i} \cong \Gamma_{p_j}$. We can use this fact to show that the coordinate change is a global quotient.

Suppose we have required charts (V_i, Γ_i, ϕ_i) $(i = 1, \dots, I)$ which cover SG_{d+1}. We take relatively compact Γ_i-invariant subsets $V_i^0 \subset V_i$ such that $(V_i^0, \Gamma_i, \phi_i)$ $(i = 1, \dots, I)$ still cover SG_{d+1}. Let $R = \bigcup_i \phi_i(\overline{V_i^0})$. Using the fact that the compact set $(K \cap SG_d) \setminus \text{Int}(R)$ does not intersect with SG_{d+1} we can show that the injectivity radius is bounded away from 0 there, in the same way as Sublemma 29.4. Let ϵ_2 be the infimum of the injectivity radius. We take $\epsilon_{p_i} < \epsilon_2/5$ and cover $(K \cap SG_d) \setminus \text{Int}(R)$ by finitely many metric balls $B_{\epsilon_{p_i}}(p_i)$. We use the exponential map to obtain an orbifold chart of $B_{\epsilon_{p_i}}(p_i)$. In the same way as the case of $d = d_0$ we can show that the coordinate changes between two new charts $B_{\epsilon_{p_i}}(p_i)$, $B_{\epsilon_{p_j}}(p_j)$ are global quotients. Moreover we may take ϵ_{p_i} so small that if $B_{\epsilon_{p_i}}(p_i) \cap \phi_i(\overline{V_i^0}) \neq \emptyset$ then $B_{\epsilon_{p_i}}(p_i) \subset \phi_i(V_i)$ and there exists a map $\{v \in T_{p_i}U \mid |v| < \epsilon_{p_i}\} \to V_i$ which is a lift of the identity map $B_{\epsilon_{p_i}}(p_i) \to \phi_i(V_i)$. We can use it to show that

[1] We need to consider the case when ϵ_{p_i} is p_i-dependent later during the proof of Proposition 29.2.

the coordinate change from a new chart to a previously defined chart $(V_i^0, \Gamma_i, \phi_i)$ is a global quotient. The proof is complete by induction. □

***Proof of Proposition* 29.2** Let $\widehat{\mathcal{U}} = (\mathfrak{P}, \{\mathcal{U}_{\mathfrak{p}}\})$ be a given good coordinate system of $Z \subseteq X$. Let $\{\mathcal{K}_{\mathfrak{p}}\}$ be its support system. For an ideal $\mathscr{I} \subseteq \mathfrak{P}$ (See Definition 9.18) we consider open subsets $U_{\mathfrak{p},i} \subset U_{\mathfrak{p}}$, $i = 1, \dots, I(\mathfrak{p})$, $\mathfrak{p} \in \mathscr{I}$ such that:

(1) $\bigcup_{i=1}^{I(\mathfrak{p})} U_{\mathfrak{p},i} \supseteq \mathcal{K}_{\mathfrak{p}} \cap s_{\mathfrak{p}}^{-1}(0)$.
(2) Let $\psi_{\mathfrak{p}}(U_{\mathfrak{p},i}) \cap \psi_{\mathfrak{q}}(U_{\mathfrak{q},j}) \neq \emptyset$ with $\mathfrak{q} < \mathfrak{p}$, $\mathfrak{q}, \mathfrak{p} \in \mathscr{I}$ or $\mathfrak{p} = \mathfrak{q} \in \mathscr{I}$, $i \geq j$ then the coordinate change from $\mathcal{U}_{\mathfrak{q}}|_{U_{\mathfrak{q},j}}$ to $\mathcal{U}_{\mathfrak{p}}|_{U_{\mathfrak{p},i}}$ is a global quotient.

We prove existence of such an open cover by induction on $\#\mathscr{I}$. The case $\#\mathscr{I} = 1$ is Lemma 29.3. We assume that we have proved the existence of such an open cover in the case $\#\mathscr{I} < d$. Suppose $\#\mathscr{I} = d$. Let $\mathfrak{q}$ be the smallest element of $\mathscr{I}$. We apply the induction hypothesis to $\mathscr{I}_- = \mathscr{I} \setminus \{\mathfrak{q}\}$ to obtain $U_{\mathfrak{p},i}$ for $\mathfrak{p} \in \mathscr{I} \setminus \{\mathfrak{q}\}$.

We apply Lemma 29.3 to $U = U_{\mathfrak{q}}$ and $K = \mathcal{K}_{\mathfrak{q}} \cap s_{\mathfrak{q}}^{-1}(0)$ to obtain an atlas $\{U_{\mathfrak{q},i}\}$ which covers $\mathcal{K}_{\mathfrak{q}} \cap s_{\mathfrak{q}}^{-1}(0)$. We take relatively compact subsets $U_{\mathfrak{p},i}^0$ of $U_{\mathfrak{p},i}$ so that (1), (2) above still holds for $\mathscr{I}_-$.

The coordinate changes among $U_{\mathfrak{q},i}$ are global quotients. The coordinate changes between $U_{\mathfrak{p},i}^0, U_{\mathfrak{p}',i'}^0$ for $\mathfrak{p}, \mathfrak{p}' \in \mathscr{I} \setminus \{\mathfrak{q}\}$ are also global quotients.

Note the open set $U_{\mathfrak{q},i}$ is the metric ball $B_{\epsilon_{p_i}}(p_i)$ in $U_{\mathfrak{q}}$. We claim that we take ϵ_{p_i} small (in each inductive step of the proof of Lemma 29.3) so that (2) above holds if we take $U_{\mathfrak{p},i}^0$ instead of $U_{\mathfrak{p},i}$. In fact, we may choose ϵ_p such that the following holds:

(*) Suppose $p \in (K \cap SG_d) \setminus \mathrm{Int}(R)$ (here the right hand side appears in the proof of Lemma 29.3) and $\psi_{\mathfrak{q}}(B_{\epsilon_p}(p) \cap s_{\mathfrak{q}}^{-1}(0)) \cap \psi_{\mathfrak{p}}(U_{\mathfrak{p},i}^0 \cap s_{\mathfrak{p}}^{-1}(0)) \neq \emptyset$.
Then $B_{\epsilon_p}(p) \subset U_{\mathfrak{p}\mathfrak{q}}$ and $\varphi_{\mathfrak{p}\mathfrak{q}}(B_{\epsilon_p}(p)) \subset U_{\mathfrak{p},i}$. Moreover the restriction of $\varphi_{\mathfrak{p}\mathfrak{q}}$ to $B_{\epsilon_p}(p)$, that is, the map $B_{\epsilon_p}(p) \to U_{\mathfrak{p},i}$, is a global quotient.

The existence of ϵ_p for each given $p \in K = \mathcal{K}_{\mathfrak{q}} \cap s_{\mathfrak{q}}^{-1}(0)$ is immediate from Definition 23.2. Therefore we can cover $K \cap SG_d$ by finitely many $B_{\epsilon_{p_i}}(p_i)$. We thus obtained the required $U_{\mathfrak{q},i}$. The induction is complete. Proposition 29.2 is the case when $\mathscr{I} = \mathfrak{P}$. □

References

[AFOOO] M. Abouzaid, K. Fukaya, Y.-G. Oh, H. Ohta, K. Ono, *Quantum cohomology and split generation in Lagrangian Floer theory*, in preparation

[AS] M. Abouzaid, P. Seidel, An open string analogue of Viterbo functoriality. Geom. Topol. **14**(2), 627–718 (2010)

[ALR] A. Adem, J. Leida, Y. Ruan, *Orbifolds and Stringy Topology*. Cambridge Tracts in Mathematics 171 (Cambridge University Press, Cambridge, 2007)

[AJ] M. Akaho, D. Joyce, Immersed Lagrangian Floer theory. J. Diff. Geom. **86**(3), 381–500 (2010)

[BH] E. Bao, K. Honda, *Semi-global Kuranishi charts and the definition of contact homology*, arXiv:1512.00580

[BF] K. Behrend, B. Fantechi, The intrinsic normal cone. Invent. Math. **128**, 45–88 (1997)

[BO] F. Bourgeois, A. Oancea, Symplectic homology, autonomous Hamiltonians, and Morse-Bott moduli spaces. Duke Math. J. **146**(1), 71–174 (2009)

[CLa] K. Chan, S.-C. Lau, Open Gromov–Witten Invariants and Superpotentials for Semi-Fano Toric Surfaces. Int. Math. Res. Not. **2014**(14), 3759–3789 (2014)

[CW1] F. Charest, C. Woodward, Floer trajectories and stabilizing divisors. J. Fixed Point Theory Appl. **19**(2), 1165–1236 (2017)

[CW2] F. Charest, C. Woodward, *Fukaya algebras via stabilizing divisors*, arXiv:1505.08146

[CT] B. Chen, G. Tian, Virtual manifolds and localization. Acta Math. Sinica. **26**, 1–24 (2010)

[CLW] B. Chen, A.-M. Li, B.-L. Wang, *Virtual neighborhood technique for pseudo-holomorphic spheres*, arXiv:math/1306.3206

[CKO] Y. Cho, Y. Kim, Y.-G. Oh, Lagrangian fibers of Gelfand-Cetlin systems, Adv. Math. **372** (2020), 107304, arXiv:1704.07213

[CFH] K. Cieliebak, A. Floer, H. Hofer, Symplectic homology. II A general construction. Math. Z. **218**(1), 103–122 (1995)

[CM] K. Cieliebak, K. Mohnke, Symplectic hypersurfaces and transversality in Gromov–Witten theory. J. Symplectic Geom. **5**, 281–356 (2007)

[Co] K. Costello, *Renormalization and Effective Field Theory*. Math. Surveys and Monographs 170 (American Mathematical Society, Providence 2011)

[DF1] A. Daemi, K. Fukaya, *Monotone Lagrangian Floer theory in smooth divisor complements: I*, submitted, arXiv:1808.08915

K. Fukaya et al., *Kuranishi Structures and Virtual Fundamental Chains*, Springer Monographs in Mathematics, https://doi.org/10.1007/978-981-15-5562-6

[DF2] A. Daemi, K. Fukaya, *Monotone Lagrangian Floer theory in smooth divisor complements: II*, submitted, arXiv:1809.03409

[Do1] S.K. Donaldson, An application of gauge theory to four-dimensional topology. J. Differ. Geom. **18**(2), 279–315 (1983)

[Do2] S.K. Donaldson, The orientation of Yang-Mills moduli spaces and 4-manifold topology. J. Differ. Geom. **26**(3), 397–428 (1987)

[Do3] S.K. Donaldson, Symplectic submanifolds and almost-complex geometry. J. Differ. Geom. **44**(4), 666–705 (1996)

[EGH] Y. Eliashberg, A. Givental, H. Hofer, Introduction to Symplectic Field Theory. GAFA 2000, Special Volume, Part II, pp. 560–673

[FJR] H. Fan, T.J. Jarvis, Y. Ruan, *The Witten equation and its virtual fundamental cycle*, arXiv:0712.4025

[Fl1] A. Floer, Morse theory for Lagrangian intersections. J. Differ. Geom. **28**, 513–547 (1988)

[Fl2] A. Floer, Symplectic fixed points and holomorphic spheres. Commun. Math. Phys. **120**, 575–611 (1989)

[FH] A. Floer, H. Hofer, Symplectic homology. I. Open sets in $\mathbb{C}^n$. Math. Z. **215**(1), 37–88 (1994)

[Fuk1] K. Fukaya, Morse homotopy, A^{∞}–category, and Floer homologies, in *Proceedings of Garc Workshop on Geometry and Topology*, ed. by H.J. Kim (Seoul National University, 1994), pp. 1–102

[Fuk2] K. Fukaya, Morse homotopy and its quantization, in *Geometry and Topology*, ed. by W. Kazez (International Press, 1997), pp. 409–440

[Fuk3] K. Fukaya, Floer homology and mirror symmetry II, in *Minimal Surfaces, Geometric Analysis and Symplectic Geometry*, Baltimore, 1999. Advanced Studies in Pure Mathematics 34 (The Mathematical Society of Japan, Tokyo, 2002), pp. 31–127

[Fuk4] K. Fukaya, Cyclic symmetry and adic convergence in Lagrangian Floer theory. Kyoto J. Math. **50**(3), 521–590 (2010)

[Fuk5] K. Fukaya, Differentiable operad, Kuranishi correspondence, and foundation of topological field theories based on pseudo-holomorphic curves, in *Arithmetic and Geometry Around Quantization*. Progress in Mathematics 279 (Birkhäuser, Boston, 2010), pp. 123–200

[Fuk6] K. Fukaya, *Answers to the questions from Katrin Wehrheim on Kuranishi structure.* Posted to the google group Kuranishi on March 21th 2012, can be obtained from https://cgp.ibs.re.kr/~yongoh/answer19.pdf

[Fuk7] K. Fukaya, Floer homology of Lagrangian submanifolds. Suugaku Expositions **26**, 99–127 (2013). arXiv:1106.4882

[Fuk8] K. Fukaya, *Lie groupoid, deformation of unstable curve, and construction of equivariant Kuranishi charts*, to appear in Publ. RIMS, arXiv:1701.02840

[Fuk9] K. Fukaya, *Unobstructed immersed Lagrangian correspondence and filtered A infinity functor*, submitted, arXiv:1706.02131

[FOh] K. Fukaya, Y.-G. Oh, Zero-loop open string on cotangent bundle and Morse homotopy. Asian J. Math. **1**, 96–180 (1998)

[FOOO1] K. Fukaya, Y.-G. Oh, H. Ohta, K. Ono, *Lagrangian Intersection Floer Theory-Anomaly and Obstruction*, preprint (Kyoto University, Kyoto, Japan, 2000)

[FOOO2] K. Fukaya, Y.-G. Oh, H. Ohta, K. Ono, *Lagrangian Intersection Floer Theory-Anomaly and Obstruction*, expanded version of [FOOO1], 2006 & 2007

[FOOO3] K. Fukaya, Y.-G. Oh, H. Ohta, K. Ono, *Lagrangian Intersection Floer Theory-Anomaly and Obstruction, Part I.* AMS/IP Studies in Advanced Mathematics 46.1 (International Press/American Mathematical Society, 2009). MR2553465

[FOOO4] K. Fukaya, Y.-G. Oh, H. Ohta, K. Ono, *Lagrangian Intersection Floer Theory-Anomaly and Obstruction, Part II.* AMS/IP Studies in Advanced Mathematics 46.2 (International Press/American Mathematical Society, 2009). MR2548482

[FOOO5] K. Fukaya, Y.-G. Oh, H. Ohta, K. Ono, Canonical models of filtered A_∞-algebras and Morse complexes, in *New Perspectives and Challenges in Symplectic Field Theory*. CRM Proceedings and Lecture Notes 49 (American Mathematical Society, Providence, 2009), pp. 201–227. arXiv:0812.1963

[FOOO6] K. Fukaya, Y.-G. Oh, H. Ohta, K. Ono, Anchored Lagrangian submanifolds and their Floer theory, in *Mirror Symmetry and Tropical Geometry*. Contemporary Mathematics 527 (American Mathematical Society, Providence, 2010), pp. 15–54

[FOOO7] K. Fukaya, Y.-G. Oh, H. Ohta, K. Ono, Lagrangian Floer theory on compact toric manifolds I. Duke Math. J. **151**(1), 23–174 (2010)

[FOOO8] K. Fukaya, Y.-G. Oh, H. Ohta, K. Ono, Lagrangian Floer theory on compact toric manifolds II : Bulk deformations. Sel. Math. New Ser. **17**, 609–711 (2011)

[FOOO9] K. Fukaya, Y.-G. Oh, H. Ohta, K. Ono, Antisymplectic involution and Floer cohomology. Geom. Topol. **21**, 1–106 (2017), arXiv:0912.2646

[FOOO10] K. Fukaya, Y.-G. Oh, H. Ohta, K. Ono, *Floer theory and mirror symmetry on compact toric manifolds*, arXiv:1009.1648v1

[FOOO10'] K. Fukaya, Y.-G. Oh, H. Ohta, K. Ono, *Floer theory and mirror symmetry on compact toric manifolds*. Astérisque 376 (2016). arXiv:1009.1648v2 (revised version of [FOOO10])

[FOOO11] K. Fukaya, Y.-G. Oh, H. Ohta, K. Ono, Toric degeneration and non-displaceable Lagrangian tori in $S^2 \times S^2$. Int. Math. Res. Not. IMRN **13**, 2942–2993 (2012), arXiv:1002.1660

[FOOO12] K. Fukaya, Y.-G. Oh, H. Ohta, K. Ono, Lagrangian Floer theory on compact toric manifolds: survey. Surv. Diff. Geom. **17**, 229–298 (2012), arXiv:1011.4044

[FOOO13] K. Fukaya, Y.-G. Oh, H. Ohta, K. Ono, *Spectral Invariants with Bulk, Quasi-morphisms and Lagrangian Floer Theory*. Memoirs of the American Mathematical Society 1254 (2019), 266pp, arXiv:1105.5123

[FOOO14] K. Fukaya, Y.-G. Oh, H. Ohta, K. Ono, Lagrangian Floer theory over integers: spherically positive symplectic manifolds. Pure Appl. Math. Q. **9**(2), 189–289 (2013), arXiv:1105.5124

[FOOO15] K. Fukaya, Y.-G. Oh, H. Ohta, K. Ono, Displacement of polydisks and Lagrangian Floer theory. J. Symplectic Geom. **11**(2), 231–268 (2013)

[FOOO16] K. Fukaya, Y.-G. Oh, H. Ohta, K. Ono, *Technical details on Kuranishi structure and virtual fundamental chain*, arXiv:1209.4410

[FOOO17] K. Fukaya, Y.-G. Oh, H. Ohta, K. Ono, Shrinking good coordinate systems associated to Kuranishi structures. J. Symplectic Geom. **14**(4), 1295–1310 (2016), arXiv:1405.1755

[FOOO18] K. Fukaya, Y.-G. Oh, H. Ohta, K. Ono, *Exponential decay estimates and smoothness of the moduli space of pseudo-holomorphic curves*, submitted, arXiv:1603.07026

[FOOO19] K. Fukaya, Y.-G. Oh, H. Ohta, K. Ono, *Kuranishi structure, Pseudo-holomorphic curve, and Virtual fundamental chain; Part 1*, arXiv:1503.07631v1

[FOOO20] K. Fukaya, Y.-G. Oh, H. Ohta, K. Ono, *Kuranishi structure, Pseudo-holomorphic curve, and Virtual fundamental chain; Part 2*, arXiv:1704.01848v1

[FOOO21] K. Fukaya, Y.-G. Oh, H. Ohta, K. Ono, Construction of Kuranishi structures on the moduli spaces of pseudo-holomorphic disks: I. Surv. Diff. Geom. **22**, 133–190 (2018), arXiv:1710.01459

[FOOO22] K. Fukaya, Y.-G. Oh, H. Ohta, K. Ono, *Construction of Kuranishi structures on the moduli spaces of pseudo-holomorphic disks: II*, arXiv:1808.06106

[FOOO23] K. Fukaya, Y.-G. Oh, H. Ohta, K. Ono, *Forgetful map and its applications to Gromov–Witten invariant and to Lagrangian Floer theory*, in preparation

[FOn1] K. Fukaya, K. Ono, Arnold conjecture and Gromov–Witten invariants for general symplectic manifolds, in *The Arnold Fest*, ed. by E. Bierstone, B. Khesin, A. Khovanskii, J. Marsden. Fields Institute Communications 24 (1999), pp. 173–190

[FOn2] K. Fukaya, K. Ono, Arnold conjecture and Gromov–Witten invariant. Topology **38**(5), 933–1048 (1999)

[Fur] M. Furuta, Perturbation of moduli spaces of self-dual connections. J. Fac. Sci. Univ. Tokyo Sect. IA Math. **34**(2), 275–297 (1987)

[Go] R. Godement, *Topologie algébrique et théorie des faisceaux* (Hermann, Paris, 1958)

[GP] T. Graber, R. Pandharipande, Localization of virtual classes. Invent. Math. **135**, 487–518 (1999)

[Gro] A. Grothendieck, *Séminaire de Géometrie Algébraic de Bois Marie 1 Revêtements étale et groupe fundamental*. Lecture Note in Mathematics 224 (Springer, Berlin, 1971)

[HS] H. Hofer, D. Salamon, Floer homology and Novikov rings, in *The Floer Memorial Volume*, ed. by H. Hofer et al. (Birkhaüser, Basel-Boston-Berlin, 1995), pp. 483–524

[HWZ1] H. Hofer, K. Wysocki, E. Zehnder, Sc-smoothness, retractions and new models for smooth spaces. Discret. Continuous Dyn. Syst. **28**(2), 665–788 (2010), arXiv:1002.3381v1

[HWZ2] H. Hofer, K. Wysocki, E. Zehnder, *Applications of Polyfold Theory I: The Polyfolds of Gromov–Witten Theory*. Memoirs of the American Mathematical Society 248 (2017)

[HWZ3] H. Hofer, K. Wysocki, E. Zehnder, *Polyfold and Fredholm Theory*, arXiv:1707.08941

[H1] M. Hutchings, Reidemeister torsion in generalized Morse theory. Forum Math. **14**(2), 209–244 (2002)

[H2] M. Hutchings, *Gluing a flow line to itself*, floerhomology.wordpress.com /2014/07/14/gluing- a-flow-line-to-itself

[IP] E.-N. Ionel, T.H. Parker, The Gopakumar-Vafa formula for symplectic manifolds. Ann. Math. **187**, 1–64 (2018)

[Ir] K. Irie, Chain level loop bracket and pseudo-holomorphic disks. J Topology **13**, 870–938 (2020), arXiv:1801.04633

[Is] S. Ishikawa, *Construction of general symplectic field theory*, arXiv:1807.09455

[Jo1] D. Joyce, *Kuranishi homology and Kuranishi cohomology*, arXiv 0707.3572v5

[Jo2] D. Joyce, *D-manifolds and d-orbifolds: a theory of derived differential geometry*, book manuscript

[Jo3] D. Joyce, On manifolds with corners, in *Advances in Geometric Analysis ALM* 21 (International Press, 2012), pp. 225–258, ISBN: 978-1571462480, arXiv:0910.3518

[Jo4] D. Joyce, A new definition of Kuranishi space, in *Virtual Fundamental Cycles in Symplectic Topology*, ed. by J. Morgan. Surveys and Monographs 237 (American Mathematical Society, 2019), arXiv:1409.6908

[Jo5] D. Joyce, *Algebraic geometry over* C^∞*-rings*. Memoirs of the American Mathematical Society 1256 (2019)

[KM] M. Kontsevich, Y. Manin, Gromov–Witten classes quantum cohomology and enumerative geometry. Commun. Math. Phys. **164**, 525–562 (1994)

[LiTi1] J. Li, G. Tian, Virtual moduli cycles and Gromov–Witten invariants of algebraic varieties. J. Am. Math. Soc. **11**, 119–174 (1998)

[LiTi2] J. Li, G. Tian, Virtual moduli cycles and Gromov–Witten invariants of general symplectic manifolds, in *Topics in Symplectic* 4*-Manifolds*, Irvine, 1996. First International Press Lecture Series 1 (International Press Cambridge, MA, 1998), pp. 47–83

[LiuTi] G. Liu, G. Tian, Floer homology and Arnold conjecture. J. Diff. Geom. **49**, 1–74 (1998)

[Li] C.-C.M. Liu, *Moduli of J-Holomorphic Curves with Lagrangian Boundary Conditions and Open Gromov–Witten Invariants for an* S^1*-Equivariant Pair*, arXiv:math/0210257

[LuT] G. Lu, G. Tian, Constructing virtual Euler cycles and classes. *International Mathematics Research Surveys, IMRS 2007*, Art. ID rym001, 220pp

[MSS] M. Markl, S. Steve, J. Stasheff, *Operads in Algebra, Topology and Physics*. Mathematical Surveys and Monographs 96 (American Mathematical Society, Providence, 2002), x+349pp

[Ma] J. Mather, Stratifications and mappings, in *Dynamical systems (Proc. Sympos., Univ. Bahia, Salvador, 1971)* 41 (Academic Press, New York, 1973), pp. 195–232

[MS] D. McDuff, D. Salamon, *J-Holomorphic Curves and Symplectic Topology*. American Mathematical Society Colloquim Publication 52 (American Mathematical Society, Providence, 1994)

[NNU] T. Nishinou, Y. Nohara, K. Ueda, Toric degenerations of Gelfand-Cetlin systems and potential functions. Adv. Math. **224**(2), 648–706 (2010)

[Oh] Y.-G. Oh, Relative Floer and quantum cohomology and the symplectic topology of Lagrangian submanifolds, in *Contact and Symplectic Geometry*, ed. by C.B. Thomas (Cambridge University Press, Cambridge, 1996), pp. 201–267

[On1] K. Ono, On the Arnold conjecture for weakly monotone symplectic manifolds. Invent. Math. **119**(3), 519–537 (1995)

[On2] K. Ono, Floer-Novikov cohomology and the flux conjecture. Geom. Funct. Anal. **16**(5), 981–1020 (2006)

[Pa1] J. Pardon, An algebraic approach to virtual fundamental cycles on moduli spaces of J-holomorphic curves. Geom. Topol. **20**, 779–1034 (2016)

[Pa2] J. Pardon, Contact homology and virtual fundamental cycles. J. Am. Math. Soc. **32**(3), 825–919 (2019), arXiv:1508.03873

[Ru1] Y. Ruan, Topological sigma model and Donaldson-type invariant in Gromov theory. Duke Math. J. **83**, 461–500 (1996)

[Ru2] Y. Ruan, Virtual neighborhood and pseudo-holomorphic curve. Turkish J. Math. **23**(1), 161–231 (1999)

[RT1] Y. Ruan, G. Tian, A mathematical theory of quantum cohomology. J. Diff. Geom. **42**, 259–367 (1995)

[RT2] Y. Ruan, G. Tian, Higher genus symplectic invariants and sigma models coupled with gravity. Invent. Math. **130**, 455–516 (1997)

[Sa] I. Satake, On a generalization of the notion of manifold. Proc. Natl. Acad. Sci. U. S. A. **42**, 359–363 (1956)

[Se] P. Seidel, *Fukaya Categories and Picard-Lefschetz Theory*. Zurich Lectures in Advanced Mathematics (European Mathematical Society (EMS), Zurich, 2008)

[Sie] B. Siebert, *Gromov–Witten invariants of general symplectic manifolds*, arXiv:dg-ga/9608005

[So1] J. Solomon, *Intersection theory on the moduli space of holomorphic curves with Lagrangian boundary conditions*, arXiv:math/0606429

[So2] J. Solomon, Involutions, obstructions and mirror symmetry, Adv. Math. **367** (2020), 107107 52p. arXiv:1810.07027

[Ta] I. Tamura, *Differential Topology*. Iwanami Lectures on Fundamental Mathematics 20 (Iwanami Shoten, Tokyo, 1983) (in Japanese)

[TF] M. Tehrani, K. Fukaya, Gromov–Witten theory via Kuranishi structures, in *Virtual Fundamental Cycles in Symplectic Topology*, ed. by J. Morgan. Surveys and Monographs 237 (American Mathematical Society, 2019), arXiv:1701.07821

[Th] R. Thom, Quelque propriétés globales des variétés différentiable. Comment. Math. Helv. **28**, 17–86 (1954)

[TX] G. Tian, G. Xu, *Gauged Linear Sigma Model in Geometric Phases*, arXiv:1809.00424

[Wu] W. Wu, On an exotic Lagrangian torus in $\mathbb{C}P^2$. Compos. Math. **151**(7), 1372–1394 (2015)

[Ya1] D. Yang, *The Polyfold-Kuranishi Correspondence I: A Choice-independent Theory of Kuranishi Structures*, arXiv:1402.7008

[Ya2] D. Yang, *Virtual Harmony*, arXiv:1510.06849

Index

K. Fukaya et al., *Kuranishi Structures and Virtual Fundamental Chains*, Springer Monographs in Mathematics, https://doi.org/10.1007/978-981-15-5562-6

C

www.ingramcontent.com/pod-product-compliance
Ingram Content Group UK Ltd.
Pitfield, Milton Keynes, MK11 3LW, UK
UKHW020141300726
14059UKWH00008B/90

* 9 7 8 9 8 1 1 5 5 5 6 1 9 *